BIOGEOGRAPHY

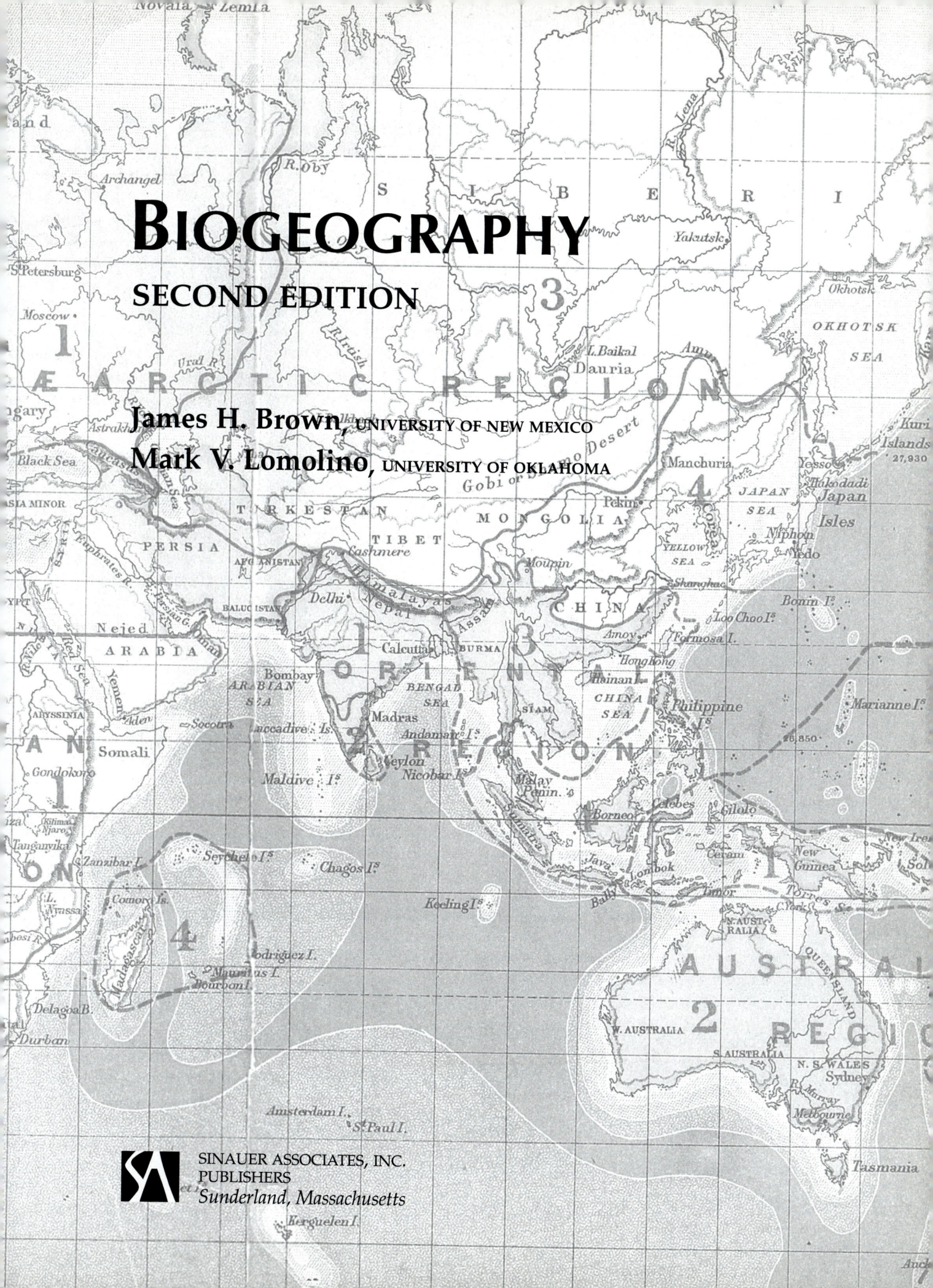

BIOGEOGRAPHY

SECOND EDITION

James H. Brown, UNIVERSITY OF NEW MEXICO

Mark V. Lomolino, UNIVERSITY OF OKLAHOMA

SINAUER ASSOCIATES, INC.
PUBLISHERS
Sunderland, Massachusetts

BIOGEOGRAPHY: SECOND EDITION
Copyright © 1998 by Sinauer Associates, Inc. All rights reserved.
This book may not be reproduced in whole or in part without permission from the publisher.
For information or to order, address:
 Sinauer Associates, Inc., P. O. Box 407
 23 Plumtree Road, Sunderland, MA, 01375 U.S.A.
 FAX: 413-549-1118
 Internet: publish@sinauer.com; http://www.sinauer.com

Library of Congress Cataloging-in-Publication Data
Brown, James H., 1942 Sept. 25–
        Biogeography / James H. Brown, Mark V. Lomolino. — 2nd ed.
              p.   cm.
        Includes bibliographical references (p.   ) and index.
        ISBN 0-87893-073-6 (hardcover)
        1. Biogeography.   I. Lomolino, Mark V., 1953–   .   II. Title.
QH84.B76   1998
578'.09—dc21   98-20356
                                                                                   CIP

Printed in U.S.A.

5 4

To our families—
wives, parents, and children—
for their love, help, and encourgement

CONTENTS

Preface   xi

Unit 1: Introducing the Discipline

Chapter 1 The Science of Biogeography   3
What Is Biogeography?   3
  Definition   3
  Relationships to Other Sciences   5
  Philosophy and Basic Principles   7
The Modern Science   9
  Doing Contemporary Biogeography   9
  Current Status   10

Chapter 2 The History of Biogeography   13
The Age of Exploration   14
Biogeography in the Nineteenth Century   17
  Four British Scientists   19
  Other Contributions in the Nineteenth Century   26
BOX 2.1 BIOGEOGRAPHIC PRINCIPLES ADVOCATED
          BY ALFRED RUSSEL WALLACE   28
The First Half of the Twentieth Century   32
Biogeography since the 1950s   33

Unit 2: The Environmental and Historical Setting

Chapter 3 The Physical Setting   39
Climate   40
  Solar Energy and Temperature Regimes   40
Winds and Rainfall   42
Soils   47
  Primary Succession   47
  Formation of Major Soil Types   48
  Unusual Soil Types Requiring Special Plant
    Adaptations   51
Aquatic Environments   53
  Stratification   53
  Oceanic Circulation   55
  Pressure and Salinity   57
  Tides and the Intertidal Zone   57
Microenvironments   59
  Small-scale Environmental Variation   59
  Colonizing Suitable Microenvironments   59

Chapter 4 Distributions of Single Species   61
The Geographic Range   61
  Methodological Issues: Mapping and Measuring the
    Range   61
  The Distribution of Individuals   64
The Distribution of Populations   65
  Population Growth and Demography   65
  Hutchinson's Multidimensional Niche Concept   67
  The Geographic Range as a Reflection of the Niche   68
  The Relationship between Distribution and
    Abundance   70
Range Boundaries   72
  Physical Limiting Factors   72
  Disturbance   81
  Interactions with Other Organisms   82
  Synthesis   91
Adaptation and Gene Flow   91

Chapter 5 The Distribution of Communities   95
Historical and Biogeographic Perspectives   95
Communities and Ecosystems   96
  Definitions   96
  Community Organization: Energetic Considerations   98
The Distribution of Communities in Space and
    Time   103
  Spatial Patterns   103
  Temporal Patterns   107
Terrestrial Biomes   110
  Tropical Rain Forest   112
  Tropical Deciduous Forest   114
  Thorn Woodland   114
  Tropical Savanna   115
  Desert   116
  Sclerophyllous Woodland   117
  Subtropical Evergreen Forest   118
  Temperate Deciduous Forest   118
  Temperate Rain Forest   119
  Temperate Grassland   120
  Boreal Forest   121
  Tundra   121

Aquatic Communities   122
   Marine Communities   123
   Freshwater Communities   126
A Global Comparison of Biomes and
   Communities   128

Chapter 6   **The Changing Earth   135**
The Geological Time Scale   135
The Theory of Continental Drift   137
   Wegener's Theory   139
   Early Opposition to Continental Drift   141
   Evidence for Continental Drift   142
BOX 6.1   STRATIGRAPHIC, PALEOCLIMATIC, AND
      PALEONTOLOGICAL DISCOVERIES THAT
      CONTRIBUTED TO THE ACCEPTANCE OF THE
      THEORY OF CONTINENTAL DRIFT   145
BOX 6.2   EXPANDING EARTH'S ENVELOPE   154
Earth's Tectonic History   157
   Tectonic History of the Continents   157
   *Tectonic Development of Marine Basins and Island
      Chains   166*
Climatic and Biogeographic Consequences of Plate
Tectonics   171

Chapter 7   **Glaciation and Biogeographic
         Dynamics of the
         Pleistocene   177**
Extent and Causes of Glaciation   177
Effects on Nonglaciated Areas   181
   Temperature   182
   Geographic Shifts in Climatic Zones   184
   Sea Level Changes in the Pleistocene   186
Biogeographic Responses to Glaciation   189
   Biogeographic Responses of Terrestrial Biotas   189
BOX 7.1   BIOGEOGRAPHICAL RESPONSES TO CLIMATIC
      CYCLES OF THE PLEISTOCENE   190
   *Dynamics of Plant Communities in the Southwestern
      United States   198*
   Aquatic Systems: Postglacial and Pluvial Lakes   200
   Biotic Exchange and Glacial Cycles   203
   Evolutionary Responses and Pleistocene Refugia   205
Glacial Cycles and Extinctions   210
   The Overkill Hypothesis   212
   *Alternative Explanations for Pleistocene
      Extinctions   214*

*Unit 3: Historical Patterns and
         Processes*

Chapter 8   **Speciation and Extinction   223**
Taxonomy   224
   Species Concepts   224

   Higher Classifications   228
Macroevolution   229
   Evolution in the Fossil Record   229
   Micro- and Macroevolution   231
Speciation   232
   Mechanisms of Genetic Differentiation   232
   Allopatric Speciation   235
   Sympatric Speciation   239
   Phyletic Speciation   243
Diversification   244
   Ecological Differentiation   244
   Adaptive Radiation   246
Extinction   249
   Ecological Processes   249
   Recent Extinctions   251
   Extinctions in the Fossil Record   253
Species Selection   255
   Processes of Species Selection   255
   Examples of Species Selection   255

Chapter 9   **Dispersal   261**
BOX 9.1   DISPERSAL VERSUS VICARIANCE:
      NO CONTEST   262
What Is Dispersal?   263
   Dispersal as an Ecological Process   264
   Dispersal as a Historical Biogeographic Event   264
Dispersal and Range Expansion   265
   Jump Dispersal   265
   Diffusion   266
   Secular Migration   273
Mechanisms of Movement   273
   Active Dispersal   273
   Passive Dispersal   276
The Nature of Barriers   279
   Physiological Barriers   281
   Ecological and Psychological Barriers   283
Biotic Exchange and Dispersal Routes   285
   Corridors   285
   Filters   285
   Sweepstakes Routes   287
   Other Dispersal Routes   289
   Dispersal Curves within and among Species   289
Establishing a Colony   291
   Habitat Selection   291
   What Constitutes a Propagule?   291
   Survival in a New Habitat   293

Chapter 10   **Endemism, Provincialism, and
         Disjunction   295**
Endemism   297
   Cosmopolitanism   299
   Classifying Endemics   300

Provincialism 302
  Terrestrial Regions and Provinces 302
BOX 10.1 ENDEMIC BIRDS AND PLANTS OF SOUTH
        AMERICA AND AUSTRALIA 306
  Biogeographic Lines 308
  Classifying Islands 312
  Marine Regions and Provinces 314
  Quantifying Similarity among Biotas 317
BOX 10.2 SIMPLE SIMILARITY INDEXES USED BY VARIOUS
        AUTHORS TO ESTIMATE BIOTIC SIMILARITIES 318
Disjunction 320
  Patterns 320
  Processes 322

Chapter 11 The History of Lineages 325
Classifying Biodiversity 326
  Evolutionary Classifications 326
  Phylogenetic Systematics 328
BOX 11.1 THE BASIS OF HENNIG'S PARADIGM: A HYPO-
        THETICAL EXAMPLE OF CLADOGENESIS 330
BOX 11.2 HOW TO CONSTRUCT A CLADOGRAM:
        A HYPOTHETICAL EXAMPLE 331
  Molecular Systematics 333
  Limitations of Phylogenetic Classifications 334
The Fossil Record 335
  Limitations of the Fossil Record 336
  Biogeographic Implications of Fossils 337
Toward a Historical Synthesis? 343

**Chapter 12 Reconstructing Biogeographic
            Histories 345**
Early Efforts: Determining Centers of Origin 346
  Concepts and Criteria 346
  An Example: Sea Snakes 346
  Critical Issues 349
Panbiogeography and Vicariance Biogeography 350
  Croizat's Panbiogeography 350
  Hennig's Progression Rule 352
  Vicariance Biogeography 352
Modern Historical Biogeography 357
  Approaches 357
  The Hawaiian Example 360
  Assessment and Prospects 364

**Unit 4: Contemporary Patterns and
        Processes**

**Chapter 13 Island Biogeography: Patterns
            in Species Richness 369**
Historical Background 369
BOX 13.1 INDEPENDENT DISCOVERY OF THE EQUILIBRIUM
        THEORY OF ISLAND BIOGEOGRAPHY 372

Island Patterns 372
  The Species-Area Relationship 372
BOX 13.2 INTERPRETATIONS AND COMPARISONS OF CON-
        STANTS IN THE SPECIES-AREA RELATIONSHIP:
        AN ADDITIONAL CAUTION 376
  The Species-Isolation Relationship 376
  Species Turnover 377
The Equilibrium Theory of Island Biogeography 379
  Strengths and Weaknesses of the Theory 382
  Tests of the Model 384
  Additional Patterns in Insular Species Richness 390
  Nonequilibrium Biotas 394
Krakatau Revisited 401

**Chapter 14 Island Biogeography: Patterns
            in the Assembly and Evolution
            of Insular Communities 407**
Assembly of Insular Communities 408
  The Selective Nature of Immigration 408
  Establishing Insular Populations 410
  The Selective Nature of Extinction 411
  *Patterns Reflecting Differential Immigration and
      Extinction* 414
  Patterns Reflecting Interspecific Interactions 419
Evolutionary Trends on Islands 429
  *Flightlessness and Reduced Dispersal Ability on
      Islands* 429
  Evolution of Body Size on Islands 434
BOX 14.1 TIME DWARFING ON THE "ISLAND CONTINENT"
        OF AUSTRALIA 441
The Taxon Cycle 444

**Chapter 15 Species Diversity in
            Continental and Marine
            Habitats 449**
Measurement and Terminology 449
  Species Richness and Diversity Indexes 449
  Scales of Diversity: Alpha, Beta, and Gamma 450
The Latitudinal Gradient 450
  Patterns 451
  Processes 458
Other Diversity Patterns 461
  Peninsulas 461
  Elevation 463
  Aridity 464
  Aquatic Environments 465
  Associated Patterns 468
Causes of the Patterns 471
  Nonequilibrial Mechanisms 472
  Equilibrial Mechanisms 475
  Toward a Synthetic Explanation? 484

Unit 5: Biogeography and Conservation

Chapter 16 Continental Patterns and Processes 487

Single-Species Patterns 488
Ecogeographic Rules 488
Geographic Variation in Life History and Population-Level Characteristics 492
Multispecies Assemblages 494
Areography: Sizes, Shapes, and Overlaps of Ranges 494
Macroecology: Assembly of Continental Biotas 500
Relationships between Local and Regional Diversity 506
Biotic Interchange 507
The Great American Interchange 508
Lessepsian Exchange: The Suez Canal 513
Maintenance of Distinct Biotas 514
Barriers between Biogeographic Regions 514
Resistance to Invasion 514
Avian Migration and Provincialism 516
Divergence and Convergence of Isolated Biotas 518
Divergence 518
Convergence 520
Overview 527

Chapter 17 The Status of Biodiversity 533

The Biodiversity Crisis and the Linnaean Shortfall 534
Geographic Variations in Biodiversity 537
Terrestrial Hot Spots 538
Hot Spots in the Marine Realm 542
The Geography of Extinctions 544
The Prehistoric Record of Extinctions 544
The Historical Record of Extinctions 545
Species Introductions: The Ecology and Geography of Invasions 547
BOX 17.1 SPREAD AND IMPACT OF ZEBRA MUSSELS IN NORTH AMERICAN FRESHWATERS 550
Current Patterns of Endangerment 556
Habitat Loss and Fragmentation 558
Applications of Biogeographic Theory 564
Designing Nature Reserves 564
Predicting the Effects of Global Climatic Change 567
Biodiversity Surveys 568

Chapter 18 Applied Biogeography: Single Species 573

The Biogeography of Humanity 573
Human Origins and Colonization of the Old World 574
Conquering the Cold: Expansion to the New World 577
Conquering the Oceans: The Island Biogeography of Humanity 579
Lessons from the Biogeography of Humanity 586
Applied Biogeography: Focal Species Patterns and Approaches for Conserving Biodiversity 586
Patterns of Range Collapse 588
Patterns of Distribution among Insular or Fragmented Habitats 598
Species Responses to Global Climatic Change 601

Chapter 19 Biogeography for the Twenty-first Century 613

Technological Advances 613
New Data 613
Analytical Methods 614
Contributions of Technology 615
Conceptual Advances 616
New Theory 617
Synthetic and Interdisciplinary Studies 618
Biogeography: Past, Present, and Future 620
Applications 621
Human Ecology 621
Management and Conservation 622

Glossary 625
Bibliography 637
Index 675

PREFACE

The first edition of *Biogeography* by J. H. Brown and A. C. Gibson was published in 1983 and went through several printings. It was widely used as a textbook for courses in biogeography, zoogeography, and phytogeography, and as a general synthesis and reference work for these fields. In recent years, however, *Biogeography* became seriously out of date. Biogeography has grown and changed rapidly in the last 20 years. Several factors have contributed. Biogeography lies at the interfaces of several different scientific disciplines: ecology, evolution, systematics, paleobiology, geography, and the physical earth sciences. All of these fields have themselves seen important advances in the last few decades. They have contributed to the factual and conceptual underpinnings of biogeography and have stimulated many areas of interdisciplinary research. Advances in computing and other technologies, such as remote sensing, geographic information systems, and geostatistics, have provided new ways to obtain, analyze, and interpret large quantities of spatially referenced data at geographic scales. Practical concerns about the detrimental impacts of the growing human population on the earth and its environment have lead to an explosion of interest and research on global change and conservation biology. Biogeography is central to these issues because it focuses on geographic distributions of organisms and spatial patterns of biological diversity.

We hope that the first edition of *Biogeography* contributed in some small measure to the increased interest in the field. This second edition aims to emulate the first edition in being a broad, integrative, synthetic, and comprehensive text and reference book. We aim to provide balanced coverage of the whole discipline of biogeography, to integrate ecological and historical approaches, to emphasize general concepts and illustrate them with empirical examples, and to draw those examples from a wide variety of organisms, environments, and geographic areas. Optimal breadth and integration are difficult to achieve, however, and some of our personal biases will be apparent. We can write with greater enthusiasm and authority about the conceptual questions that we find most interesting, and about the kinds of organisms and geographic areas that we know best.

This second edition has been completely rewritten. The book has been totally reorganized, as indicated by changes in the number, order, titles, and contents of chapters; two chapters on applications of biogeography to conservation have been added. All but a few paragraphs and sentences have been changed. The conceptual and factual content has been updated by rewriting the text and substituting and adding many new figures, tables, and references. The result, we hope, is a book that conveys both the present exciting state and the enormous future prospects of the discipline of biogeography.

In addition to those who contributed to the first edition, many additional people—far too many to recognize individually—have helped to improve and produce this book. We are especially indebted to Bruce Patterson and Evan Weiher, who read the entire text and made countless helpful suggestions. Larry Heaney, Dawn Kaufman, and Paul Martin commented on selected chapters. Many readers of the first edition, from undergraduate students to eminent scientists, made suggestions that have influenced this second edition. Alix Ohlin of the University of New Mexico, and the editors and staff at Sinauer Associates, especially Andy and Nan Sinauer and Norma Roche, made herculean efforts to produce a high-quality book on a tight schedule. We thank our families, students, and colleagues for their help and patience while we were writing. Finally, we are indebted to all biogeographers and other scientists, from the earliest workers to our contemporaries, for their contributions to the discipline. It is their individual, collaborative, and cumulative contributions that make biogeography so exciting to teach and to study.

James H. Brown

Mark V. Lomolino

April, 1998

UNIT 1

Introducing the Discipline

CHAPTER 1

The Science of Biogeography

Living things are incredibly diverse. There are probably somewhere between 5 million and 50 million kinds of animals, plants, and microbes living on earth today. Of these, fewer than 2 million have been formally recognized as species and described in the scientific literature. The remainder are represented by specimens in museums waiting to be described, or by individuals in nature waiting to be discovered. Additional untold millions, probably billions, of species lived at some time in the past but are now extinct; only a small fraction of them have been preserved as fossils.

Nearly everywhere on earth, from the frozen wastes of Antarctica to the warm, humid rainforests of the tropics, from the cold, dark abyssal depths of the oceans to the near-boiling waters of hot springs—even in rocks several kilometers beneath the earth's surface—at least some kinds of organisms can be found. But no single species is able to live in all these places. In fact, most species are restricted to a small geographic area and a narrow range of environmental conditions. The spatial patterns of global biodiversity are a consequence of the ways in which the limited geographic ranges of the millions of species overlap and replace each other over the earth's vast surface.

What Is Biogeography?

Definition

Biogeography is the science that attempts to document and understand spatial patterns of biodiversity. It is the study of distributions of organisms, both past and present, and of related patterns of variation over the earth in the numbers and kinds of living things.

A science can be characterized by the kinds of questions its practitioners ask. Some of the questions posed by biogeographers include the following:

1. Why is a species or higher taxonomic group (genus, family, order, and so on) confined to its present range?
2. What enables a species to live where it does, and what prevents it from colonizing other areas?

3. What roles do climate, topography, and interactions with other organisms play in limiting the distribution of a species?
4. How do different kinds of organisms replace each other as we go up a mountain, or move from a rocky shore to a sandy beach nearby?
5. How did a species come to be confined to its present range?
6. What are the species' closest relatives, and where are they found? Where did its ancestors live?
7. How have historical events, such as continental drift, Pleistocene glaciation, and recent climatic change, shaped the species' distribution?
8. Why are the animals and plants of large, isolated regions, such as Australia, New Caledonia, and Madagascar, so distinctive?
9. Why are some groups of closely related species confined to the same region, and others found on opposite sides of the world?
10. Why are there so many more species in the tropics than at temperate or arctic latitudes? How are isolated oceanic islands colonized, and why are there nearly always fewer species on islands than in the same kinds of habitats on continents?

The list of possible questions is nearly endless, but in essence we are asking: How are organisms distributed, over the surface of the earth and over the history of the earth? This is the fundamental question of biogeography. It has always intrigued scientists and laypersons who were curious about nature. Only within the last few decades, however, have scientists begun to call themselves biogeographers and to focus their research primarily on the distributions of living things. Not surprisingly, biogeographers have not yet answered all the questions listed above. They have, however, learned a great deal about where different kinds of organisms are found and why they occur where they do. Much of this progress has been made in just the last few decades, stimulated in large part by exciting new developments in the related fields of ecology, genetics, systematics, paleontology, and geology, as well as by technological developments.

Biogeography is a broad field. To be a complete biogeographer, one must acquire and synthesize a tremendous amount of information. But not all aspects of the discipline are equally interesting to everyone, including biogeographers. Given different biases in their training, their biogeography courses and writings tend to be uneven in coverage. A common specialization is taxonomic—for example, **phytogeographers** study plants and **zoogeographers** study animals, and within these categories one finds specialists in groups at all taxonomic levels. Although viruses and bacteria play crucial roles in ecological communities and in human welfare, microbial biogeography is poorly known and rarely discussed. Some biogeographers specialize in **historical biogeography** and attempt to reconstruct the origin, dispersal, and extinction of taxa and biotas. This approach contrasts with **ecological biogeography**, which attempts to account for present distributions in terms of interactions between organisms and their physical and biotic environments. **Paleoecology** bridges the gap between these two fields, investigating the relationships between organisms and past environments; it uses data on both the biotic composition of communities (abundance, distribution, and diversity of species) and abiotic conditions (climate, soils, water quality, etc.) derived from fossilized remains preserved in ancient sediments, ice cores, tree rings, and other places. Different biogeographers have emphasized different methods for understanding distributions: some approaches are primarily descriptive, designed to document the ranges of particular living or extinct organisms, whereas others are mainly conceptual, devoted to building and testing theoretical models to account for distribution patterns. All of these approaches to the subject are valid and valuable, and discounting or overemphasizing any division or specialization is counterproductive and unnecessary. Whereas no researcher or student can become an expert in all areas of biogeography, exposure to a broad spectrum of organisms, methods, and concepts leads to a deeper understanding of the science. As we hope to show, the various subdisciplines contribute to and complement each other, unifying the science.

Relationships to Other Sciences

Biogeography is a synthetic discipline, relying heavily on theory and data from ecology, population biology, systematics, evolutionary biology, and the earth sciences. Consequently, we do not want to draw sharp lines between biogeography and its related subjects, as some authors have attempted to do. For example, various authors have recommended that paleontology (the study of fossils and extinct organisms) and ecology be divorced from biogeography; this would make biogeography largely a descriptive, mapmaking endeavor. It would deprive biogeography of its central role as a synthetic discipline that not only has its own theoretical and empirical approaches, but also readily incorporates conceptual and factual advances from many other sciences.

Biogeography is a branch of biology, and not surprisingly, a good knowledge of biology is an important starting point. This is why our treatment devotes considerable space to reviewing and developing the ecological and evolutionary concepts that are used throughout the book (Chapters 4, 5, 8, 9, and 10). In addition, one must be acquainted with the major groups of plants and animals and know something about their physiology, anatomy, development, and evolutionary history. These topics are not the subjects of separate chapters, but are integrated throughout the text, usually by the device of using different kinds of organisms with distinctive biological characteristics to illustrate biogeographic patterns, processes, and concepts. For example, it is only by knowing several critical features of their biology that we can understand why amphibians and freshwater fishes have only rarely crossed even modest stretches of ocean to colonize islands, whereas birds and bats have done so much more frequently.

Naturally it is important to know some geography. The locations of continents, mountain ranges, deserts, lakes, major islands and archipelagoes (groups of islands), and seas, during the past as well as the present, are indispensable information for biogeographers, as are past and present climatic regimes, ocean currents, and tides. To remind the reader of the major geographic features of the earth, we have illustrated many of them on the colored maps in the front endpapers of this book. The major patterns of abiotic variation, in climate, soils, and oceanographic and limnological conditions, are described and explained in Chapter 3.

We must also remember that the earth is dynamic, and that we need to understand the causes of the changes that are continually occurring on all temporal and spatial scales. We present a brief history of the earth and explain many of these changes in Chapters 6 and 7. It may be hard to envision, but the landscapes and habitats that are so familiar to us today were completely different only 15,000 years ago. (This may seem like a long time, but *Homo sapiens* and most other contemporary species are millions of years old and show little evidence of change within the last 15,000 years.) At that time, three of the present North American Great Lakes were covered with vast sheets of glacial ice,

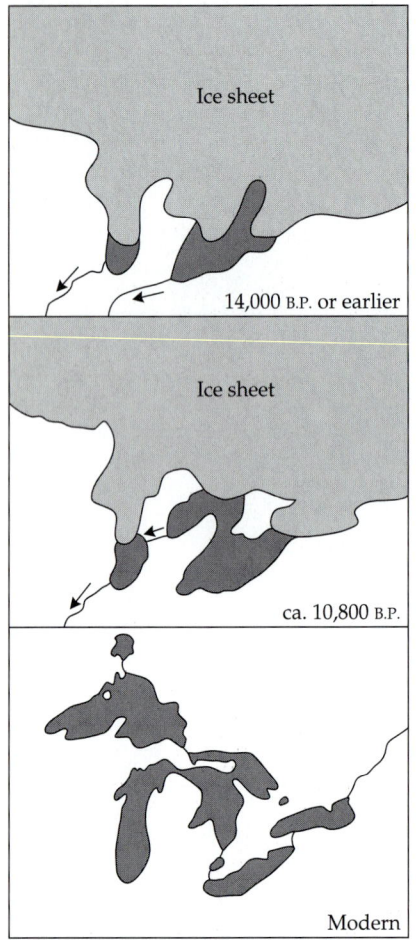

Figure 1.1   Changes in the geography of the Great Lakes region of North America from about 14,000 years ago to present. During the full glacial period of the latest Pleistocene, most of the area was covered by a great continental ice sheet, which melted about 10,000 years ago, leaving the lakes in their present configuration. Imagine the effects of these changes on the distributions of plants and animals, both freshwater and terrestrial, in this region. (After Hutchinson 1957.)

and the other two lakes, antecedents of contemporary Lake Erie and Lake Michigan, had markedly different shapes and were interconnected (Figure 1.1). In the southwestern part of North America, woodlands and forests grew where there are now deserts; Death Valley, now the driest place on the continent, was filled with a gigantic freshwater lake. Imagine how different the distributions of temperate North American plants and animals must have been.

On longer time scales the changes have been even greater. Up until the 1970s most scientists believed that the sizes, shapes, and locations of the continents and oceans had remained fixed over time. Then advances in plate tectonics demonstrated that throughout the earth's history, large blocks of crust have moved over its molten mantle, carrying land and oceans to new locations, breaking them apart and reassembling them in new configurations. With the realization that the drifting continents and oceans took their animals and plants with them, modern biogeographers have had to revise their interpretations of the distributional histories of many groups of organisms. Reflect for a moment on the history of North America. For most of its history, North America has been connected to, or in close proximity to, western Europe, and the final separation of these two landmasses occurred only 60 million years ago. At about the same time, the part of northwestern North America that we now know as Alaska first established a land connection with Siberia in northeastern Asia; the resulting Bering landbridge served intermittently as an important highway for biotic exchange until 10,000 years ago. Also, as North America drifted away from Europe, it moved closer to South America, finally making contact at the Isthmus of Panama about 3.5 million years ago. While this new land connection provided an important corridor of biotic exchange between the formerly isolated continents, it simultaneously isolated the tropical Atlantic and Pacific oceans from each other, initiating divergence of many marine forms on opposite sides of this new barrier to dispersal.

Knowledge of such historical events is essential for understanding the distributions of many organisms, such as the many kinds of mammals that North America shares with both Eurasia and South America. Similarly, several groups of living fishes, insects, birds, earthworms, and certain extinct plants and reptiles are shared among South America, Africa, and Australia. These distributions make sense when we realize that at least until 135 million years ago, while these groups were evolving and expanding their ranges, all three continents were joined as part of a single giant landmass called Gondwanaland. For these reasons, it is important for biogeographers to study earth history, and to keep abreast of new developments in plate tectonics (see Chapter 6).

The occurrence of structurally and functionally similar organisms in geographically isolated parts of the earth is not simply a consequence of history, however. Contemporary climatic patterns also play a major role. For example, distinctive vegetation occurs throughout the world in isolated regions where **Mediterranean climates** prevail. Rainfall is low in these regions, and most of it occurs during the mild winter months; summers are dry and hot. Places sharing this climatic regime are found around the Mediterranean Sea, in coastal central Chile, in southwestern Australia, in the Cape Region of South Africa, and in coastal Baja California and southern California in North America (Figure 1.2). The distinctive semiarid plant communities of these regions, variously called chaparral, matorral, macchia, maquis, or fynbos by local people, or sclerophyllous scrub by scientists, look superficially similar, and the plants share adaptations to the distinctive climate and to periodic wildfires. However, most of the plants in each region belong to different genera and even families, showing that they have evolved from different ancestors and

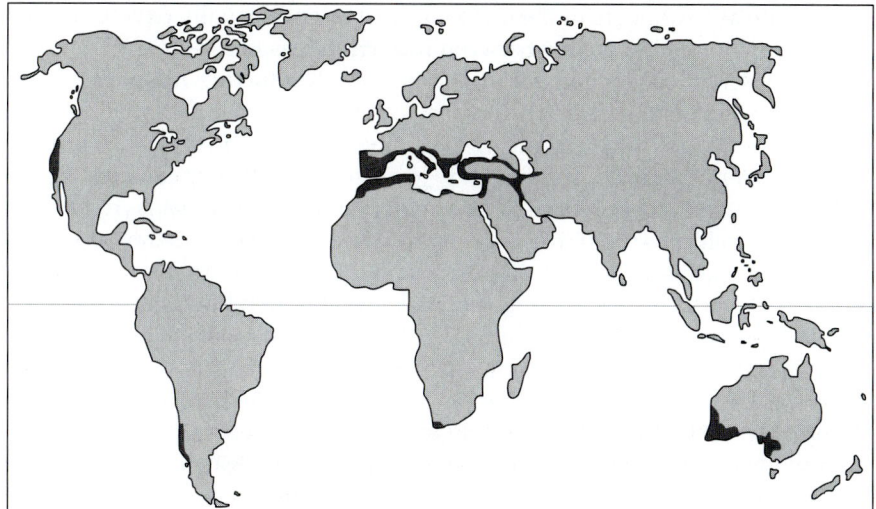

Figure 1.2 Worldwide distribution of regions with Mediterranean climates and vegetation. Note that these regions tend to occur in predictable locations: on the western coasts of continents between 30° and 40° latitude. Although these regions are widely separated, they support superficially similar plant life, because unrelated lineages have independently evolved similar morphological and physiological adaptations to this distinctive climatic and fire regime. (After Thrower and Bradbury 1977.)

that their similar forms and functions are the result of convergent evolution to adapt to similar environments. Knowledge of the world's climatic patterns is therefore essential for understanding the distributions and adaptations of different kinds of plants and animals.

Philosophy and Basic Principles

Most people have a vague, misleading impression of what science is, how scientists work, and how major scientific advances come about. Scientists try to understand the natural world by explaining its enormous diversity and complexity in terms of general patterns and basic laws. Philosophers and historians of science, viewing its progress with 20/20 hindsight, often suggest that it is possible to give a recipe for the most effective way to conduct an investigation. Unfortunately, as most practicing scientists know, scientific inquiry is much more like working on a puzzle or being lost in the woods than like baking cookies or following a road map. There are numerous mistakes and frustrations. Luck, timing, and trial and error play crucial roles in even the most important scientific advances. Some important discoveries, such as Alfred Wegener's evidence for continental drift, are long ignored or even totally rejected by other scientists. While the progress of science owes much to such admirable human traits as intelligence, creativity, perseverance, and precision, it is also retarded by equally human but less admirable characteristics such as prejudice, jealousy, short-sightedness, and stupidity. This is not meant to imply that science has not made great advances in our understanding of the natural world. It has. But like most human activities, the progress of science has followed a much more complex and circuitous path than is usually portrayed in textbooks. As we shall see, beginning in the next chapter, biogeography is no exception.

In essence, scientists proceed by investigating the relationships between pattern and process. **Pattern** can be defined as nonrandom, repetitive organization. The occurrence of pattern in the natural world implies causation by some general **process** or processes. Science usually advances by the discovery of patterns, then by the development of mechanistic explanations for them, and finally by rigorous testing of those theories until the ones that are necessary and sufficient to account for the patterns become widely accepted.

Traditional treatments of the philosophy of science usually devote considerable space to distinguishing between **inductive reasoning**—reasoning from specific observations to general principles—and **deductive reasoning**—reasoning from general constructs to specific cases. Several influential modern philosophers, especially Popper (1968), have strongly advocated so-called **hypothetico-deductive reasoning**. Any good scientific theory contains logical assumptions and consequences, and if any of these can be proven wrong, then the theory itself must be flawed. The hypothetico-deductive method provides a powerful means of testing a theory by setting up alternative, falsifiable hypotheses. First, an author puts forth a new, tentative idea, stated in clear, simple language, that can be tested and potentially falsified by means of experiments or observations. After the statement has withstood the severest empirical tests, it can be considered to be supported or corroborated, but by hypothetico-deductive logic, a theory can never be proven true, only falsified.

New general theories or paradigms are the ultimate source of most major scientific advances, and most of them have been arrived at by inductive methods. The theory of evolution by natural selection, the double helical structure of DNA, and the equilibrium theory of insular biogeography all were derived largely by assembling factual data, recognizing a pattern, and then proposing a general mechanistic explanation. Although it might be safe to say that theories usually arise by inductive methods and are tested by deductive ones, often the actual conduct of scientific research is much more complex. Empirical observations and conceptual models are played back and forth against each other, theories are devised and modified, and understanding of the natural world is acquired slowly and irregularly. This is particularly true of biogeography, in which new empirical observations and theoretical paradigms have frequently overturned existing dogma. Thus, for example, although Alfred Wegener's ideas of continental drift were rejected for many decades, by the 1970s the data and theory of tectonic plate movements became overwhelmingly convincing. Biogeographers had to revise historical explanations for distributions that had been based on a static distribution of continents and oceans.

Unlike much of contemporary biology, biogeography usually is not an experimental science. Questions about molecules, cells, and individual organisms typically are most precisely and conveniently answered by artificially manipulating the system. In such experiments, the investigator searches for patterns or tests specific hypotheses by changing the state of the system and comparing the behavior of the altered system with that of an unmanipulated control. It is impractical and often impossible to use these techniques to address many of the important questions in biogeography and the related disciplines of ecology and evolutionary biology. Historical evolution and historical biogeography are, as their names imply, history; they produce no exact predictions about the future.

A few biogeographers have used experimental techniques to manipulate small systems, such as tiny islands, sometimes with spectacular success [e.g., Simberloff and Wilson (1969)]. However, most important questions involve such large spatial and temporal scales that experimentation would be impractical or unethical. This methodological constraint does not diminish the rigor and value of biogeography, but it does pose major challenges. Other sciences, such as astronomy and geology, face the same problems. As Robert MacArthur once observed, Copernicus, Galileo, Kepler, and Newton never moved a planet, but that did not prevent them from advancing our understanding of the motion of celestial bodies. Wallace, Darwin, and Hooker used the patterns of animal and plant distributions that they observed in their world travels to develop important new ideas about evolution and biogeography. Islands have had a great influence on these and numerous subsequent biogeographers, ecologists, and evolutionists because they represent natural experiments—replicated natural systems in which many factors are held relatively constant while others vary from island to island. Despite the difficulty of performing artificial experiments, it is possible to develop and rigorously evaluate biogeographic theories by the same logical procedures used by other scientists: searching for patterns, formulating theories, and then testing assumptions and predictions independently with new observations.

In dealing with historical aspects of their science, most biogeographers make one critical assumption that is virtually impossible to test: they accept the principle of **uniformitarianism** or **actualism**. Uniformitarianism is the assumption that the basic physical and biological processes now operating on the earth have remained unchanged throughout time because they are manifestations of universal scientific laws. The principle of uniformitarianism is usually attributed to the British geologists Hutton (1795) and Lyell (1830), who realized that the earth was much older than had been previously supposed, and that its surface was constantly changing. In this same spirit, one of Darwin's great insights was the recognition that changes in domesticated plants and animals over historical time through selective breeding resulted from the same process as natural changes in organisms over evolutionary time.

As noted by Simpson (1970), acceptance of uniformitarianism has never been universal, in part because some authors have attached additional meanings to the concept. Some have assumed that the term implies that the average intensities of processes have remained approximately constant over time, and that both geological and biological changes are always gradual. Neither of these amendments is acceptable. We know that some historical events have been of great magnitude, but so infrequent that they have never been observed in recorded human history. For example, contrary to science fiction movies, no humans were living 65 million years ago when dinosaurs roamed the earth. Consequently, no one observed the collision of the earth with an asteroid that presumably caused the mass extinction that eliminated the dinosaurs and several other groups of terrestrial and marine organisms (see Chapter 8). Yet there is abundant evidence, preserved in ancient rocks, for this event. Scientists expect that the intensity of forces will vary from time to time and place to place; only the nature of the laws that control these processes is timeless and constant.

To avoid these unfortunate connotations associated with uniformitarianism, we will follow Simpson in adopting the term *actualism*, which is conceptually similar to methodological uniformitarianism (Gould 1965). Historical biogeographers in particular use this principle to account for present and past distributions, assuming that the processes of speciation, dispersal, and extinction operated in the past by the same mechanisms that they do today. This premise has become an accepted tool for interpreting the past and predicting the future. The most serious problem in using this principle is discovering which timeless processes have operated to produce different patterns.

The Modern Science

Doing Contemporary Biogeography

Biogeography differs from most of the biological disciplines, and from many other sciences, in several important respects. We have mentioned one of them above: biogeography is, for the most part, a comparative observational science rather than an experimental one because it usually deals with scales of space and time at which experimental manipulation is impossible. Thus most of the inferences about biogeographic processes must come from the study of patterns: from comparisons of the geographic ranges, genetics, and other characteristics of different kinds of organisms or the same kinds of organisms living in different regions, and from observations of differences in the diversity of species and the composition of communities over the earth's surface. But while it is difficult for biogeographers to manipulate these systems in controlled experiments, it is possible for them to study the effects of natural and human-caused perturbations. Thus, for example, we have learned a great deal about the colonization of oceanic islands by monitoring the buildup of biotas after volcanic eruptions, as on Krakatau in the East Indies and Surtsey in the North Atlantic. There is also much to be learned from the changes being caused by modern humans, who have altered habitats, connected previously isolated areas and fragmented previously continuous ones, and introduced exotic species and caused the extinctions of native ones.

Another way that biogeography differs from most other sciences is that it is usually dependent on data collected by many individuals working over large areas for long periods of time. Much of biogeography involves studying the geographic ranges of species or the species composition of regional biotas. Doing either of these accurately usually requires reliance on the previous fieldwork of many persons who have collected and identified specimens, deposited them in museum collections, and published the information in the scientific literature. Although being a biogeographer carries the sometimes deserved connotation of adventurous collecting expeditions to exotic places, most practicing biogeographers obtain more of their data from museums and libraries than from their own fieldwork. In this respect biogeography is unlike many other disciplines, in which individual scientists or groups of collaborating investigators collect most or all of their primary data themselves.

Finally, as mentioned above, biogeography is typically a synthetic science. Biogeographers work at the interfaces of several traditional disciplines: ecology, systematics, evolutionary biology, geography, paleontology, and the "earth sciences" of geology, climatology, limnology, and oceanography. Much of their best work comes from putting together data or theories from two or more of these disciplines. This might seem to require such exceptionally broad training as to be intimidating to the beginner. However, while some breadth of knowledge is desirable, the interdisciplinary nature of biogeography also means that individuals with diverse backgrounds and interests can make important contributions—sometimes on their own, and more often through collaboration with specialists from other areas. In fact, biogeography is accessible to almost anyone with curiosity and motivation. Large research grants, modern laboratory facilities, and even sophisticated statistical techniques, while often desirable, are not required to do state-of-the-art biogeographic research. All that is required is a good idea and access to a library, museum collection, or other source of data on distributions of organisms. This means that even beginning students can do original research. The authors of this book usually require that each student in our biogeography courses do an original research paper. Typically this involves coming up with an interesting question, finding a source of appropriate data in the literature or a museum, and doing the analyses to try to answer the question. Some of the work discussed in this book began as student research projects in our courses.

Current Status

Biogeography has a long and distinguished history. Many of the greatest scientists of their eras were biogeographers, including the great evolutionary biologists of the nineteenth century, Charles Darwin, Alfred Russel Wallace, and Joseph Dalton Hooker, as well as some of the most eminent scientists of the twentieth century, George Gaylord Simpson, Ernest Mayr, Robert MacArthur, and Edward O. Wilson. While these giants did not usually refer to themselves as biogeographers, their comparative studies of distributions and diversity provided much of the raw material for their theoretical and empirical contributions, which have laid the foundations for much of contemporary evolutionary biology and ecology. In the next chapter, we place the contributions of these and other influential scientists in the context of the history of biogeographic research.

It is only recently, however—within the last few decades—that biogeography has emerged as a widely respected science with its own identity. Although there were a few earlier influential works, especially in plant geography (e.g., Wulff 1943; Cain 1944), it was really not until the 1950s and 1960s that major books and papers with the word "geography" in their titles began to appear regularly (e.g., Hesse et al. 1951; Croizat 1952, 1958; Ekman 1953; Simpson 1956, 1965; Dansereau 1957; Darlington 1957, 1965; MacArthur and Wilson 1963, 1967; Nelson 1969; Neill 1969; Udvardy 1969). It was about a decade later, in 1973, that the first issue of the *Journal of Biogeography*, the first scientific periodical devoted exclusively to the discipline, was published. Since then, however, the journal has grown rapidly in size, circulation, and reputation. So great was its success that in 1991 its editors created a second journal, *Global Ecology and Biogeography Letters*, for rapid publication of short articles of topical interest. One quantitative measure of the coalescing interest in the field is provided by a computerized search of *Biosis* for occurrences of the word "biogeograph…" in titles, abstracts, or keywords at 10-year intervals: 33 in 1975, 358 in 1985, and 1238 in 1995. Classes and textbooks in biogeography were virtually nonexistent before the 1980s, but are now part of the curriculum at most major universities and many smaller institutions.

The emergence of biogeography as a vigorous modern science can be attributed to several coincident and interacting developments. One is the transformation of the discipline—and of its image in the eyes of other scientists—from a largely descriptive science closely allied to traditional taxonomy to a conceptually oriented discipline concerned with building and testing biogeographic theory. The introduction of new and mathematical theory into ecology, evolution, and systematics has spilled over to trigger major conceptual progress in both ecological and historical biogeography (see Chapters 2, 11, 12, and 13). Contemporary advances in the earth sciences, especially the theory of plate tectonics and a wealth of data from the fossil record, have also been influential.

A second stimulus to the progress of modern biogeography has been the development and application of new technology. Computers have allowed the compilation and manipulation of enormous quantities of data on truly geographic scales: distributional records and other information on organisms, as well as data on climate, landforms, geology, soils, limnological and oceanographic conditions, vegetation, land uses, and other human activities. Advances in computer hardware and software have made possible the development and application of powerful new techniques, including simulation modeling, geographic information systems (GIS), and a whole battery of statistical methods such as multivariate analyses and geostatistics. Satellites, submersible vessels, and ground-based data collection systems, usually automated and computerized, have provided a wealth of new information on the environment of the earth. Other technologies, including radioisotope, stable isotope, and molecular biological techniques, have permitted increasingly accurate reconstructions of the history of both the earth itself and its organisms. Biogeography, being one of the most synthetic sciences, has readily adopted new technologies from many other disciplines and has made rapid use of the new kinds of information that they can provide.

A final stimulus to the emergence of a vigorous modern science of biogeography has come from the need to understand and manage the impact of human beings on the earth. A major wave of environmental concern and activism developed in the late 1960s and early 1970s. Initially, much of the focus was on local responses to fairly small-scale problems: seeking alternatives to persistent pesticides; alleviating local air, water, and soil pollution; setting aside reserves of relatively unspoiled habitat; and saving local populations of endangered species. By the 1980s, however, it was becoming increasingly apparent that modern humans have been transforming the earth on regional and global, as well as local, scales. We now know that every place on earth—from the polar ice caps to the abyssal ocean depths to the upper atmosphere—has been substantially changed by the activities of our ever-growing population and our increasingly industrial and technological society. These changes include increases in atmospheric carbon dioxide and other greenhouse gasses, depletion of stratospheric ozone, extinctions of species and losses of other components of biodiversity, and conversion of natural landscapes and ecosystems to agricultural fields and human habitations. With the realization of the global scale of human influences has come a need for biogeographic expertise to measure and predict the effects of these changes on organisms and ecosystems, and to find ways to manage, mitigate, or minimize their most damaging effects. One fortunate result has been increasing respect and funding for biogeographic research.

In less than two decades, biogeography has gone from being a rather esoteric and unappreciated science, whose practitioners usually called themselves ecologists, systematists, or paleontologists rather than biogeographers, to a vigorous and respected science whose practitioners are using the latest conceptual advances and technological tools to say important things about the present and future state of the earth and its living inhabitants.

CHAPTER 2

The History of Biogeography

Biogeography has had a long history, one that is inextricably woven into the development of evolutionary biology and ecology. The problems of distribution and variation on geographic scales were matters of primary interest to evolutionary biologists, including the distinguished "fathers" of the field such as Lamarck, Darwin, and Wallace. The field of ecology, a relatively young offspring of this lineage, grew out of attempts to explain biogeographic patterns in terms of the influence of environmental conditions and interactions among species. But the origin of these three fields is ancient, dating back well before the Darwinian revolution. Indeed, Aristotle was contemplating many of the questions we ponder today:

> But if rivers come into being and perish and if the same parts of the earth are not always moist, the sea also must necessarily change correspondingly. And if in places the sea recedes while in others it encroaches, then evidently the same parts of the earth as a whole are not always sea, nor always mainland, but in process of time all change. (*Meteorologica*, ca. 355 B.C.)

Aristotle offered this prophetic view of a dynamic earth in order to explain variation in the natural world. He was one of the earliest of a long and prestigious line of scientists to ask the same questions: Where did life come from, and how did it diversify and spread across the globe?

Despite Aristotle's great insights, answering these questions would require a much more thorough understanding of the geographic and biological character of the earth. Thus the development of biogeography, evolution, and ecology was tied to the age of exploration. As we shall see below, the early European explorers and naturalists did far more than just label and catalogue their specimens. They immediately, perhaps irresistibly, took to the task of comparing biotas across regions and developing explanations for the similarities and differences they observed. The comparative method served these early naturalists well, and by the eighteenth century the study of biogeography began to crystallize around fundamental patterns of distribution and geographic variation. We here trace the development of biogeography from the age of exploration to its current status as a mature and respected science.

Many, if not all, of the themes central to modern biogeography (see Table 2.1) have their origins in the pre-Darwinian period. This is not to say that biogeography has not advanced tremendously in the past few decades—only that modern biogeographers owe a great debt to those before them who shared the same fascination with, and asked the same types of questions about, the geography of nature. As Newton once replied when asked how it was that he had developed such unparalleled insights in physics, "If it appears that I have been able to see more than those who came before me, perhaps it is because I stood on the shoulders of giants." Like physics, biogeography has a history of "giants," visionary scientists each building on the collective knowledge of those who came before them.

The Age of Exploration

It is hard for us to appreciate that roughly 250 years ago, biologists had described and classified only 1% of all the plant and animal species we know today. Biogeography was essentially founded and rapidly accelerated by world exploration and the accompanying discovery of new kinds of organisms. Biologists and naturalists of the eighteenth century were largely driven by a calling to serve God. The prevailing belief was that the mysteries of creation would be revealed as these naturalist/explorers developed more complete catalogues of the diversity of life. Up until the mid-eighteenth century, the prevailing world view was one of stasis—the earth, its climate, and its species were immutable. However, as the early biogeographers (then called naturalists or simply geologists) returned with their burgeoning wealth of specimens and accounts, two things became clear. First, biologists needed to develop a standardized and systematic scheme to classify the rapidly growing wealth of specimens. Second, it was becoming increasingly obvious that there were too many species to have been accommodated by the biblical Noah's ark. It was just as difficult for these early biologists to explain how animals and plants, now isolated and perfectly adapted to dramatically different climates and environments, could have coexisted at the landing site of the ark before they spread to populate all regions of the globe.

One of the most ambitious and visionary of these eighteenth-century biologists was Carolus Linnaeus (1707–1778). He believed that God spoke most clearly to man through the natural world, and felt that it was his task to methodically describe and catalogue the collections of this divine museum. Toward that end, Linnaeus developed a scheme to classify all life: the system of binomial nomenclature that we continue to use today. Linnaeus also set his energies to the task of explaining the origin and spread of life. Like his contemporaries, he believed that the earth and its species were immutable. He realized that the challenge was not just in explaining the number of species,

Table 2.1
Persistent themes in biogeography

1. *Classifying geographic regions based on their biotas.*
2. *Reconstructing the historical development of biotas, including their origin, spread, and diversification.*
3. *Explaining the differences in numbers as well as types of species among geographic areas.*
4. *Explaining geographic variation in the characteristics of individuals and populations of closely related species, including trends in morphology, behavior, and demography.*

but in explaining patterns of diversity as well. The rapidly growing list of species included organisms adapted to environments ranging from the moist tropics to deserts, forests, and tundra. Given that species were immutable, how could they have spread from a single site (Noah's landing) to become perfectly adapted to such diverse environments? Linnaeus's solution, although perhaps naive in retrospect, was logically sound. He hypothesized that life had originated, or survived the biblical deluge, along the slopes of Mount Ararat, a high mountain near the border of Turkey and Armenia where the ark was said to have landed (Figure 2.1). At successively higher elevations was a series of environments ranging from deserts to alpine tundra. Linnaeus reasoned that each of these elevational zones harbored a different assemblage of species, each immutable but perfectly adapted to that environment. Once the Flood receded, these species migrated down from the mountain and spread to eventually colonize and inhabit their respective environments in different regions of the globe.

Comte de Buffon (1707–1788) was a contemporary of Linnaeus, but his studies of living and fossil mammals led him to a very different view of the origin and spread of life. Buffon noted two problems with Linnaeus's explanation. First, he observed that different portions of the globe, even those with the same climatic and environmental conditions, were often inhabited by distinct kinds of plants and animals. The tropics, in particular, contained a great diversity of unusual organisms. Second, Buffon reasoned that Linnaeus's view of the spread of life required that species migrate across inhospitable habitats

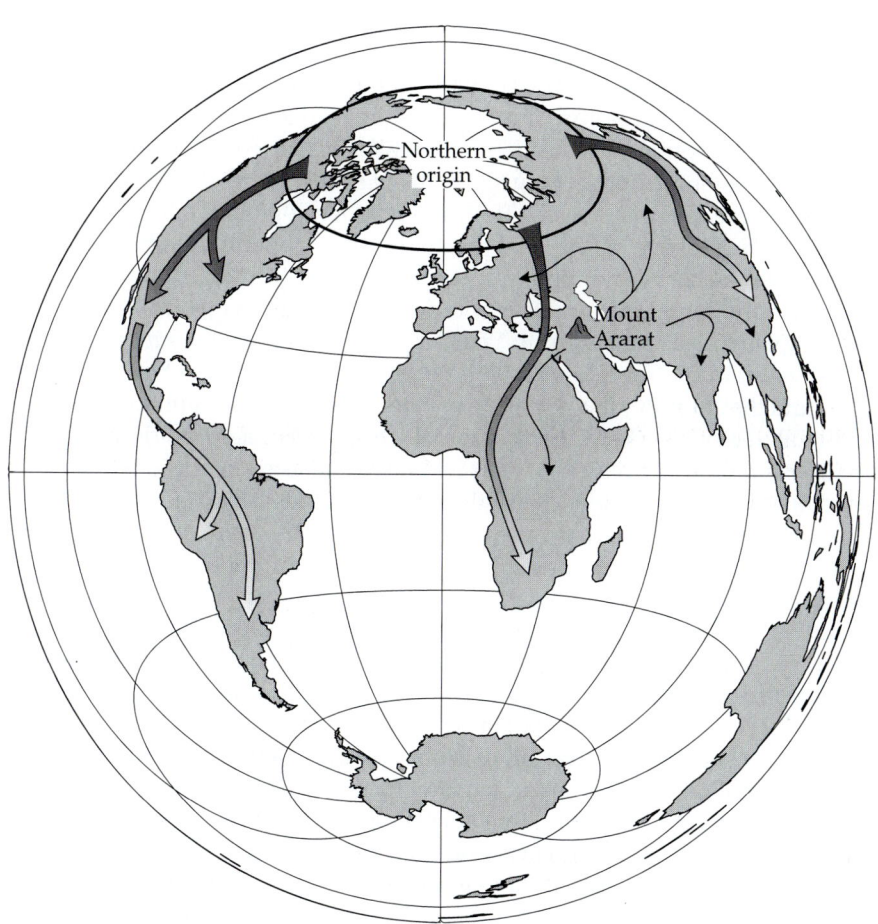

Figure 2.1 Two early hypotheses proposed to account for the diversity and distributions of terrestrial organisms. Both scenarios assume that there was just one center of origin for all life forms. Linnaeus hypothesized that terrestrial plants and animals survived the biblical Flood along the slopes of Mount Ararat, near the present-day border of Turkey and Armenia, and spread to suitable environments from that point (thin arrows). Buffon, on the other hand, hypothesized that species originated in a region much farther to the north and then spread southward, adapting or "evolving" as they colonized climatically and ecologically diverse landmasses in both the New and Old Worlds (thick arrows).

following the Flood. Species adapted to montane forests, for example, would have had to migrate across deserts before they could colonize deciduous and coniferous forests to the north. If species were immutable and therefore incapable of adapting to new environments, then their spread would have been blocked by these environmental barriers. Buffon, therefore, hypothesized that life originated not on a mountain in the temperate regions of Eurasia, but on a landmass in the far north, in an earlier period when climatic conditions were more equable (Figure 2.1). He speculated that this northern landmass was relatively continuous with both the New and Old Worlds; thus, when climates cooled, life forms could have migrated south to their current locations. During this migration, the populations of the New and Old Worlds were separated and became increasingly modified, until tropical biotas of the New and Old Worlds shared few if any forms. While Buffon's explanation may now seem fanciful, it provided two key elements of what would become central parts of modern biogeographic theory: the ideas that climates and species are mutable. Moreover, Buffon's observation that environmentally similar but isolated regions have distinct assemblages of mammals and birds became the first principle of biogeography, known as **Buffon's law**.

From 1750 to the early 1800s, many of Buffon's colleagues continued to explore the diversity of nature and to write catalogues and general syntheses of their work. One of the most prominent naturalist/collectors of this period was Sir Joseph Banks, who, during a 3-year voyage around the world with Captain James Cook on the *Endeavor* (1768–1771), collected some 3600 plant specimens, including over 1000 species not known to science (i.e., in addition to the 6000 species described by Linnaeus in his *Species Plantarum*). The efforts of Banks and many other naturalist/explorers resulted in two important developments. First, they affirmed and generalized Buffon's law. Second, they developed a much more thorough understanding of and appreciation for the complexity of the natural world. Banks and his colleagues discovered some interesting exceptions to Buffon's law—specifically, cosmopolitan species. Moreover, they noted other biogeographic patterns, which in their own right would become major themes as biogeography developed.

In the latter part of the eighteenth century, Johann Reinhold Forster (1729–1798) made many important contributions to phytogeography and to biogeography in general. In his account of his circumnavigation of the globe with Captain Cook, published in 1778, he presented one of the first systematic world views of biotic regions, each defined by its distinct plant assemblages. He found that Buffon's law applied to plants as well as to mammals and birds, and to all regions of the world, not just the tropics. Forster also described the relationship between regional floras and environmental conditions, and how animal associations changed with those of plants. He provided some important early insights into what was to become island biogeography and species diversity theory. He noted that insular (island) communities had fewer plant species than those on the mainland, and that the number of species on islands increased with available resources (island area and variety of habitats). Forster also noted the tendency for plant diversity to decrease from the equator to the poles, a pattern that he attributed to latitudinal trends in surface heat on earth.

In 1792, another German botanist, Karl Willdenow (1765–1812), wrote a major synthesis of plant geography. He not only described the floristic provinces of Europe, but offered a novel explanation for their origin. Rather than one site of creation (or survival during the biblical deluge), Willdenow suggested that there were many sites of origination—mountains that in ancient times were separated by global seas. Each of these mountain refuges was inhabited by a distinct assemblage of locally created plants. As the Flood receded, these plants spread downward to form the floristic regions of the world.

Ironically, it is not Willdenow, but one of his students, Alexander von Humboldt (1769–1859), who is generally viewed as the father of phytogeography. Humboldt was able to advance the insights of his mentor and add many of his own from his experiences in the New World tropics. He was a great naturalist, and was keenly aware that fundamental laws of nature could be discovered through the study of distributions. After studying the works of Latreille on arthropods and Cuvier on reptiles, Humboldt further generalized Buffon's law to include plants as well as most terrestrial animals. In an essay published in 1805, Humboldt noted that the floristic zonation that Forster described along latitudinal gradients could also be observed at a more local scale along elevational gradients. After conducting many floristic surveys in the Andes, Humboldt concluded that even within regions, plants were distributed in elevational zones, or **floristic belts**, ranging from equatorial tropical equivalents at low elevations to boreal and arctic equivalents at the summits.

Thus, Forster, Willdenow, and Humboldt all observed that plant assemblages were strongly associated with local climate. One of Humboldt's friends and colleagues, Swiss botanist Augustin P. de Candolle (1778–1841), added another fundamentally important insight: not only are organisms influenced by light, heat, and water, but they compete for these resources as well. Candolle emphasized the distinction between biotic provinces or regions (which he termed "habitations") and local habitats ("stations"). Later, he returned to Forster's observations on insular floras, adding that while species number is most strongly influenced by island area, other factors, such as island age, volcanism, climate, and isolation, also influence floristic diversity. Thus, some of the key elements of ecological biogeography and modern ecology were established by the early 1800s. Moreover, in an essay published in 1820, Candolle appears to be the first to write about competition and the struggle for existence, a theme that would prove central to the development of evolutionary and ecological theory.

Biogeography in the Nineteenth Century

By the early 1800s, the first three themes of biogeography were well established. Biogeographers were studying the distinctness of regional biotas, their origin and spread, and the factors responsible for differences in the numbers and kinds of species among local and regional biotas. Buffon's observations on mammals from the New and Old World tropics had been generalized to a law applying to most forms of life. But with this increased appreciation of the complexity and geographic variation of nature, the challenges for biogeography were becoming even greater. Scientists had made great progress in describing fundamental biogeographic patterns, including many that we continue to study today, but explanations for those patterns were wanting. How was it, for example, that isolated regions with nearly identical climates shared so few species? Causal explanations would have to include factors accounting for the differences as well as the similarities among isolated regions. Explanations for Buffon's law would also have to account for the exceptions to it.

During the next century, the legacy of Buffon's work would be evidenced by a long and continuing succession of explanations for his law, which would lead from a view of a static earth populated by immutable, cosmopolitan species to one in which the earth, its climate, and its species were dynamic. This view would prove essential for classifying biogeographic regions based on their biota (Theme 1 in Table 2.1) and for reconstructing the origin, spread, and diversification of life (Theme 2). Other biogeographers of the nineteenth century would focus their energies on fundamental patterns of species richness (Theme 3). It was already well established that in nearly all taxa, the number of local species (at Candolle's "stations") tended to increase with area and to decrease with distance from the equator or up a mountain. But why? To summarize, we see that nearly two centuries ago, biogeographers (then called naturalists or simply geologists) had described many of the patterns we study today, explored the generality of those patterns, and developed causal explanations. For the most part, however, they fell short on this last challenge. Their understanding of the mutability of the earth and its species still lagged far behind their rapidly increasing knowledge of the earth's geological structure, climatic patterns, and biological diversity. For the field to come of age, and to develop more rigorous, testable explanations for its fundamental patterns, it would have to await three important advances of the nineteenth century:

1. A better estimate of the age of the earth (many early biogeographers were working with an estimate of only a few thousand years)
2. A better understanding of the dynamic nature of the continents and oceans (i.e., continental drift and plate tectonics)
3. A better understanding of the mechanisms involved in the spread and diversification of species—specifically, dispersal, vicariance, extinction, and evolution

To make these advances, biogeography had to draw on new discoveries in geology and paleontology. During the nineteenth century, Adolphe Brongniart (1801–1876) and Charles Lyell (1797–1875; Figure 2.2), regarded as the fathers of paleobotany and geology, respectively, concluded that the earth's climate was highly mutable. Both men used the fossil record to infer conditions of past climates (Ospovat 1977). They found that many life forms adapted to tropical climates once flourished in the now temperate regions of northern Europe. Lyell also documented that sea levels had changed, and that the earth's surface had been transformed by the lifting up and eroding down of mountains. This, he argued, was the only way to account for the existence of marine fossils on mountain summits. Lyell also provided incontrovertible evidence for the process of extinction. Many fossil forms, once dominant and presumably perfectly adapted to existing climatic conditions, had perished and left no further trace in the fossil record. The causal agent, again, was inferred to be climatic variation and associated changes in sea level. Lyell, however, held firm to the belief that, although extinctions occurred, species were not mutable, at least not to the point that new species arose from existing ones. He also believed that, despite episodes of extinctions, the diversity of the earth, on a grand scale, remained relatively constant. He resolved this apparent contraction by suggesting that each episode of extinction was followed by an episode of creation, establishing a new set of species perfectly adapted to the altered climatic conditions. Not only were there many sites of creation, but there were many periods of creation as well!

Lyell saw in the layers of rock and their fossils undeniable evidence that the earth's surface and its biota were dynamic. He argued, moreover, that these great changes resulted from physical processes such as mountain building and erosion, processes that had operated continually throughout the history of the earth. This theory of uniformitarianism replaced earlier catastrophic explanations for the changes in the earth, its landforms, and its inhabitants. It permitted new thinking about the dynamics of living systems because, after all, their physical and biotic parts are historically inseparable. Moreover, Lyell and other geologists such as James Hutton (1726–1797) realized that, given the gradual nature of these geological processes, the earth must be much older than just a few thousand years. Only with an ancient earth could they account for the for-

Figure 2.2 Charles Lyell, often regarded as the "father of geology," strongly influenced the development of biogeography in the nineteenth century, largely through his *Principles of Geology*, first published in 1830. (Courtesy of the Council of the Linnaean Society of London.)

mation and erosion of mountains, the submergence of ancient landmasses, and the migrations or replacements of entire biotas that were documented in the fossil record.

As we shall see, the acceptance of uniformitarianism and the antiquity of the earth was essential to Darwin's and Wallace's theory that organic diversity results from the gradual yet persistent effects of natural selection operating over thousands of generations. Ironically, for most of his long and prestigious career, Lyell rejected the idea that species, like the geological features he studied, were the results or "creations" of physical forces that acted throughout time. Despite his insistent denial of what we now regard as the incontrovertible fact that species arise from other species, Lyell played a key role in the development of biogeography, largely through his treatise entitled *Principles of Geology*. In this massive work, he discussed at great length not only the geological dynamics and antiquity of the earth, but also the geography of terrestrial plants and animals and the distributions of marine algae.

Four British Scientists

Perhaps most prominent among the nineteenth-century naturalists were four British scientists: Charles Darwin, Joseph Dalton Hooker, Philip Lutley Sclater, and Alfred Russel Wallace (Figure 2.3). Although many others produced important works during this period, these four British scientists are responsible for major advances in biogeography and evolutionary biology. They all

(A)

(C)

(B)

(D)

Figure 2.3 Four British scientists who, in the mid-nineteenth century, revolutionized our understanding of the history of the earth and the distributions of its organisms: (A) Charles Darwin, (B) Joseph Dalton Hooker, (C) Philip Lutley Sclater, and (D) Alfred Russel Wallace. (A from Darwin 1859; B from Turill 1953; C from Goode 1896; D from McKinney 1972.)

studied the works of Linnaeus, Buffon, Forster, Candolle, and Lyell. They shared similar experiences as naturalists and explorers, traveling to distant archipelagoes, high mountains, and tropical and temperate regions of the New and Old Worlds. They also shared a common goal: to account for the diversity of life, including the origin, spread, and diversification of biotas. Therefore, in retrospect, it is not surprising that they featured so prominently in the development of biogeography and evolutionary biology, or that they developed great mutual respect and lasting friendships. The correspondence between these four scientists is in many cases as informative and insightful as their formal papers. In it they reveal their shared preoccupation with what we now call biodiversity and their conviction that the key to understanding the natural world was to study patterns of distribution. In a letter to Hooker in 1845, Darwin referred to the study of geographic distribution as "that grand subject, that almost keystone of the laws of creation."

With a copy of the first volume of Lyell's *Principles of Geology* in hand, young Charles Darwin set sail in 1831 on a 5-year surveying voyage aboard HMS *Beagle*, on which he served as a naturalist and gentleman companion for its captain, Robert Fitzroy (Figure 2.4). His travels would take him around the world, visiting islands of the Atlantic, Pacific, and Indian Oceans, and exploring the tropical, pampas, and Andes regions of South America. He studied geology, native plants and animals, indigenous peoples, and domesticated animals in an attempt to understand the order of life. From his diary and collections of spec-

Figure 2.4 The route of HMS *Beagle* (1831–1836). Darwin's voyage on the *Beagle* was instrumental in the development of his theory of natural selection and the origin of species.

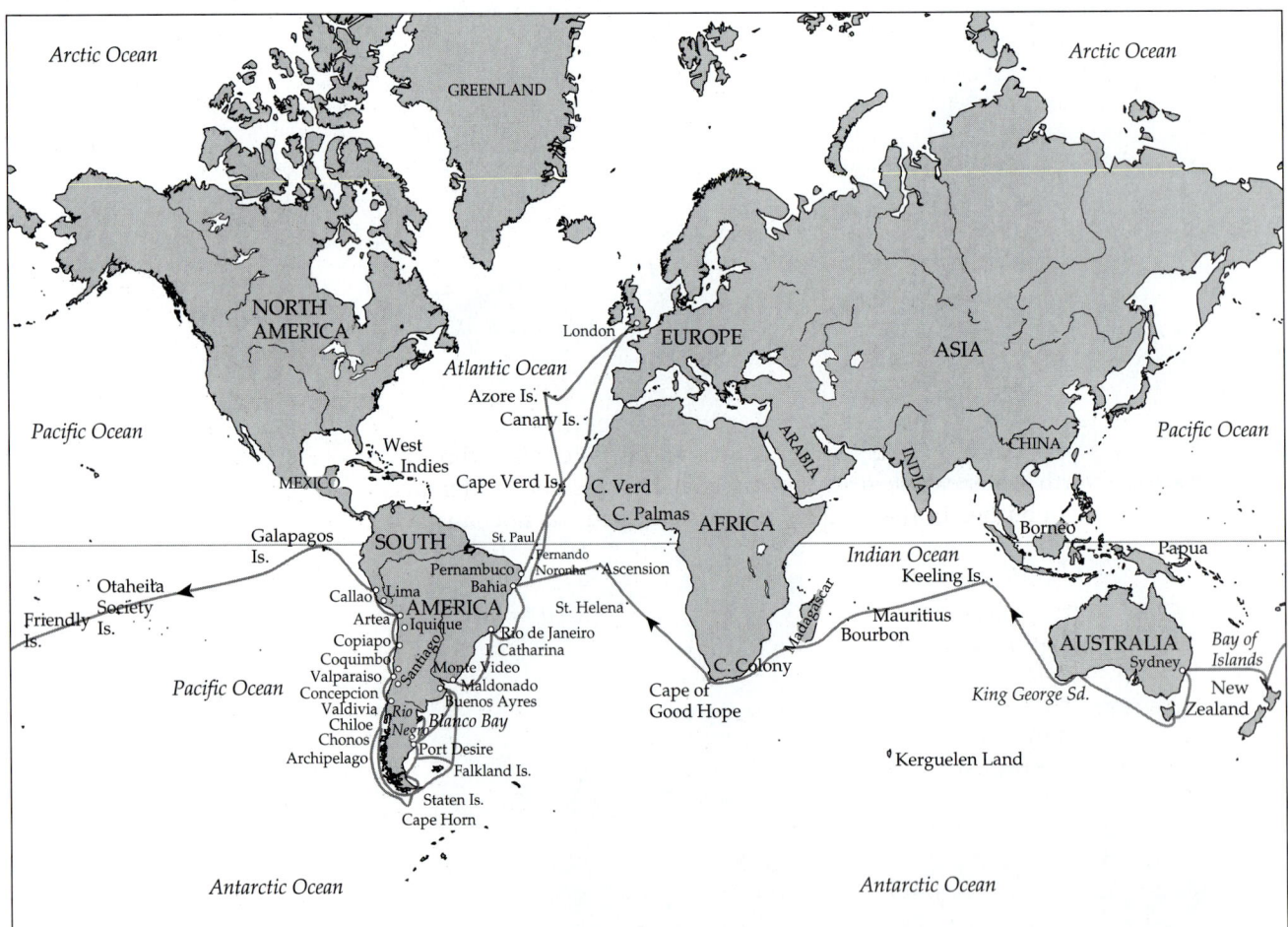

imens, he later published a fascinating account (1839) of his adventures and observations during the voyage of the *Beagle*. Darwin was intrigued and perplexed by the patterns he observed: the fossils of extinct mammals in Argentina, the presence of seashells at high elevations in the Andes, and the occurrence of unique forms of life on islands. The patterns of variability in the Galápagos Archipelago, where different forms of tortoises and finches inhabited different islands, suggested to him the idea that geographic isolation facilitates inherited changes within and between populations. On his return to England, Darwin developed his theory of evolution, invoking natural selection as the primary mechanism by which new forms of life arose and are still arising today. This theory ranks as one of the most important scientific advances of all time and is woven into all aspects of biogeography.

The writing and eventual publication of Darwin's theory of evolution by natural selection is itself an interesting story that has been the subject of much review. Darwin drafted a manuscript on the subject in 1845, but withheld the idea from print for 15 years while he continued to amass evidence to support his theory. He was finally forced to publish when he learned that another brilliant scientist, Alfred Russel Wallace, had written a manuscript that expounded the identical theory, developed independently, but based on similar observations of the natural world. A paper by Darwin and one by Wallace were read together before the Linnaean Society of London in 1858; the following year, Darwin's great book, *The Origin of Species*, was published and became an immediate best-seller.

It is difficult to overemphasize Darwin's contribution to the field of biogeography. He, along with Wallace, provided the basis for understanding changes in the adaptations and distributions of organisms across time as well as space. He proposed that the diversification and adaptation of biotas resulted from natural selection, while the spread and eventual isolation and disjunction of biotas resulted from long-distance dispersal. Darwin argued this latter point perhaps more passionately and more convincingly than anyone in the history of biogeography. His arguments were drawn not only from inferences based on the distributions of isolated biotas, but also on ingenious "experiments" on dispersal and colonization. Through these studies Darwin was able to show that seemingly unlikely events, such as dispersal of seeds embedded in mud clinging to the feet of birds, were the most likely means by which land plants had colonized oceanic islands.

Darwin's arguments on dispersal threatened to overturn the long-held static view of biogeography. In the mid-nineteenth century this older view was championed by Louis Agassiz (1807–1873), a Swiss-born naturalist who trained most of North America's leading zoologists and geologists (Dexter 1978). In two papers published in 1848 and 1855, Agassiz argued that not only were species immutable and static, but so were their distributions, with each remaining at or near its site of creation. However, as a result of Darwin's arguments, which were later bolstered by those of Asa Gray and Alfred Wallace, the static view was abandoned by most biogeographers of the nineteenth century (Fichman 1977).

But the battles to be waged by the dispersalist camp had just begun. They were soon challenged by much more formidable adversaries—the "extensionists," whose ranks were no less prestigious than those of the dispersalists, and included such respected scientists as Charles Lyell, Edward Forbes, and Joseph Hooker. Both camps agreed that distributions were dynamic in time and space. The extensionists, however, argued that long-distance dispersal across great and persistent barriers was too unlikely to explain distributional dynamics and related phenomena such as cosmopolitan species and disjunct distributions. Rather, they proposed that species had spread across great, but now submerged, landbridges and ancient continents (Figure 2.5). Despite Darwin's great respect for Lyell, nothing vexed him more than those extensionists who created landbridges "as easy as a cook does pancakes." In a letter to Lyell in 1856, Darwin complained of "the geological strides, which many of your disciples are taking. ... If you do not stop this, if there be a lower region of punishment of geologists, I believe, my great master, you will go there."

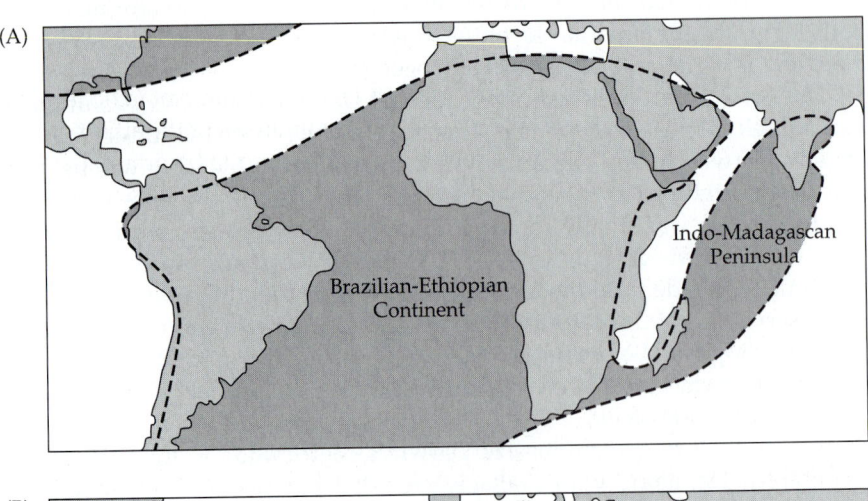

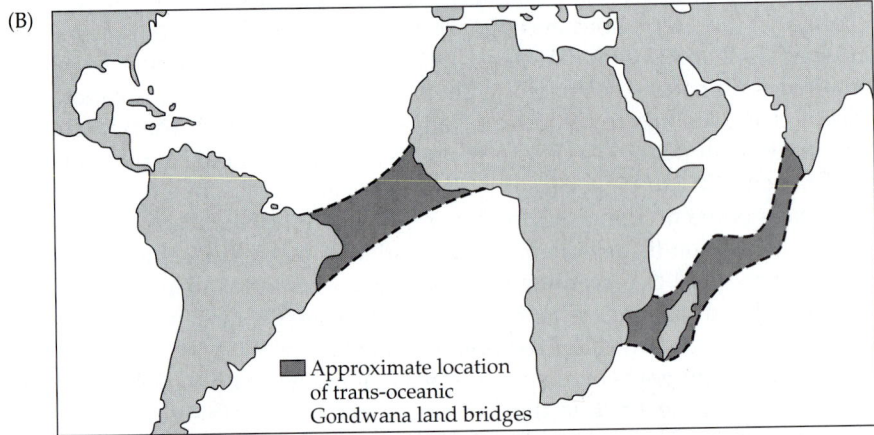

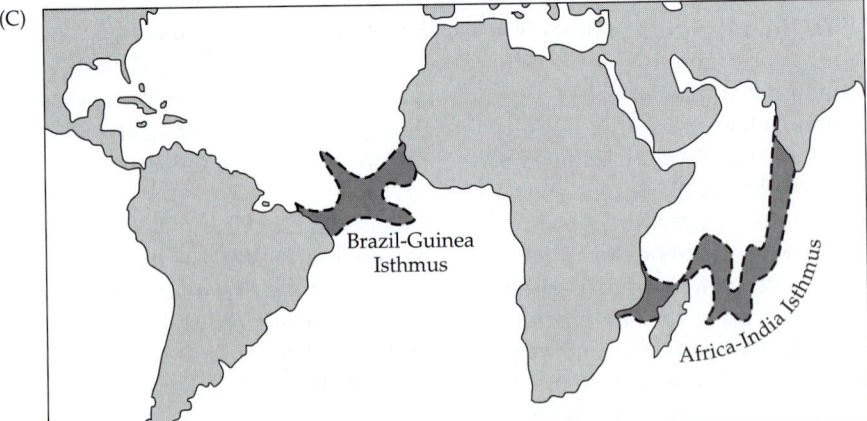

Figure 2.5 Hypothetical landbridges proposed by extensionists during the late nineteenth and early twentieth centuries to account for major disjunctions in the distributions of terrestrial organisms. (From Hallam 1967.)

Despite Darwin's persistent and passionate arguments, the extensionist camp would remain a viable and influential force throughout the latter decades of the nineteenth century. One of its greatest proponents, Joseph Dalton Hooker (Figure 2.3B), was also a good friend and admirer of both Darwin and Wallace. He was a remarkably ambitious plant collector (Turrill 1953). At the age of 22 he became the assistant surgeon and botanist on an expedition to the Southern Hemisphere led by Sir James Clark Ross. During this expedition, Hooker studied the floras of many southern landmasses, including Australia, Tasmania, Tierra del Fuego, and many archipelagoes in temperate and subantarctic waters. Hooker also traveled to Africa, Syria, India, and North America, where he studied the flora of the Rocky Mountain region with Asa Gray. In addition to his own collections, Hooker studied those of other botanists, especially those from the Galápagos Archipelago and Arctic and Antarctic regions. These collections became the raw material for his analyses of the affinities and probable origins of the flora of each of these regions, group by group.

Upon his return to England in 1843, Hooker formed a friendship with Charles Darwin, whom Hooker had admired from his account of the voyage of the *Beagle.* Within a year of their meeting, Hooker had read Darwin's manuscript on the theory of natural selection, the only person to see the manuscript before Asa Gray in 1857. Hooker shared with Darwin his ideas on the geographic distribution of plants, and was one of the few to encourage Darwin to work on, and later publish, *The Origin of Species.* In Darwin's original introduction to the book, Hooker was the only person singled out for acknowledgment. Hooker also influenced Wallace, who dedicated *Island Life* (1880) to him.

While admitting that long-distance dispersal across open oceans might account for the occurrence on remote islands of plants with easily dispersed seeds and fruits, Hooker contended that, for the most part, long-distance dispersal was an insufficient explanation for the distributions of species. He argued that the biogeographic patterns and peculiarities of the southern floras were not consistent with Darwin's dispersalist hypothesis. Rather, Hooker (1867) believed that the floristic evidence supported "the hypothesis that of all being members of a once more continuous extensive flora, ... that once spread over a larger and more continuous tract of land, ... which has been broken up by geological and climatic causes."

Hooker was correct about the affinities of Southern Hemisphere plants, although he was mistaken in his explanation. We now know that the ancestors of these plants occurred on a giant southern continent, Gondwanaland, which broke up and whose fragments began to drift apart about 180 million years ago (see Chapter 6). But the nineteenth-century geologists and biogeographers, from Lyell and Hooker to Darwin and Wallace, believed that the relative sizes and positions of the continents and oceans had changed little over geological time. So Hooker hypothesized the emergence and submergence of ancient and undiscovered continents and landbridges to account for the disjunct distributions of closely related plants. Geological evidence for these ancient landbridges never materialized. By the end of the nineteenth century, most geologists and biogeographers (including Lyell) had abandoned the extensionist doctrine.

However, we must not discount Hooker's contributions. He is still regarded as the founder of causal historical biogeography. He developed and applied many of the principles of what we now call "vicariance biogeography." In addition to emphasizing the importance of a dynamic earth and changes in climate, Hooker stressed the importance of studying insular biotas to gain insights into biogeographic processes. He confirmed the earlier observations of

Forster and others that insular floras tend to be more depauperate than those on the mainland, and noted that as island isolation increases, the number of plant species decreases, while the distinctness of the flora increases. He also observed that the most diverse floras tend to be those in lands with the greatest diversity of temperature, light, and other environmental conditions. Hooker drew an analogy between the floras of oceanic islands and those of high mountains, suggesting that both are influenced by the same processes. As we shall see, these observations and principles would be echoed by biogeographers of the twentieth century.

During most of its early history, biogeography was dominated by the contributions of botanists, from Linnaeus and Candolle to Forster and Hooker. It is no great mystery why zoogeography lagged behind phytogeography. There are many more animal species than plant species, and most animals are relatively small and difficult to collect and identify. Over 70% of all known species are animals, and most of these are insects. The search for general zoogeographic patterns had to await a better understanding of animal diversity and distributions. By the mid-nineteenth century, however, zoologists had made great strides toward these goals. In his important work on zoology published in 1866, Ernst Haeckel not only introduced the concept of ecology, but called for the recognition of biogeography (which he termed "chorology") as a new discipline, one that must incorporate the theory of natural selection.

In addition to Darwin and Hooker, two other British zoologists made major contributions to biogeography: Philip Lutley Sclater and Alfred Russel Wallace. Sclater (Figure 2.3C), a good friend of Darwin's, was an eminent ornithologist who first described 1067 species, 135 genera, and 2 families of birds. His ambitious career included service as chief executive officer of the Zoological Society of London for over 30 years, as a life member of the Royal Geographic Society of London, and as the first editor of the ornithological journal *Ibis*. Of the hundreds of articles he wrote, only a handful focused on zoogeographic patterns, but they had a major impact on zoogeography, and on biogeography in general.

In 1857, Sclater read a paper before the Linnaean Society of London entitled "On the general distribution of the members of the class Aves." He proposed a scheme that divided the earth into biogeographic regions that would reflect "the most natural primary ontological divisions of the earth's surface." As Sclater acknowledged, many others had developed biogeographic schemes, but these earlier systems had given too little regard to the flora and fauna and too much to arbitrary boundaries such as latitude and longitude. Sclater wished to do much more than just categorize each landmass with a list of species. He wanted to develop a system that would reflect the origination and development of distinctive biotas. He assumed that there were many areas of creation, and that most species must have been created in the geographic regions where they now occurred. Therefore, the ontological divisions of the globe could be identified by analyzing the similarity and dissimilarity of biotas. Sclater based his scheme on the group he knew best, birds. He acknowledged, however, that, given their impressive dispersal abilities, birds might not be the best group for such an exercise. Accordingly, he limited his work to passerines, which he believed had higher site fidelity than other birds. The system of biogeographic regions Sclater developed is illustrated in Figure 2.6. While coarse, it formed the basis for the system of six biogeographic regions we continue to use today.

It is hard to imagine how Sclater, and especially Darwin, could be overshadowed by any other zoogeographer, but they were indeed. While Sclater and Darwin spent much of their energies on other interests, biogeography was

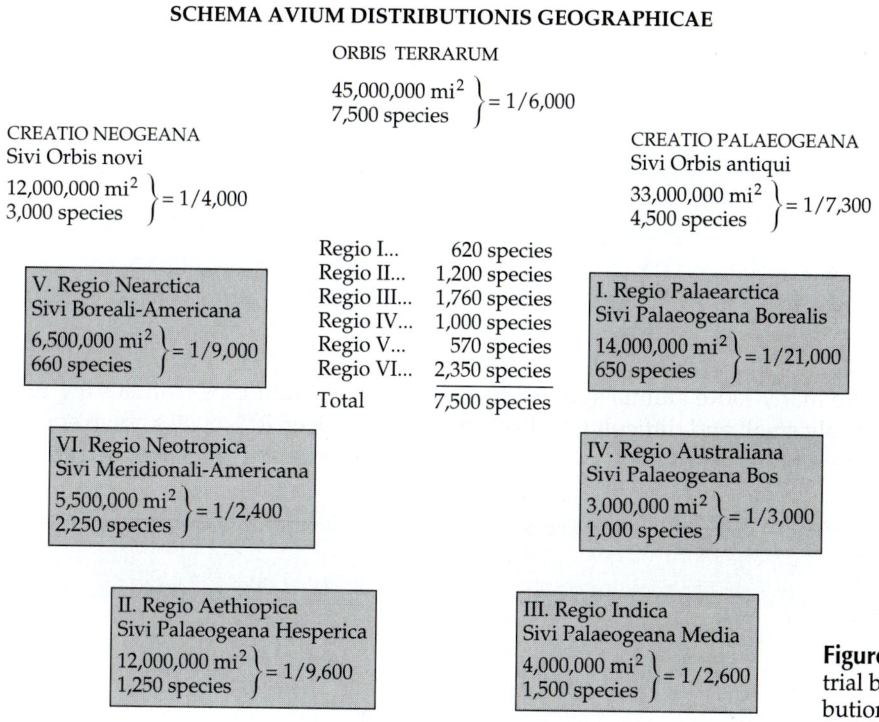

SCHEMA AVIUM DISTRIBUTIONIS GEOGRAPHICAE

ORBIS TERRARUM

$\left.\begin{array}{l}45{,}000{,}000 \text{ mi}^2 \\ 7{,}500 \text{ species}\end{array}\right\} = 1/6{,}000$

CREATIO NEOGEANA
Sivi Orbis novi

$\left.\begin{array}{l}12{,}000{,}000 \text{ mi}^2 \\ 3{,}000 \text{ species}\end{array}\right\} = 1/4{,}000$

CREATIO PALAEOGEANA
Sivi Orbis antiqui

$\left.\begin{array}{l}33{,}000{,}000 \text{ mi}^2 \\ 4{,}500 \text{ species}\end{array}\right\} = 1/7{,}300$

Regio I...	620 species
Regio II...	1,200 species
Regio III...	1,760 species
Regio IV...	1,000 species
Regio V...	570 species
Regio VI...	2,350 species
Total	7,500 species

V. Regio Nearctica
Sivi Boreali-Americana

$\left.\begin{array}{l}6{,}500{,}000 \text{ mi}^2 \\ 660 \text{ species}\end{array}\right\} = 1/9{,}000$

I. Regio Palaearctica
Sivi Palaeogeana Borealis

$\left.\begin{array}{l}14{,}000{,}000 \text{ mi}^2 \\ 650 \text{ species}\end{array}\right\} = 1/21{,}000$

VI. Regio Neotropica
Sivi Meridionali-Americana

$\left.\begin{array}{l}5{,}500{,}000 \text{ mi}^2 \\ 2{,}250 \text{ species}\end{array}\right\} = 1/2{,}400$

IV. Regio Australiana
Sivi Palaeogeana Bos

$\left.\begin{array}{l}3{,}000{,}000 \text{ mi}^2 \\ 1{,}000 \text{ species}\end{array}\right\} = 1/3{,}000$

II. Regio Aethiopica
Sivi Palaeogeana Hesperica

$\left.\begin{array}{l}12{,}000{,}000 \text{ mi}^2 \\ 1{,}250 \text{ species}\end{array}\right\} = 1/9{,}600$

III. Regio Indica
Sivi Palaeogeana Media

$\left.\begin{array}{l}4{,}000{,}000 \text{ mi}^2 \\ 1{,}500 \text{ species}\end{array}\right\} = 1/2{,}600$

Figure 2.6 Sclater's (1858) scheme of terrestrial biogeographic regions based on the distributions of birds.

Alfred Russel Wallace's life work (Figure 2.3D). Wallace, who is considered the father of zoogeography, developed many of the basic concepts and tenets of the field, combining the insights of others with his own experiences as a collector and his theory of evolution through natural selection. While Darwin and Sclater wrote only a few papers or chapters on biogeography, Wallace amassed his ideas and accounts in three seminal books: *The Malay Archipelago* (1869—dedicated to Darwin), *The Geographical Distribution of Animals* (1876), and *Island Life* (1880—dedicated to Hooker). We have summarized Wallace's contributions to the field in Box 2.1. Many of the concepts enunciated by Wallace were actually introduced by earlier scientists, but Wallace restated, documented, and interpreted them in an evolutionary context. As you read this book, you may wish to refer periodically to this box and note how many of Wallace's ideas are still being investigated by contemporary biogeographers.

Wallace collected innumerable specimens in the East Indies, and made many natural history observations on the biota of that region. He was the first person to analyze faunal regions based on the distributions of multiple groups of terrestrial animals. Wallace's analysis supported, but greatly expanded upon, Sclater's 1858 scheme. His system was based not just on birds, but on vertebrates in general, including nonflying mammals, which, because of their limited dispersal abilities, should more precisely reflect the natural divisions of the earth. Wallace developed a detailed and very precise map of the earth's biogeographic regions. His map (Figure 2.7) includes sharp divisions between regions as well as subregions, along with bathymetric divisions reflecting the isolation of different archipelagoes. A distinctive original contribution was his observation of a sharp faunal gap between the islands of Bali and Lombok in the East Indies, where many species of Southeast Asia reach their distributional limit and are replaced by forms from Australasia (Wallace 1860). This break has been called Wallace's line (see the cover of this book; Mayr 1944a; Carlquist 1965).

Figure 2.7 Wallace's (1876) scheme of biogeographic regions, which attempts to divide the landmasses into classes reflecting affinities and differences among terrestrial biotas. The regions shown are similar to those proposed by Sclater (1858; see Figure 2.6) for birds, and are still widely accepted today. Numbers identify subregions. (From Wallace 1876.)

Other Contributions in the Nineteenth Century

Other scientists of the nineteenth century were also looking for and interpreting important patterns in distributional data. Rather than considering only names and numbers of species, some of these pioneering biogeographers began to analyze geographic variation in the characteristics of individuals and populations (Theme 4 in Table 2.1). Chief among these early contributions were

GEOGRAPHICAL REGIONS, AND THE APPROXIMATE UNDULATIONS OF THE OCEAN BED.

New York: Harper & Brothers.

Stanford's Geographical Estab.t London

the generalized morphogeographic rules of C. L. Gloger (1833), C. Bergmann (1847), and J. A. Allen (1878). **Gloger's rule** holds that, within a species, individuals from more humid habitats tend to be darker in color than those from drier habitats. **Bergmann's rule** states that in endothermic (warm-blooded) vertebrates, races from cooler climates tend to have larger body sizes, and hence to have smaller surface-to-volume ratios, than races of the same species

BOX 2.1
Biogeographic principles advocated by Alfred Russel Wallace

These conclusions are summarized from Wallace's writings, and have been verified many times by researchers in the twentieth century.

1. Distance by itself does not determine the degree of biogeographic affinity between two regions; widely separated areas may share many similar taxa at the generic or familial level, whereas those very close may show marked differences, even anomalous patterns.

2. Climate has a strong effect on the taxonomic similarity between two regions, but the relationship is not always linear.

3. Prerequisites for determining biogeographic patterns are detailed knowledge of all distributions of organisms throughout the world, a true and natural classification of organisms, acceptance of the theory of evolution, detailed knowledge of extinct forms, and knowledge of the ocean floor and stratigraphy to reconstruct past geological connections between landmasses.

4. The fossil record is positive evidence for past migrations of organisms.

5. The present biota of an area is strongly influenced by the last series of geological and climatic events; paleoclimatic studies are very important for analyzing extant distribution patterns.

6. Competition, predation, and other biotic factors play determining roles in the distribution, dispersal, and extinction of animals and plants.

7. Discontinuous ranges may come about through extinction in intermediate areas or through the patchiness of habitats.

8. Speciation may occur through geographic isolation of populations that subsequently become adapted to local climate and habitat.

9. Disjunctions of genera show greater antiquity than those of single species, and so forth for higher taxonomic categories.

10. Long-distance dispersal is not only possible, but is also the probable means of colonization of distant islands across ocean barriers; some taxa have a greater capacity to cross such barriers than others.

11. The distributions of organisms not adapted for long-distance dispersal are good evidence of past land connections.

12. In the absence of predation and competition, organisms on isolated landmasses may survive and diversify.

13. When two large landmasses are reunited after a long period of separation, extinctions may occur because many organisms will encounter new competitors.

14. The processes acting today may not be at the same intensity as in the past.

15. The islands of the world can be classified into three major biogeographic categories: continental islands recently set off from the mainland, continental islands that were separated from the mainland in relatively ancient times, and distant oceanic islands of volcanic and coralline origin. The biotas of each island type are intimately related to the island's origin.

16. Studies of island biotas are important because the relationships among distribution, speciation, and adaptation are easier to see and comprehend on islands.

17. To analyze the biota of any particular region, one must determine the distributions of its organisms beyond that region as well as the distributions of their closest relatives.

living in warmer climates. The explanation for this pattern was that a lower surface area per volume ratio helps to conserve body heat in cold environments and, conversely, small size and a relatively large surface area facilitate the dissipation of heat in hot regions. Along this same line of reasoning, **Allen's rule** states that among endothermic species, limbs and other extremities are shorter and more compact in individuals living in colder climates: birds and mammals of polar regions tend to be stout with short limbs. A similar phenomenon was reported by D. S. Jordan in 1891 for ectothermic (cold-blooded) teleost fishes inhabiting marine environments. According to Jordan's law of vertebrae, as one moves farther from the equator, the vertebrae of teleost fishes become smaller and more numerous.

Despite the allure of such simple patterns, the generality and causality of these "rules" have been questioned by many biogeographers. Notwithstanding these problems, these early morpho- or ecogeographic rules were pioneering efforts in the study of geographic variation and adaptation. They stimulated the development of the field of physiological ecology and led to important observations on **allometry**—that is, how traits scale with body size.

In addition to studying geographic variation in the traits of individuals, biogeographers noted that the demographic characteristics of populations also varied across regions. In 1859, Darwin observed that within most genera, species "which range widely over the world are the most diffused in their own country, and are the most numerous in individuals"—in other words, wide-ranging species tend to occur at relatively high densities. This pattern, while not a major emphasis of early biogeography, was rediscovered and documented for a variety of taxa during the twentieth century. The study of how population-level parameters vary along geographic dimensions is now a central focus of an emerging discipline in biogeography termed **macroecology** (see Chapter 16).

Some early evolutionary "rules" were described by paleontologists who were searching for patterns in the history of life and trying to interpret the fossil record. The theory of **orthogenesis** is an example. This theory states that the evolution of a group continues in only one direction, and that this orientation is an intrinsic property of the organism and is not controlled by natural selection. Of course, this theory of evolutionary inertia was used to oppose Darwin's theory of natural selection as the agent of evolutionary change. While few, if any, modern evolutionary biologists believe in orthogenetic trends, evolution is nonetheless conservative, and is constrained by the phylogenetic history and preexisting characteristics of lineages (see Chapter 7).

A special type of orthogenesis was **Cope's rule**, which states that the evolution of a group shows a trend toward increased body size. Although there are many exceptions to this rule, it does seem that certain advantages of large size have resulted in repeated increases in size in many animal lineages. Large body size, however, seems to make species susceptible to extinction, so that large forms (such as the dinosaurs and many now extinct groups of giant birds and mammals) die out and are replaced by large representatives of new groups, which in turn evolve to a larger size (see Stanley 1973).

Simultaneously with these advances in zoogeography, phytogeographers continued to make important contributions to the development of biogeography. In the late 1800s, plant researchers in Europe began to develop novel classifications in which plant taxa were grouped according to their external architectural designs or their tolerances of abiotic stresses, such as shortages of water or excesses of salts. Contributions by the Danish workers O. Drude (1887) and E. Warming (1895) quickly led, by the early twentieth century, to the now famous classification of life forms by C. Raunkiaer (1934), who defined major types of plants based on the positions of their perennating tissues. An ecological rather than a taxonomic approach was also adopted by the great German phytogeographer A. F. W. Schimper (1898, 1903), who summarized in elaborate detail the forms and habits of plants from around the world. These works later engendered two vital areas of biogeography and ecology: plant physiological ecology, which seeks to understand how various species are adapted to the habitats in which they are found, and phytosociology, a subdiscipline of plant community ecology, which attempts to explain why certain combinations of plant species, but not others, co-occur in a given habitat (see Good 1974).

Early botanists such as Candolle and Humboldt were well aware that different types of vegetation occured at different elevations, but a zoologist, C. Hart Merriam (1894), provided one of the most valuable insights into these broad patterns. Based on his extensive field studies in southwestern North America, Merriam confirmed that elevational changes in vegetation type and plant species composition are generally equivalent to the latitudinal vegetational changes found as one moves toward the poles (Figure 2.8). He called

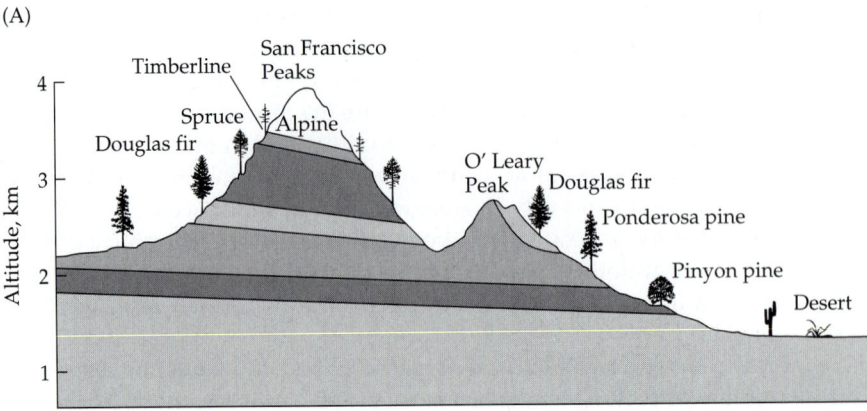

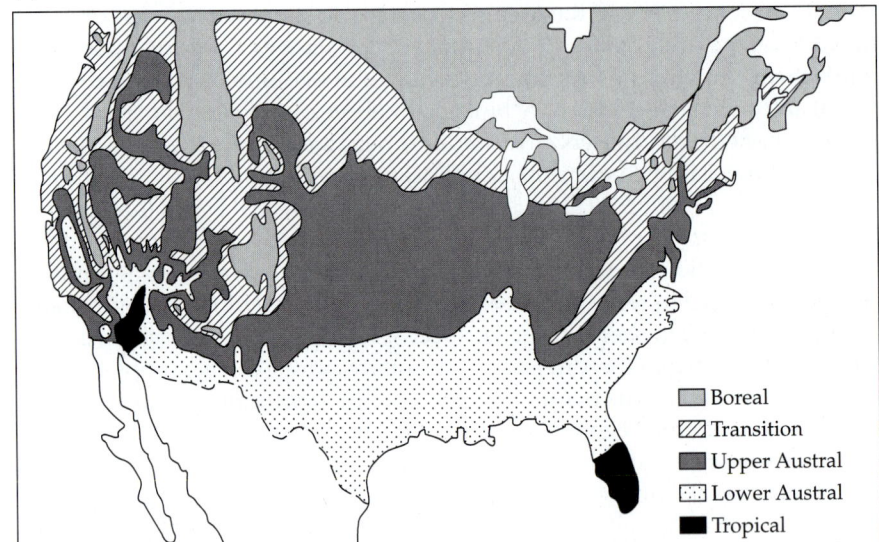

Figure 2.8   C. H. Merriam's (1890, 1894) life zones, which were based on the relationship between climate and vegetation. (A) Elevational distribution of life zones on the San Francisco Peaks of Arizona as viewed from the southeast. (B) Latitudinal distribution of life zones in North America. (A from Bailey 1996.)

these belts of similar vegetation life zones. Although Merriam was not successful in generalizing his concept of life zones to animals and to other regions of North America, he correctly concluded that elevational zonation of vegetation, like latitudinal zonation, is a response of species and communities to environmental gradients of temperature and rainfall.

In 1860 E. W. Hilgard demonstrated that climatic factors and plants are directly responsible for converting parent rock into different kinds of soils varying in pH, mineral composition, texture, and so forth. (These interrelationships are discussed in more detail in Chapters 3 and 4.) Shortly afterward, a Russian scientist named V. V. Dokuchaev recognized that each soil has a characteristic structure. These two contributions led scientists to understand that the soils of a region are governed in large part by climatic patterns, which influence the breakdown of parent materials, the growth of plants, the decomposition of organic materials, and ultimately the kinds of plants and even animals that occur there.

Some nineteenth-century naturalist/explorers began to turn their attentions to a new frontier: the oceans and their biotas. As mentioned earlier, in his *Principles of Geology*, Charles Lyell (1830) discussed the distributions of marine algae. Edward Forbes (1815–1854) produced the first comprehensive work on marine biogeography (1856), in which he divided the marine world into nine horizontal (latitudinal) regions of similar fauna ("homoizoic belts"), which he then subdivided into five zones of depth. Near the end of the nineteenth century, John Murray (1895), G. Pruvot (1896), and Arnold Ortmann (1896) published important general works on marine geography. In 1897 Philip Sclater, primarily known for his ornithological studies, published a paper on the distributions of marine mammals. Much like his earlier scheme for terrestrial zoogeographic regions (see Figure 2.6), Sclater divided the marine realm into six regions based on the distributions of geographically localized genera of marine mammals (Figure 2.9; Sclater 1897).

Despite these early contributions, just as zoogeography lagged behind phytogeography, marine biogeography would not come of age until more explorers focused on this final frontier. The present-day system of marine biogeographic "regions" was not generally accepted until the publication of John Briggs's work in the 1970s, over a century after the terrestrial system of biogeographic regions was established by Sclater and Wallace. While there remained much to learn about the marine frontier, the insights garnered from explorations of the oceans in the twentieth century would not only expand the study of biogeographic patterns to marine life, but would challenge and eventually overturn the long-held view on the permanence of the oceans and continents.

Figure 2.9   Philip Lutley Sclater's classification of biogeographic regions in the marine realm. Like his scheme for terrestrial regions, this system included six regions and was based on the distributions of geographically localized genera, in this case marine mammals.

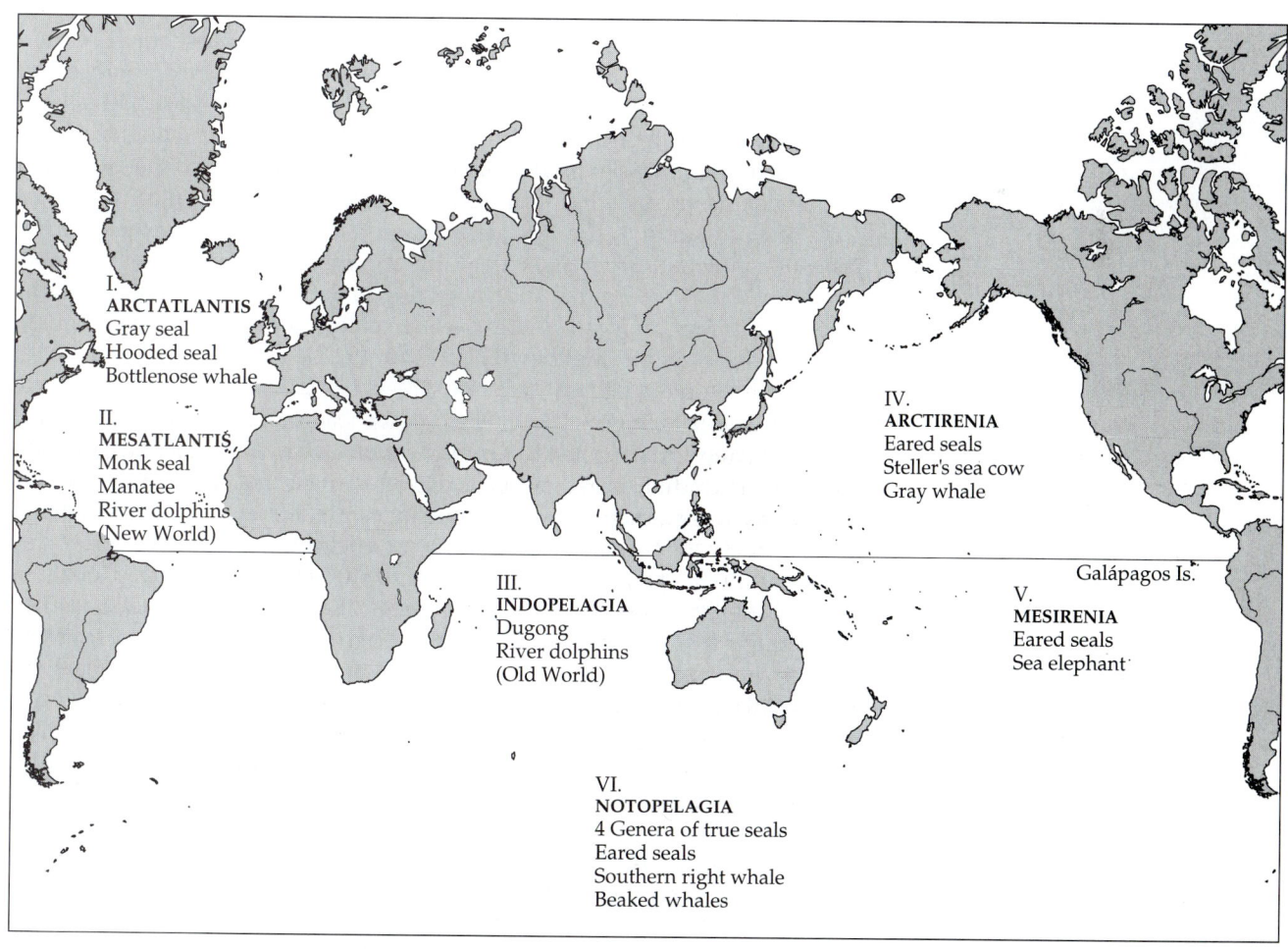

The First Half of the Twentieth Century

From 1900 through the early 1960s, several major trends in research had extraordinary effects on biogeography. Paleontology in particular deserves credit for providing new and fascinating descriptions of faunal changes on each continent. Numerous paleontologists, but especially C. Ameghino, W. D. Matthew, G. G. Simpson, E. H. Colbert, A. S. Romer, E. C. Olson, and B. Kurtén, described the origin, dispersal, radiation, and decline of land vertebrates. They showed that new groups increase in number of species, radiate to fill new ecological roles, expand their geographic ranges, and become dominant over and contribute to the extinction of older forms. Thus our present-day continental faunas have extremely long and complex histories that can be understood only by elucidating the phylogenies of the groups involved and the history of their movements between landmasses.

As mentioned earlier, explanations for how terrestrial organisms could have spread from one landmass to another were legion. Investigators proposed an incredible number of short-lived landbridges or island archipelagoes, now vanished; former continents, now sunken; or once-joined continents, now drifted apart. During this period tempers often flared as investigators debated their explanations for how groups arrived in such places as Australia, the Galápagos, and the Hawaiian Islands. Many such explanations are now rejected or considered unlikely, but these efforts infused phylogeny, paleoclimatology, and geology into biogeographic syntheses.

A common question arising out of such inquiries concerned "centers of origin." Where were the cradles of formation and diversification of various groups or biotas? Biogeographers of the early part of the twentieth century returned to the same challenges tackled by their predecessors, but with much expanded evidence and unparalleled intensity. A common goal of systematic monographs (treatments of the phylogenetic affinities and taxonomic classification of a group) was to propose a probable place of origin for the group and to describe its spread. The fossil record was spotty, so such reconstructions were often made from characteristics that included the geographic distributions of contemporary organisms, using arguments that now seem circular. Some authors were bolder and more dogmatic than others regarding the assumptions that could be used to deduce centers of origin from such limited data (see, for example, Matthew 1915; Willis 1922).

Early in the twentieth century, researchers began to investigate patterns of variation within single species. Following the lead of Bergmann and Allen on ecogeographic patterns, Joseph Grinnell, Lee R. Dice, and B. Rensch demonstrated close relationships between geographic and ecological properties of the environment and patterns of morphological variation within and among species. Subsequently, physiological and genetic variations were related to distributions in nature through the pioneering studies of Theodosius Dobzhansky on fruit flies (*Drosophila*) and J. Clausen, D. Keck, and W. Hiesey on plants. By the early 1940s evolutionary biologists were building on Darwin's synthesis to investigate patterns of geographic variation and to infer the mechanisms responsible for the origin of new species. A long list of scientists contributed to our understanding of the modes of speciation. Ernst Mayr made major contributions in the fields of systematics, evolution, and historical biogeography. Out of this work arose one unifying theme, the biological species concept, which states that a species is definable as a group of populations that is reproductively isolated from all other such groups (Mayr 1942, 1963). Moreover, Mayr's studies of patterns in the geographic distributions of species and of the underlying evolutionary mechanism now known as allopatric speciation enabled an important new synthesis in evolutionary biology and biogeography (Mayr 1944a,b, 1965a,b, 1969, 1974).

By the middle of the twentieth century, many authors had produced new and more general syntheses of biogeographic patterns for various taxa. These included studies focusing on vertebrates by Phillip J. Darlington (1957) and George Gaylord Simpson (1961); on marine zoogeography by Sven Ekman (1953) and J. W. Hedgpeth (1957); on vascular plants by Stanley A. Cain (1944) and R. Good (1947); and on island biogeography by Sherwin Carlquist (1965). Moreover, an impressive body of literature began to accumulate on ecological biogeography, which was summarized for animals by Hesse, Allee, and Schmidt (1937 and 1951) and Niethammer (1958) and for plants by Dansereau (1957). These works, and others too numerous to list here, established biogeography as a respected science that could provide insights for other fields, including evolution, ecology, and conservation biology. In a similar manner, biogeographers would continue to draw upon advances in these complementary fields as well as in geology. This led to a revitalization that would continue throughout the remainder of the twentieth century (see Watts 1979).

Biogeography since the 1950s

Four major developments have revitalized biogeography in the last 40 years: the acceptance of plate tectonics, the development of new phylogenetic methods, new ways of conducting research in ecological biogeography, and investigations of the mechanisms that limit distributions.

Until the 1960s, most biogeographers considered the earth's crust to be fixed, and without lateral movement. A theory of plate tectonics and continental drift was first introduced by Antonio Snider-Pelligrini in 1858. His radical and largely unsubstantiated theory was readily dismissed by his contemporaries. It would take another 60 years before the theory resurfaced with the arguments of a German meteorologist, Alfred L. Wegener, and an American geologist, F. B. Taylor (Taylor 1910; Wegener 1912, 1915, 1966). Wegener's theory of continental drift, published between 1912 and 1956, was based on extensive geological and some biological evidence for great movements of the continents. But again, the theory was harshly criticized and rejected by most biogeographers, including distinguished leaders such as Simpson and Darlington. Darlington, for example, argued that it was much easier to move animals than it was to move entire continents.

The theory of continental drift was accepted only in the late 1960s, when geological evidence for the process became irrefutable. Once accepted, however, this theory revolutionized historical biogeography and required scientists to rethink their explanations for many distributional patterns. Hooker had not been far off the mark—changes in the relative sizes and positions of landmasses and oceans had indeed resulted in important movements of biotas. The connections he proposed, however, rather than resulting from the emergence of mysterious continents and landbridges, were caused by great lateral movements of existing continents. Whatever one's interpretation of a distribution, the explanation eventually had to be consistent with this geological history of the earth's surface. We trace the history and modern development of plate tectonics in greater detail in Chapter 6.

Since the 1960s, biologists also have made tremendous strides toward achieving phylogenetic classifications that trace the history and relationships of taxa, thus vastly improving our understanding of how biotas are and have been related. Guidelines for reconstructing phylogenies were already available for traditional systematics (Simpson 1961) and for phylogenetic systematics or cladistics (Hennig 1950), but the issues regarding the history of lineages were crystallized in 1966 when the work of Willi Hennig was published in English. Phylogenetic research was transformed from a discipline that compared anatomical and other similarities among taxa into one in which the historical diversification of a lineage is reconstructed and the evolutionary relationships among species are quantified.

Early in the historical development of the field, biogeographers attempted to use information on geographic distributions of related species to reconstruct the histories of continents and other landmasses. In the mid-1800s, Asa Gray pioneered research on plant **disjunctions**—cases in which two closely related species are widely separated in space. Since then, biogeographers have been fascinated by disjunctions because they can reveal past land or water connections or long-distance dispersal between two regions. Interest in disjunctions as a means of evaluating past land connections has been renewed in the last 30 years, in part through the writings of L. Croizat (1958, 1960, 1964). With the availability of new phylogenetic approaches, the study of disjunct species, now called "**vicariants**," has taken a central position, particularly in historical research. Because of this renewed interest in vicariants, some of the older phylogenetic and biogeographic reconstructions are being tested and sometimes greatly revised (see Chapters 10 and 11).

Up to the 1960s, the emphasis in biogeography had been primarily an evolutionary and historical one, emphasizing the phylogenies of groups and their means of dispersing into and surviving in different regions and habitats. By the late 1950s, G. E. Hutchinson began to focus attention on questions about the processes that determine the diversity of life and the number of species that coexist in local areas or habitats. Ecologists began to emphasize the importance of competition, predation, and mutualism in influencing the distributions of species and their coexistence as ecological communities. Of all the work in ecological biogeography, perhaps the most influential was Robert H. MacArthur and Edward O. Wilson's attempt to develop a radically new theory to account for the distributions of species on islands (1963, 1967; see Chapter 13). While Agassiz's static view of distributions had long been abandoned by most biogeographers, they still assumed that species distributions within archipelagoes changed only very slowly on an evolutionary time scale. MacArthur and Wilson's equilibrium theory of island biogeography challenged this view and eventually became the new paradigm of the field. Their work changed the direction of ecological biogeography by focusing attention on a new set of questions. They asked abstract questions about patterns of distribution and species diversity, and they suggested that these patterns reflected the operation of two fundamental and opposing processes—immigration and extinction (see Chapters 14 and 15).

Questions about species diversity and coexistence, in addition to dominating the fields of ecology and theoretical biogeography, have spawned other abstract areas of inquiry, such as the extent to which the dispersal, establishment, and radiation of a lineage are **stochastic** (i.e., random) or **deterministic** (i.e., predictable if the underlying mechanisms are understood: Raup et al. 1973; Simberloff 1974b; Stanley 1979; Eldredge and Cracraft 1980). Moreover, these abstract questions have stimulated experimental testing of biogeographic concepts (e.g., Simberloff and Wilson 1969) as well as new mathematical ways of quantifying and analyzing observations (e.g., Pielou 1977a, 1979; Manly 1991; Maurer 1994; Gotelli and Graves 1996; Upton and Fingleton 1990; Cressie 1991).

Although some ecologists have emphasized the roles of interspecific interactions in influencing the distribution of species and communities (MacArthur and Connell 1966; MacArthur 1972; Whittaker 1975), others have emphasized the importance of the abiotic environment in limiting the distributions of individuals and populations and, in turn, determining the diversity of species in different regions. Important advances in techniques and instrumentation since the mid-1960s have permitted a flurry of research in this area, resulting in a tremendous amount of information that must be selectively integrated with biogeography. Advances in computer technology and related techniques, including satellite imagery, geographic information systems (GIS), and spatial statistics (geostatistics), have enabled quantum leaps in our ability to explore and analyze biogeographic patterns from local to global scales.

In summary, many great scientists have contributed to the development of biogeography. All of them, from Buffon and Candolle to Simpson, Croizat, and MacArthur and Wilson, have shared a common goal—understanding the origin, spread, and diversification of biotas. In the mid-1700s Buffon gave biogeography its first principle, while others tested its generality and broadened the field to explore a diversity of other patterns and their causal forces. As the field progressed, theories of natural selection, continental drift, and immigration/extinction dynamics provided new mechanisms to explain long-standing patterns, while new comparative and experimental methods, phylogenetic analyses, and advances in computer science and statistics provided invaluable tools. By the 1960s, biogeography had finally come of age as a rigorous and respected science. It has continued to build on those gains and flourish in subsequent decades. Since 1900, the number of publications on biogeography has increased exponentially, with most of the increase taking place in the past three decades (Figure 2.10)

Given this long list of biogeography's conceptual achievements, in themselves the seeds of whole disciplines, one can easily comprehend how it has become impossible for one person to understand and follow completely all aspects of the field. Students of biogeography can be frustrated by their inability to comprehend all the subtleties of this awesome body of knowledge—or they can be challenged and encouraged by the prospect of using biogeography as a focal point to synthesize many separate disciplines and to acquire a unique perspective on the history and distribution of life on earth.

(A)

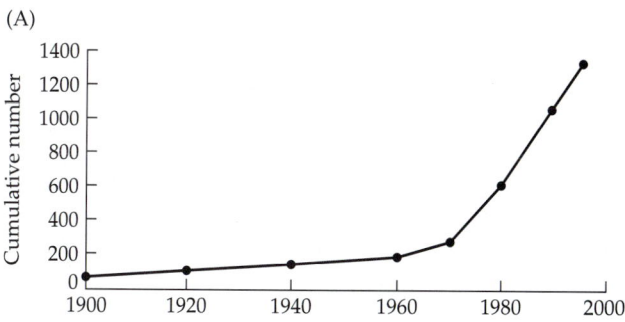

(B)

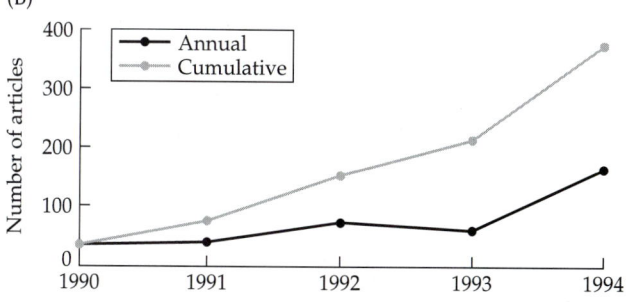

Figure 2.10 The maturity and vitality of biogeography is evidenced by the increasing number of publications in the field, especially over the last three decades, as tallied using OCLC–First Search (Online Computer Library Center, Inc.). (A) Number of books and monographs on *biogeography* found in the WorldCat database. (B) Number of articles on *biogeography* found in the Article 1st database.

UNIT 2

The Environmental and Historical Setting

CHAPTER 3

The Physical Setting

Organisms can be found almost everywhere on earth: from the cold, rocky peaks of high mountains to the hot, windswept sand dunes of lowland deserts; from the dark, near-freezing depths of the ocean floor to the steaming waters of hot springs. Some organisms even live around hydrothermal vents in the deep ocean, where temperatures exceed 100° C (but the water does not boil because of the extreme pressure). Yet no single kind of organism lives in all of these places. Each species has a restricted geographic range, in which it encounters a limited range of environmental conditions. Polar bears and caribou are confined to the Arctic, whereas palms and corals are rare outside the tropics. There are a few species, such as *Homo sapiens* and the peregrine falcon, that we call cosmopolitan because they are distributed over all continents and over a wide range of latitudes, elevations, climates, and habitats. These species, however, are not only exceptional, but are also much more limited in distribution than they appear at first glance. Humans and peregrines, for example, are absent from the three-fourths of the earth that is covered with water, and from many other places besides.

Although, as we point out in later chapters, we may need to invoke unique historical events or ecological interactions with other organisms to account for the limited geographic ranges of some species, the most obvious patterns in the distributions of organisms occur in response to variation in the physical environment. In terrestrial habitats these patterns are determined by climate and soil type. The distributions of aquatic organisms are limited largely by variation in temperature, salinity, light, and pressure.

Most geographic variation in the physical environment is regular and predictable. We all know that tropical lowlands are warm year-round, whereas the climate is colder and more seasonal at higher latitudes. Most streams and lakes contain fresh water, but the ocean is salty. Both the tops of the highest mountains and the depths of the deepest lakes and oceans are very cold. These and other climatic patterns are relatively easy to explain. To do so, we need to know the orientation of the earth with respect to the sun and the sizes and locations of major geographic features such as oceans, continents, and mountain ranges. We also need a rudimentary knowledge of the physical and chemical properties of air, water, and soil, as well as the principles of energetics and thermodynamics.

The dynamics of the earth's surface are driven by two great engines, powered by two different sources of energy. The heat stored in the earth's core at the time of the formation of the solar system is dissipated through its mantle and crust and ultimately out into space. This transfer of heat energy moves and shapes the earth's crust, shifting the positions of the crustal plates containing the continents, thrusting up mountains, and causing earthquakes and volcanic eruptions. We shall discuss these processes and some of their consequences in Chapter 6.

The other great engine is driven by the energy of the sun. Radiant energy emitted by the sun strikes the earth's surface, where it is absorbed and converted into heat, warming the surface of the land and water and the atmosphere just above them. The resulting differences in the temperature and density of air and water cause them to move over the earth's surface, both horizontally and vertically, creating the earth's major wind patterns and ocean currents. The heating of surface water also causes evaporation, and the resulting water vapor is carried by the air and redeposited as rain or snow. These processes, which are responsible for the earth's climate and for many physical characteristics of its oceans and fresh waters, are the subject of the present chapter.

Climate

Solar Energy and Temperature Regimes

Solar radiation and latitude.   Sunlight sustains life on earth. Solar energy not only warms the earth's surface and makes it habitable, but also is captured by green plants and converted into chemical forms of energy that power the growth, maintenance, and reproduction of most living things.

According to the principles of thermodynamics, heat is transferred from objects of higher temperature to those of lower temperature by one of three mechanisms: (1) conduction, a direct molecular transfer, especially through solid matter; (2) convection, the mass movement of liquid or gaseous matter; or (3) radiation, the passage of waves through space or matter. Heat flows, as radiant energy, from the hot sun across the intervening space to the cooler earth. When incoming solar radiation strikes matter, such as water or soil, some of it is absorbed, and the matter is heated. Some solar radiation is initially absorbed by the air, particularly if it contains suspended particles of water or dust (e.g., clouds), but most passes through the sparse matter of the atmosphere and is absorbed by the denser matter of the earth's surface. This surface is not heated uniformly. Soil, rocks, and plants absorb much of the radiation and may be heated intensely. Water also absorbs much solar radiation, but its heating effect is not confined to as narrow a surface layer as on land. Although air is heated to some extent by absorption of incoming solar radiation, most of the heating of air occurs at the earth's surface, where it is warmed by direct contact with warm land and water, by latent heat released by the condensation of water, and by long-wave infrared radiation emitted from the surfaces of warm objects such as leaves and bare soil.

The angle of incoming radiant energy relative to the earth's surface affects the quantity of heat absorbed. The most intense heating occurs when the surface is perpendicular to incident solar radiation, for two reasons: (1) the greatest quantity of energy is delivered to the smallest surface area; and (2) a minimal amount of radiation is absorbed or reflected back into space during passage through the atmosphere because the distance it travels through air is minimized (Figure 3.1). This differential heating of surfaces at different angles to the sun explains why it is usually hotter at midday than at dawn or dusk,

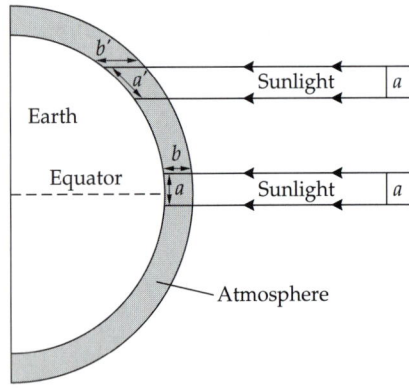

Figure 3.1   Average input of solar radiation to the earth's surface as a function of latitude. Heating is most intense when the sun is directly overhead, so that incoming solar radiation strikes perpendicular to the earth's surface. The higher latitudes are cooler than the tropics because the same quantity of solar radiation is dispersed over a greater surface area (*a'* as opposed to *a*) and passed through a thicker layer of filtering atmosphere (*b'* as opposed to *b*).

Equinox

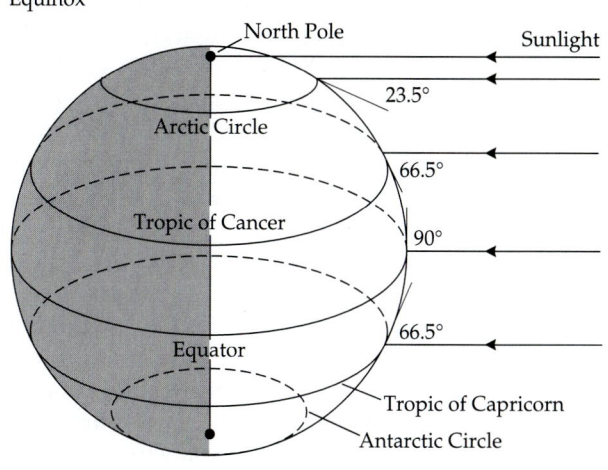

Summer solstice

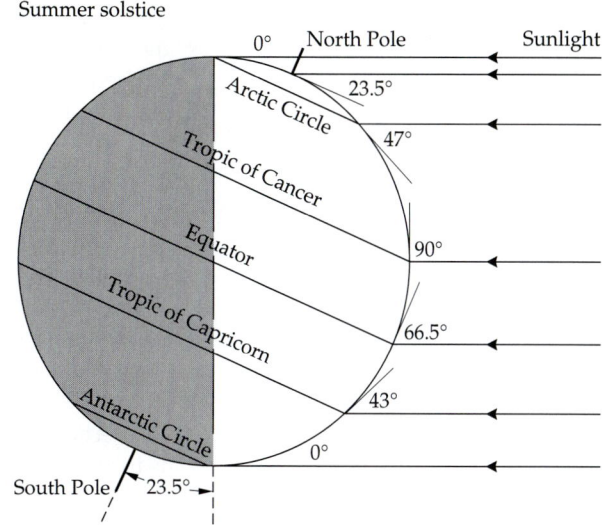

Figure 3.2 Seasonal variation in day length with latitude is due to the inclination of the earth on its axis. At the equinoxes, the sun is directly overhead at the equator, and all parts of the earth experience 12 hours of light and 12 hours of darkness each day. At the summer solstice, however, the 23.5° angle of inclination causes the sun to be directly over the Tropic of Cancer, while the Arctic Circle and areas farther north experience 24 hours of continuous daylight; at the same time all regions in the Southern Hemisphere experience less than 12 hours of daylight per day, and the sun never rises south of the Antarctic Circle.

why average temperatures in the tropics are higher than at the poles, and why south-facing hillsides are warmer than north-facing ones in the Northern Hemisphere (and the reverse in the Southern Hemisphere).

Because the earth is tilted 23.5° from vertical on its axis with respect to the sun, solar radiation falls perpendicularly on different parts of the earth during an annual cycle. This differential heating produces the seasons. The seasons are also characterized by different lengths of day and night. Only at the equator are there exactly 12 hours of daylight and darkness every 24 hours throughout the year (Figure 3.2). At the spring and fall equinoxes (March 21 and September 22, respectively) the sun's rays fall perpendicularly on the equator, equatorial latitudes are heated most intensely, and every place on earth experiences the same day length. At the summer solstice (June 22), sunlight falls directly on the Tropic of Cancer (23.5° N latitude), and the Northern Hemisphere is heated most intensely, experiences longer days than nights, and enjoys summer, while the Southern Hemisphere has winter. At the winter solstice (December 22), the sun shines directly on the Tropic of Capricorn (23.5° S latitude), and the Southern Hemisphere enjoys its summer while the Northern Hemisphere has winter, cold temperatures, and long nights. The seasonality of climate increases with increasing latitude. At the Arctic and Antarctic Circles, 66.5° latitude, there is one day each year of continuous daylight when the sun never sets, and one day of continuous darkness, each at a solstice. Although every location on the earth theoretically experiences the same amount of daylight and darkness over an annual cycle, the sun is never directly overhead at high latitudes. Considerable solar radiation is absorbed during the long summer days, however. Temperatures in excess of 30° C are commonly recorded in Alaska. The warmest days typically are in July, after the summer solstice, because of the time lag required to heat up the earth's surface.

The cooling effect of elevation. The processes just described account for seasonal and latitudinal variation in temperature, but it remains to be explained why it gets colder as we ascend to higher altitudes. The fact that Mount Kilimanjaro, nearly on the equator in tropical East Africa, is capped with permanent ice and snow seems to be in conflict with our intuitive expectation. Mountain peaks are nearer the sun, so why are they cooler than nearby lowlands? The answer lies in the thermal properties of air. The density and pressure of air decrease with increasing elevation. When air is blown across the earth's surface and forced upward over mountains, it expands in response to the reduced pressure. Expanding gases undergo what is called **adiabatic cooling,** losing heat energy as their molecules move farther apart. The same process occurs in a refrigerator as freon gas expands after leaving the compressor. The rate of adiabatic cooling of dry air with increasing elevation is about 10° C per km, so long as no condensation of water vapor and cloud formation occurs.

Higher elevations are also colder because the less dense air allows a higher rate of heat loss by radiation back through the atmosphere. Water vapor and carbon dioxide in the atmosphere retard such radiant heat exchange and produce the so-called **greenhouse effect**. These gases act like the glass in a greenhouse: they allow the short wavelengths of incoming solar radiation to pass through, but trap the longer wavelengths emitted by surfaces that have been warmed by the sun. The resulting warming effect is pronounced in moist lowland areas, where water vapor in the air retards cooling at night. In contrast, mountains and deserts typically experience extreme daily temperature fluctuations, because there is little water vapor in the air to prevent heat loss by radiation to the cold night sky.

Winds and Rainfall

Wind patterns.   Differential heating of the earth's surface also causes the winds that circulate heat and moisture. As we have already seen, the most intense heating is at the equator, especially during the equinoxes, when the sun is directly overhead. As this tropical air is heated, it expands, becomes less dense than the surrounding air, and rises. This rising air produces an area of reduced atmospheric pressure over the equator. Denser air from north and south of the equator flows into the area of reduced pressure, resulting in surface winds that blow toward the equator (Figure 3.3). Meanwhile, the rising equatorial air cools adiabatically, becomes denser, is pushed away from the equator by newly warmed rising air, and eventually descends again at about 30° N and S latitude (the horse latitudes). This vertical circulation of the atmosphere results in three convective "Hadley cells" in each hemisphere, with warm air ascending at the equator and at about 60° N and S latitude, and cool air descending at about 30° N and S and at the poles. These circulating air masses produce surface winds that typically blow toward the equator between 0° and 30° and toward the poles between 30° and 60°. In the upper atmosphere between the Hadley cells are the jet streams, high-speed winds blowing approximately parallel to the equator.

The surface winds do not blow exactly in a north-south direction; they are deflected toward the east or west by the **Coriolis effect**. Although the Coriolis effect is often called the *Coriolis force*, it is not a force, but a straightforward consequence of the law of conservation of angular momentum. Every point on the earth's surface makes one revolution every 24 hours. Because the circumference of the earth is about 40,000 km, a point at the equator moves from west to east at a rate of about 1700 km h⁻¹. Points north or south of the equator move at a slower rate (remember that the lines of longitude converge at the poles). Consider what happens at the equator if you shoot a rocket straight upward. Where does it come down? Right where it was launched; the rocket travels not only up and down, but also eastward at a rate of 1700 km h⁻¹, the same rate as the earth moving beneath it. Now suppose the rocket is propelled northward away from the equator. It continues to travel eastward at 1700 km h⁻¹, but the earth underneath it moves ever more slowly as the rocket travels farther north, and consequently its path appears to be deflected toward the

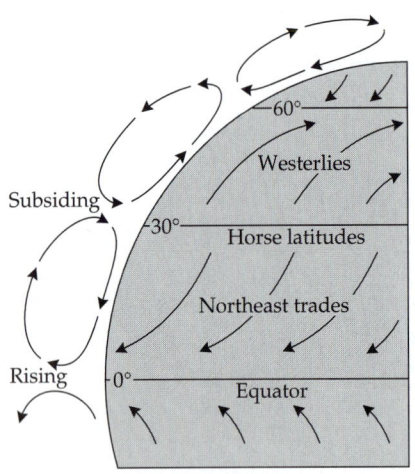

Figure 3.3   Relationship between vertical circulation of the atmosphere and wind patterns on the earth's surface. There are three convective Hadley cells of ascending and descending air in each hemisphere. As the winds move across the earth's surface in response to this vertical circulation, they are deflected by the Coriolis effect, producing easterly trade winds in the tropics and westerlies at temperate latitudes.

right. The Coriolis effect describes this tendency of moving objects to veer to the right in the Northern Hemisphere and to the left in the Southern Hemisphere. The winds approaching the equator from the horse latitudes are deflected to the west, and are therefore called northeast or southeast **trade winds**. (Winds are described based on the direction of their *sources*.) Winds blowing toward the poles between about 30° and 60° N and S latitude, called the **westerlies**, are deflected to the east (Figure 3.3). These winds naturally were very important to commerce in the days of sailing ships, when both the westerlies and the trade winds, or "trades," got their names. Ships coming to the New World from Europe traveled south to the Canary Islands and Azores in tropical latitudes to intercept the trades before heading westward, but they returned to Europe at higher latitudes with the westerlies behind them.

The surface winds, influenced by the Coriolis effect, initiate the major ocean currents. The trade winds push surface water westward at the equator, whereas the westerlies produce eastward-moving currents at higher latitudes. Responding to the Coriolis effect, these water masses are deflected toward the east or west, and the net result is that the ocean currents move in great circular gyres, clockwise in the Northern Hemisphere and counterclockwise in the Southern Hemisphere (Figure 3.4). Warm currents flow from the tropics along eastern continental margins; as these water masses reach high latitudes, they are cooled, producing cold currents down the western margins.

Precipitation patterns. By superimposing these patterns of temperature, winds, and ocean currents, we can begin to understand the global distribution of rainfall. We will also need some additional background in physics. As air warms, it can absorb increasing amounts of water vapor evaporated from the land and water. As it cools, it eventually reaches the **dew point**, at which it is

Figure 3.4 Main patterns of circulation of the surface currents of the oceans. In each ocean, water moves in great circular gyres, which move clockwise in the Northern Hemisphere and counterclockwise in the Southern Hemisphere. These patterns result in warm currents along the eastern coasts of continents and cold currents along the western coasts. Note the Pacific equatorial countercurrent, the small current along the equator that flows from west to east opposite to the gyres, and which strengthens in some years to cause the El Niño phenomenon.

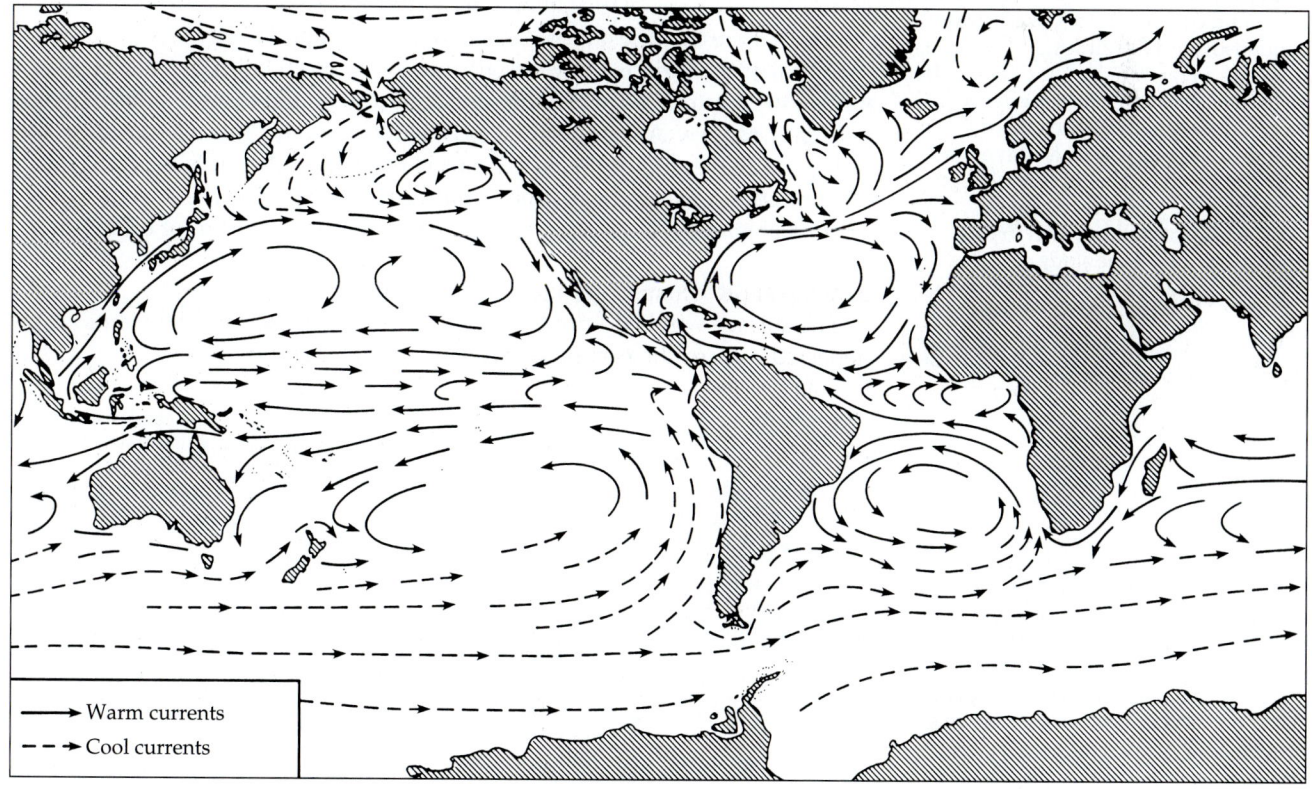

→ Warm currents
- - → Cool currents

saturated with water vapor. Further cooling then results in condensation and the formation of clouds. When the particles of water or ice in clouds become too heavy to remain airborne, rain or snow falls. In the tropics, the cooling of ascending warm air laden with water vapor produces heavy rainfall at low and middle elevations, where rain forests and cloud forests occur. Rainy seasons in the tropics tend to fall when the sun is directly overhead and the most intense heating occurs. The tropical grasslands of Kenya and Tanzania in East Africa, which lie virtually on the equator but at higher elevations than rain forests, experience two rainy seasons each year, corresponding approximately to the equinoxes, and two dry seasons, corresponding to the solstices. In contrast, the area around the Tropic of Cancer in central Mexico has only one principal rainy season, in the summer. Most tropical regions have at least one dry season.

At the horse latitudes, where cool air descends from the upper atmosphere, two belts of relatively dry climate encircle the globe. Descending air warms, and can therefore absorb more moisture, drying the land. In these belts lie most of the earth's great deserts (including the Mojave, Sonoran, and Chihuahuan in southwestern North America, the Sahara in North Africa, and the Arid Zone in central Australia), and adjacent to these deserts are regions of semiarid climates and grassy or shrubby vegetation. Within these belts, the seasonality of climate is very marked on the western sides of continents, which experience **Mediterranean climates**. Parts of coastal California, Chile, the Mediterranean region in Europe, southwestern Australia, and southernmost Africa have dry, usually hot summers and mild, rainy winters (see Figure 1.2). In winter, when the land tends to be cooler than the ocean water, the westerly winds bring ashore moisture-laden air, condensation occurs, and fog and rain result. In summer, when the land is warmer than the ocean, the westerlies blowing inland from the cold offshore currents are warmed, able to hold more water vapor, and result in dry conditions on land. The effects of cold currents are even more pronounced in localized regions of western South America and southwestern Africa, where they contribute to the formation of coastal deserts that are the driest areas on earth (Amiran and Wilson 1973).

Several of the deserts between 30° and 40° N and S latitude are located not only on the western sides of continents, but also on the eastern sides of major mountain ranges. As westerly winds blow over the mountains, they are cooled, until eventually the dew point is reached and clouds begin to form. Condensation releases heat—the latent heat of evaporation—so that wet air cools adiabatically at a slower rate than dry air: 6° C per km of elevation, as

Figure 3.5 Factors causing rain shadow deserts. (A) Air blowing over a mountain cools as it rises, water vapor condenses, and the air loses much of its moisture as rain on the windward side, so that the leeward side experiences warm, dry winds. (B) The rate of change in air temperature with elevation is affected by the presence of condensed water vapor, resulting in warmer, drier conditions on the leeward side than at the same elevation on the windward side. (After Flohn 1969.)

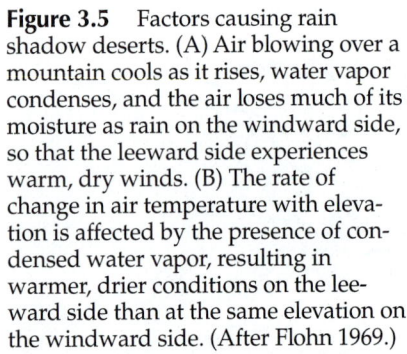

(A)

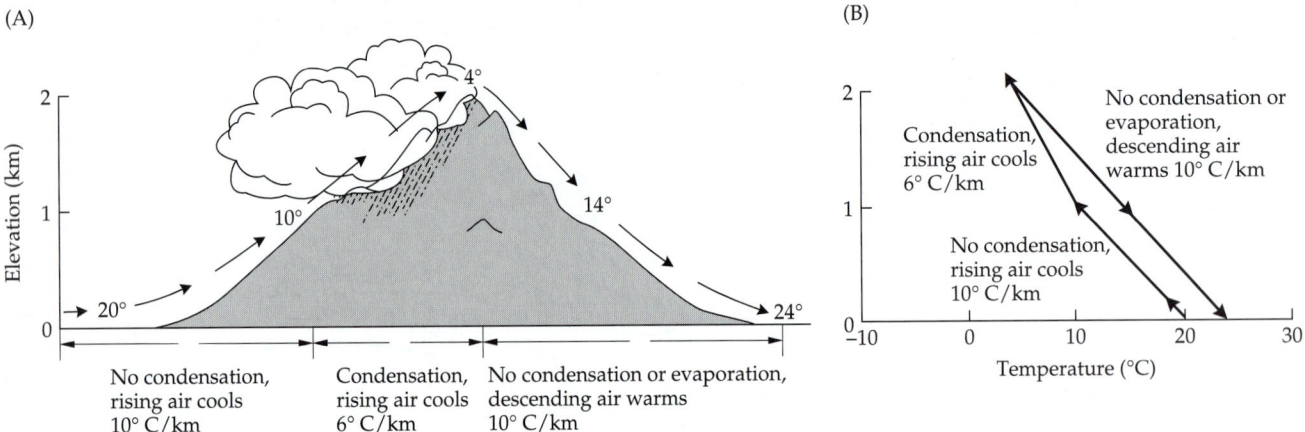

(B)

opposed to 10° C per km for dry air. As the air continues to rise and cool, most of its moisture falls as precipitation on the western side of the mountain range. When the air passes over the crest and begins to descend, the remaining clouds quickly evaporate, and the dry air warms at the higher rate. This **rain shadow** effect causes the warm, dry climates found on the eastern sides of temperate mountains (Figure 3.5). Thus, for example, the Sierra Nevada in California has lush, wet forests of giant sequoias and other conifers on its western slopes, but arid woodlands of piñons and junipers on its eastern slopes; a bit farther east, with an elevation below sea level, lies Death Valley, the driest place on the North American continent. Similarly, the Monte Desert is in the rain shadow on the eastern side of the Andes in Argentina.

These global patterns of temperature and precipitation frequently are summarized in climatic maps like that in Figure 3.6. Such maps are useful, but can be misleading, because they fail to show the small-scale patterns of spatial and temporal variation that influence the abundance and distribution of organisms.

Small-scale spatial and temporal variation. The same processes that we have just described on a global scale can also produce great climatic variation on a local scale. The effect of mountains is particularly great, as we can illustrate with several examples. From Tucson, Arizona, it is only 25 km by a paved road to the top of Mount Lemmon, at 2800 m elevation, in the Santa Catalina Mountains. But the climate and the plants at the summit are far more similar to those in northern California and Oregon, 1500 km to the north, than to those in the desert just below (Table 3.1). Similarly, the spruce-fir forests on the

Figure 3.6 Major climatic regions of the world. Note that these regions occur in distinct patterns with respect to latitude and the positions of continents, oceans, and mountain ranges. (After Strahler 1973.)

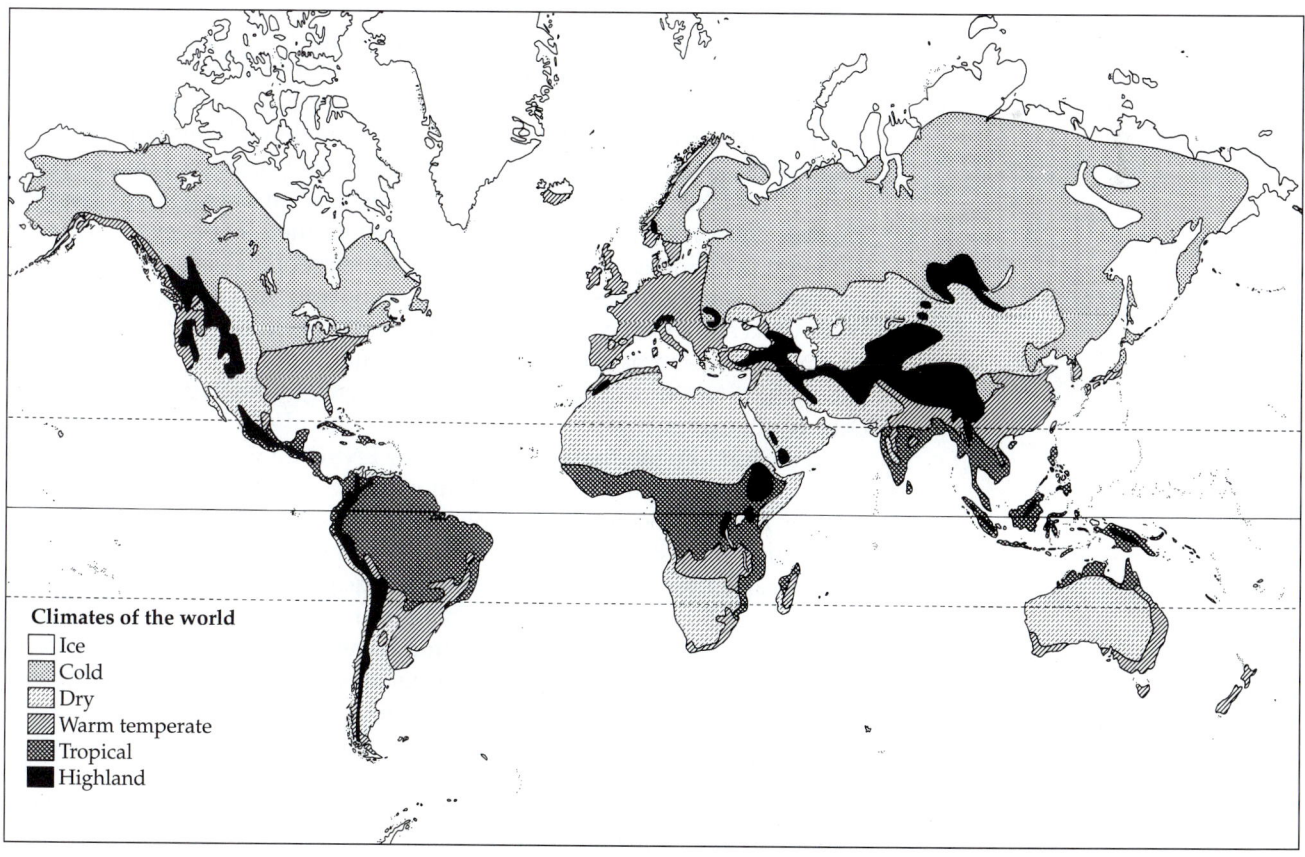

Climates of the world
- Ice
- Cold
- Dry
- Warm temperate
- Tropical
- Highland

Table 3.1
The influence of elevation on climate

Site	Elevation (m)	Temperature (°C)				Mean annual precipitation (cm)
		Mean January	Mean July	Lowest	Highest	
Tuscon, Arizona	745	10.8	30.7	−9.4	46.1	27.3
Mt. Lemmon, Arizona	2791	2.3	17.8	−21.7	32.8	70.0
Salem, Oregon	60	3.2	19.2	−24.4	40.0	104.3

Source: Data from U.S. Weather Bureau.

Note: Two of the sites are near one another in Arizona; the third site is in Oregon. Note that the climate of the high-elevation site in Arizona, Mt. Lemmon, is much more similar to that of Salem, Oregon, 1700 km to the north, than to that of Tucson, only 25 km away but 2000 m lower in elevation.

summit of the Great Smokey Mountains in Tennessee are more similar to the boreal forests of northern Canada than to the deciduous forests in the valleys below. Puerto Rico, which lies in the Caribbean Sea at 18° N latitude, is about 150 km long and 50 km wide, and has a central mountainous backbone rising to about 1000 m. The lowlands on the northern and eastern sides are lush and tropical, but much more rain falls at higher elevations on the northeastern slopes, and this is where the best-developed rain forests are found. So much moisture is lost as the northeast trade winds traverse the mountains that the southwestern corner of Puerto Rico is extremely dry; the cacti and shrubby vegetation that occur there remind a visitor of the deserts and tropical thorn forests of western Mexico (Figure 3.7). Even more dramatic are the combined effects of the cold Humboldt Current and the rain shadow cast by the Andes in Peru and Chile, where up to 10 m of precipitation per year drenches the tropical rain forests on the eastern slope, while there may several years in sucession with no rain at all in the Atacama Desert on the western slope.

There are also year-to-year and longer-term temporal variations in climate. We have recently learned that the entire global system of moving air masses, ocean currents, and patterns of precipitation fluctuates on a 5- to 7-year cycle. The fluctuations appear to be initiated by events in the vast tropical Pacific Ocean (although similar events occur in the tropical Atlantic). This pattern is called the **El Niño Southern Oscillation**, or **ENSO** for short. We are uncertain about its initial cause—perhaps variation in the output of solar radiation or intrinsic fluctuations in the atmosphere-ocean system. But whatever the ultimate cause, the pattern of tropical ocean circulation changes. While the primary ocean currents are the great hemispheric gyres mentioned above, close inspection of Figure 3.4 will show a small current running west to east right along the equator. It is called the **equatorial countercurrent** because it runs in the opposite direction to the gyres. It is usually small, as the figure suggests, but in some years it becomes much stronger, and pushes warm water away from the equator up the coasts of North and South America. As the westerly winds pass over this warm water, they pick up moisture and carry it onto the adjacent continents, causing heavy precipitation in the winter, when the land is colder than the offshore waters. This phenomenon is called **El Niño** (literally, "little boy" in Spanish), because the resulting rains tend to fall around

(A)

(B)

Figure 3.7 Comparison of vegetation on opposite sides of the central mountain range on the tropical island of Puerto Rico. (A) On the northeastern side, which receives the moisture-laden trade winds, lush rain forests occur. (B) In marked contrast, the southwestern side lies in a rain shadow, has a hot, dry climate, and has cacti and other plants typical of desert regions. (A courtesy of E. Orians; B courtesy of A. Kodric-Brown.)

Christmas, when Hispanic cultures celebrate the birth of the Christ child. El Niño years are the only times that it rains in the extremely arid coastal deserts of South America. The seemingly lifeless Atacama Desert bursts into bloom as plants that have survived as seeds dormant in the soil germinate, grow, and reproduce. Seabirds and other marine organisms along the Pacific coast and in the Galápagos Islands suffer wholesale reproductive failure and mortality due to the unusual rain and reduced upwelling.

Other kinds of temporal variation can also have important biogeographic consequences. For example, a hurricane may pass over a Caribbean island only once in a century on average, yet these rare, unpredictable storms wreak incredible devastation. Hurricanes probably are one of the primary causes of disturbance on Caribbean islands. Such large, infrequent storms may increase or decrease biodiversity; while they can inundate tiny islets, causing extinction of some terrestrial animals and plants, they also clear space in forests and coral reefs, facilitating the continued existence of competitively inferior species.

Soils

Primary Succession

Except for the polar ice caps and the perpetually frozen peaks of the tallest mountains, almost all terrestrial environments on earth can and do support life. Areas of bare rock and other sterile substrates created by volcanic eruptions or other geological events are gradually transformed into habitats capable of supporting living ecological communities by a process called **primary succession**. This process involves the formation of soil, the development of vegetation, and the assembly of a complement of microbial, plant, and animal species.

We cannot understand the distribution of soils without a knowledge of the role of climate and organisms in successional processes. The type of vegetation covering a region depends primarily on three ingredients: climate, type of soil, and history of disturbance. For example, three distinct vegetation types (temperate deciduous forest, pine barrens, and salt marsh) occur in northern New Jersey in close proximity to one another, but on different soil types (Forman 1979). Moreover, if a mature stand of deciduous forest is destroyed, as at the hands of humans or by natural fire, it is not reestablished immediately. Instead, certain plant species colonize the area and are in turn replaced by later colonists, beginning with weedy pioneer species and continuing until the mature or **climax** vegetation is reestablished. This process is called **secondary succession**. Throughout this process, both the microclimate and the soil of the site also change, becoming more favorable for some species and less favorable for others.

Soil is formed by the weathering of rock and the accumulation of organic material from dead and decaying organisms. The process by which new soil is formed from mineral substrates is usually long and complicated. Physical processes, such as freezing and thawing, and water and wind erosion, break down the parent rock material. Organisms also play key roles: lichens hasten the weathering of rock; decaying corpses of plants, animals, and microbes add organic material; the activities of roots and microbes alter the chemical composition of the soil; and burrowing animals mix and aerate it. Totally organic soils (**histosols**), such as peat, form in certain unusual environments.

The rate of soil formation varies widely, depending largely on the nature of the parent material and the climatic setting. The formation of shallow soils may take thousands of years in Arctic and desert regions, where temperature and moisture regimes are extreme (e.g., McAuliffe 1994). For example, soils only a few centimeters deep cover much of eastern Canada, where the retreat of the last Pleistocene ice sheets left bare rock only about 10,000 years ago. In other cases, especially when soils are formed from sand, lava, or alluvial materials in regions with warm, moist climates, primary succession can be amazingly rapid. In 1883, the small tropical island of Krakatau in Indonesia experienced an explosive volcanic eruption that exterminated all living things and left only sterile volcanic rock and ash. Organisms rapidly recolonized Krakatau from the large neighboring islands of Java and Sumatra, and by 1934, only 50 years after the eruption, 35 cm of soil had been formed, and a lush tropical rain forest containing almost 300 plant species was rapidly developing (Docters van Leeuwen 1936).

Formation of Major Soil Types

Anything we write about soils must be a gross oversimplification, because both the classification and the distributions of soils are very complex, even controversial. Visit the vast flat plains of the United States or the Ukraine and you will find just one or a few soil types distributed for as far as the eye can see, but in other geographic regions, especially mountainous areas, soil maps are mosaics that look like complicated abstract paintings (Figure 3.8). Great Britain has a series of unusual organic soil types formed in cold, wet environments, plus soils greatly modified by centuries of human activities.

We can begin to appreciate the diversity and distribution of soils by studying the four major processes that produce the primary, or zonal, soil types. These so-called **pedogenic regimes** are those that typically occur in habitats characterized by temperate deciduous and coniferous forests (**podzolization**), tropical forests (**laterization**), arid grasslands and shrublands (**calcification**), and waterlogged tundra (**gleization**).

Podzolization occurs at temperate and subarctic latitudes and at high elevations where temperatures are cool and precipitation is abundant. In such climates plant growth may be substantial, but the low temperatures inhibit microbial activity, so organic matter, called **humus**, accumulates. As the

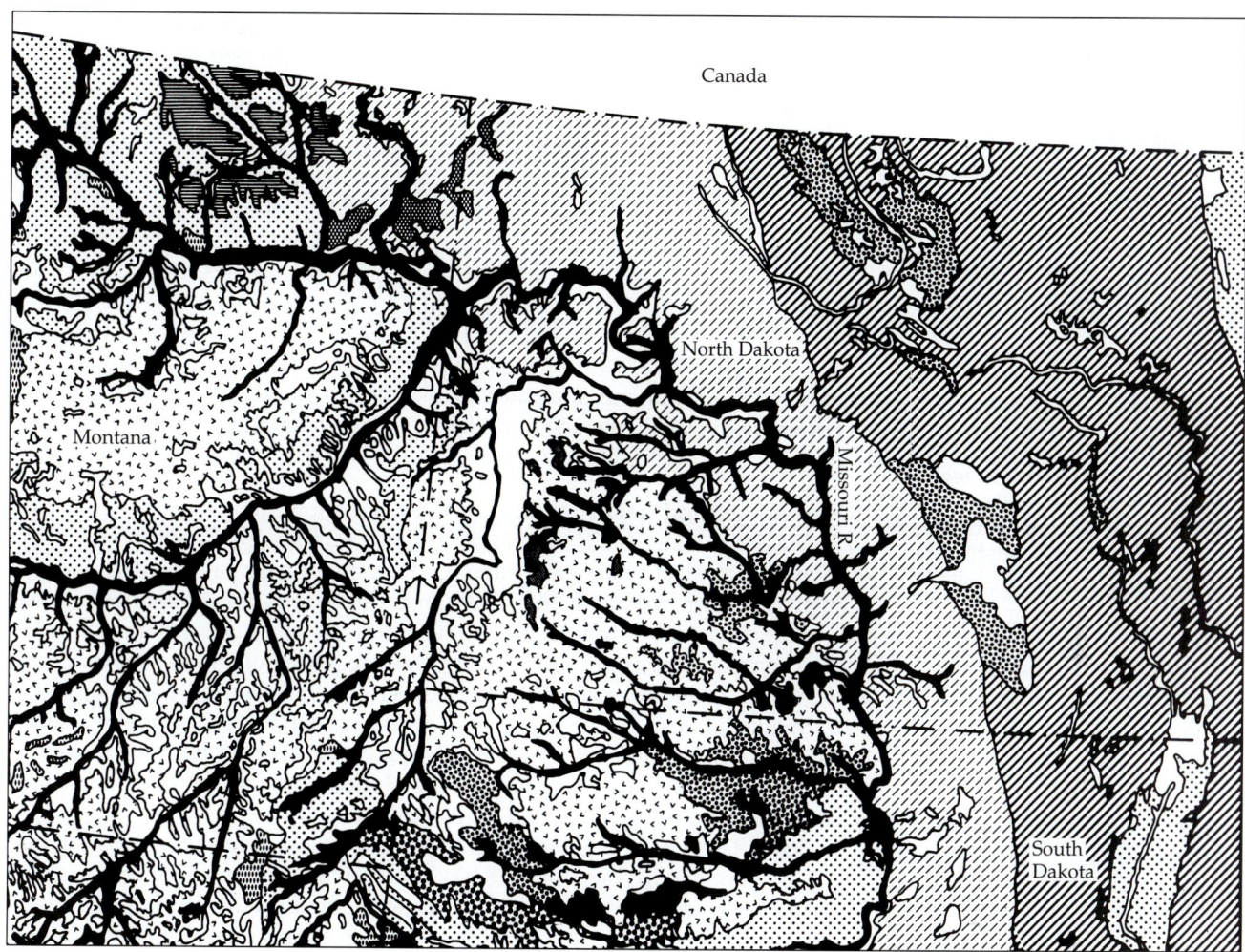

Canada

North Dakota

Montana

Missouri R.

South Dakota

Figure 3.8 Soil map for the Great Plains and adjacent Rocky Mountains of North Dakota, Montana, and adjacent states. Note that relatively homogeneous soils occur over extensive areas of the plains (right), but a much more heterogeneous mosaic of soil types occurs in the topographically and geologically diverse mountainous areas.

humus decays, organic acids are released and carried downward (**leached**) through the soil profile by percolating water. The hydrogen ions of these acids tend to replace cations that are important for plant growth, such as calcium, potassium, magnesium, and sodium, which are removed by leaching from the soil (Figure 3.9A). This process leaves behind a silica-rich upper soil containing oxidized iron and aluminum compounds, but few cations. Coniferous forests, which thrive in such acidic conditions, are a characteristic vegetation on podzolic soils.

In the humid tropics, which experience high temperatures and heavy rainfall, microbes and other organisms rapidly break down dead organic material, so little humus can accumulate. In the absence of organic acids, oxides of iron and aluminum precipitate to form red clay or a bricklike layer (laterite). The heavy rainfall causes silica and many cations, such as potassium, sodium, and calcium, to be leached out of the soil (Figure 3.9B), leaving behind a firm and porous soil with very low fertility. In some areas, if the tropical forest cover is removed, the organic material and its bound nutrients are easily lost, and the intense equatorial sun bakes the exposed lateritic soils hard, retarding secondary succession and making the area unsuitable for agriculture.

Calcareous soils typically occur in arid and semiarid environments, particularly in regions where thick layers of calcium carbonate were deposited beneath ancient shallow tropical seas. Where rainfall is relatively low, so that

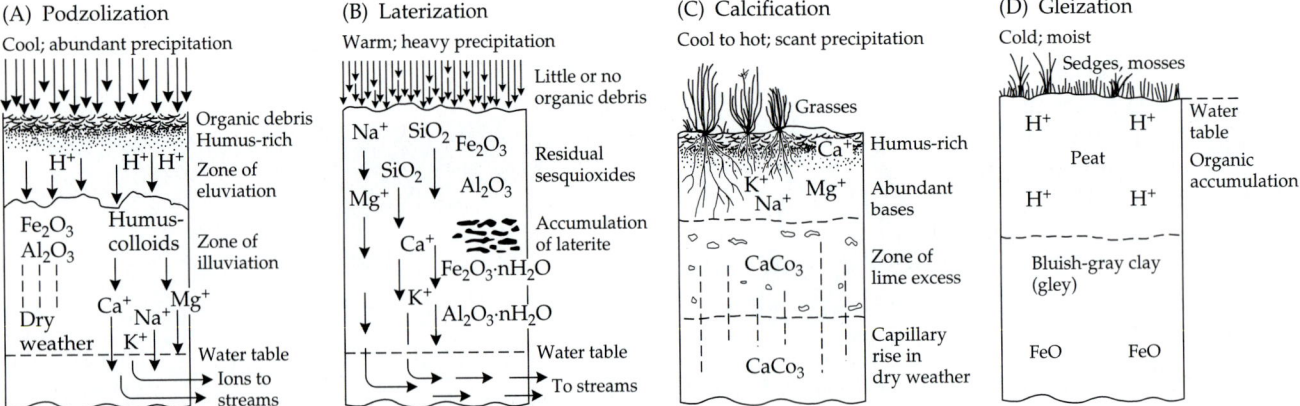

Figure 3.9 Schematic representations of the four major pedogenic regimes, showing the resulting soil profiles: (A) podzolization, (B) laterization, (C) calcification, and (D) gleization. (After Strahler 1975.)

evaporation and transpiration exceed precipitation, cations are generally not leached out. Instead, they are carried downward through the soil profile to the depth of greatest water penetration, where they precipitate, forming a layer rich in calcium carbonate (Figure 3.9C). In desert soils, the scanty rainfall penetrates only a short distance below the surface, where it leaves behind a rock-like layer of calcium carbonate, called **caliche**. In regions where precipitation is higher, water and roots penetrate deeper into the soil profile, leading to the formation of deep, fertile soils rich in organic material and essential nutrients, such as potassium, nitrogen, and calcium. Such soils are typical of tallgrass and shortgrass prairie habitats, although little of the former remains because these soils are so highly prized for agriculture.

In cold, wet polar regions, gleization is the typical process of soil formation. At the permanently wet (or frozen) surface, where the low temperatures and waterlogged conditions prevent decomposition, acidic organic matter builds up, sometimes forming a layer of peat that can be several meters thick (Figure 3.9D). Below this organic upper layer, an inorganic layer of grayish clay, containing iron in a partially reduced form, typically accumulates. While few nutrients are lost through leaching, the highly acidic conditions cause nutrients to be bound up in chemical compounds that cannot be used by plants. Thus gley soils typically support a sparse vegetation of acid-tolerant species.

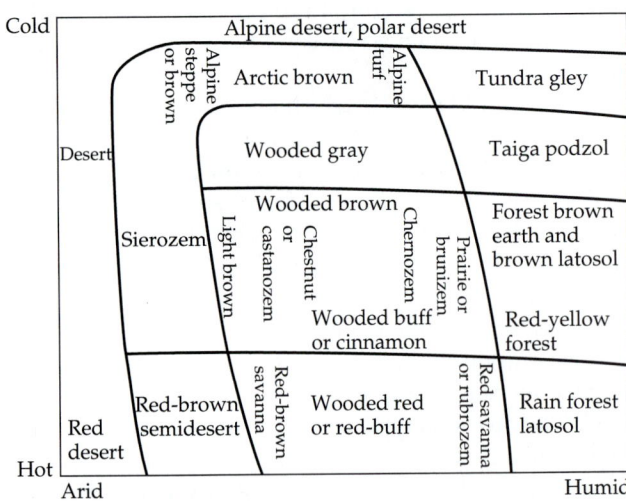

Figure 3.10 Schematic diagram depicting the relationships between major soil types and climate, showing that different combinations of temperature and precipitation cause the formation of distinctive soil types. (From Whitaker 1975.)

The above descriptions represent four idealized cases of soil formation processes. Given the complex variation in parent material and climate over the earth's surface, pedogenic regimes vary in complex but predictable ways. The processes of soil formation and soil types described above occur where the chemical composition of the parent material is typical of the common rock types: sandstone, shale, granite, gneiss, and slate. The soils that are derived from these "typical" rocks are called **zonal soils**. A simplified summary of the relationship between climate and zonal soil type is given in Figure 3.10. The global distribution of zonal soil types (Figure 3.11) can also be compared with the global climate map (Figure 3.6) to demonstrate the close relationship between soils and climate.

Unusual Soil Types Requiring Special Plant Adaptations

In addition to such zonal soils, there are unusual soil types derived from parent material of unusual chemical composition. Certain rock types, such as gypsum, serpentine, and limestone, contain unusually high amounts of some compounds and little of others. Serpentine, for example, is particularly deficient in calcium, and gypsum contains an excess of sulfate. Few plant species can tolerate such **azonal soils**, and the low-diversity plant communities that do grow on such soils have special physiological adaptations for dealing with their unusual chemical composition.

One example of a soil type that requires special adaptations by plants is **halomorphic soil**, which contains very high concentrations of sodium, chlorides, and sulfates. Halomorphic soil typically occurs near the ocean in estuaries and salt marshes, and in arid inland basins where shallow water accu-

Figure 3.11 World distribution of major soil types. Note the close correlation of these soil types with the climatic zones shown in Figure 3.6, reflecting the influence of temperature and precipitation on soil formation.

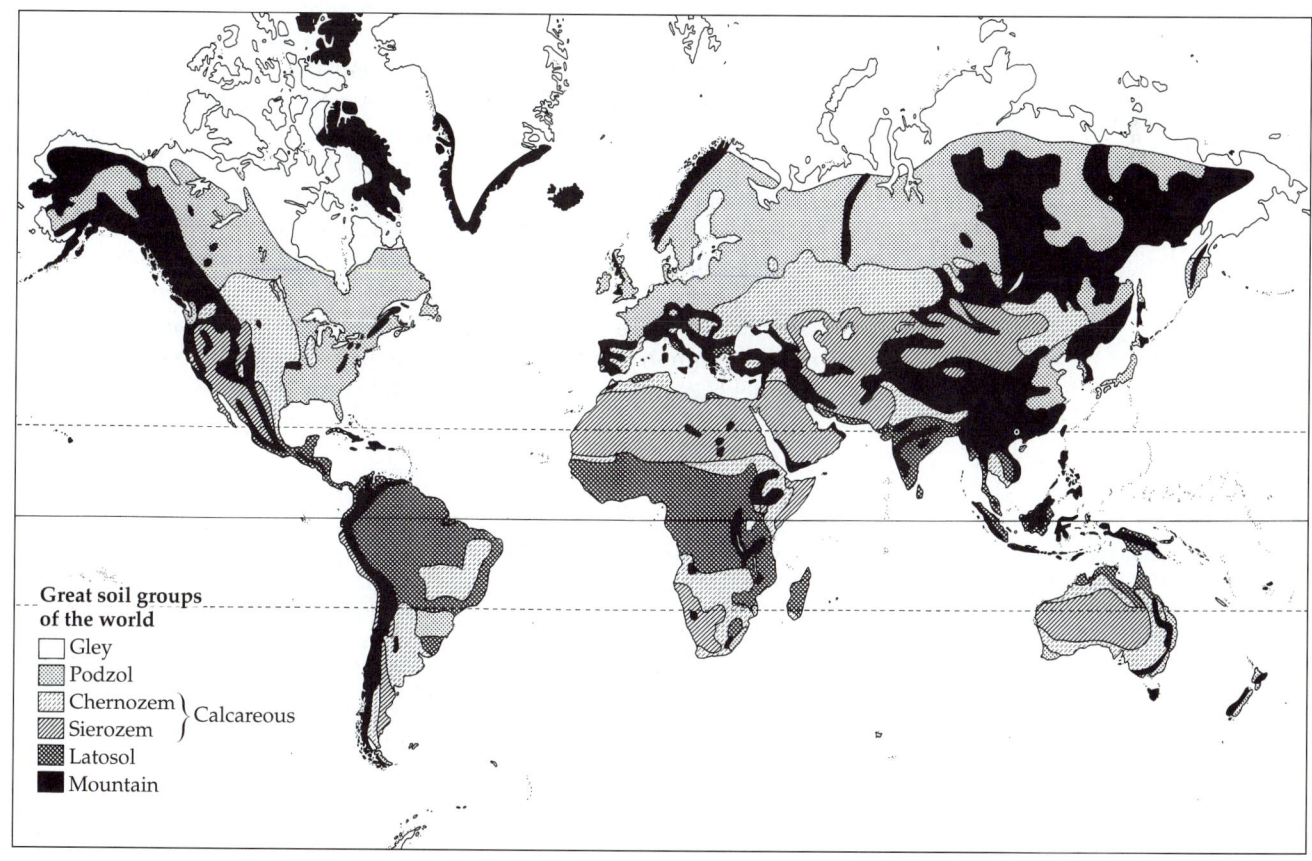

Great soil groups of the world
- ☐ Gley
- Podzol
- Chernozem ⎫
- Sierozem ⎬ Calcareous
- Latosol
- Mountain

mulates and evaporates, leaving behind high concentrations of salts. A small number of specialized **halophytic** ("salt-loving") plant species grow in such areas. They include a variety of taxonomic and functional groups, each of which has special adaptations for dealing with the problem of maintaining osmotic and ionic balance in these environments. Some species of mangroves and grasses excrete salts from specialized cells in their leaves, whereas pickleweeds and ice plants store salts in special cells in their succulent leaves.

As mentioned above, highly acidic soil conditions cause essential nutrients, especially nitrogen and phosphorus, to be bound in compounds that plants cannot use. Pitcher plants, sundews, Venus's flytraps, and other insectivorous plants can grow in highly acidic soils or other environments where nutrients are severely limiting. These plants obtain their nitrogen and phosphorus by capturing living insects, digesting them, and assimilating the nutrients. A less spectacular adaptation to acidic and other nutrient-poor soils is evergreen vegetation (Beadle 1966). Because nutrients are lost when leaves are dropped, and more minerals must then be taken up by the roots to produce new leaves, plants can use limited nutrients more efficiently by retaining their leaves for longer periods. In mesic temperate climates, where the predominant vegetation is usually deciduous forest, it is common to find evergreens growing on acidic and nutrient-poor soils. Examples are the pine barrens of the eastern United States and the *Eucalyptus* forests of Australia (Daubenmire 1978; Beadle 1981).

In addition to their chemical composition, the physical structure of soils can influence the distribution of plant species and the nature of vegetation. In arid regions, for example, the size and porosity of soil particles affect the availability of the limited moisture to plants by affecting the runoff, infiltration, penetration, and binding of water. Thus, even within a small region of uniform climate, differences in soil texture can cause large differences in vegetation. A striking example is provided by the **bajadas,** or alluvial fans, of desert regions (Figure 3.12). These interesting geological formations are made up of sediments carried out of mountains by infrequent but heavy flooding of the canyons. As the floodwater gradually loses energy, it deposits sediments in a gradient, dropping large, heavy rocks at the mouth of the canyon and small sand- and clay-sized particles at the bottom of the fan. The resulting bajada shows a corresponding gradient in water availability and vegetation (Bowers and Lowe 1986). Cacti predominate on the coarse, rocky, well-drained soils high on the bajada, where water is available only for short periods during and after rains. These succulents can take up water rapidly through their extensive shallow roots and store it in their expandable tissues. Shrubs and grasses are much more common farther down the bajada, where their roots can extract the water held on and among the smaller soil particles.

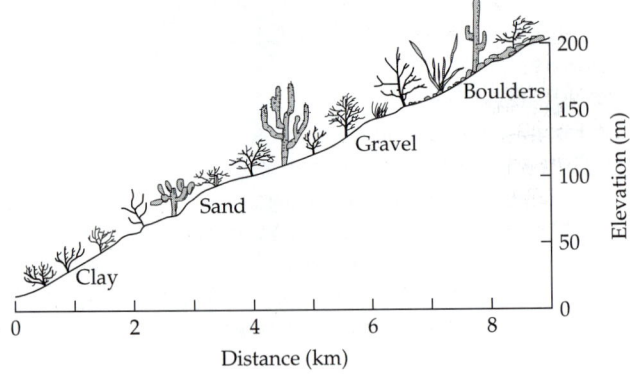

Figure 3.12   Schematic representation of the local elevational distribution of soil particle size and vegetation on a desert bajada on the Sonoran coast of the Gulf of California (Sea of Cortez). At the upper end of the alluvial fan, where large boulders have been deposited, the vegetation is dominated by cacti and other succulents that can take up water rapidly, before it percolates below the root zone. At the lower end, where water infiltration is poor and the existing water is tightly bound by fine clay particles, the vegetation consists of sparse, shallowly rooted shrubs. The greatest water availability, productivity, and species diversity occur at intermediate elevations, where the soils are sandy, infiltration is high, and water is not tightly bound by soil particles.

A somewhat similar situation occurs along the coast of the Gulf of Mexico in the southeastern United States. The uplands have coarse, sandy, well-drained soils that support a drought-tolerant coniferous woodland/savanna vegetation. In contrast, the lowlands have accumulated fine, water-retaining soils, and this, as well as their proximity to the water table, allows them to support a much more mesic vegetation. Thus a person interested in the factors influencing plant distributions and community composition at this local to regional scale must pay particular attention to how subtle characteristics of soil structure affect the runoff, infiltration, and retention of rainwater.

Although we have concentrated here on the relationship between soils and vegetation, soils also affect the distributions of animals, both indirectly, by controlling which plant species are present, and directly, through the effects of the chemical and physical environment on their life cycles. Many kinds of mammals, reptiles, and invertebrates are restricted to particular types of soils that meet their specialized requirements for burrowing and locomotion. For example, in North American deserts, lizards of the genus *Uma*, the kangaroo rat *Dipodomys deserti*, and the kangaroo mouse *Microdipodops pallidus* are all restricted to dunes and similar patches of deep, sandy soil. Another set of species, including chuckwallas (*Sauromalus obesus*), collared lizards (*Crotophytus collaris*), and rock pocket mice (*Chaetodipus intermedius*), show just the opposite habitat requirement, being restricted to rocky hillsides and boulder fields.

Aquatic Environments

As anyone who has ever tried to keep tropical fish knows, warm and relatively stable temperatures are essential for their survival and reproduction. Salinity, light, inorganic nutrients, pH, and pressure also play key roles in the distributions of aquatic organisms. Like terrestrial climates, the physical characteristics of water often exhibit predictable patterns along geographic gradients, which can be understood with a basic background in physics.

Stratification

Thermal stratification. When solar radiation strikes water, some is reflected, but most penetrates the surface and is ultimately absorbed. Although water may appear transparent, it is much denser than air, and its absorption of radiation is rapid. Even in exceptionally clear water, 99% of the incident solar radiation is absorbed in the upper 50 to 100 m, and this absorption occurs even more rapidly if many organisms or colloidal substances are suspended in the water column. Longer wavelengths of light are absorbed first; the shorter wavelengths, which have more energy, penetrate farther, giving the depths their characteristic blue color.

This rapid absorption of sunlight by water has two important consequences. First, it means that photosynthesis can occur only in surface waters where the light intensity is sufficiently high (the photic zone). Virtually all of the primary production that supports the rich life of oceans and lakes comes from plants living in the upper 10 to 30 m of water. Along shores and in very shallow bodies of water, some species, such as kelp, are rooted in the substrate. These plants may attain considerable size and structural complexity, and may support diverse communities of organisms. In the open waters that cover much of the globe, however, the primary producers are tiny, often unicellular algae, called **phytoplankton**, which are suspended in the water column. **Zooplankton**, tiny crustaceans and other invertebrates that feed on phytoplankton, migrate vertically on a daily cycle: up into the surface waters at night to feed and down into the dark, deeper waters during the day to escape their visually hunting fish predators.

Second, the rapid absorption of sunlight by water means that only surface water is heated. Any heat that reaches deeper water must be transferred by convection or by currents. Consequently, deep waters are characteristically cold, even in the tropics. The density of pure water is greatest at 4° C, and declines as its temperature rises above or falls below this point. This unusual property of water is significant for the survival of many temperate and polar organisms because it means that ice floats. Ice provides an insulating layer on the surface that prevents many bodies of water from freezing solid. The presence of salts in water lowers its freezing point, and some organisms are therefore able to exist in unfrozen water below 0° C (de Vries 1971).

A more general consequence of the relationship between water density and temperature is that water tends to acquire stable thermal stratification. When solar radiation heats the water surface above 4° C, the warm surface water becomes lighter than the cool deeper water, and so tends to remain on the surface, where it may be heated further and become even less dense. In tropical areas and in temperate climates during the summer, the surfaces of oceans and lakes are usually covered by a thin layer of warm water. Unless these bodies of water are shallow, the deep water below this layer is much colder (sometimes near 4° C). The change in temperature between the surface layer and deeper water is called a **thermocline** (Figure 3.13). Mixing of the surface water by wave action determines the depth of the thermocline and maintains relatively constant temperatures in the water above it. In small temperate ponds and lakes that do not experience high winds and heavy waves, the thermocline is often so abrupt and shallow that swimmers can feel it by letting their feet dangle a short distance. In large lakes and oceans, where there is more mixing of surface waters, the thermocline is usually deeper and less abrupt.

Tropical lakes and oceans show pronounced permanent stratification of their physical properties, with warm, well-oxygenated, and lighted surface water giving way to frigid, nearly anaerobic, and dark (aphotic) deep water. Oxygen cannot be replenished at great depths where there are no photosynthetic organisms to produce it, and the stable thermal stratification prevents mixing and reoxygenation by surface water. Only a relatively few organisms can exist in these extreme conditions. The feces and dead bodies of organisms living in the surface waters sink to the depths, taking their mineral nutrients with them. The lack of vertical circulation thus limits the supply of nutrients to

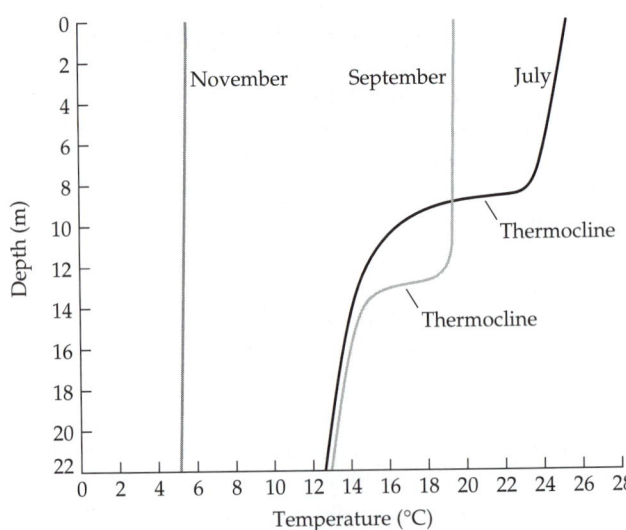

Figure 3.13 Vertical temperature profiles of Lake Mendota, Wisconsin, at different dates from summer through fall, showing the loss of thermal stratification as the lake cools. In July, the thermocline is pronounced and shallow. In September it is less pronounced and deeper, and by November it has disappeared, allowing surface and deep waters to mix in the fall overturn. (After Birge and Juday 1911.)

the phytoplankton in the photic zone. Consequently, deep tropical lakes are often relatively unproductive and depend on continued input from streams for the nutrients required to support life.

Overturn in temperate lakes.   The situation is somewhat different in temperate and polar waters. Deep lakes, in particular, undergo dramatic seasonal changes: they develop warm surface temperatures and a pronounced thermocline in summer, but freeze over in winter. Twice each year, in spring and fall, the entire water column attains equal temperature and equal density, the temperature/density stratification is eliminated, and moderate winds may then generate waves that mix deep and shallow water, producing what is called **overturn** (see Figure 3.13). This semiannual mixing carries oxygen downward and returns inorganic nutrients to the surface. Phosphorus and other mineral nutrients may be depleted during the summer, when warm temperatures allow algae to grow and reproduce at high rates; overturn replenishes these nutrients, stimulating the growth of phytoplankton. Temperate lakes, such as the Great Lakes of North America, are often quite productive and support abundant plant and animal life, including valuable commercial fisheries. However, abnormally high nutrient inputs—often due to runoff from agricultural fields and discharges of inadequately treated sewage—can cause excessive production, rapid algal growth, depletion of oxygen, fish kills, and other environmental problems.

Oceanic Circulation

The vertical and horizontal circulation of oceans is more complicated than that of lakes, in part because oceans are so vast, extending through many climatic zones, and in part because salinity affects the density of water. Salts are dissolved solids carried into the oceans by streams and concentrated by evaporation over millions of years. The presence of salts in water increases its density, causing swimmers to experience greater buoyancy in the ocean than in fresh water. Varying salinity and density have important effects on ocean circulation. Rivers and precipitation continually supply fresh water to the surface of the ocean, and this lighter water tends to remain at the surface. If you have ever flown over the mouth of a large, muddy river, such as the Mississippi, you may have noticed that its water remains relatively intact, flowing over the denser ocean water for many kilometers out to sea. In polar regions, the input of fresh water to the ocean from rivers and precipitation generally exceeds losses from evaporation, but the reverse is true in the tropics. This pattern creates a somewhat confusing situation, because warm tropical surface water tends to become concentrated by evaporation and to increase in density, counteracting to some extent stratification owing to temperature. On the other hand, cold polar water, which would be expected to show little stratification, may become somewhat stabilized as low-density fresh water accumulates on the surface.

Vertical circulation occurs in oceans, but the rates of water movement are so slow that a water mass may take hundreds or even thousands of years to travel from the surface to the bottom and back again. Areas of descending water tend to occur at the convergence of warm and cold currents in polar regions, where the colder, denser water sinks under the warmer, lighter water. Areas of rising water, called **upwelling**, are found where ocean currents pass along the steep margins of continents. This happens, for example, along the western coast of North and South America, where there is little continental shelf and the land drops sharply offshore. As the Pacific gyres sweep toward the equator along these shores, the Coriolis effect and, in tropical latitudes, the easterly trade winds, tend to deflect the surface water offshore, and water wells up from the depths to replace it. Because upwelling, like overturn in lakes, returns nutrients to the surface, productivity tends to be high in areas of upwelling (see Figure 5.30B). Probably the greatest commercial fishery in the world is located in the zone of upwelling off the coasts of Chile and Peru.

Surface currents, such as the great hemispheric gyres (see Figure 3.4), are relatively shallow and rapidly moving, so that they tend to form discrete water masses, each of which has a characteristic salinity and temperature profile distinct from those of neighboring water masses. Some organisms with limited capacity for locomotion may drift in currents for long distances without leaving a single uniform water mass. Organisms that can move actively to overcome the currents must also be able to tolerate the contrasting physical environments in different water masses.

Although oceanographers have recognized the existence of distinct water masses within the oceans for many years, modern technology has revealed the extent of spatial heterogeneity in shallow ocean waters. For example, investigators from the Woods Hole Oceanographic Institution have studied the physical environment and the biota of Gulf Stream rings (Wiebe 1976, 1982; Lai and Richardson 1977). These rings are small masses of cold or warm water that have broken away from the southern or northern edges of the Gulf Stream to drift through water of contrasting temperature in the North Atlantic. They can be readily seen on infrared satellite images that show sea surface temperatures (Figure 3.14). These rings not only have physical environments that are strik-

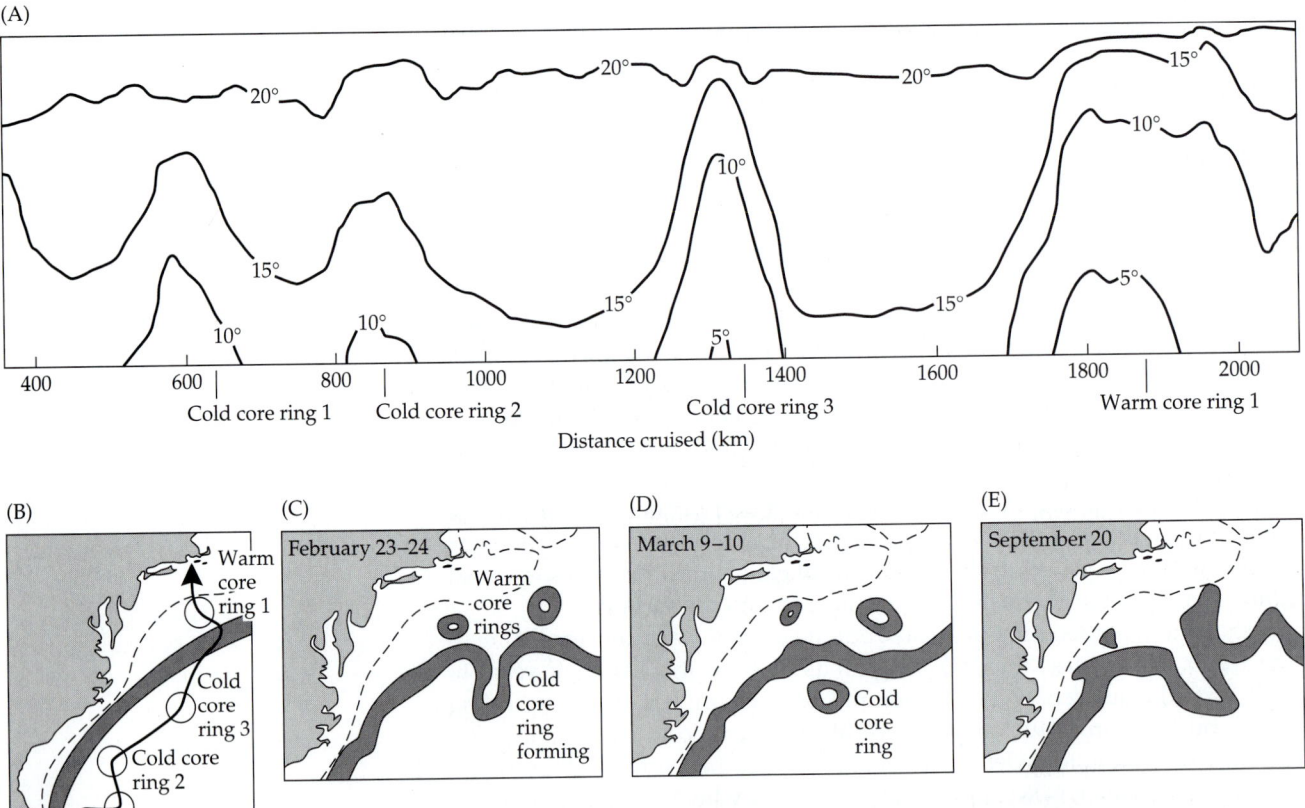

Figure 3.14   Small-scale spatial and temporal heterogeneity of surface waters in the North Atlantic Ocean is caused by small water masses, called rings, that split off from the Gulf Stream (dark shaded line). (A) Temperature/depth profile recorded by an oceanographic vessel that traveled through several rings, as indicated by the line on the map (B). (C–E) Changes in water surface temperatures as mapped by infrared satellite imagery, showing the formation, movement, and disappearance of rings. (After Wiebe 1982.)

ingly different from their surroundings, but also contain a unique biota that can persist in these special conditions far from its normal distribution in the Gulf Stream. The possible roles of these floating warm- or cold-water eddies in trans-Atlantic dispersal, both now and in the past, are intriguing.

Pressure and Salinity

Pressure and salinity vary greatly among aquatic habitats. These variations have major effects on the distributions of organisms because special physiological adaptations are necessary to tolerate the extremes. As every scuba diver knows, water pressure increases rapidly with depth. It becomes a major problem for organisms in the ocean, where the deepest areas are up to 6 kilometers below the surface. Pressure increases at a rate of about one atmosphere (about 1.5 mega-Pascals) for every 10 m of depth. In the abyssal depths, pressures are more than 200 times greater than at the surface. Organisms adapted to living in surface waters cannot withstand the pressures of the deep sea, and vice versa.

Variation in salinity is relatively discontinuous. The vast majority of the earth's water is in the oceans and is therefore highly saline (greater than 34 parts per thousand of solutes). In contrast, freshwater lakes, marshes, and rivers, which account for less than 1% of the earth's waters, contain very few dissolved salts. Habitats of intermediate or fluctuating salinity, such as salt marshes and estuaries, constitute only a tiny fraction of the earth's aquatic habitats. Consequently, most aquatic organisms are physiologically adapted and geographically restricted either to fresh water, where the physiological problem is obtaining sufficient salts to maintain osmotic balance, or to salt water, where the problem can be eliminating excess salt. Only a few widely tolerant (**euryhaline**) organisms have the special physiological mechanisms required to survive in the widely fluctuating salinities of estuaries and salt marshes.

Tides and the Intertidal Zone

We can learn a great deal about the factors determining the distributions of organisms by studying environmental gradients: both gradual changes, such as variation in light and pressure with depth in lakes and oceans, and rapid changes, such as the variation in temperature in the cooling outflow of a hot spring. One of the steepest, best-studied, and most interesting environmental gradients occurs where the ocean meets the land. Along the shore is a narrow region that is alternately covered and uncovered by seawater. It is called the **intertidal zone** because it experiences a regular pattern of inundation and exposure caused by tides.

Sir Isaac Newton explained how the gravitational influences of the moon and sun interact to cause the global fluctuations in sea level that we call tides. The entire story is complicated, but the main pattern and its mechanism are simple. The tides are flows of surface waters. They occur in response to a net tidal force, which reflects a balance between the centrifugal force of the spinning earth and the gravitational forces of the moon and sun (Figure 3.15). Because the gravitational force exerted by an object is equal to its mass divided by the square of its distance, the smaller but nearer moon has a greater effect than the sun.

Most shores typically experience both a daily and a monthly tidal cycle: there are two high and two low tides every 24 hours, and there are two periods of extreme tides each month, corresponding to the new and full moons (Figure 3.16). During these periods, the moon and sun are in the same plane as the earth, and their gravitational effects are additive, causing high-amplitude or **spring tides**, with the highs occurring at dawn and dusk and the lows near

Figure 3.15 (A) Schematic representation of how the centrifugal force of the spinning earth and the gravitational force of the moon cause the tides. On the side of the earth closest to the moon, the gravitational force (white) is stronger than the centrifugal force (gray), and the net tidal force (black) tends to draw surface water toward the moon. On the side opposite the moon, the gravitational force is weaker than the centrifugal force, and the net tidal force tends to draw water away from the moon. In between these extremes, the gravitational and centrifugal forces are balanced, and there is essentially no net tidal force. (B) The movement of surface waters in response to these tidal forces.

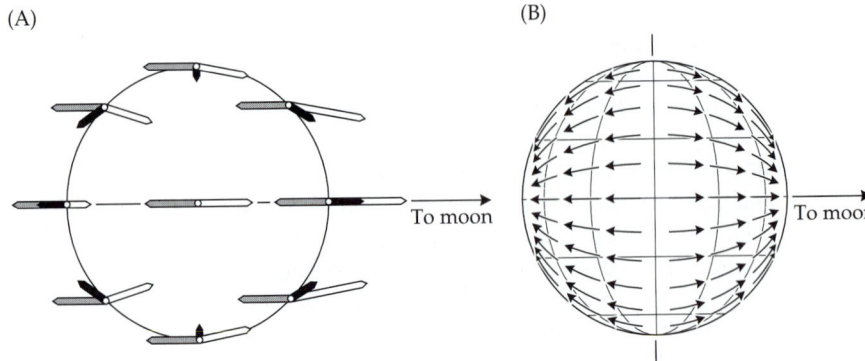

noon and midnight. During the quarter moons, the sun and moon are at right angles to each other (from the perspective of earth), and their gravitational effects tend to cancel each other, resulting in low-amplitude or **neap tides**.

A distinct community of plant and animal species lives in the intertidal zone. Nearly all aspects of the lives of these organisms are dictated by the cyclical pattern of inundation by seawater at high tide and exposure to desiccating conditions at low tide. Most species are confined to a very narrow zone of tidal exposure, so that their distributions form thin lines running horizontally along the shore. As we shall see in Chapter 4, the narrow ranges of species in the intertidal zone (typically only a few centimeters or meters), and the ease with which critical environmental conditions can be manipulated experimentally, have produced a wealth of information about the factors limiting the distributions and regulating the diversity of species.

Figure 3.16 A tide calendar for the northern Gulf of California (Sea of Cortez), showing the typical pattern of tides due to the gravitational influences of the moon and sun. Note that there are two high and two low tides each day. There are also two periods of low-amplitude (neap) and high-amplitude (spring) tides each month; the latter correspond to the times of the new and full moons, when the gravitational forces of moon and sun are aligned. (Courtesy of D. A. Thomson.)

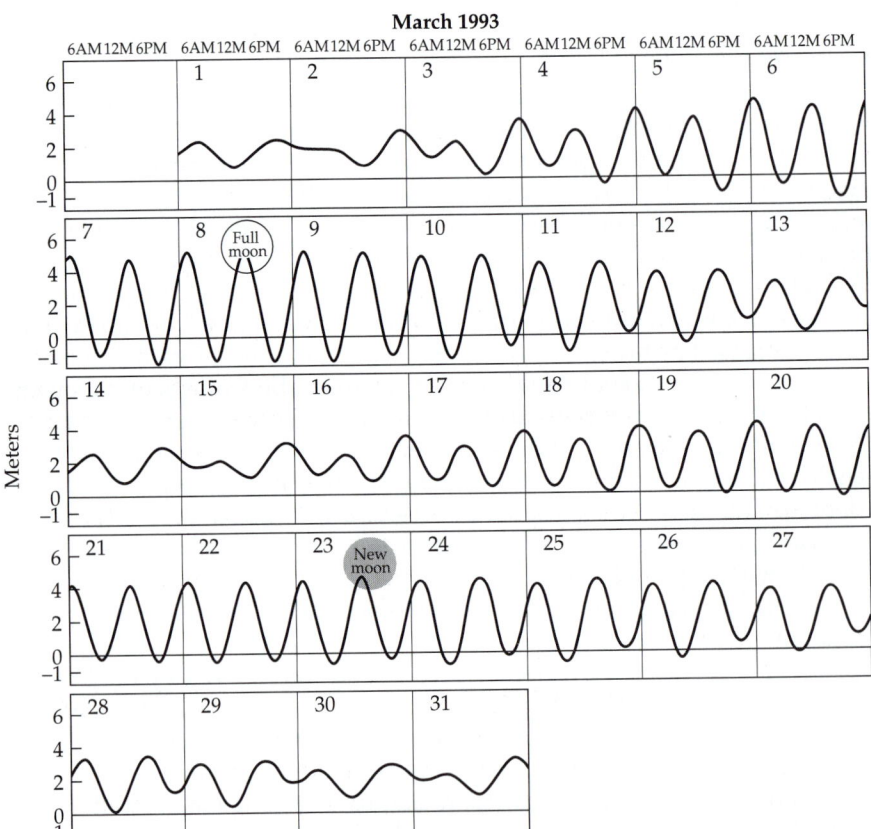

Microenvironments

Small-Scale Environmental Variation

It would be misleading to end here without a word of qualification. In this chapter we have been concerned primarily with global patterns of variation in abiotic environments that influence the distributions of organisms. Upon closer inspection, however, these patterns tell us surprisingly little about the actual conditions experienced by an organism living in a particular region. This point was best stated by the observer who noted that the climatic data recorded by the National Weather Service measures accurately only the climate experienced by the spiders living in the shelters that house the recording instruments!

Microenvironmental variation is less of a problem in aquatic habitats than on land because the physical properties of water tend to prevent the occurrence of abrupt small-scale changes. But even there, changes can be very abrupt, and conditions can go from favorable to intolerable in distances of just a few centimeters or meters. Examples include the rapid changes in temperature at thermoclines and around hydrothermal vents (see Chapter 6), and in salinity in estuaries where rivers enter the ocean.

In terrestrial habitats, the climates of small places, called **microclimates**, may bear little relationship to large-scale climatic patterns. On the one hand, two organisms living only a few centimeters apart may live in radically different physical environments—humid or arid, hot or cold, windy or protected. On the other hand, by selecting appropriate microenvironments, individuals can be distributed over a wide range of latitudes and elevations and still experience virtually identical physical conditions. Examples of both situations abound. Lizards are conspicuous elements of most desert faunas because they are active during the day and are able to tolerate the hot, dry conditions. The same deserts, however, may also be inhabited by frogs and toads, which spend most of their lives buried in the cool, relatively moist soil, emerge to feed only on rainy or humid nights, and possess adaptations for breeding in ephemeral ponds that form after occasional heavy rains. Perhaps the best examples of organisms that live in similar physical environments over a wide geographic range are internal parasites and microbial symbionts of birds and mammals. The same species may occur in tropical rain forests and arctic tundras, but still live in virtually identical, homoeostatically regulated environments within the bodies of their hosts.

The most distinctive microenvironments are typically small and widely dispersed sites. The capacity of organisms to exploit specific microenvironments depends largely on their mobility or vagility (regardless of whether they are actively or passively transported), body size, special physiological properties, and behavioral selectivity. We can readily imagine how mobile animals can seek out and settle in a particular habitat, but we should keep in mind that plants also may have adaptations that result in effective microhabitat selection. For example, many species have seeds that are attractive to certain kinds of animals, which disperse them to favorable microsites. Many seeds also require specific cues for germination that indicate the presence of favorable environmental conditions.

Colonizing Suitable Microenvironments

In order to live in isolated localities and microclimates, organisms must be able to get to them. Many plants, invertebrate animals, and microbes accomplish this during special dispersal stages of their life cycles (e.g., as seeds, eggs, or spores) that are adapted to be carried long distances and to tolerate extreme

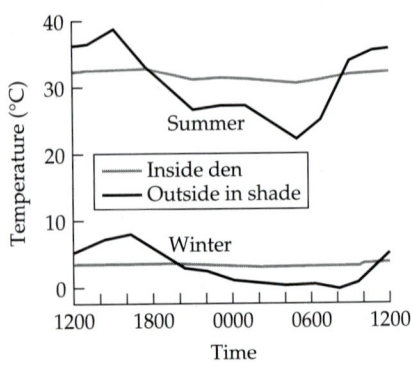

Figure 3.17 Temperatures inside (black lines) and outside (gray lines) the den of a bushy-tailed woodrat (*Neotoma cinerea*), a deep crack between large boulders in the high desert of southeastern Utah, during midsummer (above) and midwinter (below). Because the den, where the animal spends most of its time, experiences much less variation than the macroclimate outside, it affords vital protection from stressfully high and low temperatures in summer and winter, respectively. (After Brown 1968.)

environments while in transit. Often, however, their arrival at a suitable microsite is largely a matter of chance. In contrast, many animals are able to use their sophisticated sensory and locomotor systems to seek out isolated microenvironments. To demonstrate both active and passive dispersal, it is only necessary to create a small artificial pond and observe the rapidity with which it is colonized both by zooplankton (copepods and other small crustaceans), which disperse passively as resistant eggs, and by large insects (diving beetles and dragonflies), which fly long distances to actively seek out suitable sites for colonization.

Some distinctive microenvironments are so isolated that the specialized organisms that inhabit them cannot disperse among them. How, then, were they originally colonized? Some organisms colonized them in the past when bridges of suitable habitat existed between them, or the intervening areas were at least not so extensive and inhospitable. Examples include the fishes of isolated lakes, which require freshwater connections in order to disperse. Still other microenvironments are so inaccessible that their biotas include many unique species that have evolved in situ, diverging from ancestral forms that occurred in neighboring habitats. Examples include many of the highly differentiated, blind, unpigmented cave animals that have evolved in each cave system from surface-living ancestors. Similarly, many of the unique plants that inhabit isolated pockets of serpentine soils have been derived from species that occurred on the surrounding zonal soils.

While many organisms select microclimates that are appropriate for their lifestyles, some are able to create their own microenvironments. Many kinds of small mammals dig burrows or build other structures that provide favorable microsites in otherwise inhospitable environments. Animals that use subterranean burrows experience nearly constant favorable microclimates year-round because the sun's heat is not conducted more than a few decimeters through the soil, and the humidity stays near 100%. Desert-dwelling woodrats (*Neotoma* spp.), for example, make their dens in deep rock crevices or in burrows beneath "houses" they construct by piling up large quantities of sticks, stones, and other debris. These dens provide relatively stable and moderate temperatures and high humidities, even when conditions just outside are lethal (J. Brown 1968) (Figure 3.17). These structures are used for shelter not only by their woodrat owners, but also by a large number of other organisms, including not only spiders, scorpions, insects, lizards, snakes, and mice, but also many kinds of fungi and bacteria (see, for example, Reichman 1985; Seastedt et al. 1986; Hawkins and Nicoletto 1992).

The fact that many organisms are found only in particular microenvironments has important consequences for our understanding of geographic patterns. On the one hand, it means that some species may have much broader geographic ranges than we would have predicted from a cursory comparison of their physical tolerances and climatic patterns. On the other hand, careful studies have shown that the local distributions of many organisms are highly patchy, because within their geographic ranges, they are confined to microsites that provide very specific environmental conditions. In the next chapter, we consider in more detail how different kinds of environmental conditions limit the local distributions and geographic ranges of individual species.

CHAPTER 4

Distributions of Single Species

The Geographic Range

The proposition that each species has a unique **geographic range** is central to all of biogeography. Biogeographers study many phenomena—locations of occurrence of individual organisms, shifts in the local or regional distribution of a population, present and past distributions of higher taxa or **clades** (lineages of species descended from a common ancestor; see Chapter 11), patterns of biodiversity—but the ecological processes and historical events that have shaped the ranges of species are directly relevant to nearly all of them.

Methodological Issues: Mapping and Measuring the Range

If the geographic range is a basic unit of biogeographic investigation, how do we define and measure it? At first glance, this seems straightforward. Field guides and more technical systematic publications on regional floras or faunas are filled with **range maps**. These maps are seemingly easy for researchers to prepare, and equally easy for other scientists to use as sources of data for biogeographic studies. Before we start using range maps to illustrate biogeographic patterns and processes, however, we should critically consider just what they tell us.

There are three basic kinds of range maps: outline, dot, and contour. **Outline maps** usually depict the range as an irregular area, often shaded or colored, within a hand-drawn boundary (Figure 4.1). The boundary line presumably defines the limits of the known distribution of the species, but its accuracy can vary widely depending on how well the distribution is actually known and how precisely the author has incorporated this information into the map. Often, the author will use his or her knowledge of the organism to make educated guesses about the probable distributional limits when adequate data are not available.

Dot maps plot points on a map where the species has been recorded (Figure 4.2). Dot maps are often prepared as part of a taxonomic study of a species, and the dots show localities where verified museum specimens have been collected. Such maps convey both more and less information than outline maps. On the one hand, they accurately depict known records of the species' distribution. On the other hand, locations of specimens or other records, such as

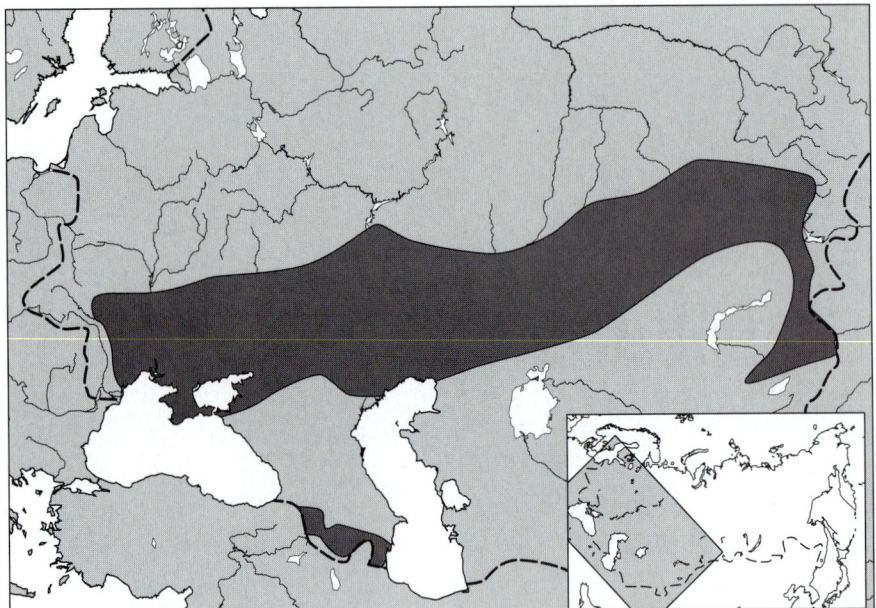

Figure 4.1   An example of an outline map of the geographic range of a species —in this case the endangered butterfly *Zegris eupheme*, which occurs in southwestern Asia. The outer boundary has been drawn by hand to include the localities where the species is known to occur. (From Borodin et al. 1984.)

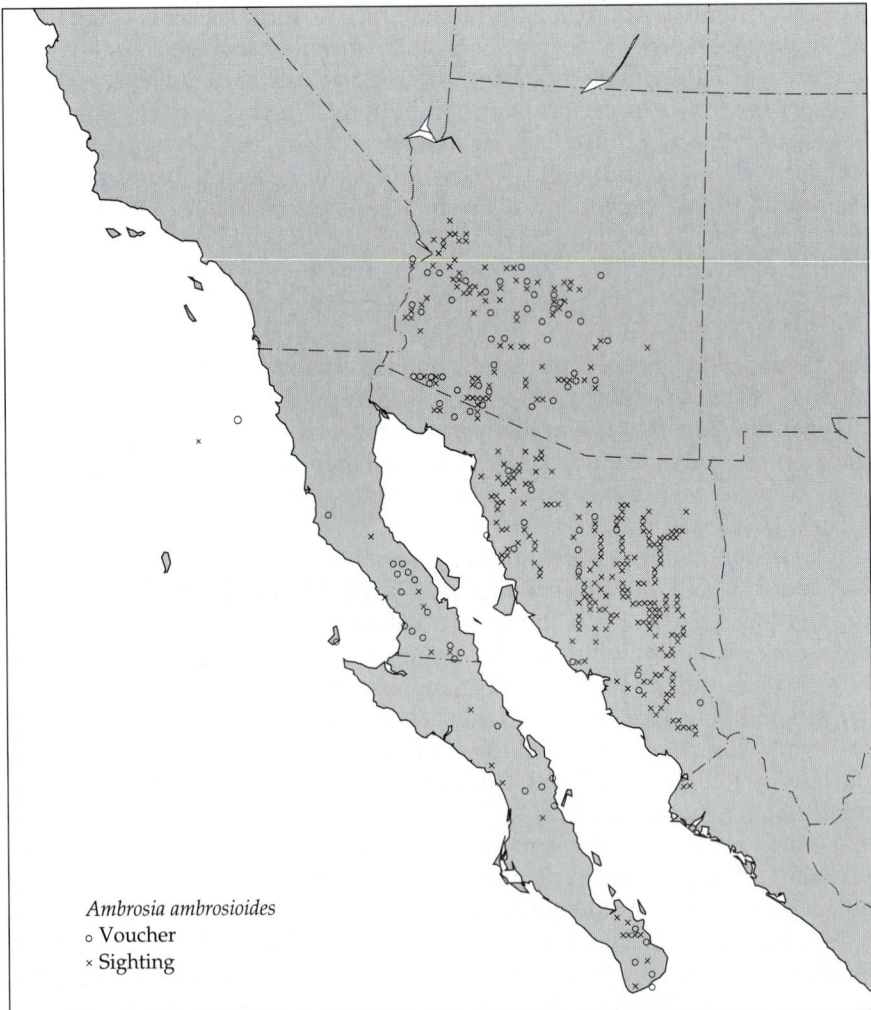

Ambrosia ambrosioides
○ Voucher
× Sighting

Figure 4.2   An example of a dot map of the geographic range of a species— in this case the Sonoran Desert plant *Ambrosia ambrosioides*. Each circle represents a locality where someone has documented the presence of the species by collecting a voucher specimen and depositing it in an herbarium. Each cross represents an additional record based on a sighting and identification of the plant in the field. (From *Sonoran Desert Plants* by Raymond M. Turner, Janice E. Bowers, and Tony L. Burgess. Copyright © 1995 by the Arizona Board of Regents. Reprinted by permission of the University of Arizona Press.)

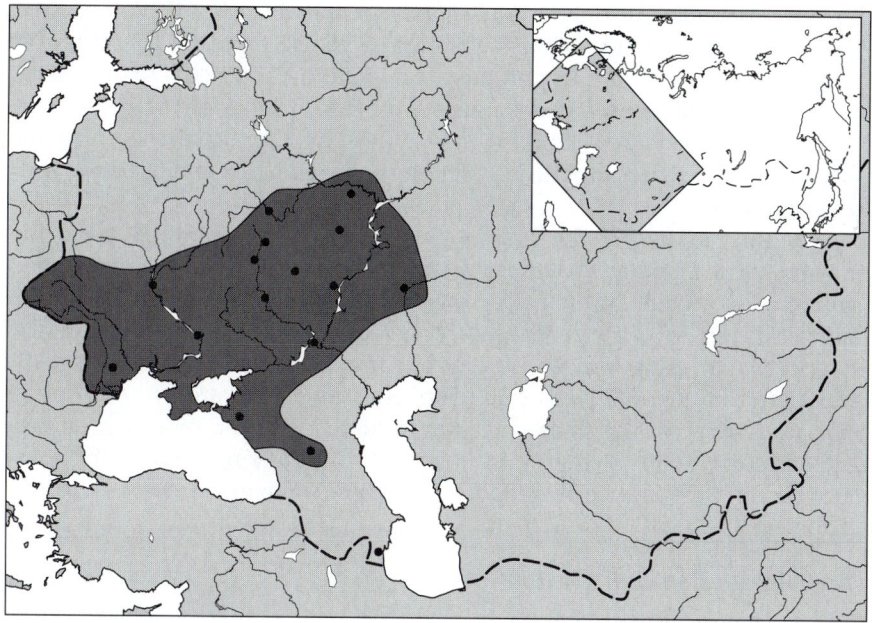

Figure 4.3 An example of a combination dot and outline map of the geographic range of a species—in this case the endangered butterfly *Zerynthia polyxena*, which is restricted to a small area north of the Black Sea in southern Eurasia. Each dot represents a locality where the species has been recorded. A line has been drawn by hand to include the outermost dots, thereby enclosing the known geographic range. (From Borodin et al. 1984.)

sightings of bird species, can represent only an infinitesimal fraction of the actual places where individuals of most species live at present or occurred in the past. While a small minority of species with tiny ranges are known to be restricted to just one or a small number of highly localized sites, the documented records of occurrence of most species represent only a small sample of their actual distribution. So a disadvantage of dot maps is that they do not extrapolate beyond the relatively few sampled locations to make inferences about the potential distribution of the species. Sometimes, however, the author draws a free-form line around the peripheral location records, creating a combination dot and outline map (e.g., Figure 4.3).

Recently, investigators have obtained sufficient information on abundance within the geographic ranges of some species to produce **contour maps** (Figure 4.4). Because they use contour lines or other graphical techniques to indicate variation in density, these maps convey much more information than either outline or dot maps. Contour maps should be interpreted with caution,

Figure 4.4 An example of a contour map of the geographic range of a species—in this case the winter range of the blue jay, *Cyanocitta cristata*, showing geographic variation in abundance. (A) Each contour line, or isocline, indicates a 20th-percentile class of relative abundance. (B) A three-dimensional landscape depicting relative abundance. The data on abundance come from the North America Christmas Bird Counts. The raw data from these census counts (number of birds seen per hour per field party) have been entered into a computer program, which averaged and smoothed them to estimate abundance between the actual census localities in order to draw the maps. (From Root 1988a.)

(A)

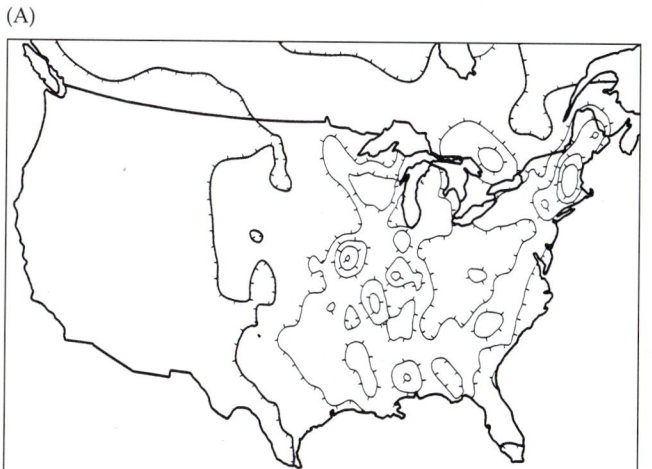

(B)

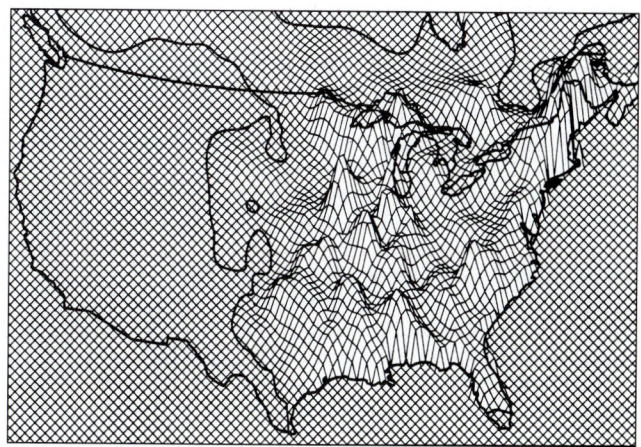

however, because information on abundance usually is available for only a limited number of fairly widely separated localities. Typically, computer programs and a statistical technique called **kriging** are used to interpolate between the data points and generate a three-dimensional landscape depicting variation in abundance within the range. So let the user beware; even contour maps are highly oversimplified depictions of the complex and dynamic structure of ranges.

Nevertheless, despite the limitations of all three kinds of range maps, they are invaluable summaries of biogeographic information. Although their precision may be limited, they usually provide a reasonably accurate and unbiased large-scale picture of the geographic range of a species. They can be compiled, computerized [using software such as geographic information systems (GIS)], and analyzed quantitatively. It is relatively straightforward, for example, to measure such variables as the area, northernmost and southernmost latitude, and easternmost and westernmost longitude of ranges from maps and to use such data in comparative biogeographic studies. You will see many such applications in this book.

The Distribution of Individuals

Even the best map, however, can convey only a highly simplified and abstracted picture of the geographic distribution of a species. The real units of distribution are the locations of all the individuals of the species. A map depicting these locations would be impossible to prepare for most kinds of organisms, but we can get some idea of what it would look like from aerial photographs in which we can identify individuals of certain conspicuous species (Figure 4.5). Rapoport (1982) prepared a map of the distribution of a distinct, easily recognizable palm (*Copernicia alba*) from aerial photographs along a transect through part of its range in Argentina (Figure 4.6). As we can see from both the sample aerial photograph and Rapoport's data, the distribution is complex, with individual plants occurring in clumps separated by gaps. As the edge of the range is approached, the individuals tend to be more sparsely distributed, the clumps smaller, and the gaps between them larger.

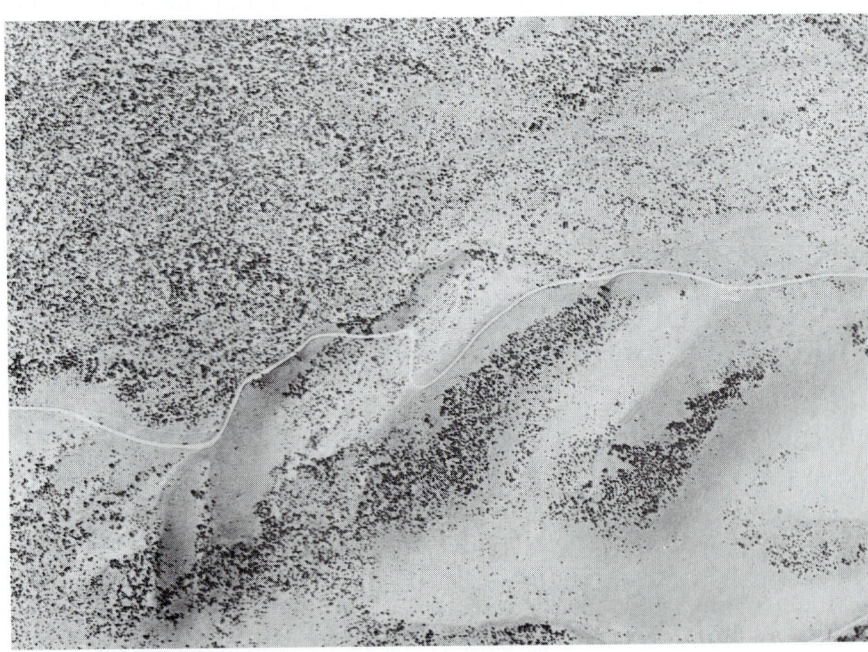

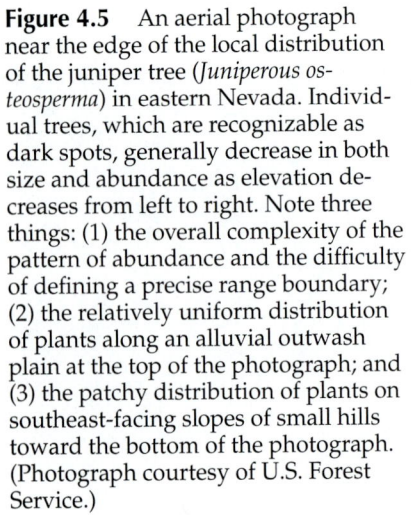

Figure 4.5   An aerial photograph near the edge of the local distribution of the juniper tree (*Juniperous osteosperma*) in eastern Nevada. Individual trees, which are recognizable as dark spots, generally decrease in both size and abundance as elevation decreases from left to right. Note three things: (1) the overall complexity of the pattern of abundance and the difficulty of defining a precise range boundary; (2) the relatively uniform distribution of plants along an alluvial outwash plain at the top of the photograph; and (3) the patchy distribution of plants on southeast-facing slopes of small hills toward the bottom of the photograph. (Photograph courtesy of U.S. Forest Service.)

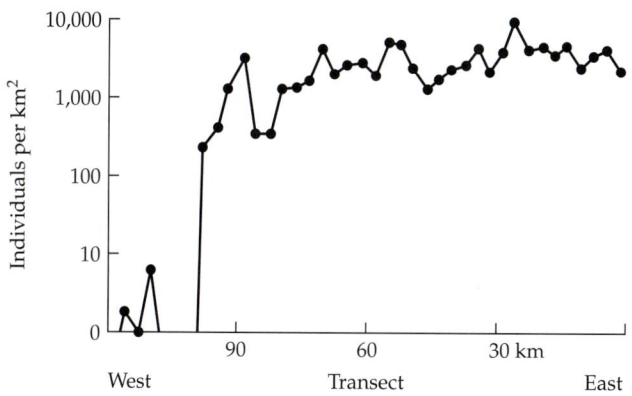

Figure 4.6 Abundance of the palm tree *Copernicia alba* along a 3 km wide transect from west to east through the edge of its geographic range. The data were taken from aerial photographs, on which individual palms were easily recognized by their distinctive shapes. Note that near the edge of the range, abundance tends to be low and the distribution of trees tends to be patchy (as indicated by values of zero abundance.) (From Rapoport 1982.)

The clumpy-gappy distribution of individuals across the landscape means that any kind of map of the species range is not only an abstraction, it is a scale-dependent abstraction. Imagine that we superimposed a grid on an aerial photograph such as Figure 4.5. Whether an individual will be found in an individual grid cell will depend on the size of that cell: the larger the cell, the higher the probability of its containing at least one individual. This exercise reveals that the edge of the range is also a scale-dependent abstraction. Although in Figure 4.5 there are no individuals in the lower right-hand corner, the exact definition of the distributional boundary will depend on the scale at which we connect the locations of individuals to draw an edge. Furthermore, in addition to the relatively obvious range boundaries, there are "holes" within the range where no individuals occur. Therefore, Rapoport (1982) has likened the range to a slice of Swiss cheese. But this is an oversimplification, because the sizes and locations of the areas where no individuals are considered to occur will also depend on the spatial scale of analysis. Perhaps the best representation of the effect of spatial scale on the perceived distribution of a species is still Erickson's (1945) classic maps of the distribution of the shrub *Clematis fremontii* (Figure 4.7).

Even such a faithful depiction of the distribution of a species as an aerial photograph fails to capture another critical feature of the geographic range, however, because it represents only a snapshot in time. The distribution of any species is dynamic, and any accurate depiction of its range should in theory be constantly updated to reflect the changes that occur as individuals are born, move, and die, and as populations colonize new areas and go locally extinct in parts of their former range. Andrewartha and Birch (1954), for example, documented large shifts in the geographic ranges of several species of insects in Australia. They showed diagrammatically how the apparent range boundary varies as local populations episodically go extinct and then are recolonized from other areas (Figure 4.8). Despite the seemingly static nature of most published range maps, such expansions and contractions are always occurring in response to both natural environmental variation and human activities (see below and Chapter 18).

The Distribution of Populations

Population Growth and Demography

The size of a range, the location of its boundaries, and shifting patterns of abundance within those boundaries reflect the influence of environmental conditions on the survival, reproduction, and dispersal of individuals and the

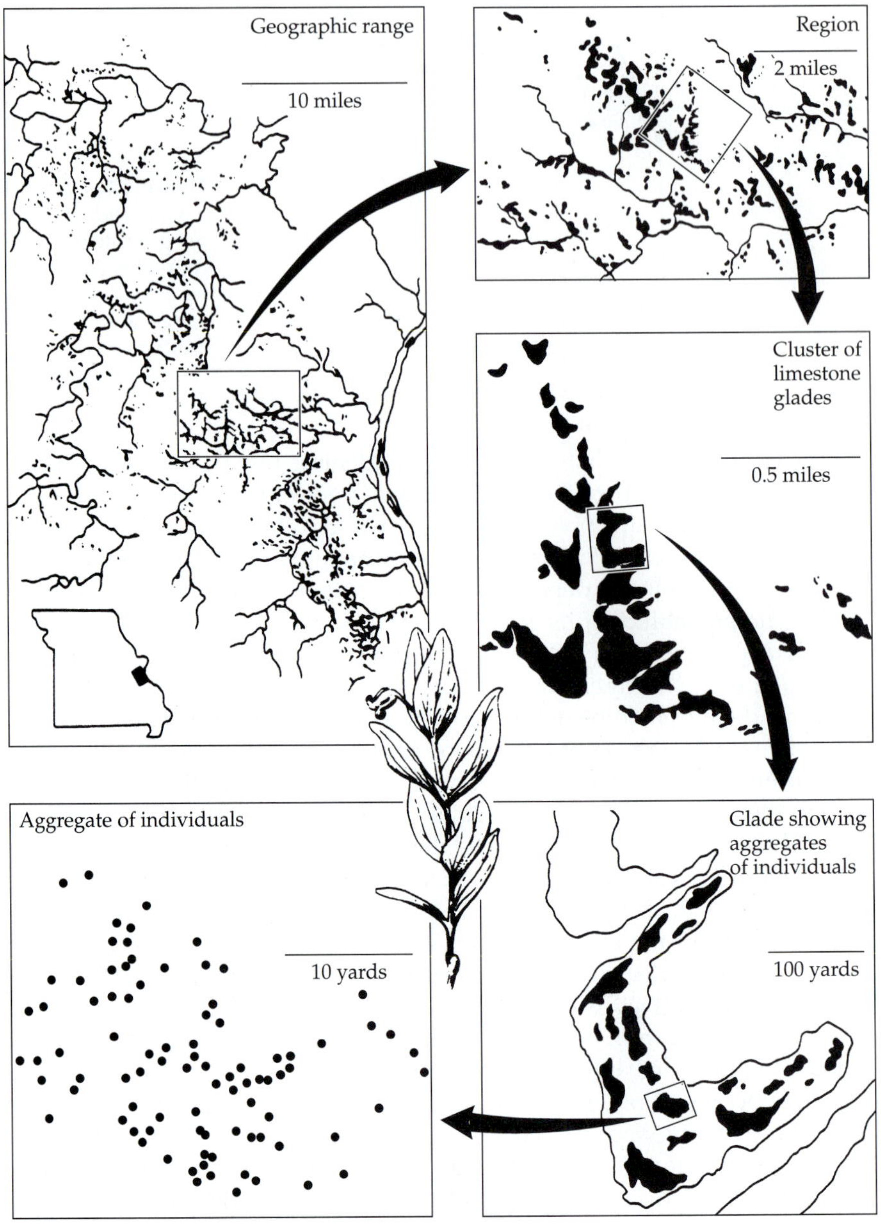

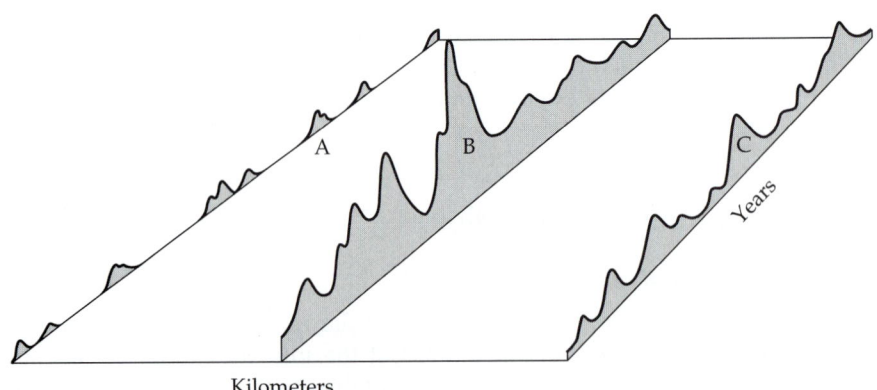

Figure 4.7   Erickson's classic depiction of the distribution of the shrub *Clematis fremontii*, within the state of Missouri in the central United States, on different spatial scales. The largest scale shows the geographic range based on known collecting localities. Successively smaller scales show the distribution of populations. The smallest scale shows the dispersion of individual plants within a single local population. Note that at all scales the distribution is patchy, and that areas where plants are found are separated by uninhabited areas. (From Erickson 1945.)

Figure 4.8   A schematic diagram showing how the abundance and distribution of a hypothetical organism might vary in time and space. Shown are fluctuations in abundance over many years at three different localities (A–C) separated by distances of several kilometers. Note that all three populations fluctuate. At locality A, which is presumably at the margin of the local or geographic range of the species, only a few individuals are intermittently present, indicating repeated episodes of local extinction and recolonization. (From Andrewartha and Birch 1954.)

dynamics of populations. In 1798, in his *Essay on the Principle of Population*, Thomas Malthus showed that all kinds of organisms have the inherent potential to increase their numbers exponentially. A population increases when the combined rates of birth and immigration exceed the combined rates of death and emigration. We can express this concept mathematically as

$$r = b + i - d - e \quad (4.1)$$

where r is the per capita rate of population growth (if r is positive, the population increases; if r is negative, it decreases), b and d are the per capita birth and death rates, respectively, and i and e are the respective per capita rates of immigration from and emigration to other populations. Given unlimited resources and favorable environmental conditions, a population will grow continuously at its maximum possible r. It will increase its numbers as described by the equation

$$dN/dt = rN \quad (4.2)$$

where dN/dt is the rate of change in numbers of individuals, N, with respect to time, t, and r is the population growth rate, as above. We call this **exponential growth**, and we can describe its rate in terms of the time interval required for the population to double its numbers. If it kept growing exponentially, any species would eventually cover the earth with its own kind. The time required would depend on r; bacteria and houseflies would require only a few years, whereas slowly reproducing trees and elephants, with their lower values of r, would take a few thousand years. The global human population has been growing at a nearly exponential rate for the last several thousand years (Figure 4.9). Malthus recognized, however, that because resources ultimately limit growth, and because many environments are unsuitable, no organisms actually continue to increase indefinitely at such exponential rates.

Hutchinson's Multidimensional Niche Concept

In 1957 Evelyn Hutchinson developed the concept of the multidimensional **ecological niche** to conceptualize how environmental conditions limit abundance and distribution. Hutchinson's view of the niche was a modification of the earlier niche concepts of Grinnell (1917) and Elton (1927; see also James et

Figure 4.9 The estimated growth of the human population over the last 10,000 years. Note the almost continuously increasing, near-exponential shape of the curve as *Homo sapiens* not only increases in local abundance but also spreads over most of the earth. All populations have the capacity to increase exponentially so long as environmental conditions are not limiting. Colonizing exotic species typically show similar near-exponential growth rates during a period of rapid range expansion. (From Desmond 1965.)

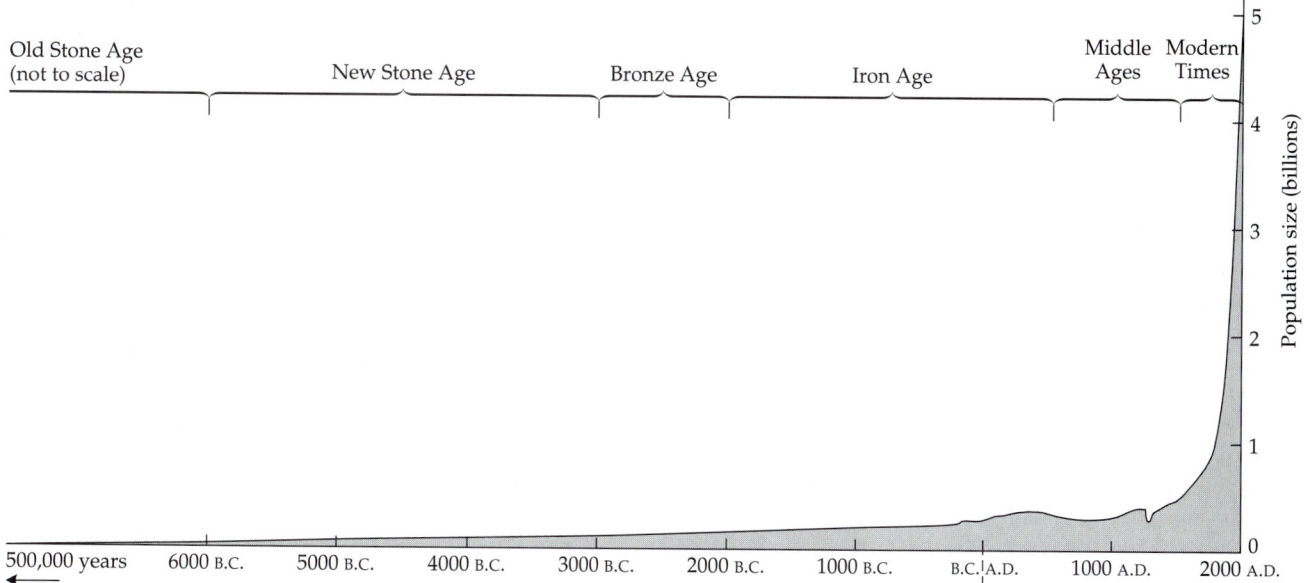

(A)

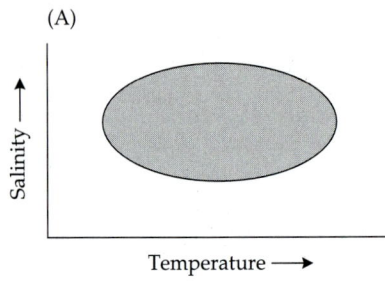

(B)

(C)

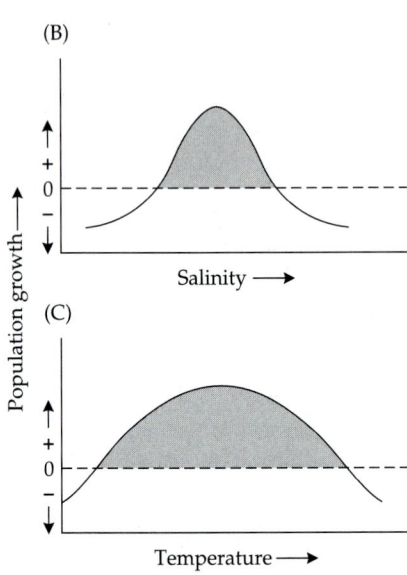

Figure 4.10   (A) A diagram of two dimensions—temperature and salinity—of the niche of a hypothetical aquatic species. The shaded area represents the combinations of these two variables—a broad range of temperatures but a narrow range of salinities—under which individuals can survive and reproduce or a population can increase. In (B) and (C), population growth rate is plotted as a function of each variable, with the horizontal dashed line showing the zero value used to plot the elliptical niche space in (A).

al. 1984; Schoener 1988). Hutchinson realized that over a period of time and over its geographic distribution, every species is limited by a number of environmental factors. He conceptualized the species' environment as a multidimensional space, or "hypervolume," in which the different axes or dimensions represent different environmental variables. The niche of the species represents the combinations of these variables that allow individuals to survive and reproduce and populations to maintain their numbers.

This concept sounds intimidating, and it is indeed a bit abstract. But the basic idea is very simple (Figure 4.10). Imagine the effects of just two environmental variables—say, temperature and salinity—on some aquatic species. If all other conditions are favorable, individual performance and population growth will be limited by the joint effects of these two variables. We can plot out the range of conditions under which population growth will be negative, zero, or positive, and the space inside the zero growth contour, or isocline, represents the niche space. In reality, it is almost certain that our aquatic organism would also be limited by other variables: perhaps dissolved oxygen concentration, a competing species, and a predator. Each of these variables would represent another dimension of the niche, causing the niche space to be a multidimensional volume, which is hard to visualize or draw, but not too hard to imagine. It is easy to see that the niche of every species is unique, because each species differs at least slightly, and sometimes greatly, from all others in the combinations of environmental conditions required for the survival and reproduction of its individuals and the growth of its populations.

The Geographic Range as a Reflection of the Niche

The geographic range of a species can be viewed as a spatial reflection of its niche: the species occurs where environmental conditions are suitable, and is absent from areas where one or more essential resources or necessary conditions is missing. The boundaries of the range, and the pattern of abundance within these boundaries, constantly shift as local populations grow, decline, colonize, and go extinct in response to changing environmental conditions.

One of the earliest studies of the ecological niche of a species is still one of the most complete. Joseph Connell (1961) studied the environmental factors that limit the range of a barnacle, *Chthamalus stellatus*, on the rocky coast of the Isle of Cumbrae in Scotland. As mentioned in Chapter 3, the intertidal zone is that narrow strip of coastline that is alternately inundated by seawater and then exposed between tides. Species are typically restricted to a narrow range of exposures within the intertidal zone. Connell used some elegantly simple field experiments to characterize important variables of the niche of *C. stellatus*, and to explain its distribution in the uppermost portion of the intertidal zone (Figure 4.11). He showed that the upper edge of the species range is set by the ability of the barnacles to tolerate the physiological stress of desiccation while exposed during low tides. The lower edge is set by interactions with other intertidal organisms, primarily by competition with another barnacle species, *Balanus balanoides*, and secondarily by predation by a snail, *Thais lapillus*. The power of the experimental method is illustrated by the effect of removing *B.*

Figure 4.11 Diagrammatic representation of the effects of interspecific competition and other factors on the distribution of the barnacle *Chthamalus stellatus* in the intertidal zone on rocky shores in Scotland. The diagrams on the left show the species distribution; the width of each bar indicates the population density at that elevation. Larvae settle over a wide range, but many die before reaching maturity, leaving adults confined to a much narrower zone. The diagrams on the right indicate the effects of three mortality-causing factors: desiccation between tides, A, which sets the upper limit of distribution; predation by the snail *Thais*, B, and competition from the barnacle *Balanus balanoides*, C, which together set the lower limit. The width of each bar shows the strength of each effect at that elevation. (After Connell 1961.)

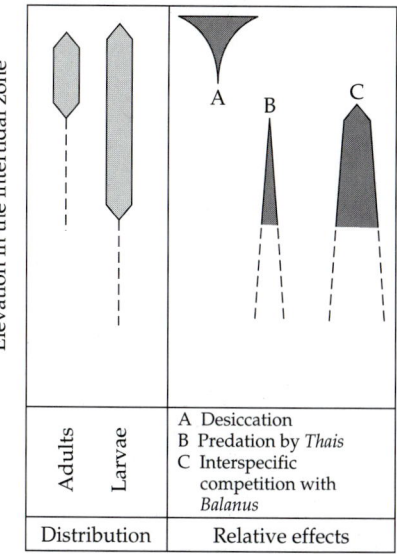

balanoides from small patches of shore: on those plots where its competitor was absent, *C. stellatus* extended its range lower into the intertidal zone.

Connell's study pioneered the use of field experiments in ecology. It is also a classic for demonstrating how three niche variables—exposure to desiccation, competition with another barnacle species, and predation by a snail—can largely explain the limited distribution of *C. stellatus* on the rocky shores of the Isle of Cumbrae. Note, however, that these are almost certainly not the only niche variables affecting the distribution of this barnacle: for example, some other factor(s) presumably accounts for the absence of *C. stellatus* from the sandy and muddy substrates that occur only a short distance from Connell's study site.

While the multidimensional environmental niche provides a conceptual framework for understanding how environmental limiting factors influence both the geographic ranges and the local population densities of species, niche variables alone are inadequate to account for all patterns of distribution and abundance. Three complications will be mentioned briefly here, and then considered further in later chapters. First, it is too simplistic to assume that environmental conditions are equally favorable for a species at all localities where it occurs. Some localities may be so favorable that birth rates exceed death rates; these localities can serve as "**source habitats**," producing surplus individuals that migrate out to other areas (Figure 4.12; Pulliam 1988). Some other localities may be so unfavorable that death rates exceed birth rates, but they may still be inhabited if they act as "**sink habitats**" and receive a sufficient supply of immigrants to maintain a local population [refer to Equation (4.1) and note the terms *i* and *e*, giving the contributions of immigration and emigration to the population growth rate]. One might expect that some of the areas near the border of a species' range would be sink habitats, and that their environmental conditions would be so marginal that they would not be able to sustain populations in the absence of immigration. *Cakile edentula*, an annual plant that lives on coastal sand dunes, provides an example of this pattern. The small proportion of individuals growing in exposed seaward sites have the highest growth rates and produce the majority of seeds, but most plants occur in unfavorable inland sites, where storms have carried large numbers of dispersing fruits (Keddy 1982).

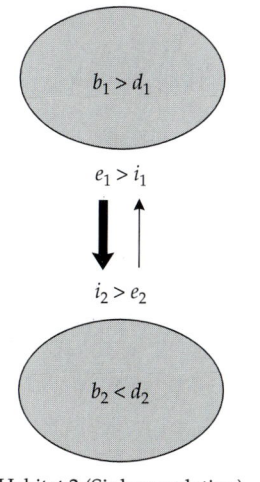

Habitat 1 (Source population)

$b_1 > d_1$

$e_1 > i_1$

$i_2 > e_2$

$b_2 < d_2$

Habitat 2 (Sink population)

Figure 4.12 Diagrammatic representation of source and sink habitats, showing the relative magnitude of the four processes that determine the growth and persistence of their populations. In the source population (Habitat 1), the birth rate (b_1) is greater than the death rate (d_1), but the population does not increase. Instead, the "excess" individuals disperse, resulting in a higher rate of emigration (e_1) than immigration (i_1). The opposite situation results in the sink population (Habitat 2), which is able to persist despite having a lower birth (b_2) than death rate (d_2) because the rate of immigration (i_2) is enough higher than the rate of emigration (e_2).

Second, just as there may be sites where environmental conditions are unfavorable, but that are nonetheless inhabited, there may also be favorable localities that are uninhabited. Some ecologists would say that there is an unfilled niche for the species in such places, but most prefer to define the niche as a characteristic of organisms (i.e., of species) rather than of places. As mentioned in Chapter 3, a species is likely to be absent from many places where it could live. Often this is because such favorable sites are isolated from inhabited areas by some combination of distance and intervening areas with unfavorable environmental conditions, so that individuals have not been able to disperse to these locations. This situation is common, as demonstrated by the large number of exotic species that have been able to become established in new habitats when humans have transported them over the intervening barriers (see Chapter 18). Biogeographers often invoke "history" to account for situations in which species are absent from apparently suitable areas. Indeed, to understand why a species occurs where it does and not elsewhere, it is necessary to understand the history both of the apparently favorable places and the barriers between them, and of the species itself. We will return to consider these problems in Chapters 9, 10, and 12.

Finally, some places may be inhabited only intermittently. Local populations increase and decrease—sometimes to local extinction—as environmental conditions fluctuate or as stochastic events affect their growth. If the habitat is patchy, the species may consist of a series of isolated populations separated by uninhabited areas. When a species population is subdivided in this way, it is said to be a **metapopulation** comprised of multiple **subpopulations**. In such cases, some of the subpopulations are likely to go extinct intermittently, especially if they are sink populations. Even source populations in favorable patches of habitat, however, may go extinct by chance. New subpopulations are also likely to be founded by immigration to uninhabited patches, including those vacated by previous local extinction events (e.g., Gilpin and Hanski 1991). We will return to consider the implications of such metapopulation and source/sink dynamics in Chapters 9, 13, and 18. For the moment it is sufficient to stress that they are especially likely to occur on the periphery of species ranges and to contribute to dynamic shifts in range boundaries.

The Relationship between Distribution and Abundance

The contour maps discussed above (see Figure 4.4) imply that there is considerable variation in abundance within a species' range. In fact, most published contour maps underestimate the magnitude of this variation because of the statistical and graphical methods used to interpolate between data points and construct the maps. The real spatial abundance patterns of nearly all species are extremely heterogeneous—what statisticians would call **clumped** or **aggregated** (Brown et al. 1995). That is, compared with a random distribution of individuals across the landscape, some places have many more individuals, and others have many fewer, or none at all.

Although the complications of source/sink dynamics and history discussed above are relevant here, most of this spatial variation in abundance presumably reflects the extent to which the local environment meets the niche requirements of a species. Each species tends to be most abundant where all niche parameters are in the favorable range, and to be rare or absent where one or more environmental factors are strongly limiting (Brown 1984; Hengeveld 1990; Lawton et al. 1994; Brown et al. 1995).

Common species are typically several orders of magnitude more abundant at some sites than at others (Figure 4.13). Thus, for example, only a single red-eyed vireo (*Vireo olivaceus*) was recorded on more than 200 of the standardized

Figure 4.13 Variation in the abundance of two common songbird species, (A) the red-eyed vireo (*Vireo olivaceus*) and (B) the Carolina wren (*Thryothurus ludovicianus*), among hundreds of census routes distributed throughout their geographic ranges in eastern North America. The numbers of birds counted on each standardized census route of the North American Breeding Bird Survey are plotted in rank order, so that routes with one bird are on the left and routes with the maximum number of birds recorded on any route are on the right. Note that for both species, fewer than 5 birds were recorded on the vast majority of census routes, but more than 100 birds were counted on least one route. This highly clumped pattern of abundance is characteristic of most birds as well as many other organisms. (From Brown et al. 1995.)

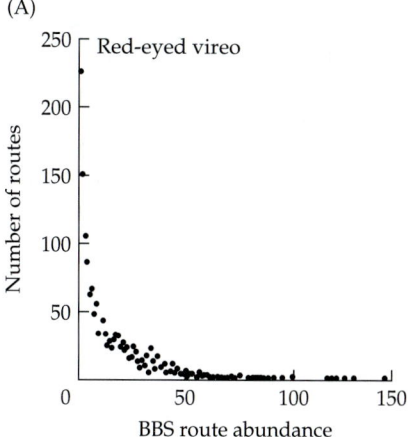

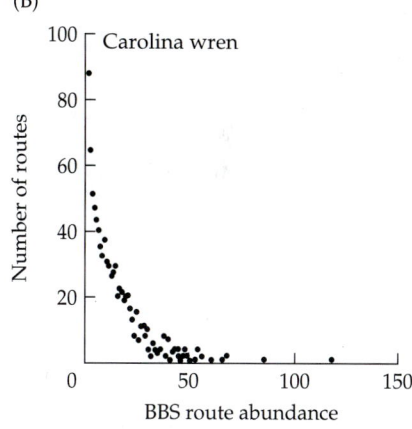

census routes of the North American Breeding Bird Survey, but more than 100 individuals were counted on 6 routes. One consequence of this highly clumped distribution pattern is that the majority of individuals of a species actually occur in a very small proportion of its geographic range. For the majority of common songbirds in eastern North America, more than half of the individuals occurred at fewer than 20% of the sites within their geographic ranges. Of course, rare species may be uncommon throughout their ranges (Rabinowitz et al. 1986; Gaston 1994), but they too typically have patchy distributions and are absent from many, presumably unfavorable, localities.

There is also variation in abundance and distribution over time, and most of it presumably reflects temporal variation in niche parameters. The fluctuations of Australian insect populations in response to climatic variation documented by Andrewartha and Birch (1954) are excellent examples (see Figure 4.8). The migratory locusts of the Old World provide additional, perhaps even more dramatic, examples (Waloff 1966; Albrecht 1967; White 1976). Source populations of these grasshoppers persist in limited regions, called outbreak areas, where conditions are suitable for their continued survival and reproduction. During periods when weather and food supplies are particularly favorable, these populations increase fantastically, change their morphology and behavior, aggregate into huge swarms, and migrate outward from the outbreak areas to forage over an enormous area. Such plagues of both the African migratory locust (*Locusta migratoria*) and the red locust (*Nomadacris septemfasciata*) have occurred two or three times in the last century, sweeping over most of southern Africa, an area of millions of square kilometers and more than 1000 times the size of the outbreak area from which they originated (Figure 4.14).

Such fluctuations do not occur only in insects in arid regions. In the tundra and taiga (coniferous forest; see Chapter 5) of northern North America, Europe, and Asia, several species of voles and lemmings (mouselike rodents, subfamily Microtinae) fluctuate over several orders of magnitude in abundance over a 3- to 4-year period (see Finerty 1980; Lidicker 1988). Unlike locusts, however, these rodents have very limited dispersal abilities, so their geographic ranges do not change very much during these cyclical fluctuations, but their local patterns of habitat use may shift considerably. Some northern birds, such as snowy owls that feed on these rodents and crossbills that feed on conifer seeds, also show wide fluctuations in abundance. Associated with

Figure 4.14 The temporally shifting range of the red locust, *Nomadacris septemfasciata*, in Africa. The small black areas are the source habitats at the core of the range. These outbreak areas are the only places known to sustain permanent populations. Enormous population increases and geographic expansions begin in these areas, and in favorable years the locusts can spread into sink habitats throughout the invasion area (gray area), which includes about half of the continent. (After Albrecht 1967.)

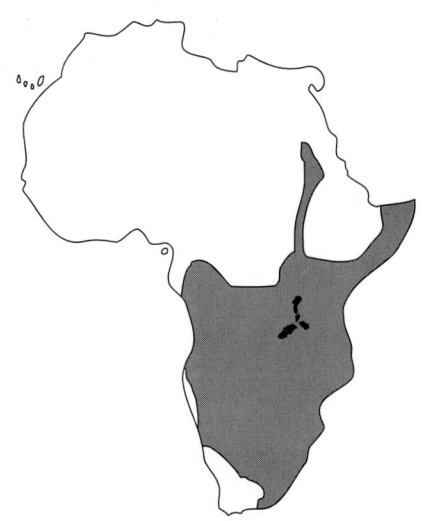

Snowy owl

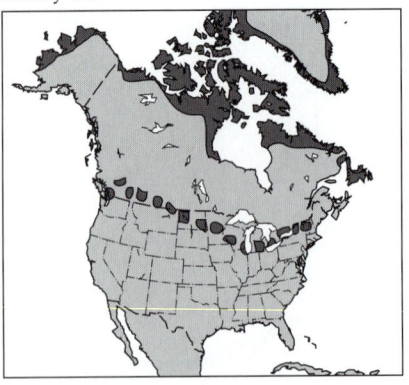

White-winged crossbill

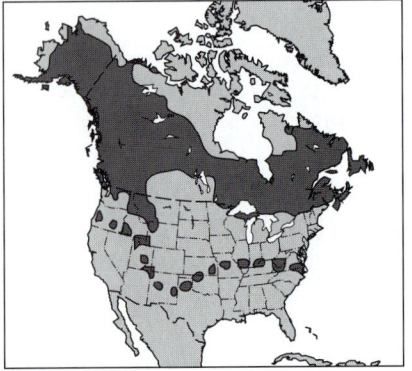

Common redpoll

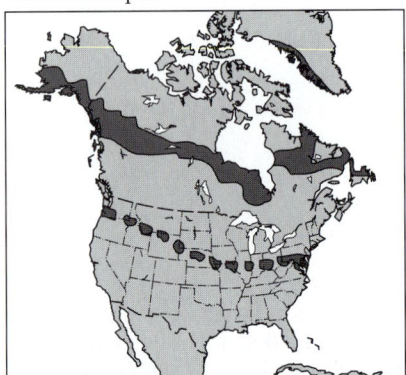

Figure 4.15   The winter ranges of three bird species, the snowy owl (*Nyctea scandiaca*), white-winged crossbill (*Loxia leucoptera*), and common redpoll (*Carduelis flammea*), that normally winter at high latitudes (dark gray area), but in years of food shortage disperse far to the south (to the dotted line), greatly expanding their ranges.

the population dynamics of these birds are large shifts in their winter ranges, which may shift hundreds of kilometers to the south in years of low food availability (Figure 4.15; Bock and Lepthian 1976).

Superimposed on all of this spatial and temporal variation in abundance are certain spatial patterns. We will illustrate and comment briefly on three of them (Figure 4.16). First, abundance is autocorrelated in space—which is a technical way of saying that abundances tend to be more similar at sites that are closer together. This is just what we would expect if abundance reflects the suitability of the local environment and if the niche variables also exhibit spatial autocorrelation. Second, abundance often varies systematically over the geographic range, from relatively high numbers near the center to zero at the boundaries. In particular, localities near the edges of the range tend to have consistently low populations, whereas sites near the center of the range can have a wide range of abundances, from very low to very high. This pattern is consistent with the idea that the boundaries of ranges occur where one or more niche factors become unfavorable. Third, an exception to the previous pattern is that abundance may be high near an edge of a range that is set by a coastline. Again, this is what would be expected if a range boundary is determined by just one niche variable that abruptly becomes unfavorable.

Range Boundaries

So far we have talked about niche dimensions and limiting factors only in abstract terms. What are these environmental variables that cause the boundaries of a species' range by limiting its abundance and distribution? Our previous discussion of the multidimensional environmental niche, spatial and temporal variation in abundance, and Connell's study of barnacle distributions in the intertidal zone would suggest that the locations of range boundaries are set by multiple environmental factors, and that these can include both physical conditions (**abiotic** factors) and effects of other organisms (**biotic** factors).

Unfortunately, there have been few studies as comprehensive as Connell's. Most investigators have studied the effects of just one or few factors in a small area near one edge of a geographic range. This is understandable, because the geographic ranges of most species are large, and the scientific expertise of most ecologists is limited (i.e., physiological ecologists study the effects of different kinds of abiotic stresses, and community ecologists study the different kinds of interactions among species). Nevertheless, we are left with a literature that provides many examples of single limiting factors, but little overview and synthesis (but see Gaston 1994; Brown et al. 1996). So first let's examine some of these examples, and then try to see whether they reveal any general patterns and processes.

Physical Limiting Factors

Many widespread species appear to be limited in at least part of their geographic ranges by physical factors, such as temperature regime, water availability, and soil and water chemistry. For example, many Northern Hemisphere plants and animals become increasingly restricted to low elevations and south-

(A)

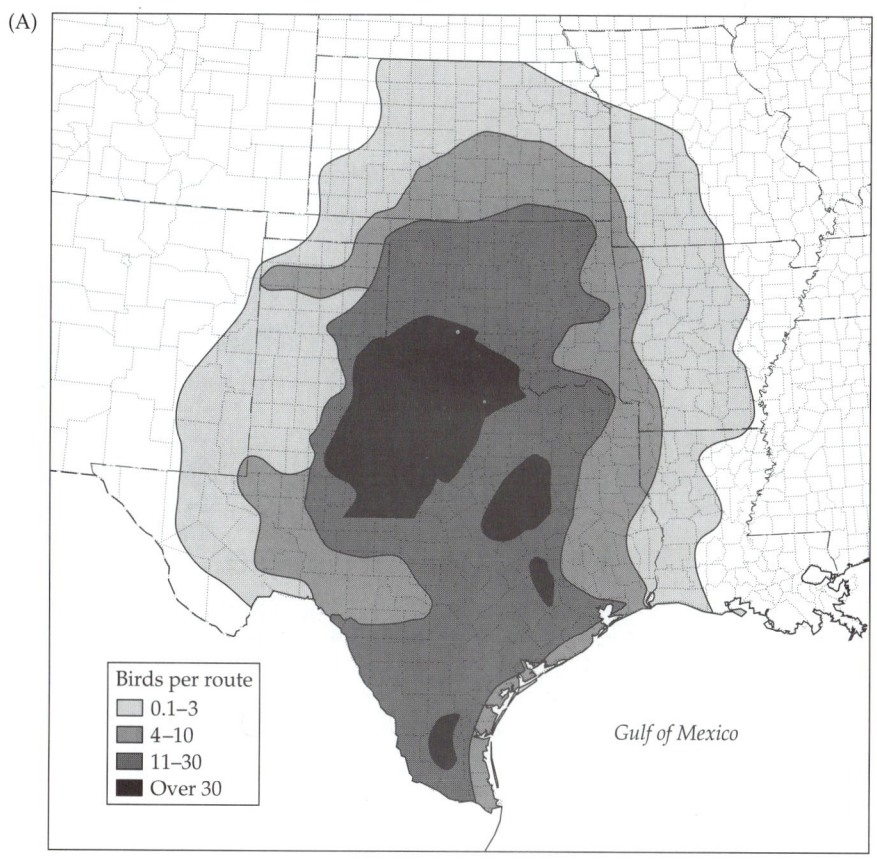

Birds per route
- 0.1–3
- 4–10
- 11–30
- Over 30

Gulf of Mexico

(B)

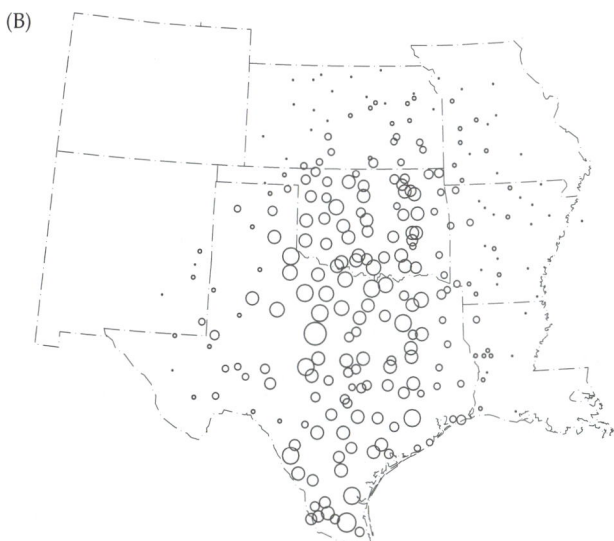

Figure 4.16 Variation in abundance across the breeding range of the scissor-tailed flycatcher, *Muscivora forficata*. Using data from the North American Breeding Bird Survey, spatial variation in local population density is depicted in two ways: (A) a contour map produced by averaging and interpolating data from census routes (courtesy of D. Bystrak), and (B) actual values for the individual census routes as shown by circles of proportionate diameter (courtesy of D. Mehlman). Both plots illustrate three common features of patterns of abundance within many geographic ranges: (1) autocorrelation, such that nearby areas tend to have similar abundances; (2) maximal abundances in one or more areas usually located near the center of the range; and (3) generally low abundances at the edge of the range, except where the boundary is set by a coastline (compare with Figure 4.4).

facing exposures as they approach the northern limits of their ranges, suggesting that their distributions are determined by ambient temperature. An excellent example is afforded by the giant saguaro cactus (*Carnegiea gigantea*) as it approaches the northern edge of its range in the Sonoran Desert of southern Arizona (see Figure 4.17 and the more detailed discussion below). Such correlations provide only circumstantial evidence, however, and do not necessarily indicate direct causal relationships. The species in question might, for example,

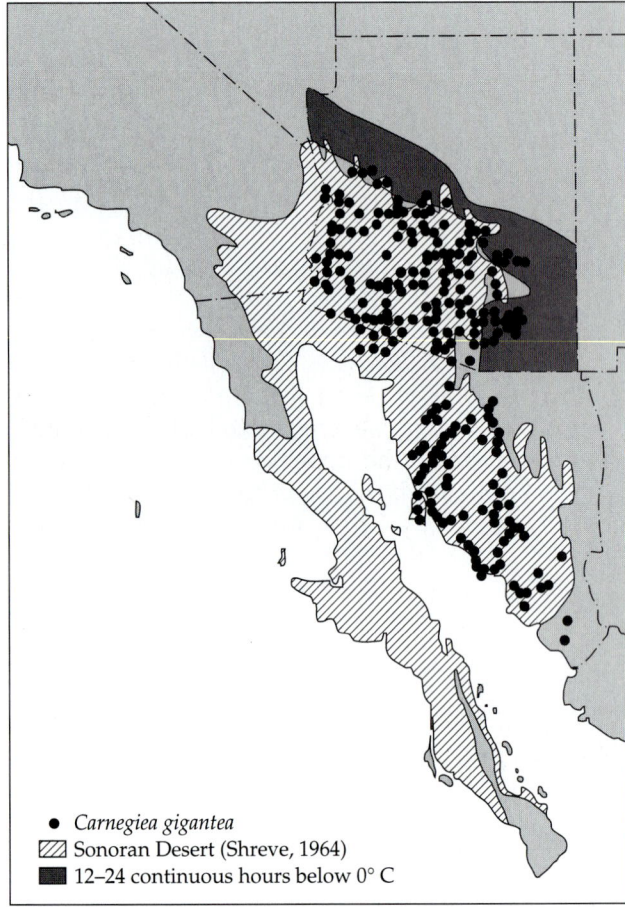

Figure 4.17 The distribution of the saguaro cactus (*Carnegiea gigantea*) in relation to winter temperature regime. This cactus, like many other Sonoran Desert plants, is intolerant of prolonged freezing. Note the close correspondence between the northern limit of the saguaro, the northern boundary of the Sonoran Desert, and the region where temperatures remain below 0° C for more than 12 hours. (Data from Hastings and Turner 1965; Hastings et al. 1972).

● *Carnegiea gigantea*
▨ Sonoran Desert (Shreve, 1964)
■ 12–24 continuous hours below 0° C

be limited not by their inability to tolerate low temperatures, but by competition from other species that are superior competitors in cold climates.

It might seem an easy matter to investigate the distributional limits of a particular species in order to identify the limiting factor and discover its mechanism of action on the organism. In 1840 Justus von Liebig suggested that biological processes are limited by that single factor that is in shortest supply relative to demand, or for which the organism has the least tolerance. At one time ecologists accepted this idea so completely that they called it **Liebig's law of the minimum** and tried to identify *the* single factor that limited the growth of each population. Now, however, as implied by Hutchinson's multidimensional niche concept and demonstrated by numerous empirical studies, we realize that Liebig's concept is too simplistic. Abundance and distribution are influenced by multiple and interacting environmental variables.

For example, many temperate and arctic birds and mammals appear to be limited by their inability to tolerate cold temperature regimes in the winter. Often this is not because they simply cannot survive at such low temperatures (Dawson and Carey 1976; Root 1988b,c), but rather because these temperature regimes increase the energy requirement for thermoregulation beyond the food supply available in the environment. In this case food supply and temperature interact to limit distributions, and increasing the food supply enables populations to inhabit colder climates. A number of bird species, including the cardinal, tufted titmouse, and mockingbird, have expanded their ranges far northward in eastern North America (e.g., Boyd and Nunneley 1964). These birds are year-round residents, and a new reliable winter food supply, provided by backyard bird feeders, has almost certainly contributed to their expansion into colder environments.

Another problem in determining the causes of distributional limits is the difficulty of identifying the mechanisms by which environmental factors affect the growth of populations. Cold, for example, is not a single variable, and different aspects of low temperature regimes limit different populations in different ways. The adults of some plant species may be killed by critically low short-term temperatures, such as those experienced on a single, exceptionally cold winter night. Other species may be more susceptible to damage from prolonged freezing. Still other species may be limited by cold climates not because they cannot withstand low winter temperatures, but because the summer growing season is too short to allow for sufficient growth and reproduction.

As mentioned above, one of the best-documented examples of cold temperatures limiting the upper elevational and latitudinal distribution of a species is provided by the saguaro cactus (Figure 4.17). This giant multi-armed columnar cactus, which may reach 15 m in height and 200 years of age, is a conspicuous part of the landscape in much of the Sonoran Desert of southern Arizona and northwestern Mexico. Although it lives where winter nighttime frosts are not infrequent, the saguaro is extremely sensitive to temperatures below –7° C and to prolonged freezing. Individuals are often killed by frost damage to their tissues, especially by destruction of the growing shoot tips. Young saguaros are more susceptible to frost damage than adults, but seedlings typically become established under the canopy of small desert trees, which provide the young cacti with a protective microclimate for the first few decades of their lives (Nobel 1980b). These "nurse trees" shield the young saguaros from the cold night sky and prevent their freezing in much the same way that frost damage to tomato plants can be prevented by covering them at night with paper or plastic—the loss of heat by infrared radiation to the sky is retarded by the nurse trees. Before they reach reproductive age, the saguaros grow above their nurse trees, often killing the trees in the process; but by then they are large enough not to be affected by overnight frosts. Nobel (1978, 1980a) has studied the thermal relations of stems and shoot apices using computer simulations and direct field measurements. The results show that the large stem diameter of the saguaro enables it to maintain higher minimal temperatures of its apical buds, and thus to have a more northern distribution than related species of columnar cacti.

Steenburgh and Lowe (1976, 1977) studied populations of saguaros at Saguaro National Monument outside Tucson, Arizona, near the northeastern and upper elevational limit of the species range. Extensive mortality of both young and adult plants occurred as a result of exceptionally low temperatures in January of 1937, 1962, 1971, and 1978. The 1971 freeze killed about 10% of individuals and severely injured an additional 30%; many of the injured cacti died during the next few years as a result of microbial infections that started at the site of the frost damage (Figure 4.18). These observations of episodic winter kill, together with the close correspondence between the northern and eastern boundaries of the species range and areas that experience below-freezing temperatures for more than 12 hours at a time (see Figure 4.17), suggest that low temperatures directly limit the distribution of saguaros in this region.

The distributions of many plant species appear to be limited by low temperatures interacting with other environmental conditions, such as water availability and soil chemistry. Hocker (1956) studied the distribution of loblolly pine (*Pinus taeda*) in the southeastern United States and concluded that the northern and western edges of the range were set by low tempera-

(A)

(B)

(C)

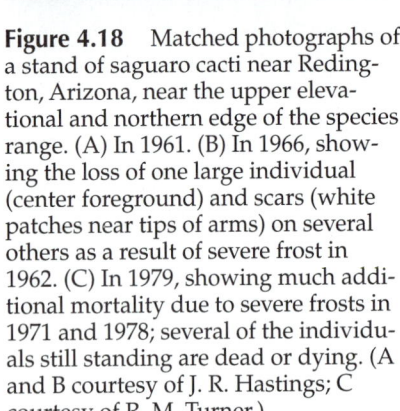

Figure 4.18 Matched photographs of a stand of saguaro cacti near Redington, Arizona, near the upper elevational and northern edge of the species range. (A) In 1961. (B) In 1966, showing the loss of one large individual (center foreground) and scars (white patches near tips of arms) on several others as a result of severe frost in 1962. (C) In 1979, showing much additional mortality due to severe frosts in 1971 and 1978; several of the individuals still standing are dead or dying. (A and B courtesy of J. R. Hastings; C courtesy of R. M. Turner.)

tures in concert with low soil moisture. He suggested that this resulted from the inability of the roots to take up sufficient water to replace the quantities lost by evaporation when environmental temperatures were low. Shreve (1922) noted that the upper elevational limits of many desert plant species appear to be determined largely by temperature, because extensive mortality occurs in the populations at the highest elevations during exceptionally cold winters. However, many of these same species, such as the ocotillo (*Fouquieria splendens*), occur at much higher elevations, and tolerate substantially colder winter temperatures, on limestone soils than on granitic and other zonal soil types (Shreve 1922; Whittaker and Niering 1968).

Many investigators have attempted to determine the cause of **timberline**, the upper elevational limit of trees on mountains. The large-scale geographic position of timberline seems to be related to the mean or maximum temperature during the warm months of the growing season, as reflected in its variation with respect to latitude and elevation (Figure 4.19), although it is influenced locally by other factors, such as snow depth, wind, and energy balance (Daubenmire 1978; Stevens and Fox 1991). At timberline, established trees often live for a long time, but their growth is very slow, and successful reproduction and seedling establishment are rare. Bristlecone pines (*Pinus longaeva*), which often grow just below timberline on arid mountains in the southwestern United States, are the oldest known living things, some individuals being more than 3000 years old. They grow slowly and produce exceptionally hard, dense wood that is highly resistant to decay (Figure 4.20). The annual growth rings of living and dead bristlecones provide a valuable record of the climatic history of the Southwest, because the width of each ring is proportional to the suitability of the growing season in the year it was laid down (Fritts 1976). In some places dead bristlecones form a "fossil timberline" above the timberline of living trees. Apparently this area was once favorable for tree growth, but climatic changes during the last few thousand years killed the trees and caused a contraction in their elevational range (La Marche 1973, 1978).

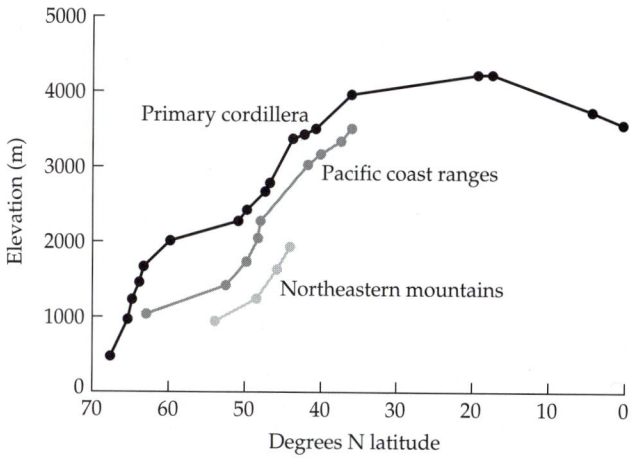

Figure 4.19 Relationship of timberline to elevation and latitude in three different major mountain chains in North America. In general, timberline increases in elevation with decreasing latitude, reflecting the influence of increasing temperature. Note, however, that the relationship is different in each mountain chain, due to other factors such as the length of the summer growing season. Note also that along the primary cordillera (the chain extending from the northern Rocky Mountains to Panama), there is essentially no change in the elevation of timberline within the tropics and subtropics, between about 35° N latitude and the equator. (After Daubenmire 1978.)

Unlike plants and sessile animals (such as the barnacles discussed earlier), most animals can move to seek out favorable microenvironments, and thus can avoid the most stressful abiotic conditions. Nevertheless, even highly mobile animals such as fishes can be limited directly by physical factors such as environmental temperatures. Pupfishes of the genus *Cyprinodon* are extremely eurythermal and euryhaline; that is, they can tolerate a wide range of temperatures and salinities (Brown 1971c; Brown and Feldmeth 1971; Soltz and Naiman 1978). Species of this genus occur in rigorous physical environments, including shallow streams and marshes in deserts and small pools in tidal flats, estuaries, and mangrove swamps, where temperature and salinity may fluctuate widely. Some populations even inhabit hot springs, although they generally cannot tolerate temperatures in excess of about 43° C (which is still amazingly high for a fish). Some pupfishes occur in the cooler outlets of hot springs whose temperatures exceed lethal limits. The local distribution of one such population of *C. nevadensis* near Death Valley, California, is limited directly by temperature (Brown 1971c). Fish occur in all waters cooler than about 42° C, including small pools only a few centimeters away from much hotter water (Figure 4.21). Occasionally individuals stray or are frightened into lethally hot water and die instantly—a clear demonstration of the thermal niche axis! Less frequently, rapidly changing weather conditions cause lethally hot water to flow far downstream from the source, trapping thousands of fish in pools where they are killed when the temperatures rise to over 43° C.

As in many organisms, adult pupfishes are tolerant of a wider range of physical conditions than are the early developmental stages. Eggs of *C. nevadensis* develop normally only in water between 20° and 36° C, although adults of this species can withstand temperatures between 0° and 42° C (Shrode 1975). Eggs are also less tolerant than adults of extreme salinity. Consequently, adult pupfishes can be found in sink habitats where reproduction is impossible, so long as they can immigrate from nearby source microhabitats that are suitable for egg development. On the other hand, species of *Cyprinodon* are conspicuously absent from some cold springs and other habitats

Figure 4.20 A bristlecone pine (*Pinus longaeva*) growing near timberline in the White Mountains of California. As is typical of individuals at the upper elevational edge of the range, this one has much dead wood and a highly contorted growth form. (Courtesy of A. Kodric-Brown.)

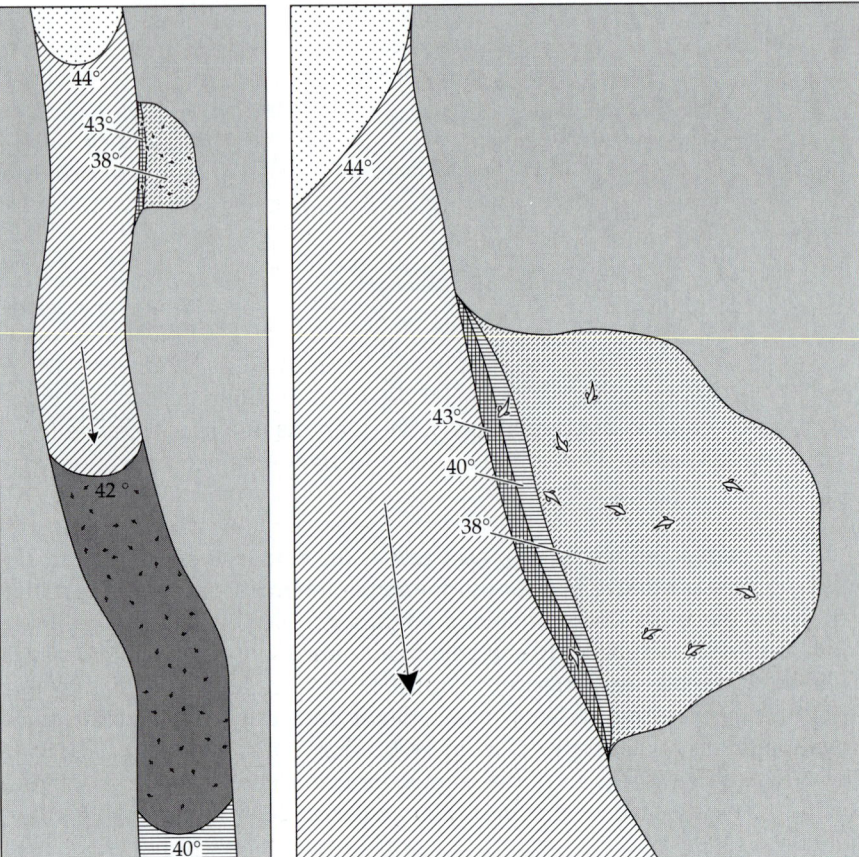

Figure 4.21 Temperature limits the local distribution of a desert pupfish (*Cyprinodon nevadensis*) in the outflow of a hot spring near Death Valley, California. The fast-flowing main channel is above the lethal temperature of 43° C; fish are trapped, but able to survive, in the cooler side pool (enlarged at right). (From Brown 1971c.)

where the adults can grow and survive, but there are no microclimates suitable for the earlier stages.

In addition to temperature, other physical and chemical factors, such as moisture, light, oxygen, pH, salinity, and soil and water elements, limit the distributions of animals, either singly or in interaction with one another. A simple example of such an interaction is the effect of temperature and oxygen concentration on many fishes and aquatic invertebrates. As water increases in temperature, its capacity to hold oxygen and other dissolved gases decreases. The resulting combination of high temperature and low oxygen concentration is very stressful to many fishes and aquatic invertebrates, because high temperatures cause elevated metabolic rates and increased demand for oxygen.

A few additional examples will suffice to illustrate distributional boundaries that can be attributed, at least in part, to such factors as moisture, salinity, and soil chemistry. Many terrestrial plants are limited by low soil moisture at the drier edges of their ranges, just as they are by low temperatures at the colder margins. In nearly all vascular plants, photosynthetic rates decline as soil moisture decreases. Plants can compensate for decreased water uptake through the roots by closing their stomates and reducing transpiration from the leaves, but rates of photosynthesis are reduced concomitantly.

Plants have diverse anatomical, physiological, and phenological adaptations that enable them to grow in a wide range of temperature, moisture, and light regimes (Bazzaz 1996). For example, species that can grow in full sunlight on dry soils (**xerophytes**) show many specialized mechanisms for keeping their stomates open despite low levels of water in their leaves. In contrast, species from wetter and more shaded environments (**mesophytes**) typically close their stomates when subjected to drought and temperature stress, and without evaporative cooling, their leaves suffer high, often fatal, heat loads. On the other hand, xerophytes typically have relatively low rates of photosynthesis when abundant water is available, and are intolerant of shade. A consequence of this trade-off is that mesophytes are physiologically incapable of growing on dry soils, whereas xerophytes can grow where there is little moisture, but are competitively excluded from wetter soils by shading from mesophyes that have higher growth rates (see Odening et al. 1974) (Figure 4.22). These physiological findings provide a mechanistic basis for the conclusion of Shreve (1922) and other early plant ecologists that in gradients of increasing aridity, the limits of plant distributions are determined largely by inability to tolerate low soil moisture. Widespread diebacks in drought years are commonly observed in local populations at the margins of a species range (e.g., Sinclair 1964; Westing 1966). Similar kinds of trade-offs between photosynthetic rate and ability to tolerate low nutrient levels, high salinity, extreme pH, or high concentrations of toxic minerals probably account, at least in part, for the failure of many otherwise widespread plant species to occur locally on soils with these characteristics, while species with special adaptations to these soil types are often restricted to them (Whittaker 1975).

These kinds of physical and chemical factors also limit animal distributions. Because of their osmoregulatory physiology, the vast majority of freshwater fishes and invertebrates are intolerant of salinities even approaching the concentration of seawater, whereas marine species cannot survive in fresh water. One consequence is that salt marshes and estuaries are inhabited by neither freshwater nor marine species, but rather by specialized euryhaline species that can tolerate great, often daily, fluctuations in salinity caused by tides, floods, and storms.

Only a few kinds of specialized organisms occur in those lakes and springs that are even saltier than seawater. Great Salt Lake in Utah, for example, has a salinity approximately seven times that of seawater. It contains no fish and only two macroscopic invertebrates, the pelagic brine shrimp (*Artemia salina*)

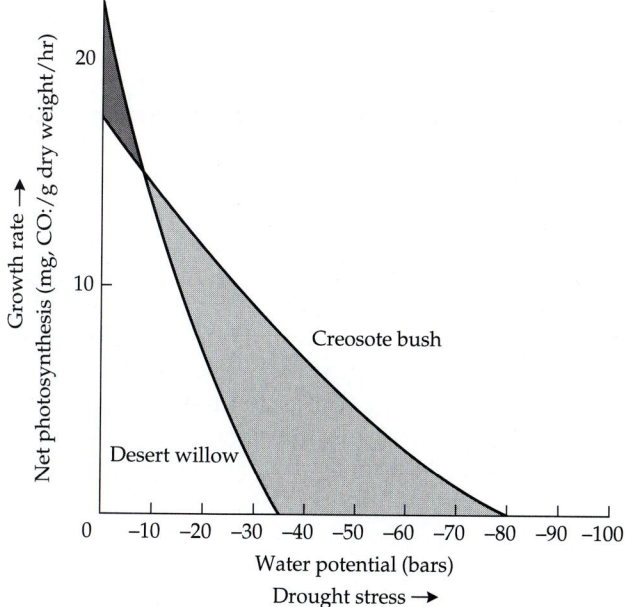

Figure 4.22 Trade-offs between growth rate and drought tolerance in two species of desert shrubs: creosote bush (*Larrea tridentata*), which grows in some of the driest North American deserts, and desert willow (*Chilopsis linearis*), which has an overlapping geographic range, but is more mesophytic, occuring in microhabitats along watercourses where the soil is permanently moist. Note that at higher soil moisture levels (light gray region), desert willow has the higher net photosynthetic rate and is able to grow faster, shade, and competitively exclude creosote bush, but when soils are drier (dark gray region), creosote bush has the higher growth rate. (After Odening et al. 1974.)

and the benthic brine fly (*Ephydra cinerea*). Several fish species and numerous invertebrates inhabit the freshwater streams and marshes that empty into the lake. These animals have abundant opportunities to extend their ranges and colonize the lake, but they are prevented from doing so by their inability to tolerate its high salinity. This is dramatically illustrated by the widespread mortality that occurs when occasional windstorms push salt water from the lake into the adjacent marshes.

(A)

(B)

(C)

Figure 4.23 Effect of a natural lightning-caused wildfire on savanna vegetation near Elgin, Arizona. (A) A few days after the fire in June 1975. (B) Several months later, September 1975. (C) Eleven years later, February 1986. Note that while the live oak trees lost most of their leaves as a result of the fire, only a few, such as the one in the center foreground, were killed or suffered major, lasting damage. Fires kill invading shrubs and some trees, maintaining this open, grassy habitat. (Photographs courtesy of R. M. Turner.)

Disturbance

Another class of factors that influences both local and geographic distributions of many organisms includes fires, hurricanes, volcanic eruptions, and other agents of sudden widespread disturbance and destruction. These natural disasters are capable of completely destroying habitats and their inhabitants. In regions where such disturbances occur with some frequency, however, they are a natural part of the environment (Figure 4.23). Many species are dependent on these periodic disturbances for their continued existence, and there is a regular pattern of colonization and replacement of species following a disturbance, which ecologists call secondary succession. For example, jack pines (*Pinus banksiana*) in eastern North America and lodgepole pines (*P. contorta*) in the Rocky Mountains have closed cones that require the heat of periodic forest fires to release their seeds. Somewhat similar cyclical successional processes occur in the chaparral shrublands of coastal California and the coniferous forests of the Rocky Mountain region, which are swept by periodic fires; on the forested islands and offshore coral reefs of the Caribbean region, which are occasionally but inevitably decimated by hurricanes; and in intertidal habitats throughout the world, which are subjected to heavy wave action and sand scouring during major storms.

On the other hand, periodic natural disturbances may be sufficiently severe and frequent to prevent the expansion of some species into areas where they could otherwise survive. In most grasslands, lightning-caused fires are a natural part of the environment. At forest/grassland boundaries, where soil moisture is high enough for woody vegetation to become established, frequent fires may prevent trees and shrubs from extending their ranges (see Figure 4.23). Artificial fire suppression within the last 200 years has contributed to the expansion of forest and shrubland at the expense of prairie and other grassland habitats along the eastern margin of the Great Plains of central North America (e.g., Beilmann and Brenner 1951; Hartnett et al. 1996). A similar phenomenon occurs along the drier margins of the Great Plains in southwestern North America, where fire influences the boundary between arid grassland and desert shrubland. Fire suppression, along with livestock grazing, has played a major role in **desertification**—the degradation of grassland to shrubland—in the southwestern United States, northern Mexico, and other arid regions throughout the world (Figure 4.24; Johnston 1963; Bahre 1995). In many of these places where natural fires once burned unchecked over enormous areas, prescribed burns must now be used as a management tool to preserve grassland and prevent the invasion of woody vegetation and exotic species. The tiny reserves of tallgrass prairie in the north central United States, for example, are too small to experience a natural frequency of lightning-caused fires. Now, prescribed burns are essential to prevent the local extinction of native prairie plant species.

At smaller scales, other causes of disturbance, especially biological ones, can be equally important. Typically, such small-scale disturbance has the effect of removing the dominant plants or sessile animals, allowing fast-growing but competitively inferior species to colonize. This creation and filling of gaps results in a patchwork of microsuccessional stages that in total supports many species. In many tropical and temperate forests, gaps in the canopy caused by the falling of single trees, or even large limbs, create a sunny microclimate on the forest floor that is essential for the existence of certain understory species and for the establishment of the seedlings of some canopy trees (e.g., tulip tree, *Liriodendron tulipifera*: Picket and White 1985). On rocky intertidal shores, such as those of northwestern Washington, the predatory starfish *Pisaster ocraceous* removes the competitively dominant mussel, *Mytilus californianus*, creat-

Figure 4.24   Contraction of grassland habitat in southern Texas between 1860 and 1960. These changes, reconstructed from historical records, are due largely to fire suppression, livestock grazing, and invasion of woody vegetation, especially mesquite (*Prosopis*). (After Johnston 1963.)

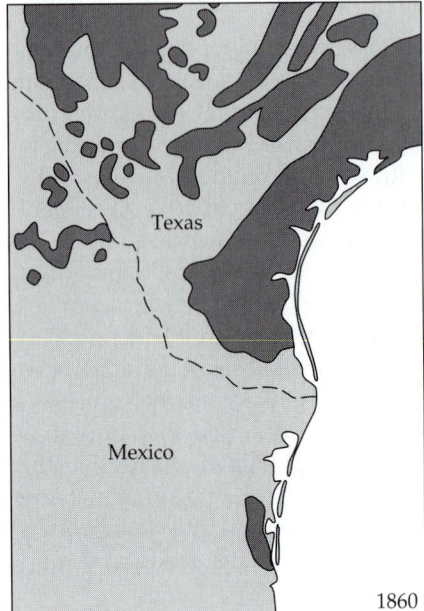
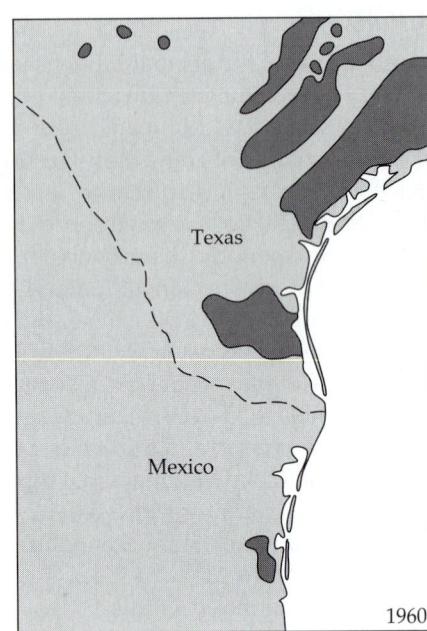

ing gaps in the otherwise continuous mussel beds that rapidly growing algae and invertebrates can colonize (Paine 1966). On some exposed shores, logs banging against the mussel beds can have a similar effect (Dayton 1971). In North American tallgrass and shortgrass prairies, the soil disturbance caused by the activities of mammals may be as important as fire for the maintenance of a diverse grassland community. The wallows of bison, the burrows of badgers, and the mounds of pocket gophers all provide patches of bare, sunny soil that are essential for the establishment and persistence of certain plant species (e.g., Platt 1975; Reichman and Smith 1985; Inouye et al. 1987).

Interactions with Other Organisms

In many cases, geographic distributions are not limited directly by physical factors. Botanical gardens and zoos provide perhaps the most dramatic evidence that individuals can survive, grow, and even reproduce under a much wider range of physical conditions than they encounter anywhere in their natural geographic range. The fact that many plants can thrive in botanical gardens, suburban landscapes, or agricultural fields—but only if they are protected from competing plants, animal herbivores, and microbial pathogens—demonstrates the importance of interspecific interactions in limiting distributions.

There are three major classes of interspecific interactions: competition, predation, and mutualism. All of these can influence the dynamics of populations and limit the geographic ranges of species.

Competition.   **Competition** is a mutually detrimental interaction between individuals. Organisms that share requirements for the same essential resources necessarily compete with each other, and suffer reduced growth, survival, and reproduction if the resources are in sufficiently short supply. Plants may compete for light, water, nutrients, pollinators, or physical space, whereas animals most frequently compete for food, but also compete for shelter, nesting sites, or mates, or for living space that contains these resources. These interactions may be purely **exploitative**, so that individuals use up resources and make them unavailable to others. Alternatively, competition may involve some form of **interference**, in which aggressive dominance or active inhibition is used to deny other individuals access to resources. For example, some plants, such as the black walnut (*Juglans nigra*), and some sessile marine animals, such as bryozoans and corals, use a form of chemical warfare called **allelopathy** to defend space from competitors.

There is much circumstantial evidence that competition limits geographic ranges. There are many examples of ecologically similar, closely related species that occupy adjacent but nonoverlapping geographic ranges. Five species of large kangaroo rats (*Dipodomys*) are found in desert and arid grassland habitats in the southwestern United States and northern Mexico, but their geographic ranges do not overlap (Figure 4.25). Two species, *D. ingens* and *D. elator*, have isolated, or disjunct, ranges, but *D. spectabilis* shares an extensive border with *D. deserti* in the west and with *D. nelsoni* in the south. Although such cases strongly suggest that competition limits distributions by preventing coexistence, they are subject to alternative explanations, and often there is no direct evidence of competitive interactions occurring on the boundaries.

Better evidence for the limiting effects of competition comes from "natural experiments" in which one species, simply by chance, is absent from regions that are apparently suitable for it. If a second species has expanded its range to include habitat types that are normally occupied by the first species, this

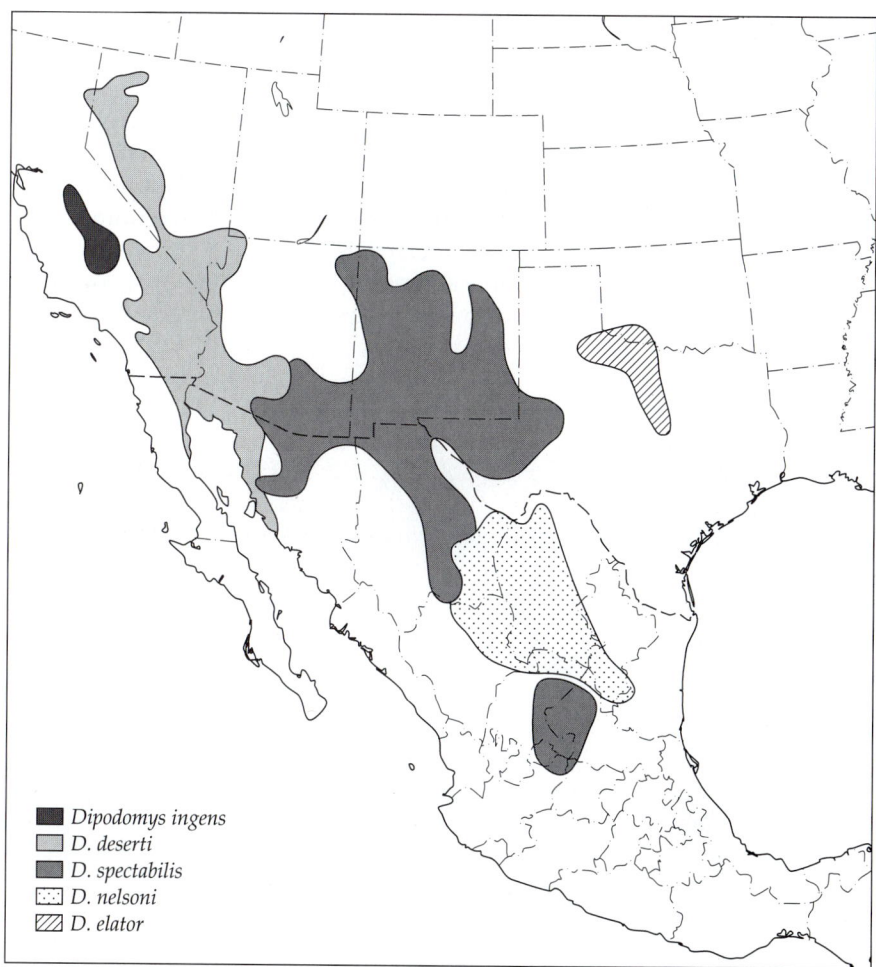

Key:
- ■ *Dipodomys ingens*
- ▨ *D. deserti*
- ■ *D. spectabilis*
- ⠿ *D. nelsoni*
- ▨ *D. elator*

Figure 4.25 Nonoverlapping geographic ranges of five species of large kangaroo rats (*Dipodomys*) in southwestern North America. These rodents are similar in their ecology, and the fact that their ranges frequently come into contact, but rarely overlap, suggests that they limit one another's distributions through competition. (After Bowers and Brown 1982.)

implies that in other areas where the first species is present in these habitats, the second is limited by competition. The forests and shrublands of the western United States are inhabited by more than 20 species of chipmunks of the genus *Eutamias*. In the mountains of the Southwest, two or three species typically occur in the woodlands and forests, but they are segregated by habitat and elevation. *Eutamias dorsalis* inhabits the open, xeric woodlands at lower elevations, and is replaced in the denser coniferous forests at higher elevations by a species of the *E. quadrivittatus* group. A third species, *E. minimus*, is sometimes present in the spruce and fir forests and above-timberline habitats on the highest peaks. There are at least 24 isolated desert mountain ranges where appropriate habitats seem to be present, but one of these species is absent, apparently because it either never colonized or became extinct sometime in the past. In every case, regardless of which species is absent, the remaining species has expanded its range to include all forested habitats from the edge of the desert to the timberline (Patterson 1981; Figure 4.26). Such examples of niche and range expansion are particularly convincing evidence of competition when, as in the present case, similar distributional shifts have occurred independently in several different places.

The mechanisms of competitive interaction among these chipmunk species have been investigated in some detail. Brown (1971a) placed feeding stations and observed behavioral interactions in the narrow zone where the ranges of *E. dorsalis* and *E. umbrinus* come into contact. He concluded that *E. dorsalis*, the more aggressive and terrestrial species, was able to exclude *E. umbrinus* from open woodlands, where chipmunks have to do most of their traveling on the ground, by aggressively defending patchy food resources. However, in denser forests, where food is harder to defend because it is more abundant and the chipmunks can travel through the trees, *E. umbrinus* wins out because *E. dorsalis* wastes excessive time and energy on fruitless chases. Chappell (1978) studied a more complex situation where the ranges of several species come into contact in the Sierra Nevada of California. He also found that these mutually exclusive distributions could be attributed primarily to the influence of habitat on the outcome of aggressive interactions.

Such "natural experiments" implicating competition in limiting geographic ranges are by no means confined to small mammals. Diamond (1975) gives

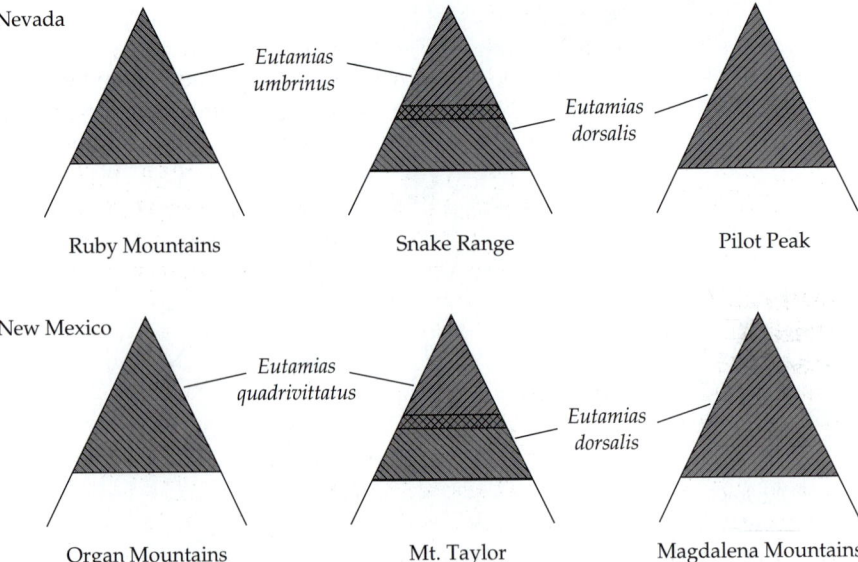

Figure 4.26 Diagrammatic representation of elevational distributions of chipmunks (*Eutamias*) on mountain ranges in the southwestern United States. On most ranges two species are present and their ranges overlap only slightly. On some ranges, however, only a single species occurs, in which case its range has expanded to include nearly all habitats and elevations normally occupied by both species. Such natural experiments are good evidence of competitive exclusion, which in this case has been confirmed by field studies.

several examples of niche and range expansion by birds in the absence of competing species on isolated mountain ranges within New Guinea and on nearby islands. Schoener (1970) and Roughgarden (1996; see also Roughgarden et al. 1983; Pacala and Roughgarden 1985) describe several cases among *Anolis* lizards on islands of the West Indies. Perhaps the best example in plants is the segregation of two species of cattails along a gradient of increasing water depth in marshes in central North America. In an elegant series of reciprocal transplant experiments, Grace and Wetzel (1981) showed that *Typha latifolia* was little affected by its congener, *T. angustifolia*. In the absence of its competitor, however, *T. angustifolia* could grow over the entire gradient.

Competition need not involve similar species to limit distributions. In fact, the kinds of pairwise interactions between closely related, ecologically similar species described above probably account for only a small part of the competition experienced by most species. For example, distantly related plants of different growth forms often experience intense, asymmetrical competition (Johansson and Keddy 1991). We described above how stressful physical conditions and fire limit the distributions of many plant species and vegetation types. The other side of the story is that the plants that are the least tolerant of abiotic stresses (drought, fire, and flooding) are usually superior competitors (Keddy and MacLelan 1990). For example, while some combination of fire, grazing, and drought may prevent trees and shrubs from invading grasslands, these woody plants are superior competitors where these conditions are not too severe. They not only can prevent establishment by grassland species, but during the last two centuries, as fire frequency has been reduced and grazing regimes altered, these woody plants have aggressively invaded grasslands and replaced herbaceous species (Steinauer and Collins 1996) (see Figure 4.24). Such asymmetrical niche relationships, in which one species or an entire functional group of species is limited by competition while the other is restricted by its inability to withstand disturbance or physical stress, appear to be extremely common among many kinds of organisms—recall Connell's study of barnacles.

Predation. **Predation** can be defined as any interaction between two species in which one benefits and the other suffers. According to this definition, relationships between herbivores and their food plants, parasites and their hosts, and Batesian mimics and their models would also be classified as predation. Predator/prey interactions can limit the distribution of either participant because, on the one hand, predators may depend on particular prey for food or other benefits necessary to support their own populations, whereas on the other hand, predators may limit prey populations by killing or damaging individuals.

When predators are highly specific, it is obvious that their distributions must depend in part on the availability of appropriate prey. It is hardly surprising that the geographic ranges of many specific parasites and herbivores correspond almost precisely with those of their animal or plant hosts. Thus the distribution of the checkerspot butterfly *Euphydryas editha* in coastal California is limited to the immediate vicinity of the patches of serpentine soil to which its host plant, *Plantago hookeriana*, is restricted (Ehrlich 1961, 1965). Of course, even highly specific predators may range less widely than their hosts, because in some areas their populations are limited by factors other than the availability of suitable prey.

It is much more difficult to document cases in which the distributions of prey populations are limited by their predators. Some of the best examples are artificial in that they involve introductions by humans of predators into regions where they did not originally occur. In some cases these introductions were deliberate attempts to control pest species. Two conspicuously successful examples of such biological control involve drastic reductions in plant populations by introduced herbivores. The prickly pear cactus (*Opuntia stricta*) was introduced into Australia in the mid-1800s to serve as an ornamental garden plant. By the early 1900s the cactus had escaped from cultivation and had become a serious pest on grazing lands. In 1926 Australian scientists introduced a moth (*Cactoblastis cactorum*) whose larva is a specific feeder on *Opuntia* in its native Argentina. By 1940 *O. stricta* had been effectively checked as a pest species in eastern Australia, although small patches of cacti and local populations of the moth remain (Dodd 1959). Unfortunately, many other species of exotic plants are invading Australian habitats, and the search for suitable agents of biological control is ongoing.

Similarly, Klamath weed (*Hypericum perforatum*) was introduced into northwestern North America from Eurasia in about 1900 and became an agricultural pest, but was subsequently controlled by the introduction of a specific leaf-eating beetle (*Chrysolina quadrigemina*) from its native habitat. In the southern part of its range, Klamath weed persists only in small populations along roadsides and in shady areas, but beetle populations do not do well in colder climates, and the weed is more widely distributed in British Columbia (Huffaker and Kennett 1959; Harris et al. 1969). In the cases of both the prickly pear cactus and Klamath weed, specific herbivores have drastically reduced plant populations, resulting in very limited local distributions, but they have not greatly altered the distributions of the weeds on a larger geographic scale.

Perhaps the best examples of complete elimination of prey populations from parts of their geographic ranges by voracious but nonspecific predators are provided by artificial introductions of large predatory fishes into certain freshwater habitats. Many of the small native fishes of the southwestern United States have suffered great reductions in their geographic ranges and complete extinction of local populations as a result of the introduction of large predatory game fishes (especially largemouth black bass, *Micropterus salmoides*) into their habitats (Miller 1961a). The native fishes are not adapted to large generalist predators because they have evolved in isolated lakes, streams, and springs for thousands of years. Zaret and Paine (1973) documented a similar example, the extinction of at least seven species of native fishes in Lake Gatun, Panama, following the introduction of the predatory fish *Cichla ocellaris*. Many of the approximately 300 endemic species of cichlid fishes have declined following the introduction of the Nile perch (*Lates niloticus*) into enormous Lake Victoria in central Africa. The lake trout (*Salvelinus namaycush*) is widely distributed over northern North America, but is not adapted to the presence of lampreys of the genera *Petromyzon* and *Entosphenus*, which are voracious predators. Niagara Falls formerly prevented *Petromyzon* from entering the upper Great Lakes, which supported large populations of lake trout. Construction of the Welland Canal, however, enabled the lamprey to colonize these lakes. The result has been a precipitous decline of lake trout in the upper Great Lakes despite a major effort to control the lamprey and save this valuable commercial fishery.

In the previous examples, the effects of predators on prey populations are particularly clear because we have been able to observe responses to artificial introductions of the predators. Without this historical perspective, however, we would be hard put to infer the extent to which the prey are limited by their predators. Today, most remaining patches of *Opuntia* in Australia are not infested with *Cactoblastis*. Similarly, it is difficult to observe black bass preying on the pupfishes of the southwestern United States, because the native fishes have already been extirpated from most waters where bass are present. It is likely that many prey populations are limited, at least in part, by their predators, but it is difficult to obtain convincing evidence without performing manipulative experiments.

While the effects of predators have been studied for decades, the effects of parasites and disease-causing microbes received much less attention from ecologists and biogeographers until quite recently. It is apparent, however, that they can also limit both local abundances and geographic distributions. There are an increasing number of examples in which domesticated plants and animals have been able to expand their ranges into previously uninhabitable areas following the elimination or control of their parasites or pathogens. Likewise, there are examples of range contraction following the human-aided invasion of new parasites or diseases. One example of the latter is the influence of avian malaria on native Hawaiian birds. Ever since the first humans arrived on the islands, this spectacular endemic avifauna has been suffering extinctions, range contractions, and population reductions (see Chapter 18). An initial wave of extinctions, presumably due primarily to hunting and habitat destruction, followed the arrival of the Polynesians about 1000 years ago. More extinctions followed European settlement, which began less than 200 years ago. The Europeans not only caused additional habitat destruction and brought in exotic avian competitors and mammalian predators, but also initiated biological warfare. Along with the exotic birds they brought in as domesticated fowl, game birds, and pets—and sometimes deliberately released— came avian malaria (*Plasmodium*). This parasite has become pandemic at lower elevations, where its persistence and transmission is favored by warm temperatures, availability of mosquito carriers, and relatively resistant exotic avian hosts. While it is difficult to tease out the multiple interacting influences of all the factors affecting the native bird species, avian malaria has clearly played a major role in the extirpation of most endemic Hawaiian birds from the lower elevations, even in areas where relatively undisturbed habitat remains (Warner 1968; van Riper et al. 1986).

Mutualism. The third class of interspecific interactions is **mutualism**, in which each species benefits the other. Examples of mutualistic associations are provided by plants and their animal pollinators and seed dispersers, corals and the photosynthetic zooxanthellae (algae) that live in their tissues, ants and aphids, and cleaner fishes and their hosts. Compared with competition and predation, mutualism has been little studied by ecologists and biogeographers (Boucher et al. 1984; Boucher 1985; Fleming and Estrada 1993). Much remains to be learned about the effects of these mutually beneficial associations on the abundance and distribution of the participating populations.

When mutualistic relationships are obligate for at least one, and especially for both, of the partners, then the interaction must have a major influence on distributions. Some plant populations are dependent on the services of specific pollinators for sexual reproduction. For example, red clover (*Trifolium pratense*) did poorly after being introduced into New Zealand until its pollinator, the bumblebee (*Bombus* spp.), was also introduced (Cumber 1953; Free 1970). Janzen (1966) studied the association between ants of the genus *Pseudomyrmex* and trees and shrubs of the genus *Acacia* in the New World tropics. Although many species of *Acacia* have no ants associated with them, and some species of *Pseudomyrmex* may not be dependent on acacias, the relationship is apparently obligate for numerous species. The trees provide the ants with enlarged thorns in which to build their nests and with specialized foods rich in sugars, oils, and proteins. In return, the ants attack herbivorous insects and vertebrates that attempt to feed on the trees, and clear away surrounding vegetation, reducing

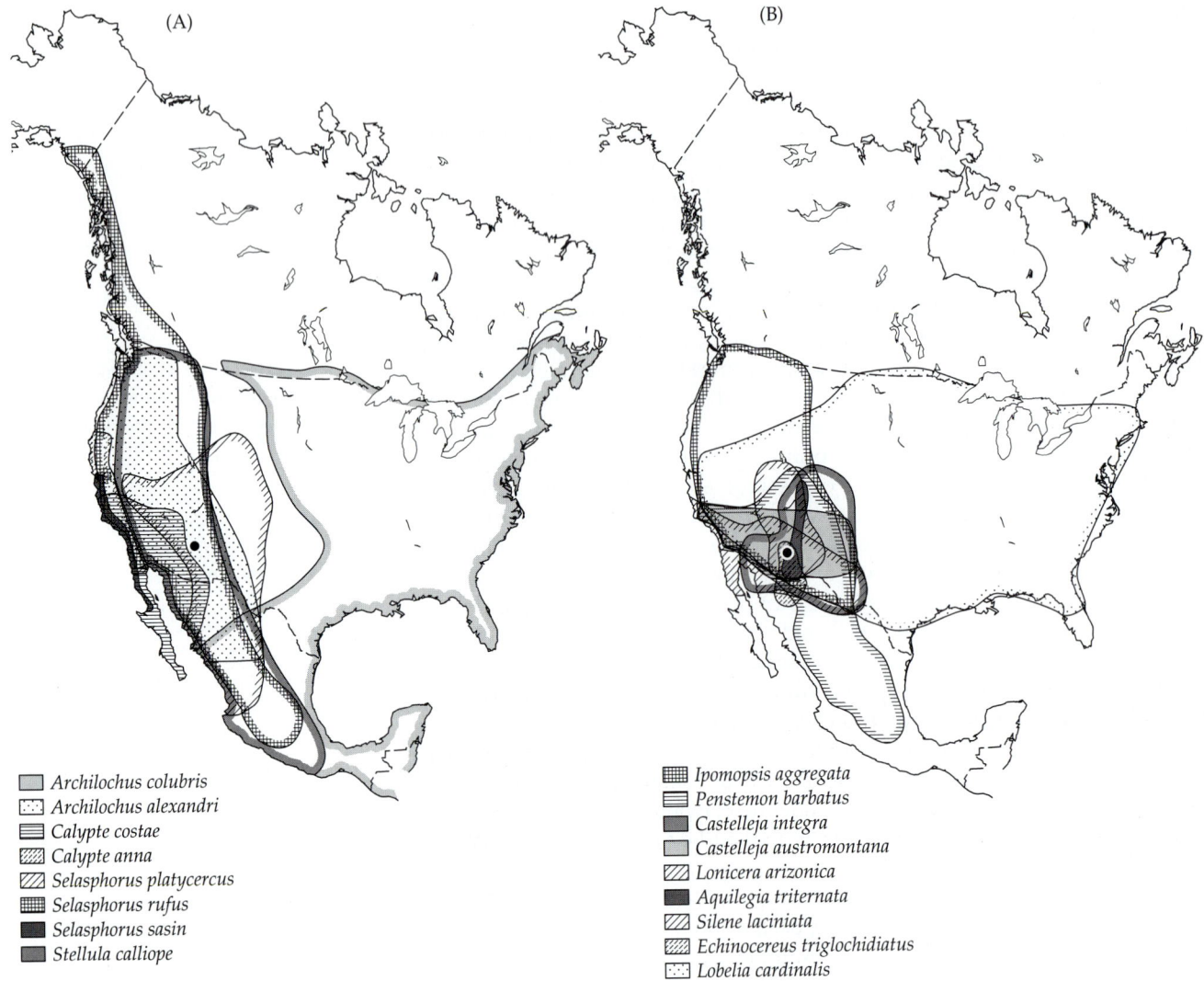

Archilochus colubris
Archilochus alexandri
Calypte costae
Calypte anna
Selasphorus platycercus
Selasphorus rufus
Selasphorus sasin
Stellula calliope

Ipomopsis aggregata
Penstemon barbatus
Castelleja integra
Castelleja austromontana
Lonicera arizonica
Aquilegia triternata
Silene laciniata
Echinocereus triglochidiatus
Lobelia cardinalis

Figure 4.27 Distributions of temperate North American hummingbirds and of some of the plants they pollinate while foraging for nectar. (A) Breeding ranges of the eight species of hummingbirds that occur substantially north of the United States/Mexican border. (B) Geographic ranges of the nine species of red tubular flowers commonly visited by hummingbirds at just one site in Arizona (circle). It would be too confusing to plot the ranges of the approximately 130 species of flowers used by these hummingbirds throughout their breeding ranges. Note that despite their close mutualistic relationships, there is little relationship between the geographic ranges of the specific plants and their pollinators. (From Kodric-Brown and Brown 1979.)

competition from other plants. Such coevolved specializations apparently have made these two mutualists so dependent on each other that they have virtually identical ranges.

These examples of close ecological and biogeographic associations may be exceptional, however. Often, relationships among mutualists, at least at the species level, are not so obligate, and the influence of the interaction on geographic distributions is not so apparent. Usually at least one of the partners does not absolutely require the mutualistic service of the other, or several different species can supply the service. Thus, most plant/pollinator and plant/seed disperser relationships are not obligately species-specific—at least outside the tropics (Waser et al. 1996).

Figure 4.27 shows the relationship between the geographic ranges in North America of eight hummingbird species and some plant species with red tubular flowers that they feed upon and pollinate. There is no correspondence between the ranges of particular pairs of hummingbird and plant species, but the distributions are such that, no matter where they occur, the hummingbirds have flowers to feed upon, and the plants have hummingbirds to pollinate them (Kodric-Brown and Brown 1979). A similar relationship is seen in the

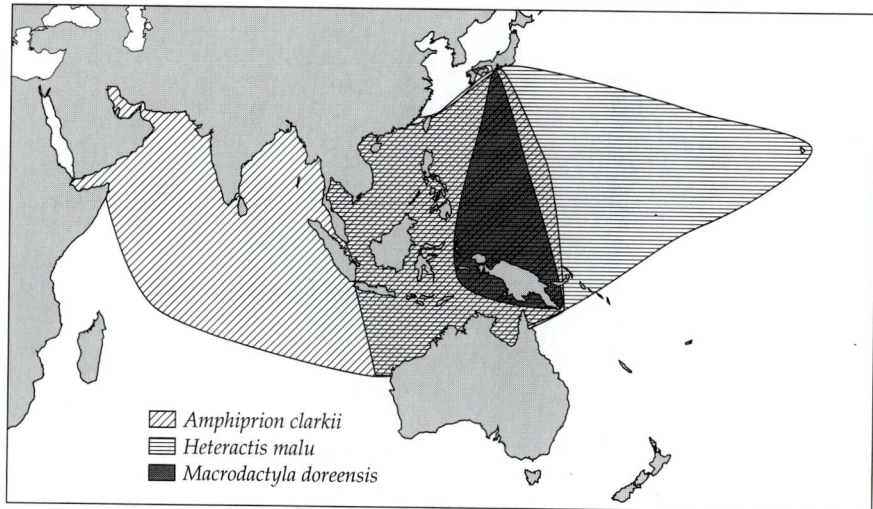

Figure 4.28 Distribution of the Indo-Pacific clownfish (*Amphiprion clarkii*) and two sea anemones (*Heteractis malu* and *Macrodactyla doreensis*) that serve as its hosts. *Amphiprion clarkii* is the only fish that is mutualistic with *H. malu*, but it uses other anemones, including *M. doreensis*. As with the hummingbirds and flowers shown in Figure 4.27, even relatively closely coevolved mutualists may be able to switch partners, and therefore do not necessarily have coincident ranges. (Unpublished data courtesy of D. Dunn.)

ranges of clownfishes (*Amphiprion* spp.) and the sea anemones with which they are associated (Figure 4.28). While all clownfishes appear to require the protection of anemones, and some kinds of anemones may be unable to exist without clownfishes, the particular species are much more interchangeable.

Multiple interactions.   In addition to those cases in which it is possible to isolate the limiting effect of one species on the distribution of another, there are undoubtedly many situations in which ranges are structured by more diffuse biotic interactions. Such limits may be the result of different, interacting effects of several species (see the discussion of competition between woody and herbaceous plants above). MacArthur (1972) noted that the southern limits of the ranges of many North American bird species apparently could be attributed neither to climate (because the climate becomes more equable at lower latitudes), nor to habitat (although it might be important for some species), nor to any particular species of competitor or predator. He suggested that for many of these species, the limiting factor must be diffuse competition from an increasing number of tropical species. He noted, for example, that of 202 land bird species that breed in Texas, only 29 also breed in Panama, but that Panama has a total of 564 breeding land bird species. One of the few species that breeds in both the United States and Panama is the yellow warbler (*Dendroica petechia*), a small, insectivorous foliage gleaner. In the United States the yellow warbler is abundant in a wide variety of shrubby and forested habitats, but in Panama it is restricted to mangrove swamps and small offshore islands (Figure 4.29). Because the forests of Panama contain many species of highly specialized foliage-gleaning birds, whereas mangroves and islands have few such species, MacArthur attributed the distribution of the yellow warbler to diffuse competition—the combined negative effects of competition with many other bird species.

In such isolated cases, however, there is little hard evidence that competition alone is responsible for the observed distributional patterns. Other factors, such as predators, parasites, diseases, and mutualists, could, and very likely do, play major roles. Recall the example of avian malaria limiting the lower elevational distributions of several native bird species in the Hawaiian Islands. In the absence of historical and epidemiological evidence for the role of this parasite, it would be easy to mistakenly attribute the restricted distributions of native species solely to competition with introduced birds. Bertness (1989,

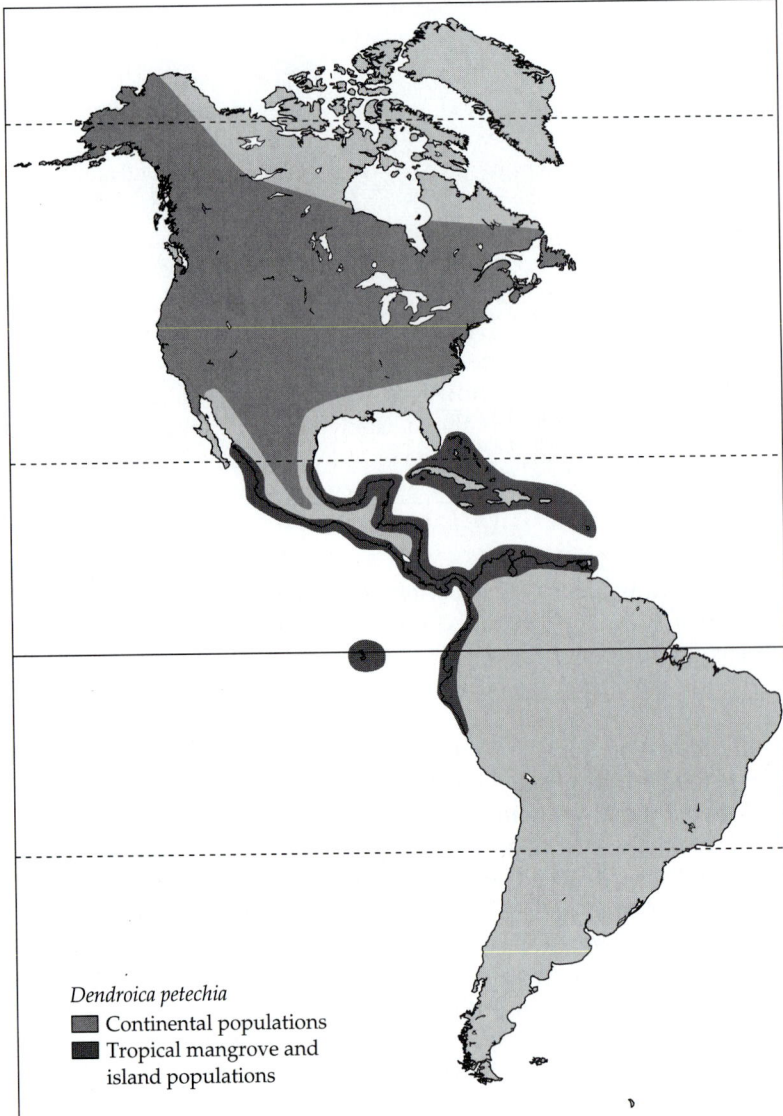

Figure 4.29   Geographic range of the yellow warbler (*Dendroica petechia*). This passerine bird is widely distributed over temperate North America, but in the tropics it is restricted to coastal mangrove swamps and islands. MacArthur (1972) suggested that diffuse competition from the many species of small insectivorous birds in most habitats on the mainland of tropical America accounts for the limited distribution of the yellow warbler in the tropics.

Dendroica petechia
 Continental populations
 Tropical mangrove and
   island populations

1991, 1993) suggests that in abiotically stressful environments, such as salt marshes and intertidal zones, mutualistic interactions may have at least as much influence on the abundance and distribution of species as competition and predation.

In the last few decades, ecological studies have documented a variety of complicated interactions involving many species. There is a large literature on indirect interactions, "bottom-up" and "top-down" chain reactions in food webs, and the ramifying effects of "keystone species" (e.g., Carpenter 1988; Schmitt 1987; Strauss 1991; Wootton 1992; Jones and Lawton 1994). The vast majority of these studies have used manipulative experiments on small spatial and temporal scales to show how such complex interactions can influence abundances and other ecological characteristics of species within local ecological communities. While there is every reason to suspect that these kinds of multispecies interactions also influence the distributions and diversity of species on geographic scales, rigorous documentation is usually lacking. The increased emphasis on larger-scale phenomena and nonexperimental methods should lead to a better understanding of the biogeographic consequences of these complex interactions.

Synthesis

Having presented numerous examples of the kinds of abiotic and biotic factors that limit the geographic ranges of species, it is appropriate for us to draw some general conclusions. First, however, a note of caution is in order: Despite all of the examples given above, it may not be productive to search for single limiting factors and simple explanations for the geographic distributions of most species. Not only may a single species be limited by different factors in different parts of its range, but even in one local area, several factors may interact in complex ways to prevent expansion of populations.

Given this complexity, however, there is a hint of one pervasive pattern, which has been alluded to above: Many species appear to be limited on one range margin by abiotic stress and on the other by biotic interactions. This pattern is particularly apparent in studies of distributions along environmental gradients, as of sessile organisms in the intertidal zone and plants along terrestrial gradients of temperature and moisture. An interesting feature of many such examples is that on the margin limited by physical stress there are relatively few other species, whereas on the margin limited by biotic interactions the diversity of other organisms is much higher. This pattern suggests an interesting relationship between species diversity and geographic range limitation: Where conditions are physically stressful for one species, they are likely to be unfavorable for other organisms as well, and vice versa at the other end of the gradient where abiotic conditions are more favorable. This is not surprising. What is more interesting is the suggestion that biotic interactions tend to become limiting when physical conditions are less severe, and that the higher the diversity of other species in the environment, the more likely that some of these will be sufficiently powerful enemies, whether competitors, predators, parasites, or diseases, to limit the range by excluding the species from areas where it otherwise could occur. A special application of this concept is the conjecture of Dobzhansky (1950) and MacArthur (1972) that biotic interactions are more likely to limit abundances and distributions in the tropics, while abiotic stresses are more likely to be limiting at higher latitudes. We will consider the implications of these relationships in Chapter 15, when we discuss the correlates and causes of geographic patterns of species diversity.

Adaptation and Gene Flow

In our discussion of ecological niches and geographic range limits, we have glossed over the issue of geographic variation within species. The niche of a species is not constant in either space or time. Since natural selection is a universal process that tends to increase the capacity of individuals to survive and reproduce, we would expect populations to adapt to their local environments. Such adaptation might be evidenced by geographic variation within species in physiology, morphology, or behavior (see below). It might also be reflected in adaptive changes in niche relationships over time. We might expect peripheral populations to be able to adapt to the environmental factors that limit their ranges, increase in density, and colonize adjacent areas (see Baker and Stebbins 1965). Lewontin and Birch (1966) describe an apparent example in the Australian fruit fly, *Dacus tryoni*. During the last century this species has expanded its range several hundred kilometers to the south along the eastern edge of the continent. These flies are limited by low temperatures, and their expansion has been accompanied by adaptation of the peripheral populations for increasing cold tolerance. But clearly not all species are increasing their ranges in this fashion. A historical perspective suggests that over the last 60 million years or so, the distributions of some species and higher taxa have indeed increased, but that these expansions have been almost equally matched by contractions in the ranges of other organisms. As a result, there appears to have been little change in global biodiversity during this period.

Why don't the peripheral populations of all species adapt to local conditions, resulting in a continual expansion of the range on all margins? In some cases the reason is obvious: There are fundamental constraints that cannot be easily overcome by local adaptation. Terrestrial life forms along coastlines, for example, cannot simply evolve adaptations for aquatic life and invade the marine realm. This kind of adaptation did happen in the past, as the terrestrial ancestors of penguins and whales invaded the sea, but it took millions of years and required that these animals give up their capacity for living on land. But such cases hardly explain why species do not expand their ranges along ecological gradients. Why don't local populations adapt to deal with the limiting physical stress or biological enemy? Why don't they expand the species range by evolving to tolerate just a bit more cold, aridity, predation, or competition?

Genetic and ecological processes seem to interact to limit the capacity of peripheral populations to adapt to local conditions. The exchange of genes among populations, called **gene flow**, may prevent local populations from acquiring and maintaining the combinations of genes necessary for continual adaptation. Gene flow is caused by the migration of individuals or gametes between populations, which bring with them genes for traits that have been selected for different local environments. Sufficiently high rates of gene flow can swamp a local population with genes from outside, effectively preventing adaptation to local conditions.

The critical question in any particular case is whether gene flow is high enough to overwhelm natural selection, prevent the continual adaptation of peripheral populations, and thereby preclude expansion into new areas of ever more extreme environments. The answer to this question for many organisms is not clear, and it has been the subject of much debate among population geneticists and evolutionists. Certainly, many species show adaptive genetic changes in response to variation in their environments over their geographic ranges. The deer mouse (*Peromyscus maniculatus*), whose enormous geographic range encompasses most of the North American continent, has been the subject of many genetic and ecological studies. There is a great deal of geographic variation in the color and shape of deer mice, which is reflected by the subdivision of the species into many geographic races or subspecies (see Figure 8.5). Classic studies by Lee Dice and his students showed that much of this variation reflects adaptation to local environments. Coat color tends to match the local soil color because of strong selection by owls and other predators against contrastingly colored individuals (Dice and Blossom 1937, 1947). Compared with individuals from desert and grassland habitats, animals from forest populations tend to have longer feet and tails because they use these appendages in climbing (Horner 1954). Populations in regions where contrasting habitat types come into close proximity, however, show lower levels of such morphological and behavioral specialization because of the diluting influence of gene flow (Thompson 1990).

These and similar studies of other organisms support theoretical models that predict that even modest levels of gene flow may be sufficient to preclude local adaptation (Wright 1978). It follows that gene flow from more central populations could prevent the adaptation and expansion of peripheral populations. If gene flow could be reduced or eliminated, however, adaptation could proceed, and further range expansion might be possible. Evidence in support of this conjecture comes from studies of the genetics of Old World burrowing rodents of the genus *Spalax*. Differences in chromosome number among populations of *S. ehrenbergi* appear to have been important in facilitating their colonization of the most arid deserts of the Middle East by reducing gene flow from populations adapted to more mesic areas (Wahrman et al. 1969; Nevo and Bar-El 1976; Figure 4.30). Gene flow between populations with different chromosome numbers and configurations is reduced because when individuals with different karyotypes breed, problems with pairing and separation of chromosomes reduce the viability and fertility of their offspring (see also Chapter 8). Patton (e.g., 1969, 1972, 1985) describes other examples of how local and regional ecological conditions, natural selection, and genetic barriers to gene flow interact to influence the distributions of small mammals.

Chromosomal rearrangements also have been implicated in limiting the distributions of plants. Populations at the geographic margins of a species range, as well as those inhabiting extreme soil types or climates, often are characterized by major chromosomal changes, especially polyploidy (Stebbins 1971b). Although these rearrangements of genetic material may themselves confer specific adaptations, their general effect is to reduce the frequency of crossing with other populations and thereby reduce or completely block gene flow, permitting adaptation to the local environment and facilitating colonization of new areas.

Ultimately, of course, the capacity of populations to adapt to new environments and to colonize new habitats is limited. Although some species are more widely distributed and more tolerant of varying conditions than others, there are no superorganisms that occur everywhere. Adaptation inevitably involves trade-offs and compromises. In order to tolerate the physical conditions and deal with the biotic interactions in some environments, a species must sacrifice its ability to do well in other habitats. Using a combination of "common garden" and breeding experiments on yarrow (*Achillea millefolium*), Clausen et al. (1940, 1947, 1948) elegantly demonstrated the interacting roles of genetic isolation, local selection, and trade-offs in determining the degree of adaptive differentiation of local populations and allowing this species to have a wide distribution along an elevational gradient in the Sierra Nevada of California (Figure 4.31). Gene flow can prevent populations from adapting to dif-

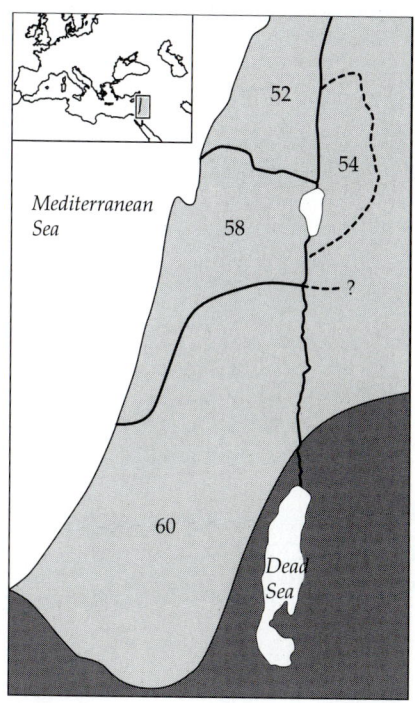

Figure 4.30 Distribution of karyotypes of the burrowing rodent *Spalax ehrenbergi*. The number of chromosomes increases along a gradient of increasing aridity from north to south (and also west to east) in the Middle East. Note that these "chromosomal races" replace each other with virtually no overlap, suggesting that reduced gene flow between them has facilitated adaptation and expansion into increasingly arid environments. (After Nevo and Bar-El 1976.)

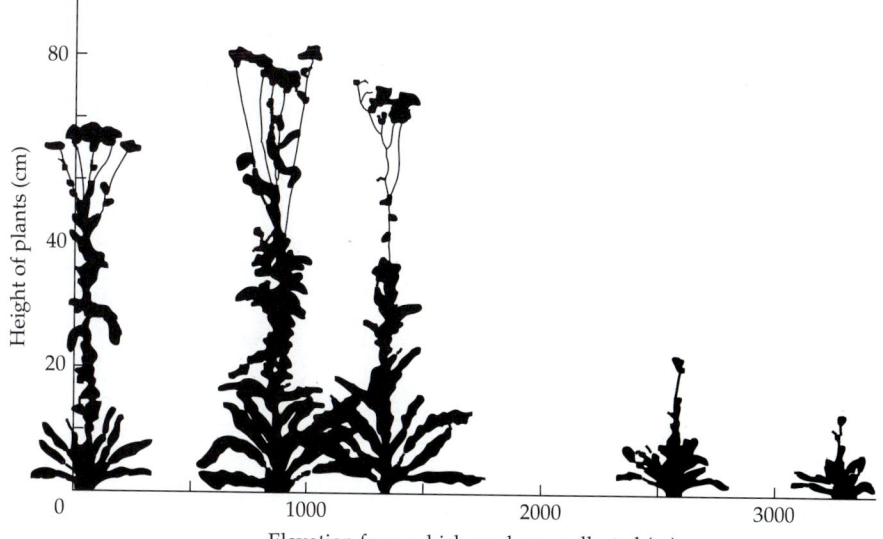

Figure 4.31 Morphological differentiation of the plant *Achillea millefolium* along an elevational gradient in the Sierra Nevada of California. Note the distinctive characteristics of different populations of the plant from different places along the gradient. Since these plants were all grown in the identical environment of a "common garden," we can infer that the differences among them are genetic. Presumably these differences reflect local adaptation, due to a combination of natural selection and reduced gene flow. (After Clausen et al. 1948.)

ferent local environments, but when gene flow between populations is interrupted, as it is during the process of speciation (see Chapter 8), then over evolutionary time, populations can and often do diverge in response to natural selection, adapt to widely different conditions, and expand their ranges to occupy new habitats and geographic areas.

CHAPTER 5

The Distribution of Communities

No living thing is so independent that its abundance and distribution are unaffected by other species. The physical and chemical composition of the earth's surface has been drastically altered by the activities of organisms, which, among other things, have created the oxygen in the atmosphere and contributed to the development of soil. Organisms vary greatly, however, in the extent to which they are dependent on other organisms. Some autotrophic organisms, such as certain kinds of algae and lichens, not only make their own food from sunlight, carbon dioxide, water, and minerals, but also inhabit extremely rigorous physical environments, such as hot springs and bare rocks, where they may encounter and interact directly with few, if any, other species. Heterotrophic organisms, which include all animals, fungi, and some non-photosynthetic vascular plants, cannot make their own food and are dependent on autotrophs for usable energy. Even many photosynthetic plants are directly dependent on specific kinds of other organisms, such as nitrogen-fixing bacteria and mycorrhizal fungi, which make available essential mineral nutrients, and insects and vertebrates, which pollinate flowers and disperse fruits and seeds.

Historical and Biogeographic Perspectives

Species occur together in complex associations called ecological communities. In the early twentieth century, as the new science of ecology was becoming firmly established, some of its foremost practitioners debated the nature of the relationships among species that determine the organization of communities. F. E. Clements (1916) suggested that a community could be regarded as a type of superorganism with its own life and structure as well as its own spatial and temporal limits. According to this view, individual organisms and species could be analogized to the cells and tissues of an organism, and the process of secondary succession could be likened to the growth and development of an individual. In contrast to Clements's concept of the community as a discrete and highly integrated unit, H. L. Gleason (1917, 1926) viewed a community as merely the coexistence of relatively independent individuals and species in the same place at the same time. Focusing primarily on plants, Gleason pointed out that the occurrence of species in an area depends primarily on their individual capacities to immigrate to and to grow in the local environment.

As is often the case with such debates, both sides made some important points. Communities do have certain properties, analogous to those of individuals, that can be measured and studied. These properties include photosynthesis (**primary productivity**) and metabolism (**respiration**), as well as more complex processes associated with the transfer and use of energy and nutrients and with changes in species composition and habitat during succession (e.g., Odum 1969, 1971). Some species are interdependent because they require others as mutualists, prey, or hosts (see Chapter 4). On the other hand, plant ecologists such as J. T. Curtis (1959) and R. H. Whittaker (1967, 1975; Whittaker and Niering 1965) amassed much data showing that the abundances and distributions of most vascular plant species vary in time and space as if they were independent of those of other members of their communities (McIntosh 1967).

This question is of particular interest to biogeographers: To what extent are species distributed together as interdependent communities as opposed to being distributed essentially independently of one another? If communities are integrated sets of species that are adapted to one another and tolerant of similar physical environments, then we might expect communities to be distributed as discrete units.

Even the casual observer will notice that certain kinds of plants tend to occur together in particular climates to create distinctive vegetation types. Ecologists and biogeographers refer to these as **life zones**, **ecoregions**, or **biomes**, and recognize that specific kinds of animals and microorganisms are associated with these vegetation formations. For example, a broad band of coniferous forest, sometimes referred to facetiously as the "spruce-moose biome," extends around the world at high latitudes in the Northern Hemisphere. Not only similar vegetation, but also many of the same species and genera of plants, animals, and microbes, are distributed over the Old World from Scandinavia to Siberia and across the New World from Alaska to Nova Scotia and Newfoundland. These organisms are generally adapted to living with one another and in similar physical environments. Their ranges extend southward together where mountain ranges provide appropriate habitats at high elevations. In North America these coniferous spruce-fir forests extend southward along the Appalachian Mountains in the east and along the Cascades, Sierra Nevada, and Rocky Mountains in the west. Several kinds of organisms are restricted to these coniferous forest habitats and have similar geographic distributions (Figure 5.1). We should be cautious, however, about inferring from such patterns that communities represent discrete, highly integrated ecological and biogeographic units, because there are alternative explanations, including historical ones (see Chapters 10 and 11).

Communities and Ecosystems

Definitions

Communities and ecosystems, the highest levels of ecological organization, are rather arbitrarily defined. As stated above, a **community** consists of those species that live together in the same place. The member species can be defined either taxonomically or on the basis of more functional ecological criteria, such as life form or diet. Ecologists study two fundamental properties of communities. **Community structure** refers to static properties, including the diversity, composition, and biomass of species. **Community function** includes all the dynamic properties and activities that affect energy flow and nutrient cycling (e.g., photosynthesis, interspecific interactions, decomposition). The

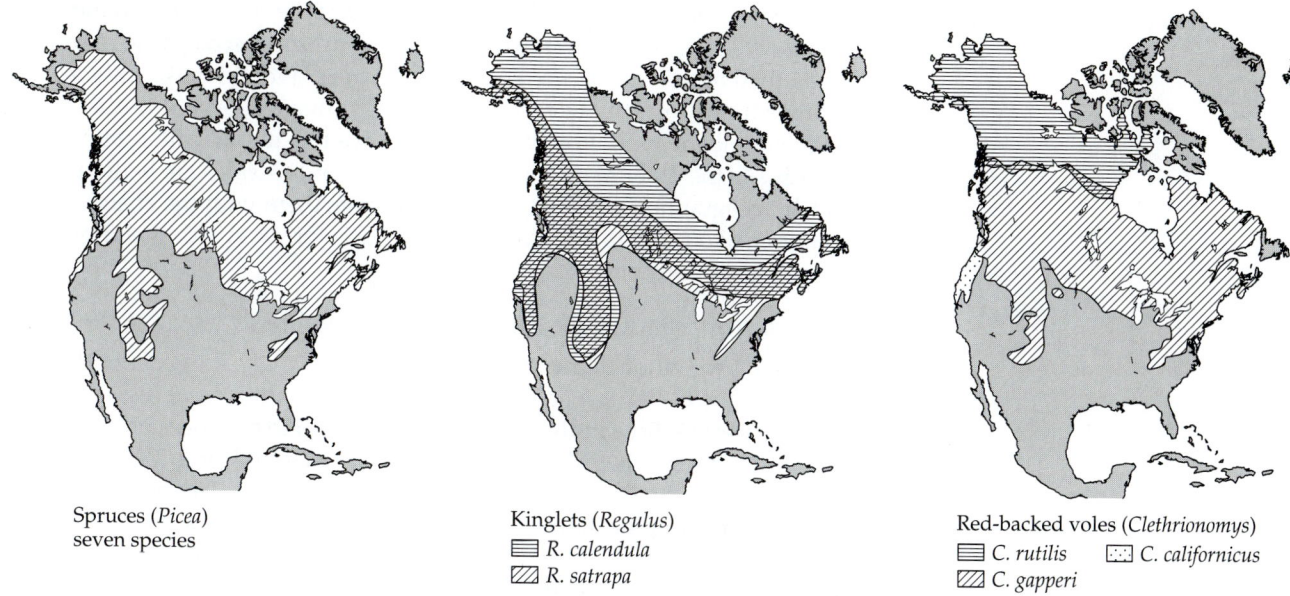

Spruces (*Picea*)
seven species

Kinglets (*Regulus*)
▤ *R. calendula*
▨ *R. satrapa*

Red-backed voles (*Clethrionomys*)
▤ *C. rutilis*   ▦ *C. californicus*
▨ *C. gapperi*

Figure 5.1 Three groups of organisms inhabiting coniferous forests exhibit similar geographic ranges in North America. These seven species of trees (not shown individually), two species of birds, and three species of mammals are typical inhabitants of the coniferous forests that spread across the northern part of the continent and extend southward at high elevations in the mountains.

place where member species occur can be designated either by arbitrary boundaries or on the basis of natural topographic features. Thus we can speak of the fish community in some ecologist's 2-hectare study area on a coral reef on the north shore of Jamaica; of the grazing community (which would include representatives of several invertebrate phyla as well as some fish species) on the entire reef, which might extend uninterrupted over many square kilometers; or of the entire community of all coral reef-inhabiting organisms in the Caribbean Sea.

An **ecosystem** includes not only all the species inhabiting a place, but also all the features of the physical environment. Because ecosystem ecologists are interested primarily in the exchange of energy, gases, water, and minerals among the biotic and abiotic components of a particular ecosystem, they often try to study naturally confined areas where the input and output of energy and materials are restricted and hence easier to control or monitor (e.g., Odum 1957; Teal 1957, 1962; Hutchinson 1957, 1967, 1975, 1993). Small, relatively self-contained ecosystems are often called **microcosms** because they represent miniature systems in which most of the ecological processes characteristic of larger ecosystems operate on a reduced scale. A sealed terrarium is a good example of an artificial microcosm, and a small pond is an example of a natural one.

Few ecosystems are as isolated and independent as they might appear. The largest and only complete ecosystem is the **biosphere**, which encompasses the earth. The interdependence of all ecosystems is vividly demonstrated by the widespread impact of human activities. Acid rain falls on virgin forests and is carried into pristine lakes far from the source of pollution. Deforestation, especially the cutting of tropical forests, and the burning of fossil fuels are changing the composition of the atmosphere, and perhaps altering the climate, throughout the world.

Many ecologists are actively investigating the structure and function of communities and ecosystems. As is usually the case in a rapidly developing field, many of the important questions remain unanswered, and some tentative conclusions are highly controversial. Because many of these unresolved issues are relevant to biogeography, this situation can be both challenging and frustrating for those trying to account for plant and animal distributions. On the one hand, biogeographers have the opportunity to use their own methods and data to make important advances in community ecology. Researchers with a broad biogeographic perspective are in a position to make unique contributions to the unraveling of the interacting ecological processes that influence the numbers and kinds of species that live together in different parts of the earth. On the other hand, until these ecological processes are better understood, biogeographers will be unable to integrate them with the effects of historical events to provide a solid conceptual basis for interpreting and predicting distribution patterns.

Community Organization: Energetic Considerations

In Chapter 4 we emphasized that each species has a unique ecological niche that reflects the biotic and abiotic environmental conditions necessary for its survival. How are these individual niches organized to produce complex associations of many species? This is the ultimate question of community ecology, and one for which we can advance only tentative answers. Two characteristics, body mass and trophic status, especially influence the roles species play in communities.

If you sought to take a single measurement that would provide the most information on the biology of an organism, you would be best advised to measure its body mass. Body mass has more important physiological and ecological consequences than any other character. The larger an organism, the more energy it requires for maintenance, growth, and reproduction. The rate of energy uptake and expenditure of animals at rest, the **basal metabolic rate** (m), varies with body mass (M) according to the relationship $m = cM^{0.75}$. Although the value of the constant (c) varies somewhat among taxonomic groups, the exponent, 0.75, is very general (Hemmingsen 1960; Calder 1984; Peters 1983; Figure 5.2). The ecological consequences of this simple formula are profound. Because the exponent is positive, total energy requirements increase with body

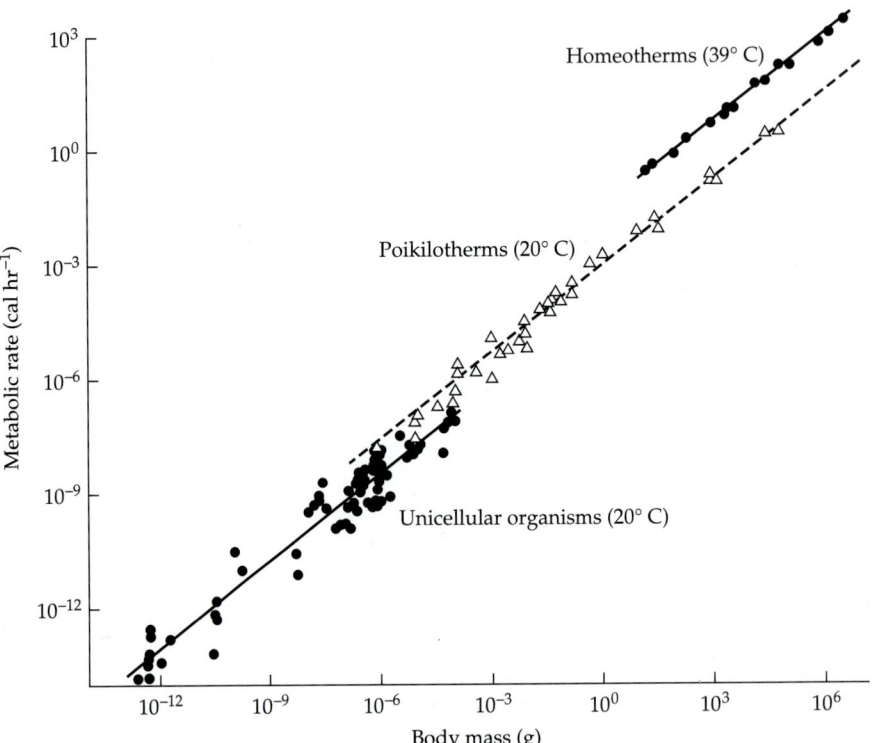

Figure 5.2 Relationship between metabolic rate (m) and body mass (M) for a wide variety of organisms, from unicellular forms to poikilothermic (cold-blooded) animals to homeothermic birds and mammals. Note that the axes are on a logarithmic scale, so that the relationship is described by a power function of the form $m = cM^{0.75}$, where the constant (c) varies slightly among the three different groups, but the exponent (slope) of 0.75 is remarkably constant. (After Hemmingsen 1960.)

size. It takes about 14,000 times more energy to maintain a 5000 kg elephant than a 15 g mouse. The same applies to all organisms: a mature redwood tree uses much more energy than a strawberry plant. On the other hand, because the exponent is less than 1, the metabolic rate per unit of mass of a small organism is greater than that of a larger one. It requires about 25 times more energy to maintain a gram of mouse than a gram of elephant. Apparently for this reason, almost all rate processes (e.g., cardiac rate and respiration frequency in vertebrates) are accelerated in small organisms, which are more active and have higher reproductive rates and shorter life spans than large ones.

Although metabolic rate and energy requirements increase as less than linear functions of body mass, storage capacities (e.g., energy stored as fat, water volume, mineral stores in bone tissue) increase in proportion to mass. Therefore, all else being equal, larger organisms have greater capacities to withstand prolonged stresses such as starvation, dehydration, and subfreezing temperatures. Body size also has important ecological consequences because it influences the scale at which organisms use the environment. Because small organisms require fewer resources per individual than large ones, they can use smaller areas, be more specialized, and still maintain population densities high enough to avoid extinction. Small organisms are better able than large ones to respond to what Hutchinson (1959) termed the mosaic nature of the environment—the spatial heterogeneity or patchiness of the environment that is most pronounced on a small scale. Consequently, small organisms have been able to divide the environment more finely, so that any geographic area contains a greater number of small-bodied species than large ones (Van Valen 1973a; May 1978; Brown and Maurer 1986) (Figure 5.3). Consider the tremen-

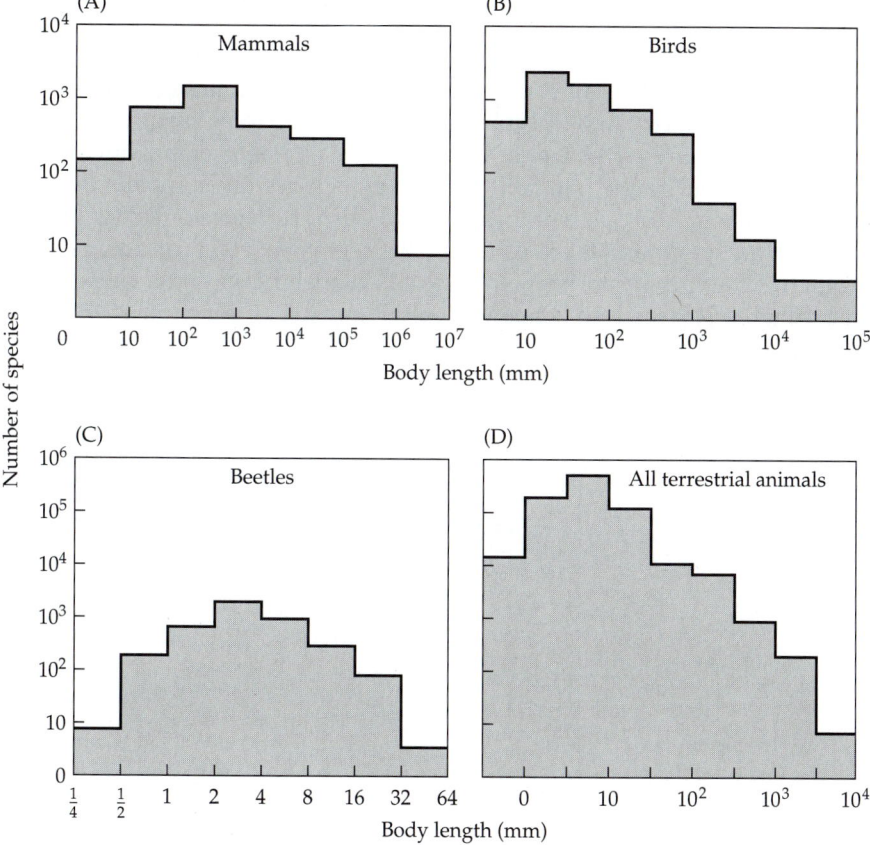

Figure 5.3 The frequency distribution of body size among species for several different kinds of organisms: (A) terrestrial mammals of the world; (B) land birds of the world; (C) beetles of Great Britain; (D) all terrestrial animals of the world (approximately). Note that the axes are on a logarithmic scale, so that small species are much more numerous than large ones. This pattern is very general, and accounts for the obvious fact that insects are much more numerous (in numbers of species as well as individuals) than vertebrates in terrestrial environments. (After May 1978.)

dous diversity of insects as compared with terrestrial vertebrates (about 750,000 versus 24,000 known species, respectively). A biogeographic correlate of this pattern is that, at least among animals, large organisms are constrained to have broad geographic ranges. The reason for this should be obvious: because large individuals require more space (McNab 1963; Schoener 1968b), the carrying capacity of the environment for these animals is low, and only large areas can support sufficient numbers of individuals to maintain the species over evolutionary time. In contrast, small areas can support large populations of small organisms, which consequently have a low probability of extinction. Small species may or may not have very restricted ranges (Figure 5.4; see also Chapter 16).

The **trophic status** of organisms, or how they acquire energy, also influences the role they play in community structure. Outside of a handful of fascinating exceptions (e.g., chemosynthetic communities of geothermal vents on the ocean floor), all of the energy used by living things ultimately comes from the sun. Autotrophic green plants use solar radiation, carbon dioxide, water, and minerals to synthesize organic compounds, and produce oxygen as a by-product. These organic compounds are not only used by the plants for making structures and fueling basic metabolism, but are also the sole source of energy for the heterotrophic organisms in the community. Solar energy, trapped in organic molecules, is transferred from one species to another and gradually used up as herbivores eat plants, carnivores eat herbivores, and so on. These heterotrophic species oxidize organic compounds to obtain usable energy, and in the process consume oxygen and release carbon dioxide.

The unidirectional paths of energy flow between species and through communities are termed **food chains**. The different links in a food chain are called **trophic levels** (Figure 5.5). The first level contains green plants, or **primary producers**; the second, **herbivores**, or **primary consumers**; the third, **carnivores**, or **secondary consumers**; and so on. At the ends of food chains are decomposers, or **detritivores**, mostly bacteria and fungi, which break down the remaining organic matter and release the inorganic minerals into the soil or water, where they can be recycled and again taken up by plants.

Although each tiny packet of energy follows a linear path along a food chain, the actual trophic relationships among species are complex. Because most consumers feed on several species and are in turn consumed by several kinds of predators and decomposers, the aggregate paths of energy flow through a community form interconnected branching patterns called **food webs** (Figure 5.5; see also Schoener 1989; Pimm 1991; Winemiller 1990). As

Figure 5.4   The relationship between area of geographic range and body mass among North American terrestrial mammals. Note that although there is much variation, the areas of the smallest ranges increase with increasing body mass and are smaller for herbivores than for carnivores. Open circles indicate species that actually have larger ranges than indicated because they range into Central America (only the North American area of the distribution has been measured). (From Brown 1981.)

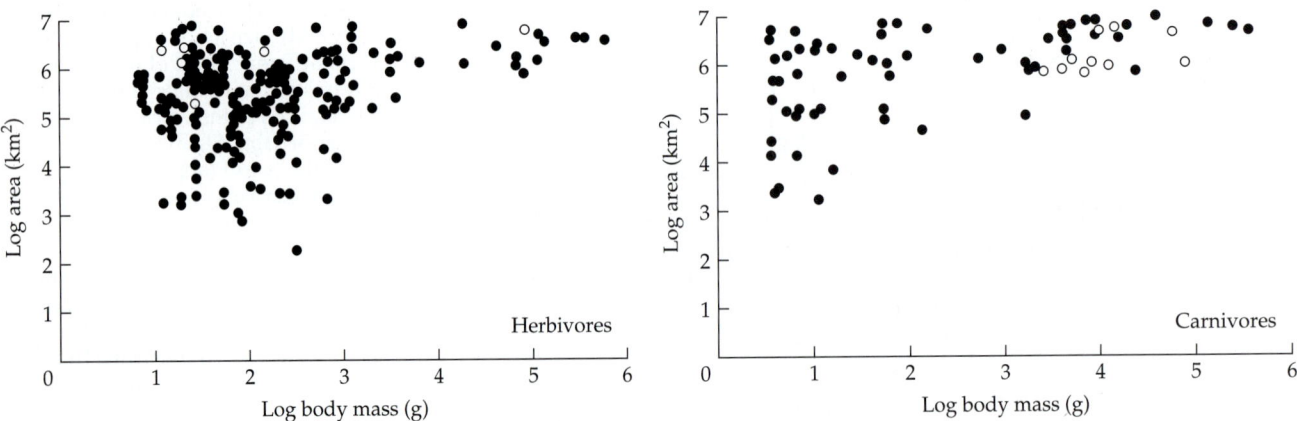

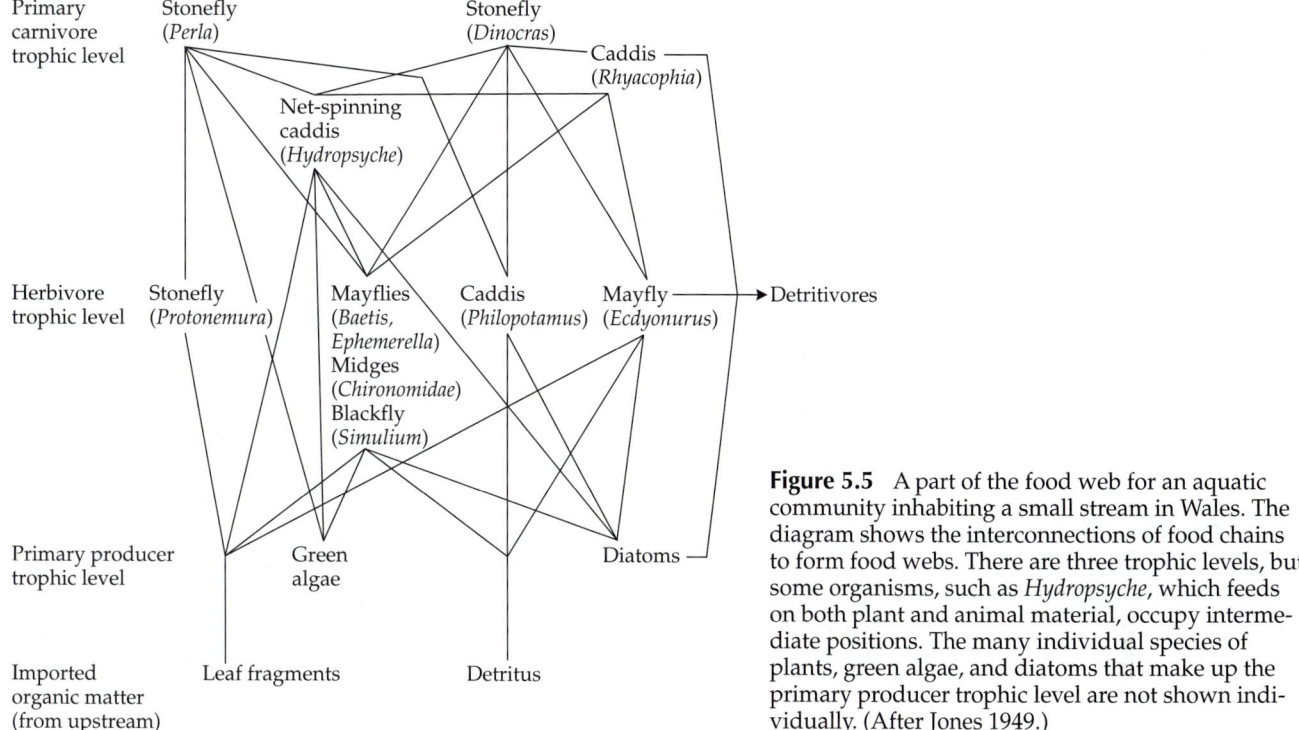

Figure 5.5   A part of the food web for an aquatic community inhabiting a small stream in Wales. The diagram shows the interconnections of food chains to form food webs. There are three trophic levels, but some organisms, such as *Hydropsyche*, which feeds on both plant and animal material, occupy intermediate positions. The many individual species of plants, green algae, and diatoms that make up the primary producer trophic level are not shown individually. (After Jones 1949.)

energy is distributed in this fashion through a community, it obeys the laws of thermodynamics, which have important implications for community organization. Simply stated, the **first law of thermodynamics** says that energy is neither created nor destroyed, but can be converted from one form to another. During photosynthesis, for example, solar energy is converted into high-energy bonds in the organic molecules of plants, which in turn can be consumed by animals and transformed into ATP to support muscular contractions. The **second law of thermodynamics** states that as energy is converted into different forms, its capacity to perform useful work diminishes, and the disorder (entropy) of the system increases. Organisms use the energy stored in organic molecules to perform the work of moving, growing, and reproducing. As organic compounds are oxidized to provide energy for these activities, much of the energy is dissipated as heat. In fact, most organisms are very inefficient. Most animals are able to incorporate only 0.1% to 10% of the energy they ingest into the organic molecules of their own bodies; the remaining 90% to over 99% is lost as heat.

One ecological consequence of these thermodynamic properties of organisms is that substantially smaller quantities of energy are available to successively higher trophic levels. This can be shown diagrammatically by portraying the community as an ecological pyramid (Figure 5.6). In any community, the green plants have the highest rates of energy uptake, and they usually comprise the largest number of individuals, the greatest number of species, and the greatest **biomass** (the total quantity of living organic material). Successively higher trophic levels tend to have less than 10% the rate of energy uptake of the level below them, and usually contain proportionately lower biomass and fewer individuals and species (Lindeman 1942).

The availability of energy must decrease substantially with each successive trophic level in accordance with the laws of thermodynamics, but there are exceptions to the other patterns. The distribution of biomass, individuals, and

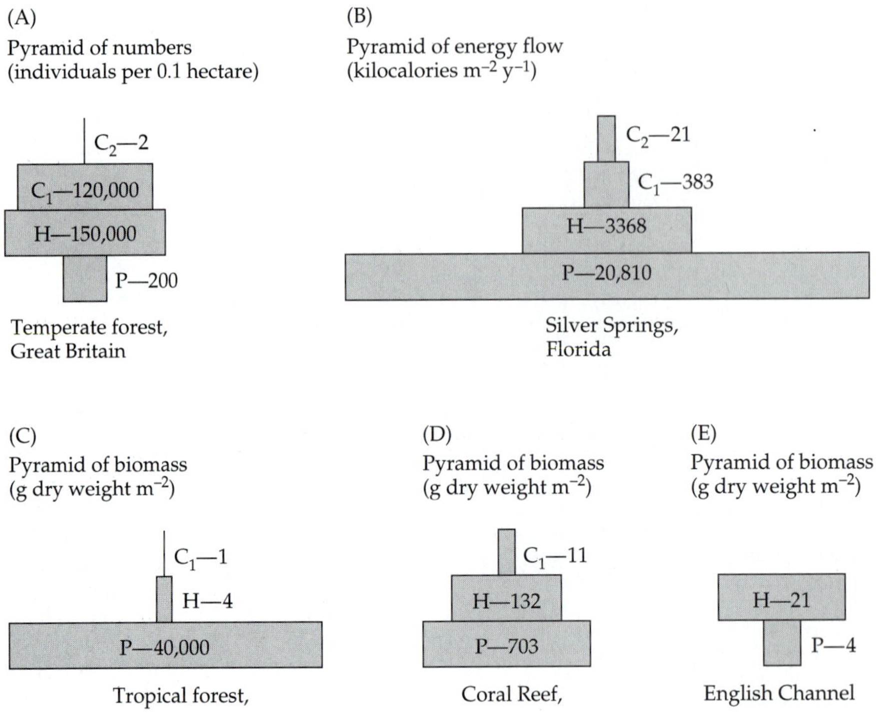

(A)
Pyramid of numbers
(individuals per 0.1 hectare)

C_2—2
C_1—120,000
H—150,000
P—200

Temperate forest,
Great Britain

(B)
Pyramid of energy flow
(kilocalories m^{-2} y^{-1})

C_2—21
C_1—383
H—3368
P—20,810

Silver Springs,
Florida

Figure 5.6   Ecological pyramids of energy flow, biomass, and number of individuals for several different communities. Note that three of the pyramids (B–D) are regular, with each successive trophic level reduced compared with the one below, but two (A, E) are not. The pyramid of individuals for the temperate forest (A) and the pyramid of biomass for the English Channel (E) have fewer and a lower biomass of primary producers than herbivores. P, primary producers; H, herbivores; C_1, primary carnivores; C_2, secondary carnivores. (After Odum 1971.)

(C)
Pyramid of biomass
(g dry weight m^{-2})

C_1—1
H—4
P—40,000

Tropical forest,

(D)
Pyramid of biomass
(g dry weight m^{-2})

C_1—11
H—132
P—703

Coral Reef,

(E)
Pyramid of biomass
(g dry weight m^{-2})

H—21
P—4

English Channel

species among trophic levels depends on how energy is acquired and used by the species in the community. In some communities, such as those inhabiting caves (Poulson and White 1969) and the dark depths of ocean trenches, there are no autotrophic organisms, and all energy is imported in organic form, usually as dead material called **detritus**. Deciduous forests of eastern North America and western Europe may have inverted pyramids of abundance and species richness—that is, fewer individuals and species of primary producers than of primary consumers. Here, most of the energy used by plants is monopolized by a relatively small number of large trees, whereas the energy used by herbivores is allocated among many individuals and species of small insects (Elton 1966; Varley 1970) (Figure 5.6A). Inverted pyramids of biomass also are possible and, in fact, quite common in some aquatic ecosystems (Figure 5.6E). In the open ocean, the photosynthetic rates of phytoplankton may be so high that these tiny producers can support a biomass of consumers far exceeding their own. This is possible because the consumers depend on *rates* of energy transfer (i.e., molecules metabolized per unit of time) and not simply biomass per se. Phytoplankton populations are so productive that they can replace themselves many times per day and thus support high rates of herbivory by zooplankton. However, consistent with the first law of thermodynamics, pyramids of energy flow can never be inverted.

Because the **carrying capacity** (measured in units of usable energy) of any area is lower for successively higher trophic levels, the organisms occupying these levels exhibit predictable characteristics that affect their ecological roles and their geographic distributions. Not only are there fewer species of carnivores than of herbivores and plants, but the carnivores also tend to be larger and more generalized than herbivores. Carnivores usually have to be large enough to overpower their prey. They also tend to feed on several prey species, and to have relatively broad habitat requirements and wide geographic distributions (see Figure 5.4; Rosenzweig 1966; Van Valen 1973a; Wilson 1975; King and Moors 1979; Zaret 1980; Brown 1981; Gittleman 1985). The cougar or mountain lion (*Puma concolor*), for example, is one of the top carnivores in the New World. It weighs 50 to 100 kg and takes a variety of prey, mostly mammals ranging in size from rabbits to elk. The cougar also has the widest geographic distribution of any American mammal. It ranges from Alaska to the southern tip of South America and, before European settlement, ranged from the Atlantic to the Pacific coasts on both continents. The cougar inhabits tropical rain forests, coniferous and deciduous hardwood forests, shrublands, and deserts. Other top carnivores, such as the wolf (*Canis lupus*) and jaguar (*Panthera onca*), also have broad geographic and habitat distributions, whereas herbivores of comparable size tend to have more restricted ranges.

Parasites are an exception to these patterns, but they still illustrate the fundamental importance of energetic relationships in community organization. Parasites are usually much smaller and more highly specialized than their hosts. Consequently, individual parasites need not consume large numbers of host individuals to meet their energy requirements, and parasites can be more numerous than their hosts. One or more parasites usually inhabit a host for an extended period, often for the entire life span of either parasite or host. Parasites are often highly specialized to infect only one or a few, often taxonomically related, host species. Although these adaptations have allowed parasites to maintain geographic ranges and numbers of individuals roughly comparable to those of their hosts in the trophic level below them, many species have had to solve the problem of finding and infecting sparsely distributed hosts. The elaborate, highly specialized life cycles of many parasites appear to be largely adaptations for locating and infecting appropriate hosts. Price (1980) and Hawkins (1994) provide interesting accounts of some special features of parasite ecology, biogeography, and evolution.

These kinds of energetic considerations lead to the prediction that the capacities of habitats or geographic areas to support many individuals and diverse species of organisms should ultimately depend on their total productivity. Everything else being equal, the higher the fixation rate of sunlight into organic material, the more usable energy should be available to be subdivided among individuals, species, and trophic levels. As we shall see, productivity varies greatly among different habitats, depending on such factors as climate, soil type, water availability, and the influence of human activity (Rosenzweig 1968; Jordan 1971; Lieth 1973; Whittaker and Likens 1973). In general, the predicted relationship between productivity and diversity is observed. Widespread, highly productive habitats such as tropical rain forests and coral reefs are renowned for their great diversity of specialized species. In contrast, small, isolated areas, such as small islands, and widespread, unproductive habitats, such as boreal forests and tundra, contain fewer and for the most part more generalized species. We will examine such patterns of species diversity in more detail in Chapters 13 and 14.

The Distribution of Communities in Space and Time

Spatial Patterns

Because species are often adapted to the same physical conditions as other ecologically similar species, we might expect biotic communities to be distributed as discrete units, with rapid turnover of species as we move along an environmental gradient from one community to the next. On the other hand, because of competition, we might expect ecologically similar species to adapt to different niches, and thus to be distributed more or less independently of each other across space. How do we resolve this dilemma?

It is obvious that abrupt changes in the environment, such as those found at the shore of a lake or the edge of a forest, will usually be accompanied by rapid transitions between two different associations of species. If the environmental discontinuity is rapid and severe, most of the species living on each side will be limited almost simultaneously when they encounter inhospitable conditions at the border. If the two habitat types are not too dissimilar, however, the edge, or **ecotone,** may actually contain more species than either pure habitat type. Species from either side of the boundary may be able to mix and occur together in the narrow area where the two environments meet. If the transition zone is fairly productive and not too narrow, it may even support its own assemblage of organisms uniquely adapted to live there. This is especially likely if the boundary is not simply an ecotone, with conditions intermediate between those on either side, but an environment in which conditions are unique or fluctuate back and forth between those found on either side. The most dramatic example of such a community is provided by the intertidal zone. Special adaptations are required to withstand periodic inundation in seawater followed by exposure to a desiccating terrestrial environment. Along a narrow strip of shore live entire communities of such plants and animals, which show their own small-scale patterns of association and segregation within the vertical gradient of tidal exposure (e.g., Connell 1961, 1975; Menge and Sutherland 1976; Lubchenco and Menge 1978; Lubchenco 1980; Souza et al. 1981; Peterson 1991; Menge et al. 1994).

How are species distributed along environmental gradients, such as the relatively abrupt gradient in exposure within the intertidal zone or the more gradual gradient in climatic conditions caused by latitudinal or elevational changes? Some of the most thorough studies of such patterns have been those of Robert Whittaker and his colleagues on the distribution of tree species on mountainsides in the United States. Their data can be used to evaluate five alternative hypotheses (see Whittaker 1975; Figure 5.7):

1. Groups of species exhibit similar ranges along the gradient and are distributed as discrete communities ("superorganisms" in the Clementsian sense) with sharp boundaries between them. This pattern could be caused by competitive exclusion between dominant species and by other species evolving to coexist with the dominants and with one another (Figure 5.7A).
2. Individual species abruptly exclude one another along sharp boundaries, but most species are not closely associated with others to form discrete communities (Figure 5.7B).
3. Much as in hypothesis 1, species form discrete communities, but replacement of communities along the gradient is gradual. This could happen if groups of species evolved to coexist with one another, but competitive exclusion did not cause rapid replacement of species along the gradient (Figure 5.7C).
4. Individual species gradually appear and disappear, seemingly independently of the presence or absence of other species. Species neither competitively exclude each other nor associate to form discrete communities, and species replacement along the gradient is random (Figure 5.7D).
5. The ranges of most species are nested within the ranges of a few dominant species that are **overdispersed** (nonoverlapping) along the gradient. Thus, species distributions may appear highly ordered at geographic scales (with little overlap among assemblages), but random at local scales (Figure 5.7E).

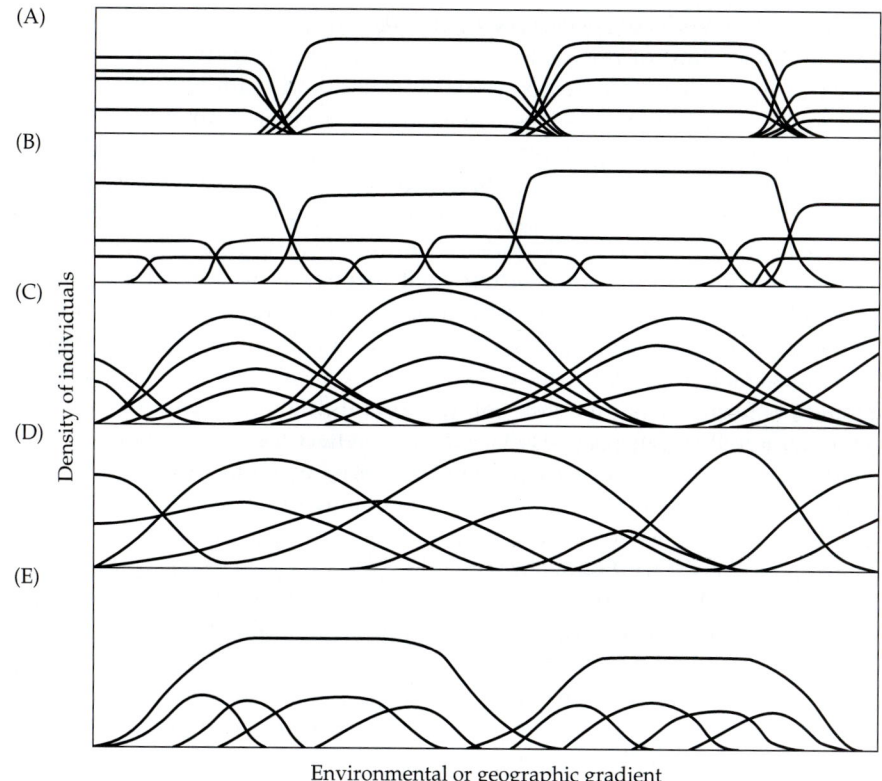

(A)

(B)

(C)

Density of individuals

(D)

(E)

Environmental or geographic gradient

Figure 5.7 Five hypothetical patterns of distributions of species along an environmental or geographic gradient. (A) Species are distributed as discrete communities that replace each other abruptly. (B) Species are not segregated into communities, but some sets replace each other abruptly. (C) Species are distributed as discrete communities, which gradually replace each other. (D) Species behave as if they are independent of each other, neither associating in discrete communities nor replacing each other abruptly. (E) Most species are nested within the ranges of a few dominant species, but otherwise occur independently of each other. (After R. H. Whittaker 1975.)

Whittaker and his associates (e.g., Whittaker 1956, 1960; Whittaker and Niering 1965) sampled several sites at varying elevations on mountains. The sample sites were chosen so as to keep soil type, slope, and exposure as constant as possible, but to allow natural variation in temperature and rainfall. The results of two such studies, shown in Figure 5.8, clearly support the hypothesis that species are distributed as if they were independent of one another, showing no evidence of either abrupt replacements that might be attributed to competitive exclusion or association of species to form discrete communities.

It should be pointed out, however, that Whittaker's methods of collecting and analyzing data may have contributed to his obtaining the patterns shown in Figure 5.8. By censusing relatively large plots and then averaging the results to obtain his species abundance curves, Whittaker would have tended to miss

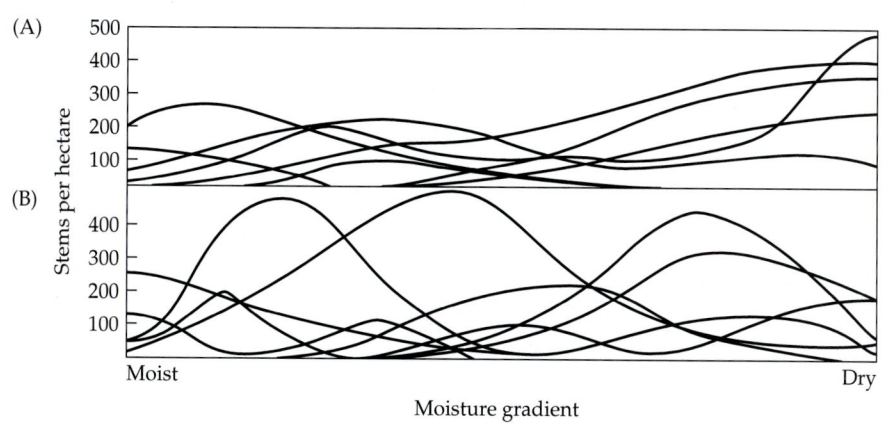

(A)

500
400
300
200
100

Stems per hectare

(B)

400
300
200
100

Moist   Dry

Moisture gradient

Figure 5.8 Distributions of tree species along two moisture gradients: (A) in the Siskiyou Mountains, Oregon, 760 to 1070 m elevation; (B) in the Santa Catalina Mountains, Arizona, 1830 to 2140 m elevation. Note that the species replace one another gradually and seemingly independently of one another. The species have narrower elevational ranges along the steeper moisture gradient in Arizona. (After Whittaker 1967.)

abrupt replacements of species owing to competitive exclusion and mediated by the distribution of microsites in a spatially heterogeneous environment. Yeaton (e.g., Yeaton 1981; Yeaton et al. 1981) has made careful analyses of the elevational distributions of pines (*Pinus*) in western North America. He found many instances of abrupt replacements by species of similar growth form, apparently as a result of interspecific competition (Figure 5.9; see also Shipley and Keddy 1987). For example, digger pine (*P. sabiniana*) reaches its upper elevational limit and is abruptly replaced by ponderosa pine (*P. ponderosa*) at approximately 840 m on southeastern-facing slopes of the Sierra Nevada near Yosemite National Park. Digger pine is at its greatest density and growth rate in the mesic environments at higher elevations just before it is supplanted by ponderosa pine, suggesting that it is competitive exclusion, not inability to tolerate the physical environment, that determines the upper limit of its range. Thus, although Whittaker's results accurately reflect the typical distribution of many plant species along environmental gradients, abrupt replacements by competing species can undoubtedly occur in many cases in which ecologically similar or closely related species come into contact (e.g., chipmunks along elevational gradients; see Figure 4.26).

Careful studies of highly coevolved species at different trophic levels, such as parasites and their hosts or plants and their pollinators, reveal many instances in which such pairs of species have, as we might expect, virtually identical ranges. Most of the species in a community, however, do not interact strongly with each other. The number of possible pairwise interactions between species rapidly increases with the number of species (S) as ($S^2 - S$)/2. If a community contained only 50 species, each species could interact with each of the 49 others, resulting in a total of 1225 possible direct pairwise interactions. Clearly each species cannot be finely adapted to all of the other species with which it coexists. On the other hand, an organism may be involved with just a few other species in strong competitive, predator/prey, or mutualistic interactions that have important influences on its abundance and distribution. In some cases a single **keystone species** may strongly influence community structure through its direct as well as indirect effects (Terborgh 1986; Paine 1974; Menge et al. 1994; Cox et al. 1994).

Interactions among three or more species also may be quite common, and may sometimes be as influential as two-way interactions. For example, many species may share a common limiting resource. While the effects of each pairwise interaction may be minor, the collective effects of such **diffuse competition** may strongly influence how many and which species can coexist (MacArthur 1972; Diamond 1975). Even where competition is intense, preda-

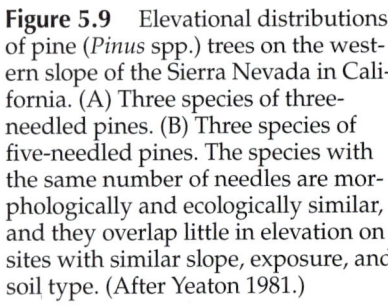

Figure 5.9   Elevational distributions of pine (*Pinus* spp.) trees on the western slope of the Sierra Nevada in California. (A) Three species of three-needled pines. (B) Three species of five-needled pines. The species with the same number of needles are morphologically and ecologically similar, and they overlap little in elevation on sites with similar slope, exposure, and soil type. (After Yeaton 1981.)

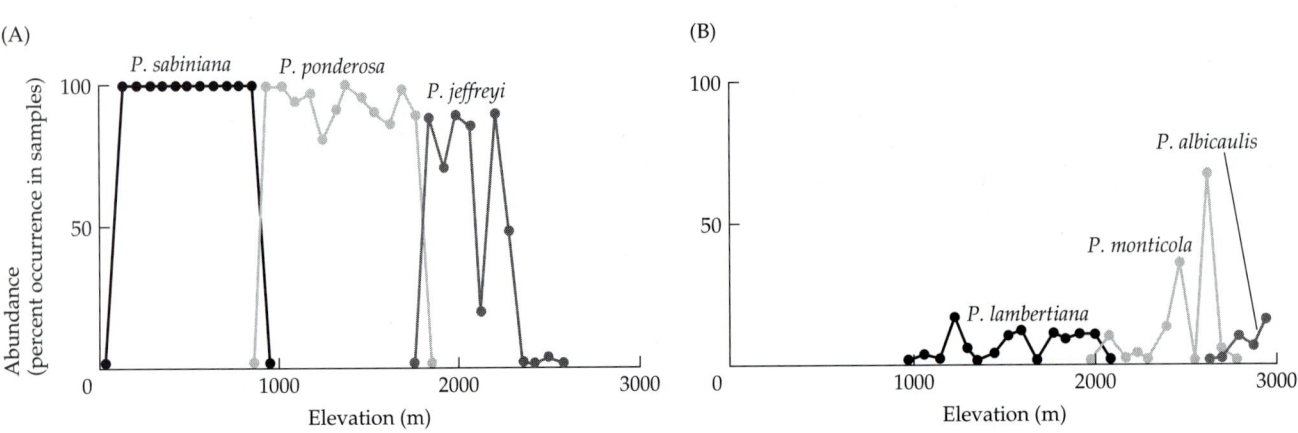

tors can allow the coexistence of otherwise incompatible competitors by preying most heavily on the most abundant prey species. Changes in the population densities of such predators and other keystone species may have pervasive effects that cascade through the community to affect species across many trophic levels (Carpenter et al. 1985, 1987; Vanni and Findlay 1990; Holt 1994). In response to the extirpation of sea otters from some regions of their historical range, populations of sea urchins (invertebrate herbivores) increased to the point at which they overgrazed aquatic plants such as kelp, reducing them to a fraction of their original biomass (see Estes and Duggins 1995). On land, the extirpation of wolves and other large predators in Eurasia and North America has also resulted in the ecological release of their herbivorous prey. Populations of deer (*Odocoileus* spp.) and other browsing mammals have increased to the extent that they have become pests. As well as entering suburbs and cities, these animals have denuded the understory of temperate forests, removing important cover and nesting material and decreasing the breeding success of native songbirds.

Temporal Patterns

Individual species and entire ecological communities replace each other over time as well as in space. These changes are of great interest both to ecologists, who study relatively short time scales, and to paleontologists, who study very long ones. Many ecologists, especially those studying sessile organisms, have devoted much effort to the study of succession (Cowles 1899, 1901; Cooper 1913; Clements 1916; Beckwith 1954; Monk 1968; Odum 1969; Drury and Nisbet 1973; Horn 1974, 1975, 1981; Connell and Slayter 1977; Usher 1979; McIntosh 1981; Miles 1987; Facelli and Pickett 1990; Farrell 1991; McCook 1994). **Succession**, as we saw in Chapter 3, is progressive change in community structure and function over ecological time. Succession is a normal process in any ecosystem in which disturbances repeatedly eliminate entire communities from patches of local habitat. If all soil and organic material is removed, as by a glacier or a volcanic eruption, the process is termed *primary succession*. On the other hand, if a disturbance such as a fire or storm removes most of the living organisms but leaves the soil intact, the process is termed *secondary succession*.

Despite decades of research, succession theory remains a hotly debated topic among ecologists. Much of the controversy echoes the Clementsian-Gleasonian debate over community organization; indeed, much of Clements's work was devoted to the study of succession. Here, the question is whether communities replace each other as interdependent units or as collections of independent species over *temporal* gradients (rather than spatial gradients, as in Figure 5.7). Clements's view, which dominated successional theory for most of the twentieth century, held that succession was deterministic, predictable, and convergent. Pioneering species colonize a site and then modify it to the point that another, better adapted set of species can invade, outcompete, and replace the first. This process is then repeated through a progressive sequence of communities, or **sere**, until a relatively stable "climax" community is established. According to the Clementsian school, successional seres are characterized by an increase in biomass, complexity, and stability (Odum 1969). The climax community and the particular sere are largely determined by climate and soil conditions, and therefore are predictable and characteristic of a given region (or biome).

In contrast to this orderly, deterministic view of succession, the Gleasonian camp observed an impressive diversity of successional seres and alternative climaxes, even for a given region. These ecologists argued that succession may not be the orderly process envisioned by Clements. Rather than being driven largely by species interactions and autogenic modification of local conditions, succession may simply reflect the idiosyncratic capacities of independent species to disperse, establish themselves, and survive at a local site with a particular combination of environmental conditions.

This debate, while nearly a century old, is not likely to be resolved soon. Until recently, succession was largely regarded as a botanical phenomenon. Whether deterministic or stochastic, it was viewed as a sequence of progressive changes in *plant* communities. Plant ecologists acknowledged that animals exhibited similar successional dynamics, but felt that they did so only because of their dependence on plants. This view, of course, is problematic. Grazers, granivores, pollinators, and decomposers, whether animals, bacteria, fungi, or protists, all influence plant community structure. Indeed, the search for the driving force, or forces, of ecological succession may require ecologists to look outside their taxonomic biases and study communities defined by functional, rather than taxonomic, criteria.

As with all debates between two such extreme points of view, it is likely that the answer to this one lies somewhere in between. Both the purely deterministic Clementsian camp and the stochastic Gleasonian camp agree that within a region of similar soil and climate, succession is a directional process that tends to result in communities characterized by a predictable set of dominant species.

Paleoecologists and historical biogeographers also study temporal changes in communities and biotas, albeit over larger geographic areas and longer time periods. Regional biotas are strongly influenced by the major shifts in climate and soil conditions that have occurred throughout history. Paleobotanists have often used analyses of fossil pollen to document the reestablishment of eastern deciduous forest communities in response to the climatic and geological changes that followed the retreat of the last glaciers in North America (Figure 5.10; see Davis 1969, 1976; see also Bernabo and Webb 1977; Peng et al. 1995). Martin and his colleagues (e.g., see Betancourt et al. 1990) analyzed pack rat middens (plant material collected by these rodents and preserved in caves in rock crevices) to reconstruct the dynamics of vegetation in arid regions of western North America over the past 40,000 years. Such studies have revealed that the retreat of the glaciers and the associated climatic changes resulted in major shifts in terrestrial vegetation far removed from the ice sheets. Communities, however, did not behave as perfectly interdependent units. Apparently, rates of invasion depended in large part on mechanisms of seed dispersal and other aspects of the life history that differed markedly among species. While some groups of species shifted in concert, others became associated with different communities, while still others were unable to adapt and went extinct (see Chapters 6 and 11).

Paleoecologists such as those cited above study the dynamics of biotas over the span of millennia. Other paleobiologists focus on the major upheavals that have occurred on an evolutionary, or geological, time scale (i.e., on the order of millions of years). The taxonomic specialization of most paleontologists and the fragmentary nature of the fossil record pose formidable challenges for those attempting to reconstruct the changes in communities that have occurred over geological time (Gray et al. 1981; Behrensmeyer et al. 1992). Much attention has been focused on **mass extinctions**, episodes of relatively abrupt (on the span of a few million years) replacement of virtually entire biotas. Although the exact causes of these catastrophic changes are still hotly debated, many, if not all, must have been triggered by drastic environmental perturbations. In many ways these biotic upheavals are comparable to the succession that occurs after ecological disturbances, or to the abrupt spatial

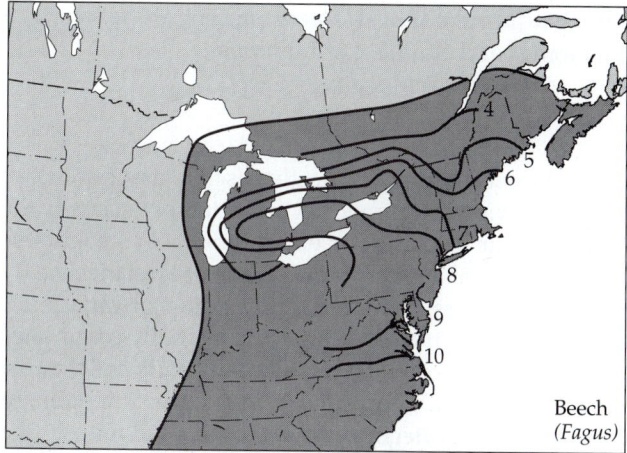

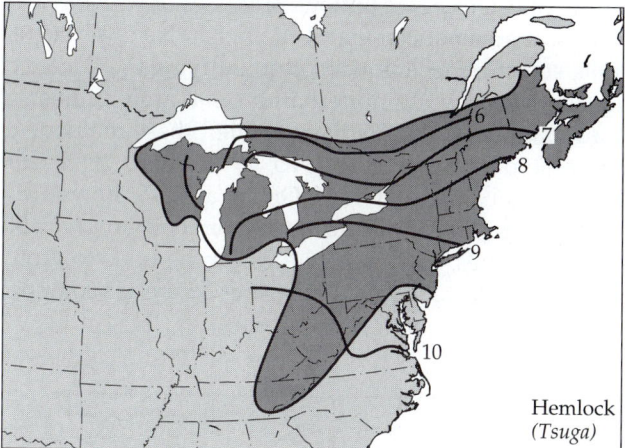

Figure 5.10   Reconstruction from fossil pollen records of the recolonization of North America by two tree species, beech (*Fagus*) and hemlock (*Tsuga*), after the last Pleistocene glaciation. The numbered lines indicate the fronts of each species' range at 1000-year intervals B.P., showing the progressive northern migration of each species. Note that the migration of these two trees was quite different, although the northern borders of their present ranges (darker areas) are virtually identical. (After Bernabo and Webb 1977.)

replacement of communities across major environmental discontinuities. These catastrophic events were spectacular, but they were the exceptions— these short pulses of extinction and evolutionary change were interspersed with extremely long spans of more gradual changes in regional biotas.

Throughout geological time, many species have colonized, speciated, and become extinct relatively independently of one another, in a pattern much like that of the spatial replacement of many contemporary species observed along gradual physical gradients. As evidence of this pattern we can point to the existence in modern communities of recently evolved species alongside "living fossils" (e.g., the ginkgo, horseshoe crab, and coelacanth), forms that have survived virtually unchanged for hundreds of millions of years. However, we must caution against sweeping generalizations. While much of the fossil record suggests independent shifts of species, other evidence may reflect interdependence among species. Mutualism, widespread today (Janzen 1985), must also have been common in the distant past. We know from research on contemporary communities that the loss of a keystone species, or the invasion and establishment of an exotic species, can cause major changes and even wholesale reorganization of communities. It is therefore likely that the evolution, range shifts, and extinctions of such keystone species contributed to the dynamics and upheavals of regional biotas in the past (e.g., see Petuch 1995). The fossil record, riddled as it is with gaps and mysteries, also may hold many insights into the forces structuring ecological communities, both past and present.

Before delving any further into the past, we now present an overview of contemporary terrestrial biomes and aquatic communities.

Terrestrial Biomes

The fact that communities do not represent perfectly discrete associations of species in either time or space obviously complicates any attempt to classify the communities of species that occur in particular environments or geographic regions. Where physical and geographic changes are abrupt, it is relatively easy to recognize distinct community types, but where environmental variation is gradual, we are faced with the problem of dividing the essentially continuous variation in species composition, life form, and other community traits into a discrete number of arbitrary categories. The human mind seems to require, or at least to depend heavily upon, such categories (e.g., the stages of mitosis, the geological periods, the biogeographic regions) even when we are well aware that they represent artificial divisions of continuous processes or variables.

In the last few decades, mathematicians and biologists have developed sophisticated multivariate statistical techniques for quantifying the degree of similarity (or difference) between two samples based on a large number of variables. Many plant and animal ecologists have applied these methods to ecological and biogeographic data (see Pielou 1975, 1979; Smith 1983; Cornelius and Reynolds 1991; Birks 1987; Bailey 1996; Omi et al. 1979; Rowe 1980; Robinove 1979; McCoy et al. 1986). If comparable measurements are available for a large number of communities, these techniques can be used to group them into hierarchical clusters reflecting their similarities in species composi-

Figure 5.11   The world distribution of the major terrestrial biomes. Note that the locations of these vegetation types correspond closely to the distribution of climatic regimes and soil types (see Figures 3.6 and 3.11). Several different vegetation types (e.g., tropical deciduous forest and savanna) have been grouped together in some cases so that the general zonal pattern of biomes can be observed.

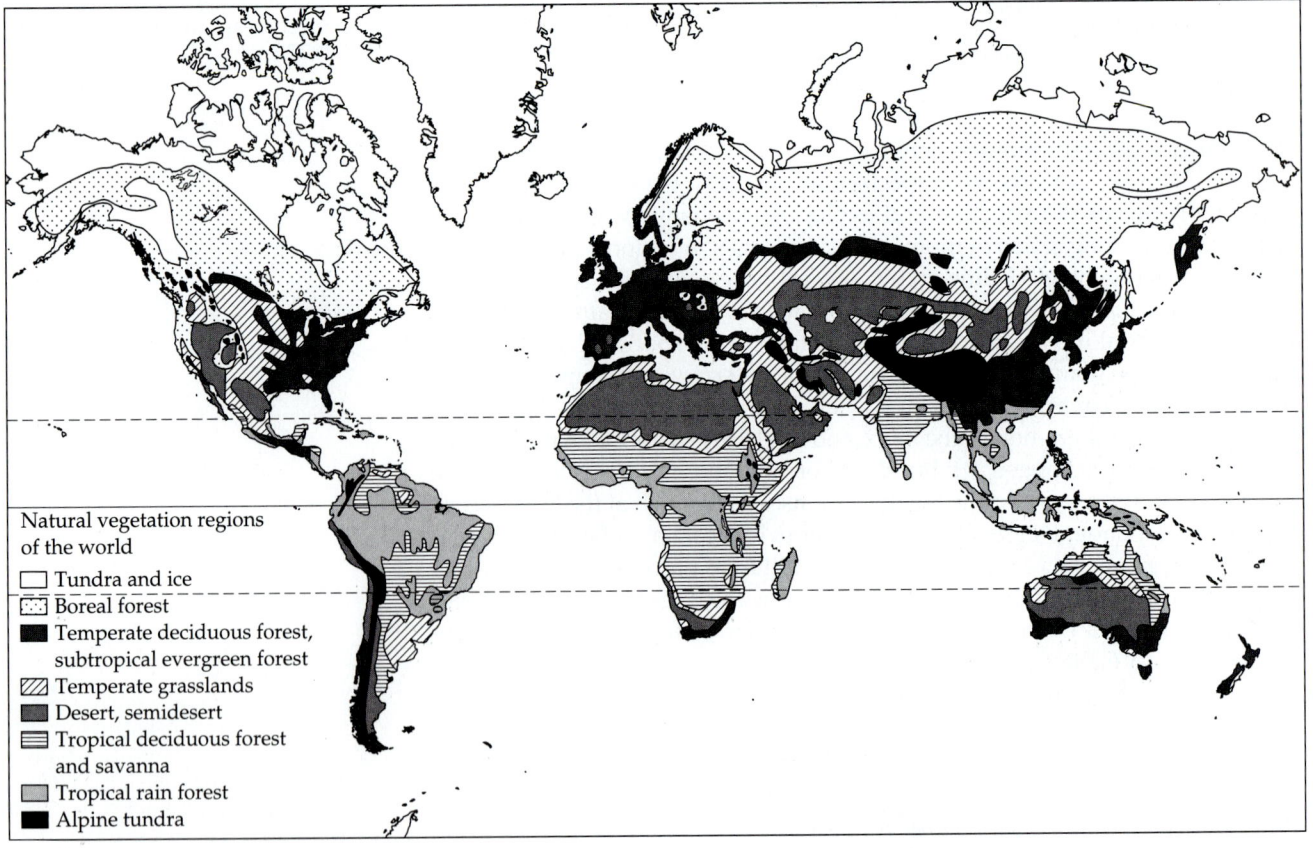

tion, life form, or other attributes of interest. As we shall see in Chapters 10 and 11, these statistical methods are extremely useful for detecting quantitative patterns of floral and faunal resemblance that may suggest the influence of historical geological events or contemporary ecological conditions. Unfortunately, this approach has not been used systematically to classify community types on a worldwide basis.

There has, however, been no lack of attempts to produce simpler, less objective groupings of community types. Beginning with the pioneering classifications of Schouw (1823) and the early phytogeographers, continuing with Merriam's (1894) life zones (see Figure 2.8) and Herbertson's (1905) natural regions, and extending to the contemporary concept of biomes, ecologists and biogeographers have almost without exception classified terrestrial communities on the basis of the structure or physiognomy of the vegetation (Figure 5.11). Implicit in all these classifications is the recognition that the life forms of individual plants and the resultant three-dimensional architecture of the vegetation reflect the predominant influence of climate and soil on the kinds of plants that occur in a region. Some authors, such as Holdridge (1947) and Dansereau (1957), have attempted to depict these relationships more quantitatively, showing the fairly tight relationships between ranges of climatic variables (such as temperature and precipitation) and specific vegetation types (Figure 5.12) (Whitaker 1975; see also Leith 1956). Similar climatic regimes do tend to support structurally and functionally similar vegetation in disjunct areas throughout the world. Often these similarities result from convergence —that is, unrelated plant species in geographically isolated regions have evolved similar forms and similar ecological roles under the influence of similar selective pressures (see the discussion on convergence of geographically isolated communities in Chapter 16).

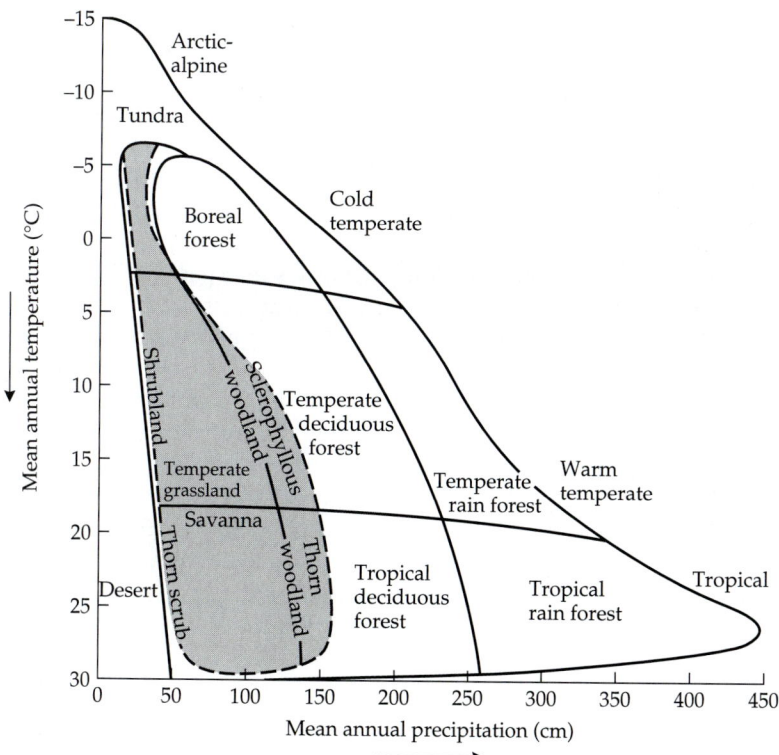

Figure 5.12   A climograph, which is a simple diagram quantifying some aspects of the relationships between climate and vegetation types. (After Whittaker 1975.)

There are almost as many different classifications of vegetation types as there are textbooks in ecology and phytogeography. In general, most biogeographers recognize six major forms of terrestrial vegetation:

1. **Forest:** a tree-dominated assemblage with a fairly continuous canopy
2. **Woodland:** a tree-dominated assemblage in which individuals are widely spaced, often with grassy areas or low undergrowth between them
3. **Shrubland:** a fairly continuous layer of shrubs, up to several meters high
4. **Grassland:** an assemblage in which grasses and forbs predominate
5. **Scrub:** a mostly shrubby assemblage in which individuals are discrete or widely spaced
6. **Desert:** an assemblage with very sparse plant cover in which most of the ground is bare

We recognize 12 common terrestrial biomes, whose geographic distributions are mapped in Figure 5.11. Note that the occurrence of these biomes corresponds approximately to the distribution of climatic zones (Figure 3.6) and soil types (Figure 3.11, Table 5.1). These latitudinal and elevational patterns reflect the fact that vegetation is highly dependent not only on the local climate and the underlying soil, but also on the influence of regional climate and topography on soil formation (see Chapter 3). We now consider the characteristics of the major principal biomes, and then conclude this chapter with a global comparison of their salient features.

Tropical Rain Forest

Tropical rain forests (Figure 5.13) are the richest and most productive of the earth's terrestrial biomes, covering just 6% of its surface, but harboring about 50% of its species. Hundreds of tree species may occur in just a few hectares of tropical rain forest, and here one also finds the world's highest diversity of arboreal insects and other invertebrates. The diversity of terrestrial and flying vertebrates is no less impressive. This enormous diversity of species sets the stage for an incredible complexity of biotic interactions.

Tropical rain forests are found at low elevations at tropical latitudes (chiefly 10° N to 10° S) where rainfall is abundant (over 180 cm annually; Figure 5.14). Although most tropical rain forests receive some precipitation throughout the year, rainfall tends to be seasonal. Temperatures are nearly uniform year-round (typically over 18° C), and vary less seasonally than diurnally. The dominant plants are large evergreen trees that form a closed canopy at 30 to 50 m. The architecture of the trees is often convergent, featuring buttressed bases and smooth, straight trunks, but the height and shape of the crowns can be highly variable. The evergreen leaves also tend to be convergent in form, robust and broad with smooth edges. There may be several levels of trees below the uppermost canopy, and palms and other distinctive plants typically occur in the understory. On the trees in the upper layers grow numerous **lianas** (woody vines) and **epiphytes** (orchids, ferns, and in the New World, bromeliads), and the leaves may be covered with **epiphylls** (thin layers of mosses, lichens, and algae). Very little light penetrates the dense, multilayered canopy to reach the forest floor, which is usually surprisingly open and devoid of vegetation. Annual plants are conspicuously absent.

Figure 5.13   Tropical rain forest. La Selva, Costa Rica. (Courtesy of E. Orians.)

Table 5.1
Relationship between climatic zone, soil type, and vegetation communities

Zonal soil type[a]	Zonal vegetation
Latisols (Oxisols)	Evergreen tropical rain forest (selva)
Latisols (Oxisols)	Tropical deciduous forest or savanna
Chestnut, brown soils, and sierozem (Mollisols, aridisols)	Shortgrass
Desert (Aridisols)	Shrubs or sparse grasses
Mediterranean brown earths	Sclerophyllous woodlands
Red and yellow podzolic (Ultisols)	Coniferous and mixed coniferous-deciduous forest
Brown forest and gray-brown podzolic (Alfisols)	Coniferous forest
Gray-brown podzolic (Alfisols)	Deciduous and mixed coniferous-deciduous forest
Podzolic (Spodosols and associated histosols)	Boreal forest
Tundra humus soils with solifluction (Entisols, inceptisols, and associated histosols)	Tundra (treeless)

Source: Bailey (1996).

[a]Names in parentheses are soil taxonomy orders (USDA Soil Conservation Service 1975).

Lowland tropical rain forests are the most diverse and productive of the major terrestrial biomes. As a result of the high temperatures and high humidity, decomposition of dead organic matter occurs so rapidly that little litter accumulates on the forest floor or in the soils. Many trees have adapted to this environment by developing extensive horizontal root mats, mostly within the upper 20 cm of soil, to capture the nutrients released when detritus decomposes. Mycorrhizal fungi, which facilitate the uptake of nutrients, are also

Figure 5.14 Global distribution of tropical and temperate rain forests. Temperate rain forests (indicated by arrows) are limited to regions along the west coasts of continents and large islands at the mid-latitudes, whereas tropical rain forests are much more broadly distributed between 10° north and 10° south latitude.

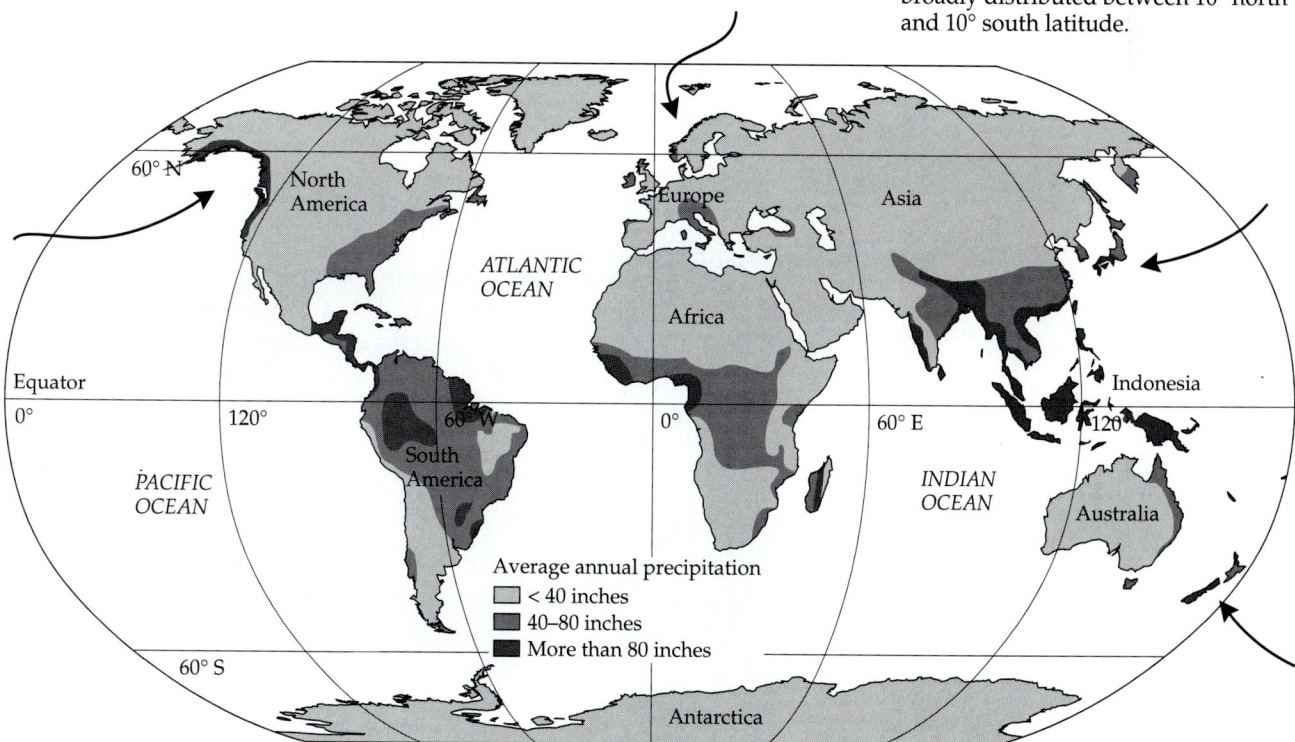

Figure 5.15   Tropical deciduous forest in the dry season when most plants are leafless. Santa Rosa National Park, Costa Rica. (Courtesy of E. Orians.)

closely associated with these root mats. These adaptations, combined with the rapid leaching rates characteristic of tropical soils, are responsible for the paradox of tropical rain forests: one of the world's most productive systems grows on some of its poorest soils.

Tropical Deciduous Forest

Tropical deciduous forests (Figure 5.15) usually occur in hot lowlands outside the equatorial zone (between 10° and 30° latitude) where rainfall is more seasonal, and the dry season is more pronounced and more extensive, than in regions of tropical rain forest. Compared with tropical rain forest, the canopy is lower and more open, and because more light reaches the ground, there is often more understory vegetation. To conserve water, many of the trees and understory plants shed their leaves during the long dry season, although much flowering and fruit maturation may occur at this time.

The dominant vegetation is often called **rain-green forest** because the forest trees leaf out during the first heavy rains following the dry season. The most luxuriant form is the **monsoon forest**, the layperson's "jungle." Monsoon forest, especially well developed in southern Asia, has many large leaves and dense undergrowth rich in bamboos. These areas are frequently drenched with torrential rainstorms, and some are among the rainiest, but most seasonal, habitats in the world.

Thorn Woodland

Tropical and subtropical **thorn woodlands** (Figure 5.16) are low arborescent vegetation types that grow in hot semiarid lowlands. The dominant plants are small spiny or thorny shrubs and trees. Members of the genus *Acacia* and other legumes (Fabaceae) are common in these biomes on all continents. Succulents, such as cacti (Cactaceae) in the New World and convergent forms of the genus *Euphorbia* (Euphorbiaceae) in Africa, are often abundant. Most

Figure 5.16   Tropical woodland near Lake Chircro, Zimbabwe. (Courtesy of E. Orians.)

plants lose their leaves during the prolonged dry season, but the trees leaf out and a dense herbaceous understory develops during the rainy season.

Thorn woodlands are often found on drier sites adjacent to tropical deciduous forests. As the climate becomes even drier along a gradient, thorn woodlands give way to **thorn scrub**. A minimum of 30 cm of annual rainfall usually is necessary to establish a thorn scrub, and there is usually a 6-month dry season with virtually no rainfall.

Tropical Savanna

Tropical savannas (Figure 5.17) are biomes dominated by a nearly continuous layer of xerophytic perennial grasses and sedges and scattered with fire-resistant trees or shrubs. Savannas usually occur at low to intermediate elevations at intertropical latitudes (primarily between 25° N and 25° S). They are characterized by marked seasonality of precipitation, with one or two rainy seasons followed by intense droughts. These weather patterns are largely driven by seasonal shifts in the **intertropical convergence zone**, the zone of most intense solar radiation and convergence of trade winds from the north and south (see Figure 3.3). As the sun shifts between tropical latitudes, the intertropical convergence zone shifts as well, passing twice over the equator, but just once over the higher latitudes of the tropics. As a result, equatorial savannas experience two rainy seasons, while those near the limits of the tropics experience a single, longer rainy season. Thus, annual rainfall varies from 30 to 160 cm. Most savannas, however, are strongly influenced by three common factors: seasonally intense precipitation, fire during the dry season, and migratory or seasonal grazing. These dynamic forces combine to make the savanna one of the most spatially and temporally heterogeneous biomes on earth.

The most extensive savannas are found in intertropical Africa, where they support the most abundant and diverse community of large grazing mammals in the world, as well as a variety of large carnivorous mammals. While easily overlooked, the diversity of smaller herbivores and carnivores living on savannas is no less impressive—typically many times that of their larger counterparts. Savannas have also played an important role in the development of human civilizations. Native peoples inhabiting the savannas of Africa and other continents followed the natural migrations of ungulates, depending on them for food and clothing. Eventually these early "pastoralists" domesticated

Figure 5.17 Tropical Savanna, Champagne Ridge, Buffalo Springs National Reserve, Kenya. (Courtesy of E. Orians.)

some species, such as cattle, horses, donkeys, and camels, and moved their herds along with the natural grazers to exploit seasonal shifts in the productivity of the savanna's native communities.

Desert

Hot deserts and semideserts (Figure 5.18) occur around the world at low to intermediate elevations, especially in the belts of dry climate from 30° to 40° N and S (the horse latitudes) between the humid tropics and the mesic temperate biomes. Rainfall is not only scanty (often less than 25 cm per year) and seasonal, but also highly unpredictable. The evaporative potential of hot desert climates is so strong that most plants have special adaptations to take up, store, and prevent the loss of water. The key feature common to deserts is that the amount of rainfall is far less than the evaporative potential and is highly unpredictable within, as well as among, years. Even regions with well over 30 cm of average annual precipitation may be dominated by desert vegetation under these circumstances. Some extremely arid regions may not experience any rainfall for several years in a row, and therefore have little or no perennial vegetation. In less arid regions, the dominant vegetation consists of widely scattered low shrubs, sometimes interspersed with succulents (cacti, euphorbs, yuccas, and agaves). Where shrubs predominate, the vegetation is usually called **desert scrub**.

Desert plants posses a variety of adaptations to withstand periods of drought and capitalize on the short, unpredictable, but sometimes heavy rains. Following the rains, ephemeral forbs and grasses may grow rapidly to carpet the normally bare ground. Many plants, especially the succulents, are able to swell and store water, and often possess extensive shallow root systems, which act like inverted umbrellas to capture rainfall before it penetrates the soil. Animals also exhibit a variety of adaptations to arid environments, including the ability to derive all the water they need from seeds and other foods, or remain seasonally or diurnally inactive until environmental conditions become more equable. Some large desert mammals are able to store substantial amounts of water in their tissues, while birds and bats can avoid the most stressful seasons by migration.

Figure 5.18   Desert vegetation near Puertecitos in northern Baja California, Mexico. Note that the vegetation, which consists almost entirely of small shrubs, is extremely sparse. A few small trees of palo verde (*Cercidium microphyllum*) grow in the sandy bottom of the dry watercourse in the center, where the plants in general are larger and denser than on the surrounding rocky hillsides. (Photograph by J. R. Hastings.)

Despite their hardy appearance, desert ecosystems are quite fragile. They show very poor resilience and, once disturbed, may take centuries to recover. On the other hand, overgrazing by livestock and diversion of water for agriculture and other uses have converted otherwise more mesic systems into deserts, albeit atypical ones.

Sclerophyllous Woodland

Sclerophyllous woodlands and chaparral (Figure 5.19) occur in mild temperate climates with moderate winter precipitation but long, usually hot, dry summers. Sclerophyllous woodlands may also occur in regions with moderate precipitation, but whose sandy soils have little water-holding capacity. This biome includes a broad variety of xeric woodlands ranging from piñon-juniper woodlands and pine barrens to sandhill pine woodlands, sandpine scrub, and pine flatwoods. The dominant plants have **sclerophyllous** (hard, tough, evergreen) leaves. Sclerophyllous woodlands can be tall, open forests that receive over 100 cm of annual rainfall, like the eucalypt woodlands of southwestern Australia, or shorter woodlands that experience less rainfall, like the oak and conifer woodlands of western North America (with evergreen *Quercus*, *Juniperus*, and *Pinus* species).

(A)

(B)

Figure 5.19   Sclerophyllous woodland. (A) Chaparral, Mendocino County, California. (B) Fynbos, Cape of Good Hope, South Africa. (A and B courtesy of E. Orians.)

Figure 5.20   Subtropical evergreen forest, Queensland, Australia. (Courtesy of E. Orians.)

Those areas that receive less than 60 cm of rainfall per year tend to have low, shrubby vegetation. Sclerophyllous scrublands, called **chaparral**, **matorral**, **maquis**, **fynbos**, or **macchia**, are characteristic of Mediterranean climates (see Chapter 3). Much of their land surface is covered with a dense and almost impenetrable mass of evergreen vegetation only a few meters high. Fires frequently sweep through these habitats, burning off the aboveground biomass and apparently playing a major role in preventing the establishment of trees. The shrubs resprout from their root crowns to reestablish the vegetation.

Subtropical Evergreen Forest

Subtropical evergreen forests (Figure 5.20), some of which have been called **oak-laurel forests** or **montane forests**, are common in subtropical mountains at intermediate elevations. These broad-leaved forests cover extensive areas of China and Japan, disjunct areas in the Southern Hemisphere, and much of the southeastern United States. These areas may receive as much as 150 cm of annual rainfall evenly distributed throughout the year, but subtropical evergreen forests cannot occur where the mean annual temperature is much below 13° C or where severe frosts occur (Wolfe 1979).

Most of the dominant species are dicotyledons with broad, sclerophyllous evergreen leaves, such as laurels (Lauraceae), oaks (*Quercus*, Fagaceae), and magnolias (Magnoliaceae) in the Northern Hemisphere and southern beeches (*Nothofagus*) in the Southern Hemisphere. The canopy is not usually well stratified, and understory plants, especially mosses, may be exceedingly common. A number of temperate broad-leaved deciduous trees occur in these forests, and as the climate becomes colder, broad-leaved evergreens are gradually replaced by deciduous trees or conifers.

Temperate Deciduous Forest

Temperate deciduous forests (Figure 5.21) grow throughout temperate latitudes almost wherever there is sufficient water to support the growth of large

Figure 5.21   Temperate deciduous forest, near Moscow, Russia. (Courtesy of E. Orians.)

trees. They are also called **summer-green deciduous forests** because they have a definite annual rhythm—the trees are dormant and leafless in the cold and snowy winter and leaf out in the spring. Temperate broad-leaved deciduous forests are extremely variable in their structure and composition across eastern North America, western Europe, and parts of eastern Asia. In otherwise arid southwestern North America, similar vegetation occurs along permanent watercourses (**riparian deciduous woodland**). The height and density of the canopy and the importance and composition of the understory vary greatly depending on local climate, soil type, and the frequency of fires. The diversity and coverage of understory plants can be quite high, especially during spring before the trees leaf out. As a result of their extensive accumulation of organic matter and the high capacity of their soils to hold water, temperate deciduous rain forests are much less prone to fire than many other biomes.

In many parts of the Northern Hemisphere, temperate deciduous forests are located next to other arborescent communities, especially temperate ever-green forests, sclerophyllous woodlands, and coniferous forests. Consequently, many phytogeographers recognize a long list of hybrid associations between these communities, such as mixed evergreen-deciduous forest. The trees of temperate forest climax communities grow slowly, and most forests have been significantly affected by logging over the last few centuries.

Temperate Rain Forest

Temperate rain forest (Figure 5.22) is an uncommon but interesting biome found along the western coasts of continents where precipitation exceeds 150 cm per year and falls during at least 10 months (see Figure 5.14). Cool temperatures predominate year-round, but these regions are always above freezing, and they experience much fog and high humidity, permitting the growth of large evergreen trees. The moderately dry season during the summer inhibits the growth and dominance of deciduous trees. Cool temperatures account for the absence of any true tropical plants, such as palms, and the relatively low number of tree species. Temperate rain forests do not have many kinds of lianas, but the epiphyte diversity is high, consisting of mosses,

Figure 5.22 Temperate rain forest of the Olympic Peninsula, Washington State. (A) Riparian corridor of Olympic National Forest. (B) The world's largest sitka spruce tree which has a circumference of 17.6 m (58 ft) and is approximately 1000 years old. (Photographs by Mark V. Lomolino.)

(A)

(B)

lichens, epiphyllous fungi, and some ferns. These cool and moist forests are largely uninfluenced by fire. Therefore, while growth rates are relatively slow, temperate rain forests are renowned for possessing some of the world's oldest and largest trees. Their canopies tend to be closed, with many dead standing trees, or "snags," while the humid understory is often covered with lush mats of mosses, ferns, and lichens. Decomposition rates tend to be slow, again due to the relatively low temperatures. Combined with the slow growth rates of the dominant tree species, this means that forest development requires many centuries of relatively stable climatic conditions.

The best example of a temperate rain forest in North America is the spectacular forest on the Olympic Peninsula in Washington, which receives nearly 200 cm of annual precipitation (rain plus fog). This is a coniferous forest, composed of huge spruces and firs with narrow leaves. In contrast, the temperate rain forests of the Southern Hemisphere are dominated by large trees with broad evergreen leaves, such as *Agathis*, *Eucaplyptus*, *Nothofaugus*, and *Podocarpus*; the lower strata contain large tree ferns (mostly *Dicksonia* and *Blechnum*).

Temperate Grassland

Temperate grasslands (Figure 5.23) are situated both geographically and climatically between the deserts and the temperate forests. While they are broadly distributed between 30° and 60° latitude, they are most extensive in the interior plains of the Northern Hemisphere. The climate of these regions is markedly seasonal, with substantial annual variation in both temperature and rainfall. The vegetation is confined to a single stratum, which is dominated by grasses, sedges, and other herbaceous plants. Vegetation height tends to vary directly with precipitation, both factors decreasing from the tall grasslands, or **prairie** (**veldt** of South Africa, **puszta** of Hungary, **tallgrass prairie** of North America, or **pampas** of Argentina and Uruguay), to shortgrass plains, or **steppe,** in colder latitudes and desert grasslands adjacent to warm arid regions. Even in relatively moist tallgrass prairies, drought, fire, and heavy grazing pressures combine to prevent the establishment of woody plants and favor the dominance of herbaceous plants.

The dominant grasses are perennials with **basal meristems** (growth tissue located in the soil), which make them tolerant to defoliation; indeed, in these

Figure 5.23 Temperate grassland, the Tallgrass Prairie Preserve in Northern Oklahoma. (Courtesy of Bruce Hoagland.)

Figure 5.24   Boreal coniferous forest, near Eagle Trail, Alaska. (Courtesy of E. Orians.)

species, vegetative growth is stimulated by fire and grazing. Although grasslands are typically dominated by just a few grass species, they actually harbor a surprising diversity of both plants and animals. Grasses, which may account for over 90% of the biomass, typically constitute less than 25% of the plant species in grassland ecosystems. Despite the variable and sometimes luxuriant layer of vegetation above the surface, most of the grassland biomass lies belowground in the extensive root systems of the perennial plants. The ratio of belowground to aboveground biomass varies with annual precipitation, ranging from less than 2:1 for arid grasslands to 13:1 for tallgrass prairies (Wiegert and Owen 1971). Accordingly, grassland soils tend to have high accumulations of organic material, which in turn supports a rich diversity of soil invertebrates and microbial decomposers. Many vertebrate species, especially some rodents, are completely **fossorial** (burrow-dwelling), while others are cursorial grazers, consuming as much as two-thirds of the aboveground production.

Because of their deep, fertile soils, many temperate grasslands have been converted to agricultural uses. As a result of cultivation or desertification, natural grasslands, which once covered approximately 40% of the earth's surface, have now been reduced to about half of their presettlement range.

Boreal Forest

Boreal forests (also called **taiga** or "**swamp forest**") occur in a broad band across northern North America, Europe, and Asia in regions with cold climate and adequate moisture (Figure 5.24). At high elevations, this biome also extends well southward into the temperate latitudes. For example, boreal forests extend down the cordilleras of western North America all the way to southern Mexico; much of highland Mexico is covered by boreal forest.

Boreal forests, although often thick, are typically dominated by just a few species of coniferous trees, such as spruce (*Picea*), fir (*Abies*), and larch (*Larix*). Because of the cool temperatures and waterlogged soils, decomposition rates are relatively slow, resulting in the accumulation of peat and humic acids, which render many soil nutrients unavailable for plant growth. The acidic soils combine with the relatively cool temperatures to limit the diversity and productivity of the few tree species able to survive these stressful conditions. Often the canopy is not dense, and a well-developed understory of acid-tolerant shrubs, mosses, and lichens may be present in the most mesic sites.

Tundra

Tundra (Figure 5.25) is a treeless biome found between the boreal forest and the polar ice cap and at high elevations on tall mountains (**alpine tundra**). Even more than the boreal forest, the tundra is characterized by stressful environmental conditions. Temperatures remain below freezing for at least 7 months of the year, precipitation is often less than in many hot deserts, and tundra soils tend to be even more nutrient-limited than those of the boreal forest. Soils are also saturated with water because of slow evaporation rates and the presence of **permafrost** (a frozen, impermeable layer of soil that lies at a depth of a meter or less in the summer). Consequently, the primary productivity, biomass, and diversity of the tundra are lower than that of almost all other terrestrial biomes.

Arctic, antarctic, and alpine tundra are all covered with a single dense stratum of vegetation, usually only a few centimeters or decimeters in height. The

Figure 5.25   Tundra, Denali National Park, Alaska. (Courtesy of E. Orians.)

dominant plants tend to be dwarf perennial shrubs, sedges, grasses, mosses, and lichens. Despite their generally low productivity during the rest of the year, tundra plants exhibit bursts of productivity during the short growing season. The lush vegetation is then heavily grazed by migratory or nomadic ungulates, including caribou (*Rangifer tarandus*), muskoxen (*Ovibos moshatus*), and Dall sheep (*Ovis dalli*). Other important herbivores of the tundra include geese (*Branta* spp.), ptarmigan (*Lagopus*), and small mammals (voles and lemmings), whose populations fluctuate dramatically in a complex interaction with the plant community.

Found above the timberline on mountaintops in the equatorial zone is **tropical alpine scrubland**, with vegetation taller than that of the arctic tundra. The dominant plants are tussock grasses and bizarre, erect rosette perennials with thick stems. These vegetation types are found at elevations above 3300 m in the Andes (**paramo**) in South America, on the upper slopes of the highest mountains in East Africa, and on mountaintops in New Guinea.

Just like its hot desert counterpart, the tundra is a fragile system. Oil exploration and other human activities are major threats to this delicate environment. Once these activities disturb the permafrost, natural communities may take many decades to recover.

Aquatic Communities

Marine and freshwater ecologists and biogeographers do not classify aquatic communities into categories analogous to those used for terrestrial biomes. For one thing, the relatively simple arrangement of sessile plants growing on a land surface is not comparable to the three-dimensional diversity of the water column. Terrestrial habitats are essentially two-dimensional, in the sense that organisms do not remain permanently suspended above the soil surface. To the extent that a third dimension is present, it is formed by the vertical growth of sessile plants and by arboreal and flying animals. The three-dimensional organization of aquatic communities is very different. On the one hand, a well-developed structure of attached, vertically growing organisms is absent from most aquatic habitats, although there are obvious exceptions, such as kelp forests, coral reefs, and the submersed vegetation of lakeshores. On the other hand, many aquatic organisms spend much or all of their lives suspended in the third dimension, either drifting passively or swimming actively in the water column.

The physical factors that vary in time and space to affect the abundance and distribution of aquatic organisms are also quite different from those that determine both the terrestrial climate and the organization of terrestrial communities. Because of the high specific heat of water, temperature varies less on a daily, seasonal, and latitudinal basis in aquatic environments than in terrestrial ones. On the other hand, variations in pressure, salinity, and light are important in aquatic systems. Tidal cycles, which fluctuate bimonthly with the phases of the moon, are more important to many marine shore communities than are daily or seasonal cycles (see Figure 3.16).

Oceanographers, limnologists, and aquatic ecologists have developed classification systems for marine and freshwater communities. Like the division of terrestrial communities into biomes, these systems use arbitrary groupings that break up a continuous spectrum of biological associations into a number of convenient categories. Salinity, depth, water movement, and nature of the substrate are the physical characteristics that most influence the abundance and distribution of aquatic organisms and are most often used in classifying aquatic communities.

The first major division of aquatic systems is into marine and freshwater communities. On biological as well as geographic grounds, the earth's bodies of water can be divided into the oceans, which form a huge interconnected water mass covering over 70% of the earth's surface, and the comparatively tiny, highly fragmented lakes, ponds, rivers, and streams, which together cover only a small fraction of the remaining surface. The oceans and these bodies of fresh water differ greatly in salinity. The salt concentration of the oceans varies slightly around 35 parts per thousand, whereas even the hardest fresh waters have salinities of less than 0.5 parts per thousand. As stressed in Chapter 4, this difference in salinity can have dramatic effects on distributions. Only a tiny fraction of aquatic organisms can live in both salt and fresh water, so salinity effectively divides aquatic communities into two nonoverlapping groups. Because of this strong dichotomy, marine and freshwater ecosystems have largely been studied independently by different groups of ecologists (**oceanographers** and **limnologists**, respectively), who have developed different classifications of communities. These classification schemes are best considered separately.

Marine Communities

Compared with terrestrial and freshwater environments, the ocean is large and essentially continuous. Organisms live everywhere in the ocean, but the abundances and kinds of life vary greatly depending on the local physical environment. Perhaps the most important features are light, temperature, pressure, and substrate. The current system of biogeographic regions of the marine realm was not generally accepted until the early 1970s (Figure 5.26A), roughly a century after Sclater (1858) and Wallace (1876) proposed the system of terrestrial biogeographic regions that we continue to use today. This system is primarily based on water temperature; therefore, the biogeographic regions of the marine realm encompass broad latitudinal zones and tend to be elliptical due to circular oceanic currents (compare Figures 5.26A and 5.26B).

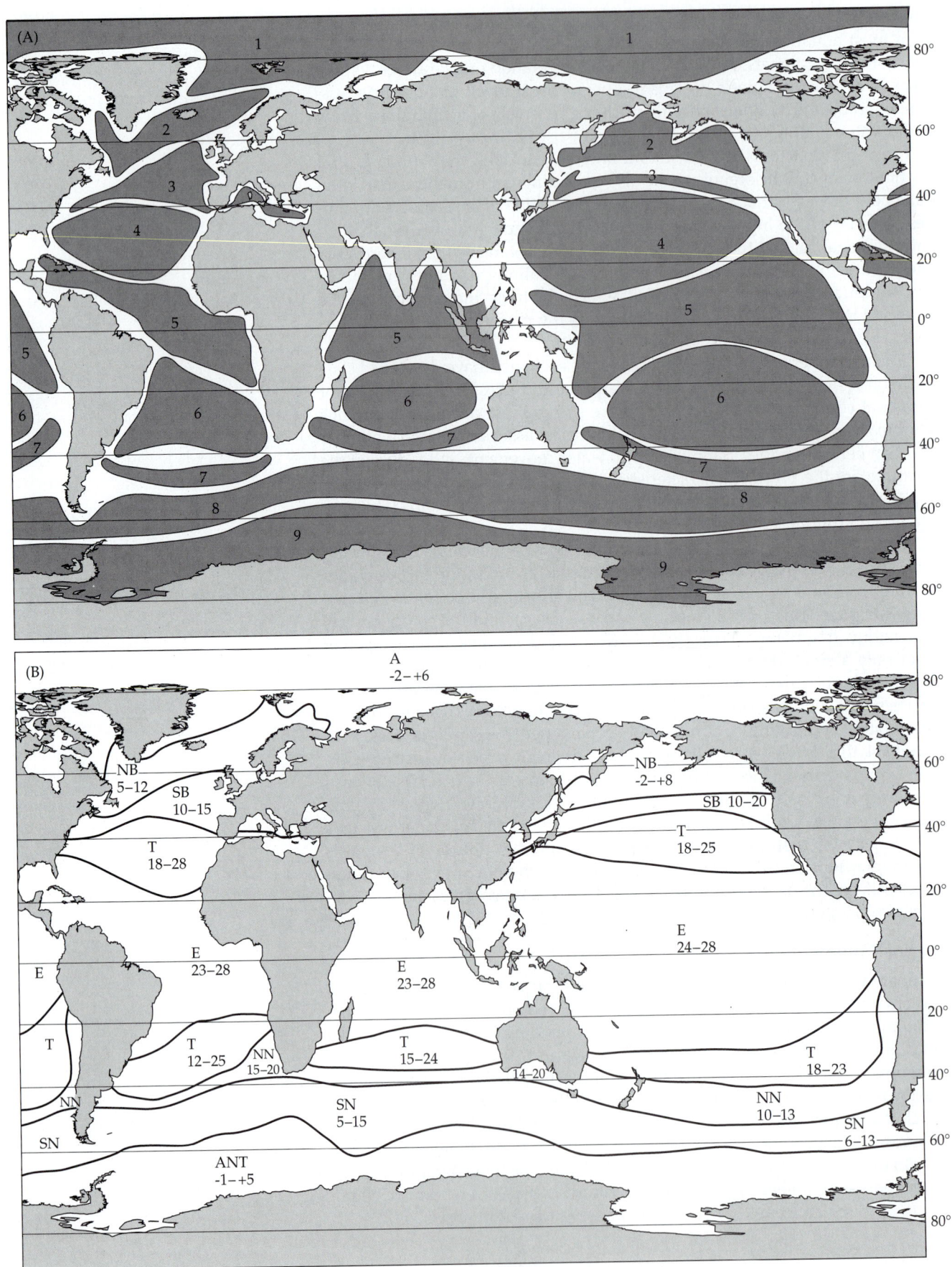

◀ **Figure 5.26** Biogeographic and climatic regions of the world's oceans. (A) Biogeographic regions: 1, arctic; 2, subarctic; 3, northern temperate; 4, northern subtropical; 5, tropical; 6, southern subtropical; 7, southern temperate; 8, subantarctic; 9, antarctic. (B) Climatic regions based on mean monthly water temperatures: A, arctic; NB, northern boreal; SB, southern boreal; T, tropical waters; E, equatorial region; NN, northern notal; SN, southern notal; ANT, antarctic. (After Rass 1986.)

Within each of these regions, the ocean can be divided into two vertical zones—the **photic zone** and the **aphotic zone**—based on the penetration of sunlight (Figure 5.27; see Chapter 3). Because sunlight is gradually absorbed by water with increasing depth, the boundary between these zones is somewhat arbitrary, but is usually set where light penetration is reduced to between 1% and 10% of incident sunlight. The depth of the photic zone increases from coastal waters, where light rarely penetrates more than 30 m because of organisms and inanimate particles suspended in the water column, to the open ocean, where it may extend to a depth of 100 m or more.

The significance of this zonation by light is, of course, that photosynthesis can occur only in the photic zone. Essentially all of the organic energy that sustains marine life is produced in this shallow surface layer of the ocean. Most organisms in the aphotic zone obtain their energy by consuming organic material that is produced in the photic zone and reaches deep water as feces and dead bodies that sink to the bottom. In the late 1970s, however, oceanographers discovered entire flourishing communities of organisms, including

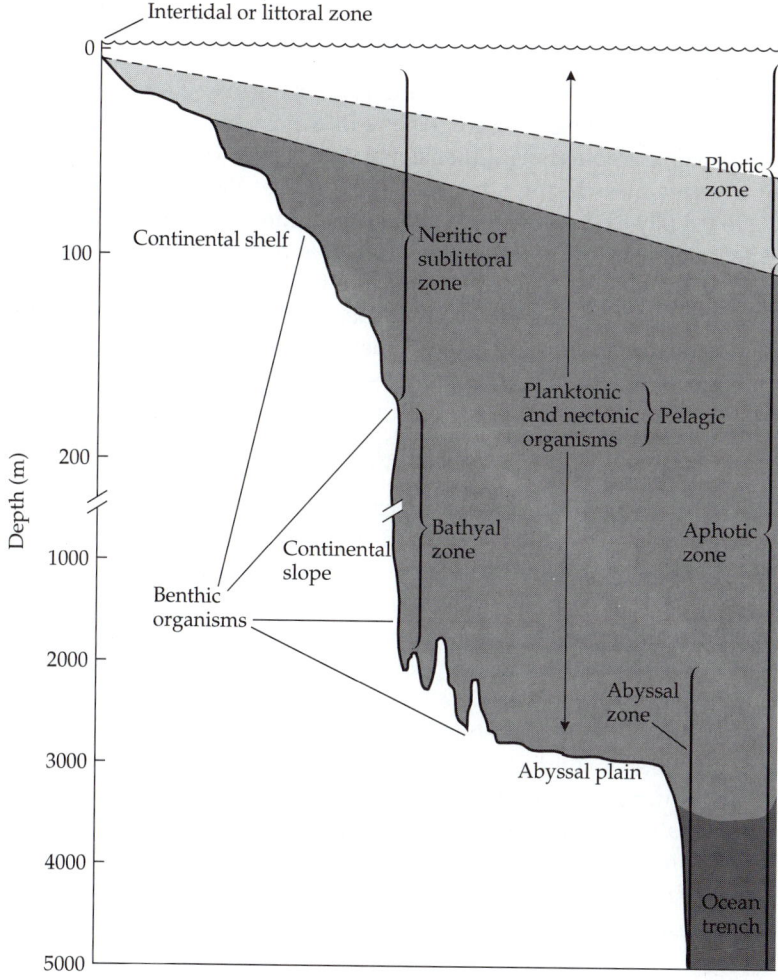

Figure 5.27 Division of marine communities into major zones. Note that these zones are based primarily on water depth, light levels, and relationships between organisms and substrates.

unique kinds of worms, mussels, and crabs, that do not depend on material from the photic zone. These communities live in highly localized areas such as those on the otherwise barren slopes of the Galápogos rift (eastern tropical Pacific) where submarine hot springs emit hydrogen sulfide. Chemosynthetic bacteria obtain energy by oxidizing the hydrogen sulfide, and these unusual autotrophs serve as the base of the food chain for the community (Jannasch and Wirsen 1980; Karl et al. 1980).

Marine communities are also classified into another set of zones on the basis of **bathymetry** (i.e., the depth and configuration of the ocean bottom) (Figure 5.27). The shallowest zone is the intertidal or **littoral zone**, which occurs on the shore where sea meets land. Although it is inhabited almost exclusively by marine organisms, the intertidal zone is actually an ecotone between land and ocean. Beyond the intertidal zone is the **neritic** or **sublittoral zone**, encompassing waters of a few meters to about 200 m deep that cover the continental shelves. At the edges of the continental plates is a region of highly varied relief, called the **bathyal zone**, with the marine equivalent of mountainsides and canyons, in which the waters rapidly drop away to the great ocean depths. The **abyssal zone**, which constitutes most of the ocean, covers extensive areas in which the water ranges in depth from 2000 to more than 6000 m. Deep ocean waters provide some of the most constant physical environments; they are continually dark, cold (4°C), subject to enormous pressures, and virtually unchanging in chemical composition.

The organisms that inhabit the oceans are often classified as either **benthic** or **pelagic**, depending on whether they are closely associated with the substrate or distributed higher in the water column (see Figure 5.27). Benthic communities vary greatly in composition depending on the nature of the substrate. On hard substrates, attached benthic organisms often form a three-dimensional structure that varies in complexity from low crusts and turfs of algae and sessile invertebrates to tall "forests" of kelp and coral. On soft sandy or muddy substrates, there is often a comparable three-dimensional complexity, but it is formed by burrowing invertebrates that live beneath the surface. Pelagic (open water) organisms are usually divided into two groups, **plankton** and **nekton**. The former consists of primarily microscopic organisms that float in the water column. The plankton typically includes many simple plants, or **phytoplankton**, such as diatoms, and tiny animals, or **zooplankton**, such as small crustaceans and the larvae of many invertebrates and fishes. The nekton comprises the actively swimming animals, including fishes, whales, and some large invertebrates, which usually occupy higher trophic levels than planktonic organisms.

Freshwater Communities

Freshwater communities are widely distributed as small, isolated lakes, ponds, and marshes, sometimes connected by long, branching streams and rivers. These environments are usually divided into two categories: **lotic** or running-water habitats, such as springs, streams, and rivers, and **lentic** or standing-water habitats, such as lakes and ponds. Lotic habitats are often divided into **rapids** (or "**riffles**") and **pools**. In the former, the velocity of the water is sufficient to keep the water well oxygenated and the substrate clear of silt. Stream rapids are usually inhabited by organisms that live on the surface of the rocky substrate or swim strongly in the current. Pools are characterized by deep, slowly moving water and silty, often poorly oxygenated bottoms. Swimming animals are common in stream pools, and many of the benthic species burrow in the substrate. Although some of the organic material in streams is manufactured in place by benthic plants or phytoplankton, most of it is washed in from the surrounding watershed.

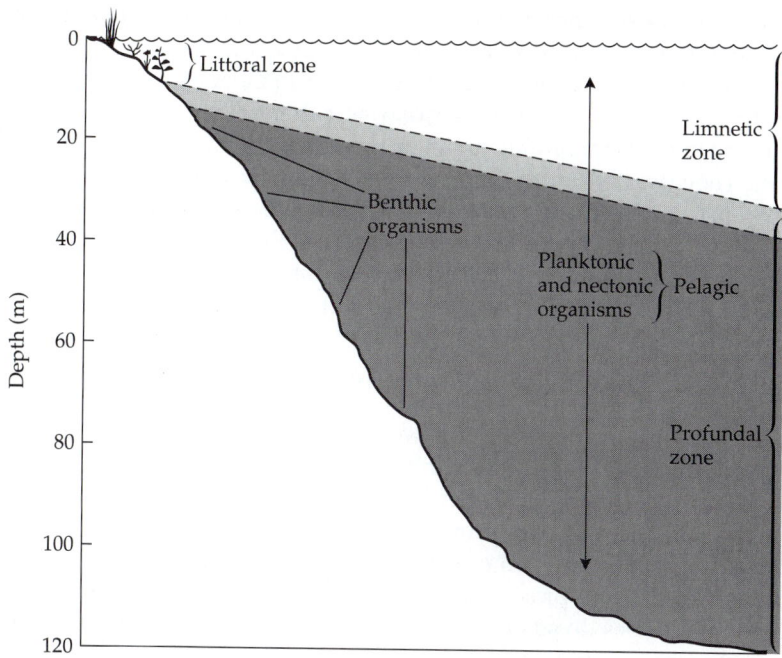

Figure 5.28 Division of freshwater lentic habitats into major zones. This classification scheme is similar to that used for marine environments because it is based on variation in light penetration, water depth, and relationships between organisms and substrates. Nevertheless, somewhat different terms are used.

Lentic habitats are often divided into zones reminiscent of those of the oceans, although somewhat different terms and meanings are applied (Figure 5.28). The littoral zone consists of shallow waters where light penetrates to the bottom and rooted aquatic vegetation may be present. The offshore waters are divided into a surface **limnetic zone**, where light penetrates sufficiently for photosynthesis to occur, and a deep **profundal zone** beyond the depth of effective light penetration. Lakes can be highly productive, supporting extensive food webs based on photosynthesis by attached vegetation in the littoral zone and phytoplankton in the limnetic zone. Productivity is limited largely by the availability of inorganic nutrients, such as phosphorus, which wash in from surrounding watersheds and, in temperate and subarctic lakes, are seasonally replenished from the organic material on the lake bottom when thermal stratification disappears and the waters overturn (see Chapter 3). Temperate lakes are often classified as either **eutrophic** or **oligotrophic**. Eutrophic lakes are shallow lakes that are highly productive because light penetrates almost to the bottom and vertical circulation of the water column occurs each spring and fall, returning limiting nutrients to the surface and oxygen to the depths. Oligotrophic lakes are characterized by low nutrient input and tend to be deeper than eutrophic lakes. In fact, many oligotrophic lakes are so deep that little or no vertical circulation occurs. As a result, productivity is relatively low, despite the high water clarity of many oligotrophic lakes.

The preceding classification scheme omits a number of freshwater communities. **Swamps (marls)** and **marshes (moors)** are two very common types of wetlands that tend to develop on mineral soils and are distinguished primarily by whether their dominant vegetation is woody or herbaceous, respectively. Some other freshwater communities are much rarer, but are interesting because they represent atypical physical environments that pose special problems for the few organisms that are able to live there. Examples of such harsh environments are hypersaline lakes (undrained lakes, such as Great Salt Lake in North America, the Dead Sea in the Middle East, and the Aral Sea in Central Asia, which are much more saline than seawater), caves (which admit no light and contain communities supported entirely by imported organic matter), and hot springs (among the most physically rigorous of all environments).

Another important group of communities occurs in those areas where fresh water, the sea, and the land meet. The **estuaries**, **salt marshes**, and (in tropical regions) **mangrove swamps** that occur at these sites are highly productive ecosystems. They usually contain only a few species that can tolerate the physical rigor of successive exposure to fresh water, seawater, and the terrestrial climate. On the other hand, the circulation provided by the tides and the input of nutrients from the rivers, and ultimately the land, often permits these habitats to support high biomasses and densities of individuals.

A Global Comparison of Biomes and Communities

In concluding this discussion of biomes, we must sound a note of caution. The biomes described above, and their distributions as shown on the map in Figure 5.11, indicate only the general kind of climax vegetation that we would expect to find in a region, based primarily on its climate. However, someone visiting many of the areas on the map might have difficulty finding good stands of the vegetation typical of these biomes, and might find some unexpected vegetation types. In some cases, this might be the result of secondary succession occurring in response to natural disturbances. More often, it is caused by human destruction of the original vegetation and modification of the landscape. For example, tallgrass prairie, the most productive temperate grassland, once covered much of Illinois and Iowa and stretched from Saskatchewan to Texas. Now most of this area has been converted to agriculture, and only a few stands of native prairie, totaling less than 5% of its original area, remain, mostly in a small number of preserves (see Steinauer and Collins 1996). Perhaps the most significant change now occurring is the rapid destruction of virgin lowland tropical rain forests, whose boundaries are evershrinking, giving way to successional communities of reduced biotic diversity and diminished economic value.

In some places, local variations in topography or soil type cause types of vegetation to be found in regions where one would not predict them based on the general map. For example, throughout temperate grasslands, sclerophyllous scrublands, tropical thorn scrub, and deserts, there are galleries of riparian forest vegetation along permanent streams. Such diversity of vegetation types contributes greatly to the overall biotic richness of a region, because the distributions of many other plants and numerous animal species are strongly influenced by the dominant vegetation. Bird species distributions, and the diversity of bird communities, for example, are highly dependent on vegetation structure (MacArthur and MacArthur 1961), and the riparian deciduous woodlands in the deserts of the southwestern United States support exceptionally high bird species diversity (Carothers and Johnson 1974).

Given the great diversity of world biomes and communities, it may be instructive to conclude this overview with a global comparison of their salient features (Table 5.2). **Net primary productivity** (NPP) is a measure of the rate at which solar energy is converted to plant tissue, typically expressed as mass produced per unit of surface area (e.g., $g\ m^{-2}\ yr^{-1}$). NPP is one of the most fundamental and important measures of community function, as it represents the energy available to maintain the biomass and diversity of almost all forms of life. As we have seen, biomes and communities vary markedly in temperature, precipitation, nutrient availability, and many other factors that influence primary productivity. Consequently, the world's biomes vary markedly in NPP, biomass, and diversity. Tropical rain forests are renowned for their high pro-

Table 5.2

Net primary production and biomass of major kinds of biomes and marine communities

Biome/Community	Area (10⁶ km²)	Net primary production per unit area (g m⁻² yr⁻¹)	Total net primary production (10⁹ MT yr⁻¹)	Mean biomass per unit area (kg m⁻²)
Terrestrial and Freshwater				
Tropical rain forest	17.0	2000.0	34.00	44.00
Tropical deciduous forest	7.5	1500.0	11.30	36.00
Temperate rain forest	5.0	1300.0	6.40	36.00
Temperate deciduous forest	7.0	1200.0	8.40	30.00
Boreal forest	12.0	800.0	9.50	20.00
Savanna	15.0	700.0	10.40	4.00
Cultivated land	14.0	644.0	9.10	1.10
Woodland and shrubland	8.0	600.0	4.90	6.80
Temperate grassland	9.0	500.0	4.40	1.60
Tundra and alpine meadow	8.0	144.0	1.10	0.67
Desert scrub	18.0	71.0	1.30	0.67
Rock, ice, and sand	24.0	3.3	0.09	0.02
Swamp and marsh	2.0	2500.0	4.90	15.00
Lake and stream	2.5	500.0	1.30	0.02
TOTAL TERRESTRIAL AND FRESHWATER	149.0	720.0	107.09	12.30
Marine				
Coral reefs and Algal beds	0.6	2000.0	1.10	2.00
Estuaries	1.4	1800.0	2.40	1.00
Upwelling zones	0.4	500.0	0.22	0.02
Continental shelf	26.6	360.0	9.60	0.01
Open ocean	332.0	127.0	42.00	0.003
TOTAL MARINE	361.0	153.0	55.32	0.01
WORLD TOTAL	510.0	320.0	162.41	3.62

Source: After Whittaker and Likens (1973).

ductivity, and indeed, they are clearly the most productive of the terrestrial biomes. However, as we see in Figure 5.29A, some aquatic systems rival tropical forests in NPP. In aquatic systems, productivity tends to be highest in shallow-water environments, where high photosynthetic rates are favored by the relatively high levels of sunlight and nutrients.

These trends in NPP among biomes are paralleled by trends in biomass (Figure 5.29B). Again, the most productive biomes and communities tend to support the highest density of living tissue. Aquatic communities, however, exhibit consistently lower biomass than their terrestrial counterparts. Phytoplankton, the smallest plants, account for approximately 90% of the NPP of aquatic communities. Unlike terrestrial macrophytes, more than a third of whose biomass is photosynthetically inactive tissue, phytoplankton are extremely efficient. Under optimal conditions, solar energy is rapidly converted into phytoplankton tissue and in turn made available to aquatic consumers and decomposers. As mentioned earlier, this high turnover rate, which supports extensive food webs, explains why pyramids of biomass are inverted for some lake and ocean ecosystems (see Figure 5.6E).

We caution that the above estimates of productivity and biomass are averages over space and time. At finer scales, each of these biomes and communities is remarkably heterogeneous. Each is composed of a collection of successional and disturbance stages, all influenced to varying degrees by seasonal changes. Temperate grasslands, meadows, and deserts, for example, can exhibit impressive bursts of productivity during the growing season.

Figure 5.29 Comparisons of (A) net primary productivity and (B) biomass among terrestrial and aquatic communities (see Table 5.2). Despite their relatively low biomass, aquatic communities often rival tropical rain forests in productivity on a per area basis.

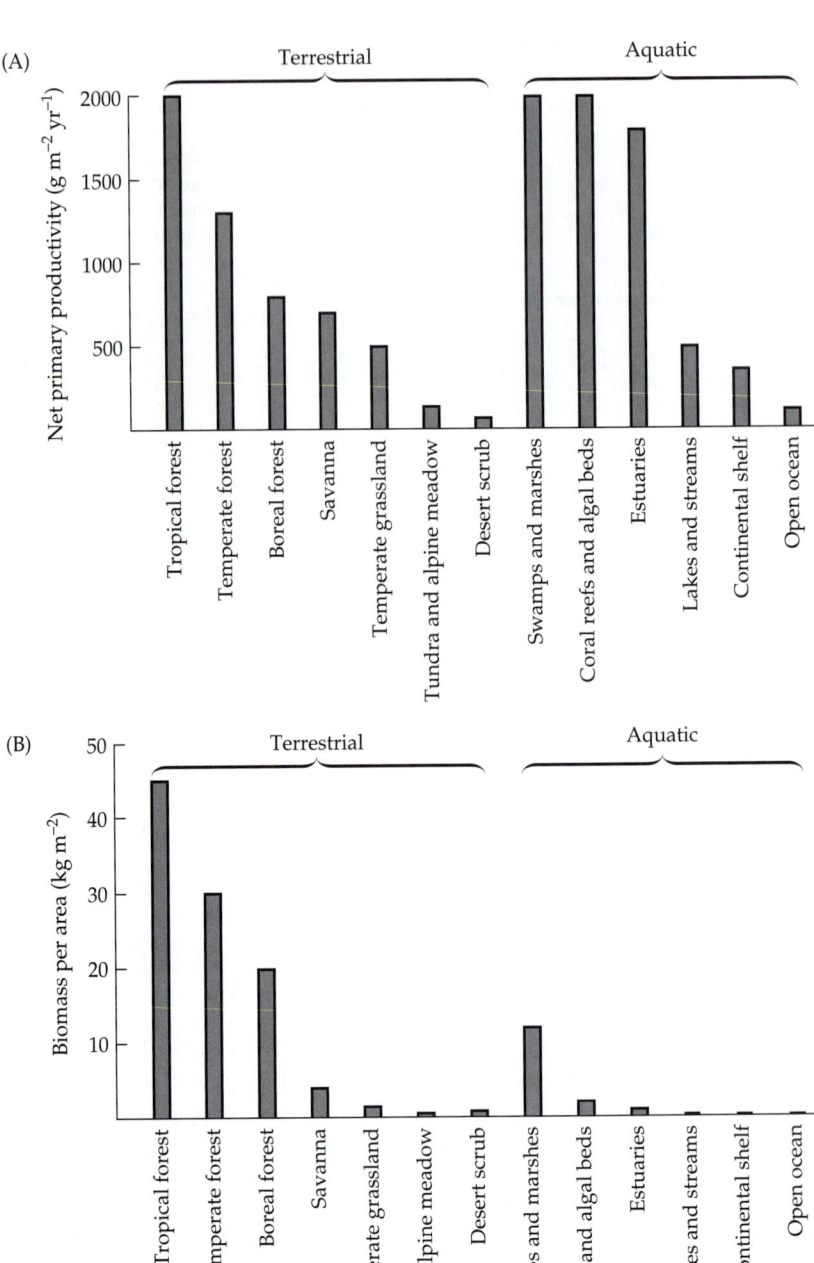

We should also bear in mind that up to now we have been comparing biomes and communities on a productivity per area basis. Let us now study these systems from a different perspective—a global view. If an astronaut were able to view the earth through a biologically sensitive lens, one that could distinguish productivity levels, it would look something like Figure 5.30. Both terrestrial and aquatic ecosystems exhibit pronounced latitudinal effects, but other patterns differ markedly between them. On land, NPP tends to be strongly correlated with precipitation and temperature. Thus, except for the arid regions along the horse latitudes, terrestrial productivity tends to be highest in tropical and subtropical regions and to decrease as we move toward the poles. In the oceans, however, phytoplankton productivity tends to be most

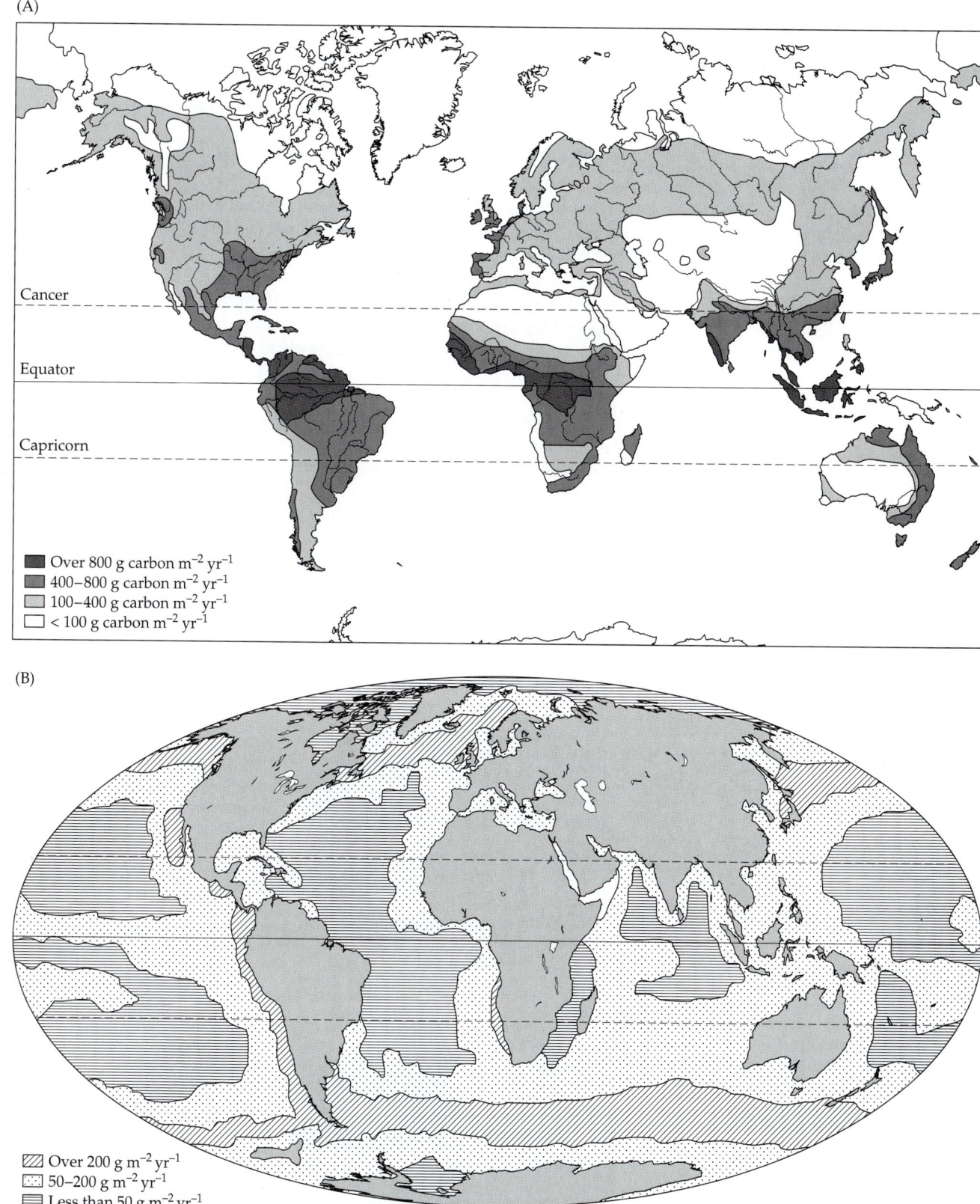

Figure 5.30 Productivity profiles for the world's landmasses (A) and oceans (B). In both realms, productivity profiles reflect geographic patterns in temperature, sunlight, and available nutrients. (A after Cox and Moore 1985; B from Tivy 1993, based on Bunt 1975.)

strongly limited by the availability of dissolved nutrients, especially phosphorus and nitrogen. Consequently, hotspots of marine productivity coincide with areas of nutrient input from the discharge of large rivers along continental shelves or upwelling from the nutrient-rich depths of the ocean. As we can see in Figure 5.30B, ocean upwelling tends to occur along the west coasts of continents and at higher southern latitudes.

Reviewing Figure 5.11 makes it quite clear that the world distribution of biomes and communities is far from uniform. In fact, the earth's surface is dominated by open ocean, one of the least productive ecosystems per unit of surface area (Figure 5.29A). Open oceans, however, cover nearly three-fourths of the earth's surface (Figure 5.31A). Consequently, when all of the open oceans are totaled, we find that they account for over one-fourth of the earth's primary production, a proportion equal to or perhaps slightly exceeding that of all tropical forests (Figure 5.31B). On the other hand, some of the most productive communities on a per area basis, such as coral reefs and estuaries, contribute only a minor fraction to the earth's total primary production.

Finally, our global view, while generally accurate and informative, is really just a snapshot in time. The distribution of biomes and the productivity profile of the earth (see Figure 5.30) have changed dramatically throughout geological time. Since the breakup of Pangaea, approximately 180 million years ago, continental drift has caused major changes in global temperatures, precipitation patterns, prevailing winds, and ocean currents. The positions of the continents with respect to latitude and solar radiation have changed substantially as they have drifted, sometimes from one hemisphere to the other (see Chapter 6). These environmental changes have resulted in major shifts in biomes, including the extinction of once dominant ancient biomes and communities

Figure 5.31 Comparisons among total area (A) and total primary production (B) of the world's biomes. Tropical rain forests are so efficient (on a per area basis) at transforming solar energy into plant tissue that even though they cover just 4% of the earth's surface, they account for roughly one-fourth of the earth's total primary production. On the other hand, even though open ocean communities are relatively inefficient at fixing solar energy, because they cover nearly three-fourths of the earth's surface, they rival tropical rain forests in total primary production.

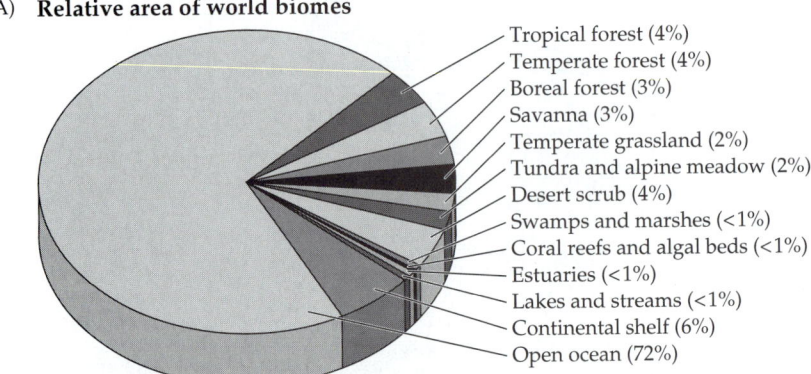

(A) **Relative area of world biomes**

- Tropical forest (4%)
- Temperate forest (4%)
- Boreal forest (3%)
- Savanna (3%)
- Temperate grassland (2%)
- Tundra and alpine meadow (2%)
- Desert scrub (4%)
- Swamps and marshes (<1%)
- Coral reefs and algal beds (<1%)
- Estuaries (<1%)
- Lakes and streams (<1%)
- Continental shelf (6%)
- Open ocean (72%)

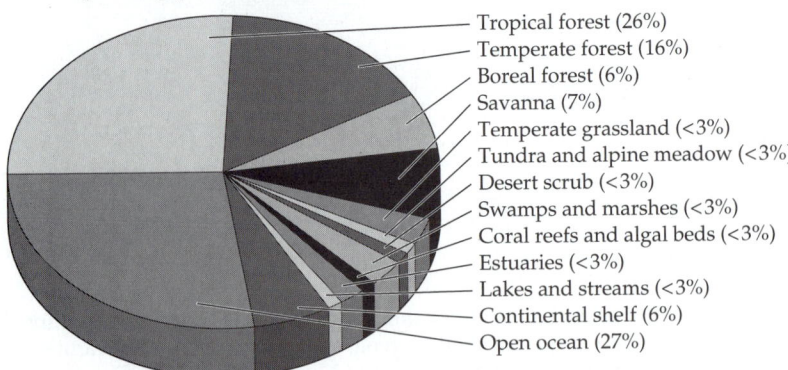

(B) **Global primary production of the earth's biomes**

- Tropical forest (26%)
- Temperate forest (16%)
- Boreal forest (6%)
- Savanna (7%)
- Temperate grassland (<3%)
- Tundra and alpine meadow (<3%)
- Desert scrub (<3%)
- Swamps and marshes (<3%)
- Coral reefs and algal beds (<3%)
- Estuaries (<3%)
- Lakes and streams (<3%)
- Continental shelf (6%)
- Open ocean (27%)

(Behrensmeyer et al. 1992; Erwin 1993). Even as recently as 18,000 years ago—only an instant ago on the geological time scale—most of the northern land-masses were covered by glaciers, often a kilometer or more thick. Warm deserts and some other biomes, once relatively rare and limited to lower latitudes and lower elevations, have since expanded at the expense of others. As we shall see in subsequent chapters, these historical changes often leave a lasting imprint on biogeographic patterns.

We can learn much from this history lesson and, just as important, apply it to the future. Even within ecological time—decades to hundreds of years—human activities have caused significant changes in the global climate and the distribution of biomes. Given recent advances in our ability to monitor and model these changes, we can apply some of these lessons from the past to develop a prospective view of tomorrow's biogeography. This is the focus of the final chapters.

CHAPTER *6*

The Changing Earth

The earth's surface has changed continually during the history of life. Continents have moved, seas have expanded and contracted, mountain ranges have risen and been eroded away, islands have appeared and disappeared, and glaciers have advanced and retreated. In addition, the fossil record reveals that the earth's climate has experienced profound changes. The width of the tropics has varied, influencing global patterns of vegetation and associated animal life. The positions of the equator and poles have not changed, but because continents have moved, regions that are tropical or polar today were not always so in the past. A sound knowledge of these past physical changes is essential for understanding the influence of historical events on both past and present distribution patterns.

The Geological Time Scale

Anyone studying historical biogeography needs to be familiar with the time scale used to date the history of the earth (see Table 6.1). Early geologists recognized that each layer in a stratigraphic column contained a unique assemblage of fossils, characteristic of a single time span, and that these assemblages could therefore be used to correlate the ages of rock strata in one locality with those in distant localities. The most reliable fossils for use in such correlations were wide-ranging species whose lifestyles were largely independent of small-scale patchiness in the environment, especially those that were freely dispersed among marine habitats by currents. These forms are called **index** or **guide fossils**. Examples are the planktonic, calcareous foraminiferans and siliceous radiolarians (phylum Protozoa) of the Cenozoic era; the chitinous colonial graptolites (phylum Hemichordata), floating animals of the Ordovician and Silurian periods; and the swimming ammonoid cephalopods (phylum Mollusca) of the Mesozoic era, whose buoyant calcareous shells probably drifted with currents after death. On the continents, the most widespread fossils are found in the coal beds of the Carboniferous period, which are characterized by numerous distinctive vascular plants.

Correlations of fossils around the world gave only relative estimates of the ages of various rocks and fossils; early scientists had no specific knowledge of the actual dates of rocks. Alfred Russel Wallace (1880), for example, accepted an estimate of 400 million years for the absolute age of the earth. The discov-

Table 6.1

The geological time scale (all numbers are in million of years B.P.)

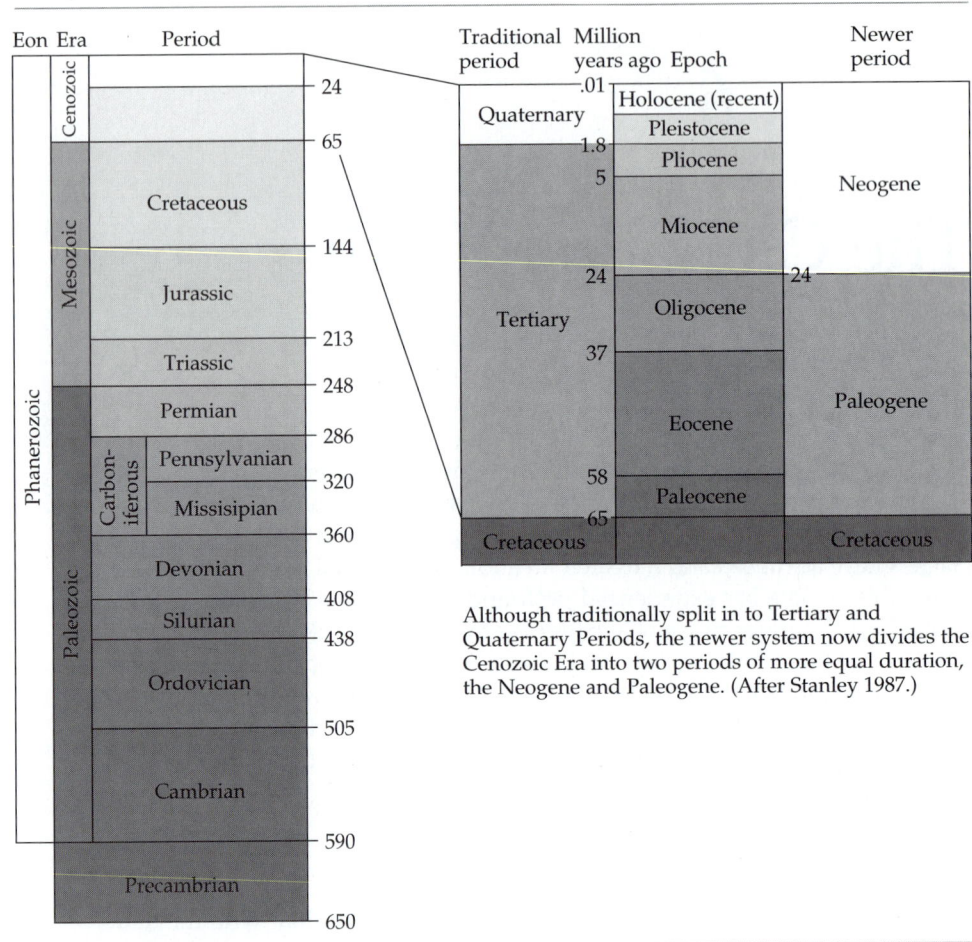

Although traditionally split in to Tertiary and Quaternary Periods, the newer system now divides the Cenozoic Era into two periods of more equal duration, the Neogene and Paleogene. (After Stanley 1987.)

ery in the twentieth century of radioactive materials finally led to more exact dating procedures. Radioactive elements are unstable, and decay through a series of intermediate unstable products to stable atoms; during this disintegration process, atomic particles are released. The rate of decay, which is independent of environmental factors, can be quantified and expressed as a **half-life**, the amount of time needed for half the radioactive material to decay to the stable element. By calculating the ratio of radioactive element to stable end product, one can determine, within limits, the age of a sample. Thus, by using radioactive isotopes of uranium and thorium, whose stable end product is lead, scientists have pushed back the estimate of the earth's age to 4.6 billion years. The oldest known fossils have been dated at 3.5 billion years before the present (B.P.).

The potassium-argon method is another valuable technique for dating Phanerozoic rocks. Radioactive potassium ($^{40}K^{19}$) decays to stable calcium ($^{40}Ca^{20}$) and the inert gas argon ($^{40}Ar^{18}$); the half-life of potassium 40 is 1.31 billion years. A major problem with this technique is that argon gas will escape from rock heated above 300° C—as it would be during metamorphism—and is therefore not wholly reliable. Measurement of the decay of rubidium 87 (^{87}Rb) to strontium 86 (^{86}Sr) is used mainly for dating rocks older than 100 million years. For very recent material, **radiocarbon dating** is extensively used. Carbon 14 decays to carbon 12 at a fairly rapid rate (half-life 5730 ± 30 years). After 50,000 years, so little radiocarbon remains that detection is very difficult. Consequently, the radiocarbon dating method is presently useful only in late Quaternary research.

Because many trees form an annual ring each growing season, analysis of tree growth rings in temperate latitudes is a reliable method for dating fossils formed within the last 10,000 years, and the results compare closely with carbon 14 values. In fact, tree rings can be used to calibrate radiocarbon dating. Paleoclimatic reconstructions are also possible because the width of a fossil growth ring is correlated with the length of the growing season and the availability of water (Fritts 1976).

These methods, in combination, have been used to estimate the times of major geological and evolutionary events during the past 600 million years, known as the Phanerozoic eon (see Table 6.1). The geological time scale is hierarchical, with each division among **eons**, **eras**, **periods**, or **epochs** marking transitions among geological strata and embedded fossil assemblages. Given the great difficulty of dating such ancient events, it should not be surprising that precise dates are not universally accepted. For example, the scale accepted in 1971 divided the Cretaceous period into 12 equal epochs of 6 million years each, but radioisotope dating has revealed that some epochs were longer and others quite short (Baldwin et al. 1974). Reexamination of Triassic deposits may eventually lead to a drastic shortening of that period and to increased time spans for the Jurassic and Permian. Accurate dates within the Mesozoic are crucial because the early evolution and dispersal of major lineages of land vertebrates and seed plants, as well as the extinctions of certain marine groups and radiations of others, occurred during that era.

Traditionally, the Cenozoic era was divided into the Tertiary and Quaternary periods. However, the Tertiary covered some 63 million years, while the Quaternary lasted just 1.8 million years. This traditional scheme is often replaced by a newer one that divides the Cenozoic into two periods of more similar duration: the Paleogene (65 to 24 million years B.P.) and the Neogene (24 to 0.01 million years B.P.; see Table 6.1).

The Theory of Continental Drift

No contribution to biogeography has had more of an impact than the theory of **continental drift**. This theory developed from a highly speculative idea in the early 1900s to a well-established fact by the 1960s. Simply defined, the theory of continental drift states that continents and portions of continents have rafted across the surface of the globe on the weak, viscous upper mantle beneath the earth's crust. Thus the earth's crust is not composed of fixed ocean basins and continents, as was once supposed, but instead is a changing landscape in which once distant lands are now in juxtaposition, and others once attached are now widely separated.

The evidence in favor of crustal movements is conclusive, and within the last few decades the theory of continental drift has given rise to a respected science. Today's more comprehensive theory, referred to as **plate tectonics**, explains the origin and destruction of the earth's plates as well as their lateral movement or drift. Throughout the long history of the field, however, biogeographers, from Candolle to Lyell, Darwin, and Wallace, insisted on the fixity of the continents and great oceans. Lyell, Cuvier, and others had long ago established that the land and sea exhibited great vertical fluctuations throughout the geological record. But the notion that great masses of the earth's crust could drift, collide, and separate like ice on a partially frozen river must have been a tough sell indeed.

Scholars have searched diligently to determine who first proposed the theory of continental drift. Various authors have attributed the germ of the idea to early writers—for example, to Sir Francis Bacon (1620), although Bacon never really discussed the subject. Lyell, realizing that fossils found in Europe indicated that a tropical climate had once prevailed in that region, suggested that the earth experienced cycles of global climatic change. These changes were triggered by vertical shifts in the earth's crust, which rose above sea level in one region of the world while sinking in another. According to Lyell, when elevated landmasses were concentrated near the equator, global warming occurred. At other times, landmasses were concentrated near the poles, resulting in what Lyell called earth's "great winter," which was accentuated by glacial episodes. Lyell's theory of geoclimatic cycles, albeit novel, had little if any impact on the field. Lyell insisted that, while the earth's crust shifted vertically during these great cycles, the continents (plates) changed little in size, shape, or position relative to one another (Figure 6.1). Lyell, one of the most persuasive and revered scientists of his time, never made this theory one of his great crusades. As we shall see, his model bore only a trivial resemblance to the modern theory of continental drift and plate tectonics.

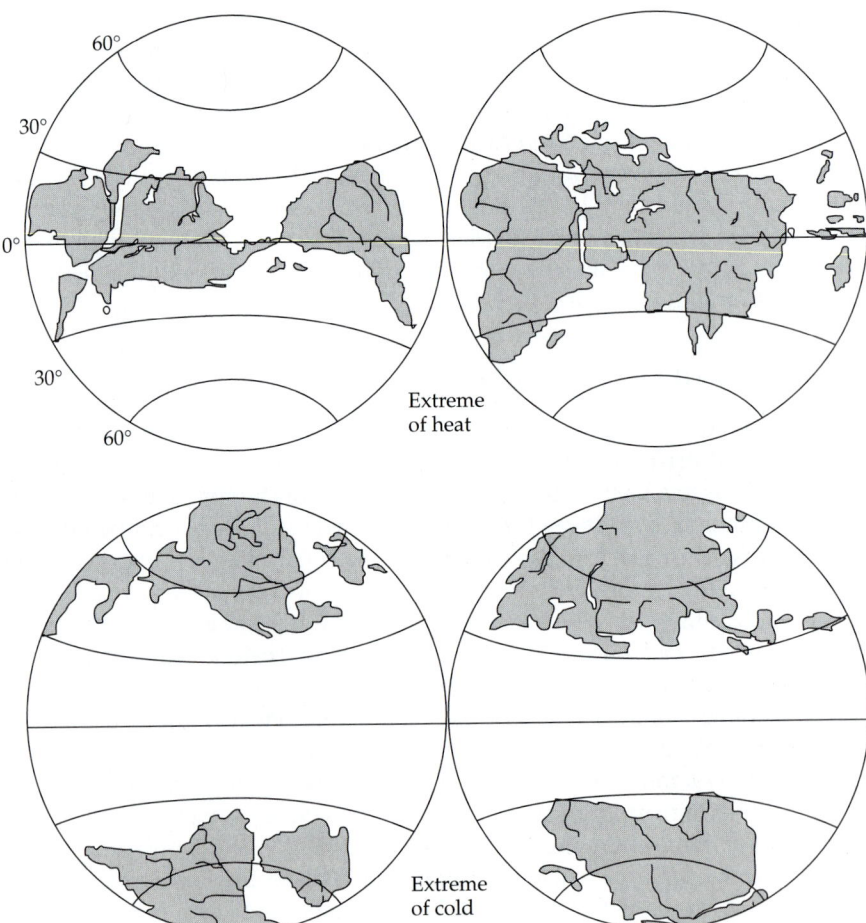

Figure 6.1 Lyell's theory of cycles of global climatic change. In the third edition of his *Principles of Geology*, Lyell suggested that although they did not change in shape or position relative to one another, the continents shifted in concert across the globe. This, he argued, could account for the paleontological evidence of major shifts in climates.

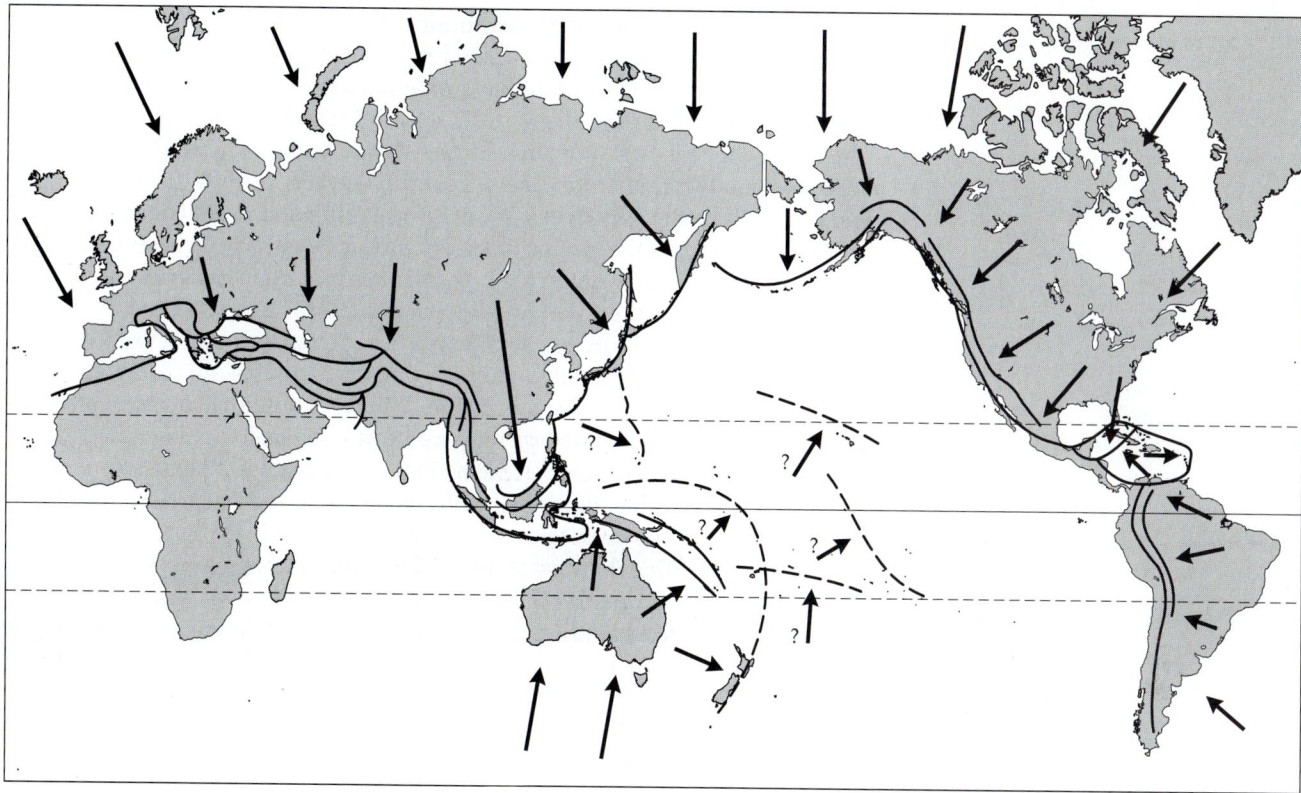

Figure 6.2 An early model of crustal movement proposed by Taylor (1910). This scenario suggests a general drift of the major landmasses toward the equator.

Figure 6.3 Alfred Wegener (1880–1930), who developed the ideas leading to the modern science of plate tectonics and its confirmation of continental drift. (From Schwarzbach 1980; photo courtesy of Deutsches Museum, Munich.)

On the other hand, one of Lyell's contemporaries, Antonio Snider-Pelligrini (1858), may have been the first to demonstrate the geometric fit of the coastlines of continents on opposite sides of the Atlantic Ocean and to argue cogently that they once formed a supercontinent that subsequently split apart. Yet throughout the nineteenth century and into the early decades of the twentieth century, geologists had little understanding of the past relationships of continents. In 1908, F. B. Taylor, an American geologist, presented a model (privately published in 1910) in which the continents were hypothesized to move, distorting crustal materials into mountain ranges and island chains (Figure 6.2). Taylor postulated that moving continents form mountains at their forward margins and leave oceans behind them. Although Taylor did not correctly perceive the directions of continental movements, his ideas were innovative, and later research would show that crustal upheaval and continental movement are indeed intimately related.

Wegener's Theory

Alfred L. Wegener (Figure 6.3), a German meteorologist, conceived and championed the theory of continental drift, which he presented with admittedly scanty evidence at first, but which anticipated much of our current knowledge. Wegener developed his ideas on continental displacement in 1910, independently of Taylor, while observing on a world map the congruence of opposite coastlines across the Atlantic. In January 1912, Wegener unveiled his working hypothesis, with supporting evidence, in two oral reports, published later that year (1912a,b). These observations were expanded into his classic book, *Die Entstehung der Kontinente und Ozeane* (*The Origin of Continents and Oceans*) (1915).

Wegener's theory on horizontal continental movements not only discussed all the continents, but also synthesized evidence from many disciplines: geology, geophysics, paleoclimatology, paleontology, and biogeography. The strongest attribute of the theory was its integration of these many types of phenomena for the first time. In addition to the geometric fit of the continents, Wegener pointed to the now incontrovertible evidence that the earth's crust had fluctuated markedly in elevation. If the great landmasses could move vertically, why couldn't they move laterally? Indeed, what would keep them from doing so? Wegener also noted the alignment of mountain belts and rock strata on opposite sides of the Atlantic. As Lyell and others had observed much earlier, the coal beds of North America and Europe indicated that both of these landmasses were once situated over the tropical latitudes. Similarly, Wegener observed that glacial deposits, or "**tillites**," in what are now subtropical Africa and South America suggest that they were once displaced poleward. In addition, the tillites of North America seemed continuous with those of Europe. Many anomalous biogeographic patterns, such as the occurrence of extant marsupials in South America and Australia, were easily explained by the theory that these continents were connected during the Permian period.

Like many revolutionary ideas, Wegener's ideas were ignored by most scientists and ridiculed by others. At first, his conclusions were accepted by just a few geologists and biogeographers, most notably those in the Southern Hemisphere. After all, the idea of ancient connections and biotic affinities among the southern biotas was not new to students of those biotas. J. D. Hooker had suggested it in 1853 after reviewing distribution patterns of plants among the southern continents and archipelagoes:

> Enough is here given to show that many of the peculiarities of each of these the three great areas of land in the southern latitudes…is agreeable with the hypothesis of all being members of a once more extensive flora, which has been broken up by geological and climatic causes. (1853)

The second edition (1920) of Wegener's treatise received some attention, albeit negative, when it was criticized by several prominent geologists, but wide knowledge of the theory came only after the third edition (1922) was translated into five languages, including English (1924). A fourth edition (1929), the one generally used now, contained more information, but throughout the various editions, the substance of Wegener's ideas remained the same. The following are some of his pertinent conclusions.

1. Continental rocks, called **sial**, are fundamentally different, less dense, thicker, and less magnetized than those of the ocean floor (basaltic rocks, called **sima**). The lighter sialic blocks, the continents, float on a layer of viscous, fluid mantle.

2. The major landmasses of the earth were once united as a single supercontinent, **Pangaea**. Pangaea broke into smaller continental plates, which moved apart as they floated on the mantle. The breakup of Pangaea began in the Mesozoic, but North America was still connected with Europe in the north until the late Tertiary or even the Quaternary (Figure 6.4).

3. The breakup of Pangaea began as a rift valley, which gradually widened into an ocean, apparently by adding materials to the continental margins. The midoceanic ridges mark where opposite continents were once joined, and the ocean trenches formed as the continental blocks moved. The distributions of major earthquake centers and regions of active volcanism and orogeny (mountain building) are related to the movements of these blocks.

Upper Carboniferous

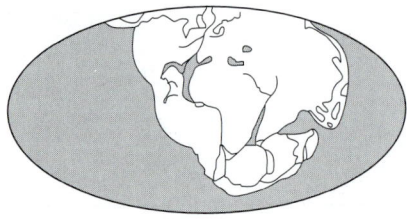

Eocene

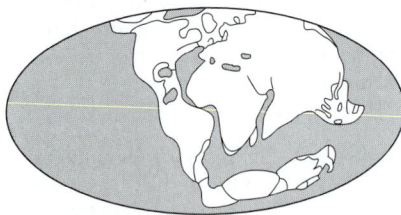

Lower Quaternary

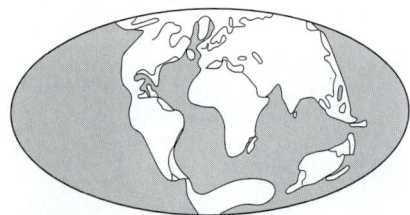

Figure 6.4   Wegener's (1929) model of continental drift, showing how he envisioned the continents, initially united in one giant landmass, to have moved apart during the Mesozoic and early Tertiary. In Wegener's time, the geological epochs and periods were thought to have been more recent than has been indicated by modern dating methods. Nevertheless, comparison with Figure 6.15E and Figure A in Box 6.2 shows that Wegener's view was extremely similar to current reconstructions of continental movement. (After Wegener 1966.)

4. The continental blocks have essentially retained their initial outlines, except in regions of mountain building, so the manner in which the continents were once joined can be seen by matching up their present margins. When this is done, similarities in the stratigraphy, fossils, and reconstructed paleoclimates of now distant landmasses demonstrate that those blocks were once united. These patterns are inconsistent with any explanation that assumes fixed positions of continents and ocean basins.

5. Rates of movement for certain landmasses range between 0.3 and 36 meters per year, the fastest being Greenland, which may have separated from Europe only 50,000 to 100,000 years ago.

6. Radioactive heating in the mantle may be a primary cause of block movement, but other forces are probably involved. Whatever the causal processes, they are gradual and not catastrophic.

Early Opposition to Continental Drift

Wegener's ideas were clearly prophetic, and form the basis of our modern theory of continental drift and plate tectonics. Yet, despite his cogent and persistent arguments, continental drift was not generally accepted until the early 1960s, some 50 years after he and Taylor first published their ideas. Why was the theory resisted for so long? The history of this debate serves as an important lesson in the nature of scientific revolutions. Strong criticism of Wegener's theory arose as soon as translated volumes made it available to most geologists and biogeographers in the mid-1920s. Some scientists resisted the new idea because it conflicted with their preconceived ideas of fixed continents and a solid earth, and because it was proposed by a meteorologist—a man who was not part of the geological establishment.

Other scientists opposed Wegener's theory on much more objective and defensible grounds. Although it would eventually prove to be one of science's most important paradigms, Wegener's theory suffered from at least four shortcomings. First, the nature of scientific revolutions is such that new theories that challenge long-held paradigms are resisted until the evidence clearly shows them to be more parsimonious. The prevailing attitude in biogeography and paleontology was first expressed by Alexander du Toit: "Geological evidence *almost entirely* must decide the probability of this hypothesis" (1927, p. 118). Wegener's theory included too many assumptions that remained unsubstantiated by the available geological evidence. Paleontologists in particular were unconvinced by the biogeographic and fossil evidence marshaled to support the continental drift model. For example, in 1943, G. G. Simpson published an analysis of past and present mammalian distributions to show how these data fit alternative scenarios of past intercontinental connections. After correcting prevalent errors in the literature concerning these distributions, he pointed out that most of the known Cenozoic patterns could be explained without invoking continental drift. It was difficult, using paleontological data, to discriminate between alternative theories such as stable continents with periodic flooding, ancient landbridges, and continental drift. Although Wegener's thesis was plausible, the theory of continental drift would not be accepted until enough unambiguous new evidence was collected to make it the most parsimonious explanation for geological and biogeographic patterns.

A second shortcoming of Wegener's theory is that it contained many factual errors. The theory aroused skeptical interest in many fields, and scientists in each discipline recognized major factual errors in Wegener's presentations. These inconsistencies had to be resolved. Even du Toit, Wegener's strongest proponent, who published two books (1927, 1937) in favor of the theory, had to concede unquestionable errors by Wegener. For example, Wegener proposed that the plates move at an incredibly high rate, as rapidly as 36 meters per year. His theory might have been much more palatable if his estimates were closer to what we now believe to be true—that is, rates of 2 to 12 cm per year. Wegener's error, however, is understandable, as he and his colleagues were working with what they believed to be a much younger earth than we now assume, and were limited to very crude methods of measuring displacement rates. Third, in addition to correcting these errors, Wegener and his followers would have to gather much more evidence from a variety of disciplines in geology and biogeography to test their model.

Finally—and this was perhaps more critical than any of the other short-comings—Wegener's theory lacked a plausible mechanism. How could the plates—rigid, enormous masses of rock—move about, and what force or forces could drive such movements? Wegener's insights in this area, however, are largely overlooked by science historians. In fact, he discussed potential causal mechanisms at length in the 1929 revision of his book (see Wegener 1929, pp. 167–179). Still, he was very conservative, and justifiably so:

> The theory [of continental drift] is still young and still often treated with suspicion . . . It is probable, at any rate, that the complete solution of the problem of the driving forces will still be a long time coming, for it means the unraveling of a whole tangle of interdependent phenomena, where it is often hard to distinguish what is cause and what is effect.

Wegener discussed three potential driving or "displacing forces." Not surprisingly, given that he was a meteorologist and an astronomer, two of these involved celestial phenomena (the effects of centrifugal forces on the earth's surface; the combined effects of the gravitational fields of the earth, moon, and sun). The third force Wegener discussed, convective currents of molten rock beneath the earth's crust, is indeed the same ultimate force assumed by today's model of plate tectonics. Wegener admitted that his ideas on causal mechanisms were speculative and that "the problem of forces which have produced and are producing continental drift is still in its infancy" (p. 179). Speculation, however, may be a normal and perhaps an essential part of scientific revolutions. As Darwin wrote in a letter to Wallace (1857), "I am a firm believer that, without speculation, there is no good and original observation."

Soon after completing the fourth and final revision of his book, Wegener set out on an expedition in 1930 to Greenland to document its purportedly rapid movement and, by so doing, verify his theory of continental drift. In fact, his view of a dynamic earth with rapidly drifting landmasses was largely stimulated by his earlier expeditions to Greenland in 1906 and 1912. It is one of science's great and tragic ironies that, as Pascual Jordan observed, Wegener "died in a snowstorm on the very Greenland expedition he undertook to verify his theory."

Evidence for Continental Drift

Providing convincing evidence for Wegener's theory was to take nearly five decades of research conducted by many teams of geologists, paleontologists, and biogeographers. Even Snider-Pelligrini's (1858) early observations on the fit of the continents were not generally accepted for nearly a century. From the first time Wegener proposed the supercontinent of Pangaea, his opponents criticized the liberties he took to achieve a "good fit." A good reconstruction was finally achieved when S. W. Carey (1955, 1958b), an Australian geologist, used plasticene shapes of landmasses sliding over a globe. Nonetheless, the fit was not widely accepted until three researchers (Bullard et al. 1965) combined computer mapping techniques and statistical analyses to test continental fits. Their analyses showed that the continents do fit together if one uses the submarine contours of the continental shelves to delineate the margins of the continental plates (see also Hallam 1967).

Other stratigraphic, paleoclimatic, and paleontological evidence gradually accumulated to support the theory of continental drift. Along with these important discoveries (summarized in Box 6.1), some of the most compelling evidence in favor of the theory of continental drift was provided by marine geologists.

Marine geology.   After World War II, a second generation of scientists, who had not been directly involved in the initial debates, made some important discoveries about ocean basins and rock magnetism that encouraged reexamination of the ideas and evidence advanced by Wegener and du Toit. When Wegener proposed his ideas, very little was known about the structure of the ocean floor. On the basis of loose samples obtained by dredging, geologists suspected that the ocean floor was composed of basalt (sima, consisting primarily of silicon and magnesium), but no one had actually taken core samples of the deep basins. Sialic continental rocks (composed largely of silicon and aluminum), however, were well known. Echo soundings from several transoceanic expeditions had portrayed ocean bottoms as smooth structures (abyssal plains) lying 4 to 6 km beneath the ocean surface. A midoceanic ridge was known only in the Atlantic Ocean. Finally, deep cuts in the ocean floor, known as trenches, had been found on the ocean sides of island arcs, and were known to display unusual gravitational properties.

Oceanographic research was just beginning to accelerate before the outbreak of World War II, when charting ocean topography became a practical goal. During the war, Herman H. Hess, a marine geologist who was using an echo sounder as he sailed aboard a U.S. troop transport, discovered some flat-topped submarine volcanoes 3000 to 4000 m high. Peaked submarine volcanoes, called **seamounts**, had been previously identified. The new structures, which Hess later named **guyots** in honor of a Princeton geologist, were thought to be volcanic islands that had formed above the ocean surface, were later truncated by wave action, and finally sunk to 1 or 2 km beneath the waves (see Figure 6.6). Guyots are common in the northern and western Pacific Ocean, and as we shall see shortly, they figured prominently in the development of models of continental drift.

Following the war, marine exploration blossomed due to generous funding by the Allied navies. Important discoveries were made using new deep-sediment piston corers and explosive charges. From the samples they obtained with these new techniques, geologists learned that under recent sediments, all ocean floors are composed of basalt, and that this basement is young, at a maximum dating back only to the Jurassic (150 million years B.P.). Thus the oceans are considerably younger than the continents, whose ancient foundations, called **cratons** or **Precambrian shields**, are older than 1 billion years!

By the mid-1950s a team of scientists had recognized that the submarine mountain ranges bisecting the oceans are really segments of a continuous global system 65,000 km long (Figure 6.5). This system is marked by a central rift valley, which is closely associated with a zone of frequent shallow earthquakes. New instruments measured remarkably high temperatures in these rifts, suggesting that molten mantle material was being released there. Scientists began to interpret the midoceanic ridges as zones where the oceans expand, establishing the concept of **seafloor spreading** (Figure 6.6).

Located far from the ridges, oceanic **trenches** are so deep that until the 1950s, most knowledge of them was obtained by taking soundings. Oceanic trenches are V-shaped troughs about 10 km deep (Figure 6.6). Through the use of seismic refraction techniques, marine geologists learned that the earth's crust is extremely thin in the trenches, and the heat flow beneath the trenches

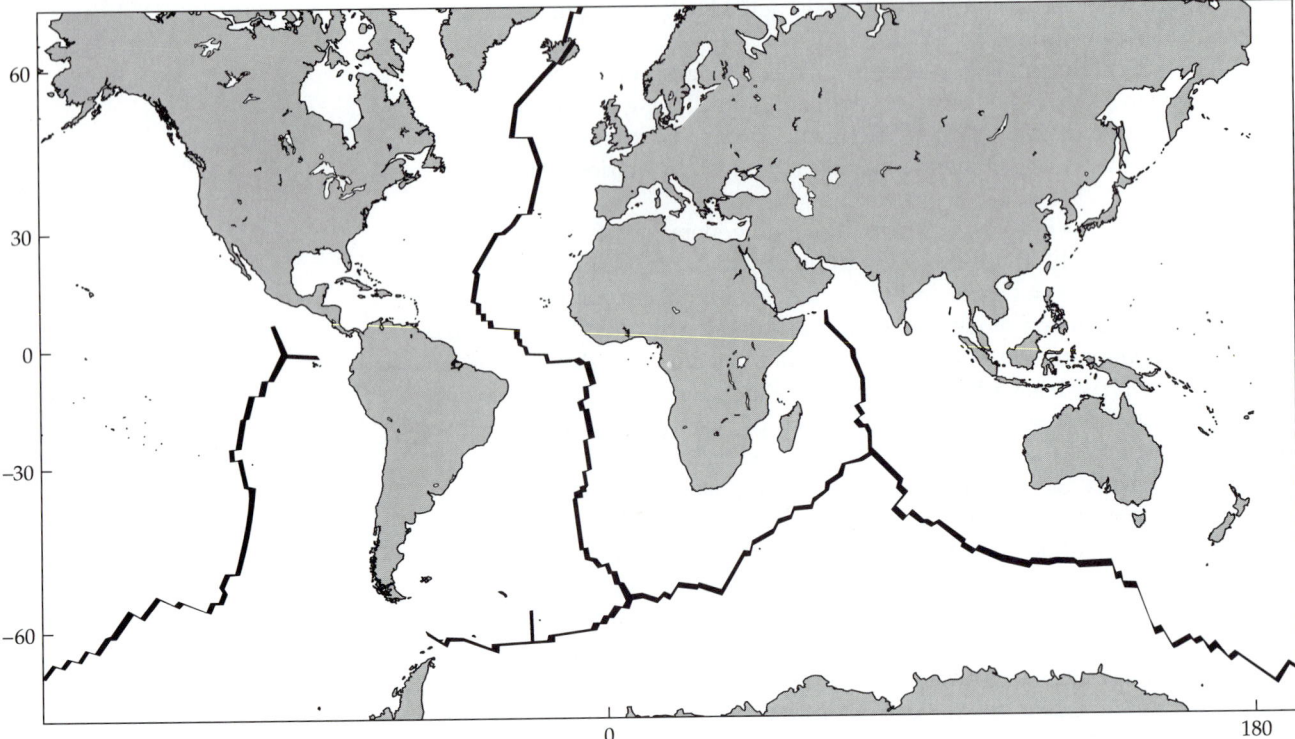

Figure 6.5   The global system of midoceanic ridges, which mark regions of seafloor spreading. (After Scotese et al. 1988.)

Figure 6.6   A highly simplified model of seafloor spreading that depicts how oceanic plates are pushed apart at the spreading center by the upwelling of magma from the mantle, which causes the plates to slide away from the midoceanic ridges over the viscous asthenosphere. Magma may also produce volcanic islands near the spreading center, but as a point on the plate is displaced from the ridge, it also descends to 4 to 6 km below sea level, and the islands become submerged. These submerged volcanic structures (seamounts or guyots) eventually disappear into an oceanic trench where the oceanic plate meets another plate. In the case illustrated, the heavier oceanic plate descends beneath the lighter continental plate, which causes the metamorphosis of the surface material on the oceanic plate into ophiolites and their deposition on the continent, the consumption of the volcanic islands, and eventually the remelting of the plate itself. The asterisks indicate the epicenters of earthquakes resulting from the contact of the two plates (the Benioff zone).

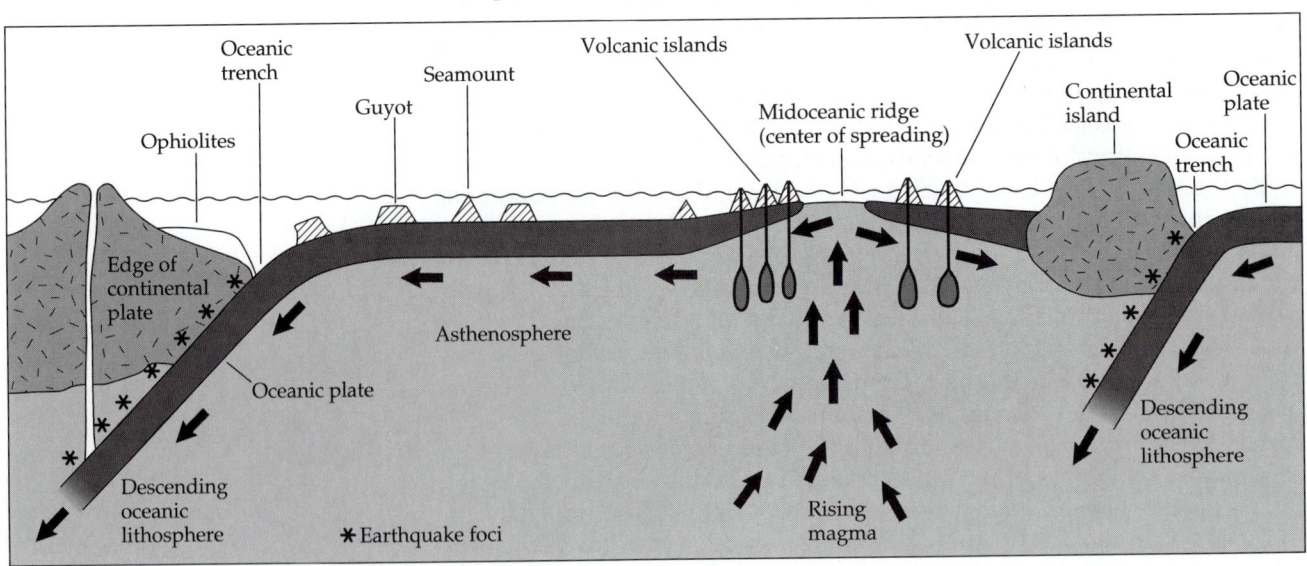

Box 6.1

Stratigraphic, paleoclimatic, and paleontological discoveries that contributed to the acceptance of the theory of continental drift

By the 1970s, stratigraphic, paleoclimatic, and paleontological data provided strong support for the theory of continental drift. In particular, many pieces of evidence came together to support the matchup of continents—particularly the southern continents—in their former positions as part of Pangaea.

Stratigraphic evidence. Topographic features, including mountains, oceanic ridges, and island chains, along with specific rock strata (e.g., Precambrian shields and flood basalts) and fossil deposits, were found to be aligned along Wegener's hypothesized connections of the now fragmented portions of Gondwanaland (Figure A) (Hurley 1968; Hurley and Rand 1969). In addi-

tion, on each of the now isolated southern continents, rocks from the late Paleozoic and early Mesozoic contained the same stratigraphic sequence (see Allard and Hurst 1969): glacial sediments, coal beds, and sand dunes and other desert deposits, all overlain by a layer of volcanic rock.

Paleoclimatic evidence. All of the continents in the Southern Hemisphere have late Paleozoic glacial deposits (tillites) in their southernmost regions. Moreover, as glaciers move, they scour the underlying rocks, leaving deep scratches that mark the direction of their movements. If we plot these glacial lines on a map with the southern continents in their current positions, the pat-

terns appear quite confounding (Figure BI). Not only do the glaciers appear to have been situated in what are now some relatively warm latitudes, but many appear to have risen out of the sea. This perplexing anomaly disappears when the same glacial lines are plotted on a reconstruction of Gondwanaland as it was during the Permian period (Figure BII).

Paleontological evidence. The late Paleozoic glacial deposits of the southern continents are covered with Permian rocks bearing the so-called *Glossopteris* flora (Schopf 1970a,b). These arborescent gymnosperms are presumed to have been adapted to (and therefore indicative of) temperate climates because they had deciduous leaves and conspicuous growth rings in their wood (Schopf 1976). When the occurrence of the *Glossopteris* flora is plotted on a map of Pangaea (Figure B), the points circumscribe a discrete region of Wegener's Gond-

Figure A (I) Distributions of Precambrian shields (hatched areas), illustrating how they match up if the continents are reassembled as they were before the breakup of Pangaea. (After P. M. Hurley and J. R. Rand 1969.) (II) Although substantially isolated today, flood basalts also serve as evidence of the previous connections of the southern continents in the former Gondwanaland. (After Storey 1995.)

(I)

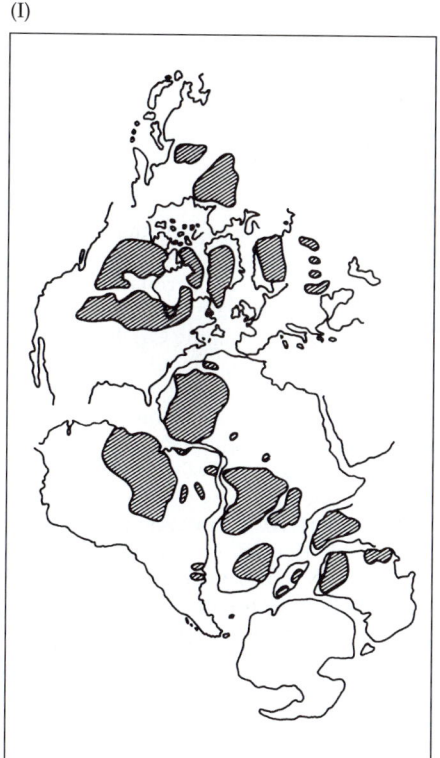

(II)

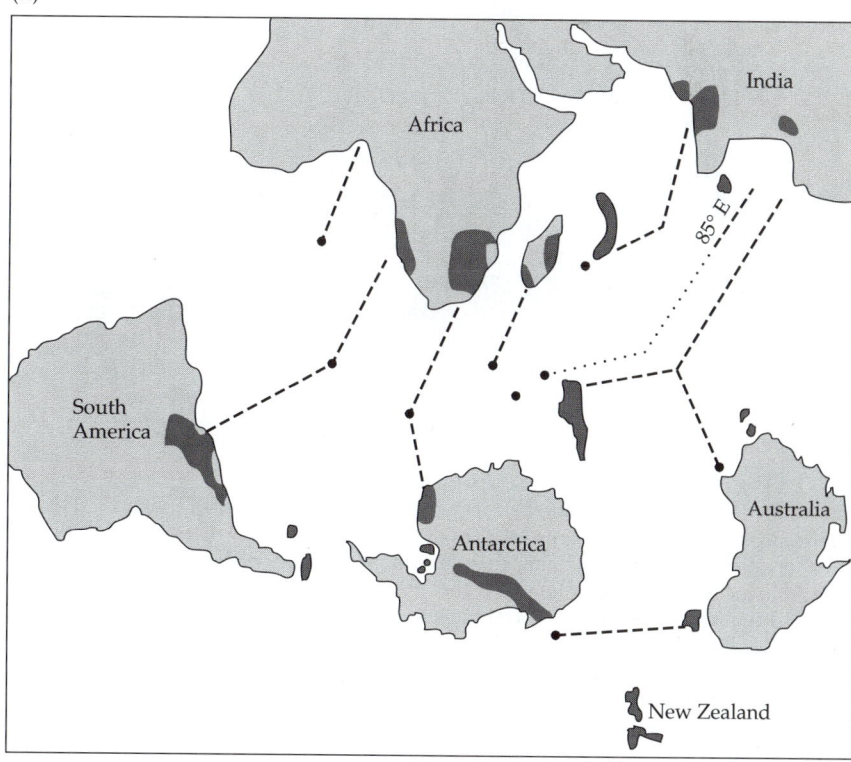

Box 6.1 *(continued)*

wanaland, thought to be correlated with the margins of the glaciers.

In 1969 D. H. Elliot and E. H. Colbert unearthed the first tetrapod fossils found in Antarctica (Elliot et al. 1970) from mudstone and volcanic sandstone of the early Triassic. Additional finds provided convincing evidence that many of these bones belonged to *Lystrosaurus*, a mammal-like reptile also found in rocks of similar age in the Karoo of southern Africa and the Panchet Formation of southern India. The reconstruction of Gondwanaland explained many such biogeographic anomalies in the distributions of extant as well as fossil assemblages. Many vertebrates underwent major radiations during the Permian, when the proximity of the continents allowed their rapid spread across what are now isolated landmasses. Like basaltic rocks, glacial tillites, and fossil assemblages, many of these extant forms exhibit disjunct distributions. The extensive breaks in their current ranges are consistent with the former connections of Pangaea and its subcontinents (Figure C).

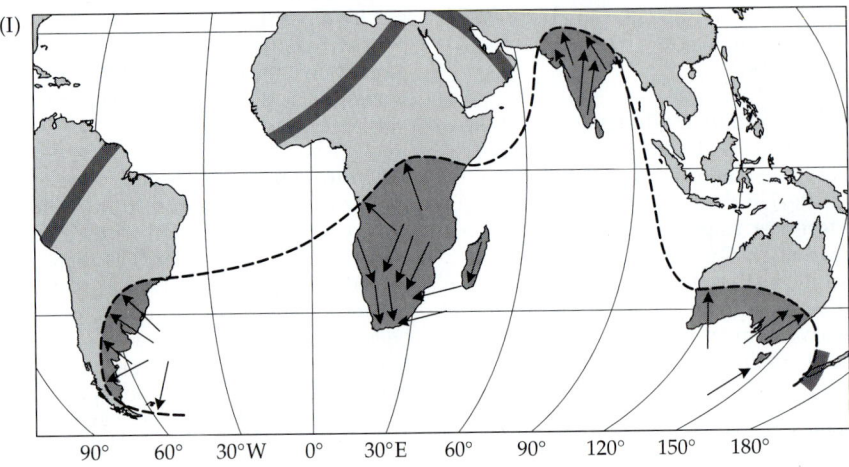

(I)

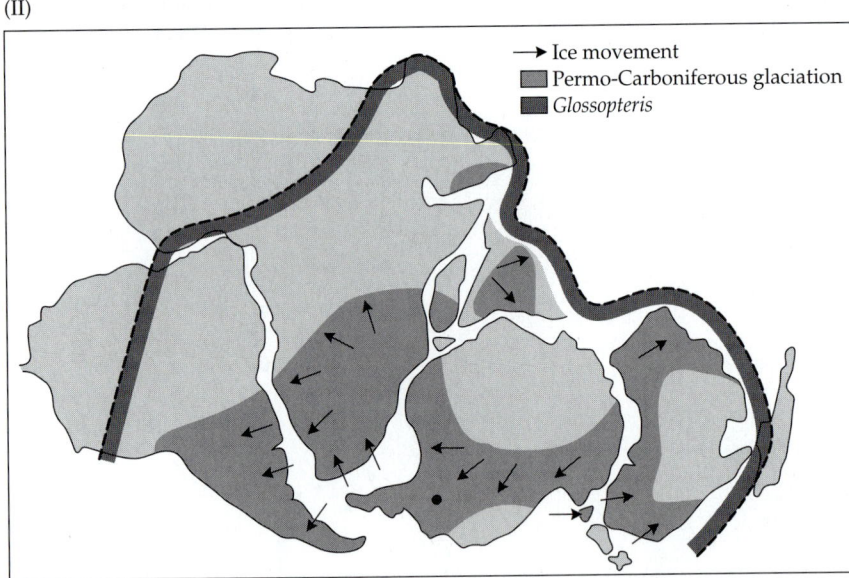

(II)

→ Ice movement
▨ Permo-Carboniferous glaciation
▨ *Glossopteris*

Figure B　Two lines of paleontological evidence for continental drift found on the southern continents. Glaciers carved lines in the underlying rock material, marking their location and direction of movement (arrows). The *Glossopteris* flora (or "southern beeches") included several groups of plants that grew along the margins of the glaciers. (I) The origin and directions of glacial movement (shaded area with arrows) and the distributions of *Glossopteris* fossils (darker shading) are difficult to explain based on the current positions of the southern continents because they imply that glaciers moved from oceans onto land. (II) These patterns, however, are consistent with reconstructions of Gondwanaland as it was during the Permian period. (I after Stanley 1987; II after Windley 1977.)

Figure C　The disjunct distributions ▶ of some living taxa suggest that their ancestral forms radiated across Gondwanaland in the Permian period. (I) Southern temperate beetles of the tribe Migadopini of the family Carabidae. (II) Fishes of the superfamily Galaxioidea. These fishes are restricted to nontropical waters in the Southern Hemisphere. (III) Plants of the family Proteaceae. This group is found on all of the southern continents, but barely reaches the Northern Hemisphere. (IV) Clawed aquatic frogs of the family Pipidae. This family is comprised of two subfamilies, the Pipinae in tropical South America and the Xenopinae in tropical Africa, suggesting a common ancestor that was once distributed in western Gondwanaland. (I after Darlington 1965; II after Berra 1981; III after Johnson and Briggs 1975; IV after Savage 1973.)

(I)

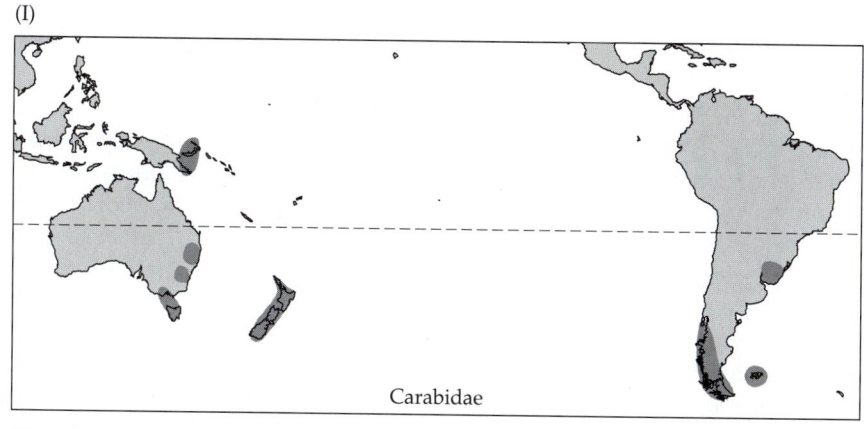

Carabidae

(II)

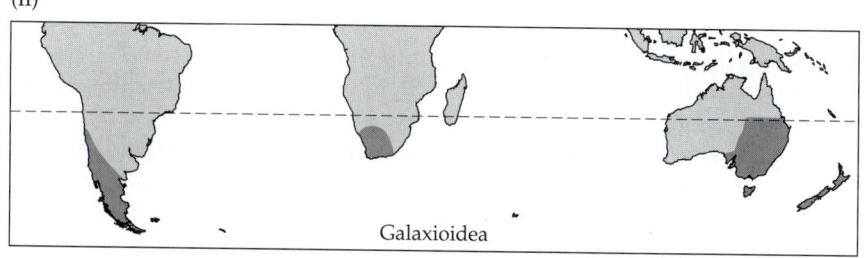

Galaxioidea

(III)

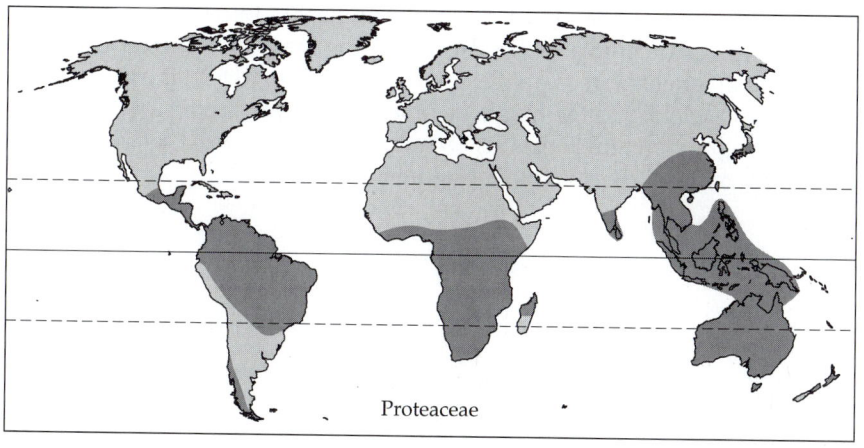

Proteaceae

(IV)

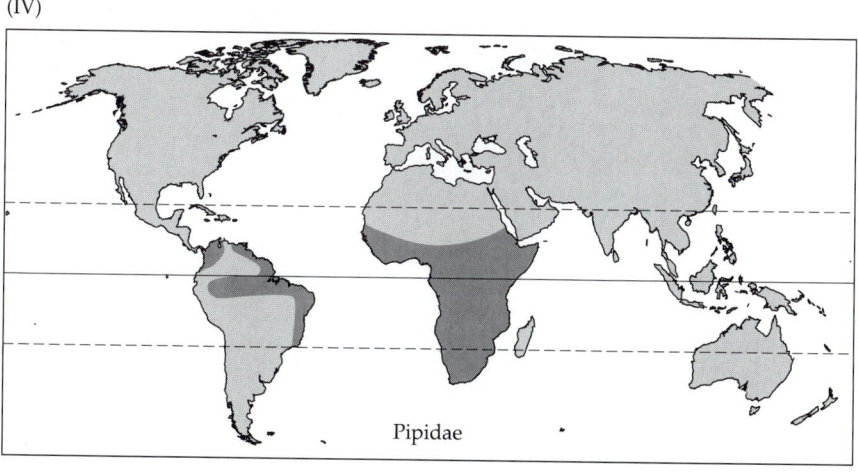

Pipidae

is half that found in the abyssal plain, implying that heat is consumed in the trenches. Gravity measurements in the trenches are lower than in any other place on earth. Geologists therefore postulated that the crust is pulled downward into these trenches and its material reincorporated into the mantle.

Around the Pacific Ocean is a belt of volcanism and earthquake activity known as the "Ring of Fire." It was proposed that the subduction (downward movement) of crustal materials in the trenches was the direct cause of these violent geological events. Hugo Benioff (1954) provided the first convincing evidence for this hypothesis. By plotting the positions and depths of earthquake epicenters in the vicinity of trenches, he demonstrated that the epicenters closest to trenches are shallow and those farther away are progressively deeper. The epicenters are aligned along a zone dipping downward at about 45° behind the trench, indicating that the earthquakes are caused as the cold, rigid crustal slab descends into the mantle (see also Calvert et al. 1995). These zones are now termed **Benioff zones**.

Paleomagnetism and the emergence of a mechanism.   Studies of **Paleomagnetism** provided additional evidence for seafloor spreading. Paleomagnetism refers to the orientation of magnetized crystals at the time of mineral formation—that is, when molten rock solidifies. Rocks containing iron and titanium oxides become magnetized as they solidify and cool, and this magnetization is reflected in their crystalline structure, which remains "frozen" in the rock, oriented as a fossil compass in the direction and declination of the then-prevailing magnetic field. This high-temperature magnetization, referred to as **remnant magnetism**, is very stable unless the rock is reheated to extremely high temperatures (the **Curie point**). Hence, by measuring the direction and declination of remnant magnetism in cooled lavas, it is possible to determine the relationship of any landmass to the magnetic poles at the time the rock was formed, and by triangulation and computer techniques, to reconstruct the positions of landmasses relative to one another (Figure 6.7).

Figure 6.7   (A) The earth acts as a great bar magnet. Because its magnetic fields are oriented toward its core as well as poleward, latitudinal position can be read as declination in a compass needle. The same phenomenon also influences the orientation of crystals during the formation of magnetically active rock, thus recording the latitudinal position of the rock when it was formed. (B) Such paleomagnetic information can be used to reconstruct the positions and movements of the continents, such as the movements of Gondwanaland relative to the South Pole during the Paleozoic era. Gondwanaland, including some present-day equatorial regions, drifted over the South Pole twice, in the late Ordovician and in the late Devonian. (From Stanley 1987.)

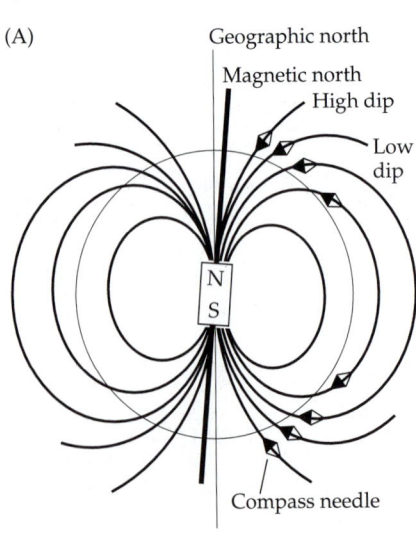

(A)

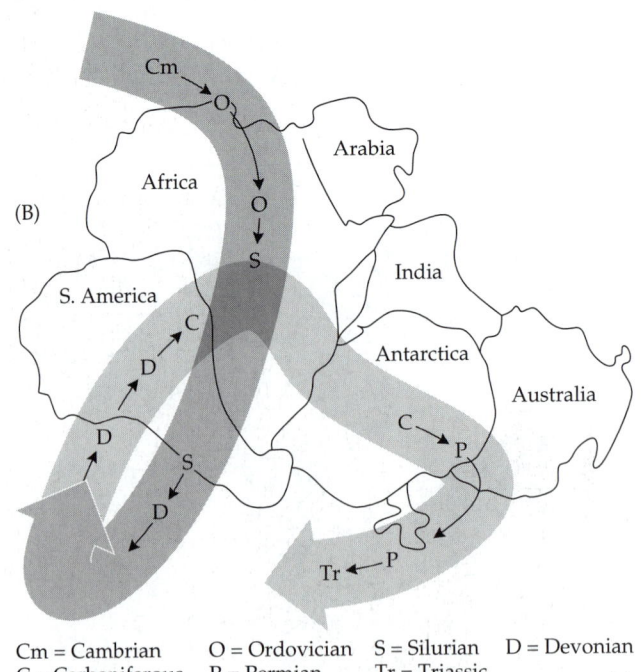

(B)

Cm = Cambrian   O = Ordovician   S = Silurian   D = Devonian
C = Carboniferous   P = Permian   Tr = Triassic

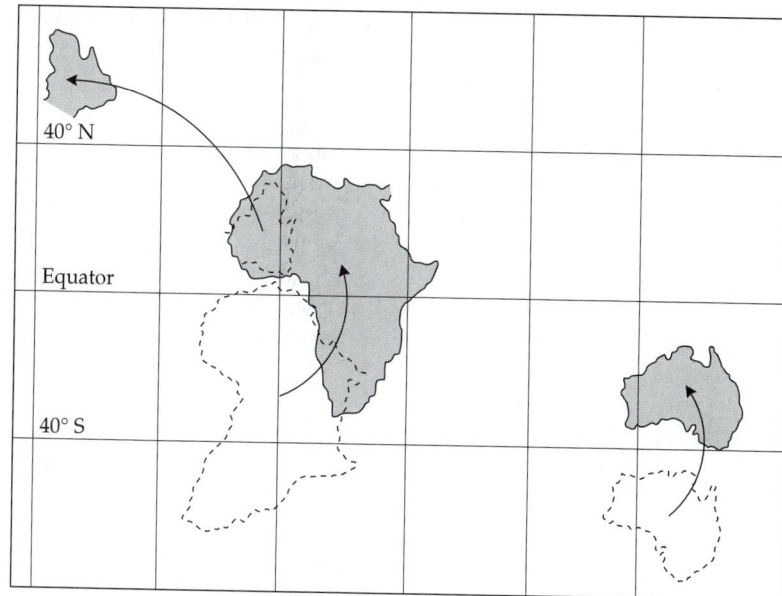

Figure 6.8   Shifts in the orientation and latitudinal positions of Labrador, Africa and Australia between the Triassic (200 million years B.P.; dashed outline) and the present (solid outline) are revealed by paleomagnetism. (After Pielou 1979.)

In the early 1950s the British physicist P. M. S. Blackett invented a new supersensitive magnetometer (magnetic detector) that could be used to determine continental orientation throughout geological history. First, the magnetometer was used to show that the British Isles had rotated 34° clockwise since the Triassic (Clegg et al. 1954). A major breakthrough came when other British scientists (Creer et al. 1954, 1957; Runcorn 1956) analyzed geological strata in Europe and North America and provided strong evidence that the two continents had once been joined, but later drifted apart. Subsequent studies around the world reaffirmed the necessity of continental movements to explain the existing paleomagnetic patterns (Irving 1956, 1959; Runcorn 1962) (Figure 6.8).

At the beginning of the twentieth century in central France, Bernard Bruhnes (1906) first discovered **magnetic reversals** when he found lavas that were magnetized in a direction opposite that in recently formed ones. Such patterns reflect reversals of the earth's magnetic field, which occur every 10^4 to 10^6 years. Since then, many investigators have found geological evidence of these reversals. On the ocean floor, alternating patterns of normally and reversely magnetized basalt appear as **magnetic stripes**, which retain their spacing and shapes for long distances (Figure 6.9; see Cox 1973).

Marine geologists were the first to perceive the significance of magnetic stripes for continental drift theory. Two seminal discoveries were provided in the early 1960s by Frederick Vine and Drummond Matthews and by Herman Hess. Vine and Matthews (1963) discovered several important properties of ocean floors:

1. Basaltic rocks at the midoceanic ridges have normal field (present-day) magnetic properties.

2. The widths of the alternating magnetic stripes on the opposite sides of a ridge are often roughly symmetrical, and the stripes are generally parallel to the long axis of the ridge.

3. The banding pattern of any one ocean closely matches that of the others, and the ocean patterns correspond approximately to reversal timetables from terrestrial lava flows.

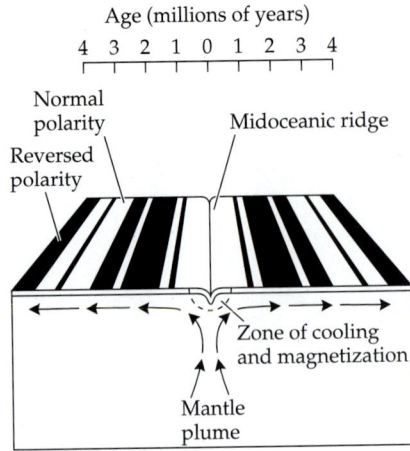

Age (millions of years)

4 3 2 1 0 1 2 3 4

Normal polarity

Reversed polarity

Midoceanic ridge

Zone of cooling and magnetization

Mantle plume

Figure 6.9 During seafloor spreading, reversals in the earth's magnetic field are recorded as the magnetically sensitive, iron-rich crust cools. Differences in the widths of the magnetic stripes reveal differences in the duration of these polarity episodes and in the rate of seafloor spreading over time and among regions. (From Stanley 1987.)

Herman Hess proposed the first model of seafloor spreading to account for major tectonic events (see Figure 6.6). This synthesis was presented at Princeton University in 1960, but was not published for general readership until 1962. By that time, R. S. Dietz (1961) had published similar but less detailed accounts of global continental movements. Hess and Dietz hypothesized that the oceans are formed by addition of material and spreading at the midoceanic ridges. Moving away from the ridges, basalt along the ocean floor increases in age and is marked by magnetic stripes, which record the polarity of the prevailing magnetic field. The stripes tend to be highly symmetrical on opposite sides of a ridge. In contrast, differences in stripe widths indicate that the rate of seafloor spreading varies over time, and that it is not uniform across different oceans, or even different parts of the same ocean.

With this model of seafloor spreading, Wegener's theory finally had a plausible mechanism. The initial theory of continental drift was now included in a more general theory, plate tectonics, which included not just the lateral movements of the continents and ocean basins, but their origin and destruction as well.

In Great Britain, most geologists who studied global tectonics were converted to continental drift theory by 1964, convinced especially by the soundness of Hess' model and the new synthesis. Acceptance in North America, however, lagged behind by several years. Meanwhile, widely circulated articles appearing in *Scientific American*, *Science*, and *Nature* did much to make the entire scientific community aware of the rebirth of Wegenerism. Young scientists became aware of the latest evidence and helped to create a wave of acceptance following the mid-1960s. The theory of plate tectonics is now firmly established as a unifying paradigm for much of geology and biogeography.

The current model.   The theory of plate tectonics remains an active and exciting area of research. Following its general acceptance, geologists, paleontologists, and biogeographers were freed to focus their attention on more intriguing questions relating to the temporal and spatial patterns of plate dynamics and potential causal mechanisms.

The lateral movements of the plates result from a complex interaction among the earth's crust, the underlying mantle, and the core, which is the site of the intense heat that drives plate movement (Figure 6.10). The plates are roughly 100 km thick and are composed of a relatively thin, rigid layer of crust, which adheres to the upper layer of the mantle, the **lithosphere**. The mantle also includes a deeper, more fluid layer, the **asthenosphere**, which is composed primarily of molten material. There is an emerging consensus suggesting that plate movements are caused by a combination of forces, including **ridge push**, **mantle drag**, and **slab pull** (Kerr 1995). Ultimately, all of these forces are generated by heat and convective forces deep in the earth. Ridge push occurs at the midoceanic ridges, where **magma** (molten rock) upwells from the asthenosphere to the surface. It is believed that the parent rock of the mantle is partially melted and the basaltic portion is then brought to the surface. The addition of basaltic magma at the center of the ridge causes the older rocks on either side to spread—to be literally pushed apart. Thus, ridge push is the cause of seafloor spreading in Hess' model (see Figure 6.6).

Mantle upwelling is part of a convective cell system that also includes lateral flow of the mantle beneath the plates and downward flow of cooler rock toward the earth's core (Figure 6.10). The lateral flow and the friction between the mantle and the overlying plate create a dragging force much like that of a conveyor belt (i.e., mantle drag). However, much, if not most, of the drifting force may be generated at **subduction zones**, where dense oceanic plates sink deep into the magma, eventually contributing to the convective gyre of molten rock. As the

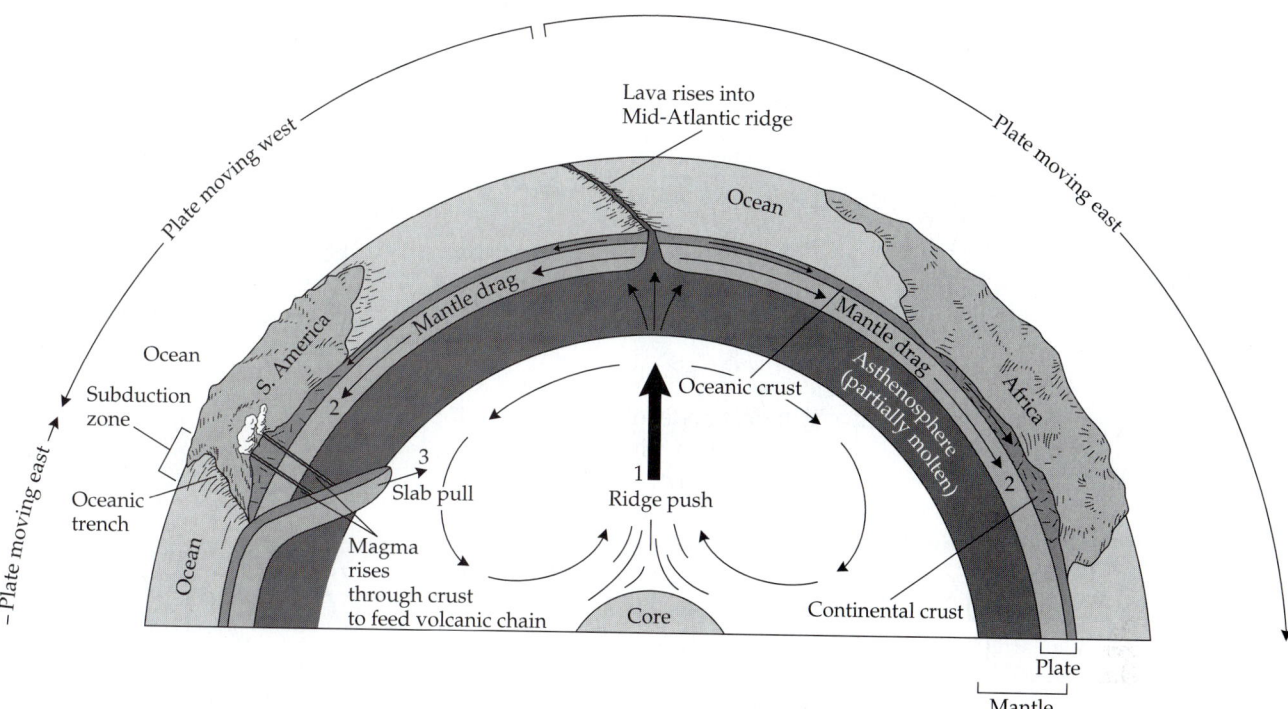

leading edge of the subducted plate descends, it pulls the rest of the plate laterally toward the subduction zone (see Kerr 1995 and references therein).

The relative importance of these three forces—ridge push, mantle drag, and slab pull—appears to vary markedly across tectonic regions and geological periods. Computer modeling conducted by Lithgow-Bertelloni and Richards (1995) suggests that slab pull may at times account for over 90% of the net tectonic forces. On the other hand, slab pull cannot account for the movement of those continental plates, such the South American, that lack subduction zones (actually, the Pacific Plate is being subducted under the western edge of the South American Plate). Obviously, this central issue—the underlying mechanisms of plate tectonics—will remain an active and hotly debated question for some time (see Box 6.2).

Despite the remaining uncertainties about their ultimate causal mechanisms, geologists have developed a sound understanding of plate configurations, plate movements and interactions, and related phenomena, including earthquakes and volcanism. While the number of plates has varied continually throughout geological time, 16 major plates are currently recognized (Figure 6.11). These plates range in size from the Gorda, which is roughly 750 km², to the Pacific Plate, with an estimated area of over 100 million km². The rate of lateral movement also varies markedly among the plates, with some, such as the western portions of the Pacific Plate, drifting as rapidly as 5 cm per year, while others appear fixed. The biogeographic relevance of plate tectonics becomes immediately evident when we compare plate configurations with Wallace's map of biogeographic regions (compare Figures 6.11 and 2.7): biogeographic regions are defined by biotas and portions of the earth's crust that share both evolutionary and tectonic histories, with representative assemblages on each plate evolving in isolation from those on other plates.

While the geological history of the earth's plates must be profoundly complex, plate boundaries take three basic forms: **spreading zones**, **collision zones**, and **transform zones**. Other tectonic phenomena, including earthquakes, volcanism, and the formation of mountains and island arcs, are closely associated with these interactions among plates. As noted above, midoceanic ridges mark

Figure 6.10 The current model of plate tectonics includes the possibility that at least three forces may be responsible for crustal movements: (1) ridge push, or the force generated by molten rock rising from the earth's core through the mantle at the midoceanic ridges; (2) mantle drag, the tendency of the crust to ride the mantle much like boxes on a conveyor belt; and (3) slab pull, the force generated as subducting crust tends to pull trailing crust after it along the surface. (After Stanley 1987.)

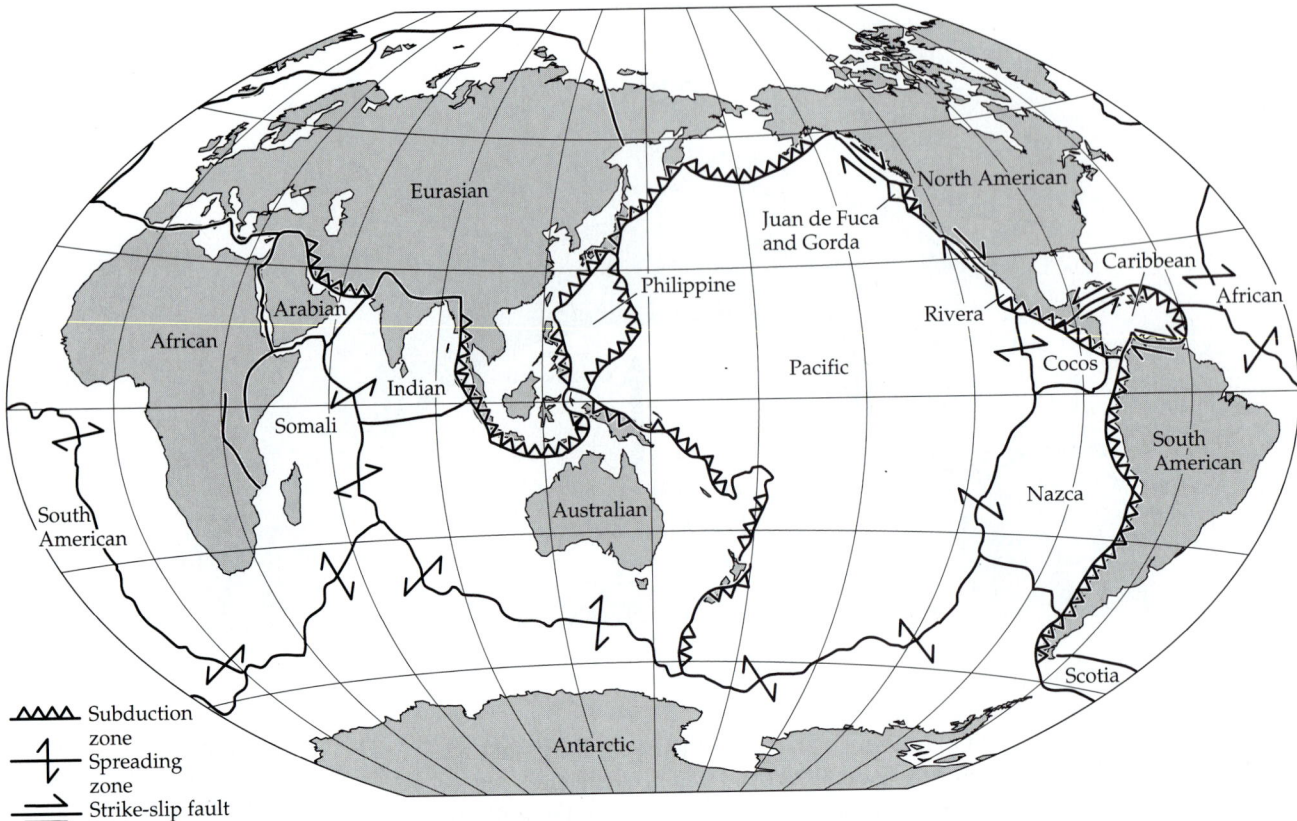

Figure 6.11   The earth's major tectonic plates.

the sites where two plates are drifting apart. Spreading zones, however, are not confined to oceanic plates. On the continents, plates also diverge to form **rift zones**, which in the past created the Red Sea and the great, deep lakes of the Baikal rift zone and the East African Rift Valley. The latter region remains a tectonically active area marked by frequent earthquakes. Rather than continually and gradually sliding away from spreading centers, plates tend to resist moving until tectonic forces finally exceed some threshold, creating powerful bursts of movement, interspersed with relatively long periods of stasis. Spreading zones also are marked by volcanic activity where magma rises to the surface (see Figure 6.6). Along ridge systems, volcanoes may become emergent as oceanic islands. Once formed, such an island is eventually carried away from the ridge and down a slope to the abyssal plain. This movement decreases the elevation of the island relative to sea level and eventually draws it beneath the surface, making it into a submarine seamount. In classic models, wave action wears an island flat to form a guyot; however, some evidence suggests that flat-topped guyots may actually be formed that way without ever having been emergent (see Figure 6.6). If the seafloor spreading model is correct, the ocean floor and its associated chains of volcanic islands, seamounts, and guyots should be youngest at the ridges and oldest in the subduction trenches. Beginning with Tarling (1962) and Wilson (1963b), various researchers have shown that these predictions are correct (Figure 6.12).

Far away from the spreading zone, along a plate's leading margin, it will often collide with another plate. If the two plates are of roughly equal density, their collision will cause violent uplifting and the formation of mountain ranges along the plate boundary. This is what occurred when the Indian Plate collided with the Eurasian Plate to form the Himalayas during the early Tertiary period

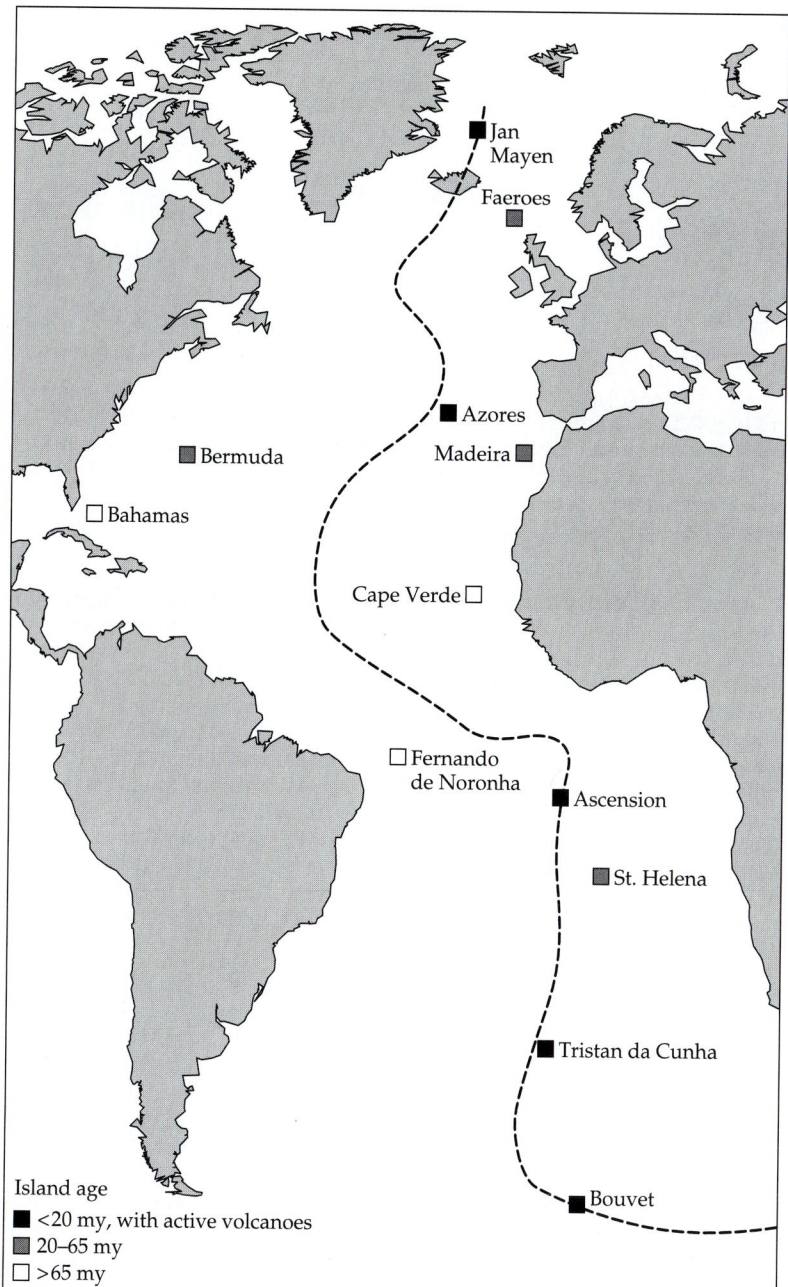

Island age
- ■ <20 my, with active volcanoes
- ▦ 20–65 my
- ☐ >65 my

Figure 6.12 Because new crust is added along the spreading zones of midoceanic ridges, the age of islands and other crustal features tends to increase with increasing distance from these ridges, as demonstrated by islands at different distances from the mid-Atlantic ridge (dashed line). (From Pielou 1979.)

(approximately 45 million years B.P.). More often, relatively dense oceanic plates collide with and sink beneath the lighter continental plates to form a subduction zone and a deep oceanic trench (Figures 6.10 and 6.13). Again, the plates tend to undergo long periods of stasis followed by violent episodes of movement and resulting earthquakes. As the oceanic plate is drawn deep into the molten layer of the mantle, its accumulation of relatively wet sediments is heated, causing mantle plumes to rise to the surface and form a ring of volcanoes far upstream from the superficial layers of the subduction zone. Extensive subduction zones are found along the western margins of North and South America. As oceanic plates are subducted over geological time, seamounts and guyots are scraped

Box 6.2
Expanding earth's envelope

The theory of continental drift and plate tectonics has weathered some eight decades of skepticism, criticism, and at times even ridicule to finally become embraced as a unifying paradigm of geology and biogeography. While most scientists now accept the general tenets of the theory, a few have proposed modifications or alternatives, some of which seem quite radical. Here, we highlight just three of these alternative views.

At least as early as the mid-1970s, geologists and biogeographers hypothesized the existence of an ancient continent, "Pacifica," which lay east of New Zealand and Australia before the Permian (see Nur and Ben-Avraham 1977; Melville 1981; Nelson and Platnick 1981). During the Mesozoic, this continent supposedly fragmented and dispersed, and its remnants became embedded along the growing margins of other continents. The terranes around the current margins of the Pacific Ocean (Figure 6.14) were thought to be remnants of Pacifica. However, subsequent analyses indicated that these terranes were not of continental origin, but rather accretions of island chains, guyots, and seamounts. Further, reconstructions of seafloor spreading and crustal movements over the past 180 million years indicated that the purported remnants of Pacifica could not have diverged from a single "homeland."

Based primarily on biogeographic evidence, including the disjunct distributions of many plant species (see Figure 10.19), Humphries and Parenti' (1986) suggested a more substantial modification of the current model of plate tectonics. In their view, many disjunctions,

Pre-Pangaea (>300 B.P.)   **Pangaea (225 B.P.)**   **Recent**

Figure A   A possible explanation for the amphitropical distributions of many plants (see Figure 10.19). This hypothesis suggests a pre-Pangaean supercontinent in which austral and boreal forms were located adjacent to one another. These landmasses may then have separated and drifted toward opposite poles before the formation of Pangaea. Finally, Pangaea split north to south to provide the modern pattern of amphitropical disjunctions. (After Cox 1990.)

including **amphitropical distributions** (see Figure 10.19), in which species are restricted to boreal and austral regions, are consistent with the existence of a global, pre-Pangaean supercontinent during the early or mid-Paleozoic. The arrangement of landmasses of that supercontinent may have been quite different from that of Pangaea, with today's austral landmasses occupying northern latitudes adjacent to the then equatorial, now boreal landmasses (see Figure A). This supercontinent then

broke up, and its landmasses migrated to approximate their current relative positions during the late Carboniferous. While this hypothesis would explain many biogeographic anomalies, existing geological and paleomagnetic data are not consistent with the hypothesized movements of this pre-Pangaean landmass (see Cox 1990).

Perhaps the most radical alternative to the current model of plate tectonics is the expanding earth theory. This theory was first proposed in the 1950s to account for

Figure B   Terella model illustrating the expanding earth hypothesis. (I) According to this alternative theory of plate tectonics, for most of earth's 4.5 billion year history it remained relatively small and its surface was covered with a densely packed mass of continental plates (forming Pangaea)—devoid of oceanic plates or a world ocean (Panthalassa). (II) Earth's expansion and the concomitant development and expansion of oceanic basins is hypothesized to occur sometime during the Carboniferous and Permian periods (360–248 million years B.P.), thus initiating the breakup of Pangaea and its subcontinents. (III–V) The hypothesized expansion of the earth continued throughout the late Mesozoic and early Cenozoic Eras, with oceanic expansion occurring at the mid-oceanic ridges (dotted lines in IV and V)—i.e., just as hypothesized for the currently accepted model of plate tectonics.

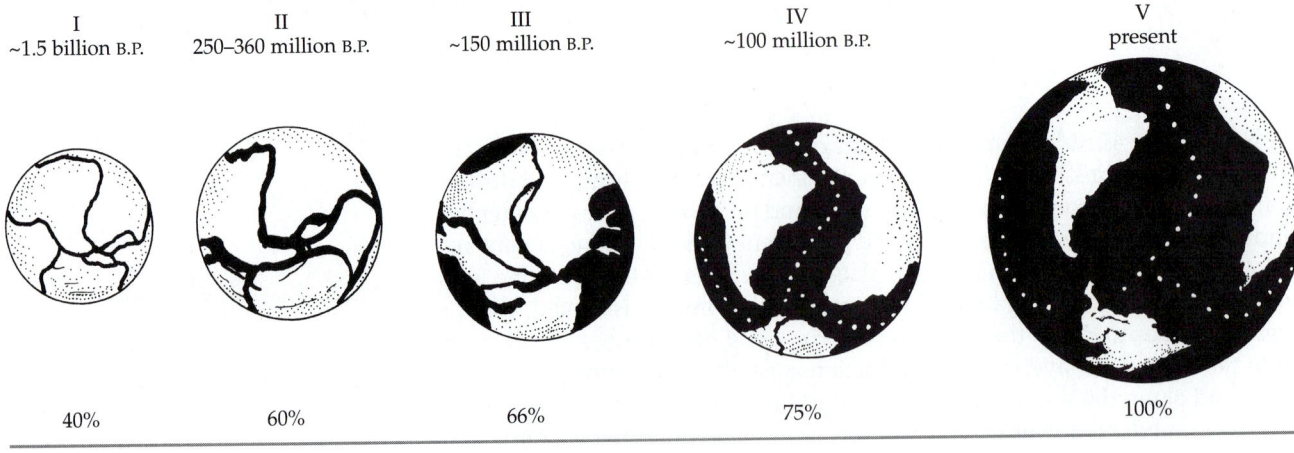

I	II	III	IV	V
~1.5 billion B.P.	250–360 million B.P.	~150 million B.P.	~100 million B.P.	present
40%	60%	66%	75%	100%

the same features addressed by the Wegenerians: the fit of the continents, their breakup since the Permian, the relative ages of continental and oceanic crust, and the alignment of various geological and paleontological features across the oceans (Egyed 1956, 1957; Carey 1958). The expansionists proposed that the earth's volume is not constant, but has increased—perhaps even doubled—since the Permian (see also Kremp 1992; Egyed 1956). As a result of this expansion, the earth's crust—all continental at that time—was ripped apart, and new, oceanic crust filled the expanding gaps. Perhaps the most controversial, yet intriguing, feature of the expansion

theory is that it implies that the gravitational "constant" is not—that is, that the expansion is a result of weakening of the earth's gravitational field. Not surprisingly, these provocative ideas have attracted attention from physicists (see Jordan 1971). Figure B summarizes Kremp's (1992) reconstruction of major geological features associated with the earth's hypothesized expansion since the Permian period.

Taken together, the above hypotheses lie far outside the mainstream views of plate tectonics and related phenomena. In most cases, they are not supported by the available information, are less parsimonious than the accepted paradigm,

and are often championed by scientists outside the primary fields. This may sound familiar: when the theory of continental drift was first proposed by Wegener, a meteorologist, it received much the same notoriety, and many decades passed before what at first seemed a very radical theory was embraced by the scientific establishment.

Thus, while the earth's plates are unlikely to make great progress in their movements during the next century, views of our ancient earth and its tectonic machinery may experience great leaps and revolutions, perhaps embracing some of what we now view as radical hypotheses, including some yet to be described.

---

onto the continental plates, forming coastal mountain ranges such as the Olympics of Washington State. Thus, subduction zones are marked with parallel bands of earthquakes and volcanism near their active edges and mountain ranges and accumulations of marine sediments, or **terranes**, which increase in age as we move upstream from the active zone (Figure 6.14) (see also Coney et al. 1980; Calvert et al. 1995).

Finally, plates of roughly equal density may slide and grind against each other, without either subducting, to form a **transform zone**. Not surprisingly, transform zones are characterized by high seismic activity as plates grind against each other. In southern California, the North American and Pacific

Figure 6.13   A subduction zone, where a relatively heavy oceanic plate slides beneath a lighter continental plate. Here, seafloor spreading causes the Gorda Plate to be subducted beneath the continental crust of the Pacific Northwest region of North America. This, in turn, results in faults and terranes (accumulations of oceanic material; see Figure 6.14) at the subduction zone and volcanic activity deep beneath the continental plate, forming the Cascade Range.

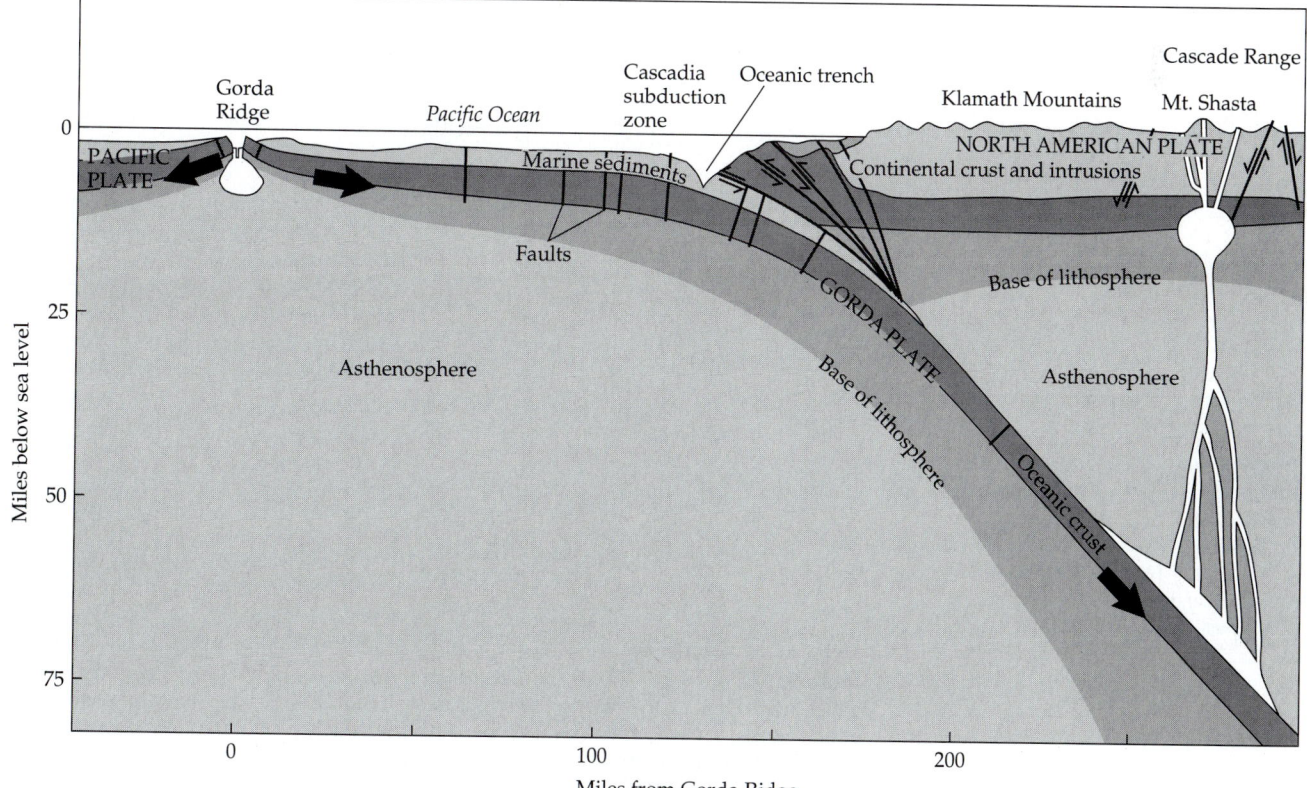

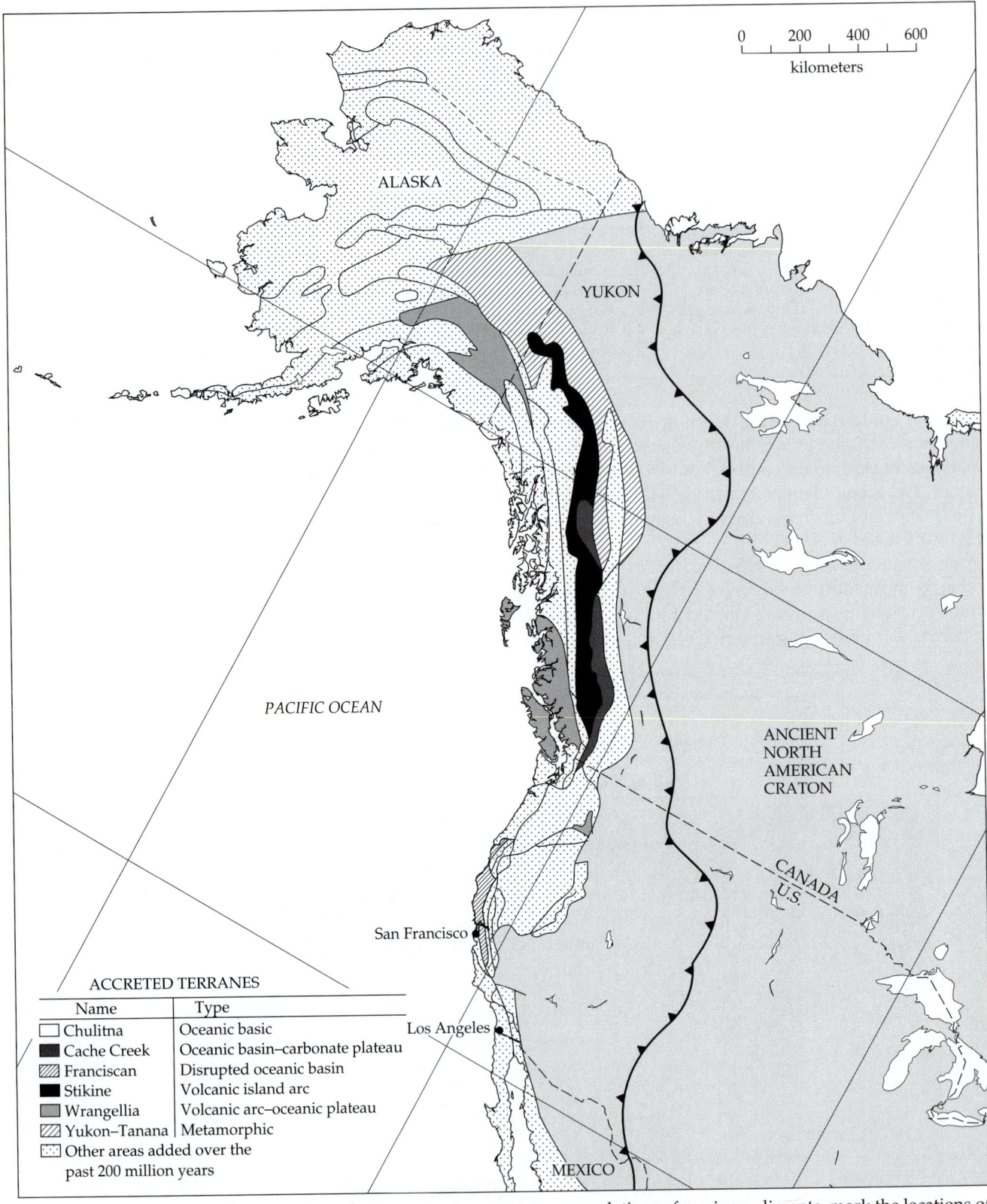

0   200   400   600

kilometers

ALASKA

YUKON

PACIFIC OCEAN

ANCIENT
NORTH
AMERICAN
CRATON

CANADA
U.S.

San Francisco

Los Angeles

MEXICO

ACCRETED TERRANES

Name	Type
Chulitna	Oceanic basic
Cache Creek	Oceanic basin–carbonate plateau
Franciscan	Disrupted oceanic basin
Stikine	Volcanic island arc
Wrangellia	Volcanic arc–oceanic plateau
Yukon–Tanana	Metamorphic
Other areas added over the past 200 million years	

Figure 6.14   Terranes, or accumulations of marine sediments, mark the locations of earlier subduction zones along the west coast of North America. Over the past 200 million years, oceanic plates slipped beneath the relatively buoyant continental plates, but islands, seamounts, and other superficial features were scraped off and added to the continental plate. (After Jones et al. 1982.)

Plates grind past each other at the San Andreas fault. While this may be the most notorious example, transform faults are quite common. Some are closely associated with major plate boundaries (e.g., those along the margins of the Scotia and Antarctics Plates, along the southern tip of South America), while others are found at great distances from plate boundaries (e.g., the Altyn Tagh fault of western China).

Earth's Tectonic History

The tectonic processes described above have recurred throughout the earth's 4.5-billion-year history, and will certainly continue to modify the configuration of its surface. It is hard to overemphasize the effects of plate tectonics on biogeographic patterns of virtually all organisms. The origin, spread, and radiation of many taxa took place when the earth's surface was dramatically different from its current profile. These processes, along with the many extinctions evidenced in the fossil record, were often associated with the collision, separation, or destruction of continents or oceanic basins, which altered opportunities for biotic exchange or affected climate on regional to global scales.

In the following sections, we summarize the current view of earth's tectonic history, focusing on the dynamics of its continents and marine basins during the Phanerozoic. As you can imagine, the detective work involved in this 600-million-year reconstruction, while fascinating, is a great challenge. Over the past three decades, many teams of geologists and biogeographers have taken up this challenge. Paleomagnetic data can show us where the landmasses were at any given time, and **suture zones** show where past oceans disappeared and landmasses were welded together (Burke et al. 1977; Dewey 1977). While there remains disagreement over many details and the precise timing of particular events, a general consensus seems to have emerged about the salient features associated with the formation and ultimate disintegration of the great continent of the Permian period, Pangaea (Figure 6.15).

Tectonic History of the Continents

Gondwanaland, Laurasia, and the formation of Pangaea.   Supporters of the theory of continental drift originally believed that prior to the Mesozoic, all landmasses had been continuously united in a supercontinent, Pangaea (Figure 6.15B–D), which occupied one-third of the earth's surface. Now geophysicists tell us that Pangaea was a more temporary structure that probably existed as a single unit only during late Paleozoic and early Mesozoic time (Figure 6.15D). Its northern half, Laurasia, had a very complex early history that was substantially different from that of the southern half, Gondwanaland.

Gondwanaland included the foundations of present-day South America, Africa, Madagascar, Arabia, India, Australia, Tasmania, New Guinea, New Zealand, New Caledonia, and Antarctica. It was, by far, the most ancient of the Pangaean landmasses, forming some 650 million years B.P. during the Precambrian era. From this time through the middle Ordovician period (about 475 million years B.P.), Gondwanaland remained a relatively continuous supercontinent in the Southern Hemisphere, although it drifted substantially between the equator and the South Pole (see Figure 6.7B). If one examines the history of Queensland, Australia, now at 12° S latitude, one discovers that it was located near the North Pole in the Proterozoic (1 billion years B.P.), at the equator in the Silurian, and at 40° S in the Mesozoic (Embleton 1973). The great antiquity of Gondwanaland accounts in part for the similarity of the now isolated biotas of the Southern Hemisphere.

(A) Late Silurian (425 million years B.P.)

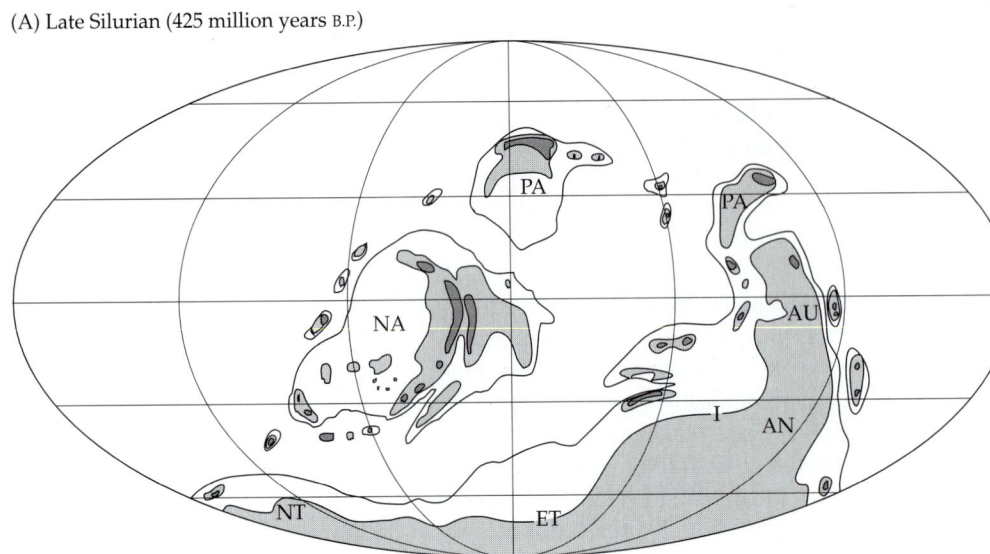

(B) Late Devonian (363 million years B.P.)

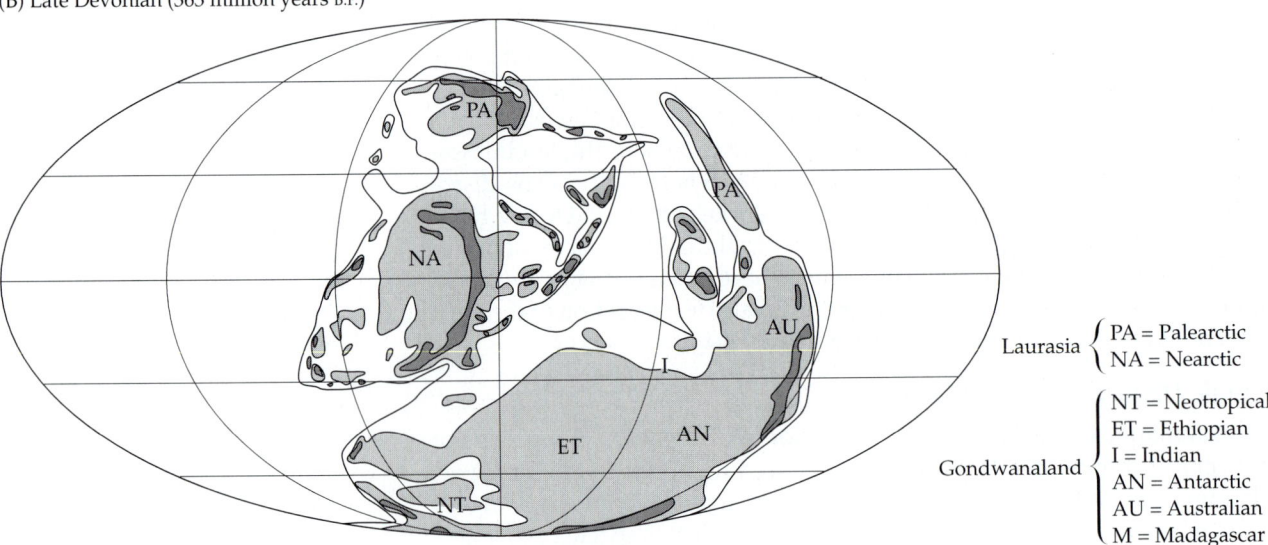

(C) Late Carboniferous (306 million years B.P.)

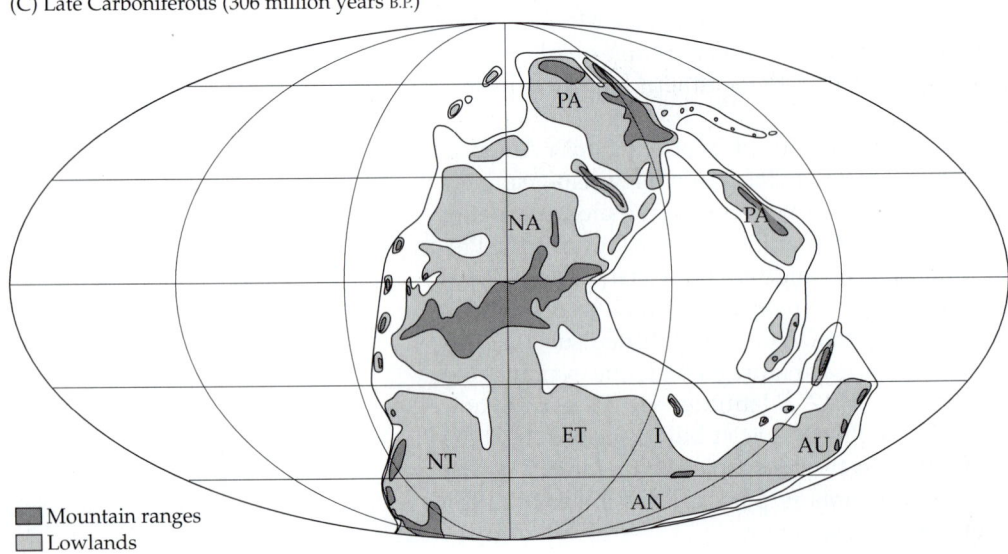

Figure 6.15 Continental drift and paleotopography of the earth's landmasses from the late Silurian (425 million years B.P.) to the present. [After Chris Scotese of the PALEOMAP project, University of Texas at Arlington; for information on digital images and other PALEOMAP products, contact via internet (www.scotese.com), email (chris@scotese.com), or toll free (888) 288-0160.]

Laurasia $\begin{cases} \text{PA = Palearctic} \\ \text{NA = Nearctic} \end{cases}$

Gondwanaland $\begin{cases} \text{NT = Neotropical} \\ \text{ET = Ethiopian} \\ \text{I = Indian} \\ \text{AN = Antarctic} \\ \text{AU = Australian} \\ \text{M = Madagascar} \end{cases}$

- ▪ Mountain ranges
- ▪ Lowlands
- ▫ Continental shelf

(D) Late Permian (255 million years B.P.)

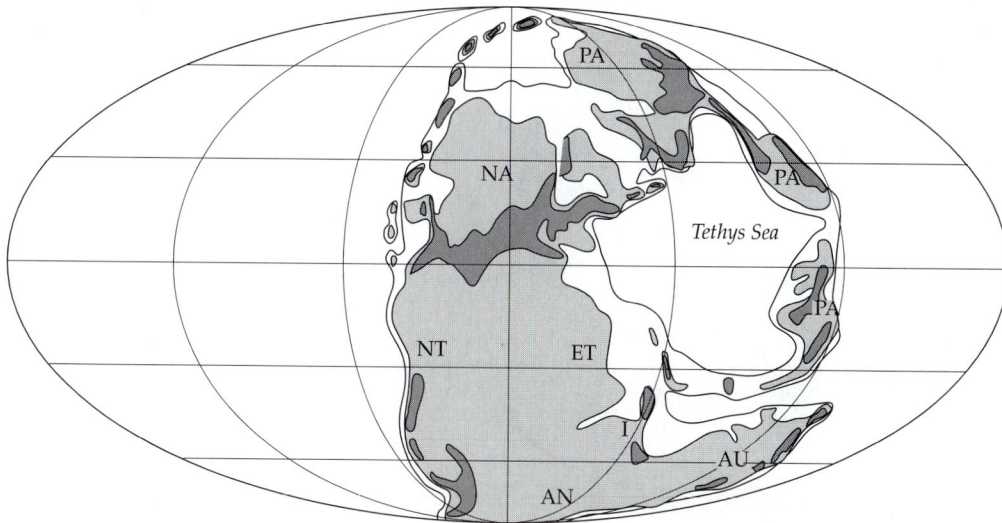

(E) Early Jurassic (195 million years B.P.)

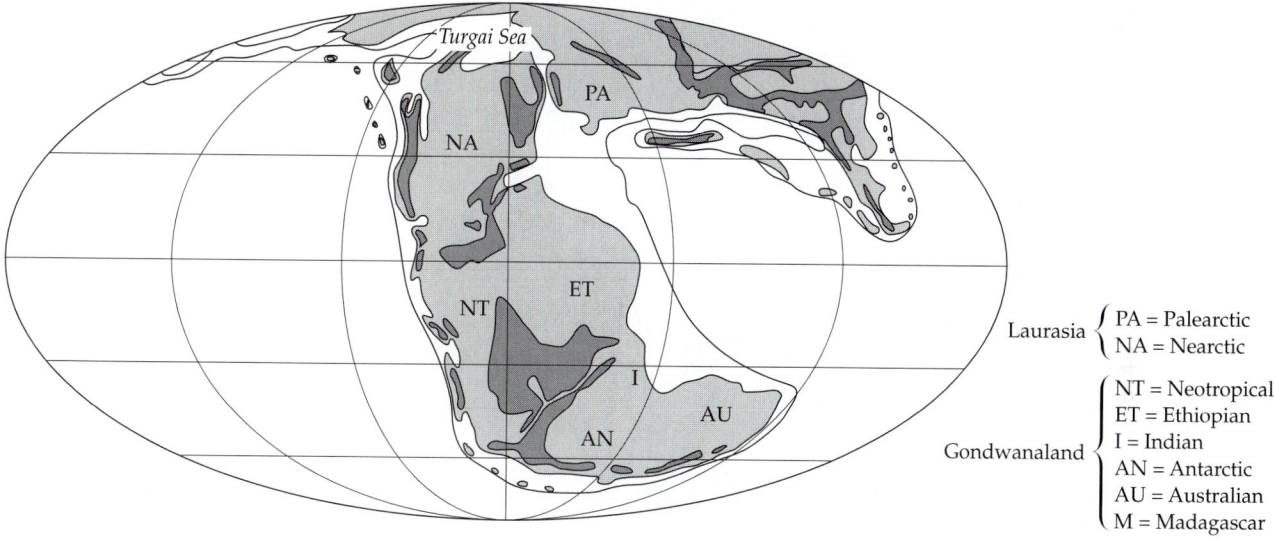

Laurasia $\begin{cases} \text{PA = Palearctic} \\ \text{NA = Nearctic} \end{cases}$

Gondwanaland $\begin{cases} \text{NT = Neotropical} \\ \text{ET = Ethiopian} \\ \text{I = Indian} \\ \text{AN = Antarctic} \\ \text{AU = Australian} \\ \text{M = Madagascar} \end{cases}$

(F) Late Jurassic (152 million years B.P.)

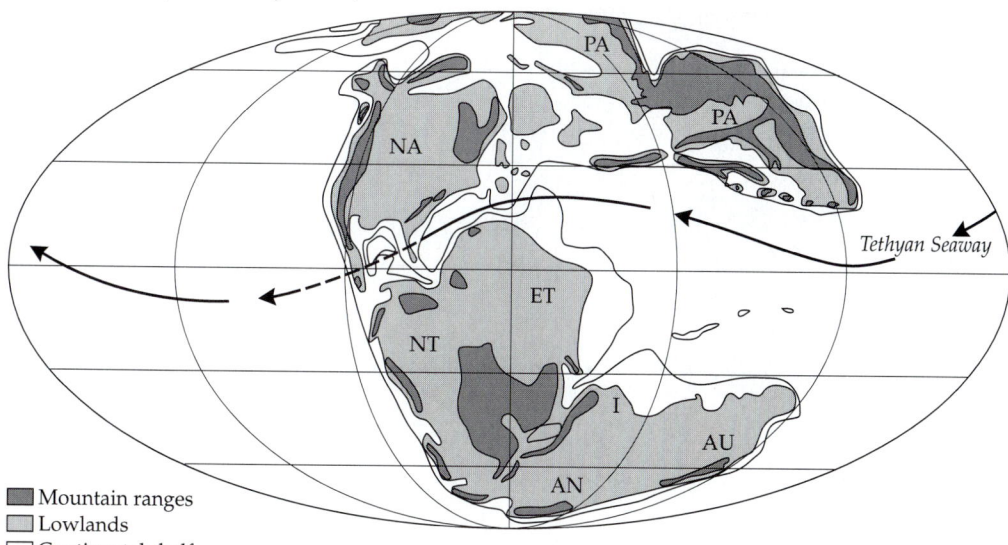

■ Mountain ranges
▨ Lowlands
□ Continental shelf

(G) Early Late Cretaceous (94 million years B.P.)

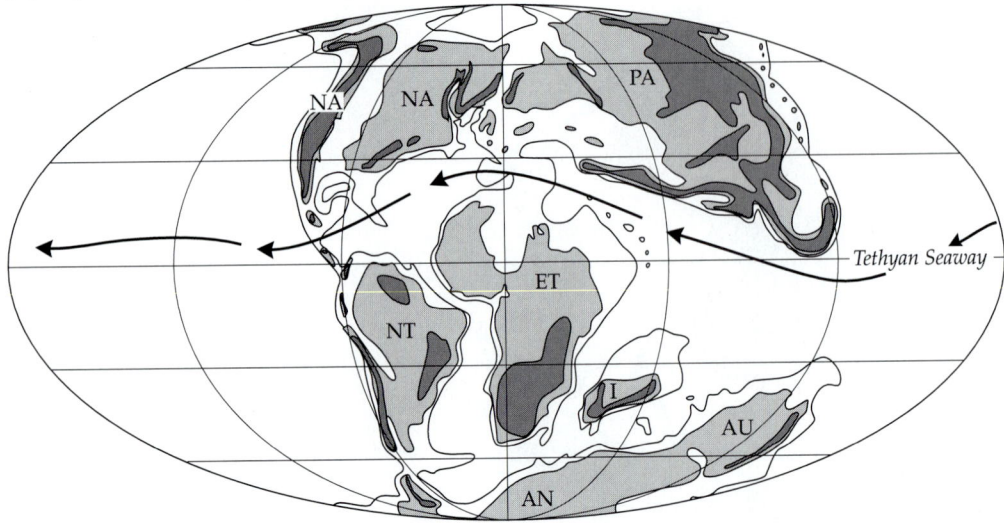

(H) Middle Eocene (50 million years B.P.)

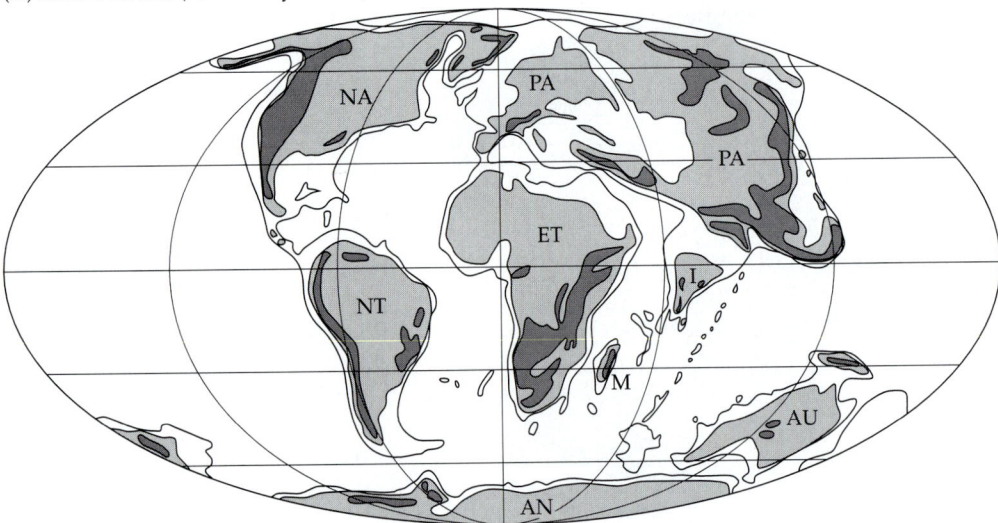

(I) Middle Miocene (14 million years B.P.)

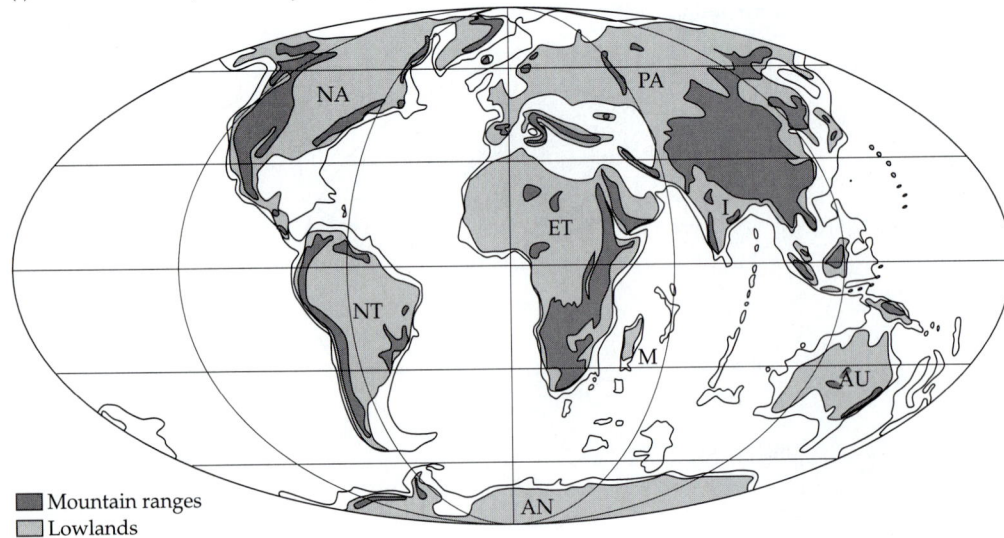

■ Mountain ranges
▨ Lowlands
☐ Continental shelf

(J) Modern World

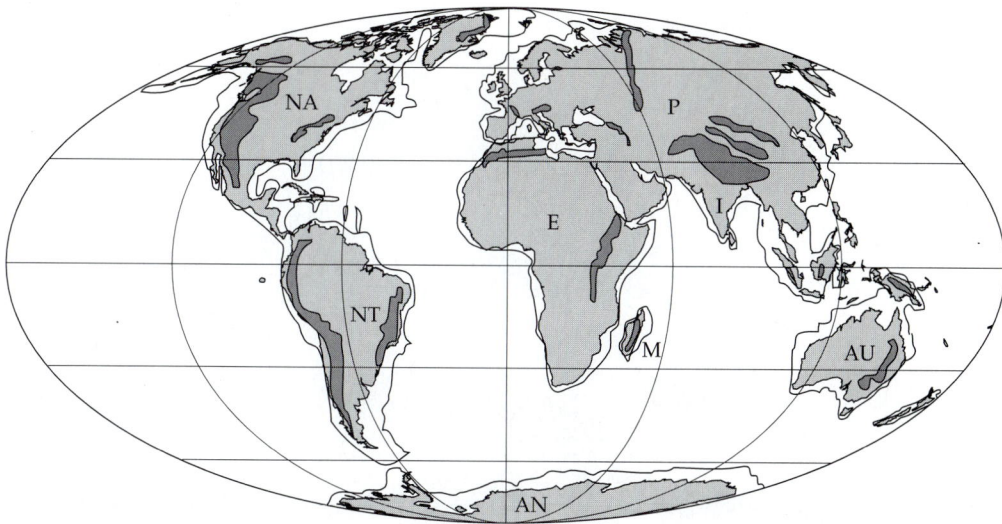

(K) Future (+150 million years)

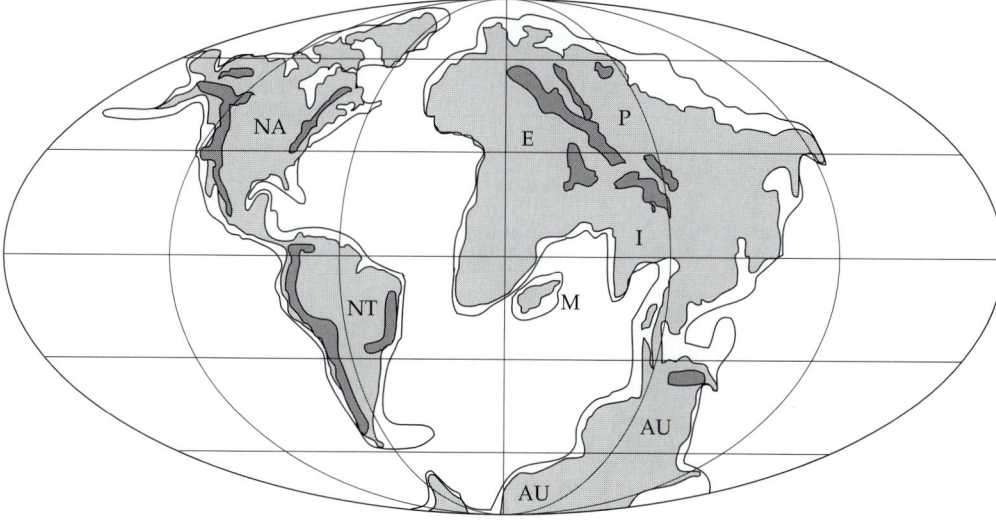

(L) Future (+250 million years)

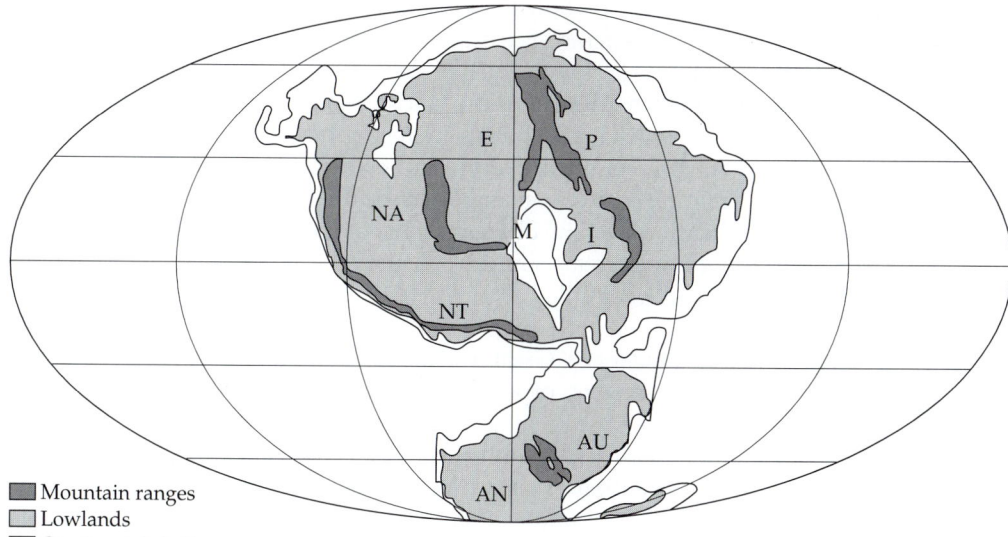

■ Mountain ranges
▨ Lowlands
□ Continental shelf

Figure 6.16 The land-masses that constituted the Old Sandstone Continent of the late Devonian (365 million years B.P.). Note that reefs covered most of the region that eventually formed North America, and that regions that now lie in temperate zones (such as Germany and the Appalachians) once lay near the equator. (From Stanley 1987.)

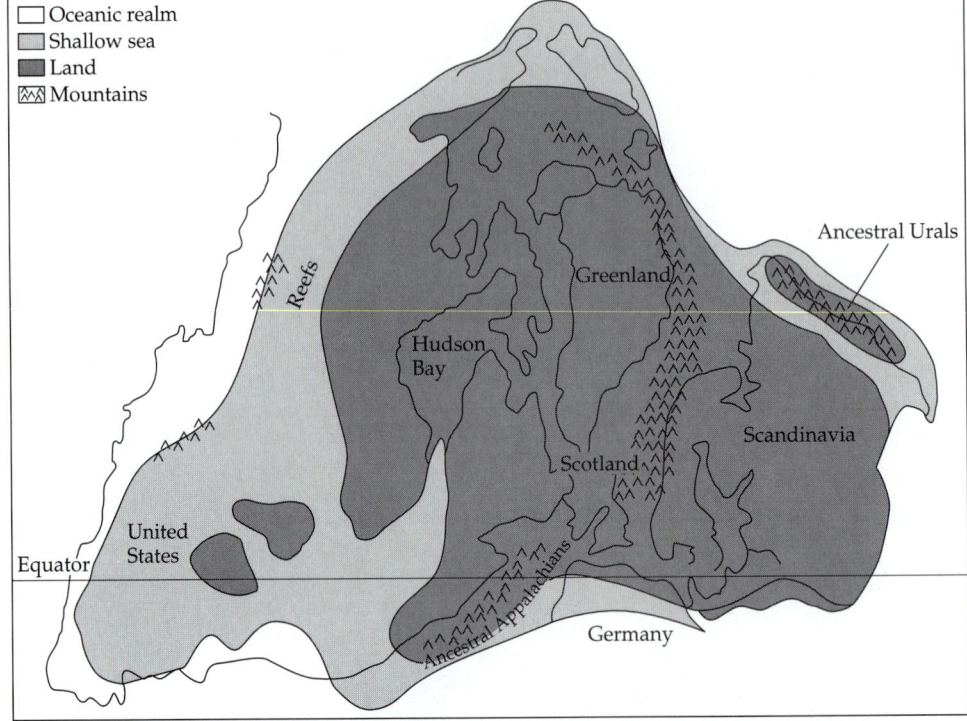

In contrast to its ancient southern counterpart, the history of Laurasia is much more recent. Pre-Laurasian landmasses remained isolated until the early Devonian (about 400 million years B.P.), when the precursors of Northern Europe, which once occupied subantarctic latitudes, drifted northward across the equator to collide with the precursors of North America and the northern part of Siberia, forming the "Old Sandstone Continent" (Figure 6.16). The precursors of China and southern Siberia remained isolated in subtropical latitudes of the Northern Hemisphere, far from Gondwanaland, which was then situated near the South Pole. During the Carboniferous period, Gondwanaland began drifting northward, and eventually united with the Old Sandstone Continent about 300 million years B.P. Later, during the early Permian (about 270 million years B.P.), western Asia collided with Europe to form the Ural Mountains and unite the bulk of Pangaea. At this time, eastern Asia still remained isolated as several large islands, which partially divided the world's oceans and encircled the Tethys Sea along the eastern margin of Pangaea (Figure 6.15D). By the late Permian, these landmasses consolidated, and Pangaea formed a great, continuous landmass stretching from pole to pole.

Although this fact is often overlooked, coincident with the formation of Pangaea, the marine basins were united to form one great ocean, **Panthalassa.** Thus, the Permian was a time of great connectivity and exchange among both terrestrial and marine biotas, many of which had evolved in isolation for hundreds of millions of years.

The breakup of Pangaea.   Great evolutionary radiations of most terrestrial families and many orders, especially among vertebrates, seed plants, and insects, occurred after the breakup of Pangaea and Panthalassa. In fact, the diversification of many taxa may have been a direct outcome of this breakup and the resultant isolation of terrestrial and marine biotas. As the continents fragmented and drifted toward different latitudes and climates, reduced gene flow and altered selective pressures allowed rapid speciation and radiation of their rafting biotas. Again, the history of tectonic events provides invaluable insights for interpreting biogeographic patterns, both past and present.

The breakup of Laurasia and its rifting from Gondwanaland. The breakup of Pangaea was actually initiated in the early Jurassic (about 180 million years B.P.) when the Turgai Sea (Figure 6.15E) expanded southward from the Arctic to split Asia from Euramerica (i.e., eastern from western Laurasia). By the mid-Cretaceous (about 100 million years B.P.), a shallow sea separated western North America from Europe and eastern North America, dividing Laurasia into three landmasses (Figure 6.15F–G). By this time, active seafloor spreading had begun to open the Atlantic, and the remaining narrow landbridges between Laurasia and Gondwanaland (near Central America and North Africa) were broken, transforming the Tethys Sea into a circum-equatorial seaway.

By the late Cretaceous (75 million years B.P.), a land connection between the Laurasian landmasses had been reestablished across the Bering Strait (Figure 6.15G–H). Thus, at about the same time that one land connection between eastern North America and western Europe was severed, a new one formed between western North America and Asia, permitting continued exchange of organisms. This connection, **Beringia**, persisted through most of the Cenozoic, although an uninterrupted landbridge was not continually present (see Chapter 8). Transient connections among the remnants of Laurasia (between North America and Europe) also existed across the Greenland-Scotland Ridge during the early Tertiary (i.e., the Thulian and DeGeer routes). Paleontological evidence indicates that these connections across the North Atlantic were important corridors for the migration of animals and plants (McKenna 1972a,b; Sclater and Tapscott 1979; Barron et al. 1981). Earlier, biotic exchange also occurred across transient land connections that often linked the British Isles with Greenland and Canada during the early Cretaceous (Hallam and Sellwood 1976). As a result of this long history of exchange between the Nearctic and Palearctic regions, they share many species of plants and animals, and have often been coupled as a superregion, the **Holarctic**.

Other areas of the Nearctic have a much more recent history. As North America drifted away from Europe, its southeastern coast, including land from New Jersey to Yucatán, Mexico, became submerged to form a great sedimentary basin. The Atlantic Gulf Coast was under water from the late Jurassic until the Eocene, so all the forms living there today are recent colonists.

The breakup of Gondwanaland. Roughly simultaneously with the initial breakup of Laurasia, the landmasses of Gondwanaland began to separate. According to one recent review by Storey (1995), the separation of Gondwanaland proceeded in three stages (Figure 6.17). First, during the early Jurassic (about 180 million years B.P.), rifting opened the Somali, Mozambique, and Weddel Seas between eastern and western Gondwanaland. This initial rifting was followed by seafloor spreading, which united these relatively narrow seas and split Africa and South America from the remainder of Gondwanaland (Madagascar, India, Australia, Antarctica, and New Zealand) about 160 million years B.P. At this time, Madagascar and India were adjacent to the northeastern margin of Africa, while Australia was located to the south along the eastern margin of Antarctica in a position rotated approximately 90° clockwise from its present position (Figure 6.17B).

During the second stage of the breakup, Madagascar and India were separated from the Antarctica-Australia landmass by a narrow seaway (about 130 million years B.P.). Active seafloor spreading had begun by about 125 million

Figure 6.17   Stages in the breakup of Gondwanaland. Bold lines show zones of seafloor spreading. (After Storey 1995.)

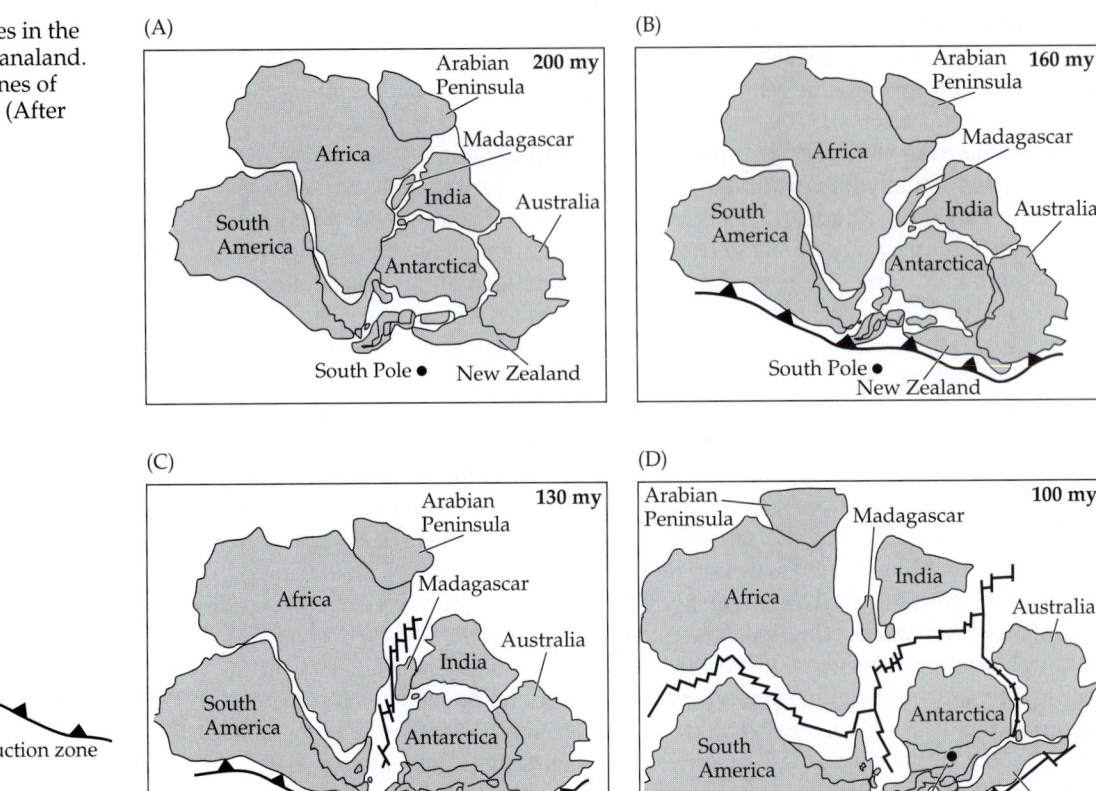

Subduction zone

Active seafloor spreading

years B.P., further isolating India and Madagascar from Antarctica and Australia (Figure 6.17C). Also at about this time, South America and Africa rifted apart, thus forming four Gondwanan landmasses (South America, Africa, Madagascar-India, and Antarctica-Australia-New Zealand). Finally, during the mid-Cretaceous (about 100 million years B.P.), Australia and New Zealand rifted from Antarctica, and extensive seafloor spreading continued to disperse the remnants of Gondwanaland. Later, during the Paleocene (about 60 million years B.P.), the Mascarene basin expanded to separate Madagascar and the Seychelles archipelago from India (Figure 6.15G–H). Throughout this period, India's migration northward was very rapid, averaging 15 cm per year. At about 60 million years B.P., India collided with Eurasia, initiating the uplift of the Himalayas (see Figure 6.15H–I; Butler 1995).

Much farther to the south, New Zealand retained a connection to Antarctica and what is now the eastern margin of Australia until about 90 million years B.P. New Zealand broke away, first from Australia and then from Antarctica, about 80 million years B.P., drifting farther east of Australia during the development of the Tasman Sea. While other remnants of Gondwanaland were drifting toward the lower latitudes, Antarctica continued its slow poleward journey. By the Miocene, Antarctica was situated over the South Pole, triggering the development of the great polar icecap.

History of Central America and the Antilles.   No matter what brand of historical biogeography you practice, the history of Mesoamerica and the Caribbean region is a complex and intriguing case study. The reasons for this are obvious. First, biogeographers need to know how much opportunity there was for the exchange of organisms between the Nearctic and Neotropical regions following the breakup of Pangaea. Second, we need to know whether the rich biota of the Caribbean islands, especially the Greater Antilles, arrived there exclusively by long-distance dispersal, as suggested by Darlington (1938) and many others, or whether the islands are remnants of an ancient landmass that at one time was continuous with the mainland (i.e., either North or South America).

Following the formation of Pangaea, the precursors of North and South America remained connected until about 160 million years B.P. The ancient connection between Mexico proper and South America was apparently severed by the late Jurassic (approximately 150 million years B.P.). During the early Cretaceous (120 to 140 million years B.P.), a chain of volcanic islands began to emerge along the eastern edge of the Caribbean Plate. By the late Cretaceous, this plate had drifted eastward so that this archipelago, called the Proto-Antilles, was in a position to serve as a stepping-stone route for limited exchange of some terrestrial organisms between the Nearctic and Neotropical regions (Figure 6.18). The Proto-Antilles continued to drift eastward along the leading edge of the Caribbean Plate to eventually form the Greater Antilles, the Caribbean Mountains of Venezuela, and the ABC (Aruba, Bonaire, and Curaçao) islands off the coast of Venezuela. Some small fragments of the Proto-Antilles may also have been displaced to the region of the Lesser Antilles. By Eocene times (58 million years B.P.), the core of the Greater Antilles had achieved their present positions.

The formation of the current landbridge between North and South America was an important event for New World biogeography. Waves of migration across this landbridge, called the **Great American Interchange**, profoundly affected the diversity and composition of the Nearctic and Neotropical faunas (see Chapter 16). Like the Proto-Antilles, the Central American landbridge may have first emerged during the late Cretaceous (80 to 65 million years B.P.) as a chain of islands far west of its current position. It then drifted eastward with the boundary between the Cocos and Caribbean Plates (Briggs 1994; Figure 6.18). At the end of the Cretaceous (65 million

Figure 6.18   One possible reconstruction of tectonic events in the Caribbean. Central America first formed as an archipelago in the Pacific Ocean during the early Cretaceous (120 to 140 million years B.P.), then continued to drift eastward along with the Caribbean Plate and the Proto-Antilles. The Central American archipelago eventually drifted to its position between the two continents by the mid-Miocene (20 million years B.P.), but did not form a complete landbridge until the late Pliocene (around 3.5 million years B.P.). (After Briggs 1994.)

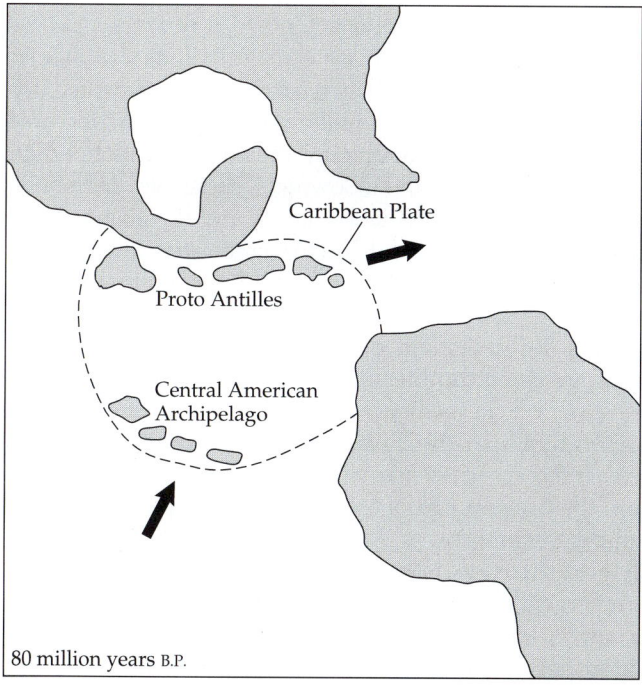

Caribbean Plate

Proto Antilles

Central American Archipelago

80 million years B.P.

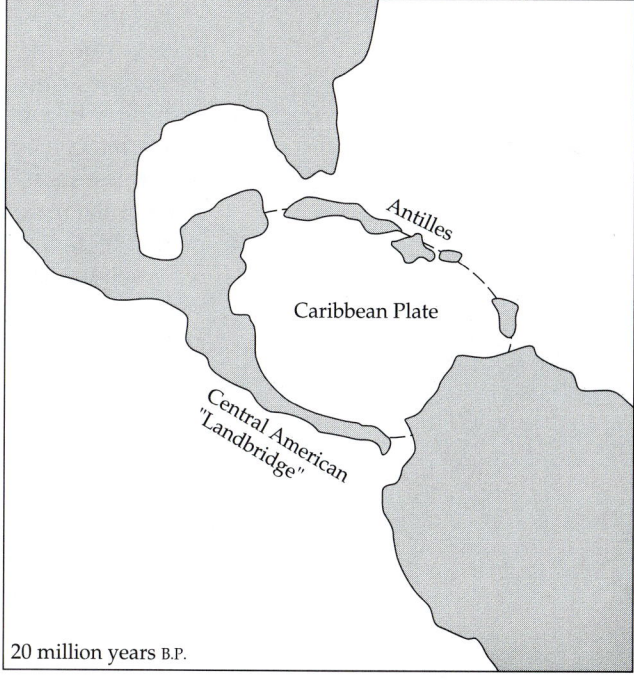

Antilles

Caribbean Plate

Central American "Landbridge"

20 million years B.P.

years B.P.), the sea level fell substantially to expose a continuous, albeit narrow, landbridge. This scenario remains hypothetical, but it is consistent with the available biogeographic evidence, which suggests isolation of marine invertebrates on either side of the Central American rise simultaneous with the Paleocene exchange of some mammals, lizards, frogs, and freshwater fishes between Neotropical and Nearctic regions (see Gayet et al. 1992; Briggs 1994). This hypothesized Cretaceous-Paleocene landbridge, if it ever existed as a continuous strip of land, must have quickly submerged beneath rising waters during the late Paleocene.

The final emergence of Central America in the Neogene was produced by convergence of the Cocos, Nazca, and Caribbean Plates. By the late Miocene (10 to 5 million years B.P.), the Central American archipelago provided a stepping-stone route for the dispersal of a variety of terrestrial organisms. Approximately 3.5 million years B.P., the archipelago finally fused to form the current Central American landbridge between the Nearctic and Neotropical regions (see Robinson and Lewis 1971; Coney 1982).

Tectonic Development of Marine Basins and Island Chains

We commented earlier on the unfortunate tendency of biogeographers to ignore the tectonics of the marine realm, except as it relates to the biogeography and paleontology of terrestrial biotas. Yet the marine realm, in addition to creating major barriers to dispersal for terrestrial organisms and influencing climates on continents and islands, constitutes most of the biosphere. As we noted earlier, when the earth's landmasses were consolidated to form Pangaea, its marine basins also were joined to form one world ocean, Panthalassa. As the Pangaean landmasses split apart, seafloor spreading created new marine basins at the expense of Panthalassa, which was fragmented and isolated by the drifting continents. Just as terrestrial climates are strongly influenced by adjacent oceans, changes in the positions of continents changed paleocurrents dramatically, strongly influencing the temperatures and basic chemistry of marine waters. Often throughout the earth's history, these climatic changes have produced glacial cycles, which lower sea levels during their peak phases and raise sea levels to flood the continents during glacial minima.

Again, a sound understanding of its unique tectonic history is essential for anyone studying the biogeography of the marine realm. While continental plates have drifted, broken apart, and united, they have neither been created nor destroyed. In contrast, the ocean floor is continually created at the midoceanic ridges and consumed at the trenches. Consequently, most of the present ocean floor is less than 50 million years old. In the following sections we summarize some of the major events associated with the formation and disintegration of marine basins and shallow, epeiric seas, and discuss associated phenomena, including the formation of island chains and the dynamics of marine currents.

Epeiric seas.   **Epeiric seas**, also called **epicontinental seas**, are formed when sea levels rise and the oceans flood continental plates. We happen to be living today in one of the driest periods in the history of the earth. Anyone who has hunted for fossils in the interior of the United States or Eurasia, however, knows that vast seas covered these areas many times, leaving thick deposits of limestone, formed from the calcium carbonate of ancient coral reefs and the fossilized remains of marine organisms.

Epeiric seas usually act as barriers to terrestrial organisms, subdividing a landmass into smaller emergent regions. In the early Cretaceous (about 50 million years B.P.), for example, epeiric seas split Australia into three subcontinents. Such events had important biogeographic consequences for biotic differentiation and extinction (see Chapters 7 and 10). Some of the more interesting examples since the Paleozoic are the marine transgression in North America from the Gulf of Mexico as far west as Arizona, and the subdivision of mainland Africa (Cooke 1972) (see Figure 6.15H–I). The mid-American seaway dried up by the Paleocene (about 65 million years B.P.), while the Turgai Sea dried up during the Oligocene (about 30 million years B.P.), again allowing biotic exchange. India is one of the few large regions that has had relatively little change in its coastal outline throughout the Phanerozoic.

On a smaller scale, we can find important changes in river systems. In South America, the Amazon had a complex history, changing its drainage several times, from westward in the early Cenozoic prior to the uplift of the Andes to its present drainage eastward into the Atlantic. River systems such as the Mississippi and the St. Lawrence in the United States also have changed course many times. In the western United States, the Great Basin, which includes much of Utah, Nevada, and eastern California, now has no drainage to the ocean, only numerous alkali sinks and dry lake and river beds. However, during wetter periods over the last million years, there were connections to both the Colorado River in the south and the Columbia River (via the Snake River) in the north.

Formation of the Mediterranean and Red Seas.   After being freed from North America and South America, Africa began to swing counterclockwise toward Eurasia and closed the formerly extensive Tethys Sea. A bridge was formed between Asia and Africa through Arabia following their collision in the middle Tertiary (35 million years B.P.), which not only created the Zagros Mountains of Iran, but also confined the Mediterranean Sea along its eastern margin. As Africa approached southern Europe, a number of deformations were initiated, including the counterclockwise rotation of Italy, which produced its characteristic diagonal orientation (McElhinny 1973a; see also Hsü 1972; Biju-Duval and Montadert 1977; Bureau de Recherches Géologiques et Minières 1980a,b). By the early Miocene (24 million years B.P.), the Straits of Gibraltar had closed, causing desiccation and a 6% increase in salinity in the Mediterranean.

While the Mediterranean formed through the closure of a once expansive, circum-equatorial seaway, the Red Sea formed de novo through the rifting of continental landmasses. The Red Sea, lying at the eastern margin of the Mediterranean, provides a recent example of a marine basin created by rifting, which began approximately 35 million years B.P. (Figure 6.19). Seafloor spreading then accelerated during the early Miocene (21 to 25 million years B.P.). The Straits of Gibraltar rifted apart during the late Miocene to open the Mediterranean Sea to the Atlantic. Rifting proceeded more rapidly in the south, resulting in an angular separation of the African and Arabian Plates hinged along their northern juncture. This created a large basin continuous with the Indian Ocean to the south, but isolated from the Mediterranean. The rift system that created the Red Sea seems continuous with Africa's Great Rift Valley. If current trends continue, the tectonic events creating the Red Sea may be repeated in eastern Africa, with its great lakes eventually connecting to form an extensive inland seaway continuous with the Indian Ocean. The biota of eastern Africa would then drift in isolation, much like that of India during its northward migration from Gondwanaland, providing a fascinating natural biogeographic experiment.

Dynamics of the Pacific Ocean.   Early articles on continental drift focused mainly on the continental plates, but soon considerable attention was focused

Figure 6.19   Tectonic activity in the Great Rift Valley system of eastern Africa. Arrows indicate directions of plate movements; rates of movements are in cm per year. The Rift Valley and its Great Lakes formed as the African Plate separated and rotated away from the Somali Plate. (After Jordan 1971).

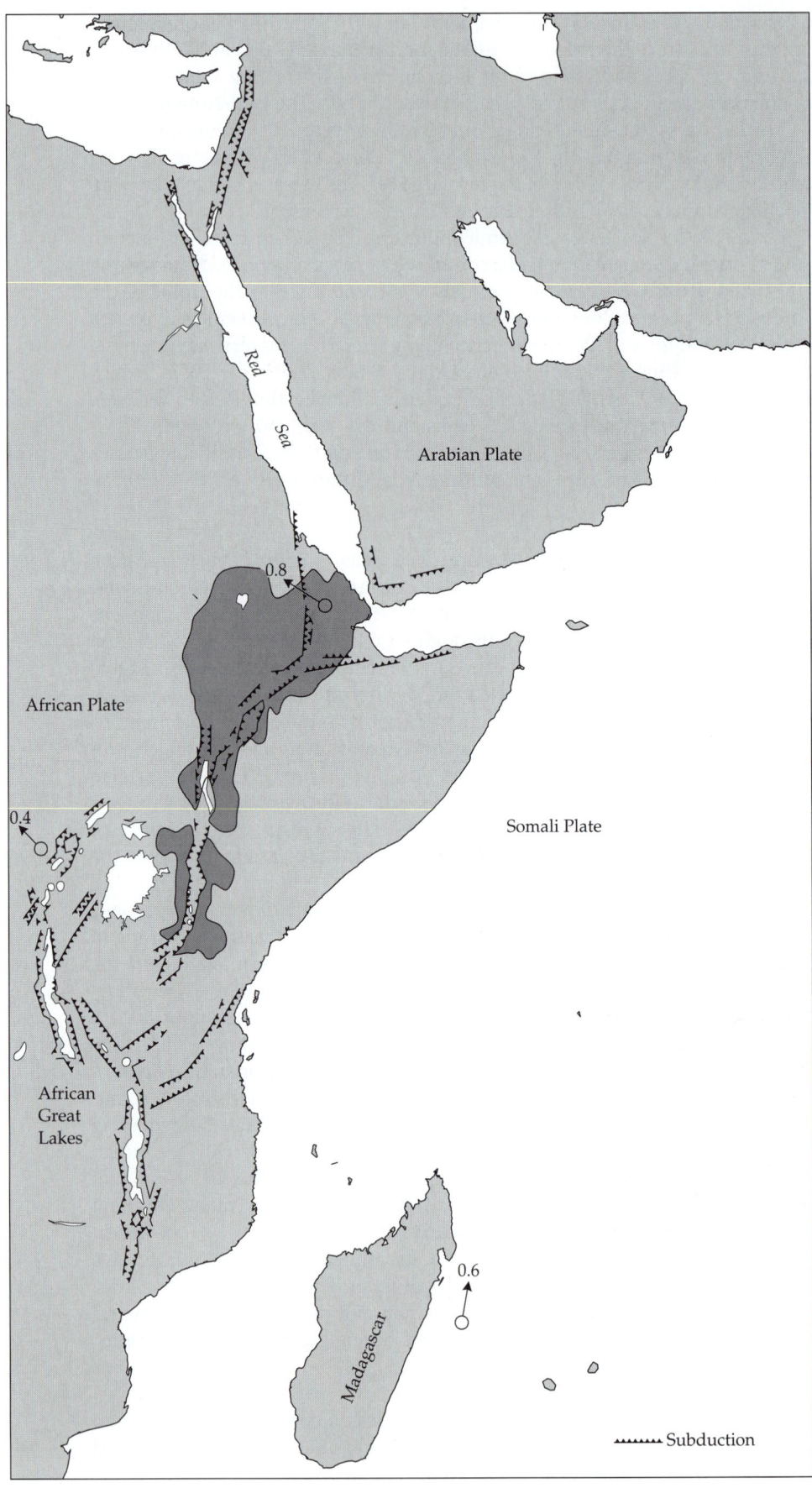

on the oceanic plates in the Pacific basin, which contain relatively little emergent land. It may surprise you to learn that our largest ocean, the Pacific, has been getting smaller; it once was part of the continuous global ocean, Panthalassa, which covered two-thirds of the globe when Pangaea existed.

Hot spots and triple junctions. In the basic model of seafloor spreading, volcanic islands are formed either at midoceanic ridges in files perpendicular to the ridge, or behind oceanic trenches as parallel arcs of islands (see Figure 6.6). However, in the Pacific Ocean, there is no single central ridge, and many of the islands are not associated with ridges or trenches. The Hawaiian Islands are a perfect example: this narrow, linear island chain is located far from any ridge or trench. The oldest emergent island is Midway, and the youngest is Hawaii, which remains volcanically active. J. Tuzo Wilson (1963a) proposed a unique mechanism to explain this pattern, which he called a **hot spot**: a fixed weak spot in the mantle at which magma is released. As the oceanic plate passes over this hot spot, volcanoes are produced at the surface, causing the formation of islands (Figure 6.20). The spot is very narrow, so the islands form in a chain, which is linear as long as the plate moves strictly in one direction. The youngest islands in the chain are emergent, and the progressively older ones gradually become submerged (see also Jarrard and Clague 1977).

One nice feature of Wilson's model was that it explained the occurrence of most islands in the eastern Pacific. The Hawaiian Islands, many of which are still emergent, are really only the most recent part of a long series extending to the Aleutian Trench, including the submerged Emperor seamount chain (Mor-

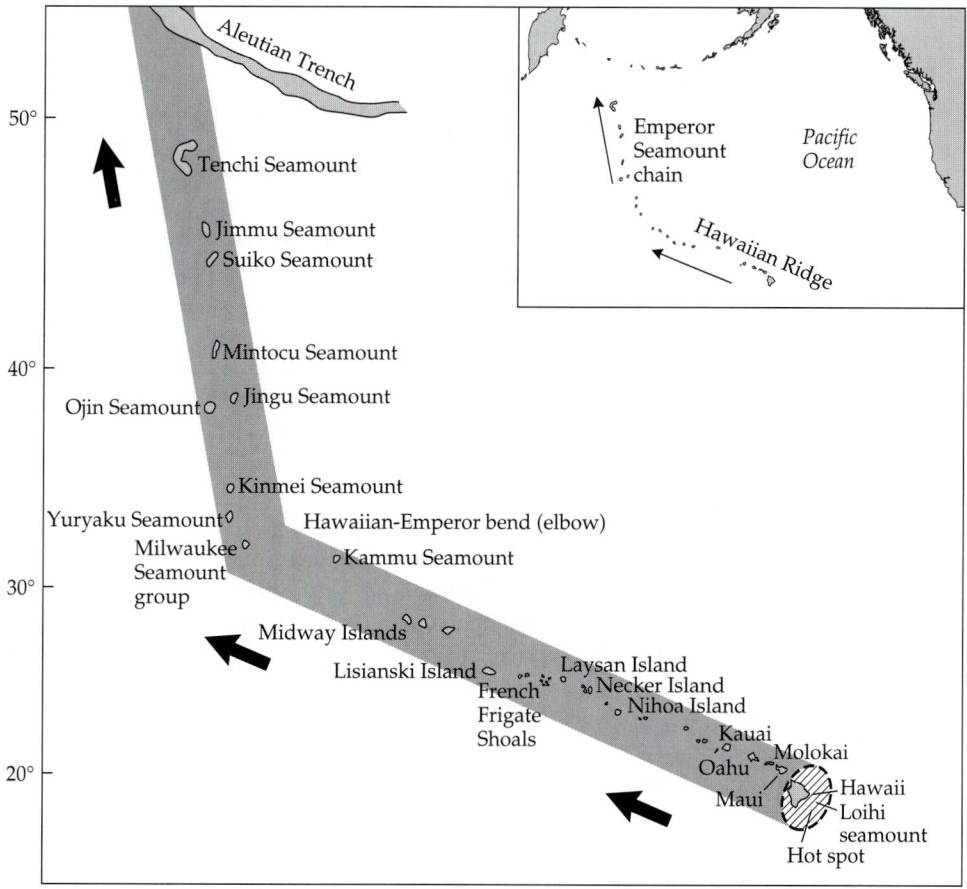

Figure 6.20 A hot spot has produced the volcanic Hawaiian Islands and the Emperor seamount chain in the Central Pacific. The hot spot, presently located near the Hawaiian Deep, is presumably responsible for the intense volcanic activity on the island of Hawaii. Several authors agree that these islands were created as their locations on the Pacific Plate successively drifted over the hot spot. The older islands to the north were formed first; they have disappeared beneath the surface to become seamounts, and the oldest have disappeared into the Aleutian Trench. The sharp bend between the Emperor and Hawaiian chains occurred when the drifting oceanic plate changed direction about 44 million years B.P. (After Jackson et al. 1972.)

gan 1972a; Dalrymple et al. 1973). The seamount closest to the trench is naturally the oldest (about 80 million years B.P.). Morgan (1972a,b) noted that three very similar chains of emergent and then submerged volcanoes occur in the Pacific basin: the Tuamoto Archipelago (including Easter Island) and the Line Islands chain, the Austral chain and the Marshall-Gilbert-Turalu island chain, and a chain from the Cobb Seamount opposite Washington through the Gulf of Alaska. Another interesting quality of three of these four chains is that the recent islands are all roughly parallel; then there is a sharp bend (elbow), so that all the older sections are parallel again (see Figure 6.20). This is the pattern you would expect if a rigid plate started moving in a different direction as it passed over a stationary hot spot. The Hawaiian-Emperor elbow is dated at 44 million years B.P. Consequently, we can use the distance and orientation of the chains to reconstruct their movements through geological time.

Volcanic islands may also be formed at a **triple junction**, a place where three plates rest against each other to form a complex of trenches (McKenzie and Morgan 1969). Volcanoes are often associated with triple junctions. Because each plate has its own rate of subduction into the trench, the triple junction and its associated volcanic activity can shift position through time, forming an **island arc**, or archipelago. An example of an archipelago formed in this manner is the Galápagos, which originated near the intersection of the Nazca, South America, and Cocos Plates.

Inferred history of the Pacific basin.   Using the data on hot spots and triple junctions, many researchers have attempted to reconstruct the early history of the Pacific basin. Most such reconstructions recognize six ancient plates. The Phoenix Plate was located in the southern Pacific and disappeared due to the enlargement of the Pacific Plate, and the Kula Plate has disappeared under Alaska and Siberia. A triple junction existed at the intersection of the Kula, Pacific, and Farallon Plates, which migrated to come into contact with the North American Plate; at 150 million years B.P., this junction was just east of the Hawaiian hot spot. Most of the Farallon Plate has been subducted under western North America, and is responsible for much of the mountain building there, but small pieces of this plate remain in the vicinity of the Pacific Northwest (the Juan de Fuca Plate and its southern extension, the Gorda Plate) and opposite central mainland Mexico (called the Rivera Plate). The Pacific Plate also became fused with a section of mainland Mexico to become Baja California, while its northward movement began the formation of the Gulf of California. This gulf has been spreading over the last 4 million years. The convergence of the Cocos and Caribbean Plates has produced the present mountainous backbone of Central America, and the meeting of the Nazca Plate and South America induced the formation of the Andean Cordillera, mostly since the Oligocene.

Paleoclimates and paleocirculations.   As discussed in Chapter 3, global circulation patterns of wind and ocean currents are controlled not only by the equatorial to polar gradients of solar radiation and temperature, but also by the relative proportions and distributions of land and water. Given the great shifts, splittings, collisions, submergences, and emergences of landmasses that occurred, oceanic currents, and thus terrestrial climates, must have changed dramatically through the Phanerozoic eon, strongly affecting both marine and terrestrial biotas.

During the Permian, gyres and surface currents in the Tethys Sea and throughout Panthalassa were replaced by circulation through a warm, tropical, circum-equatorial Tethyan Seaway, which formed during the late Cretaceous (Figure 6.21A–B). Later, as Australia separated from Antarctica during

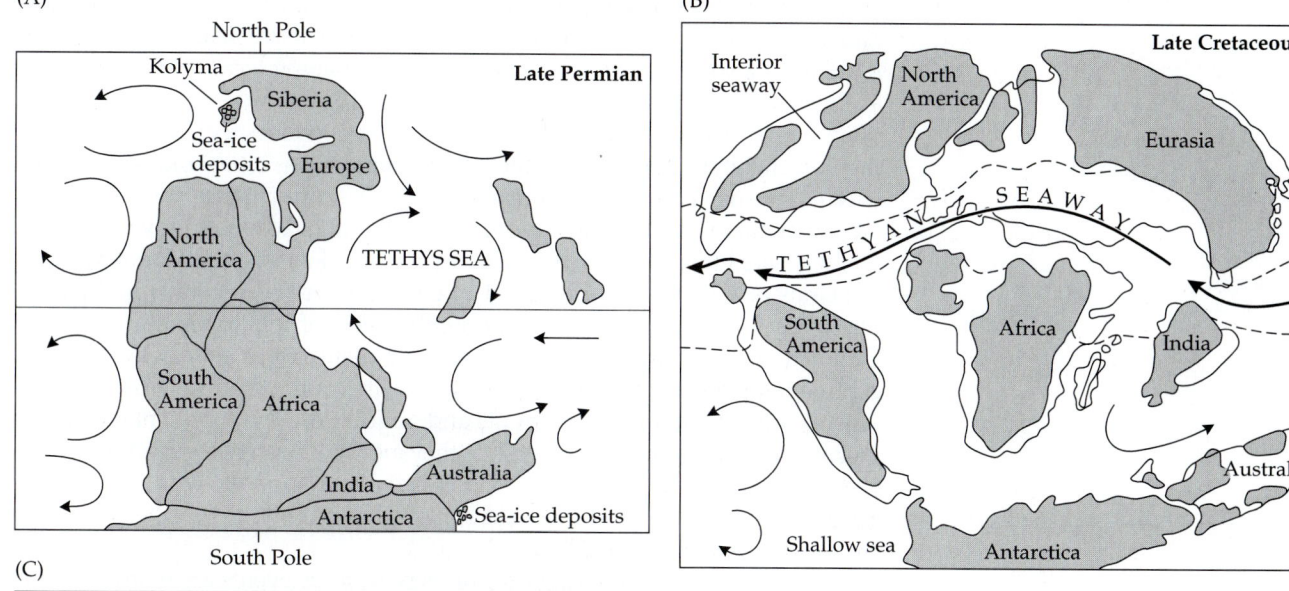

(C)

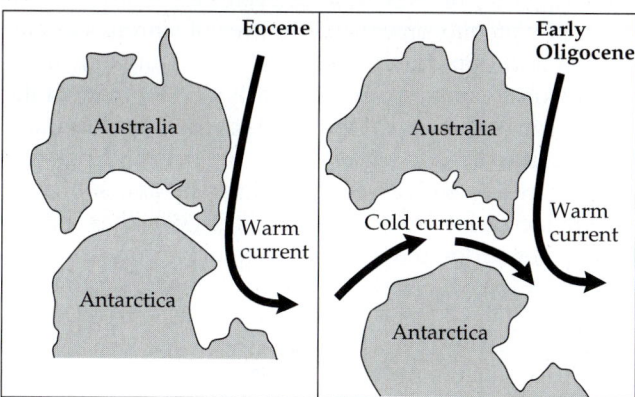

Figure 6.21   Continental drift strongly influenced paleocirculations and paleoclimates. (A–B) As the result of the breakup of Pangaea, surface currents throughout Panthalassa, the world ocean, were replaced by flow through the warm, circum-equatorial Tethyan seaway. (C) In a similar manner, the separation of Australia from Antarctica between the Eocene and early Oligocene epochs established the cool, circum-polar Antarctic seaway. These and many other shifts in paleocirculations dramatically altered the thermal regimes of marine and terrestrial environments, and are often linked to heavy extinctions of stenothermal species. (From Stanley 1987.)

the late Eocene (about 40 million years B.P.), cold currents became established between those two continents, deflecting the relatively warm South Pacific gyre, which had until then moderated the water temperatures and coastal environments of southeastern Australia and northeastern Antarctica (Figure 6.21C). As it continued to separate from Australia during the Oligocene, Antarctica drifted farther poleward, and the cold currents strengthened. The warm circum-equatorial currents of the late Cretaceous were thus interrupted by drifting continental landmasses and replaced by cold circum-antarctic currents, setting the stage for regional extinctions of thermally intolerant marine invertebrates and for global cycles of glaciation and climatic change.

Climatic and Biogeographic Consequences of Plate Tectonics

Plate tectonics, perhaps more than any other phenomenon, has had profound effects on the biogeographic patterns of both terrestrial and marine biotas. Over geological time, new plates have arisen and expanded at the expense of more ancient ones, which have been subducted and consumed within the deeper layers of the earth's mantle. Plates have varied dramatically in shape and size and, during their existence, have split apart, slid against each other, or collided to feed the mantle's convective cycle.

The surface features of the plates have also varied tremendously over time and space. Many plates were often divided by extensive shallow seas, which harbored a great diversity of marine life while isolating terrestrial biotas. Also, as illustrated in the series of maps in Figure 6.15, extensive mountain ranges often served as barriers to lowland species or as corridors for those adapted to montane habitats. Even during the late Permian, when most landmasses were united, extensive mountain ranges along the convergence zone between Gondwanaland and the Old Sandstone Continent of Laurasia may have effectively isolated their respective biotas. As we noted earlier, the Himalayas and the Urals also formed as the result of collisions between continental plates, effectively isolating many species on either side of this topographic barrier. Other mountain ranges formed on oceanic plates were subducted beneath continental plates, which caused great buckling of the land surface (as in the Andes) or the welding of seamounts and guyots onto the continental plate (as in the Olympic Mountains). Finally, other mountain ranges, including island chains, formed as a result of volcanic activity associated with subduction zones or as plates drifted over hot spots in the mantle (as in the Hawaiian island chain; see Figure 6.20).

The biogeographic consequences of these tectonic events are manyfold, but most stem from the effects of plate movements on the area of landmasses and marine basins, on isolation and opportunities for biotic exchange, or more indirectly, on regional and global climates. Since the early development of the field, biogeographers have been well aware of the effects of area, isolation, and latitudinal position on diversity. Across a great variety of taxa, time periods, and ecosystems, diversity increases with area, while diversity and similarity among biotas decrease with increasing isolation. As Hallam (1983) has shown, the diversity of ammonites, a once dominant but now extinct group of marine

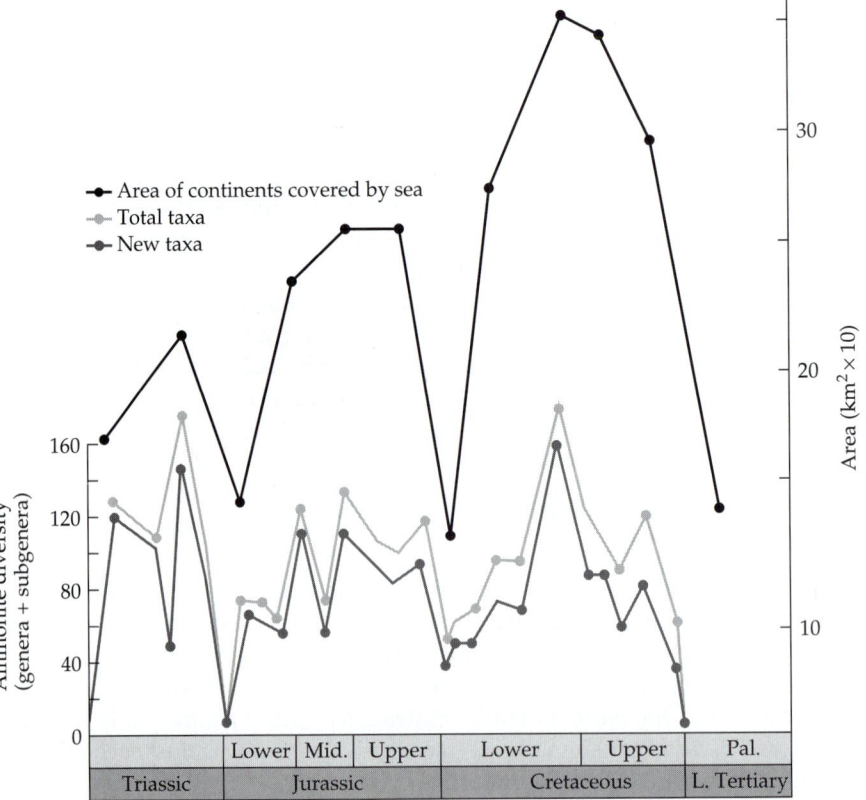

Figure 6.22 Many extinction events in the marine realm may be linked to changes in sea level, which dramatically altered the area of the continents covered by shallow epeiric seas, the source for the majority of marine fossils. The diversity of ammonites, once dominant marine invertebrates, appears to be directly correlated with the area of the earth's epeiric seas. (From Hallam 1983.)

Figure 6.23   Seafloor spreading in the North Atlantic caused separation and divergence of shallow-water marine invertebrates during the Mesozoic and Cenozoic. As the North Atlantic widened and the coastal environments on either side became more isolated, the taxonomic similarity of their biotas decreased. Simpson's coefficient of similarity = 100 × (the number of taxa in common/the number in the sample with the smallest number of taxa). EJ, early Jurassic; MJ, mid-Jurassic; LJ–EK, late Jurassic to early Cretaceous; EK, early Cretaceous; LK, late Cretaceous; P, Paleogene; N, Neogene. (From Hallam 1994.)

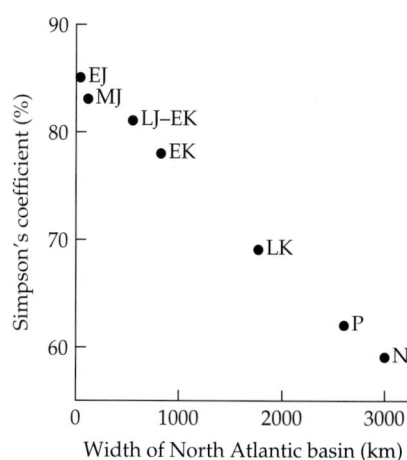

invertebrates, was strongly correlated with the area of epeiric seas (Figure 6.22). Other paleontological work with marine invertebrates shows that when shallow-water marine communities on either side of the North Atlantic drifted apart, their similarity (i.e., the proportion of shared species) decreased (Figure 6.23) (Fallaw 1979).

Precipitation patterns, paleocurrents, wind, and temperature also must have varied tremendously as continents split, converged, or drifted between the poles. On a global scale, the proportion of landmasses at different latitudes has been nearly reversed since the Silurian (425 B.P.), when most landmasses were situated south of the equator (Figure 6.24). When landmasses drifted over the poles, they triggered cycles of glaciation and changes in sea level (Table 6.2). The glacial cycles of the Pleistocene are now familiar to most biologists, but these events were dwarfed by the magnitude of some earlier glaciations and fluctuations in sea level. Similarly, the formation of Pangaea isolated most of its interior land area from the buffering effects of the ocean, increasing temperature fluctuations and aridity and, in turn, causing the drying of shallow seas.

We have tried to summarize the major tectonic, climatic (including currents and sea levels), and biotic events of the Phanerozoic in Table 6.2. We admit

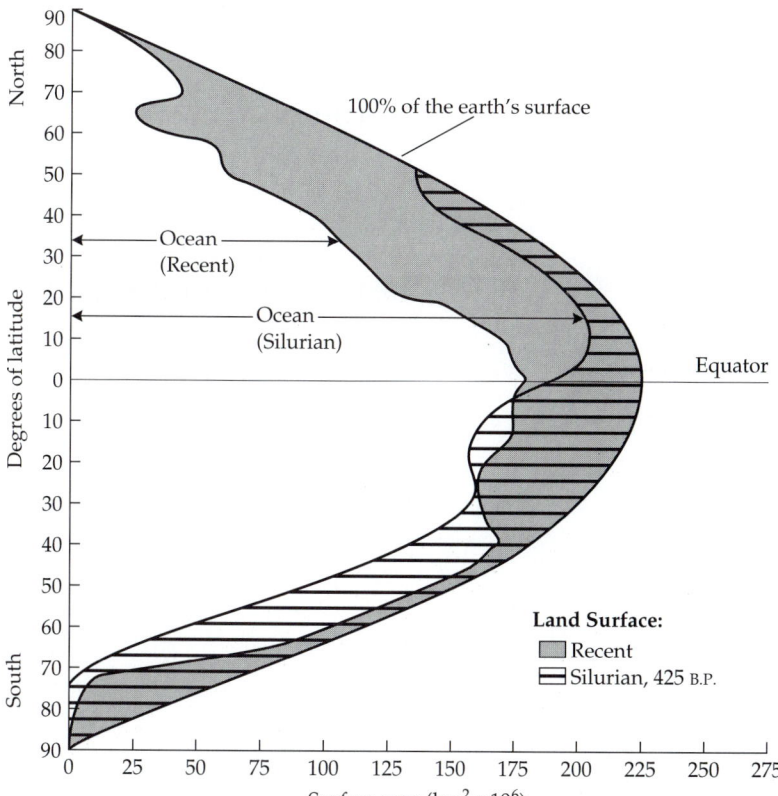

Figure 6.24   Continental drift has substantially altered the latitudinal distribution of landmasses and, in turn, has caused dramatic changes in regional and global climates. (After Rosenzweig 1995.)

Table 6.2
Summary of the tectonic, climatic, and biotic events of the Phanerozoic

MY B.P.	Period	Tectonic events	Climatic/marine events
	Neogene	Central American Archipelago emerges Red Sea forms	Repeated (about 20) glacial events Aridity peak

24		Africa collides with Eurasia	Climate cooling, drop in sea level
			  [Tethys seaway closed, circum-antarctic
			  seaway opens]

	Paleogene	India collides with Eurasia, Himalayas uplifting Madagascar and Seychelles separate from India Beringia forms	Warming of global climate
65			Tethys circum-equatorial seaway opens
	Cretaceous	Separation of Gondwanaland is complete	Global transgression of shallow seas
		India and Madagascar drift from Antarctica-Australia	Shallow seas recede
		Proto-Antilles emerge between Mexico and Colombia	
144		N. and S. America separate	Global transgression of shallow seas
	Jurassic	E. and W. Gondwanaland rifting	
213		Laurasia rifting	
	Triassic	Formation of Pangaea and Panthalassa	Minor glacial event, sea levels drop, shallow seas dry
248			
	Permian		Glaciers wane in Gondwanaland
286			
Carboniferous	Pennsylvanian		
	Mississippian		
360			Glacial event in Gondwanaland
	Devonian	Old Red Sandstone Continent of Pre-Laurasia forms	
408			
	Silurian	Australia near the equator	
438			Glacial event in Gondwanaland Glaciation
	Ordovician	Most landmasses located in southern or equatorial latitudes	Circum-arctic current established
505			Lowering of sea level
	Cambrian		
590			
	Precambrian		
650		Gondwanaland forms	Glacial event

Source: Sea levels after Hallam, 1994.

Table 6.2 *(continued)*
Summary of the tectonic, climatic, and biotic events of the Phanerozoic

Global sea level *(shading marks cooler periods)*	**Biotic events**

Extinctions: plankton, marine invertebrates, land mammals
Mass biotic exchange between Nearctic and Neotropics

Extinctions: plankton, marine invertebrates, land mammals

Extinctions: plankton, marine invertebrates, marine reptiles; dinosaurs
Flowering plants become dominant on land

Extinctions: marine invertebrates

Extinctions: marine invertebrates, dinosaurs
Archaeopteryx ——————

Extinctions: marine invertebrates   —— Earliest birds

Proto-Aves ——————

Earliest mammals evolve
Extinctions: sea-floor protozoans, marine invertebrates, mammal-like reptiles

Mammal-like reptiles evolve

Earliest reptiles evolve

Vertebrates invade land
Extinctions: plankton, marine invertebrates, primitive fishes

Extinctions: marine invertebrates, including reef builders

Earliest fishes
Extinctions: marine invertebrates (trilobites and brachipods)

Great diversification of marine life

First multicellular organisms

that this account is far from complete or precise, but it should be instructive. While the mass extinction at the end of the Cretaceous appears to have been caused by an asteroid impact, others appear to have been associated with tectonic events. The great Permian event, which marked the extinction of over 90% of the earth's species, is associated with the formation of Pangaea and the concurrent drop in sea level and the drying of shallow seas. The challenge of establishing cause and effect across geological time remains a daunting one indeed, but we should note that Cuvier himself attributed the mass extinctions to great changes in relative sea level. We should, however, add a note of caution: Mass extinctions, as spectacular and intriguing as they appear, tend to attract a disproportionate, and perhaps unwarranted, amount of attention. As Stanley (1987) noted, the victims of the major mass extinctions include only a small minority of all extinct species: "most of the global turnover of species was very gradual, not catastrophic." On the other hand, tectonic events also set the stage for major radiations of new species, which are evident throughout the fossil record. As we noted earlier, many terrestrial vertebrates evolved and spread across the relatively continuous landmasses of Pangaea during the late Permian. Their major radiations, however, did not take place until Pangaea split, isolating its biotas, which were then free, if they survived, to diversify under different selective pressures and in the absence of gene flow. Few contemporary scientists doubt that tectonic, climatic, sea level, and biogeographic phenomena are causally related (see Fischer 1984).

Our extensive discussions of plate tectonics, ecology, and evolution in the early chapters of this book are a recognition of the fundamental importance of these disciplines to biogeography. Again, all of these disciplines are intricately related, and any distinction we attempt to make is surely an arbitrary one. In the next chapter we take a closer look at the relationship between climate and biogeographic events during the past 2 million years.

CHAPTER 7

Glaciation and Biogeographic Dynamics of the Pleistocene

Chapter 6 explored the biogeographic consequences of events occurring across a broad span of the geological record—the entire Phanerozoic. Here we focus on a thin slice of this record, the Pleistocene and Holocene epochs (i.e., the late Neogene), which include only the past 2 million years of earth's history. While the earth's tectonic plates were relatively stable during this short period, global and regional climates experienced many upheavals, which in turn had profound effects on most biotas. Because these events were so recent, biogeographers and paleoecologists can capitalize on a variety of paleontological methods and data not available to those studying more ancient periods. These methods include analyses of tree rings, pack rat middens, pollen deposits, accumulations of thermally sensitive indicator species in ice cores (e.g., foraminiferans, radiolarians, and coccoliths), lithological evidence (including tillites and evaporites), and the depths of fossilized coral reefs, as well as radiocarbon dating. The resulting high-resolution reconstructions of the late Neogene provide us with an intriguing opportunity to explore the relationship between climate and biogeographic patterns of current as well as ancient biotas.

By devoting a full chapter to the Pleistocene, we recognize that events during that time had profound effects on today's distributions of plants and animals. By making this the second of two chapters on earth history, however, we are reminding the reader that these more recent changes have not erased the patterns resulting from earlier events. We first review some of the major climatic events of the Pleistocene, then discuss their biogeographic consequences, including range shifts and biotic exchanges, adaptations and evolution, and extinctions. In addition to elucidating the interrelations between historical and ecological biogeography, studies of the climatic and biotic dynamics of the Pleistocene may also provide invaluable insights for predicting the biogeographic responses of contemporary biotas to anticipated anthropogenic changes in global climates (see Chapters 17 and 18).

Extent and Causes of Glaciation

As we noted in the previous chapter, glacial events were often associated with the positioning of large landmasses over or near the poles. Thus, as a result of the drifting of continents during the Phanerozoic, the earth has undergone several periods of extensive glaciation (see Table 6.2). Throughout most of the

Mesozoic and early Cenozoic eras, however, the global climate was relatively warm and equable, with little variation across seasons or latitudes. The glacial-interglacial cycles that characterized the late Neogene were truly unprecedented episodes in the biogeographic and evolutionary history of most biotas, many of which developed and radiated subsequent to the breakup of Pangaea.

During the Pleistocene, the earth experienced several **glacial-interglacial cycles** during which continental glaciers advanced and retreated. These glaciers were incredibly massive sheets of ice. They were often 2 to 3 km thick, and their mass was so great that it deformed the underlying lithosphere. At their maximum extent, these ice sheets covered up to a third of the earth's land surface. During these periods of glacial maxima the climates of many unglaciated temperate regions were cooler and wetter than those of today; therefore, glacial maxima in now arid regions are also referred to as **glacio-pluvial** (ice-rain) periods. In contrast, tropical regions tended to be drier during glacial maxima. At the height of the most recent glacial period, ice sheets in the Northern Hemisphere extended from the Arctic southward to cover most of North America and Central Asia to approximately 45° N latitude (see Figure 7.1). These glacial periods alternated with **interglacial** periods, when the climate warmed and the ice sheets retreated. As the recent paleontological record tells us, these events had profound effects on both terrestrial and marine biotas.

Most authors focus on Pleistocene glaciation episodes in the Northern Hemisphere because over 80% of the glacial ice occurred there (Figures 7.1 and 7.2). In the Southern Hemisphere (except in Antarctica, where glaciers began

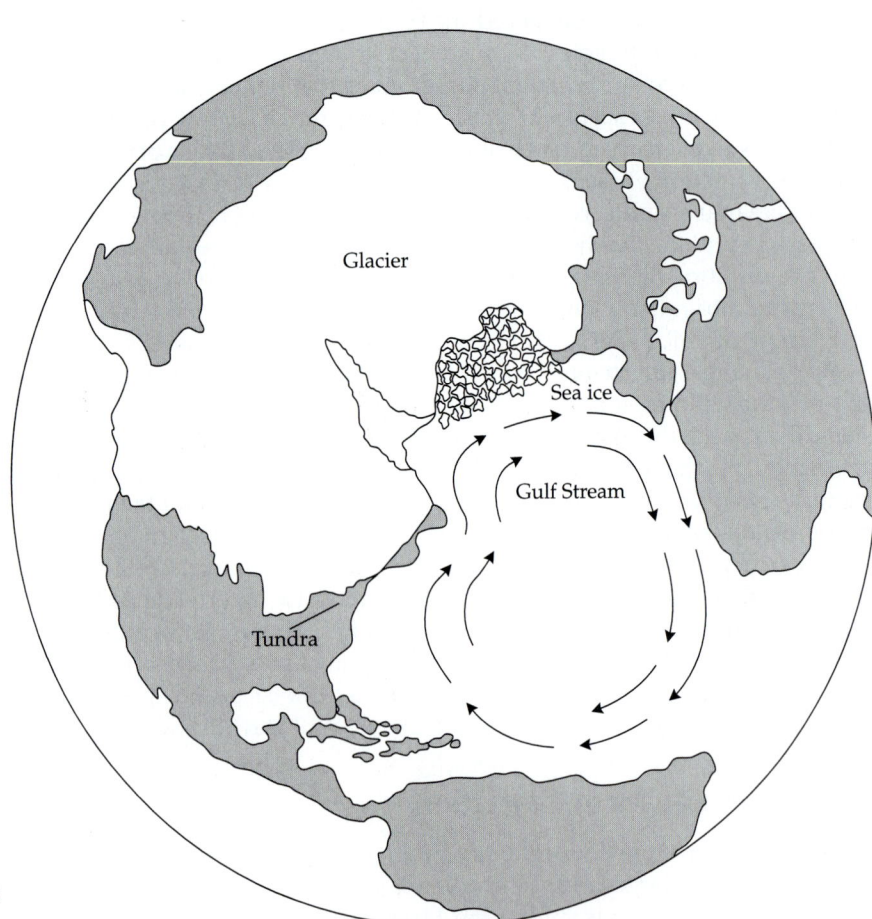

Figure 7.1   Glaciation of the Northern Hemisphere during the most recent glacial maximum (18,000 years B.P.). The north-flowing waters of the Gulf Stream tended to warm the North Atlantic, but this same oceanic gyre cooled coastal regions of southern Europe and Africa as it flowed southward. (After Stanley 1987.)

18,000 years B.P.

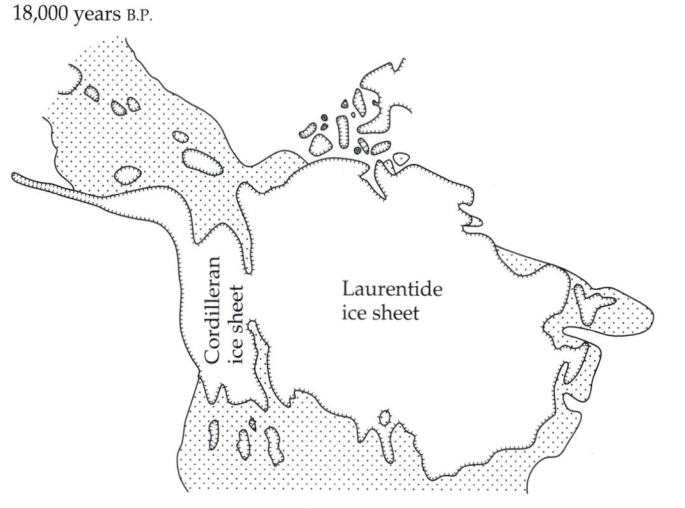

13,000 years B.P.

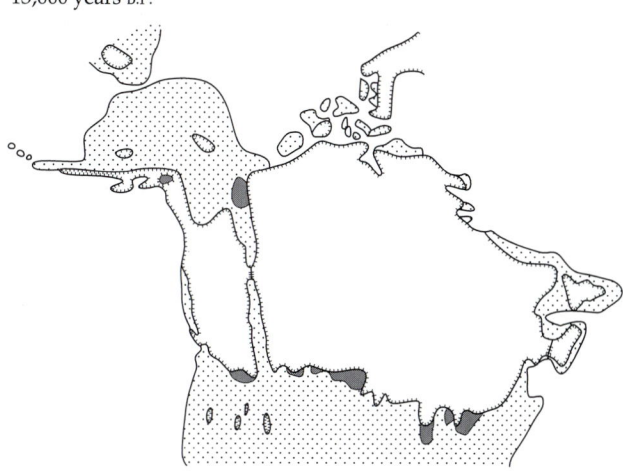

10,000 years B.P.

7,000 years B.P.

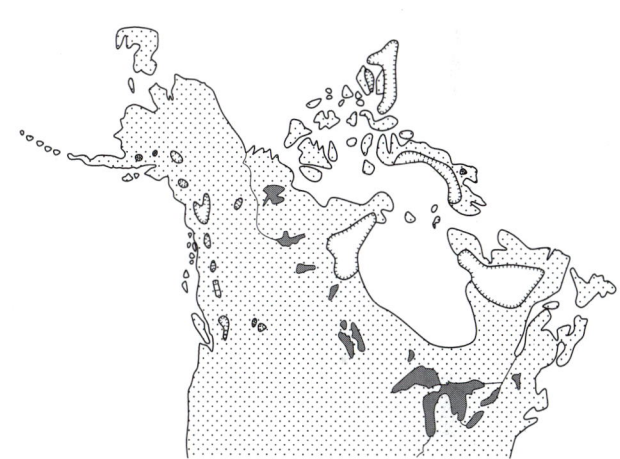

Figure 7.2 The sequence of glacial recession in North America since the most recent glacial maximum (18,000 years B.P. to present). White areas with etched borders represent glaciers; stippled areas, unglaciated land surfaces; shaded areas, lakes. Black areas in the present map represent land surfaces currently covered with glaciers. (After Pielou 1991.)

Present

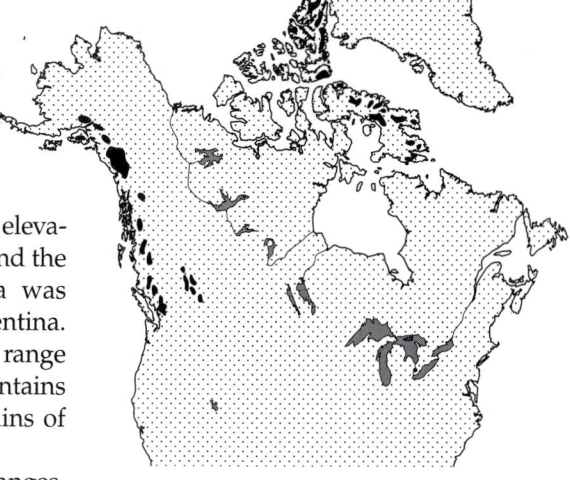

to form in the Miocene), glaciation was mostly confined to high elevations at high latitudes, such as the Central Plateau of Tasmania and the New Zealand Alps (see Flint 1971). The Andean Cordillera was glaciated, but the greatest ice coverage was in Chile and Argentina. Mainland Australia was unglaciated except for a small mountain range in Victoria, and Africa lacked glaciation except in the Atlas Mountains of the extreme northwestern corner and in the highest mountains of eastern Africa.

Before turning to the biogeographic effects of these climatic changes, we first focus on their causes. The glacial events of the Pleistocene, unlike those before them, were not caused by the drifting of continents. In the past, scientists attributed these events to changes in the output of solar radiation. However, while solar output has varied by as much as 25–30% during earth's 4.5-billion-year history, it has changed little during the Phanerozoic (the last 0.6 billion years: Gates 1993). Instead, the climatic reversals of the Pleistocene were caused by changes in the interception and absorption of solar radiation by the earth's surface due to changes in its orbit (Figure 7.3) (Gates 1993; Muller and MacDonald 1995). Three characteristics of the earth's orbit around the sun change over time, each with a characteristic periodicity. These

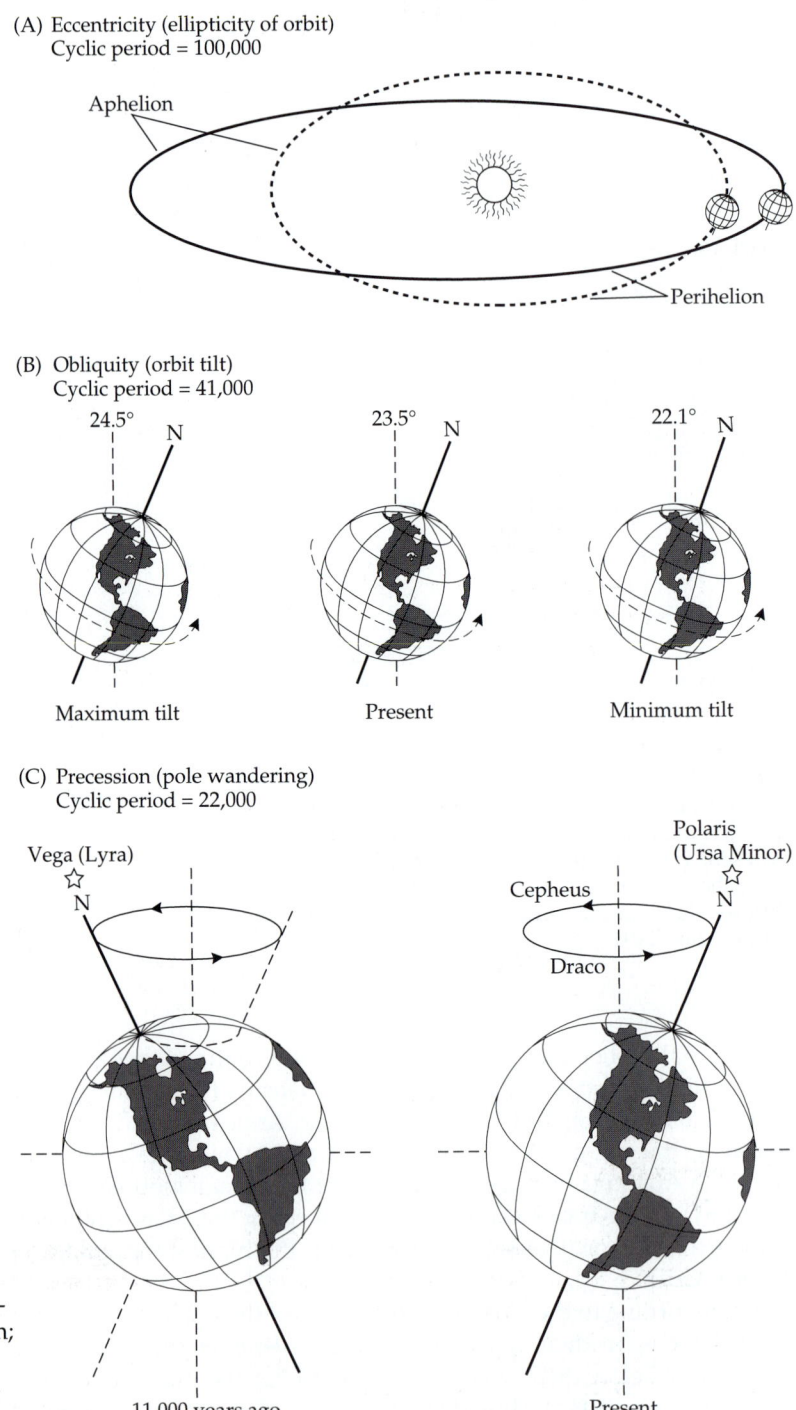

Figure 7.3   Milankovitch cycles are periodic changes in the eccentricity, obliquity, and precession of the earth's orbit. Each of these changes influences the earth's interception of solar radiation; therefore, these cycles may have been largely responsible for the glacial cycles of the Pleistocene. (After Gates 1993.)

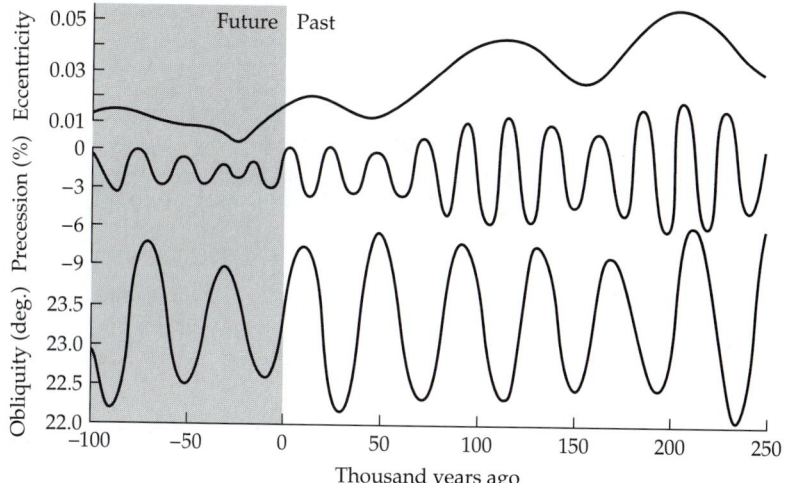

Figure 7.4 Past and predicted changes in Milankovitch cycles. Note that while the periodicities of these orbital phenomena remain relatively constant, their amplitudes vary substantially. (After Gates 1993.)

changes are called **Milankovitch cycles** after their discoverer. First, the earth's orbit is not perfectly circular, but varies in ellipticity, or "**eccentricity**," with a period of 100,000 years. Second, the tilt of the earth on its axis—its **obliquity**— varies from 22.1° to 24.5° with a 41,000-year period. Finally, the earth's orientation, or **precession**, wanders, with the axis of the North Pole shifting from one "north star" (presently Polaris of Ursa Minor) to another (Vega of Lyra) with a periodicity of approximately 22,000 years. The combined effects of these cyclical changes in the earth's orbit cause substantial fluctuations in the amount of solar energy striking the earth, ultimately combining in a complex fashion to create the glacial-interglacial cycles and climatic reversals of the Pleistocene (Figure 7.4).

To further complicate this picture, it seems that transitions between glacial and interglacial periods were influenced by some remarkable feedback effects. During the initial stages of glaciation, for example, developing fields of ice and snow increased reflectance (**albedo**) over large areas of the earth. This phenomenon reduced the effective solar heating of the earth and further increased cooling rates, causing more rapid glaciation. The reverse process, deglaciation, also seems to be influenced by feedback and secondary driving forces. It appears that glaciers melted much more quickly than they formed—too quickly to be accounted for solely by the relatively gradual changes in the earth's orbit. Instead, glacial melt seems to have been driven in part by global changes in the concentration of greenhouse gases, especially carbon dioxide and methane. Apparently, global warming accelerated the natural biogenic release of these gases into the atmosphere. In fact, this natural greenhouse warming may have accounted for as much as half of the 4.5° C temperature rise since the last glaciation (see Figure 7.6; Gates 1993).

Effects on Nonglaciated Areas

The steep thermal latitudinal gradient from frozen poles to warm equator that now characterizes our global climate is a relatively recent phenomenon. Throughout most of the Phanerozoic, equatorial climates were tropical, as they are today, but latitudinal gradients in temperature were less pronounced, and terrestrial climates were generally more equable. Beginning in the Miocene, global climates gradually began to cool and become drier. The extension of the Antarctic ice sheet and the intensification of oceanic and atmospheric

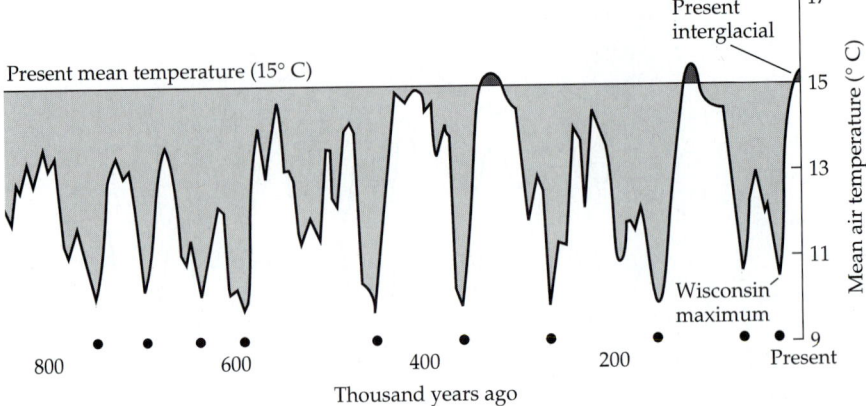

Figure 7.5 Estimated average global air temperature over the past 850,000 years, inferred from oxygen isotope measurements from ice cores. Note that the current interglacial represents one of the warmest periods in this record, and that interglacials, in general, tend to be very limited in duration. Circles mark glacial maxima. (After Gates 1993.)

circulation during the mid-Miocene (15 million years B.P.) established a strong latitudinal thermal gradient that intensified during interglacial periods. During glacial maxima of the Pleistocene, average air temperatures were 4° to 8° C cooler than during interglacial periods (Figure 7.5), and cooling appears to have been more pronounced in the lower latitudes.

Due to the high heat capacity of water, the global temperature of the oceans varied by just 2° or 3° C between glacial and interglacial periods. On the other hand, atmospheric and oceanic circulation changed dramatically, and climatic zones shifted substantially in latitude and altitude with each climatic flush. In mountainous regions, snow lines shifted as much as 1000 m in elevation between glacial and interglacial periods (Behrensmeyer 1992). Before exploring some of their biogeographic and evolutionary consequences, we summarize some of the major climatic changes associated with glacial-interglacial cycles.

Temperature

Investigators can estimate paleotemperatures by several qualitative methods or, more quantitatively, by determining oxygen isotope ratios in marine fossils or ice cores. Most marine exoskeletons are composed of calcite ($CaCO_3$), made by combining water and carbon dioxide. In water, the two common isotopes of oxygen are ^{16}O and the heavier ^{18}O. The lighter isotope (^{16}O) evaporates more rapidly, especially during warmer periods. Thus, with some sophisticated calibrations and analysis, the ratio of ^{18}O to ^{16}O locked in the shells of marine organisms or in ice cores can serve as a paleothermometer.

The resultant reconstruction of global temperatures during the past 2 million years reveals that the earth underwent at least ten major glacial periods (i.e., periods during which global temperatures were at least 4° C below the current, interglacial temperature). At these times, glaciers in the Arctic expanded into the lowlands and mountains of middle latitudes in Eurasia and North America (see Figure 7.1). Prior to these recent, high-resolution reconstructions of global temperatures, geologists and paleobiologists recognized just four or five glacial cycles during the Pleistocene, each identified by geological evidence of their southernmost extent in North America or Europe (Table 7.1).

As Figure 7.5 indicates, glacial conditions prevailed throughout most of the Pleistocene, with intermittent interglacials accounting for less than 10% of this epoch. A closer look at the thermal record for the Pleistocene and Holocene reveals that climatic change can be remarkably rapid (Figure 7.6). Since the peak of the most recent glaciation, the Wisconsin in North America or the Würm in Europe, global temperatures have increased by approximately 4.5° C,

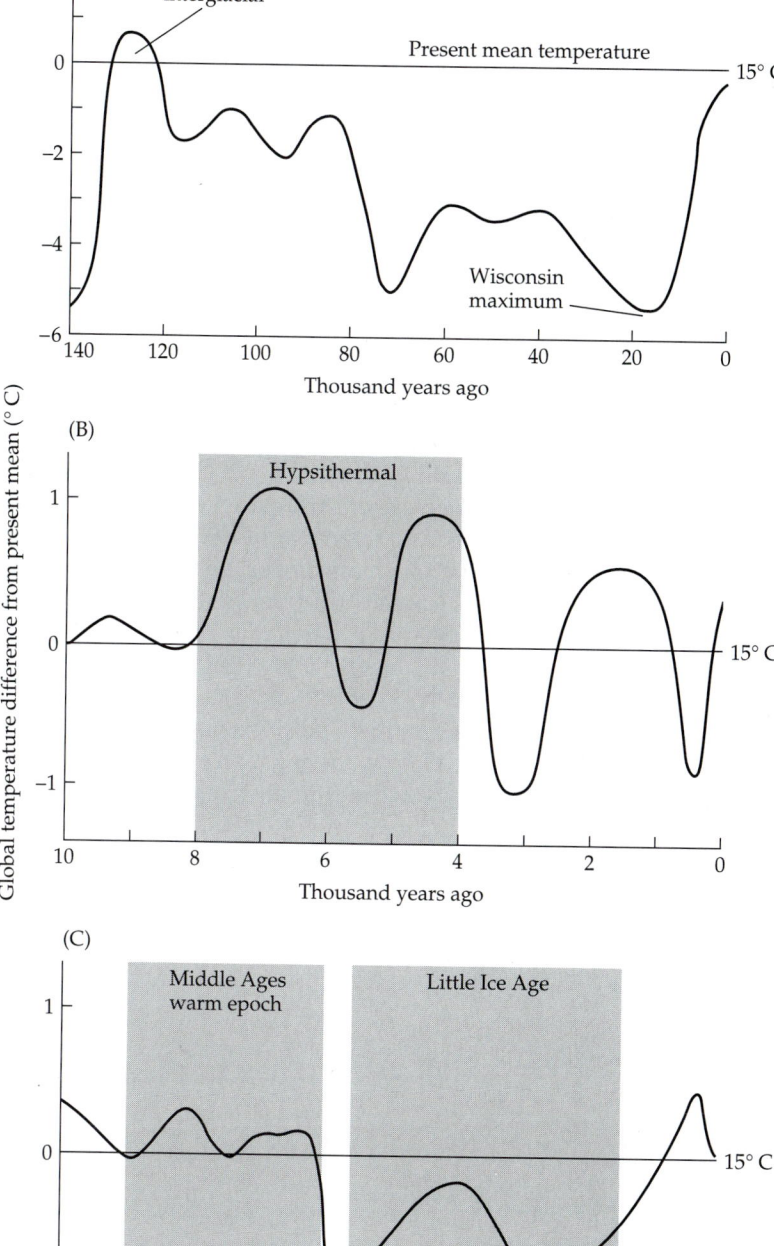

Figure 7.6 Records of global temperatures over three temporal scales reveal that climatic change can occur very rapidly. (A) Mean global ocean temperatures over the last 140,000 years show that the last two shifts from glacial maxima to interglacials (4 to 5°C increases in temperature) occurred over a span of just a few thousand years. (B) Air temperatures from Eastern Europe during the past 10,000 years and (C) mid-latitude air temperatures from the Northern Hemisphere over the past 1000 years show that more recent thermal shifts, while not as dramatic, were also quite rapid, sometimes shifting a degree or two in just a few centuries or even decades. (After Gates 1993.)

with most of that increase occurring within a span of just 4000 years (Figure 7.6A). Using higher-resolution reconstructions (Figure 7.6B), we see that temperatures have varied substantially even during the Holocene. Global temperatures from about 8000 to 4000 years B.P.—a period called the hypsithermal—were often 0.5° to 1° C warmer than current temperatures. Following this

Table 7.1
Sequence of glacial and interglacial stages in various regions of the Northern Hemisphere during the last 600,000 years

		North America	European Alps	British Isles	Northern Europe	Russia (European)
Time →	Fourth glaciation	Wisconsin	Würm	Devensian	Mecklenburgian	Waldai
	Interglacial	Sangamon	Riss/ Würm	Ipswichian	Neudeckian	Mikulino
	Third glaciation	Illinoian	Riss	Polandian	Moscow	
	Interglacial	Yarmouth	Mindel/Riss	Helvetian	Odintzovo	
	Second glaciation	Kansan	Mindel	Wolstonian	Saxonian	Dnieper
	Interglacial	Aftonian	Günz/Mindel	Hoxnian	Norfolkian	Lichwin
	First glaciation	Nebraskan	Günz	Anglian	Scanian	Oka

Source: After West (1977).

relatively warm period, global temperatures dropped by 2° C in just 500 years. Finally, global temperatures have varied significantly even over the past millennium (Figure 7.6C). After the relatively warm Middle Ages (1100 to 1400 A.D.), temperatures cooled rapidly, and much of the world experienced "Little Ice Ages," the most recent occurring between 1650 and 1850 A.D. (Davis 1986; Gates 1993).

Geographic Shifts in Climatic Zones

The Pleistocene glacial-interglacial cycles involved shifts not just in temperatures, but in entire climatic regimes. During the most recent glacial period, most nonglaciated regions experienced declines in air temperatures ranging from 4° to 8° C (Figure 7.7). In general, climatic zones shifted toward the equa-

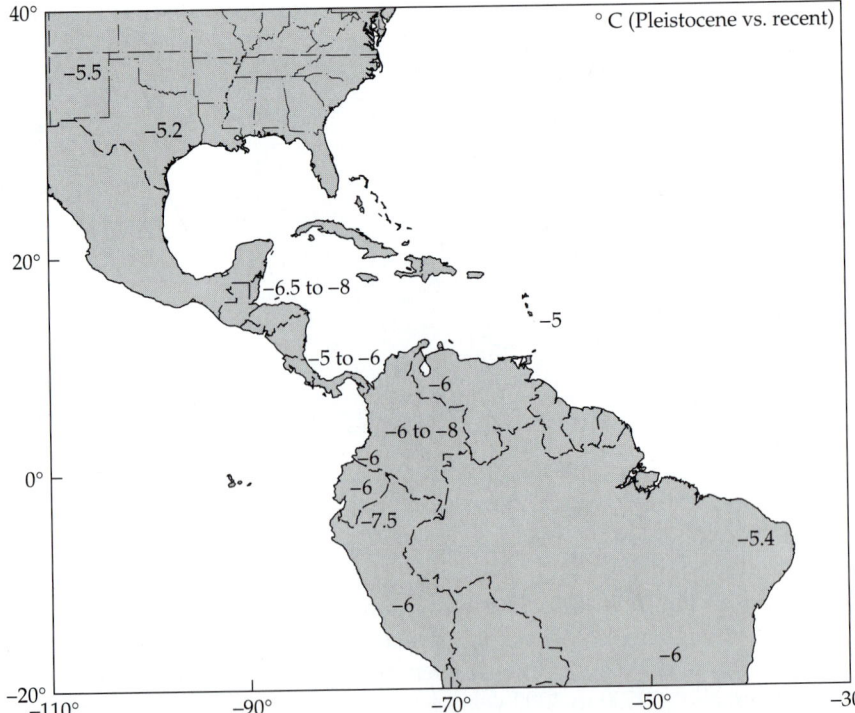

Figure 7.7 Glacial cycles of the Pleistocene influenced regional climates far from the edges of the glaciers. Temperatures over much of North and South America, for example, ranged from 4° to 8° C cooler during the Wisconsin. (After Stute et al. 1995.)

tor during glacial periods and poleward during interglacials. The pattern of shifts, however, was complicated by the configurations and positions of land and sea, and by the combined effects of glaciers, open ocean, and land on atmospheric circulation patterns.

Glaciers were so massive that they greatly reduced the flow of cool polar air masses to nonglaciated regions. Even air masses that were not effectively blocked by the glaciers would have undergone substantial adiabatic warming as they descended 2 to 3 km, creating steep thermal gradients at the glacial margins. Thus, despite the generally cooler conditions, glacial winters were less severe, while glacial summers were cooler and less subject to the heat waves that characterize contemporary climates, especially in temperate regions. While cold and dry conditions characterized the glacial climates of many tropical regions and some temperate ones, such as Europe, other regions, including southwestern North America, were wetter during the glacial periods (see Van der Hammen et al. 1971; Wells 1979; Spaulding and Graumlich 1986). Recent studies suggest that this pattern of cooler and more equable climates and less pronounced thermal latitudinal gradients during glacial maxima was true of marine environments as well (see D'Hondt and Arthur 1996).

An important, but often overlooked, effect of glacial cycles is that as thermal zones shifted, novel combinations of temperature, prevailing winds, ocean currents, and precipitation created climatic zones that lack any contemporary analogues. The glaciers caused dramatic changes in prevailing winds and ocean currents, which in turn strongly influenced regional climates. During the Wisconsin, the jet stream of North America was split and diverged around the glacier, and an anticyclonic (clockwise) circulation pattern was established over the Laurentide ice sheet (Figure 7.8). As a result, while the average global temperature of the oceans cooled by just 2° to 3° C, the surface temperatures of waters in the North Atlantic cooled by as much as 10° C (see also Figure 7.12). Along the southwestern edge of the glacier, relatively dry, easterly winds from the interior caused desiccation of lakes in the American Northwest. Farther south, the prevailing westerly winds brought saturated oceanic air masses, causing elevated water levels in lakes of the American Southwest.

Paleoclimatic data also reveal a strong association between glacial events and monsoonal circulation and precipitation. Monsoons are driven by differ-

Figure 7.8 (A) The Laurentide ice sheet (18,000–9000 years B.P.) caused major changes in prevailing winds, including the jet stream and westerlies over North America, which strongly influenced regional climates. (After Gates 1993.) (B) As relatively cool winds descended the 2 to 3 km high faces of the glaciers, they were adiabatically warmed, thus moderating climatic conditions adjacent to the glaciers.

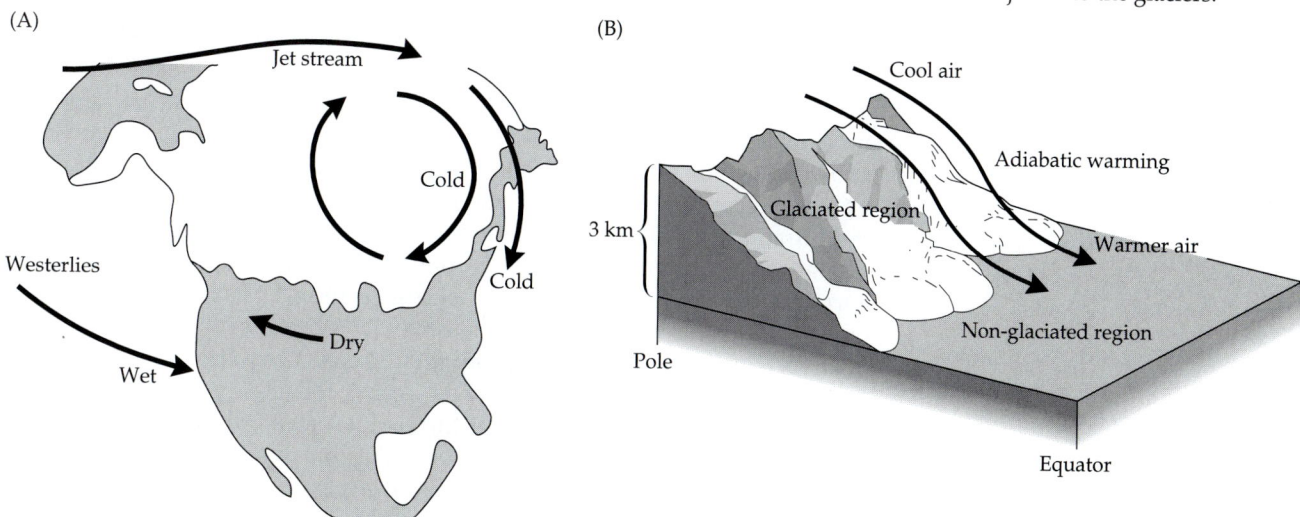

(A)

Jet stream

Cold

Cold

Westerlies

Dry

Wet

(B)

Cool air

Adiabatic warming

3 km

Glaciated region

Warmer air

Pole

Non-glaciated region

Equator

ential solar heating of land and sea during hot summers in tropical and subtropical latitudes. Because of its lower heat capacity, land heats much more rapidly than water. Ascending air masses over land create convective cycles that draw oceanic air masses inland, causing heavy summer precipitation—monsoon rains. Pollen records and global circulation models indicate that the strongest monsoonal events of the Pleistocene coincided with periods of peak summer solar radiation (roughly 126,000, 104,000, 82,000, and 10,000 years B.P.: Gates 1993). Conversely, monsoons were weakest during glacial maxima, often resulting in aridification of otherwise moist tropical regions.

In summary, the world we see outside our windows is not typical of the conditions that animals and plants lived and evolved under for the past 2 million years. The current very warm period was perhaps equaled during only 10% of the Pleistocene. While climatic zones tended to shift in latitude and altitude with glacial cycles, the specific nature of these zones varied markedly, often resulting in novel combinations of temperature, precipitation, and atmospheric or oceanic circulation. Not only were climatic zones altered in both character and location, but they frequently were displaced to occupy new soil regimes. Finally, while most climatic zones shifted toward the equator, tropical climates had little opportunity for geographic shifts. The inevitable reduction of what we now view as tropical climatic zones raises an intriguing question: How is it that the biotas adapted to these zones not only survived the cooler climates of glacial cycles, but maintained or developed a diversity unparalleled by that of any other region on earth? In the next section and in later chapters we return to this fascinating paradox, along with other biogeographic consequences of glacial events. First, however, we need to focus on one of the most pervasive and most important events associated with glacial cycles: regional and global changes in sea level.

Sea Level Changes in the Pleistocene

Throughout the Pleistocene, sea levels have fluctuated dramatically on both global and regional scales. **Eustatic** changes are global fluctuations in sea level resulting from the freezing or melting of great masses of sea ice. In contrast, **isostatic** changes occur when portions of the earth's crust rise or sink into the less buoyant asthenosphere, causing local or relative changes in sea level even when global levels remain unchanged. In some areas of the high northern latitudes, glaciers have caused crustal downwarping by over 300 m. As we shall see, both eustatic and isostatic changes during the Pleistocene strongly influenced the distributions and diversity of biotas.

During the most recent glacial maximum, nearly one-third of all land in the Northern Hemisphere was covered with thick glaciers, and great fields of sea ice occurred in both polar regions (see Figure 7.1). In consequence, a large volume of water was removed from the ocean, probably exceeding the equivalent of 50 million km^3 of ice, more than the present-day ice sheets in Antarctica and Greenland combined. However, the resulting 100 m drop in sea level during the Wisconsin glaciation was relatively mild in comparison to some earlier glacial maxima, when sea levels dropped by well over 160 m. Similarly, the current interglacial is not the warmest one on record. Sea levels during some of the most intense earlier interglacials may have been as much as 70 m higher than current levels. The total amplitude of sea level changes during the Pleistocene may have exceeded 230 m. Thus, despite the relative stability of the tectonic plates during this thin slice of the geological record, the earth's biogeographic profile may have been transformed as much as during any period in its history. Even the lowering of the sea level by 100 m during the early Holocene exposed a considerable area of the continental shelf and allowed landbridges to form between biogeographic regions (Figure 7.9). The biogeographic relevance of these events is difficult to overstate. Wallace's line, for example, which separates the Oriental and Australian biogeographic regions, coincides with the division between the glacial landbridges of the Sunda Shelf and Australia-New Guinea-Tasmania.

In Figure 7.10 we summarize global, or eustatic, changes in sea level. For many regions, however, isostatic changes have been just as large, with one

(A)

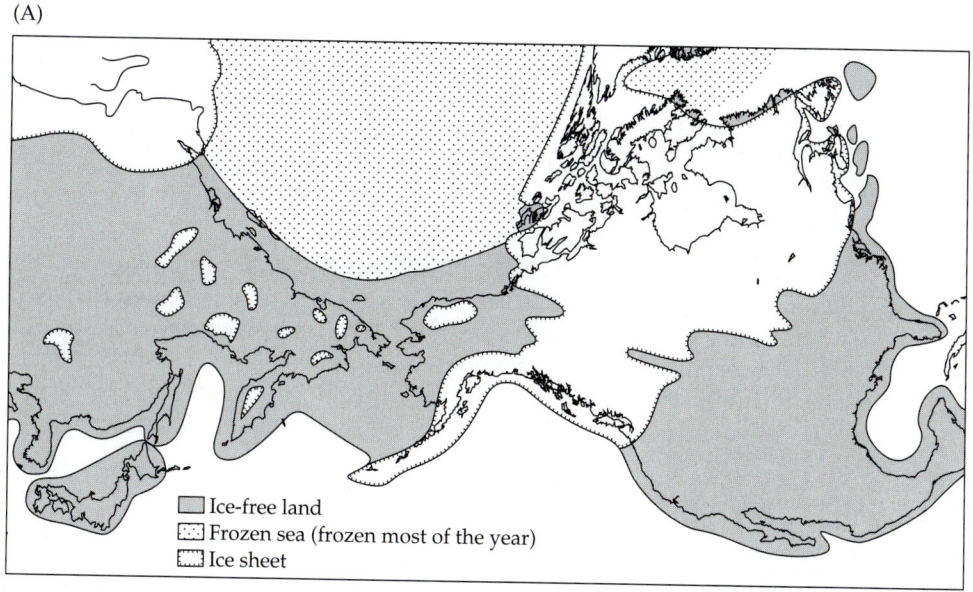

- ◼ Ice-free land
- ▦ Frozen sea (frozen most of the year)
- ▦ Ice sheet

(B)

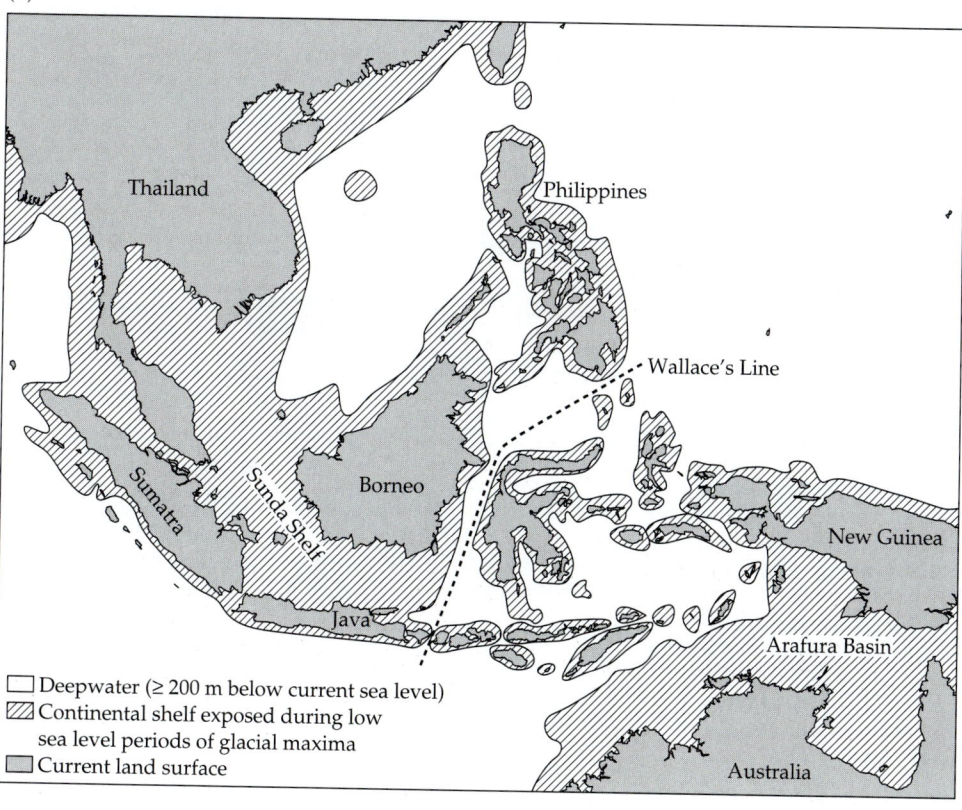

- ☐ Deepwater (≥ 200 m below current sea level)
- ▨ Continental shelf exposed during low
  sea level periods of glacial maxima
- ◼ Current land surface

Figure 7.9 Glaciation during the Pleistocene resulted in the lowering of sea levels by 100 to as much as 160 m below their current levels. As a result, many terrestrial regions and associated biotas now isolated by oceanic barriers were connected during glacial maxima. (A) Beringia connected North America and Asia. (After Pielou 1991). (B) Many islands of Indonesia were connected to mainland Asia and Australia, respectively. Wallace's line, marking a division between the biotas of Southeast Asia and Australia, coincides with the division between these glacial landmasses.

Figure 7.10   Changes in global sea level during the past 140,000 years based on oxygen-isotope data (black line) from analysis of benthic foraminifers found in deep sea cores of the Caribbean, and raised coral reefs (gray line) in New Guinea. In addition to these global (or eustatic) changes in sea levels, regional sea levels may vary substantially as the earth's crust rises and sinks in the asthenosphere. Such isostatic fluctuations in sea level can occur even when global levels remain unchanged. (After Hopkins et al. 1982.)

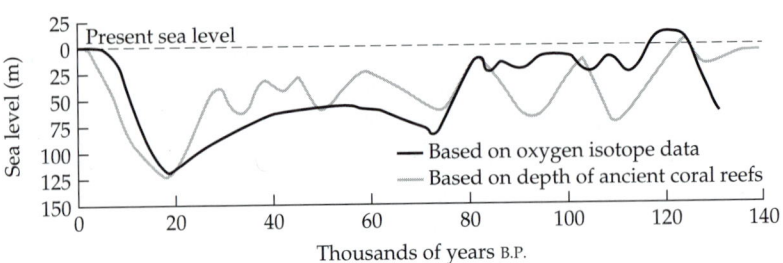

important difference: isostatic changes tend to lag far behind eustatic changes. It takes many centuries for downwarped crust to rebound after glacial recession. Consequently, as temperatures rise, rising seas often spill over onto the downwarped regions of continents, creating extensive shallow seas. During the early Holocene (about 12,000 to 10,000 years B.P.), for example, the Saint Lawrence Valley and Great Lakes of North America were inundated with marine waters from the Atlantic (Figure 7.11A). This saltwater corridor provided a dispersal avenue for shoreline flora of the Atlantic coast (e.g., beach plants including *Ammophila breviligulata, Cakile edentula,* and *Euphorbia polygonifolia;* coastal bog species such as *Xyris caroliniana;* aquatic macrophytes such

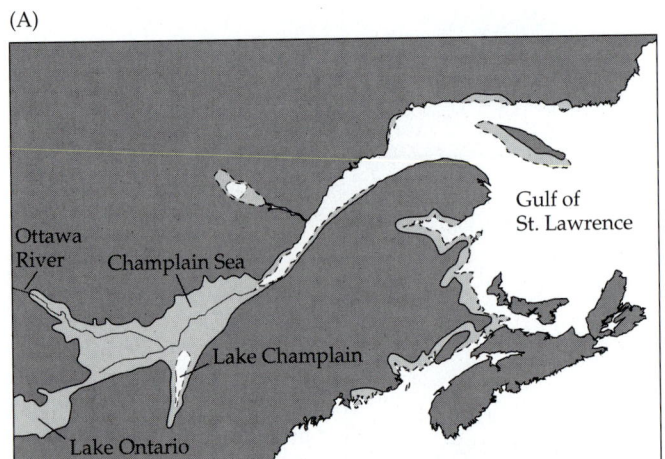

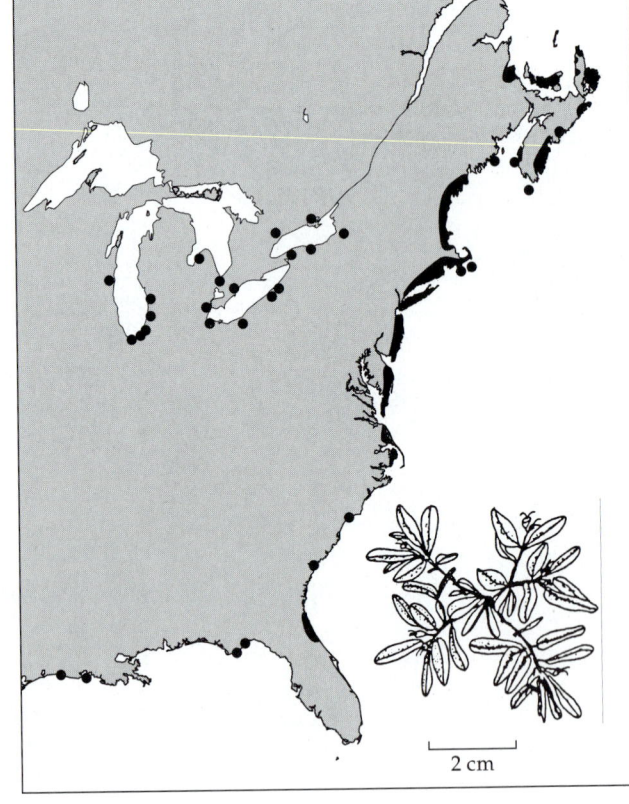

Figure 7.11   During the most recent glacial retreat, sea levels rose relatively rapidly, while previously glaciated land surfaces were slow to rebound from the downwarping force of the massive glaciers. As a result, marine waters spilled over onto these low-lying regions of the continents, creating shallow seas and providing saltwater corridors for the migration of plants and animals. (A) About 10,000 years B.P., shallow seas covered the Saint Lawrence River valley, Lake Champlain, and the Ottawa River, forming the Champlain Sea (the dashed line depicts the modern coastline; shading, the area covered by shallow seas.) (B) The existence of such transient corridors explains the disjunct distributions of many shallow-water, marine, and estuarine species, such as the seaside spurge (*Euphorbia polygonifolia;* distribution indicated by dark areas), along the coastlines of eastern North America and the Great Lakes. (After Pielou 1979.)

as *Utricularia purpurea*: Pielou 1991). Subsequent upwarping by as much as 275 m drained this corridor and transformed it into the freshwater Saint Lawrence River, which drains the Great Lakes. Events such as this may account for the disjunct ranges of many species adapted to coastal environments (Fig 7.11B).

In summary, glacial events altered dispersal routes among nonglaciated regions far from the ice sheets and long after the glaciers receded. During glacial maxima, the once continuous ecosystems of northern regions were often isolated on widely scattered pockets of unglaciated land situated along the margins of the ice sheet, while other, long-isolated biotas mixed across formerly submerged landbridges. Although these landbridges provided dispersal routes for terrestrial biotas, they created barriers for marine life. The glacial cycles thus created great, alternating waves of biotic exchange among both terrestrial and marine biotas.

Biogeographic Responses to Glaciation

The biogeographic dynamics of Pleistocene biotas were triggered by three fundamental changes in their environments:

1. Changes in the location, extent, and configuration of their prime habitats
2. Changes in the nature of climatic and environmental zones
3. Formation and dissolution of dispersal routes

The responses of biotas, long adapted to relatively stable and equable climates, also were of three types:

1. Some species were able to "float" with their optimal habitat as it shifted across latitude or altitude.
2. Other species remained where they were and adapted to altered local environments.
3. Still other species underwent range reduction and eventual extinction.

As we shall see, the environmental stresses of the glacial-interglacial cycles triggered all three of these responses. The specific nature of biogeographic responses varied considerably among species and geographic regions, but we have summarized some of their salient features in Box 7.1. We emphasize that, while such a synopsis may be useful, a richer understanding of these truly complex and fascinating events can be developed only by thoroughly exploring individual case studies.

Biogeographic Responses of Terrestrial Biotas

Many of the trends summarized in Box 7.1 are illustrated in the vegetative dynamics of the most recent deglaciation (see Figures 7.13–7.19). Climatic changes associated with the glacial maximum caused the general expansion of steppes, savannas, and other open-canopied terrestrial ecosystems at the expense of closed ecosystems, especially tropical rain forests. In general, biomes have shifted from 10° to 20° in latitude between glacial and interglacial periods, yet have tended to occupy the same relative positions across latitudes or elevations (i.e., a sequence from tundra and boreal forest to savanna and tropical rain forest). This is because the climatic belts of the earth (see Figure 3.6), then as now, created zonal patterns of vegetation, although the zones were compressed during glacial episodes. In the marine realm, latitu-

Box 7.1
Biogeographic responses to climatic cycles of the Pleistocene

1. The gradual period of cooling during the mid-Cenozoic was followed by repeated and dramatic climatic reversals during the Pleistocene.

2. Communities and coevolved assemblages of plants and animals that may have persisted for tens of millions of years during the equable Mesozoic were disrupted, with many species responding independently of one another based on their particular physiological tolerances, life history strategies, and dispersal abilities.

3. Many species were able to track the geographic shifts of their prime climates and habitats, but they typically lagged behind, often by centuries, sometimes by millennia.

4. Vegetation zones tended to shift toward the equator (or lower elevations) during glacial periods and toward the poles (or higher elevations) during interglacials, but the shifts were complicated and strongly influenced by geographic features (e.g., mountains, ocean basins, prevailing winds, and proximity to the ice sheet).

5. In general, open-canopied biomes (tundra, savannas, grasslands, and prairies) expanded during glacial maxima at the expenses of closed biomes (i.e., forests). These trends were reversed during periods of global warming, but again, the rates of shifts varied substantially among biomes, as did the particular species composition of each biome and community.

6. Despite substantial variation among regions, glacial climates tended to be dry as well as cool. On the other hand, glacial warming resulted in flooding of coastlines, submergence of landbridges, introgression of marine waters onto land, and formation of extensive shallow seas and great, post-glacial lakes and rivers.

7. On land, climatic zones changed dramatically, not only in location and areal coverage, but also in their characteristic nature (i.e., combinations of temperature, seasonality, precipitation patterns, and soil conditions). As a result, the Pleistocene events created novel environments, fostering development of novel communities, while other communities disappeared.

8. Although there was much variation within taxonomic groups, plants tended to shift more slowly than animals. The geographic dynamics of species during the Pleistocene created many isolated populations, in some cases promoting evolutionary divergence and diversification of certain biotas.

9. Many plants and animals that were unable to track their shifting environments were able to remain in situ by adapting to the altered conditions.

10. The remaining species, unable to shift or adapt, went extinct. During the initial cycles of climatic reversals, extinctions were much more common among plants than ani-

mals. This may have been a consequence of the comparatively limited ability of plants to disperse and the decoupling of associations among plants and between plants and animals that served as pollinators, parasites, and herbivores.

11. In contrast, until the most recent glacial cycles, animal extinctions were relatively few, and many groups, especially the large herbivores and carnivores, underwent major radiations.

12. The tables were turned, however, during the more recent glacial cycles, which witnessed waves of extinctions of many animals, especially larger ones, while comparatively few plants suffered extinctions. It appears that the initial climatic reversals may have "weeded out" most of the intolerant plants, leaving behind those more capable of dispersing with, or adapting to, climatic reversals.

13. During the most recent glacial cycle, large mammals may have become too specialized on the now waning glacial habitats (especially steppes and savannas). Alternatively, these "megafaunal" extinctions may have resulted from biotic exchanges associated with glacial events, which, again, decoupled coadapted groups of species or introduced novel competitors and predators, including *Homo sapiens* (see Chapters 17 and 18).

dinal shifts in isotherms and biogeographic patterns have tended to be substantial in the mid-latitudes (35° to 55°), but relatively minor at lower latitudes (Figure 7.12).

Shifts in climatic zones and biomes, however, are complicated by currents and topographic features, including mountain ranges, large rivers, and other bodies of water. In Europe, the southward shift of some biomes during the most recent glacial maximum was blocked by the Alps, Pyrenees, and Mediterranean Sea (Figure 7.13). In contrast, the north-south-running rivers and mountain ranges of North America facilitated extensions of high-latitude biomes deep into subtemperate and subtropical latitudes. During the Wisconsin maximum (about 18,000 years B.P.), boreal forests and tundra penetrated deep into the interior of the continent along the Mississippi River valley and

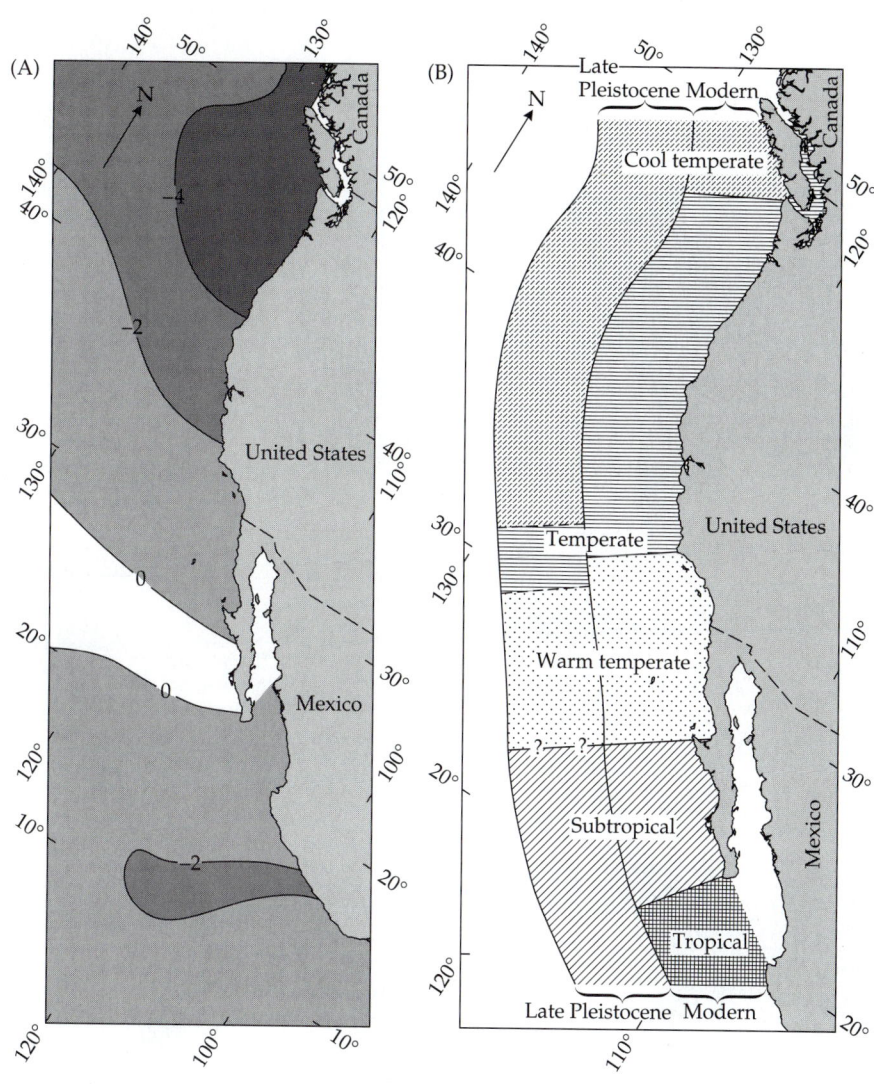

Figure 7.12 Shifts in (A) February sea surface temperatures (temperatures are in °C; minus sign indicates cooler temperatures during the Pleistocene) and (B) marine biotic provinces along the west coast of North America between 18,000 years B.P. and the present. Note that sea surface temperatures and nearshore marine provinces have remained relatively stable in the waters off southern California and northern Baja, while those in other regions have varied substantially during the same period. (After Fields et al. 1993.)

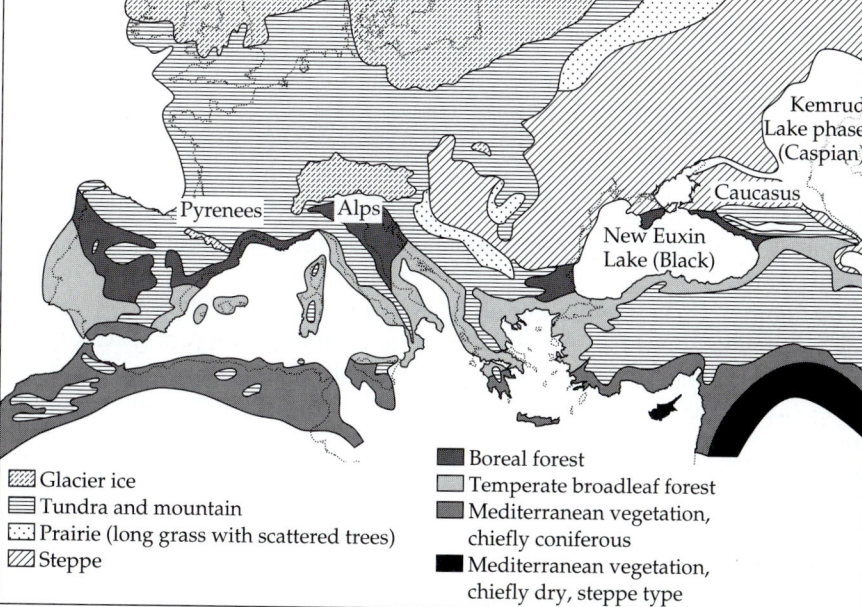

Figure 7.13 Zones of vegetation in Europe during the Würm glacial maximum (18,000 years B.P.). Major vegetation types were shifted southward of their present locations by 10° to 20° of latitude, and the precursors of the Black Sea and Caspian Sea were interconnected. Mountain ranges that probably blocked latitudinal shifts of biomes are shown. (After Flint 1971.)

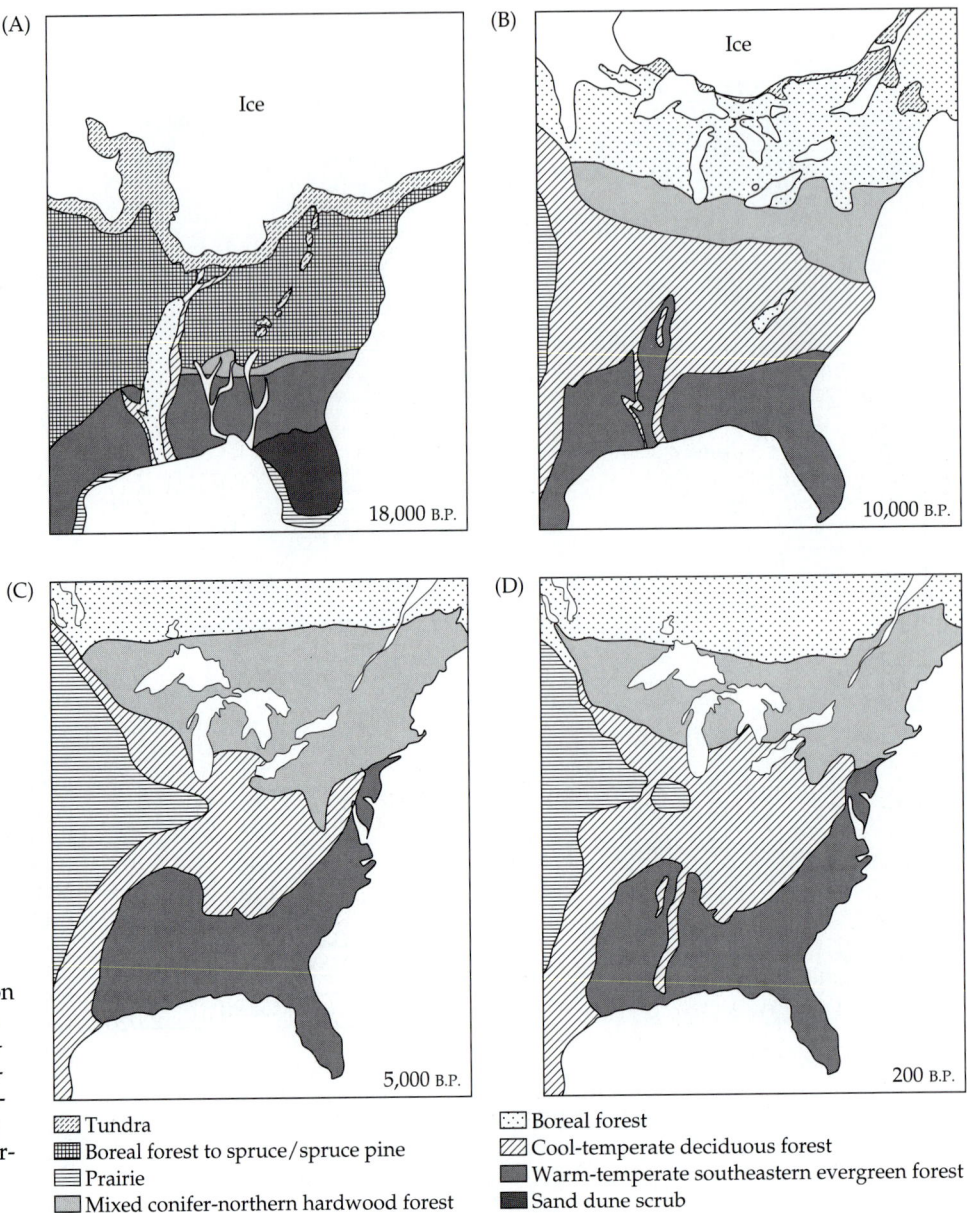

Figure 7.14 Shifts in vegetation zones of eastern North America during the most recent deglaciation. Note the great southern expansion of boreal forest and tundra along the Mississippi Valley and Appalachian Mountains during the Wisconsin (A). (After Gates 1993).

Tundra
Boreal forest to spruce/spruce pine
Prairie
Mixed conifer-northern hardwood forest

Boreal forest
Cool-temperate deciduous forest
Warm-temperate southeastern evergreen forest
Sand dune scrub

the Appalachian Mountains (Figure 7.14A). Similar extensions of boreal forests and other northern biomes occurred along the Rocky Mountains of western North America.

In Figure 7.15, pollen profiles of the Andes in Colombia show how each of the vegetation zones has shifted upward since the most recent glacial maximum (see Flenley 1979). The lower tropical, sub-Andean, and Andean forest types (including the tropical rain forest) have much wider elevational amplitudes now than they did 14,000 years ago. Subparamo has remained about the same, but paramo and superparamo are elevationally compressed. Because the upper vegetation types occur closer to the narrow mountain peaks, the total area covered by these biomes has fluctuated greatly, causing extinctions when populations were isolated in restricted montane areas. In these mountainous regions, elevational shifts ranged from 150 to 1500 m between glacial and interglacial periods, and were typically much more rapid than latitudinal shifts.

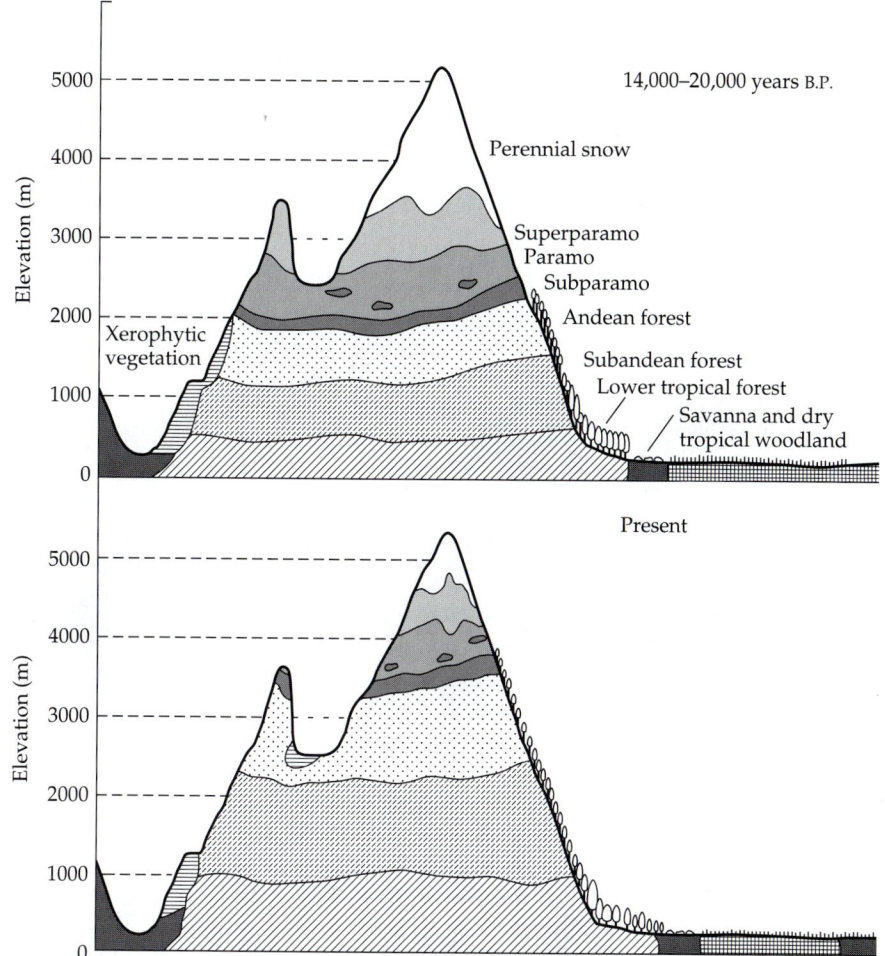

14,000–20,000 years B.P.

Perennial snow

Superparamo
Paramo
Subparamo

Andean forest

Subandean forest
Lower tropical forest
Savanna and dry
tropical woodland

Xerophytic
vegetation

Present

Figure 7.15 Elevational shifts in vegetation zones in the eastern Cordillera of the Andes in Colombia in response to climatic change following the most recent glacial maximum. Note that while all zones tended to shift in concert, the upper zones became narrower as they shifted upward in response to global warming. (After Flenley 1979a.)

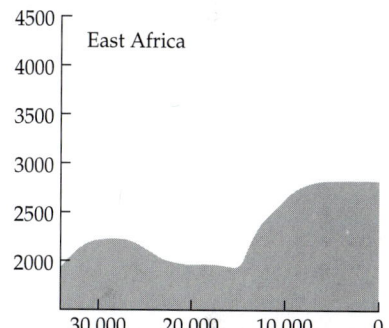

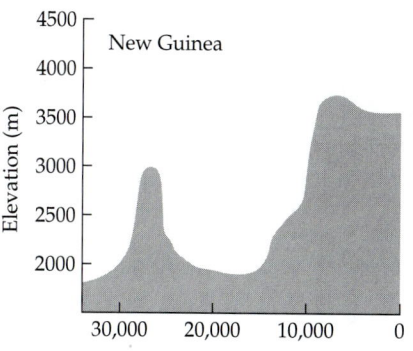

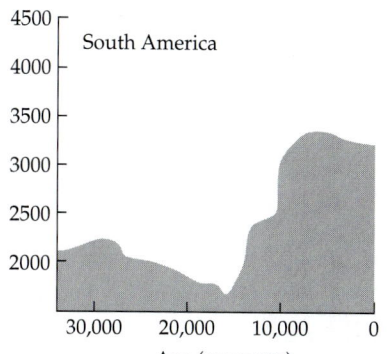

Figure 7.16 tracks the probable upper elevational limit of subtropical forests in three equatorial regions (East Africa, New Guinea, and South America). Each graph suggests several drastic elevational shifts for these forests during the last 34,000 years. The curves are not identical, but in each case the upper elevational limit of tropical forest began to decrease gradually beginning about 29,000 to 27,000 years B.P., but then reversed and increased sharply about 16,000 to 15,000 years B.P. Flenley (1979) also noted that in Africa, South America, Indo-Malaya, and New Guinea, montane vegetational zones were depressed by about the same amounts (1000 to 1500 m) at roughly the same times.

Just south of New Guinea, Queensland, in northeastern Australia, exhibited an exceptional and rapid change from sclerophyllous woodland to rain forest

Figure 7.16 Shifts in the upper elevational limits of tropical forests in three different regions (South America, East Africa, and New Guinea) during the past 33,000 years. Note that these elevational shifts were synchronous, although the exact upper elevational boundary was somewhat different in each region. (After Flenley 1979a.)

between the late Pleistocene and 6000 years B.P. (Figure 7.17). The latter point helps to illustrate how difficult it is to extrapolate trends in terrestrial paleoclimatology and to understand vegetational history merely by studying present-day vegetation. Again, the responses to glacial conditions were not uniform across the globe. While African and Amazonian rain forests contracted in response to glacial aridity, those of Sumatra remained intact (see Flenley 1979; Maloney 1980).

Geographic shifts in response to the climatic changes of the Pleistocene were even more complex for individual species. Rather than responding as discrete units to shifts in climates, biomes and communities often disinte-

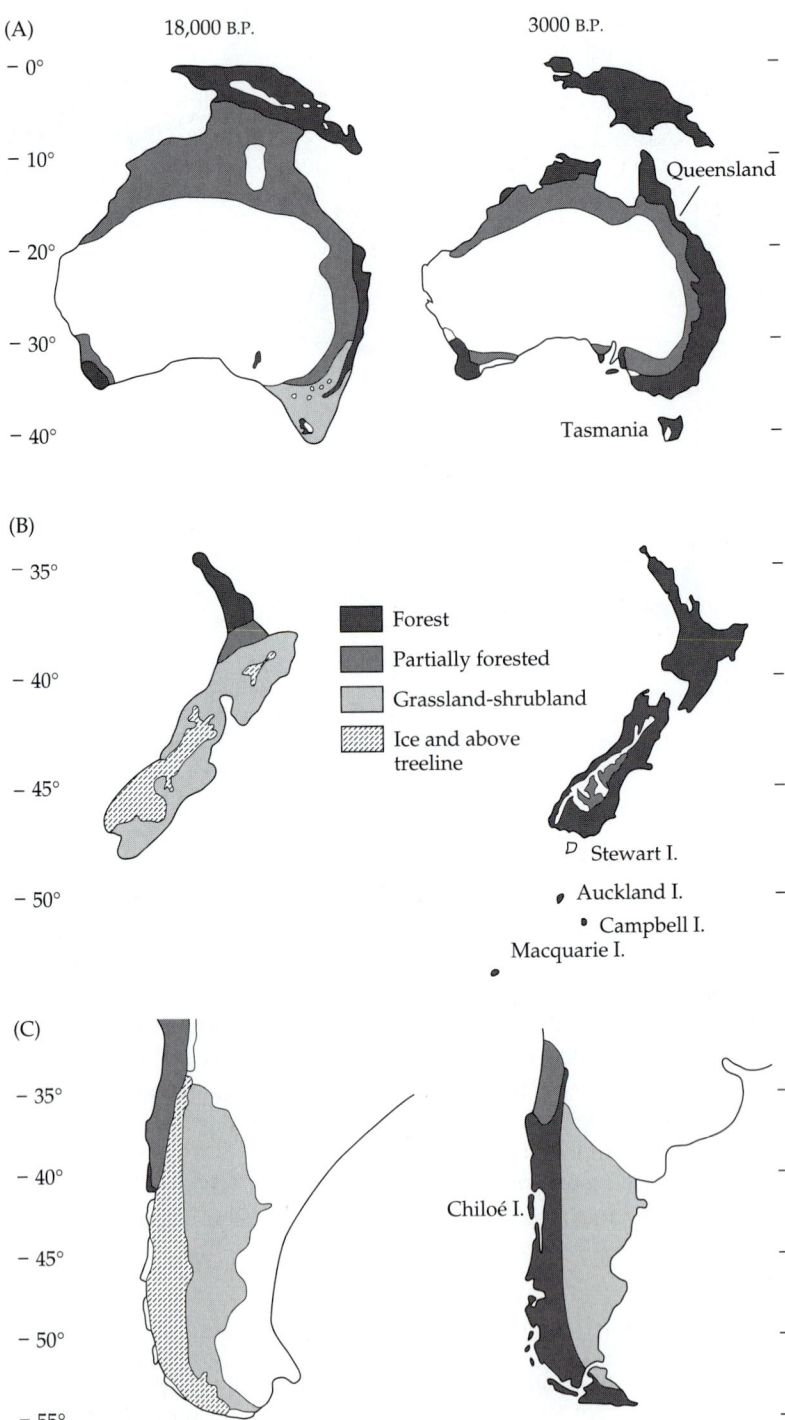

Figure 7.17 Comparisons of distributions of vegetation zones during the most recent glacial maximum (18,000 years B.P.) and 3,000 years B.P. (i.e., before significant disturbance by humans) for three regions of the Southern Hemisphere: (A) Australia; (B) New Zealand; (C) southern South America. (After Markgraf et al. 1995.)

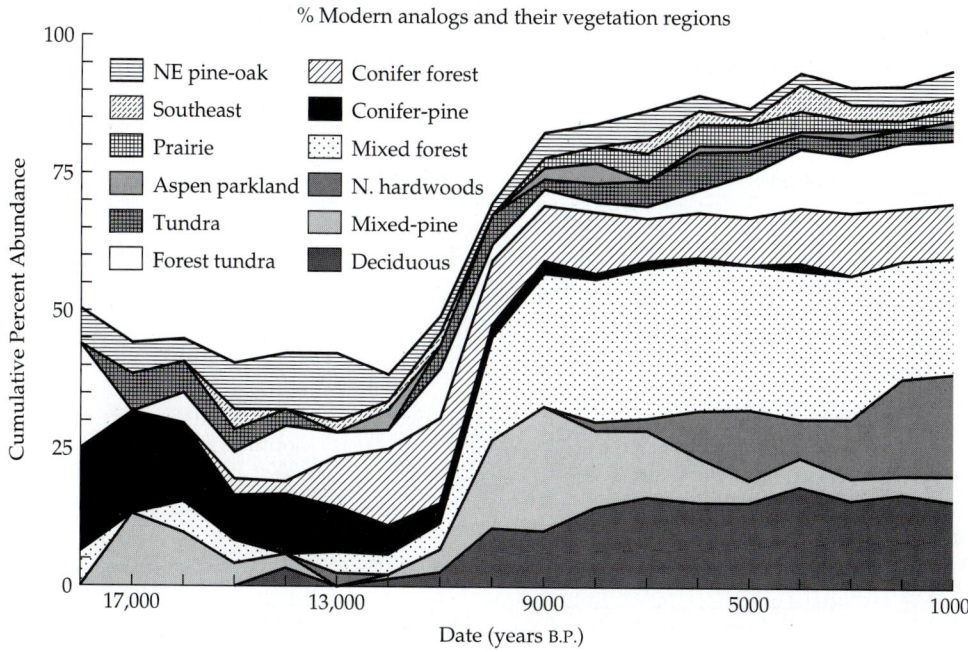

Figure 7.18 Variation in the relative abundance of different types of vegetative communities in North America since the most recent glacial maximum (based on samples of fossil pollen). Note that the most rapid changes occurred between 14,000 and 10,000 years B.P. (After Shane and Cushing 1991.)

grated, with many, if not most, species responding in an individualistic manner. Shifts in species ranges were influenced by both extrinsic and intrinsic factors. Extrinsic factors included climate, soils, prevailing winds, ocean currents, and topographic features such as those discussed above. The colonization of deglaciated areas of North America by beech provides an exemplary case study of the influence of these factors. According to Margaret Davis (1986), after beech became established along the southeastern shore of Lake Michigan (about 7000 years B.P.), it took another 1000 to 2000 years to reach the far side of the lake. Birds, including the now extinct passenger pigeon (*Ectopistes migratorius*), may have played an important role in carrying the seeds of beech and many other species across such geographic barriers. The potential for range shifts also may have been strongly influenced by other interspecific interactions, especially competition, predation, and parasitism (Figure 7.18).

Individualistic responses to climatic change among species derive largely from their considerable differences in ability to respond to these extrinsic factors. Physiological tolerances, behavior, life history strategies, intrinsic rates of increase, and dispersal abilities all vary markedly among species, even those occupying the same community. Pollen analyses and other paleontological records document considerable interspecific variation among tree species in rates of migration (Table 7.2). Average rates of range extension during the Holocene in North American trees varied from 100 to 400 meters per year. As a result, some species now recognized as dominants may be relatively recent arrivals in contemporary communities. The American chestnut (*Castanea dentata*), for example, was a dominant species in oak-chestnut forests of the Appalachian Mountains for over 5000 years, but it was a relatively recent invader of similar forests in Connecticut (arriving about 2000 years B.P.). In western North America, lodgepole pine (*Pinus contorta* var. *latifolia*) has extended its range at about 200 meters per year, reaching southern Alaska just a few centuries ago, and may still be expanding its range northward (Figure 7.19A).

Rates of range expansion can vary considerably within, as well as among, species. At the end of the Wisconsin in North America, for example, white spruce (*Picea glauca*) migrated northward along both the eastern and western

Figure 7.19 Rates of migration of trees following glacial recession varied substantially among species, and were strongly influenced by extrinsic factors such as topographic features and prevailing winds. (A) Inland varieties of the lodgepole pine (*Pinus contorta* var. *latifolia*) have expanded their range northward over the past 12,000 years, and may be continuing this range expansion in modern times. The dots indicate northern range boundaries at various times based on pollen samples. (B) Northward range expansion of the white spruce (*Picea glauca*) during the retreat of the Laurentide glacier was strongly influenced by the prevailing winds (hollow arrows). The species moved rapidly once it reached the western edge of the glacier, where north-flowing winds aided its range expansion. Dots indicate locations of sample points. The long arrows indicate slow migration; the short arrows, rapid migration. (After Pielou 1991.)

Table 7.2
Average rates of Holocene range extensions of trees (m/yr) following glacial recession

Species	North America	European mainland	British Isles
Pines	300–400	1500	100–700
Oak	350	150–500	350–500 (50 near northern limit)
Elm	250	500–1000	550 (100 near northern limit)
Beech	200	200–300	100–200
Hazelnut	—	1500	500
Alder	—	500–2000	500–600 (50–150 near northern limit)
Basswood	—	300–500	450–500 (50–100 near northern limit)
Ash	—	200–500	50–200
Spruce, Larch Balsam Fir, Maples, Hemlock and Hickory	200–250	—	—
Chestnut	100	—	—

Source: North American data after Davis 1981; European mainland and British Isles data after Huntley and Birks 1983.

Note: For most of these species, the rate of range extension slowed as they approached their northern limits.

margins of the retreating Laurentide glacier. From 14,000 to 7000 years B.P., it migrated along the eastern edge of the ice sheet at about 300 m per year. In contrast, its migration along the western edge of the ice sheet, where north-flowing winds aided its dispersal, was nearly an order of magnitude faster (Figure 7.19B).

Species varied not only in their rates of range expansion, but in the directions as well, with many species shifting as much in longitude as in latitude (Figure 7.20) (e.g., see Graham 1986; Graham et al. 1996). In a sim-

(A)

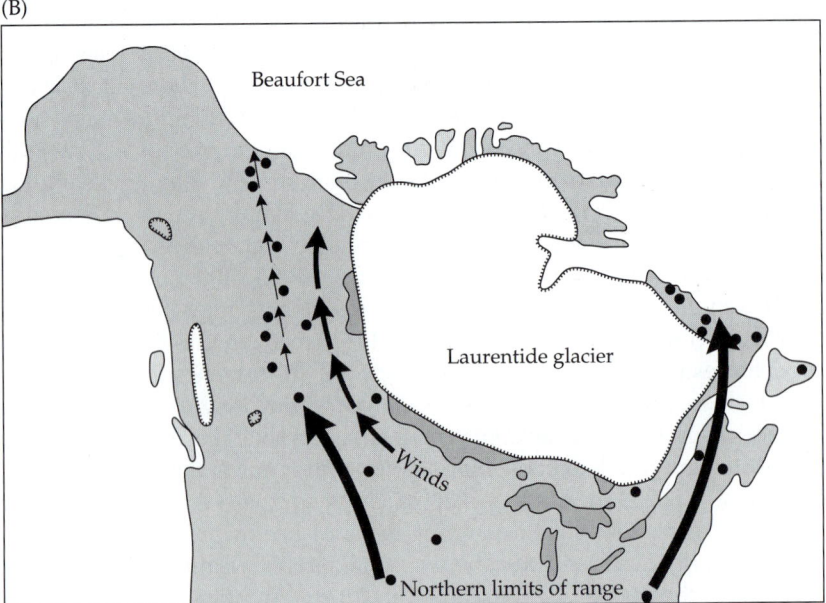

(B)

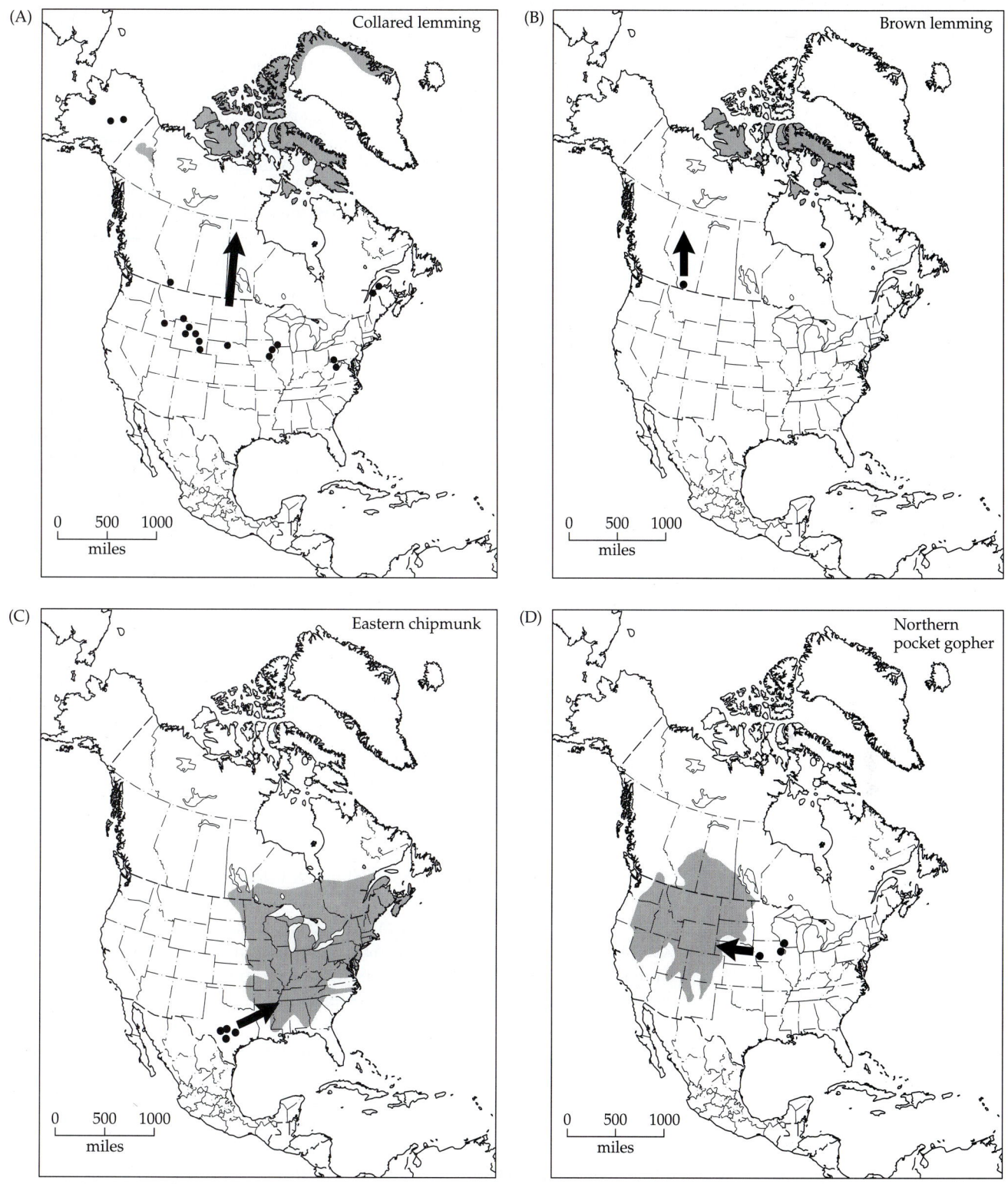

Figure 7.20   Geographic range shifts in four species of rodents during the Holocene. Shaded areas represent the present range; dots indicate locations of late Pleistocene fossils. Note that these range shifts differed in extent, with collared lemmings (*Dicrostonyx*) shifting farther northward than brown lemmings (*Lemmus*), and in direction, with the above species shifting northward, while eastern chipmunks (*Tamias striatus*) shifted toward the northeast and northern pocket gophers (*Thomomys talpoides*) shifted toward the west. (After Graham 1986.)

ilar manner, many plants and animals inhabiting mountainous regions exhibited individualistic shifts in elevation (e.g., see Jackson and Whitehead 1991). The overall result of these individualistic responses was a great reshuffling of communities from one climatic reversal to the next. Again, animals tended to shift much more rapidly than plants, apparently influenced more strongly by vegetative structure than by abundances of particular plant species. Thus, glacial reversals often caused decoupling of communities. In the subtemperate latitudes of the American West (30° to 40° N latitude), for example, glacial and contemporary communities typically share fewer than one-third of their plant species. While biomes shifted in a predictable manner, the species composition of each system varied significantly, often with communities of one glacial stage having no comparable analogue in the next (Figure 7.21).

Not only were species decoupled and reshuffled among communities, but communities and biomes may never have reached equilibrium with their Pleistocene climates (Davis 1986; but see Wright, 1976; Webb 1987). Animals, and especially plants, lagged far behind climatic zones, which shifted at least an order of magnitude more rapidly (Gates 1993). In fact, the plant communities of the Holocene are often viewed as ephemeral collections of species that repeatedly disassociated and reassociated with each other as climates changed (Behrensmeyer et al. 1992).

Dynamics of Plant Communities in the Southwestern United States

Many of the biogeographic dynamics discussed above can be illustrated by detailed accounts of vegetative shifts in arid regions of the southwestern United States. Because the deserts of the Northern Hemisphere lie between about 30° and 40° N latitude (see Chapter 3), they were not covered by ice sheets or mountain glaciers. Yet these and other regions far removed from the glaciers were still strongly influenced by the glacial cycles of the Pleistocene. In fact, if it were possible to visit these desert regions 12,000 to 18,000 years ago, such a visitor would be rather surprised to find substantially different climates and vegetation.

Because dry deserts are notoriously poor environments for fossilization, it has been difficult to reconstruct the biotic history of deserts. Is a particular desert old or young? Did deserts move closer to the equator during glacio-pluvial periods, or did they stay where they were, but with greatly contracted areas? Are the desert communities seen today the same ones that existed before and during the Pleistocene? Where did desert vegetation survive during the mesic pluvial periods? Many of these questions remain largely unanswered, but they are the focus of much current research. Traditionally, the history of deserts has been inferred mainly from theoretical paleoclimatological models, reinforced wherever possible by fossils, especially pollen records from the sediments of pluvial lakes. Neither provides a very reliable account of the small-scale history of deserts.

In recent decades, a new type of fossil data has been used to reconstruct the vegetational history of one complex region: the desert zones of the semiarid and arid southwestern United States (e.g., Betancourt et al. 1990; Grayson 1993). Pack rats (*Neotoma*) are abundant rodents in xeric habitats. They hoard plant materials in large caches, which are sometimes protected in caves or under rock ledges. The remains of these caches become solid structures, called **middens**, and persist for thousands of years if kept dry. They are excellent sources of plant fossils because they provide a relatively complete and often quite continuous sample of the plants growing within 100 m of the rats' den during its occupation. By collecting pack rat middens at different elevations and locations and dating the materials using radiocarbon methods, researchers can reconstruct the shifts in elevation and composition of vegetation types and deduce past climatic regimes (i.e., those back to 40,000 years B.P.).

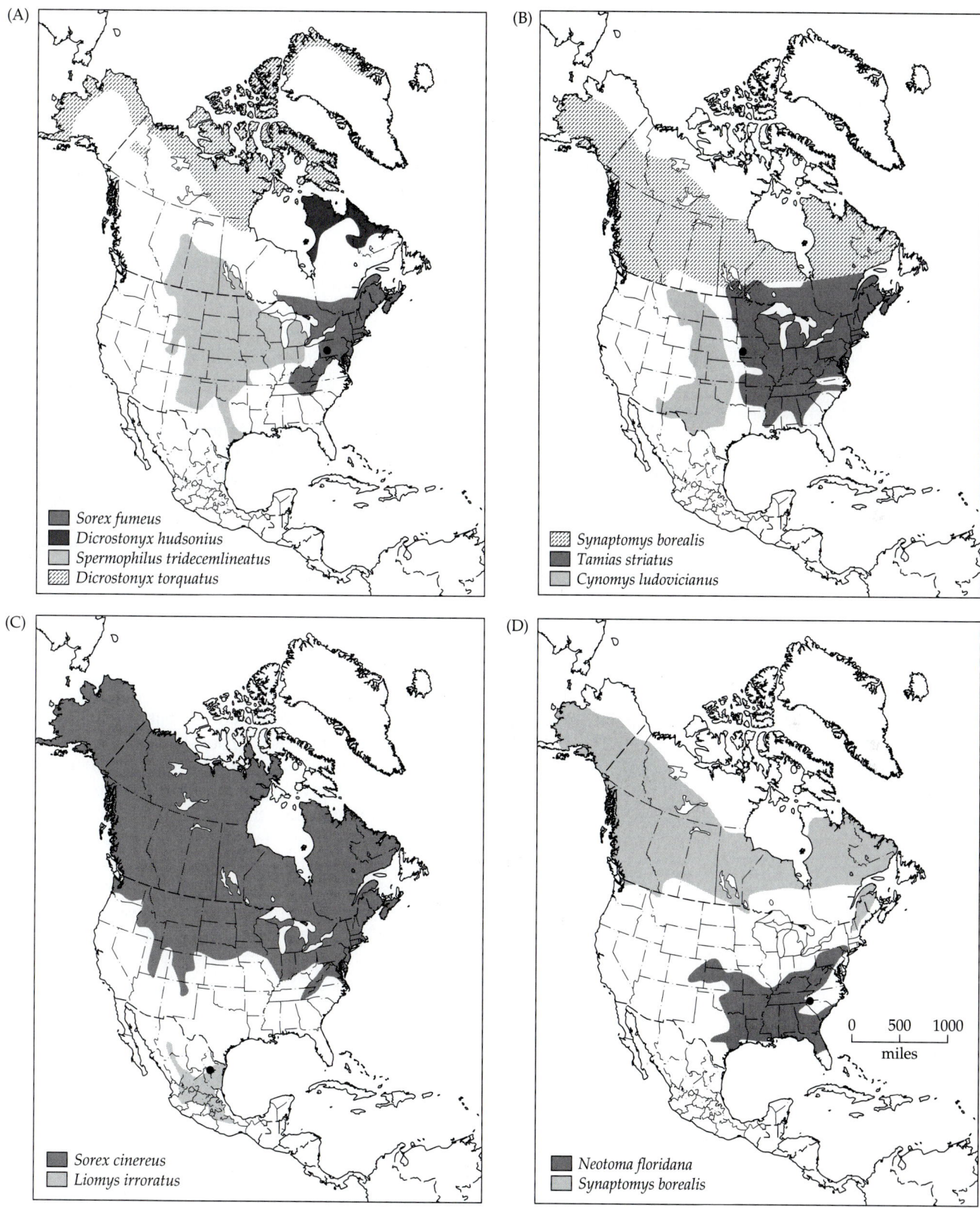

Figure 7.21   As a result of the differences in extent and direction of range shifts during the Holocene, species that co-occurred during the most recent glacial maximum often exhibit disjunct ranges today. In each map, the black dots indicate coincident occurrences of the named species during the late Pleistocene, while the shaded areas indicates their current ranges. (After Graham 1986.)

Data from pack rat middens are continuing to be synthesized (Van Devender 1977; Van Devender and Spaulding 1979; Wells 1979; Betancourt et al. 1990; see also Fritts 1976), but some general trends are apparent. It is clear that there have been major changes in the vegetation at all elevations throughout the arid Southwest within the last 20,000 years. In general, climates in this region were substantially cooler and wetter during the most recent (Wisconsin) glacial maximum. During this period, vegetation zones were displaced as much as 500 to 1000 m below their present limits (Figure 7.22). The change to the present configuration of climates and vegetation types occurred primarily within the last 8000 to 12,000 years.

In general, pack rat middens reveal a history consistent with that found in pollen records and geological studies of pluvial lakes, but they provide a much more detailed picture of vegetation changes. It is now clear that plant communities did not simply move as entire entities up and down the mountains. Rather, they changed dramatically in composition. For example, during the most recent glacial maximum (21,000 to 15,000 years B.P.), most plant species in the Grand Canyon occurred 600 to 1000 m lower than at the present time. This finding indicates a cooler, wetter climate (Cole 1982). Many of the species that inhabited areas along the rim of the Grand Canyon, however, are no longer found in similar communities in the same region today. In fact, the community contained several species that are presently found in northeastern Nevada and northwestern Utah, at least 500 km to the north (Cole 1982).

Many of the plants now dominant in the coniferous forests of the intermountain West (e.g., ponderosa pine and piñon pine) were rare and restricted in glacial times. On the other hand, species that were much more widespread 10,000 to 30,000 years ago are now narrowly distributed or no longer occur in this region.

Aquatic Systems: Postglacial and Pluvial Lakes

Glacial (cryogenic) lakes.   No other force of lake formation can compare with the glacial activity of the Pleistocene (Hutchinson 1957; Wetzel 1975). As glaciers melted, many shallow-water marine systems were decimated by rising waters. Throughout the world, great volumes of meltwater created rivers and lakes of such magnitude that they dwarf their contemporary analogues. Modern Lake Superior, the world's largest freshwater lake, is less than a fourth the size of Glacial Lake Agassiz, which covered approximately 350,000 km^2 during the early Holocene. Like Lake Agassiz, most of these postglacial lakes developed between 11,000 and 12,000 years B.P. and peaked in extent

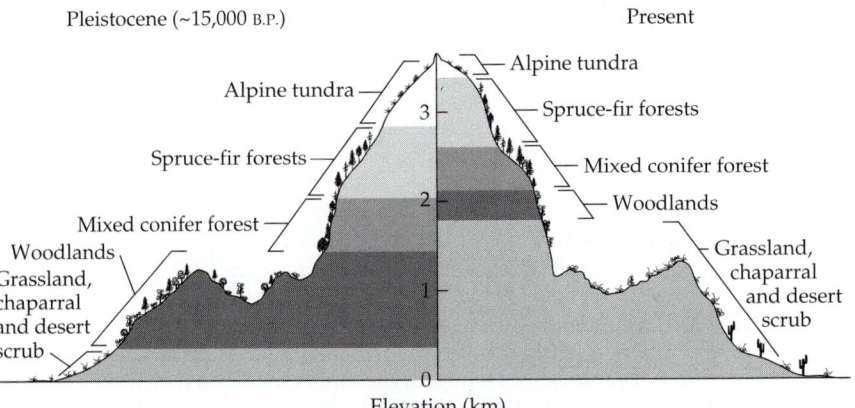

Figure 7.22   Elevational shifts in vegetation zones in the mountainous region of the American Southwest near southern Arizona. (After Lomolino et al. 1989; Merriam 1890.)

Pleistocene (~15,000 B.P.)

Present

Alpine tundra

Alpine tundra

Spruce-fir forests

Spruce-fir forests

Mixed conifer forest

Mixed conifer forest

Woodlands

Woodlands

Grassland, chaparral and desert scrub

Grassland, chaparral and desert scrub

Elevation (km)

about 10,000 to 9000 years B.P. While many of them were long-lived, their demise was often catastrophic and spectacular. These lakes were typically dammed along at least one side by the remnants of a receding glacier (Figure 7.23A). As the glaciers melted, these ice dams eventually gave way in explosive outbursts. For example, postglacial Lake Missoula, equivalent in size to modern-day Lake Ontario, emptied its 2000 km^3 volume of water in less than two weeks (McPhail and Lindsey 1986, cited in Pielou 1991).

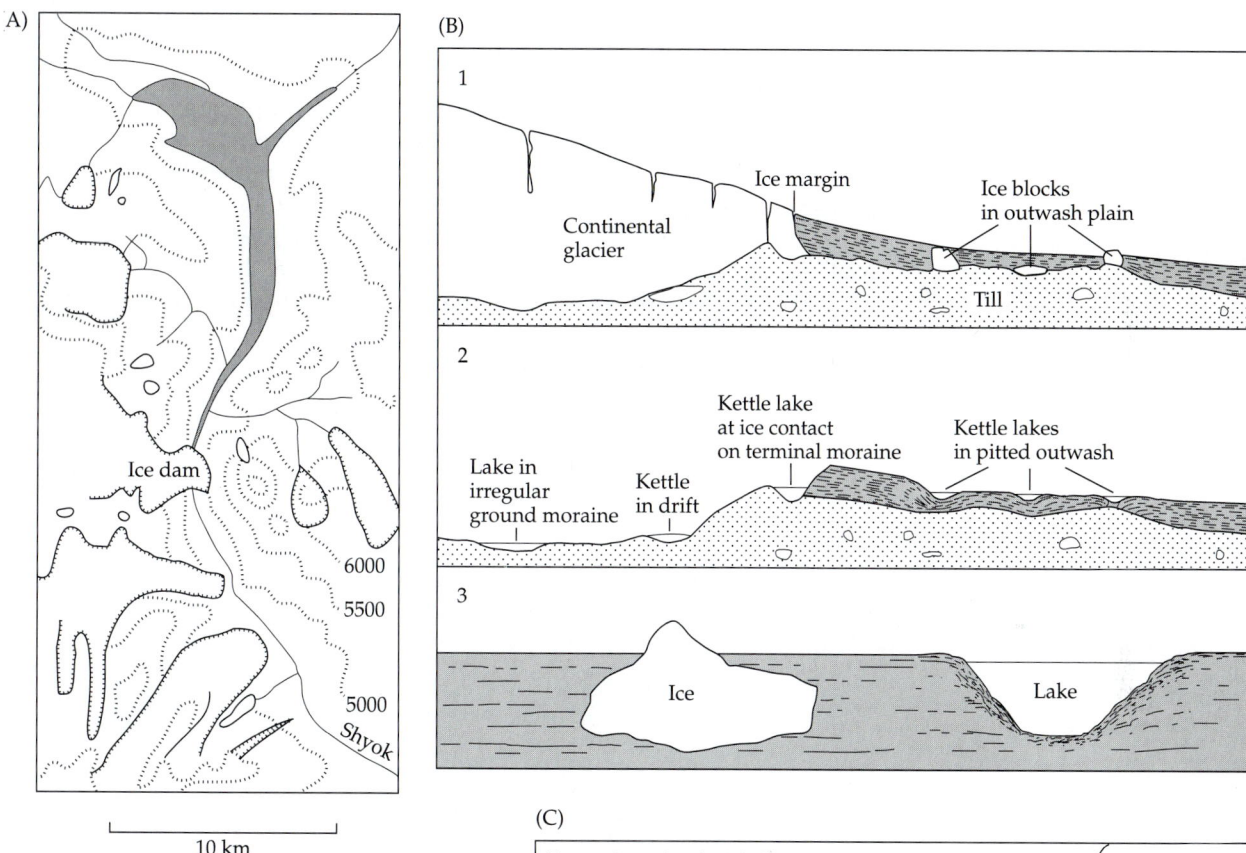

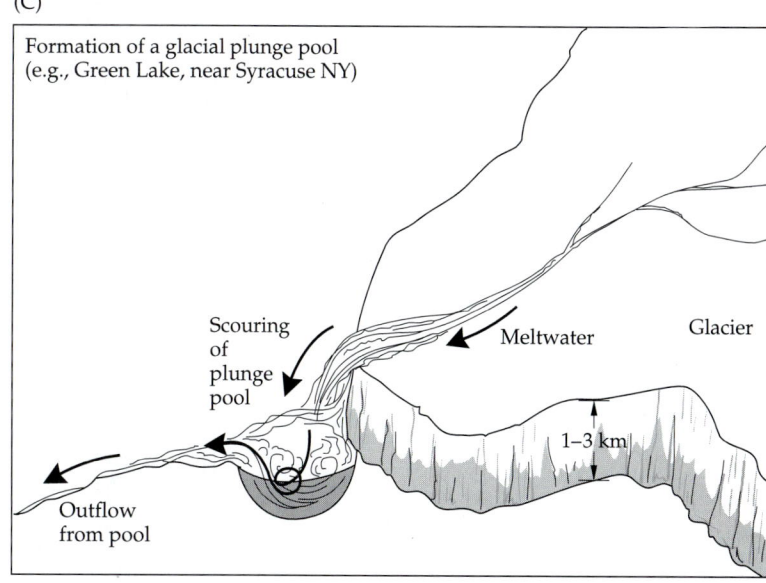

Figure 7.23 Many of today's lakes were formed by glacial activity. (A) Shyok Ice Lake along the Western margin of the Himalaya Mountains, one of many examples of a postglacial lake formed when retreating glaciers acted as dams, with glacial meltwater accumulating in valleys carved by previous glacial activity. (B) Formation of kettle lakes: 1, retreating ice sheet leaves stagnant ice blocks on the outwash plain; 2, lakes are formed in the outwash and in the till by melting blocks of ice; 3, a large, partly buried ice block melts, causing irregular sliding of the outwash originally covering the sides of the block. (C) Formation of a glacial plunge pool lake. Glacial meltwater sometimes formed large rivers that flowed over the surface of the retreating glacier, then plunged as much as 3 km down the face of the glacier to carve out a roughly circular lake basin below. (A and B after Hutchinson 1957.)

Retreating glaciers often left great masses of ice in their wake. These enormous blocks of ice often persisted for many centuries, leaving their imprints as prairie potholes in temperate regions of the Holarctic. The largest blocks of ice created deep and persistent depressions that later filled to form **kettle lakes** (Figure 7.23B). The legacy of glacial recession can also be seen in the relatively deep, roughly circular lakes called **plunge pools**, which were formed by glacial meltwaters that flowed over the surface of the glacier before plunging off its edge to carve a basin in the earth some 2 to 3 km below (e.g., Green Lake near Syracuse, New York; Figure 7.23C). Countless other lakes were formed by the scouring action of glaciers and meltwater, which formed basins, many of them dammed by the deposition of rock debris along glacial moraines (e.g., Lake Mendota and Devil's Lake of Wisconsin and the Finger Lakes of New York).

Pluvial lakes in arid regions.   Glacial cycles affected plant and animal communities in regions far withdrawn from glaciers, including those that are now dominated by deserts. During the pluvial periods large freshwater or saline lakes formed in these regions because of a combination of low evaporation rates (primarily due to lower temperatures) and high precipitation rates ("pluvial" means "rainfall"). Few biogeographers appreciate the size and number of pluvial lakes that existed in places that now have desert climates. A splendid

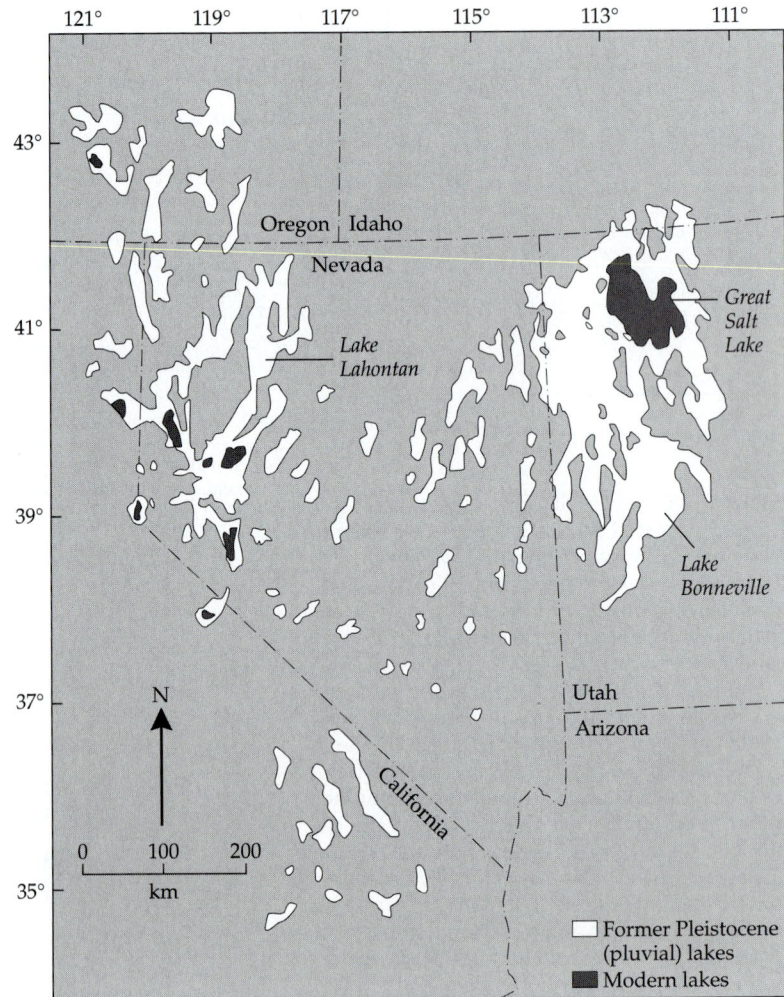

Figure 7.24   Distribution of pluvial lakes in western North America during the most recent (Wisconsin) glacial maximum. During glacio-pluvial periods most of the arid region of the continent experienced a wetter and cooler climate, and lakes and marshes filled what are now desert valleys. (After Elias 1997; Benson and Thompson 1987.)

area in which to study ancient pluvial lakes is the southwestern United States. This desert region has a **basin-and-range topography**, in which many low, flat areas (**basins**) are interrupted by isolated mountain ranges. During pluvial times, many of these basins filled with water (Figure 7.24). The largest such water body was Lake Bonneville in Utah and parts of Nevada and Idaho. At times it contained fresh water and drained northward into the Snake and Columbia Rivers. In the Middle Wisconsin, this lake was 330 m deep, had an area exceeding 50,000 km^2 (slightly smaller than present-day Lake Michigan), and supported a freshwater community that included cutthroat trout (*Salmo clarki*) and other fish. The present Great Salt Lake is a small remnant of Lake Bonneville. Nevada, southern Oregon, eastern California, southeastern Arizona, and southwestern New Mexico had numerous large and small pluvial lakes. Yet most of these lakes had evaporated by 10,000 years ago. One of these lake basins, Death Valley, with the lowest elevation in North America (–93 m), now contains perhaps the most extreme desert on the continent.

Pluvial lakes were also present in what are now deserts on other continents. Their remnants—saline lakes and dry lake beds—are abundant in the Atacama Desert of northern Chile, the Monte of Argentina, many areas in interior Australia, the region of the Dead Sea in the Middle East (ancient Lake Lisan), the Kalahari Desert of southern Africa, and many places in arid and semiarid Asia (Flint 1971). The western portion of the Sahara Desert was still relatively mesic until 5000 years ago. Perhaps the most remarkable of these remnants is Lake Chad. This lake, now just 16,000 km^2, was once 950 km long and covered over 300,000 km^2 (the present size of the Caspian Sea), including a significant portion of the southern Sahara Desert. Lake Chad remained at this maximum size from 22,000 to 8500 years B.P.

The disappearance of pluvial lakes during the Holocene had several profound biogeographic effects. It caused the wholesale extinction of many plants and animals living in or around these bodies of water. In addition, the dissection of large lakes into smaller, isolated units led to the vicariant speciation of surviving forms, as with the pupfishes (*Cyprinodon*) of the southwestern United States (Miller 1961b; Smith 1981).

Biotic Exchange and Glacial Cycles

Figures 7.15 and 7.22 demonstrate how downward shifts of high-altitude vegetation (primarily forests) during glacial maxima could have created avenues of dispersal: such a lowering would allow a plant or animal species, previously isolated on individual peaks, to cross ridges and migrate along mountain ranges. This mechanism is precisely the one invoked by many authors to account for the intracontinental dispersal of vegetation types and associated animals. When followed by isolation as mesic biomes contracted, it could result in taxonomic disjunctions (Simpson 1975; Vuilleumier and Simberloff 1980).

Glacial cycles may have had similar effects on the distributions of many marine organisms (Figure 7.25). In the case of cold-water stenothermal species, relatively warm tropical waters serve as an effective physiological barrier to dispersal, limiting their distributions to the middle or high latitudes of the Northern or Southern Hemisphere. During glacial maxima, however, cooling of marine waters could have allowed range expansion into the lower latitudes. Subsequent rewarming of tropical waters during interglacials could have again caused range contraction and possibly bipolar distributions of these species (Figure 7.25C).

As we noted earlier, eustatic and isostatic changes in sea level greatly altered opportunities for biotic exchange for both terrestrial and marine biotas (see Figure 7.9). Greatly reduced sea levels during the Wisconsin created

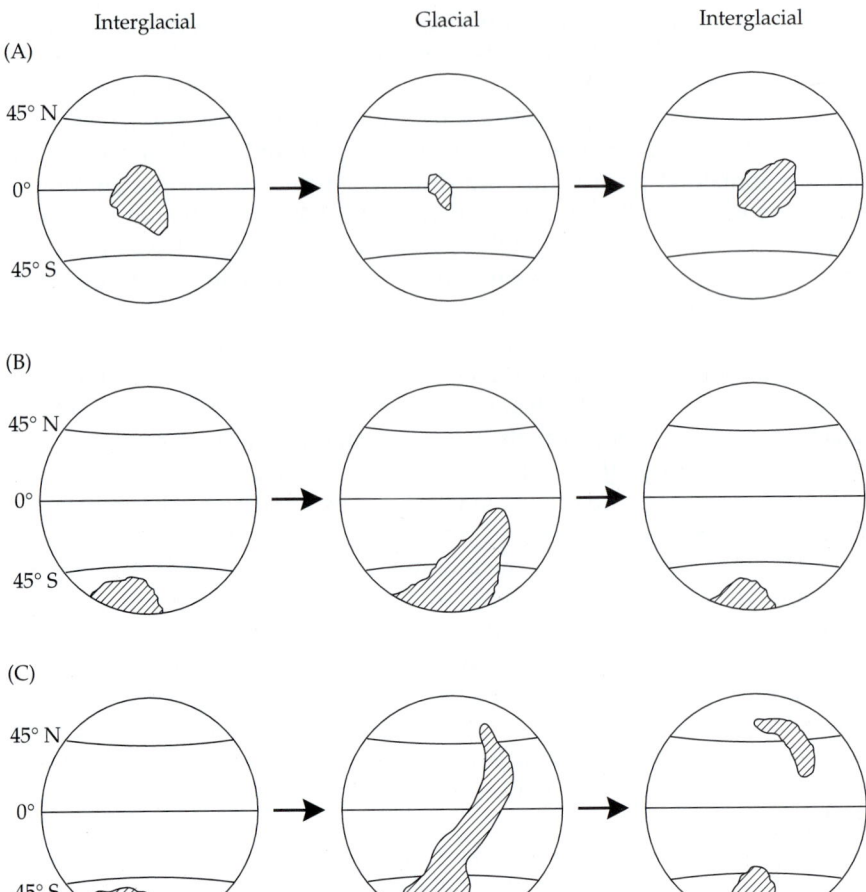

Figure 7.25 Potential range expansions and contractions of hypothetical warm stenothermal (A) and cool stenothermal (B and C) marine taxa. During a glacial period, cool stenothermal taxa could penetrate from one hemisphere to the other (middle diagram in C) and could possibly achieve bipolar distributions when waters rewarm and their ranges contract toward the poles. (After Crame 1993.)

extensive landbridges, such as Beringia (connecting Siberia and North America), the Sunda Shelf (connecting Malaysia, and Indonesia), and the Arafura Sea and Bass Straits (connecting New Guinea, Australia, and Tasmania). While these landbridges eliminated and fragmented marine biotas, they served as important dispersal corridors for terrestrial organisms. In most cases, however, biotic exchange was asymmetrical, with more species migrating from larger (species-rich) to smaller areas than vice versa (e.g., from Siberia to Alaska; from Southeast Asia to the "islands" of the Sunda Shelf; from Australia to Tasmania). Biotic exchange of terrestrial organisms across Beringia (see below) contributed significantly to the similarity between Nearctic and Palearctic biotas. Our own species used this glacial landbridge to colonize North America from Siberia.

Similarly, biotic exchange of marine organisms often tended to be asymmetrical, depending on the size and diversity of each species pool and the ocean currents and other factors influencing dispersal. While Beringia served as a dispersal corridor for terrestrial organisms during glacial maxima, the Bering Strait was an important corridor for the dispersal of marine life during interglacial periods. Again, more species migrated from the larger, more species-rich region—that is, from the Pacific basin northward. During the late Cenozoic, 125 species of marine invertebrates invaded the Arctic-Atlantic region from the Pacific, while no more than 16 species colonized in the reverse direction (Durham and MacNeil 1967; see also Vermeij 1991).

In Chapter 6 we discussed the tectonic events that ultimately formed a Central American landbridge between North and South America (about 3.5 million years B.P.). The resultant waves of biotic exchange between Nearctic and Neotropical biotas, referred to as the **Great American Interchange**, were made possible not just by tectonic events, but by eustatic changes and vegetative shifts associated with glacial cycles. During glacial maxima, the lowering of sea levels increased both the area and the elevation of the Central American landbridge. Perhaps just as important, the relatively dry conditions that prevailed during glacial maxima caused savannas to expand toward the equator and form a continuous habitat corridor for the migration of many species adapted to these open habitats (savannas and shortgrass prairie; see Webb 1991). We shall return to the profound effects of these waves of biotic interchange in Chapter 16.

Evolutionary Responses and Pleistocene Refugia

As we shall see in Chapter 8, evolutionary divergence is intimately related to biogeographic dynamics and the resultant mixing or isolation of gene pools. Given the repeated shifting and fragmentation of species ranges during the Pleistocene, glacial cycles may have had profound evolutionary effects. Here we focus on the biogeographic and evolutionary responses of Pleistocene biotas confined to two types of hypothesized refugia: Neotropical forest refugia and glacial refugia, or nunataks.

Neotropical Pleistocene refugia.   Lowland tropical rain forests (see Chapter 5) are awe-inspiring structures, teeming with an amazing diversity of life. Traditionally such places were regarded by many investigators as stable refuges that had persisted virtually unchanged for many millions of years while climates became harsh and fluctuating elsewhere. However, more recent lines of evidence suggest that these rain forests have experienced marked climatic changes, so that they have varied greatly in distribution within the last 40,000 years, sometimes being much more restricted than at present.

Within the Amazon basin, which contains the most extensive continuous rain forest in the world today, researchers have found centers of high species richness and relatively high endemism for both plants and animals. Haffer (1969, 1974, 1978, 1981), for example, identified six principal areas to which about 150 species of birds are narrowly restricted (Figure 7.26). Frequently these areas are inhabited by several subspecies of the same species, or by several similar species of the same superspecies. This pattern is repeated in families of reptiles, amphibians (Vanzolini and Williams 1970; Dixon 1979; Lynch 1979), woody plants (Vuillemier 1970; Prance 1973, 1978, 1982), and butterflies (Brown, K. S. 1982), although a greater number of centers can be identified for plants and butterflies.

These insular patterns of species richness and endemism in what had seemed to be homogeneous habitats suggested to Haffer and others that these centers were **refugia**—islands of lowland rain forest that persisted during glacial maxima. Haffer and his colleagues developed a model of **cyclical vicariance** —also known as the **speciation pump model**—to attempt to account for the amazing diversity of tropical rain forests. According to their model, populations of wide-ranging species dependent on tropical forests became isolated as their prime habitat became fragmented during glacial maxima (Figure 7.26). These relatively dry glacial periods lasted for up to 100,000 years, thought to be long enough for significant evolutionary divergence among isolated sister populations in different refugia. As the glaciers receded, these isolated rain forests expanded and eventually reconnected to form the continu-

(A)

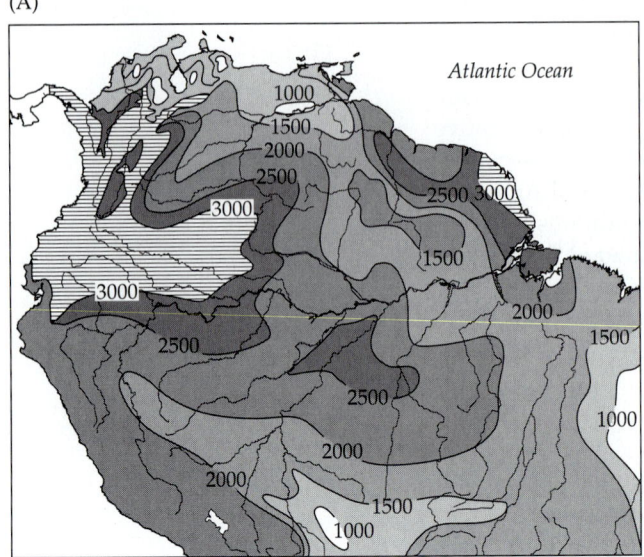

(B)

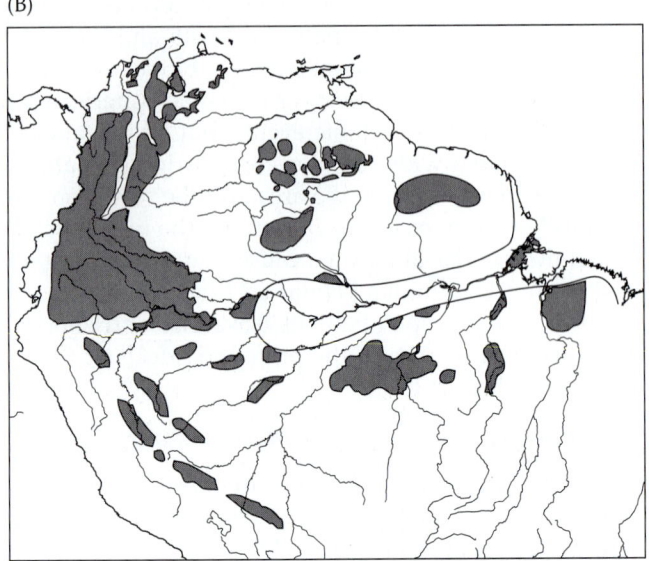

(C)

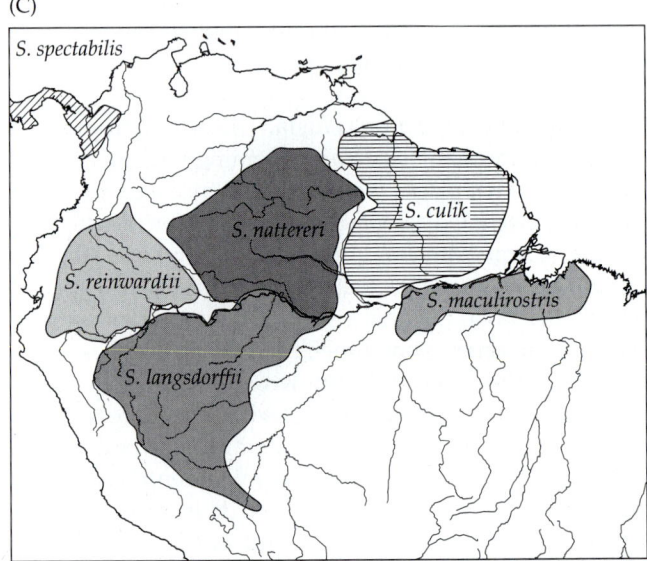

Figure 7.26   Haffer's Pleistocene refugium hypothesis was proposed to account for the relatively high species richness and endemicity of Amazonian plants and animals. (A) Contemporary patterns of rainfall were used to identify the locations of Pleistocene rain forest refugia (numbers indicate rainfall in mm per annum). (B) Regions receiving more than 2500 mm of rainfall annually were postulated to have served as forested refugia during the Pleistocene. (C) The limited geographic distributions of a number of species groups (such as the toucanets, *Selenidera* spp., in this example) seemed consistent with the postulated distributions of Pleistocene refugia, and with the hypothesis that these refugia served as speciation pumps. (After Terborgh 1992.)

ous forest of Amazonia. Populations from the various refugia likewise expanded their ranges and came into contact with related populations from adjacent refugia. Haffer postulated that their divergence during their isolation may have been reinforced by mechanisms promoting reproductive isolation in these zones of secondary contact. However, complete reproductive isolation between many close relatives apparently was not achieved during the most recent glacial cycle. In insects, lizards, and birds, zones of secondary contact are regions of frequent hybridization between closely related forms.

Although Haffer's refugia hypothesis provided an interesting explanation for the high diversity and relatively high endemism of tropical rain forests, over the past decade or so, many scientists have questioned its validity based on the following objections (see Prance 1982). First, Haffer and his colleagues may have greatly overestimated the reduction of tropical rain forests during glacial maxima. Because cooling was relatively uniform across Amazonia,

Neotropical rainforests and their associated populations may not have been as fragmented and isolated as hypothesized (see Colinvaux 1989; Colinvaux et al. 1996). Second, molecular data and other genetic and phenotypic data indicate that many of the endemic forms are much older than the hypothesized refugia (i.e., older than 18,000 years B.P.; see Cracraft and Prum 1988; Hackett and Rosenburg 1990; Hackett 1993; Marshall and Lundberg 1996). Third, there appears to be little overlap in centers of endemism among different taxa (see Beven et al. 1984). Finally, as interesting and provocative as it was, Haffer's hypothesis is not the most parsimonious explanation for the high diversity and endemism of today's relatively continuous tropical rain forest. Tropical forest ecosystems are much more spatially heterogeneous than once assumed. It now appears that the hypothesized "refugia" tend to be areas of relatively high elevation, which have distinctive abiotic environments and vegetation.

Despite these objections, Haffer's hypothesis served an important purpose by stimulating an increased emphasis on understanding the effects of glacial cycles on biotas far removed from the glaciers. Although Haffer's original model now seems tenuous, it is possible that during an earlier, more severe glacial period, Neotropical forests were sufficiently fragmented and isolated to promote biological diversification.

Nunataks: Glacial refugia.   **Nunataks**, refugia that persisted within or adjacent to ice sheets, are a popular topic in Pleistocene phytogeography. Nunataks could have developed along the continental periphery of a glacier, internally as isolated pockets along a mountain range (for example, on the tops of the highest peaks), along a coastline where bluffs are too steep for glaciers, or between adjacent ice sheets. Geologists have identified several possible "internal" nunataks. The Laurentide ice sheet covered most of northeastern and north central North America. The most famous nonglaciated region within this ice sheet is the so-called **driftless area** in southern Wisconsin and adjacent Illinois and Iowa, an elliptical area that was bypassed by the glacial front. The Cordilleran glacier complex occurred in westernmost Canada, extending from the Canadian Rockies and adjacent southern Alaska (through the Aleutians) to the coast, and including small portions of northern Washington, Idaho, and Montana. A number of narrow nunataks may have occurred between the Laurentide and Cordilleran ice sheets and in intermountain valleys (Figure 7.27).

Numerous phytogeographers have used plant distributions as evidence for other northern nunataks, but the response to these, especially from geologists, has not always been favorable. Reminiscent of the hypothesized Neotropical refugia, these nunataks are alleged to have occurred where investigators have found small, isolated pockets of arctic-alpine vegetation that are especially rich in species, including rare and endemic forms. Examples include two alleged nunataks in Norway, which contain some interesting disjunctions. The Lapland rosebay (*Rhododendron lapponicum*), for example, occurs in Greenland and in the Arctic and alpine regions of the New World and Asia, but nowhere else in Europe besides these two areas. Pielou (1979) has discussed the many problems in interpreting these areas as nunataks, the most serious being that it is difficult to demonstrate that they were not areas of recolonization following the retreat of the ice sheets at the end of the Pleistocene.

In contrast to the questionable importance of these relatively small, internal nunataks, those along the periphery of glaciers may well have provided refugia for a great many species. For example, at least three large refugia persisted during the Wisconsin in North America: the expansive iceless areas of Beringia, the coastal regions of the Pacific Northwest, and Nova Scotia (see

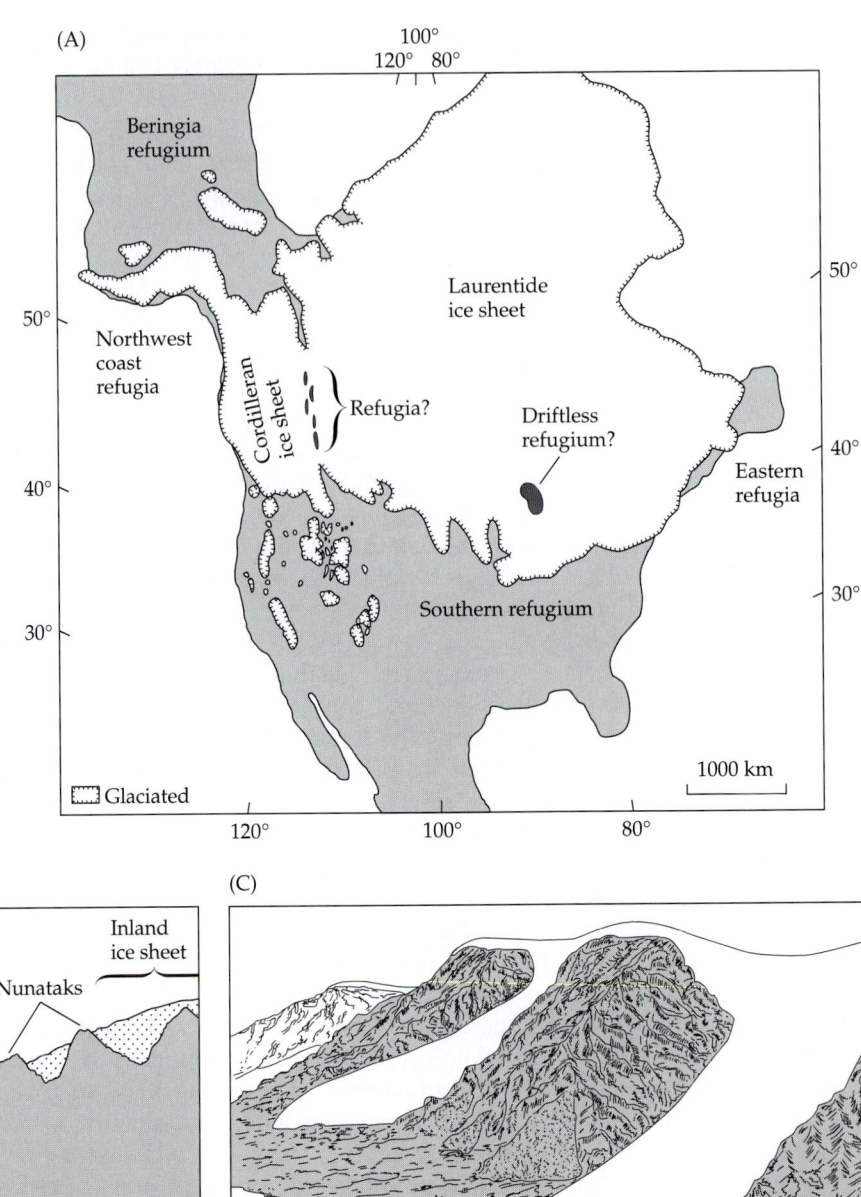

Figure 7.27 (A) Even during the Wisconsin glacial maximum, ice-free refugia may have occurred between the Laurentide and Cordilleran ice sheets and in a region called the driftless area. (B, C) Ice-free areas in the mountainous regions along the Pacific coast may also have served as refugia, and possibly as migration corridors for plants and animals, during full glacial conditions. (A after Rogers et al. 1991; B and C after Pielou 1991.)

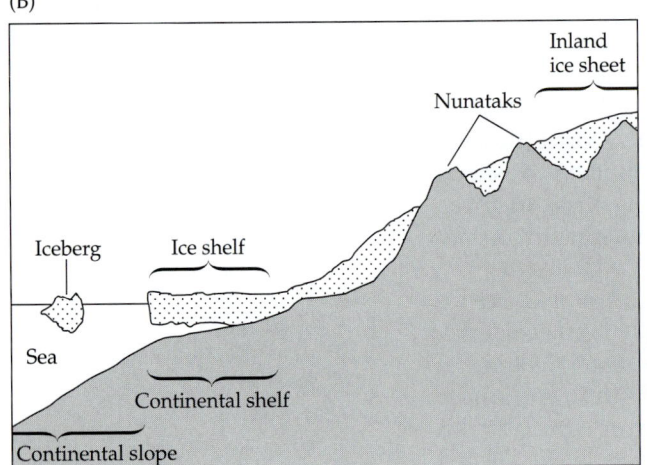

Figure 7.27). Many species that were unable to disperse with their shifting habitats persisted, and in some cases diversified, in these refugia. Thus, many Holarctic biogeographic patterns bear witness to the importance of these glacial refugia.

The thesis that Beringia was a Pleistocene refugium was clearly presented by the phytogeographer E. Hultén (1937) (Figure 7.28). Hultén's work on this subject is a classic, one of the best efforts to document plant distributions on a broad scale. His exact distribution maps, which stimulated an important creative synthesis, should be required study for all biogeographers. Although Hultén's idea was ahead of its time, several lines of evidence helped to con-

Figure 7.28 The high frequency of endemic plant species in central Alaska was used by Hultén (1937) to advance the idea that a large part of Beringia remained unglaciated and served as a refugium for arctic forms during the Pleistocene. Each contour line indicates the number of narrow endemics present in that region.

firm that Beringia was vegetated during the Wisconsin glacial maximum. First, Colinvaux (1981, 1996) and others analyzed fossil pollen records and showed that mesic tundra was widespread in the northern and central portions of the landbridge at the height of the Wisconsin glacial period (see Elias et al. 1996). Second, Hopkins and Smith (1981) presented rather strong evidence for the occurrence of deciduous dicotyledonous trees and larch (*Larix*, Pinaceae) in the Yukon at this same time. Finally, Weber et al. (1981) and others excavated late Pleistocene mammalian fossils (aged from 40,000 years B.P.) in the interior of Alaska, finding a great diversity of large ungulates (woolly mammoth, woolly rhino, barren-ground and forest muskox, mountain sheep, steppe antelope, reindeer, horse, and camel; see also Hopkins et al. 1982; Guthrie 1990). Paradoxically, these grazing mammals occupied Beringia during a period when its vegetative cover and the productivity of their preferred forage were much lower than in other contemporary communities. The waves of extinctions that followed the Wisconsin glacial maximum, when ungulate forage was increasing, remain an equally controversial and intriguing paradox, and are attributed, perhaps justifiably, to humans, as we shall see below. While it may take many years to solve this mystery, it is clear that the Bering landbridge, at times 1000 km wide, served as an important refuge and dispersal corridor for many Holarctic plants and animals, including humans.

On the other hand, the formation and development of glaciers fragmented once continuous ranges of northern species, promoting their evolutionary divergence in isolated refugia. As we have just noted, Beringia was the largest of these glacial refugia in North America, and was occupied by a high diversity of animals during the most recent glacial maximum. Smaller refugia, including those along the coast of the Pacific Northwest (see Figure 7.27), supported fewer species. In general, however, regions formerly within glacial refugia still tend to harbor a higher diversity of animals then those in formerly glaciated regions.

In the absence of gene flow among refugia, populations diverged genetically, often to the specific level in small mammals or to the subspecific level in larger species. Examples of such divergence between populations in Beringia and those in regions south of the glaciers include northern and southern redbacked voles, tundra and arctic shrews, and arctic and Columbian ground squirrels (see Hoffman 1971; Nadler et al. 1973; Nadler et al. 1978). Examples in large mammals include the divergence of moose (*Alces alces gigas*) and Beringian Dall's sheep from their southern counterparts (Peterson 1955; Korobytsina et al. 1974). In contrast, the smaller refugia of the Pacific Northwest not only supported fewer species, but their populations underwent less divergence than those inhabiting Beringia. Shrews, deer mice, voles, marmots, chipmunks, brown bears, and ermine have exhibited modest divergence, especially those isolated on islands of the Queen Charlotte Archipelago (see Banfield 1961; Hoffmann et al. 1979; Hoffman 1981; Riddle 1996). Finally, anthropological data also bears strong witness to the legacy of glacial refugia. Biogeographic patterns in the dental traits, genetic markers, blood proteins, and linguistic diversity of native Americans appear to parallel those found in other mammals. Diversity in these traits (for example, number of languages per unit of area) tends to remain highest where refugia occurred, and higher in larger refugia (i.e., higher south of the ice sheets than in Beringia, and in Beringia than in the Northwestern refugia; see Rogers et al. 1991).

In summary, Pleistocene refugia, especially the largest and most isolated ones, provided important opportunities for genetic divergence. Few of their endemic forms, however, spread as rapidly into deglaciated areas as their southern counterparts (i.e., those persisting south of the ice sheets). On the other hand, the high endemicity of former glacial refugia may derive, at least in part, from the inability of many other species to invade these sites.

Glacial Cycles and Extinctions

As we noted earlier, most late Neogene plant extinctions occurred during the initial glacial cycles, between 5 and 0.7 million years B.P. Those plants that survived these glacial periods did so by virtue of their ability to disperse along with their climatic zones or to adapt to new zones and environments. Some groups, including those of the arid land regions of North America, underwent extensive radiations during glacio-pluvial periods. It appears that by the later glacial cycles of the Pleistocene, vegetative communities had come to be dominated by plants that were preadapted to climatic changes. Thus, while vegetative communities were reshuffled and shifted substantially in latitude and elevation, relatively few plant extinctions occurred during the most recent glacial cycles.

Mass extinctions of marine invertebrates during the late Neogene also appear to be associated with glacial cycles. As with plants, bivalve extinctions in the Mediterranean and North Seas were heaviest during the earlier glacial cycles, with most occurring in the Pliocene between 3.2 and 3.0 million years B.P.—that is, coincident with the earliest evidence of glaciation in those regions. Additional pulses of extinctions continued through the early Pleistocene, but not into the mid- or late Pleistocene. It appears that the victims were primarily stenothermal species. By the mid-Pleistocene, these species had been "weeded out," and as a result, marine biotas became dominated by eurythermal species, which were little affected by fluctuations in water temperatures during subsequent glacial cycles (see Raffi et al. 1985).

The pattern of terrestrial vertebrate extinctions during the Pleistocene is quite different. On display in American natural history museums are skeletons and models of many large mammals that dominated American faunas until as recently as 20,000 years ago, but are now extinct (Figure 7.29). Gone from North America are most of its large herbivores, such as mastodons, mammoths, camels, llamas, horses, tapirs, ground sloths, and cave bears, as well as species of ungulates related to contemporary deer, bison, and pronghorn antelope (Martin and Wright 1967; Kurtén and Anderson 1980). Gone also are

(A)

(B)

(C)

(D)

(E)

|← 17 ft. →|

Figure 7.29 The mass extinction of terrestrial vertebrates that occurred in North America during the late Pleistocene and early Holocene included the loss of a highly disproportionate number of large mammals and birds, often referred to as the Pleistocene megafauna.(A) Mammoths (*Mammuthus* spp.); (B) ground sloths (*Megalonyx* spp.); (C) sabertooth cats (*Smilodon* spp.); (D) giant bison (*Bison latifrons*); (E) teratorns (*Teratornis* spp.) These groups are either without modern day analogues in North America (mammoths and ground sloths) or they have been replaced by much more diminutive forms (sabertooth cats replaced by mountain lions and smaller felines; giant bison by American bison; teratorns by condors and vultures). (E after Pielou 1991.)

many of the large predators that hunted those herbivores, including hyenas, dire wolves and other canids, saber-toothed tigers, and even lions. Large birds, especially raptors and scavengers, were also disproportionately subject to extinctions during this period. These casualties included the teratorns of North America, believed to be the largest flying birds that ever lived (the wingspan of *Teratornis incredibilis* was approximately 5 m, as compared with 3 m in today's California and Andean condors; see Steadman and Martin 1984; Steadman 1987).

So many of these fossils have been unearthed from Pleistocene beds that paleontologists could not help but be impressed by the disappearance of this North American megafauna at the end of the Pleistocene. The record of this mass extinction prompted scientists in the early nineteenth century to search for a specific cause for the wholesale destruction of the North American megafauna. Were the extinctions sudden or gradual? Did other landmasses experience similar waves of extinction, and if so, were they synchronous across regions? Did small animals and plants become extinct at the same time? Were the Pleistocene extinctions caused by climatic and geological changes, or did intense hunting by humans result in the extirpation of these large beasts? These are questions that must be answered if we are to understand faunal change and biogeography in the Pleistocene.

The Overkill Hypothesis

The prehistoric, or Pleistocene, **overkill hypothesis** states that humans were responsible for the mass extirpation of large herbivorous mammals (over 50 kg), and the carnivores and scavengers dependent upon them, after the Wisconsin glaciers had retreated. This is an old hypothesis, but is also one that has been most clearly presented as a straightforward explanation with potentially

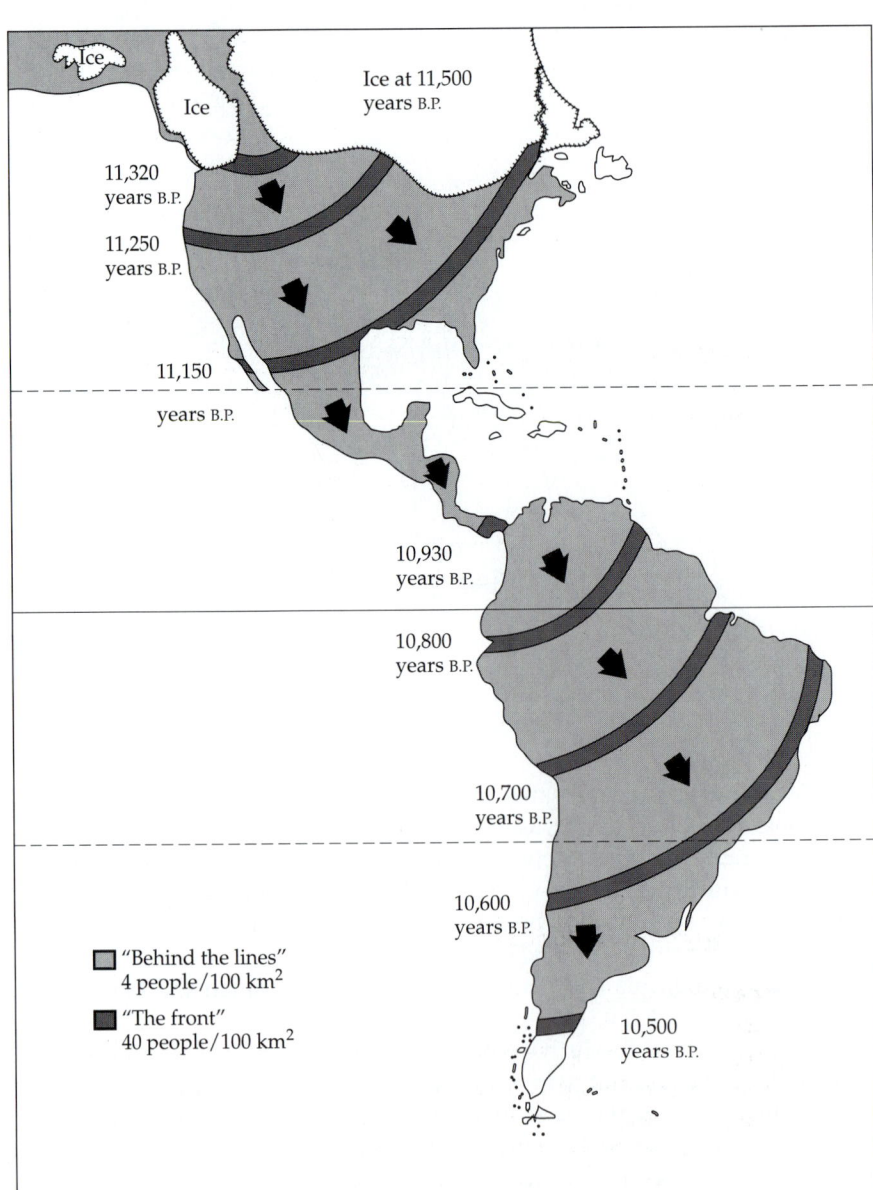

Figure 7.30   The temporal sequence of advancing populations of human big-game hunters correlates well with the progressive extinction of large Pleistocene mammal species. According to the Pleistocene overkill hypothesis, sophisticated hunters crossed Beringia and expanded southward, maintaining a relatively dense population by subsisting on large mammals. Human populations may have colonized the America's well ahead of these dates, but their population densities, technology, and ability to cause significant ecological disturbance were very limited in comparison to the more sophisticated hunting societies that followed. (After Martin 1973.)

falsifiable assumptions and predictions (Martin 1967, 1973, 1995). Let us consider this hypothesis in detail for comparison with alternative explanations.

The overkill model suggests that a population of aggressive and skillful human hunters entered North America during the late Wisconsin by crossing Beringia from Asia. Once these hunters colonized America, they spread southward and eastward through North America and into South America, killing large animals as they went (Figure 7.30). The native American animals lacked adequate defensive behaviors to outwit or elude their new predators. The abundant food supplies obtained from their hunts permitted human populations to remain high and in constant need of new and massive food sources. Behind this trail of carnage, there were no more waves of mammalian immigrants from Asia to replace those species that became extinct. Most of the large mammals that survived were those that had spread to the New World from the Old World since the evolution of Pleistocene humans, so they were presumably already adapted to human hunters.

The evidence supporting this scenario is of several types. First, fossil evidence shows that prehistoric humans and large mammals coexisted in the Americas and that the people hunted the extinct herbivores. Arrow points in carcasses and remains of massive animal kills are clearly demonstrated. Second, late Wisconsin extinctions in North America were nonrandom in that many more large and very large mammals than smaller ones became extinct during the period from 12,000 to 10,000 years B.P. (Figure 7.31). Third, as noted above, immigrants from Beringia and Eurasia, including caribou, moose, deer, and Dall's and bighorn sheep, fared much better than native species (Figure 7.32; Kurtén and Anderson 1980). Fourth, extinctions of large mammals appear to have begun in the north and proceeded rapidly and systematically southward (compare to Figure 7.30). Finally, when the dates of the last known occurrences of species are compared with a computer simulation of southward human migration (assuming high human population densities), the two appear to coincide rather closely (Mosiman and Martin 1975).

The Pleistocene overkill model could be tested and possibly falsified by showing that many different types of animals and plants became extinct at the same time, that extinctions were under way before humans arrived, that aggressive human hunters coexisted with large mammals for long periods, that human populations were never at high densities, or that comparable extinctions on other continents did not correspond with an invasion by ecologically significant human societies.

Figure 7.31   Pleistocene extinctions of mammals tended to be highly nonrandom, disproportionately affecting relatively large taxa. (A) Extinction rates among mammalian herbivore genera of different body sizes in North America, South America, Europe, and Australia during the late Pleistocene. (B) Distribution of body weights (on a log scale) of extant eutherian omnivore and herbivore taxa, excluding primates, in Europe, Canada, East Africa, and Thailand. The dashed line represents the frequencies of relatively large mammal taxa (the megafauna) in these biotas before the Pleistocene extinctions. (A after Owen-Smith 1988; B after Caughley 1987.)

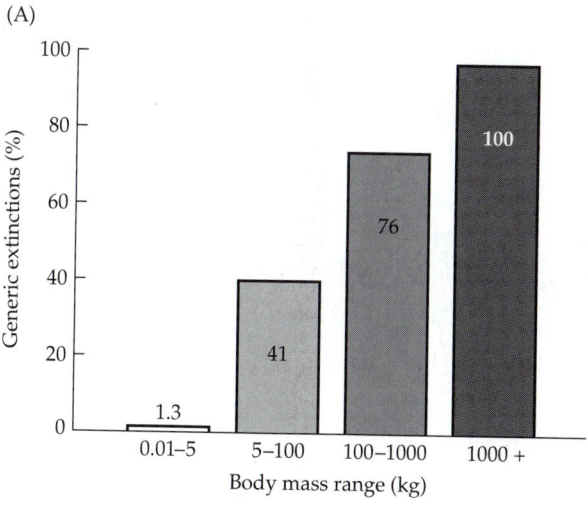

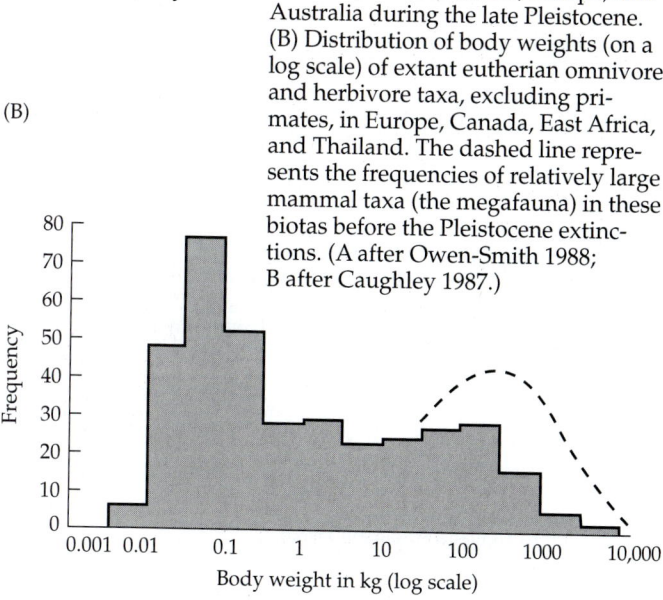

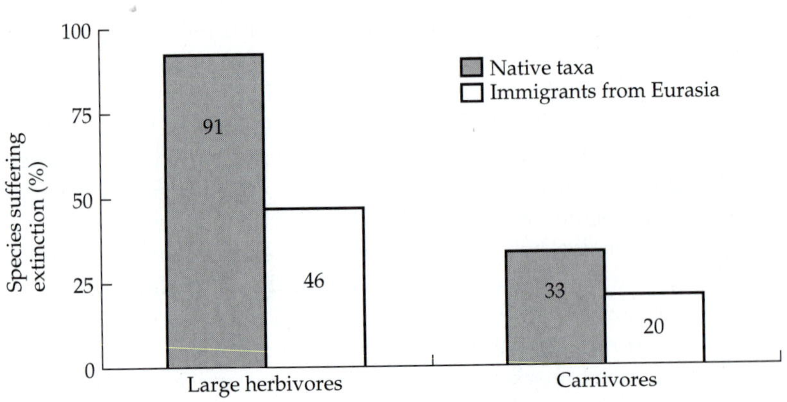

Figure 7.32 Extinction rates among native and immigrant large herbivores and carnivores in North America during the late Wisconsin. (After Kurtén and Anderson 1980.)

Alternative Explanations for Pleistocene Extinctions

As with any controversial theory in biogeography, there are several alternative explanations to account for Pleistocene extinctions of mammals. The overkill hypothesis, if correct, paints a rather brutal and disparaging picture of the early human pioneers in North America. Some authors feel that these colonists have been given a "bum rap"—they may have been instrumental in reducing prey population sizes, but extinctions were already occuring in response to climatic shifts at the end of the Ice Age. They point to other groups of organisms, such as raptors and large scavenging birds, that experienced high rates of extinction at the same time as large herbivores (Grayson 1977). However, as we noted earlier, such extinctions are entirely consistent with the overkill hypothesis, which predicts a loss not just of the megafauna, but of dependent predators and scavengers as well (see Owen-Smith 1988).

One observation that has always been puzzling is that the North American fauna did not disappear until after the Wisconsin glaciers retreated. Hence the late Pleistocene extinctions cannot be related directly to glaciation, cold climate, or any other catastrophic geological event, such as flooding. Nevertheless, many researchers contend that climatic changes were the direct cause of the extinctions, either through increased aridity (Guilday 1967) or by decreased equability (Slaughter 1967; Axelrod 1967).

An excellent discussion of Pleistocene extinctions by Kurtén and Anderson (1980) explains why paleontologists generally prefer a climatic explanation. Pleistocene extinctions of mammals were not restricted to the period of 12,000 to 10,000 years B.P., but were part of a fairly continuous series of episodes during the latest Cenozoic (Table 7.3). The Pliocene Blancan extinction (mainly

Table 7.3
Extinction of North American mammals during the last 4 million years

Animal size	Blancan (3.5–1.8)	Irivingtonian (1.8–0.7)	Rancholabrean (0.7–0.01)	Extinction total	Surviving species	Percent extinct
Small (1–907 g)	97	55	29	181	166	52%
Medium (908 g–181 kg)	31	25	33	89	50	64%
Large (182–1730 kg)	5	12	35	52	16	76%
Very large (> 1730 kg)	1	2	5	9	1	90%

Source: After Kurtén and Anderson (1980).

Note: Duration of periods is in million years B.P.

between 3.3 and 2.4 million years B.P.) resulted in the disappearance from North America of at least 125 mammalian species, of which three-fourths were animals smaller than 1 kg in body mass. During that time, aridity increased, grasslands replaced forests, and many forest dwellers and browsers died out. Following that depletion of the fauna, surviving grazers and rodents underwent evolutionary radiation, and small carnivores also increased in abundance and diversity. In the Irvingtonian extinction of the Pleistocene (1.8 to 0.7 million years B.P.) only 89 taxa disappeared, 80% of which were small or medium-sized (< 180 kg). Extinction rates during the Pleistocene were fairly low and constant until the late Wisconsin, when many small, medium, and large animals disappeared. However, as stated earlier, a highly disproportionate number of large mammals became extinct in the late Wisconsin as compared with other episodes (see Figure 7.31A).

The cause of the late Pleistocene extinctions remains one of the most important and intriguing mysteries of our field. In a wonderful essay on the nature and causes of historic extinctions, Jared Diamond (1984) called on one of the world's greatest detective minds to help solve the mystery: Sir Arthur Conan Doyle—alias Sherlock Holmes. In "Silver Blaze," Holmes called attention to "the curious incident of the dog in the night-time." When the dim-witted Inspector Gregory observed that "the dog did nothing in the night-time," Holmes remarked that "that was the curious incident"—it indicated that the stables pet was familiar with the "intruder." The decisive clues to the causes of Pleistocene extinctions also may be the "dogs that did nothing in the night-time," namely, the species and biotas that survived while others became extinct.

Champions of the overkill hypothesis call our attention to four such "curious incidents." First, they ask, why did the North American megafauna diversify when climatic conditions seemed least favorable during the Wisconsin, only to suffer so many extinctions when climates warmed and environmental conditions became more favorable? Second, why did the megafaunal extinctions not occur during an earlier glaciation? Third, why were most groups of small animals spared from mass extinctions while their larger counterparts were devastated? Finally, why didn't the megafaunal species of Africa suffer extinctions comparable to those among biotas on other large continents (Figure 7.33)?

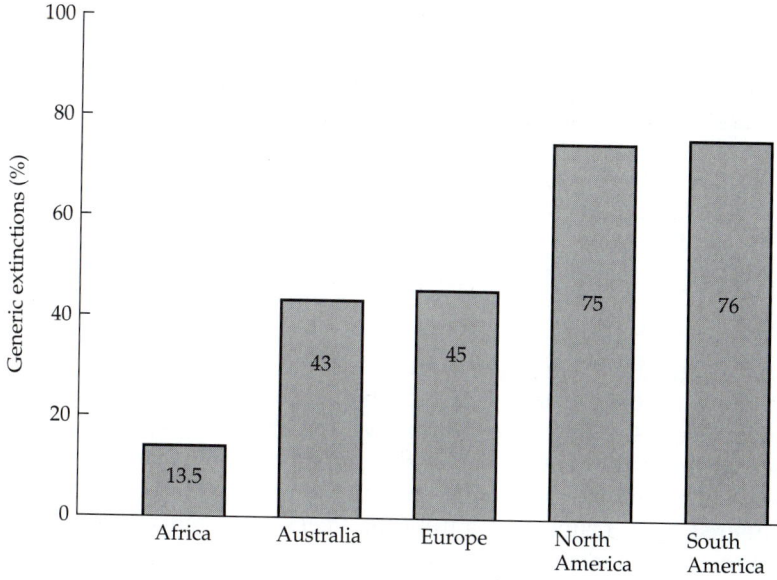

Figure 7.33 Extinction rates among mammalian herbivore genera with medium to large body sizes (> 5 kg) on different continents during the late Pleistocene. (After Owen-Smith 1988.)

1000–2000 kg

100–1000 kg

10–100 kg

1–10 kg

Figure 7.34   Selective extinctions of large, megafuanal mammals in Australia during the Pleistocene and Holocene. Shown are outline drawings of species known to occur in these regions when they were colonized by aboriginal humans. Species suffering extinction during the Pleistocene and early Holocene are shown in black, while those that became extinct or endangered following European colonization are shaded (open outlines indicate extant, nonendangered species). (Information courtesy of T. F. Flannery, art based on original drawings by Tish Ennis, Australian Museum, Sydney.)

Some have argued that familiarity is the key to the last question: *Homo sapiens* has had such a long history in Africa that, at least until modern times, native African societies have had little impact on the megafauna that coevolved with them. Proponents of the overkill hypothesis note that Africa may be the exception that proves the rule. In fact, the megafauna of most landmasses suffered mass extinctions during the Pleistocene, but they were far from synchronous. A mass extinction of the Australian megafauna for exam-

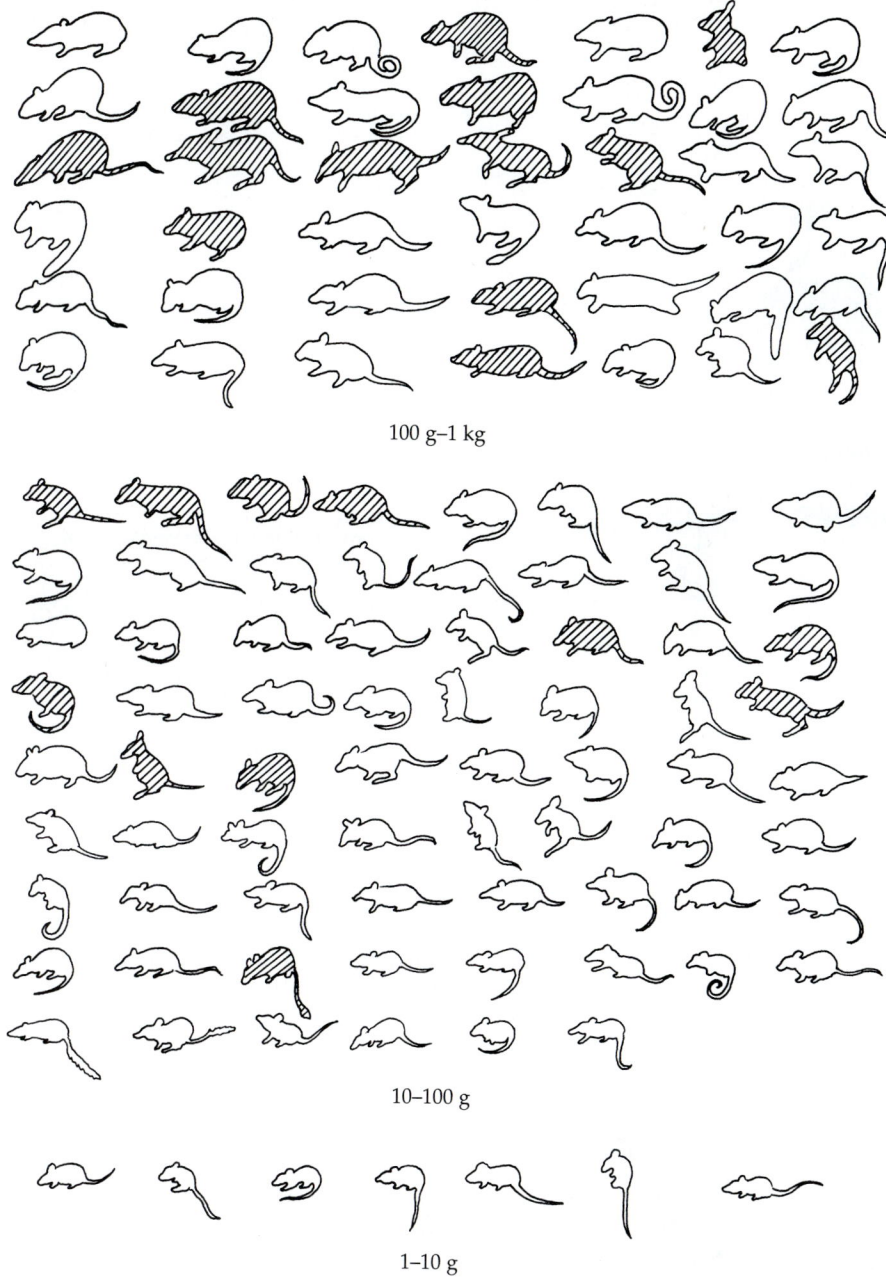

100 g–1 kg

10–100 g

1–10 g

ple, occurred roughly 35,000 years B.P. (Figure 7.34), while mass extinctions on Madagascar and the Galápagos Islands have occurred within the last 1000 years (Figure 7.35). These chronologies are difficult to explain according to the climate-based hypothesis, since climatic reversals were synchronous across the globe. On the other hand, the timing of these waves of megafaunal extinction is coincident with invasions by aggressive hunting societies, increased human population levels, extensive use of fire, and other activities that significantly modified native ecosystems.

Despite these arguments, champions of the climate-based hypothesis continue to take issue with the overkill hypothesis. Webb and Barnosky (1989), for example, recognized six major episodes of Neogene extinctions of North American mammals, each thought to be coincident with rapid climatic changes. But

Afroeurasia: No major episodes of extinction during past 100,000 years, although some losses occurred.

Meganesia: Humans arrive 40,000–60,000 years B.P.; major extinction episode follows, but extends to circa 15,000 years B.P. (or later?).

Americas: Ecologically significant human populations arrive 12,500 years B.P.; major extinction episode terminates circa 10,500 years B.P., few extinctions thereafter.

Mediterranea: Humans arrive 10,000 years B.P.; major extinction episode follows and terminates circa 4,000 years B.P., few extinctions thereafter.

Antillea: Humans arrive 7000 years B.P.; major episode of extinction follows, but extends to circa A.D. 1600 (or later?).

Madagascar: Humans arrive 2000 years B.P.; major episode of extinction follows and terminates circa A.D. 1500, few extinctions thereafter.
Mascarenes (Ms): Humans arrive A.D. 1600; major episode of extinction follows and terminates circa A.D. 1900.
New Zealand: Humans arrive 800–1000 years B.P.; major episode of extinction follows and terminates circa A.D. 1500.

Commander Islands (C): Humans arrive A.D. 1741; Steller's sea cow extinct by A.D. 1768.
Wrangel Island (W): Humans arrive ?; mammoths survive to 4000 years B.P.
Galápagos Islands (G): Humans arrive A.D. 1535; modern-era extinctions only.

Figure 7.35   The geography and chronology of Pleistocene extinctions may be correlated with major episodes of human colonization. Extinction episodes during the Pleistocene were relatively minor in regions with a long history of human occupation, but severe and coincident with colonization by ecologically significant human cultures elsewhere (see also Chapter 18). (After MacPhee and Marx 1997; based on drawings by Clare Flemming.)

why, then, ask the advocates of the overkill hypothesis, were these extinctions not synchronous across the continents, and why were they not recorded for all of the ten or so major climatic reversals of the Pleistocene?

As we can see, this debate is far from over, and it is likely that the pendulum of consensus will swing many times between champions of the competing theories. In fact, it is likely that the true causes of the Pleistocene extinctions involve a multitude of factors, including anthropogenic as well as climate-driven changes in native ecosystems, physiological effects of altered climates, and biotic exchanges and resultant changes in interspecific interactions, including the appearance of new predators, competitors, diseases, and parasites (see MacPhee and Marx 1997; Flannery 1994). In other words, many combinations of ecological or evolutionary factors (anthropogenic and otherwise) might have contributed to the waves of Pleistocene extinctions, and hunting by early human colonists may merely have delivered the final and fatal blow to some species. Unfortunately, this explanation is difficult to evaluate, even if it is correct.

Regardless of its cause or causes, the elimination of the majority of large mammals and birds has been one of the most important events in the relatively recent history of terrestrial biotas. Not only did these animals themselves experience reductions in their populations and geographic ranges ending in their extinction, but their disappearances may also have had important effects on other species. The herbivores that did survive were faced with fewer potential competitors, and the surviving carnivores had to make do with fewer prey. Parasites, scavengers, and mutualists of the extinct species either switched to new associates or became extinct themselves (see Owen-Smith 1988). For example, Steadman and Martin record high extinction rates among carrion-feeding birds at the end of the Pleistocene, including eagles, vultures, teratorns, and condors. Finally, it is known from fossil feces (**coprolites**) that the extinct large herbivores consumed large quantities of certain extant plant species. To what extent has release from such herbivory contributed to the shifts in plant species ranges and the changes in the distribution of vegetation types that are known to have occurred within the last 20,000 years? It is perhaps as important to investigate these questions as it is to solve the riddle of the extinctions of the Pleistocene megafauna. To paraphrase G. G. Simpson, our ability to understand and effectively curtail the ongoing wave of extinctions may well depend on our ability to learn the lessons of these prehistoric extinctions (see Chapters 17 and 18).

Historical Patterns and Processes

CHAPTER *8*

Speciation and Extinction

Species have histories that are somewhat analogous to the lives of individuals. New species originate through the multiplication of old ones (**speciation**); they survive for varying periods of time, during which they may or may not leave descendants; and they die (**extinction**). There are, however, important differences between the histories of individuals and of species. Whereas the births and deaths of most individuals are discrete, easily recognizable events, it is often difficult to determine when a species is fully formed. Rather than becoming extinct abruptly when the last individual dies, some species may disappear by evolving into a new kind of organism that is recognized as a different species.

There is also a difference between individuals and species in the branching patterns of lineages (Figure 8.1). Individuals of sexual organisms have two parents, four grandparents, and so on, so the lineages of individuals are "**reticulate**," not only branching out but also joining and rejoining as we trace them through time. We think of strictly asexual organisms—in which each individual has only a single parent, and therefore lineages show simple splitting patterns—as exceptions to this general rule. The evolution of species lineages (**cladogenesis**) in most kinds of organisms is a simple branching process similar to asexual reproduction: a single ancestor gives rise to one or more descendants. We think of organisms that commonly exhibit reticulate cladogenesis as being exceptions to this general rule. Some vascular plants and other kinds of organisms, for example, in which two or more parental species may hybridize to produce a single descendant species, regularly exhibit reticulate evolution.

Like related individuals, related species possess particular combinations of traits, at both the structural and the molecular levels, that have been inherited from their common ancestors. If a lineage has been diverse and successful, one can usually infer that the traits that make the group distinct are **adaptations**—that is, that they represent innovative solutions to a problem that limited survival and reproduction in other populations. It is not necessarily true, however, that all new traits of an evolutionary lineage are adaptive and have arisen as a direct result of natural selection.

Further, although a particular lineage may flourish for a period of time and radiate to produce a diverse group of species, this success is usually ephemeral on a geological time scale. Eventually the rate of extinction exceeds

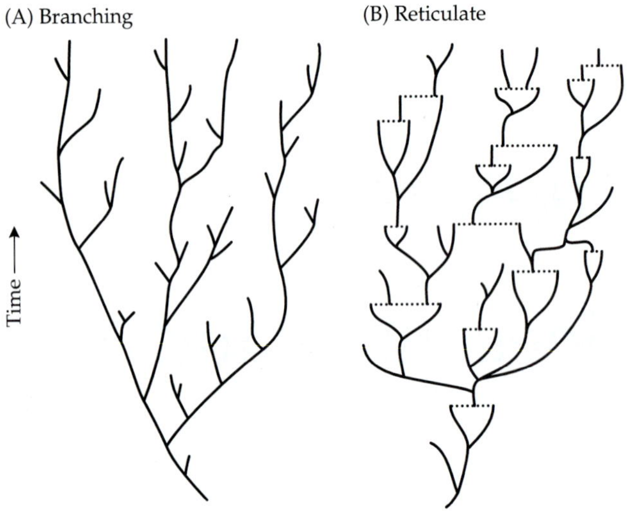

(A) Branching

(B) Reticulate

Time ⟶

Figure 8.1 Schematic representation of two patterns of of diversifying lineages: (A) simple branching evolution due to speciation events and (B) reticulate evolution, resulting from a combination of lineage splitting due to speciation and lineage rejoining due to hybridization. Hybridization and reticulate evolution are assumed to be rare in vertebrates and most other animal groups, but they are known to be common in many kinds of higher plants.

the rate of speciation, the diversity of the group decreases, and all or nearly all of its representatives go extinct. These repeated episodes of speciation, radiation, and extinction have occurred many times during the history of the earth to produce its variety of living and extinct organisms. While the vast majority of the species that have ever lived went extinct without leaving descendants, some lineages obviously survived. Every species living today has an unbroken line of descent extending all the way back to a single common ancestor at the origin of life (see Chapter 11).

Taxonomy

Species Concepts

Before we can talk meaningfully about the evolutionary process, we must define our terms. This requires a brief digression into the subject of taxonomy. **Taxonomy** is the discipline that assigns names to organisms and classifies biological diversity. It describes fundamental units, called **species**, and gives them **scientific names**, Latinized binomials: first a genus and then a species epithet (e.g., the scientific name for humans is *Homo sapiens*). It arranges these species in a hierarchical classification scheme that aims to reflect the historical evolutionary pattern of their ancestry and descent.

Morphological species. If the fundamental unit of biogeography is the geographic range of a species, then what is a species? This is a difficult question. Even the taxonomists who describe species and the systematists who try to reconstruct the history of evolutionary lineages do not agree on how species should be defined. The classic definition of a species is the **morphological species** concept. This concept recognizes that each species usually is morphologically distinguishable from its closest relatives. However, the criteria for determining in which traits, and by how much, a population must differ from others to warrant recognition as a distinct species vary from group to group.

Biological species. Since the new synthesis in evolutionary biology early in the twentieth century, many taxonomists have preferred to employ the **biological species** concept, which defines a species as a population of organisms that is actually or potentially reproductively isolated from other populations (e.g., see Mayr 1963; Futuyma 1986; Otte and Endler 1989). When a population is **reproductively isolated**, it constitutes a separate evolutionary lineage that is prevented by geographic or biological barriers from interbreeding with other populations. It is free to follow its own course in response to the genetic processes and environmental influences that cause evolutionary change.

As straightforward as it may sound, the criterion of reproductive isolation is difficult to employ. It cannot be applied to fossil organisms; the best one can do is to determine whether the morphological gaps between specimens are as large or larger than those between living species that are reproductively isolated. The biological species concept is unwieldy for extant groups as well. Some organisms are exclusively asexual, so the reproductive isolation criterion cannot be applied to them—or every individual would be a separate biological species! Asexual organisms include not only a number of microbes, but also some rotifers, mollusks, arthropods, vertebrates (fishes and lizards), and vascular plants (apomictic and obligately selfing forms), so this creates no small problem. There are also many examples of **hybridization**: interbreeding that produces viable and fertile offspring, but which occurs between populations that remain distinct genetic and evolutionary units. This is especially a problem in vascular plants, in which hybridization is common between species and genera and has been known to occur even between plants in different families (Levin 1979; Grant 1981; Briggs and Walters 1984; Wyatt 1992). Between the two extremes of undifferentiated populations that do not interbreed and highly differentiated ones that interbreed freely are many cases in which the level of actual or potential interbreeding is difficult to determine, usually because the populations in question are spatially isolated and somewhat differentiated. Only in a few such cases have mating tests actually been conducted, because of the time and expense involved and the fact that many organisms cannot easily be reared in the laboratory. Usually, as in the case of fossil species, it is left for the taxonomist to make an educated guess about whether individuals would have interbred fertilely to form a single population if the isolated populations had come back into contact.

Other species concepts.   Two new species concepts have been proposed in an effort to remedy some of these problems. One is the **evolutionary species** concept (Wiley 1981), which would recognize each independent evolutionary lineage as a species. It differs from the biological species concept in that actual, rather than potential, genetic isolation is the criterion for recognition of a species. Every population that is sufficiently isolated by either genetic or biogeographic barriers so that the current level of gene flow from other populations (see Chapter 4) is too low to prevent its genetic divergence would be considered a species. The evolutionary species concept also takes account of the fact that cladogenetic evolution can be reticulate: populations may become isolated from one another and begin to diverge, but then the isolation breaks down, hybridization occurs, and the populations merge. Such populations would be recognized as independent units while they were isolated and as one species after they rejoined.

The other new concept is the **phylogenetic species** concept (Cracraft 1989), which would recognize as a species any group of organisms in which all individuals share a unique **derived**, or **apomorphic**, characteristic—one not present in their ancestors or relatives. This concept defines species in a manner consistent with the theory and practice of phylogenetic systematics, the discipline that reconstructs the history of evolutionary lineages (which will be described more fully in Chapter 11). The phylogenetic species concept is similar to the morphological species concept in that it uses unique distinguishing characteristics to identify species. Because of the increasing use of molecular techniques for phylogenetic reconstruction, however, these characteristics are often molecular genetic traits rather than morphological ones.

Both the evolutionary and phylogenetic species concepts get around certain problems with the biological species concept (Wiley 1981; Cracraft 1989; Avise and Ball 1990; Baum and Donoghue 1995). But both concepts also create new problems. In particular, they are difficult for most practicing taxonomists to use when describing species, and for all kinds of biologists to use to identify the organisms they are studying. The application of either new concept would result in the breaking up of many currently recognized species, now defined on the basis of morphological or biological species concepts, into multiple evolutionary or phylogenetic species. This would be especially true of **polytypic** species: species that are composed of multiple geographic races or subspecies. The use of these concepts would thus substantially inflate estimates of biodiversity. The evolutionary species concept would often require even more guesswork than the biological species concept to estimate which populations are currently sufficiently isolated (i.e., have sufficiently low rates of genetic exchange with other populations) to be undergoing independent evolution. If applied rigorously, the phylogenetic species concept could result in nearly every local population being recognized as a separate species, because each one is likely to have some seemingly unique molecular genetic variant. Both of these concepts, like the biological species concept, would require that every individual of obligately asexual organisms be considered a separate species, because each is a separate evolutionary unit and each will almost certainly have some unique new mutation in its genetic material.

Recently, one of us (Brown 1995) suggested that the difficulties of defining species reflect fundamental problems that stem from the way in which organisms vary in their characteristics. Variation among organisms in most traits is **discontinuous**: there are clumps separated by gaps. In other words, some individuals are very similar to one another, but are considerably different from other such clumps of individuals. Figure 8.2 illustrates this pattern using variation in beak size among the Darwin's finches that occur on several islands of the Galápagos. This "clumpy-gappy" organization of biological diversity is apparent in nearly all characters of organisms, not only in morphology and genetics, but also in physiology, ecology, and behavior. It reflects something very basic in the genetic and evolutionary processes through which both individuals and populations inherit characteristics from their ancestors. At the individual level, these processes are essentially the same in all organisms, but at the population level, as we saw above, they are fundamentally different in sexual and asexual organisms: the former have two parents and thus reticulate patterns of inheritance, whereas the latter have only one parent and thus only a simple branching pattern of ancestry and descent.

All species concepts are attempts to capture important features of this clumpy-gappy organization of life in some more formal scheme that enables us to recognize and assign names to the major clumps. Doing this is far from straightforward in practice, however, for all of the reasons mentioned above. While the same clumps are often recognizable on the basis of distinctive morphological, genetic, physiological, behavioral, and ecological characteristics, sometimes they differ in some of these features, but not others. The clumps are not of the same size, and they are separated by gaps of varying magnitudes. Over evolutionary time, clumps break apart and come together again as populations diverge in isolation and merge through hybridization. Any scheme that tries to force these complex patterns of variation into rigid categories will have the kinds of problems that plague all of the current species concepts.

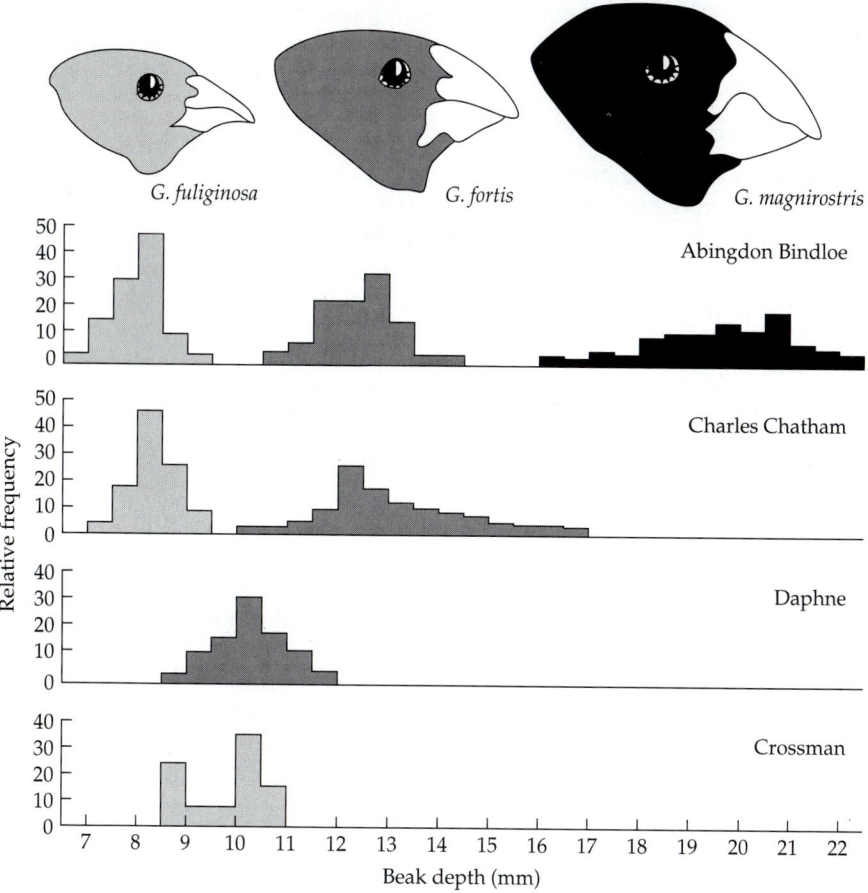

G. fuliginosa *G. fortis* *G. magnirostris*

Abingdon Bindloe

Charles Chatham

Daphne

Crossman

Beak depth (mm)

Relative frequency

Figure 8.2 The distribution of one trait, beak depth, in several populations of Darwin's finches inhabiting different islands of the Galápagos. This discontinuous variation, especially within a single island, illustrates the "clumpy-gappy" distributions of traits that most species concepts try to capture. The pattern of interisland differences also illustrates *character displacement;* that is, the fact that species tend to be more different from one another where they occur together—presumably reflecting adaptive divergence to avoid competition—than where they do not coexist with closely related forms. (After Lack 1947.)

The best that can probably be done is to use a combination of all of these concepts: one that recognizes the discontinuous organization of biological diversity, allows us to recognize the major clumps as "species," and is practical to apply when identifying organisms in the field or laboratory. It is important to remember that all species concepts have been invented by humans to simplify and classify biological diversity. Many of the entities that are currently recognized as species are not such discrete, well-defined natural units as any of the species concepts imply. However, we can take some comfort from the fact that specialists on most taxonomic groups share a common idea of how different populations must be in order to be classified as separate species.

Subspecies and ecotypes.   Taxonomists, evolutionary biologists, and ecologists recognize that some well-differentiated populations probably do not warrant recognition as distinct species using any species definition. These populations are variously called **subspecies**, **geographic races**, **varieties**, or **ecotypes**. *Subspecies* is a formal term usually reserved for populations that are morphologically, and presumably genetically, distinct. Typically a subspecies is given a trinomial Latin name, such as *Peromyscus maniculatus gracilis*, in which the last part is the subspecies designation. The other terms are applied more loosely and informally. Plant ecologists use the term *ecotype* to refer to a distinct population that occurs in a particular habitat type. Unlike subspecies and geographic races, which normally have nonoverlapping geographic ranges, two or more ecotypes may occur together in the same local area, so long as they are restricted to different habitats. Typically ecotypes have distinctive morphological and physiological traits, which may reflect genetic adaptations to their unique environments.

Higher Classifications

Having considered the problem of how species are defined, let's briefly turn to the equally vexing problem of how species are arranged in a taxonomic classification scheme. The great eighteenth-century naturalist Carolus Linnaeus developed the hierarchical scheme for classifying organisms that is still used today. In the Linnaean classification system, species are grouped into genera, genera into families, families into orders, and so on (Table 8.1). In Linnaeus's time, when species were thought to have been individually created by God, this classification was simply meant to reflect the fact that species differ in the degree to which they resemble one another.

Now that we know that all living things are descended, through a complex historical pattern of branching lineages, from a common ancestor, most taxonomists want their hierarchical classification schemes to reflect this history of ancestor-descendent relationships. The discipline of **phylogenetic systematics** has had considerable success in developing methods for reconstructing the evolutionary history of lineages using molecular and other kinds of data. We will discuss phylogenetic systematics and its applications to biogeography extensively in later chapters (especially Chapters 11 and 12).

For the moment, it is sufficient to know the levels of the taxonomic hierarchy and to understand what these imply. The major levels of classification are given in Table 8.1. Additional levels are sometimes recognized by taxonomists working on particular groups—examples include superspecies, subgenera, infraorders, and so on. It is usually easy to figure out from their names where these levels are located in the hierarchy.

The hierarchical classification scheme implies that the taxa grouped together at each level are most closely related to one another. Thus, for example, species within the same genus are more closely related to one another than to species in any other genus, as are genera within the same family, and so on. By *more closely related*, we mean that they more recently shared a common ancestor. Because organisms inherit most of their characteristics from their ancestors, this usually also means that the taxa within a group share similar traits, including morphological, physiological, genetic, ecological, and behavioral characteristics. One thing that the classification scheme does not

Table 8.1
The primary units in the Linnaean system of hierarchical taxonomic classification of living things, illustrated by the classification of the human species, *Homo sapiens*

Kingdom	Animalia
Phylum	Chordata
Class	Mammalia
Order	Primates
Family	Hominidae
Genus	*Homo*
Species	*sapiens*

imply, however, is that all of the taxa at the same level of classification are equally closely related. Thus, if three families are placed in an order, two may have split from their common ancestor more recently than the third. In some groups, such as certain vascular plants and vertebrates, species in the same genus may have diverged only within the last few million years, whereas congeners in other groups may have split tens of millions of years ago. In Chapter 11 we will consider in some detail how systematists reconstruct phylogenetic relationships and what these genealogies imply.

Before leaving the subject of taxonomy, we must mention one recent and radical development. Because of all the problems in defining species and higher taxonomic categories, Michael Donoghue and others have begun calling for abandonment of the Linnaean system of classification and the substitution of a more flexible and quantitative scheme. It is hard to know where this development will lead. On the one hand, most systematists and other biologists recognize that cladogenetic evolution is far too complex for all of its intricacies to be captured by the simple Linnaean scheme. On the other hand, most practicing taxonomists and systematists seem reluctant to abandon a system that has worked well for 250 years—at least until someone can propose an alternative system that is both practical and conceptually superior.

Macroevolution

Recent studies of changes in organisms observed in the fossil record have given us new insights into the patterns and processes of evolution. Until the 1970s most biologists assumed that evolution usually proceeds gradually through the successive incorporation of relatively small genetic changes as populations respond to environmental change in space and time. However, many eminent evolutionary biologists (e.g., Simpson 1953; Mayr 1963) emphasized that evolutionary rates can vary widely, even within the same evolutionary lineage. Mayr and others also suggested that major, rapid changes can occur during the process of speciation.

The prevailing view of evolution was the product of the "new synthesis" (see Chapter 2) that dominated evolutionary thought for most of the twentieth century. The new synthesis was concerned primarily with how the characteristics of populations change as a result of the genetic mechanisms of natural selection, mutation, genetic drift, and gene flow. Such changes have come to be referred to as **microevolution**. Microevolutionists have traditionally studied contemporary organisms, either observing experimentally induced changes in laboratory populations of organisms such as microbes and fruit flies, or drawing inferences from comparative studies of populations in the field.

Evolution in the Fossil Record

In the 1970s a group of young paleontologists that included Niles Eldredge, Stephen J. Gould, and Steven M. Stanley began to emphasize the kinds of evolutionary change that are recorded in fossil remains. Because of the long time frame typically involved, these changes tend to be large, and include speciation and extinction events. The paleobiologists referred to these large changes as **macroevolution**. They argued that the fossil record suggests a view of evolution different from the prevailing microevolutionary view. They suggested that macroevolution occurs primarily as a result of two processes, which they called punctuated equilibrium and species selection (Eldredge and Gould 1972; Stanley 1979).

The fossil record implies that the evolution of a lineage usually consists of long periods of virtually no change (**stasis**) interspersed with relatively brief periods of rapid change, which often appear to be associated with speciation

Figure 8.3 "Punctuated" equilibrium in the evolution of fossil mollusks in the Lake Turkana Basin in eastern Africa. The diagram depicts the reconstructed history of shell morphology in several genera. Dotted lines indicate inferred changes during periods when no fossils have been recovered. Note that most lineages exhibited long periods of virtual stasis followed by rapid, substantial change. The latter, punctuational events often occurred either (1) virtually simultaneously in several different lineages, suggesting major environmental changes, such as shifts in lake level owing to climatic change (confirmed by other evidence); or (2) in association with speciation events, which often left one species virtually unchanged while the other diverged substantially. (After Williamson 1981.)

events. This pattern of highly variable rates of evolution is referred to as **punctuated equilibrium**. It is supported by data from the fossil record, especially that of marine and freshwater invertebrates, in which new fossil species often appear abruptly and then persist for millions of years with virtually no morphological change. Evidence of rapid change associated with speciation events comes, for example, from P. G. Williamson's (1981) work on the molluscan fauna of the Lake Turkana Basin in eastern Africa. In a stratigraphic sequence representing several million years of fossil deposition, Williamson found several lineages of mollusks that showed long periods of virtually no change in shell morphology, punctuated by rapid shifts (Figure 8.3). The changes took place within a time interval of less than 50,000 years—so rapidly with respect to the fossil record that

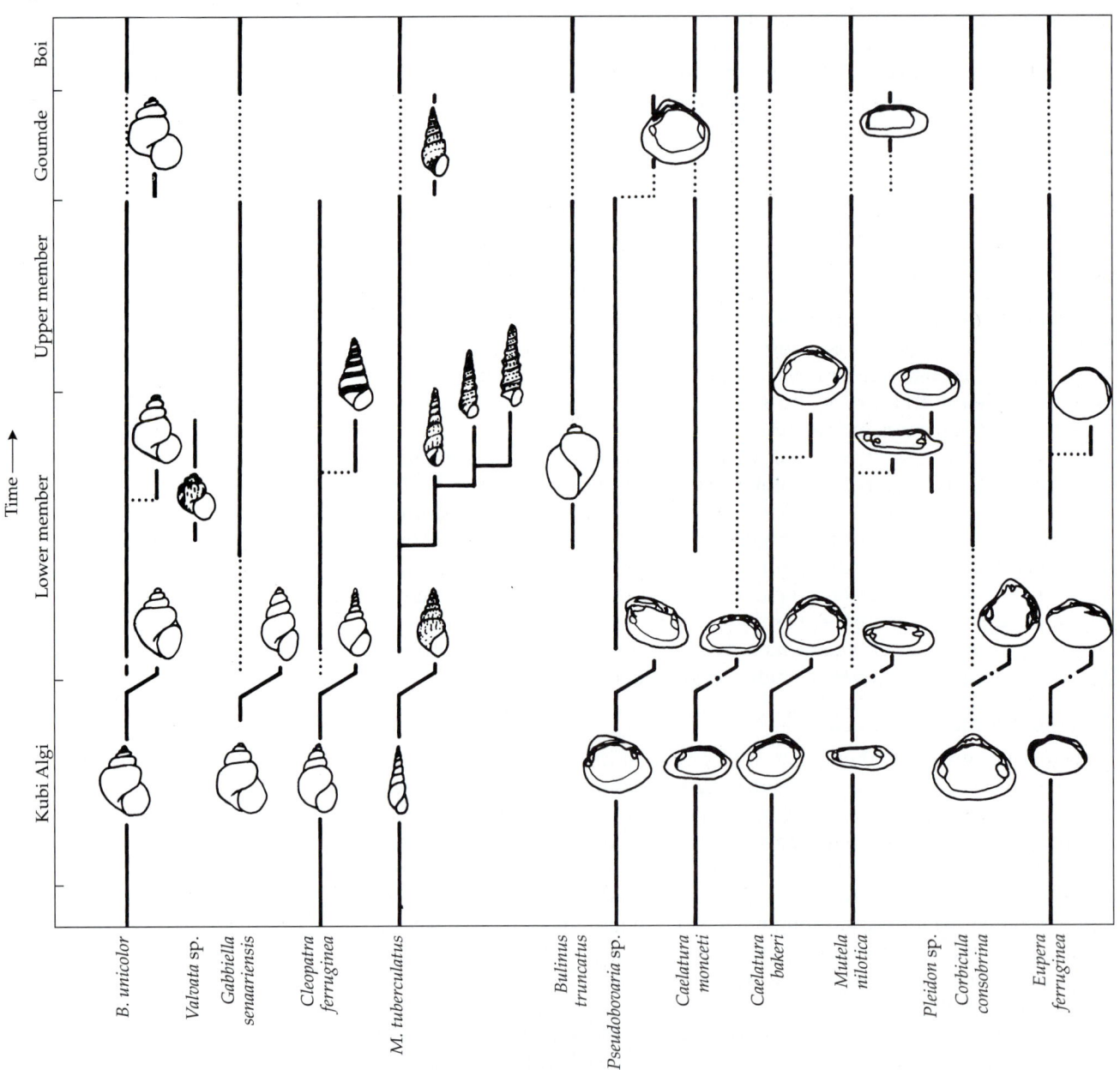

often no shells with intermediate characteristics were found. Many of these shifts were accompanied by the splitting of lineages, and they often occurred in conjunction with rapid changes in lake level.

The other macroevolutionary process evidenced in the fossil record has been called **species selection**. The idea is simple: just as natural selection is a process of evolutionary change caused by the differential survival and reproduction of individual organisms with certain heritable characteristics, so species selection is a process of evolutionary change caused by the differential survival and speciation of species with particular heritable traits. For example, in his classic study of industrial melanism, Kettlewell (1961; see below) documented microevolutionary change in the peppered moth as a result of the higher mortality of individuals that did not match their sooty-colored backgrounds. Similarly, macroevolutionists have shown that major episodes of extinction and speciation in the fossil record have been selective with respect to the characteristics of species. Thus, for example, the asteroid that struck the earth at the end of the Cretaceous period about 65 million years ago caused wholesale death and destruction, including the extinction of many species and higher taxa. But far from being random, these extinctions were highly selective. Entire lineages of large animals (e.g., dinosaurs, several other groups of giant terrestrial and marine reptiles, and several kinds of large invertebrates, including ammonites) were completely exterminated, while many lineages of small animals (e.g., insects, mammals, teleost fishes) and vascular plants survived. Similarly, in major episodes of speciation, some species with certain morphological and life history characteristics have produced many more descendant species than have other ancestral species with contrasting characteristics.

Micro- and Macroevolution

When they first aired their ideas, the macroevolutionists appeared to be using data from the fossil record to challenge the traditional view of microevolution that had come from the new synthesis. But as is often the case in such apparent controversies, neither side has been proved right or wrong; rather, we have discovered that for the most part, these apparent antagonists were actually talking about different things. In this case, the primary difference is one of scale. Microevolutionists are concerned with evolutionary changes within populations that occur as a result of the differential births, deaths, and movements of individuals with certain heritable characteristics. Such changes can occur over short periods of time and can cause geographic variation within a species (see the examples of industrial melanism, house sparrows, and clines below). Such changes are by no means always slow or uniformly continuous; however, they are rarely of sufficient magnitude to result in the formation of a new species or the extinction of an existing one. Macroevolutionists, on the other hand, are concerned with evolutionary changes in the morphological forms that they can recognize in the fossil record. These changes are often caused by the differential proliferation and extinction of species, so that they substantially alter the diversity of lineages and the composition of biotas. By comparison with the modest scale of microevolutionary change, these changes are often large, and in relation to the eons of evolutionary time over which they occur, they often appear abrupt or "punctuated."

The microevolutionary and macroevolutionary perspectives are complementary. Both are necessary for understanding the influence of geography on evolutionary processes and the influence of evolutionary processes on the distributions of organisms. We will illustrate the need for these complementary perspectives in the remainder of this chapter. We will be concerned primarily with the "macroevolutionary" processes of speciation and extinction. Yet we will show that a deep understanding of these processes also requires a microevolutionary perspective on how populations respond to environmental change in both space and time.

Speciation

Mechanisms of Genetic Differentiation

Speciation is a branching process in which different kinds of organisms originate from a single ancestral population. The magnitude of this process is staggering when we consider that all species of green land plants that have ever lived ultimately share a common ancestor in a simple green alga that lived 500 million years ago, that all vertebrates are traceable to some ancient chordate, and that millions of species of insects have evolved in the 400 million years since their ancestor first invaded land. Most of the species in all of these lineages are no longer around; they lived in the past and went extinct. How did this diversity come to be?

Genetic drift.   Populations diverge (or converge) in their genetic characteristics as a result of the microevolutionary forces of mutation, genetic drift, gene flow, and natural selection. Different forms of a gene at the same locus, known as **alleles**, are responsible for differences among individuals in their heritable traits. New alleles arise by mutation, and changes in the frequencies of alleles in populations occur primarily through genetic drift, natural selection, and gene flow. **Genetic drift** is a relatively weak force because it involves changes in the genetic constitution of a population caused solely by chance. Given sufficient time—which usually means many generations—the frequencies of alleles in a population tend to change randomly as different individuals happen to survive, mate, and produce offspring. Genetic drift has relatively little effect in large populations, but can have important influences on the evolution of small populations. How small is small is a matter of considerable discussion in evolutionary biology. Nevertheless, if new species start from small founding populations, as many island forms undoubtedly do, then genetic drift may play an important role in their initial differentiation, as we shall see below.

Natural selection.   Natural selection, on the other hand, can be a potent force for evolutionary change in both large and small populations. **Natural selection** is the change in a population that occurs because individuals express genetic traits that alter their interactions with their environment so as to enhance their survival and reproduction. Over many generations, alleles for such adaptive traits tend to increase in frequency at the expense of alleles that confer less fitness. Populations tend to diverge if there is sufficient variation in the environment to select for different characteristics to deal with different environmental conditions.

This fine-tuning of phenotypes to environmental heterogeneity can be readily documented. Above we mentioned industrial melanism. Within the last 150 years, moths in industrialized areas around cities have evolved dark color patterns to match their sooty backgrounds and are thus less visible to their avian predators (Kettlewell 1961). Rapid evolution by natural selection has also occurred in house sparrows. Since being introduced to North America from Europe less than 200 years ago, these enormously successful birds have not only spread to colonize most of the continent, but have also evolved distinct geographic races (Johnston and Sealander 1964, 1971; Johnston and Klitz 1977). This variation, in characteristics such as body size, color pattern, and physiological characteristics, is similar to that seen in the deer mouse (see Figure 8.5); it reflects different adaptations to different environmental conditions.

Gene flow. Migration, or **gene flow**, already mentioned in Chapter 4, often tends to act counter to genetic drift and natural selection to retard genetic divergence. Individuals that migrate to a new area carry their genes with them, and if they subsequently reproduce successfully, they inject their genes into the local population. Such migration, therefore, tends to have a homogenizing influence, preventing or at least retarding the development of geographically isolated and genetically differentiated populations.

Geographic variation. There is often a geographic component to genetic divergence, because both genetic drift and natural selection are facilitated, and gene flow is retarded, by geographic isolation. Genetic drift can be an important force in small, isolated populations, such as those that inhabit small outlying patches at the periphery of a species range, or those that have recently been founded by long-distance colonization. A population started by only a few colonizing individuals contains only a small random sample of the alleles present in the ancestral population. The genetic drift that occurs in these circumstances has been termed the **founder effect**. Mayr (1942) suggested that this process accounts for the apparently random differences often found among the bird populations on different islands (Figure 8.4), each of which was probably derived from a few successful colonists. Experimental evidence suggests that the founder effect can play a major role in speciation, because this initially random genetic sampling has effects on the subsequent genetic differentiation of small colonizing populations (e.g., Carson and Kaneshiro 1976; Templeton 1980a; Carson 1981).

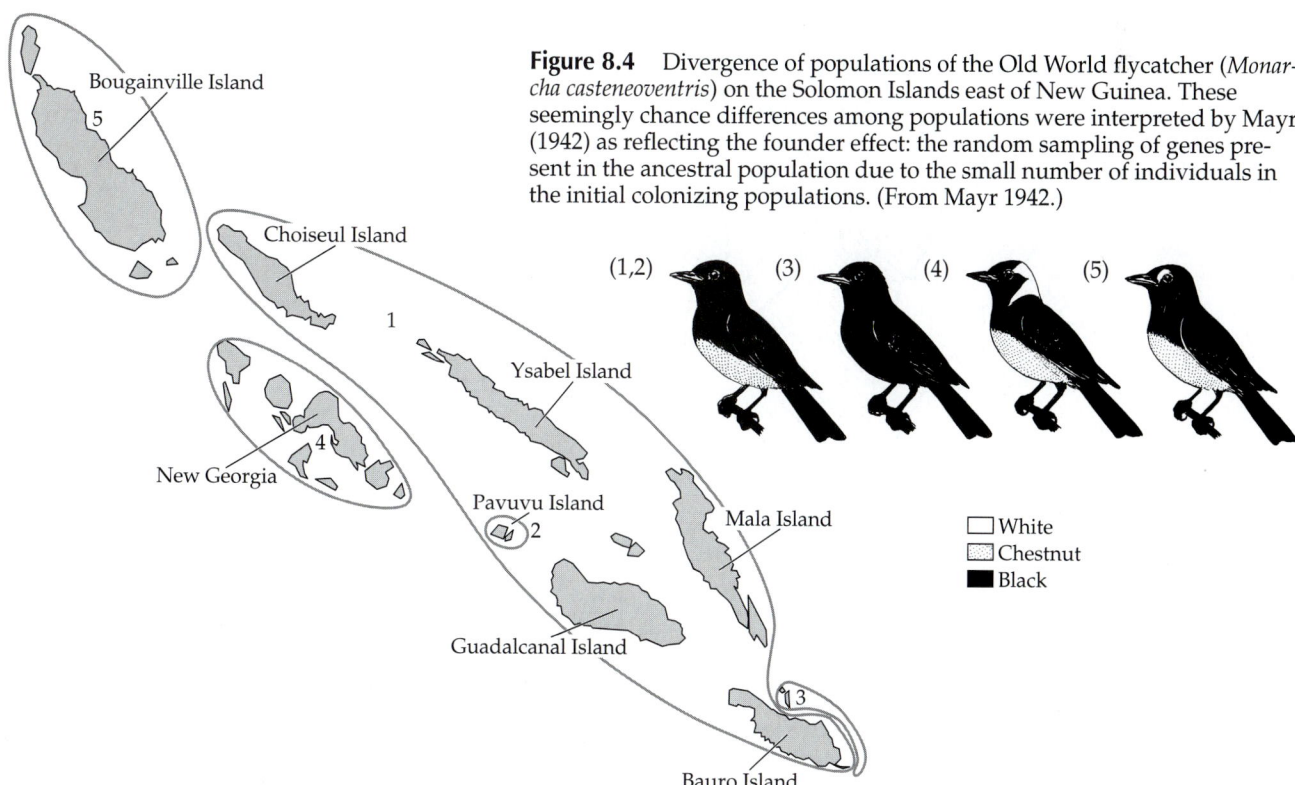

Figure 8.4 Divergence of populations of the Old World flycatcher (*Monarcha casteneoventris*) on the Solomon Islands east of New Guinea. These seemingly chance differences among populations were interpreted by Mayr (1942) as reflecting the founder effect: the random sampling of genes present in the ancestral population due to the small number of individuals in the initial colonizing populations. (From Mayr 1942.)

Bougainville Island

5

Choiseul Island

1

New Georgia

4

Ysabel Island

Pavuvu Island

2

Mala Island

Guadalcanal Island

Bauro Island

3

(1,2) (3) (4) (5)

☐ White
▨ Chestnut
■ Black

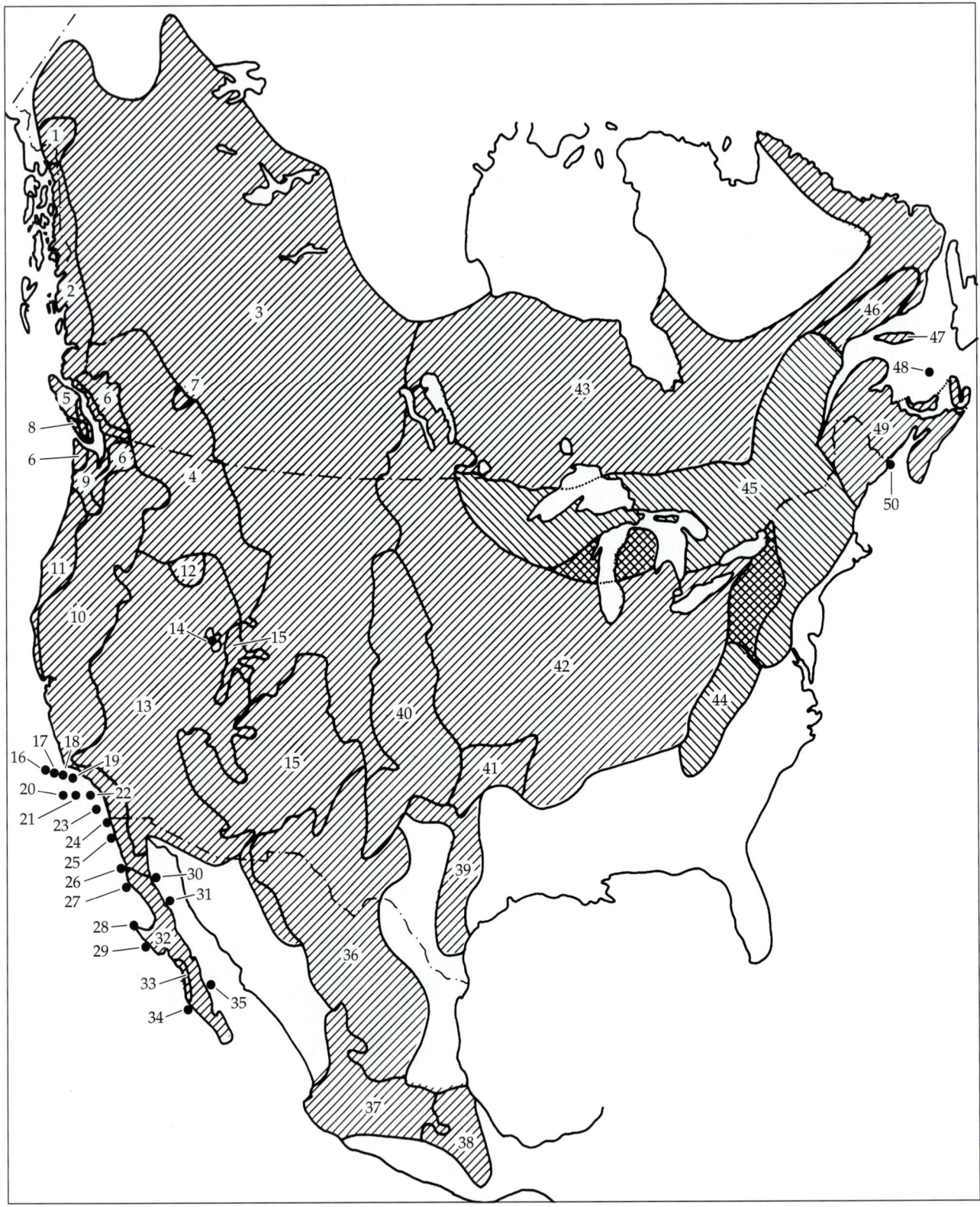

Figure 8.5   Geographic variation in the deer mouse (*Peromyscus maniculatus*) is indicated by the subdivision of the species into 50 formally recognized subspecies. Each of these geographic races has distinct characteristics, including dorsal coat color, which resembles background color and provides camouflage, and tail and hindfoot length, which are related to climbing ability and habitat structure. (After Hall 1981.)

Geographic separation of populations also facilitates genetic differentiation by natural selection. Different environmental regimes tend to select for different traits, and spatially isolated populations are likely to occur in different environments. The effects of natural selection caused by spatial environmental heterogeneity are exemplified by data on small mammals. As mentioned in Chapter 4, the coat color of the deer mouse (*Peromyscus maniculatus*) varies greatly over the wide geographic range of the species (Figure 8.5). The color of the dorsal fur closely matches the color of the soil or other substrate on which the animal is likely to be active, reflecting the consequences of selection by predators. Dice (1947) performed a classic experiment showing that owls selectively captured mice that contrasted with their background.

Patterns of geographic variation can take many forms. The term **cline** is used to describe a gradual change in one or more features along a single environmental gradient. Many birds and mammals exhibit clinal variation in clutch or litter sizes with latitude and elevation (e.g., for *Peromyscus* see Dunmire 1960; Lord 1960; Smith and McGinnis 1968; Spencer and Steinhoff 1968) (Figure 8.6). Such variation presumably reflects the adaptation of life history traits to environments that differ in temperature, seasonality, productivity, and other factors. Similarly, clines in physiological characteristics of plants and insects show that populations at progressively higher latitudes are adapted to colder temperature regimes.

Allopatric Speciation

Perhaps the simplest and most frequent process of speciation is the one that occurs when populations are geographically isolated, so that gene flow between them is mostly or entirely cut off. This process is termed **allopatric speciation**, meaning divergence that occurs in different places. The classic model of allopatric speciation by geographic subdivision (also called **geographic speciation**) was championed by Ernst Mayr (1942, 1963).

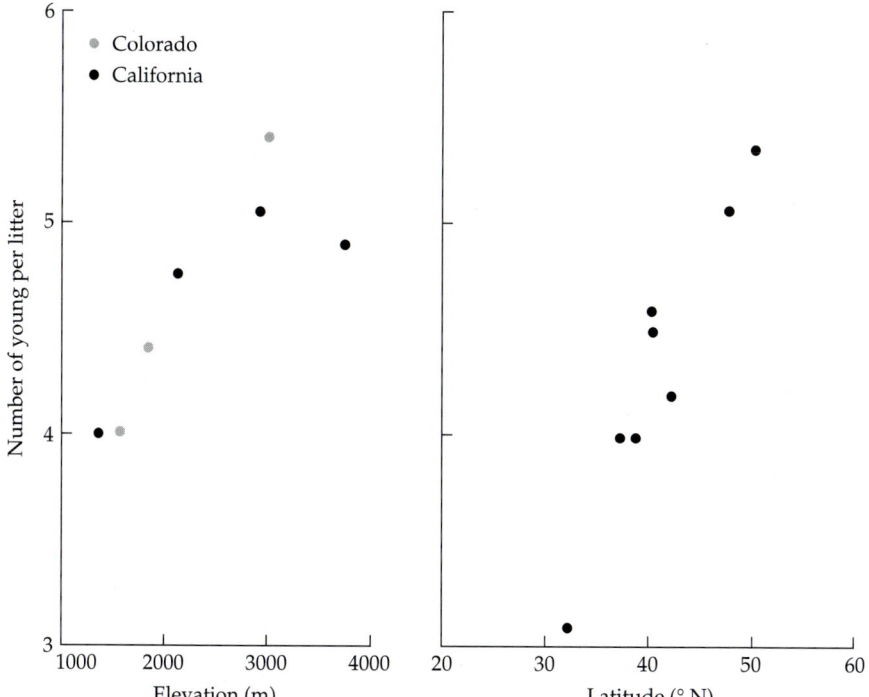

Figure 8.6   Latitudinal and elevational variation in average litter size of the deer mouse (*Peromyscus maniculatus*). Note that the number of young per litter increases with both latitude and elevation. Presumably, large litters are advantageous in colder climates, perhaps because the shorter growing seasons favor fewer litters per year with more young per litter (see also Chapter 16).

If its environment is heterogeneous, a geographically widespread ancestral population will tend to develop regional genetic differences in response to either natural selection or genetic drift. Because of barriers that limit dispersal, free gene flow from one end of the range to the other rarely occurs. Thus populations living in environmentally distinct and geographically separated regions tend to become somewhat differentiated from one another, but some gene flow maintains the genetic cohesiveness of the species. The regional races of house sparrows and deer mice, and the clinal variation in litter and clutch sizes mentioned above, are examples of such differentiation.

If, however, the regional populations become sufficiently isolated that the cohesive gene flow between them is cut off or drastically reduced, then they become independent evolutionary units. Without dispersal and gene flow, the isolated populations tend to diverge. Divergence proceeds more rapidly when substantially different environments subject the isolates to different selective pressures. For example, Darwin called attention to the morphological differences among the giant tortoises of the Galápagos, which are obviously descended from a common ancestor but presently occur on different islands. On some of the most arid islands, where treelike cacti are abundant, the endemic tortoises have evolved long necks and forelimbs and distinctively shaped shells that allow then to reach up high to feed on these plants. On wetter islands, where the tortoises feed mostly on lower vegetation, they have more generalized body forms.

Isolation: Vicariant and founder events.   We can distinguish, at least at the extremes, two ways in which such isolation can occur. At one extreme, some environmental change can create a barrier to dispersal, isolating previously connected and interbreeding populations. Rising sea levels, for example, can isolate an island on a continental shelf, tectonic events can cause part of a continent to split off and drift away, or, conversely, landmasses can drift together to isolate formerly continuous oceans. Such changes are called **vicariant events**, and they usually isolate relatively large populations (Figure 8.7A). At the other extreme, individuals may disperse across an existing barrier to colonize a previously uninhabited region. Such migrations, called **dispersal** or **founder events**, are the way in which oceanic islands and many other isolated patches of habitat come to be colonized. They typically involve a small initial population, sometimes only one or a few individuals (Figure 8.7B). As indicated above, the mechanisms and rates of initial genetic divergence may differ depending on the mode of isolation. Genetic drift may play a greater role rel-

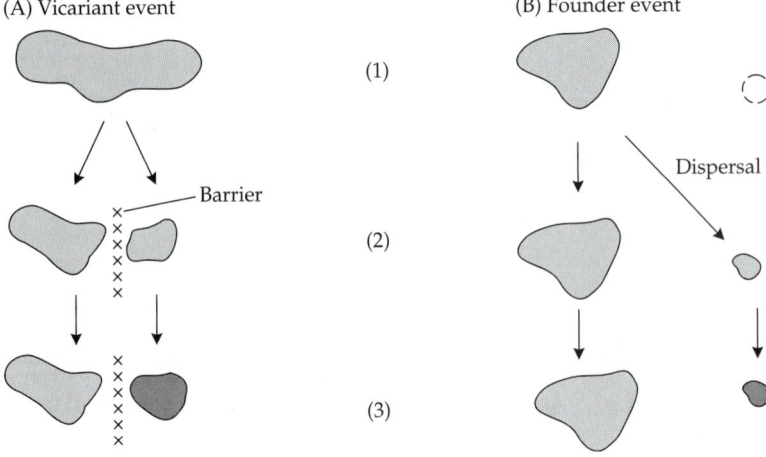

Figure 8.7   A schematic representation of two modes of allopatric speciation: (A) a vicariant event, in which part of a once widely distributed population becomes isolated due to the formation of a geographic barrier, and (B) a founder event, in which a small number of individuals disperse to found a new geographically isolated population. In both modes the speciation process involves several stages: a single ancestral population (1) gives rise to two isolated populations (2), the isolates diverge (3), and eventually become sufficiently distinct to be recognized as separate species according to one or more of the species concepts.

ative to selection, and divergence may be more rapid, at least initially, in founder events than in vicariant events (Carson 1971; Bush 1975; Templeton 1980a, 1981).

Contact and reinforcement.   Once isolates have formed and differentiated, there are several possible outcomes. First, they may come back into contact. This may occur either through the disappearance of a geographic barrier or by dispersal across it. If the populations reestablish contact, there are then three possible outcomes: (1) they may not interbreed, or may fail to produce fertile offspring, in which case reproductive isolation is complete and speciation has occurred; (2) the two populations may interbreed extensively, producing fertile, fit hybrids, so that the populations merge and their differentiation breaks down; or (3) the two populations may hybridize, but the hybrids may be less fit than the offspring of within-population matings. In this last case, selection favoring those individuals that choose mates from within their own population may be sufficiently strong to lead to reproductive isolation and the completion of the speciation process. This process of selection for mechanisms that promote within-population matings is termed **reinforcement**, and the traits that evolve to enhance reproductive isolation are called **isolating mechanisms**. A final scenario is one in which the isolated populations never do come into contact. In this case reproductive isolation may take a long tome to develop, and, as mentioned above, the isolates will cause problems for a taxonomist trying to apply the biological species concept and decide whether they should be considered different species.

While there is broad agreement that speciation can, and often does, occur through geographic isolation, there is much less consensus about the details of the process (see Paterson 1982; Otte and Endler 1989). How much do the mechanisms of speciation vary among different kinds of organisms, and even among different speciation events within lineages of closely related organisms? What are the relative frequencies of vicariant and founder events? When vicariant events are involved, how often are the isolates small, peripheral populations as opposed to large fragments of a once continuous population? How often are the initial isolates composed of only a few individuals, so that they are likely to diverge rapidly due to the founder effect and genetic drift, as opposed to large populations that are less likely to experience rapid genetic change? How important is gene flow in maintaining cohesion and preventing differentiation among populations? How important is the genetic inertia of large populations and the influence of similar selective environments in retarding divergence? Conversely, what are the roles of genetic isolation and founder events relative to divergent selective pressures in promoting divergence of isolated populations? To what extent is there active selection for "reinforcement" of isolating mechanisms, so that individuals actively avoid interbreeding with members of closely related, coexisting species? To what extent does sexual selection for mates with exaggerated traits facilitate the evolution of reproductive isolation? There is much disagreement among evolutionary and systematic biologists on the answers to these questions. We suspect that much of it stems from a well-intentioned but misplaced desire to describe one universal process of speciation. Instead of attempting this, we must recognize the enormous variation in the process due to the special characteristics of different kinds of organisms and the different historical and environmental contexts in which speciation occurs.

Examples: Allopatric speciation in the Galápagos archipelago.   Despite these uncertainties about the details, there can be little doubt that geographic isolation has been a common mode of speciation in many groups. The giant tortoises and finches that Darwin observed in the Galápagos archipelago provide examples of populations in different stages of the process. As we have seen, there are distinctly different populations of tortoises on each of the large islands. These populations have diverged from a single ancestor that originally colonized the archipelago and whose descendants subsequently dispersed to establish populations on the different islands. However, there are no cases in which two or more forms occur on the same island. Even in the unlikely event that individuals immigrated from another island and were sufficiently differentiated so that widespread hybridization did not occur, even the largest islands are probably too small and unproductive to support two species of giant tortoises. On the other hand, in Darwin's finches, the final stages of speciation are represented. Here again, all the species are believed to be derived from a single ancestral population that colonized the archipelago. In the case

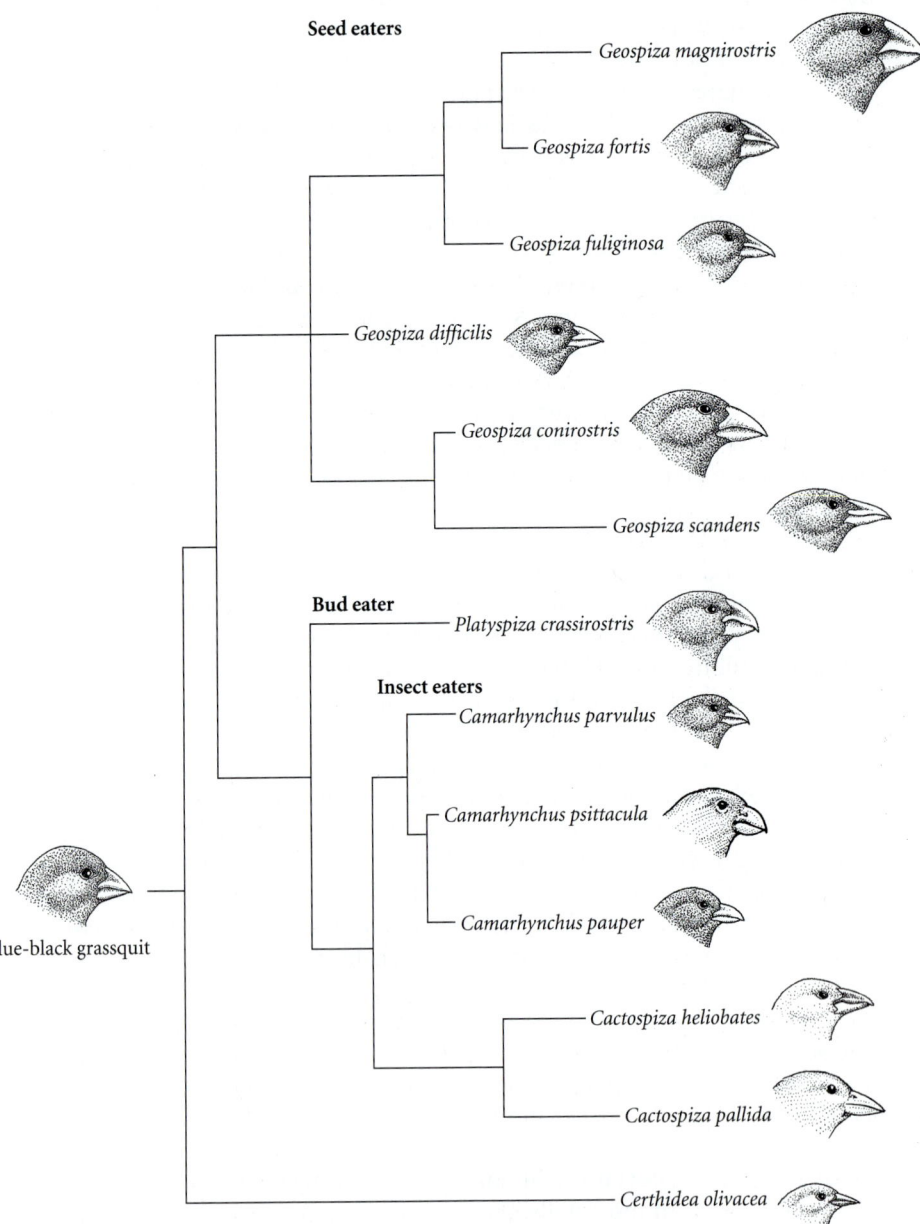

Figure 8.8 Adaptive radiation in Galápagos finches, showing the diversity of beak shapes and diets. A single ancestor, a grassquit, colonized the archipelago. Subsequent allopatric speciation events, due to repeated episodes of colonization and divergence, produced five genera and 13 species. (After Futuyma 1998, based on Purves et al. 1997; Grant 1986.)

of the finches, however, after diverging in isolation on different islands or in different habitats, some populations have successfully reinvaded already inhabited areas, so that several species (as many as 10 on certain islands) now coexist. These species have not only evolved specific mating behaviors that prevent interspecific hybridization, but have also diverged in morphology and behavior to exploit different ecological niches (Figure 8.8). The process of speciation is an ongoing one, and so there are also documented cases of inter-island colonization events and of hybridization between incompletely isolated forms (Grant and Grant 1989, 1992).

Sympatric Speciation

Although many evolutionary biologists once maintained that speciation in most organisms occurs primarily or solely as a result of geographic isolation, most now accept that speciation can, and often does, occur within spatially contiguous populations. This process is called **sympatric speciation**, meaning divergence that occurs in the same place. Two mechanisms of sympatric speciation have been proposed.

Disruptive selection. If strong selective pressures cause a population to adapt to two or more different environmental regimes or niches, they can progressively pull the population apart and eventually result in speciation. Endler (1977; see also Slatkin 1973; Rosenzweig 1978) suggested that such **disruptive selection**, acting along an environmental gradient, could gradually sharpen clinal variation until a single ancestral population fragmented into two or more species. Such speciation is often termed **parapatric** or **stasipatric** because the diverging populations are geographically distinct, but remain in contact with each other.

Bush (1975), Price (1980), and others have argued that sympatric speciation by disruptive selection may be common in certain groups of phytophagous (herbivorous) insects and animal parasites that are highly specialized for specific host species. Among these organisms, successful colonization of a new kind of host must be a rare event, but when it occurs, the colonists are immediately subjected to selection for the ability to survive and reproduce in a drastically different environment. Usually this selection pressure is intensified by counterevolution of the host to escape from or reject the parasite. Selection to meet the challenge of a new host could potentially lead to the rapid differentiation and speciation of totally sympatric populations, in which the organisms are spatially close enough to mate, but do not, often because mating occurs on the host.

Chromosomal changes. Sympatric speciation can also occur through chromosomal changes. Chance rearrangements of the genetic material of a parent during meiosis, or of an embryo during fertilization or early development, can sometimes change the number of chromosomes or the sequence of genes on chromosomes. Changes in chromosome number are of two kinds: **aneuploidy**, in which a single chromosome breaks or fuses with another to change the total number by plus or minus one; or **polyploidy**, in which an entire additional set of chromosomes is passed on, changing the number by some multiple (e.g., a doubling or tripling). In other cases, the chromosome number remains the same, but some of the genetic material is either rearranged within a chromosome (**inversion**) or transferred to another chromosome (**translocation**).

In diploid organisms, precise pairing during meiosis of the genes and chromosomes inherited from each parent usually is necessary to ensure the transmission of a complete set of genes to each gamete, and hence to produce viable offspring. Consequently, mutant individuals with new chromosomal arrangements often have impaired fertility when they mate with an individual having the original chromosomal arrangement, and may be able to reproduce only by mating with another individual having the new arrangement. For this reason, it is obviously difficult for a population with a new arrangement to become established. However, once established—especially in a small, isolated, inbred population—the new type is genetically isolated from its parental population and can diverge rapidly as a new species.

Sympatric speciation by way of polyploidy appears to have occurred frequently in some groups of organisms, especially plants (e.g., Stebbins 1971b; de Wet 1979; Lewis 1979; Briggs and Walters 1984). There are many documented ways in which polyploidy has been achieved in plants, but one such process is considered very common (de Wet 1979). Diploid ($2N$) organisms have two sets of chromosomes and produce haploid (N) gametes by meiosis. Occasionally, a female gamete is formed without undergoing meiosis, and remains diploid ($2N$). This unreduced gamete can then fuse with a haploid pollen grain (N) to produce a triploid ($3N$) plant, which will produce triploid gametes because of complications during meiosis. In the next generation, a triploid female gamete can fuse with a haploid pollen grain to yield a tetraploid ($4N$) zygote, which can survive and produce fertile offspring from diploid ($2N$) gametes, either by self-fertilization or by crossing with other rare tetraploids in the population. The resulting tetraploid population is immediately genetically isolated from the diploid population.

Polyploidy can occur either within a population, called **autopolyploidy**, or as a result of hybridization between different but usually closely related populations or species, called **allopolyploidy**. Allopolyploidy is thought by many researchers to be more common. Because the chromosomes from different species may not pair and segregate properly, interspecific hybridization often results in abnormalities in the meiotic process that can facilitate the process described above. Allopolyploids may not only arise more frequently than autopolyploids, but may also be more likely to become established. In addition to possessing a larger genome than either of the parental species, allopolyploids tend to be intermediate in their characteristics, which enables them to be superior competitors in certain habitats.

Examples: Sympatric speciation in isolated lakes.   In the decades immediately following the new evolutionary synthesis, there was a tendency to regard allopatric speciation as the general process and sympatric speciation as the exception: a process that undoubtedly occurs in some plants and possibly in a few other kinds of organisms, such as phytophagous insects and parasites, but is relatively rare. This assumption was largely due to the enormous influence and powerful arguments of Ernst Mayr (e.g., 1942, 1963), who argued for the near universality of allopatric speciation. Recently, however, the pendulum has begun to swing back as increasing evidence for sympatric speciation has accumulated. For example, as many as 70 to 80% of angiosperm plant species are now thought to be of polyploid origin (Briggs and Walters 1984).

Some of the most convincing cases of sympatric speciation involve the divergence and adaptive radiation of fishes in isolated lakes (Echelle and Kornfield 1984). There are many examples: cichlids in the Great Lakes of central Africa and in the giant Quatro Cienegas spring in northern Mexico; whitefishes in the Great Lakes of eastern North America; sculpins in Lake Baikal in Siberia; herrings in Scandinavian lakes; pupfishes in Lago Chichancanab in Yucatán; sticklebacks in lakes in British Columbia. These cases differ in when the isolation occurred and how much differentiation has taken place. At one extreme are the cichlids in Lakes Victoria, Malawi, and Tanganyika of the African Rift Valley. These fishes have been isolated in large lakes for tens of thousands to millions of years, and a few founding lineages of cichlids have branched to produce species flocks of hundreds of species (e.g., Fryer and Iles 1972; Greenwood 1974, 1984; Kaufman and Ochumba 1993). The resulting species exhibit enormous variation in morphology, much of it related to specialization for different diets and feeding modes, and in color patterns and behavior, much of it related to courtship and mating displays (Figures 8.9 and 8.10). Recent evidence indicates that shallow Lake Victoria was completely dry in the late Pleistocene (until only about 12,000 years ago; Johnson et al. 1996). The fact that it now supports a fauna of more than 300 species of endemic cichlids suggests that under conducive conditions, speciation and adaptive radiation can be extremely rapid.

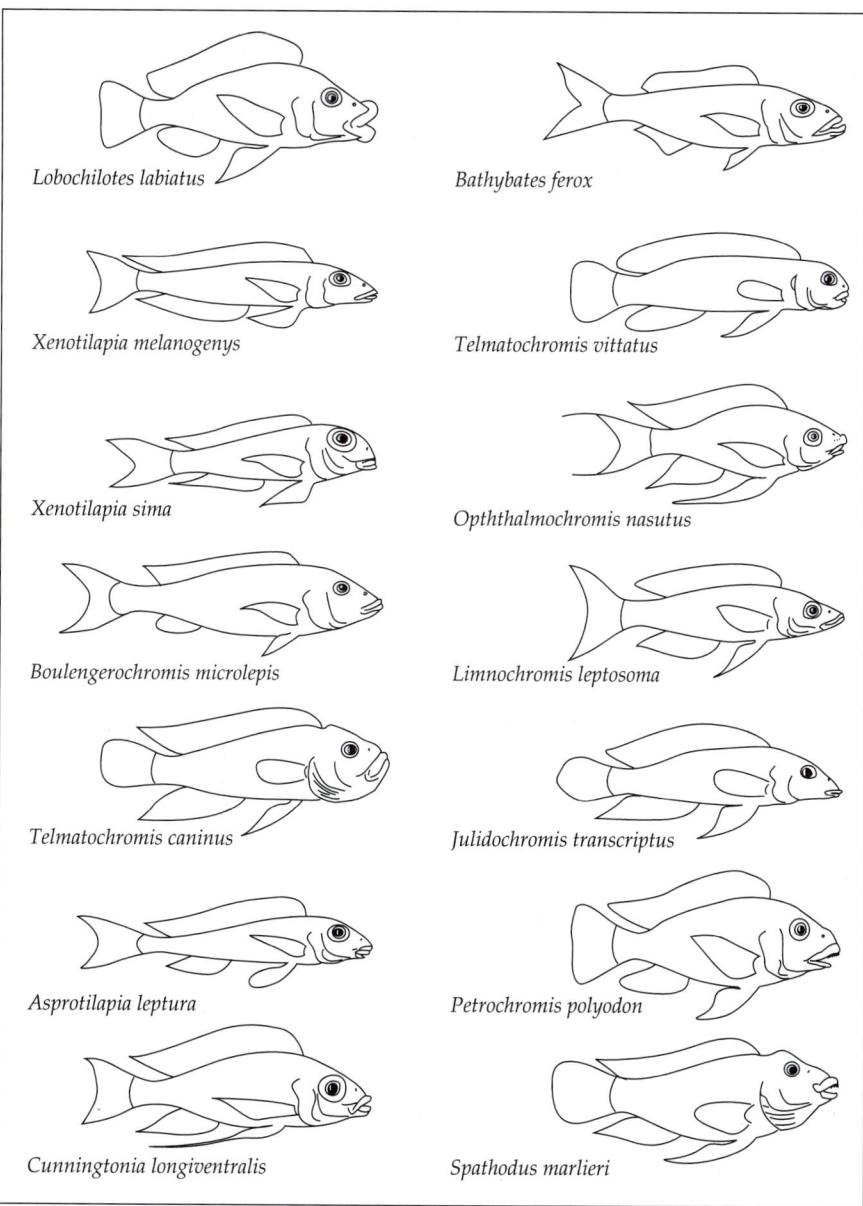

Figure 8.9 Examples of the variety of body forms resulting from the adaptive radiation of cichlid fishes in Lake Tanganyika in eastern Africa. (After Fryer and Iles 1972.)

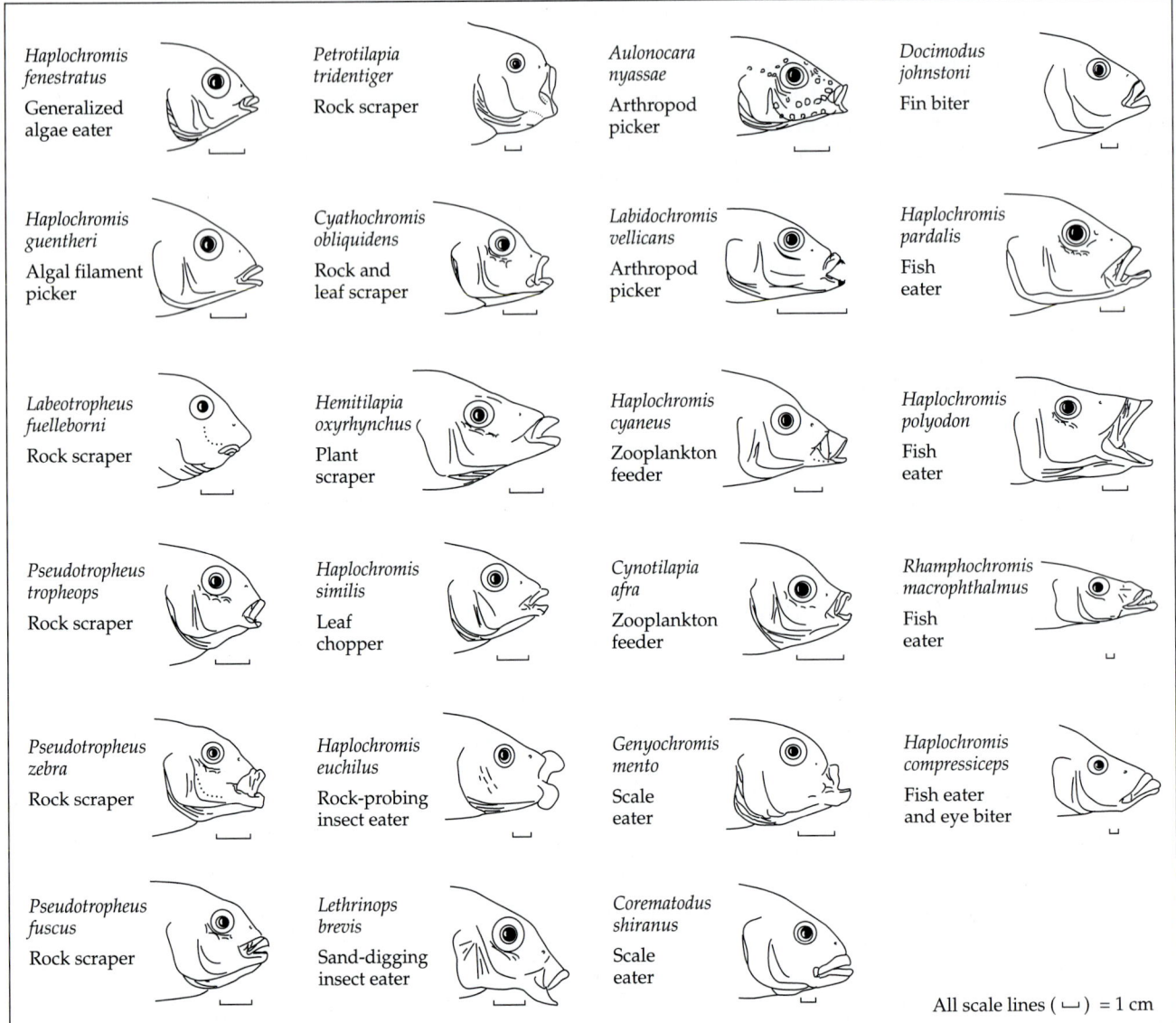

Haplochromis fenestratus
Generalized algae eater

Petrotilapia tridentiger
Rock scraper

Aulonocara nyassae
Arthropod picker

Docimodus johnstoni
Fin biter

Haplochromis guentheri
Algal filament picker

Cyathochromis obliquidens
Rock and leaf scraper

Labidochromis vellicans
Arthropod picker

Haplochromis pardalis
Fish eater

Labeotropheus fuelleborni
Rock scraper

Hemitilapia oxyrhynchus
Plant scraper

Haplochromis cyaneus
Zooplankton feeder

Haplochromis polyodon
Fish eater

Pseudotropheus tropheops
Rock scraper

Haplochromis similis
Leaf chopper

Cynotilapia afra
Zooplankton feeder

Rhamphochromis macrophthalmus
Fish eater

Pseudotropheus zebra
Rock scraper

Haplochromis euchilus
Rock-probing insect eater

Genyochromis mento
Scale eater

Haplochromis compressiceps
Fish eater and eye biter

Pseudotropheus fuscus
Rock scraper

Lethrinops brevis
Sand-digging insect eater

Corematodus shiranus
Scale eater

All scale lines (⊔) = 1 cm

Figure 8.10   Examples of the variety of head shapes, mouthparts, and feeding habits resulting from the adaptive radiation of cichlid fishes in Lake Malawi in eastern Africa. This amazing variation reflects specialization in diet due to natural selection to reduce competition and exploit ecological opportunities. (After Fryer and Iles 1972.)

At the opposite extreme are the pupfishes of Lago Chichancanab and the sticklebacks of British Columbia, which have been isolated in small lakes for only a few thousand years as a result of changing sea and lake levels since the Pleistocene (see Chapter 7). Lago Chichancanab contains five species of *Cyprinodon*, which have distinctive morphologies related to diet and also exhibit some degree of genetic and behavioral reproductive isolation (Figure 8.11) (Humphries and Miller 1981; Humphries 1984; Strecker et al. 1996). Small lakes in coastal British Columbia typically contain two forms of sticklebacks, benthic and pelagic, which exhibit morphological adaptations for swimming and feeding along the bottom or in open water, respectively (McPhail 1994). While these pupfishes and sticklebacks have diverged much less than the African cichlids, they nevertheless provide an excellent opportunity to study the roles of ecological and evolutionary processes in speciation and adaptation.

While some systematists cling to hypotheses of allopatric speciation, which typically involve changing lake levels and repeated episodes of colonization

Figure 8.11 Morphology of five forms of pupfishes (*Cyprinodon*) in Lago Chichancanab on the Yucatán Peninsula of Mexico. These forms have diverged sympatrically within the last few thousand years since the isolation of the saline lake from the sea. Genetic and behavioral studies suggest that the speciation process is not complete, and that some of these forms still interbreed to some extent. The differences in size and shape reflect dietary differences, and suggest that strong selection for trophic differentiation is driving the sympatric speciation. (After U. Strecker.)

and isolation, to explain speciation among these fishes, the simplest explanation is that differentiation has occurred sympatrically within each of these lakes. The ultimate mechanism often appears to be disruptive selection to exploit different food resources. This hypothesis is supported by the common theme of morphological and dietary divergence found among these fishes, even in the cases of the pupfishes and sticklebacks, in which reproductive isolation between some forms is only partial and hybrids are not infrequent. These cases, as well as the speciation of several kinds of animals and plants within individual islands of the Hawaiian archipelago (see Chapter 12), suggest that sympatric speciation may be much more common than most evolutionary biologists have suspected. They also suggest that disruptive selection to exploit different ecological niches may often be a sufficiently powerful force to produce speciation, even in the absence of geographic isolation.

Phyletic Speciation

In the processes of speciation that we have described, a single lineage splits to produce two or more species. There are also cases in which one ancestral species is transformed into a single descendent species by evolutionary changes over time. This process, called **phyletic speciation**, is often inferred from the fossil record. Similar organisms replace one another in successively younger deposits, often without any evidence of splitting of lineages (Figure 8.12). Each major stage in such an evolutionary sequence is called a **chronospecies**, and is given a Latin binomial for descriptive purposes. Many examples have been cited in the literature, but relatively few have been studied carefully enough that we can be confident that the chronospecies represent an unbroken series—and not, in fact, products of speciation events that were followed by rapid extinction of all but one lineage.

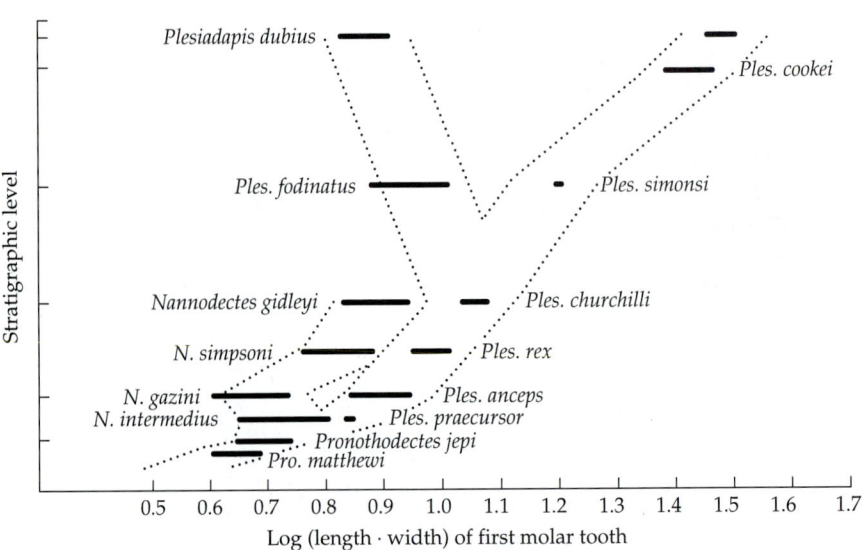

Figure 8.12 Reconstructed evolutionary history of the Plesiapidae, a primitive primate group from North America, based on changes in tooth size. The dashed lines indicate the inferred range of variation as the lineages evolved. The graph shows just two branching events, but many name changes. The latter illustrate "phyletic speciation" events—sufficiently large changes in a single lineage over time to warrant recognition as new species. (From Gingerich 1976a.)

Diversification

Ecological Differentiation

Once new species have formed, what happens to them? Immediately after a speciation event, the resulting species are often quite similar to each other. Above, we pointed out that they are most likely to diverge if they are subjected to different environments with different selective regimes. Ecological differen-

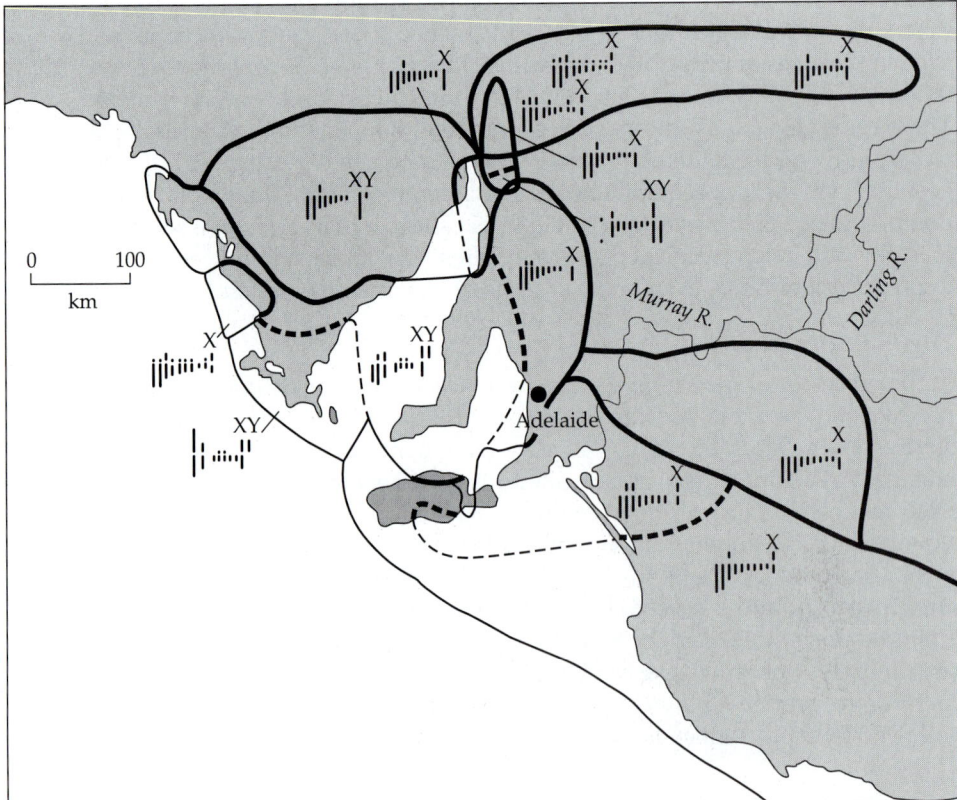

Figure 8.13 Ranges of different chromosomal forms of the grasshopper *Vandiermenella* in southern Australia. These grasshoppers are extremely similar in morphology and ecology, but their different chromosomal arrangements effectively prevent hybridization. Their contiguously allopatric, or parapatric (i.e., touching but not overlapping), distributions are typical of those of sibling species formed as a result of chromosomal changes. (After White 1978.)

tiation increases the likelihood that the species will be able to occupy overlapping geographic ranges. Thus it facilitates the buildup within a region of a biota of closely related, sympatric species. Gause (1934) showed in laboratory experiments with protozoans of the genus *Paramecium* that two species with identical resource requirements could not persist in the same environment: one species was eventually outcompeted and went extinct. Ecologists have generalized this phenomenon and termed it the principle of **competitive exclusion** (see Hardin 1960; Miller 1967; Hutchinson 1978).

A biogeographic corollary of the principle of competitive exclusion is that species that are extremely similar in their niches tend to have nonoverlapping geographic distributions, whereas species that coexist in the same area and habitat tend to differ substantially in their resource use. Perhaps the most striking example of this phenomenon is provided by so-called **sibling** or **cryptic species**: species that are genetically distinct, but extremely similar in their morphology and ecology. Examples include species that have recently formed through chromosomal changes. Sibling species often exhibit almost perfectly abutting, but nonoverlapping (i.e., parapatric), geographic ranges, as demonstrated by numerous examples in both plants and animals (Figure 8.13). Even related species that have apparently originated by allopatric speciation and are no longer extremely similar may competitively exclude each other from local habitats or extensive geographic areas. Examples include the kangaroo rats and chipmunks mentioned in Chapter 4 (see Figures 4.25 and 4.26). Several species of pocket gophers—fossorial rodents of the genera *Thomomys* and *Geomys*—have ranges in western North America that come into contact, but do not overlap. Even though the burrows of parapatric species may be only a few meters apart, the species remain separate (Figure 8.14). There is good evidence that this pattern is maintained by competitive exclusion (e.g., Miller 1964; Vaughan and Hansen 1964; Vaughan 1967).

Conversely, numerous field studies show that when closely related species do coexist in nature, they often differ substantially in their use of limiting resources (e.g., MacArthur 1958, 1972; Cody and Diamond 1975). Often these

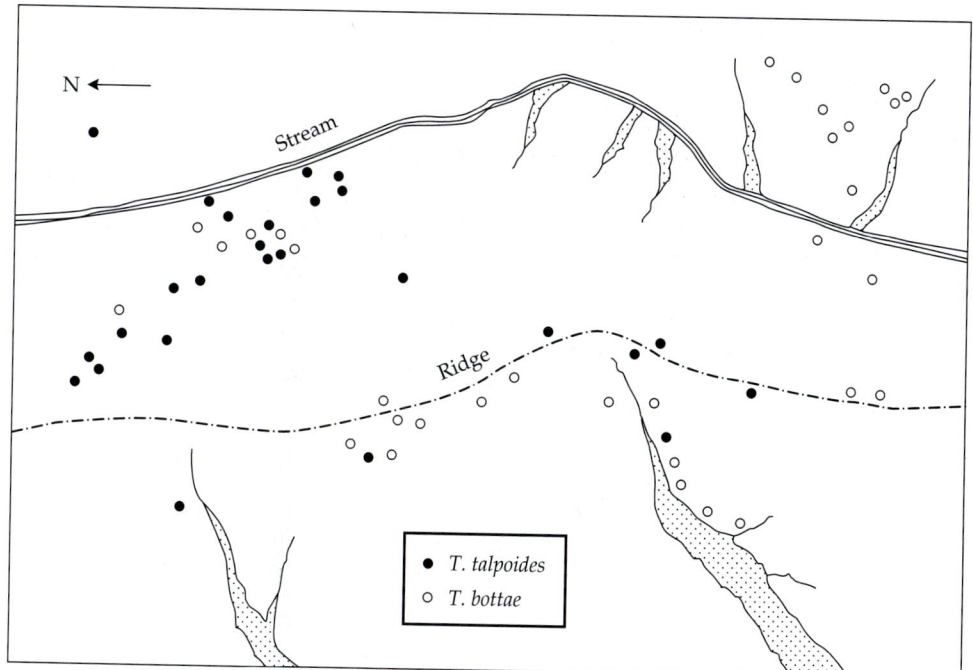

Figure 8.14   The small-scale distribution of two species of pocket gophers, as indicated by the locations of burrows of individual animals. The map encompasses approximately 1 km of a valley in the Rocky Mountains of Colorado. Note that these morphologically and ecologically similar species come into extremely close proximity, but do not overlap. (From Vaughan 1967.)

niche differences are reflected in pronounced morphological, physiological, or behavioral differences. The Galápagos finches provide excellent examples of this phenomenon. As the finches have reinvaded inhabited islands after speciating, they have diverged morphologically, behaviorally, and ecologically from other sympatric species. This process, called **character displacement** (Brown and Wilson 1956), has resulted in species being more different where they coexist than where they live allopatrically (see Figure 8.2). In these finches, character displacement is most apparent in the size of the beak, which enables coexisting forms to specialize on different kinds of foods (Lack 1947; Abbott et al. 1977; Schluter and Grant 1984; Schluter et al. 1985; Grant 1986; Grant and Grant 1989). Some intermediate phenotypes do occur, but strong selection against interspecific hybrids and other deviant phenotypes during periods of food shortage maintains the clumpy-gappy distribution of distinctive sympatric species (see Figure 8.2).

Adaptive Radiation

Adaptive radiation is the diversification of species to fill a wide variety of ecological niches. It occurs when a single ancestral species gives rise, through repeated episodes of speciation, to numerous kinds of descendants that become or remain sympatric. These coexisting species tend to diverge in their use of ecological resources in order to reduce interspecific competition. Such character displacement in response to competition obviously cannot occur if closely related species remain divided by physical barriers (i.e., vicariant), but some differentiation will still tend to occur as the allopatric species adapt to different environments.

Today, looking at the variety of living things, we can find numerous examples of successful lineages that have radiated to produce diversity at many levels. We can, for example, consider the adaptive radiation of Hawaiian honeycreepers (Drepanidinae), a group of small perching birds that probably colonized the archipelago within the last few million years (Amadon 1950; Raikow 1976; Tarr and Fleischer 1995). Or we can examine the major radiation of placental mammals, which occurred during the early Cenozoic era, after the mass extinction event that eliminated most of the giant reptiles, and produced many of the existing mammalian orders (e.g., Lillegraven 1972). In all such cases, the basic ecological and evolutionary processes are similar. New ecological opportunities are created by either an adaptive innovation or an environmental change, such as colonization of a new area or the extinction of competing species. These opportunities are exploited as an ancestral form repeatedly speciates, diversifies, and specializes to fill numerous ecological niches.

One of the most dramatic examples of adaptive radiation, mentioned above, is provided by the cichlid fishes of the lakes of East Africa, and is described in a fascinating book by Fryer and Iles (1972; see also Greenwood 1974, 1984; Meyer et al. 1990; Kaufman and Ochumba 1993; Johnson et al. 1996). The cichlids have diversified morphologically and have specialized behaviorally and ecologically to fill many different niches (see Figures 8.9 and 8.10). There are herbivores and carnivores; species with mouths and teeth adapted for catching tiny zooplankton, crushing snails, and eating other fishes whole; even forms specialized to feed just on the fins, scales, or eyeballs of other fishes.

Islands and archipelagoes provide many other examples of adaptive radiation. Madagascar, with its long history of isolation from Africa and the other southern continents (see Chapter 6), has been the site of several spectacular radiations. These include not only the well-publicized lemurs (Primates), but the less well known tenrecs (a morphologically, behaviorally, and ecologically diverse endemic family, Tenrecidae, in the mammalian order Insectivora:

Eisenberg and Gould 1970) and vanga shrikes (a similarly diverse group of perching birds belonging to the endemic family Vangidae). Among the many examples in the Hawaiian Islands (see also Chapters 12 and 14; Wagner and Funk 1995) are Hawaiian honeycreepers of the endemic subfamily Drepanidinae. There were about 33 species of these birds when Europeans first visited Hawaii, although at least 10 have since gone extinct and most of the others are endangered. They are descended from a single common ancestor, now thought to be a cardueline finch that colonized from North America, probably less than 5 million years ago (Tarr and Fleischer 1995). The lineage radiated to produce an amazing variety of species, which differ conspicuously in the sizes and shapes of their beaks (Figure 8.15) and to a lesser extent in color pattern. As in Darwin's finches and the African lake cichlids, ecological differentiation to exploit different niches, especially alternative food sources, seems to have been the primary process that led to the radiation of the honeycreepers.

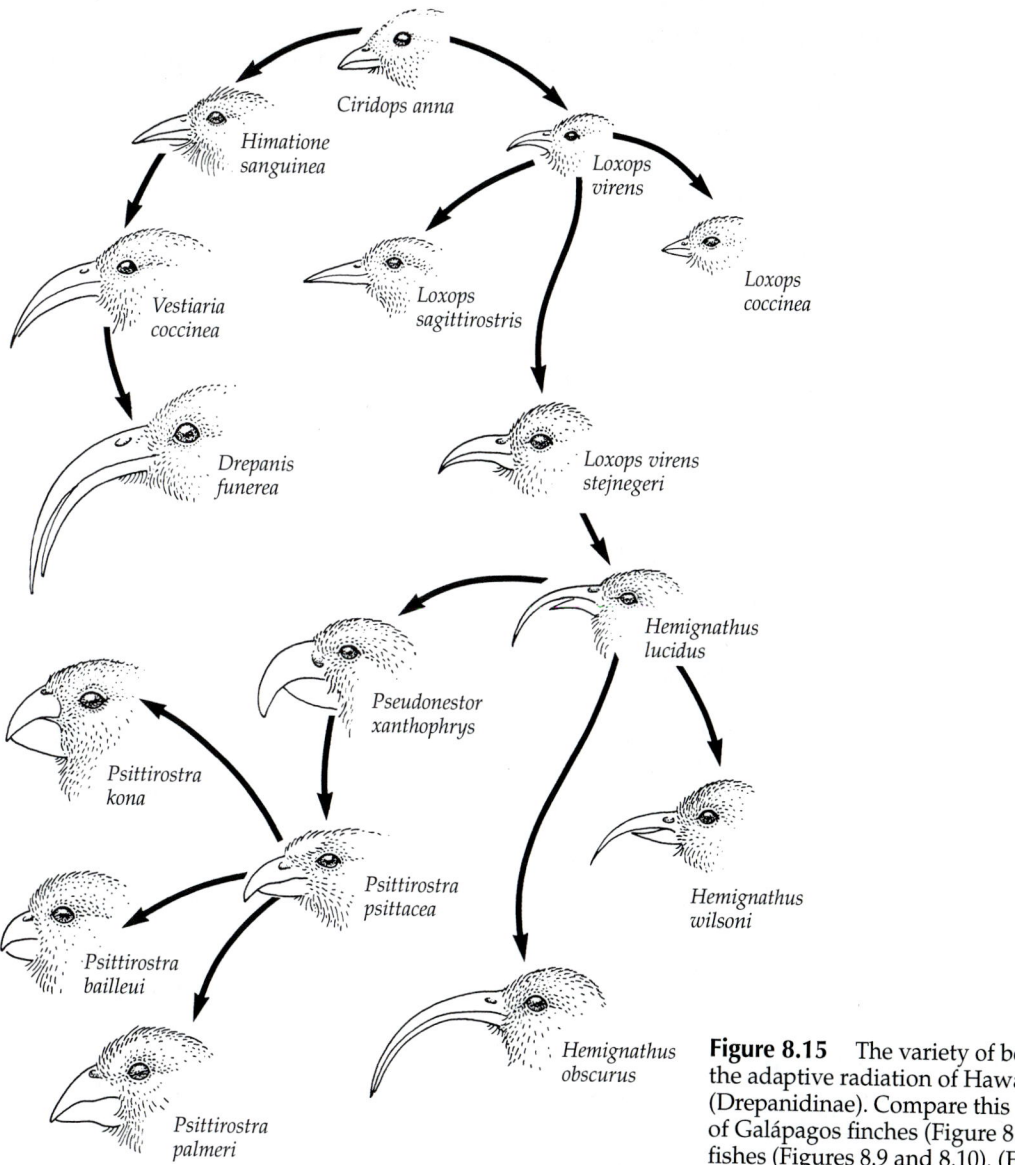

Figure 8.15   The variety of beak shapes resulting from the adaptive radiation of Hawaiian honeycreepers (Drepanidinae). Compare this group with the examples of Galápagos finches (Figure 8.8) and African cichlid fishes (Figures 8.9 and 8.10). (From Primack 1998.)

While many examples come from isolated habitats, such as islands and lakes, comparable adaptive radiations have occurred in other groups in the oceans and on continents. They have also taken place on many different temporal and spatial scales. An ancient radiation occurred in the marine realm in the cryptically colored marine anglerfishes (Lophiiformes), producing the shallow-water frogfishes (*Antennanus*), the open-ocean sargassum fish (*Histrio histrio*), the stingraylike batfishes (Ogcocephalidae), and a variety of deep-water bioluminescent forms (suborder Ceratoidea). In plants, a group that has radiated over the past 20 million years on the North American continent is the phlox family (Polemoniaceae). Within this single family, flower form and color have become amazingly variable as different species have adapted to be pollinated by hawkmoths, bees, butterflies, flies, or hummingbirds, and some have become specialized for self-pollination (Figure 8.16). Australia, the most isolated continent, is the site of many spectacular radiations: marsupial mammals, lizards of the genera *Ctenotus* and *Varanus*, and plants of the genera *Eucalyptus*, *Melaleuca*, and *Acacia*.

Whereas many genera and families exhibit a fascinating variety of ways of life, many others are extremely monotonous, differing mainly in minute structural details. Taxa that have changed only slightly over evolutionary time apparently are good at what they do and have not had the genetic flexibility or ecological opportunity to shift adaptive strategies. The wide variation in rates and degrees of divergence makes it difficult to generalize about the diversification of higher taxa and the influence of geography on adaptive radiation. The many examples of spectacular adaptive radiations on islands (including Madagascar and Australia) seem to show the importance of long historical isolation in these cases. Equally impressive radiations in less isolated settings, however, such as those of desert rodents of the family Heteromyidae in southwestern North America, the darters of the fish genus *Etheostoma* in the Missis-

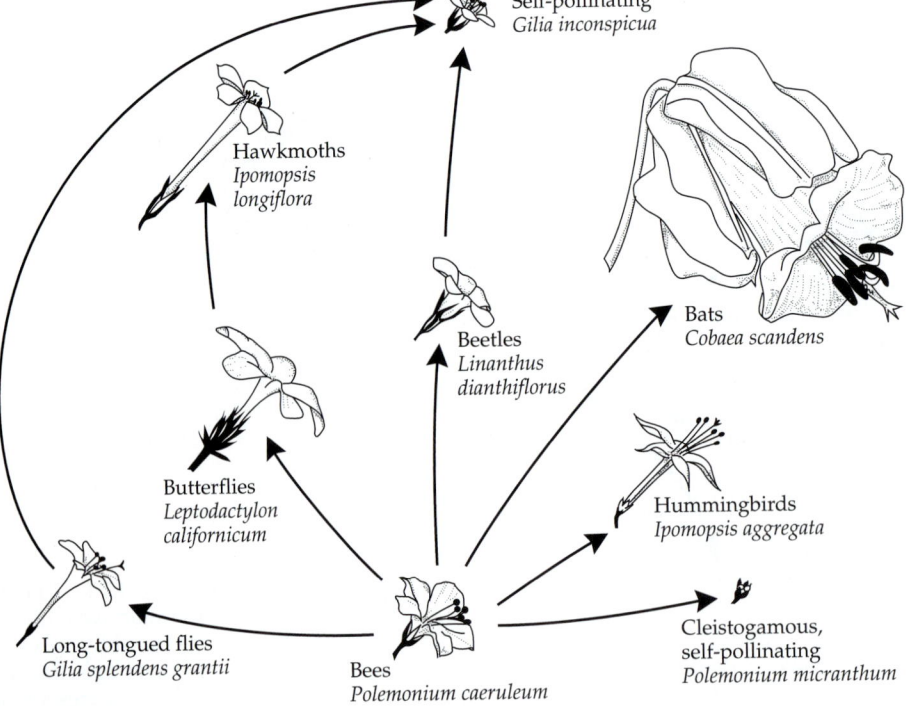

Figure 8.16 Adaptive radiation on the phlox family, Polemoniaceae, showing diversity of flower form reflecting different modes of pollination. A generalized bee-pollinated ancestor is believed to have given rise to the other forms, most coevolved for mutualistic relationships with particular kinds of animals, but some specialized for self-pollination. Plants in other families that have specialized to use the same kinds of pollinators often have convergently evolved similar flower sizes, shapes, colors, odors, and nectar and/or pollen rewards. (From Ehrlich and Holm 1963.)

Self-pollinating
Gilia inconspicua

Hawkmoths
Ipomopsis longiflora

Beetles
Linanthus dianthiflorus

Bats
Cobaea scandens

Butterflies
Leptodactylon californicum

Hummingbirds
Ipomopsis aggregata

Long-tongued flies
Gilia splendens grantii

Bees
Polemonium caeruleum

Cleistogamous, self-pollinating
Polemonium micranthum

sippi River drainage, and the spiny, succulent plants of the family Cactaceae in the arid regions of North and South America, suggest that long and nearly complete geographic isolation is not necessary. Comparative studies that are attempting to assess the influence of phylogenetic relationships and evolutionary constraints (see Chapter 11; Felsenstein 1985; Brooks and McLennan 1991; Harvey and Pagel 1991) promise to contribute importantly to our understanding of the enormous differences in rates of speciation and adaptive differentiation among lineages.

Extinction

Ecological Processes

Although all living organisms represent a continuous evolutionary lineage extending billions of years back to the origin of life, the ultimate fate of every species is extinction. This can be appreciated by taking a brief glance at the fossil record. The earth was teeming with life 100 million years ago. Both terrestrial and aquatic habitats were occupied by diverse biotas that formed complex ecological communities. However, the species and genera—and many of the families and orders—that were dominant then have been eliminated or drastically reduced by extinction, and have been replaced by new lineages. On land, dinosaurs and other reptilian groups have been replaced by birds and mammals, while ferns and gymnosperms have been largely supplanted by angiosperms. In the oceans, cephalopod mollusks have been supplanted by teleost fishes, while icthyosaurs and mesosaurs (reptiles) have been replaced by dolphins, whales, seals, and sea lions (mammals). Extinctions have apparently occurred continuously throughout the history of life, although the fossil record also catalogues occasional episodes of widespread disaster when much of the earth's biota was wiped out, apparently by rapid and drastic environmental change (Raup and Sepkoski 1982). Not surprisingly, extinctions have had a major influence not only on the kinds of organisms in existence at any given time, but also on the geographic distributions of those now extinct forms and the contemporary lineages that are descended from them.

Several authors have likened the evolutionary history of life to a continual race with no winners, only losers—those species that become extinct. This view is probably best expressed in Van Valen's (1973b) **Red Queen hypothesis**, named for the Red Queen in Lewis Carroll's *Through the Looking Glass*, who said "It takes all the running you can do to keep in the same place." The idea is that a species must continually evolve in order to keep pace with an environment that is perpetually changing, not just because abiotic conditions are shifting, but also because all the other species are evolving, altering the availability of resources and the patterns and processes of biotic interactions. Those species that cannot keep up with the changes become extinct, but others do well temporarily and speciate to produce new forms.

Van Valen points out that the probability of a species becoming extinct appears to be independent of its evolutionary age, but not of its taxonomic and ecological status. Certain taxonomic and ecological groups have consistently higher rates of extinction than others. For example, apparently due to their lower extinction rates, small and herbivorous mammals are found on more and smaller islands than large or carnivorous species (Brown 1971b; Heaney 1986; Lawlor 1986). This appears to be a general pattern (see Van Valen 1973a). Somewhat similarly, differences in diversity and duration in the fossil record among lineages of marine invertebrates are correlated with characteristics of their life history, as we shall see below.

Some researchers have developed mathematical models to predict a population's vulnerability to extinction based on its demographic characteristics (MacArthur and Wilson 1967; Richter-Dyn and Goel 1972; Leigh 1981; Gilpin and Hanski 1991). All populations experience fluctuations in size as a result of variations in environmental conditions and the activities of their enemies. When populations become very small, however, purely chance factors, such as random variations in the sex ratio, can also affect their abundance. The mathematical models show that, in general, the smaller a population becomes, the lower its ratio of births to deaths (see Equation 4.1), and the longer it remains at low numbers, the more vulnerable it is to extinction. A population with a low birth rate, especially when coupled with a high death rate, cannot recover rapidly from a temporary reduction in numbers. The long-term overall population size is probably the most important factor. The models suggest that probability of extinction increases nonlinearly as population size decreases, and becomes very high when population size becomes and remains very low—say, fewer than 100 individuals (Figure 8.17). These mathematical models are based on the intrinsic demographic characteristics of populations, but it is important to recognize that changes in extrinsic environmental conditions are likely to be the cause of the changes in those characteristics that ultimately lead to extinction.

Information on how intrinsic demographic and extrinsic ecological factors interact to cause extinction is difficult to obtain and interpret because extinctions of species, with the exception of those caused by humans, are rarely observed. Researchers have gained valuable insights, however, by studying the effects of these factors on the turnover of small, isolated subpopulations within a metapopulation (Gilpin and Hanski 1991; see also Chapters 4 and 14). One particularly well-documented example is provided by the work of Smith (1974, 1980) on pikas (*Ochotona princeps*). Pikas are small relatives of rabbits that live in rock slides and boulder fields in the mountainous regions of western North America. Smith carefully monitored pikas in the Sierra Nevada of California that had colonized the rock piles left by a mining operation. These mine tailings functioned as habitat islands for the pikas in a sea of sagebrush habitat. Smith was able to document extinctions of subpopulations by census-

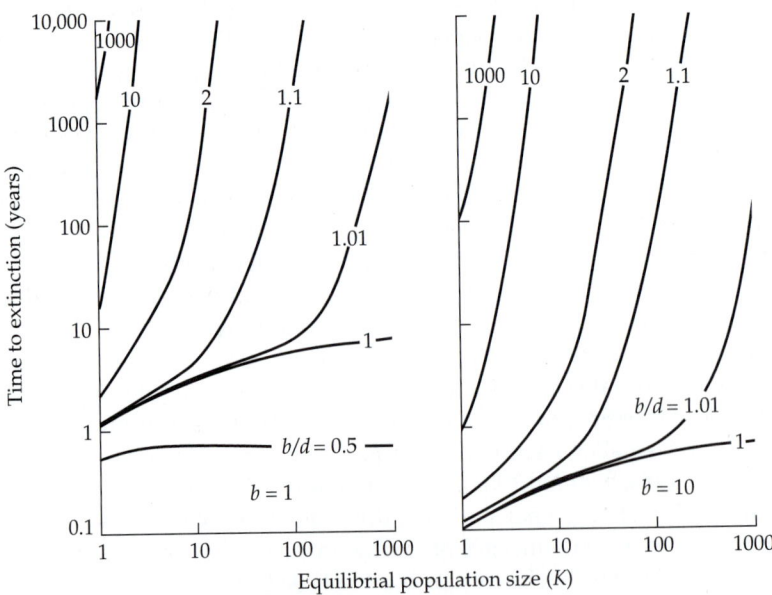

Figure 8.17   Output of a mathematical model showing how estimated time to extinction depends on two demographic characteristics of a population: equilibrial population density, or carrying capacity, (*K*) and the ratio of birth rate (*b*) to death rate (*d*). The graphs show that the probability of extinction is high (expected time to extinction is low) when populations are small and birth rates are low relative to death rates, but decreases rapidly as these demographic parameters increase. (From MacArthur and Wilson 1967.)

Table 8.2
Estimated time to extinction for pika subpopulations

Population size (number of individuals) *K*	Birth rate (per capita per year) *b*	Death rate (per capita per year) *d*	Time to extinction (years) *E*
1	0.35	0.00	2.9
2	0.35	1.63	6.9
3	0.35	2.44	46.2
4	0.35	2.84	405.1
5	0.35	3.15	3751.5

Source: Data from Smith 1974, 1980.

Note: The populations represented here occupy habitat islands of varying size (see Figure 8.17). Per capita birth and death rates were determined from the age structure of this population. A mathematical model, based on the model of MacArthur and Wilson (1967), predicts that only very small populations ($K < 3$ individuals) should have measurable rates of extinction owing to ordinary random demographic fluctuations. This prediction was confirmed by Smith's subsequent measurement of population turnover on these same islands during a 5-year period.

ing the rock piles initially in 1972 and then again in 1977. The repeat censuses documented both extinctions of previously existing subpopulations and colonizations of previously uninhabited rock piles. From data on the birth and death rates of pikas and the sizes and degree of isolation of the rock piles, Smith and colleagues developed a stochastic model that quite accurately predicted the frequency of extinction and colonization events as a function of these factors (Hanski and Smith, pers. comm.) (Table 8.2). Somewhat similarly, Lomolino (1984, 1993) and Crowell (1986) documented the extinction and immigration of small mammals on islands in the St. Lawrence River, Lake Huron, and the Gulf of Maine. Even though these studies are concerned with the turnover of small, isolated subpopulations rather than extinctions of entire species, they illustrate how both intrinsic demographic characteristics and extrinsic environmental conditions influence the extinction process.

Recent Extinctions

Over the last 200 years, humans have caused the extinctions of thousands of species. We are undoubtedly unaware of many species of microbes and small animals and plants that have disappeared, but the demise of some larger, more spectacular organisms is well documented. The recent reductions in populations, contractions of geographic ranges, and extinctions of species caused by humans will be considered in more detail in Chapter 18. We will present a few examples here, however, to illustrate some of the processes involved in extinctions, because we have much more information about some of these recent extinctions than about most of those evidenced in the fossil record.

The passenger pigeon (*Ectopistes migratorius*) was incredibly abundant in eastern North America when the first European colonists arrived. Estimates of the total population size are in the millions, perhaps billions. Because the pigeons traveled in dense flocks numbering in the thousands, feeding on beechnuts, acorns, and other abundant seeds and fruits, and nested in huge aggregations, they were very vulnerable to humans, who hunted them for food. In the 1870s, 2000 to 3000 birds were often taken in one net in a single day, and 100 barrels of pigeons per day were shipped to New York City for weeks on end. By 1890 the birds had virtually disappeared. In 1914 the last known passenger pigeon died in the Cincinnati Zoo (Pearson 1936). Similar stories could be told about the demise of the great auk, Carolina parakeet, and Steller's sea cow. Only a combination of luck and belated conservation action has prevented the whooping crane, trumpeter swan, bison, sea otter, gray whale, and northern elephant seal from suffering similar fates.

A different lesson can be drawn from the demise of the American chestnut tree (*Castanea dentata*). Along with beech, maple, oak, and hickory, the chestnut was one of the most abundant trees in the deciduous forests of eastern North America. In 1904 a pathogenic fungus (*Endothia parasitica*) was accidentally introduced, apparently from Asia, where it attacks a related but less susceptible species of chestnut. The disease spread rapidly (Metcalfe and Collins 1911), and within 40 years, mature chestnuts were virtually eliminated from their entire range. In some areas, scattered small trees that have sprouted from surviving root stock can still be found. Unfortunately, these are usually attacked by the disease and killed before they can reach reproductive size, so it is questionable whether the chestnut can much longer avoid absolute extinction. Similar cases of biological warfare due to introduced pathogens are the near-extinction of the American elm due to Dutch elm disease, of native Hawaiian birds due to avian malaria (see Chapter 4), and of native peoples on oceanic islands, and perhaps even on the American continents, due to smallpox, measles, diphtheria, and other diseases brought by European invaders.

One of the best-documented cases of local extinction of multiple species is the loss of bird species from Barro Colorado Island in Panama (see Figure 9.16). Prior to the early 1900s the island did not exist; it was simply a hill in a tract of tropical lowland forest. During the construction of the Panama Canal, the Chagres River was dammed, and the rising waters of Gatun Lake covered the adjacent lowland areas, creating Barro Colorado, an island of about 16 km². Despite its relatively large size, Barro Colorado no longer contains many of its original bird species. Because the Smithsonian Institution has long operated a biological research station on the island, its biota is well known, and some of the extinctions are well documented. Using these records, Willis (1974) calculated conservatively that at least 17 species had been lost from a total land bird fauna in excess of 150 species. More recently, by comparing the avifauna of the island with that of a mainland site of similar size and habitat, Karr (1982) concluded that at least 50 species have become extinct on the island since its isolation. Although habitat changes may have contributed to the demise of some forms, most of the missing species normally occur at low densities, and it appears that when the small populations present on the island died out, they were not replaced by colonists from across the narrow water gap.

The above examples illustrate the vulnerability of many species to environmental changes caused by modern humans (see also Chapters 17 and 18). Humans do not need to have guns, agriculture, or European culture to cause extinctions, however. Aboriginal humans almost certainly played a major role in many extinctions documented in the fossil record, as we saw in Chapter 7. On the other hand, the capacity of some species to recover from low numbers is amazing. Two species of marine mammals that inhabit the Pacific coast of North America, the northern elephant seal (*Mirouga angustirostris*) and the sea otter (*Enhydra lutris*), were hunted almost to extinction for their fat and fur, respectively. Tiny populations managed to escape detection, however, and once protected, increased rapidly to produce large, healthy populations. Another Pacific marine mammal, the gray whale (*Eschrichtius robustus*), has been increasing reassuringly following the protection of its breeding grounds off Baja California and the cessation of hunting (Rice and Wolman 1971), and was recently removed from the U.S. Endangered Species List. Several bird species that were hunted to near extinction in the late 1800s for feathers to supply the millinery industry, including trumpeter swans, sandhill cranes, and great and snowy egrets, are again abundant.

Extinctions in the Fossil Record

The fossil record provides abundant evidence of extinctions, but often provides only tantalizing clues to their causes. In many cases the fossil record is complete enough to show that many diverse species became extinct virtually simultaneously over relatively short periods of geological time. We can infer that such **mass** extinctions were caused by some drastic, widespread environmental change, but the exact nature of the perturbation and its effect on the organisms concerned is often difficult to deduce clearly. Long-standing, vigorous debates about the causes of mass extinctions mark the paleontological literature.

One of the most recent episodes of mass extinction was the disappearance of the Pleistocene megafauna of North and South America between 15,000 and 8000 years ago. The cause of this dramatic event is still the subject of controversy, as we saw in Chapter 7. Originally, paleontologists favored hypotheses invoking climatic change. The idea that aboriginal humans played a major role, advanced by Paul Martin in the 1960s and known as the "overkill hypothesis," was initially rejected by most paleontologists. Multiple lines of accumulating evidence, however, have shifted the balance, causing many scientists to conclude that humans almost certainly played a significant, and perhaps a pivotal, role in the extinctions of the megafauna (e.g., Martin and Klein 1984; Owen-Smith 1987, 1989). Examples of additional extinctions caused or threatened by modern humans will be discussed in detail in Chapter 17.

The causes of other mass extinctions that occurred even longer ago have also been highly controversial. The two most dramatic episodes saw the disappearance of the dinosaurs and many other groups of terrestrial and marine organisms at the end of the Cretaceous period, about 65 million years ago, and of even more species and higher taxa, particularly marine organisms, at the boundary between the Permian and the Triassic periods, about 225 million years ago (see Raup 1979; Raup and Sepkoski 1982). Raup estimated that 88 to 96% of all marine species then in existence went extinct at the end of the Permian. The causes most often invoked for these and other mass extinctions are large, rapid changes in the climate of the earth: either global cooling, with resulting continental glaciation and exposure of continental shelves, or global warming, with associated rising sea levels and inundation of continental shelves (e.g., Stanley 1984). But other explanations, including continental drift (e.g., Schopf 1974) and reorganization of interaction in ecological communities (Bak 1996) have had their proponents.

While the causes of the Permo-Triassic event remain poorly understood, there is now little doubt about the cause of the Cretaceous-Tertiary event. Its discovery is one of the most exciting detective stories in the recent earth sciences, rivaling the development of plate tectonic theory. It has long been apparent that the end of the Cretaceous was a time of major change. The Cretaceous-Tertiary (K-T) boundary was defined based on dramatic changes in the fossil record. Not only did major groups of previously dominant organisms (such as dinosaurs on land and ammonites in the oceans) go extinct, but there were also major changes in the abundance and species diversity of the lineages that survived (e.g., dramatic increases in birds and mammals on land and teleost fishes in the oceans; decreases in marine cephalopod mollusks). Until the early 1980s these changes were generally thought to have occurred over several million years and to have been caused by climatic change.

In 1980 three scientists at the University of California at Berkeley (Alvarez et al. 1980, 1984) suggested a radically new hypothesis: an asteroid impact. Initially, this sounded like an idea from a science fiction movie and was ridiculed by many established earth scientists. Nevertheless, the group led by Walter Alvarez presented intriguing data and a testable hypothesis. The asteroid impact hypothesis was based on the occurrence in rock strata that spanned the K-T boundary of a layer highly enriched in the element iridium (Ir). Iridium is rare in the earth's crust, but is often present in high concentrations in materials of extraterrestrial origin (meteorites and asteroids). The Alvarez group hypothesized that the iridium-enriched layer was produced by the dust injected into the atmosphere and circulated around the earth following the impact of a large asteroid. The dust presumably blocked solar radiation and resulted in rapid global cooling, which in turn caused the extinction of the dinosaurs and other lineages.

A strength of the Alvarez hypothesis was that it made testable predictions (Alvarez et al. 1984). First, it predicted that when an iridium-enriched layer occurred near the K-T boundary, it would always separate Cretaceous rocks below from Tertiary strata above. This prediction led to efforts to find stratigraphic sequences spanning the K-T boundary, to search for an iridium-enriched stratum, and to date the underlying and overlying rocks. Many independent studies of localities throughout the world confirmed the prediction: when present, the iridium-enriched layer marked the K-T boundary. Second, the hypothesis predicted that the vast majority of extinctions should have occurred right at the K-T boundary. Previous interpretations of fossil remains had suggested that some taxa disappeared millions of years before or after the K-T boundary. More accurate dating of fossil remains, however—especially with reference to the telltale iridium-enriched layer, when present—increasingly supported the hypothesis: the extinctions were concentrated during a very brief period. Third, an asteroid impact and resulting global cooling would be expected to lead to differential extinction or survival of different kinds of organisms depending on their life histories and ecologies. And indeed, the K-T extinctions were highly selective. Those groups, with capacities for body temperature regulation, such as endothermic birds and mammals, and those with life history stages that would be resistant to a brief but intense cold stress, such as seed plants, insects, and freshwater invertebrates, suffered fewer extinctions than many other kinds of organisms.

Finally, the hypothesis predicted that the crater formed by the asteroid impact might actually be discovered, and suggested that the thickness of the iridium-enriched stratum should provide clues to its location. Obtaining evidence bearing directly on this prediction initially seemed to be a long shot. No known meteorite craters had the appropriate combination of size and age to be good candidates. Further, since most of the earth's surface is ocean and most of the seafloor is young, there was a high probability that the impact occurred in the ocean and that the crater had long since been subducted into an ocean trench (see Chapter 6). However, mapping of the increasing data on the iridium-enriched stratum showed a clear geographic pattern, with the thickest layers in the Caribbean region. Then came the definitive find: the discovery of a crater—of the right size and age—on the continental shelf off the coast of Yucatán (Swisher et al. 1992).

So we now know that an asteroid did strike the earth about 65 million years ago. It appears to have caused the extinction of the dinosaurs, ammonites, and many other organisms, and in so doing, to have opened the way for mammals, teleost fishes, and other lineages to rise to their present dominance. While the major features of this event are now well documented, many pieces of the story are not. How much was the earth cooled, and how long did the cold period last? Beyond the differences in life history and ecology mentioned above, what determined which species and clades died out and which survived? Clearly, even such an unprecedented and unpredictable event as the impact of an extraterrestrial body did not cause "random" extinctions. Instead, it acted as a severe filter, eliminating many lineages, allowing others to pass through, and causing revolutionary changes in the composition of the earth's biota that have endured for 65 million years.

Species Selection

Processes of Species Selection

The historical pattern of speciation and extinction has had a major influence on the taxonomic diversity and geographic distribution of living things. Episodic events such as explosive adaptive radiations and mass extinctions have been particularly important. The aftermath of the K-T extinction event makes a point that is also illustrated by other parts of the fossil record: lineages of organisms have not been equally successful, and even the most successful lineages have been dominant during different periods of earth history and have prospered for different lengths of time. Some lineages have radiated rapidly to leave many descendants; others have survived virtually unchanged for millions of years; still others have disappeared quickly, leaving no descendants. Paleontologists and biologists have long recognized that certain lineages appear to possess particular traits that result in high speciation rates or low extinction rates, thereby leading to adaptive radiation and relative evolutionary success compared with those groups that die out or barely maintain themselves.

As mentioned in the discussion of macroevolution above, the differential survival and proliferation of species over geological time has come to be termed *species selection* (Stanley 1979), by analogy to the differential survival and reproduction of individuals (*individual selection*), which has traditionally been considered the primary mechanism of evolutionary change by natural selection. Species selection and individual selection should be viewed not as totally different biological processes, but as the consequences of generally similar ecological and genetic processes operating at different levels of biological organization. The same conditions (e.g., rapid environmental changes) that cause large differences in the birth and death rates of individuals of different genotypes and thus result in rapid evolution by individual selection are also likely to lead to differential multiplication and survival of species and thus to result in rapid evolution by species selection. Rapid evolution by individual selection is likely to be a primary cause of high speciation rates, which contribute to rapid phyletic diversification by species selection.

Examples of Species Selection

To some extent the fate of evolutionary lineages is a matter of chance and opportunity—it depends on a species with particular traits being in a favorable place at an opportune time. As emphasized above, many adaptive radiations begin when a population either colonizes a new area or evolves some new trait that substantially increases its fitness in a particular environment (Simpson 1952a). In either case, the species is suddenly presented with new ecological opportunities, which stimulate further rapid evolution and speciation. The Cenozoic radiations of mammals illustrate the influence of both of the above factors. During the late Mesozoic the ancestors of modern placental mammals acquired several new traits, including higher metabolic rates, improved temperature regulation, and new mechanisms of nourishing their developing young. These traits represented major advances over their reptilian ancestors and over other kinds of primitive mammals, including monotremes, marsupials, and some now extinct groups. The extinction of the dinosaurs and other vertebrate groups at the end of the Cretaceous eliminated their rivals and provided them with new ecological opportunities. Immediately afterward, in the early Cenozoic, a relatively few lineages of placental mammals radiated explosively (Figure 8.18). They not only "replaced" the extinct reptilian groups, but also diversified at the expense of other mammalian lineages, such as the "primitive" docodonts and the rodentlike multituberculates, that had survived the K-T mass extinction.

The radiation of placental mammals occurred throughout the world except in Australia and, for most of its Cenozoic history, South America. On these isolated continents, surrounded by ocean barriers, egg-laying monotremes, pouched marsupials, and other kinds of primitive mammals persisted. In the absence of the dominant reptiles, some of them underwent their own radiations. South America subsequently was colonized by many groups of placental mammals from North America, and many of the original South American taxa became extinct (see Chapter 16). Australia has remained isolated right up to the present. Two groups of egg-laying monotremes (duck-billed platypus and echidna) survived, and the marsupials underwent an extensive radiation, giving rise to many families and diverse morphological and ecological types. These include a sand-burrowing "mole" (Notorydtidae), small to medium-sized insectivores and carnivores, some called marsupial "mice," "cats," and "wolves" (Dasyuridae), numbats (Myrmecobiidae), possums, gliders, and koalas (Phalangeridae), bandicoots (Peramelidae), wombats (Phascolomiidae), and a wide variety of wallabies and kangaroos (Macropodidae). Prior to the

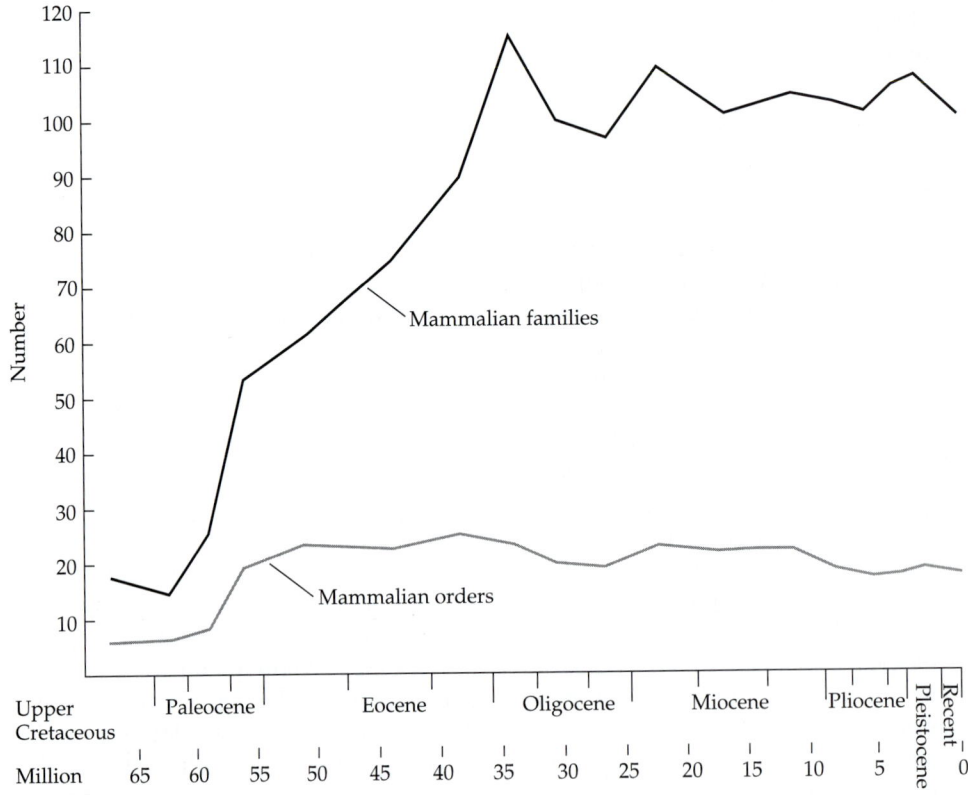

Figure 8.18 The "explosive" radiation of placental mammals during the Cenozoic, illustrated by the rapid increase in number of families. This radiation occurred after the K-T mass extinction event as mammals diverged and specialized to take advantage of ecological opportunities presented by the extinction of the dinosaurs and other groups of previously dominant reptiles. (After Lillegraven 1972.)

Pleistocene and the arrival of humans, dogs (dingoes), and other mammals that the humans imported, only two orders of placental land mammals, rodents and bats, managed to colonize Australia from Asia across water barriers. The rodents diversified dramatically, after at least two different colonization events, to produce about 80 species, most belonging to endemic genera. These include water rats (*Hydromys*), desert hopping mice (*Notomys*), and stick nest builders (*Leporillus*), which are ecologically and morphologically similar to North American muskrats (*Neofiber*), kangaroo rats (*Dipodomys*), and pack-rats (*Neotoma*), respectively (Keast 1972a, b, c).

In the absence of a historical event that eliminated competitors or provided access to a new area, radiations often were slower, and the buildup of diversity took longer. Many such radiations apparently occurred when some evolutionary innovation gave an ancestral species an advantage over coexisting organisms. Despite its apparent superiority, however, it usually took considerable time for the founding lineage to speciate, diversify ecologically, and supplant other groups of organisms that were already present. An example is provided by the neogastropods, a group of specialized, predatory marine snails that originated in the Cretaceous, 130 million years ago, and has gradually diversified to become a dominant invertebrate group. Another is provided by the angiosperm plants, which also originated in the Cretaceous. They possessed a number of structural and functional innovations in reproductive biology that, among other things, allowed them to evolve mutualistic associations with the animals that pollinate their flowers and disperse their seeds. Despite these advantages, however, it took until the mid-Cenozoic—about 100 million years—for the angiosperms to largely supplant the previously dominant ferns and gymnosperms (Figure 8.19; Niklas et al. 1983; Knoll 1986; Taylor 1996).

The importance of evolutionary innovations, or at least particular combinations of adaptive traits, is illustrated by macroevolutionary and biogeographic patterns in mollusks. There is an interesting relationship between speciation rate, geographic distribution, and mode of larval dispersal among these groups (Jackson 1974; Hansen 1980; Jablonski and Lutz 1980; Jablonski 1982). Some groups have planktotrophic larvae—juvenile stages that drift passively in the ocean—whereas others brood their offspring or have other specializations that do not involve their young being dispersed in the plankton. The mode of larval dispersal can be determined for fossil as well as for recent forms because an individual's original larval shell is preserved as part of its adult shell, which is readily fossilized. Molluscan species with planktotrophic larvae tend to have larger geographic ranges (Figure 8.20A) and lower speciation rates than related species whose young rarely disperse far from their par-

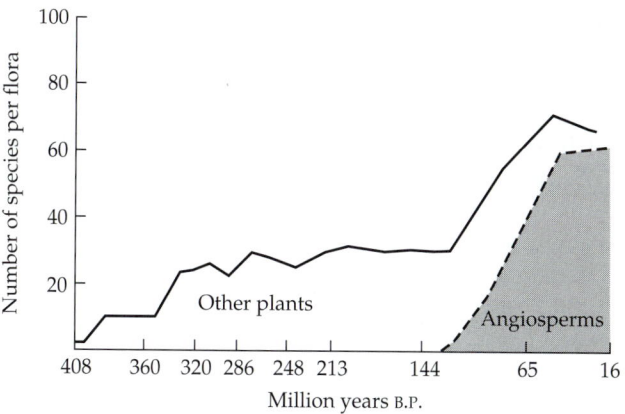

Figure 8.19 The relatively gradual radiation of angiosperms as illustrated by changing composition of fossil floras representing past communities. Although innovations in reproductive biology apparently gave the angiosperms advantages over the previously dominant gymnosperms, the increase in the former and decrease in the latter took over 100 million years. The angiosperms ultimately attained much higher diversity than their gymnosperm progenitors, although some gymnosperms, such as conifers and cycads, still survive today. (After Knoll 1986.)

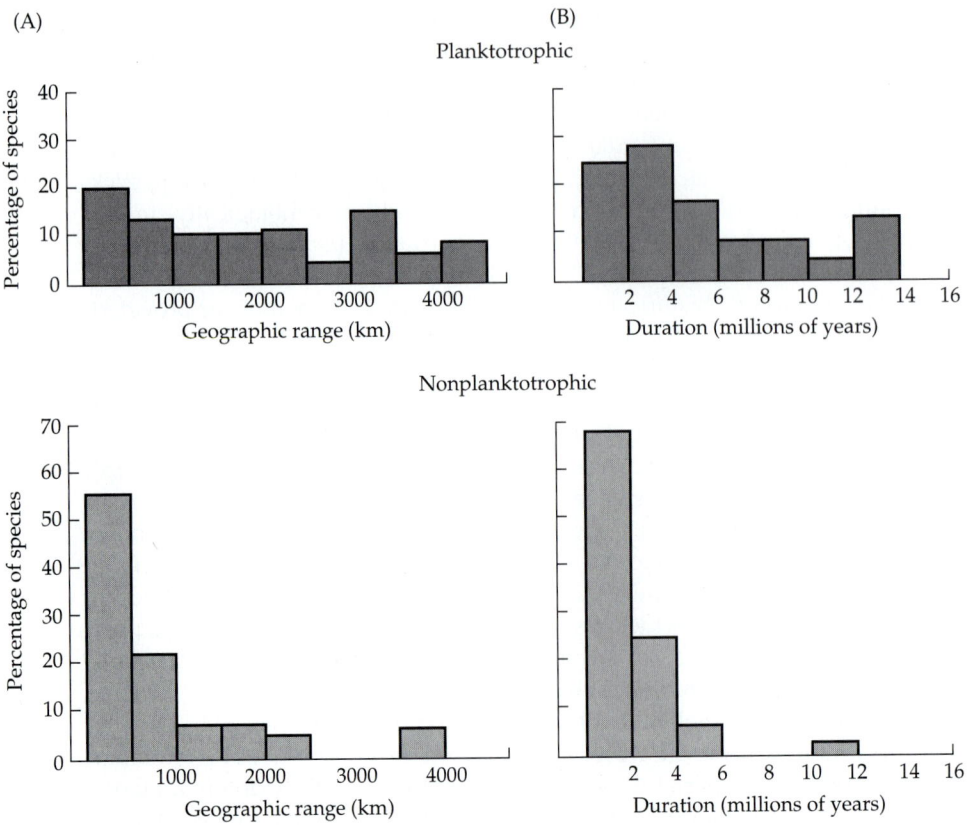

Figure 8.20   Relationship between mode of larval dispersal, extent of geographic range (A), and survival time in the fossil record (B) of late Cretaceous mollusks on the east coast of North America. Note that species whose larvae can disperse long distances in the plankton (planktotrophic species) tended to have larger geographic ranges and lower extinction rates (longer survival times) than those species whose larvae settle close to their parents (nonplanktotrophic species). (After Jablonski 1982.)

ents. Limited dispersal of offspring presumably reduces gene flow, enables populations to adapt to local conditions, and thus facilitates genetic differentiation and allopatric speciation. On the other hand, because small, specialized populations are also particularly vulnerable to extinction, nonplanktotrophic species tend to persist in the fossil record for shorter periods of time than do those with planktotrophic larvae (Figure 8.20B).

As implied by this example, patterns of species extinction, which depend in part on the characteristics of the organisms themselves, can influence the evolutionary histories of lineages by species selection. Over a period of 600 million years, one group of marine bottom-dwelling invertebrates, the clams (Mollusca, class Bivalvia), has been replacing a phylum, the brachiopods (Brachiopoda) (Figure 8.21). Because these two groups have superficially similar morphology, feeding habits, and habitat requirements, it had long been thought that the clams were supplanting the brachiopods by competitive exclusion (Elliot 1951). In fact, species of the two groups may compete significantly where they occur together, but careful examination of the fossil record indicates that the increase of clams and the dramatic decline of brachiopods was also facilitated by mass extinctions that eliminated many more taxa of brachiopods than of clams (Gould and Calloway 1980). Valentine and Jablon-

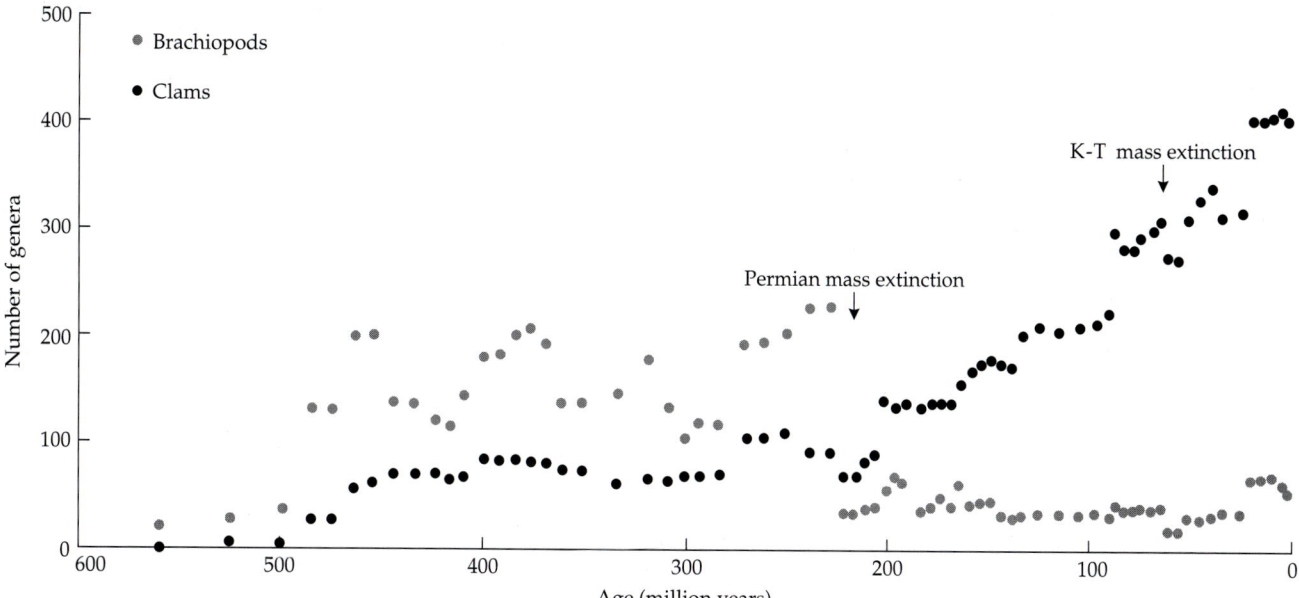

Figure 8.21   The "replacement", over a 600-million-year period, of brachiopods (phylum Brachiopoda) by clams (phylum Mollusca, class Bivalvia). This shift in dominance has often been attributed to competition because the two groups are generally similar in morphology and ecology. While competition may have played an important role, so did several mass extinction events: after both the Permian and the K-T mass extinctions, brachiopods declined in diversity, while clams increased. (After Gould and Calloway 1980.)

ski (1982) suggest that the brachiopods are differentially susceptible to catastrophic extinction because they do not have planktotrophic larvae, but instead brood their young or produce larvae that spend only a brief time in the plankton. As noted above, marine invertebrates with nonplanktotrophic larvae tend to have restricted geographic distributions and high extinction rates. Brachiopods have largely been eliminated from shallow-water habitats, where they are exposed to environmental fluctuations, but some forms persist in deep waters, where their mode of reproduction is less disadvantageous.

The fossil record documents the origination and extinction of lineages at all taxonomic levels, from species to phyla. Many of these lineages radiated to produce considerable morphological and ecological diversity, and some spread throughout the world. But success was almost invariably fleeting: most groups declined in diversity and ultimately went extinct. Sometimes their fates were due primarily to chance events, such as genetic drift, random extinctions of small populations, vicariant isolation due to plate tectonics, or the collision of an asteroid with the earth. Often, however, their biological attributes, shaped by the microevolutionary force of natural selection and the macroevolutionary force of species selection, played key roles. The more we learn about the history of life, the more we appreciate the influence of the nonrandom biological processes that have shaped the patterns of speciation and extinction.

CHAPTER 9

Dispersal

Nature has been defined as a principle of motion and change, and it is the subject of our inquiry. We must therefore see that we understand the meaning of motion for if it were unknown, the meaning of nature too would be unknown.

(Aristotle, Physics, III, 1)

There are three fundamental processes in biogeography: evolution, extinction, and dispersal. All the biogeographic patterns that we study derive from the effects of these processes. While Aristotle may have been unaware of the importance, or even existence, of evolution and extinction, his writings clearly speak to the role of movements.

Throughout the development of biogeography and its maturation as a respected discipline, the relative importance of movement, or **dispersal**, has been a subject of great debate. As described in Chapter 2, Charles Darwin was one of the most passionate and persuasive champions of its importance. Dispersalists, who included Darwin, Alfred Wallace, and Asa Gray, argued that species disjunctions can best be explained by long-distance dispersal across existing barriers. Their antagonists in this debate were the extensionists, including Charles Lyell, Edward Forbes, and Joseph Hooker, who argued that disjunctions were the result of movements along ancient corridors that subsequently disappeared. The point of contention was not whether dispersal occurred, but whether range expansion and disjunction required the crossing of a barrier.

Evidence for the extensionists' great landbridges never surfaced, but a new mechanism of dispersal emerged in the early-twentieth century: continental drift. No longer was it necessary to propose such rare and unlikely events as long-distance dispersal to account for disjunctions. Species could simply ride on the continents as they split and migrated across the surface of the earth. This splitting of once continuous populations would serve as a vicariant event and promote divergence of the now isolated populations. The dispersalist/

extensionist debate has thus been replaced by one that is no less heated and contentious: a debate between dispersalists and vicariance biogeographers (Box 9.1; see also Chapters 10, 11, and 12). The earlier debate has been transformed into one that focuses on the relative importance of long-distance dispersal versus plate tectonics, orogeny, and other vicariant events in shaping biogeographic patterns: in other words, did dispersal occur before or after barriers formed?

Finally, some biogeographers have questioned whether dispersal, over any time frame, has had any lasting influence on biogeographic patterns, past or present. This view, in perhaps its most extreme form, is summarized by **Bejerinck's law**: "Everything is everywhere, but the environment selects" (Sauer 1988). Proponents of this theory suggest that, given enough time, dispersal is inevitable for many, if not most, species. Therefore, regardless of their differences in dispersal ability, all species exhibit geographically limited ranges simply because they differ in their abilities to adapt to different environments. This view may have been derived from a misinterpretation of G. G. Simpson's suggestion that even exceedingly unlikely events become more likely if enough trials are performed. Increased likelihood, however, does not mean eventual certainty of the dispersal of all species to all points across the globe. Of course, the other extreme view—that dispersal is the only force influencing isolated communities—is just as problematic.

Box 9.1
Dispersal versus vicariance: No contest
by C. A. Stace

It is true that few new ideas or approaches are not controversial, and that controversy is a healthy and possibly essential part of scientific progress. Nevertheless, it is most unfortunate and not in the least beneficial that the advocacy of cladistic methods in taxonomy and biogeography has been so bitterly and vindictively pursued and opposed, and that criticism on both sides has so often been aimed as much at individuals as at ideas. As usual when such emotions surface, polarization of views and extremist stances emerge. It is a fair bet in such cases that neither stance is fully sustainable, and I believe that 'cladistic biogeography' provides and excellent example of such an outcome.

At the center of the argument is whether a disjunct distribution of a taxon was caused by a *dispersal* event (migration by the taxon across a barrier from A to B) or a *vicariance* event (erection of a barrier between A and B, both of which were already occupied by the taxon)–in other words, whether the occupation of both areas followed or preceded the erec-

tion of the barrier. The most obvious and strongest spatial barriers are mountain ranges and oceans, the formation of which are today reasonably well understood and can usually be at least approximately dated. The idea of continental drift, first forwarded in 1912, was not generally accepted until the 1960s, and so it was not until after that date that vicariance was widely championed as the major causative element in disjunction, despite the publication of Croizat's views in the 1950s (Nelson and Platnick 1981).

The most widely used argument against dispersal solutions to distribution patterns 'is that they lack a testable definition of relationship', and so 'they are irrefutable' (Humphries and Parenti 1986). The corollary, of course, is that they are also unprovable. Both are true, and it is indeed a severe handicap to any scientific hypothesis that it is not capable of objective testing, preferably by using it to predict outcomes whose reality can be observed. It would be preferable to adapt the hypothesis in some way so as to make it testable, or it might be better to

abandon it altogether. *But it still could be the case that it is the correct hypothesis. Lack of means of proving that an idea is true is not an indication that it is false, nor even a reason to believe that it might be false. It is, however, a reason to consider alternative ideas that might be more testable* [italics added].

There can surely be no doubt that long-range dispersal has played and does still play a large part in determining distribution patterns. This is nowhere better demonstrated than in the colonization of new oceanic volcanic islands that *de novo* were barren but that become vegetated. But there are many other dispersal events that explain distribution patterns far better than do vicariance events. The colonization of northern Europe after glaciation has been much debated, especially that into Britain and Ireland. Indeed, some of the geological facts as well as the biological events are still disputed (Devoy 1985). Rose (1972) demonstrated that most species arrived in Britain before the Channel formed *c.* 7500 years B.P., probably along the Somme Valley to Sus-

This debate over the biogeographic relevance of dispersal is, of course, a critical one. However, as with most debates, the truth lies somewhere between the extremes. The relative importance of dispersal, extinction, and evolution probably varies from one biotic group and region to another. As Simpson (1980) put it, "A reasonable biogeographer is neither a vicarist nor a dispersalist, but an eclecticist." Before we can assess the relative importance of dispersal, vicariant events, and selection, however, it is essential that we more fully explore the nature of dispersal

What Is Dispersal?

All organisms have some capacity to move from their birthplaces to new sites. The movement of offspring away from their parents is a normal part of the life cycle of virtually all plants and animals. Often dispersal is confined to a particular stage of the life history. Higher plants and some aquatic animals are sessile as adults, but in their earlier developmental stages are capable of traveling long distances from their natal sites. Mobile animals can shift their locations at any time during their lives, but many settle in one place and confine their activities to a limited home range for long periods of time.

Dispersal should not be confused with **dispersion**, an ecological term referring to the spatial distribution of individual organisms within a local population.

Box 9.1 *(continued)*

---

sex/Dorset axis. The rare continental species of Kent are mainly light-seeded Orchidaceae and Orobanchaceae that are probably more recent arrivals across what is now the narrowest disjunction. In other words, most of the British/French disjunctions are explicable by a vicariance event, but some by a dispersal event.

The Irish/North American disjunction has been no less debated. The small group of species that are considered amphi-Atlantic natives (e.g. *Eriocaulon aquaticum*) can be equally argued as the result of dispersal or of vicariance events. Other species are more doubtfully native, but some are certainly aliens (e.g. *Juncus tenuis*, probably *Hypericum canadense*) whose arrival in western Ireland has been by unknown means. Thanks to the increasingly large numbers of amateur bird-watchers on the look-out for rare species, we now know that many birds arrive in the British Isles direct from America each year. Lest anyone doubts the likelihood of dispersal across the Atlantic, let him consider the case of *Juncus planifolius*, quite recently found in natural habitats in western Ireland with its nearest native stations in Chile. Whatever the numbers of amphi-Atlantic species explicable by vicariance events, some are *only* explicable by dispersal events.

As a third example, the tropical family Combretaceae (*c.* 100 species in America, *c.* 250 in Africa) contains only three amphi-Atlantic species, all of which are coastal (two of them mangroves). Similarly the single species of the family common to Africa and Asia is also a mangrove. The most likely explanation is that these maritime species have been able to occupy both sides of the Atlantic due to their dispersal across it, not that these three alone of hundreds of species have remained indistinguishable for 180 million years.

I think the evidence is overwhelming that *both* dispersal and vicariance have played major roles in producing disjunct distribution. To deny either is futile. In each case the evidence needs to be weighed by whatever means are available, and it is to be expected that the relative importance of the two types of events will vary from situation to situation, for example from Hawaiian Islands to Australia/South Africa/South America. Nor is it certain that the two processes are always so different. Geographic disjunctions mostly do not arise overnight, but gradually, and there is no reason to believe that plants did not migrate to fresh areas at about the same time that these areas were becoming isolated. Post-glacial recolonization might well come into this category. This phenomenon, the wholesale migration of complete floras and faunas, also demonstrates how preposterous is the notion held by some cladistic biogeographers that if many different groups of organisms show the same patterns of disjunction, then dispersal events are not likely to be the cause....

...We have a very long way to go before we can be confident of our understanding of disjunct distributions and the interpretation these provide for biogeographic phenomena. When we do get close to the truth I feel absolutely certain that the explanations will be as diverse as the phenomena, and that it will be quite impossible to consider either dispersal events or vicariance events insignificant.

Source: Stace, C. A. (1989). *Journal of Biogeography* 16: 200–201. Reprinted with permission of Blackwell Science Ltd.

Dispersal as an Ecological Process

Plants and animals have evolved an incredibly diverse array of dispersal mechanisms (see Carlquist 1974, 1981; Udvardy 1969; Gadow 1913; Krebs 1978; Sauer 1988; Den Boer 1977; Stenseth and Lidicker 1992). Yet in all cases dispersal is basically an ecological process that is an adaptive part of the life history of every species. Natural selection favors individuals that move a modest distance from their natal site. A new location is always likely to be more favorable than the individual's exact birthplace, in part because intraspecific competition between parent and offspring and among siblings is reduced, and in part because the environment, and hence the quality, of the natal site is always changing. On the other hand, as distance from the natal site increases, habitats become more dissimilar and, as a result, would-be colonists are less likely to be well adapted to their new habitats.

Dispersal as a Historical Biogeographic Event

The role of dispersal in biogeography is very different from its role as a demographic phenomenon. Although dispersal occurs continually in all species, most of it does not result in any significant change in their geographic distributions. As pointed out in Chapter 4, the geographic ranges of most species are limited by environmental factors and remain relatively constant over ecological time. Biogeographers are concerned primarily with the exceptions: those rare instances in which species shift their ranges by moving over long distances. This occurs so infrequently that we seldom see it happening, and even less often are we able to study it. Usually biogeographers must look at dispersal as a historical process, and must infer the nature and timing of past long-distance movements from indirect evidence, such as the distributions of living and fossil forms. Making such inferences is a monumental task. The distribution of every taxon reflects a history of local origin, dispersal, and local extinction extending back to the very origin of life. Patterns of endemism, provincialism, and disjunction of geographic ranges (see Chapter 10) indicate that the dispersal of some groups has been so limited that their histories are indeed reflected in the distributions of their living and fossil representatives. However, reconstructing dispersal history usually requires that we deal with a process operating so infrequently that its effects may often appear to be highly random and idiosyncratic.

The problem of dealing with rare but important events is not unique to biogeography. In many ways the role of successful dispersal in biogeography is analogous to that of beneficial mutations in evolution. Beneficial mutations provide the raw material for evolutionary change, but they occur essentially randomly and so infrequently that they are difficult to observe and study (see Elena et al. 1996). Every inherited feature of an organism has a history comprising a series of unique genetic changes. Trying to reconstruct the history of the mutations that resulted, for example, in a reptilian scale evolving into a feather is conceptually similar in many ways to attempting to reconstruct the biogeographic events that led to the present disjunct distribution of large predaceous birds (e.g., ospreys, hawks, kestrels, and owls) in Africa, Australasia, New Zealand, and South America.

We draw this analogy to make two points. First, long-distance dispersal events may be infrequent and somewhat stochastic, but that does not mean that they are unimportant. On the contrary, these movements are among the most important of the events that have shaped present distributions. Second, we cannot afford to ignore the role of dispersal in biogeography just because it is difficult to study. One of the great challenges in understanding the geography and evolution of life is to develop ways of evaluating the influence of rare but important events such as beneficial mutations, asteroid impacts, and long-distance dispersal to archipelagoes as remote as the Galápagos.

Dispersal and Range Expansion

In order to expand its range, a species must be able to (1) travel to a new area, (2) withstand potentially unfavorable conditions during its passage, and (3) establish viable populations upon its arrival. Biogeographers often distinguish among three kinds of dispersal events that can have this result, primarily by the relative rates of movement and range expansion involved (e.g., Platnick and Nelson 1978; Pielou 1979). These three mechanisms of range expansion are called jump dispersal, diffusion, and secular migration (Pielou 1979).

Jump Dispersal

There is abundant evidence that many species have undergone long-distance dispersal, also known as **jump dispersal**. Anyone who has built a small pond in the backyard cannot help but be impressed by the rate at which populations of aquatic insects, snails, other invertebrates, vascular plants, and algae become established there. The same process of colonization occurs on a much larger geographic scale. When a volcanic eruption blasted the Indonesian island of Krakatau in 1883, it covered both what was left of Krakatau and the neighboring island of Verlaten with a deep blanket of ash, obliterating all life. Biological surveys, primarily of birds and plants (Figure 9.1), documented the rapidity with which new populations of organisms became established on the island (Docters van Leeuwen 1936; Dammermann 1948; see also Brown et al. 1919). By 1933, only 50 years after the eruption, Krakatau was once again covered with a dense tropical rain forest, and 271 plant species and 31 kinds of birds, as well as numerous invertebrates, were recorded on the island. Where did these organisms come from, and how did they get there? In the case of Krakatau, the answer to the first question is relatively clear: they dispersed across the water gap from the large islands of Java and Sumatra, which lie 40 and 80 km,

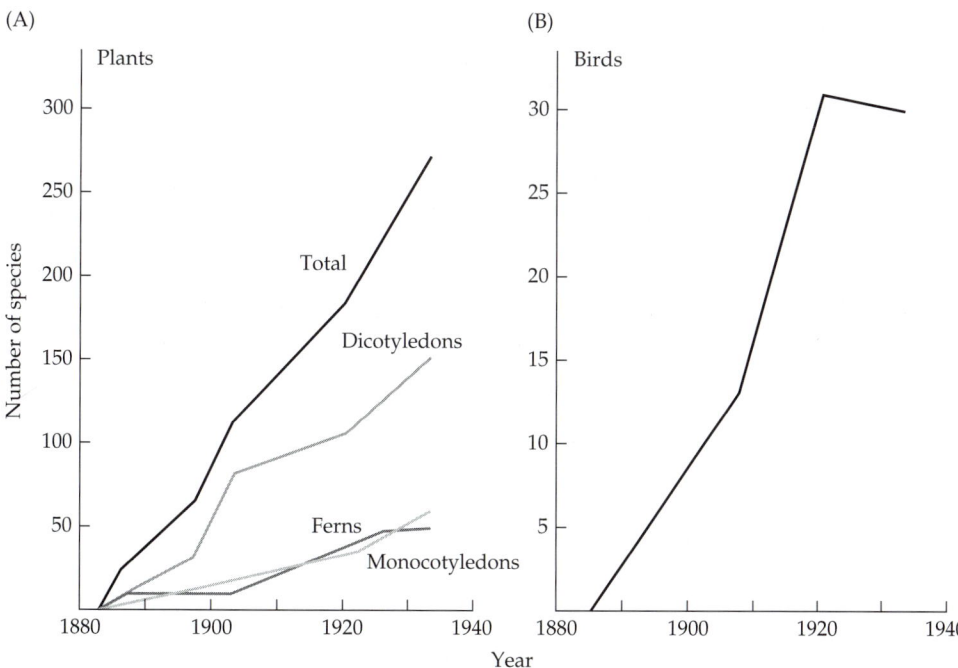

Figure 9.1 Rapid recolonization of the island of Krakatau by (A) plants and (B) birds after all life on the island was destroyed by a volcanic eruption in 1883. Several biological surveys recorded the arrival of colonizing species, which probably traveled across at least 40 km of ocean. (After MacArthur and Wilson 1967.)

respectively, from Krakatau (see the discussion of the colonization of Krakatau in Chapter 13).

The case of Krakatau is unusual only in that the creation of completely virgin habitat was so dramatic that the sources of colonizing organisms could be easily identified and their immigration carefully documented. The same processes have occurred over much longer distances and time periods for other archipelagos. The Galápagos and Hawaiian archipelagoes lie far out in the Pacific Ocean, 800 km west of Ecuador and 4000 km west of Mexico, respectively. These oceanic islands are actually the tops of volcanoes that arose from the ocean floor. Although they have never been much nearer to continents than they are today, they have acquired diverse biotas from propagules that have dispersed across the ocean.

Some authors continue to downplay the importance of long-distance dispersal in favor of alternative explanations (e.g., the ancient landbridges of the extensionists). Nevertheless, there is undeniable evidence that many groups of organisms have reached distant islands by traveling through the air or floating on the sea (Ridley 1930; J. M. B. Smith et al. 1990; J. M. B. Smith 1994; Vagvolygi 1975; Baur and Bengtsson 1987; Enckell et al. 1987). We also have evidence, both from recent occurrences and from historical patterns, that other species have traveled equally long distances over continental areas (McAtee 1947; Cruden 1966). Moreover, long-distance dispersal may have a strong selective component. Those species that have special adaptations for long-distance travel, or that send out more than one immigrant to found a population, have been especially successful at colonizing islands (see Chapters 13 and 14). Both the Galápagos and the Hawaiian Islands have land snails and bats as well as numerous species of trees, insects, and birds. In addition, giant tortoises and native rats inhabit the Galápagos. Noteworthy in their absence from these islands are **nonvolant** (nonflying) mammals, amphibians, freshwater fishes, and other forms poorly adapted for dispersal across open oceans.

Long-distance dispersal has at least three important consequences for biogeographers. First, it can be used judiciously to explain the wide and often discontinuous distributions of many taxa of animals, plants, and microbes. Second, it accounts in part for both the similarities and the differences among biotas inhabiting similar environments in different geographic areas. As we have seen, the ability to disperse successfully over distance and over habitat barriers varies in a predictable manner among different kinds of organisms (e.g., bats and birds versus amphibians and nonvolant mammals). However, because chance plays such an important role in the successful dispersal and establishment of colonists, there is a certain degree of taxonomic randomness (or **stochasticity**) in the composition of biotas. Finally, it emphasizes the importance of the many changes that have occured as expanding human civilizations have aided the long-distance transport of species to the most remote points of the globe. We shall return to the biogeographic and ecological effects of anthropogenic dispersal in Chapter 17.

Diffusion

In comparison to jump dispersal, **diffusion** is a much slower form of range expansion that involves not just individuals, but populations. Whereas jump dispersal can be accomplished by just one or a few individuals within a short period of their life span, diffusion typically is accomplished over generations by individuals gradually spreading out from the margins of a species' range. These two mechanisms of range expansion are, however, closely related, as diffusion often follows the jump dispersal of a species into a distant, but uncolonized, region of hospitable habitat.

An excellent example of both kinds of range expansion is provided by the cattle egret, *Bubulcus ibis* (Crosby 1972). This small heron originally was native to Africa, where it inhabits tropical and subtropical grasslands in association with large herbivorous mammals, foraging for insects and other small animals that are flushed by the grazing herbivores. In the late 1800s cattle egrets colonized eastern South America, having dispersed under their own power across the South Atlantic Ocean (Figure 9.2). During the succeeding decades the immigrants and their descendants thrived and spread throughout much of the New World, finding abundant food and habitat as a result of the clearing of the tropical forests, primarily for grazing livestock. The cattle egret has now expanded its breeding range northward to the southern United States and has colonized all of the major Caribbean islands.

Other well-known examples of diffusion in birds include the range expansions of starlings (*Sturnus vulgaris*) and house sparrows (*Passer domesticus*) after their intentional introductions into North America (Figure 9.3A–B). Similarly, the American muskrat (*Ondatra zibethica*, Figure 9.3C) rapidly expanded its range after introduction into Europe, and opossums (*Didelphis virginiana*) and armadillos (*Dasypus novemcinctus*, Figure 9.3D), both natives of South America, continue to expand their ranges northward through North America. Cases of diffusion of invertebrates are perhaps all too common, but among them, the most notorious and problematic are the spread of Africanized "killer" bees (*Apis mellifera*) northward following their introduction into Brazil, and of fire ants, which are continuing their relatively rapid expansion into the southern portion of the United States. In aquatic ecosystems, many "weeds,"

Figure 9.2 Colonization of the New World by the cattle egret, *Bubulcus ibis*. This heron crossed the South Atlantic from Africa under its own power, becoming established in northeastern South America by the late 1800s. From there it dispersed rapidly, and it is now one of the most widespread and abundant herons in the New World. (After Smith 1974.)

(A)

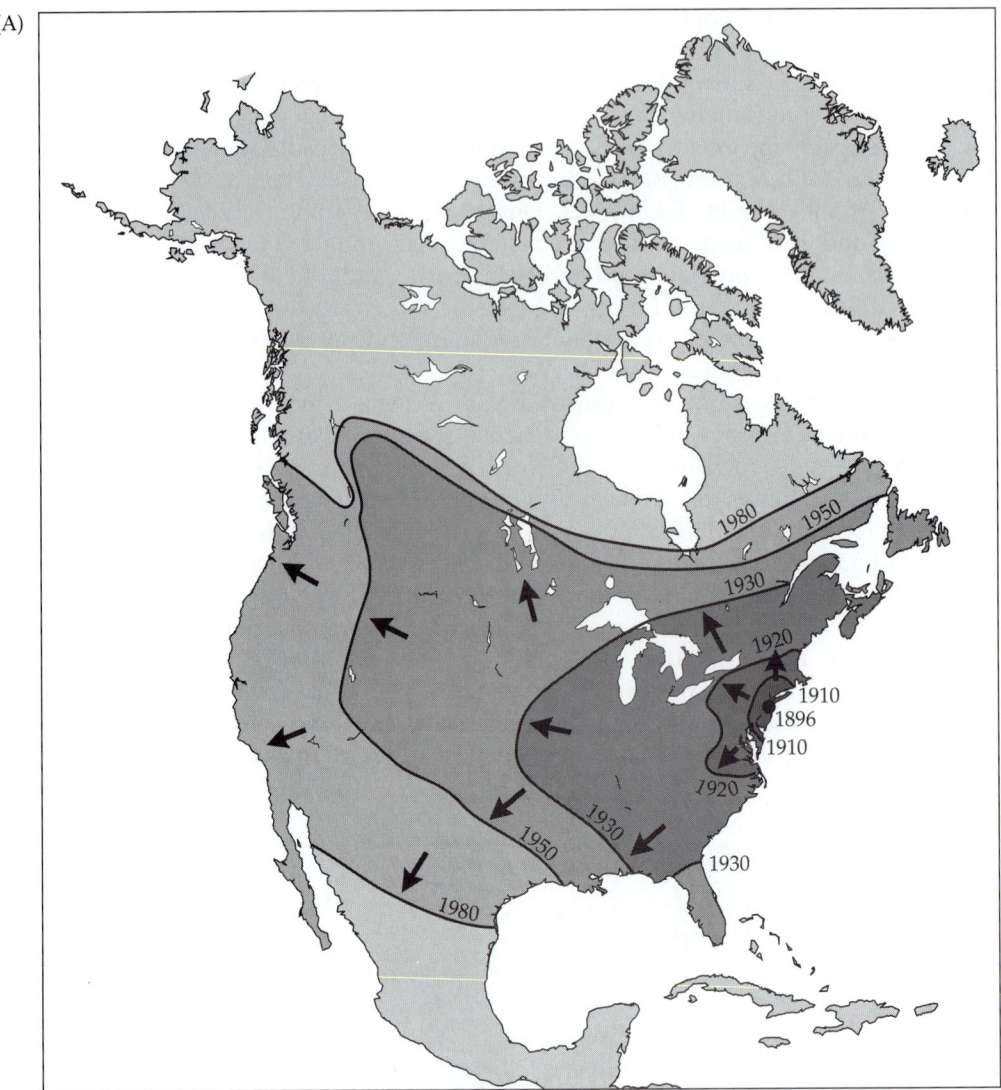

(B)

Figure 9.3   Range expansions of selected species of animals: (A) European
starling (*Sturnus vulgaris*) in North
America. (B) House sparrow (*Passer domesticus*) in North America. Note that
the current populations derive from at
least three introductions, including
those of 1852 in New York, 1871 in California, and 1873 in Utah. (C) American
muskrat (*Ondatra zibethica*) in Europe
since its 1905 introduction near Prague.
(D) Nine-banded armadillo (*Dasypus
novemcinctus*) in the southern United
States. (E) European rabbit (*Oryctalagus
cuniculus*) in Australia. (F) Red fox
(*Vulpes vulpes*) in Australia. (A compiled from various sources; B after
Lowther and Cink 1992; C after Van
den Bosch et al. 1992; D after Taulman
and Robbins 1996; E after Stoddart and
Parer 1988; F after Dickman 1996.)

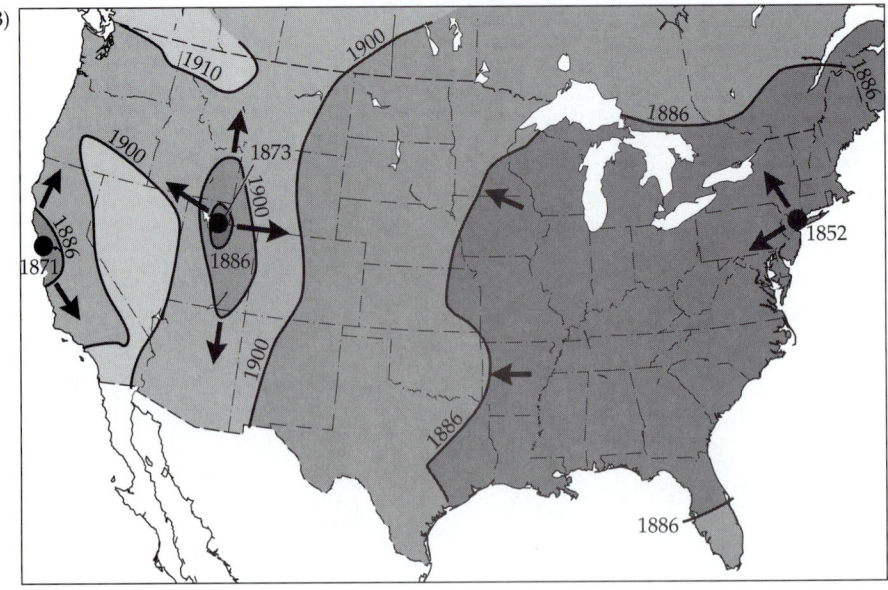

(C)

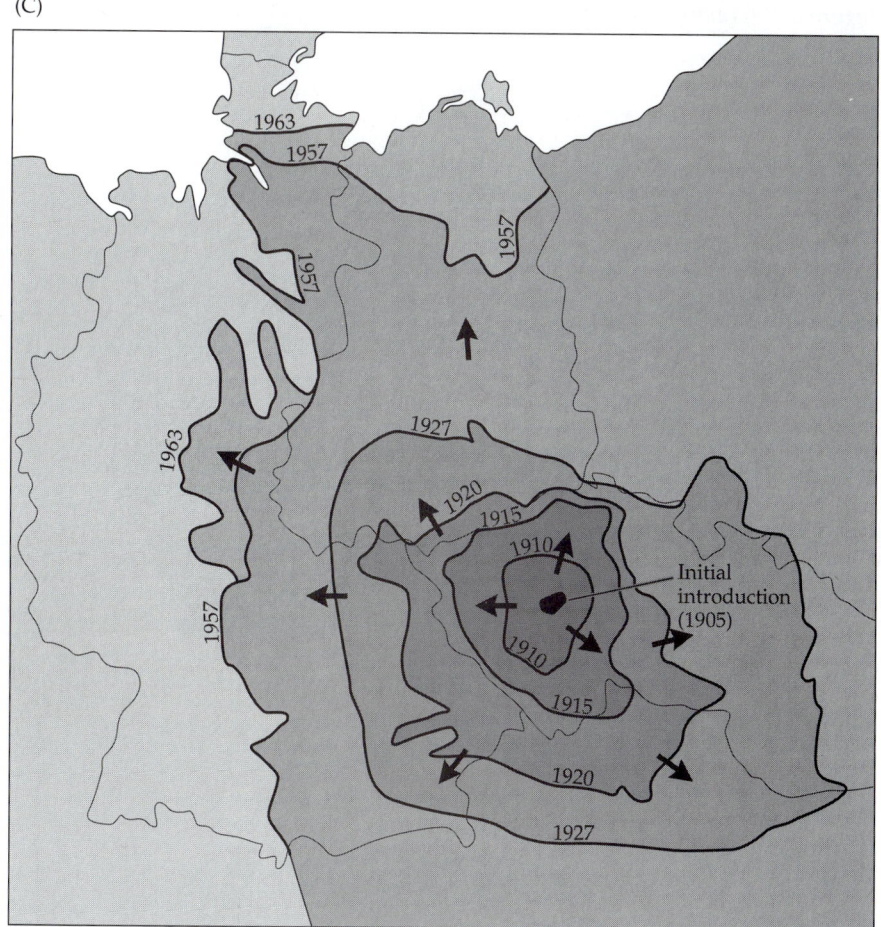

(D)

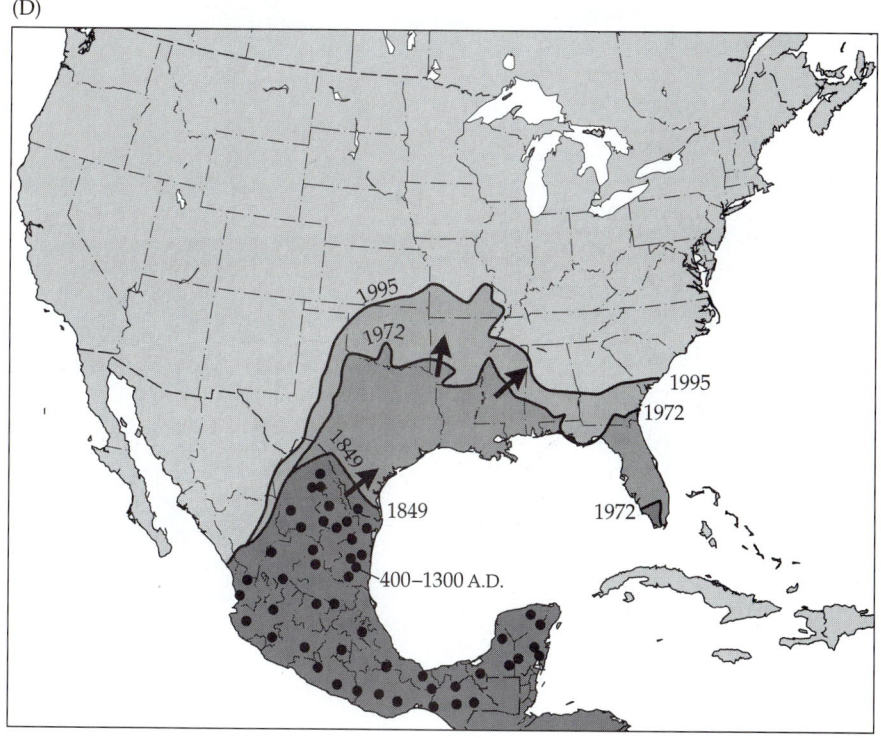

Figure 9.3 (*continued*)

(E)

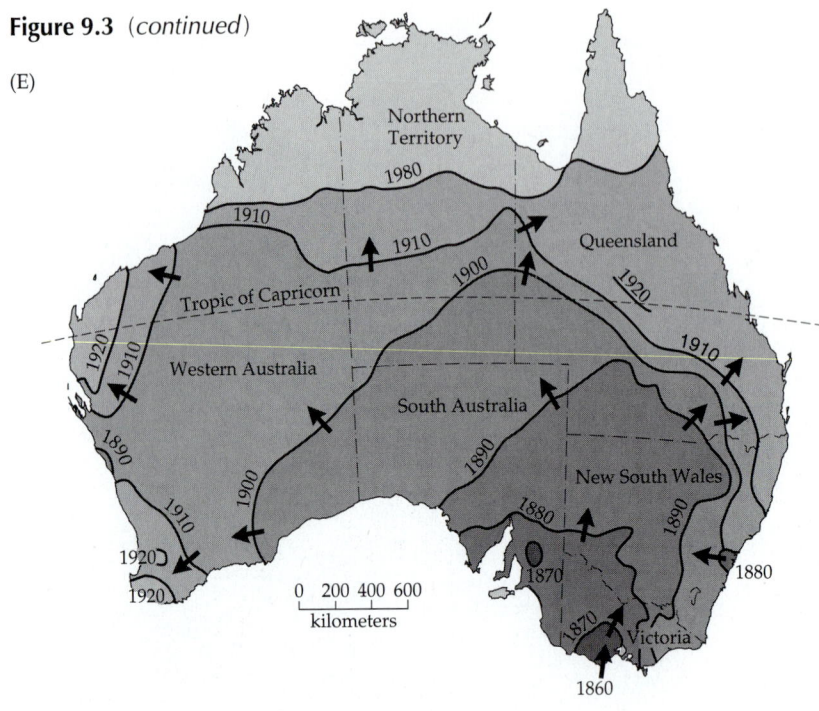

(F)

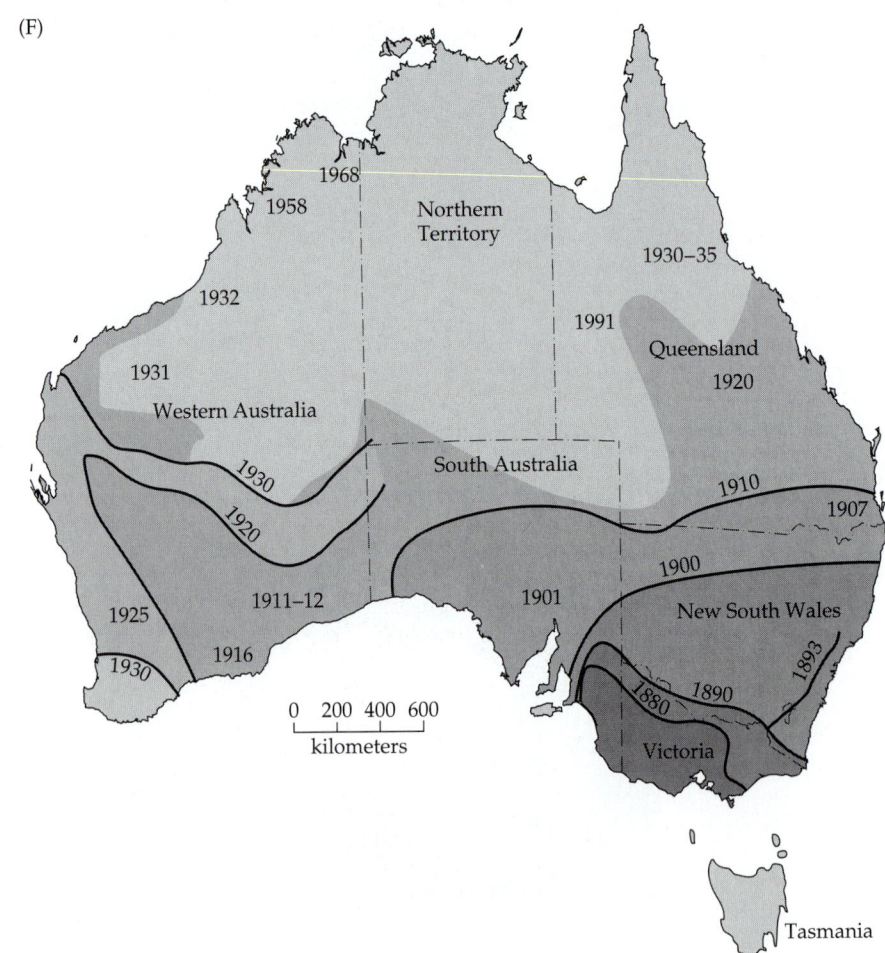

such as water hyacinths (*Eichhornia crassipes*), water fern (*Salvinia molesta*), and Hydrilla (*Hydrilla verticilatta*; Figure 9.4), and many animal pests, including zebra mussels (*Dreissena polymorpha*; see Box 17.1) have exhibited rapid range

(A)

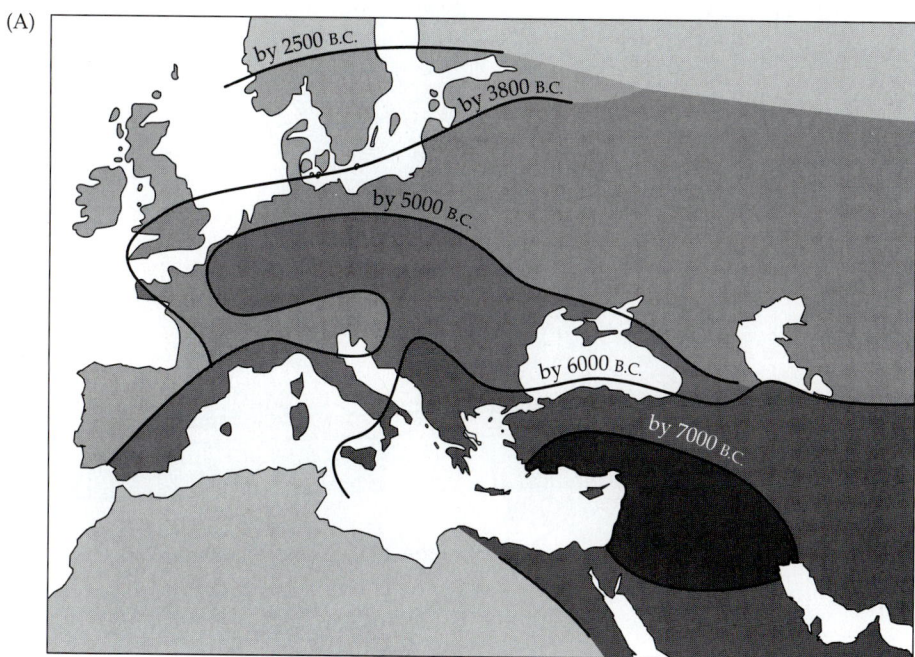

(B)

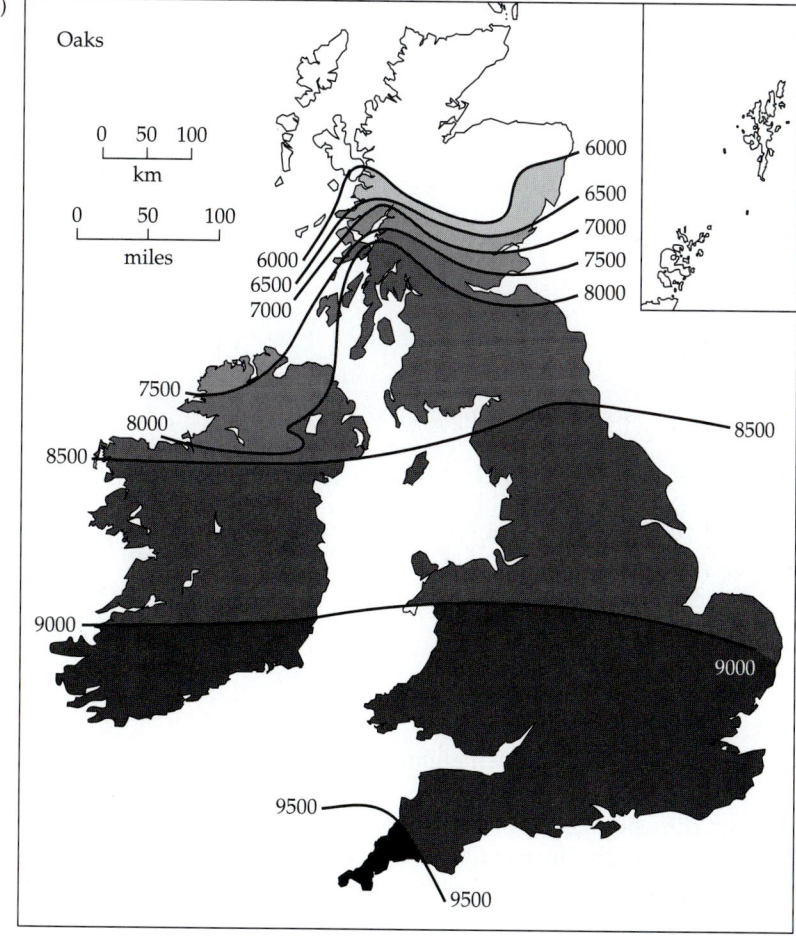

Figure 9.4 Range expansion in selected plants. Maps show the spread of (A) "Fertile Crescent" crops across western Eurasia; (B) oaks (*Quercus* spp.) in Great Britain (numbers indicate years B.P.); (C) elm (*Ulmus* spp.) in Great Britain (numbers indicate years B.P.); (D) purple loosestrife (*Lythrum salicaria*) in North America. (A after Diamond 1997; B and C after Birks 1989; D after Thompson et al, 1987.)

Figure 9.4 *(continued)*

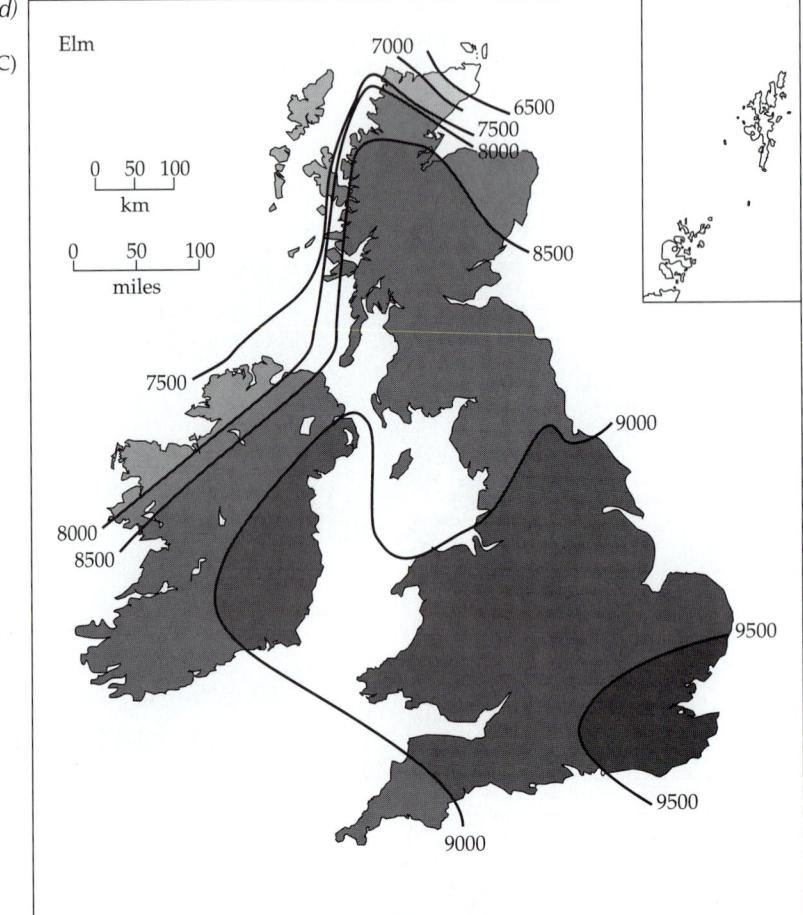

(C)

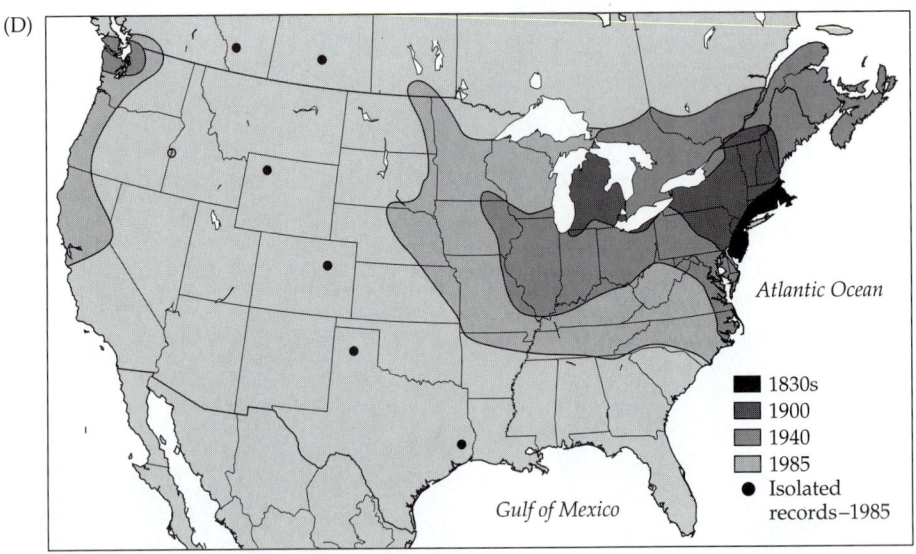

(D)

expansion following their accidental introduction into lakes and rivers outside their original range.

Although diffusion is an inherently complex process, it typically proceeds in three stages. Initially, invasion and range expansion may be very slow, and may require repeated dispersal events and adaptation to the characteristics of the ecosystems being invaded. Once an invasive species becomes established, however, its geographic range often expands at an exponential rate (Van den Bosch et al. 1992). Eventually, range expansion slows when the species encounters physical, climatic, or ecological barriers, which also tend to distort the shape of the species' geographic range. At this stage, geographic ranges may remain relatively stable unless environmental conditions become substantially altered or the species somehow crosses the barrier and reinitiates the invasion sequence.

Secular Migration

In contrast to these relatively rapid forms of range expansion, **secular migration** occurs so slowly—on the order of hundreds of generations—that species have ample opportunity to evolve en route. Although Herbert Louis Mason appears to have coined the term in 1954, the concept of secular migration is an old one, dating back to the early development of biogeography (Pielou 1979). During the eighteenth century, Buffon hypothesized that most life forms originated in the northern regions of the Old and New Worlds (see Chapter 2). From there, they migrated southward and, once isolated, became modified such that the biotas of the New and Old World tropics share very few forms (i.e., Buffon's law).

While Buffon's explanation for the diversification and spread of terrestrial life forms now seems fanciful, the fossil record provides convincing evidence of evolutionary divergence during range expansion. For example, some now extinct forms of camels spread southward through North America and eventually across the newly formed Central American landbridge during the Pliocene epoch. Extant products of this secular migration include the guanaco (*Lama guanicoe*) and vicuña (*Vicugna vicugna*) of South America. Similarly, camels and horses spread from the Nearctic region into the Old World. In both cases, the descendants persisted although their ancestors were extirpated from their North American homeland.

In Chapter 4 we described several cases in which a species had gradually expanded its geographic range by colonizing new regions at the boundaries. Most such cases that have been documented in written records involve shifts in response to habitat modification by humans, possibly accompanied by adaptations of the species to the new niches. Presumably the same processes—habitat change and adaptation to new conditions—were responsible for historical expansions that cannot be attributed to human influence, such as the invasion of South America by many groups of North American mammals during the Pliocene (see the discussion of the Great American Interchange in Chapter 16).

Mechanisms of Movement

Active Dispersal

Organisms can disperse either actively, moving under their own power, or passively, being carried by a physical agent, such as wind or water, or by other organisms. The terms **vagility** and **pagility** are sometimes used to denote the ability of organisms to disperse by active or passive means, respectively.

Only a few animals have the capacity to travel long distances under their own power. Of these, strong fliers, such as many birds, bats, and large insects (e.g., dragonflies, some lepidopterans, beetles, and bugs), have the greatest capability for long-distance dispersal. Many of these animals regularly travel hundreds or thousands of kilometers during their seasonal migrations, which are a normal part of their annual life cycles. When stressed or aided by favorable winds, some of these same animals can cover comparable distances during a single flight.

A few examples of normal migratory routes demonstrate the potential of dispersal by flight. The golden plover (*Pluvialis dominica*), a medium-sized shorebird weighing about 150 g, breeds in the Arctic and winters in southern South America, southern Asia, Australia, and the islands of the Pacific. Migrating individuals regularly fly nonstop from Alaska to Hawaii, a distance of 4000 km (Figure 9.5) (Dorst 1962). The ruby-throated hummingbird (*Archilochus colubris*), one of the smallest birds at 3.5 g, regularly commutes twice a year nonstop across the Gulf of Mexico, a distance of 800 km, en route between its breeding grounds in the eastern United States and its wintering grounds in southern Mexico (Dorst 1962).

In an insightful yet often overlooked paper, Joseph Grinnell (1922) discussed "accidental" (extralimital) occurrences of birds. In addition to noting that these seemingly rare events are actually commonplace for many species, Grinnell concluded that "the continual wide dissemination of so-called accidentals ... provided the mechanism by which each species spreads [i.e., diffusion] or by which it travels from place to place when this is necessitated by shifting barriers [i.e., jump dispersal]." Extralimital sightings are well documented for many taxa. Every year a few individuals of bird species native to Europe are seen in eastern North America, and vice versa, usually following severe North Atlantic storms. These "rare" events, although difficult to study, may play a major role in determining biogeographic patterns. For example, bats of the genus *Lasiurus* (body weight 10–35 g) migrate from northern North America to winter in the Neotropics (Findley and Jones 1964). The species of this bat that is native to the Hawaiian Islands undoubtedly is derived from migrating individuals that went astray.

Figure 9.5   Migratory routes of the golden plover (*Pluvialis dominica*), a shorebird that breeds in the Arctic and winters in temperate regions of the Southern Hemisphere. Each year these birds fly prodigious distances, often crossing huge expanses of open ocean.

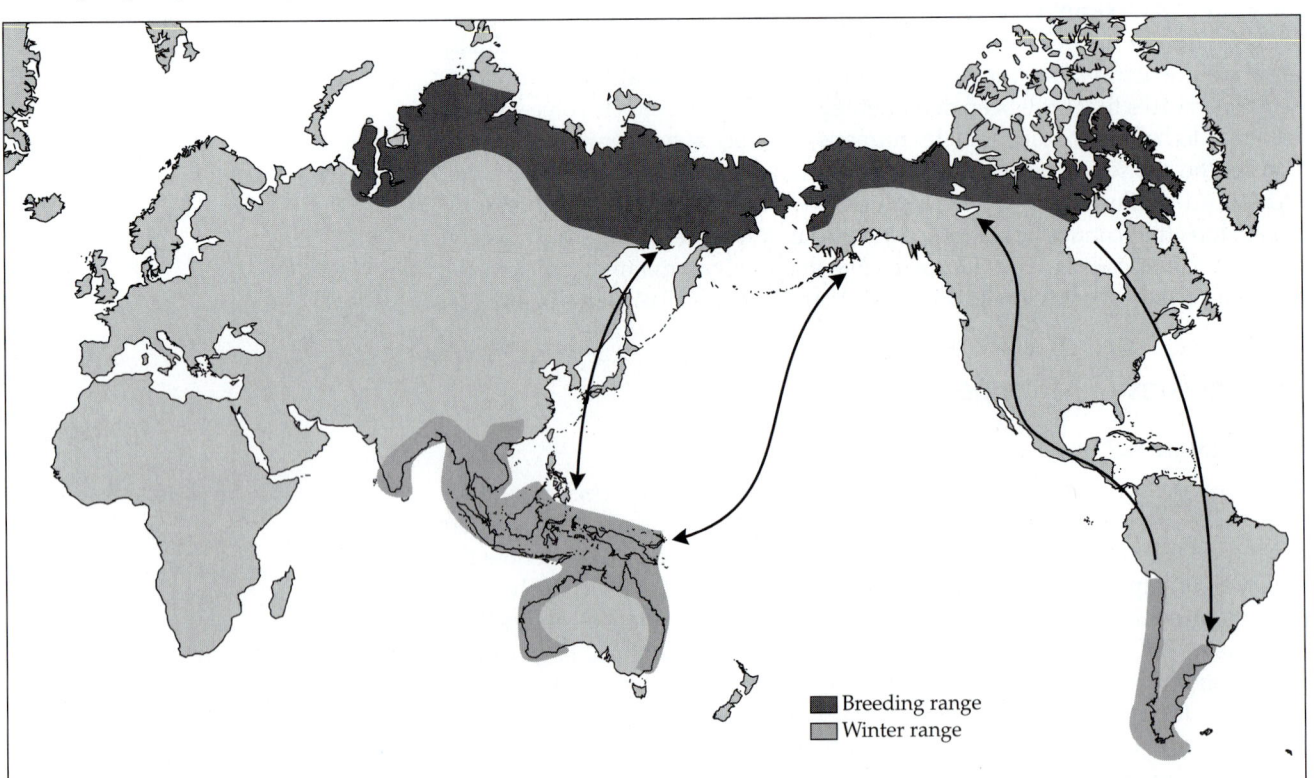

Monarch butterflies (*Danaus plexipus*) and some dragonflies (*Anax*) migrate distances comparable to those flown by many songbirds, from southern Canada and the northern United States to the southern United States and central Mexico (Figure 9.6). Individual monarch butterflies may fly as far as 375 km in 4 days and 4000 km in their lifetimes (Urquhart 1960; Brower 1977; Brower and Malcolm 1991). Occurrences of individuals in Cuba and other regions of the Caribbean may indeed represent fatal "accidents," but they might also be the equivalent of genetic "hopeful monsters." These misguided few, or their descendants, may someday found a new race or species of monarchs.

In contrast to these volant species, only a few nonvolant animals, such as some of the larger mammals, reptiles, and fishes, are able to disperse substantial distances by swimming or walking. Active dispersal by these means is generally less effective than flight because the animals are forced to swim or walk through unfavorable intervening habitats, whereas flying animals can simply vault barriers. Nevertheless, some large animals, especially aquatic ones such as whales, sharks, predaceous fishes, and sea turtles, have wide and often discontinuous geographic distributions produced in part by active dispersal.

Stories about swimming terrestrial vertebrates have appeared repeatedly in the literature, and as absurd as some reports may seem, we cannot deny that such incidents occasionally occur. A well-documented case involves elephants, which enjoy being in water. However, who would think that an elephant would

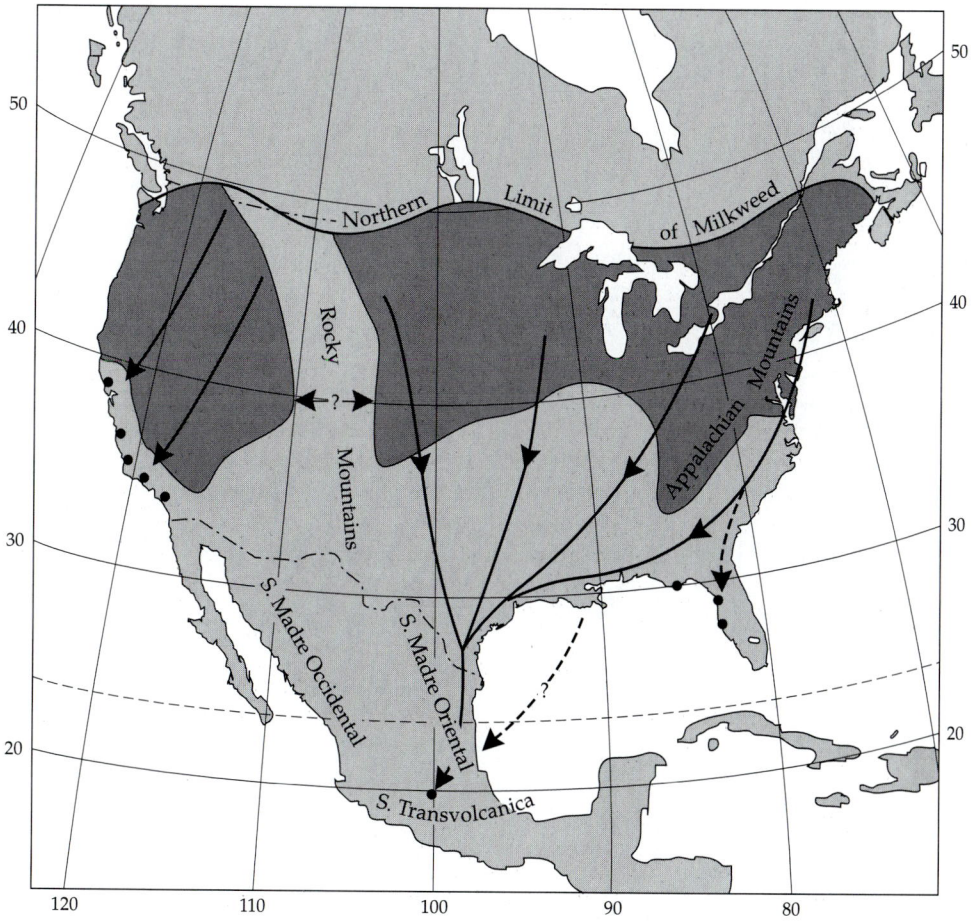

Figure 9.6 Fall migration routes of eastern and western populations of the monarch butterfly (*Danaus plexippus*). Most, if not all, overwintering individuals of the eastern populations gather in spectacular winter congregations in montane forests of the Transvolcanic region of Mexico. During the summer, the northern limits of this species are coincident with the northern limits of its host plants, milkweeds (*Asclepias*). (From Brower and Malcolm 1991.)

be found swimming in the ocean, with its trunk serving as a snorkel? Odd as it may seem, such observations are well authenticated. A cow and her calf not only were photographed voluntarily swimming in fairly deep ocean water up to 50 km to an island off the coast of Sri Lanka, but were also followed and timed, showing that they made a return trip to Sri Lanka (Johnson 1978, 1980, 1981). Johnson and others believe that elephants are able to traverse narrow straits between islands and mainlands in search of new food supplies and, therefore, that their presence is not a reliable indicator of a solid land connection. Other large vertebrates, such as tigers and terrestrial snakes, have been spotted over a kilometer from land, but in many of these cases the animals were probably not there by choice, having been washed out to sea during a torrential storm.

Passive Dispersal

As impressive as active dispersal may seem, the vast majority of organisms disperse largely or solely by passive means. In any plant community, for example, we can easily observe the movement of **diaspores** (seeds, spores, fruits, or other plant propagules) away from the parent plant (Figure 9.7). The wind carries seeds and fruits that have attached wings, hairs, or inflated processes. Birds and mammals consume fleshy and dry diaspores, scattering some of them during feeding and distributing others later with their feces. The seeds, fruits, and spores of some plants become attached to the feathers or fur

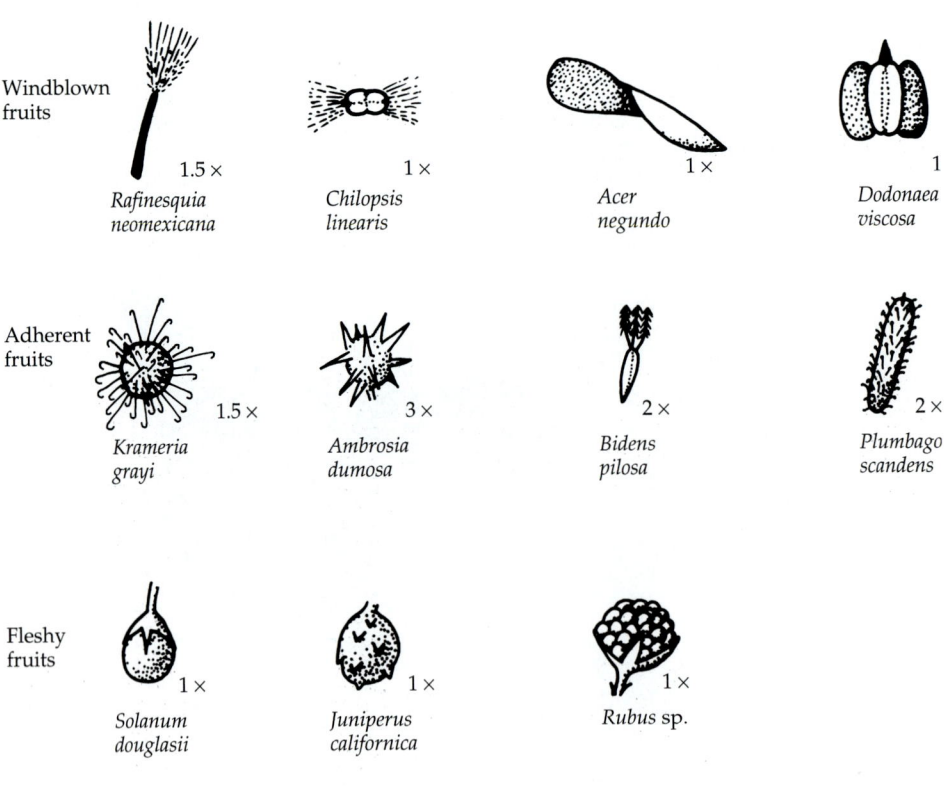

Figure 9.7   A variety of diaspores from North American plants designed to enable seeds to disperse from the mother plant. Seeds that are carried by wind often bear hairlike (*Rafinesquia, Chilopsis*) or papery wings (*Acer, Dodonaea*). Seeds that hitchhike on the bodies of animals have barbs (*Krameria, Bidens*), sharp spines (*Ambrosia*), or sticky glands (*Plumbago*). Seeds meant to be eaten and later defecated by birds and mammals are enclosed in fleshy fruits (*Solanum, Juniperus, Rubus*). In an unusual case, seeds of the dwarf mistletoe (*Arceuthobium*) are explosively discharged from a fleshy base.

Windblown fruits

1.5 ×   *Rafinesquia neomexicana*

1 ×   *Chilopsis linearis*

1 ×   *Acer negundo*

1 ×   *Dodonaea viscosa*

Adherent fruits

1.5 ×   *Krameria grayi*

3 ×   *Ambrosia dumosa*

2 ×   *Bidens pilosa*

2 ×   *Plumbago scandens*

Fleshy fruits

1 ×   *Solanum douglasii*

1 ×   *Juniperus californica*

1 ×   *Rubus* sp.

Dehiscent fruit

3 ×   *Arceuthobium americanum*

of animals and ride as hitchhikers until they are dislodged accidentally or by grooming. Some diaspores are explosively released over short distances, whereas others simply fall to the ground at the base of the parent plant. On the ground, ants, rodents, and birds compete for the fallen seeds and may carry them considerable distances, in some cases many meters. Finally, flowing water, tides, and ocean currents may displace diaspores and carry them long distances (Stebbins 1971a; Pijl 1972).

These means of passive dispersal are used not only by plants, but also by animals, fungi, and microbes. Some of them are obviously more effective than others. The distance passively traveled in a generation by a seed, spore, or other disseminule depends, of course, on the velocity, distance, and direction of movement by the dispersal agent. Aerial dispersal by winds, or by birds and bats, is especially successful in moving disseminules over long distances. Animal transporters, however, may be nomadic or migratory, leaving their original habitats and covering great distances, or they may be territorial residents that rarely leave their home ranges. Wind tends to have local effects except in violent storms, during which organisms can be widely disseminated. Jet streams are also alleged to be important dispersal agents, especially for tiny spores (Gressitt 1963; Clagg 1966). Such so-called **aerial plankton** also includes tiny mites, spiders, and insects that do not necessarily have special stages adapted for dispersal.

Even passive mechanisms for dispersal are not necessarily completely haphazard or random. Aerial invertebrates, for example, occupy different strata above the land surface, so it is likely that they differ in their tendency to be wind-dispersed (Figure 9.8). In addition, all else being equal, small organisms tend to be more effectively transported by wind than large organisms (e.g., see Vagvolygi 1975). But all else is not equal. The structural features that promote long-distance dispersal differ markedly among species and taxonomic groups. For bacteria, yeast, fungi, mosses, and ferns, dispersal is effected by tiny spores that not only are readily transported by wind and water, but also are highly resistant to extreme physical environments. They are light enough to float in water or remain aloft in the air for long periods.

Many invertebrates also have small propagules. Often, fertilized eggs or other life history stages encyst to form thick-walled structures that are metabolically inactive and capable of withstanding long periods of desiccation and wide ranges of temperature. Some protozoans, rotifers, tardigrades, worms, and crustaceans disperse over long distances by such means. For example, the brine shrimp, *Artemia*, lives in highly saline pools and lakes throughout the

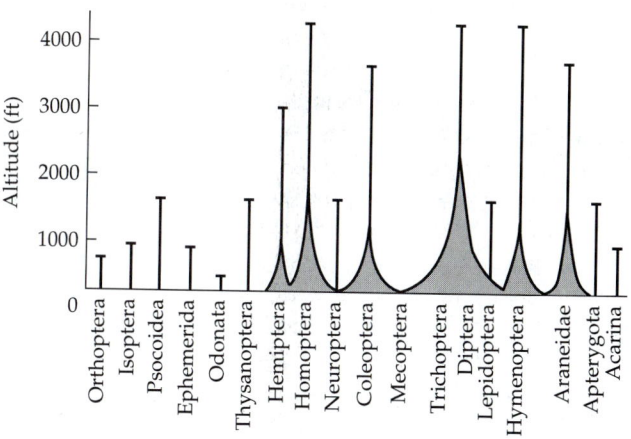

Figure 9.8 Aerial distribution of insects, spiders, and mites. Provided they can survive the rigors of high altitudes (especially low oxygen concentrations and temperatures), species found at higher altitudes should have greater dispersal abilities. (From Muller 1974.)

world. When these bodies of water dry up, the shrimp can survive for months or years as encysted fertilized eggs. When the pools refill, they resume their life cycle. These dry eggs can be purchased at pet stores, and one need only add them to salt water to rear a new generation of *Artemia*. The tiny encysted eggs are also picked up by the wind and dispersed great distances. It is apparently because of this long-distance dispersal ability that *Artemia* is found in isolated localities such as Great Salt Lake in the western United States and the Dead Sea in Israel (see Browne and MacDonald 1982).

Larger propagules are obviously heavier, and therefore require surface features to keep them aloft in wind currents. We mentioned earlier the adaptations of plants for aerial transport (see Figure 9.7); some invertebrates have equally innovative designs, including the gossamer parachutes spun by certain spiders (Araneae, Arachnida) and the long dorsal filaments of mealybugs (Hemiptera) (see Figure 9.8).

Most marine organisms have free-living juvenile stages that drift near the surface of the water. These tiny planktonic propagules—invertebrate eggs and larvae, fish larvae, algal cells, and the like—move passively with ocean currents. The distances traveled by small versus large forms may differ, but all are controlled ultimately by the circulation patterns of the oceans (see Chapter 3). Strong currents may move propagules quickly from one locality to another in a definite direction. The wind that drives oceanic circulation also disperses some species, such as the jellyfish *Vellela* (Cnidaria), which is equipped with "sails" that are oriented to carry individuals in somewhat predictable directions (Figure 9.9).

Although terrestrial vertebrates are generally too large to be carried far by the wind, some can be transported for surprising distances by water, often as passengers on mats of drifting vegetation or other debris, called "rafts." Rafts containing entire trees and carrying a large variety of organisms have been recorded (Ridley 1930; Carlquist 1965). Such large rafts are washed out to sea from large tropical rivers and may not break up for extremely long distances. Rodents, and especially lizards, apparently have managed to colonize many isolated oceanic islands by this means. Although the chance of a particular raft going ashore on a given island is admittedly very small, over sufficiently long periods of evolutionary time there is a high probability of at least a few such events occurring. As we noted earlier, it is through the chance occurrence of such individually unlikely episodes of passive long-distance dispersal that

 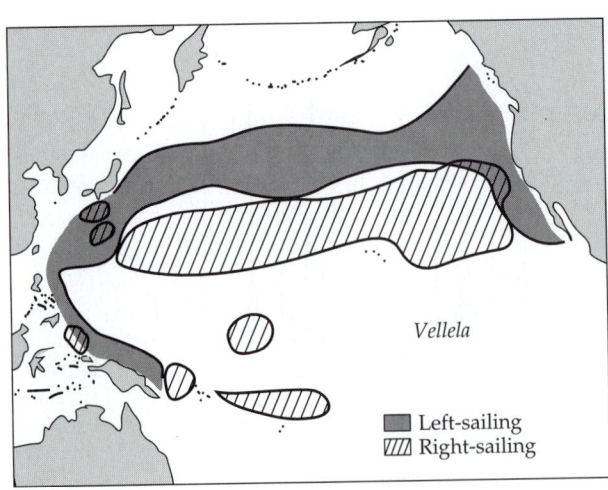 

Figure 9.9 Like their Cnidarian relative, the Portuguese man of war, individuals of the jellyfish *Velella* possess "sails," or floats, oriented so that they tend to drift either to the right or to the left. Each form is restricted to certain regions of the ocean, and it is thought that by sailing in the wind, these animals counteract their tendency to be carried off by ocean currents, thus remaining within a limited area. (After Savilov 1961.)

Sail

Tentacles

Left

Vellela

■ Left-sailing
▨ Right-sailing

Right

oceanic islands and many other isolated habitats acquire much of their distinctive biotas. A less haphazard means of dispersal by seawater is found among plants, especially those inhabiting beaches and mangrove swamps, some of which have fruits, seeds, or vegetative parts that float in salt water and remain viable even when submerged for long periods of time (Carlquist 1965, 1974; MacArthur 1972; J. M. B. Smith et al. 1990; J. M. B. Smith 1994).

A surprisingly large number of organisms are transported over long distances by other organisms. Many examples of this process, called **phoresy**, are obvious. Parasites, for example, are carried to any new areas colonized by their hosts. As Europeans explored and colonized the earth, they spread a variety of bacterial and viral diseases. In some isolated areas, the native human populations had never been exposed to these microbial pathogens and had evolved no immunity to them. Consequently, diseases such as smallpox, syphilis, and measles had devastating effects when introduced among American Indians and Pacific Islanders (McNeill 1976; Marks and Beatty 1976).

Some plants have sticky or barbed seeds or fruits that adhere to mobile animals (see Figure 9.7). Aquatic birds may transport small invertebrates in their feathers, or trapped in mud on their feet, between widely separated lakes and ponds. Other seeds—usually those found in sweet, fleshy fruits—are adapted to pass through the digestive tracts of animals. Because they are resistant to digestive juices, they are still viable when dropped in the feces; in fact, the germination of certain seeds is enhanced when they are exposed to animal digestive chemicals. Such internal transport is extremely common and is important not only in the coloniziation of isolated oceanic islands, but also in dispersal within continental regions, from one habitat to another. Birds are the usual agents, but mammals and reptiles also serve as internal seed dispersers. Similarly, insects and mycophagus mammals are important agents of dispersal of fungal spores.

A few animals are obligatory hitchhikers. Colwell (1973, 1979) and his associates have studied tiny mites that live in certain flowers (see also Athias-Binche 1993; Zeh et al. 1992; Poinar et al. 1990; Brown and Wilson 1992). These mites do not have a highly resistant stage in their life cycle and seem poorly adapted for long-distance dispersal. However, the mites occur only in flowers regularly visited and pollinated by hummingbirds, and they are transported between flowers on the birds' beaks. They crawl into the nasal openings, whose protected microclimate enables them to survive and be carried for long distances. At least two species of these specialized flower mites occur on the Greater and Lesser Antilles. Presumably, they originally colonized these islands from the continental mainland of tropical America, riding hundreds of kilometers on the beaks of dispersing hummingbirds.

The Nature of Barriers

Successful long-distance dispersal usually requires that organisms survive for significant periods of time in environments very different from their usual habitats. These unusual environments constitute physical and biological barriers that successful colonists must cross. The effectiveness of such barriers in preventing dispersal depends not only on the nature of the environment, but also on the characteristics of the organisms themselves. These characteristics, of course, vary from one taxonomic group to another, so that particular barriers may not affect all residents of a habitat equally. Thus, barriers are species-specific phenomena. Two examples will illustrate this point.

Most freshwater zooplankton have resistant stages that facilitate long-distance dispersal. Consequently, the same species may be found in widely separated localities wherever environmental conditions are similar, such as in the

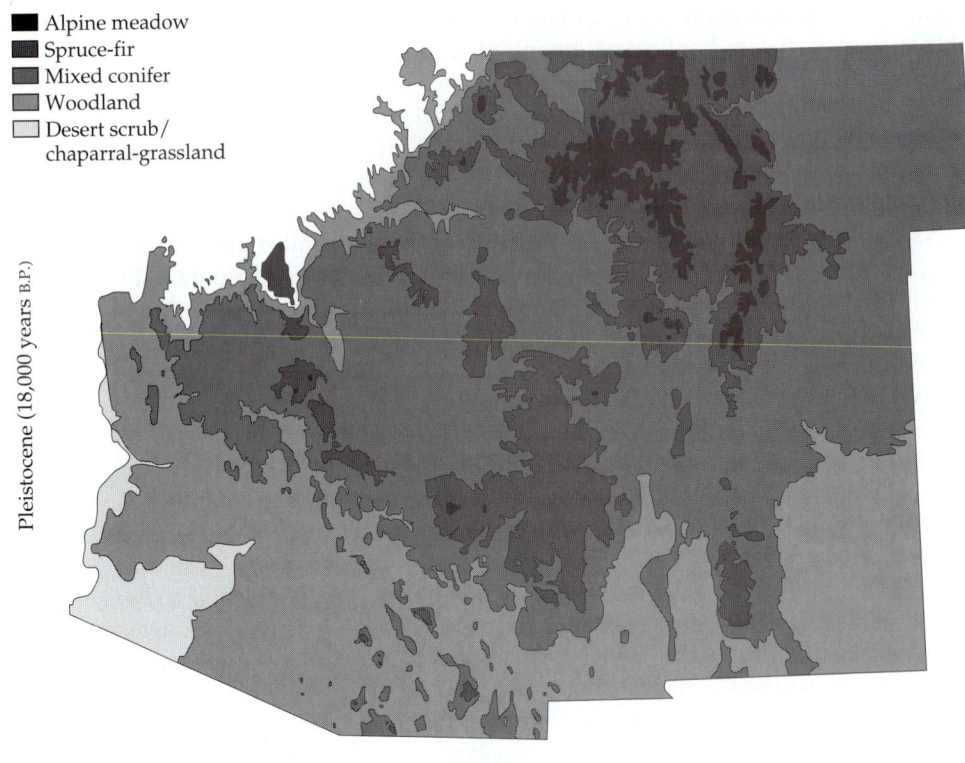

Pleistocene (18,000 years B.P.)

- ■ Alpine meadow
- ■ Spruce-fir
- ■ Mixed conifer
- ■ Woodland
- □ Desert scrub/ chaparral-grassland

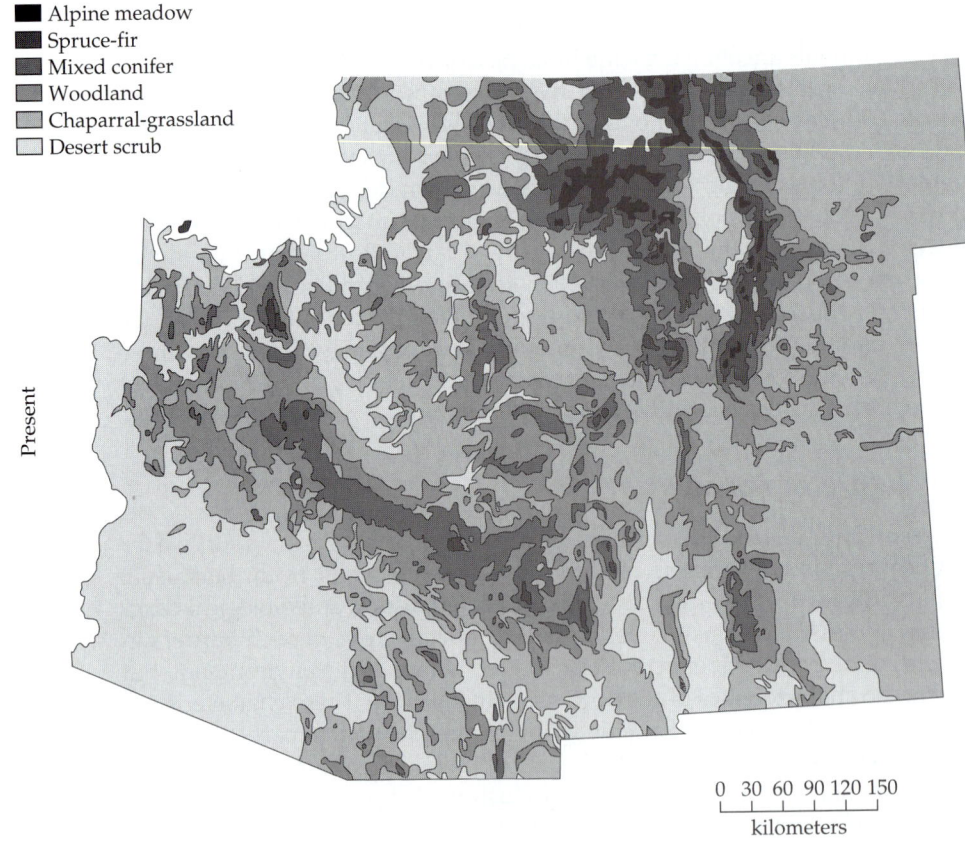

Present

- ■ Alpine meadow
- ■ Spruce-fir
- ■ Mixed conifer
- ■ Woodland
- ■ Chaparral-grassland
- □ Desert scrub

0 30 60 90 120 150
kilometers

Figure 9.10 Geographic shifts in vegetation zones of the American Southwest since the most recent glacial maximum. (Elevational shifts in these vegetation zones are depicted in Figure 7.22.) (From Lomolino et al. 1989.)

cold temperate lakes of northern North America, Europe, and Asia. On the other hand, the fishes inhabiting these same lakes appear to be unable to disperse across terrestrial and oceanic systems. Consequently, the same fish species inhabit only those lakes that have been connected by fresh water at some time in the past. Many isolated alpine lakes, which have never had such connections, support a diverse plankton fauna but have no fish at all (unless they have been recently introduced by humans).

Similarly, many of the isolated mountains of the southwestern United States are essentially islands of cool, moist forest in a sea of hot, dry desert. The desert is not a barrier to most birds and bats, which rapidly and repeatedly fly over it to colonize the montane forests. On the other hand, for many small terrestrial mammals, reptiles, and amphibians, which must disperse much more slowly on foot, the desert may represent an absolute barrier. It is likely that these animals colonized the most isolated mountains when they were connected by bridges of suitable forest and woodland habitat during the Pleistocene (Figure 9.10; see Findley 1969; Patterson 1980; J. Brown 1978; Lomolino et al. 1989).

Because the effectiveness of a particular kind of barrier depends on both the physical and biotic challenges it poses and the biological characteristics of the organisms attempting to cross it, it is difficult to make sweeping generalizations about the nature of barriers. It is usually true, however, that organisms that inhabit temporary or highly fluctuating environments are much more tolerant of extreme or unusual physical and biotic conditions than are species that are confined to permanent or stable habitats. Plants and animals from fluctuating environments are also more likely to have resistant life history stages, which not only enable them to survive periods of unfavorable conditions, but also can serve as effective propagules. Consequently, the "weedy" species that inhabit temporary or fluctuating environments are likely to be better dispersers and less limited by any kind of barrier than are species from more permanent or constant habitats.

A comparable example is provided by the pupfish *Cyprinodon variegatus*, an extremely euryhaline and eurythermal species that inhabits estuaries, tidal flats, and mangrove swamps along the eastern coast of North America from Cape Cod to Yucatán. This little fish has managed to cross hundreds of kilometers of ocean to colonize comparable habitats on many of the Caribbean islands. In contrast, the strictly freshwater sunfishes (*Lepomis*) and basses (*Micropterus*) that are so abundant and diverse in the rivers, lakes, and streams of eastern North America are not native to the Caribbean. The successful introduction of these species by humans indicates that saltwater barriers, rather than any lack of suitable habitats on the islands, had prevented their colonization.

Daniel Janzen (1967) made a related point in an insightful paper that addressed the restricted distributions of many species in mountainous areas of the tropics. He pointed out that mountain passes of a given elevation are effectively "higher" in the tropics than in temperate zones. Because temperate environments experience seasonal temperature variations, lowland plants and animals must be able to tolerate a wide range of environmental conditions, including those they would encounter at higher elevations during the summer (Figure 9.11). In contrast, because of the thermal constancy of the tropics, a tropical lowland species would never experience, and would not be adapted to withstand, the temperature regimes it would encounter if it were to travel over a high mountain pass from one lowland area to another.

Physiological Barriers

Probably the most severe barriers are presented by physical environments so far outside the range an organism normally encounters that it cannot survive

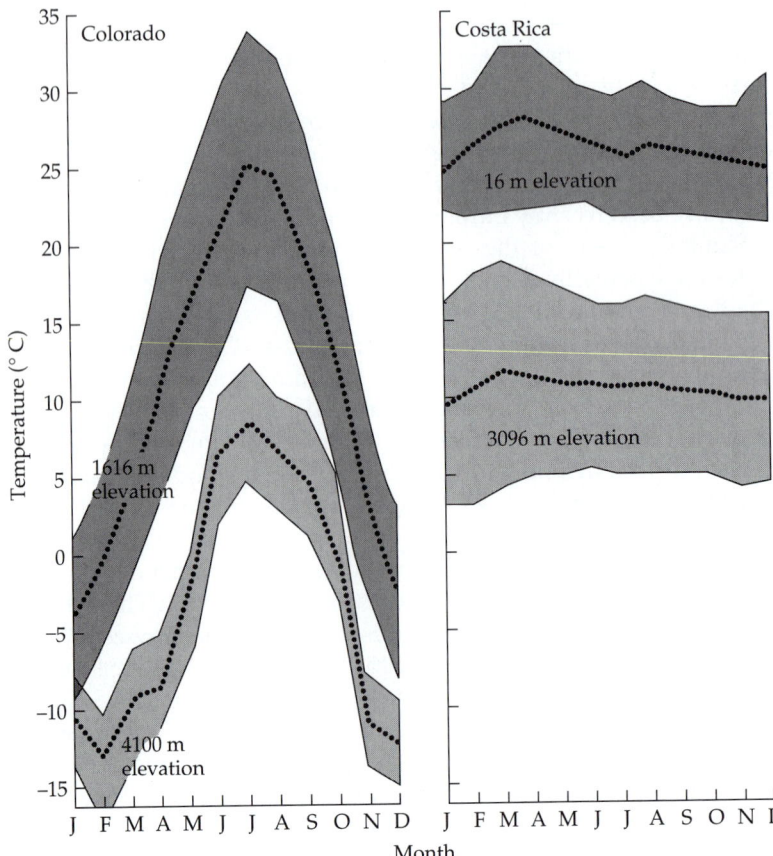

Figure 9.11   A given change in elevation tends to be a greater barrier to dispersal in the tropics than at higher latitudes, as shown by these temperature profiles. In tropical regions, sites separated by several thousand meters of elevation usually experience no overlap in temperature, whereas in the temperate zone, winter temperatures at low elevations broadly overlap summer temperatures at much higher sites. (After Janzen 1967.)

long enough to disperse across them. The vast majority of aquatic organisms live either in the oceans or in fresh water, and cannot regulate their water and salt balance sufficiently to survive more than brief exposure to the other environment. Very few freshwater fishes have successfully colonized across the oceans. The same is true of amphibians, which have permeable skins and are quite intolerant of exposure to salt water. Likewise, many terrestrial plants are unable to withstand prolonged exposure to seawater at any stage of the life cycle, including the seed. Hnatiuk (1979), however, reported that 56 of the 69 terrestrial plant species native to Aldabra Atoll in the western Indian Ocean tolerated total immersion of their seeds in seawater for 8 weeks with no inhibition of germination. This tolerance would not only facilitate the survival of these species on tiny islands subject to wave splash during heavy storms, but would also greatly increase the probability of these species surviving immersion during transport across ocean barriers, thus perhaps explaining how many of them managed to colonize the islands in the first place.

Environmental temperature regimes can also serve as physiological barriers. We have already mentioned the case of mountain passes in the tropics. A similar barrier is created by the tropics themselves, which form a band of high temperatures around the equator, isolating the cooler temperate and arctic areas toward either pole. For many cold-adapted organisms, both terrestrial and aquatic, these warm tropical climates are a major barrier to dispersal. Some groups, such as the Nearctic avian family Alcidae (auks, puffins, guillemots, and murres), are good dispersers within their climatic zone and are broadly distributed throughout one hemisphere (North or South), but have

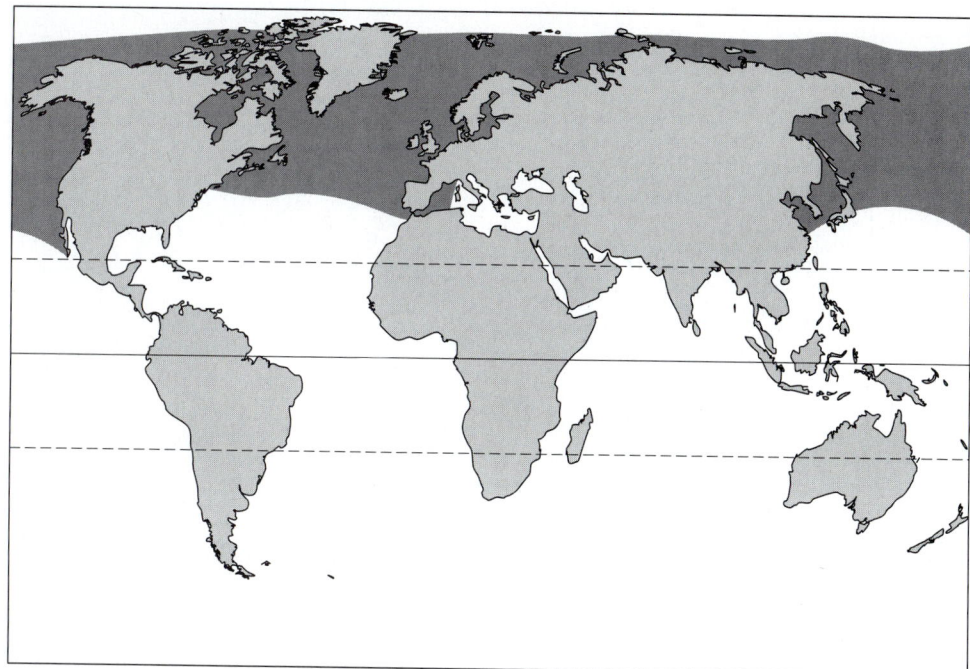

Figure 9.12 The avian family Alcidae (auks, puffins, and murres) is restricted to cooler regions of the Northern Hemisphere (shaded area), even though all of these birds are winged and most are strong fliers. The tropics appear to constitute a major barrier to the dispersal of this group, although the presence of distantly related but ecologically convergent seabirds in the Antarctic region may also prevent their colonization of the Southern Hemisphere. (After Shuntov 1974.)

not managed to cross the tropics to colonize the other hemisphere (Figure 9.12). The two hemispheres are inhabited by different forms with convergently similar morphology and ecology, such as the penguins (Spheniscidae) and diving petrels (Pelecanoididae), or the skuas (*Catharacta* spp.) and jaegers (*Stercorarius* spp.) of the Southern and Northern Hemispheres, respectively.

The nature of barriers may also vary dramatically with the seasons. For example, in temperate regions of North America, large bodies of water serve as effective barriers to the movement of many terrestrial species, including nonvolant mammals. During winter, however, these waters freeze, providing a seasonal avenue for the dispersal of winter-active species to islands or adjacent mainland sites (Figure 9.13) (Jackson 1919; Banfield 1954; Lomolino 1984, 1988, 1989, 1993; Tegelstrom and Hansson 1987).

Ecological and Psychological Barriers

Dispersing organisms must be able to survive not only the physiological stresses imposed by the environments they traverse, but also the ecological hazards. Just as predation and competition can limit the local abundances of species, they can also prevent successful dispersal. One would expect biotic interactions to be particularly important components of barriers for animals that neither move very rapidly nor disperse at a resistant stage. Although it is likely that competition and predation limit the dispersal of such organisms, we are not aware of any well-documented examples.

Surprising as it may seem, behavioral or psychological barriers appear to play a major role in preventing the long-distance dispersal of some organisms. Most organisms appear to possess mechanisms of **habitat selection**, the ability to recognize and respond appropriately to favorable environments. In some animals these traits are so well developed that they strongly inhibit active dispersal. For example, some bird species that seem perfectly capable of flying long distances are apparently unwilling to cross certain kinds of barriers. Willis (1974) has described species of antbirds (Formicariidae) that have become

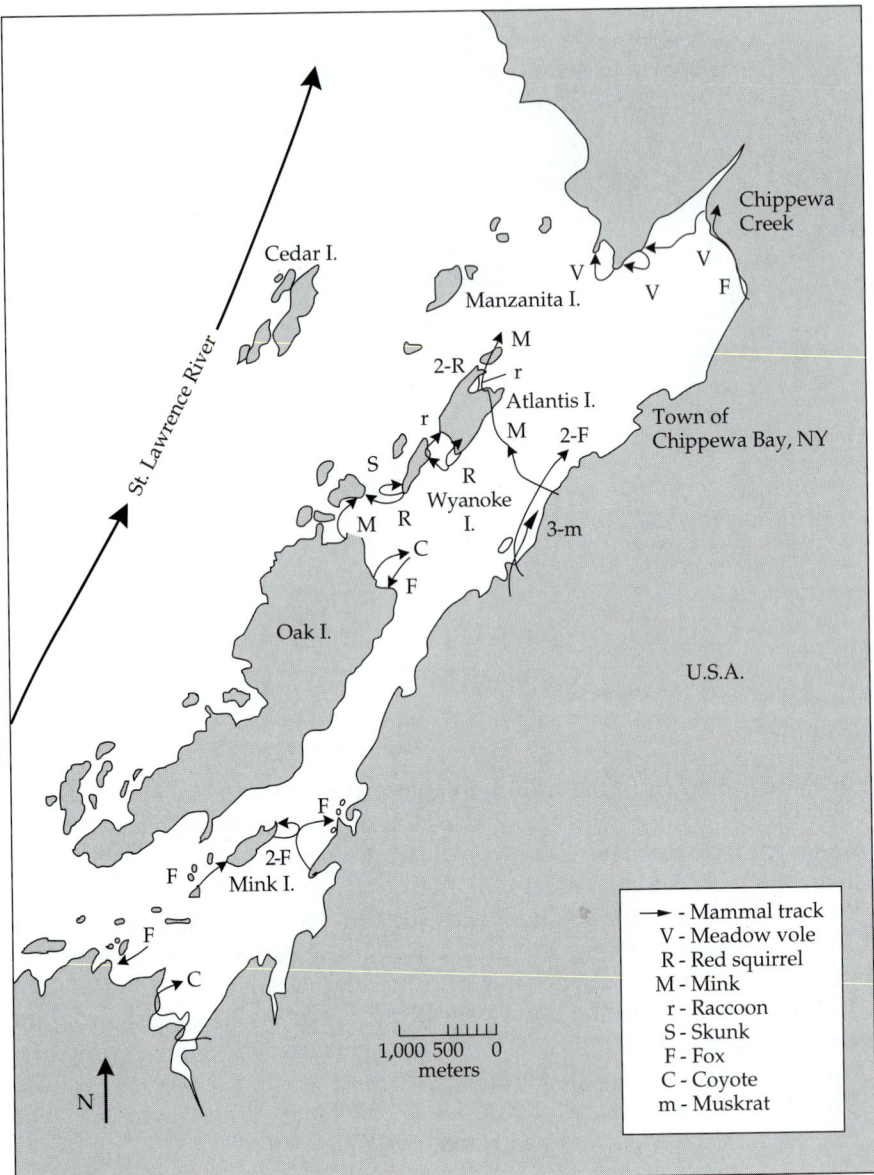

Figure 9.13 Movements of terrestrial mammals across the ice of the St. Lawrence River within the vicinity of Chippewa Bay, New York. Arrows indicate the direction of mammal tracks observed within a 24-hour period (March 1, 1979). (From Lomolino 1983, 1988.)

extinct on Barro Colorado Island and have not recolonized, even though they would have to fly only a few hundred meters across Gatun Lake. MacArthur et al. (1972), Diamond (1975b), and others have documented groups of tropical birds, such as the New World cotingas and toucans and the Old World barbets and pittas, that are strong fliers, but are repeatedly absent or poorly represented on oceanic islands, even where good habitat appears to be abundant. Often species restricted to virgin rain forest are particularly sedentary, whereas other, even closely related, species that are characteristic of second growth habitats or edges are good dispersers. Species adapted to successional habitats must continually disperse to new environments as their habitat patches undergo succession and become unsuitable. Such species are therefore much more likely to strike out across barriers and disperse successfully over long distances than are species that are restricted to very stable environments.

Biotic Exchange and Dispersal Routes

As we emphasized in the preceding section, barriers are species-specific phenomena. Even within a group of related species, a barrier to one is not likely to be a barrier to all. Now we shall broaden our perspective to look at entire biotas and the process of biotic exchange between biogeographic regions. Biogeographers often distinguish three kinds of dispersal routes based on their effects on biotic exchange. Listed in order of increasing resistance to biotic exchange, these include corridors, filters, and sweepstakes routes (Simpson 1940).

Corridors

The term **corridor** refers to a dispersal route that permits the movement of many or most taxa from one region to another (Simpson 1936, 1940; see also Udvardy 1969). A corridor therefore allows a taxonomically balanced assemblage of plants and animals to cross from one large source area to another, so that both areas obtain organisms that are representative of the other. By definition, a corridor does not selectively discriminate against any form, and must therefore provide an environment similar to that of the two source areas.

As we discussed in Chapters 6 and 7, the geological record is replete with evidence of great pulses of biotic exchange across ancient marine and terrestrial corridors. The ancient Tethyan Seaway, a circum-equatorial marine system, extended all the way from the Orient and Malaysia to westernmost Europe, separating Africa from Eurasia (see Figure 6.21B). For nearly 500 million years the Tethyan Seaway served as a dispersal highway for marine organisms, including both benthic and pelagic forms (Adams and Ager 1967). The exchange of taxa within the Tethys was not always uniform, and may have been strongly influenced by its predominantly westward flowing currents. This exchange was disrupted about 60 million years ago when Africa became connected with Asia through Arabia and India drifted northward to join Asia. Now the Mediterranean Sea has a vastly different fauna than the Indian Ocean.

Many ancient, transient connections have served as corridors for exchange among terrestrial biotas. Most recently, a great diversity of terrestrial plants and animals dispersed across the landbridges of Beringia, the Sunda Shelf, and the Tasman and Arafura Basins during the glacial maxima of the Pleistocene (see Figure 7.9). The Bering landbridge between Alaska and Siberia permitted passage in both directions, but was probably a corridor only during the first half of the Cenozoic, when a mild climate prevailed. Even then, biotic exchange across Beringia tended to be asymmetrical, with more species moving from the relatively diverse regions of eastern Asia and Siberia into Alaska.

Filters

As its name implies, a **filter** is a dispersal route that is more restrictive than a corridor. It selectively blocks the passage of certain forms while allowing those able to tolerate the conditions of the barrier to migrate freely. As a result, colonists tend to represent a biased subset of their respective species pools. Thus the biotas on the two sides of a filter share many of the same taxonomic or functional groups, but some taxa are conspicuously absent in each.

The Arabian subcontinent is a harsh filter that permits the dispersal of only a limited number of mammals, reptiles, nonpasserine ground birds, invertebrates, and xerophytic plants between northern Africa and Central Asia. Organisms that need abundant water, such as freshwater fishes, most amphibians, and forest dwellers of many taxa, are stopped by the deserts in

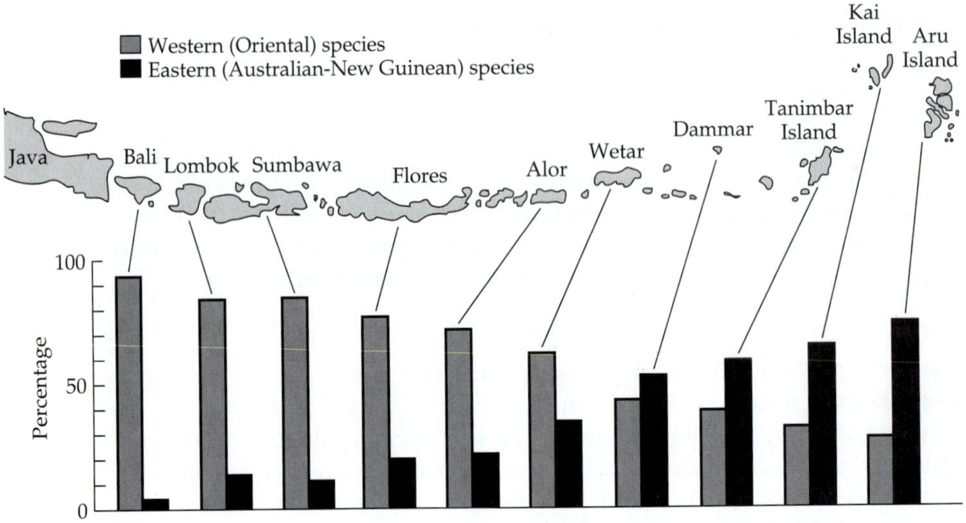

Figure 9.14   The Lesser Sunda Islands between Java and New Guinea serve as a two-way filter for the reptilian faunas of southeastern Asia and Australia. The bars quantify the decline in Oriental species and the increase in Australian species going from west to east down the island chain. (After Carlquist 1965.)

this region. This has not always been true. As mentioned above, for millions of years Arabia and Asia were widely separated by the Tethys Sea, a formidable barrier to terrestrial organisms. When the Arabian Peninsula became emergent and contacted Persia, the region intermittently served as a landbridge for movements of Asian taxa into Africa and vice versa, as in the Upper Miocene, when many African forms appeared in Asia for the first time (Cooke 1972). Thus, today's filter has been both more of a corridor and a more formidable barrier in the past, and might become either in the future.

Filters may be produced by abiotic or biotic factors. They are generally easy to identify because the number of species in certain taxa decreases with distance from the source area. Filters often form transition zones between two biogeographic regions. Figure 9.14 depicts such a transition zone: a two-way filter formed by the Lesser Sunda Islands between Java and New Guinea. Reptile species of Oriental origin decrease proportionately as one moves eastward, where Australasian groups become dominant, and, of course, the same decreasing trend is found for Australasian groups as one moves westward through Wallacea (i.e., the transition zone bisected by Wallace's line).

On the opposite side of the globe, biogeographers have discovered a terrestrial version of Wallacea. For most of the Pleistocene, assemblages of forest-dwelling terrestrial vertebrates in western North America have been isolated in two major "mainland" regions, the forests of the Rocky Mountains to the north and of the Sierra Madre to the south. The intervening habitats have, however, allowed some mixing among these regional biotas, especially during glacial maxima, when forests and woodlands expanded at the expense of more xeric habitats (see Figures 7.22 and 9.10). Thus, the xeric landscapes of the American Southwest serve as a two-way filter or transition zone between the two faunal regions (Figure 9.15; see Davis 1996; Lomolino and Davis 1997).

In Chapter 15 we discuss the biotic exchange across another terrestrial filter, the Central American landbridge. This event, termed the Great American Interchange (G. G. Simpson 1969, 1978), represents a perhaps unparalleled case study of the combined effects of dispersal, evolution, and extinction. As many have observed, however, the formation of this landbridge created a barrier to the dispersal of marine organisms. The construction of the Panama Canal reunited the Caribbean Sea and the Pacific Ocean, but biotic exchange between them has been extremely limited. The low-salinity waters flowing from Gatun Lake serve as an effective physiological barrier to most marine

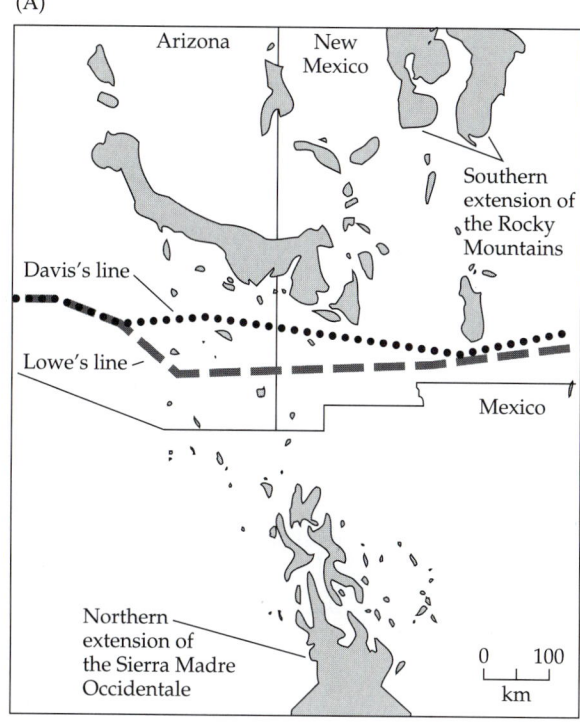

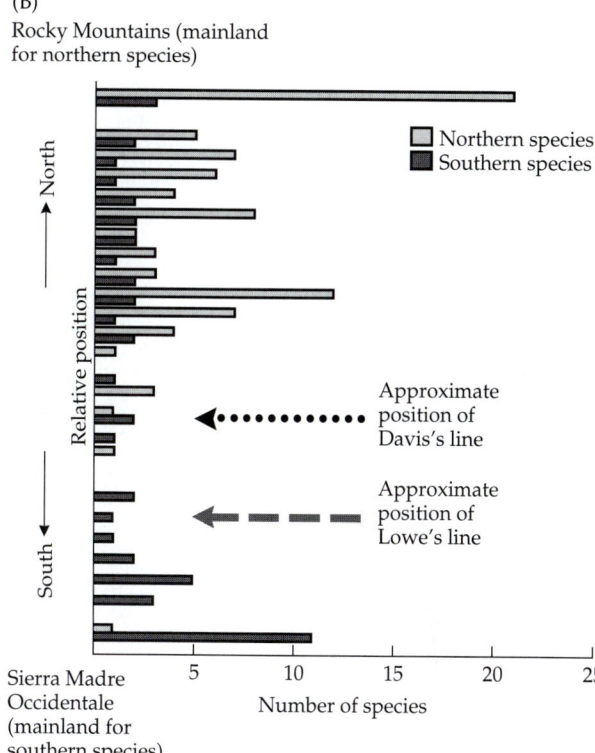

Figure 9.15 The desert "sea" of the American Southwest may serve as a transition zone, or two-way filter for forest and woodland species between the Rocky Mountain "mainland" to the north and the Sierra Madre "mainland" to the south. (A) Two biogeographic lines, Lowe's line and Davis's line, have been proposed to mark the divisions between these regional faunas. (B) The bars quantify the decline in species of nonvolant mammals as one moves away from their respective source regions. (A after Davis 1996; B after Lomolino and Davis 1997.)

animals (Figure 9.16) (Abele and Kim 1989; Hildebrand 1939; Rubinoff and Rubinoff 1977; Woodring 1966).

Sweepstakes Routes

The term *sweepstakes* was coined by G. G. Simpson to describe rare, chance dispersal from one locality to another across a major barrier. In a sweepstakes, many individuals enter the contest, but only a handful of lucky ones win prizes. In the natural world, propagules continually disperse from established populations into new areas, but only a small fraction are ever successful in founding a new population. As we pointed out earlier, the more extensive and harsh the barrier, the less chance of a propagule arriving at a new locality. Only those species or groups with features that permit long-distance jumps and tolerance of physiological hardships have any chance of arriving in a remote area. Given enough time, however, dispersal by at least some forms will probably occur. A severe barrier that results in this type of stochastic dispersal pattern is known as a **sweepstakes route**.

The seemingly chance colonization of isolated oceanic islands is the classic example of sweepstakes dispersal. This process is not as random as it may seem, however. Even when organisms cross barriers independently of one another, they may still use the same dispersal routes. Insular biotas, for example, usually have taxonomic affinities with the organisms inhabiting the nearest continent or other nearby landmass, which serves as a source of propagules. As we saw above, most of the species inhabiting the islands of the South Pacific have Indo-Malayan and, to a lesser extent, Australasian affinities, and the number of taxa shared with these landmasses decreases with increasing distance out into the Pacific (Figures 9.14 and 9.17) (Wilson 1959, 1961; Solem 1981). The mainland ancestors of most species inhabiting these remote islands

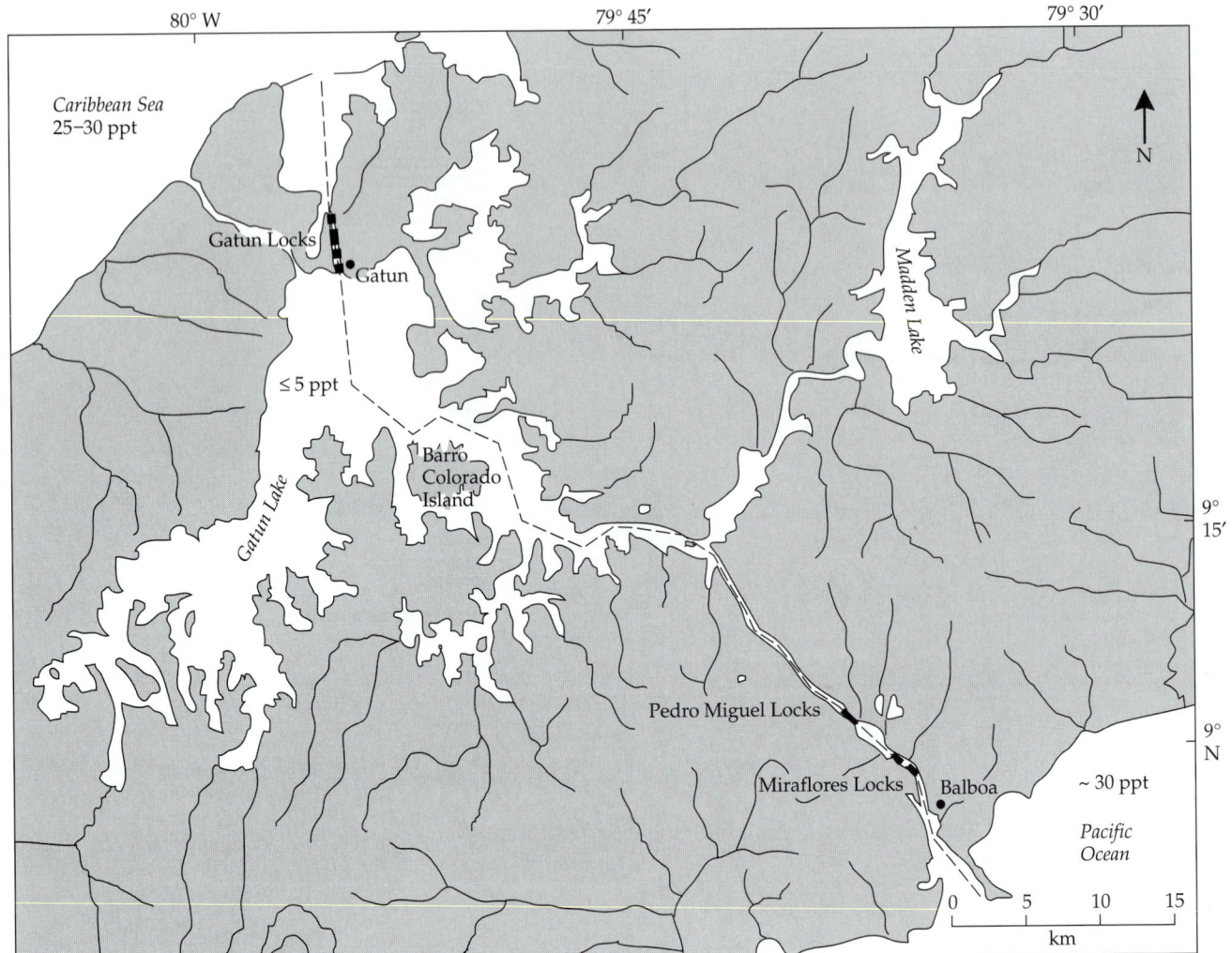

Figure 9.16 Map of the Panama Canal and associated waterways, showing salinity levels. The low-salinity waters of Gatun Lake serve as an effective physiological barrier to most marine animals. (After Abele and Kim 1989.)

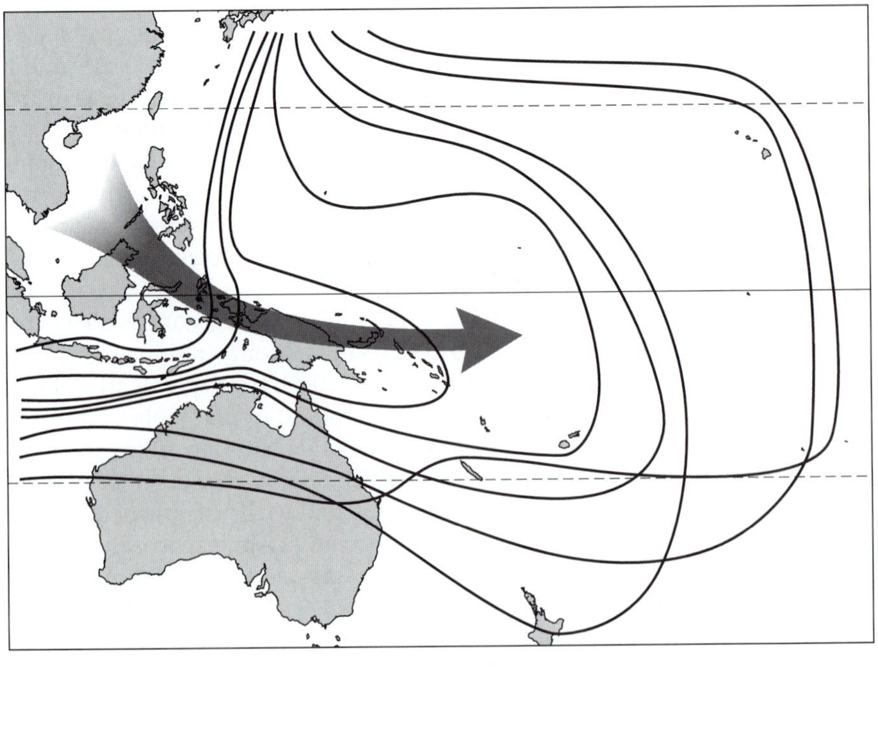

Figure 9.17 A classic example of a sweepstakes route. The lines show the limits of the distributions of eight different families of land snails in Australasia and the South Pacific. Each of these groups originated in Southeast Asia and has spread southward and eastward to a different extent. (After Solem 1981.)

were probably preadapted for long-distance dispersal—so-called "waif species." Thus, dispersal across these sweepstakes routes is random only in that it is impossible to predict which of these waif species will actually make it to any given island.

Other Dispersal Routes

When a landmass is shifted from one place to another by seafloor spreading, it carries a biota on board. These species can thus be transferred directly without crossing major barriers. India, which moved from southern Africa to Asia, has been called a "Noah's Ark" by McKenna (1973) because an assemblage of organisms rafted on it en masse to a new environment. There are other fragments of tectonic plates in southern Europe, western North and South America, and eastern Asia that may have served as arks for their respective biotas. These plate fragments also carried fossil beds associated with their former locations. The movements of these fossil beds, sometimes termed "Viking funeral ships," may complicate biogeographic analyses by mixing the remains of biotas that may never have occurred together in time.

Biologists are now eager to help earth scientists find evidence of past land connections and physical barriers. There was a time early in the twentieth century, however, when biogeographers could have made substantial contributions toward establishing the occurrence, mechanisms, and timing of continental drift, especially if our understanding of the phylogeny of organisms had been more advanced. But this did not happen, perhaps because most biogeographers were too conservative (see Chapter 6). Now we must design our reconstructions with a realistic picture of the earth's history in mind. Biogeography still can make some original contributions in these areas, however, especially by identifying barriers other than those created by drifting continents. Biological distributions generally cannot be used as proof of land connections in lieu of evidence from the rocks, but their analysis can corroborate or contradict geological data and can provide valuable clues to past events overlooked by geologists. The analysis of dispersal routes is best done separately by geologists and biogeographers, with both keeping watchful eyes out for conflicting results. Meanwhile, the burden of determining the positions of landforms and dates of tectonic changes still falls on geologists, and reconstructions of biotas, paleoecology, and paleoclimatology are projects for biogeographers.

Dispersal Curves within and among Species

For nearly all organisms, immigration rates (i.e., numbers of individuals or species arriving at a site per unit of time) tend to decline with increasing distance from the parent or source. Depending on the mechanism of dispersal, this relationship between immigration rate and distance often takes one of two general forms (Figure 9.18). If dispersal is a purely random process with a constant probability of organisms "dropping out" with each increment of distance, then the dispersal curve should take the form of a negative exponential (the dashed line in Figure 9.18B). This pattern is often found among windborne propagules and other forms that are passively dispersed.

On the other hand, for actively dispersing organisms or those dispersing on logs or rafts with normally distributed persistence times, dispersal abilities may approximate a normal or lognormal distribution. The precise form of this frequency distribution is not critical, only the fact that most individuals (or species) can disperse beyond some minimal distance, and few can travel beyond some greater distance. In such cases, immigration curves should approximate what MacArthur and Wilson (1967) termed a "normal" or **Gaussian function** (the

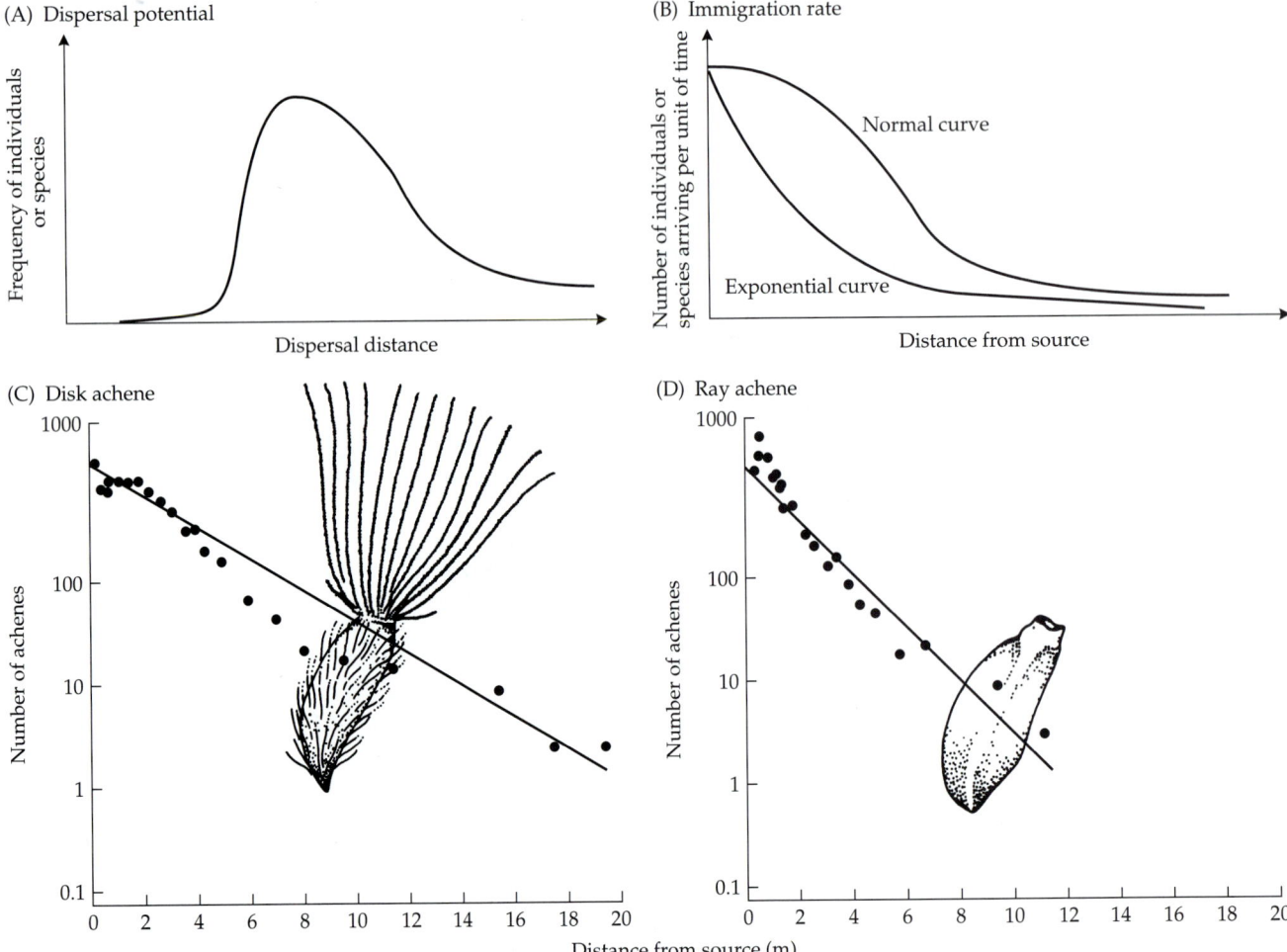

(A) Dispersal potential

Frequency of individuals or species

Dispersal distance

(B) Immigration rate

Number of individuals or species arriving per unit of time

Normal curve

Exponential curve

Distance from source

(C) Disk achene

Number of achenes

(D) Ray achene

Number of achenes

Distance from source (m)

Figure 9.18 (A) An idealized lognormal frequency distribution of dispersal potential among individuals or species. (B) Dispersal curves describe the expected decline in immigration rate as a function of increasing distance from a source region. If dispersal is a passive and random process with a constant probability of organisms dropping out with each increment of distance, then the dispersal curve should take the form of a negative exponential (dashed line). On the other hand, if dispersal is active or if organisms are carried on rafts with normally or lognormally distributed persistence times, then the dispersal curve should take the form of a "normal" function. (C and D). Seed dimorphism in the composite *Heterotheca latifolia* affects dispersal curves of two types of fruits: disk achenes, which have an attached parachute-like structure, and ray achenes that lack this dispersal structure, but have thicker fruit walls and can survive in the soil. (A and B after MacArthur and Wilson 1967; C and D courtesy of L. Venable.)

solid line in Figure 9.18B). Immigration rates should thus remain relatively high until the distance exceeds the dispersal abilities of the least vagile individuals, decline rapidly until the modal dispersal ability is approached, and then slow to asymptotically approach zero for the most distant sites.

Because the dispersal abilities of related species result from a combination of physiological, behavioral, and morphological traits, many of which are distributed as normal or lognormal functions, we expect the same patterns to hold for dispersal curves compiled across species. That is, for a given taxonomic group or pool of species, dispersal abilities should be normally or lognormally distributed, and immigration rates should also decline as a normal or lognormal function of distance.

We will return to the biogeographic relevance of dispersal curves in our chapters on island biogeography. It is important to emphasize here, however, that actual rates of dispersal or arrival at distant sites are influenced by the nature of dispersal barriers as well as by the characteristics of the species in question.

Establishing a Colony

If dispersal is to be of biogeographic significance, an organism not only must be able to travel long distances and cross barriers, but also must be able to establish a viable population upon its arrival at a new site. As Sherwin Carlquist (1965) has stated, "Getting there is half the problem." Obviously, a successful colonist is one that survives and reproduces. Several factors, including habitat selection and reproductive strategies, may play little role in dispersal per se, but may be of great importance in determining the fate of an immigrant once its journey is over.

Habitat Selection

All organisms exhibit some form of habitat selection. Highly mobile organisms actively seek out favorable environments, which they recognize either instinctively or from having learned the characteristics of their place of origin. Many passively dispersed organisms, such as the planktonic larvae of sessile marine invertebrates, will not settle unless they perceive certain sensory cues indicating a substrate that is suitable for establishment. Even the seeds and spores of plants and the cysts of invertebrates have some capacity for habitat selection. Although these resistant structures are passively dispersed, certain specific conditions of temperature, moisture, light, and other factors are usually required to break their dormancy and initiate growth and development. Unless cues indicating favorable conditions are received, the diaspore remains in the resistant stage and is capable of further dispersal.

Stanley Wecker (1963, 1964; see also Harris 1952) performed a classic study of habitat selection in deer mice (*Peromyscus maniculatus*). He worked with two subspecies that inhabit the north central and northeastern United States: *Peromyscus maniculatus gracilis*, a form primarily restricted to mature mesic forests, and *P. maniculatus bairdi*, which inhabits grasslands. The latter subspecies is of particular interest because it is an animal that originally inhabited the prairies of central North America, but extended its range eastward by colonizing agricultural cropland and old fields as the forests were cleared by Europeans (Hooper 1942; Baker 1968). In a series of experiments in which he reared mice in different environments and then tested them in a large outdoor enclosure that was half old field and half forest, Wecker showed that each subspecies tended to wander and explore until it found its appropriate habitat, whereupon it would establish residence (Figure 9.19). The mice used genetically inherited information, supplemented by early experience in their natal habitat, to select the correct environment. Because forest and field habitats form a temporally fluctuating patchy mosaic throughout the northeastern United States, this precise habitat selection has probably been important to the successful dispersal of deer mice into these successional habitats.

What Constitutes a Propagule?

In addition to finding the right kind of habitat upon arrival, successful colonists must be able to reproduce and establish a viable new population. The size and composition of the dispersing unit are of critical importance in this regard. MacArthur and Wilson (1967) used the term **propagule** to define the unit necessary to found a new colony.

Colonization is a major problem for the many higher plants and animals that reproduce sexually, because reproduction normally requires the participation of two individuals of different sexes or mating types. Since long-distance dispersal is a rare event, it is relatively unlikely that potential mates will arrive sufficiently close to each other in space and time to find each other and repro-

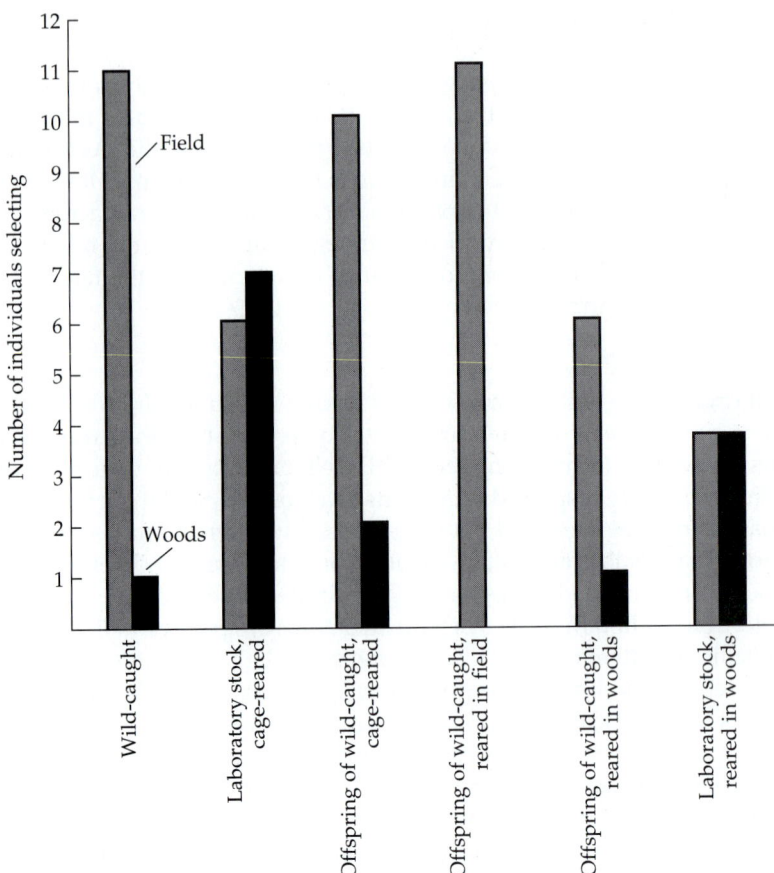

Figure 9.19   Habitat selection by the field (grassland) subspecies of the deer mouse, *Peromyscus maniculatus bairdi*, depends on both inherited ability and early experience. Field-caught individuals and their offspring, especially if they were exposed to field habitat at the time of weaning, spent most of their time in the field end of a large enclosure that covered a field/forest boundary. Many generations in the laboratory or early experience in the woods diminished this preference. (After Wecker 1963.)

duce. In contrast, asexually reproducing organisms are often successful colonists, because a single individual can found an entire population by fission, budding, or some other asexual means (see Browne and MacDonald 1982). Although many microbes and lower plants and animals are capable of sexual reproduction, many of them can also reproduce asexually. Many of the crustaceans that constitute the freshwater zooplankton can go through many generations of asexual reproduction after emerging from encysted zygotes (which are produced by a sexual generation that occurs periodically during the life cycle). Many plants also are capable of asexual reproduction by vegetative growth via rhizomes, stolons, root crowns, or bulbs.

Even among the normally obligately sexual higher plants and animals, there are some species with reproductive patterns that make them much better colonists than others. A large number of plant species are both hermaphroditic and self-compatible; that is, the same individual has both male and female flower parts and can produce viable seeds by self-fertilization. This is true of many species, such as tumbleweed (*Salsola iberica*), that we normally think of as weeds. Dandelion (*Taraxacum*), another common weed, is **apomictic**; it can produce viable seeds from unfertilized ovules. In all of these kinds of weeds a single seed can serve as a propagule. Animals also exhibit a marvelous diversity of reproductive strategies. A few species of fish, amphibians, and reptiles are parthenogenetic: the entire population of these species consists solely of females that produce female offspring asexually (White 1973; Cuellar 1977). Some fishes can also change their sex, depending on the sex of other members of their local population.

Although these reproductive patterns do not necessarily mean that the organisms will be good colonists, they appear to facilitate successful colonization in some species. At least three genera of gecko lizards, for example, contain parthenogenetic species that are distributed primarily on oceanic islands (Cuellar 1977). It takes only a single individual, drifting ashore on a floating log or arriving in an islander's canoe, to establish an entire new population. On the other hand, sexuality, if it is associated with aggressive behavior, may confer a competitive advantage to some invading lizard species over asexual natives (Petren et al. 1993; Case et al. 1994).

Some sexual species have reproductive cycles and social behaviors that facilitate simultaneous colonization by several individuals. In animals with internal fertilization, a propagule can consist of a single female with stored sperm or developing embryos. In plants in which several seeds are normally retained and dispersed in a fruit, the fruit can be the unit of dispersal even in self-incompatible, obligately sexual species. Some birds, which have virtually none of the characteristics listed above, are good dispersers and successful colonists not only because they can fly long distances, but also because they normally travel in flocks, which obviously increases their chances of establishment in new locations. Thus we find that fruit pigeons (*Ducula*), medium-sized, highly social birds that often travel in flocks far out to sea, are widely distributed in the tropical Pacific (Diamond 1975b) and were even more widespread before extinctions caused by aboriginal humans (see Chapters 17 and 18; Steadman 1995).

Survival in a New Habitat

Because long-distance dispersal almost inevitably means colonizing a habitat that is not exactly like the one in which the colonists originated, successful colonists must be able to survive physical stresses and biological hazards to which they may not be adapted. Thus, organisms from highly fluctuating and unpredictable environments probably tend to be good dispersers not only because they can tolerate stressful conditions encountered en route, but also because they are well prepared to meet the unknown physical and biotic challenges they must face after they arrive. The biological hazards posed by competitors and predators (including parasites and diseases) should not be underestimated. Organisms from large, continuous areas with diverse ecological communities, such as productive continental habitats, tend to be relatively successful in establishing populations in small, isolated habitats containing few species, such as oceanic islands. Insular species, on the other hand, are rarely successful in invading mainland habitats, presumably because they are not adapted to cope with the variety of threats posed by a diverse array of competitors and predators.

When small numbers of individuals are involved in colonization, as is almost always the case in long-distance dispersal, chance becomes very important in determining the course of events. Colonies may fail, for example, because a freak storm or an unlikely predator kills just one or two individuals. Thus, despite this incredible diversity of preadaptations for long-distance dispersal, even if a viable propagule arrives in an apparently suitable environment, successful colonization is not assured: in fact, most colonizations fail. Most human-assisted introductions of exotics also fail to establish breeding populations. Therefore, successful long-distance dispersal typically is a very rare phenomenon. Yet we know that even events as rare as dispersal across sweepstakes routes have strongly influenced the distributions of most, if not all, biotas.

CHAPTER *10*

Endemism, Provincialism, and Disjunction

The most pervasive feature of geographic distributions is the fact that they have limits. No species is completely cosmopolitan, and most species and genera, and even many orders and families, are confined to restricted regions, such as a single continent or ocean. Some distributions are even more limited. The minute redfinned blue-eye, the sole species in the fish genus *Scaturiginichthys*, lives in one tiny spring, containing at most a few thousand liters of water, in arid western Queensland, Australia (Ivantsoff et al. 1991). The avian family Opisthocomidae contains only a single species, the weird hoatzin (*Opisthocomus hoazin*), a nearly flightless, leaf-eating bird that occurs only in a small region of northern South America. Even some diverse taxa have very restricted distributions. The avian family Furnariidae (ovenbirds), with more than 50 genera and 200 species, is confined to South and Central America plus several neighboring islands. Several large fish families, including the catfish eels (Plotosidae) and archerfishes (Toxotidae), are confined to the Indo-Pacific region. In South and Central America there are over 50 families and subfamilies of flowering plants that occur nowhere else (Table 10.1); many of these are small taxa, but they also include several extremely diverse families. The cacti (Cactaceae) and bromeliads (Bromeliaceae) would be endemic to the New World if one species of each had not recently crossed the ocean by natural means (long-distance dispersal) and become established in Africa.

Endemics are not distributed randomly, but tend to be concentrated in certain regions. Australia, southern Africa, Madagascar, New Zealand, and New Caledonia, for example, contain a large percentage of endemic species and numerous endemic higher taxa, whereas other regions, such as Europe, northern North America, and the southern Atlantic Ocean, share much of their biotas with other areas. Different groups of plants and animals tend to show similar patterns of endemism, occurring not only in the same ocean or on the same continent or island, but also in the same localities and habitats within those regions, a phenomenon that is called **provincialism**. These coincident distributions of endemics often do not correspond precisely to the present boundaries of continents and oceans, and they certainly do not always coincide with obvious characteristics of abiotic and biotic environments. As pointed out in Chapters 4 and 18, the many successful introductions of species by humans demonstrate that species can thrive in regions far from their native

Table 10.1
Families and subfamilies of angiosperms endemic or nearly so to the Neotropical regions

Aextoxiacaceae	Luzuriagoideae of Liliaceae
Agdestioideae of Phytolaccaceae	Malesherbiaceae
Alstroemerioideae of Liliaceae	Marcgraviaceae
Alvaradoroideae of Simaroubaceae	Mayacaceae[a]
Asteranthoideae of Lecythidaceae	Misodendraceae
Bixaceae	Morkilliodeae of Zygophyllaceae
Brunelliaceae	Neotessmannioideae of Tiliaceae
Calyceraceae	Nolanoideae of Solanaceae
Caryocaraceae	Pakaraimoideae of Dipterocarpaceae
Catopherioideae of Lamiaceae	Pellicieroideae of Theaceae
Columellioideae of Saxifragaceae	Peridiscaceae
Cyclanthoideae of Cyclanthaceae	Phytelphantoideae of Arecaceae
Cyrillaceae	Picramnoideae of Simaroubaceae
Dictyolomatoideae of Rutaceae	Plocospermatoideae of Loganiaceae
Duckeodendroideae of Solanaceae	Quillajeoideae of Rosaceae
Eremolepidaceae	Quiinaceae
Goetzeaceae	Rapateaceae[a]
Gomortegaceae	Rhabdodendroideae of Rutaceae
Goupioideae of Celastraceae	Siparunoideae of Monimiaceae
Gyrocarpoideae of Hernandiaceae (except *Gyrocarpus americanus*)	Stegnospermataceae
Halophytaceae	Styloceratoideae of Buxaceae
Henriquezioideae of Rubiaceae	Tetralicoideae of Tiliaceae
Houmiriaceae[a]	Theophrastoideae of Myrsinaceae
Lactoridaceae (Juan Fernández Island)	Thurnioideae of Juncaceae
Lecythidoideae of Lecythidaceae	Tovarioideae of Capparaceae
Ledocarpaceae	Tropaeolaceae
Lennoaceae	Vellozioideae of Velloziaceae
Leonioideae of Violaceae	Vivianiaceae
Lithophytoideae of Verbenaceae	Vochysiaceae[a]
Lophophytoideae of Balanophoraceae	

[a]Taxa with African taxa.

habitats. These introductions emphasize the unique influences of historical events in determining where organisms occur today. In many cases, the spread of taxa from the regions in which they evolved has been blocked by barriers to dispersal. In other cases, one species of a once widespread group has persisted in a limited area after its representatives in other regions have become extinct.

Disjunct distributions provide a dramatic exception to this general pattern of provincialism. **Disjunctions** are those cases in which two or more closely related taxa occur in widely separated regions but are absent from intervening areas. Disjunct distributions reflect past events: the disjunct forms either dispersed long distances over geographic barriers, were carried to distant sites aboard crustal plates as they drifted apart, or are the surviving remnants of a once widespread taxon. Usually disjuncts are morphologically similar and inhabit similar environments. The evergreen southern beeches (*Nothofagus*), for example, grow in wet, cool temperate forests in South America, New Zealand, New Caledonia, New Guinea, New Britain, and Australia (Figure 10.1). Disjuncts are not always similar, however. The flightless ratite birds (order Struthiformes) have a disjunct Southern Hemisphere distribution similar to that of the southern beeches and other groups that once occurred on Gondwanaland. Most ratites, such as the ostriches of Africa, emus of Australia, and rheas of South America, inhabit open, relatively arid habitats. The cassowaries of New Guinea and tropical Australia, however, occur in tropical rain forests. The ecologically most divergent of the ratites are the strange,

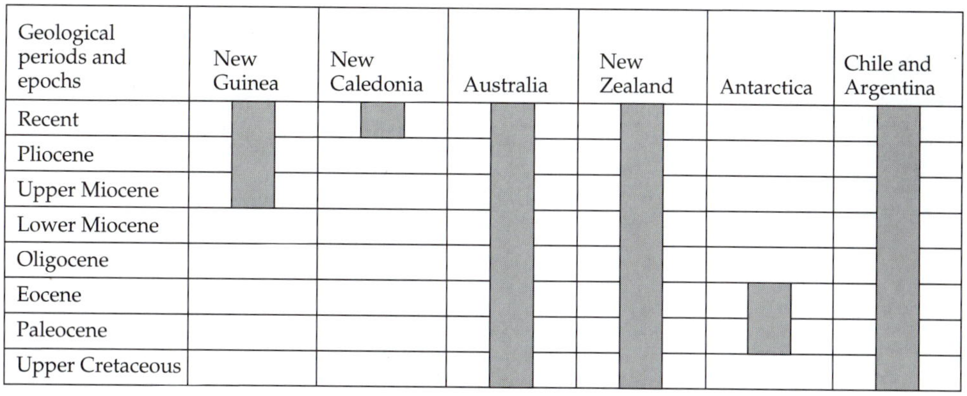

Geological periods and epochs	New Guinea	New Caledonia	Australia	New Zealand	Antarctica	Chile and Argentina
Recent	▓	▓	▓	▓		▓
Pliocene	▓		▓	▓		▓
Upper Miocene	▓		▓	▓		▓
Lower Miocene			▓	▓		▓
Oligocene			▓	▓		▓
Eocene			▓	▓	▓	▓
Paleocene			▓	▓		▓
Upper Cretaceous			▓	▓		▓

Figure 10.1   Recent and fossil distribution of the Southern Hemisphere beeches, *Nothofagus* (Fagaceae). This distribution is both relictual and disjunct, with representatives of the genus presently occurring on many of the widely dispersed fragments of the ancient southern continent of Gondwanaland. Also note that the fossil record of the genus in some of these places extends back to the Mesozoic, and that fossil localities also include Antarctica. As with many other Gondwanan relicts, no representatives of the genus presently occur on Africa, Madagascar, or India. (After Schlinger 1974.)

"mammal-like" kiwis, which live in wet temperate forests of New Zealand, are nocturnal, feed on earthworms, and nest in burrows.

In this chapter we consider general patterns of endemism, provincialism, and disjunction, as well as some of the conceptual and methodological problems in analyzing and interpreting these patterns. Methods for reconstructing the biogeographic histories of endemics and disjunctions using information on phylogenetic relationships and present distributions will be introduced in Chapters 11 and 12.

Endemism

Because the term **endemic** simply means occurring nowhere else, organisms can be endemic to a geographic location on a variety of spatial scales and at different taxonomic levels. Organisms can be endemic to a location for two different reasons: because they originated in that place and never dispersed, or because they now survive in only a small part of their former range.

Taxonomic categories are hierarchical, so the distributions of lower taxa within that of a higher taxon are also organized in a hierarchical fashion. An order, for example, contains a nested set of families, genera, and species that represent the historical branching pattern of a single evolutionary lineage. Similarly, the geographic range of an order contains within its boundaries the ranges of all of its families, genera, and species in a cumulative series. For this reason, the lowest taxonomic categories, species and genera, tend to be more narrowly endemic than the higher taxa, such as families and orders, of which they are members. Figure 10.2 provides an example. The rodent family Heteromyidae, containing the pocket mice and their relatives, is endemic to western North America, Central America, and northernmost South America. Each of the five genera of heteromyids have more restricted ranges; the genus *Microdipodops* (kangaroo mice), for example, occurs only in the Great Basin of the western United States. The genus *Dipodomys* (kangaroo rats) has a much wider distribution, and its species vary greatly in their ranges, from *D. ordii*, which occurs in most of the desert and arid grassland regions of the western

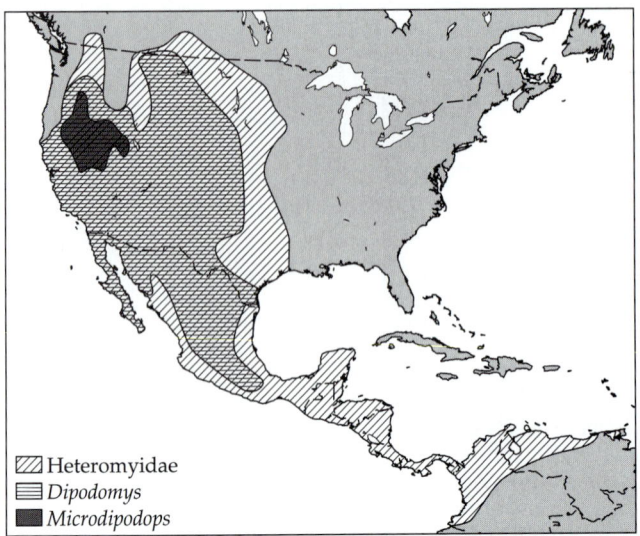
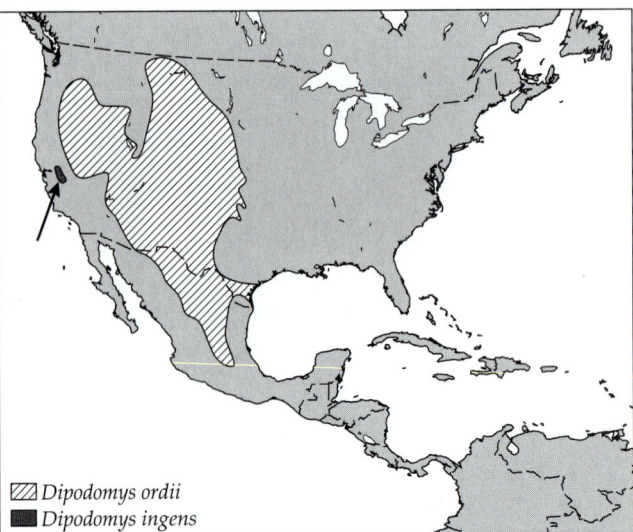

Figure 10.2 Hierarchical patterns of endemism in the rodent family Heteromyidae, which includes the kangaroo rats and pocket mice. The entire family is endemic to the New World, ranging from southern Canada to northern South America. The distributions of its five genera vary in extent, from the kangaroo rats (*Dipodomys*), which range over most of western North America, to the kangaroo mice (*Microdipodops*), which are endemic to the Great Basin. Similarly, the range of just one species, *D. ordii*, encompasses most of the range of its genus, whereas *D. ingens* is endemic to the San Joaquin Valley of California.

United States, to endangered *D. ingens*, which is endemic to an area of a few thousand square kilometers in the San Joaquin Valley of California.

Many species and genera, and even some families and orders, are entirely restricted to tiny islands or equally small patches of habitat. The entire population of the Devil's Hole pupfish (*Cyprinodon diabolis*) numbers fewer than 600 individuals, which are confined to a spring pool measuring 20 by 3 meters in the Mojave Desert of Nevada, just east of Death Valley. Some remarkable plant endemics occur in California, especially on the Channel Islands off the southern coast. The only population of the distinctive shrub *Munzothamnus* (Asteraceae) lives on the tiny island of San Clemente. The entire known population of a species of mountain mahogany, *Cercocarpus traskae* (Rosaceae), consists of four individuals growing in one small canyon on Santa Catalina Island. Also on Santa Catalina, San Clemente, Santa Rosa, and Santa Cruz islands lives the Catalina ironwood, *Lyonothamnus* (Rosaceae), an elegant evergreen tree whose fossils are known from the mainland, including some from Death Valley. Even some flying organisms can have narrow ranges. The todies (Todidae), for example, are a family of tiny birds entirely restricted to a few West Indian islands.

Of course, isolated islands, such as Madagascar and New Zealand, are famous for their endemics. Some of the endemic groups that have radiated on Madagascar are mentioned in Chapter 8. Perhaps the most famous endemic is the tuatara (*Sphenodon punctatus*) of New Zealand, a lizardlike reptile that is the sole surviving representative of the order Rhynchocephalia. Widespread on continents in the Mesozoic, it persists only on a few small islands that have not been reached by introduced rats and cats. Not so lucky was the Stephen Island wren (*Xenicus lyalli*), which occurred only on one small islet in the strait between North and South Islands of New Zealand. The original size of its population is unknown, because the only individuals known to science were "collected" by the lighthouse keeper's cat, which apparently single-handedly exterminated the entire species.

Cosmopolitanism

In contrast to such narrow endemics are **cosmopolitan** taxa, organisms that are widely distributed throughout the world. No species, genus, or family is truly cosmopolitan, although our own species, *Homo sapiens*, comes close. And some plant, animal, and microbe species are now widely distributed because humans have intentionally or inadvertently introduced them throughout the world. Terrestrial organisms that have achieved nearly worldwide distributions by natural means include the peregrine falcon (*Falco peregrinus*), the diverse plant genus *Senecio* (groundsel), and the bat family Vespertilionidae (Figure 10.3). Although they do not occur on Antarctica and on some remote islands, these and other exceptionally widespread taxa are often said to be cosmopolitan. Numerous minute animals and plants, such as protozoans, algae, and fungi, also have extremely broad ranges because their resistant life stages are dispersed widely by water or wind.

Even in the sea, where any organism could, theoretically, swim around the world unimpeded by land barriers, there are relatively few species or genera that are actually found in all the oceans. The most notable exceptions are certain whales and some invertebrates that have been widely dispersed by ships. On the other hand, some kinds of freshwater organisms are surprisingly widely distributed, especially considering that other groups, such as fish and mollusks, contain some of the most narrowly distributed endemics. Most freshwater plant families, and even genera and species, tend to have broad ranges. Some species, including many duckweeds (Lemnaceae), some aquatic ferns (*Azolla, Salvinia, Marsilea*), water milfoil (*Myriophyllum*), and hornwort (*Ceratophyllum*), are nearly cosmopolitan. Similarly, several genera and even species of freshwater zooplankton (e.g., the water flea, *Daphnia*, and other tiny invertebrates, such as rotifers and tardigrades) are distributed worldwide, or nearly so. The key to these wide distributions appears to be dispersal (see Chapter 9). Freshwater plants are dispersed largely by wading and swimming birds, which carry the seeds or plantlets from one pond or lake to another. The

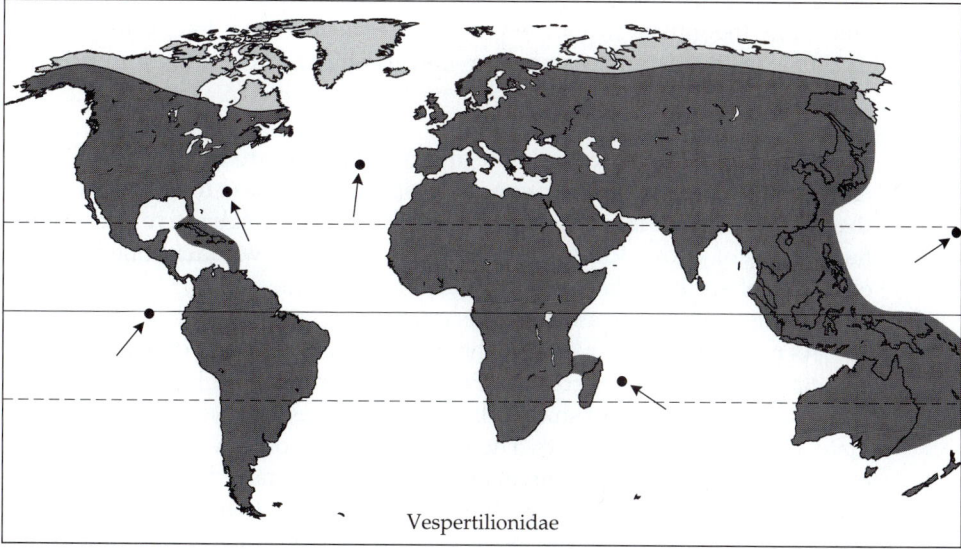

Vespertilionidae

Figure 10.3   The nearly cosmopolitan distribution of the bat family Vespertilionidae. Representatives of this group occur on all of the continents except Antarctica and have colonized isolated archipelagoes such as Hawaii, the Galápagos, and the Azores. (After Koopman and Jones 1970.)

widely distributed freshwater zooplankters have resistant life history stages, often fertilized eggs, that are capable of surviving long periods of desiccation as they are transported by water birds or blown like dust in the wind.

Few species are truly cosmopolitan. However, many families and genera with exceptionally wide distributions contain at least some species that also have very large geographic ranges, indicative of their broad ecological tolerances and their capacities for dispersing long distances with or without human assistance. On the other hand, many high-ranking taxa, such as orders, classes, and families, are essentially cosmopolitan, because the ecological diversity within these groups is broad enough to include forms that can exist in most terrestrial or aquatic habitats, and also because these groups are old enough to have had historical opportunities to colonize most parts of the world.

Classifying Endemics

By place of origin.   The origins of endemics are indicated by a variety of terms. An **autochthonous** endemic is one that differentiated where it is found today, whereas an **allochthonous** endemic originated in a different location from where it currently survives. Prime examples of allochthonous endemics are **relicts** or **epibiotics**, organisms such as the tuatara that were once widespread but are now confined to a very small region.

By taxonomy or geography.   There are two kinds of relicts, taxonomic and biogeographic. **Taxonomic relicts** are the sole survivors of once diverse taxonomic groups, whereas **biogeographic relicts** are the narrowly endemic descendants of once widespread taxa. Often the two categories coincide, especially for organisms called **living fossils.** One example, already mentioned, is the tuatara of New Zealand. Another is the monito (*Dromiciops gliroides*), a primitive marsupial that superficially resembles some opossums and is restricted to southern beech forests of Chile and Argentina. Still another is the coelacanth (*Latimeria*), known only from the deep waters of the tropical Indian Ocean. This "primitive" fish is the only living member of the lobe-finned fishes, the crossopterygians, a group that was widely distributed in freshwater habitats as well as in oceans and shallow epicontinental seas in the Paleozoic, and which gave rise to the amphibians. An example of a plant relict is the ginkgo (*Ginkgo biloba*, Ginkgoales), a gymnosperm native to a small region in eastern China, the sole survivor of a group that was quite diverse in the Mesozoic. Now the ginkgo is a widely distributed ornamental tree, valued for its unusual, aesthetically pleasing form and its ability to tolerate drought, poor soil, and air pollution.

By age.   The terms **paleoendemic** and **neoendemic** are used to identify old and recently formed endemic species, respectively. One can see immediately that the use of such terms requires judgments, usually subjective ones, about the origins of endemics. In the previous paragraph we mentioned several examples of ancient relicts, all of which are paleoendemics.

Also easy to identify are some very recent endemics, those of Quaternary age. As pointed out in Chapter 7, the Pleistocene was a time of great geological, climatic, and biogeographic change. Within just the last 10,000 years, many species ranges have shifted dramatically in response to the warming climate and retreating ice sheets. The ranges of many once widespread species, especially those of cool mountain climates, have contracted, so that now only small, isolated populations are found. An example is the bristlecone pine (*Pinus longaeva*), which is now restricted to arid, rocky microenvironments just below timberline on a few isolated mountains in the Great Basin of California and Nevada, but was widely distributed at lower elevations during the most recent glacial period (Figure 10.4, see also Figure 4.20). The saltbush genus *Atriplex* is widely distributed in western North America, but several endemic polyploid forms occur in distinctive habitats surrounding Great Salt Lake. Since they occur in areas that were covered by the waters of Pleistocene Lake Bonneville, they have almost certainly formed within the last few thousand years (Stutz 1978). On the other hand, many desert-dwelling species of plants and animals have expanded their distributions greatly since the beginning of the Holocene. At the same time, boreal forest and tundra species have reinvaded large areas of northern North America and Eurasia that were previously glaciated. Species restricted to these newly colonized areas, whether their ranges are large or small, are neoendemics.

As pointed out in Chapter 4, the restriction of a taxon to a particular geographic range is a consequence of both historical events and ecological processes. Ecological explanations must be invoked to explain the present range limits of an endemic. The survival and extinction of its local populations affect its ability to persist in particular localities, and dispersal processes affect its ability to colonize (or recolonize after local extinction) favorable but isolated localities (see Chapter 9). Abiotic and biotic limiting factors determine the dis-

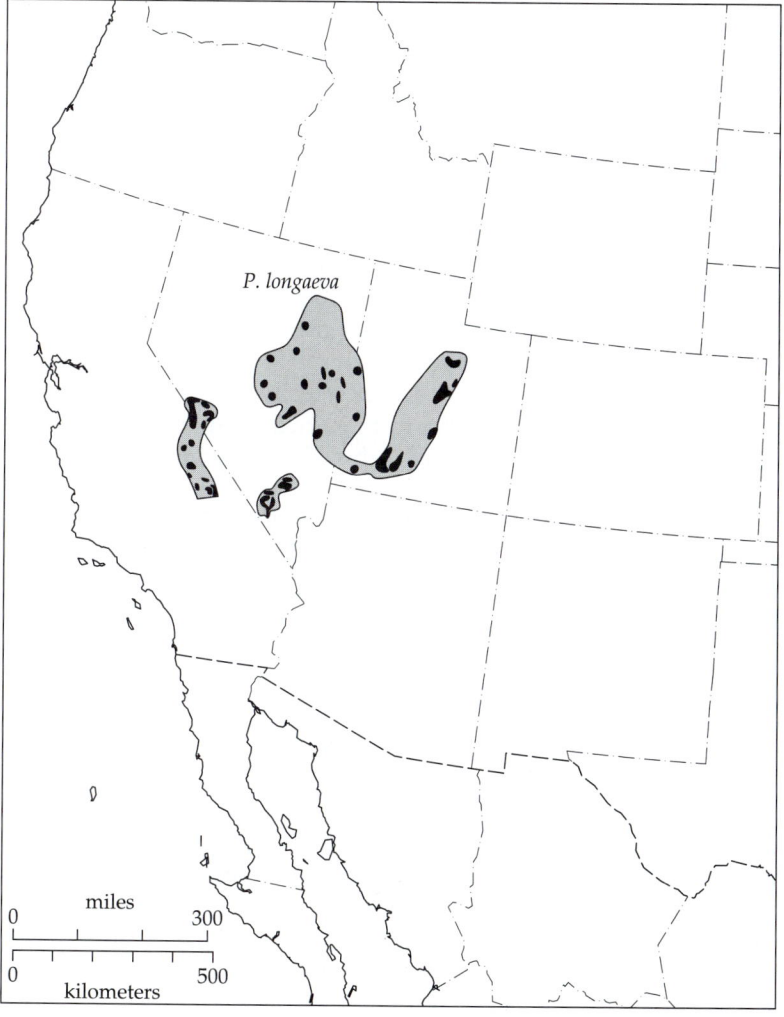

Figure 10.4 Current (black) and late Pleistocene (shaded) distributions of the bristlecone pine, *Pinus longaeva*. This species is now restricted to a few high mountain localities in the Great Basin, but only a little more than 10,000 years ago it occurred at lower elevations and was much more widespread. For a photograph of this species, see Figure 4.20. (After Wells 1983.)

tribution of habitable places within the geographic range and prevent the species from expanding at the periphery of its range (see Chapter 4). Together, dispersal ability and limiting factors determine the nature of the barriers that prevent the species from occurring in favorable but distant areas.

On the other hand, historical events must be invoked to explain how the taxon became confined to its present range and to reconstruct the geographic origin, spread, and contraction of the taxon. As one looks for the influence of historical events, such as the formation of barriers by drifting continents, changing sea levels, and glaciation, it must also be kept in mind that there were other taxon-specific events, such as relatively stochastic long-distance dispersal and extinction of small populations, going on at the same time. Further, geological, climatic, and other environmental changes cause the expansion and contraction of the ranges of many different species, allowing new combinations of organisms to come into contact and to limit each other's distributions through biotic interactions. Thus, while investigators may search for satisfying, simplistic explanations for the origins of endemics, what they often find is a complex picture, explained in part by past earth history and in part by past and present ecological processes—sometimes mixed in ways that are hard to disentangle.

Provincialism

Terrestrial Regions and Provinces

When the ranges of organisms are examined closely, it can be seen that endemic forms are neither randomly nor uniformly distributed across the earth, but instead are clumped in particular regions. Three patterns are apparent. First, the most closely related species tend to have overlapping or adjacent ranges within restricted parts of continents or oceans. Second, completely unrelated higher taxa—for example, certain plant and animal orders and classes—often show similar patterns of endemism. Third, a small but significant number of taxa have markedly disjunct ranges, with species living in widely separated areas on different continents or islands. The first two patterns make it possible for biogeographers to identify circumscribed regions of the earth's surface that share common, taxonomically distinctive biotas. The third pattern encourages us to search for historical explanations for how these organisms came to be distributed among such widely separated regions.

Provincialism was one of the first general features of land plant and animal distributions noted by such famous nineteenth-century biologists as the phytogeographers Schouw (1823) and de Candolle (1855) and the zoogeographers Sclater (1858) and Wallace (1876). As soon as biologists traveled among different continents, they were impressed by the differences in their biotas. The limited distributions of distinctive endemic forms suggested a history of local origin and limited dispersal. As discussed in Chapter 2, much biogeographic research has been devoted to identifying centers of origin, areas where one or more groups of organisms originated and began their initial diversification (e.g., Udvardy 1969; Nelson and Platnick 1981; see also Chapter 12). There have been complementary efforts to find evidence of historical barriers that blocked the exchange of organisms between adjacent regions, and of historical corridors that allowed dispersal between currently isolated regions. The result has been a division of the earth into a hierarchy of regions reflecting patterns of faunal and floral similarities. In order of decreasing size, the common subdivisions are usually referred to as **realms** or **regions**, **subregions**, **provinces**, and **districts**.

The largest units are the most general. As mentioned in Chapter 2, the division of the earth into biogeographic regions recognizes that these areas contain endemic and closely related taxa in many different groups. These large units have often been divided into hierarchically nested subunits on the basis of distinctive endemic biota. A classic example of a biogeographic region is Australia, whose terrestrial fauna is typified by marsupials (kangaroos, koalas, wombats, possums, etc.) rather than placental mammals, and whose flora contains unique eucalypts (Myrtaceae) and proteads (Proteaceae). Within the continent, biogeographers have classically recognized two major subregions. One, the Eyrean Subregion, named for the giant Lake Eyre Basin, encompasses the entire central two-thirds of the continent. It is a vast arid and semiarid area without major mountain ranges or other internal barriers. The other subregion comprises the wetter fringe of Australia, and is further subdivided into three parts (Figure 10.5). The northern portion, called the Torresian Province, is a warm tropical belt that contains animals and plants with close affinities to those in New Guinea and sometimes also Southeast Asia. Their isolation is relatively recent, often only about 10,000 years old. During glacial episodes of the Pleistocene, when low sea levels created a landbridge across the present Torres Strait, many of these organisms ranged continuously from Australia to New Guinea. The southeastern portion of Australia, including Victoria, Tasmania, and some of the surrounding islands, is called the Bassian Province, and is inhabited by animals and plants adapted to cool, mesic temperate climates. This is the location of the southern beech forests, which contain *Nothofagus* and other relicts dating back to the ancient southern continent of Gondwanaland. The other province, called Westralia, includes the southwestern corner of the continent, where great numbers of endemic forms reside. Many of the taxa shared between the Mediterranean climatic regions of Australia and South Africa, such as certain groups in the plant family Proteaceae, occur in Westralia.

The Australian Region is one of the six large units used by Sclater to describe the world distributions of bird families and genera (see Figure 2.6). These were essentially the same six zoogeographic regions later adopted by Wallace in his major treatise on zoogeography. In the Northern Hemisphere, landmasses north of the tropical zone are called the **Holarctic**, which is composed of the **Nearctic** (North America) and **Palearctic** regions (Eurasia and northernmost Africa). The remaining regions are primarily tropical: the **Neotropical** (Central and South America and the West Indies), the **Ethiopian** (Africa south of the Sahara, and Madagascar), and the **Oriental** (Southeast Asia and the adjacent continental islands). The faunas of oceanic islands in the Pacific basin are anomalous in this classification scheme because they contain a small number of taxa from adjacent continents and have relatively few unique groups.

The majority of articles and books on the distributions of organisms now use these six regions as fundamental descriptors of animal distributions, and this classification scheme has become one of the primary empirical foundations of biogeography. This is a tribute to the early zoogeographers who created it, who were working with woefully incomplete information on distributions and using classifications that did not accurately reflect phylogenetic relationships. It is also a testimony to the clarity and generality of the distributional patterns of terrestrial organisms.

The Sclater-Wallace classification scheme is basically similar to those used for plants, but there are a number of significant differences (Figure 10.6). Disregarding for the moment the fact that phytogeographers call the provinces of vertebrates regions, and the districts provinces, there are other important dif-

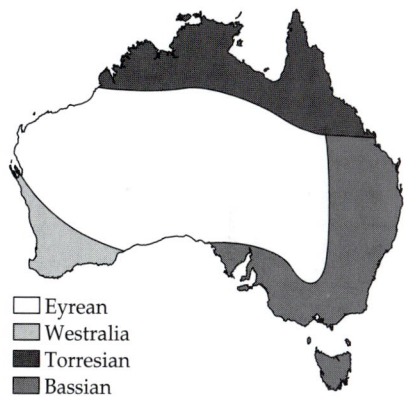

Eyrean
Westralia
Torresian
Bassian

Figure 10.5 Biogeographic provinces of Australia. Australia is typically divided into two subregions: the Eyrean Subregion, which includes the entire arid center of the continent, and the strip of moister, mostly forested environments around the northern, eastern, and southwestern periphery. The latter is divided into three provinces, designated the Torresian, the Bassian, and Westralia, respectively. All three provinces are rich in endemics and in groups with representatives on other southern continents. The Bassian province shares southern beeches (*Nothofagus*) with New Zealand, New Caledonia, and South America, while Westralia shares several disjunct plant groups with the Mediterranean climatic region of South Africa.

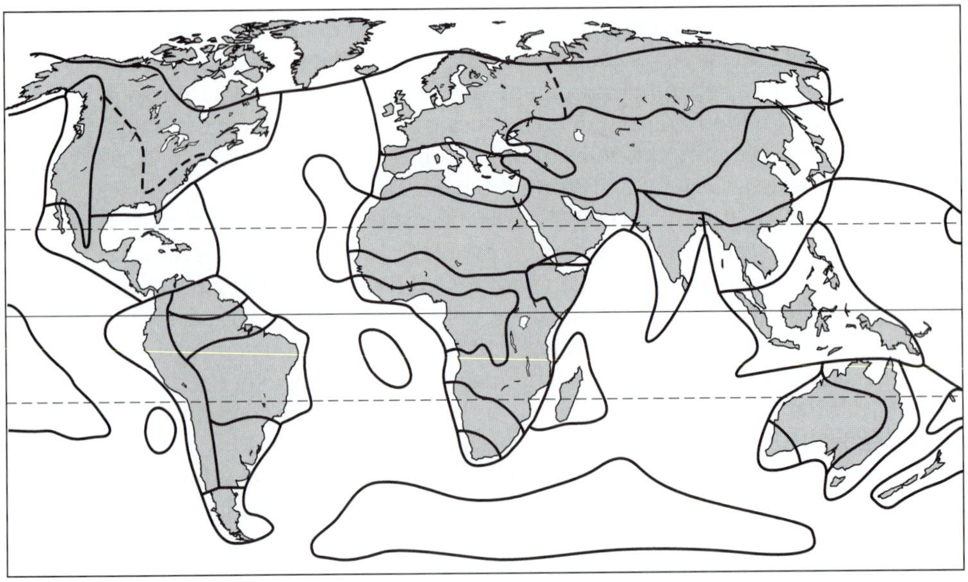

Figure 10.6 Division of the world into biogeographic regions and provinces based on the distributions of land plants. While the major regions have been subdivided into provinces, comparison with Figure 2.7 shows the close correspondence between the divisions used for plants and for animals. (After Good 1974.)

ferences that reflect the distinctive characteristics of plants. Vegetation types are tightly restricted by abiotic factors, such as temperature and rainfall. Consequently, regions of endemism are more clearly defined by climatic and other physical barriers for plants than for animals, which are often able to surmount climatic barriers by dispersal or by physiological and behavioral adaptations to tolerate stressful conditions. Hence the tip of South Africa, which has a Mediterranean climate, is a very distinctive phytogeographic region. This small area has an exceedingly rich flora, and about 90% of its species are endemic (Dyer 1975; Goldblatt 1978). In contrast, most animal groups of southernmost South Africa are not restricted to that zone and are not spectacularly diverse (Werger and van Bruggen 1978). South America has been divided into numerous provinces based primarily on the distributions of plant taxa. As elsewhere, these provinces are characterized by particular combinations of temperature, precipitation, and soil. However, Cabrera and Willink (1973), who defined the provinces shown in Figure 10.7, pointed out that distinctive animal species also are largely restricted to or exceptionally abundant in these regions.

As implied above, the distinctive floras and faunas used to characterize regions and provinces have been shaped by both ecological and historical factors. Continental taxa have been stranded on certain landmasses both because their ranges have been limited by abiotic and biotic factors (see Chapter 4) and because major barriers to dispersal have restricted them to those landmasses for millions of years (see Chapters 6 and 12). To avoid complicating our general discussion here, we shall illustrate patterns of endemism in two groups, land birds and angiosperm plants, in two regions, Australia and South America, in Box 10.1. Despite the enormous differences in lifestyle and dispersal ability between birds and plants, it is readily apparent that these historically isolated southern continents are centers of endemism.

By itself, however, the observation that these regions contain endemic taxa is of limited interest. Even randomly selected areas would be expected to contain endemics, given the fact that most taxa have limited distributions. For example, Beadle (1981) has observed that Australia has about one-fifth as many endemic plant families as South America, but that it is also about one-fifth the size. The recognition of formal biogeographic provinces implies not

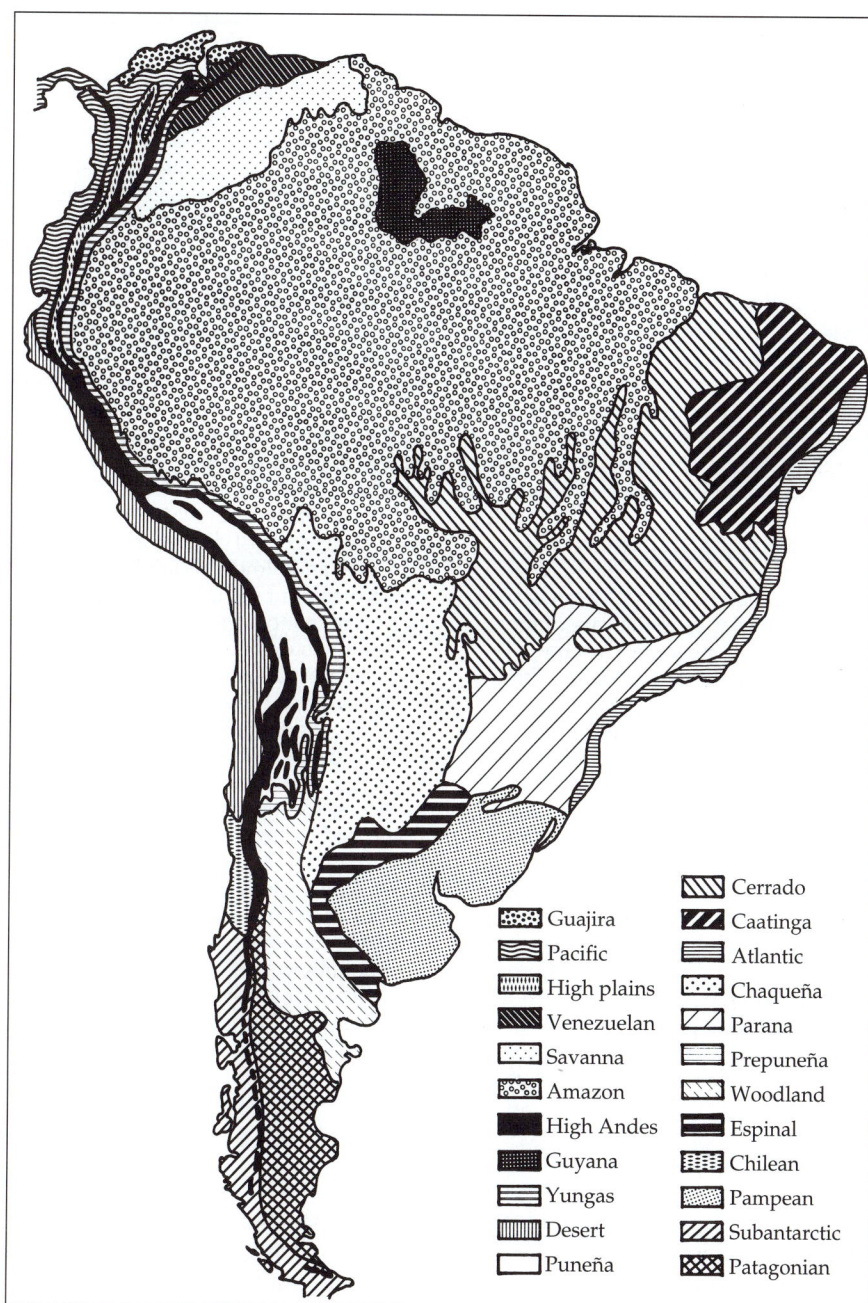

Figure 10.7 Division of South America into biogeographic provinces based primarily on the distributions of land plants. Note that the configuration of the provinces corresponds closely to geological and climatic features, such as the Andes, showing the influence of climate and soil on the distributions of plant groups and vegetation types. (After Cabrera and Willink 1973.)

Legend:

- Guajira
- Pacific
- High plains
- Venezuelan
- Savanna
- Amazon
- High Andes
- Guyana
- Yungas
- Desert
- Puneña
- Cerrado
- Caatinga
- Atlantic
- Chaqueña
- Parana
- Prepuneña
- Woodland
- Espinal
- Chilean
- Pampean
- Subantarctic
- Patagonian

only that each province contains distinctive endemic taxa, but also that the biota within each province is more homogeneous than between adjacent areas. While this is what would be expected if the provinces have long been isolated by some combination of unique ecological conditions and barriers to dispersal, most biogeographic units have not been quantitatively analyzed to show that they do indeed contain relatively homogeneous assemblages. Much quantitative characterization of patterns of endemism still remains to be done. Nevertheless, at least three lines of evidence suggest that many of the biogeographic provinces do indeed indicate the pervasive influence of geography, geology, and climate on the historical origin, diversification, and spread of lineages—and that the legacy of this history is preserved in the distributions of contemporary taxa.

Box 10.1
Endemic birds and plants of South America and Australia

South America and Australia share some important features of earth history and biogeography. Both were parts of the giant southern-hemisphere continent of Gondwanaland that drifted apart in the Mesozoic, and both were completely isolated island continents for most of the Tertiary. By the late Tertiary, however, continental drift had carried the northern, tropical parts of both South America and Australia into close proximity to large northern continents that were once part of the giant continent of Laurasia. South America joined with North America at the very narrow Central American isthmus about 3.5 million years B.P. Australia and New Guinea, while part of the same plate and joined to each other by land connections for most of their history, drifted close to southeastern Asia (especially when the Sunda Shelf was exposed in the Pleistocene), but never established land connections with it.

Both South America and Australia contain many endemic taxa. Here we use patterns of endemism in angiosperm plants and birds to illustrate some features of their biogeography that reflect the long histories of isolation of these continents. Much attention has been given to the few groups, such as southern beeches (*Nothofagus*) and giant flightless birds (ratites), that are shared between the two continents and are undoubtedly Gondwanan relics. Many

of the endemic groups, however, are greatly differentiated, and their nearest presumed relatives do not necessarily occur on other southern continents. At the level of families and subfamilies, approximately 60 plant taxa are endemic, or nearly so, to South and Central America, and more than 40 are endemic or nearly so to Australia and New Guinea. The "nearly endemic" taxa have one or a few close relatives that occur in a different biogeographic region, apparently as a result of fairly recent long-distance dispersal: to Africa in the case of the South American Houmiriaceae, Mayacaceae, Rapataceae, and Vochysiaceae and to Melanesian or Polynesian islands in the case of the Australian Balonpaceae, Corynocarpaceae, Davidsoniaceae, and Trimeniaceae.

Despite large differences in their capacities for dispersal and other traits, similar patterns of endemism occur in birds. About 25 families and subfamilies containing a total of about 700 species are endemic, or nearly so, to South and Central America. This compares with approximately 30 families and subfamilies containing more than 500 species that are endemic, or nearly so, to Australasia. While no South American lineages of birds apparently have, like plants, recently dispersed across the Atlantic to Africa, several groups of Aus-

tralian birds, including the Loriidae (lorikeets), Pachycephalinae (whistlers), Meliphagidae (honeyeaters), and Artamidae (wood swallows), have colonized Pacific islands.

Another pattern of endemism seen in both plants and birds is exhibited by a large number of additional taxa that are New World endemics. Many of the families and subfamilies included in the above analysis are not restricted to South America, but have spread northward to colonize at least some of the Central American isthmus. In addition, many more taxa are most diverse in South America, and almost certainly originated there, but include several species or genera that have spread varying distances into subtropical and even temperate North America. These include the plants (such as those whose ranges are shown in Figure 10.19) that are arid habitat disjuncts and the main groups of birds (e.g., flycatchers, vireos, wood warblers, and tanagers; see Chapter 16) that breed in temperate North America but migrate to winter in the tropics of Central and South America. Thus the South American and Australian plants and birds illustrate the point that often (but not always: see text and Chapter 12) very different groups of organisms show strikingly similar patterns of endemism.

Perhaps the most convincing evidence is the correspondence between the hierarchy of phylogenetic and taxonomic categories and the hierarchy of biogeographic regions, with the clearest patterns occurring at the largest scales. Thus, higher taxa (orders and families) tend to have fairly broad distributions within a continental landmass, presumably reflecting the ancient origin and long confinement of major lineages, while progressively less differentiated forms (genera and species) tend to be confined to smaller areas within those regions. As pointed out earlier, nested patterns are required by the hierarchical classifications of both organisms and biogeographic regions, so this apparent pattern could be an artifact. Nevertheless, the relationships between the two hierarchies suggest that the progressively finer biogeographic subdivisions reflect a history of less ancient and severe barriers isolating more recently differentiated lineages (more on this in Chapter 12).

The second line of evidence that biogeographic subdivisions reflect the long-standing influence of geological events and climatic patterns is that the locations of provinces and their boundaries, determined independently for different groups of organisms, tend to coincide. We have already described, above and in Chapter 2, how the division of the earth into the six major biogeographic regions was originally developed for birds by Sclater, then generalized to terrestrial animals by Wallace, and finally extended to land plants and, by implication, all terrestrial organisms. The same kind of generality across different taxa often applies to the finer divisions as well. The North American deserts, for example, have long been divided into four provinces: the Great Basin, Mojave, Sonoran, and Chihuahuan (Figure 10.8). This division reflects the distributions of many endemic plants and animals, which are typically restricted to only one or a subset of the four deserts but are fairly wide-ranging within the desert(s) where they occur. These patterns of endemism can in turn be attributed to the unique histories and environments of the deserts. On the one hand, during the glacial periods of the Pleistocene, the climate of southwestern North America was cooler and wetter than at present, and the deserts contracted to lowland basins isolated by barriers of grassland and woodland at higher elevations. On the other hand, even during the interglacial periods, such as the present one, when these barriers were removed, the distinctive precipitation and temperature regime of each desert limited distributions and prevented wholesale mixing of the desert biotas. The result is that each desert has distinctive endemic plants and animals. For example, the saguaro, *Carnegia gigantea*, and four other species of large columnar cacti, the Sonoran Desert toad (*Bufo olivarius*), the regal horned lizard (*Phrynosoma solare*), Bendire's thrasher (*Toxostoma bendirei*), and Bailey's pocket mouse (*Chaetodipus baileyi*) are but a few of the many plant and animal species that are endemic to the Sonoran Desert of southern Arizona and northwestern

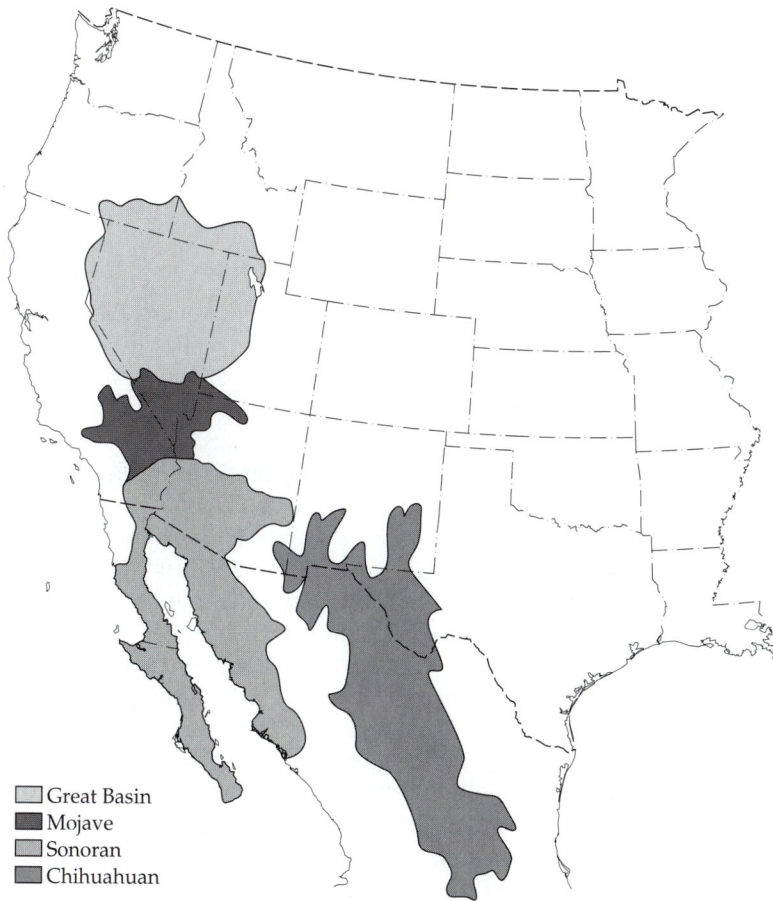

□ Great Basin
■ Mojave
□ Sonoran
▨ Chihuahuan

Figure 10.8   The division of arid North America into four desert provinces: the Chihuahuan, Sonoran, Mojave, and Great Basin. The first three are relatively hot, low-elevation deserts, distinguished primarily by an east-west gradient of seasonality of precipitation: rain falls in summer in the Chihuahuan, in both summer and winter in the Sonoran, and in winter in the Mojave. The Great Basin, due to its higher latitude and elevation, is a cold desert or shrub-steppe. Each province has distinctive endemic plants and animals.

Mexico. Such congruence in patterns of endemism across different lineages of organisms suggests that they have responded similarly to historical geological and climatic events and to the geographic variation in important geological, topographic, and climatic features of the earth.

Freshwater organisms include forms with both some of the most cosmopolitan and some of the most narrowly endemic distributions. The former, as mentioned above, tend to have life history stages that can withstand desiccation and are readily dispersed. The latter tend to lack such resistant stages and to be equally intolerant of exposure to air and to seawater. As a consequence of their limited dispersal abilities, some freshwater groups, such as fishes and mollusks, show clear patterns of endemism that correspond to the Sclater-Wallace biogeographic scheme. Furthermore, these same taxa typically show pronounced patterns of provincialism within continents as well. Each major river drainage or lake basin has distinctive, well-differentiated endemic forms (e.g., for fishes, see Figure 10.9; Hocutt and Wiley 1986; Mayden 1992). Thus, each of the world's great river systems, such as the Amazon, Congo, Volga, and Mississippi, has a diverse fish fauna that contains many endemic species and genera. And such endemism is found even at much smaller spatial scales. Each of the great lakes of the Rift Valley of Africa, such as Victoria, Tanganyika, and Malawi, has been a major center of endemic speciation, and supports unique lineages containing many closely related species of both fish and mollusks. These same groups are spectacularly diverse not only in the Mississippi River itself, but also in many of its tributaries. Especially in the Ozark and Appalachian highlands, even the smaller drainages contain endemic forms of darters (*Etheostoma*), shiners (*Notropis*), and clams. Giant Lake Baikal in Siberia contains not only endemic fishes and invertebrates, but also a unique seal (*Phoca sibirica*) that colonized sometime during the Pleistocene when the now landlocked lake drained into the Arctic Ocean. Because of the inability of most freshwater fishes and certain invertebrates to disperse across either land or sea, their distributions probably reflect historical events more faithfully than do those of most other organisms. These groups offer fruitful systems for research on speciation and diversification, as well as on many aspects of historical biogeography.

Biogeographic Lines

Another observation that reflects the pervasiveness of provincialism is the rapid turnover of taxa at the boundaries between regions. Traditionally, biogeographers have drawn lines to define fairly precisely the limits of regional biotas. Although these biogeographic lines are usually derived for one taxon at a time, they often coincide with geological or climatic barriers that have prevented the dispersal of many kinds of organisms. A prime example of such a sharp boundary is Wallace's line, drawn between the Indonesian islands of Borneo and Sulawesi (Celebes) and Bali and Lombok to mark the boundary between the Oriental and Australian Regions. Not all taxa show distributional boundaries corresponding precisely to Wallace's line, and other lines have been described to accommodate them (Figure 10.10). Nevertheless, two things about Wallace's line are noteworthy. First, most of the other lines deviate only slightly from the boundary described by Wallace. Second, Wallace's line corresponds almost exactly with the outer limit of the Sunda Shelf, the part of the continental shelf of Southeast Asia that was intermittently exposed by lowered sea levels during the Pleistocene. Thus, although Wallace did not realize this, his line corresponds to the deep water marking the limit of historical land connections among the major East Indian islands and between them and the Southeast Asian mainland. Lydekker's line (Figure 10.10), just west of New Guinea, cor-

Figure 10.9 Biogeographic provinces for North American freshwater fishes. Note that most of these provinces correspond to major drainage basins, such as the Great Lakes, Mississippi, and Colorado. The two numbers given are the numbers of families and species, respectively, and the values in parentheses are the percentages of the species that are endemic to each province. (After Burr and Mayden 1992.)

responds to the edge of the Sahul Shelf, which was exposed in the Pleistocene as part of the supercontinent that included both Australia and New Guinea. It marks the northwestern limit of the ranges of many species of the Australian Region. In between these two lines are islands surrounded by deep water that were not connected by Pleistocene landbridges to either Southeast Asia or Australia-New Guinea, and which contain a mixture of species from both regions that have managed to disperse across the water barriers.

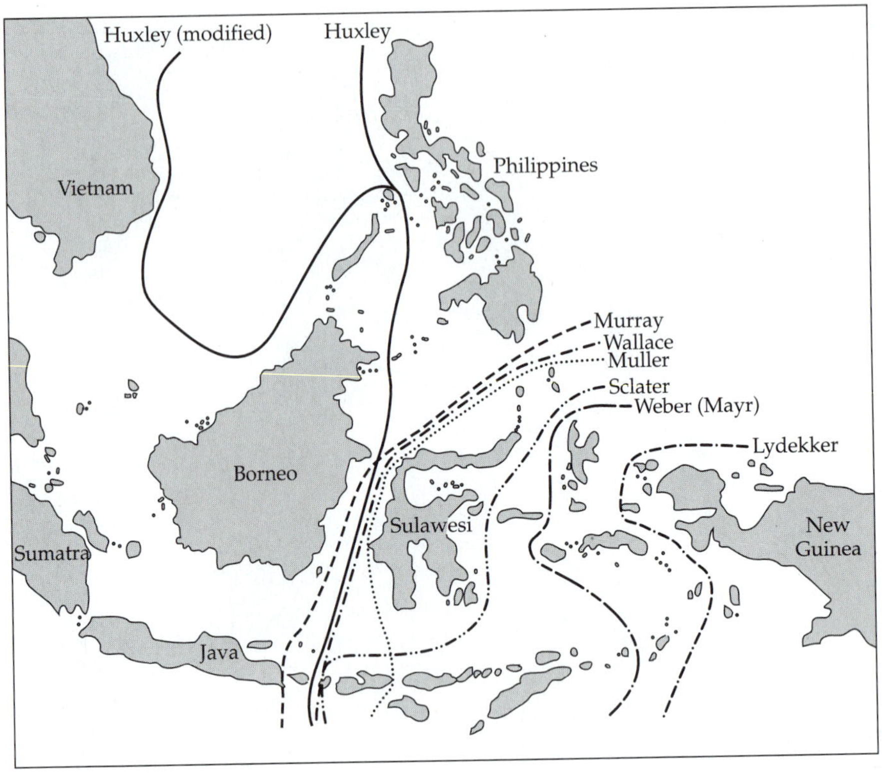

Figure 10.10 The biogeographic lines drawn by Wallace and later biogeographers to mark the boundary between the Oriental and Australian regions. The locations of the various lines indicate that different taxa have dispersed to different degrees into the islands of the East Indies from their continents of origin. The approximate limits of the continental shelves are given by Lydekker's line in the east and Huxley's modified line in the west, although the Philippines have a complex tectonic history (see Figure 10.13).

Boundaries between other biogeographic regions are not as distinct as Wallace's line. Such boundaries usually reflect a combination of two interrelated phenomena. First, the historical geological and climatic isolation of the regions was often not so discrete, causing the apparent boundary to be blurred. Second, different groups of organisms responded somewhat differently to the same historical events and climatic patterns, resulting in a lack of consistency across taxa. These two factors have conspired to make it difficult to draw a single line defining the boundary between the Nearctic and Neotropical regions. The effort to draw this boundary illustrates the influence of both historical and ecological factors on the phenomenon of provincialism.

The Isthmus of Panama is the location of the historical separation between the two regions. It marks where the last marine barrier separating the formerly isolated continents of North and South America closed about 3.5 million years ago. When the Central American landbridge became available, many organisms took advantage of ecological opportunities and dispersed along it (see Chapter 16). Because the climate and habitats of Central America and southern Mexico are tropical, there was much biotic exchange between this area and tropical regions in northern South America. Thus, even though Central America and southern Mexico have many endemic forms, they are usually recognized as a subregion of the Neotropics. But the problem of where to draw the line remains. Often it is placed at the Isthmus of Tehuantepec, which marks both the approximate northernmost extension of tropical rain forest and a major lowland gap in the mountain cordilleras running the length of western North America. By no means all Neotropical taxa have their northern limit there, however. Figures 10.11 and 10.12, which plot the northernmost and southernmost limits of the ranges of terrestrial mammals and freshwater fishes, respectively, illustrate the difficulty of defining the line separating the Nearctic and Neotropical regions. Each family has colonized southward or

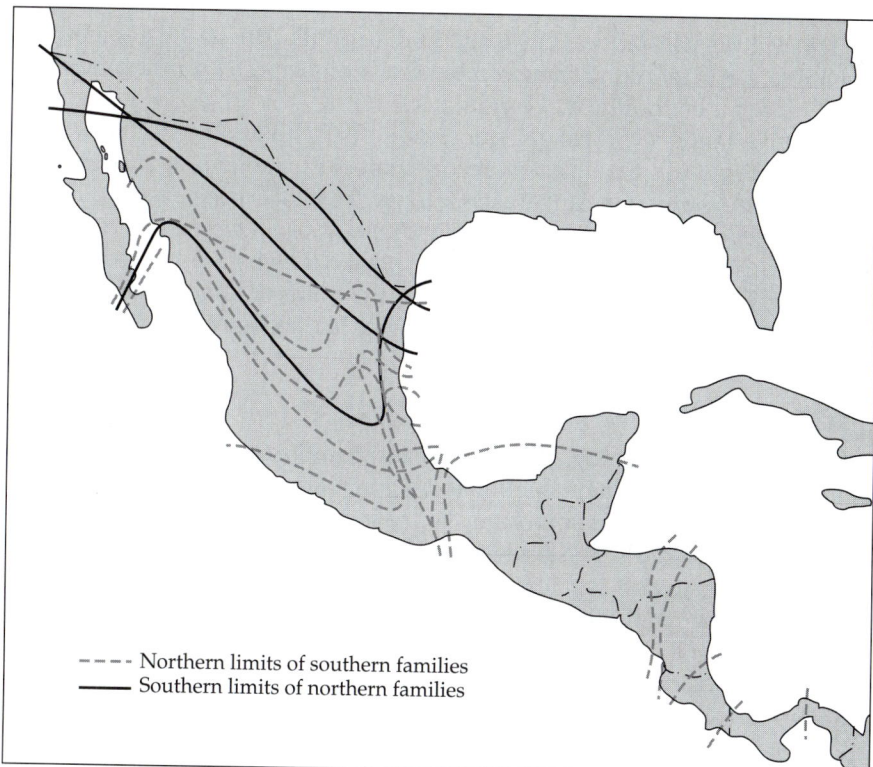

Figure 10.11 Northern limits of the ranges of Neotropical mammal families (dashed lines) and southern limits of the ranges of Nearctic families (solid lines). Note that the transition between these historically isolated mammal faunas occurs over a wide area, making it difficult to draw a definitive line separating the two biogeographic regions. Similarly complex patterns occur in nearly all taxa (e.g., see Figure 10.12 for fishes).

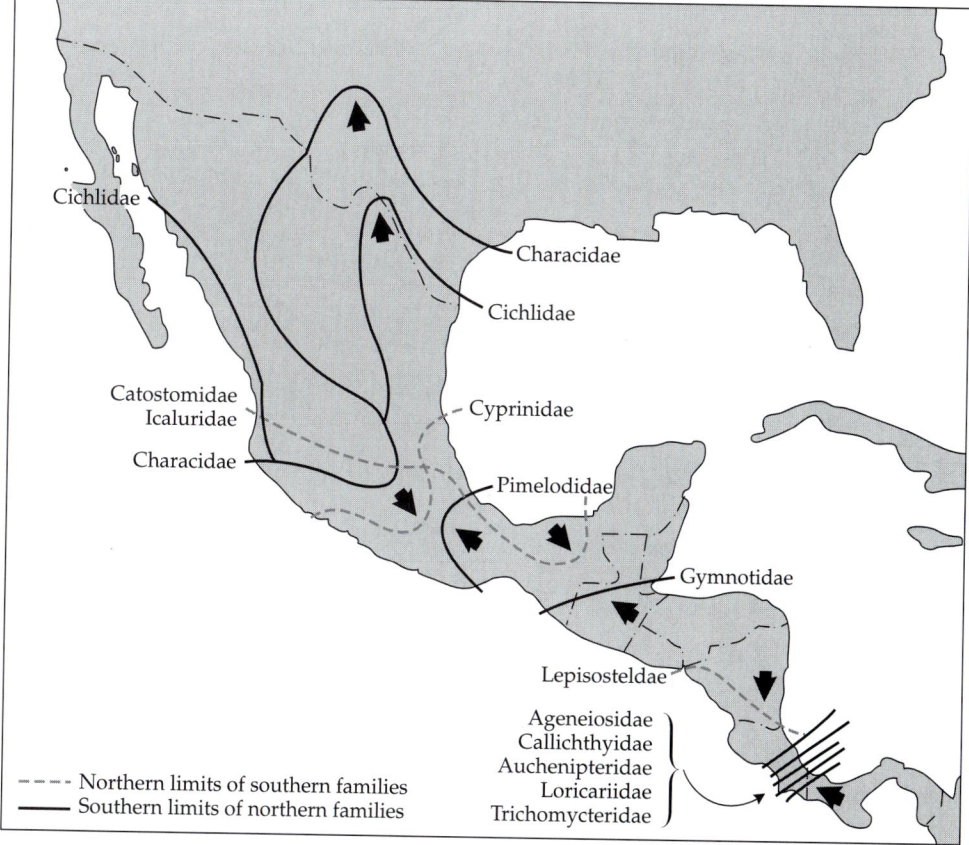

Figure 10.12 Distributional limits of freshwater fish families of South American (dashed lines) and North American (solid lines) origin. Only two species of obligately freshwater fishes of South American origin have reached the United States, and North American forms extend no farther south than Costa Rica. Note that, as in the case of the mammals shown in Figure 10.11, no single line can be drawn to separate unambiguously the Neotropical and Nearctic faunas. (After Miller 1966.)

northward from its historical origin in North or South America in a somewhat different pattern, reflecting some combination of its capacity for dispersal and its tolerance for environmental conditions.

While the tendency of range boundaries of species and higher taxa to be coincident facilitates the objective recognition of biogeographic provinces, a more interesting question can be asked: To what extent are the distributions of taxa a reflection of the history of the landmasses or water masses in which these organisms now live, and to what extent can they be used to reconstruct the history of geographic changes on the earth's surface? We will consider the relationships between phylogenetic, biogeographic, and earth history in much more detail in Chapter 12.

Classifying Islands

Since Wallace (1876, 1880), biologists have attempted to classify islands as either continental or oceanic (see Table 10.2). Initially, such classifications were made on the basis of the composition of the insular biotas. As we shall see in Chapter 13, both kinds of islands typically have substantially fewer species than are found in comparable habitats on nearby continents. **Continental islands**, as their name implies, support plants and animals that are closely related to forms on the nearby mainland; these forms are often so little differentiated as to be considered populations of the same species. Further, continental islands usually have kinds of organisms that are poor over-water dis-

---

Table 10.2
Some biogeographically interesting islands classified according to their modes of origin

FULLY OCEANIC ISLANDS

Totally volcanic islands of fairly recent origin that have emerged from the ocean floor and have never been connected to any continent by a landbridge.

Midoceanic island chains or clusters formed from hot spots (HS) or along fracture zones (FZ) within an oceanic plate

Austral-Cook Island chain (HS or FZ); Carolines (HS); Clipperton Island (FZ); Galápagos Islands (FZ, Carnegie Ridge); Hawaiian Islands (HS); Kodiac Bowie Island chain (HS); Marquesas (HS); Society-Phoenix Island chain (HS, but some contribution by the Tonga Trench)

Island arcs formed in association with trenches

Aleutians (may have been part of the Bering Landbridge in the Cenozoic); Lesser Antilles; Lesser Sunda Islands; Marianas; New Hebrides; Ryukyus (may have been associated with neighboring islands); Solomons; Tonga and Kermadec

Islands formed at presently spreading midoceanic ridges

Ascension Island; Azores (some islands have continental rocks); Faeroes; Gough Island; Tristan da Cunha

CONTINENTAL ISLANDS

Formed as part of a continent and subsequently separated from the landmass. Some of these have added oceanic material since they were formed.

Islands permanently separated from the mainland since the split was initiated (time of final separation in parenthesis)

Greater Antilles (80 million years B.P.); Kerguelen Island (Upper Cretaceous); Madagascar (ca. 100 million years B.P.); New Caledonia (ca. 50 million years B.P.); New Zealand (80–90 million years B.P.); Seychelles (65 million years B.P.); South Georgia (45 million years B.P.)

Island groups with connections of some islands, but not others, to the mainland

Canary Islands

Islands most recently connected to some mainland in the Pleistocene by land or an ice sheet

British Isles; Ceylon; Falklands; Greater Sunda Islands; Japan and Sakhalin; Newfoundland and Greenland; New Guinea (with Australia); Taiwan; Tasmania

persers, such as large-seeded plants, amphibians, and nonvolant terrestrial mammals. Thus, they tend to have representative samples of the floras and faunas of nearby continents. These two features suggest recent land connections between continental islands and the mainland.

In contrast, **oceanic islands** typically have biotas of lower taxonomic richness, not only at the species level but at higher taxonomic levels as well. Often these forms are insular endemics that are well differentiated from their apparent relatives on nearby continents. Often, an oceanic island, or an archipelago of many such islands, supports a group of species obviously more closely related to one another than to any continental form. This suggests that not only evolutionary divergence, but also speciation, has occurred within the island or archipelago. Further, the limited taxonomic richness of oceanic islands is strongly biased in favor of kinds of organisms that are relatively good over-water dispersers, such as plants with small wind- or bird-dispersed seeds and flying animals. Some endemic island forms have become so differentiated that they have lost their structures and capacities for dispersal (e.g., parachutes, hooks, and awns have been lost from seeds; reduction and loss of wings have occurred in insects and birds; see Chapter 14 and Carlquist 1965, 1974). Their phylogenetic affinities, however, suggest that these organisms lost their dispersal abilities secondarily, after their ancestors had colonized the islands. Together, these characteristics of their biotas suggest that oceanic islands have had a long history of isolation, perhaps without any previous land connections to continents.

From Wallace's time until the 1960s, this division between continental and oceanic islands, based on their biotic composition, appeared to work well. It was presumed to reflect geological history in a straightforward way: continental islands, located on the continental shelves, were presumed to have had recent land connections to the mainland during periods of lowered sea levels, whereas oceanic islands, located in deeper waters, were presumed never to have been connected to the mainland, but to have pushed been up from the sea floor, usually as a result of volcanism. As knowledge of plate tectonics and insular geology accumulated, however, this clear dichotomy between continental and oceanic islands was called into question. Some islands, isolated from all continents by oceans hundreds of kilometers wide and hundreds of fathoms deep and long thought to be oceanic, were found to contain continental (sialic or andesitic) rocks. And as tectonic plate movements were reconstructed, it was realized that these islands had once been part of continental plates. Madagascar, New Zealand, and New Caledonia, for example, were recognized as fragments of the ancient supercontinent of Gondwanaland, and the islands of the Greater Antilles were discovered to be the remnants of the ancient Caribbean Plate.

In the wake of this accumulating geological evidence for ancient continental connections, the biological and biogeographic evidence of isolation was reexamined for these islands (as was the case for continents; see Chapter 6). Many elements of their floras and faunas may indeed be descendants of over-water colonists (e.g., Lack 1976). Often, however, their biotas contain telltale elements, ignored or misinterpreted by earlier biogeographers, that are clearly relictual survivors of ancient continental floras and faunas. Examples include the southern beeches (*Nothofagus*) and freshwater galaxioid fishes of New Zealand and New Caledonia, and the flightless ratite birds, the elephant birds of Madagascar and the moas and kiwis of New Zealand, all now recognized as remnants of the Gondwanan biota.

There are also examples of the converse situation for presumably continental islands. On the one hand, increasing information on bathymetry has permitted much more accurate estimation of the timing and extent of past land connections. Of particular importance is the depth of the water between island and continent in relation to Pleistocene fluctuations in sea levels. The 100 m depth contour is now known to mark the approximate sea level during the most recent glacial period (see Chapter 7). Islands isolated by shallower depths can usually be assumed to have had their continental land connections severed within the last 10,000 years. This assumption is generally consistent with the biogeographic evidence, such as the occurrence of a diverse and only moderately differentiated terrestrial mammal fauna on the islands of Great Britain and of the Sunda Shelf (e.g., Java and Sumatra), which had extensive land connections with Europe and Southeast Asia, respectively, during the latest Pleistocene.

On the other hand, we cannot not simply assume that about 10,000 years ago, all continental islands had their mainland connections severed, and that this stopped all biotic interchange. Just as with oceanic islands, both the geological and biogeographic histories of continental islands can be complex. One example concerns the more isolated islands of the East Indies, especially Borneo and the Philippines (Heaney 1985, 1986; Heaney and Rickart 1990). Both geological and biogeographic evidence suggest that these islands had past land connections to the islands of the Sunda Shelf and to the mainland of Southeast Asia, but not during the latest Pleistocene (Figure 10.13). Thus, these islands are separated by lower sea levels, and contain much more differentiated endemic land plants, terrestrial mammals, and other organisms, than other islands on the Sunda Shelf.

Also, just because most continental islands have not had land connections with the mainland within the last 10,000 years does not necessarily mean that there has not been more recent biotic interchange. Forms with well-developed capabilities for over-water dispersal, such as bats and some birds, tend to be especially well represented, even on tiny islands, and to exhibit virtually no evidence of evolutionary divergence. For example, the continental island of Trinidad, separated from Venezuela by two shallow straits, both less than 25 km wide, is famous in birdwatching circles for its scarlet ibises. These hardly constitute an isolated population, however, because many of the birds commute on a daily basis to feed on the South American mainland. While this is an extreme example, many kinds of organisms have continued to disperse from the mainland after the isolation of continental islands by rising sea levels at the end of the Pleistocene. The rescue effect of such over-water immigration (see Chapter 13) has tended both to recolonize the islands after extinction events and to prevent extinction and divergence of the insular populations.

Marine Regions and Provinces

Numerous biogeographers have attempted to define regions and provinces in the oceans. Some of these classifications, however, are really ecological characterizations based on such criteria as water temperature, depth, and substrate. Such classifications often emphasize vertical rather than horizontal divisions of the three-dimensional marine realm—and understandably so, because there is much less difference among the marine biotas in different oceans than among terrestrial biotas on different continents. Not only are the oceans more interconnected than the continents, but many marine forms have tiny planktonic eggs or larvae that are passively and widely dispersed by ocean currents. Most marine plants and animals are cosmopolitan at the familial level, and many even at the generic level. Consequently, it is difficult to detect patterns in the distributions of marine taxa that clearly reflect the histories of the present water masses. This is especially true of open-water, or pelagic, organisms. The

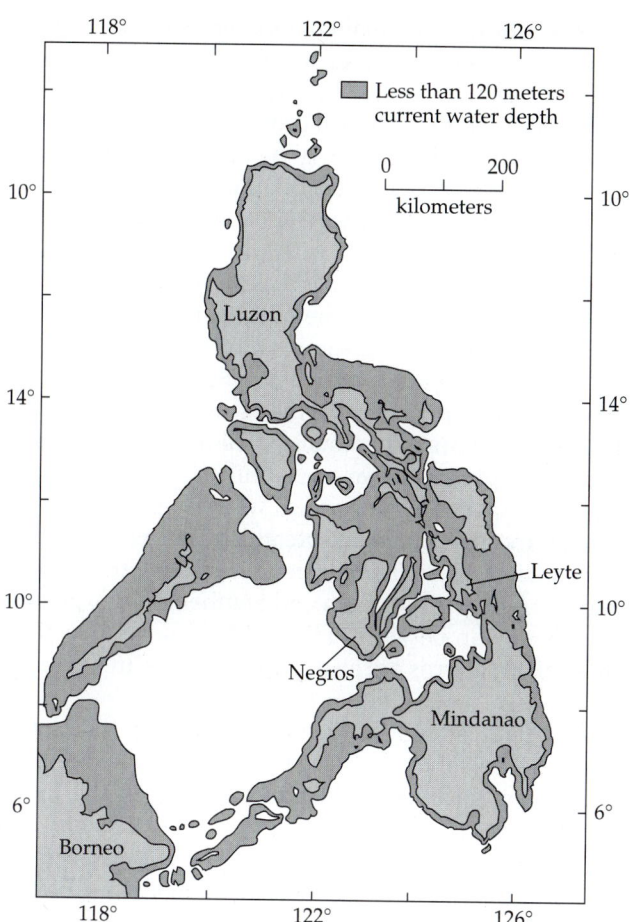

Figure 10.13   Map of the eastern Sunda Shelf and the Philippines, showing the apparent extent of land above sea level in the late Pleistocene. Note that while many of the currently isolated islands (lighter shading) were connected and the straits between others were much narrower than at present, there were still no land connections between the Philippines and Borneo (to the southwest) or the rest of the exposed Sunda Shelf, and thus to the Southeast Asian mainland. (After Heaney 1986.)

benthic plants and animals that live close to, on the surface of, or buried in marine substrates tend to have more restricted distributions, and consequently, to exhibit more provincialism.

Part of the problem in reconstructing the histories of oceanic biotas is the very long and complicated history of most oceans. By studying fossil marine invertebrates of the Paleozoic and Mesozoic, we can see that during certain periods there was very little provincialism in some, if not the majority of, taxa. For example, in the Silurian, many genera of even benthic forms, such as brachiopods, gastropods (Boucot and Johnson 1973), and graptolites (Berry 1973), had nearly cosmopolitan distributions. On the other hand, the majority of phyla contain distinctive taxa that, at least during some periods, were restricted in distribution to some early ocean, such as the Tethys (Hallam 1973a; Gray and Boucot 1979). So marine invertebrates show an accordion effect, with ranges expanding and contracting to produce alternating stages of provincialism and cosmopolitanism. This fact should cause us to be cautious in searching for legacies of ancient (i.e., pre-Cenozoic) earth history in the distributions of present-day marine organisms.

Some history is preserved, however, in the distributions of certain taxa, especially those benthic forms with limited dispersal ability. An obvious horizontal pattern is latitudinal variation in species diversity and composition. Several biogeographers have described marine provinces distributed latitudinally along continental coastlines (Figure 10.14; e.g., Valentine 1966; Pielou 1977b; Horn and Allen 1978) that are separated by zones of rapidly changing species composition. However, these latitudinal patterns appear to be primarily the result of environmental gradients in temperature and correlated abiotic limiting factors, rather than a legacy of unique historical events (see Chapter 15). Thus, the locations of the boundaries between provinces have shifted latitudinally with changing ocean temperatures during the Pleistocene (Valentine 1961, 1989; Valentine and Jablonski 1992), and the taxonomic composition of local assemblages has changed as species have expanded and contracted their ranges (Enquist et al. 1995).

There is, however, marine provincialism that more clearly reflects the influence of tectonic and oceanographic history on the origin and distribution of lineages. Warm tropical oceans have served as significant barriers to dispersal for many cold-adapted temperate and arctic organisms. So the high-latitude seas represent centers of endemism and speciation for groups such as fishes, seabirds (penguins and auks), and marine mammals (whales and seals). Other polar groups do not exhibit such provincialism, apparently because they have frequently dispersed across the equator in deep water, where temperatures are also very cold (an obvious impossibility for air-breathing marine birds and mammals). Some marine organisms exhibit amphitropical disjunctions: ranges that include the cold waters of both the Northern and Southern Hemispheres, but not the warm tropical oceans in between. Figure 10.15 shows two such examples for whales, but similar patterns are also found in fishes and invertebrates.

Some shallow-water and coral reef organisms show significant differentiation among different oceans or coastal regions (see Figure 2.9). Coral reef fish faunas are quite variable in composition, so that certain areas in the Indo-Pacific and the Caribbean not only are centers of high species diversity, but also contain a number of endemic genera (Briggs 1974). There is also significant endemism at the generic level among the corals and mollusks. Fossils indicate that many genera of these invertebrate groups were widely distrib-

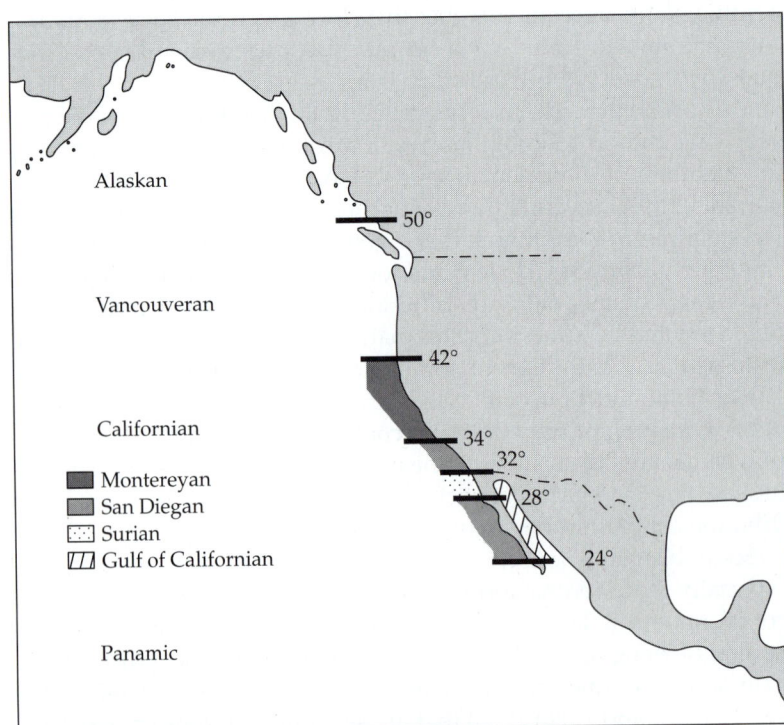

Figure 10.14   Map of the Pacific coast of western North America, showing the latitudinal boundaries of the marine provinces that have been designated for shallow-water benthic marine organisms. (From Southern California Coastal Water Research Project 1973.)

(A)

(B)

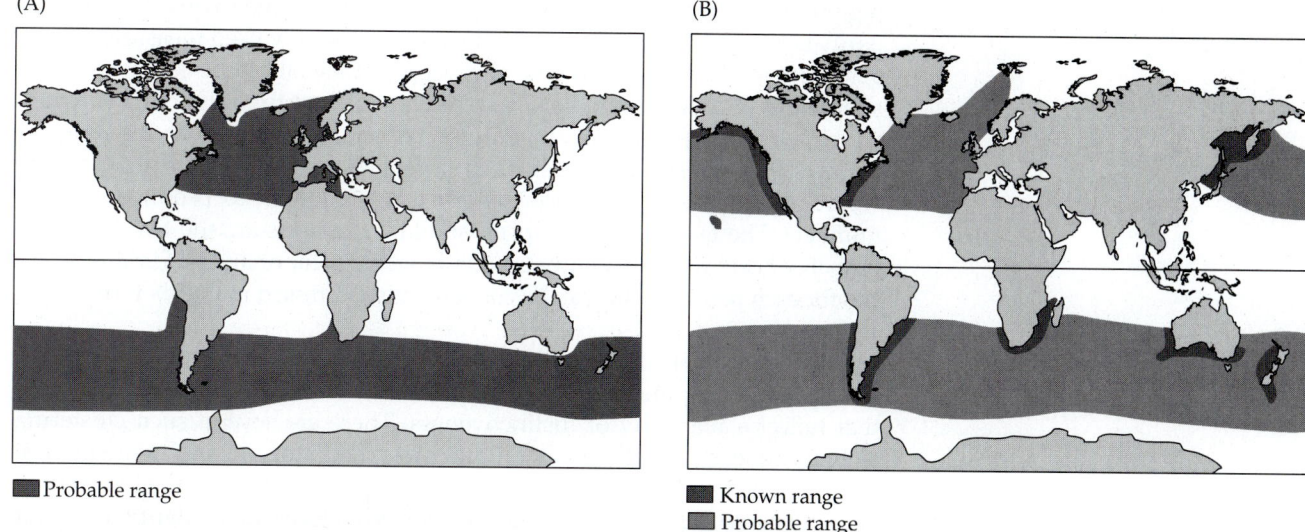

■ Probable range

■ Known range
■ Probable range

Figure 10.15 Amphitropical distribution of the long-finned pilot whale, *Globicephala melas*, (A) and the right whale, *Eubalaena glacialis*, (B). Warm tropical waters are powerful barriers to dispersal, isolating populations of these cetaceans in the Northern and Southern Hemispheres, and apparently preventing the pilot whales from colonizing the North Pacific. (After Martin 1990.)

uted in the Mesozoic, which means that many of today's narrowly restricted forms are relicts of once circumtropical forms (Vermeij 1978). Recent information suggests that many shallow-water marine organisms, even some of those with planktonic larval stages, do not disperse as far as was once believed (Gaines and Bertness 1992).

On the other hand, we are just beginning to learn about the biology and distribution of deep-water benthic organisms, including those inhabiting hydrothermal vents (see Chapter 3) and the vast abyssal plains that constitute the majority of the ocean floor. At least some of these organisms seem to show remarkably little differentiation or endemism, suggesting that they may have enormous powers of dispersal. Clearly, the relationship between marine biogeography and historical tectonic events will be a fruitful area for future research as we explore one of the earth's last frontiers.

Quantifying Similarity among Biotas

The early biogeographers and many of their successors defined biogeographic provinces and regions subjectively. This does not necessarily mean that their classifications are unreliable; in fact, the human brain has an exceptional capacity for recognizing patterns. The biogeographic regions first defined subjectively by Sclater almost certainly summarize real patterns that could be defined objectively by modern quantitative analyses. During the last several decades, however, quantitative techniques have been applied increasingly to systematics and biogeography in order to make the process of classifying organisms and biogeographic regions more rigorous, objective, and repeatable.

In principle, quantitative techniques are simple. First, the items to be classified are described using objective criteria. In the case of biogeographic studies, the data usually consist of a complete list of the relevant taxa that occur at a specific site or within a given area. Sometimes other data, such as average abundance or area of geographic range, are available for each taxon for each region. Several mathematical techniques can then be used to quantify the similarity between each pair of biotas. The most commonly used measures are the similarity indexes shown in Box 10.2, because they all can be computed from simple presence-absence data. These indexes differ primarily in the extent to which they incorporate taxa that are present in both regions, in the range of values they can assume, and how they behave mathematically (that is, how variations in presence-absence patterns are combined to produce a single number). The Jaccard and Simpson similarity indexes are the two that have probably been most frequently used in biogeographic analyses. Similarity values for each pair of biotas can be conveniently expressed in matrix form (Table 10.3).

Once the similarity between each pair of biotas has been computed, a quantitative clustering method is typically used to divide the biotas into groups that reflect a hierarchy of distinctiveness. There are several such clustering techniques, each of which has unique procedures for making mathematical computations and can give somewhat different results. Once the biotas have been clustered, all that remains is to decide what levels of similarity to use for designating different biogeographic ranks, such as provinces and subprovinces, so that the results can be plotted with exact boundaries on a map. Given a suitable data set, the similarity indexes can be calculated and the clustering procedure performed rapidly using a computer.

Most applications of quantitative methods to define biogeographic provinces have been limited to a single group of organisms within a limited geographic region. Examples include the division of Australia into ten provinces based on avian distributions (Kikkawa and Pearse 1969); analyses of similarities among mammalian faunas on the different landmasses (Flessa et al. 1979; Flessa 1980, 1981); and the division of the North American deserts into a hierarchy of biogeographic regions based on plant distributions (McLaughlin 1986, 1989). Such studies illustrate the practicalities of applying quantitative methods to biogeographic problems, but are so limited in taxonomic scope that their results may not be general.

Independent analyses of the distributions of several groups in the same region are potentially more informative. Connor and Simberloff (1978) compared the similarities among the various Galápagos Islands for both land birds and land plants, and found both similarities and differences between the two groups that presumably reflect the influences of such factors as the geological

---

Box 10.2
Simple similarity indexes used by various authors to estimate biotic similarities

Jaccard	$\dfrac{C}{N_1+N_2-C}$	Second Kulczynski	$\dfrac{C(N_1+N_2)}{2(N_1 N_2)}$	Braun-Blanquet	$\dfrac{C}{N_2}$
Simple matching	$\dfrac{C+A}{N_1+N_2-C+A}$	Otsuka	$\dfrac{C}{\sqrt{N_1 N_2}}$	Fager	$\dfrac{C}{\sqrt{N_1 N_2}}-\dfrac{1}{2\sqrt{N_2}}$
Dice	$\dfrac{2C}{N_1+N_2}$	Correlation ratio	$\dfrac{C^2}{N_1 N_2}$		
First Kulczynski	$\dfrac{C}{N_1+N_2-2C}$	Simpson	$\dfrac{C}{N_1}$		

Note: A, absent in both units compared; C, present in both units; N_1, total present in the first unit; N_2, total present in the second unit (when the first unit contains the fewer taxa). After Cheatham and Hazel 1969.

Table 10.3

A matrix of similarity coefficients (Simpson index) between the mammalian faunas of various regions

	North America	West Indies	South America	Africa	Madagascar	Eurasia	SE Asian Islands	Philippines	New Guinea	Australia
North America		40	55	8	9	19	8	9	6	6
West Indies	67		33	11	9	11	7	7	9	11
South America	81	73		3	7	4	7	6	4	3
Africa	31	27	25		30	25	21	27	17	12
Madagascar	38	27	35	65		32	26	22	22	17
Eurasia	48	27	36	80	69		75	64	25	14
SE Asian islands	37	20	32	82	63	92		73	30	18
Philippines	40	20	32	88	50	96	100		26	18
New Guinea	36	21	36	64	50	64	79	64		46
Australia	22	20	22	67	38	50	61	50	93	

Source: After Flessa et al. 1979. Courtesy of The Geological Society of America.
Note: Values above the diagonal are similarities at the generic level; values below the diagonal are similarities at the familial level.

histories and ecological environments of the islands and the dispersal capabilities and population biologies of the organisms. An ambitious and early use of quantitative methods was Holloway and Jardine's (1968) analysis of the distributions of butterflies, birds, and bats in the Indo-Australian area (Figure 10.16). They measured similarities among different islands at the species level, and made inferences about past trends of dispersal and speciation. They con-

(A)

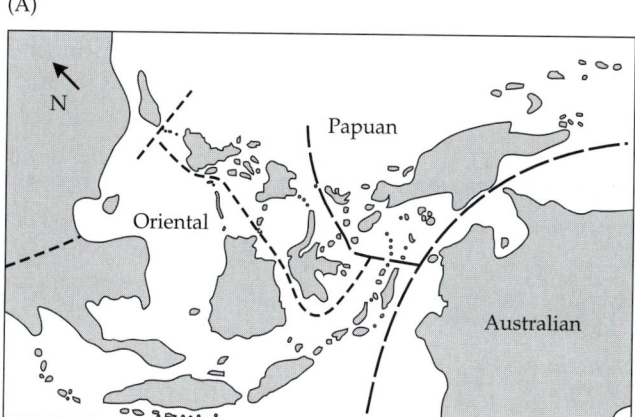

(B)

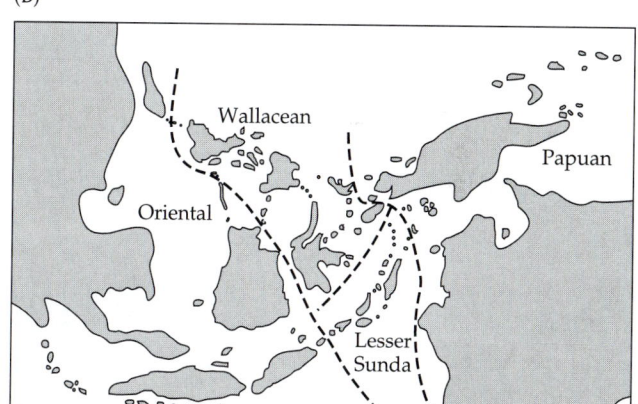

(C)

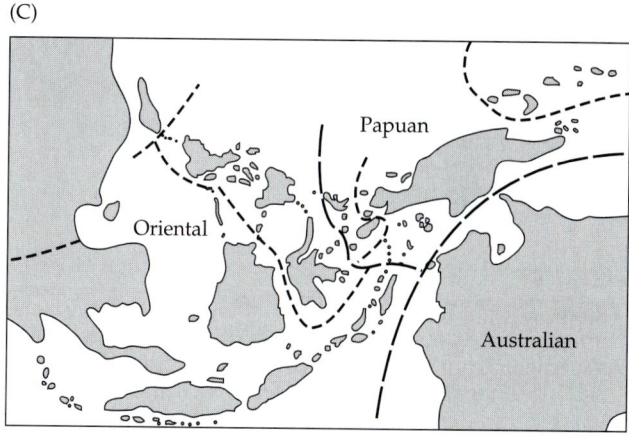

Figure 10.16 Biogeographic regions and subregions in the Indo-Australasian area defined for (A) birds, (B) bats, and (C) butterflies, based on the quantitative analyses of Holloway and Jardine (1968). Note that the divisions for birds and butterflies are virtually identical, but those for bats are quite different. Also compare the lines separating these provinces with those drawn to separate the Oriental and Australian regions in Figure 10.10. (From Holloway and Jardine 1968.)

cluded that all three of these groups of flying organisms dispersed predominantly eastward out of Southeast Asia, but that bats showed a different pattern of faunal differentiation than birds and butterflies. Subsequently, Nelson and Platnick (1981) reanalyzed the original data of Holloway and Jardine, applied the alternative approach of phylogenetic biogeographic analysis (see Chapter 12), and made somewhat different interpretations.

Interestingly, the use of similarity indexes and clustering methods to define biogeographic provinces has declined since the 1970s. There appear to be two reasons for this. First, as indicated above, at levels finer than the Wallace-Sclater regions, the provinces defined in this way for one group usually do not generalize well to other taxa. This should not be surprising. Different kinds of organisms originated and spread during different historical periods, when the earth's geography and climate were different, and their patterns of speciation and dispersal have been affected by their different intrinsic characteristics, such as dispersal modes and life histories, and by different environmental factors, such as biogeographic corridors and barriers. Second, efforts to more rigorously characterize and interpret the legacy of biogeographic history have seen methods based on biotic similarity measures largely supplanted by methods based explicitly on the distributions of organisms with respect to their phylogenetic relationships. We will examine this approach in the next two chapters.

Disjunction

Patterns

Disjunctions are those distributions in which closely related organisms live in widely separated areas. There are many classic examples. Many related forms occur on some combination of the Southern Hemisphere landmasses of Aus-

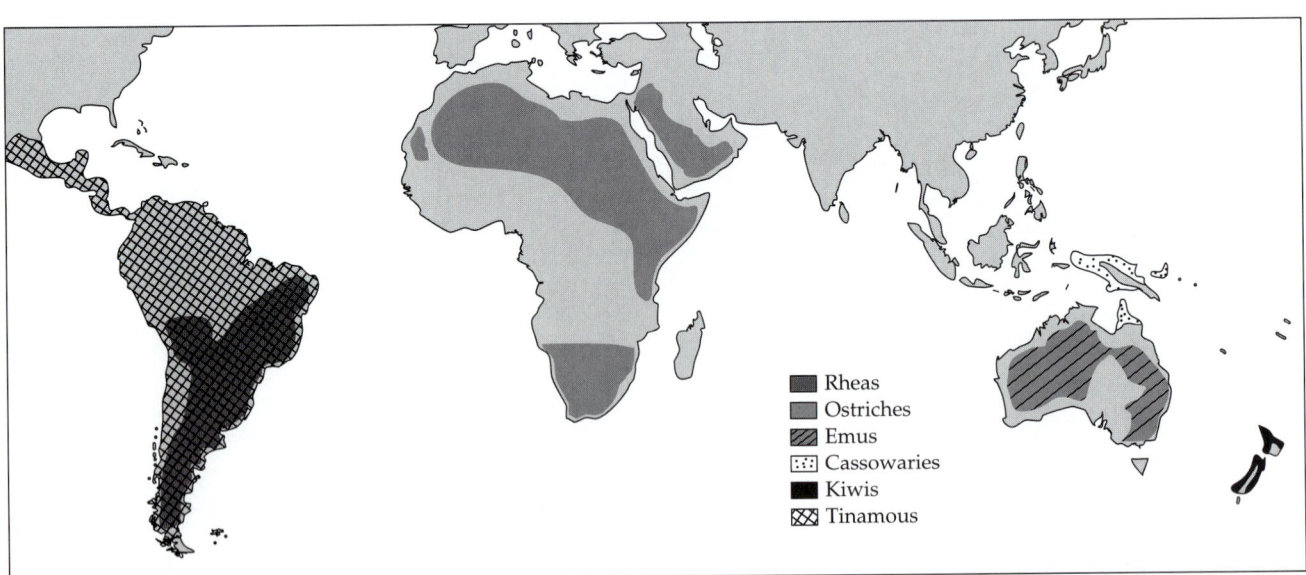

Figure 10.17 The disjunct distribution of the surviving members of the bird lineage that includes the tinamous and flightless ratites: ostriches, emus, cassowaries, kiwis, and rheas. The current widely disjunct distribution of this group reflects their original occurrence on Gondwanaland and their subsequent isolation as the ancient southern continent fragmented and drifted apart. Additional evidence of the historical Gondwanan distribution of this group comes from fossils of extinct giant flightless moas on New Zealand and elephant birds on Madagascar.

tralia, Africa, South America, and sometimes New Zealand and Madagascar: these disjuncts include southern beeches (*Nothofagus*), trees and shrubs of the genus *Acacia*, several groups of mayflies, lungfishes, galaxoid fishes, clawed frogs (family Pipidae), ratite birds (emus, cassowaries, ostriches, rheas, tinamous, kiwis, and extinct moas and elephant birds; Figure 10.17), and marsupial (pouched) mammals. Lungless salamanders (family Plethedontidae) are broadly distributed in the wetter parts of the New World; the only representative in the Old World, the genus *Hydromantes*, occurs in southern Europe (France, Italy, and Sardinia), but also occurs in California (Figure 10.18). Many of the common genera and even species of plants in the deserts of southwestern North America, including creosote bush (*Larrea*), mesquite (*Prosopis*), and paloverde (*Cercidium*), also occur in the desert regions of southern South America, but not in the extensive tropical and montane areas in between. Other areas are shown in Figure 10.19.

There are disjuncts on smaller scales as well. Plants provide some of the best examples, perhaps because they often have very specific environmental requirements and are able to persist for long periods in small patches as long as conditions are suitable. The sweet gum tree, *Liquidambar*, occurs in the deciduous forests of the eastern United States and also in a series of isolated localities in central and southern Mexico. The Black Belt and Jackson Prairies in the southeastern United States are isolated grassland habitats that occur on unusual soils. They contain a number of disjunct plants whose nearest populations occur far to the west on the Great Plains (Schuster and McDaniel 1973). Some wetland plants found along the Atlantic coastal plain of eastern Canada also occur as isolated populations well inland in central Ontario (Wisheu and Keddy 1989).

Animals also exhibit small-scale disjunctions. Like plants, many kinds of insects have been able to disperse to and persist in widely separated areas that meet their narrow niche requirements. A vertebrate example is the lizard genus *Uma*, which occurs only on isolated sand dunes in the southwestern North American deserts and not in the intervening habitats (Figure 10.20). The checkerspot butterfly, *Euphydryas editha*, occurs in many isolated populations widely dispersed across western North America (Ehrlich et al. 1975). Interestingly, many of the isolates near the southern and lower elevational limits of

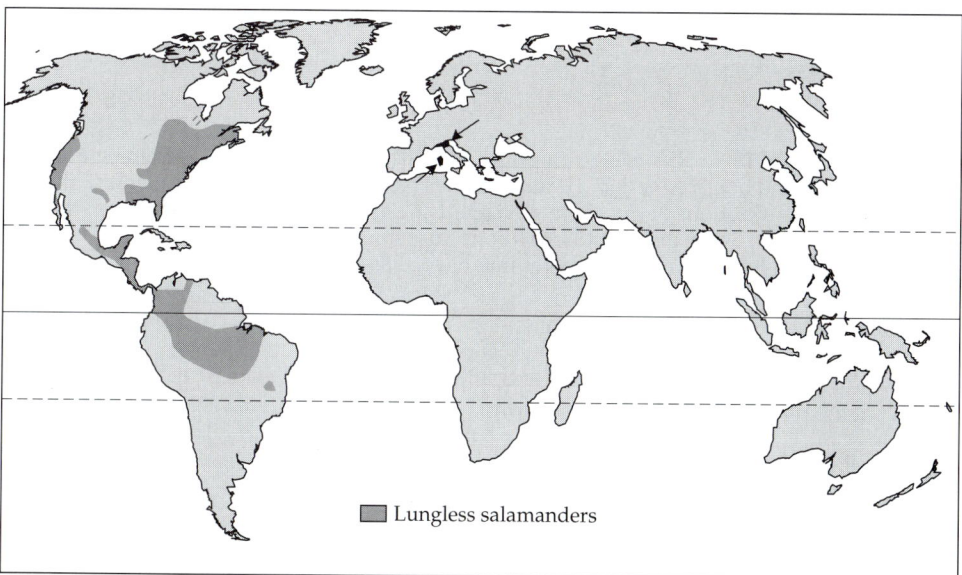

Lungless salamanders

Figure 10.18 The disjunct distribution of lungless salamanders (Plethedontidae). Note that this ancient group has many isolated relicts in North America, and even one in the Mediterranean region of France, Italy, and Sardinia. (After Wake 1966.)

(A) *Nama dichotomum*

(B) *Phacelia crenulata*

(C) *Phacelia magellanica*

(D) *Agoseris heterophylla*, North America
Agoseris coronopifolia, South America

(E) *Sanicula crassicaulis*

(F) *Bowlesia incana*

(G) *Osmorhiza depauperata* (obtusa)

(H) *Osmorhiza chilensis*

Figure 10.19   Examples of amphitropical disjunct distributions of plant species in North and South America. These examples represent only a small fraction of the disjunct distributions of closely related plant species, mostly in arid regions, on the two continents. These disjunct distributions raise interesting questions about the historical geological and climatic events that have allowed these plants to disperse across the tropics.

the species' range have gone extinct in the last few decades, apparently due to global warming (Parmesan 1996).

Processes

Since the early nineteenth century, explaining disjunct distributions has been a major goal of biogeography. These distributions raise an obvious challenge: If the organisms are indeed closely related, then how did they get from where their common ancestor occurred to their current widely separated ranges?

There can be only three answers: (1) their ancestors occurred on pieces of the earth's crust that were once united, but have subsequently split and drifted apart; (2) their ancestors were once broadly distributed, but populations in the intervening areas have gone extinct, leaving isolated relics; (3) at least one lineage has dispersed a long distance from the area where its ancestor(s) originally occurred.

Examples of all three mechanisms have been found. The Southern Hemisphere disjuncts mentioned above are all vicariant relics of forms that once occurred on Gondwanaland and were isolated when that ancient supercontinent broke up and its fragments drifted apart. Examples of disjunctions caused by the extinction of intervening populations include, at an intercontinental scale, the camels and tapirs, which now occur only in South America and Eurasia, but are found in North America as fossils, which clearly indicate that they had a much wider and apparently continuous range during the Pleistocene. Many of the organisms that exhibit smaller-scale disjunctions within North America and other continents are relics left by the extinction of intervening populations due to changes in climate or other environmental conditions since the Pleistocene. Many fishes found in now isolated desert springs and mammals restricted to the cool, mesic habitats on isolated mountain ranges had wider, more continuous ranges in the Pleistocene (see Chapter 13). Some disjuncts are the result of long-distance dispersal, either by natural means, such as the cattle egrets that a few decades ago crossed the South Atlantic Ocean to colonize South America from Africa; or by human transport, such as the exotic European house sparrows and rabbits that are now established in such outposts as Australia, New Zealand, and southern South America.

There are also examples of forms once thought to be disjuncts that are now known not to be close relatives at all. Thus, "porcupines," rodents characterized by hairs modified to form distinctive defensive quills and by certain features of the skull, were once thought to be disjuncts between the New and Old World. These animals occur in Africa, southern Europe and Asia, and North and South America, but are absent from all of temperate Eurasia. Subsequent systematic research showed, however, that New and Old World porcupines are not closely related; they arose independently from primitive ratlike rodents, retaining unspecialized features of the skull and evolving quills by convergence. Similarly, the anteaters of the tropical New World and the pangolins or scaly anteaters of tropical Africa and Asia were once thought to be related disjuncts and were placed in the same order, Edentata. These animals share several characteristics, especially anatomical features of the skull and tongue, that are related to their exclusively insectivorous diet, but these have arisen by convergence. These "anteaters" actually are unrelated members of two different orders, Edentata (which includes the New World anteaters, armadillos, and sloths) and Pholidota (which includes only the Old World pangolins).

The revelation that some organisms with seemingly disjunct distributions are not in fact close relatives underscores the need for good systematics. Patterns of cosmopolitanism, provincialism, and disjunction raise important questions in historical biogeography. We want to know how the origin, diversification, and spread of a particular kind of organism was related to and influenced by the geographic, geological, climatic, and biotic features of the earth at the times when those processes were occurring. But unless we have an accurate assessment of the phylogenetic relationships among organisms, we do not even know which biogeographic questions to ask. So before we proceed further with our effort to reconstruct and interpret past distributions, a consideration of phylogenetic systematics is in order. That is the subject of the next chapter.

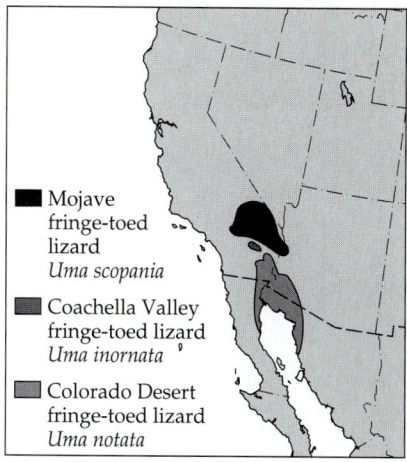

■ Mojave fringe-toed lizard *Uma scopania*

■ Coachella Valley fringe-toed lizard *Uma inornata*

■ Colorado Desert fringe-toed lizard *Uma notata*

Figure 10.20 The disjunct distribution of the lizard genus *Uma*, which is restricted to dunes of wind-blown sand in the desert region of southwestern North America. Many other plants and animals that are restricted to specific kinds of spatially isolated habitats, such as mountaintops, lakes and springs, and patches of unusual soil types, show similar patterns of small-scale disjunct distributions.

CHAPTER *11*

The History of Lineages

All features of life on earth, including the distributions of species and higher taxa, have been influenced by "history." It is important to recognize that the characteristics of contemporary organisms have been shaped by two kinds of events in the past. We will call these the **history of place** and the **history of lineage** (Brown 1995). The history of place is the history of the earth itself: the changes in geography, geology, climate, and other environmental characteristics that are *extrinsic* to the particular organism or group of organisms being studied. We like to think of the history of place as the history of the environmental template, including the other kinds of organisms, that each organism has experienced during its own unique evolutionary history. Thus, the history of place includes the geological history of the earth, changes in soils, climate, and oceanographic and limnological conditions, and the varying composition of biotic communities. Past environments, by influencing the abundance, distribution, and adaptive evolutionary changes of its ancestors, may have influenced a contemporary species or lineage in many ways. Even the most distantly related kinds of organisms that lived together in the past at a particular place, and hence experienced some features of a common environment, share some history of place. Sharing a history of place does not depend on sharing ancestors.

The history of lineage is the series of evolutionary changes that have occurred in the *intrinsic* characteristics of an individual organism, species, or higher taxon. These are changes in heritable characteristics, derived from and constrained by the characteristics of the organisms' ancestors. Only to the extent that organisms share common ancestors do they share a history of lineage. The history of place strongly affects the history of lineage, because characteristics of past environments (e.g., geology, climate, and biotic composition) undoubtedly influenced the survival, distribution, and diversification of all lineages that occurred in those places. But the converse is less true. The history of lineage influences the history of place only to the extent that the activities of particular organisms altered the past environment for themselves and other lineages. While many kinds of organisms substantially modify their environments—the building of reefs and atolls by corals and the modification of the climate of the Amazon basin by rain forest trees are just two examples—usually these influences are diffuse and cannot be attributed to just one lineage or taxonomic group.

Chapters 6 and 7 described many important events in the history of place. This chapter focuses on the histories of lineages, and how we reconstruct and interpret them.

Classifying Biodiversity

As pointed out in Chapter 2, humans have long recognized that some kinds of organisms are more similar to one another than others, and they have tried to capture these patterns of similarity by naming and classifying living things. Aboriginal people have names for the plants and animals in their environments, and these names typically recognize degrees of difference. The present scientific system for naming organisms and arranging them in a hierarchical classification scheme dates back to the eighteenth-century Swedish naturalist Linnaeus. Of course, in his time, it was believed that each kind of plant and animal had been specially created by God. So the classification scheme simply reflected the perceived degree of similarities and differences among them, especially in morphological characteristics.

Evolutionary Classifications

In the nineteenth century, the triumph of Darwinism replaced special creation with evolution. It was recognized that all living things are related through a past history of ancestry and descent, and that the similarities among different kinds of organisms are due to the inheritance of common characteristics from common ancestors. Since then, taxonomists and systematists have endeavored to develop methods of classification that ever more accurately reflect evolutionary history. As pointed out in the discussion of species definitions in Chapter 8, the initial classification schemes were still based largely on morphological characteristics. However, as evolutionary biology revealed more about the patterns and processes of evolutionary change, information from paleontology, developmental biology, population biology, genetics, and other disciplines was incorporated.

By the 1960s, the "new synthesis" in evolutionary biology had led to the widespread adoption of so-called "evolutionary classifications" (e.g., Mayr 1942; Simpson 1945, 1961a; Dobzhansky 1951; Rensch 1960). In these schemes, the classifications were intended to reflect the branching tree of ancestry and descent, and the hierarchical taxa (i.e., kingdoms, phyla, orders, families, genera, and species) were intended to reflect both the pattern of descent from common ancestors and the degree of divergence. These schemes ran into problems, however, because these two features of the evolutionary process did not always follow the same patterns, much less occur at the same rates. This made it difficult to incorporate both relative time of lineage splitting and rate of change since the split into one consistent classification. This problem is illustrated by the classic case of the crocodilians. The crocodiles, alligators, and caymans have retained many traits of their reptilian ancestors, such as scales, ectothermy, simple teeth, and quadrupedal locomotion, but they are descended from archosaurs, a group of extinct reptiles that also gave rise to modern birds. So a classification based on branching sequence would have to place the crocodilians closer to birds than to other living reptiles, whereas one based on rate of divergence would leave them with the reptiles. Prior to the 1960s, classifications of vertebrates unanimously considered crocodiles to be reptiles (Figure 11.1).

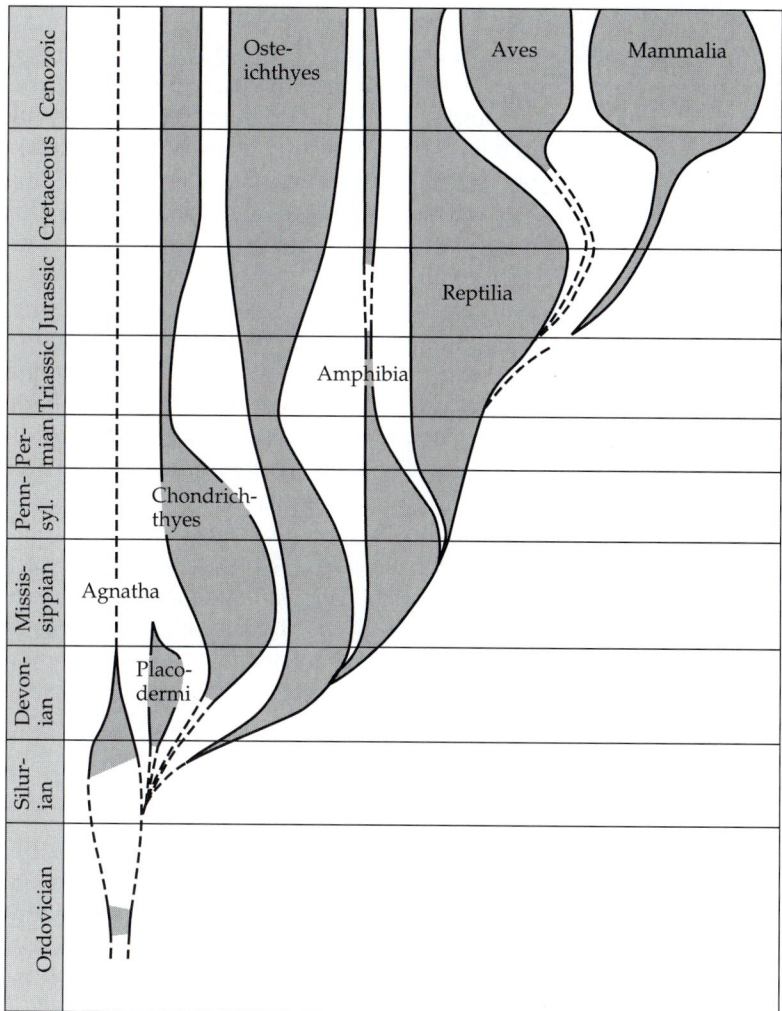

Figure 11.1 An evolutionary classification and reconstructed phylogeny of living vertebrates from the classic work of A. S. Romer. The thickness of the branches indicates the approximate diversity of the groups through time. The degree (rate) of evolutionary divergence from a common ancestor as well as the time of the split are taken into account in grouping organisms into taxa. Note that the cartilaginous fishes (Chondrichthyes), bony fishes (Osteichthyes), amphibians, reptiles, birds (Aves), and mammals are all recognized as equivalent units and given class rank. Compare this figure with the phylogenetic reconstruction or cladistic classification in Figure 11.2. (After Romer 1966.)

This example illustrates another problem with evolutionary classifications: they tended to be highly subjective. Different systematists, assigning different importances to branch points versus evolutionary rates and using their own schemes for choosing and weighting characters, produced different classifications for the same groups—and then argued about which one best reflected the "true" evolutionary relationships. In response, some systematists tried to make the process of classification more objective and quantitative. The result was an approach called **numerical taxonomy** (Sneath and Sokal 1973), which was very influential for a brief period in the 1960s and early 1970s. Many numerical taxonomists wanted to abandon the effort to create explicitly evolutionary classifications. As an alternative, they advocated purely "phenetic" classifications, obtained by clustering taxa into hierarchical units based on quantitative analyses of their phenotypic traits. This trend in systematics paralleled and occurred contemporaneously with the approach of defining biogeographic provinces quantitatively (see Chapter 10). Many of the same people were involved in both endeavors, and they used similar techniques for clustering units based on the degree of similarity. Both approaches were short-lived because they were rapidly replaced by new and better methods for reconstructing the evolutionary histories of organisms and their distributions.

Phylogenetic Systematics

In the mid-1960s, two breakthroughs occurred that would revolutionize systematics and have an enormous influence on historical biogeography. One was the publication in English of a comprehensive monograph by Willi Hennig (1966), a German systematist then little known outside the world of insect taxonomy. In this work (first published in German in 1950), Hennig laid out a revolutionary new method for determining the phylogenetic relationships among organisms. Although it is objective and repeatable, the method relies more on logic than on numerical measurement.

Hennig's paradigm.   Hennig's genius was to recognize the logical consequences of a feature of the evolutionary process that had been known for a long time. Organisms are very complex systems; they comprise many parts and processes that must interact in many ways to produce a functional individual that can survive and reproduce. The kinds of changes that can occur over generations of evolution—and also during the ontogenetic development of individual organisms—are highly constrained. New structures and functions are almost never created de novo. Instead, they are obtained by modifying already existing structures and functions. The history of these changes is recorded in the similarities and differences in the complex characteristics of related organisms—in the extent to which the characteristics of their common ancestors have been modified by subsequent additions, losses, and transformations. From this simple insight, Hennig developed an elegantly simple logical method for reconstructing the history of a lineage and assessing the relative degree of relationship among its taxa.

Assumptions and definitions.   Hennig's method is based on several readily defensible assumptions and operational definitions. First, its goal is to reconstruct the branching pattern that depicts the evolutionary sequence of ancestry and descent. Each branch in this evolutionary tree is required, by definition, to be **monophyletic**. This means that the taxa on each branch all share a common ancestor, and that the branch includes all descendants of that common ancestor. A monophyletic lineage is called a **clade**. The goal of phylogenetic systematics is to produce a branching diagram, called a **cladogram**, that accurately reconstructs the sequence of past speciation events through which the common ancestor gave rise to its descendants (Figure 11.2).

It is important to recognize that these cladistic reconstructions, and the cladistic classifications that are based on them, avoid the problem of trying to incorporate in one scheme both the relative time of branching and the rate of evolutionary divergence since the branching. They do so by *explicitly assuming that the cladogram depicts only the sequence of branches*. Consequently, phylogenetic classifications are often quite different from the older evolutionary classifications (for example, compare Figures 11.1 and 11.2). Note that a cladogram of the vertebrates does not recognize "fishes" as a taxonomic category, because the common ancestor of what we call "fishes" was also the common ancestor of all other vertebrates. Using similar logic, birds are placed with the reptiles, because birds and crocodilians share a more recent common ancestor than either does with the other lineages of "reptiles."

Another set of assumptions and definitions is required to characterize ancestor-descendant relationships. If the characteristics of organisms are derived with modification from those of their ancestors, this implies that when we look back on the evolution of a trait, it will be seen to have undergone a series of transformations from more ancestral, or **plesiomorphic**, character states to more derived, or **apomorphic**, character states (Box 11.1). Cladistic

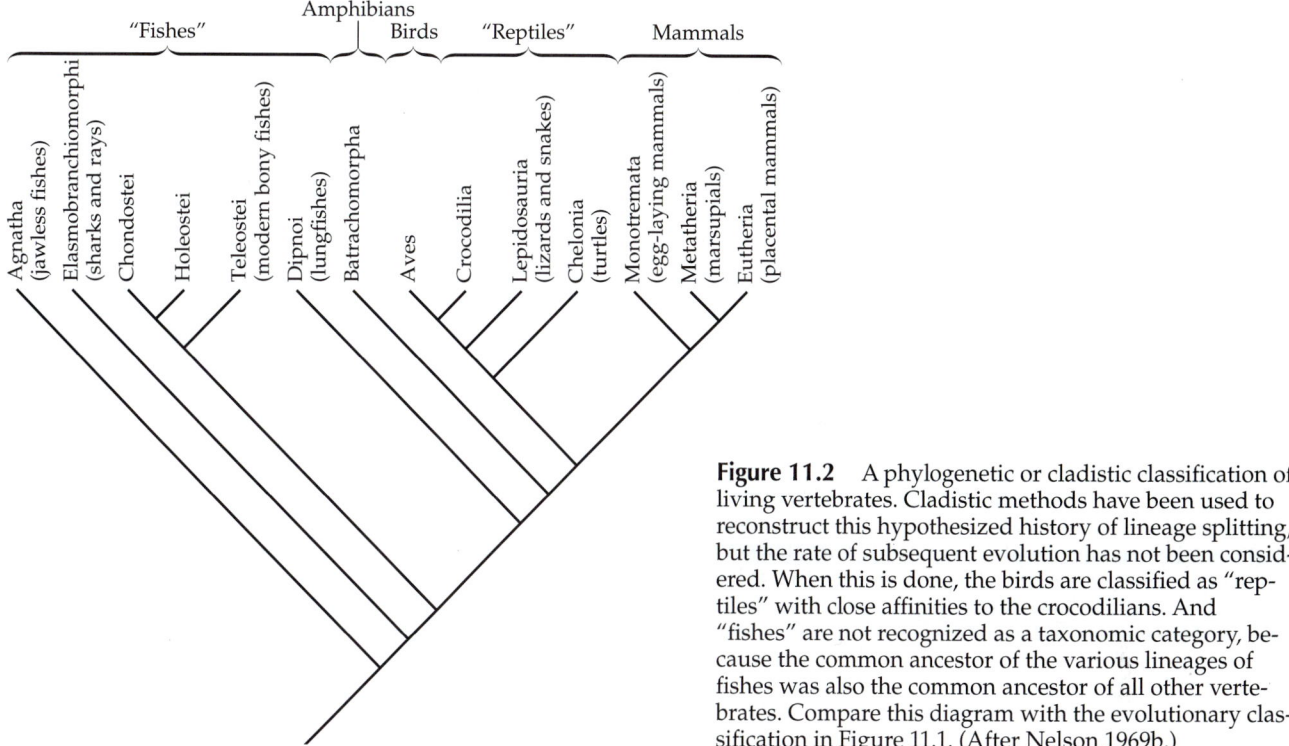

Figure 11.2 A phylogenetic or cladistic classification of living vertebrates. Cladistic methods have been used to reconstruct this hypothesized history of lineage splitting, but the rate of subsequent evolution has not been considered. When this is done, the birds are classified as "reptiles" with close affinities to the crocodilians. And "fishes" are not recognized as a taxonomic category, because the common ancestor of the various lineages of fishes was also the common ancestor of all other vertebrates. Compare this diagram with the evolutionary classification in Figure 11.1. (After Nelson 1969b.)

reconstructions are based solely on shared apomorphic traits (**synapomorphies**), because it is these that indicate the sequence of changes from the ancestral condition. Organisms are expected to share some plesiomorphic traits (**symplesiomorphies**) with their relatives because they will have been inherited from the common ancestor. For the purposes of phylogenetic reconstruction, they are relatively uninformative. For example, most vertebrates have two eyes and two pairs of appendages, but these shared plesiomorphic characteristics are not very useful in classifying the vertebrate groups.

To reconstruct the history of ancestry and descent from the transformation of characters, it is necessary to assume that independent evolution of the same character state occurs very infrequently—at least much less frequently than the transition from some common ancestral state to a unique derived state. This assumption itself makes implicit assumptions about the kinds of changes that are possible, or at least likely, in a complex integrated system. It suggests that care be taken in the selection of characters so as to avoid those, such as body size, that are simple traits known to be subject to reversible and independent evolution.

It is next necessary to deduce the direction of change. How do we know which character states are apomorphic and which are plesiomorphic? The standard procedure is to identify and make comparisons with an **outgroup**, a taxon closely related to but not a member of the clade being analyzed. The assumption is that the outgroup shares many plesiomorphic character states with the ancestor of the clade. As the clade evolved, those plesiomorphic states were often modified to apomorphic ones (sometimes repeatedly). So it is possible to order the sequence of changes that most likely occurred to produce the derived character states of each species in the clade. Because of its key role in polarizing transitions in character states, it is important to choose the outgroup with care, doing one's best to ensure that it is closely related to, but not a member of, the lineage being classified.

Box 11.1
The basis of Hennig's paradigm: A hypothetical example of cladogenesis

The following example and diagram will illustrate Hennig's insight into how the branching pattern of phylogeny leaves its legacy in the derived characteristics of contemporary forms. Assume that we have five contemporary species, numbered 0–4. We will designate species 0 as an outgroup, closely related to but outside the lineage of the four species whose lineages we will follow. We will follow the evolution of six characters, lettered A–F.

The outgroup and the common ancestor that it shares with the clade (located at the basal branch point, or node, of the tree) have the ancestral, or plesiomorphic, state of each character. When a character evolves to a new derived, or apomorphic, state, we indicate this with a prime sign (e.g., A → A').

Now let's start with the ancestor, which has character states ABCDEF, and follow its evolution. It speciates, and one lineage gives rise to the outgroup without any change in the six characters, so species 0 also has states ABCDEF. After this first split, the second lineage evolves

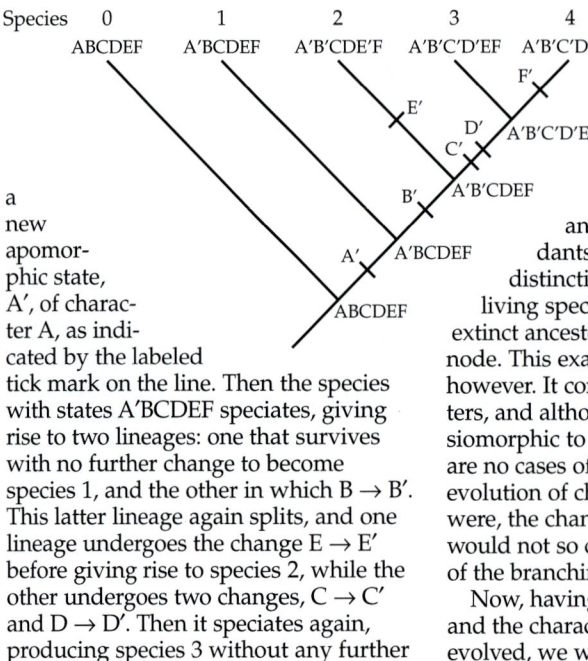

a new apomorphic state, A', of character A, as indicated by the labeled tick mark on the line. Then the species with states A'BCDEF speciates, giving rise to two lineages: one that survives with no further change to become species 1, and the other in which B → B'. This latter lineage again splits, and one lineage undergoes the change E → E' before giving rise to species 2, while the other undergoes two changes, C → C' and D → D'. Then it speciates again, producing species 3 without any further change and species 4 with the change F → F'.

This example illustrates how changes in the states of characters, from plesiomorphic to apomorphic, accumulate as the lineage branches and each ancestor gives rise to descendants. These character states distinctively characterize not only living species, but also their now extinct ancestors represented by each node. This example is highly simplified, however. It considers only a few characters, and although they change from plesiomorphic to apomorphic states, there are no cases of reversal or independent evolution of character states. If there were, the changes in these characters would not so clearly preserve evidence of the branching of the lineage.

Now, having dictated how this lineage and the characters of its species have evolved, we will see in Box 11.2 how this phylogeny can be reconstructed by cladistic analysis.

Procedures.   Armed with these operating assumptions and definitions, it is conceptually straightforward to generate a cladogram (Box 11.2). Basically, this is done by working "backward" with respect to the direction of evolution. Starting with the terminal taxa (the "twigs" of the tree), the pair that shares the most derived character states is joined by a branch point, or **node**. This process is repeated until all the branches have been joined at the common ancestor (the "trunk" of the tree). A cladogram is typically depicted as the kind of diagram shown in Figure 11.2. The terminal "twigs" represent the taxa being classified—sometimes species, but often higher-level groups such as genera or families. The outgroup serves to "root" the cladogram by specifying the plesiomorphic character states. The little tick marks throughout the diagram illustrate the transitions in character states, and these are sometimes coded to indicate which specific changes occurred in which traits (Figure 11.3). Each node represents the most recent common ancestor of all the species above that node. A historical speciation event is inferred to have caused the branching in the lineage. Because we know the reconstructed history of the character transitions, we can predict the characteristics (character states) of each ancestral species at each node. Initially, cladograms were constructed based largely or entirely on morphological traits, but as other kinds of characteristics have been documented by biological studies, they have often been incorporated into phylogenetic analyses.

Because the logic of Hennig's paradigm is so simple, it is conceptually straightforward to program a computer to generate a cladogram from a data set of taxa (including a specified outgroup) and their character states. Commercially available software is now widely used for phylogenetic reconstruc-

Box 11.2
How to construct a cladogram: A hypothetical example

Here we will use the characters of the hypothetical species whose cladogenetic evolution we followed in Box 11.1 to reconstruct the phylogenetic branching pattern by applying the techniques of cladistic analysis.

We begin with a group of four species, numbered 1–4, with six characters, lettered A–F, each of which can exhibit one of two states. The first thing we need to do is to determine which character states are ancestral, or plesiomorphic, and which are derived, or apomorphic. To do this, we need to identify an outgroup, a species closely related to species 1–4 but whose lineage diverged at an earlier time. In this case, the outgroup is species 0, and it has character states ABCDEF. Thus, we assume that the character states without prime signs are plesiomorphic and that those with prime signs are apomorphic.

Now, all that remains is to cluster our species by the number of apomorphic characters that they share. The easiest way is to prepare a matrix table like the one below. The first row gives the character states for each species. The other entries, in bold type, show the number of shared apomorphic (synapomorphic) characters for each pair of species, followed by those character states in parentheses.

From the table it is apparent that species 3 and 4 share the largest number of derived character states, 4 (A'B'C'D'), so these species should share the most recent common ancestor, and can be connected by a dichotomous branch:

Species 3 Species 4
A'B'C'D'EF A'B'C'D'EF'

Next, we ask which species shares the largest number of apomorphies with this pair. The answer is species 2, which shares two apomorphic states, A'B', with species 3 and 4, so we can connect it with a branch that splits earlier (lower in the tree):

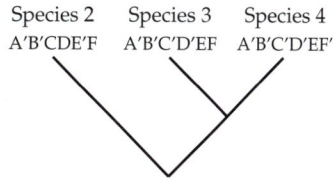

Species 2   Species 3   Species 4
A'B'CDE'F   A'B'C'D'EF A'B'C'D'EF'

Finally, we ask about the last species, species 1, and find that it shares only one apomorphy, A', with the others, so we connect it with an earlier branch. And we can put in the outgroup, Species 0, splitting at a still earlier node, if we wish. This gives us the following cladogram:

Species 0   Species 1   Species 2   Species 3   Species 4
ABCDEF   A'BCDEF   A'B'CDE'F   A'B'C'D'EF   A'B'C'D'EF
                                          F'
                                  E'
                              D'   A'B'C'DEF
                              C'
                          B'   A'B'CDEF
                      A'   A'BCDEF
                  ABCDEF

Note that this is the same tree whose historical branching pattern we followed in Box 11.1. This exercise shows how we were able to reconstruct the history of this lineage knowing only the character states of the species and, by using an outgroup, which states were plesiomorphic or apomorphic.

Now derive a cladogram yourself for the following outgroup and four species with characters designated as before: species 0 (outgroup): GHIJKL; species 1: G'H'I'JKL; species 2: G'H'IJKL; species 3: G'HIJ'K'L; species 4: G'HIJ'K'L'. Make a table of shared apomorphic character states and place species on branches based on the number of synapomorphies. You should obtain the cladogram shown at right.

Again, we emphasize that these are simplified hypothetical examples. They were constructed to illustrate the elegantly simple logic that is the basis for the cladistic method of phylogenetic reconstruction. In these examples, all of the character changes are internally consistent, so the branching of each lineage can be explained by a unique cladogram. In real cases, more characters would be used, and some would have more than one apomorphic state. Further, there might well be cases of reversal or independent evolution of character states, although these should be rare. If we allow for these possibilities, it is possible to derive several alternative tree diagrams. It is then necessary to use some formal procedure, such as parsimony or maximum likelihood analysis, to determine which one should be preferred. No matter what procedures are used, it is essential to remember that a cladogram is only a hypothesis about the likely course of cladogenesis: evolution could actually have followed one of the less likely alternative pathways.

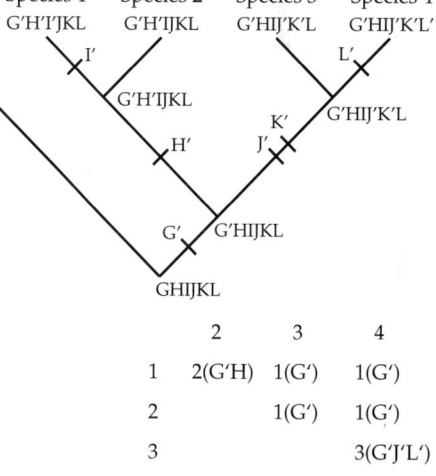

Species 0   Species 1   Species 2   Species 3   Species 4
GHIJKL   G'H'I'JKL G'H'IJKL   G'HIJ'K'L   G'HIJ'K'L'
                                          L'
                          I'
                              G'H'IJKL   G'HIJ'K'L
                          H'   J'
                                  K'
                      G'   G'HIJKL
                  GHIJKL

	2	3	4
1	2(G'H)	1(G')	1(G')
2		1(G')	1(G')
3			3(G'J'L')

Character state	Species 0 ABCDEF	Species 1 A'BCDEF	Species 2 A'B'BCDE'F	Species 3 A'B'C'D'EF	Species 4 A'B'C'D'EF'
Apomorphies shared with:					
Species 1	0	—	1(A')	1(A')	1(A')
Species 2	0	1(A')	—	2(A'B')	2(A'B')
Species 3	0	1(A')	2(A'B')	—	4(A'B'C'D')
Species 4	0	1(A')	2(A'B')	4(A'B'C'D')	—

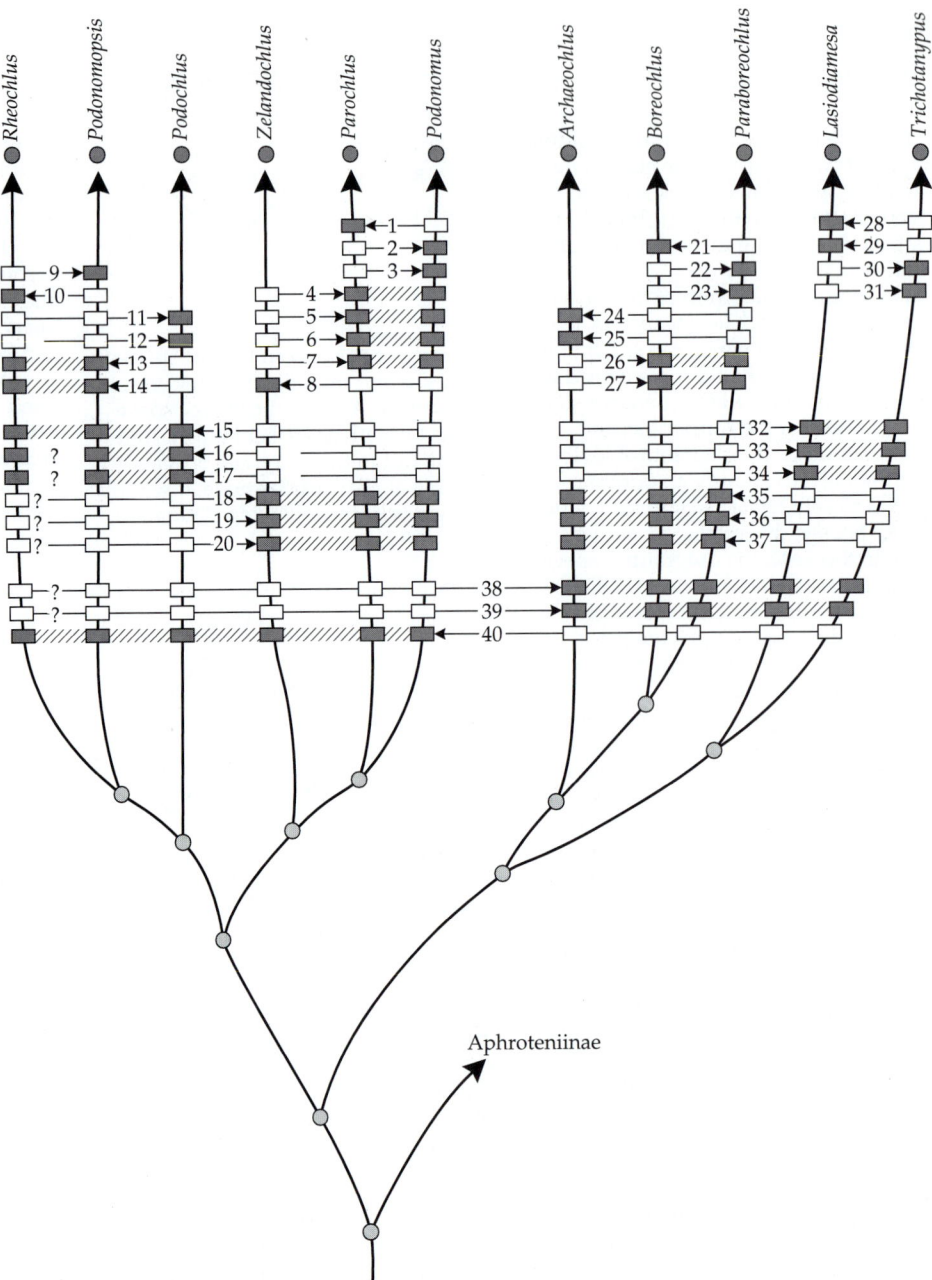

Figure 11.3 A cladogram, or hypothesized reconstruction of phylogenetic relationships, for the genera of midges in the subfamily Podonominae. The circles represent living genera (solid circles, above) and their presumed common ancestors (open circles, below). The numbered rectangles show the patterns of shared derived character states used to generate the cladogram. Each number refers to a different character. Ancestral, or plesiomorphic, states are represented by open rectangles; derived, or apomorphic, states are represented by solid rectangles; shared derived (synapomorphic) character states are joined by shaded bars; and transitions from an ancestral to a derived state are indicated by arrows. Thus, for example, the genera *Lasiodiamesa* and *Trichotanypus* are joined in one clade that shares the common ancestor indicated by the open circle farthest to the right; this lineage differs from other clades because the two genera and their presumed common ancestor share the same unique derived states of characters 32, 33, and 34. (After Brudin 1965.)

tions (e.g., PAUP: Swofford 1993). In practice, however, there are enough complications that anyone seriously attempting phylogenetic classification and reconstruction should be thoroughly familiar with current systematic theory and practice. Just to mention some of the issues involved: (1) many characters and their various states may be statistically correlated and functionally interdependent, so that they cannot be considered independent traits; (2) different algorithms, such as parsimony or maximum likelihood, can be used to estimate the probabilities of character state transitions and thus to derive the cladogram; (3) clear resolution of all branches is often difficult or impossible, so that the cladogram may contain unresolved polychotomies rather than just dichotomous branches; and consequently, (4) more than one tree topology (branching pattern) may be equally or nearly equally consistent with the available data (Figure 11.4).

Molecular Systematics

The second breakthrough in systematics that began in the 1960s was the widespread use of techniques and data from molecular genetics. In 1953, Watson and Crick elucidated the molecular structure of DNA, the molecule that encodes and transmits genetic information. It was not long before biologists were devising ways to decode DNA and make comparisons of molecular genetic differences among individuals, species, and higher taxa. Systematists soon realized that just as the structural and functional characteristics of organisms are legacies of phylogenetic evolution, so too are the molecules that code for these characteristics. Transformations in the genetic code are preserved in the sequence of bases in DNA molecules. Further, while the number of variable independent morphological traits available for phylogenetic analyses is often limited, the genome contains an enormous number of potentially useful characters. And while statistical correlation and functional interdependence among morphological traits creates problems of analysis and interpretation, the rapidly accumulating knowledge of molecular genetics and the growing data bases of molecular traits offer ways to address these problems (e.g., see Hillis and Moritz 1990; Avise 1994; Hillis 1996).

Molecular systematics initially used relatively simple and unsophisticated measures of genetic similarity, such as immunological reactions and DNA hybridization, which provided a single quantitative estimate of genetic similarity between two organisms, or protein electrophoresis, which detected differences in the products produced by specific genes (e.g., Sibley 1970; Sibley and Ahlquist 1990). As the discipline has evolved, the methods have become progressively more sophisticated and standardized. Now, most studies are based on comparisons of nucleotide sequences, obtained either directly by

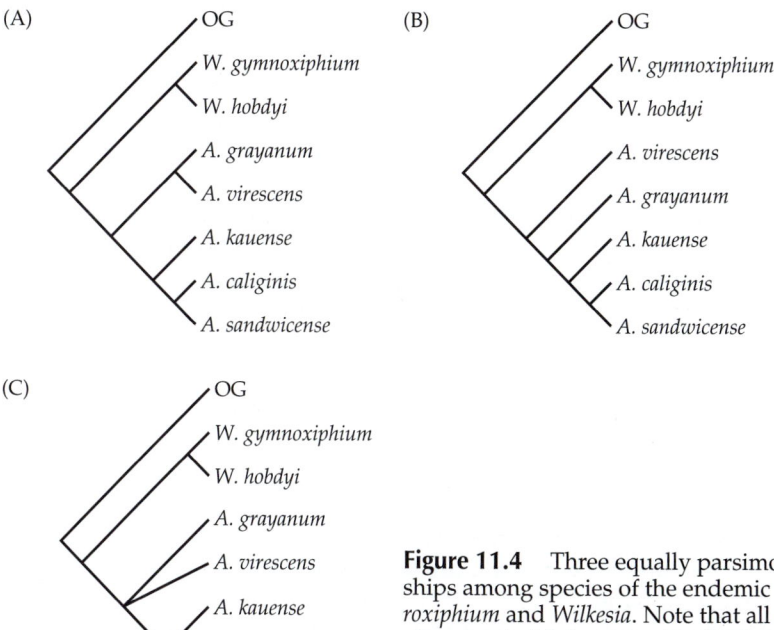

Figure 11.4   Three equally parsimonious cladograms depicting phylogenetic relationships among species of the endemic Hawaiian silverswords, plants of the genera *Argyroxiphium* and *Wilkesia*. Note that all of these phylogenetic reconstructions agree on the placement of *A. kauense, A. calignis,* and *A. sandwicense,* but hypothesize different relationships for the other two species, *A. virescens* and *A. grayanum.* OG indicates the outgroup used in this analysis. (From Funk and Wagner 1995.)

sequencing particular genes or indirectly by using restriction enzymes that "cut" the DNA molecule at particular sequences. Since different genes are now known to evolve at different rates, the molecular systematist can choose a gene likely to offer the desired degree of resolution: a fast-evolving one for working out relationships among recently derived, closely related species, or a more conservative one for resolving the more ancient basal branches of more distantly related lineages.

Initially, most studies in molecular systematics simply clustered taxa into a hierarchical scheme based on quantitative measures of genetic similarity. Increasingly, however, explicitly cladistic methods for analyzing and interpreting molecular data have been developed. These methods require some sophistication, because there are many kinds of possible changes in DNA sequences (i.e., inversions, insertions, and deletions of one or more nucleotides, as well as simple point mutations from one base to another), which differ in their probabilities of occurring repeatedly or reversibly. The rate of progress has been amazingly rapid. The technology for sequencing genes is being streamlined and automated, and improved computer algorithms and statistical methods are being developed for constructing cladograms from molecular data.

Limitations of Phylogenetic Classifications

All of these advances mean that more and better information on phylogenetic relationships is becoming available to assist biogeographers in their efforts to reconstruct the distributional histories of organisms (see Riddle 1996; Chapter 12). It is important, however, that biogeographers understand the limitations of cladistic classifications and reconstructions. First and foremost, a cladogram is a *hypothesis* about phylogenetic relationships. It is only as valid as the assumptions, data, and analytical methods used to construct it. There is some probability that any given cladogram contains errors. Often the main branches of a lineage are fairly well resolved, but there may be considerable uncertainty about the exact configuration of the terminal branches. Computer programs, such as PAUP, that construct cladograms from data sets provide one kind of estimate of uncertainty: they not only construct a consensus, or most parsimonious, tree, but also indicate the number of other tree topologies that are as consistent, or almost as consistent, with the data (see Figure 11.4). There are other kinds of uncertainty as well: for example, different data bases containing information on different characters, or different computer algorithms, can generate different tree topologies.

Further, there are some deeper conceptual problems with cladistic reconstructions. Most cladistic methods assume a branching rather than a reticulating pattern of evolutionary diversification. While this assumption may often be justified, in some kinds of organisms, such as plants, hybridization is common (see Chapter 8). Hybridization may sometimes occur between quite distantly related species in different genera and even families. And some microbes—plastids and viruses—are known to insert their own DNA into host genomes and to transfer genetic material between different hosts. It is easy to imagine how rejoining of branches due to hybridization, or lateral transfer of genetic traits among branches by microbes, could complicate efforts to reconstruct phylogenetic relationships—and how such reticulating evolutionary events, if they go undetected, could cause errors in the construction and interpretation of cladograms.

A final problem with phylogenetic reconstruction concerns the interpretation of cladograms in a historical context. Most cladograms, and all those based on molecular data, include only contemporary organisms. So what do we do with the fossil record? Some doctrinaire cladists would ignore it altogether, but most systematists would like to have ways to include fossil taxa in their reconstructions. For one thing, this would allow them to assign approximate dates to some of the branches. Problems remain, however, because a fossil species might be either an ancestor somewhere along a line of descent leading to contemporary forms, or a representative of a side branch that went extinct without leaving any living descendants (Figure 11.5). This problem brings up another feature of cladograms that systematists and biogeographers must bear in mind. While each node in a cladogram represents a past speciation event, not all speciation events—and none of the extinction events—are likely to be represented in a cladogram. The vast majority of organisms that ever lived went extinct without leaving any currently living descendants. Thus, any cladogram derived for contemporary taxa is likely to be a highly oversimplified representation of the actual pattern of phyletic evolution, missing many branches representing undetected speciation and extinction events.

The Fossil Record

The task of reconstructing the evolutionary histories of organisms has by long tradition been carried out by two different groups of scientists. We have just seen how systematists or taxonomists have been attempting to classify organisms into hierarchical groups (taxa) that summarize patterns of phenetic branching and evolutionary relationships. At the same time, paleontologists and paleobiologists have been studying the evolutionary histories of organ-

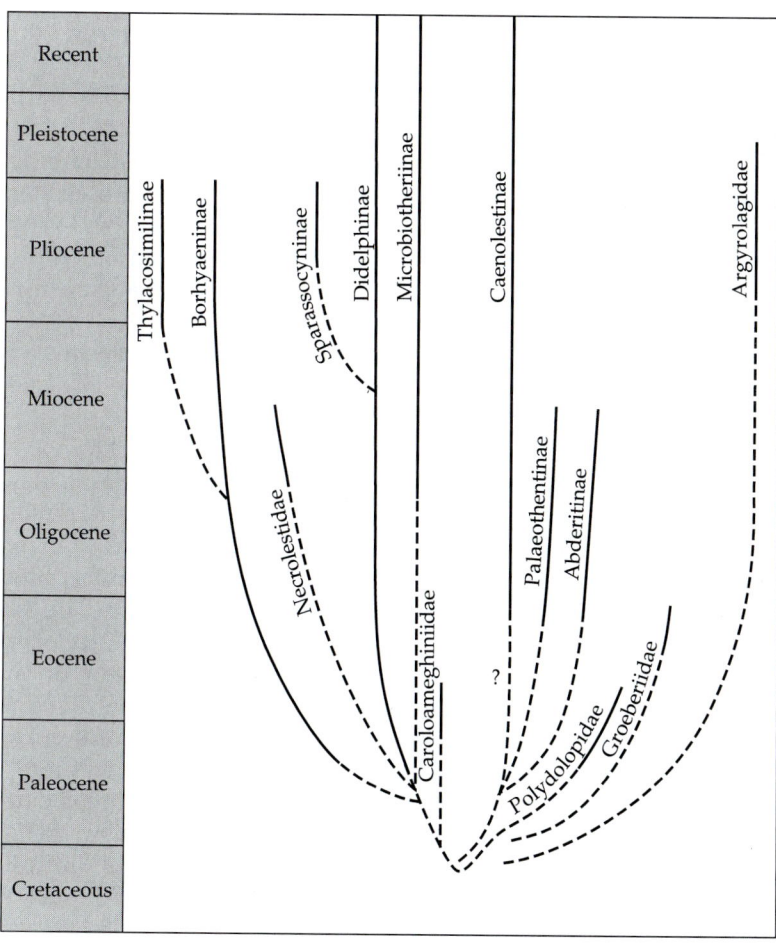

Figure 11.5 A phylogenetic hypothesis for the family-level relationships and evolutionary history of South American marsupials. The fossil record shows that this group had a wide radiation on the isolated South American continent during the early Cenozoic, but only three families have survived to the present. The solid lines show the known time spans of the families in the fossil record; the dashed lines show their inferred durations and relationships to other families. This hypothetical reconstruction uses a modification of cladistic methods, based on detailed information on the living forms supplemented by limited data on morphology and dates obtained from the fossils. A classification that employed molecular and strict cladistic techniques would show only the relationships among the three extant families, missing much of the rich history of this group of South American mammals. (After Patterson and Pascual 1972.)

isms directly from the fossil record. **Fossils**, the preserved remains of organisms that lived at various times in the past, provide the most direct factual evidence of evolution.

Limitations of the Fossil Record

Although it might seem more straightforward to rely on the fossil record than on constructed cladograms, the fossil record has its own set of practical and logical problems. The known fossil record is incomplete, and its interpretation can be difficult. There are many uncertainties in dating fossils, assigning them to taxonomic groups, understanding how the living organisms were structured and how they functioned, and reconstructing their paleoenvironments.

Only a minute fraction of all the organisms and species that have ever lived have been found as fossils. **Taphonomy**, a subdiscipline of paleontology, is concerned with the processes by which remains of living things become fossilized, and the ways in which these processes can bias the fossil record or cause problems of interpretation (e.g., Behrensmeyer et al. 1992). Animals lacking hard tissues and plants without durable chemicals in their cell walls are disproportionately poorly represented in the fossil record because they are readily decomposed. Some organisms are known only from their distinctive chemical remains, such as the limestone formed from coral reefs and the coal and oil formed from undecomposed biomass in ancient swamps and shallow seas. Some animals, especially soft-bodied forms, are known only from their tracks or burrows, called **trace fossils**. Of those fossils that are formed, many are destroyed by erosion or tectonic processes, and many others are inaccessible, hidden away in deep strata. The relatively recent fossil record is the most complete (Figure 11.6), one reason why so much more is known about organisms and their environments during the Pleistocene (see Chapter 7) than during earlier periods of earth history (see Chapter 6).

Even when organisms have been fossilized and those fossils have been discovered, problems of interpretation remain. Most fossils are deposited under water and preserved in sediments. Remains may have been transported by wind or by the currents of streams, rivers, lakes, and oceans and deposited far from where the organisms originally lived. A striking example is provided by the remains of recently deceased fish that are frequently washed up on the shore of Great Salt Lake. A naive observer might easily conclude that these fish live in the lake. In fact, the reverse is true: the fish die in the lake. They are killed when storms push the hypersaline water of the lake into the freshwater marshes and streams where they live, and their bodies float into the lake and are eventually washed ashore.

One consequence of such long-distance transport of remains before fossilization, therefore, is that species that occur together in fossil deposits may not have actually coexisted locally in the same communities. Pleistocene fossils of small vertebrates from river bottom and cave deposits often represent more species than occur together in single habitats today, probably in part because they have been collected from a larger area. It is easy to imagine how a river could collect sediments from a large basin, but it might be thought that a dry cave would contain fossils only of the animals from its immediate vicinity. Many of the small vertebrate fossils preserved in caves, however, were brought in by owls, which regurgitated the bones of prey captured in their territories, which may have been several square kilometers in area and included a variety of habitat types. So, the take-home message seems to be a disappointing one: the fossil record is biased in favor of easily preserved organisms, temporally and geographically uneven, still greatly unexplored, and sometimes difficult to interpret.

(A) Area

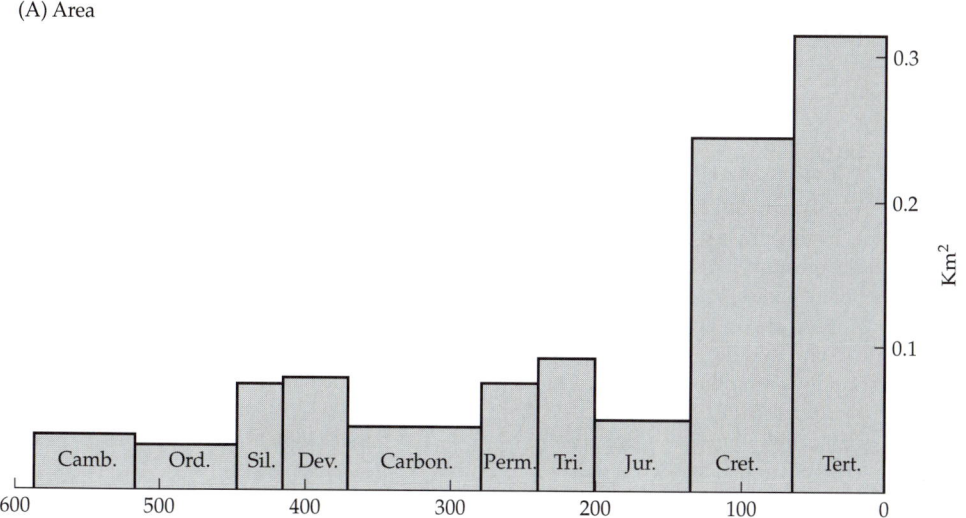

(B) Volume

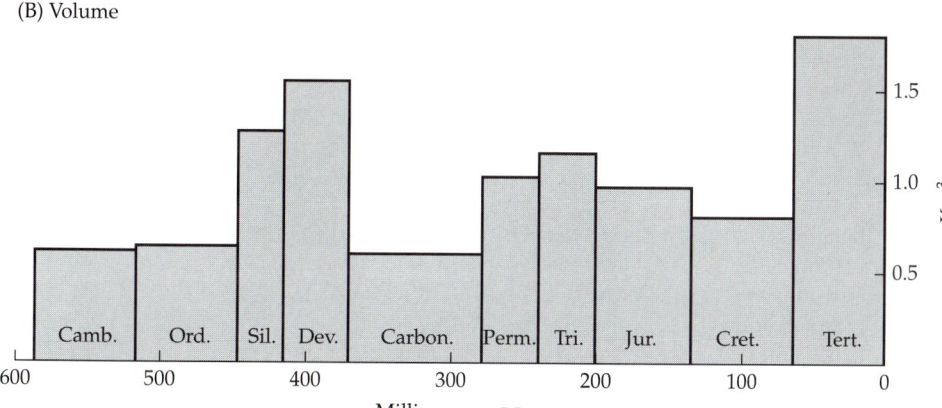

Million years B.P.

Figure 11.6 Estimated area (A) and volume (B) of potentially fossil-bearing sedimentary geological strata of varying ages (here measured as area or volume of rock per year of the past preserved). Note that the area of exposed rocks decreases with age, introducing an important sampling bias into the fossil record. Although some ancient strata are extensive and rich in fossils, in general there is simply less opportunity to find fossils from older geological periods. (After Raup 1976.)

Biogeographic Implications of Fossils

Despite these limitations, a great deal of information about the history of life on earth can only or best be obtained from fossils. Only fossils record with certainty the kinds of organisms that occurred at particular times and places in the past. Fossils provide the ages when different taxonomic lineages lived (Table 11.1) and when they occurred in particular areas. Because of the incomplete nature of the record, however, the organisms may have been present for unknown lengths of time before and after their known fossils were preserved. Fossils document some of the enormous diversity of prehistoric life, and some of the important events in the histories of lineages: speciations and radiations, expansions and contractions of ranges, and extinctions.

Fossil histories of lineages.   Fossils are sources of several kinds of invaluable historical information. For one thing, they can provide insights into the history of a lineage that cannot be obtained from cladistic reconstructions based solely on recent material. They can supply data on the characteristics and character states of earlier representatives of the lineage, some of which may have been the ancestors of contemporary forms. But even when the fossils are not on the direct branches leading to living representatives, they can show the unique combinations of traits that these organisms possessed. Thus, while recent analyses suggest that *Archaeopteryx* may not have been, as was once thought,

Table 11.1
Oldest known fossils of selected taxa

Taxon	Oldest known undisputed fossil	Age (10^6 yr)	Number of extant species
Vascular plants (Tracheophyta)	*Cooksonia*	Pridolian, U. Silurian (405)	232,000
Ginkgoads (Ginkgoales)	*Sphaerobaiera*	U. Permian (240)	1
Angiosperms (Annonopsida)	*Retimonocolpites* (and others)	Barremian, L. Cretaceous (117)	220,000
Sunflower family (Asteraceae)	*Echitricolporites*	L. Miocene (20)	21,000
Crabs (Brachyura)	*Eocarcinus*	L. Jurassic (180)	4,500
Insects (Insecta)	Protorthoptera	Namurian, L. Pennsylvanian (320)	800,000
Lampreys (Petromyzones)	*Mayomyzon*	M. Pennsylvanian (290)	31
Anurans (Salienta)	*Vieraella*	L. Jurassic (180)	2,500
Turtles (Testudines)	*Proganochelys*	U. Triassic (200)	230
Birds (Aves)	*Archaeopteryx*	U. Jurassic (150)	8,600
Mammals (Mammalia)	*Kuehneotherium* (and others)	U. Triassic (200)	4,100
Monotremes (Prototheria)	*Obdurodon*	Miocene (20)	3
Bats (Chiroptera)	*Icaronycteris*	L. Eocene (50)	860

Note: Number of extant species cannot be predicted from age of oldest fossil or projected age group.

a direct ancestor of living birds (Feduccia 1996), its exquisitely preserved Jurassic fossils show the special combination of avian and reptilian traits found in a bird that lived 150 million years ago. Knowing the characteristics of earlier representatives permits tests of cladistic assumptions about the kinds and directions of change in character states that occurred during the evolution of a lineage.

In addition, fossils can be dated (see Chapter 6) to provide fairly precise information on when prehistoric representatives of a lineage showing particular degrees of differentiation lived on earth. Such dating allows cladograms, which depict only relative times of lineage branching, to be calibrated with respect to real time. This calibration is invaluable, because the timing and rates of diversification have been very different among different plant and animal groups, and even among different subclades within the same group. Such information can have important implications for biogeography. For example, in a fossil deposit dated about 15 million years B.P. from the mid-Cenozoic of the Magdalena River Basin in Colombia, South America, Lundberg et al. (1986) found the exquisitely preserved remains of a fish that was identified as a living species, *Colossoma macropomum*, which currently occurs in the Orinoco and Amazon basins. The discovery of fossils of this and other fishes on the western side of the Andes, where they no longer occur, allowed minimum dates to be assigned to events of speciation and extinction as well as to dispersal between now isolated watersheds (see also Lundberg and Chernoff 1992).

Past distributions.   Another unique biogeographic contribution of fossils is their ability to document localities of past occurrence. There are countless examples of fossil representatives of lineages being found far outside the current geographic range of the group, and often such occurrences provide key answers to interesting biogeographic puzzles. Of course, it must be borne in mind that when the fossils were deposited, the environment may have been very different than it is today. Some fossils have been transported on drifting crustal plates to locations far from where the organisms originally lived. Thus, for example, the discovery in Antarctica of an ever-growing number of fossils, including the plants *Nothofagus* and *Glossopteris* (Schopf 1970a, 1970b, 1976), the reptiles *Lystrosaurus*, *Thrinaxodon*, and *Procolophodon* (Elliot et al. 1970), and marsupials (Woodburne and Zinmeister 1982; Goin and Carlini 1995; Woodburne and Case 1996), not only provides increasing evidence of the rich biota that continent once shared with Gondwanaland, but also indicates that

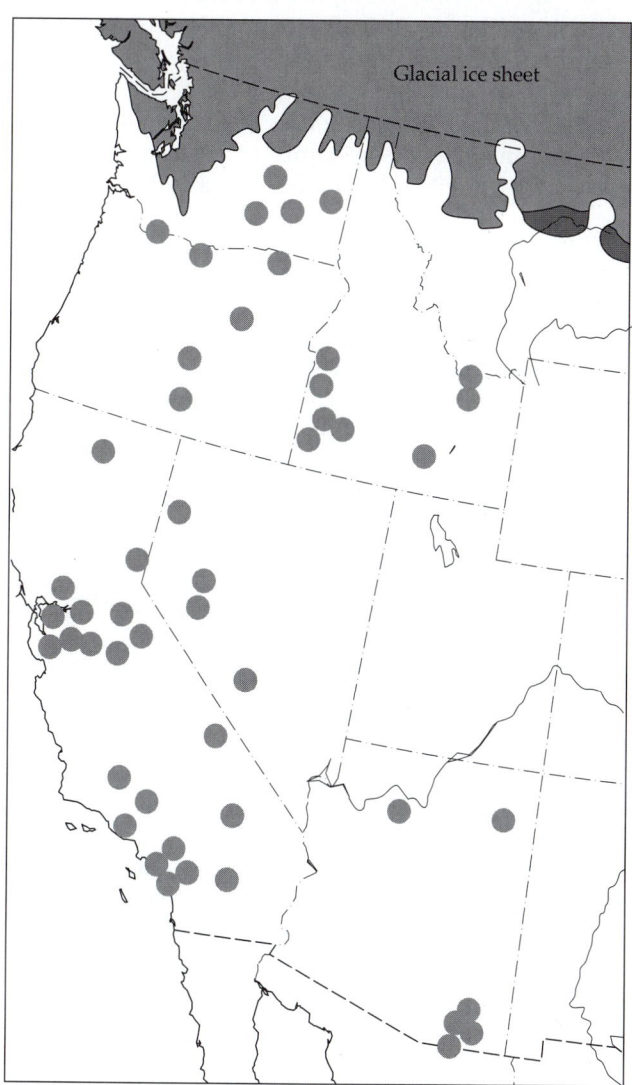

Figure 11.7   Fossil localities of camels (family Camelidae) from the Pleistocene in western North America. The records show that these two kinds of mammals were widely distributed south of the continental ice sheet (shaded area) until they went extinct abruptly—probably due to human influences—less than 20,000 years ago. The extinctions left a disjunct distribution of camelids with guanacos and vicanas in South America and dromedary and batrician camels in the Old World. (From Hay 1927.)

Antarctica did not always occur at polar latitudes and was once much warmer than it is today.

Fossils show that many now narrowly endemic forms are relicts of much more diverse and widely distributed lineages. Examples of such fossils include a member of the avian family Todidae, now endemic to the Greater Antilles, from the Oligocene of Wyoming (Olson 1976) and a platypus, a representative of the order of egg-laying mammals, Monotremata, now restricted to Australia, from the Paleocene of Argentina (Pascual et al. 1992). Cenozoic fossils of tapirs and camels from temperate North America show how intervening populations went extinct to create the present disjunct distributions between South America and Asia (Figure 11.7). The extent of the historic range contractions that can be documented using the fossil record can be impressive. We know that some "living fossils," such as the superficially lizardlike tuatara (*Sphenodon punctatus*), now found only on a few small islands of New Zealand (see Chapter 10), were members of once widespread groups. Another spectacular example is the fish family Ceratodontidae, now represented by a single species, *Neoceratodus fosteri*, in Queensland, Australia, but known from fossils found on all the continents except Antarctica (Figure 11.8). Many cladistic or phylogenetic biogeographers are still attempting to reconstruct past geographic distributions of lineages without using fossil evidence (see Chapter 12), but as new fossil finds are made, they promise to provide strong evidence to support or reject these phylogeographic hypotheses.

Paleoecology.   Finally, fossils and the other materials that are preserved with them provide invaluable information on the nature of past environments. Some of this information comes from the physical and chemical composition

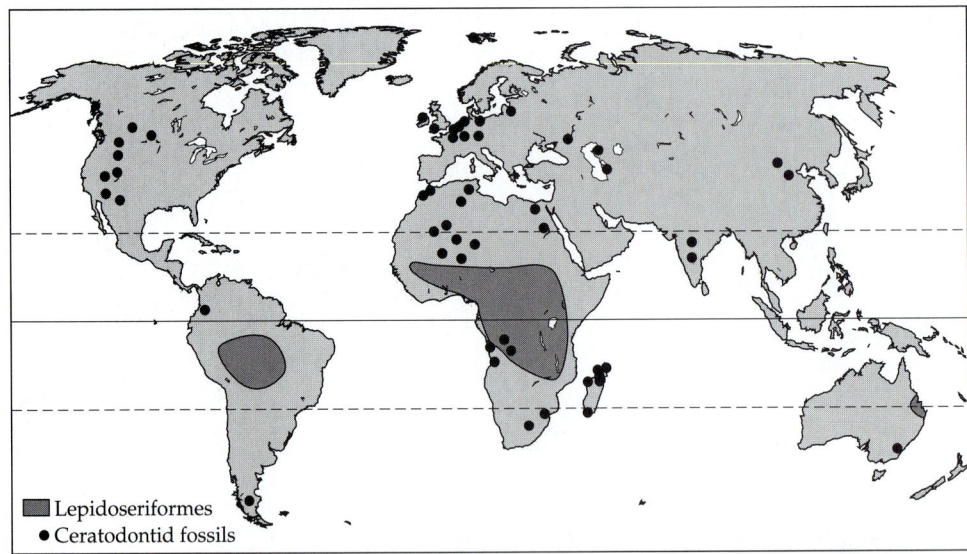

Figure 11.8   Information from the fossil record is essential for interpreting biogeographic history. The shaded areas show the current relictual distribution of the lungfishes (Lepidoseriformes) on three southern continents. The circles show fossil localities for one of the three lungfish families, Ceratodontidae. This family contains just one surviving species (*Neoceratodus fosteri*), which is restricted to Queensland, in northeastern Australia, but was distributed nearly worldwide in the Mesozoic. Accurate dating of the fossils is obviously crucial to determining how tectonic events, such as the breakup of Gondwanaland, affected the distribution of the lungfishes. (After Sterba 1966; Keast 1977a.)

of fossil-bearing rocks. Their particle size and structure indicates the nature of the substrate and the surrounding geological formations where the remains were preserved. Their chemical composition can indicate whether the environment was marine, freshwater, or terrestrial, anoxic or oxygen-rich, and can often give some idea of the temperature. For example, the fossilized remains of ancient coral reefs form distinctive limestone rocks, and since reef-building corals require sunlight for their symbiotic algae and water temperatures above about 20°C, these strata indicate the past occurrence of shallow tropical and subtropical seas (Figure 11.9). The orientation of crystals due to paleomagnetism indicates the latitude of liquid volcanic rocks when they were cooling. Mapping of the past positions of the drifting continents and oceans relative to the equator has enabled paleobiologists to determine that the latitudinal gradient of species diversity is an ancient feature of the biogeography of the planet, dating back at least 100 million years (see Chapter 14; Stehli 1968; Stehli et al. 1969; Stehli and Wells 1971; Crane and Lidgard 1989).

Remains of organisms preserved together in the same strata can often provide abundant information on the climate, vegetation, and biotic communities when these organisms were alive. Perhaps the fossils most informative about paleoecology are **catastrophic death assemblages**. The lives and environment of the people inhabiting the Roman cities of Pompeii and Herculaneum are frozen in time, exquisitely preserved by the ash spewed out by the eruption of Mount Vesuvius in 79 A.D. Similarly, many soft-bottom marine assemblages were preserved in situ when they were covered by large quantities of sediment. Individuals were trapped in their exact locations, providing an invaluable record of the depth and orientation of burrows, the size and age structure of populations, and the species composition and spatial relations of communities. Extensive fossil deposits in the Niobrara River basin of Nebraska preserve a catastrophic death assemblage of the terrestrial mammals that inhabited the area during the Miocene, about 4 million years ago.

Pleistocene assemblages provide a wealth of paleoecological information. Not only is there a large quantity of well-preserved fossil material, but the fact that most of the plants and animals were very similar to their living descendants enables us to make strong inferences about their behavior and ecology. Not only do data on oxygen isotope ratios provide information on global, regional, and local temperatures (see Chapter 7), but the kinds of organisms and their characteristics also allow for reconstructions of past climates. Fossil pollen from lakes and bogs and plant fragments from woodrat middens provide so much information on the occurrences and abundances of still extant plant species that it is possible to reconstruct the structure and composition of Pleistocene plant communities and the climatic regimes in which they lived. From plant and animal fossil deposits, it is also possible to document the shifts in geographic ranges and changes in composition of communities as a consequence of the advances and retreats of glaciers and the associated changes in climate during the Pleistocene. For example, since the maximum of the Wisconsin glacial period about 20,000 years ago, some small mammals have shifted their ranges hundreds of kilometers, and not always consistently in a northward direction, as would be expected if the climate had simply warmed up uniformly. As a consequence, Pleistocene communities contained combinations of species strikingly different from contemporary communities (see Chapter 7; see also Graham 1986; Graham et al. 1996).

Fossils can provide convincing evidence of the role of biotic interactions in the past. As discussed in Chapter 8, many researchers have used shifts in dominance over time between certain functionally similar but distantly related groups of organisms to infer that competition has played a major role in the

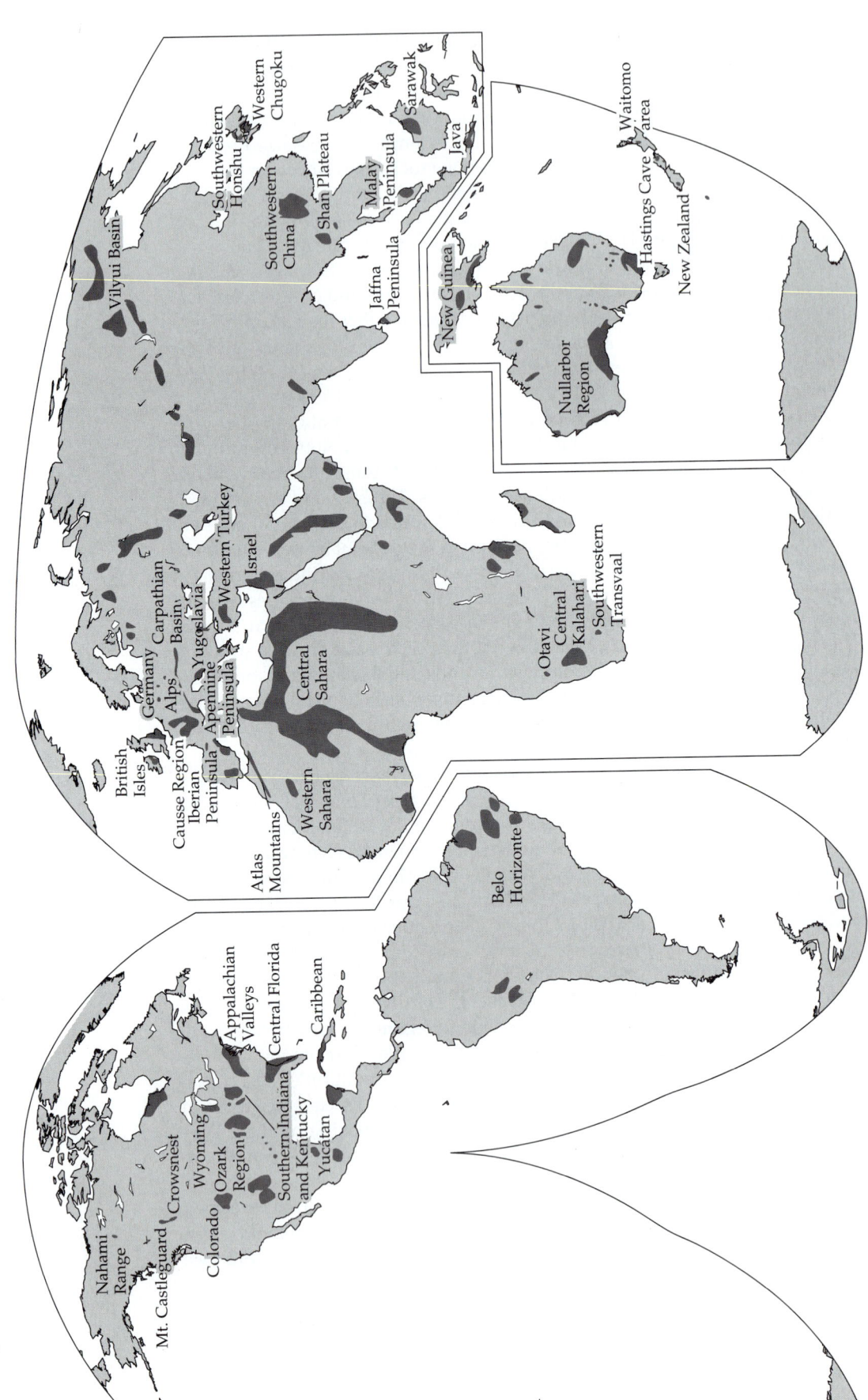

Figure 11.9   Global distribution of limestone, indicating past locations of shallow marine waters and coral reefs. Note that many of these limestone formations occur far from contemporary coastlines, sometimes near the centers of continents. (From Snead 1980.)

history of these lineages: brachiopods and mollusks (Elliot 1951; but see Gould and Calloway 1980), gymnosperm and angiosperm plants (Knoll 1986), and multituberculate and rodent mammalian herbivores (Simpson 1953; Lilligraven 1972). Pollen profiles and woodrat middens provide evidence of successional dynamics and reorganization of past plant communities in response to shifts in climate (e.g., Davis 1986, 1994; Delcourt and Delcourt 1977, 1994, 1996; Woods and Davis 1989). Vermeij (1974, 1978) has used patterns of co-occurrence of different groups of carnivorous and herbivorous mollusks and evolutionary trends in shell form to document the history of the coevolutionary arms race between predators and prey throughout most of the Phanerozoic. But such coevolution proceeds slowly, at least in mollusks. The drill holes made by predatory snails in the shells of their prey are clearly visible on well-preserved fossils. The sizes and locations of 100,000-year-old drill marks are identical to those found on contemporary shells, indicating that there has been no detectable change in predatory behavior over that period (Tull and Bohning-Gaese 1993). Fossils are also providing incontrovertible evidence of the role of predation by aboriginal humans in the extinctions of other vertebrates. The correspondence between the time of arrival of humans and wholesale extinctions of native birds and mammals on islands and continents shows the effect that an invading predator can have on its prey, and demonstrates that humans have been having a major influence on global biodiversity for thousands of years (see Chapter 17).

Toward a Historical Synthesis?

We conclude this chapter by noting that knowledge of the history of lineages is being advanced rapidly by research in two areas. On the one hand, systematists have been using the techniques of cladistics and molecular genetics to reconstruct phylogenetic trees. On the other hand, paleontologists and paleobiologists have been using new finds and better interpretations of fossil remains to document what kinds of organisms lived in particular places at different times in the history of the earth. Unfortunately, there has been too little communication between these two groups of scientists, and too little synthesis of their findings. In part, this is because the scientists themselves historically came from different backgrounds—systematists from biology and paleontologists from geology—and because they present their results in different journals and meetings. But in part, it is because they have traditionally treated each other as rivals and with suspicion, rather than as mutualists who can bring different backgrounds, techniques, and perspectives to a common endeavor: the formidable task of reconstructing the history of life on earth. Given the importance of this endeavor, and the central role that it must play in historical biogeography, let us hope that in the future there will be more mutualism and less competition, more synthesis and less parochialism than in the past.

CHAPTER *12*

Reconstructing Biogeographic Histories

Traditionally, the goal of historical biogeography has been to develop and evaluate hypotheses about how the distributions of organisms changed as the evolutionary histories of lineages played out over the changing surface of the earth. A corollary endeavor, which has long been a component of biogeographic research, has been the development of hypotheses about earth history based on the geographic distributions of organisms.

The realization that historical tectonic events have caused changes in the positions and relationships of landmasses and water masses over geological time, together with recent advances in phylogenetic systematics, has stimulated a revitalization of historical biogeography. If the climate of the earth and the positions of continents had remained fixed over time, this would drastically limit the kinds of hypotheses that could be advanced about worldwide patterns of endemism, provincialism, and disjunction. For example, there would be only one likely explanation for the disjunct distribution of a terrestrial organism between South America and Australia: the ancestors originally occurred on one of those continents, and individuals dispersed long distances across extensive oceanic barriers to colonize the other. Application of the theory and data of plate tectonics, however, provides an alternative hypothesis: the ancestors originally inhabited Gondwanaland, and the present occurrences are vicariant relicts created by the fragmentation of that ancient supercontinent. Of course, there are more complicated versions of both hypotheses, perhaps involving past occurrences of the lineage on the northern continents and extinction of those populations, but some combination of over-water dispersal and continental drift is required to explain the current disjunction. Using information from phylogenetic reconstruction, plate tectonics, and the fossil record, it should be possible to erect and evaluate appropriate hypotheses. How definitive the answers will be will depend on the quantity and quality of the evidence. For example, if it could be shown that the ancestors evolved after the breakup of Gondwanaland, then some long-distance, over-water dispersal would be required to explain the disjunction, but additional information would be required to reconstruct the most likely dispersal path.

Early Efforts: Determining Centers of Origin

Concepts and Criteria

To appreciate the enormous recent advances in historical biogeography, it is instructive to consider the state of the discipline just a few decades ago. Since the time of Buffon and de Candolle (see Chapter 2), biogeographers have realized that the distributions of organisms have shifted over time. This realization led naturally to efforts to determine the birthplace, or **center of origin**, of each taxon. Two general questions motivated the search for centers of origin. First, investigators wanted to know whether certain geographic regions have served as cradles for the evolution of new kinds of organisms, and of the special features that allowed these lineages to be successful. Second, biogeographers wanted to understand how biotas have been assembled: where taxa started from, what routes they followed to disperse around the world, and what factors produced present patterns of endemism, provincialism, disjunction, and diversity. These are still the goals of historical biogeography, but fortunately, the methods used to infer the origins and movements of lineages have improved greatly.

Initially, biogeographers tried to develop simple rules for determining the center of origin for any taxon. For example, Adams (1902, 1909) listed ten criteria that could help to identify such centers. Some authors insisted categorically on using a single criterion to the exclusion of others. Matthew (1915), for instance, followed Buffon in believing that centers of mammalian origin were in the Holarctic. He argued that new, successful forms arose in response to the challenges of temperate climates; these forms eventually supplanted their progenitors and other lineages, forcing them into peripheral habitats, then to lower latitudes, and eventually into the Southern Hemisphere. Thus Matthew thought that the center of origin is where the derived forms reside. This is in direct opposition to the idea that the center of origin is where the primitive forms live today, sometimes called the progression rule (Hennig 1966). According to this rule, an ancestral population remains at or near the point of origin and derived forms disperse outward. In actuality, either explanation might be correct for different groups, because displacement of one form by another depends in part on how they speciate, disperse, and interact with their biotic and abiotic environments.

A thorough evaluation of the problem of determining centers of origin was made by the phytogeographer Stanley Cain (1944). He listed the thirteen criteria in use at the time (Table 12.1), and argued convincingly that none by itself could reliably identify the center of origin. Some authors had claimed, for example, that the center of origin should be where the greatest number of species in a group resides. This assumption is, of course, invalid if the majority of forms inhabit a region of secondary radiation, as do the heaths of South Africa (*Erica*), with 605 species in the Cape Region (Baker and Oliver 1967), or the hundreds of species of *Drosophila* on the Hawaiian Islands (see Chapter 14). Is the location of a primitive form or the earliest fossil an absolute criterion? No, said Cain, because primitive forms often survive in isolated regions containing few competing species and located far from their original ranges. Examples of such relicts are described in Chapter 10.

An Example: Sea Snakes

To give the flavor of a center of origin scenario, let's examine one case in which a number of these criteria point to a similar conclusion. Consider the sea snakes (Hydrophiidae) of the Indo-Pacific region, a group whose fossil record

Table 12.1

Criteria used and abused for indicating center of origin of a taxon

1. *Location of greatest differentiation of a type (greatest number of species)*
2. *Location of dominance or greatest abundance of individuals (most successful area)*
3. *Location of synthetic or closely related forms (primitive and closely related forms)*
4. *Location of maximum size of individuals*
5. *Location of greatest productiveness and relative stability (of crops)*
6. *Continuity and convergence of lines of dispersal (lines of migration that converge on a single point)*
7. *Location of least dependence on a restricted habitat (generalist)*
8. *Continuity and directness of individual variation or modifications radiating from the center of origin along highways of dispersal (clines)*
9. *Direction indicated by geographic affinities (e.g., all Southern Hemisphere)*
10. *Direction indicated by the annual migration routes of birds*
11. *Direction indicated by seasonal appearance (i.e., seasonal preferences are historically conserved)*
12. *Increase in the number of dominant genes toward the centers of origin*
13. *Center indicated by the concentricity of progressive equiformal areas (i.e., numerous groups are concentrated in centers, and numbers decrease gradually outward)*

Source: After Cain (1994).

is unknown. Sea snakes are venomous marine predators closely related to the Elapidae—the cobras, kraits, coral snakes, and mambas. Over 50 species are known from tropical and subtropical coastal habitats along reefs in the western Pacific and Indian oceans, including some brackish inlets and rivers; several species have entered and adapted to freshwater habitats in the Philippines and the Solomon Islands (Figure 12.1; Dunson 1975; Heatwole 1987). The species with the widest distribution, which includes the geographic range of the entire family, is the pelagic form *Pelamis platurus* (Figure 12.2), which ranges across the tropical Pacific to the west coast of the Americas, from Mexico to Ecuador. Determining a center of origin for sea snakes is made easier

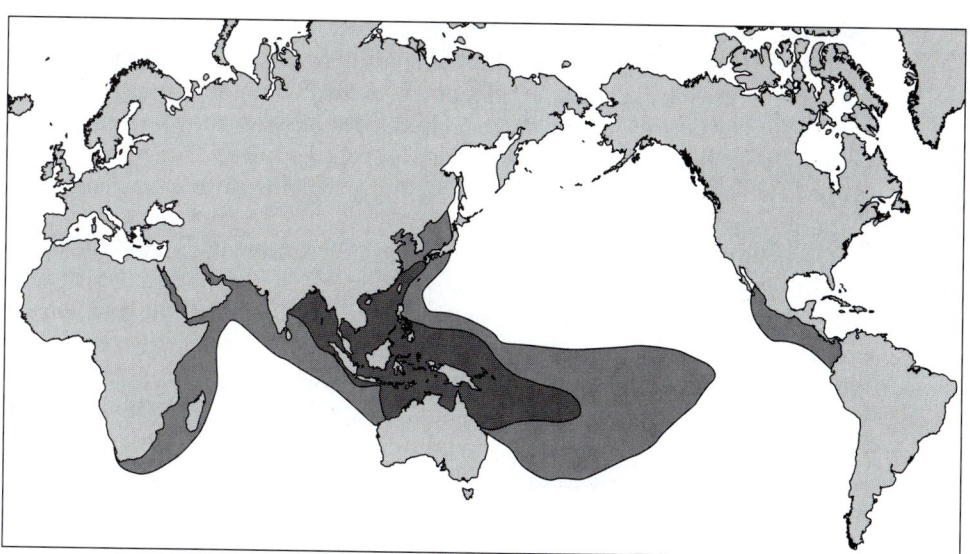

Figure 12.1   Map showing the current distribution of sea snakes (family Hydrophiidae). All but one species are confined to the densely shaded area. The exceptional, very wide-ranging species, *Pelamis platurus*, occurs in this region, but also in the more lightly shaded areas. The concentration of species and lineages in the Indo-Pacific oceans and the occurrence of the seemingly most closely related outgroup in swamps in Australia led Cogger (1975) and others to hypothesize that the sea snakes spread from a center of origin in the Australia-New Guinea region. (From Heatwole 1987.)

(A)

(B)

Figure 12.2   Two species of sea snakes. (A) *Pelamis platurus* is pelagic, and is the most widespread species in the group. This individual was washed ashore on a beach in western Mexico during a storm. (B) *Laticauda laticauda* is representative of a distinctive group often classified as a separate tribe or subfamily. This specimen was photographed in shallow water on the island of Fiji. (A courtesy of R. E. Brown; B courtesy of W. A. Dunson.)

because they possess several useful attributes: (1) inability to tolerate cold water (below 20° C) has created an absolute barrier, confining them to the Indian and Pacific oceans; (2) they appear to be a young family (of Cenozoic age); (3) their closest relatives (the Elapidae mentioned above) overlap with them in distribution; and (4) the lines of morphological specialization appear to be relatively unbroken, thus eliminating problems of tracing geographic patterns obscured by major extinctions.

Cogger (1975) concluded that the Hydrophiidae originated in the Australia-New Guinea region. This is the region of highest hydrophiid diversity, with more than 30 species. Here too live their putative elapid ancestors (*Rhinoplocephalus* and *Drepanodontis*), swamp-inhabiting Australian snakes, as well as the most primitive and unspecialized hydrophiids (McDowell 1969, 1972, 1974). *Ephalophis*, the putative missing link, shares several traits with the elapids and is a notably weak swimmer (Dunson 1975). Most genera of Hydrophiidae are classified in the *Hydrophis* group, and most of these taxa are Australian. Genera that range to the northwest of Australia often have at least one representative in the Australian coral reefs as well. Finally, the genus *Lati-*

cauda (Figure 12.2), which appears to be an early offshoot of the family (a separate tribe or subfamily), occurs around New Guinea. [Cogger's reconstruction is not universally accepted, however; see Voris (1977) for a somewhat different interpretation of the same systematic and geographic data.]

Scenarios like the above are just that: stories concocted to explain most or all of the known facts about a group that might be relevant to its history. In a sense, such scenarios are hypotheses, because many specific parts of the story can potentially be falsified. For example, all it would take to reject the above scenario would be the discovery of a fossil along the west coast of Africa or a phylogenetic analysis showing that the sea snakes are more closely related to the New World pit vipers of the genus *Bothrops* (fer-de-lance and relatives) than to the Old World elapids. On the other hand, there is precious little direct evidence to support the above scenario for sea snakes. In particular, the story makes no reference to a rigorous phylogenetic analysis or to fossil evidence, and it is not linked to any specific events in earth history. This is not to say, however, that when more and better evidence becomes available, all or parts of this scenario will not be supported; that remains to be determined.

Critical Issues

Our reasons for going into so much detail about the center of origin concept are three. First, inferring centers of origin using some or all of the criteria in Box 12.1 was common practice in historical biogeography until quite recently, up until the 1970s and 1980s—and it is still practiced, at least implicitly. When other evidence is not available, and a group is diverse and widely distributed in one region but has a few outliers in another, the former region is often assumed to have been inhabited for a long period and the latter region to have been more recently colonized. For example, several clades of small birds, including the hummingbirds (Trochilidae), flycatchers (Tyrannidae), wrens (Troglodytidae), tanagers (Thraupinae), blackbirds (Icterinae), and sparrows (Emberizinae), that occur in temperate North America but are much more diverse in the Neotropics are often assumed to have originated in South America and subsequently to have colonized North America as the two continents drifted into proximity during the Cenozoic.

Second, scenarios about centers of origin and directions of subsequent dispersal have historically been used to make sweeping generalizations about the innovations that triggered the success of lineages, and about global patterns of diversity. We have mentioned that first Buffon (1761) and then Matthew (1915) argued that new lineages and new innovations arose predominantly as adaptive responses to the climatic rigors of the temperate zones, especially in the Holarctic, and then spread southward to thrive and diversify in the milder tropics. The exact opposite position was advanced by Darlington (1959b), who argued that new lineages and new innovations were more likely to have originated in the tropics and spread toward the poles. He suggested that the greater diversity of species and higher taxa at low latitudes meant an increased probability that some of these "evolutionary experiments" would be successful.

These questions about the origins of lineages and their unique attributes remain of great inherent interest (see Chapter 15). Biogeographers, systematists, and ecologists still debate whether the tropics are the source or the last refuge, or both, of the distinctive combinations of traits that characterize lineages. A related example is the discussion among marine paleobiologists of whether rates of origination of new lineages and extinction of existing ones have been higher in more abiotically stressful onshore or more benign offshore environments (e.g., Jablonski et al. 1983; Miller 1989; Jablonski and Bottjer 1990, 1991).

Third, we now have much better methods and data to use in reconstructing biogeographic histories. Currently there are signs of an emerging consensus in historical biogeography, but for the last few decades the field has been wracked by discord and controversy (e.g., see Nelson and Rosen 1981), and some of this divisiveness persists. To appreciate the evolution of the ideas that have been put forward, some of which seem silly in retrospect, it is helpful to see what their proponents were reacting against. It is clear that historical biogeography, based largely on studies of single taxa and plagued by flawed concepts of centers of origin and dogmatic ideas about dispersal (see Chapter 2), was in need of more rigorous approaches.

Panbiogeography and Vicariance Biogeography

Croizat's Panbiogeography

As is often the case with revolutions, the rapid changes in historical biogeography, which took place largely in the 1970s, were triggered by a few individuals with radical ideas. And like the bizarre evolutionary changes in organisms that have occurred on islands, novel and divergent ideas often seem to arise in isolated circumstances. Thus it was that Leon Croizat, an unorthodox plant biologist who labored for most of his career in relative obscurity in Venezuela, provided the spark that ignited the revolution in historical biogeography. Croizat (1952, 1958, 1960, 1964), like many phytogeographers before him, recognized that the present limited distributions of narrowly endemic, disjunct taxa are the relicts of ancestral taxa, and often of entire floras, that were more broadly distributed in the past. Unlike most previous researchers, who typically studied just one taxon, Croizat compared the distributions of different groups. He realized that very distantly related kinds of organisms often exhibited similar disjunctions, and that these distributional patterns were legacies of historical events that influenced many different lineages. Amassing data on disjunctions from all over the world, Croizat developed an approach that he termed **panbiogeography**. He argued that each pattern of multiple disjunctions reflected the fragmentation of a biota that originally inhabited interconnected regions. He plotted the ranges of narrowly endemic species on a map, and then drew lines, called tracks, connecting the distributions of the most closely related taxa. Croizat then superimposed the maps for multiple taxa and noted where the positions of their tracks coincided. He called the resulting tracks multiple or generalized tracks (Figure 12.3). He inferred that such tracks indicate historical connections—that they were pathways taken by the isolated fragments of a formerly continuous biota. By plotting all the generalized tracks on a map, he could theoretically recreate the way in which regional biotas had developed in time and spread over space.

When first published, Croizat's ideas received little favorable response. For one thing, the basic concept of historical subdivision and isolation of formerly widespread biotas was hardly new. For another, Croizat arrogantly dismissed alternative views while zealously promoting his own. For a third, some of his ideas and examples were so extreme that many of his contemporaries did not take his work seriously. His books contained many technical errors, and some of his purported disjunctions were based on questionable systematics. He categorically rejected dispersal over long distances and across barriers to account for the generalized tracks because he reasoned that "dispersal must be orderly and continuous in time and space" (1958). He argued that long-distance dispersal could not have been a major mechanism of geographic range expan-

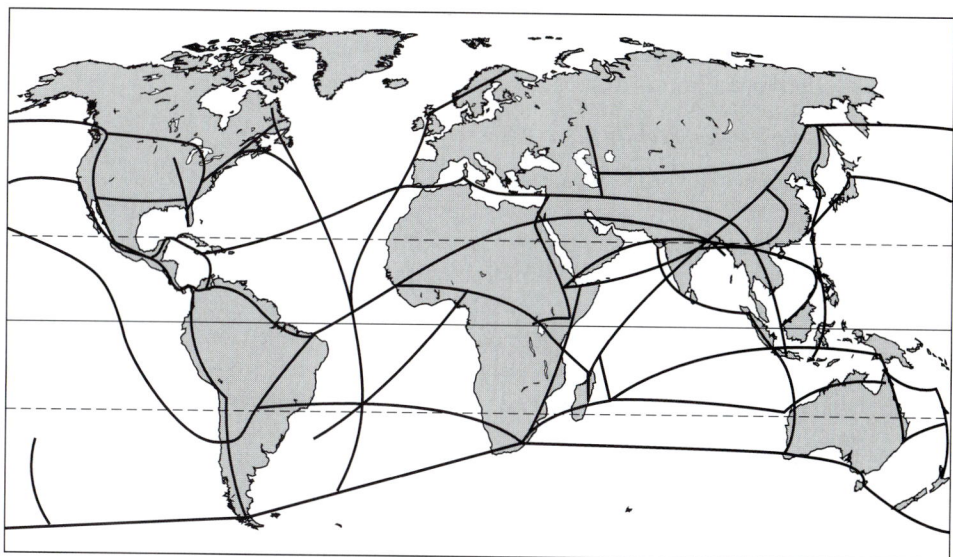

Figure 12.3   A composite drawing of the major and moderately important "tracks" drawn by Croizat (1952, 1958, 1960, 1964) to show hypothesized past connections between the distributions of terrestrial organisms that currently occur in widely separated regions. These tracks were derived by plotting the distributions of endemic species within several taxa and then drawing lines to connect regions that share disjunct endemics in several taxonomic groups. The tracks radiating out from Madagascar, for example, show that its endemics have close affinities with disjuncts in Africa, India, Australia, and the East Indies. While many of Croizat's tracks correspond to past land connections, others are highly problematic. Note, for example, the lines connecting northeastern North America and northwestern Europe with islands and continents in the South Atlantic Ocean.

sion, even for the colonization of distant oceanic islands. And while continental drift would seemingly have provided a mechanism to explain many of the observed disjunctions, Croizat, in his earlier writings, denounced Wegenerism (see Chapter 5). Rather than moving continents and oceans, he kept their locations fixed and invoked ephemeral land and water bridges to explain most of the tracks connecting disjunctions, such as those among the southern continents.

Today, except for a few zealous disciples of panbiogeography (e.g., Craw 1982, 1988; Heads 1984), most biogeographers recognize Croizat's approach as being flawed by its unrealistic and idiosyncratic assumptions (Nelson and Platnick 1981; Patterson 1981; Parenti 1981; Humphries and Parenti 1986; Page 1990; Morrone and Crisci 1995). Nevertheless, Croizat made an important contribution to the field. He emphasized the need for formalized methods of inferring past distributions from present ones. His search for distributional patterns that hold across many different kinds of organisms was a major departure from much previous work, which had been content to confine itself to a single taxon and to propose ad hoc events and taxon-specific mechanisms to account for its distribution. While he went to the extreme of insisting that his methods be applied uniformly across all organisms without regard for their dispersal mechanisms and times of divergence, Croizat was reacting to the lack of rigor in the methods that many of his predecessors and contemporaries had been using to infer centers of origin, mechanisms of dispersal, and directions of range expansion, and thus to develop historical scenarios to explain contemporary distributions. Croizat rightly pointed out that even though these narratives may be correct, many of them are ad hoc, untested constructs of human imagination. To materially advance the science, they needed to be framed and tested as rigorous hypotheses.

Hennig's Progression Rule

As intimated above, a weakness of Croizat's work was that it lacked a solid systematic and phylogenetic foundation. If purported disjuncts are not in fact close relatives, there is obviously no point in hypothesizing past connections between them. Without information on the relative timing of lineage branching, it is difficult to reconstruct the historical sequence of the formation and severing of geographic connections. It was Willi Hennig (1966), who developed the conceptual foundations of phylogenetic systematics (see Chapter 11), who also emphasized the importance of phylogenetic relationships in reconstructing biogeographic history. He emphasized that the hierarchical pattern of ancestor-descendant relationships among the taxa in a clade also has a spatial history. The unique sequence of lineage branching must reflect a unique sequence of geographic range dynamics, and vice versa.

Hennig's view of the interrelationship between phylogeny and biogeography is embodied in his concept of a **progression rule**. He noted that as lineages diverge from their ancestors, they acquire new traits, largely by the evolution of ancestral or plesiomorphic character states into derived or apomorphic ones. Similarly, he suggested that as the evolving lineage disperses away from its common place of origin, it takes with it the historical pattern of progressive character transformations. Selection for new character states and combinations of traits in the newly colonized areas would reinforce this trend. As a consequence, Hennig predicted, the geographically most displaced taxa in a lineage would exhibit the most derived character states, while the taxa remaining nearest to the site of divergence would show ancestral traits.

Like Croizat's panbiogeography, Hennig's progression rule has its critics. In particular, it implies that taxa with derived characteristics are most likely to be found in peripheral regions, far distant from their place of origin. This prediction is seemingly falsified by the many examples of "living fossils" and other organisms with primitive features that persist as paleoendemics in extreme environments and isolated locations far from their places of origin (see Chapter 10). As we will see shortly, the distributions of many Hawaiian organisms also do not obey a simple progression rule. It is easy to imagine a mechanism that would cause the rule to be violated, such as organisms evolving in the area of origin undergoing selection for new, derived traits. This might happen if the place of origin were also an area of high species diversity, since ecological interactions with other organisms might select for evolutionary innovations. Tropical forests and other areas of high diversity, for example, may have high rates of speciation and character evolution (see Chapter 17). More important than the ultimate fate of Hennig's progression rule, however, is his emphasis on incorporating phylogenetic information into studies of biogeographic history.

Vicariance Biogeography

It remained for others, especially G. Nelson, N. Platnick, and D. Rosen from the American Museum of Natural History, to take historical biogeography to the next stage. They combined Hennig's approach to phylogenetic systematics with Croizat's approach of analyzing patterns of endemism in multiple taxa, and began to develop rigorous methods of reconstructing the biogeographic history of lineages. Their approach, which came to be called **vicariance bio-**

geography, was closely tied to cladistic methods of phylogenetic reconstruction, and was built on basically the same premises.

Concept and methods. The vicariance approach developed by the "American Museum group" is summarized in a number of publications, perhaps most clearly by Platnick and Nelson (1978), Nelson and Platnick (1981), and Wiley (1981, 1988). Platnick and Nelson noted that historical explanations for disjunct distributions of related organisms fall into two classes: dispersal hypotheses (which they called **dispersal biogeography**), in which the organisms are assumed to have migrated across preexisting barriers, and vicariance hypotheses (vicariance biogeography), in which the formation of new barriers is assumed to have fragmented the ranges of once continuously distributed taxa. Extant distributions are usually inadequate not only to determine whether a given barrier was in existence before or after the migration of a particular taxon to its present disjunct areas, but also to reconstruct the direction of the migration or the sequence of barrier formation. If, however, the organisms inhabit three or more disjunct areas, the techniques of cladistic systematics (see Chapter 11) can be used to determine the sequence of branching in the lineage (Figure 12.4).

If we can assume that the speciation events in a lineage were caused by geographic isolation, then the phylogenetic relationships within the lineage also indicate the relative times of initial spatial separation of the now disjunct groups. If we can further assume that these ancient geographic separations have been preserved, and are reflected, in the current distributions of the species of the lineage, then the cladistic phylogeny provides not only a phylogenetic hypothesis about the historical ancestor-descendant relationships among the taxa, but also a biogeographic hypothesis about the historical relationships among geographic localities. Both phylogenetic and biogeographic reconstructions are hypotheses about the historical branching of a taxonomic lineage in time and space. The biogeographic reconstruction is called an **area cladogram** because of its precise logical analogy to a cladistic reconstruction of phylogenetic relationships.

Figure 12.4 shows a hypothetical example of an area cladogram. Three taxa are distributed in three disjunct areas. The arrangement of the diagram indicates the relationships among the taxa as revealed by cladistic analysis: taxa 2 and 3 are more closely related to each other than either is to taxon 1. From this we can infer that taxa 2 and 3 share not only a more recent common ancestor,

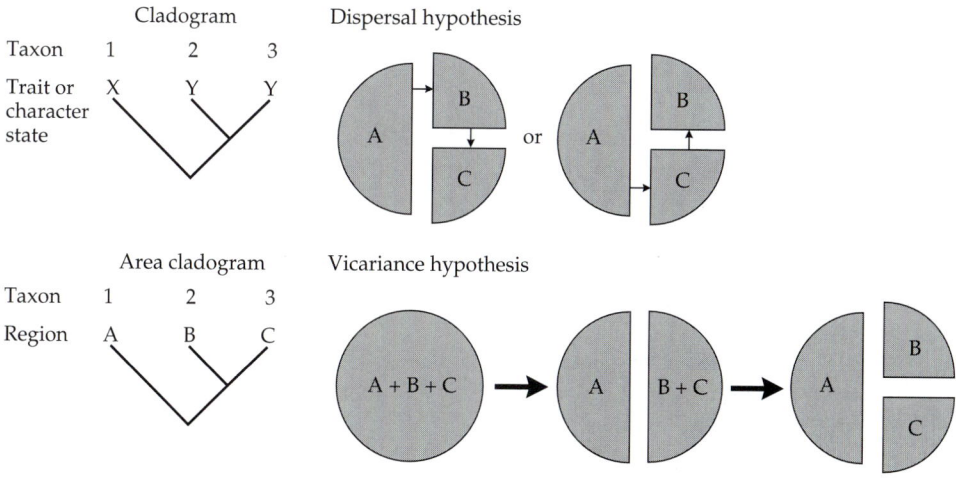

Figure 12.4 A hypothetical cladogram showing reconstructed phylogenetic relationships among three taxa (numbers) and corresponding dispersal and vicariance hypotheses to account for their current distributions among three areas (letters). Note that even without any extinction or multiple colonization events, there are two dispersal hypotheses that account equally well for the known phylogenetic and area relationships, but only one vicariance hypothesis.

but also a more recent common geographic range, than either does with taxon 1. We can now advance two kinds of biogeographic hypotheses that are consistent with these relationships. A dispersal hypothesis would have the ancestral population inhabiting area A, with a propagule dispersing to area B and then, after some additional time, another propagule dispersing to area C. Because taxa 2 and 3 are symmetrically related to each other, the colonization order could also have been A → C → B, but dispersal from either B or C to A is inconsistent with the phylogenetic diagram. A vicariance hypothesis would have the ancestral taxon inhabiting all three areas, A, B, and C, followed by the formation of a barrier to isolate B and C (still interconnected) from A, and finally by the formation of a barrier to isolate B from C.

How would we test these hypotheses? Initially, the American Museum group wanted, like Croizat, to reject out of hand any hypothesis that called for long-distance dispersal. Platnick and Nelson (1978) and others tried to justify this seemingly arbitrary procedure by suggesting that dispersal hypotheses are extremely difficult to falsify, even with information from the fossil record. They argued that an episode of long-distance, barrier-crossing dispersal is an event of low probability and predictability. Each episode is likely to represent an independent event in which a propagule of one taxon crosses the barrier and colonizes the isolated region; several different kinds of organisms are unlikely to disperse together or simultaneously. Further, if a species could disperse across a barrier once, presumably it could have done so repeatedly, and this would invalidate the assumption that the historical separation created by the original allopatric speciation event has been preserved. It is easy to show that if we allow for the possibility of repeated episodes of colonization and extinction, a wide variety of area cladograms can result, and most of them will not preserve the geographic history of past speciation events (Figure 12.5).

The American Museum group argued that a vicariance hypothesis is easier to falsify than a dispersal hypothesis. Provided that barriers, once formed by a vicariance event, cannot be crossed, a vicariance hypothesis makes explicit predictions about the historical connections among areas. In particular, the formation of a new barrier is likely to isolate populations of many different kinds

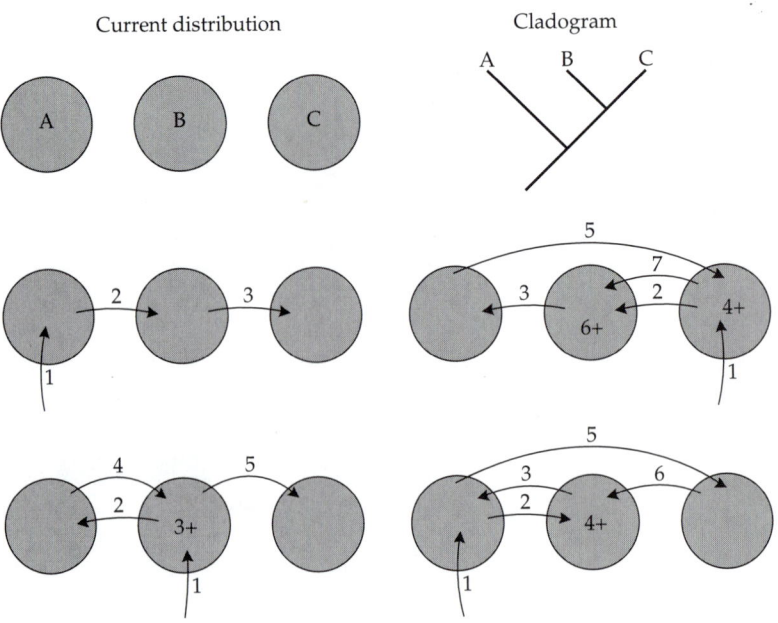

Figure 12.5 Hypothesized sequences of colonization and extinction events that are all consistent with a cladogram in which taxa B and C are more closely related to each other than either is to A. The numbers show the sequence of events; arrows indicate colonizations, and crosses indicate extinctions. In the sequence at the lower left, for example, (1) the ancestor colonizes the center, (2) a population disperses to the left, (3) the original population in the center goes extinct, (4) dispersal from the left recolonizes the center, and (5) dispersal from the center colonizes the right. The four sequences shown here, only a subset of the possible sequences that are consistent with the cladogram, demonstrate how repeated colonization and extinction events can erase the evidence of earlier biogeographic history.

of organisms at approximately the same time. Therefore, a vicariance hypothesis would be supported if multiple taxa exhibit similar patterns of endemism, as Croizat emphasized. A vicariance hypothesis would be falsified if it was incompatible with data on the geological and climatic history of the earth or with information from the fossil record on past distributions of the lineage(s).

In the case of our hypothetical example in Figure 12.4, clear geological evidence that the barriers between the areas were in existence before the taxa occurred there would falsify a vicariance hypothesis. The occurrence of a fossil of taxon 2 or 3 in area A would falsify the vicariance area cladogram, as would a fossil of 3 in B or of 2 in C. Perhaps the strongest test, however, could be conducted by constructing area cladograms for other extant groups with endemic taxa in areas A, B, and C. If all the lineages were in existence at the same time, one would expect them to have responded similarly to the sequential fragmentation of their ranges. Therefore, they should exhibit similar or congruent area cladograms, reflecting similar responses to the same sequence of barrier-forming geological and/or climatic events. Of course, there is some probability that area cladograms will be congruent by chance alone. These probabilities can be determined, however, because they depend on the number of areas and taxa being considered. As the number of areas, and especially the number of independent taxonomic groups, included in the analysis increases, the probability of observing congruent cladograms simply by chance diminishes rapidly. Calculating these probabilities may not be as simple as it may seem, but it is possible, at least in principle, to test the degree of congruence of area cladograms in a rigorous statistical framework (Simberloff et al. 1981a,b).

Examples. A number of studies applied the vicariance approach, beginning in the late 1970s. Rosen (1975, 1978), of the American Museum group, demonstrated the empirical application of their methods with two analyses of distributions of organisms in the West Indies.

Other early studies focused on one of the most obvious test cases, the breakup of Gondwanaland. As the continents drifted apart, they permanently isolated many freshwater and terrestrial organisms, providing a clear case of barrier formation isolating previously widely distributed populations. Further, the relative timing of the separation of the various plates has been increasingly well documented by research in plate tectonics (see Chapter 6). So it is possible to make clear predictions: multiple taxa with endemic representatives on the southern continents, whose ancestors are presumed to have been present on Gondwanaland and to have been isolated by its breakup, should exhibit congruent area cladograms.

Many analyses of Southern Hemisphere distributions, beginning with Brundin's (1965, 1966, 1967, 1972) classic studies of midges, supported these predictions (Cracraft 1980, 1981; Humphries 1981; Humphries and Parenti 1986). In many groups, closely related taxa occur in New Zealand, South America, and Australia, and more distantly related forms, if present, are found in Africa and Madagascar. The interpretation of this pattern is complicated, however, by the fact that many organisms that exhibit it occur primarily in cool, mesic forests and other habitats that are well represented in New Zealand, South America, and Australia, but not on the other southern continents. There are many variations on and exceptions to the above pattern (Figure 12.6). More exceptions undoubtedly will be found, especially as studies are extended to taxa found in other kinds of habitats. For example, many plants endemic to the Mediterranean climatic region of southwestern Australia have their closest relatives in southern Africa (see Chapter 10). Still other

(A) Silphid beetles

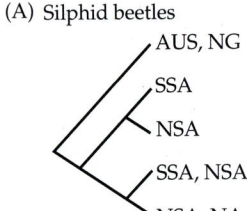

(B) Diamesine midges

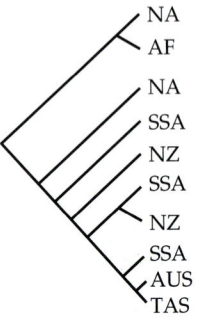

(C) Metallicine beetles

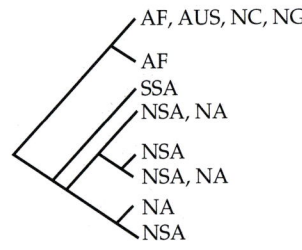

(D) *Nothofagus* and *Fagus* beeches

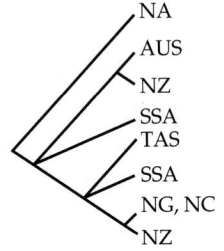

Figure 12.6   Area cladograms for various taxa with endemic forms inhabiting some of the southern landmasses. AF = Africa; AUS = Australia; NC = New Caledonia; NG = New Guinea; NA = North America; NSA = northern South America; SSS = southern South America; TAS = Tasmania. Repetition of geographic codes within the same cladogram implies that different groups within the clade had different histories. Although there is only one geological history of breakup of Gondwanaland, none of these area cladograms exactly reflects that sequence of tectonic events. Instead, the area cladograms hypothesize a different geographic history for each clade. For example, two of the clades have no representatives in Africa, and the other two show completely different affinities for their African representatives. (After Crisci et al. 1991.)

groups, such as some plants, freshwater fishes, and mammals, show strong affinities between South America and Africa (e.g., Bremer 1993).

What about the geological evidence? Current evidence suggests that the first rifting separated Africa and South America from the remainder of Gondwanaland, and that Antarctica, New Zealand, and Australia were the last landmasses to be separated (see Chapter 6). Thus, the distributions of many taxa among the southern continents provide clear evidence of the vicariant events that accompanied the breakup of Gondwanaland. Many of these organisms are much more closely related to one another than to any forms on the northern continents. It is questionable, however, whether historical biogeographic relationships are sufficiently well preserved in the phylogenetic affinities of contemporary organisms to show the precise sequence of tectonic events (Figure 12.6).

Pursuing Hennig's ideas (1966), Daniel Brooks (1985, 1988; Brooks and Bandoni 1988; Brooks and McLennan 1991) suggested that comparisons of phylogenies and area cladograms of parasites and their hosts should provide powerful tests of the vicariance approach. Since parasite and host presumably share a long history of close association, they would be predicted to have been affected by the same historical barrier-forming events, to have speciated concurrently, and therefore to exhibit congruent area cladograms. Therefore, it should be possible to use the occurrence of parasite taxa as "characters" of their hosts for reconstructing both phylogenetic and biogeographic histories. Brooks assembled several area cladograms, mostly for freshwater fishes and their helminth parasites, that generally supported these predictions. In a somewhat similar vein, Hafner and Nadler (1991) compared cladograms of pocket gophers (burrowing rodents) and their chewing lice parasites (Figure 12.7). They found a relatively high degree of congruence, although there were important exceptions: one species of louse had apparently managed to colonize a different genus of gopher than all of its close relatives, and two closely related gopher species were each parasitized by a pair of louse species, one each from two distantly related lineages.

These and other studies (e.g., Page 1991, 1993a,b, 1994; Hafner and Page 1995) of parasites and hosts illustrate some of the potential contributions and pitfalls of the vicariance approach. On the one hand, parasites and hosts must have coexisted at least temporarily for either cospeciation or host shifts to occur, so these cladograms can suggest hypotheses about biogeographic history. On the other hand, it is not clear how much of the congruence among cladograms reflects coevolution, and how much of it simply reflects similar but independent evolutionary patterns because of a shared history in the same region. Even if two groups of organisms did not interact closely, we might expect to observe similar patterns of speciation and diversification if they lived in the same region and were similarly affected by its geography and geologi-

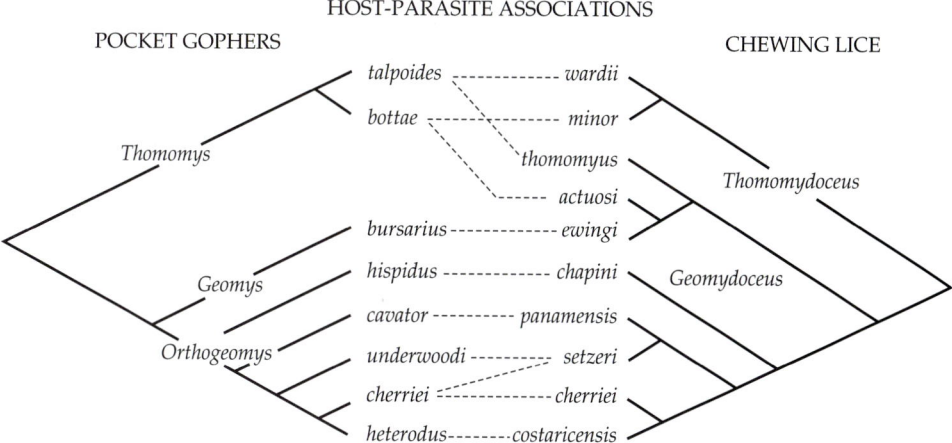

Figure 12.7 Cladograms for a group of chewing lice (right) and their pocket gopher hosts (left) show a substantial degree of congruence. While similarities in the branching patterns suggest that speciation may occur coincidently in both lineages, differences in the pattern imply that the parasites are also able to colonize distantly related hosts (e.g., such switching seems necessary to account for the occurrence of three closely related louse species on two distantly related gopher genera: *Geomydoceus actuosi* and *G. thomomyus* on *Thomomys*, and *G. ewingi* on *Geomys*. Since the chewing lice are obligate, specific parasites on these rodents, the two cladograms have important historic biogeographic implications: for example, parasite and host must live in the same place for cospeciation or host switching to occur. (From Hafner and Nadler 1991.)

cal history. So, it is still not clear just what can be inferred about either mechanisms of speciation or biogeographic history from comparing phylogenies of closely interacting species.

The vicariance biogeography advocated by the American Museum group represented an important advance over Croizat's panbiogeography, but it had some of the same weaknesses. In its quest for analytical rigor, it made some highly restrictive and unrealistic assumptions, including the following: (1) all speciation occurs by geographic isolation (see the discussion of sympatric speciation in Chapter 8); (2) long-distance dispersal does not occur (or at least cannot be incorporated into the cladistic/vicariant framework; see above and examples of long-distance dispersal in Chapter 9); and (3) the geographic isolation that produced past speciation events is preserved in the spatial configurations of current geographic ranges (note the magnitude of relatively recent changes in distributions described in Chapters 7 and 17). These assumptions are so restrictive, and so obviously violated by many kinds of organisms in many geographic regions, that it is little wonder that historical biogeography went through a very contentious period.

Modern Historical Biogeography

Approaches

Over the last 15 to 20 years, American Museum–style vicariance biogeography has largely been replaced by a broader approach to reconstructing biogeographic history. Most current research retains three key features of vicariance biogeography: (1) its emphasis on rigorous logic and hypothesis testing; (2) its close interrelationship with phylogenetic systematics; and (3) its use of area cladograms. It departs from vicariance biogeography largely in its willingness to incorporate more kinds of data, including fossils and geological evidence, and a wider range of hypotheses and mechanisms, including long-distance dispersal. These changes are highlighted by the fact that many historical biogeographers now refer to their work as phylogeography (Riddle 1996), cladistic biogeography (Humphries and Parenti 1986; Humphries et al. 1988), or simply historical biogeography, rather than vicariance biogeography. While there remain serious disagreements among some of the leading investigators, there now seems to be much less of the strident divisiveness that was present during the early days of vicariance biogeography.

While it may be too soon to claim that a conceptual consensus and a unified approach are emerging, research seems to be coalescing around several distinct, but mutually reinforcing, themes (Cracraft 1982, 1988; Humphries and Parenti 1986; Humphries et al. 1988; Page 1990; Crisci et al. 1991; Morrone and Crisci 1995). One is the incorporation of state-of-the-art phylogenetic and molecular systematics. The methods of phylogenetic reconstruction have advanced rapidly in the last few decades (see Chapter 11), and historical biogeographers have been quick to incorporate the new methods and the resulting data into their research. While many early studies in vicariance biogeography were potentially compromised by flawed or poorly resolved phylogenetic reconstructions, recent studies have typically been based on much stronger cladistic hypotheses. Further, knowledge of rates of molecular evolutionary change can now be used to estimate the absolute, not just the relative, times of speciation events (e.g., Riddle 1996).

Such advances are resolving some important biogeographic problems. For example, the Pleistocene refugium hypothesis for the speciation and diversification of Amazonian rainforest organisms has been cast into doubt by molecular studies purporting to show that even those taxa (some trees, butterflies, frogs, and birds) that exhibit superficially similar patterns of endemism within the Amazon basin speciated at drastically different times in the past (Heyer and Maxon 1982; Lynch 1988; Silva and Patton 1993; see also Endler 1982; Bush 1994; Patton et al. 1994; Bates et al., in press). Other studies of patterns of diversification in North American terrestrial vertebrates are similarly casting doubt on the old idea that many contemporary species formed as a result of isolation during the most recent glacial-interglacial cycles of the Pleistocene. For example, sister taxa of desert rodents once thought to have diverged during or since the most recent (Wisconsin) glacial period are turning out to have been effectively isolated in geographically separated desert basins for a much longer period (Riddle 1995 1996). In both the Amazon basin and North American desert cases, the time of many speciation events has been pushed back by more than an order of magnitude, from 10,000–20,000 to as much as several million years B.P.

Another theme characteristic of much modern historical biogeography is the increasing incorporation of information from paleontology and geology. Many phylogenetic systematists and vicariance biogeographers ignored important information about fossils and earth history because such data did not fit conveniently into their tightly constructed logical paradigms for historical reconstruction. They were justly criticized for producing flawed reconstructions that could have been corrected, or at least improved, by incorporating known facts about the past. Some systematists and biogeographers are now actively collaborating with paleobiologists and geologists to understand the relationships between the histories of places and the histories of the lineages that have evolved in those places. Efforts to use molecular data to estimate the times of past cladogenetic events are being complemented by attempts to identify the geological and climatic changes that triggered these

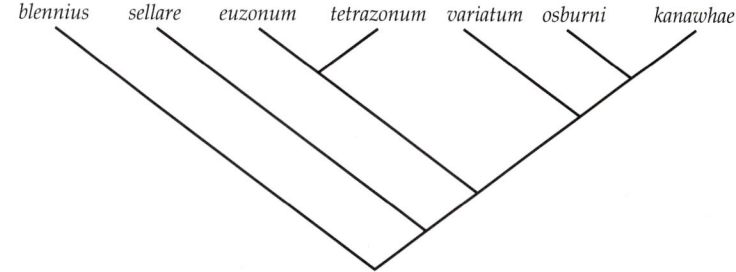

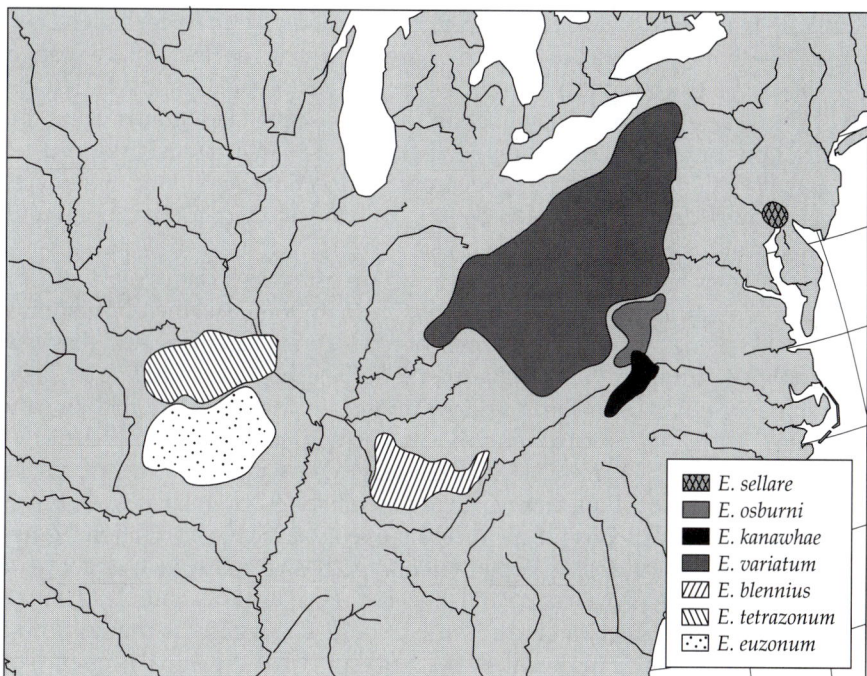

Figure 12.8 Phylogenetic and geographic relationships among several species of darters in the *Etheostoma variatum* species group. Note that there are two centers of distribution, in the Appalachian and the Ozark highlands. The species that occur in each of these areas are most closely related to each other, reflecting recent isolation and speciation events. Such patterns are typical of freshwater fishes, whose phylogenetic and biogeographic relationships typically preserve evidence of past aquatic connections among watersheds. (From Wiley and Mayden 1985.)

events. Some of the best examples are studies of the relationships between the phylogenies and distributions of freshwater fishes and the geological and hydrologic histories of river basins in temperate North America (e.g., Figure 12.8; Wiley and Mayden 1985; Mayden 1987a,b, 1992a,b; Taylor and Gotelli 1994) and tropical South America (Lundberg and Chernoff 1992).

Third, the construction and applications of area cladograms have been carefully scrutinized. This research appears to be taking two quite different directions. On the one hand, some investigators advocate treating an area cladogram as a hypothesis for the historical relationships among areas (e.g., Humphries et al. 1988; Rosen and Smith 1988; Humphries 1993). They make a precise analogy between a phylogenetic cladogram, which uses changes in many different characters to reconstruct the branching history of a lineage descended from a common ancestor, and an area cladogram, which uses the distributions of different lineages as "characters" to produce a branching diagram depicting hypothesized hierarchical historical relationships among regions (Figure 12.9). Such an area cladogram, or **consensus area cladogram,** is an explicit hypothesis about the historical connections and biotic exchanges among areas. This approach seems to be potentially flawed, however, because the analogy to a cladogram is questionable. In cladogenetic evolution, one kind of organism (an ancestor) is transformed into other kinds (its descen-

(A)

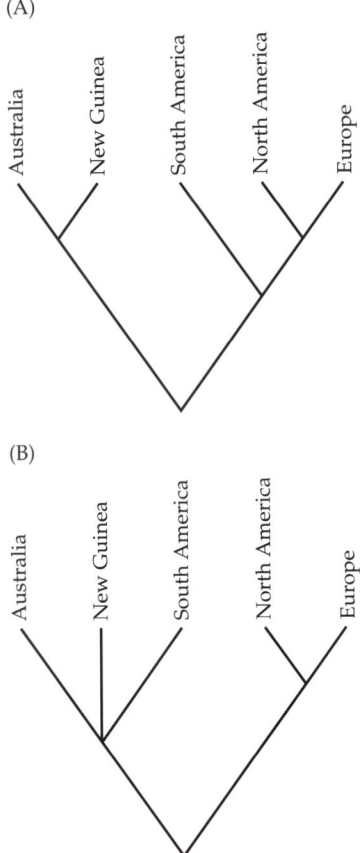

(B)

Figure 12.9 Examples of the approach of using the geographic distributions of multiple taxa as "characters" to produce an "area cladogram" or "consensus area cladogram," which is a formal hypothesis about the sequence of historical connections and biotic exchanges among areas. Here we have two representations of the relationships between some southern and northern continents: (A) for hylid frogs, ratite and galliform birds, and xylontine fishes; and (B) for *Nothofagus* and *Fagus* beeches, and podonomine and diamesine midges. Note that more than one diagram is necessary because the organisms inhabiting these landmasses did not have just one biogeographic history. In particular, they differed in the extent and timing of interchange between South and North America. (From Patterson 1981.)

dants) by a sequence of changes in traits. Hennig's paradigm uses the legacy of these changes, preserved in living (and sometimes fossil) organisms, to reconstruct the likely series of transformations and to propose a phylogenetic hypothesis. But a place is not transformed and split in a simple branching pattern by changes in either the physical environment or the distributions of organisms. Further, the assumptions of cladistics that reversals and independent evolution of character states are infrequent probably do not hold for areas. While each place on earth has experienced a unique and complex sequence of geological and environmental changes somewhat analogous to the evolutionary changes in a taxon, the distributions of the organisms that live there may have shifted back and forth repeatedly, with each change erasing, rather than adding to, the record of previous changes (as we have seen in the case of the glacial-interglacial cycles of the Pleistocene; Chapter 7). Several algorithms have been used to generate consensus area cladograms using the distributions of taxa as "characters." Whether these can accurately reconstruct the actual sequence of historical geographic connections and biotic exchanges, however, will depend crucially on the realism of their assumptions and soundness of their logic.

On the other hand, several investigators are using area cladograms simply to depict the distributions of the taxa in a clade. In this case, the cladogram is constructed using standard techniques of phylogenetic analysis, and then the geographic distributions of the taxa are plotted on the diagram (Figure 12.10). Sometimes the names of the taxa are omitted, so that only the relationship between the phylogenetic tree and the areas where the taxa occur is depicted (e.g., see Figures 12.12 and 12.13). Such a diagram provides a visual representation of the relationships among the areas inhabited by related taxa, but it should not necessarily be thought of as a biogeographic reconstruction. Sometimes it provides sufficient information to hypothesize an explicit sequence of historical interchange and isolation (speciation) events. For example, the area cladograms of the Hawaiian Islands in Figure 12.12 suggest the sequence and direction of interisland dispersal events, assuming that the actual sequence of events has not been obscured by multiple episodes of colonization and extinction. On the other hand, Figure 12.13A shows only the distributions of Hawaiian honeycreepers in relation to their pattern of phylogenetic relatedness. It does not attempt to reconstruct the direction of interisland dispersal events, because there is not enough phylogenetic or biogeographic information to infer with confidence the direction of these movements.

The Hawaiian Example

Perhaps the best way to appreciate the current state of historical biogeography is to consider in some detail one specific example: the distributions of taxa in the Hawaiian archipelago. W. Wagner and V. Funk (1995) have edited an excellent book that brings together much of the recent phylogenetic and biogeographic research on the biota of the Hawaiian Islands. Each chapter presents a cladistic and biogeographic analysis of a different group of organisms that occurs in the archipelago. In the final chapter, Funk and Wagner bring together the results of all these separate studies to address synthetic questions about the biogeographic history of the archipelago. They analyze the phylogenetic and distributional relationships of more than 20 lineages, including terrestrial invertebrates (insects and spiders), birds (honeycreepers), and flowering plants.

Historical and ecological setting.   Hawaii is the ideal historical biogeographic case study. As we saw in Chapter 6, the archipelago has a diagram-

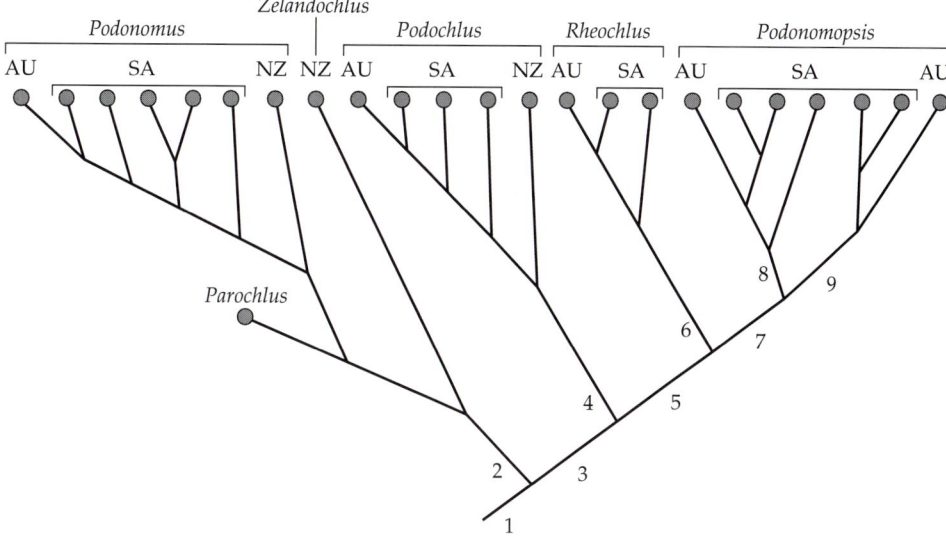

Figure 12.10 An example of the approach of generating a cladogram using traditional phylogenetic analysis and plotting the areas where the taxa presently occur. Shown is Brundin's phylogeny of the midge tribe Podomini, with the species indicated by circles, their grouping into genera indicated above the circles, and their distributions among Southern Hemisphere landmasses indicated as follows: AU = Australia; SA = South America; NZ = New Zealand. Such a diagram is implicitly a historical biogeographic hypothesis, although only one of many (but perhaps the most likely one) that might depict the actual sequence of distributional changes.

matically simple geological history. The islands, among the most isolated landmasses in the world, have been formed over a hot spot, a weak spot in the earth's mantle where magma wells up. As the Pacific Plate has drifted over the hot spot, the magma has punched through the crust to create a series of volcanoes. The hot spot is located at the southeastern end of the island chain, beneath the island of Hawaii and another volcano (Loihi Seamount) to the southeast, which is growing but still submerged below the ocean (see Figure 6.20). The formation of islands began 75 to 80 million years ago, but there may have been times since then when there was little or no emergent land. The oldest of the present major islands is Kauai, the most northwestern island, which was formed about 5.1 million years ago. The ages of the islands decrease down the chain to the southeast: Kauai, Oahu, Molokai, Maui, and Hawaii.

The islands also offer different environments. Ancient Kauai has been eroded to leave knife-edge ridges, deep valleys, and steep cliffs. Maui and Hawaii rise to more than 3000 m; at their summits are perpetually cold, sparsely vegetated habitats, and at low elevations there are rainforests on the northeastern sides and arid scrub habitats on the southwestern exposures. In addition, Hawaii has three active volcanoes and many lava flows of various ages.

Predictions. Area cladograms, of the kind that simply depict the island(s) where each taxon in a lineage occurs as a function of its hypothesized phylogenetic relationships, can address several important questions of historical biogeography. To what extent do different taxa exhibit congruent patterns? Most historical biogeographers would probably have predicted that most groups of organisms would exhibit a simple progression rule pattern of distribution, with the least differentiated taxa on the oldest island and the most derived

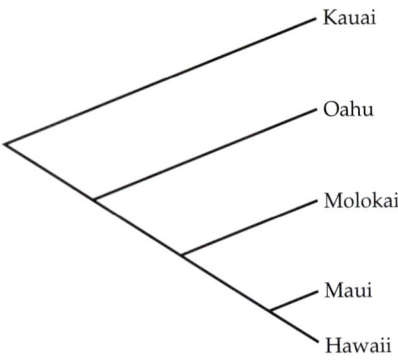

Figure 12.11   The area cladogram predicted for the Hawaiian Islands based on a simple progression rule. This is the pattern of phylogenetic relationships that would be expected if a lineage colonized each island in turn as it was formed by the emergence of a volcano from the sea.

ones on the youngest island (Figure 12.11). Area cladograms not only allow us to see whether this predicted pattern is observed, but also permit an assessment of the degree of congruence among the distributional patterns of different kinds of organisms. If the logic of Croizat and the American Museum group is sound, and different lineages have responded similarly to the geological history of the Hawaiian archipelago, then a high degree of congruence would be expected.

What is the frequency and direction of past interisland colonization events? The same logic that would yield the above progression rule would also predict that the predominant direction of colonization should be from inhabited to uninhabited and often from older to younger islands. Area cladograms should provide some assessment of the frequency of back-colonization to already inhabited and perhaps older islands.

What are the processes and geographic consequences of the speciation events that have occurred in the archipelago? Darwin's and Lack's interpretations of the differentiation of tortoises and finches in the Galápagos and Mayr's observations on variation in East Indian birds (see Chapter 8) would suggest that interisland colonization followed by divergence in isolation has been the predominant mode of speciation in archipelagoes. If this is true of Hawaii, we would predict that the most closely related taxa should occur on different, but perhaps nearby, islands.

Distributional patterns and historical processes.   So what do the results of the phylogenetic analyses show? The bottom line is that none of the simple predictions made above is supported as a general rule. Instead, the distributions of the different kinds of organisms show a rich variety of relationships with respect to their phylogenetic histories. We can present only some of this variety in a simplified form here (for the full story read Wagner and Funk 1995).

Some clades (e.g., the *Drosophila* fruit flies and certain plants) do indeed show a more or less clear progression rule, with basal taxa on older islands and progressively more derived forms on younger islands (Figure 12.12A,B). But other clades show very different patterns. One variation, found in the closely related plants *Schiedea* and *Alsinidendron*, is a series of subclades, each exhibiting its own progression rule with multiple waves of dispersal from older to younger islands (Figure 12.12C). A clear exception to any progression rule is found in the plant genus *Tetramolopium* (Figure 12.12D), in which the ancestral species clearly colonized one of the younger islands (either Hawaii or Maui) and subsequently dispersed to older ones (Oahu and Molokai), but apparently never reached the oldest (Kauai). Several other cladograms show complex patterns that clearly do not support a progression rule (e.g., the honeycreepers in Figure 12.13A, which have sister taxa on islands of contrasting age: Kauai and Hawaii), while still others are not clearly resolved, and could be interpreted to suggest two or more very different histories.

The cladograms also reveal many different patterns of colonization and speciation. The examples of progression rules mentioned above illustrate cases of dispersal from older to younger islands. But there are also many cases of colonization of older islands from younger ones. One, also mentioned above, is the genus *Tetramolopium*. Another is the Hawaiian honeycreepers, in which several recently derived taxa occur on the oldest island, Kauai (Figure 12.13A). Again, it is important to mention that for several clades, it is difficult to pinpoint the island that was first colonized, and therefore it is equally difficult to determine the direction of subsequent dispersal events that resulted in the colonization of additional islands. Such problems may be due to difficulty in

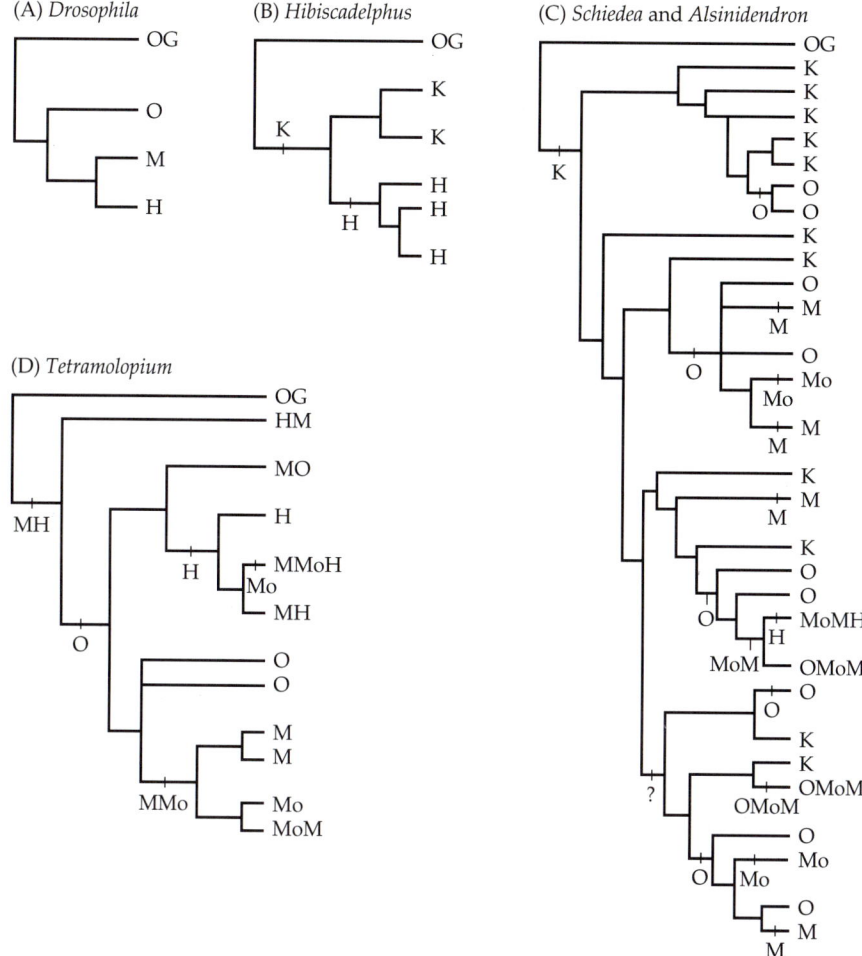

Figure 12.12 Area cladograms for four groups of Hawaiian organisms, simplified to include only the outgroup and those taxa found on the five largest islands. OG = outgroup; K = Kauai; O = Oahu; Mo = Molokai; M = Maui; H = Hawaii. Letters on the terminal branches (right) indicate present distributions; those placed on the tree indicate over-water dispersal to colonize new islands; and multiple letters for the same island without multiple colonization events indicate within-island speciation events. (A) A group of *Drosophila* fruit flies shows a progression rule, with the more derived forms occurring on progressively younger islands. (B) The endemic plant genus *Hibiscadelphus* shows a highly modified progression rule, with the more derived taxa occurring on the youngest island (Hawaii) and the ancestral taxa on the oldest (Kauai), but with multiple speciation events within these two islands, and no occurrences on the islands of intermediate age. (C) The closely related endemic plant genera *Schiedea* and *Alsinidendron*. This group comprises four subclades, each of which shows a general progression rule. Note, however, that there have also been multiple independent colonizations of the same island (e.g., Oahu six times) and speciation events within islands (e.g., especially on Kauai and Oahu). (D) The plant genus *Tetramolopium*, which is probably a fairly recent immigrant to the archipelago, shows no evidence of a progression rule. It originally colonized either Maui or Hawaii, and subsequently dispersed to Molokai and Oahu, but apparently never got established on the oldest island, Kauai. (From Funk and Wagner 1995.)

resolving the cladograms (because of too few characters or other problems), but they may also be due to unresolvable complications in the biogeographic history. For example, branches of lineages that went extinct on islands at different times in the past, and did or did not colonize other islands and leave

(A) Drepanidinae

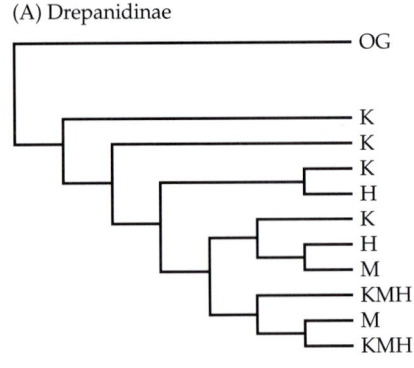

Figure 12.13 Area cladograms for two lineages of animal groups endemic to the Hawaiian archipelago and showing contrasting patterns of colonization and speciation. Localities and colonization events are coded as in Figure 12.12. (A) The Hawaiian honeycreepers of the avian subfamily Drepanidinae show many episodes of interisland colonization followed by speciation in isolation on the different islands. The direction of colonization is not known for the honeycreepers because there have been so many colonization events that the direction of dispersal often cannot be resolved from patterns of phylogenetic relatedness. (B) The cricket genus *Prognathogryllus* shows relatively few interisland colonization events, but each such event has been followed by multiple episodes of within-island speciation. (From Funk and Wagner 1995.)

(B) *Prognathogryllus*

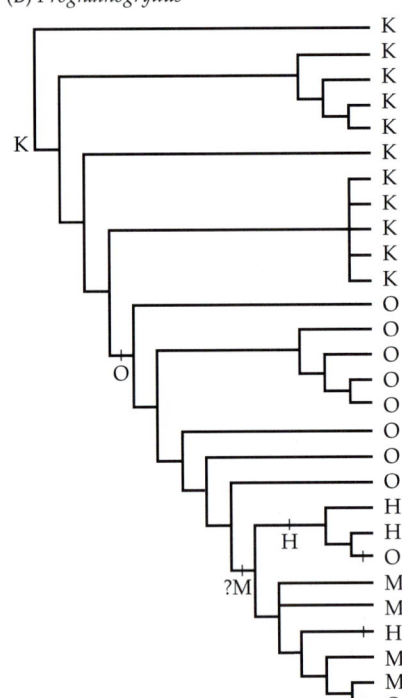

descendants there, can make it difficult to reconstruct the biogeographic history even though the phylogenetic reconstruction may be well resolved and accurate.

With respect to speciation, the cladograms do show examples of allopatric speciation presumably caused by dispersal to and differentiation on different islands. Perhaps the best example is that of the honeycreepers (Figure 12.13A). In general, the most closely related pairs of species occur on different islands, and often these islands are far from each other (e.g., Kauai and Hawaii, at the opposite ends of the archipelago). This pattern of buildup of diversity as a result of repeated episodes of colonization and speciation fits well with that seen in other groups of birds in other archipelagoes, such as the Galápagos and East Indies. On the other hand, the predominant pattern, seen in many clades of Hawaiian arthropods and plants, is one of extensive speciation and radiation within islands (Figure 12.13B). It is very similar to the pattern observed in groups of fishes and mollusks in lakes (see Chapter 8). Further, since we know the ages of the islands, we can put some minimum time on the speciation events. We can conclude that all of the within-island speciation occurred within the last 5 million years, and that some of it probably occurred (e.g., on Hawaii) within the last 500,000 years. At this point, it is important to note that just because speciation occurred within an island does not mean that spatial isolation played no role in the differentiation of the populations. All of the large Hawaiian islands have a great deal of topographic relief and habitat heterogeneity. For organisms that may disperse as poorly as some plants and invertebrates, this heterogeneity may provide not only sufficient isolation for microallopatric speciation, but also different environmental settings to select for rapid divergence. Nevertheless, the high frequency of speciation within islands, like that within lakes, raises important questions about the role of ecological and genetic processes in speciation—and especially about the relative importance of spatial isolation and divergent selection pressures.

Assessment and Prospects

The Hawaiian examples illustrate the status of contemporary historical biogeography: both its enormous promise and the many issues that still remain to be resolved. It is hard to imagine any other place on earth with such a clear, well-understood history of place as Hawaii. Yet the histories of the lineages that live in this place are complex. In retrospect, there are obvious biological reasons for this. For one thing, the different kinds of organisms differ enormously in their environmental requirements, in their abilities to disperse over both land and water barriers, and in the genetic and social structures of their populations. These differences have caused differences in colonization, extinction, and speciation processes that have left as their legacies different patterns of area cladograms. For another, just because the archipelago has a relatively simple geological history does not mean that this history has affected all organisms similarly. Indeed, different lineages have colonized the archipelago at different times during its history, and consequently have shown different patterns of subsequent speciation and dispersal. Not all species have been affected similarly by the physical barriers—terrestrial barriers within islands as well as the water barriers separating islands—that limit dispersal and promote speciation. Also, in any system as small as these islands, chance has undoubtedly played a major role in the outcomes of past events.

Given all of this complexity, it is impressive what has been learned. In fact, while many challenging questions remain, the application of modern phylogenetic systematics has revealed a great deal about the histories of the few lineages of organisms that have managed to colonize and diversify on the Hawaiian archipelago. And the story is far from over. Hawaii remains a hotbed of geological, biogeographic, evolutionary, and ecological research.

If the biogeographic history of the Hawaiian archipelago is complex, however, how much more complex is the history of distributions within a continent? If the patterns of colonization and speciation in some Hawaiian clades are difficult or impossible to reconstruct with accuracy, how much more difficult will it be to recover the biogeographic histories of lineages on continents or in the oceans?

Raising these questions is not intended as criticism. On the one hand, there are limits—practical if not conceptual ones—to how much history can be recovered. Given that we do not even know how many living species there are on earth, it seems unrealistic to think that we can ever know how many lived, say, in the early Cretaceous or late Miocene. It is even more unrealistic to believe that we can know the geographic ranges of all those organisms. Yet some of those species must have been ancestors of living taxa, and knowledge of their historical ranges may be critical to understanding current distributions.

On the other hand, we can take encouragement from the progress that has been made in just the last few decades. Enormous advances in the earth sciences have provided much information on the movements of continents and oceans, the composition of ancient biotas, and the physical, chemical, and biological characteristics of their environments. Equally great advances in systematic theory, molecular genetics, and other areas of biology have permitted increasingly detailed reconstructions of phylogenetic relationships and past distributions. Much has been learned about the histories of both places and lineages. We can be optimistic that much more will be learned, especially about the last few million years of biogeographic history. In particular, by combining information from the rich fossil record of this period with the wealth of molecular genetic data and the powerful methods of phylogenetic reconstruction, we can expect to much better understand the shifts in geographic ranges and the speciation and extinction events that have shaped the composition and distribution of the earth's contemporary biota.

Contemporary Patterns and Processes

CHAPTER 13

Island Biogeography: Patterns in Species Richness

Islands have always had a great influence on biogeography, far out of proportion to the tiny fraction of the earth's surface that they cover. The reason for this is straightforward: Islands and other insular habitats, such as mountaintops, springs, lakes, and caves, are ideal subjects for natural experiments. They are well-defined, relatively simple, isolated, and numerous—often occurring in archipelagoes of tens or hundreds of islands.

Just as conditions can be varied in artificial manipulative experiments, islands can vary in a number of environmental features (e.g., area, isolation, or presence of predators and competitors). In this way the effects of each factor on community structure can be assessed. Despite the limitations of such natural experiments, they have an important advantage over artificial manipulations in that they were established sufficiently long ago that there has been time for evolutionary response to the variables. All one needs to do is to gather the data and interpret them correctly. This, however, has not been an easy task. Basic data on insular biotas are still being compiled, and their interpretation has often been the subject of controversy. Like any other scientific study, a natural experiment must be well designed. That is, out of all the islands we could study, we must prudently select a subsample of islands that minimizes or controls variation in all factors except those central to our working hypothesis.

Of course, the challenges of designing rigorous and insightful experiments, gathering and analyzing data, and debating and testing alternative methods and ideas are common to all scientific experiments, whether natural or manipulative. As we shall see, biogeographers have utilized both approaches to gain insights into the processes influencing insular communities and, in turn, have obtained important insights into the forces structuring mainland communities as well.

Historical Background

Since the early work of Forster and Candolle during the eighteenth and early nineteenth centuries, studies of islands, mountaintops, and other isolated ecosystems have strongly influenced the development of biogeography. As we saw in Chapter 2, the influence of islands on biogeography, as well as on evolutionary biology and ecology, began in earnest in the early nineteenth century when various European nations undertook to explore, map, and study the world. Naturalists frequently accompanied these voyages of exploration. The best of these naturalists—notably, Wallace, Darwin, and Hooker—not only described and collected specimens of what they found, but also noted patterns in nature and sought explanations for them. Some of the clearest patterns were apparent among the various islands of oceanic archipelagoes. Darwin's experiences in the Galápagos, Wallace's travels in the East Indies, and Hooker's explorations in the Southern Ocean had profound effects on the thinking of these scientists, and consequently on the ideas about evolution and related areas of environmental biology that revolutionized scientific thought in the mid-1800s.

Another revolution—a much more modest one, but nevertheless a major shift in the direction of scientific thought—occurred in the mid-1900s with the integration of concepts from ecology, evolution, and biogeography. It is difficult to pinpoint the exact beginning of this endeavor, but islands again played a central role. One of the pioneers was David Lack (1947, 1976), who early in his career conducted a classic study of the evolution and ecology of Darwin's finches on the Galápagos archipelago, and, shortly before his death, investigated the distribution and ecology of birds in the West Indies. As Lack had followed Darwin to the Galápagos, another ornithologist, Ernst Mayr (1942, 1963), followed Wallace to the East Indies and returned to make major contributions to the understanding of speciation and other aspects of the evolutionary process. Another pioneer was G. E. Hutchinson (1958, 1959, 1967), who also traveled widely, but studied lakes rather than islands. In his 1959 paper, "Homage to Santa Rosalia, or why are there so many kinds of animals?" Hutchinson called attention to the problem of how to explain geographic variation in the diversity of species. This problem has remained a focus of research in ecological biogeography and community ecology right up to the present.

If any single contribution can be said to have triggered the recent revolution in ecological biogeography, however, it was Robert MacArthur and Edward Wilson's **equilibrium theory of island biogeography** (1963, 1967). This seminal work was completed when both men were young, still in their thirties. MacArthur had been a student of Hutchinson's at Yale. His doctoral dissertation (1958) was a classic study of competition and coexistence in several closely related species of warblers. After completing his degree, he did postdoctoral work in Britain with Lack and then held professorships at the University of Pennsylvania and at Princeton University. Wilson, who has spent his entire career at Harvard, began it as a systematist. Strongly influenced by Mayr, he had worked extensively on the origins and relationships of the ants of the East Indies and South Pacific. He was also a coauthor of the classic paper on character displacement (Brown and Wilson 1956). Both men had extensive experience with islands: MacArthur in the montane islands of the southwestern United States, in the West Indies, and in small islands off the coasts of Maine and Panama; Wilson in the East Indies, Polynesia, and the Florida Keys. Both men went on to have illustrious careers. MacArthur died of cancer in 1972 at the age of 42, but had already produced many theoretical papers on population and community ecology that still motivate research in those fields today. Wilson continued to work on social insects, especially ants, but his interests have shifted from systematics and biogeography, to animal behavior (Wilson 1975), and most recently, the conservation of biodiversity.

MacArthur and Wilson's equilibrium theory represented a radical change in biogeographic thought. Prior to their work, investigators had focused on historical problems and idiosyncratic approaches. The primary questions of biogeography had always been those addressed in Chapter 2: where did a particular taxonomic group of organisms originate, and how, as a result of subsequent dispersal, speciation, and extinction, did its diversity and distribution change? Obviously these questions have a historical and phylogenetic focus, and they are ad hoc questions in the sense that they are normally applied to particular taxa or specific regions.

Prior to 1960, the dominant theme of island biogeography was what is sometimes referred to as the **static theory** of islands (Dexter 1978). Basically, this theory held that insular community structure was fixed in ecological time—that is, that species composition remained unchanged unless modified by long-term evolutionary processes. According to the static theory, insular community structure resulted from unique immigration and extinction events, and species number was determined by the limited number of niches available on each island (Lack 1976). Either a species had already colonized the island in question, or it never would. Once it arrived, the species either found adequate resources for survival or failed to establish a population.

Island biogeography had no general explanations or rigorous models, mathematical or graphical, that would provide clear predictions and lead to tests among alternative explanations for observed patterns. All this was to change in a relatively short time with MacArthur and Wilson's theory. During the early decades of the twentieth century, models of dynamic equilibria had been developed to explain a variety of phenomena, ranging from chemical reactions and regulation of body temperature to gene frequencies and demography. The central idea, or **paradigm**, in all these models was the concept of a **dynamic equilibrium**—that is, that opposing forces maintain constancy in some characteristic of a system despite continual changes, or **turnover**, in its other intrinsic properties. For example, the body temperature of a bird is regulated within narrow limits, despite changes in environmental temperature and internal heat production, by opposing mechanisms of heating and cooling. Given the early development of equilibrium models in other fields, rather than marveling over MacArthur and Wilson's contributions, which truly were revolutionary, we may wonder why ecological biogeography lagged so far behind. In fact, an early form of an equilibrium theory of island biogeography was developed by Eugene Gordon Munroe in 1948, but it was ignored by his contemporaries (see Box 13.1). The following two decades witnessed a period of great growth in ecology, which had been a mathematically unsophisticated science. By the early 1960s ecological biogeography was poised for a scientific revolution, to be spearheaded by one of the world's leading mathematical ecologists, Robert MacArthur, and an equally qualified naturalist and ecologist, Ed Wilson.

MacArthur and Wilson deliberately departed from the classic ad hoc, static and historical approach and asked radically new kinds of questions. They searched for general patterns in the distributions of diverse kinds of species, independent of their phylogenetic affinities, in the hope that such patterns would have general ecological explanations rather than idiosyncratic and historical ones. Although they recognized the importance of systematics and historical geology, MacArthur and Wilson were primarily interested in patterns that might be explained without invoking unique historical events. Their approach was to focus on variations in plant and animal distributions that appeared to be correlated with the functional attributes of contemporary organisms and with measurable characteristics of their present environments. They were more impressed, for example, with the fact that birds and bats are similar both in their ability to fly and in their wide distribution on oceanic islands than with the many differences in morphological traits that reflect the long divergent evolutionary history of these two taxa.

Box 13.1
Independent discovery of the equilibrium theory of island biogeography

A correlation of this kind [between number of species and logarithm of area of an island] is as interesting as it is unexpected, for it suggests the existence of an equilibrium value for the number of species in a given island, a value which acts as a limit to the size of the fauna. The processes which determine the equilibrium value for an island of given size must be, on the one hand, the extinction of species, and, on the other hand, the formation of new species within the island, and the immigration of new species from outside it.

The above quotation does not come from one of MacArthur and Wilson's (1963, 1967) two seminal publications on the equilibrium theory of island biogeography. It was written 15 years earlier by someone else. It appears on page 117 of Eugene G. Munroe's (1948) doctoral thesis on the distribution of butterflies in the West Indies.

The earlier and independent discovery of the equilibrium theory by Munroe is more than one of those amusing incidents in the history of science. It war-rants further examination for two reasons. First, in contrast to some purported cases of prior discovery of important ideas, Munroe did not have just some vague, poorly articulated notion of species equilibrium. He clearly presented the empirical species-area relationship that stimulated his inductive discovery, investigated the generality of this pattern, and developed detailed verbal and mathematical models to explain it (Munroe 1953).

Second, given the striking similarity of the two models, it is worthwhile to ask why Munroe's discovery went unrecognized (but see Gilbert 1984) while MacArthur and Wilson's has been hailed as one of the great accomplishments of the evolutionary ecology of the 1960s. Others have been given credit for developing precursors to the equilibrium theory [including Dammermann (1948— see Thornton 1992) and LaGreca and Sacchi (1957—see Vuilleumier 1975)], but Munroe's theory was conceptually and mathematically equivalent to MacArthur and Wilson's theory (see also Wilkinson 1993).

Mathematical models
Had Munroe stopped with his doctoral thesis, the similarities to MacArthur and Wilson's later work would have been striking enough. However, Munroe developed his equilibrium theory of island biography in a subsequent paper (1953), but only the abstract was published, and this appeared 4 years after the paper was presented at a meeting in 1949. This tantalizing abstract contains only three paragraphs. The first, quoted above, points out the generality of the semi-logarithmic species-area relationship, but then Munroe goes on to present a mathematical model:

The actual form of the curves is that of a shallow sigmoid, with the equation

$$F = k'A* \left[iL/(i + kp) \right]$$

where F = number of species in the fauna at equilibrium, L = number of species in surrounding lands capable of immigrating into the island, I = the probability of any one of species actually immigrating, p = the probability of extinction of a single pair of one species, and A = the area of the island, to which the population number of each species is

Island Patterns

MacArthur and Wilson's theory was developed to explain two very general patterns in island biogeography: the tendencies for the number of species to increase with island area and to decrease with island isolation. These patterns had been well known to biogeographers since the early 1800s (see Chapter 2). Perhaps the greatest inspiration for the equilibrium theory, however, came from the more recent observation that immigrations and extinctions were relatively frequent phenomena, even in ecological time. During the twentieth century, repeated biological surveys of Krakatau and other islands that had been cleared of life by volcanic eruptions revealed that immigrations and extinctions were recurrent processes, and that island communities exhibited substantial turnover as new colonists replaced extirpated species. MacArthur and Wilson's innovation was to recognize the common themes that underlie these observations (the species-area relationship, species-isolation relationship, and turnover) and to propose a single, unifying theory to account for them.

The Species-Area Relationship

Theories, like islands, are often reached by stepping stones. The species-area curves are such stepping stones. (MacArthur and Wilson 1967, p. 8)

Box 13.1 *(continued)*

assumed to be directly proportional. (Munroe 1953, p. 53)

MacArthur and Wilson (1967, p. 26, Figure 11 and Equation 3-1) use a similar, but somewhat simpler, expression in their first mathematical model:

$$S = (IP)/(E + I)$$

where S = the equilibrium number of species, I = the initial immigration rate if the island was empty of species, P = the number of species in the species pool available to colonize, and E = the extinction rate if P species were present on the island.

Conceptual advances and scientific progress

Why didn't Monroe promote his ideas?
Despite the remarkable similarity of their ideas, Munroe's prior discovery had no apparent impact and went virtually unrecognized to this day (but see Gilbert 1984), whereas MacArthur and Wilson's later, independent development of the same concept has been enormously influential. In one sense, it is not surprising that Munroe's concept of species equilibrium remained unknown to MacArthur, Wilson, and virtually all

biogeographers and ecologists. Munroe's ideas were presented in five pages, one table, and one figure in a large unpublished doctoral dissertation devoted primarily to the systematics and descriptive biogeography of Caribbean butterflies, and in a one-page abstract in a relatively obscure regional publication. And it is not hard to understand why a young scientist with "competing interests and pressures" did not aggressively pursue ideas that apparently elicited interest from only a few colleagues. In the late 1940s, biogeography was dominated by descriptive and taxonomic approaches; this was not a propitious time for injecting mathematical theory and ecological concepts.

In another sense, however, it does seem surprising, especially with the clarity afforded by decades of hindsight, that Munroe did not make a greater effort to publicize his discovery. This is especially true considering that Munroe did not retire from productive science after receiving his degree, but went on to enjoy a distinguished career as a lepidopteran systematist. Munroe clearly devoted considerable time to developing his ideas on faunal equilibrium and recognized at least some of their important implications.

The lessons. It is unfortunate that Munroe did not get more recognition for his discovery. One of the purposes of this essay is to rectify this situation. This case is strangely reminiscent of the independent discovery of the theory of evolution by natural selection by Wallace and Darwin. And like that story, it shows that, as important as new ideas are in the progress of science, they are often not the unique inspirations of genius that are portrayed in the textbooks. On the one hand, scientific revolutions usually do depend on major conceptual innovations. On the other hand, in order for these insights to have an impact, they must be promoted cogently at a receptive stage in the development of a discipline. It is not sufficient to have a good idea; it is even more important to develop and publicize it. Munroe and MacArthur and Wilson had the same basic idea. Munroe, distracted by other interests and perhaps frustrated in his initial attempts to publish, allowed his idea to languish. MacArthur and Wilson vigorously pursued and advocated their idea, and had a major impact on their science.

(Excerpted and modified from Brown and Lomolino 1989.)

Schoener (1976) described the species-area relationship as "one of community ecology's few laws." Indeed, it is one of the most general, best-documented patterns in nature (Watson 1859; deCandolle 1855; Jaccard 1902, 1908; Brenner 1921; Arrhenius 1921; Gleason 1922, 1926). Regardless of the taxonomic group or type of ecosystem being considered, species number tends to increase with increasing area. This relationship, however, is not linear; richness increases less rapidly for larger islands (Figure 13.1).

Despite the long history of studies on this fundamental pattern, it wasn't until the 1920s that ecologists generalized it in mathematical form. In 1920, Arrhenius adapted the allometric equation (used for the scaling of morphology and metabolic processes) to describe the species-area relationship. The Arrhenius equation, more commonly referred to as the **power model**, can simply be expressed as

$$S = cA^z$$

where S = species number (or richness), c is a fitted constant, A = island area, and z is another fitted parameter that represents the slope when both S and A are plotted on logarithmic scales. Typically, this relationship is linearized by taking the log of both sides of the equation:

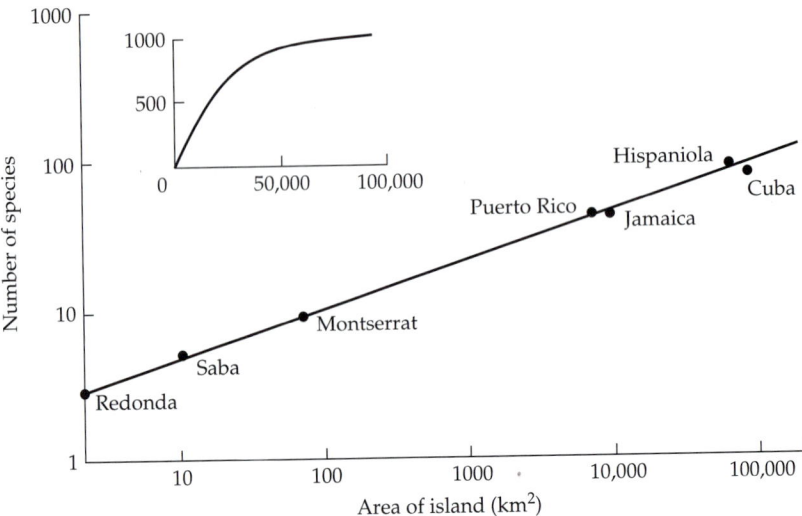

Figure 13.1   The empirical relationship between number of species, *S*, and island area, *A*, for the reptiles and amphibians of the West Indies, plotted from the original data of Darlington (1957). Note that both axes are logarithmic, and the points are well fitted by a straight line giving an equation $S = cA^z$, where *c* and *z* are fitted values. Inset depicts this relationship in arithmetic space. (After MacArthur and Wilson 1967.)

$$\log(S) = \log(c) + z \log(A)$$

The *c* and *z* values can now be easily estimated using simple linear regression of log-transformed data. In Box 13.2 we discuss the biological relevance and potential misinterpretations of *c* and *z*.

While the power model appears to be the most commonly used formula for the species-area relationship, a semi-logarithmic model also is frequently used, especially among plant ecologists. In 1922, Gleason used the following formula to study species-area relationships of plant communities:

$$S = d + k \log(A)$$

As this formula implies, *d* represents the intercept, and *k* represents the slope of the line when *S* (richness) is plotted against the log of *A* (area).

Frank Preston, an engineer by profession and a naturalist by avocation, contributed to the mathematical development of ecology during the 1950s and 1960s. Preston (1962) noted that the species-area relationship of islands was a special case of the general multiplicative increase in the number of species with an increase in the area sampled. He suggested that this was a consequence of what he termed the canonical lognormal distribution of the number of individuals among species (see also Williams 1953, 1964). In any region, only a few species are extremely common, and most are moderately or very rare. Therefore the distribution of the number of individuals among species, when plotted on a logarithmic abscissa (*x*-axis), is fairly well fitted by a normal, bell-shaped curve (Figure 13.2). Sometimes this curve is cut off on the left-hand side, and Preston suggested that this happens when the sample is so small that some of the rarest species are not observed. It is this effect that produces the species-area curve. As progressively larger areas are sampled, one obtains not only more individuals, but also more species, because some of the new individuals will be representatives of rare species that had not yet been seen. Additionally, larger areas will tend to incorporate new kinds of habitats and therefore add specialized species that are restricted to those environments.

Preston pointed out that small, isolated islands have fewer species per unit of area and higher *z* values for the species-area curve than sample areas of comparable size within large regions of continuous habitat on continents (Figure 13.3; see Schoener 1974; Sugihara 1981). The reason for this should be intuitively apparent: small, isolated islands have fewer species than compara-

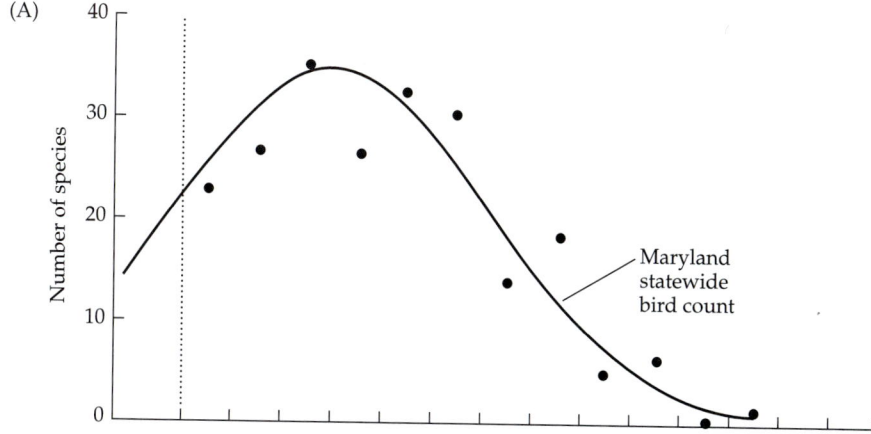

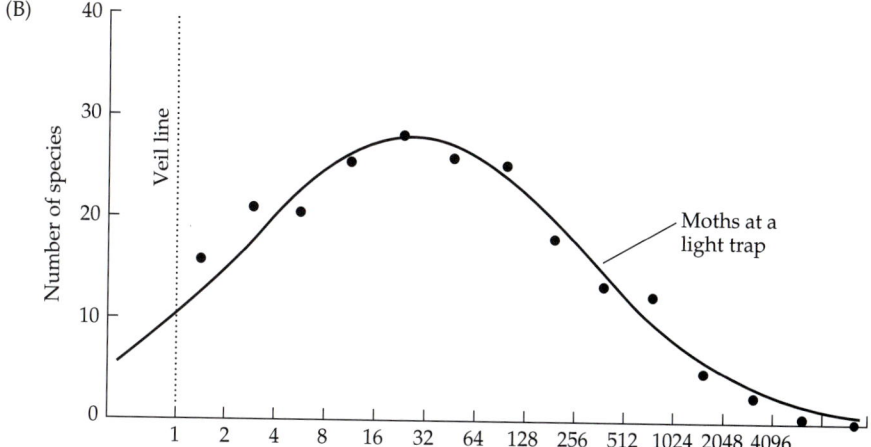

Figure 13.2 The relative abundances of species within a local biota often fit a lognormal distribution; in other words, the frequency distribution approximates a normal curve when abundance is plotted on a logarithmic scale. Note that often the left-hand tail of the distribution is cut off by what Preston called a veil line. Because the axis is logarithmic, this curve shows that every community contains more rare species than common ones. The data are from a bird census in Maryland (Preston 1957) and a count of moths at a light trap in England (Williams 1953).

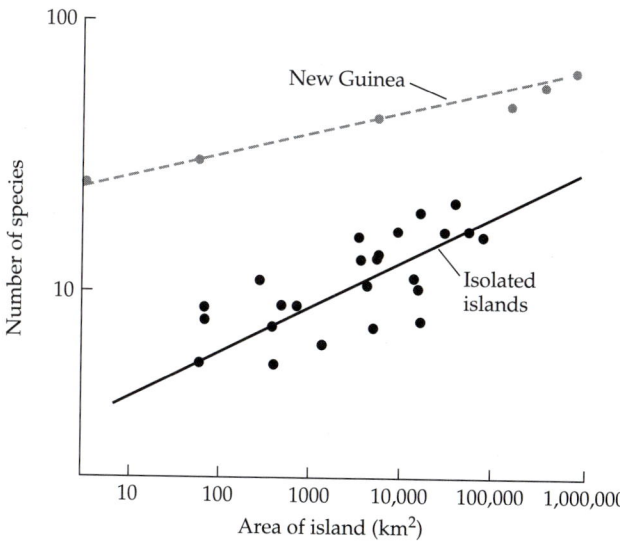

Figure 13.3 The slope of the species-area relationship in log-log space is much steeper for isolated islands than for sample areas of different sizes within a single large landmass. These data are for pomerine ants on the Moluccan and Melanesian islands (below) and in regions of increasing size on New Guinea (above). The difference between the two curves can be attributed to the greater likelihood of extinction without replacement by immigration of rare species on isolated islands. (From Wilson 1961.)

Box 13.2
Interpretations and comparisons of constants in the species-area relationship: An additional caution

The species-area relationship is one of the most important and most frequently studied patterns in biogeography and, according to Schoener, "one of community ecology's few universal regularities" (1986, p. 560). Yet the utility of the most common model describing this relationship, the power model ($S = cA^z$, where S is species richness, A is area, and c and z are fitted constants), is debatable. The strongest controversy stems not from the "fit" provided by this model, but from the biological relevance of its exponent, z (Connor and McCoy 1979; Sugihara 1981; Abbott 1983).

Many misinterpretations can result from the frequent but unfortunate reference to the z and c values as the slope and intercept of the species-area relationship, respectively. The power function, however, "intercepts" at the origin (i.e., when $A = 0$, $S = 0$). Just as important, z values represent the slopes of the relationship between log S and log A,

not species richness and area. By themselves, z values do not indicate how rapidly S increases with A. For this, we require the values of both parameters in the power model, c and z (Gould 1979).

Although this is not a mathematically startling revelation, it is not rare to see, for example, a high z value equated with a rapid increase in S with increasing A. Yet such an inference is valid only if the c values for the archipelagoes and faunas under study are equal. On the contrary, c values vary considerably (often by an order of magnitude or more among archipelagoes or taxa), whereas z values tend to be conservative (typically ranging from 0.15 to 0.35: Gould 1979, his Table 1; Wright 1981, his Table 1). This fact led Gould (1979) to suggest that we draw inferences from comparisons of c values for archipelagoes with approximately equal z values (analogous to analysis of covariance for log-transformed data).

The effects of varying one of these parameters, c or z, while holding the other constant are illustrated in Figures A and B. Note that the species-area relationship (arithmetic scale) is strongly influenced by relatively modest variation in c, but comparatively insensitive to the variations in z typical of natural communities. Furthermore, if these parameters vary simultaneously, as they surely do in nature, then the slope of the species-area relationship (again on an arithmetic scale) may actually be lower for studies reporting *higher* z values (Figure 1C of this box).

In summary, although the power model may continue to provide important insights into factors affecting species richness in isolated biotas, there remains considerable potential for misinterpretation. Studies comparing constants of the power model among archipelagoes, or comparisons with predicted values of these constants, should take measures to avoid the statistical problems discussed here and elsewhere (Connor and McCoy 1979; Gould 1979; Martin 1981).

(Excerpted and modified from Lomolino 1989.)

Effects of varying the values of c and z on the species-area relationship. The model in all cases is $S = cA^z$. (A) Effects of varying the value of c (z is held constant at 0.25). (B) Effects of varying the value of z (c is held constant at 1). (C) Effects of varying both c and z. Note that species richness increases more rapidly for the upper curve despite its substantially lower z value. (From Lomolino 1989.)

(A)

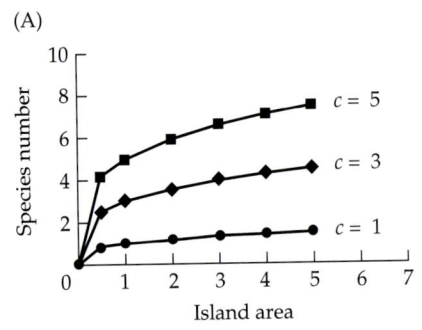

(B)

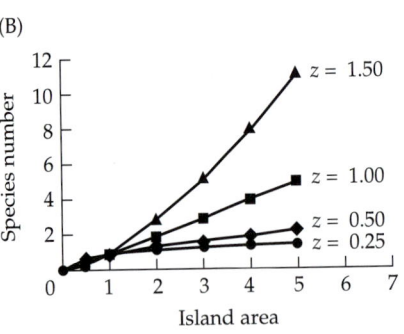

(C)

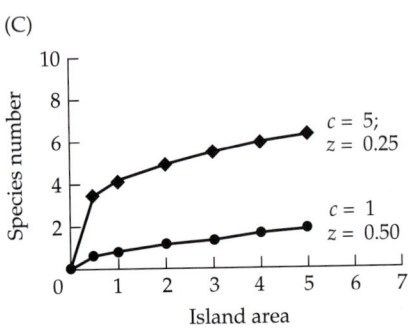

ble areas on a continent because if a species becomes too rare on an island, it is likely to become extinct, whereas on a continent its population can be sustained at low levels by the exchange of individuals between local areas. The effect of such extinctions is much more severe on small islands than on larger ones, resulting in the steeper slope of the species-area curve.

The Species-Isolation Relationship

Since the early 1800s, it has been well known that single, isolated islands far out in the ocean support fewer species than islands that are part of major arch-

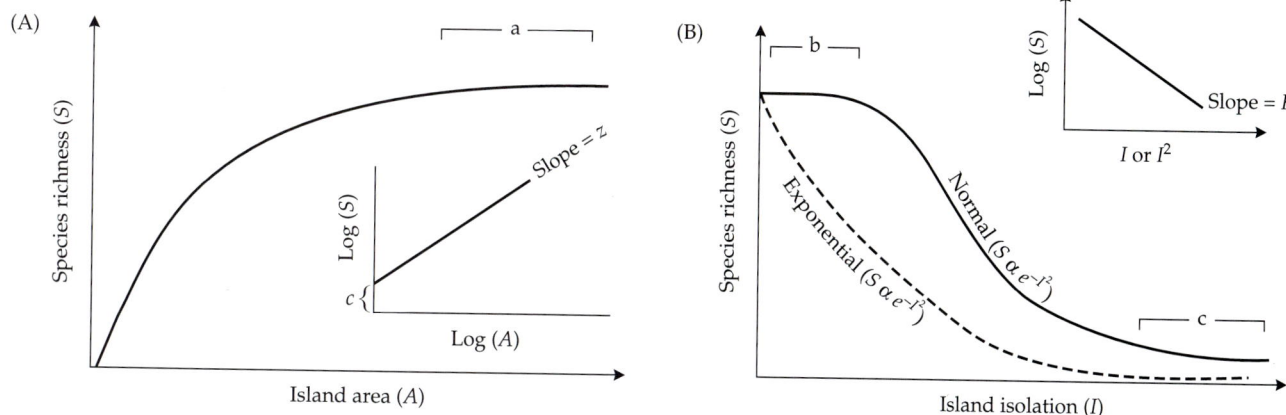

Figure 13.4   Two of the most common patterns in nature—the species-area (A) and the species-isolation (B) relationships. The log-transformed equivalents of these relationships are presented in the insets (S = species richness, A = island area, c and z are fitted constants for the power model of the species-area relationship). Note that the species-area relationship will be difficult to detect (slope near 0) if biogeographic surveys are limited to the larger islands (i.e., region a; see also Figures 13.15 and 13.16). Similarly, the species-isolation relationship will be difficult to detect if surveys are restricted to the very near or very distant islands (regions b and c; see also Figures 9.18 and 13.19A).

ipelagoes or are nearer to continents. Assuming that the decline in species richness results from a decline in dispersal rates with isolation, the form of the species-isolation relationship should be a consequence of dispersal curves for the pool of species (potential colonists from the mainland; see Chapter 9). Therefore, for a variety of taxa and ecosystems, species richness should decline as a negative exponential or sigmoidal function of isolation (Figure 13.4).

With appropriate transformations of one or both axes (i.e., richness and isolation), this relationship can be linearized to allow statistical analyses and comparisons among studies (e.g., $S = k_1 e^{-k_2(I)}$, or $S = k_1 e^{-k_2(I \cdot I)}$; where S = richness, k_1 and k_2 are a fitted constant, and I = isolation). In practice, however, the species-isolation relationship often proves much less general than the species-area relationship. This difference may derive, in part, from the tendency for many studies of islands, such as those of a single archipelago, to include a broad range in island area but only a limited range in isolation. In addition, while measures of island area are easily taken from maps, given the many possible immigration routes and sources, biologically relevant measurements of isolation are extremely challenging. Still, many studies report significant species-isolation relationships, especially when island isolation varies substantially and when the effects of island area are statistically "controlled" (e.g., by using correlation or regression analysis after accounting for the effects of area on species richness; Figure 13.5A and B).

Species Turnover

A third pattern that influenced MacArthur and Wilson was the rapidity of recolonization of the Krakatau islands. These islands are located in the Sunda Straits between the Indonesian islands of Sumatra and Borneo (Figure 13.6). A violent volcanic eruption destroyed the original island of Krakatau in 1883, leaving several remnant islands devoid of life. Recolonization, apparently from the nearby large islands of Java and Sumatra, was rapid; by 1935 a tropical rain forest, supporting numerous species of plants, birds, and other organisms, was developing. Several scientific expeditions visited the remnant islands of

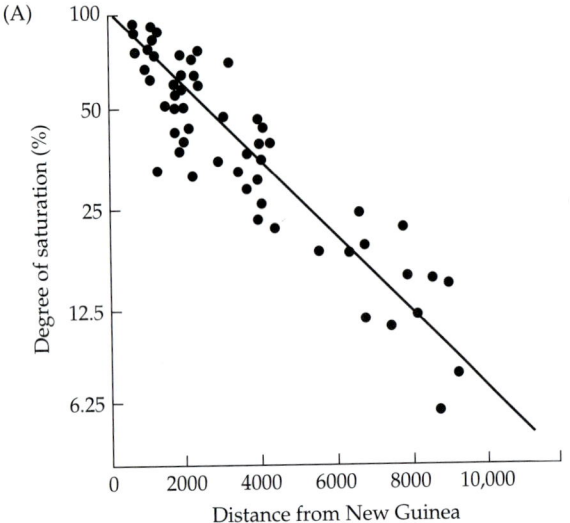

(A)

(B)

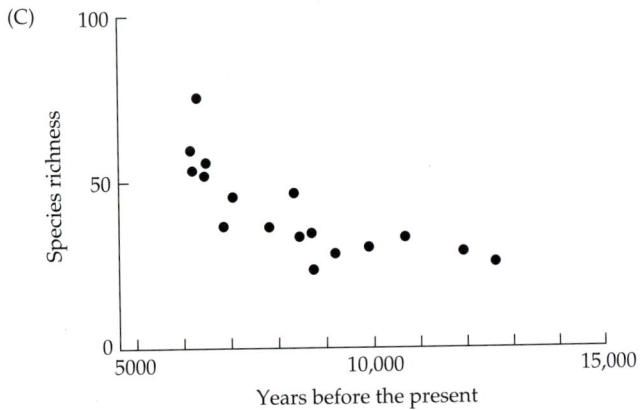

Figure 13.5 A sample of graphs illustrating the effects of isolation on the species richness of various insular biotas (see also Figure 13.22). (A) Resident land birds in the Moluccan and Melanesian archipelagoes. Here species richness is expressed as saturation, which is the species richness of isolated islands expressed as a percentage of that found on an island of equivalent size, but closer to New Guinea. (B) Nonvolant mammals of the Thousand Island region of the St. Lawrence River, New York. Here the ordinate equals the residuals about the species-area relationship (i.e., the difference between the observed insular species richness and that predicted for an island of its size based on the regression model $\hat{S} = 6.51(A^{0.305})$. (C) Species richness of lizards on landbridge islands in the Gulf of California. Here, species richness is graphed as a function of time since isolation of these landbridge islands following rising sea levels of the most recent glacial recession. (A after Diamond 1972; B after Lomolino 1982; C from Wilcox 1978.)

(C)

Figure 13.6 The Krakatau islands are located in the Sunda Straits between the Indonesian islands of Sumatra and Borneo. Rakata, Sertung, and Panjang are remnants of a much larger island that was destroyed by a massive volcanic explosion in 1883. Anak Krakatau is a relatively young island that emerged from a submarine volcanic vent in 1930. (From Whittaker and Jones 1994.)

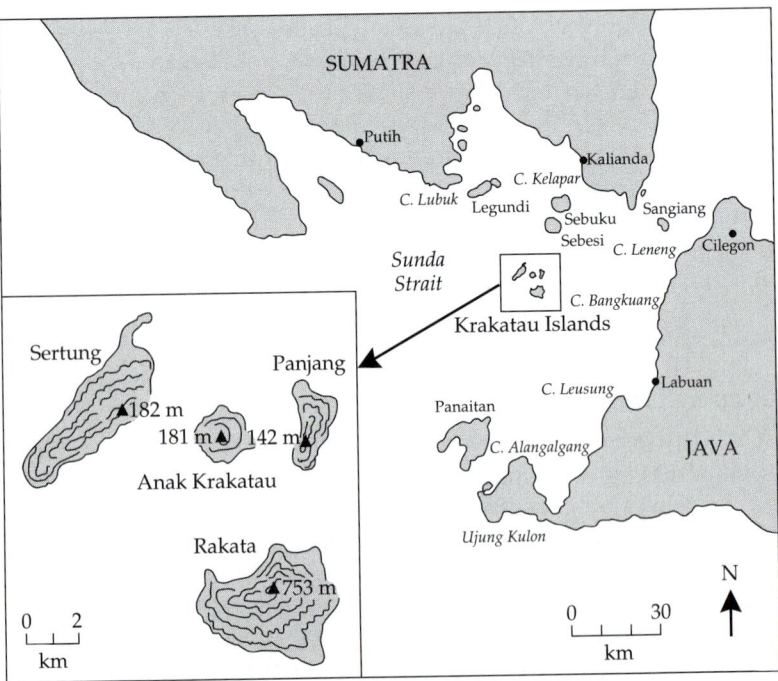

Table 13.1

Number of species of land and freshwater birds on Rakata and Sertung

Number of species found

	Rakata			Sertung		
	Nonmigrant	Migrant	Total	Nonmigrant	Migrant	Total
1908	13	0	13	1	0	1
1919–1921	27	4	31	27	2	29
1932–1934	27	3	30	29	5	34

Number of extinctions and colonizations between censuses

	Rakata		Sertung	
	Extinctions	Colonizations	Extinctions	Colonizations
1908 to 1919–1921	2	20	0	28
1919–1921 to 1932–1934	5	4	2	7

Source: After MacArthur and Wilson 1967.

Note: The number of species increased from the census of 1883 to that of 1919–1921 and then remained relatively constant despite extinction of some species and colonization of others.

Rakata and Sertung and inventoried the biota. Early censuses of the birds, which were particularly complete, are summarized in Table 13.1.

MacArthur and Wilson noted that the number of bird species on Rakata and Sertung increased rapidly until about 1920, but that after that, the total number of species remained relatively constant, despite changes in the composition of the avifauna. Species not only colonized rapidly prior to 1920, but also continued to immigrate after 1920. Some of these late arrivals became successful colonists, replacing about an equal number of species that became extinct. These offsetting colonizations and extinctions might simply have reflected successional changes in the avifauna in response to the development of a tropical forest and the concomitant elimination of open habitats, but they also suggested that continual turnover might be typical of insular biotas. Such turnover might be particularly high when, as is the case for birds on Rakata and Sertung, organisms need cross only modest barriers to reach small islands.

The Equilibrium Theory of Island Biogeography

MacArthur and Wilson produced a single theory to explain what they considered to be the three basic characteristics of insular biotas: the species-area relationship, the species-isolation relationship, and species turnover. They proposed that the number of species inhabiting an island represents a dynamic equilibrium between opposing rates of immigration and extinction. The equilibrium is termed dynamic because immigration and extinction are thought to be recurrent, opposing processes, maintaining a relatively stable species richness despite changes in species composition.

The equilibrium model can be presented graphically by plotting immigration and extinction rates as a function of the number of species present on an island (Figure 13.7). The number of species on the island (S) can range from zero to a maximum, P, the number in the pool of species that is available to colonize the island from a nearby continent or other source area. Now we can predict the shapes of the curves representing colonization and extinction rates. Let us start with an empty island. The immigration rate (defined as the rate of arrival of propagules of species not already present on the island) must decline

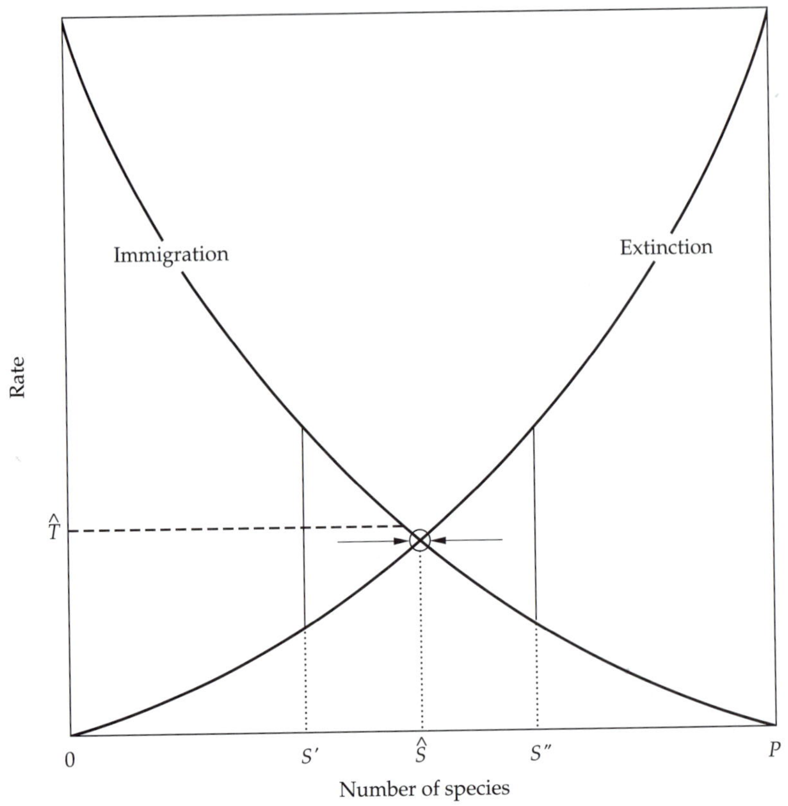

Figure 13.7 A simple model in which the number of species inhabiting an island represents an equilibrium between opposing rates of colonization and extinction. Note that the immigration rate declines and the extinction rate increases as the number of species increases from zero to P, the number in the mainland species pool. The point of intersection of the two curves represents a stable equilibrium, because if the number of species is displaced from $\hat{S}$ to either higher (S'') or lower (S') numbers, it will return (arrows).

from some maximum value when the island is empty to zero when the island contains all the species in the pool and there are no more new species to arrive. As the number of resident species grows, there remain fewer *new* species on the mainland to colonize the island. Conversely, the extinction rate (defined as the rate of loss of existing insular species) should increase from zero when there are no species present on the island to become extinct, to some maximum value when all the species in the mainland pool are inhabiting the island. Simply put, as the island fills, the number of species that can suffer extinctions increases, and therefore, extinction rates should increase accordingly.

At some number of species between zero and P, the lines representing the immigration and extinction rates cross. At this point the two rates are exactly equal, resulting in an equilibrial number of species, $\hat{S}$, and an equilibrial rate of species turnover, $\hat{T}$. This point represents a stable equilibrium because if the number of species is perturbed from this value, it should (at least theoretically) always return. For example, suppose that a natural disaster, such as a hurricane, causes the extinction of several insular species, temporarily reducing the number of species from $\hat{S}$ to S' (Figure 13.7). Then the immigration rate will exceed the extinction rate, and the island will accumulate species until it has again reached $\hat{S}$. Similarly, if S is perturbed from $\hat{S}$ to a larger number, S'', then the extinction rate will be greater than the colonization rate, and species will be lost until $\hat{S}$ is restored.

Now let us incorporate the effects of island size and isolation into this model. MacArthur and Wilson assumed that the size of an island would affect only the extinction rate. Although they recognized that a large island would provide a larger target for dispersing propagules than a small one, they reasoned that such an effect on immigration rate would be insignificant compared with the importance of island size in extinction. The population sizes of all species should decrease with decreasing island area, and as discussed in

Chapter 8, the probability of extinction increases rapidly as a population gets very small. Consequently, for a source pool biota of many species, the extinction rate should be substantially greater for a small island than for a larger one. This can be shown in the graphic model by drawing two extinction rate curves; the one for a small island is always higher than the one for a large island (Figure 13.8). Examining the intersections of these curves with the colonization curve shows immediately that the small island is predicted to have a smaller equilibrium number of species and a higher equilibrium turnover rate than the large island.

MacArthur and Wilson used similar logic to show how immigration curves would be influenced by isolation. They assumed that the distance of an island from the source pool would affect only the immigration rate. No matter what the mechanism of dispersal, if a barrier exerts a filtering effect, then the probability of an organism crossing the barrier decreases as the width of the barrier increases. This effect of isolation by distance can be incorporated into the model by drawing two immigration curves; the one for an island near a source of species is always higher than the one for a more isolated island (Figure 13.8). The intersections of these curves with the extinction curves predict that at equilibrium, near islands should have more species and higher rates of turnover than distant islands.

By combining the effects of island size and isolation in a single graph (Figure 13.8), one can see the predictions of the model. There are four intersections of immigration and extinction rate curves, one for each combination of island size, large (L) and small (S), and distance, near (N) and far (F). The number of species at equilibrium is predicted to be in the order $S_{LN} > S_{LF} \sim S_{SN} > S_{SF}$. (Note that whether a large far or small near island should have more species will depend on the exact shapes of the immigration and extinction curves.)

The model obviously explains the observations that motivated it: namely, that the number of species increases with area and decreases with isolation,

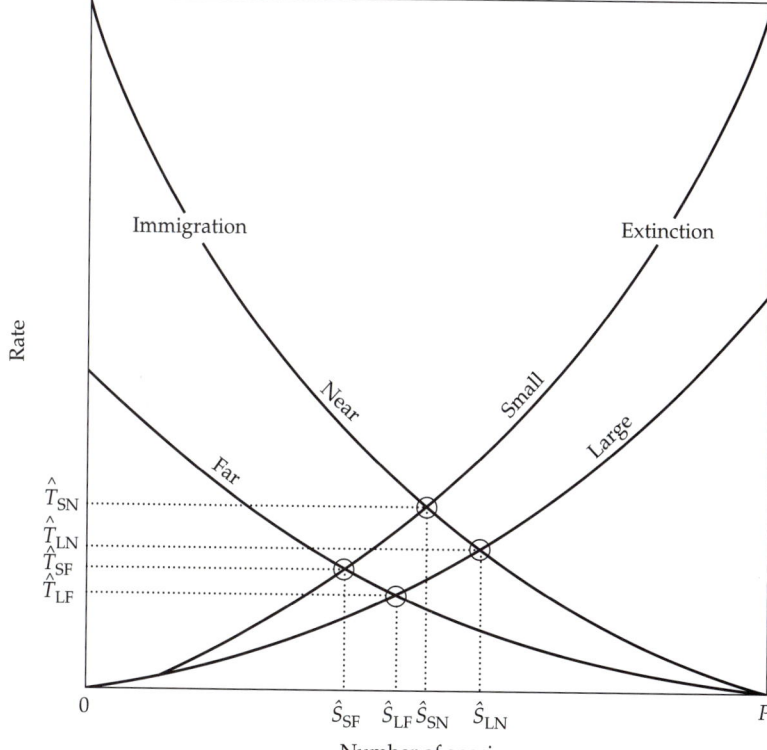

Figure 13.8 MacArthur and Wilson's (1963) equilibrium model of island biogeography, showing the effects of island size (different extinction rate curves) and isolation (different immigration rate curves) on the equilibrium number of species ($\hat{S}$) and rate of species turnover ($\hat{T}$). The intersections of the curves for islands of different combinations of size and distance can be used to predict the relative numbers of species and turnover rates at equilibrium.

and that there is a continual turnover of species. However, like any good theory, the model goes beyond what is already known to make additional predictions that can be tested only with new observations and experiments. Specifically, it predicts the following order of turnover rates at equilibrium: T_{SN} > T_{SF} ~ T_{LN} > T_{LF}. The model also predicts the relative rates at which islands of different sizes and degrees of isolation should return to equilibrium if the biota is perturbed. For example, a near island should return to equilibrium more rapidly than a distant island of the same size because it should have a higher immigration rate, but the same extinction rate.

Strengths and Weaknesses of the Theory

Like most important new ideas, MacArthur and Wilson's equilibrium theory elicited a mixed response from other scientists. It generated not only great interest and enthusiasm on the part of some investigators, but also severe skepticism and criticism from others. Certainly the theory stimulated a new wave of research in ecological biogeography. Most of these studies involved islands or other isolated habitats and were designed specifically to evaluate or elaborate on MacArthur and Wilson's model. In a review of the equilibrium theory written only a decade after MacArthur and Wilson's (1963) first seminal paper, Simberloff (1974a) cited 121 references, and the pace of research has continued to accelerate. By 1987, the 1963 paper and the 1967 monograph had been cited in over 2000 publications.

Several features of the equilibrium theory contributed to its favorable reception. Its elegantly simple graphical presentation made the essential elements of the model accessible to a wide audience, including persons with minimal mathematical training. Moreover, the equilibrium theory not only introduced stimulating new ideas that helped to bridge the gap between traditional biogeography and ecology, but presented them in the form of a model that made clear, testable predictions. Many mathematical models, especially those used in ecology, are not empirically operational—that is, they do not explicitly indicate what observations would be necessary to test the theories and reject the models. Such models can still be valuable, however, if they serve a heuristic function—if they cause one to think about a problem in new and more precise ways. MacArthur and Wilson's model certainly plays such a heuristic role, but it also indicates the kind of data that are necessary for a rigorous test. It predicts qualitative trends (increases or decreases) in numbers of species and turnover rates with island size and isolation. These predictions can be tested using simple lists of species inhabiting various archipelagoes at different times. Such lists can be compiled by original fieldwork, but the results of previous surveys of the biotas of many islands are also available in the published literature. The only other data required, the areas of the islands and distances to probable source areas, can be obtained from standard maps.

The simplicity of the theory, however, has also been the basis for much criticism. Some have argued that it is so simple as to be useless because it obscures, rather than clarifies, the patterns and processes that make island biogeography interesting (e.g., Sauer 1969; Lack 1970; Carlquist 1974; Gilbert 1980; Williamson 1989). In view of the research the theory has stimulated and the increased understanding of island distributions that has resulted (see Chapter 1), much of this criticism is unwarranted. All models are intended to help investigators understand nature by presenting a simplified, abstracted concept of a more complex reality. They inevitably sacrifice a certain amount of precision for clarity and generality. Usually new theories are presented initially in a very general, incomplete form and are corrected and refined as a result of subsequent empirical and theoretical research.

If the equilibrium theory of island biogeography survives at all, it will probably experience this fate. Critics have pointed out specific problems with several assumptions of the model, which may limit its applicability to many types of islands. Much of the recent research has served to identify and evaluate these problems and to modify the model so as to incorporate different and, it is hoped, more realistic assumptions. To their credit, MacArthur and Wilson themselves (1967) acknowledged some important weaknesses of their theory:

> First, we still know very little about the precise shape of the extinction and immigration curves, so that few numerical predictions can yet be made. Second, and more importantly, the model puts rather too simple an interpretation on the process by making an artificially clear-cut distinction between immigration and extinction … The third difficulty stems from the assumption that extinction and immigration curves have fairly regular shapes for different faunas and different islands and for different times on the same islands. When a new set of curves must be derived for a new situation, the model loses much of its virtue. Deviant cases do occur—for example the Krakatau flora. The extinction curves, furthermore, have a pronounced genetic component: rarity affects gene frequencies, and genetic as well as ecological causes of extinction act in concert.

Others were quick to echo these concerns and to point out additional shortcomings of the theory. As you will see from the concerns listed below, most of these criticisms suggest that the theory is insufficient or incomplete, yet do not take issue with its basic tenets.

1. Many insular biotas may not be in equilibrium between opposing rates of immigration and extinction. Rather, the number of species may increase or decrease over evolutionary time. This is particularly likely when immigration and extinction occur on approximately the same time scale as the geological and climatic events that create, change, and destroy islands. Here, insular communities may never reach equilibrium. An equilibrium theory may be useful for interpreting these cases, however, because it may be profitable to consider these biotas as approaching a new equilibrium following a historical perturbation. As MacArthur and Wilson (1967:19–21) observed, "a perfect balance between immigration and extinction might never be reached, … but to the extent that the assumption of a balance has enabled us to make certain valid new predictions, the equilibrium concept is useful."

2. The model assumes that the identities and characteristics of particular species can be ignored. Species are not all assumed to have identical immigration and extinction rates, but nevertheless, the resulting turnover is viewed as a highly stochastic process: new species immigrate and existing ones die out more or less at random, and only the approximate number of species remains the same. The implicit assumption that ecological processes, including interspecific interactions, do not determine *which* species can coexist on a particular island is at least technically incorrect.

3. Immigration and extinction are treated as independent processes. This is probably justified, given that immigration is defined as the arrival of propagules of new species and that secondary succession is not occurring. On the other hand, recruitment of additional individuals of species already present should tend to rescue populations on near islands from extinction. In a similar manner, area may affect immigration as well as extinction rates because larger islands may intercept more potential colonists (see the later sections on rescue and target effects).

4. It may be difficult to identify the source of an island biota without careful investigation of the systematics and historical distribution of the species that are present. Indeed, the species inhabiting a single island may be derived from several sources, including over-water dispersal from continents and other islands, past connections with other landmasses, and endemic speciation within the island itself. If insular species are acquired by immigration from multiple sources, the theory can potentially be modified to deal with this complication. For example, MacArthur and Wilson considered stepping-stone colonization, in which a species disperses from island to island down a chain of islands.

5. If insular species are derived by speciation within the island itself, then a basic assumption of the model—that species richness is affected only by immigration and extinction—is clearly violated. Again, although the original model did not do so, it can be modified to include the effects of speciation (see Figure 13.29).

6. Area provides only a very general and indirect measure of the capacity of islands to support individuals and species. Although the extent of most habitat types increases with increasing island size, so usually does the diversity of habitats. Larger islands tend to have higher mountains, more aquatic habitats, and so on, as well as larger areas of most of the vegetation types found on small islands. Consequently, some of the increase in species diversity with island size may be owing to the addition of specialists whose habitat requirements are met only on large islands. In this case a more elaborate model, which also incorporates specific habitat variables, should predict patterns of insular species diversity and distribution better than area alone (e.g., Power 1972; Johnson 1975).

These criticisms are serious, and should be kept in mind as we proceed to discuss insular distributions in this and the following chapter. We will see many cases in which the model, in its original form, is inadequate to account for observed patterns (see Williamson 1989; Haila et al. 1982; Case 1987; Haila 1986). It will also be clear that in following MacArthur and Wilson's general approach, in testing and often rejecting their theory, we have learned a great deal about the ecological and historical processes that determine the distribution of organisms among islands and other isolated habitats.

Tests of the Model

Beginning in the late 1960s, numerous authors published papers purporting to test MacArthur and Wilson's theory. Often only one or two of its predictions—namely, covariation in number of species with area or isolation—were tested. Even these incomplete tests are valuable, however, because when the predictions are not borne out by the data, they indicate that the theory in its present form cannot account for the distributions of these biotas. On the other hand, when a few observations concur with the predictions, it is unwarranted to assume, as many authors have done, that this corroborates the theory. It is possible to obtain the right result for the wrong reason because alternative explanations, based on different assumptions, can generate some of the same predictions. Before accepting a particular model as the explanation for a set of observations, it is important not only to test all possible predictions of the model, but also to evaluate the validity of the underlying assumptions and to attempt to imagine and rule out alternative explanations.

Early studies that "tested" and purported to support the equilibrium model fell into three categories, which we describe below. Most common were analyses of data on insular distributions of various taxa at one point in time. Many of these simply confirmed for more groups of organisms and more archipelagoes the relationship between number of species, island size, and island isolation pointed out by MacArthur and Wilson (e.g., Hamilton et al. 1964; Hamilton and Armstrong 1965; Johnson et al. 1968; Johnson and Raven 1973). Others described similar patterns for other insular habitats, such as mountaintops (Vuilleumier 1970) and caves (Culver 1970; Vuilleumier 1973). Frequently, they advanced ad hoc explanations to explain differences among these relationships in such things as z values and goodness of fit of the regression equation to the data points. Although these studies supported the generality of the patterns, they cannot be viewed as rigorous tests of the theory.

Estimates of turnover on landbridge islands.   In 1969, Jared Diamond reported a more direct test of one crucial prediction: continual turnover in species composition resulting from the balance of colonizations and extinctions. Diamond recensused the avifauna of the Channel Islands off the coast of southern California almost exactly 50 years after Howell (1917) had published a detailed account of the birds known to breed on each island at the turn of the century. Comparison of the two censuses revealed striking differences in species composition (Table 13.2), although there had been relatively little change in the number of species breeding on each island. In 1968 Diamond observed a number of species not known to breed there prior to 1917, and about an equal number of species present 50 years earlier that had apparently disappeared. From these observations, he concluded that at least 20 to 60% of the bird species on each island had turned over since 1917. He pointed out that actual turnover rates could well have been even higher; some species might have colonized and become extinct during the intervening 50 years. Although there were not enough islands to test rigorously for the relationship between turnover rate and either island size or isolation, Diamond noted that turnover appeared to be greatest on the islands with the fewest species.

Table 13.2
Turnover of breeding land bird species on the California Channel Islands between 1917 and 1968

Island	Area (km²)	Distance to mainland (km)	Number of species 1917	Number of species 1968	Extinctions	Introductions (by humans)	Colonizations	Turnover (%)[a]
Los Coronados	2.6	13	11	11	4	0	4	36
San Nicholas	57	98	11	11	6	2	4	50
San Clemente	145	79	28	24	9	1	4	25
Santa Catalina	194	32	30	34	6	1	9	24
Santa Barbara	2.6	61	10	6	7	0	3	62
San Miguel	36	42	11	15	4	0	8	46
Santa Rosa	218	44	14	25	1	1	11	32
Santa Cruz	249	31	36	37	6	1	6	17
Anacapa	2.9	21	15	14	5	0	4	31

Source: Diamond 1969.

[a]Turnover rate, expressed as percentage of the resident species per 51 years, is calculated as 100 (extinctions + colonizations)/(1917 species + 1968 species – introductions).

Diamond's study, widely cited as confirming or supporting the equilibrium model (e.g., MacArthur 1972), was also vigorously challenged. Lynch and Johnson (1974) pointed out that most of the thoroughly documented changes in the avifauna could be attributed directly to human influence, and some of the other apparent changes may have resulted from errors in conducting and interpreting the census. In particular, most of the extinctions involved the disappearance of large birds of prey, including the osprey (*Pandion haliaetus*), bald eagle (*Haliaeetus leucocephalus*), and peregrine falcon (*Falco peregrinus*), which were almost certainly eliminated by pesticide poisoning. Many colonizations were a result of the immigration of house sparrows (*Passer domesticus*) and starlings (*Sturnus vulgaris*), which colonized under their own power as they expanded their ranges after introduction into eastern North America from Europe. Turnover involving these latter two and probably other species had also been influenced by habitat changes caused by recent human activities. Lynch and Johnson concluded that, contrary to Diamond's claims, there was little evidence for natural turnover of breeding birds on the Channel Islands within the 50-year period.

This debate, however, was far from resolved. Jones and Diamond (1976) surveyed the birds of the California Channel Islands, especially Santa Catalina, for several years in succession, and found year-to-year changes in the breeding status of several species. Diamond and May (1976) analyzed many years of careful records of birds breeding on the small Farnes Islands off Great Britain. Several species were reported to have been present only sporadically during the years for which data were available (Table 13.3). In both of these studies, however, the "turnovers" involved species that were migratory or highly nomadic. Their presence or absence on an island is more a matter of the arrival and departure of highly mobile individuals for which a few miles of water repre-

---

Table 13.3
Year-by-year turnover of breeding bird species on the Farnes Islands, Great Britain

Number of breeding bird species	
Present per year	
Mean	5.9
Range	4–9
Exhibiting turnover	12
Present every year	4
Percent species turnover per year	
Mean	13
Range	3–22
Number of breeding pairs per species	
Of species exhibiting turnover	
Mean	0.23
Range	0–6
Of species present every year	
Mean	14.87
Range	2–60

Source: Diamond and May 1976.

Note: This table is based on 29 consecutive years of censuses. Note that the 12 species that bred in some years but not in others had very low average populations, whereas those 4 species that were present every year maintained average populations of about 15 pairs per species.

sent no significant barrier, than of the establishment and extinction of real resident populations. Terborgh and Faaborg (1973) reported apparent turnover in the avifauna of Mona, a small island west of Puerto Rico in the West Indies, as a result of comparisons between an early survey and their own subsequent census. Again, however, the island had been changed substantially by human activities, especially by the effects of introduced goats. Lack (1976) documented changes in the birds of Jamaica over more than 200 years of recorded history. On this large Caribbean island, with a resident land avifauna numbering 65 species, there have been two extinctions and one colonization, and all of these can be attributed directly to human influence (see also Abbott 1983).

All of this is not to say that organisms do not naturally colonize and suffer extinctions on oceanic islands as predicted by equilibrium theory. Rather, the data documenting such turnover of established populations are not convincing, even for birds, which are perhaps the best long-distance colonizers of all organisms.

Turnover on recently created anthropogenic islands. Human activities have fragmented and insularized once massive and continuous ecosystems across the globe. The effects of these activities on biodiversity are the focus of Chapters 17 and 18. Here, we consider the biogeographic dynamics of two anthropogenic archipelagoes created by the flooding of mountainous areas in the American tropics.

Early in the twentieth century, the Chagras River was dammed to create the Panama Canal and Gatun Lake, in turn flooding lowland areas and transforming forested hilltops into islands. Repeated biotic surveys have been conducted to trace the **relaxation** (decline toward a new equilibrium) of plant and animal communities on these islands, especially those on the largest island, Barro Colorado. Surveys conducted during the 1970s and early 1980s revealed that about 45 of the estimated 108 original species of breeding birds had disappeared (see Willis 1974; Karr 1982, 1990; Wright 1985). Because intensive surveys were not conducted until well after the island was formed, additional extinctions may have gone unrecorded. The salient observations from these studies are that extinctions and immigrations appear to be recurrent, and that turnover is lowest on the largest and most isolated islands (Figure 13.9). Numerous other studies report a similar inverse relationship between turn-

Figure 13.9 Consistent with the predictions of the equilibrium theory, turnover rates of birds on islands of Gatun Lake, Panama, tend to decrease with island area (A) and with isolation (B). Yet, because frequent immigrations can rescue otherwise dwindling populations from extinction, turnover rates may also be lower on near islands than on distant islands (e.g., the three near islands in B; see also Figure 13.13). (From Wright 1985.)

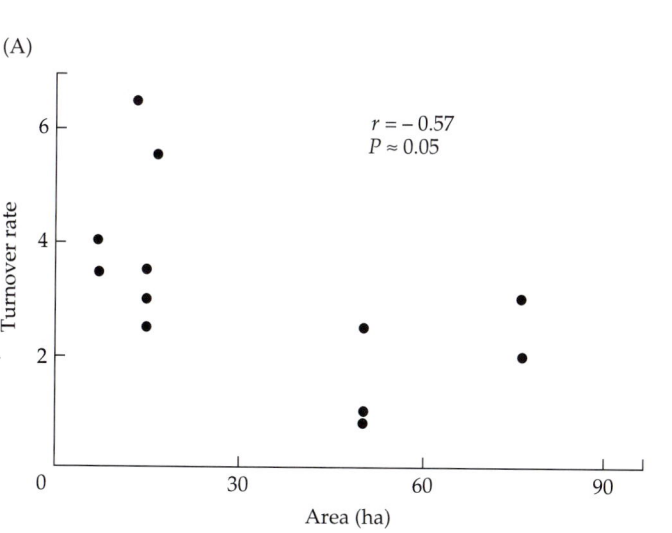

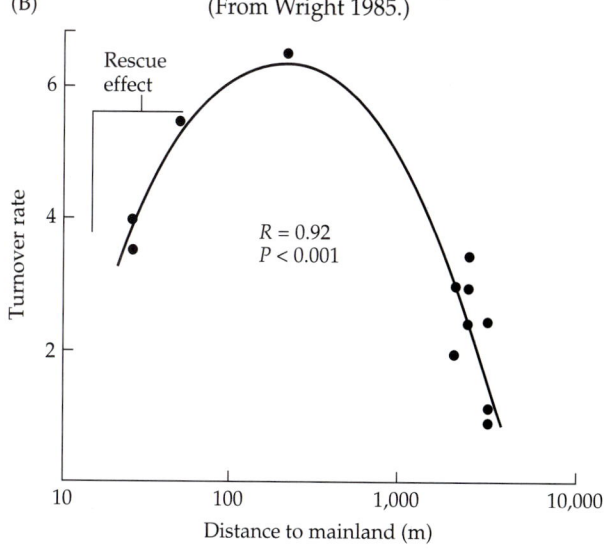

over rates and island area or isolation (Brown 1971b; Diamond 1972; Wilcox 1978, 1980; Heaney 1984, 1986; Terborgh and Winter 1980; Nilsson and Nilsson 1982). Although the evidence is limited, it appears that species richness declines most rapidly during the early stages of relaxation, immediately following isolation (MacArthur and Wilson 1967; Terborgh 1974; Wilcox 1980; see also Lovejoy et al. 1986; Robinson et al. 1992; Kattan et al. 1994; Stouffer and Bierregaard 1995). As predicted by the equilibrium theory, relaxation rates then tend to slow as species richness approaches the new equilibrium.

Recent studies by John Terborgh and his colleagues suggest that relaxation on newly created islands may be surprisingly rapid (Terborgh et al. 1997). Avian communities on relatively small (≈1 ha) islands in Lago Guri, a Venezuelan lake created during the 1980s for a hydroelectric project, may have already achieved a new dynamic equilibrium just 7 years after the islands were formed (Terborgh et al. 1997). In contrast, avian species richness still appears to be declining on the larger islands, a result that is also consistent with the predictions of the equilibrium theory. Tracking the results of these and similar opportunistic "experiments" is certain to provide additional insights into the forces structuring insular communities.

Throughout the history of biogeography, the comparative method has proved to be a powerful tool for exploring the forces influencing the distributions and diversities of biotas. In a review of 21 studies of turnover in a variety of organisms, ranging from protozoans and plants to terrestrial arthropods and vertebrates, Thomas Schoener (1983) discovered some intriguing patterns. Not only did turnover tend to be lower on larger islands, but it also decreased with the generation time of the organisms (Figure 13.10). As additional turnover studies are accumulated, reviews such as Schoener's will continue to provide important insights into the forces structuring isolated communities.

Finally, we should note that relaxation following isolation is evidence for just one, fairly noncontroversial part of the MacArthur and Wilson model. We know that small, isolated populations are especially vulnerable to extinction. The real question is whether there is a balance between the two opposing forces—are immigration rates high enough to set a new equilibrium?

Experimental defaunation.   The most rigorous early tests of MacArthur and Wilson's theory were the defaunation experiments conducted by Wilson and his student Daniel Simberloff (Simberloff and Wilson 1969, 1970; Wilson and

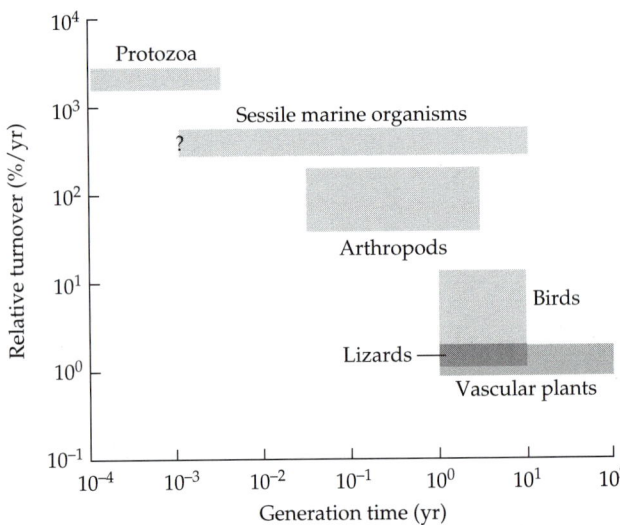

Figure 13.10 Relative turnover rates tend to be lower for organisms with longer generation times. (From Schoener 1983.)

Simberloff 1969). This study has become justly famous as an example of the use of controlled, manipulative experimentation to test theoretical models in biogeography and ecology. The basic design was simple: all species of arthropods were eliminated from tiny islets of red mangrove (*Rhizophora mangle*) in the Florida Keys, and the subsequent changes were monitored closely. Simberloff and Wilson hired an exterminator, who used methyl bromide gas to kill all insects, spiders, mites, and other terrestrial animals while leaving the mangrove vegetation virtually undamaged. This was a drastic but effective perturbation. Recolonization, monitored by careful surveys, was surprisingly rapid (Figure 13.11). Within less than a year, all but the most distant island had recovered their initial number of species. In fact, the number of species increased rapidly and appeared to overshoot the initial number before declining and stabilizing close to the initial value. Furthermore, there was a great deal of turnover, even after the number of species had stopped changing significantly. Individual species colonized and disappeared, sometimes repeatedly, during the short-term study. The high turnover rates were not surprising given the proximity of these islands to the mainland (0.002 to 1.2 km), their small size (75 to 250 m²), and the lack of intervening dry land.

Thus, Simberloff and Wilson's results strongly supported several predictions of the equilibrium model (see also Rey 1981; Strong and Rey 1982; Hockin 1982; Molles 1978). Although there were too few islands to test rigorously for the predicted relationships of initial colonization rate, equilibrium turnover rate, and equilibrium number of species to island size and isolation, nevertheless the results were generally consistent with the predictions. For example, the most isolated island (E1 in Figure 13.11) had the fewest species and the lowest rate of recolonization. Simberloff and Wilson also suggested that developing communities may pass through three and possibly four dynamic equilibria (Figure 13.12). First, species may tend to accumulate rapidly to reach a "non-interactive equilibrium," with most species occurring at relatively low population levels, perhaps too low to inhibit populations of other species. Populations may later increase to the point at which interspecific interactions can cause local extinctions of some species, resulting in a reduced, or "interactive" equilibrium. Simberloff and Wilson's studies also suggest that succession or ecological sorting may generate a subsequent increase in species richness toward an "assortative" equilibrium. Finally, on the largest and most isolated islands, species richness may continue to increase until extinction

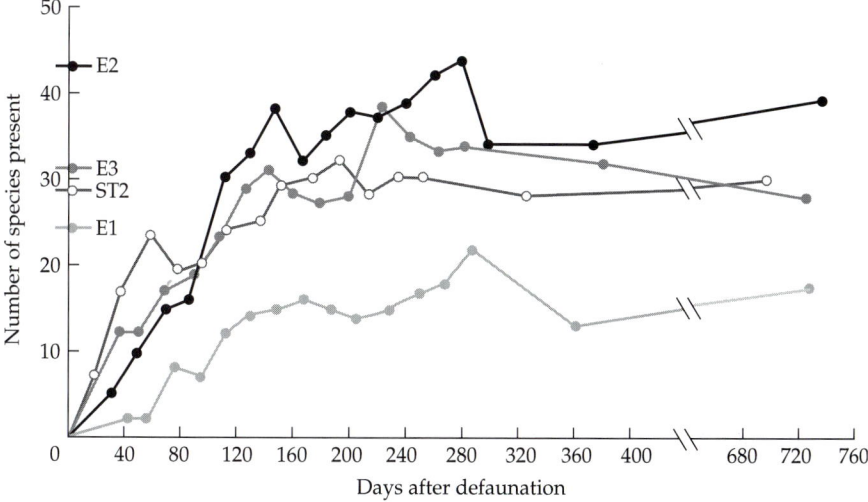

Figure 13.11
Recolonization by terrestrial arthropods of four small mangrove islands as a function of time since the fauna was removed. The initial number of species present is indicated along the vertical axis. Note that after defaunation the number of species increases rapidly, tends to overshoot the initial number, declines, and then increases gradually to approximately the initial number. Island E1, with a lower rate of colonization and a smaller number of species, was more isolated from a source of colonists than the other islands. (After Simberloff and Wilson 1970.)

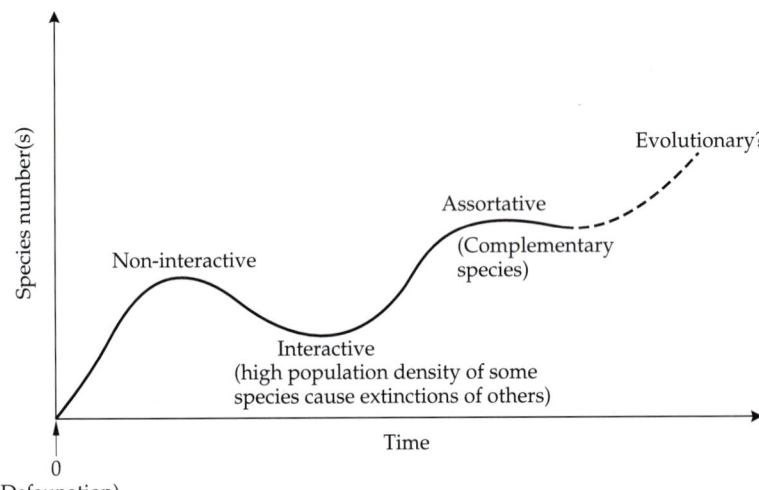

Figure 13.12 As an empty island accumulates species, its insular communities may pass through a series of equilibria reflecting demographic, ecological, and evolutionary processes. (After Simberloff and Wilson 1969, 1970.)

rates balance the combined effects of speciation and immigration ("evolutionary" equilibrium). While certainly appealing, this graphical model of dynamic equilibria has not yet been shown to be generally applicable.

Additional Patterns in Insular Species Richness

As we discussed earlier, the equilibrium theory assumes that extinction is affected only by area and that immigration is affected only by isolation. Here we consider violations of these assumptions that warrant modification of the original model.

The rescue effect. Another study of arthropods on isolated patches of vegetation points to a potentially important problem with MacArthur and Wilson's equilibrium model. Brown and Kodric-Brown (1977) censused arthropods (mostly insects and spiders) on individual thistle (*Cirsium neomexicanum*) plants growing in desert shrubland in southeastern Arizona. Although the intervening habitat may have been suitable for some of the arthropod species, the thistle plants constituted isolated patches of favorable habitat. Brown and Kodric-Brown counted the individuals and species of arthropods on the plants at 5-day intervals. Their results confirmed several major predictions of the MacArthur-Wilson model (Table 13.4). The number of individuals and species increased with plant size and decreased with increasing distance from the nearest plants. Although the arthropods did not maintain real populations on the plants, there was a dynamic equilibrium between the rates of arrival and disappearance. Defaunated thistles were reinhabited rapidly, those near other thistles more quickly than isolated plants: plants closely surrounded by others regained 94% of their original arthropod biota in 24 hours, whereas isolated plants acquired only 67% of the initial number of species in the same period. The turnover of individuals and species was higher on small plants than on large ones.

All of these results were consistent with the predictions of the equilibrium theory. However, contrary to the model's prediction, turnover rates were lower on plants in close proximity to others than on isolated ones (Table 13.4). This single exceptional result is important, because it suggests a problem with the model that may be as relevant for organisms on real islands as for arthropods on thistles. The most likely explanation for all of the results taken together is that there is an insular equilibrium maintained by opposing mechanisms of

Table 13.4
Turnover rates of arthropod species on individual thistle plants

Size–isolation category	Site 1			Site 2		
	Number of plants	Mean number of species	Mean turnover rate	Number of plants	Mean number of species	Mean turnover rate
Large–near	16	3.82	0.67	9	5.25	0.29
Large–far	7	3.78	0.78	9	4.44	0.42
Small–near	56	1.89	0.78	21	2.21	0.69
Small–far	3	1.33	1.00	11	0.80	0.91

Source: Brown and Kodric-Brown 1977.

Note: Plants were divided into objective size and isolation categories on the basis of the number of flowers and the number of other plants in the immediate vicinity. The number of species is for the second of the two censuses 5 days apart. Turnover rate equals the number of species present only in the first census plus the number of species present only in the second census, divided by the total number of species present in both censuses.

immigration and extinction, as envisioned by MacArthur and Wilson, but the factors affecting the arrival of new species are not independent of those influencing the extinction of species already present. Proximity to a source of immigrants increases the immigration rate of all species, and a continual influx of individuals belonging to species already present tends to prevent the disappearance of those species.

In the case of arthropods on thistles, this **rescue effect** is probably simply statistical: high rates of immigration reduce the probability that a species will temporarily be absent and hence recorded as a turnover. On real islands, however, immigrants may rescue populations from extinction by contributing to the breeding stock and by injecting new genetic variability to counteract the deleterious effects of inbreeding, which can be severe in small, isolated populations. In a separate study, Smith (1980) corroborated the rescue effect of immigrants on extinction rates, showing that populations of pikas (*Ochotona princeps*; rat-sized rodents similar in appearance to small rabbits) inhabiting isolated rock-slides in North America had higher turnover rates than those near a source of colonists. Other researchers have subsequently reported rescue effects for a variety of organisms (e.g., Wright 1985, Laurance 1990; see Figure 13.9B).

The rescue effect was not anticipated by MacArthur and Wilson, but it is easy to modify their model slightly to incorporate it. Drawing different extinction rate curves, as well as different colonization rate curves, for near and far islands (Figure 13.13A,B) allows us to take into account the decrease in the extinction rate on near (versus distant) islands that is due to the rescue effect. Note that when this is done, it reverses the order of the equilibrium turnover rates ($T_F > T_N$) from that predicted by the MacArthur-Wilson model, but the predicted relationship for the equilibrium number of species remains unchanged ($S_N > S_F$).

Brown and Kodric-Brown's study emphasizes the importance of testing all predictions of a model as well as critically evaluating its basic assumptions. Unless this is done, investigators risk misinterpreting data that are merely consistent with the model as corroborating evidence.

The target area effect. Just as extinction rates may be influenced by island isolation, immigration rates may be influenced by island area. Again, this is a violation of one of the assumptions of the equilibrium model, but one that requires only a slight modification. Michael Gilpin and Jared Diamond (1976) suggested that at least part of the increase in species richness with increasing

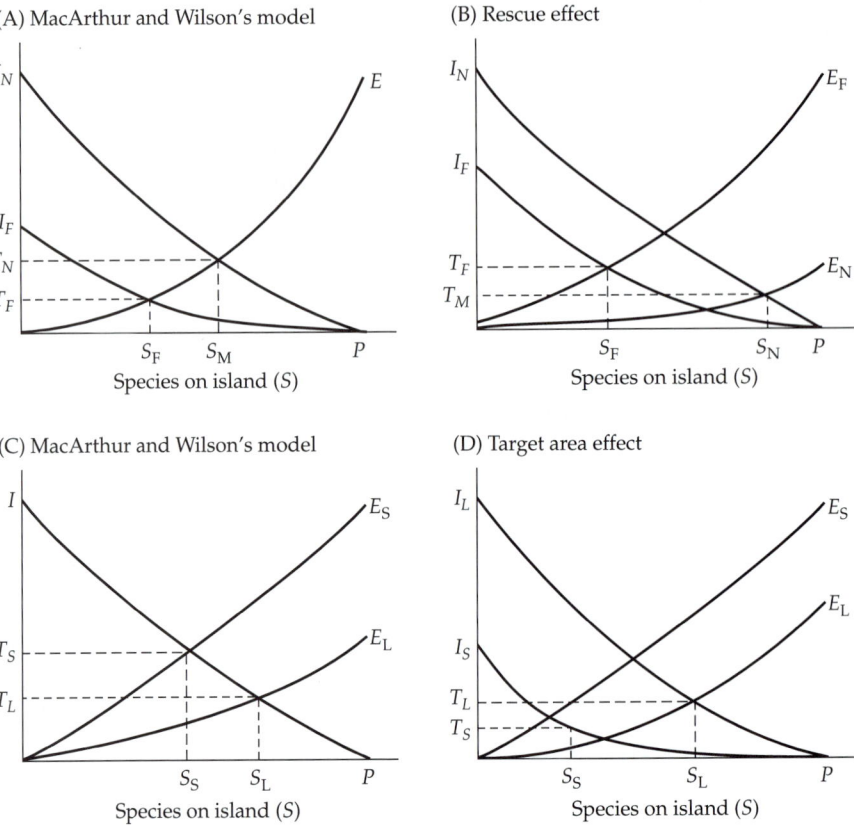

Figure 13.13 Two modifications of MacArthur and Wilson's equilibrium theory of island biogeography can be made to take observed patterns into account. The rescue effect (B) refers to the reduction in extinction (and therefore turnover) rates on near islands because recruitment of immigrants can supplement otherwise dwindling populations. The target area effect (D) refers to the tendency for immigration rates to be higher on larger islands. I = immigration rate; E = extinction rate; T = turnover rate; S = species richness; P = richness of the source or "pool" biota; S = small island; L = large island; N = near island; and F = far island. (After Gotelli 1995.)

island area may derive from a tendency for immigration rates to be higher on larger islands. That is, larger islands may serve as more effective target areas for potential immigrants because they are more likely to be seen (by active immigrators) or encountered (by passive immigrators). In his studies on recolonization of defaunated *Spartina* islands, Jorge Rey (1981) found that immigration rates of arthropods were positively correlated with island area. In a quite different system, Hanski and Peltonen (1988) found that colonization rates of shrews on islets in a Finnish lake increased with island area. While this suggests a **target area effect** for active immigrators, it is important to note that colonization includes the survival and establishment of breeding populations, as well as immigration.

Lomolino (1990) was able to conduct a more direct test of the target area hypothesis by tracking the movements of terrestrial mammals across the ice-covered St. Lawrence River during winter. The characteristics of tracks in the snow allowed the identification of species and the mapping of actual movements among islands and sites along the mainland. Immigration rates of both small and large mammals were significantly and positively correlated with island area (Figure 13.14A). Studies of plant propagules drifting onto the beaches of oceanic islands indicated that immigration rates also increase with island size (Figure 13.14B). Thus, there is evidence, albeit limited, that at least some of the species-area relationship can be attributed to the target area effect. Again, this calls for a slight modification of the equilibrium model, one that alters relative turnover rates, but not patterns in species richness (Figure 13.13C,D; see also McGuiness 1984; Hamilton et al. 1964; Connor and McCoy 1979; Coleman et al. 1982).

(A)

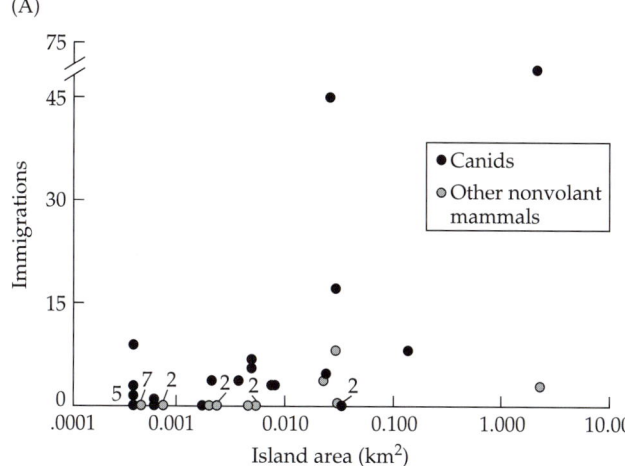

(B)

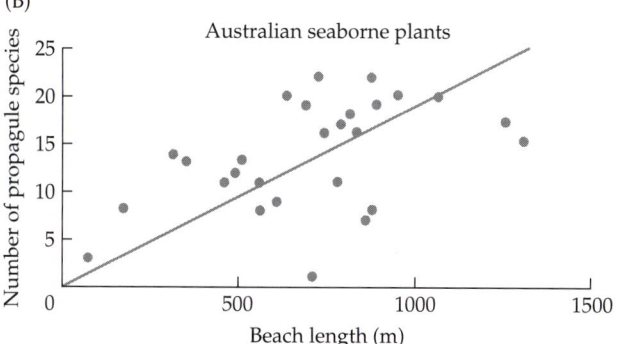

Figure 13.14   Two examples of the **target area effect**—the tendency for larger islands to attract or intercept more immigrants. (A) Nonvolant mammals dispersing across the snow-covered ice to islands of the Saint Lawrence River, which forms the border between the northeastern United States and Canada. Immigration rates increase significantly with island area for the canids (primarily coyotes and red foxes) as well as other mammals (including raccoons, weasels, red squirrels, voles, deer mice, and shrews). (After Lomolino 1990.) (B) The diversity of seaborne plant propagules found along the beaches of islands in the vicinity of northeastern Australia increases with the length of the beachfront. (After Buckley and Knedlhans 1986.)

The small island effect.   In their 1967 monograph, MacArthur and Wilson remarked that the species-area relationships of very small islands may be "truly anomalous." For example, Niering's (1963) study of higher plants on islands of the Kapingamarangi Atoll, Micronesia, indicated that species richness appears to be independent of area for the smaller islands (those of less than 3 acres; Figure 13.15). Although a few other studies also report a small island effect (see Dunn and Loehle 1988; Wiens 1962; Whitehead and Jones 1969; Woodruffe 1985), it is likely to remain a rarely reported phenomenon, as most biogeographic studies are conducted on islands large enough to maintain an appreciable diversity of species. This feature of sampling protocols does not, of course, render the small island effect insignificant. In fact, it is possible that when biogeographic surveys include many very small islands, a sigmoidal model may prove superior to the power and semi-logarithmic models, which describe monotonic relationships (Figure 13.16).

Similarly, for reasons discussed in Chapter 9, the species-isolation relationship may also be sigmoidal, this time a negative sigmoidal (see Figure 13.4B). Taken together, these patterns suggest that species richness tends to decline most rapidly across intermediate levels of area or isolation, presumably those levels that approximate the modal resource requirements or dispersal distances of the species pool. In contrast, both species-area and species-isolation relationships may be difficult to detect for extremes of island size and isolation, respectively (see Figures 13.4 and 13.16; see also Lomolino 1998). Future studies on the potential sigmoidal nature of these relationships may prove extremely valuable because, if corroborated, they may change the way we study two of the most common patterns in nature.

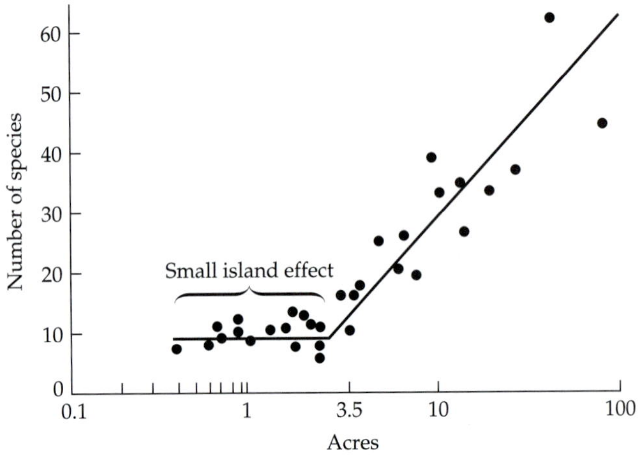

Figure 13.15   The **small island effect** refers to the tendency of the species richness of some insular faunas to remain relatively low and independent of area for the smallest islands. This pattern was found among the higher plants of the Kapingamarangi Atoll, Micronesia. (After MacArthur and Wilson 1967; data from Niering 1963.)

Nonequilibrium Biotas

Comparing observed patterns of insular species diversity with the predictions of the MacArthur-Wilson model reveals many cases in which the diversity clearly does not represent an equilibrium between contemporary rates of immigration and extinction. The number of species on these islands is not remaining approximately constant; instead, it is either increasing or decreasing steadily in response to major historical events. Ironically, as a result of the failure of some of its predictions, MacArthur and Wilson's approach to island biogeography has proved valuable for detecting and interpreting the dynamics of nonequilibrium biotas. In fact, many of the observed patterns can be explained in terms of the biota approaching a new equilibrium number of species following a historical perturbation.

Pleistocene refugia.   As pointed out in Chapter 7, Pleistocene changes in climate and sea level caused major shifts in the distributions of organisms on the continents, especially terrestrial and freshwater species. Their effects on certain insular habitats also were quite profound. In some cases, once isolated habitat islands were connected by habitat bridges, permitting a free interchange of biotas; in others, new habitat islands were created by the intrusion of formidable barriers to dispersal. The current distribution of climate, sea level, and major biome types has been approximately stable only for the last 8,000 to 12,000 years.

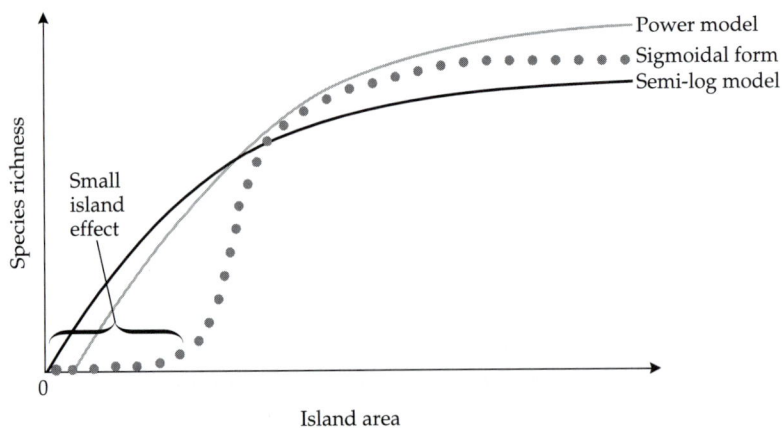

Figure 13.16   The two models most commonly used to describe the species-area relationship [i.e., the power model, where $S = cA^z$, and the semi-logarithmic model, where $S = d + k \log(A)$] typically provide a good fit. Yet the species richness of some—perhaps many—insular biotas may be best described by a sigmoidal relationship that includes the small island effect (see Figure 13.15).

The legacy of these Pleistocene perturbations is still apparent in many insular distributions. Many islands and patches of isolated habitat were formed by rising sea levels and climatic changes at the end of the Pleistocene. Sometimes the fragmentation of extensive areas of once continuous habitat left small islands oversaturated with species, and diversity began to decrease as certain species were eliminated by extinction. In other cases, completely new insular habitats were created and began to acquire species by colonization and speciation. The extent to which contemporary biotas still show the effects of these historical changes depends largely on the kinds of barriers isolating these post-Pleistocene islands and on the ability of different kinds of organisms to disperse across them.

In one of the first attempts to test the applicability of the MacArthur-Wilson model to the distributions of organisms among habitat islands, Brown (1971b, 1978) studied the small nonflying mammals, and later the birds, inhabiting isolated mountaintops in the western United States (see also Johnson 1975; Behle 1978; Grayson 1987a,b, 1993; Grayson and Livingston 1993). During the Pleistocene, the climate of this region was relatively cool and wet, and forests formed an expansive "mainland" of habitats for boreal species. Following the last glacial recession, the climate warmed and dried, and as a result, the Great Basin, which lies at about 1500 m elevation between the Rocky Mountains to the east and the Sierra Nevada to the west (Figure 13.17), developed into a vast region of desert. The now isolated mountain ranges, some rising to more than 3000 m, form islands of isolated coniferous forest and other mesic habitats surrounded by a sea of sagebrush desert.

These isolated mountaintops are inhabited by a number of boreal mammal and bird species that are restricted to the cool, moist habitats of higher elevations. The patterns of diversity of these mammals and birds exhibit both similarities and differences when compared with each other and with the biotas of

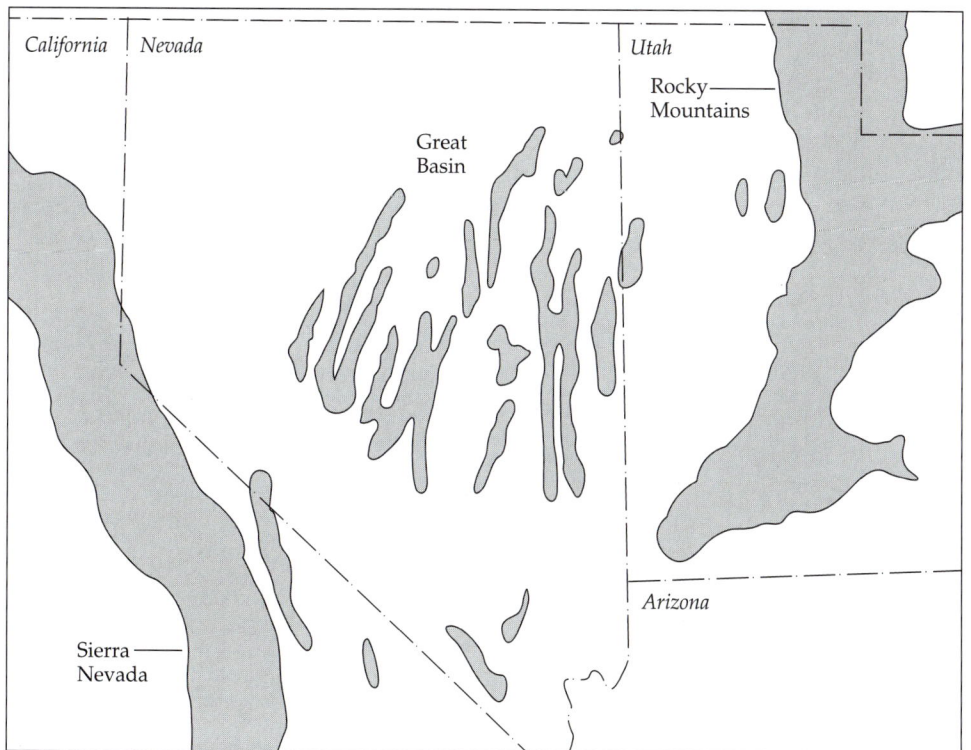

Figure 13.17 Isolated mountain ranges of the Great Basin in western North America are islands of cool, mesic forest habitat in a sea of sagebrush desert. The ranges shown, with peaks mostly higher than 3000 m, lie between two montane "mainlands": part of the Rocky Mountains to the east, and the Sierra Nevada to the west. The desert valleys between the mountains are readily crossed by birds, but they are virtually absolute barriers to small mammal dispersal. (After Barbour and Brown 1994.)

oceanic islands. The number of species inhabiting a mountaintop increases with the size (area) of the mountain range (Figure 13.18). However, there is no detectable effect of isolation by distance on the diversity of either taxon. The Rocky Mountains and the Sierra Nevada are the likely sources of colonists; they support a much higher diversity of boreal forms than the isolated mountaintops do, including all of the species that are found on the isolated mountaintops.

Brown (1971) proposed that the present boreal mammal populations of these montane islands are relicts, vicariant remnants of once widespread distributions during the Pleistocene. This model is consistent with plant fossils, which show that as recently as 10,000 to 12,000 years ago, the climate of the Great Basin was cooler and wetter, and the vegetation zones were shifted several hundred meters below their present elevations (Wells and Berger 1967; Wells 1976, 1979; Thompson and Mead 1982). This shift would have connected several presently isolated habitats across the entire Great Basin, permitting all islands to be colonized by all species for which appropriate habitat bridges existed. At the end of the Pleistocene, however, the cool, mesic habitats shrank back to higher elevations, completely isolating on the mountaintops those boreal mammal species unable to disperse across the desert valleys (see Grayson 1993). After being isolated, some insular populations became extinct, reducing the diversity of the boreal biota, especially on small mountaintops, but relictual populations of other species have survived until the present.

Brown's model proposes that in the absence of post-Pleistocene immigration, the mammalian faunas of the isolated mountaintops have been relaxing toward an equilibrium of zero species. Small islands have had high extinction rates and have lost most of their fauna in 10,000 years, whereas large islands still retain most of their original species. One additional observation supports this relaxation model. Late Pleistocene or more recent fossils of boreal species, including some forms still found on some of the larger isolated mountain ranges, have been found in the intermontane valleys and on several mountaintops, indicating that these species were indeed once present and have become extinct within the last 12,000 years (Grayson 1981, 1987a,b, 1993; Thompson and Mead 1982).

Paradoxically, birds also fail to exhibit significant species-isolation relationships because they are excellent dispersers. In contrast to the mammals, boreal bird species appear to be present on all mountain ranges where there are sufficient areas of suitable habitat (Johnson 1975; Behle 1978). Boreal birds have been observed flying across the desert valleys, which appear to pose no significant barriers to their colonization of even the most isolated mountaintops. A

Figure 13.18 Species-area relationships for the boreal resident birds (A) and small terrestrial mammals (B) inhabiting the isolated mountain ranges of the Great Basin. Birds continually recolonize these mountains, whereas the mammals are relicts of more widespread populations that existed during the Pleistocene. (After Brown 1978.)

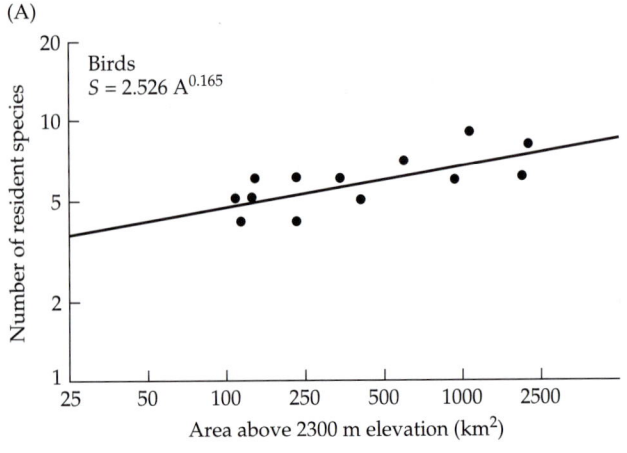

(A)

Birds
$S = 2.526 \, A^{0.165}$

Number of resident species

Area above 2300 m elevation (km²)

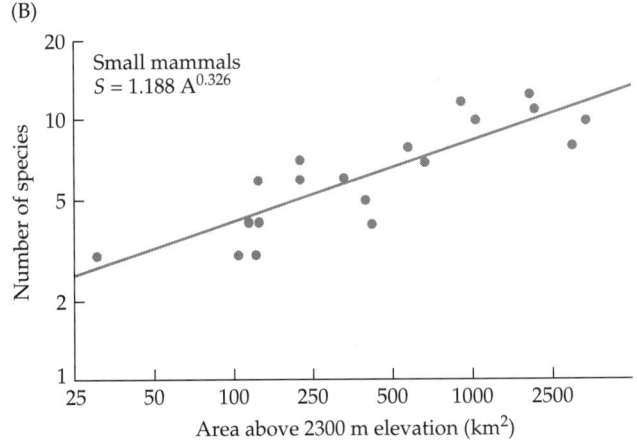

(B)

Small mammals
$S = 1.188 \, A^{0.326}$

Number of species

Area above 2300 m elevation (km²)

high immigration rate apparently continually replenishes bird populations on small islands. Thus the distribution of boreal birds may well represent an equilibrium between immigration and extinction. However, even this is not quite the sort of equilibrium predicted by MacArthur and Wilson, because the immigration and extinction processes appear to be highly deterministic instead of stochastic. The rate of colonization is so high that habitats are almost completely saturated with those species that can live there. When on rare occasions an insular population does become extinct, it probably is replaced rapidly.

Other studies also indicate that relaxation rates of supersaturated communities are inversely proportional to island area and tend to be most rapid during the period immediately following isolation (e.g., see Terborgh et al. 1997; Diamond 1972; Wilcox 1978 and 1980). It may, however, be invalid to assume that all habitat islands are inhabited by fauna in perpetual states of relaxation. Even Brown's interpretation of the Great Basin mammals may need slight modification, because some species may occasionally disperse across xeric habitats (see Grayson and Livingston 1993). Still, it is likely that most species of forest mammals in the Great Basin are completely isolated by the intervening deserts. In contrast, intermountain habitats in the American Southwest, which lie south of the Great Basin, may well represent filters, not perfect barriers, for the movements of many forest mammals (see Figure 9.10). The key difference between these two regions is that most mountain forests of the American Southwest are isolated by woodlands, not deserts. As a result, post-Pleistocene immigrations are likely for most species, and mammalian communities may have approached a dynamic equilibrium as envisioned by MacArthur and Wilson. As illustrated in Figure 13.19, the species richness of forest mammals in this region is significantly correlated with both isolation and area. In addition, intermountain dispersal and post-Pleistocene colonization has been documented for a number of these species (Figure 13.20; Lomolino et al. 1989; Brown and Davis 1995; Davis and Brown 1989; Davis and Callahan 1992; Davis and Dunford 1987).

Again, however, we caution against overgeneralizing from these studies. Patterson (1980, 1984, 1995) has provided convincing evidence that at least some of these communities, especially those on the more isolated mountains in southern New Mexico, may truly be Pleistocene relics. Just as Brown observed in the Great Basin, distant islands of the American Southwest are surrounded by deserts as well as woodlands, making post-Pleistocene immigration very unlikely, if not impossible, for many mammal species. Thus,

Figure 13.19 Species-isolation (A) and species-area (B) relationships for nonvolant forest mammals on 27 montane "islands" in the American Southwest (primarily Arizona, New Mexico, southern Colorado, and southern Utah). Because these mesic forests are isolated by "seas" of xeric woodlands and deserts, their biotic communities often exhibit biogeographic patterns similar to those of species on true islands. (After Lomolino et al. 1989.)

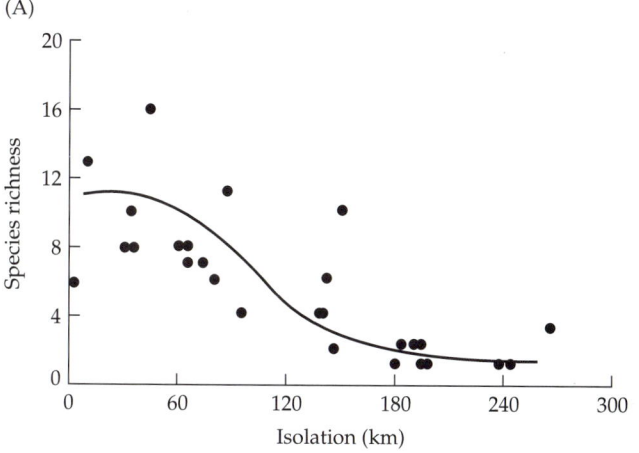

(A)

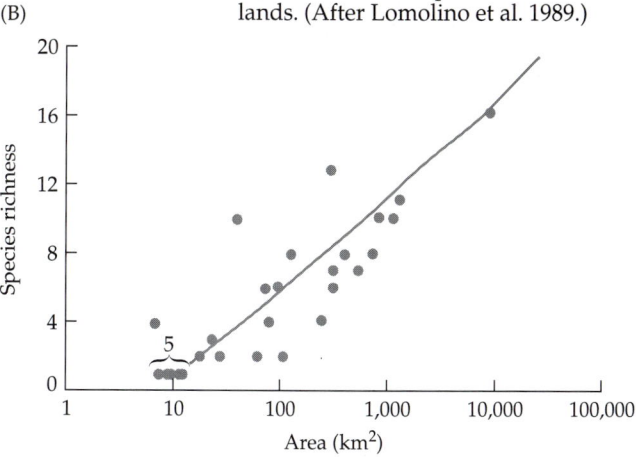

(B)

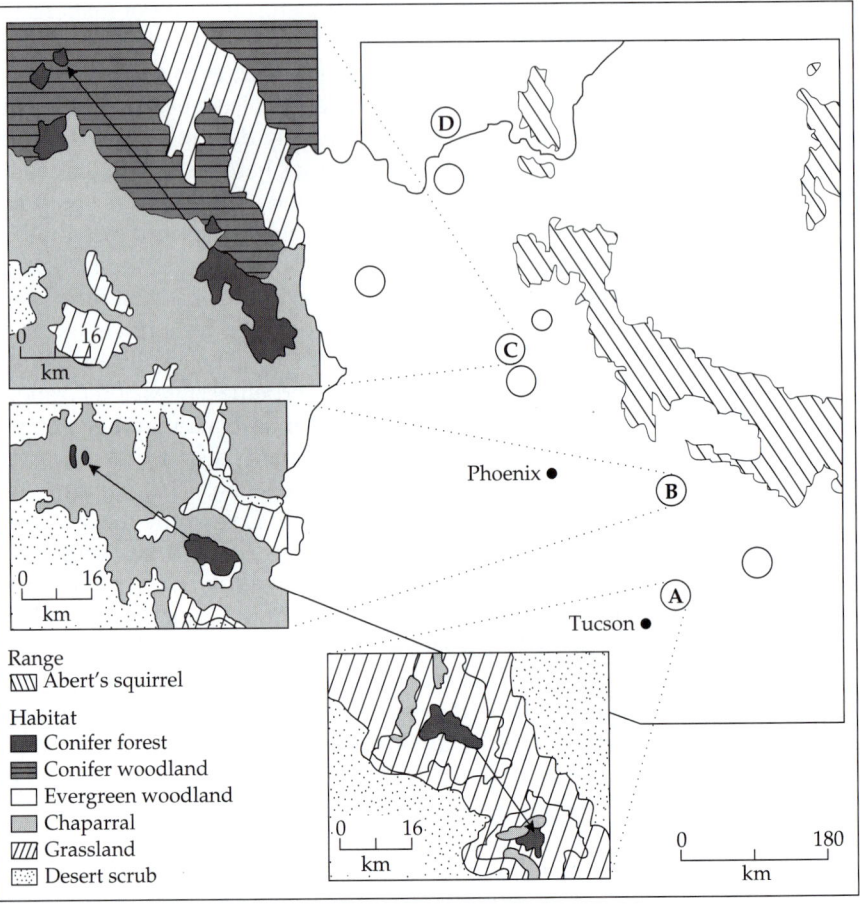

Figure 13.20 Relatively recent intermountain dispersal by Abert's squirrel (*Sciurus aberti*) in Arizona. Open circles indicate sites where the squirrels were artificially introduced between 1940 and 1945. Lettered circles indicate introductions that were followed by colonization of nearby montane forests far removed from the natural range of this species. The time between introduction and apparent colonization was approximately 20, 30, 40, and 10 years for sites A, B, C, and D respectively. The dispersal distances, 23, 29, 35, and 10 km respectively, are roughly proportional to these times. (After Lomolino et al. 1989; data from D. E. Brown and R. Davis, personal communication.)

mammalian communities on the more isolated mountains of this region may indeed be relaxing from supersaturated levels of richness in the absence of contemporary immigrations.

The take-home lesson is relatively simple: heterogeneous landscapes and species pools defy any one simple explanation for their salient patterns. Also, as we noted above, comparisons across archipelagoes also pose special challenges. Great Basin birds appear to be uninfluenced by isolation: that is, most islands may fall well within their immigration abilities. In other, more isolated regions, avian species richness declines significantly with isolation [e.g., montane birds of southern California and Baja California (Kratter 1992) and the Andes of Venezuela, Colombia, Ecuador (Vuilleumier 1970), and Argentina (Nores 1995); Figure 13.21].

Landbridge islands.   Rising sea levels since the Pleistocene have created many new islands in coastal regions. These are called **landbridge** or **continental islands** because, unlike oceanic islands, they were once part of the mainland, or at least connected to it by a bridge of terrestrial habitats. The approximate 100 to 200 m rise in sea level that occurred about 10,000 years ago inundated many landbridges and created numerous continental islands.

Diamond (1972, 1975b) found that the influence of past connections to the mainland of New Guinea is apparent in the composition of the avifauna of the landbridge islands off its coast. Islands that are separated from the mainland by water less than 200 m deep support a greater number of species than oceanic islands of comparable size and distance from New Guinea (i.e., those

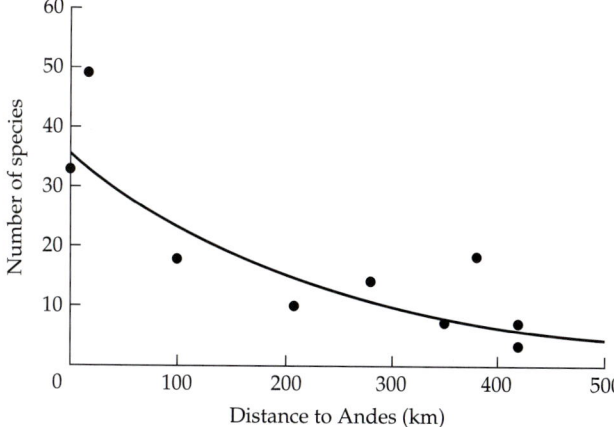

Figure 13.21 Much like that of their counterparts on oceanic islands, the species richness of birds in montane forest islands near the Andes in Venezuela, Colombia, and Ecuador decreases with isolation. (From Nores 1995.)

that arose as undersea volcanoes and have never been connected to the mainland). Although the landbridge islands lack many bird species found in comparable habitats on New Guinea, they have several kinds of birds that are found on the mainland but never occur on the oceanic islands. Of course, there are still other species that occur on both landbridge and oceanic islands. Because these birds obviously have crossed water barriers to colonize the oceanic islands, investigators cannot be sure whether their populations on the continental islands are Pleistocene relicts or whether they have been replenished by subsequent immigration (see also Lawlor 1983, 1986).

Post-Pleistocene dynamics of freshwater faunas. The present distribution of freshwater fishes in southwestern North America reflects a history of aquatic habitat connections during the cooler and wetter climate of the Pleistocene, followed by isolation and subsequent extinctions (Hubbs and Miller 1948; Miller 1948; Smith 1978). Hard as it may be to believe, only a few thousand years ago, Death Valley, in the most arid part of the North American desert, was almost completely filled by a large lake, which was supplied by a major system of permanent rivers and springs (see Figure 7.24). At least three genera of fishes inhabited this basin, and they have persisted as relictual populations in isolated springs. In the case of these fishes, there can be no doubt about the susceptibility of small, isolated populations to extinction. Many have disappeared within the last few years as humans have diverted water or introduced exotic species of competing or predatory fishes.

In most of the examples discussed so far, the biotas of insular habitats isolated since the Pleistocene are relictual. Once widespread habitats containing diverse biotas were diminished in size and became fragmented. As soon as these islands formed, they were oversaturated with species, and they have been losing species by extinction ever since. In contrast, some isolated habitats, as well as areas scoured by glaciers, remain undersaturated and continue to acquire species (see Chapter 7).

An excellent example of the latter situation is provided by the glacier-formed lakes of northern North America and Eurasia. Many lakes, including the Great Lakes and the Finger Lakes in northeastern North America, were gouged out by the advancing continental ice sheets and filled with water as the glaciers retreated (see Figure 7.23). Some groups, such as certain algae, protozoans, and invertebrates that use cysts or other effective means of long-distance dispersal, may have achieved an equilibrium between rates of colonization and extinction in these habitats over a relatively short period of time. In contrast, other less vagile animals, such as fishes (Smith 1981) and some mollusks, may never achieve an equilibrial level of species richness. The fish faunas of many of these lakes appear to be undersaturated and still gradually increasing in diversity (Barbour and Brown 1974; see also Holland and Jain 1981; Hugueny 1989; Smith 1979; Brown and Dinsmore 1988; Watters 1992; Elmberg et al. 1994; Tonn and Magnunson 1982; Browne 1981).

Only a handful of fish species can survive for more than a few minutes out of water at any stage of their life cycle, so fishes can colonize new areas only when connections of aquatic habitat are present. They usually colonize lakes from rivers and streams, but most lotic waters contain only a few species that can also be successful in lentic environments. Many glacier-formed alpine lakes lack native fishes entirely because they have never been connected by suitable habitat bridges to waters inhabited by fishes. The fact that several introduced species thrive in some of these lakes is further evidence that the absence of fishes is a consequence of barriers to dispersal. Many of those glacier-formed lakes that do contain fishes still appear to be undersaturated with species. This is particularly true of the very large ones, such as the Great Lakes in North America and Lake Baikal in the Soviet Union. There were not enough species in the rivers and streams draining these lakes to fill the niches available to fishes. Some taxa, such as the ciscoes or whitefishes (*Coregonus*) in the Great Lakes, have been diversifying by endemic speciation and adaptive radiation, but there has not been sufficient time since the Pleistocene to achieve an equilibrium between speciation and extinction.

This conclusion is supported by a comparison of species-area relationships for fishes inhabiting the glacier-formed lakes in temperate North America and the much older lakes of tropical Africa (Figure 13.22). As mentioned in Chapter 8, the relatively high diversity of Africa's Rift Valley lakes derives primarily from the spectacular endemic speciation and adaptive radiation of cichlids in the larger lakes (Fryer and Iles 1972; Greenwood 1974; Meyer 1993). Lake Victoria, the largest lake in Africa, now harbors over 300 endemic species of cichlids, yet it appears to have dried up completely during the late Pleistocene (Johnson et al. 1996). Following recolonization, speciation of these fishes must have proceeded at an extremely rapid rate indeed, possibly reaching a dynamic equilibrium before the relatively recent surge in extinctions due to introduced species and other anthropogenic activities.

Finally, the relative time scale of underlying glacio-pluvial events holds some key insights for interpreting and predicting the biogeographic dynamics of insular systems. Recall from Chapter 7 that interglacial periods are relatively short compared with glacial periods (often just 10,000–30,000 years versus 75,000–300,000 years, respectively). Thus, on a geological time scale, relaxation of relictual communities or accumulation of species in environments cleared by glaciers is likely to be a transient and incomplete phenomenon. As Gates (1993) observed, "the present interglacial should prevail for another 2,000 years, perhaps less." Paleoecologists or biogeographers of future millen-

Figure 13.22 Species-area relationships for the fishes inhabiting lakes in northern North America recently formed by glaciers (A) and much older lakes of central and eastern Africa (B). The large African lakes have acquired diverse fish faunas by endemic speciation. Although such speciation is occurring in the North American lakes, it has not yet produced high species richness in the largest lakes. (After Barbour and Brown 1974.)

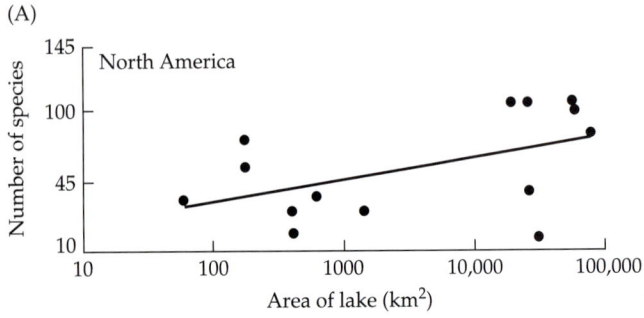

(A)

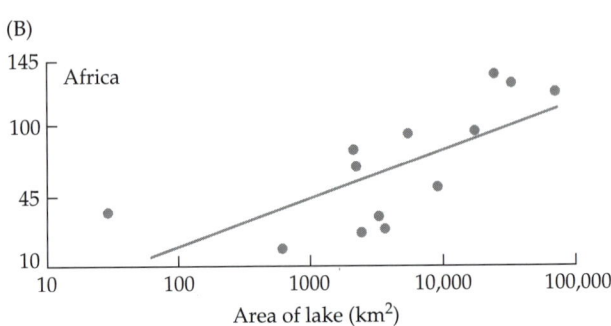

(B)

nia may well reverse our original question and ask "to what extent distribution patterns during glacial periods represent the legacy of the relatively short pulses of interglacial events."

Krakatau Revisited

We conclude this chapter with an update on the biogeographic dynamics of the Krakatau Islands (see Figure 13.6). Recall that, while the species-area and species-isolation relationships fueled the development of the equilibrium theory, biotic surveys of these islands also provided a vital spark. The Krakatau communities were found to be dynamic, experiencing frequent immigrations, extinctions, and turnover. MacArthur and Wilson's discussions were limited to surveys conducted between 1908 and 1934 (i.e., 25 to 41 years after the 1883 explosion cleared the islands of all life). Additional surveys conducted through 1990 highlight some of the strengths and weaknesses of the equilibrium theory (Thornton et al. 1990; Bush and Whittaker 1991; Thornton et al. 1993; Whittaker and Jones 1994).

First, the now century-plus record confirms the dynamic nature of insular communities. Turnover remains relatively high for most groups of plants and animals. Moreover, the general trends in immigration and extinction rates appear consistent with the basic theory: immigration rate decreases and extinction rate increases with species richness (Figure 13.23). For a number of

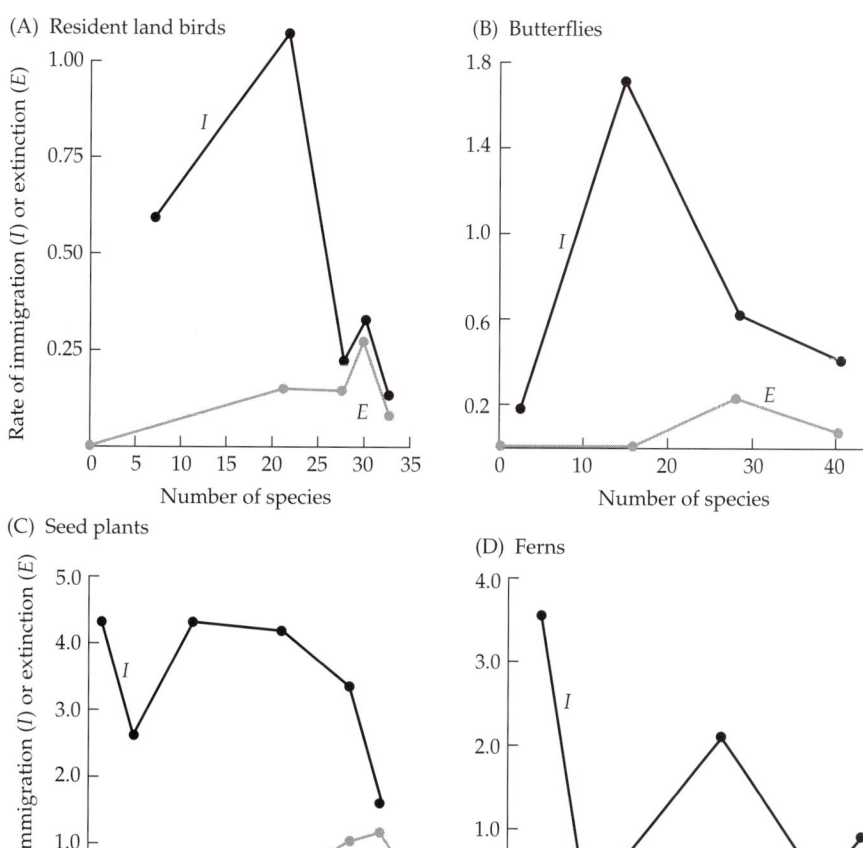

Figure 13.23 Immigration (*I*) and extinction (*E*) rates (as number of species/year) of animals and plants on Rakata, the largest of the Krakatau Islands, as a function of insular species richness (*S*). Consistent with Mac-Arthur and Wilson's equilibrium theory, immigration rate declined and extinction rate increased as the islands accumulated species. Note that at least for resident land birds, seed plants, and ferns, insular communities may have approached a dynamic equilibrium ($I \approx E$). (From Thornton et al. 1993.)

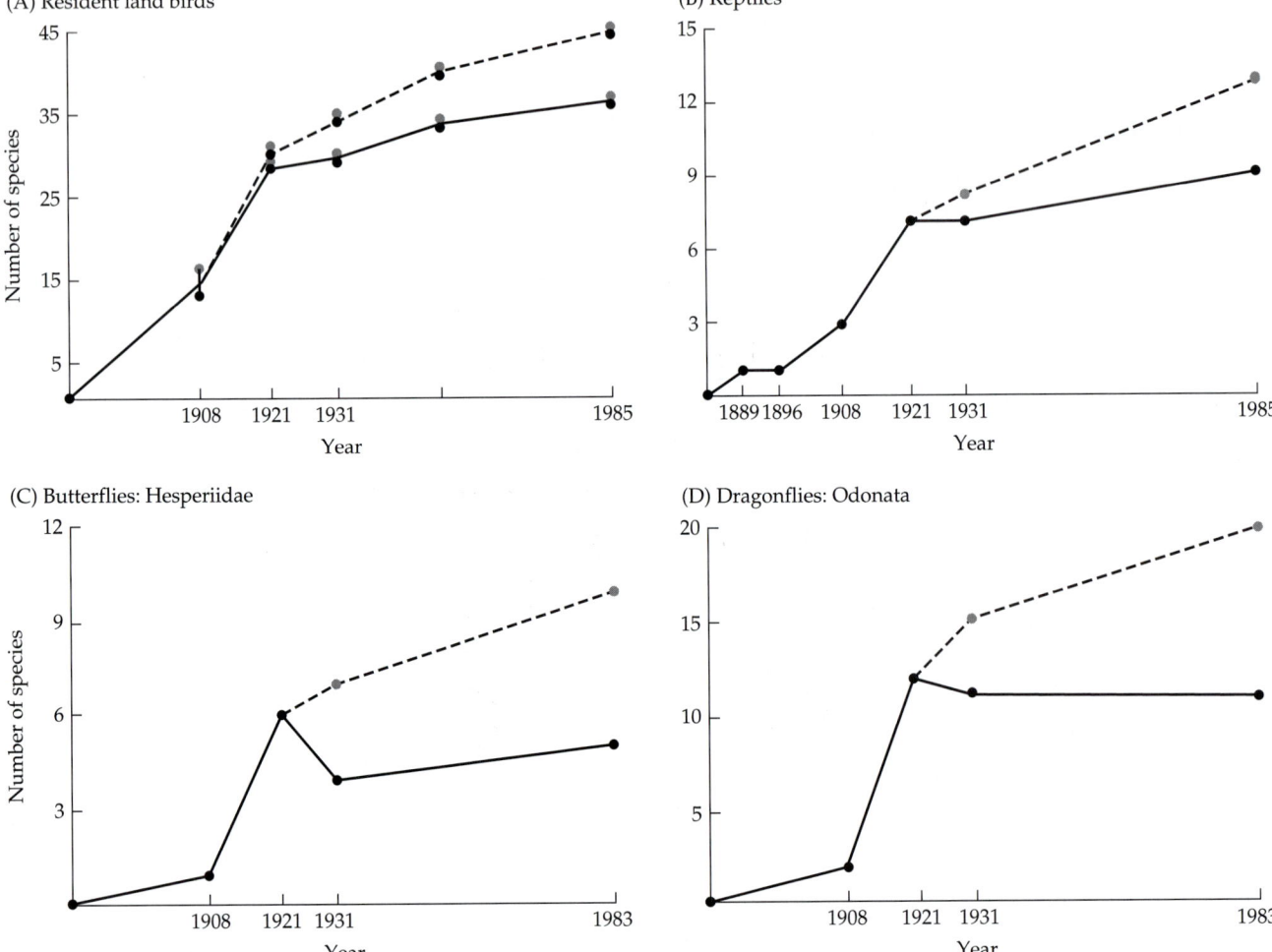

Figure 13.24 Colonization curves for various animal taxa in the Krakatau Islands: (A) resident land birds, (B) reptiles, (C) butterflies (Hesperiidae), and (D) dragonflies (Odonata). Solid lines indicate actual species numbers at the time the islands were surveyed; dashed lines indicate cumulative numbers of species detected for all surveys, combined. (From Thornton et al. 1990.)

taxonomic groups, including seed plants, dragonflies, butterflies, reptiles, and resident land birds, the rate of species accumulation has slowed substantially, and species number appears to be approaching an equilibrium (Figure 13.24).

The rate of approach toward an equilibrium varies considerably among taxa, and seems directly correlated with dispersal abilities. For example, while avian species richness on the Krakataus is only slightly below that expected based on island size, species richness of nonflying mammals is approximately one-tenth of the expected values (Figure 13.25). On Rakata, the largest island, sea-dispersed plants have accumulated much less rapidly than wind- or animal-dispersed plants, particularly during the last few decades of colonization (Figure 13.26).

While these trends confirm the basic tenets of the equilibrium theory, others reveal some important limitations of the theory. As MacArthur and Wilson (1967) anticipated, immigration and extinction curves may not be monotonic if major successional changes occur. Some species are strongly dependent on other pioneer species to create niches for them. This seems to be the case for butterflies, land birds, and ferns, whose accumulation rates increased substantially during forest development (1908 to 1931; see Figure 13.25). Interpretations of these data are, however, plagued by our inability to study immigration directly and distinguish between rates of immigration and accumulation

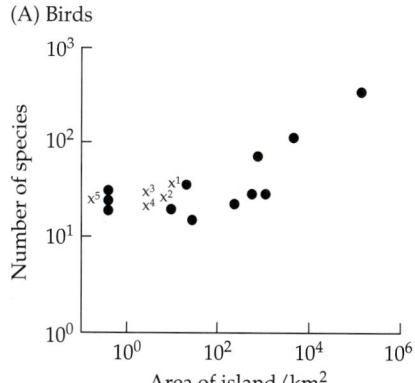

(A) Birds

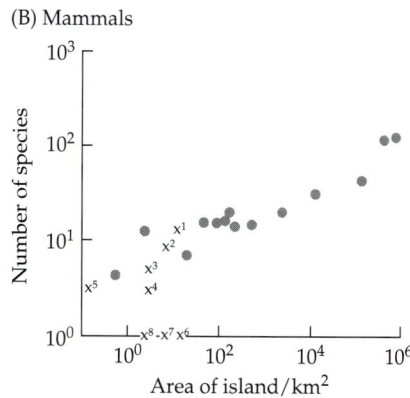

(B) Mammals

Figure 13.25 Species-area relationship for birds (A) and mammals (B) on the Krakatau Islands (Xs with superscripts) and undisturbed oceanic islands of the Sunda Shelf (solid circles). Superscripts: birds and volant mammals (bats): 1 = Rakata, 2 = Sertung, 3 = Panjang, 4 = Anak Krakatau (based on total land area), 5 = Anak Krakatau (based on vegetated area only); nonvolant mammals (rats): 6 = Rakata, 7 = Sertung, 8 = Panjang. Note that species richness of birds and bats on the Krakatau islands may have rebounded to pre-eruption levels (see also Figure 13.24), whereas richness of nonvolant mammals remains relatively low in comparison to undisturbed islands of the same area. Anak Krakatau, which emerged from the sea in 1930, still lacks terrestrial mammals. (From Thornton et al. 1990.)

of species. Recall that MacArthur and Wilson (1967) criticized their own model for "making an artificially clear-cut distinction between immigration and extinction." Immigration is defined as "the process of arrival of a species on an island not occupied by that species" (MacArthur and Wilson 1967). This says nothing about the likelihood that an immigrant will establish a breeding population, and indeed, it is likely that most immigrants fail to do so. Biotic surveys such as those conducted on the Krakataus, while insightful, measure accumulation (or establishment) rates, not immigration rates. Although immigration rates should be unaffected by ecological succession, the chance that an immigrant will become established may be strongly influenced by successional changes. In short, the apparent nonmonotonic immigration curves reported for the Krakataus' biota may be an artifact of the necessity to lump

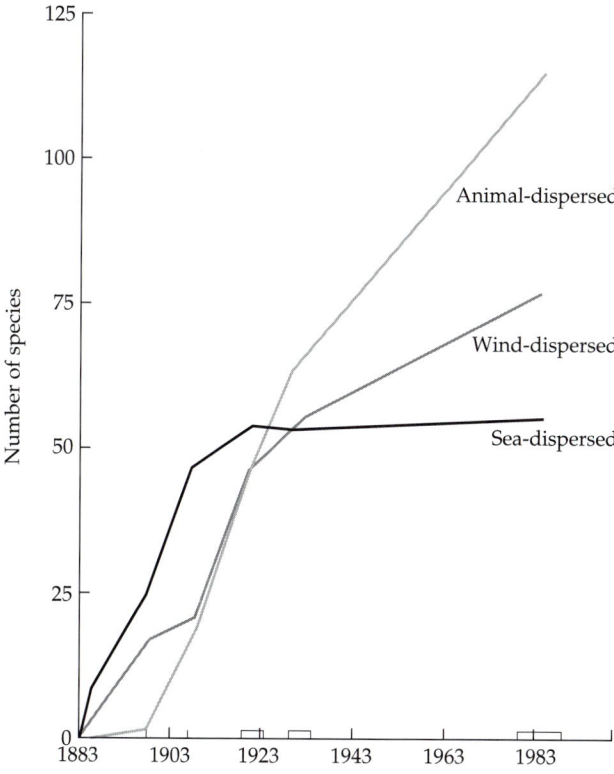

Figure 13.26 Species accumulation curves for plants on Rakata differ markedly depending on the mode of dispersal. Sea- and wind- dispersed plants were first to arrive, whereas animal-dispersed plants did not accumulate until successional changes provided suitable habitat for their animal transporters. Following 1923, however, animal-dispersed plants accumulated much more rapidly than wind- and sea-dispersed forms. (From Bush and Whittaker 1991.)

Figure 13.27 The effects of succession can be included in an equilibrium model by noting how it alters the likelihood that immigrants will survive and establish a breeding population. S = species richness, $\hat{S}$ = equilibrial species richness, I = immigration rate (number of propagules of new species arriving per unit of time), F = failure rate (number of propagules failing to establish a breeding population per unit of time), C = colonization rate ($= I - F$). (After MacArthur and Wilson 1967.)

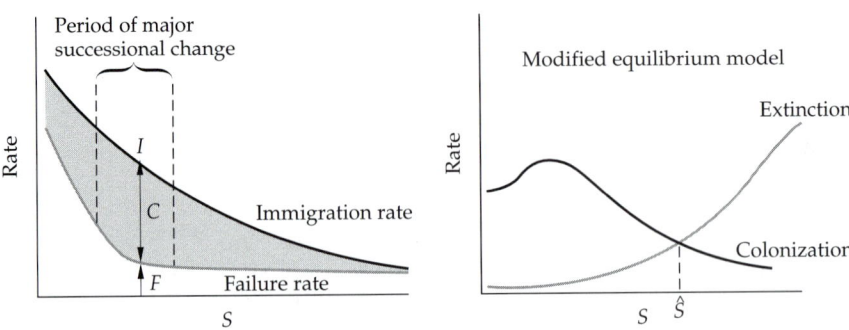

establishment (or "ecesis," *sensu* MacMahon 1987) with immigration (Figure 13.27). Interestingly, in their attempt to include successional effects in their model, MacArthur and Wilson (1967:51) lumped failure to become established with extinction, not immigration. This problem is not simply one of semantics. If the equilibrium theory is to include important ecological processes such as succession, it must explicitly address the effects of these processes on establishment as well as immigration and extinction.

Finally, Bush and Whittaker (1991) have voiced some other, more fundamental concerns, namely, that the equilibrium model does not include the effects of processes occurring at other time scales. If the approach to equilibrium is as slow as Bush and Whittaker predict, on the order of millennia, then disturbances such as hurricanes and volcanic eruptions may prevent an equilibrium for some, perhaps many, taxa (Figure 13.28). Similarly, the equilibrium model does not include speciation, which, at least on very large and isolated islands, may rival or possibly exceed immigration rates (Diamond and Gilpin 1983; Heaney 1986, 1991; Adler 1994). With just a minor modification, however, the model can include the effects of speciation as well as immigration (Figure 13.29).

What, then, is the future of the equilibrium theory, which has served as a paradigm of island biogeography for over three decades? As a heuristic model, it has served the field well indeed. Yet, as discussed above, it has many shortcomings. Given those shortcomings, the field may be primed for a major

Figure 13.28 For some insular biotas, major disturbances such as volcanic eruptions and hurricanes may occur so frequently that the insular communities seldom achieve a dynamic equilibrium as envisioned by MacArthur and Wilson. The shading in A indicates when immigration rate exceeds extinction rate (i.e., when species richness should increase). (After Bush and Whittaker 1991.)

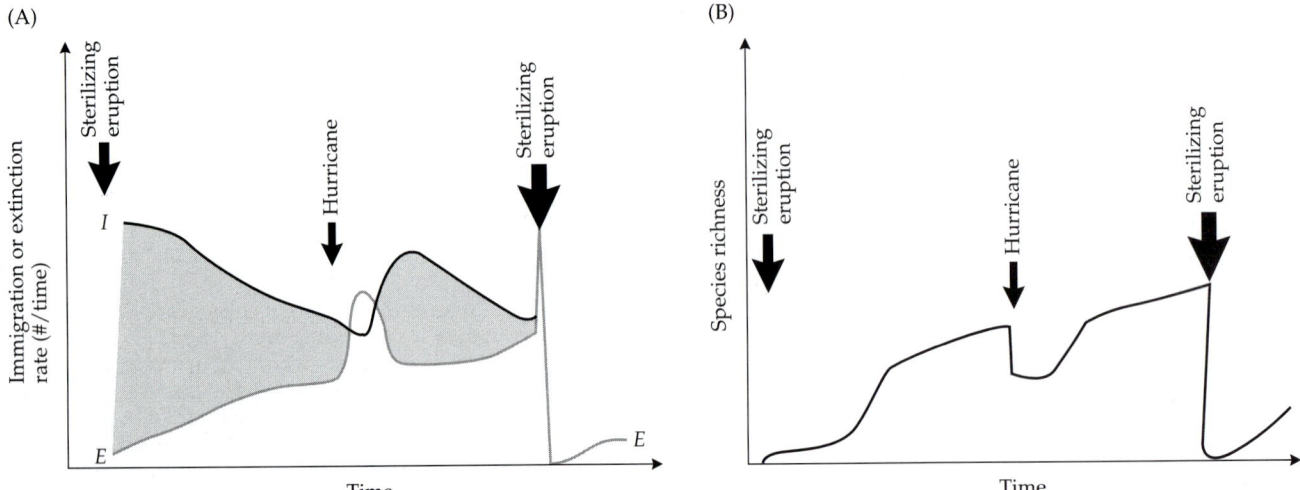

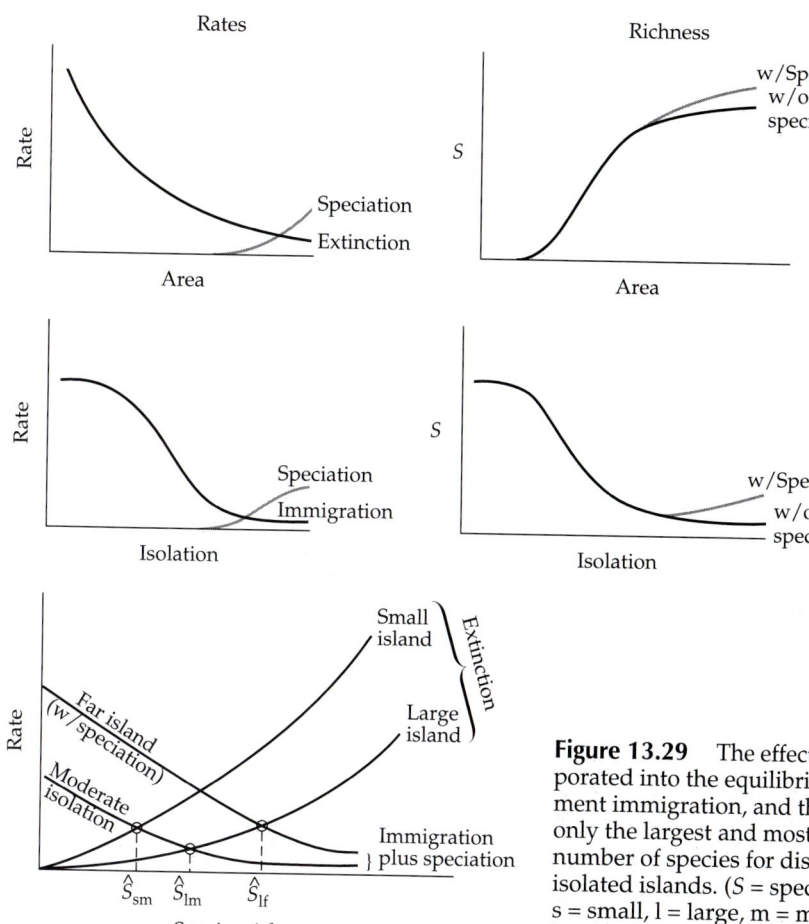

Figure 13.29　The effects of evolution on species richness can be incorporated into the equilibrium theory by noting that speciation will supplement immigration, and that speciation rates should be significant for only the largest and most isolated islands. As a result, the equilibrium number of species for distant islands may actually exceed that for less isolated islands. (S = species richness, $\hat{S}$ = equilibrium species richness, s = small, l = large, m = moderately isolated, and f = far or distant islands.)

scientific revolution in which the equilibrium theory is replaced by a radically different model. If that is the case, we hope the new paradigm serves us as well as its predecessor has. Alternatively, the equilibrium theory may slowly evolve, gradually becoming more complex and realistic by including additional processes, such as succession, disturbance, and speciation, that operate over a broad range of temporal and spatial scales. MacArthur and Wilson's model has already spawned related models of distribution patterns and persistence of isolated species (e.g., metapopulation models, incidence functions, and compensatory patterns: see Chapter 14). It is our hope that, in some way, this chapter will stimulate our readers to accept this great challenge and contribute to a major paradigm shift in biogeography.

CHAPTER *14*

Island Biogeography: Patterns in the Assembly and Evolution of Insular Communities

One of the most general patterns in biogeography is that insular communities tend to be species-poor in comparison to mainland communities. Islands are not only isolated, but also small, and thus contain a limited diversity of habitats and resources. As we learned in the previous chapter, the impoverishment of islands can be attributed to their relatively low rates of immigration and their limited ability to support populations. MacArthur and Wilson's (1967) equilibrium theory accounted for differences in species richness among islands that differ in isolation and size. Their model, however, included the simplifying assumption that species are equivalent with respect to traits influencing their dispersal ability and their ability to survive on islands.

Biogeographers often are concerned with which species occur in a place, rather than just how many. The equilibrium model, while successfully predicting very general tendencies in species richness, cannot account for patterns in species composition. Furthermore, the structure of insular communities, especially on large and isolated islands, is strongly influenced by evolution in situ. Insular "novelties," including giant flightless birds, dwarf elephants, and assemblages dominated by an astounding diversity of taxa that are insignificant components of the mainland biota have attracted attention throughout the history of biogeography. Moreover, these novelties, and related patterns in the assembly of insular biotas, have provided invaluable insights into the forces influencing community structure and evolution in all biotas.

General explanations for these intriguing patterns must recognize that differences among the species matter, because their particular physiological, morphological, behavioral, and genetic characteristics ultimately combine to influence immigration, extinction, and evolution. It should not be necessary to point out that MacArthur and Wilson knew as well as any of their colleagues that species differ in ways directly related to their potential to colonize, survive, and evolve on islands. Both of these scientists were superb naturalists and ecologists, and both produced seminal papers exploring the ecological and evolutionary consequences of differences among species (e.g., see MacArthur 1958; MacArthur and Levins 1967; MacArthur et. al. 1966; Wilson 1959, 1961).

In this chapter we focus on this same theme. First, we discuss the selective nature of immigration and extinction, and then explore how these processes can create nonrandom patterns in the composition of insular communities. After reviewing these and other ecological responses, we examine some truly fascinating evolutionary responses to insular environments.

Assembly of Insular Communities

Given that both species and islands differ with respect to factors influencing immigration and extinction, insular communities should represent nonrandom samples of their mainland source pools, samples biased in favor of the better immigrators and better survivors. On the other hand, under the null hypothesis that all species and islands are equivalent, we would expect insular communities to represent random draws from the mainland pool.

Biogeographers often use the terms **harmonic** and **disharmonic** to characterize insular biotas. Harmonic, or **balanced**, biotas are assemblages that are similar in composition to the source biota. They may have fewer species, but the proportions of species in each taxon or ecological category are roughly the same as on the mainland (Figure 14.1). Again, this is the null prediction. Where species and islands are not equivalent, we would predict some nonrandom patterns in species composition, not just differences between insular and mainland communities, but among insular communities as well. Certain types of organisms should be overrepresented on islands, and this bias should increase as islands become more isolated or smaller. As we shall see, while biogeographers have as yet failed to develop a comprehensive model of patterns in species composition, they have provided seemingly incontrovertible evidence for the selective nature of immigration and extinction—that is, for the nonrandom assembly of insular communities.

The Selective Nature of Immigration

If species vary in traits affecting their immigration potential, then they will exhibit different patterns of distribution among islands that vary in isolation. As we have seen in Chapter 9, dispersal abilities do vary, often dramatically, among species, even within a group of closely related species (Table 14.1).

To the extent that these interspecific differences are predictable, they will translate into predictable, nonrandom patterns of species distributions and community structure among islands. For example, groups with obvious adaptations for dispersing long distances over inhospitable habitats tend to be disproportionately common on islands. Birds, bats, and flying insects tend to be well represented on distant oceanic islands, whereas few if any native species of nonvolant mammals, amphibians, and freshwater fish are present there (see Carlquist 1974; J. Smith 1994; Nilsson and Nilsson 1978). Similarly, the primary means of dispersal for plants inhabiting isolated oceanic archipelagoes is birds. The predominance of avian-dispersed plants on isolated islands is probably a reflection of the relative efficiency of transport rather than sheer number of propagules. That is, while more propagules may be dispersed passively by winds and water (oceanic drift), only a tiny fraction of those potential colonists will reach land. Indeed, the chance that a randomly dispersed propagule will reach an island decreases as an exponential function of its isolation. In contrast, avian dispersal to islands typically is a nonrandom, or directed phenomenon. Therefore, plants and other avian-dispersed "hitchhikers" have a much higher probability of colonizing oceanic islands.

In an analogous fashion, mammals can be divided into two groups—bats and nonvolant forms—on the basis of their over-water dispersal ability (e.g.,

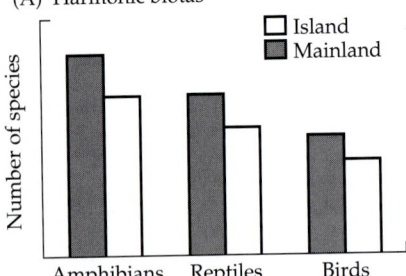

Figure 14.1 Two hypothetical examples of patterns in species composition illustrating the difference between harmonic and disharmonic biotas. In both examples, the insular communities have fewer species of each taxonomic group than the mainland biota. Disharmony refers to marked differences in the composition of insular communities from that of mainland biotas, with overrepresentation of some taxonomic or functional groups (e.g., birds in B) and scarcity of others that tend to be common elements of the mainland biota (amphibians and reptiles in B).

Table 14.1

Some examples of documented long-distance dispersal by
terrestrial animals

Reptiles	
Freshwater turtles	Possible 200 miles (322 km) to Madagascar
Snakes	500 miles (805 km) or more to the Galápagos
Lizards	1000 miles (1609 km) to New Zealand; perhaps more than 1000 miles (1609 km) for geckos
Terrestrial Mammals	
Large mammals	Perhaps 30 miles (48 km) or more[a]
Small mammals	Possibly 200 miles (322 km) to Madagascar for civets and insectivores
Rodents	500 miles (805 km) to the Galápagos
Amphibians	Perhaps 500 miles (805 km) to the Seychelles; perhaps 1000 miles (1609 km) to New Zealand
Land mollusks	More than 2000 miles (3218 km) in Polynesia (to Juan Fernandez Island)
Insects and spiders	More than 2000 miles (3218 km)

Source: Wenner and Johnson 1980.

[a]Elephants have been estimated to swim up to 48 km (Johnson 1978). Also, a semiaquatic hippopotamus occurred on Madagascar.

see Lawlor 1986). Bats have naturally colonized such distant outposts as New Zealand, New Caledonia, and the Canary and Hawaiian Islands, and they are represented by 33 genera on the Greater and Lesser Antilles. In contrast, native terrestrial mammals are completely absent from New Zealand, New Caledonia, the Canaries, and the Hawaiian archipelago, and only 26 genera are known from the Antilles (and all but 5 of those have become extinct since the late Pleistocene, owing at least in part to human activity; see Morgan and Woods 1986).

For those groups that lack flight or are otherwise incapable of being carried by winds, salt tolerance is an especially important influence on oceanic distributions (see Dunson and Mazotti 1989). Freshwater fishes and amphibians are so limited by their intolerance of salt water that their presence is often taken as an indication that an island was once connected to the mainland. Here again, there are some interesting exceptions that actually help to prove the rule. Among amphibians, the ranids and bufonids tend to be the dispersal champions (Meyers 1953). *Rana cancrivora* and *Bufo marina* are especially well adapted for oceanic transport, with high salinity tolerance in both adults and tadpoles (Inger 1954; Udvardy 1969). Consequently, these species are much more widely distributed on oceanic islands than their freshwater counterparts (see also Neill 1958).

Among the mollusks, slugs seem equivalent to freshwater fishes and hygrophylic amphibians in their intolerance of saline waters and their generally limited distributions on oceanic islands. In contrast, land snails adapted to arid conditions, and those small enough to be transported by the winds, are often significant components of oceanic biotas (Solem 1959; Vagvolyi 1975; see also Jaenike 1978).

On the mainland, animals and plants also differ substantially in their abilities to disperse to and colonize archipelagoes of freshwater systems. Strongly flying insects and groups of crustaceans, protozoans, and algae that form resistant eggs, spores, or cysts tend to have wide distributions that include lakes, springs, and streams that have never been connected to other bodies of water. Other forms, especially bivalves and weakly flying insects with short-lived terrestrial stages, are less widely dispersed. The distributions of freshwater fishes are typically so restricted that it is possible to trace historical connections between freshwater drainages by the taxonomic affinities of their fish species (e.g., Smith 1978, 1981; Hocutt et al. 1978; Rosen 1979). Bodies of water that have never had aquatic connections are devoid of fishes, although they often support diverse communities of more vagile organisms. Of course, freshwater fishes are also notoriously poor overseas dispersers. New Zealand, for example, has no native primary division freshwater fishes, but introduced trout thrive in its magnificent rivers.

Among nonvolant mammals, relatively few insectivores and small rodents are found on islands along coastal and inland bodies of water, presumably because of their limited immigration abilities (Peltonen et al. 1989; Crowell 1986). In regions where waters freeze during at least part of the year, insular distributions are strongly influenced by the ability of species to remain active and withstand subfreezing temperatures during winter. In contrast to hibernators and other seasonally inactive species, those that remain active during winter are disproportionately common on islands. Absent or surprisingly rare are chipmunks, ground squirrels, woodchucks, jumping mice, and other species unable or unlikely to utilize the winter ice cover as a seasonal avenue for migration (see Lomolino 1988, 1993b; Peltonen et al. 1989; Tegelstrom and Hansson 1987). Even among winter-active mammals, larger species generally are more capable of withstanding winter conditions, and thus are often better at dispersing across ice to colonize islands in winter.

Establishing Insular Populations

In Chapter 9 we tried to emphasize the distinction between immigration and establishment of an insular population. Obviously both are requisite phases of successful colonization, yet they are often logistically difficult, if not impossible, to study as separate phenomena. For some, and perhaps many, groups of species, ability to establish breeding populations may not be correlated with immigration ability. Or, just as confounding, these traits might be positively correlated in some species groups and negatively correlated in others. While empirical studies addressing this question remain scarce, ecological theory suggests that the former may be the case; that is, good dispersers may, in general, be preadapted to establish populations on islands and in other isolated ecosystems. Species adapted to exploit disturbed or newly created environments often combine high dispersal abilities with broad ecological tolerances. Such species should therefore be preadapted for successfully colonizing islands. These species are often referred to as r-strategists. K-strategists, on the other hand, are better adapted for efficient utilization of limited resources, especially when population levels approach the carrying capacity of the environment (K). While most species probably fall between the idealistic extremes of an r/K continuum, biogeographers have identified the characteristics of a prototypical colonizer (Table 14.2).

Empirical studies, while still very limited, tend to support most, but certainly not all, of the predictions listed in Table 14.2. For example, Baur and Bengtsson (1987) found that among land snails of the Baltic uplift archipelago, colonizing abilities were higher for species with a broader niche and for those capable of self-fertilization, a trait obviously advantageous for reproducing in newly colonized environments. In addition, colonization success was highest on islands that afforded the greatest diversity of habitats. On the other hand,

Table 14.2
Some attributes that may preadapt species to be good colonizers

| Taxon | Immigration | | Establishment | | | |
	Preferred habitats superior as points of departure	High dispersal power of propagules	Preferred habitats superior as points of arrival	Large K	Large r/λ	Unknown significance
Birds (Mayr 1965a)	?	Often fly long distances as an adaptation to scattered, temporary habitats such as fresh water	?	Often able to occupy many habitats, thus enlarging K; tendency to eat seeds, which allows larger standing populations	Tendency to clump (by traveling in small flocks)	
Ants (Wilson 1961)	Tendency to occupy ecologically "marginal" (species-poor) habitats near coast	Since marginal habitats are unstable, the species are "fugitive," with probably greater vagility	Tendency to occupy ecologically "marginal" (species-poor) habitats near coast	Large populations exist in marginal habitats; also typically able to occupy many habitats	Larger average colony size	Workers spinier, more commonly lay odor trails
Flowering plants Dipsacaceae, Asteraceae, Rubiaceae (Ehrendorfer 1965)	Weedy species, members of regular successions that constantly colonize newly opened habitats	Propagules relatively widely dispersed	Weedy species, members of regular successions that constantly colonize newly opened habitats	?	Fast individual development, large quantities of progeny, occasional autogamy	

Source: MacArthur and Wilson 1967.

Note: K, carrying capacity (as number of individuals) of the environment; r, intrinsic rate of increase of a population; λ, birth rate.

the colonization success of these species was not significantly correlated with age at maturity or other life history traits typically associated with r-strategists.

Ebenhard's (1991) review of the available information on characteristics associated with colonization success in other animals is, for the most part, consistent with theoretical predictions. For example, colonization success among introduced birds and mammals was directly dependent on propagule size and clutch size (see Crowell 1973; Mehlop and Lynch 1978; Ebenhard 1989). Numerous introduction experiments with dung beetles (Scarabaeidae) across the globe revealed that colonization success tended to be higher for species with intermediate size, rapid development, high fecundity, and broad niches (see Doube et al. 1991; Hanski and Cambefort 1991; see also Schoener and Schoener 1983; Harrison 1989).

The Selective Nature of Extinction

Despite the tremendous variation among insular environments, they all tend to be similar in that they provide limited resources, and as island size decreases, the amount of resources and their replenishment rate decreases. Thus, insular communities should be biased not just in favor of good dispersers, but also in favor of those species that require few resources to maintain their populations. Although this prediction is far from a rule, many studies are consistent with it. For example, the record of extinctions since the Pleistocene suggests that extinction has by no means been an entirely random process. Some types of species have become locally extinct much more frequently than others during the last 10,000 years, and this differential susceptibility appears to be related to their ecological characteristics. Of course, this is just what would be expected if, as suggested in earlier chapters, the ecological characteristics of species determine their carrying capacities, and in turn, these equilibrium population densities largely determine the probability of extinction (e.g., Brown 1971b, 1978, 1981; Van Valen 1973a,b). In fact, it should be possible to predict the relative abilities of species to persist on isolated land-bridge islands based on some easily measured ecological traits. Animals of large body size, carnivorous diet, and specialized habitat requirements should experience higher extinction rates than species that are smaller, herbivorous, or more generalized in habitat use.

This predicted pattern is precisely what is observed for terrestrial mammals on isolated mountaintops in western North America (Figure 14.2). Small herbivores that are habitat generalists make up the limited subset of species that are found on virtually all mountain ranges, whereas carnivores and herbivores of larger body size or specialized habitat requirements have become extinct on many montane islands and persist on only the largest ones. This pattern is particularly impressive because it is repeated on two different, geographically isolated sets of mountaintop islands in North America: those of the Great Basin (Brown 1971b) and those farther south in the American Southwest (Brown 1978; Patterson 1984).

In Chapter 7 we noted that vertebrate extinctions during the late Neogene also tended to be biased in favor of large, carnivorous, or specialist species—

Figure 14.2 All else being equal, species with relatively low resource requirements tend to be disproportionately more common on islands. The small mammal communities of mountaintop forest "islands" of the Great Basin (A) and of New Mexico and Arizona (B) both tend to be dominated by small, generalist herbivores, presumably because, in comparison to mammals with higher resource requirements, they can maintain higher populations on the same area and, therefore, are less prone to extinction. (A after Brown 1971, 1978; B after Patterson 1984.)

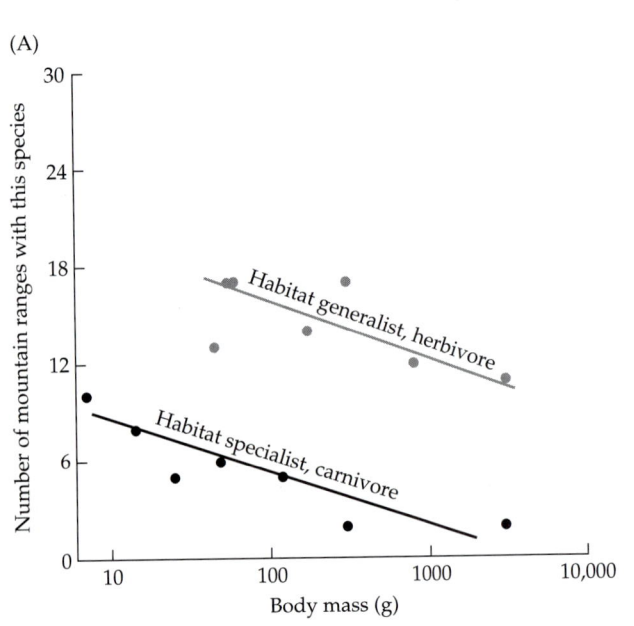

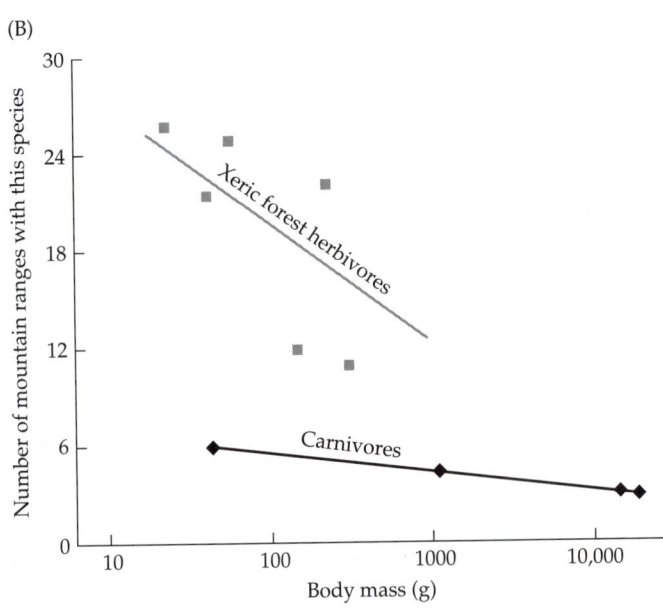

that is, species with relatively high resource demands and low reproductive rates. The record of historical extinctions and lists of currently endangered species also bear witness to the vulnerability of large animals (see Chapter 17).

Additional insights can be gained from studies of introductions of birds to hundreds of islands across the globe. Many ecologists have capitalized on these natural experiments to test some of the factors purported to influence the assembly of insular communities. Moulton and Pimm's studies of birds introduced to the Hawaiian Islands and Tahiti suggest that body size and demography strongly influenced the ability of these exotics to establish and maintain insular populations (Moulton and Pimm 1983, 1986, 1987; Moulton 1993; Pimm et al. 1988). In a similar study of recent extinctions of birds on 16 British islands, Pimm et al. (1988) observed that risk of extinction was highest for species that are migratory and tend to maintain relatively low and variable population densities. This study also revealed an interesting interaction between the effects of population size and body size. At average population sizes above seven pairs, large-bodied species exhibited a relatively high risk of extinction, while the reverse was true for populations below seven pairs. Karr's (1982) study of extinctions of birds on Barro Colorado Island, which was isolated during the construction of the Panama Canal, also produced mixed results. Extinctions of forest birds were not significantly associated with population density, but species with more variable populations were more prone to extinction.

James's (1995) recent review revealed that prehistoric extinctions in oceanic birds also were selective. Analyses of recent fossils indicated that avian richness on the Hawaiian Islands declined as much as 50 to 75% following the arrival of Polynesians some 1500 to 2000 years B.P. Again, larger, species with greater resource requirements, including raptors (eagles, hawks, and owls) and water birds (ibises, rails, geese, and ducks), were much harder hit than the relatively small passerines. Extinctions also appear to have been selective with respect to trophic strategies. Extinctions were especially prevalent among predators and terrestrial omnivores and herbivores. Faring much better were the nectarivores, which now represent the most common native birds in Hawaiian forests (Figure 14.3; James 1995; see Scott et al. 1986). Finally, James emphasized that the "extinction profile" may vary substantially among island ecosystems. For example, while prehistoric extinctions removed some 69 to 74% of the birds on Maui, recent fossil evidence indicates that avian richness on Madagascar declined by just 11 to 15% during the same period, and that

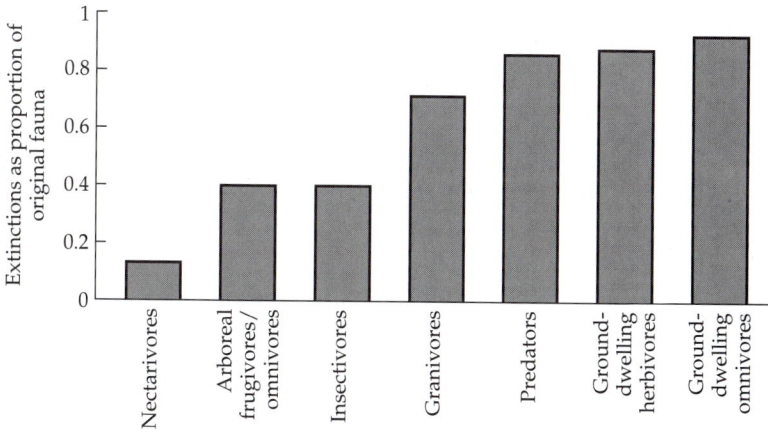

Figure 14.3 Prehistoric extinctions among birds on the Hawaiian Islands tended to be highly selective, being most prevalent among predators and ground-dwelling birds, while extinctions were relatively infrequent among nectarivores, which now represent the most common native birds on the islands. (After James 1995.)

extinctions were even more skewed toward loss of the larger species there (see also Dickerson and Robinson 1986).

In short, while many studies suggest that extinction is selective, there remains much debate over which characteristics are most closely associated with extinction. Many ecologists argue that insular survival is more strongly influenced by island attributes, including area and isolation, than by the intrinsic properties of the species (for example, see Gotelli and Graves 1996; Tracy and George 1992, 1993; Haila and Hanski 1993; Diamond and Pimm 1993; Richman et al. 1988). The geography of extinction has intrigued biogeographers throughout the distinguished history of the field, and it is likely to provide information critical to understanding and mitigating the current extinction crisis. We therefore explore this theme in depth in our discussions on biodiversity (Chapters 17 and 18).

Patterns Reflecting Differential Immigration and Extinction

Nestedness of insular communities. MacArthur and Wilson's equilibrium model has sometimes been criticized for, among other things, its assumption that all species are equivalent. Again, this criticism seems a bit unfair, because their model attempted to explain patterns in species richness, not the composition of insular communities. In addition, MacArthur and Wilson's model implicitly included some effects of interspecific differences in immigration and extinction potential among species. If each species were identical with respect to these important characteristics, immigration and extinction rates would be linear functions of species richness. Each time a species colonized an island, the immigration rate would decline and the extinction rate would increase by constant amounts. However, MacArthur and Wilson acknowledged that that the *names* of the species matter. Thus, they predicted that immigration and extinction curves should be concave.

Moreover, MacArthur and Wilson (1967:80) cited specific cases in which insular biotas tend to be biased in favor of species with superior dispersal and colonizing abilities (see Table 14.2). For example, they noted that the insects of Micronesia tend to be relatively small, a trait that facilitates their transport by winds (MacArthur and Wilson 1967; after Gressitt 1954). Similarly, Wollaston's (1877) earlier work had demonstrated that insect colonists of Saint Helena Island in the Atlantic Ocean were predominately wood borers and other forms preadapted to rafting on logs or vegetation mats. As these and many other studies demonstrated, oceans act as filters, allowing the selective passage of

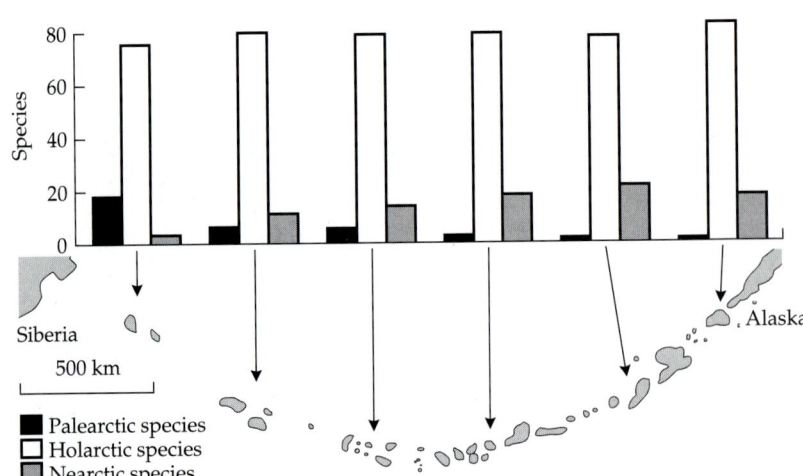

Figure 14.4 Much like the Lesser Sunda Islands (see Figure 9.14), the waters between Siberia and Alaska form an oceanic double filter, or transition zone, here characterized by the reciprocal attenuation of land plants on the Aleutian Islands. (After Williamson 1981.)

certain types of species. Often, these oceanic filters form transition zones, characterized by differences in the attenuation rates of species groups that differ in dispersal ability (Figure 14.4; see also Figures 9.14 and 9.15).

Surprisingly, the selective nature of immigration is also evident for some islands that are quite near the coast. Simberloff and Wilson's (1969, 1970; Simberloff 1978) classic defaunation experiments on invertebrates of mangrove islands not only tested some critical predictions of the equilibrium theory, but also revealed that as islands were colonized, they tended to converge on their original species composition. Leston's earlier (1957) comparisons of insect

(A)

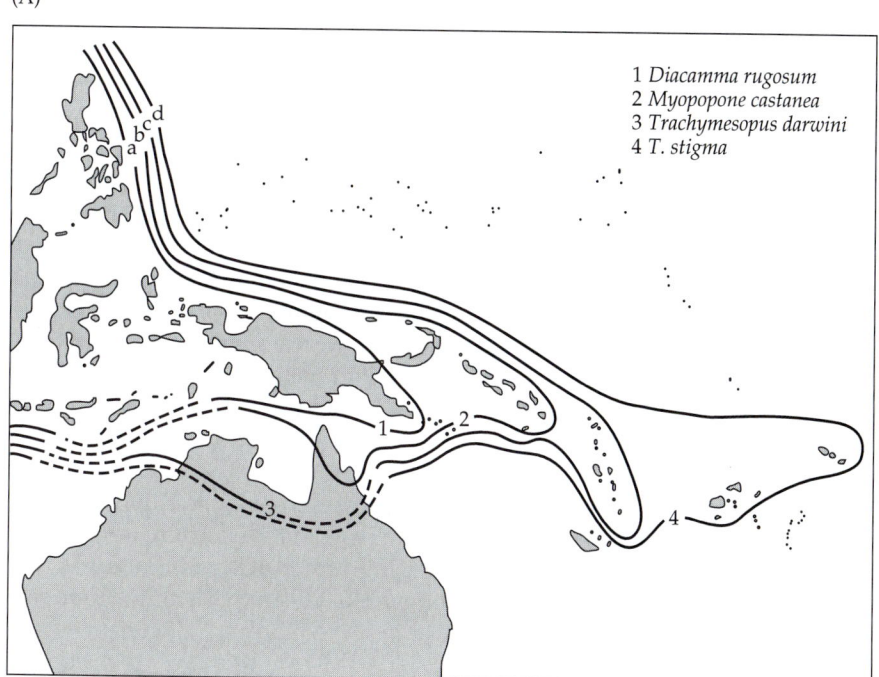

1 *Diacamma rugosum*
2 *Myopopone castanea*
3 *Trachymesopus darwini*
4 *T. stigma*

(B)

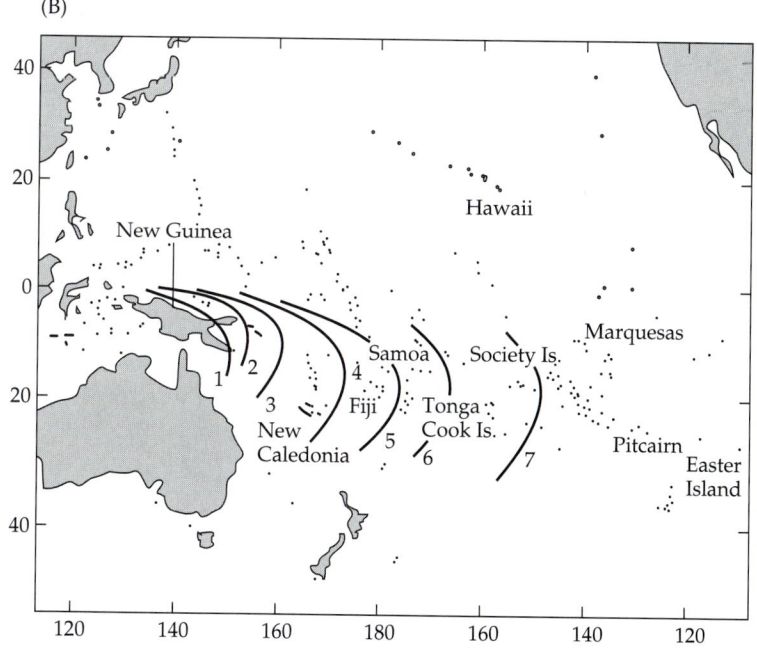

Figure 14.5 The selective nature of immigration is revealed in the geographic nestedness of many insular biotas. Both ants (A) and birds (B) of the western Pacific exhibit analogous patterns of distributions extending eastward from their sources in New Guinea and Indonesia. The lines indicate the eastern range limits of taxa. (A) Ant species: 1 = *Diacamma rugosum*; 2 = *Myopopone castanea*; 3 = *Trachymesopus darwini*; 4 = *T. stigma*. (B) Land and freshwater breeding birds: 1 = pelicans, storks, larks, pipits, birds of paradise, and nine others; 2 = cassowaries, quails, and pheasants; 3 = owls, rollers, hornbills, drongos, and six others; 4 = grebes, cormorants, ospreys, crows, and three others; 5 = hawks, falcons, turkeys, and wood swallows; 6 = ducks, thrushes, waxbills, and four others; 7 = barn owls, swallows, and starlings; beyond 7 = herons, rails, pigeons, parrots, cuckoos, swifts, kingfishers, warblers, and flycatchers. (A after Wilson 1959; B from Williamson 1981, after Firth and Davidson 1945.)

communities on isolated islands of the Atlantic and Pacific Oceans also revealed that these communities are convergent with one another and tend to form regular subsets of the mainland biota. Some taxonomic groups tend to be underrepresented on islands, while others are disproportionately common on islands in comparison with mainland areas. Just as important, these biases in species composition seem to increase with increasing isolation. This tendency for communities on species-poor islands to form regular subsets of those on species-rich islands is known as **nestedness**. Subsequent studies have confirmed that nestedness of insular biotas is an extremely common phenomenon (Wright et al. 1996), one exhibited by most animals and plants and by an overwhelming variety of islands, mountaintops, and other isolated ecosystems (Figure 14.5).

In his seminal monograph on biogeography, Darlington (1957) presented a graphical model for nestedness that he termed the **immigrant pattern** (Figure 14.6). Interestingly, Darlington's model ignored the possibility that extinction also is a selective process. His **relict pattern** predicted that, while smaller islands should have fewer species, their biotas should be random draws of the mainland pool. However, as we have discussed earlier, ability to survive on limited insular resources may differ dramatically and predictably among species, even within a group of closely related species. Therefore, regular patterns in species composition on islands may result from extinctions as well as immigrations. According to Brown's **relaxation model**, as the area of islands decreases, their communities should converge on a similar set of species, i.e., those with minimal resource requirements (Figure 14.6). Thus, across a great variety of archipelagoes, insular communities tend to form nested subsets when ordered by decreasing area or increasing isolation (or by species richness; Figure 14.7).

Although nestedness had attracted a modest degree of attention for many decades, it wasn't until 1984 that Bruce Patterson and Wirt Atmar developed a statistically rigorous approach for analyzing nested subsets (Patterson and Atmar 1986). Their important paper triggered a flurry of studies that con-

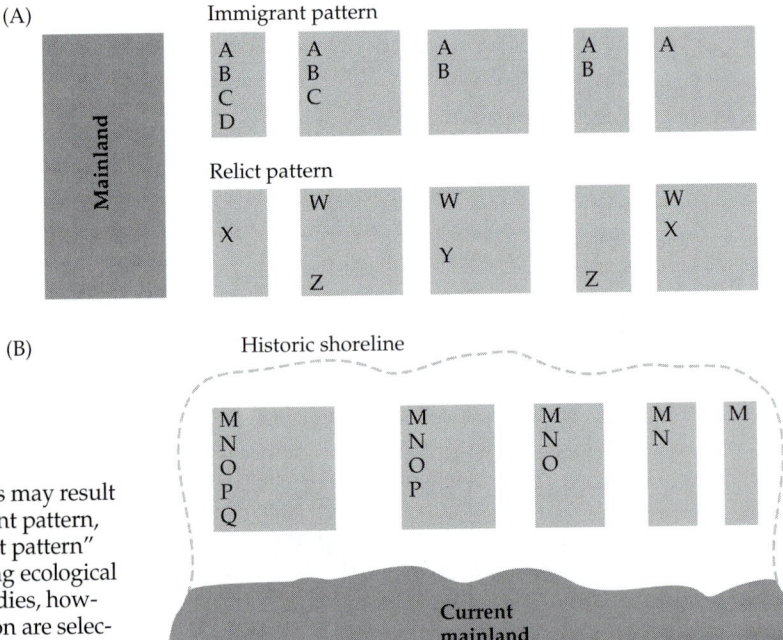

Figure 14.6   Nestedness of insular communities may result from species-selective immigration (the immigrant pattern, A) or extinction (B). Note that Darlington's "relict pattern" implies that the partial extinction of faunas during ecological relaxation might be random (A). Most recent studies, however, suggest that both immigration and extinction are selective processes. (A after Darlington 1957.)

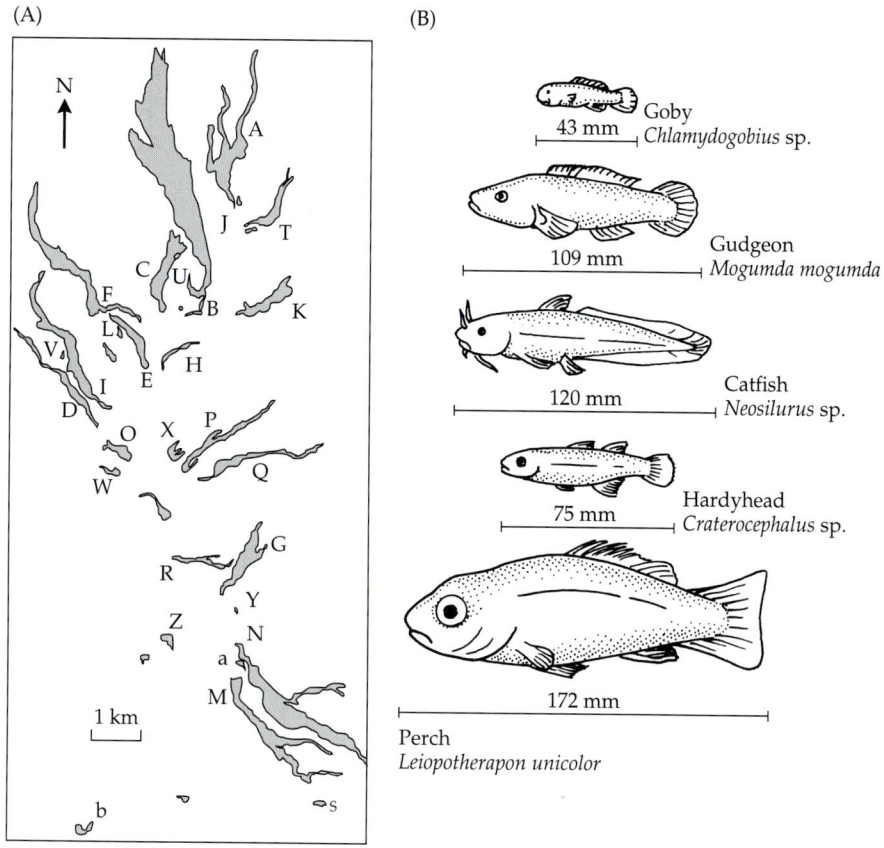

(A)

(B)

(C)

Spring (see map in part A)

Species	A B C D E F G H I J K L M N O P Q R S T U V W X Y Z a b	No. of springs
Goby	X X X X X X X X X X X X X X X X X X X X X X X X X X X X	28
Gudgeon	X X X X X X X X X X X X X X X X X X X	19
Catfish	X X X X X X X X X X X X X X	14
Hardyhead	X X X X X X X X X	9
Perch	X X X X X X X	7
Number of species	5 5 5 5 5 5 4 4 4 4 3 3 3 3 2 2 2 2 2 1 1 1 1 1 1 1 1 1	

Figure 14.7   The distributions of five fish taxa among 28 springs in the Dalhousie Basin of South Australia illustrates the nested subset pattern. Here, nestedness is nearly perfect in that, with the exception of the taxon that is absent from spring G, the faunas of the more depauperate springs are regular subsets of those found the in richer springs. (From Brown and Kodric-Brown 1993.)

firmed the generality of nestedness, provided alternative statistical approaches, and began to explore the causality of nestedness among taxonomic groups and across archipelagoes (see Patterson 1987, 1990; Cutler 1991; Wright and Reeves 1992; Patterson and Brown 1991; Atmar and Patterson 1993; Doak and Mills 1994; Simberloff and Martin 1991; Lomolino 1996; Kadmon 1995; Lazell 1983; Kitchener et al. 1980; Schoener and Schoener 1983; Cook 1995; Cook and Quinn 1995; Wright et al. 1998). As this fascinating area of research continues to mature, it is certain to contribute substantially to our understanding of the forces structuring isolated communities.

Distributions of particular species. Nestedness, while an intriguing and insightful measure of patterns among communities, provides only limited information on the distribution patterns of particular species. Again, significant nestedness indicates that species and islands differ with respect to factors influencing immigration and extinction. But how are these species actually distributed among islands that vary in both area and isolation?

A useful starting point for addressing this question may again be MacArthur and Wilson's equilibrium theory—specifically, its basic assumption that insular community structure results from the combined effects of recurrent immigrations and extinctions. Here, however, the terms *immigration* and *extinction* refer to the frequency of these events for a particular species. With the above assumption as a conceptual foundation, Levins, Hanski, and other scientists working during the past few decades have developed an entirely new theory of population biology called metapopulation theory (see Gilpin and Hanski 1991). Basically, metapopulation studies attempt to estimate the proportion of islands, or "patches," that must be occupied to ensure the survival of the interacting populations of a species—its metapopulation (see Chapter 4). Alternatively, given information on the actual proportion of patches occupied, metapopulation models can be used to estimate the time to extinction of the metapopulation.

In a similar fashion, a model of insular distributions for a particular species can be derived based on the same fundamental assumption plus two additional ones: (1) the probability or frequency of immigration decreases with increasing isolation, and (2) the frequency of extinction decreases as island area increases (see Lomolino 1986, 1997). As you can see, these assumptions are essentially equivalent to those of the equilibrium theory. Given this, we predict that the distribution of the focal species should result from the interactive, or compensatory, effects of immigration and extinction. The basic rule is that a focal species will occur on those islands where its immigration rate exceeds its extinction rate. Thus, the focal species is expected to occur on isolated islands, but only those that are large enough so that extinction rates are low enough to compensate for infrequent immigrations. Conversely, the species may be common on small islands if they are close enough to the mainland so that high immigration rates compensate for frequent extinctions. The overall result is that minimal area requirements to maintain populations of the focal species should increase as isolation increases. When this **insular distribution function** is plotted on a graph depicting island area and isolation, the intercept becomes a measure of resource requirements, while the slope is an inverse measure of the immigration ability of the focal species (Figure 14.8; see also Peltonen and Hanski 1991; Hanski 1986; Alatalo 1982; Peltonen et al. 1989).

This species-based approach to island biogeography is a relatively new and untested one. However, it complements the dynamic, ecosystem-level approach championed by MacArthur and Wilson. Here, however, the identities of the species are not lost. Indeed, many of the patterns community ecologists study result from, not in spite of, differences among species. For example, a high degree of nestedness implies that insular distributions vary in a very regular manner among species (see Figure 14.7; see also Patterson 1984). If species were equivalent, nestedness would be a rare phenomenon indeed.

The insular distribution function, in addition to providing an alternative explanation for many patterns of diversity, highlights the need to study some critical questions. As MacArthur and Wilson (1967) lamented, we still know

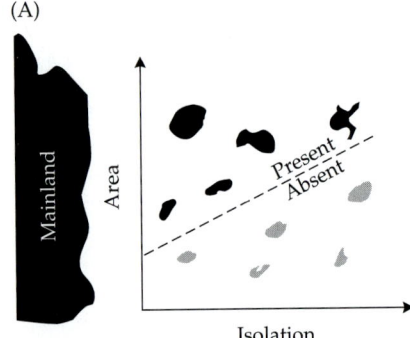

(A)

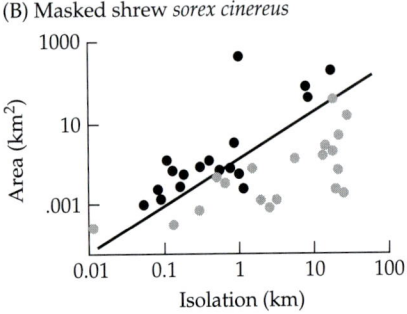

(B) Masked shrew *sorex cinereus*

Figure 14.8 Insular distributions of species should be influenced by the combined effects of immigration and extinction. Species populations may persist on small islands if they are close enough to the mainland to experience frequent immigrations. Because immigration rates should decrease with isolation, the minimum area required to maintain insular populations of a focal species should increase with isolation. The predicted pattern (A) has been observed for a variety of organisms, including small mammals (B, after Lomolino 1986 and 1998), fishes (see Figure 18.26), and butterflies (see Figure 18.27).

relatively little about interspecific differences in immigration and extinction. The general tendency for insular communities to be nested along gradients of isolation and area suggests some regular patterns of variation in immigration abilities and resource requirements, but what are those patterns? If we were to draw histograms of the frequency distributions of immigration abilities, what form would they take? Do immigration abilities and resource requirements (slopes and intercepts of the insular distribution function) covary among species? That is, do good immigrators require larger or smaller islands to maintain their populations than poor immigrators? Regardless of the fate of this particular model, these questions will persist. They are indeed challenging, but if resolved, they will certainly provide some insights fundamental to developing more comprehensive models of insular community structure.

Patterns Reflecting Interspecific Interactions

Despite their heuristic value, MacArthur and Wilson's equilibrium theory and metapopulation theory ignore the potential importance of interspecific interactions. Nestedness models such as Figure 14.6 also ignore the potential importance of interactions such as competition, which, if important, may reduce nestedness for many biotas. Modifying these general models to include the influence of one species on the distributions of others will certainly add a substantial degree of complexity. Yet, more than any other systems, islands provide compelling evidence for the importance of interspecific interactions.

Several biogeographic patterns suggest that the distributions and abundances of particular species are influenced not only by their own characteristics, but also by interactions with other species. All types of interactions, including competition, predation, parasitism, and mutualism, are probably important, but most studies of insular ecology have emphasized competition. Three patterns that have been hypothesized to be caused by interspecific competition are listed below.

1. Ecologically similar species tend to exhibit mutually exclusive distributions, seldom if ever occurring together on the same island. A corollary of this pattern is the prediction that species that do coexist should tend to be more dissimilar in ecologically relevant traits than would be expected by chance.
2. In comparison to conspecific populations on the mainland, populations on species-poor islands often should exhibit relatively high densities.
3. Insular populations tend to exhibit **ecological release**, characterized by significantly broader niches and shifts to habitats, feeding strategies, activity periods, or other characteristics that would be considered atypical for the species on the mainland.

Distributions, checkerboards, and incidence functions.   In the absence of direct experiments on the competitive exclusion of one species by another, a pattern of mutually exclusive distributions of two or more species with similar resource requirements can serve as natural evidence of competition. Archipelagoes consisting of many similar islands provide numerous examples of such patterns. Diamond (1975b) described several mutually exclusive distributions, which he called **checkerboards,** for pairs of congeneric bird species inhabiting the Bismarck Archipelago east of New Guinea. The flycatcher *Pachycephala melanura dahli,* for example, is found on 18 islands, and its congener *P. pectoralis* on 11 islands, but the two species never occur together on the same island (Figure 14.9).

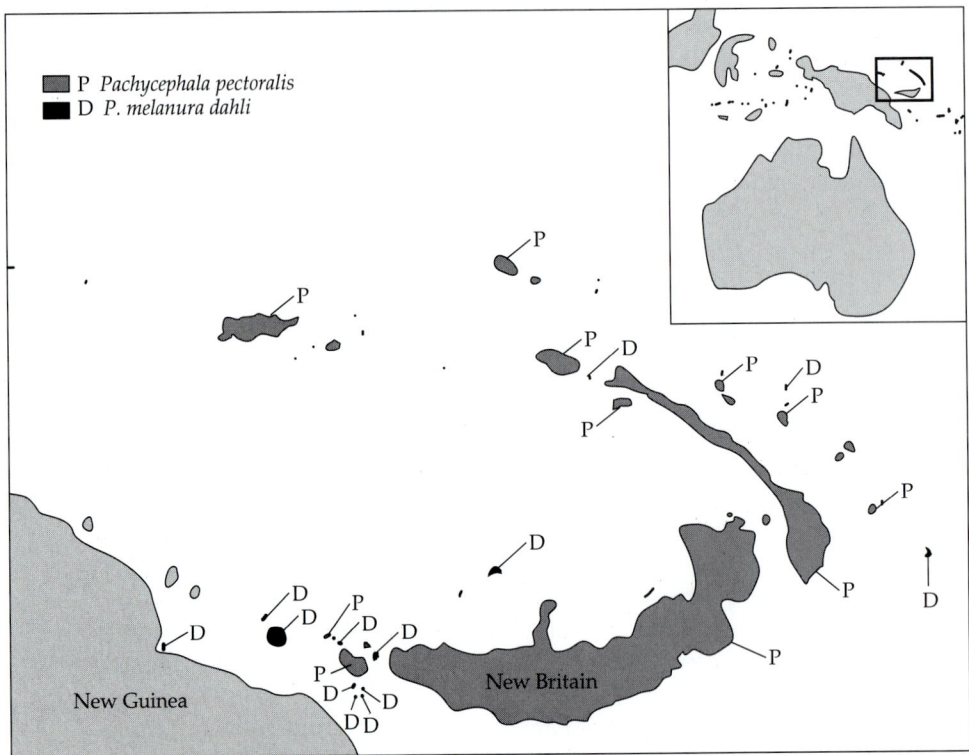

Figure 14.9   Mutually exclusive, or checkerboard, distributions of *Pachycephala* fly-catchers on the Bismarck Archipelago off New Guinea. There are two species, *P. pectoralis* and *P. melanura dahli*. Most islands have one of these species, no islands have both, and a few (especially the smallest islands) have neither. (After Diamond 1975b.)

Other negative associations resulting from interspecific competition may be more complex than simple checkerboards. Species that do not occur together on small islands may coexist on larger ones, suggesting that they competitively exclude each other where carrying capacities are low, but are able to subdivide the more plentiful resources of larger islands (see Schoener and Adler 1991). In still other cases, only certain combinations of three or more species may be able to coexist, whereas other combinations appear to be "forbidden," or at least much less common, because of competitive exclusion. Diamond (1975b) cites another example from the Bismarck Archipelago, that of cuckoo doves in the genera *Macropygia* and *Reinwardtoena*. There are four species in this group, so there are fifteen possible combinations of species that would give biotas of from one to four species (Table 14.3). Only six of these combinations are actually observed, however, suggesting that certain sets of species are incompatible (but see also McFarlane 1989; Gotelli and Graves 1996).

Sometimes diffuse competition among many unidentified species, rather than direct interactions among just a few closely related ones, has been implicated in determining insular distributions. Again, perhaps the best examples come from Diamond's (1974, 1975) studies of the birds of the Bismarck Archipelago. He noted that certain species, which he called **supertramps**, are usually found only on small islands containing few other bird species (Figure 14.10). Supertramps are also among the few species that have recolonized islands following a recent volcanic eruption. By plotting **incidence func-**

Table 14.3
Combinations of four cuckoo dove species that are present or absent in the Bismarck Archipelago

| Number of species | Observed combination | | Missing species combinations |
	Species combinations	Number of islands inhabited by this species combination	
1	A	3	N
	M	8	R
2	A, M	5	A, N
	A, R	4	M, N
	M, R	2	N, R
3	A, N, R	5	A, M, N
			A, M, R
			M, N, R
4	None		A, M, N, R

Source: Simplified from Diamond 1975b.

Note: A, *Macropygia amboinensis*; M, *M. mackinlayi*; N, *M. nigrirostris*; R, *Reinwardtoena* super-species.

tions—the proportion of islands inhabited by a given species as a function of the number of other species present—Diamond was able to quantify some interesting distributional patterns. Using such graphs, supertramps, such as the flycatcher *Monarcha cinerascens* and the honeyeater *Myzomela pammelaena*, are readily distinguished as species that appear to be excellent colonists, but poor competitors in diverse bird communities (Figure 14.10). Other species, including a starling (*Aplonis metallica*) and the incubator bird (*Megapodius*

Figure 14.10 Incidence functions for various bird species in the Bismarck Archipelago. Note that some species, called supertramps, such as *Monarcha cinerascens* and *Myzomela pammelaena*, occur on most islands with few other bird species, but are absent from islands with diverse avifaunas; other species, such as *Aplonius metallica* and *Megapodius freycinet*, are found on all but the smallest islands; and still other species, including *Accipiter brachyurus* and *Egretta intermedia*, occur on only the largest islands with many other bird species. (After Diamond 1975b.)

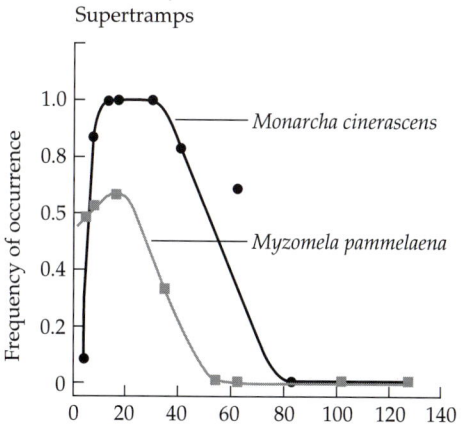
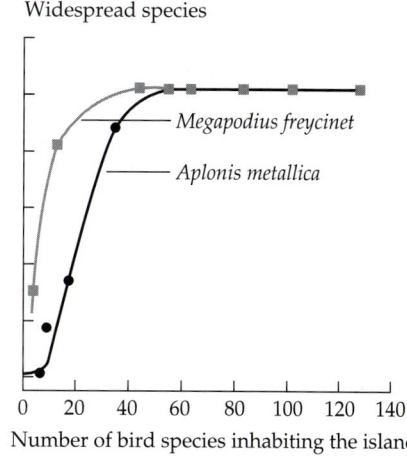
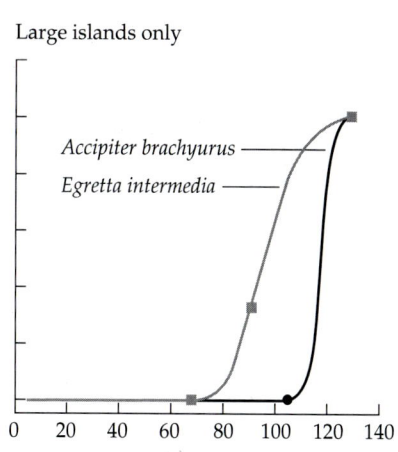

freycinet), appear to be both capable colonists and good competitors, as they are found with relatively high frequency on all but the smallest islands. Still other species, especially birds with large body sizes, specialized diets, or restricted habitat requirements, such as hawks (e.g., *Accipiter brachyurus*) and herons (e.g., *Egretta intermedia*), are restricted to the largest islands, which presumably offer more resources, but also contain more bird species (see also Hanski 1992; Peltonen and Hanski 1991; Taylor 1991).

A similar approach involves testing the prediction that ecologically similar species coexist on islands less frequently than expected by chance. Several investigators have noted that islands contain relatively few congeners, and few species that are similar in ecologically important morphological traits, such as bill size in birds (e.g., Grant 1965, 1966b, 1968, 1986; MacArthur 1972; Diamond 1973, 1975b; Lack 1973, 1976). In analyzing these patterns, the critical question is whether the differences among co-occurring insular species are greater than would be expected if the island biotas were assembled at random from the mainland species pool. Some analyses that have addressed this problem have produced results that are consistent with the competition hypothesis. For example, in the case of both Darwin's finches in the Galápagos Islands (Grant and Abbott 1980; Hendrickson 1981; Case and Sidell 1983) and hummingbirds in the Antilles (Lack 1973, 1976; Brown and Bowers 1985; Case et al. 1983), it appears that species that coexist on the same island tend to be more different in bill and wing measurements than would be expected if the species associated at random (Table 14.4). Actually, the pattern is even more dramatic than Table 14.4 shows, because when species of similar size occur on the same island, they tend to be segregated by elevation and habitat.

Earlier we mentioned Pimm and Moulton's studies on interspecific competition and success among birds introduced to Tahiti and the Hawaiian Islands (Figure 14.11; Moulton and Pimm 1983, 1986, 1987; Moulton 1993; Pimm et al. 1988). More recent ecomorphological studies on birds introduced to Bermuda provided similar results (Lockwood and Moulton 1994; see also Moulton 1993; Simberloff and Boecklen 1991). Invasion success was significantly lower when ecologically similar species were present, and surviving assemblages tended to be overdispersed in morphology.

Some patterns in insular community structure appear to be so regular that Diamond termed them **assembly rules** (Table 14.5). Each of these rules addresses the tendency of insular communities to represent nonrandom subsets of the mainland species pool. These rules have been the target of much

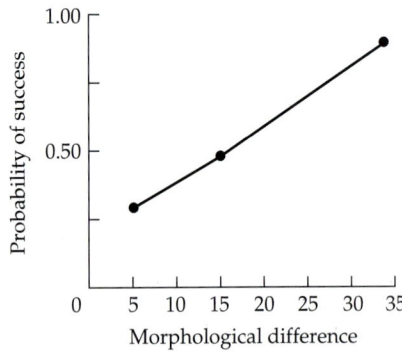

Figure 14.11 The probability that introduced species of birds will successfully establish an insular population is directly correlated with the degree to which they differ morphologically from the resident Hawaiian avifauna. Morphological difference is represented here by the difference in bill length between an introduced bird species and its most similar resident cogener. (After Moulton and Pimm 1986.)

Table 14.4

Coexistence of pairs of hummingbird species of varying size combinations on the islands of the Greater and Lesser Antilles

	Both small	One small, one large	Both large
Observed species pairs	6	27	16
Species pairs expected from a random distribution	14.3	18.7	24.3

Source: Data from Brown and Bowers, 1984.

Note: The bill length of every other species was expressed as a ratio relative to the bill length of the smallest species, and the species with ratios less than or greater than 1.8 were categorized as small or large, respectively. Note that coexisting pairs of species tend to be of different sizes; the probability of this occurring by chance is less than 0.02.

Table 14.5
Diamond's assembly rules for insular communities

1. *If one considers all the combinations that can be formed from a group of related species, only certain ones of these combinations exist in nature.*

2. *These permissible combinations resist invaders that would transform them into a forbidden combination.*

3. *A combination that is stable on a large or species-rich island may be unstable on a small or species-poor island.*

4. *On a small or species-poor island a combination may resist invaders that would be incorporated on a larger or more species-rich island.*

5. *Some pairs of species never coexist, either by themselves or as part of a larger combination.*

6. *Some pairs of species that form an unstable combination by themselves may form part of a stable larger combination.*

7. *Conversely, some combinations that are composed entirely of stable subcombinations are themselves unstable.*

Source: Diamond 1975b.

criticism, yet they have served an important purpose—refocusing our attention from relatively simple patterns in species richness to other, more challenging questions related to the forces influencing the organization of communities (see review in Gotelli and Graves 1996).

Density compensation and niche shifts on islands. If competition influences the organization of communities by limiting species to only part of their fundamental niches, one would expect populations to expand their niches and increase in density on islands where they interact with few other species. Both of these phenomena, niche expansion (or shifts) and increased densities, appear to be relatively common among insular populations. Crowell (1962) was the first to focus attention on these patterns by comparing bird populations on Bermuda (900 km east of North Carolina in the Atlantic Ocean) with those of similar habitats on the North American mainland. He not only found niche shifts among these birds, but also observed that three species on Bermuda maintained a combined population density at least as great as that of the entire avian community on the mainland (see also Diamond 1970a, 1973, 1975; MacArthur et al. 1972, 1973; Kohn 1978; Yeaton 1974; but see also Faeth 1984; Martin 1992).

This latter phenomenon, termed **density compensation**, is often attributed to relatively low levels of interspecific competition, but it may just as easily result from the paucity of insular predators. Mice and shrews inhabiting islands along the coasts and within freshwater systems of North America often exhibit relatively high densities along with substantial niche shifts (e.g., see Hatt 1928; Bishop and Delaney 1963; Grant 1971; Crowell and Pimm 1976; Lomolino 1984). At least in some of these cases, predation is implicated as an important factor influencing the niche dynamics and densities of these mammals. For example, in many regions of North America, short-tailed shrews (*Blarina brevicauda*) often prey on small rodents such as meadow voles (*Microtus pennsylvanicus*) and deer mice (*Peromyscus* spp.), especially on the nestlings and juveniles. In the absence of these and other predators, voles and deer mice often exhibit significant ecological release, occupying habitats considered atypical for their species, often at relatively high densities (Figure

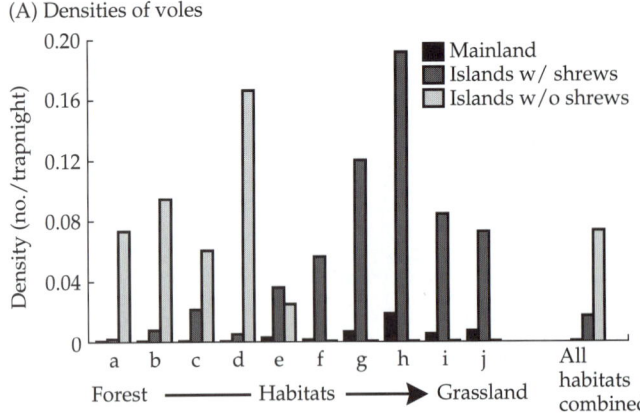

(A) Densities of voles

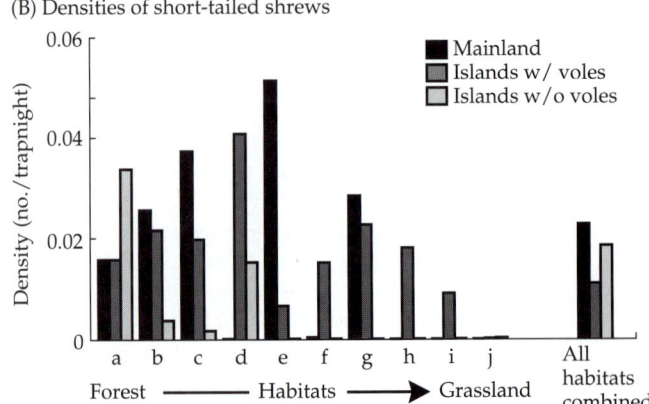

(B) Densities of short-tailed shrews

Figure 14.12   Insular populations often exhibit niche shifts, exhibiting relatively high densities and occupying habitats considered atypical for the species on the mainland. Such shifts may result from the paucity of competitors or predators on islands. (A) In the Thousand Island region of the Saint Lawrence River, meadow voles (*Microtus pennsylvanicus*) exhibit significant niche shifts on islands lacking one of their predators, the short-tailed shrew (*Blarina brevicauda*). (B) Conversely, on islands lacking meadow voles, the distribution of short-tailed shrews contracts to habitats where the availability of alternative prey items (primarily macroinvertebrates in forest soils) remains relatively high.

14.12A). In the absence of short-tailed shrews, meadow voles, typically grassland specialists, expand their niches to occupy all available habitats at relatively high densities. Experimental introductions of short-tailed shrews onto islands of the Saint Lawrence River confirmed that these shrews prey heavily on juvenile voles and, if their populations persist, can cause extinction of insular vole populations (Lomolino 1984). Interestingly, short-tailed shrews also exhibit insular niche shifts. On islands where voles are absent, they retreat from meadows to concentrate on more heavily forested sites where earthworms and other alternative prey items are more abundant (Figure 14.12B).

The role of release from predation may be much more important than originally believed. In an elegant field experiment, Schoener and Spiller (1987) tested whether excessive densities of spiders on small Bahamian islands resulted from predatory release. The primary predators of these spiders are lizards (*Anolis* spp. and *Ameiva* spp.). On islands lacking lizards, spider densities were about ten times as high as on islands where those predators were present. Schoener and Spiller tested the predatory release hypothesis by removing lizards from 84 m² enclosed plots and monitoring spider densities on experimental and control plots for just over a year. Spider density rapidly increased in the absence of lizards, reaching a level approximately 2.5 times that of the control plots by the end of the experiment (Figure 14.13).

Schoener and Spiller's study is just one in a long and distinguished series of studies on the ecology and evolution of West Indian lizards. No field biologist can visit a small Caribbean island without being impressed by the incredible abundance of *Anolis*. These small reptiles seem to be everywhere, from the ground to the tops of the tallest trees, from disturbed habitats along roadsides and in cities to pristine native forests. Indeed, these lizards are much more abundant on most islands than they are anywhere on the tropical American mainland. E. E. Williams of Harvard University and his students T. W. Schoener, G. C. Gorman, J. Roughgarden, B. Lister, and R. Holt, and their students in turn, have studied the evolution, ecology, and biogeography of the

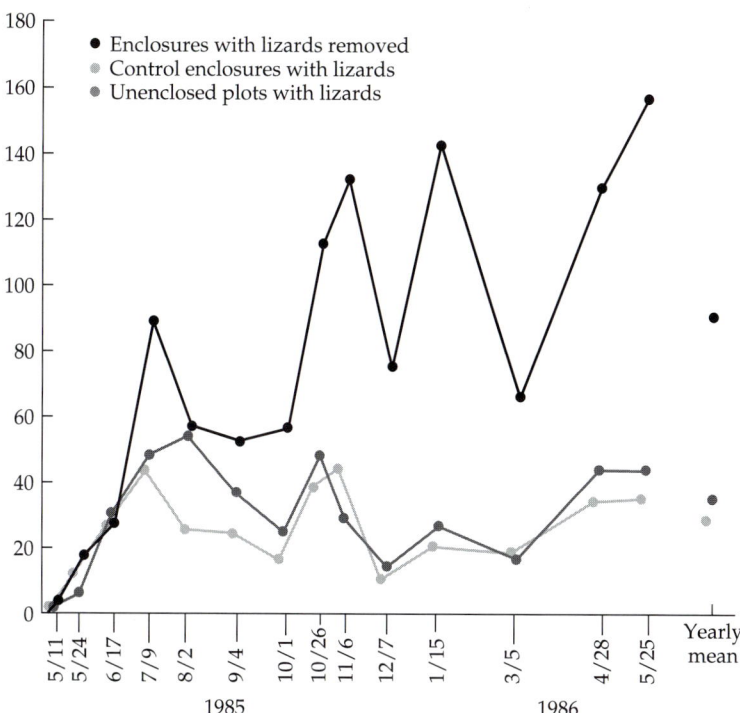

Figure 14.13   On Bahamian islands lacking predatory lizards, spiders often exhibit predatory release, reaching densities about ten times as high as on islands inhabited by these lizards. Experimental removal of lizards from enclosed plots demonstrated that population densities of spiders are strongly influenced by predation. (From Schoener and Spiller 1987.)

Caribbean *Anolis* in great detail. Although many distributional patterns have been well documented, the underlying causal processes have proved more difficult to demonstrate convincingly.

Anolis is but one of several important lizard genera on the mainland of tropical America, but it is by far the most abundant genus of vertebrates on the Caribbean islands. On the large islands of the Greater Antilles, a few colonizing species have given rise by endemic speciation and adaptive radiation to a diverse *Anolis* fauna. Hispaniola, the second largest and ecologically most diverse island, has at least 35 species, which were probably derived from four separate invasions (Williams 1976, 1983). These species occupy a variety of ecological niches. They range from tiny insectivores to large carnivores, with head and body lengths ranging from 40 to more than 200 mm. They are morphologically, physiologically, and behaviorally specialized for distinctive habitats and microenvironments, from sunny sites to deep shade, from open ground and rocks to grasslands and scrub forests to different layers in the complex vegetation of mature tropical and montane forests (Figure 14.14). In contrast, the small islands of the Lesser Antilles each have only one or two generalized species (Roughgarden and Fuentes 1977; Roughgarden et al. 1983). When two species coexist on a small island, they differ in body size, prey size, and habitat; but when only one species is present, it is intermediate in size, takes a wide range of prey, and occupies virtually all habitats (Figure 14.15). Clearly, the fundamental niche of an *Anolis* species that has evolved in the absence of congeners is very broad. Some of the observed niche expansion of *Anolis* on small islands may represent behaviorally mediated ecological responses to the absence of competing species. Nevertheless, most of the niche shifts in Caribbean species represent evolutionary adaptations to communities containing different numbers and kinds of other species, and these are reflected in morphological, physiological, and behavioral changes (see Roughgarden 1995).

sh
Pine zone
f
ri
in
c
e
rm
Montane broadleaf-shade
ri
al
d
cy
se
ri
eu
ch
Montane-open
d
cy
se
Lowland mesic
d/b
wh
o
Lowland arid
Elevation
eu *eugenegrahami*
f *fowleri*
sh *shrevei*
ri *ricordii* group
in *insolitus*
c *christophei*
e *etheridgei*
rm *rimarum*
ch *chlorocyanus*
al *aliniger*
d *distichus*
c *cybotes* group
se *semilineatus*
d/b *distichus/brevirostris*
o *olssoni*
wh *whitemani*
Shade
Sun

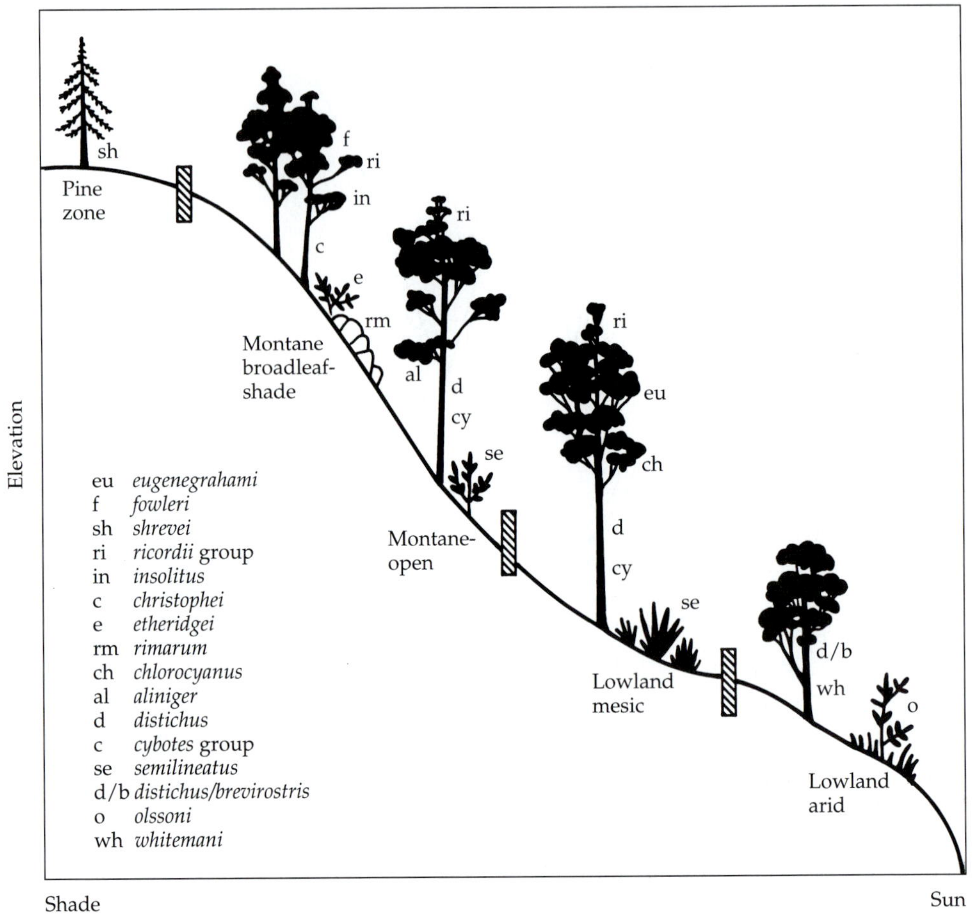

Figure 14.14 Diagrammatic representation of the habitats occupied by different *Anolis* species on the northern part of Hispaniola. Their niches differ in elevation, vegetation type, perch height, and position on a gradient from sunlight to shade. The species indicated here, only a fraction of the at least 35 species that inhabit the island, have been produced largely by speciation and adaptive radiation within the island. (From Williams 1983.)

Jamesby
Petit Tobac
Petit Bateau
Palm
Union
Marie Galante
Saba
St. Croix
St. Kitts
Barbada
St. Maarten
St. Eustatius
Grenada
St. Vincent
8
10
12
14
16
Jaw length (mm), adult males

Variation in *Anolis* densities and niche characteristics between islands and between habitats within islands has also been investigated, especially by Schoener (1968a, 1975), Lister (1976a,b), and Holt (1994). They have shown that when a species occurs on an island or in habitats where there are few other lizards, increases in niche breadth often parallel increases in density (Figure 14.16). Again, it is tempting to attribute these patterns of density compensation and niche expansion to a release from competition with other *Anolis* species, but this assumption may be unwarranted. The small islands where the lizards show the most spectacular increases in density contain not only fewer *Anolis* species, but also fewer species of all terrestrial animals. Thus the *Anolis* populations are potentially released from competition not only with congeners, but also with insectivorous arthropods (e.g., spiders), birds, frogs, and lizards of other genera, and from predation by birds and other lizards. It is also possible that small islands support higher standing stocks of insect prey. Holt studied

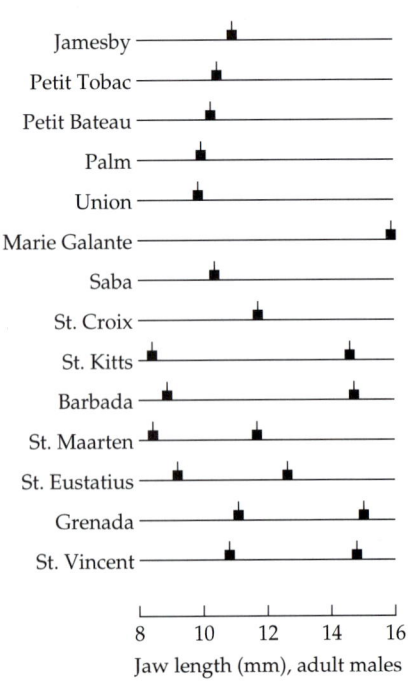

Figure 14.15 Body sizes of *Anolis* lizards of the Lesser Antilles. Note that all of these islands have either one or two species. When two species co-occur, they tend to be displaced in size, whereas when only one species is present, it tends to be of intermediate size. (Data from Roughgarden 1974; Roughgarden and Fuentes 1997; Roughgarden et al. 1983.)

Anolis on islands off Trinidad and concluded that direct competition with other *Anolis* or other lizard species could not account for the observed density changes. He suggested that the absence of predatory birds might be the most important of the various factors contributing to increased *Anolis* density and niche breadth on small islands.

Thus, while ecological release and high densities among insular populations are often attributed to reduced competition, they are more generally a consequence of the depauperate nature and reduced intensity of interspecific interactions in general.

Density overcompensation. One puzzling aspect of insular density compensation is that the total population densities of a few species inhabiting a small island may actually exceed the combined densities of a much greater number of species of the same taxon occupying similar habitats on the mainland. This phenomenon, called **density overcompensation** or **excess density compensation**, is fairly well documented, especially for birds on small oceanic islands. In fact, Crowell's (1962) studies, mentioned above, revealed that just 10 species of small passerine birds on Bermuda together maintained a population not just equal to, but about 1.5 times greater than the combined densities of 20 to 30 species on the mainland. Subsequent studies by MacArthur et al. (1972) on the Pearl Islands south of Panama, by Diamond (1970b, 1975b) on the Bismarcks and other archipelagoes north and east of New Guinea, and by Emlen (1978) on the Bahamas have described similar patterns. Case (1975) documented density overcompensation among the lizards on the islands in the Gulf of California compared with the mainland of Baja California and Sonora.

Several explanations have been proposed for such overcompensation, and for density compensation as well:

1. Because bird species of large body size tend to be absent from small islands, the same resources can support substantially larger populations of small birds. Note, however, that Crowell's (1962) study found that biomass, as well as density, was substantially higher among insular birds than among mainland birds.
2. Bird densities reflect release from competition not only with missing bird species, but also with other taxa, such as mammals and amphibians, which use similar foods and other resources, but are even less common on oceanic islands than birds because they are poorer overwater dispersers.
3. Inflated population densities on islands reflect the absence or paucity of predators and parasites, which also tend to be poorly represented on oceanic islands (MacArthur et al. 1972; Grant 1966; Case 1975; George 1987; McLaughlin and Roughgarden 1989).
4. Oceanic islands are more productive, at least in terms of foods and other resources required by small birds (see Case 1975).
5. On oceanic islands, renewable food resources are harvested at rates closer to their maximum sustained yields than on mainlands, where intense competition leads to overexploitation and lower productivity of resources.
6. Populations can become more finely adapted to their local environment, and hence attain higher densities, on isolated islands than on continents, where extensive gene flow among populations occupying different habitats tends to prevent specialization for more efficient use of local resources (Emlen 1978, 1979).

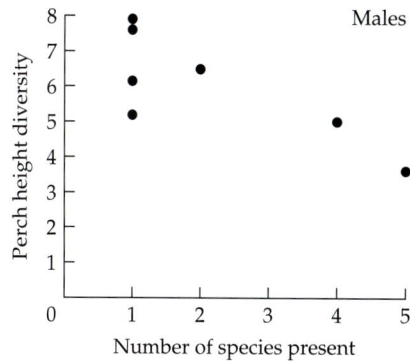

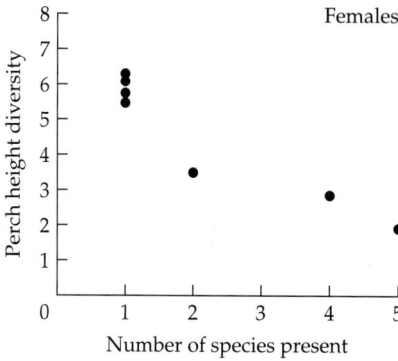

Figure 14.16 Relationship between the diversity of heights of perches used by males and females of *Anolis sangrei* and the number of other *Anolis* species occurring in the same habitat on the same island. Note than when *A. sangrei* is the only species present, it uses a wide variety of perch heights, but the breadth of this niche dimension contracts and *A. sangrei* is excluded from arboreal habitats as it encounters an increasing number of coexisting congeneric species. (After Lister 1976a.)

7. The surrounding waters act as a fence, preventing emigration of individuals into marginal habitats (the **fence effect**; see Krebs et al. 1969; Emlen 1979; MacArthur et al. 1972).

As is often the case with the most general patterns in ecology and biogeography, these patterns may result from a combination of convergent mechanisms, rather than just one unique, overriding one.

Patterns of combined densities of insular species, including compensation and overcompensation, may be summarized in a graphical model (Figure 14.17A; Wright 1980). We again emphasize that, like their mainland counterparts, insular communities often are Gleasonian, with species responding sometimes uniquely to one another and to environmental conditions. That is, while the overall pattern exhibited by a particular insular community may be density compensation, some of its species are likely to exhibit atypically low populations, while densities of others are compensatorily high. For example, Crowell (1962) found that common crows, house sparrows, and starlings were extremely rare in all but farmland habitats on Bermuda, while catbird, cardinal, and white-eyed vireo populations were 3 to 30 times higher in comparison to the same habitats on the mainland. Similarly, in a study of Caribbean

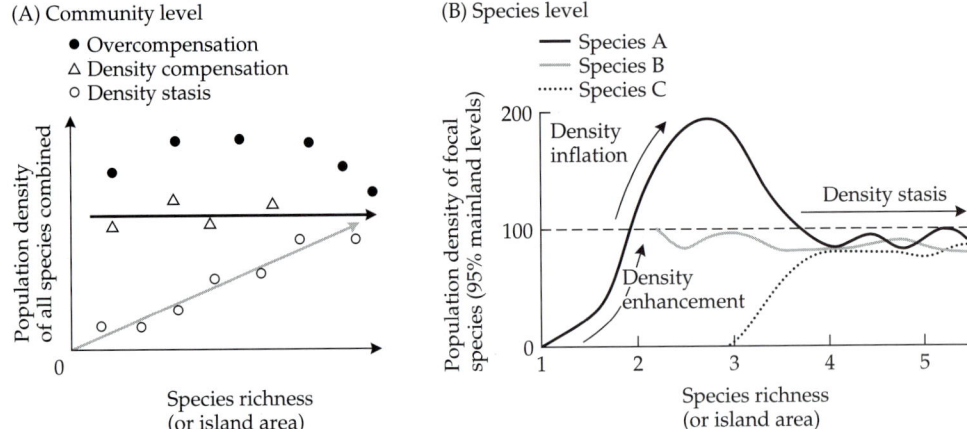

Figure 14.17   (A) Insular population densities (number of individuals per unit of area) may exhibit one of three qualitatively different responses to variations in species richness (or island area). If population densities are not influenced by competition, predation, and other interactions among species, then the densities of all species combined should increase in proportion to the total number of species (density stasis). On the other hand, if interspecific interactions do limit insular populations, they may exhibit density compensation or overcompensation on species-poor islands (i.e., populations of one or a few insular inhabitants may equal or exceed those of a much richer assemblage of species). (B) Population densities of individual species also may exhibit three qualitatively different responses to differences in species richness (or island area) among islands. Density enhancement (*sensu* Jaenike 1978): on very small, species-poor islands, population levels of a species may fall below those of its conspecifics on the mainland, but increase as island area increases and environmental conditions become more suitable. Density inflation: population levels may continue to increase, with increasing area, above their mainland levels, until interspecific interactions begin to regulate population densities of the focal species. Density stasis: at intermediate to high levels of species richness, many species may regulate one another's densities such that few, if any, exhibit any consistent trend of population density with species richness or island area. (A after Wright 1980.)

insects, Janzen (1973) found that generalists, especially the Homoptera, tended to exhibit density compensation on islands, while other insects did not.

A species-based alternative to Wright's model (Figure 14.17B) introduces an additional density pattern termed **density enhancement** (see Jaenike 1978). Population densities of some, and perhaps most, species on very small islands may indeed be much lower than their typical densities of the mainland, but increase with island area. Larger islands not only support more species, some of which may be mutualists or commensals, but also tend to provide less harsh, more stable, and otherwise more favorable habitats for particular species. Alternatively, population densities of at least some species may remain relatively stable over a broad range of species richness—a pattern that Williamson (1981) termed **density stasis**. In fact, the same species may exhibit all three patterns—density enhancement on tiny islands, density inflation on islands of intermediate size, and density stasis on larger, more species-rich islands (see Figure 14.17B, species A). Perhaps density inflation also varies systematically among species: highest for supertramps and lowest for those species restricted to the richest, and largest, islands. We admit that these are largely untested predictions, but the pertinent data seem to be available, at least for some insular communities.

In summary, studies on the ecological responses of insular populations have been, and will certainly continue to be, fertile ground for research on the forces influencing community structure. The comparative approach enables identification of some intriguing patterns, and variation among islands provides the opportunity for natural experiments, while their isolation, small size, and simplicity often enable manipulative experiments in a controlled, but realistic, arena.

Evolutionary Trends on Islands

Flightlessness and Reduced Dispersal Ability on Islands

> As with mariners shipwrecked near a coast, it would have been better for the good swimmers if they had been able to swim still further, whereas it would have been better for the bad swimmers if they had not been able to swim at all and had stuck to the wreck. (Darwin 1859, p. 177)

With this one metaphor, Darwin explained a truly fascinating paradox of insular biotas: many insular forms, especially those on the most isolated oceanic islands, have little or no ability to disperse to other islands. As many have remarked, there is no greater anomaly in nature than a bird that cannot fly (Darwin 1859). Yet, flightless birds and insects are relatively common on many oceanic islands, as are other animals and plants with little ability or propensity for dispersal. And herein lies the paradox: how could these relatively sedentary forms have colonized such remote ecosystems?

During Darwin's day, the extensionists may have used these observations to bolster their argument that ancient landbridges once connected even the most isolated archipelagoes with the continents. As we saw in Chapter 2, nothing "vexed" Darwin more than these post hoc scenarios of the earth's dynamism. Instead, he offered an explanation embodied in the above quote. There was no doubt, at least in Darwin's mind, that these peculiar insular forms were the products of evolution, and that their ancestors had arrived via long-distance dispersal.

Unfortunately, Darwin attributed some of these patterns to what was termed the "law of disuse." For example, he believed "that the nearly wingless condition of several birds, species which now inhabit several oceanic islands, tenanted by no beast of prey, has been caused by disuse." He acknowledged, however, that the flightlessness of many insular forms, such as 200 of the 550 described beetles of Madeira Island, was "wholly, or mainly due to natural selection." The key point is that the selective forces operating on a potential immigrant may be entirely different from those operating on its insular descendants. In Darwin's metaphor, selection during immigration favors "good swimmers," but once they (or their descendants) have reached the "wreck" (island), selection then favors "bad swimmers"—those less likely to swim off or be carried away by winds and lost at sea.

Birds.   Flightlessness and reduced dispersal ability have evolved repeatedly in a variety of birds and insects (see Carlquist 1974). Derived flightlessness is reported in at least eight orders of birds (Struthioniformes, Anseriformes, Psittaciformes, Strigiformes, Columbiformes, Gruiformes, Ciconiiformes, and Passeriformes) and in most oceanic archipelagoes. On New Zealand, some 25 to 35% of the terrestrial and freshwater birds are—or were, in the case of extirpated forms—flightless. Similarly, 24% (20 species) of Hawaii's endemic birds were flightless. The long list of flightless birds includes a cormorant of the Galápagos (*Phalacrocorax harrisi*) and many flightless giants, including the dodo (*Raphus cucullatus*) of Mauritius, the solitaires (*R. solitarius* and *Pezophaps solitariaon*) of Réunion and Rodriguez, the elephant birds (Aepyornithidae) of Madagascar, and the moas (*Pachyornis* spp.) of New Zealand (see review by McNab 1994a). Flightlessness is especially common among the rails, having evolved independently within at least 11 groups of rails (Diamond 1991). In fact, Olson (1973) has speculated that flightlessness in rails may evolve over a time span of decades to centuries, rather than millennia. While this may seem astounding, paleontological evidence suggests that, before human colonization, most Pacific islands were inhabited by a flightless rail species (see Steadman 1986 and 1989; Steadman and Olson 1985).

Currently, the most widely accepted explanation for the evolution of flightlessness in birds involves selective pressures associated with the absence of predators and limited resources on islands (McNab 1994b; Diamond 1991). Under reduced pressure from predators, and for that matter, from competitors as well, insular populations would be likely to undergo ecological release. Evolutionary responses could include changes that would conserve energy, such as reduced size overall, or a reduction in metabolically expensive tissues, such as the otherwise large flight muscles. On the mainland, such selection to conserve energy is of course countered by the selective advantage of being able to escape ground-dwelling predators. On many islands, however, this is not the case, as "reduction in flight muscle brings great energy savings with little penalty" (Diamond 1991). If these selective factors were combined with natural, heritable variation in the mass of flight muscles, their relative size and ability to power flight would atrophy over generations. Thus, at least for the rails, flightlessness was probably achieved primarily by reduction in mass of flight muscle, with perhaps a modest reduction in overall mass as well.

Evolution of flightlessness in other birds is often associated with an increase in body size (Morton 1978; Livezey 1993). This suggests an alternative path to the loss of flight: many insular birds may have simply outgrown their wings. In the absence of mammals and other large terrestrial vertebrates, selective pressures may have promoted increased body size overall, without a compensatory increase in flight muscle mass (see the following section on the evolution of insular body size). The result was moas, dodoes, solitaires, and elephant birds—avian giants no more capable of flight than we are. Indeed, these and other flightless insular birds may have converged on niches left vacant by mammals—shifting their diets, developing more drab, cryptic coloration, and increasing in size so as to eventually lose their power of flight (see Baker et al. 1995; Livezey 1993; Trewick 1996).

Insects and other invertebrates.   The flightless beetles of Madeira Island in the Atlantic Ocean are just one of the countless groups of insular insects whose powers of flight have been lost, or greatly reduced. In addition to Coleoptera, flightlessness has been reported in insular Lepidoptera, Diptera, Hymenoptera, Orthoptera, and Homoptera, on Madeira and in many other archipelagoes. Like many of the avian examples discussed above, the wetas (Orthoptera) of New Zealand may represent another fascinating case of evolutionary convergence on a mammalian niche (Figure 14.18). In the absence of ground-dwelling mammals, wetas have lost the power of flight, increasing in size and occupying the ground-dwelling, or "geophilous," niche otherwise occupied by rodents.

While the evolution of flightlessness in insects may have involved energy conservation (Roff 1986, 1990), it is more likely that other factors were much more important. In contrast to birds, mammals, and other relatively large animals, insects perceive a much more coarse-grained environment. Insects may live out their entire life cycle in just a small patch of a bird's home range. Thus, site fidelity, or what Darlington (1943) termed **precinctiveness**, may be at a premium. Because more distant patches or habitats tend to be more dissimilar than those closer to their natal sites, selection may favor offspring or adults that limit their movements—in Darwin's terms, those that "stick to the wreck." Precinctiveness, however, applies at a much finer scale than in Darwin's metaphor: individuals do not have to be blown off the island to suffer lower fitness, only to a less favorable habitat.

Of course, such advantages are not unique to insects on islands. Many mainland landscapes also offer patchy, yet relatively stable, environments where energy conservation and site fidelity may increase fitness. Accordingly, flightlessness is quite common among insects on the mainland as well as on islands. Flightlessness, or reduced capacity and propensity for flight, also tends to be more common among insects of montane environments than among those living at lower elevations (Figure 14.19; see Roff 1990; Darlington

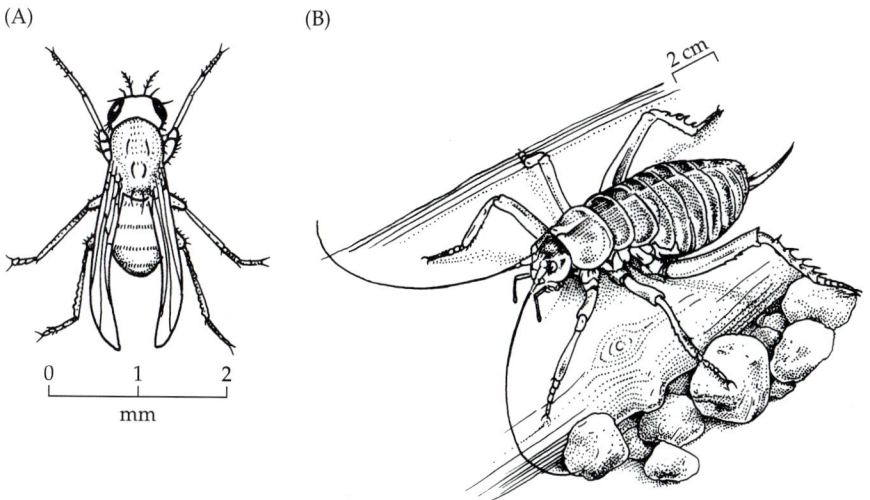

(A)   (B)

0   1   2
mm

2 cm

Figure 14.18   Two examples of flightlessness in insular insects: (A) *Scaptomyza frustulifera*, a drosophilid from Tristan da Cuhna, and (B) the Stephens Island weta, *Deinacrida rugosa*, a flightless orthopteran from Stephens and Mana Islands, Cook Strait, New Zealand. The weta may be an ecological equivalent of rodents. It has been extirpated from the New Zealand mainland and many offshore islands (which lacked native nonvolant mammals) by introduced rats and other ground-dwelling mammals. (A after Williamson 1981; B after Collins and Thomas 1991.)

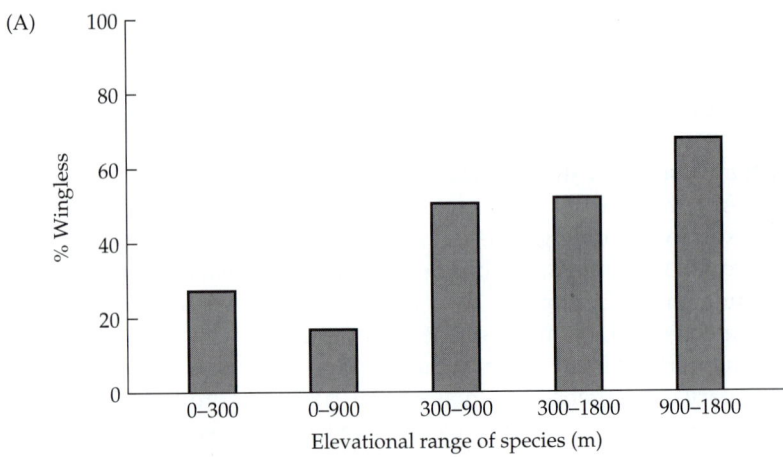

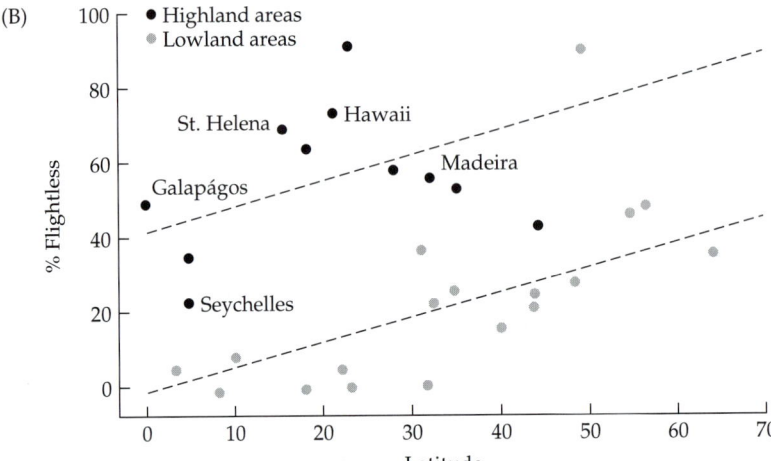

Figure 14.19 Flightlessness in insects is a common phenomenon among both insular and mainland forms, and tends to be more common in mountainous regions and at higher latitudes. (A)Wingless coleopterans of Madeira Island. (B) Flightless carabids on oceanic islands. (After Roff 1990.)

1943; Barlow 1994; Hawkins and Lawton 1995). Again, the characteristics of biotas inhabiting these habitat islands seem to be convergent with those inhabiting true islands.

Other animals and plants.   Land snails, while of course never achieving the power of flight, demonstrate sometimes astounding abilities to disperse great distances, especially the smaller species. As we mentioned in Chapter 9, ability to be dispersed by wind is inversely correlated with body size. As a result, oceanic islands tend to be colonized by a highly disproportionate number of microsnails (Vagvolyi 1975; see also Zimmerman 1948). Following colonization, however, many of these forms have evolved larger body sizes, with a concomitant decrease in their dispersal abilities.

Reduced dispersal ability also is remarkably common in insular plants. In his comprehensive summary on this subject, Carlquist (1974) wrote the following:

> While working with evolutionary phenomena in the Hawaiian Islands and other Polynesian islands, I observed types of fruits and seeds that seemed incongruously poorly adapted at dispersal. The floras of these islands must have arrived by long-distance dispersal, yet during evolution on the island areas various groups of plants have lost their dispersibility.

Many of the changes that result in reduced dispersal ability in plants are associated with shifts from adaptations for colonizing habitats along the beachhead to traits better suited for living in interior, mesic forests. Precinctiveness appears to be just as important for insular plants as it is for insular insects. Accordingly, both seeds and fruits tend to become heavier, less buoyant, and less resistant to seawater, and structures that facilitate air lift or attachment to animals (spines, wings, and hooks) are greatly reduced (Figure 14.20; Carlquist 1974). Morphological changes associated with reduced dis-

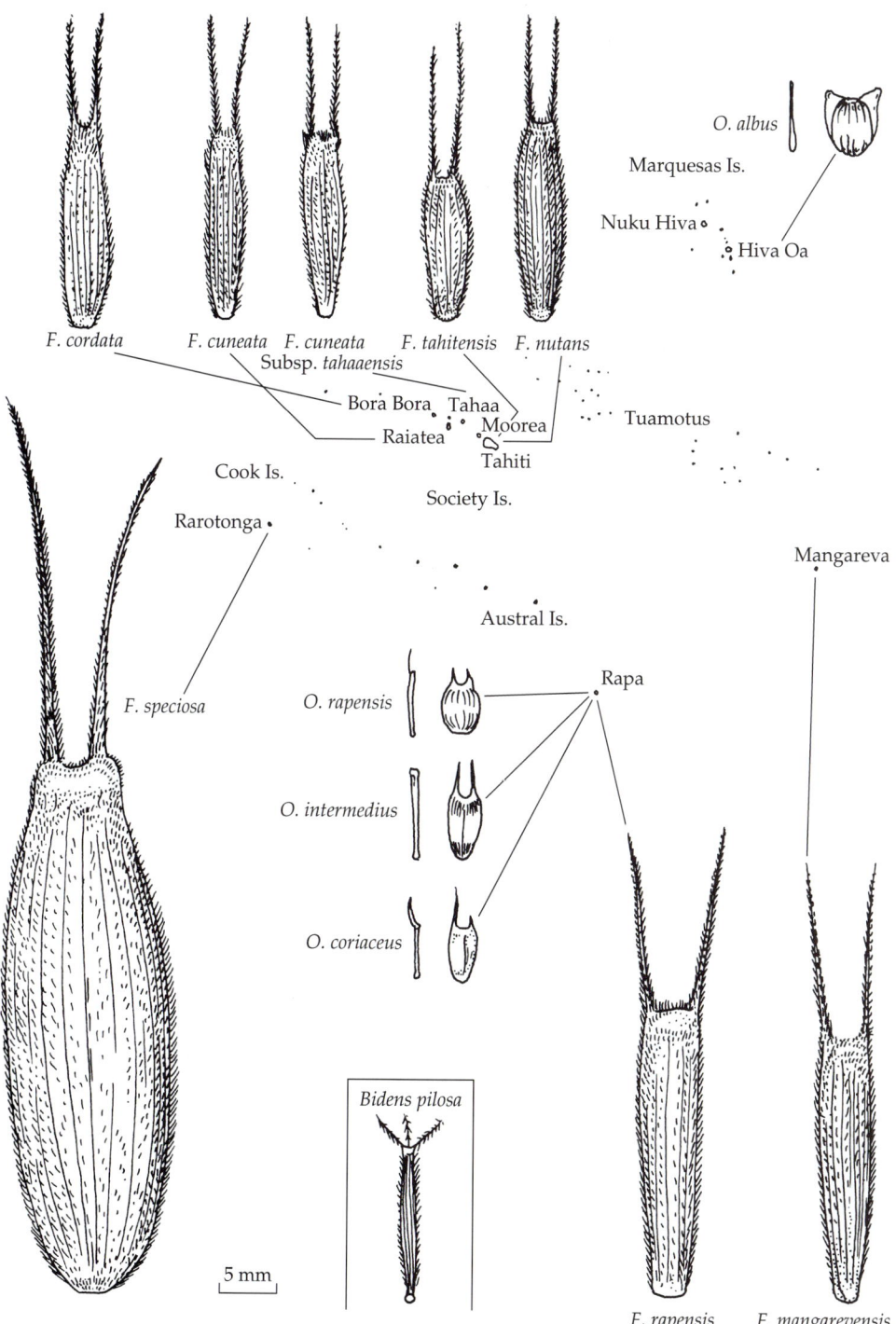

Figure 14.20 On islands, the seeds and fruits of many plant species exhibit traits reflecting reduced dispersal ability, including fruits and seeds that are heavier, less buoyant, and less resistant to seawater, and reduction in spines, wings, and structures that would otherwise facilitate lift or attachment to animal dispersers. Fruits of insular forms of *Fitchia* and *Oparanthus* (Asteraceae) are shown; *Bidens pilosa*, which exhibits a more typical mainland morphology, is shown in the inset for comparison.

persal ability in insular plants may evolve very rapidly, perhaps in less than a decade (see Cody and Overton 1996).

Many species of plants that are typically herbaceous on the mainland evolve woodiness and large size on isolated islands (Darwin 1859:392), resulting, for example, in sunflowers the size of oak trees. This pattern is as widespread as it is impressive, occurring on islands throughout Polynesia and Melanesia and on New Zealand, Juan Fernández, the Desventuradas, the Galápagos, the California offshore islands, the Canary Islands, Madeira, the Cape Verde Islands, the Azores, the West Indies, Madagascar, the Comoros, and Saint Helena (see Carlquist 1974). The cause of such evolutionary novelties in insular plants is probably complex, but may stem from the disharmonic nature of insular biotas. In comparison to those of large trees and other woody plants, the propagules of small herbaceous plants are typically more easily transported by winds and water currents. Thus, isolated islands are colonized by a highly disproportionate number of small herbaceous species. Just as important, isolated islands often lack large herbivorous mammals, including browsers and folivores, which tend to feed most heavily on woody plants. Under these conditions, herbaceous plants often undergo ecological release and evolve to occupy the large woody plant niche left vacant by their mainland competitors. As we shall see in the following section, these types of responses to the depauperate and disharmonic nature of insular communities are common among animals as well as plants.

Finally, many cases of reduced dispersal ability among insular plants and insects may result from a type of ecological lottery that characterizes most islands. Given that islands are depauperate by nature, plants that depend on particular animals for dispersal (exozoochory), or animals that depend on larger animals for dispersal (phoresy), may find themselves stranded. Luck may still be with them, however, if they can adapt and find a niche somewhere on the "wreck" and stick to it.

Evolution of Body Size on Islands

Flightlessness, of course, is not the only way in which a species can respond to limited resources and a paucity of interspecific interactions. As we have mentioned above, those species that adapt to these insular conditions and "stick to the wreck" often undergo remarkable changes in body size (Figure 14.21). These seemingly divergent changes, ranging from dwarfism in elephants to gigantism in earwigs, seem as chaotic as they are fantastic. But there is order to this variation. Gigantism and dwarfism may simply reflect differences in selective pressures among different species.

Many insular species exhibit trends toward larger size. In the absence of competitors and predators, selective pressures shift so as to favor intraspecific competition and exploitation of a greater breadth of resources. Under these conditions, large size has a number of selective advantages:

1. Larger individuals can exploit a greater diversity of resources. Larger predators can feed on large as well as small prey; larger squirrels and other granivores can crack large as well as small nuts.
2. Given that they can acquire sufficient resources to survive and reproduce, larger individuals can produce larger litters or clutches.
3. Although larger individuals require more resources, they also tend to dominate in territorial battles or other intraspecific interactions associated with competing for those resources.
4. Finally, larger individuals have relatively greater energy and water reserves and, therefore, greater ability to withstand famine or drought.

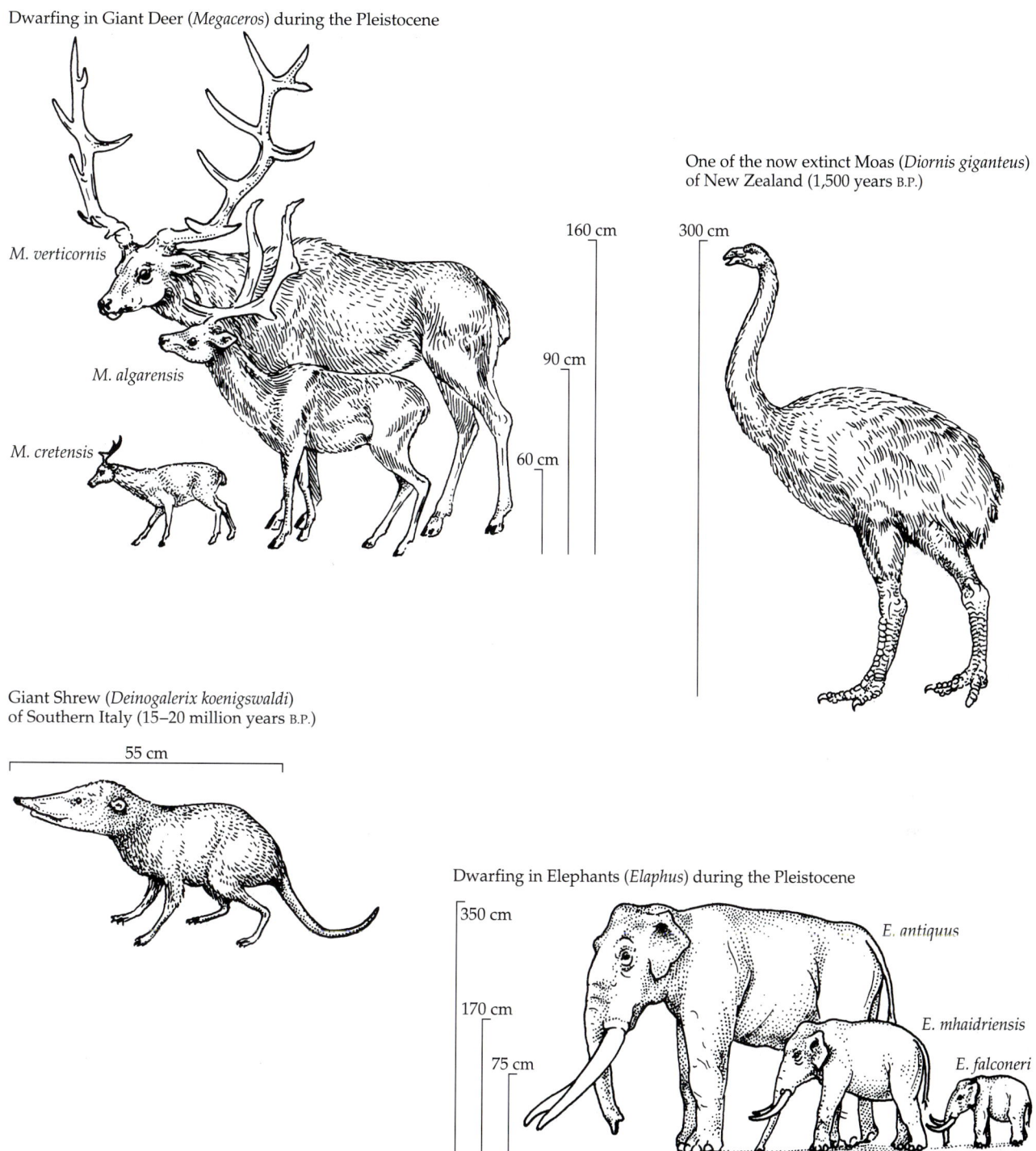

Dwarfing in Giant Deer (*Megaceros*) during the Pleistocene

M. verticornis

M. algarensis

M. cretensis

160 cm

90 cm

60 cm

One of the now extinct Moas (*Diornis giganteus*) of New Zealand (1,500 years B.P.)

300 cm

Giant Shrew (*Deinogalerix koenigswaldi*) of Southern Italy (15–20 million years B.P.)

55 cm

Dwarfing in Elephants (*Elaphus*) during the Pleistocene

350 cm

170 cm

75 cm

E. antiquus

E. mhaidriensis

E. falconeri

Figure 14.21 Insular forms of mammals often exhibit markedly divergent morphological trends, ranging from dwarfism in typically large mammals such as elephants and ungulates to gigantism in shrews, mice, and many birds.

Given all these advantages of being large, why are not all insular forms giants? Body size on islands—or for that matter, body size in general—represents a compromise between benefits and costs. Being small on islands also has its advantages:

1. Smaller individuals require fewer resources to survive and reproduce. This should be especially important on small islands where resources are in short supply.
2. Smaller individuals, because they require fewer resources, tend to be more specialized and more efficient at assimilating nutrients and energy.
3. Smaller individuals can exploit small shelters and refugia from predators, and avoid stressful environmental conditions, that larger forms cannot.

In addition to the above selective pressures—ecological release (promoting gigantism) and resource limitation (promoting dwarfism)—insular body size also may be influenced by selection for better immigrators: in Darwin's terms, selection for the good swimmers for the ability to swim still farther. If body size is somehow associated with immigration ability, then the founders of an insular population may represent a biased subset of the mainland population. For example, among active immigrators, larger individuals should have higher vagilities, and founding populations may be made up of relatively large individuals. On the other hand, we might expect just the opposite in passive immigrators, in which small individuals are more easily dispersed by wind or water. The effects of immigrant selection can persist, however, only if the trait in question—body size—has a genetic basis (see Falconer 1960; Roth and Klein 1986). Immigrations also must be frequent enough to counter the effects of genetic drift and opposing postcolonization selection, or these selective forces must be weak or convergent with the effects of immigrant selection.

Given that insular populations may be influenced by a combination of selective forces favoring gigantism or dwarfism, we might well expect trends in insular body size to be chaotic, lacking any recognizable pattern across taxonomic groups or trophic categories. As we will see, however, insular faunas often exhibit some remarkably consistent patterns of variation in body size, suggesting that at least one of the above selective forces varies in some systematic way among species.

Insular mammals. "Pygmy mammoth" is, of course, an oxymoron, but one familiar to most evolutionary biologists and, indeed, to many of the lay public as well. Paleontological studies have documented the ability of these now extinct relatives of today's elephants to colonize and adapt to island environments. The fossil record reveals that these species were once widespread throughout Malaysia and on selected islands of the Mediterranean, the California Channel, and the Arctic Ocean (see Roth 1990; Sondaar 1977; Hooijer 1976).

One of the most remarkable cases of insular dwarfism was recently described by Vartanyan and his colleagues (see Vartanyan et al. 1995; Martin 1995). Although the woolly mammoth (*Mammuthus primigenius*) was widespread across Siberia and northern Europe during the last Ice Age, its geographic range contracted rapidly during the onset of the current interglacial. Its demise may have resulted from the contraction of its primary habitat, steppe/grasslands, and hunting by aboriginal humans, who expanded their ranges northward as the glaciers receded. The mammoth's range contracted northward during the glacial recession (20,000 to 10,000 B.P.) until it survived on just one small refuge, Wrangel Island, which lies in the East Siberian Sea just north of the Arctic Circle. Here, provided with remnants of its primary habitat and some isolation from its human predators, mammoths persisted well into the current interglacial, perhaps as recently as 2000 B.P.

An additional key to the persistence of mammoths on Wrangel Island was a marked and relatively rapid development of dwarfism [see also Lister's (1989) example of rapid development of dwarfed forms of the red deer on Jersey Island during the previous interglacial]. Just as their relatives became dwarfed on other islands, the Wrangel Island mammoths decreased in size from 6 tons to just 2 tons over roughly 5000 years. These changes no doubt reduced the resource requirements of the animals, effectively increasing their carrying capacity and prolonging their persistence by many centuries, if not millennia.

While mammoths and some other large mammals tend to be dwarfed on islands, many small mammals exhibit gigantism on islands. In 1964, J. Bristol Foster published an important synthesis that revealed that these patterns were in fact features of a very general pattern, which later became known as the **island rule**. According to Foster (1963, 1964), different groups of mammals tend to exhibit different trends in insular body size. Carnivores, including canids and felids, exhibit dwarfism, while rodents tend to exhibit gigantism (Table 14.6). Rather than representing any taxonomic bias, however, the island rule can be more simply described as a graded trend from gigantism in smaller species to dwarfism in larger species. This pattern is evident within as well as among taxonomic groups (Figure 14.22A), and therefore may reflect predictable changes in the relative importance of selective forces influencing body size on islands (Figure 14.22B).

Small mammals are relatively poor immigrators. In addition, they are more likely to be the smaller members of a guild of species, or prey that escape predators by seeking the protection of small refugia (see McNab 1971; F. Smith 1992). In either case, these relatively small species tend to increase in size in the absence of predators and larger competitors. Thus, immigrant selection and ecological release, selective forces favoring gigantism, should be most important for the smaller species of mammals. On the other hand, larger species, such as mammoths, deer, and wolves, require more energy to survive and reproduce. Thus, resource limitation and concomitant selection for dwarfism should be most important for these species. The combined effects of these selective forces, which vary in intensity with the body size of the species, result in the graded trend in insular body size illustrated in Figure 14.22.

Some biogeographers have scrutinized an additional feature of this pattern. If the island rule results from a balance between forces promoting gigantism in the smaller species and dwarfism in the larger species of insular mammals, where is the fulcrum? That is, is there an "optimal" insular body size—above

Table 14.6

Taxonomic patterns in the relative sizes of insular mammal subspecies as compared with their relatives on the mainland

	Number of Subspecies		
	Smaller	Same	Larger
Marsupials	0	1	3
Insectivores	4	4	1
Lagomorphs	6	1	1
Rodents	6	3	60
Carnivores	13	1	1
Artiodactyls	9	2	0

Source: Foster 1964.

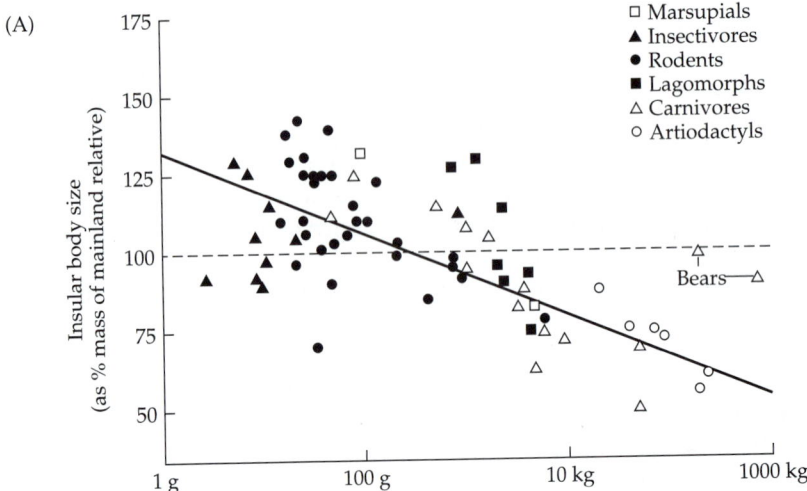

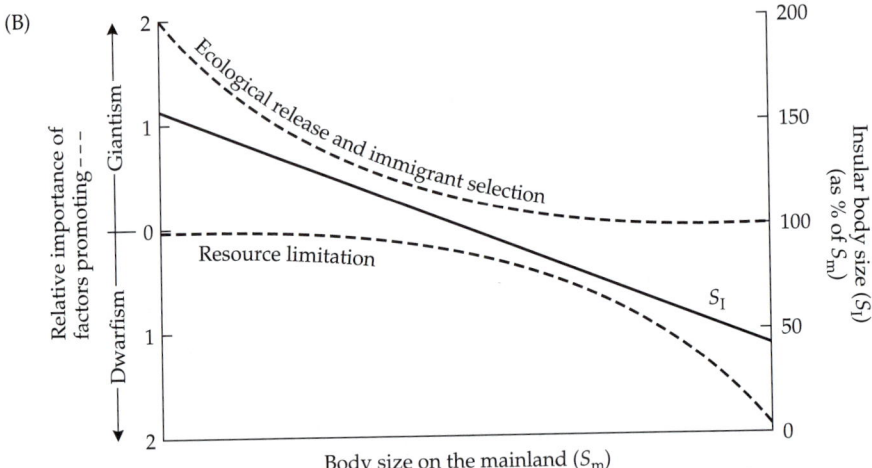

Figure 14.22 (A) The island rule refers to the graded trend from gigantism in small species of mammals (e.g., mice and voles) to dwarfism in large species (e.g., canids, ungulates, and elephants). (B) This pattern may result from the combined effects of ecological release and selection for better immigrants, which tends to select for gigantism in small mammals, and resource limitation, which tends to select for dwarfism, especially in larger species. Solid line represent trend (regression line) for mammals in general. S_I = insular body mass, expressed as a percentage of the body mass of that species on the mainland; S_m = mainland body mass. (After Lomolino 1985.)

which species tend to decrease, and below which they tend to increase in size? In fact, the hypothesized insular fulcrum can be estimated empirically by simple linear regression. Based on this analysis, the so-called optimal size for insular mammals is approximately 250 grams, roughly the size of a red squirrel (see Maurer et al. 1992). Brian Maurer and his colleagues have argued that optimal body size is actually a function of land area. Bolstering their argument, they report that the median body size of land mammals ranges from 85 g for species in North American to 216 g in Australia and 231 g in Madagascar, a trend converging on the hypothetical optimal size for smaller islands (Maurer et al. 1992).

Like most other biogeographic patterns, these observed trends in insular body size in mammals leave us with many intriguing questions. While the relatively simple model shown in Figure 14.22B successfully accounts for the island rule, it remains largely untested. How much of the residual variation in Figure 14.22A can be accounted for by factors in addition to the body size of the ancestral species? The model predicts, for example, that the relative body sizes of insular mammals should increase with island area and isolation, and that they should be higher for herbivores than for carnivores. Consistent with the former prediction, the body sizes of tri-colored squirrels and fruit bats of

Southeast Asia increase with island area (Figure 14.23; Heaney 1978; McNab 1994b; see also Krzanowski 1967). In addition, the body sizes of at least six species of small North American mammals tend to increase with island isolation (Figure 14.24). Wrens of the British and Scottish Isles also appear to exhibit a similar trend of increasing body size with isolation (Berry 1964). The potential influence of diet on the evolution of insular body size is suggested by the relative sizes of lagomorphs (rabbits and hares) and artiodactyls (deer and related species), which tend to fall above the general trend in Figure 14.22A, while insular canids and felids tend to fall below the general trend. Similarly, rodents exhibit stronger tendencies toward gigantism than do insectivores (shrews and moles). On the other hand, insectivores seem anomalous in that they fail to exhibit the expected graded trend from gigantism to dwarfism. Bears also appear to be exceptional in that they exhibit only a modest degree of dwarfism despite their relatively large size and carnivorous habits (see rightmost triangles in Figure 14.22A; see also Gordon 1986).

In summary, nonvolant mammals exhibit a very regular pattern in insular body size, but one with some perplexing outliers and exceptions. In addition, secondary patterns, including the relationships between body size and island area and isolation, merit much more comprehensive and rigorous tests to evaluate their validity. Perhaps the most promising strategy for future studies will be to focus on the exceptions and on groups with dramatically different vagilities and resource requirements. Bats, for example, have much higher vagilities and greater resource requirements than nonvolant mammals of similar mass. Thus, bats should be less influenced by immigrant selection and more strongly affected by resource limitation. Consistent with this prediction, a review by Krzanowski, published some three decades ago, indicated that bats tend to be dwarfed on islands (74 cases of relatively small size versus 32 cases of relatively large size: Krzanowski 1967). Finally, just like their insular counterparts, rodents on isolated mountaintops of Europe seem to exhibit gigantism (Carlquist 1974; Zimmermann 1950), but again, more rigorous tests are needed to determine whether the island rule applies to montane mammals. The generality of the island rule and its corollaries, including the size-isolation and size-area relationships, the effects of trophic strategies, and parallel patterns for mountaintops and other isolated ecosystems, remain promising areas for future studies (see Box 14.1).

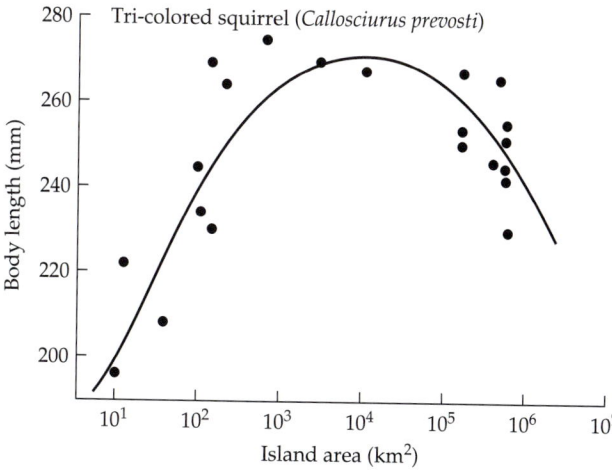

Figure 14.23 Mean body size of subspecies of the tri-colored squirrel (*Callosciurus prevosti*) as a function of island area. (After Heaney 1978.)

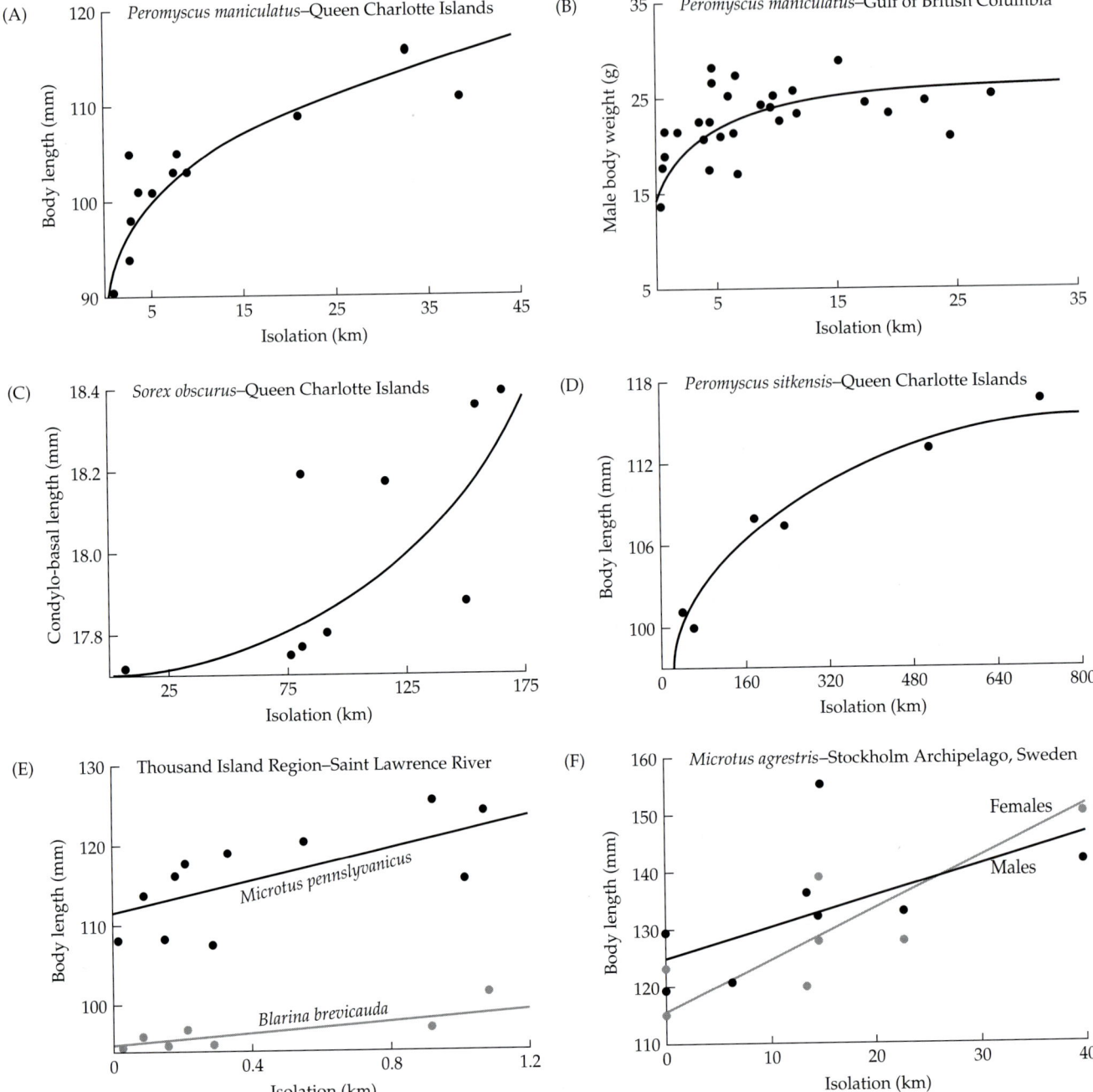

Figure 14.24   The body sizes of insular forms of a number of small mammals tend to increase with island isolation. (A–E after Lomolino 1983, 1984; F after Ebenhard 1988.)

Other insular vertebrates.   In comparison to other insular faunas, birds and reptiles represent unrivaled subjects for ecological and evolutionary studies. In comparison to mammals, birds and reptiles are more widely distributed on islands, and are generally easier to observe and collect. Some studies suggest that insular populations of birds and reptiles exhibit trends in body size similar to those seen in mammals. For example, like tri-colored squirrels and fruit bats, rails and ducks tend to increase in body size on larger islands, suggesting

Box 14.1
Time dwarfing on the "island continent" of Australia

Evolutionary trends in the divergence of insular forms from their mainland ancestors in body size, such as those illustrated in Figure 14.22 and summarized in Table 14.7, are inferred from comparisons of the respective forms in a snapshot in time. Inferences regarding these evolutionary patterns are based on the assumption that insular and mainland forms shared a common ancestor, but then became isolated by some vicariant or dispersal event after which the populations were free to evolve in isolation and respond to the different selective pressures of their respective environments (i.e., mainland or insular).

Occasionally, paleobiologists can piece together a more comprehensive picture of evolutionary trends in the morphological characteristics of a lineage through time. Perhaps most notable among these trends is Cope's rule, which refers to the tendency for body size to increase during the early evolutionary history of a lineage. In a recent and truly intriguing account of evolutionary trends in Australian mammals, Tim Flannery (1994) reported that, like their counterparts on the larger continents, Australian mammals tended to increase in body size during the past 15 million years. Flannery, however, also reported a startling reversal of this trend during the late Pleistocene. During the last 40,000 years, body sizes of most large Australian marsupials (i.e., > 5 kg) have actually decreased. This phenomenon, labeled **time dwarfing** by Flannery, appears to be a very general one: the very few apparent exceptions include just three extant species of wombats (*Lasiorhinus krefftii*, *L. latifrons*, and *Vomatus ursinus*) and possibly humans. Flannery also discovered that the degree of dwarfing of Australian marsupials during the late Pleistocene was greatest for the largest forms (see the figure on this page). Species less than 5 kg in mass failed to exhibit any detectable degree of dwarfism, while those ranging in size from 5 to 300 kg exhibited a graded trend in dwarfism much like the trend for insular mammals illustrated in Figure 14.22. The largest of Australia's extant mammals, red and grey kangaroos (*Macropus rufus* and *M. giganteus*), are quite small in comparison to their late Pleistocene ancestors, perhaps just 70 to 75% of their body length and roughly half their mass. The largest of Australia's marsupials, including a diverse assemblage of species ranging from the hippolike diprotodons, weighing in at nearly 2000 kg, to marsupial rhinos, marsupial lions, short-faced kangaroos, five wombats, and seven kangaroos, remained relatively unchanged in body size, but suffered extinction (see the figure).

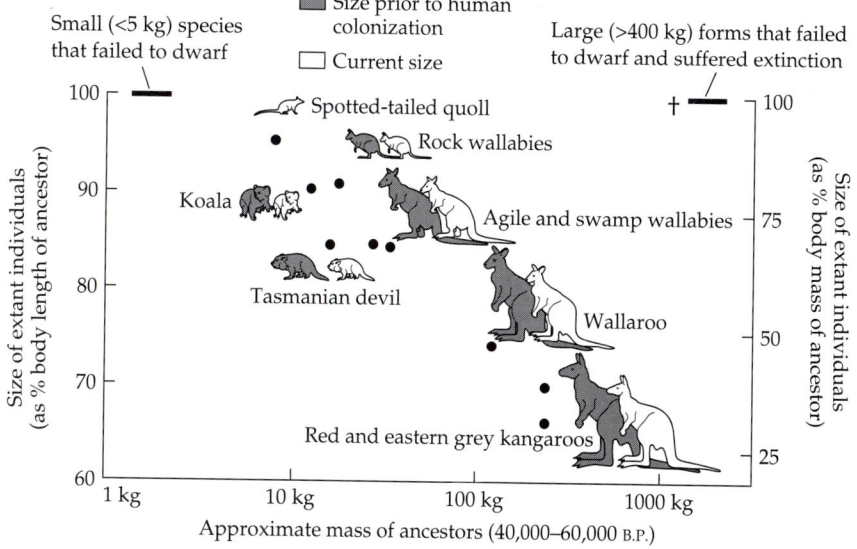

Explanations for the time dwarfing of Australia's marsupials are the same as those proposed to account for the megafaunal extinctions during the Pleistocene: climatic change and overkill by humans (see Chapter 7). Proponents of the climatic change hypothesis argue that time dwarfing is a response to a presumed reduction in forage quality, which may have been associated with climatic change during the most recent glacial recession. Yet the quality of forage for large herbivorous marsupials began to decline as early as 15 million years B.P., when most regions of Australia became increasingly arid. The body size of Australia's largest herbivores, the diprotodons, actually *increased* during this period. Further, opponents of the climatic change hypothesis argue that this explanation ignores the fact that there were some 17 glacial cycles during the past 2 million years. Why should marsupials start to dwarf just during the last one, and why just those weighing 5 to 300 kg?

Flannery and many other paleoecologists have suggested that time dwarfing may have resulted from hunting by aboriginal humans, which appears to have concentrated on the larger species (those > 5 kg) and may have culled the largest individuals. The onset of time dwarfing in Australia is coincident with the arrival of humans, roughly 40,000 to 60,000 B.P. According to this version of the overkill hypothesis, aboriginal hunting pressures either resulted in extinctions or, by decreasing the fitness of relatively large individuals, caused the dwarfism of surviving species.

The intriguing pattern of time dwarfing serves as another compelling case for the need to explore the deep history of biogeographic patterns. The existing assemblage of Australian mammals is in many ways not representative of those that occurred there prior to the arrival of humans. Inferences based solely on the limited diversity and relatively small size of the extant species may allow only an incomplete, and possibly misleading, view of the forces influencing the geographic variation of nature.

that resource limitation may influence insular body size in these species (McNab 1994a,b; Weller 1980). In 1966, Michael Soulé reported that the average body size among insular populations of the lizard *Uta stansburiana* increases as the number of resident competitor species decreases (Figure 14.25), suggesting that, just as has been postulated for insular rodents, these lizards tend to exhibit gigantism in the absence of competitors (see also Case and Bolger 1991).

In addition, different families and orders exhibit different evolutionary tendencies, much as Foster first described for mammals. Gigantism, for example, is common in insular iguanids, herbivorous lizards, whiptails, and tiger snakes, while rattlesnakes, rails, and ducks tend to be dwarfed on islands. The data presented in Table 14.7 suggest that insular birds may follow a pattern similar to the island rule, with smaller species—namely, wrens—exhibiting gigantism while larger forms, especially ducks and rails, tend to exhibit dwarfism. Yet this pattern, if indeed it is valid, is far less general for birds and reptiles than it is for mammals. Just as troubling, various authors have reported different trends for the same group of species (see Table 14.7). These difficulties may derive, paradoxically, from the wealth of information available for insular reptiles and birds. Measurements of body size have been taken from populations over a great variety of islands, which differ substantially in area, isolation, climate, and other factors that may influence the evolution of insular body size. Moreover, these islands are often much more isolated than those inhabited by mammals, thus making it more difficult to identify the source populations suitable for comparative studies.

To understand this latter difficulty, consider the case of the Wrangel Island mammoths. Compared with almost all other mammals—except, of course, their Pleistocene ancestors—they seemed gigantic. Yet we know that they underwent substantial and relatively rapid dwarfism during the early Holocene. Now consider the extant reptilian "mammoths." The land tortoises of the Galápagos in the Pacific and Aldabra Island in the western Indian Ocean are indeed the largest extant tortoises, but are they larger than their ancestors? As recently as 500 years ago, "giant" tortoises occurred on at least 20 islands in the Indian Ocean, including Madagascar, which harbored two species. According to Arnold (1979), the supposed ancestor of the Aldabra Island tortoise is a Madagascan species that was at least 115 cm in length, larger than the biggest Aldabra species, which was 105 cm. One Madagascan species reached 122 cm in body length, rivaling the extant giant tortoises, while the two species of Rodriguez Island were just 85 cm and 42 cm. As Williamson (1981) observed, we cannot be certain whether large size evolved

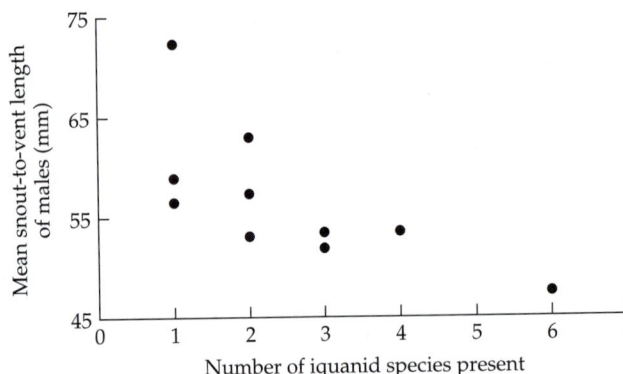

Figure 14.25   The lizard *Uta stansburiana* tends to exhibit increased body sizes on islands lacking its mainland competitors. (After Soulé 1966.)

Table 14.7
Some apparent trends in insular body size in reptiles and birds

Taxon	Apparent trend	Reference
Reptiles		
Lizards		
Most herbivorous lizards	Gigantism	Case 1978
Iguanids	Gigantism	Case 1978
Teiids (whiptails)	Gigantism	Case 1978
Lacertids (geckos)	Gigantism	Case and Bolger 1991
	Dwarfism	Case 1978
Varanids (monitor lizards)	Dwarfism	Mertens 1934
	Gigantism	Case and Schwaner 1993
Snakes		
Tiger snakes	Gigantism	Schwaner and Sarre 1990
Crotalids (rattlesnakes)	Dwarfism	Case 1978
Tortoises	Gigantism	Hooijer 1951
	Dwarfism	Arnold 1979
Birds		
Wrens	Gigantism	Williamson 1981
Fruit pigeons	No changes, or occasional gigantism	McNab 1994a,b
Rails	Dwarfism	
Ducks	Dwarfism	Weller 1980
Ratites	?Gigantism	
	Dwarfism	Wallace 1857; Greenway 1967
Raptors	Occasional gigantism	Azzaroli 1982

after the islands were colonized, or whether it simply made dispersal easier for the initial colonist from Madagascar.

In an analogous fashion, cassowaries are large by most avian standards, but as Wallace noted in 1857, the New Guinea form (*Casuarius bennetti*) is small compared with its relative on the Australian mainland (*C. casuarinus*; body lengths of 52 versus 65 inches, respectively). Emus (*Dromaeius novaehollandiae*) of the small islands of Bass Strait are much smaller than those of Tasmania and the Australian mainland (Greenway 1967). Even the monitor lizards, including the famous Komodo dragon, have proved to be a perplexing case study in insular evolution. The Komodo is, after all, the world's largest living lizard, measuring up to 3 m in length and tipping the scales at up to 150 kg (Auffenberg 1981). Even when considering just extant forms, however, the trend for monitors is equivocal, with some insular forms tending toward gigantism while others appear to be dwarfs (see Table 14.7). Yet all these forms are tiny when compared with a monitor that inhabited Australia during the Pleistocene—a 7 m, 600 kg behemoth (Rich 1985). Until sound phylogenies are available, we must admit the possibility that the Indonesian "dragons" are actually insular dwarfs!

The above difficulties, while challenging, are probably not insurmountable. The influence of island characteristics on trends in insular body size can be statistically controlled, so that we can better focus on the influence of differences among species (e.g., trophic category or size of presumed ancestor). In addition, phylogenetic analyses can be used to estimate the most likely ancestral state—in this case, the relative size of the immediate mainland ancestor (see Pianka 1995; Miles and Dunham 1996).

There are, however, some other reasons to expect reptiles and birds to differ from mammals. Granted, reptiles and birds must respond to some of the same selective pressures that shaped the evolution of insular mammals—especially those associated with isolation, ecological release, and resource limitation. Yet there remain some important differences among these groups of vertebrates. Because they are ectothermic, reptiles typically require less than one-tenth the resources needed to maintain similar-sized endotherms (i.e., birds and mammals; see McNab, 1994a,b). Thus, resource limitation may be relatively unimportant for reptiles, except for larger forms and those restricted to relatively small islands.

It is likely that the evolution of insular body size in reptiles and birds is strongly influenced by the disharmonic nature of insular communities—specifically, the absence or paucity of medium-sized to large mammals. With the large herbivore and carnivore niches vacant, birds and reptiles have often evolved features convergent with those of the absent mammals, including large size. For example, several scientists have suggested that the giant monitor of Australia was an ecological equivalent of saber-toothed cats (Pianka 1995; Auffenberg 1981; Losos and Greene 1988).

Considering the various factors discussed above, including the relatively high diversity and broad distributions of insular reptiles and birds, the low energy requirements of reptiles, and insular conditions favoring ecological release and evolutionary convergence on mammalian niches, it is not surprising that body size trends in these groups seem so complex. Despite all these difficulties—or, rather, because we can identify them—we are optimistic that an interpretable signal, perhaps one rivaling the island rule, will emerge from more comprehensive studies of birds, reptiles, and other insular biotas.

While the research program we outline is ambitious, it may well provide us with a better understanding of a long-standing and fascinating evolutionary problem. Case's (1978) earlier work and Pianka's (1995) and McNab's (1994a,b) more recent syntheses are excellent examples of this approach. Finally, as ambitious as this call for a more general theory of insular body size seems, it pales in comparison with the theme of the concluding section of this chapter, the taxon cycle, which attempts to explain many of the ecological and evolutionary phenomena discussed in this chapter.

The Taxon Cycle

Oceanic barriers severely limit the immigration of propagules, but those that manage to gain a foothold have a high probability of success. This infrequent, but continual, establishment of colonists drives a pattern of ecological and evolutionary change in island biotas that E. O. Wilson (1959, 1961) termed the **taxon cycle**. Insular species evolve through a series of stages from newly arrived colonists, indistinguishable from their mainland relatives, to highly differentiated endemics, which ultimately become extinct (Table 14.8). Although this process can be prevented by sufficient gene flow via immigrants to prevent insular differentiation, it is rightly termed a "cycle" because once insular populations begin to differentiate and adapt to island life, they appear to be doomed to extinction and replacement by new colonists from the mainland.

When Wilson originally described the taxon cycle, he described what appeared to be stages in the expansion and taxonomic differentiation of ants on the islands of Melanesia, north and east of New Guinea in the western Pacific (see Figure 14.5A). He also described successive changes in the niches of these ants as they evolved through the taxon cycle. Those species that are

Table 14.8
Stages of the taxon cycle

Stage I The initial stage of colonization and establishment of populations of a species across an archipelago before its insular populations have differentiated from one another or from the source population on the mainland. In this stage, the species has a relatively continuous range across the archipelago, but exhibits little geographic variation. Such "invaders" are often broad-niched species from marginal habitats on the mainland, and are therefore preadapted for marginal habitats along beachfronts and other habitats occupying the island's periphery.

Stage II The insular populations have differentiated to the point at which they may represent endemic subspecies, or even species. The populations have invaded and become adapted to habitats within the island's interior, and often exhibit associated changes such as reduced dispersal ability, shifts in body size, and increased specialization. Some insular populations, however, have become extinct, so that the range of the taxon has contracted and its distribution is now spotty.

Stage III Differentiation and range contraction have continued to the point that the taxon now comprises just a few relictual, endemic species whose populations are highly specialized and restricted to interior habitats.

Stage IV The ranges of the relictual populations have contracted further, both within and among islands of the archipelago. As a result of their extreme specialization, perhaps hastened by competition with new invaders (stage I species), relictual populations disappear from the archipelago, presumably to be replaced by other, stage III species.

Source: After Wilson 1959 and 1961.

good over-water colonists typically occur in coastal or disturbed habitats in New Guinea, and recently arrived, undifferentiated populations are found in similar habitats on the Melanesian islands. As they differentiate in isolation, however, the ants also change their ecological requirements and expand into interior habitats, such as native forests (Figures 14.26 and 14.27). They are replaced by a new wave of colonists occupying the beaches and disturbed habitats. Highly differentiated endemic forms, representing the last stage of the taxon cycle, typically are restricted to a narrow range of environments, usually rain forest or montane forest deep in the interior of the islands.

Thus, even before co-developing what would prove to be a paradigm of ecological biogeography—the equilibrium theory—Wilson presented a theory that could account for patterns associated with colonization, ecological release, niche shifts, character release and displacement, reduced dispersal ability, shifts in insular body size, range contraction, endemicity, and extinction. Later,

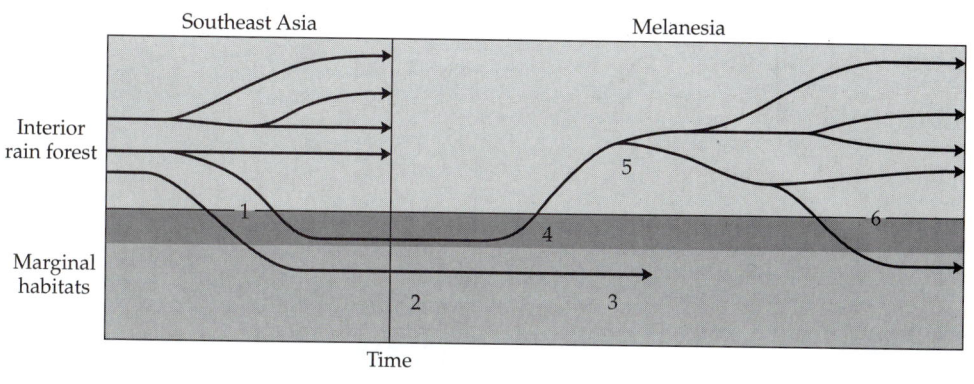

Figure 14.26 Ecological changes accompanying the taxon cycle in the ants of Melanesia. Species of forest ancestry that have secondarily invaded marginal (usually disturbed or coastal) habitats in Southeast Asia (1) tend to be good dispersers and to colonize similar habitats on the islands of Melanesia (2). These populations become extinct fairly rapidly (3), or else they invade interior rain forest habitats (4), where they may undergo differentiation and adaptive radiation (5). Sometimes they give rise to forms that secondarily invade marginal habitats and disperse in stepping-stone fashion to more distant islands (6). (After Wilson 1959.)

Figure 14.27 A graphic model of the processes suggested to be involved in the insular taxon cycle. Numbers are as in Figure 14.26. The cycle is driven by the colonization of generalized species, which evolve to become more specialized, sacrificing competitive ability and experiencing reduced population density within particular habitats. Pushed further along the habitat gradient by superior competitors, species in the terminal stages are forced to specialize further, but evolutionary constraints prevent them from becoming well adapted to these new niches and hence from increasing in density and competitive ability. Consequently, they evolve into rare endemics and ultimately become extinct.

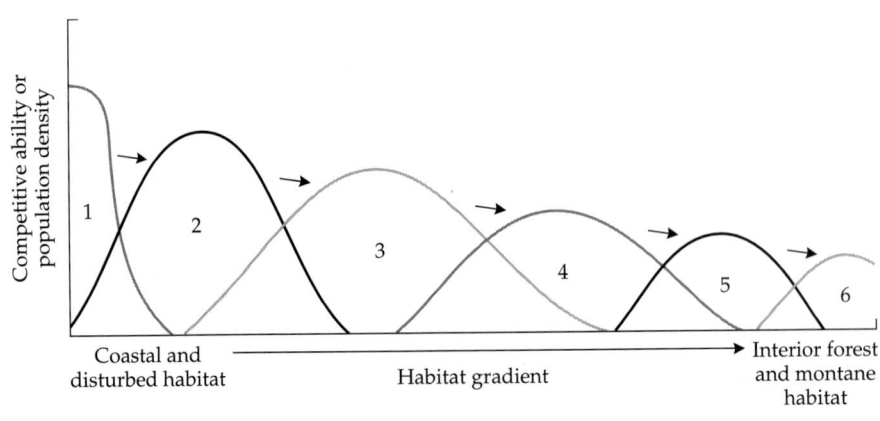

Figure 14.28 Stages of the taxon cycle as illustrated by birds of the West Indies. Stage I, a widespread, undifferentiated species that presumably has recently colonized from South America, represented here by the flycatcher *Tyrannus dominicensis*. Stage II, a widespread form with well-differentiated races (letters) on different islands, represented by the finch *Loxigilla noctis*. Stage III, with well-differentiated races on only a few islands, represented by the warbler *Dendroica adelaidae*. Stage IV, a narrowly endemic species, represented by the finch *Melanospiza richardsoni*, confined to Saint Lucia. (From Ricklefs and Cox 1972.)

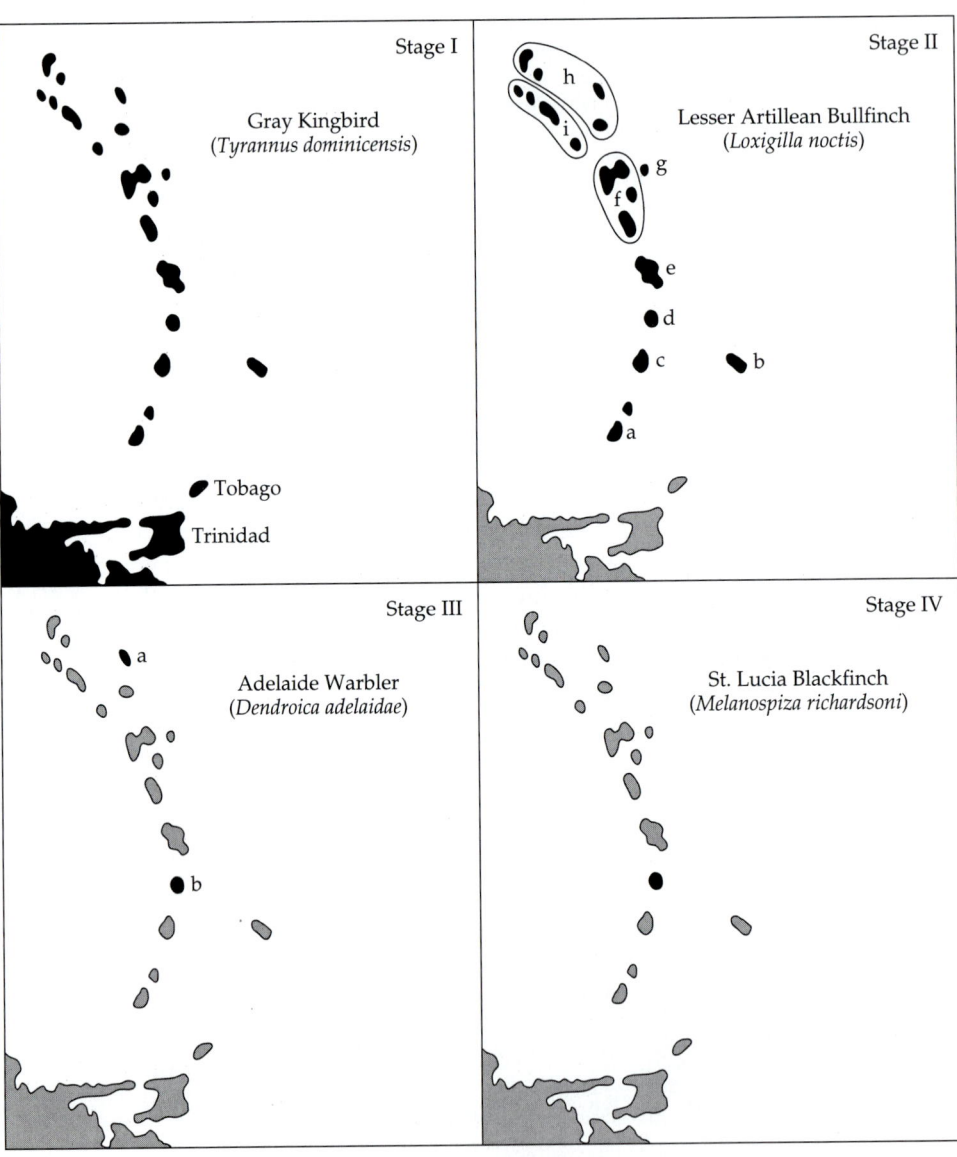

Ricklefs and Cox (1972, 1978) reported ecological and evolutionary shifts in West Indian birds that showed successive stages of what they interpreted as a taxon cycle (Figure 14.28). Similarly, Roughgarden and his colleagues used the theory of taxon cycles to explain the evolution of body size in West Indian anoles (Figure 14.29; Roughgarden and Pacala 1989; Roughgarden et al. 1987; Roughgarden 1992; see also Miles and Dunham 1996). Other researchers, however, have offered alternative and seemingly more parsimonious explanations for observed patterns in the body sizes of these lizards (Losos 1992; Losos et al. 1993; Taper and Case 1992). In fact, Wilson's theory has fallen far short of achieving the status of a paradigm. For the past three decades, it has attracted surprisingly little attention from most biogeographers and ecologists, and many have dismissed it as an interesting but failed attempt. Such declarations, however, seem unfortunate and premature. Given its ability to integrate distributional, ecological, and evolutionary phenomena, Wilson's theory of taxon cycles merits far more attention and more rigorous assessment before we evaluate its true worth as pyrite or gold.

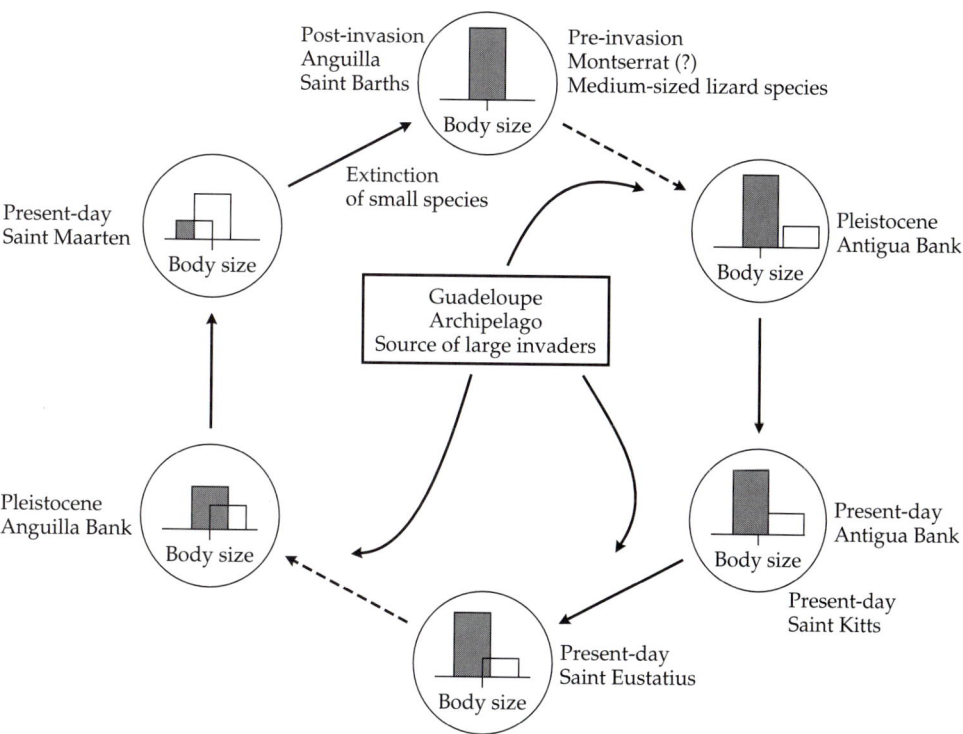

Figure 14.29   A proposed taxon cycle for *Anolis* lizard species in the northern Lesser Antilles of the Caribbean. An island with just one resident species (gray bar) evolves a "characteristic" or "optimum" body size (top circle, bars indicate relative population density). This island is then invaded by a second and larger species (open bar), which then evolves toward the size of the original resident, which, in turn, becomes smaller. The range of the initial resident begins to contract until it eventually becomes extinct. The "invader" now becomes the specialized resident, and its size and range continue to decrease as the island is again colonized by another, larger species to reinitiate the cycle. (From Roughgarden et al. 1989.)

CHAPTER *15*

Species Diversity in Continental and Marine Habitats

The same kinds of questions we have asked about the distributions and diversity of species on islands and in other isolated habitats can also be asked about the biotas of continents and oceans. What determines the numbers and kinds of species that occur together in one place, and why do the numbers and kinds of species vary from one place to another? The taxonomic identities and affinities of species depend, of course, on the history of phyletic lineages, which in turn have been greatly influenced by historical geological and climatic events (see Chapters 6, 7, 11, and 12). In this chapter, we will be concerned primarily with describing and attempting to explain large-scale patterns in species diversity on the continents and in the oceans.

Measurement and Terminology

Species Richness and Diversity Indexes

For many biogeographers, it is sufficient to be concerned with **species richness**, the number of species in a census or sample from a local area or geographic region. The rare species may be as interesting and important as the common ones; in fact, many taxa of interest to both historical and ecological biogeographers are both highly restricted in their distributions and uncommon where they do occur (see Chapter 4). To compare the similarity in species diversity or taxonomic composition of two regions, it is often most useful, as well as most convenient, simply to compare lists of species, which can usually be obtained from systematic works and faunal and floral surveys. Assuming that the samples are relatively complete, or that the sampling effort for each region is at least comparable, species counts can provide simple, relatively unambiguous measures of the biological diversity of different geographic areas.

Many ecologists, on the other hand, are interested primarily in the organization of natural communities. To them, the relative abundances of species and other characteristics indicative of their ecological roles may be of more concern than the number of species present. Rare species may be important components of ecological systems (see Gaston 1994; Kunin and Gaston 1997), but ecologists often deemphasize them and focus on species that dominate the

449

community in terms of abundance, biomass, energy use, cover, or some other estimate of "importance." Consequently, ecologists have adopted several measures, or indexes, of species diversity that give greater weight to the dominant species. Many such diversity indexes have been proposed, and several are used frequently in the ecological literature. These indexes are defined, and their uses, strengths, and weakness discussed, in several reviews (e.g., Pielou 1975; Whittaker 1977; Ludwig and Reynolds 1988; Magurran 1988; Rosenzweig 1995). Although a thorough understanding of these indexes is important for interpreting many ecological studies, a detailed treatment of them is beyond the scope of this book. For our purposes, it is sufficient to know that in most natural communities, nearly all of the approximately 20 indexes commonly used to quantify species diversity are highly correlated with species richness. For this reason, we will use the term *species diversity* somewhat loosely, and synonymously with species richness.

Scales of Diversity: Alpha, Beta, and Gamma

Like most other ecological patterns, species diversity varies with the spatial scale on which it is studied. The continuum of spatial scales is commonly divided into three convenient categories (Whittaker 1975, 1977; Cody 1975; Wilson and Schmida 1984; Magurran 1988; Ricklefs and Schluter 1993). **Alpha diversity** refers to the species richness of a local ecological community—that is, the number of species recorded within some standardized area, such as a hectare, a square kilometer, or some naturally delineated habitat patch. **Beta diversity** refers to the change (or turnover) in species composition over a relatively small distance, often between recognizably different but adjacent habitat types. An example of beta diversity would be the difference between two distinct communities on a mountain slope, for example, between lowland rain forest and montane evergreen forest in the tropics, or between deciduous and coniferous forest in a temperate region. Several methods have been used to measure beta diversity (see Wilson and Schmida 1984; Ricklefs and Schluter 1993), but their goals are the same: to express the overall turnover in species composition. **Gamma diversity** refers to the total species richness of a large geographic area, such as a biome or continent. Gamma diversity reflects the combined influences of alpha and beta diversity, so it will be highest in regions, such as the Amazon basin, where there are many species within local communities and a high turnover of species among habitats and across the landscape.

In most of the following discussion we will be concerned primarily with patterns in alpha diversity. The primary data available for numerous taxa and a variety of geographic regions are species counts for small areas. These counts usually are published either by systematists describing the results of biotic surveys or by ecologists summarizing censuses or samples of local habitats. Changes in species richness and composition across the landscape, however, are also of great interest to biogeographers. We shall discuss these patterns in beta and gamma diversity in the last part of this chapter and in the following chapter.

The Latitudinal Gradient

One of the most striking characteristics of life on earth is the gradient of increasing species diversity from the poles to the equator. The earliest explorer-naturalists noticed that the tropics teem with life, the temperate zones have fewer kinds of animals and plants, and the Arctic and Antarctic are stark and barren by comparison (Wallace 1878). This pattern is a general one in that

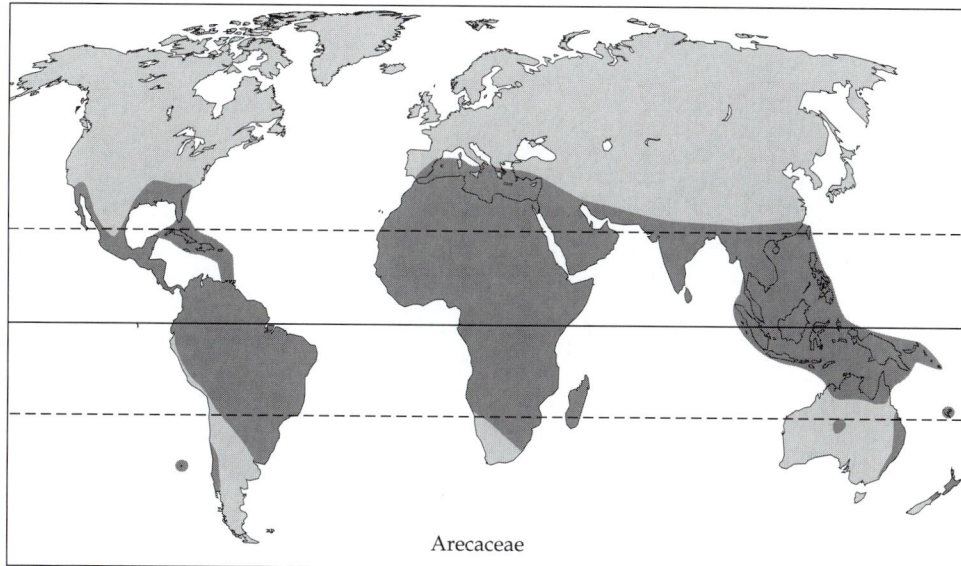

Figure 15.1 Pantropical distribution of the palm family, Arecaceae. The primary factor limiting the distribution of these plants is temperature. Although a few palms can tolerate freezing, none can withstand extreme cold. Where palms are found outside the tropics, they invariably occur in areas with moderate climates, such as New Zealand and southeastern Australia. (After Good 1974; data from Moore 1973.)

it holds true not only for organisms as a whole, but also for most major taxa (classes, orders, and families) of microbes, plants, and animals. There are exceptions: penguins (Spheniscidae) and seals (Phocidae) reach their greatest diversity at high latitudes, and ichneumonid wasps (Ichneumonidae) and coniferous trees (Pinaceae) are most diverse in the temperate zones. For every such exception, however, there are numerous groups that are confined, or nearly so, to the tropics; the New World fruit bats (Phyllostomidae), the Indo-Pacific giant clams (Tridacnidae), and the palms (Arecaceae; Figure 15.1) are just a few examples.

Patterns

Beginning with classic studies by Alfred Fischer (1960) and George Simpson (1964), several authors have quantified latitudinal diversity gradients (for a recent review, see Rosenzweig 1995). For groups of organisms whose distributions are well known, this can be done by counting the species in local areas of approximately equal size and drawing contour maps of species richness.

The diversity patterns for North American land birds and mammals are shown in Figure 15.2. The most striking feature that emerges from such maps is, of course, the rapid increase in diversity from the Arctic to the tropics. Interestingly, however, the rate of change of species richness with latitude is not the same in the two taxa. Not only are there more birds than mammals in every region, but bird species richness increases about twelvefold in the 60° of latitude shown, whereas mammalian diversity increases only eightfold. There are also differences among the different families of birds. Flycatchers (Tyrannidae) show a typical gradient of decreasing species richness from the tropics to the Arctic, but the sandpipers (Scolopacidae) are one of the exceptional groups that exhibit the opposite pattern (Figure 15.3).

Table 15.1 summarizes the latitudinal patterns of species richness in a variety of taxa. Although all of these groups decrease in species richness with increasing latitude, the rate of change varies from group to group. It is difficult to show the exceptional patterns in such a table. One exceptional group that has been particularly well studied is the plant family Pinaceae—the conifers (pines, spruces, firs, and their relatives). These plants are restricted to the Northern Hemisphere and achieve their highest diversity at mid-latitudes in

(A)

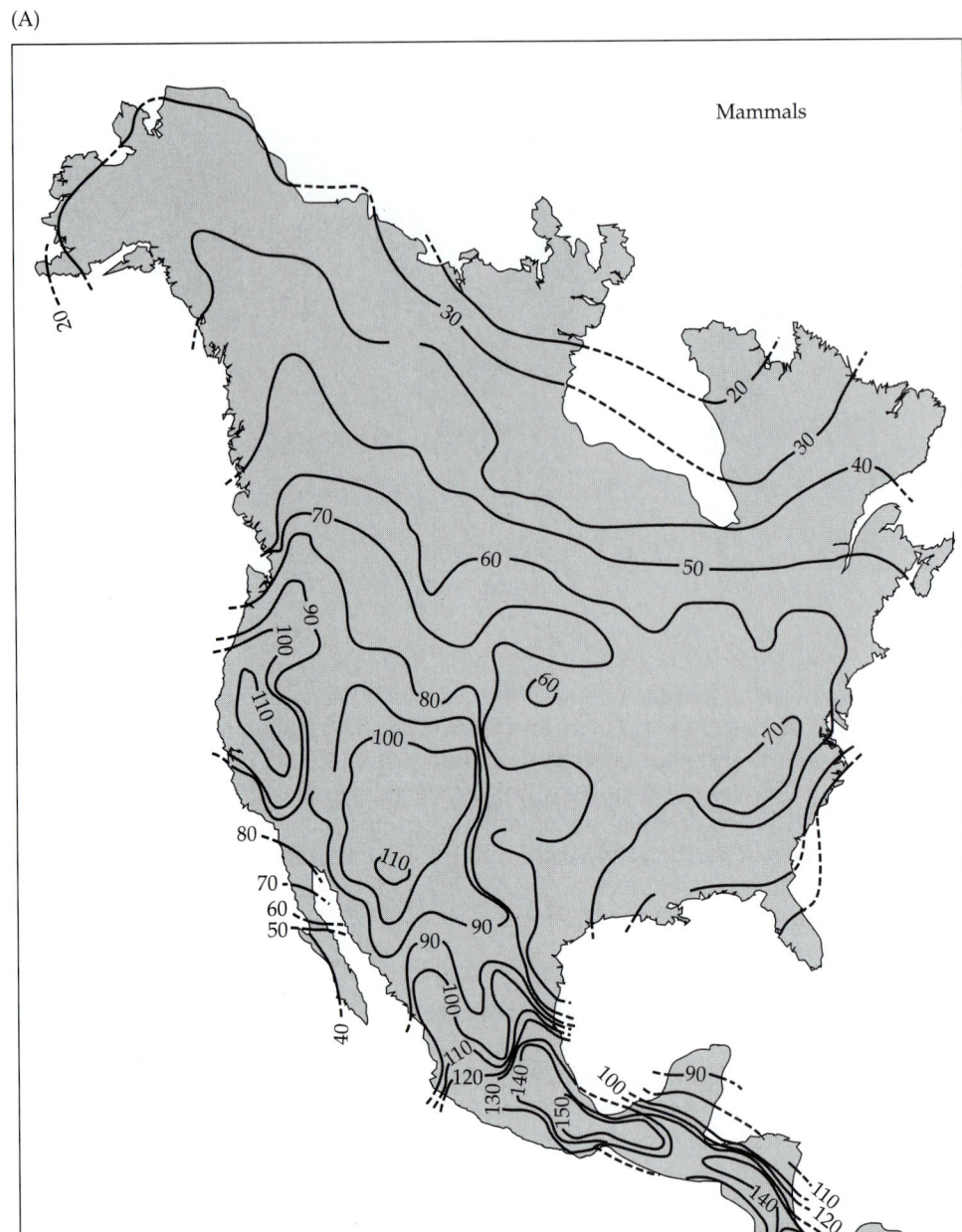

Mammals

Figure 15.2 Geographic variation in the species richness of North American terrestrial mammals (A) and breeding land birds (B). These data were compiled by dividing the continent into grid squares, tallying the numbers of species whose geographic ranges overlapped each square, and then interpolating and smoothing the data to produce the smooth isopleths depicting diversity. Note the similar patterns in both groups, with numbers of species increasing toward the tropics and also showing high values in the diverse habitats in the mountainous regions of western North America. (After Simpson 1964 and Cook 1969.)

both North America and Eurasia (Figure 15.4; Stevens and Enquist 1998). The Ichneumonidae, a large family of parasitic wasps, are also most diverse in temperate regions, with the greatest species richness in North America occurring at about 40° N latitude (Owen and Owen 1974; Janzen 1981; see also Hespenheide 1978).

Latitudinal trends in species richness are also apparent from ecological studies of small areas of terrestrial habitat. Perhaps the most dramatic trend is in numbers of forest tree species. A single hectare of tropical rain forest in South or Central America may contain 40 to 100 different kinds of trees (Richards 1957; Anderson and Brown 1980; Hubbell and Foster 1986), whereas

(B)

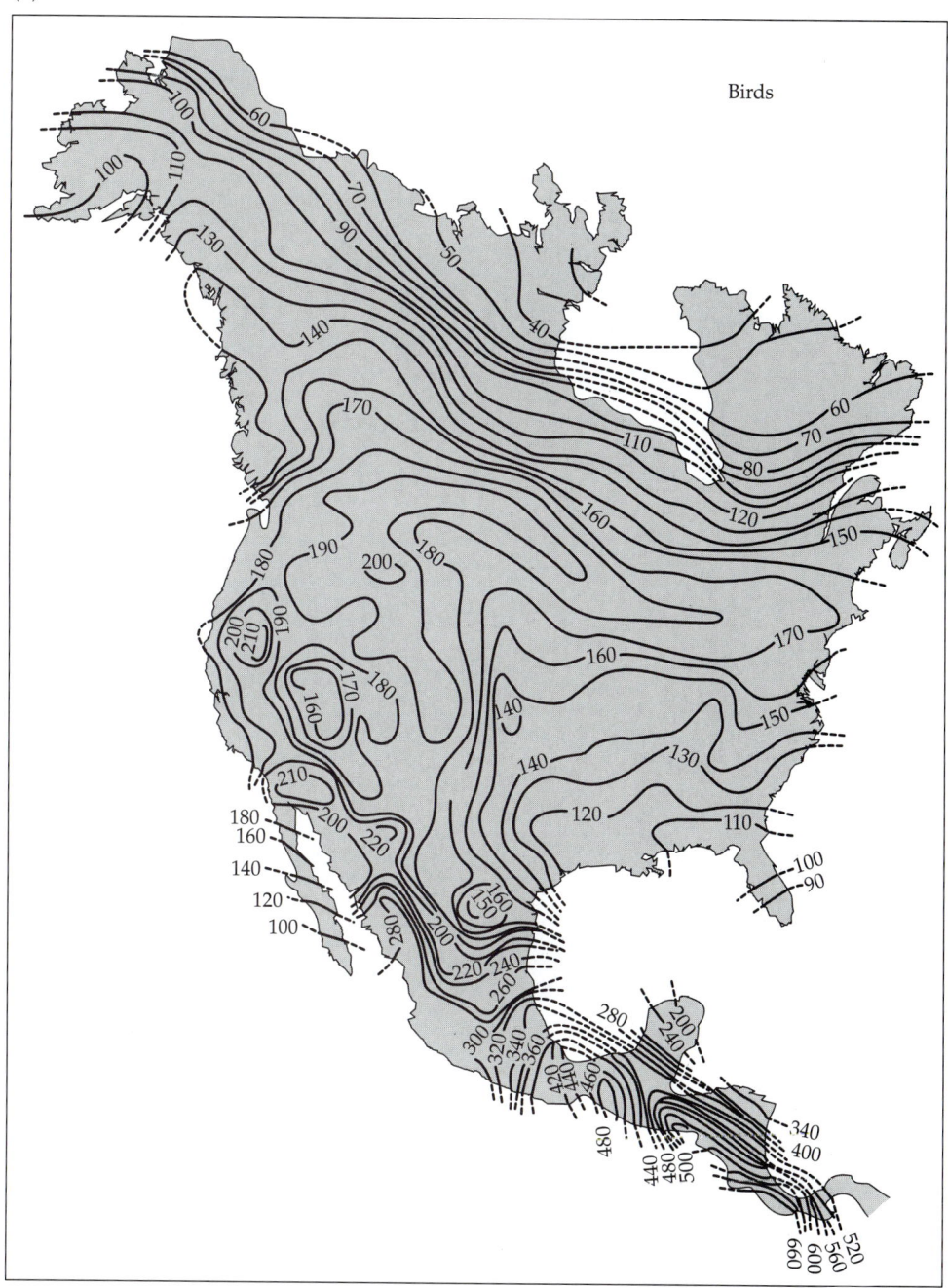

Birds

a hectare of deciduous forest in eastern North America is likely to contain 10 to 30 species, and a hectare of coniferous forest in northern Canada, only 1 to 5 species (Braun 1950; Glenn-Lewin 1977; Latham and Ricklefs 1993; Scheiner and Rey-Benayas 1994). Because many insect species are specific herbivores and pollinators of individual plant species, it is expected that many insect taxa will show comparable patterns when the data become available (see Otte 1976; Erwin 1983, 1988). When Fleming (1973) compiled data on numbers of mammal species in North American forest habitats, he found that 15 to 16 species occur in boreal coniferous forests in Alaska, 31 to 35 species inhabit deciduous forests in the eastern United States, and 70 species are found in both wet and

(A)

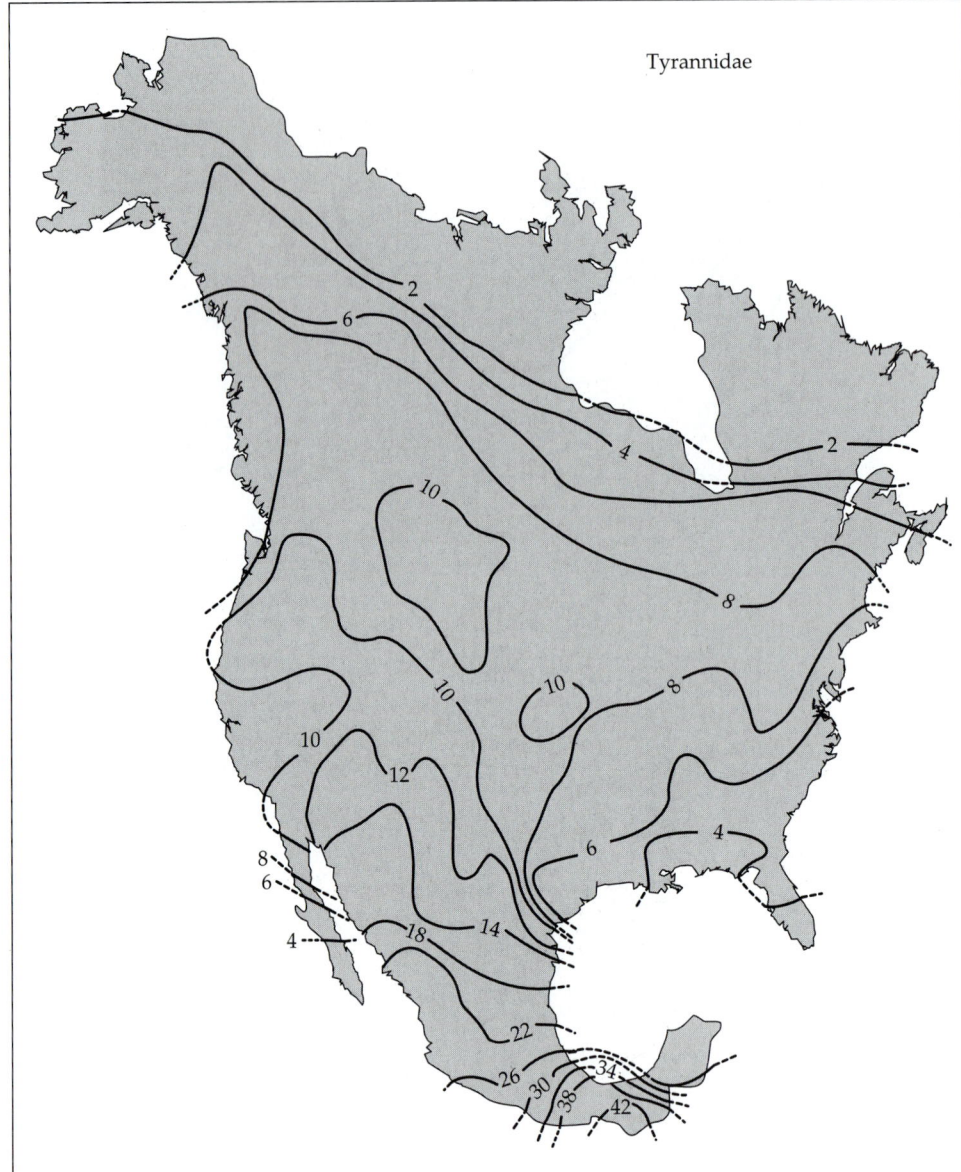

Tyrannidae

Figure 15.3 Geographic variation in the species richness of two families of birds in North America: (A) flycatchers (Tyrannidae) and (B) sandpipers (Scolopacidae). These maps, compiled from maps of the birds' breeding ranges, show contrasting patterns. The flycatchers, typical of most groups, increase in diversity toward the tropics, whereas the greatest numbers of sandpiper species breed in tundra habitats at high latitudes. (After Cook 1969.)

dry tropical forests in Panama. As implied by Table 15.1, the latitudinal gradient in land mammal diversity can be attributed largely to the rapidly increasing numbers of bat species at lower latitudes. Nonflying mammals of certain orders—such as Primates (monkeys) and Edentata (armadillos, sloths, and anteaters)—and families (especially of rodents) that are specialized for arboreal life or have insectivorous or frugivorous diets also contribute importantly to the composition of the diverse tropical communities (Willig and Selcer 1989; Willig and Sandlin 1989; Kaufman 1995; Kaufman and Willig, in press). Although the data are somewhat more sketchy, similar latitudinal patterns of diversity also characterize Old World forest mammals.

Clear latitudinal gradients of species diversity, qualitatively similar to those seen on the continents, also occur in the oceans. Figure 15.5 and Table 15.1 illustrate the patterns in some marine taxa that have been fairly well studied

(B)

Scolopacidae

(see also Stehli 1968; Stehli et al. 1969; Taylor and Taylor 1977; Turner 1981). Sanders (1968) found about five times as many species (100 versus 20) of bottom-dwelling annelids and bivalve mollusks in the Bay of Bengal (20° N latitude) as in comparable areas off Cape Cod, Massachusetts (40° N latitude). There is even a pronounced latitudinal gradient of diversity in the abyss (Figure 15.6; Stuart and Rex 1994). This gradient is of particular interest because the abiotic environment at such great depths is relatively constant throughout the world. These examples reinforce the less quantitative observation that tropical coral reefs support the most diverse marine communities, whereas the cold oceans and shores of polar latitudes have fewer species.

It was once thought that freshwater organisms were an exception to the latitudinal patterns of species richness observed in other habitats (Patrick 1961, 1966). Now it appears that this is not the case (Clarke 1992; France 1992; Rex et

Table 15.1
Latitudinal gradients of species richness in various taxonomic groups

Taxon	Region	Latitudinal range	Range in species (or generic richness)	Source
Land mammals	North America	8–66° N	160–20	Simpson 1964; Wilson 1974
Land mammals	South America	54° S–13° N	26–189	Kaufman and Willig, in press
Bats	North America	8–66° N	80–1	Simpson 1964; Wilson 1974
Bats	South America	54° S–13° N	2–138	Kaufman and Willig, in press
Breeding land birds	North America	8–66° N	600–50	Cook 1969; MacArthur 1972
Reptiles	United States	30–45° N	60–10	Kiester 1971
Lizards	Northern Hemisphere	20–67° N	20–1	Arnold 1972
Lizards	United States	30–45° N	15–1	Schall and Pianka 1978
Snakes	Northern Hemisphere	20–67° N	25–2	Arnold 1972
Amphibians	United States	30–45° N	40–10	Kiester 1971
Anurans	Northern Hemisphere	20–67° N	15–1	Arnold 1972
Marine fishes	California coast	32–42° N	229–119	Horn and Allen 1978
Papilionid butterflies	New World	0–70°	80–3	Scriber 1973
Papilionid butterflies	Old World	0–70°	50–4	Scriber 1973
Sphingid moths	South America	0–50° S	100–2	Schreiber 1978
Ants	South America	20–55° S	220–2	Darlington 1965
Termites	Old World	37–3° S	2–60	Collins 1989
Calanoid crustacea	Atlantic coast of North America	25–50° N	80–10	Fischer 1960
Gastropod mollusks	Atlantic coast of North America	25–50° N	300–35	Fischer 1960
Bivalve mollusks	Atlantic coast of North America	25–50° N	200–30	Fischer 1960
Deep-sea snails	North Atlantic	77–0° N	3–32	Stuart and Rex 1994
Permian brachiopods	Northern oceans	10–55° N	80–40	Stehli et al. 1969
Hermatypic corals (genera)	Western Pacific	0–30°	45–0	Stehli and Wells 1971
Hermatypic corals (genera)	Northwestern Atlantic	20–40° N	20–0	Stehli and Wells 1971
Planktonic foraminifera	World oceans	0–70°	16–2	Stehli 1968
Trees	Eurasia	45–70° N	12–2	Silvertown 1985
Orchids	New World (South America)	0–66° S	2500–15	Dressler 1981

al. 1993; Cotgreave and Stockley 1994; Oberdorff et al. 1995). In a quantitative study of stream insects, Stout and Vandermeer (1975) usually found 30–60 species at tropical American sites compared with 10 to 30 species in the temperate United States. The number of freshwater fish species clearly decreases with increasing latitude in North America (see Figure 10.12), and a similar trend holds for lakes around the world (Barbour and Brown 1974). Despite this pattern, there is a great deal of variation owing largely to differences in the sizes and histories of individual bodies of water, as might be expected from

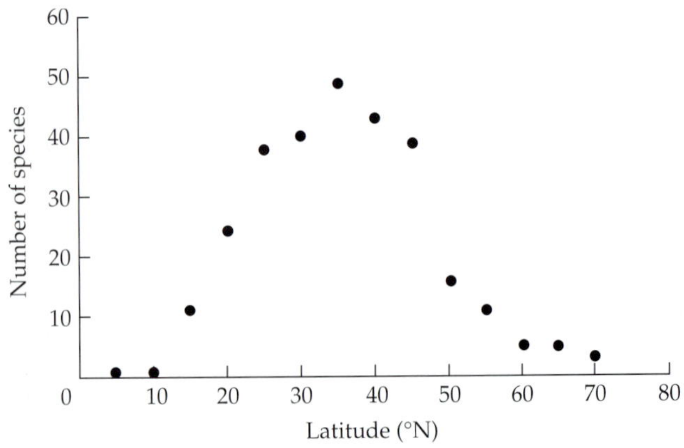

Figure 15.4 Latitudinal variation in species richness in the pines (genus *Pinus*). Like other members of the conifer family, Pinaceae, pines are confined to the Northern Hemisphere, and are most diverse in temperate climates at intermediate latitudes. (From Stevens and Enquist 1998.)

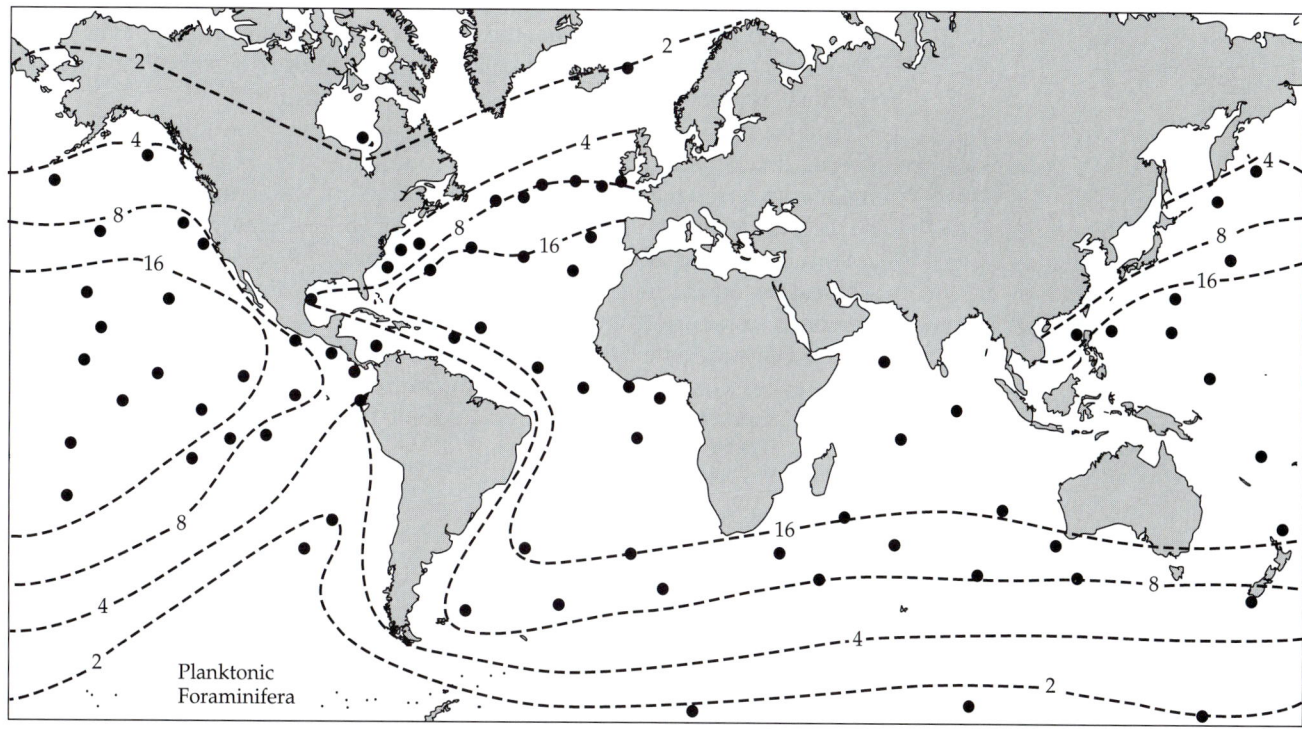

Figure 15.5 The worldwide pattern of species richness in planktonic foraminifera shows a latitudinal gradient that is fairly typical for marine organisms. Dots represent locations of samples used to draw the isopleths. Corals, mollusks, and fishes are other marine groups that are most diverse in tropical waters. (After Stehli 1968.)

our discussion in Chapter 12. The highest diversity of lacustrine fish species occurs in the Great Lakes of the African Rift Valley, where Lakes Malawi, Tanganyika, and Victoria contain hundreds of species in just one family, Cichlidae (Fryer and Iles 1972; Meyer 1993). With respect to rivers, although the exact numbers are still uncertain, there seems to be little doubt that the Amazon supports more fish species than any other river system in the world. This is true both for the entire river basin (gamma diversity) and within small areas of relatively uniform habitat (alpha diversity). As documented in recent scientific

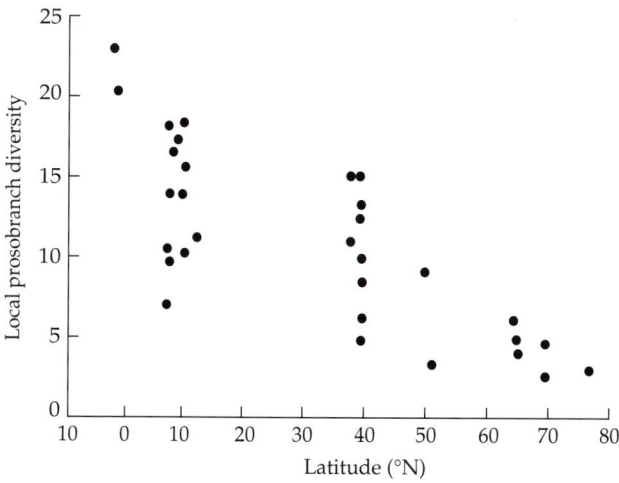

Figure 15.6 Species diversity of prosobranch (gastropod) mollusks as a function of latitude in the abyssal depths of the North Atlantic Ocean. Samples were collected using a dragged sled trap, and species diversity was measured using Hurlbert's expected number of species (one of the diversity indexes). Note two things: first, there is considerable variation in diversity, even among samples collected at comparable latitudes; presumably this reflects the spatial heterogeneity of abyssal environments. Second, despite this variation, there is a clear tendency for diversity to increase toward the equator. (From Stuart and Rex 1994.)

studies and television specials, the Amazon fish fauna includes a substantial component that moves into seasonally flooded forests to exploit the unusual habitats and food resources there, including even fruit.

Finally, it is becoming increasingly clear that the latitudinal gradient of species diversity is an ancient feature of our planet. Remains of plants and animals from terrestrial habitats in the tropics are not well represented in the fossil record. Nevertheless, Crane and Lidgard (1989) estimate that there has been a latitudinal gradient of land plant diversity for at least 110 million years. There is a good fossil record of tropical marine organisms, such as brachiopods, corals, and single-celled foraminiferans and diatoms. These fossils also indicate that the latitudinal gradient of species diversity has been in existence at least since the Mesozoic, for more than 250 million years in brachiopods and 70 million years in planktonic foraminifera (Stehli et al. 1969).

Processes

Latitudinal gradients in species diversity have intrigued ecologists, systematists, and biogeographers for many years. The observation that they occur in nearly all kinds of organisms living in a wide range of terrestrial, freshwater, and marine habitats, and have been in existence for hundreds of millions of years, stimulates the search for an equally general explanation. Clearly this pattern cannot be attributed to the influence of unique events in the histories of just a few lineages. Instead, it must reflect some pervasive way in which the geography of the earth has influenced the diversification of living things. More than 25 hypotheses have been advanced to explain it; the most frequently cited ideas are listed in Table 15.2 (see also Pianka 1966; Brown 1988; Rohde 1992; Rosenzweig 1992, 1995).

Several things about these hypotheses warrant comment. For one thing, they are not mutually exclusive. Many of the proposed factors could have operated together, either sequentially or concurrently, to influence current latitudinal patterns of diversity. This is not only theoretically possible, it is highly likely. Many of the environmental factors and mechanistic processes listed in Table 15.2 are known to vary in a systematic way with latitude; examples include the area of both the earth's surface overall and major habitat types,

---

Table 15.2
Processes that have been hypothesized to account for geographic patterns of species diversity

Hypothesis or theory	Mechanism of action
Historical perturbation	Habitats that have experienced historical changes are undersaturated because of inadequate time for species to colonize and adapt.
Productivity	The greater the availability of usable energy, the larger the number of species that can be supported, and the greater the specialization of coexisting species.
Harshness	Small, isolated, ephemeral, or physically extreme habitats have lower colonization rates or higher extinction rates than large, continuous, permanent, and physically equable habitats.
Climatic stability	A fluctuating environment may preclude specialization or increase the extinction rate, whereas in a constant environment the species can specialize on predictable resources and persist when rare.
Habitat heterogeneity	Diverse physical habitat structure permits finer subdivision of limiting resources and hence greater specialization.
Interspecific interactions	Competition, predation, or mutualism promote coexistence and specialization.

solar energy input and climate, and productivity. Many of these factors and mechanisms have been shown to influence diversity on smaller scales when manipulative ecological experiments or careful comparative analyses have controlled for the effects of many other, potentially interacting variables; examples include historical perturbations, productivity, structural habitat heterogeneity, and interspecific interactions. Thus, as tempting as it may be to seek one grand unifying hypothesis, it is important to realize that the latitudinal diversity gradient might result from multiple, mutually reinforcing processes.

Most of the proposed hypotheses can be divided into two classes: equilibrial and nonequilibrial. Nonequilibrial hypotheses imply that the patterns of diversity we see today reflect the lingering influence of perturbations in the past. Thus some of the low diversity at high latitudes may reflect the influence of the glacial-interglacial cycles of the Pleistocene (see Chapter 7; Fischer 1960; Fischer and Arthur 1977; Stanley 1979). The paucity of some kinds of organisms at high latitudes may reflect high extinction rates due to their inability to migrate away from advancing ice sheets or to small populations being isolated in unglaciated areas. Some of those species that survived may not yet have recolonized previously glaciated areas. For example, tree species diversity in high-latitude forests has been increasing over the last 10,000 years as individual species with different dispersal rates have colonized northward in response to the retreat of the glaciers and the warming of the climate (Bernabo and Webb 1977; Davis 1986). Whitefishes in the North American Great Lakes, sticklebacks in lakes in British Columbia, and pupfishes in Lago Chichancanab in Yucatán are currently speciating in lakes formed as a result of changes in climate and sea level in just the last 10,000 years (see Chapter 8).

Equilibrial hypotheses, on the other hand, imply that the pattern we see today reflects the adjustment of the biota to current geological, climatic, and oceanographic conditions. Thus the climatic stability and productivity hypotheses imply that the major features of the diversity gradient reflect latitudinal variation in the input of solar radiation to the earth's surface. While there has been quantitative fluctuation in solar radiation over time (which has caused, among other changes, the glacial-interglacial cycles of the Pleistocene; see Chapter 7), there has always been a pronounced gradient of decreasing temperature and increasing seasonality from the equator to the poles.

The following thought experiment is a useful way to visualize the distinction between equilibrial and nonequilibrial hypotheses. Imagine that each year for the next 100 million years, every place on earth experienced exactly the same environmental conditions that it did in, say, 1998. Would there still be a latitudinal diversity gradient? Of course we will never know, but most of us would have difficulty imagining that the polar ice caps would ever support as many kinds of organisms as currently live in the Amazon basin. If so, this would suggest that no perturbations on the scale of years to millennia, including Pleistocene glaciations, can account entirely for the latitudinal diversity gradient.

The proposed hypotheses also differ in the level of causal explanation that they offer. Primary hypotheses seek to explain diversity in terms of some underlying abiotic characteristic of the earth. Unless the latitudinal gradient is due purely to chance (see Colwell and Hurtt 1994), it must ultimately be due to some characteristic(s) of the earth's physical-chemical template. Otherwise, why is high diversity centered around the equator, one of the earth's most prominent geographic features? One feature of the earth that varies directly with latitude is its relationship to the sun, which is used to define the equator and other latitudinal landmarks such as the tropics of Cancer and Capricorn, the Arctic and Antarctic Circles, and the North and South Poles. While these are just places on the globe, the influence of latitude on abiotic and biotic characteristics of the earth is due to the input of solar radiation, which is most intense and seasonally unvarying at the equator (see Chapter 3). Most hypotheses that invoke productivity, climatic stability, or environmental harshness are based on one or more correlates or consequences of the latitudinal pattern of solar radiation.

Another feature of the earth that is unambiguously related to latitude is its surface area. Because of the way that latitude is mapped onto the spherical earth, its surface area decreases from the equator to the poles, as indicated by the lines of longitude converging at the poles. But this relationship is tricky, because the equator is defined by the relationship of the earth to the sun. In both a mechanistic and a statistical sense, the rate and seasonal pattern of solar radiation input is the independent variable and surface area is a dependent variable, so the area hypothesis can be considered a secondary one. That does not mean, however, that area has not influenced the latitudinal diversity gradient. The importance of area to the diversity of isolated habitats is immediately apparent from the species-area relationship (see Chapter 13). As we shall see, it is also almost certainly an important part of the explanation for latitudinal gradients within the continents and oceans (Rosenzweig 1995).

Although some authors have tried (e.g., Pianka 1966; Brown and Gibson 1983; Brown 1988; Rohde 1992), it is hard to divide the hypotheses unambiguously into primary, secondary, and so on. Are hypotheses such as productivity or harshness primary or secondary? Clearly, they depend directly on characteristics of the earth's abiotic template in relation to the input of solar radiation, but they also depend on characteristics of organisms. Productivity reflects the capacity of organisms to convert solar energy into biomass and biologically usable energy. Harshness is a relative term, which reflects the capacities of organisms to tolerate physical and chemical stresses of their environment.

Some hypotheses, however, are clearly secondary, because they rely on purely biological characteristics and make no direct reference to the earth's abiotic template. Secondary hypotheses include those that invoke varying degrees of specialization and strengths of biotic interactions. For example, it is often suggested that tropical organisms are more specialized or have narrower niches, thereby allowing more species to coexist within a local community than at higher latitudes. It has also been claimed that predation (including herbivory, parasitism, and disease) is more intense in the tropics, thereby reducing competition and dominance and allowing more species to coexist. Hypotheses about competition have been argued both ways: that competition is weaker in the tropics because of the specialization and/or predation just mentioned, or that it is stronger as a consequence of the large number of coexisting species with overlapping resource requirements. Note that such secondary hypotheses are in a sense incomplete, because they do not explain *why* degree of specialization or interaction strength is related to latitude. Unless it is pure chance, there must be something about the geography of the earth—presumably some consequence of the differential input of solar radiation—that causes these biotic relationships to vary in some predictable way with latitude.

A final level of hypotheses invokes speciation and extinction rates. It is often suggested that rates of speciation are higher, rates of extinction are lower, or both, in the tropics than at higher latitudes. As we have seen in our exploration of island biogeography (see Chapter 13), any complete understanding of species richness must take into account the opposing forces that regulate diversity: speciation and colonization on the one hand and extinction on the other. It should also be clear, however, that the relationship of origination and extinction processes to latitude must be mediated through the abiotic and biotic mechanisms we have been discussing here. Ultimately, speciation and extinction processes must reflect in some way the differential input of solar radiation to the earth's surface.

At this point we will leave temporarily our consideration of the hypotheses that have been proposed to explain the latitudinal diversity gradient. Having put them on the table, and explored their relationships to the earth's geography and to one another, we will resist for the moment the temptation to evaluate them. We will return to these hypotheses after examining some of the other geographic patterns of species diversity. As we shall see, some of these are almost as general as the latitudinal gradient, and they offer additional insights into the processes that account for variation in organic diversity over the planet.

Other Diversity Patterns

Peninsulas

A pattern of diversity in terrestrial environments is the tendency of species richness to decrease along peninsulas with increasing distance from their continental attachments (Taylor and Regal 1978; Milne and Forman 1986; Means and Simberloff 1987). The generality and the causes of this pattern have been much debated, however. The situation is usually complicated by the effects of correlated variation in environmental factors and possible historical influences. Several factors in addition to distance from a continent typically vary along a peninsula (e.g., Milne and Forman 1986; Means and Simberloff 1987). For example, the Florida and Baja California peninsulas in North America, the Malay and Indochinese peninsulas in Southeast Asia, and the Cape York Peninsula in northeastern Australia all extend into lower latitudes from their continental connections, so distance from the continent is confounded by a gradient in latitude and by many correlated environmental conditions. And if contemporary environments vary along peninsulas, then it is likely that past environments did as well. Consequently, patterns of diversity may be due in part to historical events that have affected the development of peninsular biotas through processes of colonization, speciation, and extinction.

To appreciate these complications, consider two of the peninsulas just mentioned: Florida and Baja California. Both show a **peninsula effect** of decreasing species richness in certain taxonomic groups, and both have superficially similar geographic settings, extending from the temperate landmass of the North American continent southward into the subtropics. In most respects, however, Florida and Baja California are very different; they offer stark contrasts in environmental setting and biogeographic history. Florida currently has a uniformly low elevation, well-drained sandy soils, and a relatively hot, dry climate. This distinctive abiotic environment has probably limited the number of temperate North American organisms that have been able to colonize far down the peninsula. During periods of higher sea level in the Pleistocene, much of the peninsula was submerged, leaving a low, arid island. The historical influence of reduced habitat area and isolation by marine and terrestrial barriers has prevented the colonization of Florida by many taxa that occur in similar arid lowland habitats in tropical America. That this is a historical effect is evidenced by the many exotics of West Indian or other tropical American origin that have become established in southern Florida once they were introduced. Thus, in taxa with limited capacity for over-water dispersal, such as reptiles and amphibians, there is a pronounced gradient of decreasing species diversity from the base to the tip of the Florida peninsula (Means and Simberloff 1987). In other groups, such as butterflies, tropical forms have been able to colonize the southern tip of the peninsula, and there is little, if any, gradient of diversity (Brown and Opler 1990).

Equally varied patterns are seen in Baja California (Figure 15.7), even though it has a very different contemporary environment and biogeographic history. Perhaps most importantly, it is mountainous. On the one hand, this means that there is a great deal of environmental heterogeneity, and that most of the peninsula has been above sea level throughout the Pleistocene. On the other hand, it means that the diverse habitats of the peninsula currently are, and historically have been, small and isolated. The montane areas are effectively islands, isolated by long distances and inhospitable low desert or marine barriers from comparable, more species-rich environments in the southwestern United States and northern Mexico. Even the different kinds of desert habitats, characterized by particular soils and climates, often occur as small, isolated patches. Thus, the desert-dwelling heteromyid rodents show a dramatic decline in species richness from the base to the tip of the peninsula (Taylor and Regal 1978; Figure 15.7A). In other groups of desert and montane organisms, however, the historical influence of diverse and patchy environments is reflected in both high species diversity and a relatively high degree of endemism. Scorpions, for example, attain their highest diversity near the center of the peninsula, where habitat heterogeneity is highest (Due and Polis 1986; Figure 15.7B). Finally, the Cape region at the southern tip of the peninsula is tropical and only semiarid, but it is also highly isolated from comparable environments on the mainland of western Mexico. It has reasonably high levels of both species richness and endemism in some groups, especially plants and insects. In these groups—butterflies are an excellent example (Brown 1987; Figure 15.7C)—there is often a second peninsular peak in diver-

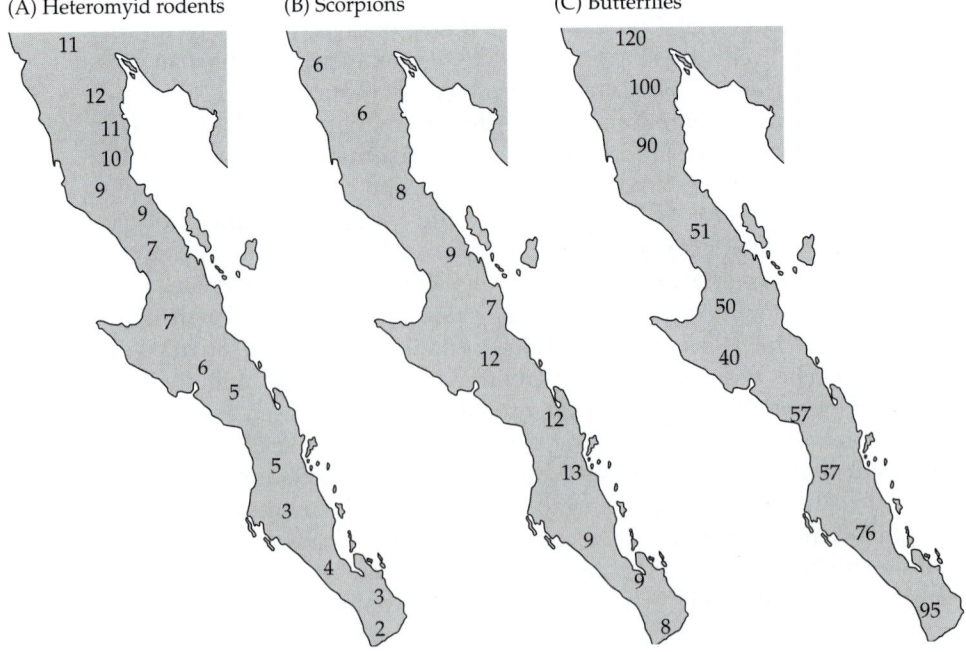

Figure 15.7 Variation in the species richness of three different groups of organisms in Baja California, showing the variety of patterns observed on this peninsula. The diversity of heteromyid rodents (kangaroo rats and pocket mice, A) decreases down the peninsula from base to tip; that of scorpions (B) does not show a clear pattern, but may peak near the center of the peninsula; and that of butterflies (C) is distinctly bimodal, with peaks near the base and tip and the lowest diversity in the center. (Data from Taylor and Regal 1978; Due and Polis 1986; Brown 1987.)

sity in the Cape region. Nevertheless, the tip of the peninsula has much lower species richness than comparable habitats on the continental mainland.

The variation in diversity along peninsulas presents, on a small scale, many of the same problems in interpreting patterns and processes as variations in continental species richness. On the one hand, the patterns are usually relatively clear, and there is a general tendency for diversity to decrease with increasing distance from the continental landmass. On the other hand, position along a peninsula is often associated with variation in contemporary environmental conditions and with the legacy of historical events. These complications make it difficult to distinguish correlation from causation and arrive at simple mechanistic explanations for the observed patterns.

Elevation

In terrestrial habitats, variation in species diversity along gradients of elevation is almost as general and striking as latitudinal variation. Just as the number of species decreases in progressively cooler climates from the tropics to the poles, it decreases in the cooler environments encountered as we ascend mountains. Although this pattern has long been obvious to ecologists and biogeographers (see Chapter 2), it is not as well documented quantitatively as the latitudinal gradient of diversity. This is in part because there have been fewer censuses of species on a sufficiently small scale along elevational transects. However, it is also in part because elevational gradients of temperature are often confounded by variations in available moisture. For example, in otherwise arid regions of the horse latitudes, moisture availability increases with increasing elevation, and the high mountains typically have cool mesic forests, alpine tundra, and even permanent snowfields at the highest elevations. To analyze the influence of any one factor, it is desirable to find "natural experiments" in which that factor is the only important variable and other conditions remain relatively constant. Unfortunately, this is not always possible. For reasons explained in Chapter 3, both temperature and precipitation tend to vary systematically with both elevation and latitude.

Despite the paucity of quantitative studies, the general decline in species diversity with decreasing temperatures at higher elevations is as obvious as the latitudinal pattern. The lowland tropics of the Amazon basin support a rich rain forest biota, but the peaks of the nearby Andes, covered with bare rock, ice, and snow, are as barren of life as the polar ice caps. The effect of temperature is most unambiguous in elevational gradients in mesic regions, where moisture is not a confounding factor. Yoda (1967) quantified the dramatic decline in tree species richness with increasing elevation in the Himalayas. Subsequently, species richness in many different taxa has been shown to vary inversely with elevation in mesic tropical regions (Figure 15.8; see also Kikkawa and Williams 1971; Terborgh 1977; Heaney 1991; Heaney et al. 1989; Fernandes and Lara 1993; Daniels 1992; Patterson et al. 1996). Whittaker (1960, 1977) has found similar patterns for trees on mesic mountains in North America (Figure 15.9; see also Stenthouraskis 1992).

There are exceptions to this general rule (Rahbek 1995). As with the latitudinal gradient, certain restricted taxonomic or ecological groups do not conform to the general pattern. Herbaceous plant species sometimes reach their greatest richness at intermediate elevations on temperate mountains (Figure 15.9; Whittaker 1960, 1977), and orchids (Orchidaceae) attain their greatest diversity on tropical mountainsides in the low-stature cloud forests that occur considerably higher than lowland rain forests (Dressler 1981). We might question how general the pattern is, because many different kinds of organisms show a peak in diversity at intermediate elevations. On desert mountains,

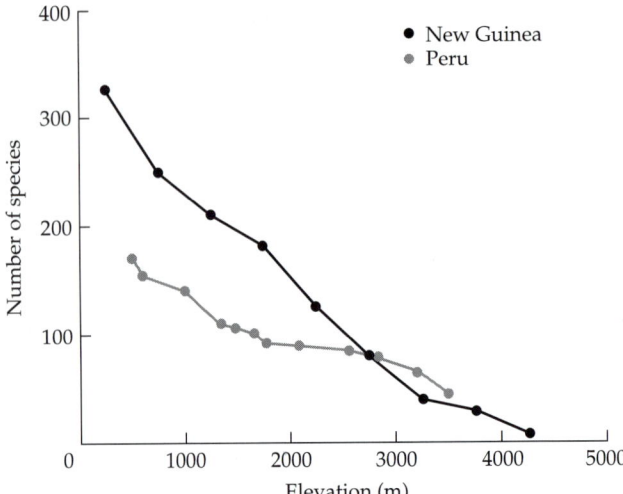

Figure 15.8 Variation in bird species richness with elevation in New Guinea (black circles) and on the eastern slope of the Andes in Peru (gray circles). The data from each region were collected using somewhat different field methods, so they are not exactly comparable. Note, however, that in both regions, diversity declines fairly continuously with increasing elevation. (After Kikkawa and Williams 1971; Terborgh 1977.)

such as those of the southwestern United States, the diversity of nearly all kinds of organisms shows this pattern (e.g., for plants, see Figure 15.10; Whittaker and Niering 1965). In this case, the pattern can be attributed to the fact that the elevational gradient is really two different gradients. Temperature, length of the growing season, and other related variables decrease with increasing elevation, but precipitation and overall moisture availability increase. These two gradients act in opposition to affect diversity, with the result that conditions are "most favorable," and the most species occur, at the intermediate elevations. It is by no means only in arid regions, however, that there is a peak in diversity at intermediate elevations (Rahbek 1995).

Aridity

Species diversity also decreases with diminishing water availability, from mesic forests to more arid shrublands and grasslands to xeric deserts. Deserts are so named because they appear barren and lifeless in comparison with more mesic grasslands, shrublands, and forests. Again, this pattern is obvious, but it is not well documented. It too is easily confounded by spatial scale and environmental heterogeneity. At a large scale, most arid regions are not all

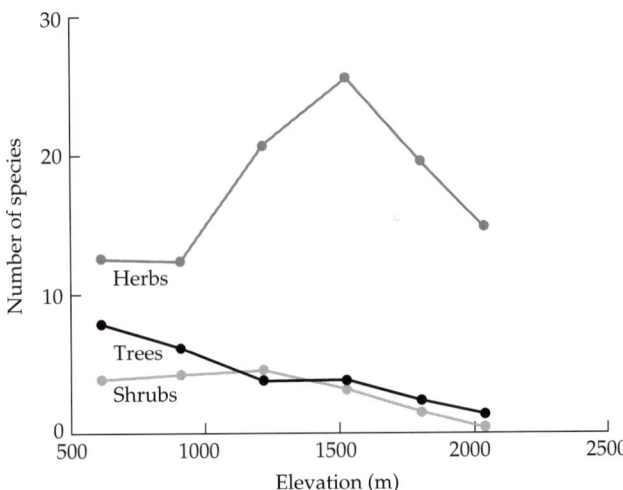

Figure 15.9 Variation in the species richness of three plant life forms, herbs, shrubs, and trees, as a function of elevation in the Siskiyou Mountains of California and Oregon. Note that herbs and shrubs both show a mid-elevational peak in diversity, whereas trees exhibit a continuous decline in species richness from low to high elevations. (Data from Whittaker 1960.)

Figure 15.10 Variation in plant species richness with elevation on a desert mountain range, the Santa Catalina Mountains in southern Arizona. Note the conspicuous mid-elevational peak in diversity, which is apparently due to the combined limiting effects of extreme aridity at low elevations and low temperature and short growing season at high elevations. (Data from Whittaker and Niering 1975; after Brown 1988.)

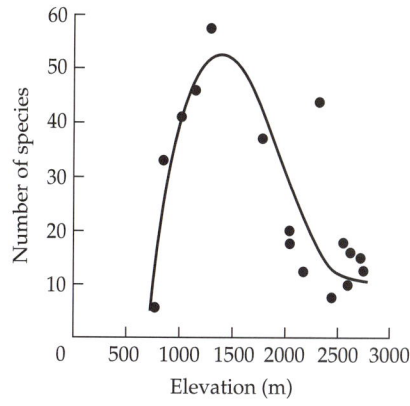

desert. In the generally arid southwestern United States, states such as Arizona and New Mexico have many more species of terrestrial plants and animals than comparably sized states such as Georgia and Alabama in the mesic Southeast. But, as implied above, much of the diversity in the Southwest is due to the mountainous terrain and the mesic environments found at higher elevations. Even at low elevations, the large permanent rivers flowing from the mountains to the sea are associated with riparian deciduous forests, meadows, and marshes, each contributing many species and distinctive higher taxa to the biota. Similarly, the high diversity of most taxa in eastern Queensland, in northeastern Australia, is largely due to the fact that this region is so topographically and environmentally diverse; it contains a variety of habitats, ranging from arid shrubland and savanna to temperate (at higher elevations) and tropical rain forest.

It is also likely that the relationship between moisture availability and species diversity has a peak at some intermediate value. The wettest habitats, including swamps and marshes as well as seasonally flooded forests and grasslands (e.g., the llanos of northern South America and the flooded forests of the Amazon basin), often have fewer species than drier surrounding habitats. We shall return to this observation shortly when we consider the concepts of environmental favorability and harshness.

Aquatic Environments

Some of the patterns of species richness found in water are remarkably similar to those found on land. We have already mentioned that small lakes, like small islands, contain fewer species than large ones, and that comparably sized lakes, rivers, and streams at high latitudes support lower diversity than those in the tropics. Other patterns of diversity, however, suggest that aquatic systems differ in important ways from terrestrial ones. While there are many examples of positive correlations between productivity and diversity in the terrestrial realm, as we shall see below, productive waters do not necessarily have high species diversity. Some productive marine habitats, such as coral reefs, are also rich in species. On the other hand, some small productive areas of ocean water, such as local regions of upwelling and enriched sites near rivers or estuaries, often contain relatively few species, especially of plankton. McGowan and his associates (McGowan and Walker 1979; Hayward and McGowan 1979) studied the plankton of the north central Pacific, the most extensive area of uninterrupted water on earth. They found that this region is rich in species, even though the water is unproductive and the density of individual planktonic organisms is usually low. Many of the species present are extremely widely distributed, ranging thousands of kilometers across the region. Several studies of temperate lakes have shown an inverse relationship between productivity and planktonic diversity (Figure 15.11; Whiteside and Harmsworth 1967), although this relationship is confounded by the sizes of the lakes. Small lakes tend to be shallow and to experience much vertical mixing of water, which substantially increases their productivity. Thus highly pro-

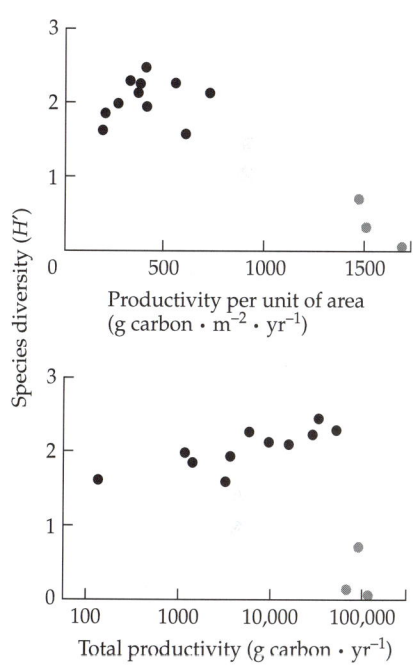

Figure 15.11 The relationship between species diversity (here measured in terms of one of the commonly used diversity indexes, H') of chydroid cladoceran zooplankton and primary productivity for 14 lakes in Indiana. The gray circles represent three heavily polluted lakes with exceptionally high productivity and low diversity. Note that for the remaining lakes, there is no correlation when productivity is expressed per unit of surface area, but a positive relationship when productivity is expressed for the lake as a whole. These data suggest a positive effect of productivity on diversity if one assumes that it is the productivity of an entire lake that determines its capacity to support zooplankton species. (Data from Whiteside and Harmsworth 1967; plotted by D. H. Wright.)

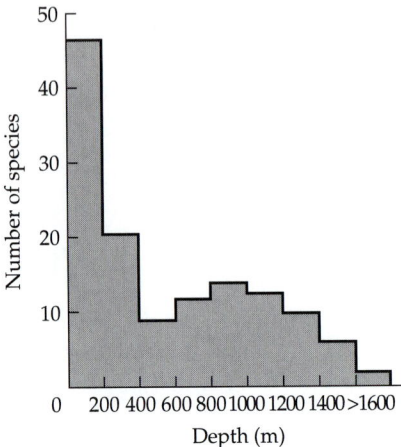

Figure 15.12 Species richness as a function of depth for benthic marine fishes in the Pacific Ocean off the Oregon coast. Note that unlike the marine invertebrates in Figure 15.13, which are most diverse at intermediate depths on the continental slope, these fishes have their highest diversity in the shallow waters of the continental shelf. (Data from Day and Pearcy 1968.)

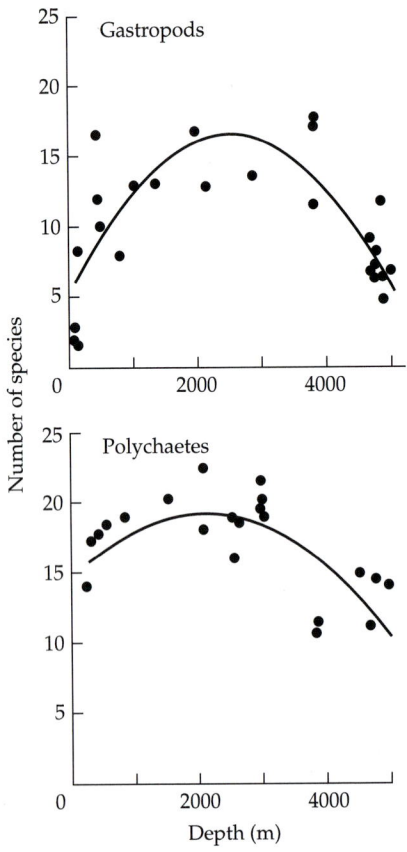

ductive marine and freshwater environments tend to be small and patchy, whereas unproductive regions are usually larger and more extensive.

At first glance, variation in species richness with water depth may appear reminiscent of the pattern of variation with elevation on land. For example, the cold, dark, oxygen-poor abyssal depths of tropical and temperate lakes contain fewer species in most taxonomic groups than do the shallow surface waters. Some large lakes, including the Great Lakes and the Finger Lakes in northeastern North America, Lake Baikal in Siberia, and several of the large lakes of central Africa, are more than 100 m deep, but few organisms are found in the deepest waters. In Chapter 8 we described the great diversity of fishes, especially in the family Cichlidae, that inhabit the Great Lakes of central Africa. Although Lake Malawi and Lake Tanganyika are 704 and 1470 m deep, respectively, their hundreds of species of fish are most diverse in the surface waters, and none are found below about 200 m. Their lower depths apparently are inhabited by only a few kinds of anaerobic invertebrates and bacteria.

The general pattern of decreasing diversity with increasing depth probably also holds true for the oceans. Marine plants, including both phytoplankton and larger (usually attached) algae, are confined to the photic zone (see Chapter 3), which rarely extends below 30 m. Pelagic animals, both passively floating plankton and free-swimming nekton, are most diverse in surface waters. Although the diversity of pelagic forms decreases rapidly with depth, many kinds of bizarre deep-sea fishes and some unusual invertebrates inhabit the abyssal depths. Bottom-dwelling animals exhibit more complex patterns, but many are again reminiscent of those seen in terrestrial organisms on elevational gradients. Along the continental margins, the diversity of benthic fishes typically appears to be highest in shallow waters (Figure 15.12). Benthic marine invertebrates, however, often show a pronounced peak in species richness at intermediate depths (Figure 15.13; Vinogradova 1962; Sanders 1968; Rex 1981). Alpha diversity in substrate samples shows that invertebrate diversity is low on the shallow continental shelves, reaches a maximum near the top of the continental slope, and then decreases with increasing depth.

Sanders (1968) performed classic studies on the diversity of benthic bivalve mollusks and polychaete worms collected by dredging in the oceans. Although he used a somewhat unconventional method to express his data (Figure 15.14), his results exhibit two general trends. First, within a region, he found consistently higher species richness on the continental slopes than on the shallower continental shelves inshore. Sanders (1968, 1969; Slobodkin and Sanders 1969) attributed this pattern to the temporal stability of the deepwater sites, which do not experience the pronounced seasonal and long-term fluctuations in temperature and other factors that are characteristic of shallow waters. Second, he found substantial geographic variation in diversity: he found higher species richness in the more productive waters of the Pacific Northwest than at comparable latitudes in the Atlantic off New England, and even higher richness in the tropical waters of the Bay of Bengal (Figure 15.14).

Figure 15.13 Variation in species richness with depth in two groups of benthic marine organisms: gastropod mollusks (above) and polychaete worms (below). These humped-shaped relationships suggest that diversity is low in shallow coastal waters, highest in the heterogeneous terrain of the continental slope, and low in the abyssal depths. This is probably the typical pattern for samples of small areas (see also Figure 15.14), but recent data suggest that the abyssal plain may have high diversity at larger scales. (After Rex 1981.)

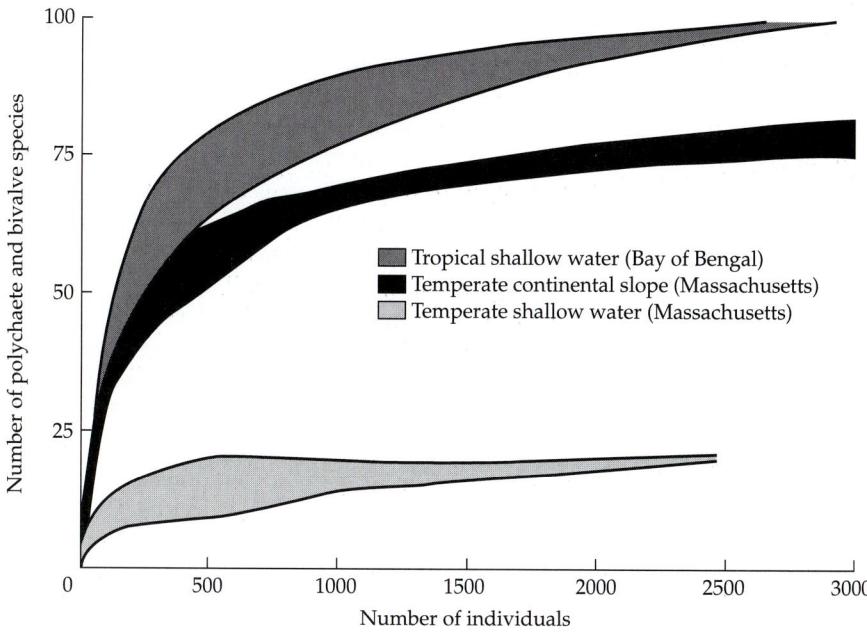

Figure 15.14 Sanders's data on species diversity in benthic marine bivalve mollusks and polychaete worms. Cumulative numbers of species are plotted against the numbers of individuals counted, so that diversity is expressed as a curve that approaches an asymptote when the entire biota has been sampled. The thickness of the curve shows the amount of variation among replicate samples. Note that the tropical locality has a higher diversity than the temperate one at a comparable depth. In the temperate localities, as in Figure 15.13 (but unlike the fishes in Figure 15.12), there is much higher diversity on the continental slope than in the shallower waters of the continental shelf. (After Sanders 1968.)

Most of these patterns of diversity in freshwater and marine systems are not too difficult to reconcile with those seen on land.

Recent explorations of the ocean depths, however, may challenge some of our ideas about diversity. Most of the ocean floor constitutes the vast abyssal plains that lie at depths greater than 2000 meters. This environment has unusual historical and ecological characteristics. Except for its gradual creation at mid-oceanic ridges and its consumption in trenches (see Chapter 6), the abyss, at a large scale, is perhaps the earth's most constant as well as its most spatially extensive environment. This vast area is uniformly dark, cold, and unproductive. Aside from the scientifically fascinating but quantitatively insignificant thermal vents (see Chapter 6), the sole energy source available to abyssal organisms is the incompletely decomposed organic material produced by photosynthesis at the surface that drifts down through the water column. As might be expected with such a limited energy supply, the abundance, biomass, and species richness per unit of area of the abyss does indeed appear to be very low.

On the other hand, there appears to be a great deal of spatial heterogeneity in both the environment and the organisms of the abyss. Much of this heterogeneity is abiotic, ranging from subtle variations in topography and substrate composition to the dramatic differences in temperature and chemical composition between the hydrothermal vents and the surrounding abyssal waters. Some of the heterogeneity, however, is biotic, or at least biogenic. Some reflects the effects of disturbance by bottom-feeding and burrowing animals, and some is due to variation in the kinds and spatial and temporal distribution of food resources, which range from tiny particles to entire whale carcasses. As a consequence of all this heterogeneity, both beta diversity and overall gamma diversity in the abyssal ocean depths may be surprisingly high (see Grassle 1991; Grassle and Maciolek 1992; Snelgrove et al. 1992, 1996).

As mentioned earlier (see Figure 15.6), there also appears to be a clear latitudinal diversity gradient on the abyssal plain. It seems likely that this gradient is due to resource availability, a consequence of higher productivity in warmer surface waters. Much more remains to be learned about life in the deep sea, and the findings are likely to have a major impact on our ideas about global patterns of diversity and the mechanisms that have caused them.

Smaller-scale patterns of diversity have been thoroughly documented for benthic organisms in intertidal and shallow subtidal habitats. The intertidal zone presents an interesting gradient in both abiotic conditions and biological diversity. For some reason that is probably important but not well understood, nearly all organisms living in the intertidal zone are of marine ancestry. The species diversity of these plants and animals declines dramatically with increasing elevation on the shore. So this intertidal gradient of decreasing diversity is also a gradient of increasing physical stress for these organisms, due to exposure to solar radiation, high temperatures, and desiccating conditions during the increasing periods of exposure between tides (see Chapter 3 and the discussion of Connell's studies of intertidal barnacles in Chapter 4).

Another common observation by marine and intertidal ecologists is that soft, unstable substrates such as mud and especially sand support many fewer species than nearby hard, permanent substrates such as rock and coral. Rocky intertidal habitats in the temperate zone and coral reefs in the tropics support particularly diverse communities, in which sessile plants and animals extend above the surface and crevice-dwelling and burrowing forms live beneath the surface, creating a complex, three-dimensional structure. In contrast, sandy beaches and mudflats throughout the world usually appear virtually lifeless, although the latter especially may support reasonable numbers of burrowing forms. It has been suggested that these differences can be attributed at least in part to variations in the frequency and extent of perturbations (e.g., Stanley 1979). Hard substrates, and the organisms that are firmly attached to them, are relatively resistant to buffeting from severe storms, which can virtually destroy soft substrates and temporarily eliminate most of their inhabitants.

Associated Patterns

Before returning to the hypotheses proposed to explain these observed gradients of species diversity, it is important that we look at two additional patterns. While they are not themselves gradients of species diversity, these patterns have important implications for diversity.

Rapoport's rule.   Rapoport (1982) observed that within widespread species of North American mammals, the areas of the geographic ranges of subspecies or geographic races showed an interesting relationship with latitude: the taxa had larger ranges at higher latitudes. Stevens (1989) showed that the same pattern holds at the level of species: there is a positive correlation between the range of latitudes over which a species occurs and the latitude of the center of its range (Figure 15.15). The area of the geographic range also tends to increase with the latitude of the center of the range, although the pattern is messier when plotted this way (Figure 15.16; Brown 1995). Stevens showed that this pattern, which he called **Rapoport's rule**, is found in many different kinds of organisms, including mammals, birds, mollusks, and plants. Rapoport's rule also has its analogue in elevational gradients in terrestrial environments and in

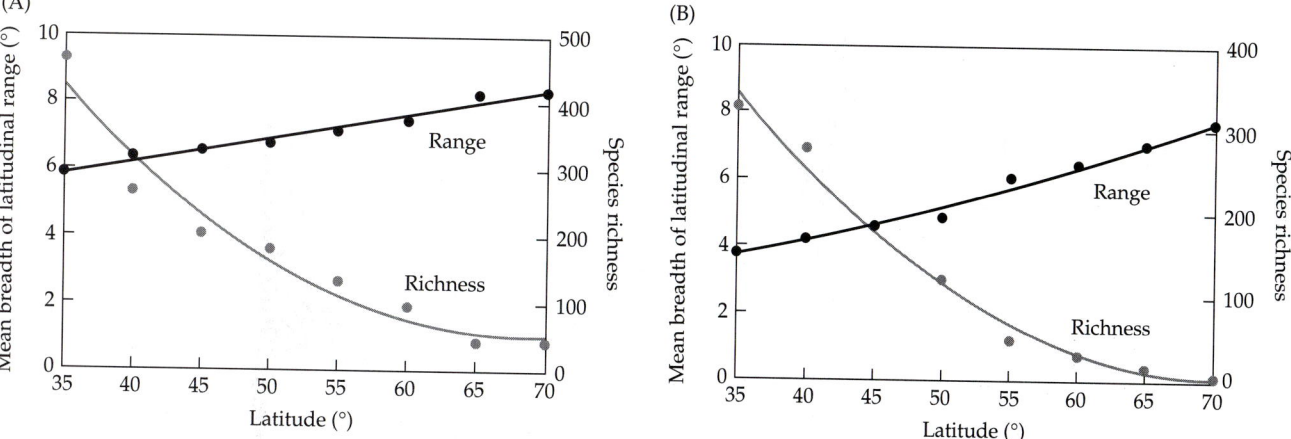

Figure 15.15 Two examples of Rapoport's rule in marine mollusk species along the Pacific Coast of North America (A), and in tree species in the continental United States and Canada (B). As species richness declines with increasing latitude, the remaining species tend to have geographic ranges that extend over a broader range of latitudes. This method of plotting Rapoport's rule (after Stevens 1992, 1989) depicts the average breadths of the latitudinal ranges of the species whose ranges are centered on 10° latitudinal bands. (After Brown 1995.)

depth gradients in the marine realm (Figure 15.17; Stevens 1992, 1996; but see Rhode et al. 1993, 1996). Over and over again, along geographic gradients, as species diversity increases, the range of the gradient inhabited by each species tends to decrease.

One important implication of Rapoport's rule is that alpha and beta diversity appear to be positively correlated. Along major geographic gradients, as the diversity of species within a habitat or small sample area increases, the

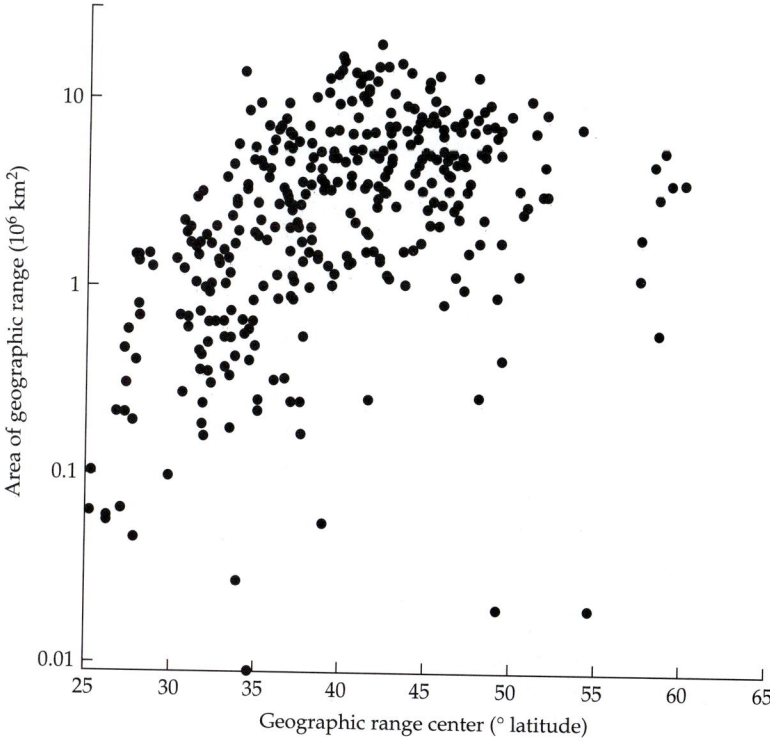

Figure 15.16 Variation in size of geographic range with latitude for North American breeding land birds. Although this method of plotting the data appears to show much more variation than in Figure 15.15, it also provides an example of Rapoport's rule: increasing range size with increasing latitude. (After Brown 1995.)

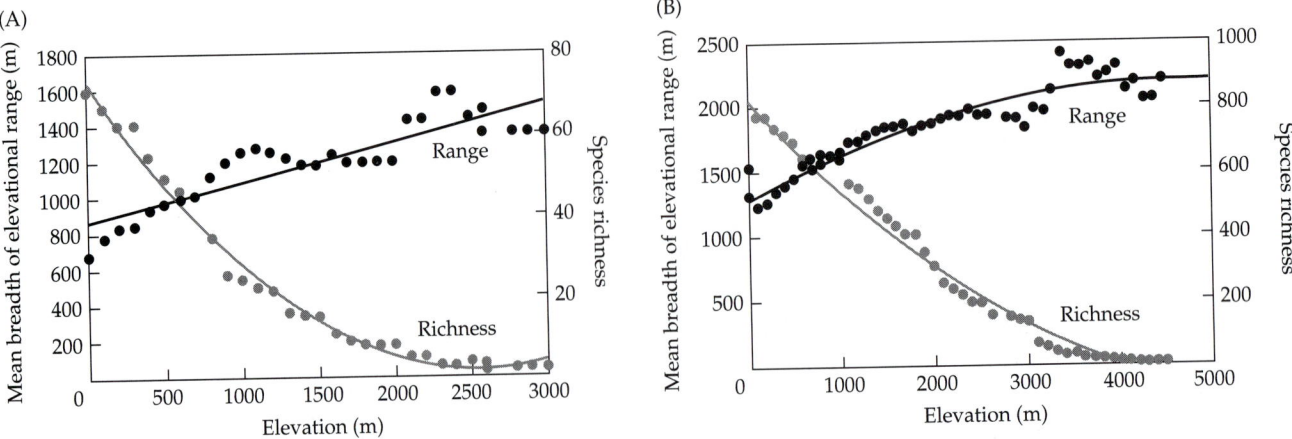

Figure 15.17   The elevational equivalent of Rapoport's rule: species tend to be distributed over a wider range of elevations as elevation increases. This pattern is seen both in tree species in Costa Rica (A) and in bird species in Venezuela (B). Compare with the latitude patterns in Figure 15.15. (After Stevens 1992.)

turnover of species between habitats and across the geographic landscape also tends to increase. This latitudinal pattern of beta diversity is shown in Figure 15.18, where we plot the proportion of the terrestrial mammal species in the regional species pool that were recorded in intensive censuses of small patches of uniform habitat. As the number of species in the regional pool increases, the absolute number of species occurring in a patch also increases, but the proportion of the species in the regional pool that occur in a patch decreases.

Dominance-diversity relationships.   A second pattern associated with latitudinal diversity gradients concerns the relative abundances of species. Above, we briefly mentioned that most indexes of species diversity combine two variables, the total number of species and the relative abundances of the species, in a single numerical value. A better representation of the distribution of abundances among species can be obtained by plotting the number of individuals or population density of each species, expressed as a proportion of all the individuals present, as a function of that species' abundance rank (Figure 15.19). Usually the proportional abundance variable is scaled logarithmically, and sometimes, especially for plants, some other measure of "dominance," such as cover, energy use, or biomass, is used instead of abundance. When such curves are compared for tropical and temperate areas, the tropical ones are typically much flatter (Hubbell 1979). This means that there tends to be less "dominance" in tropical habitats: not only are there many more rare species, but even the most common species account for a relatively small proportion of the total sample. The same trend of decreasing community dominance with increasing species diversity is observed for gradients of elevation. There is reason to suspect, therefore, that the inverse relationship between community dominance and species richness is a general phenomenon.

Thus, we see that the latitudinal gradient, and very likely the other major gradients of species diversity as well, may not be simply gradients of varying species richness. Other important characteristics of assemblages exhibit correlated variation. Not only do alpha and gamma diversity increase with decreasing latitude, but beta diversity increases, geographic range size decreases, and the distribution of abundances among species becomes more equitable. Any

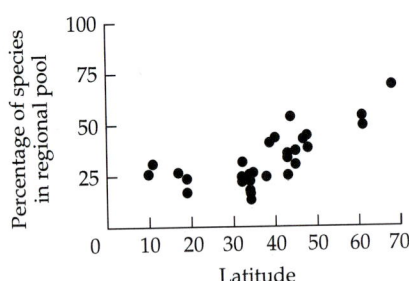

Figure 15.18   The relationship between beta diversity and latitude for North American terrestrial mammals. In this case, beta diversity is expressed as the percentage of the species in the regional species pool that occur in small patches of relatively uniform habitat; low percentage values indicate high turnover of species among different habitat types, and thus high beta diversity. (Data compiled by D. M. Kaufman.)

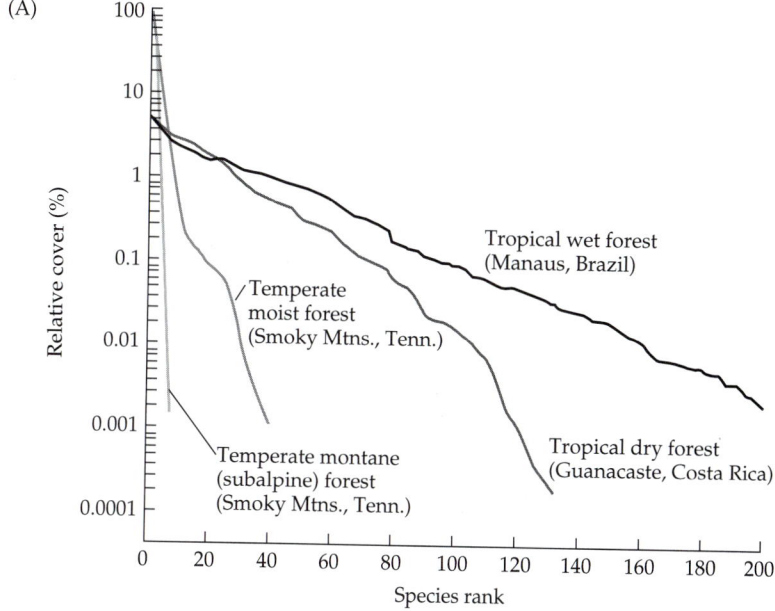

(A)

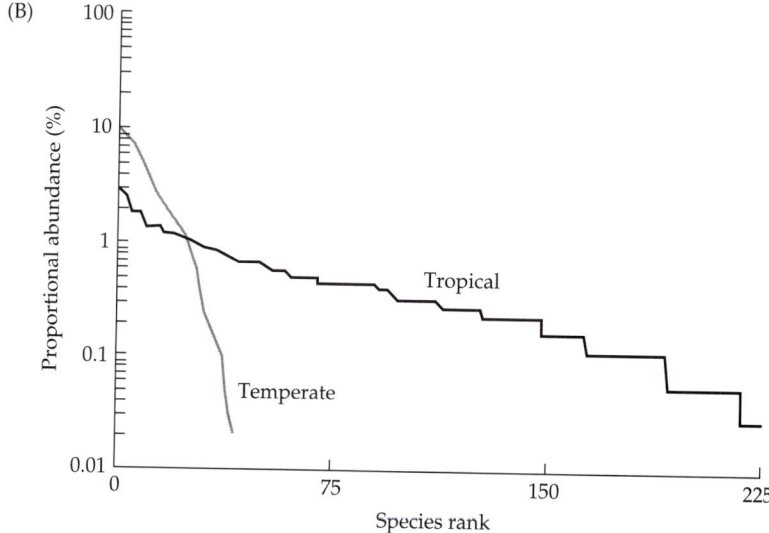

(B)

Figure 15.19   Dominance of tree species (A) and bird species (B) at sites of varying latitude. Here the dominance of each species (relative cover for trees; proportional abundance for birds) is plotted on a logarithmic scale in rank order, and the points are connected to form a curved line. Note that in the temperate habitats the most abundant species typically accounts for at least 10% of all individuals, and abundance declines rapidly with rank. In tropical habitats there is much less dominance; that is, abundance is much more equally distributed among species. (A after Hubbell 1979; B, data from Hubbard Brook, New Hampshire, and Manu National Park, Brazil, compiled by D. M. Kaufman.)

reasonably general explanation for the latitudinal and other major diversity gradients should presumably take account of all these relationships.

Causes of the Patterns

In discussing the major geographic patterns of species richness, we have already mentioned some of the hypotheses advanced to account for them. Here we shall consider these hypotheses in greater detail. Because many of these proposed mechanisms are not necessarily mutually exclusive, it is probably more realistic to ask how they might contribute to causing the observed patterns than to attempt to support or refute each one individually.

Although there is little disagreement about the existence of most of the patterns described above, the same cannot be said of the proposed explanations. One reason for the controversy is that most investigators have focused on testing a single hypothesis to account for just one pattern—most often the latitudinal gradient. Perhaps taking a broader perspective and developing a conceptual framework that includes many patterns and processes (see Pianka 1966, 1978) will eventually lead to better understanding and more consensus.

Nonequilibrial Mechanisms

Glaciation and climatic change.   Several investigators (e.g., Fischer 1960; Fischer and Arthur 1977; Stanley 1979) have stressed the importance of historical geological and climatic changes in causing latitudinal and other gradients of diversity. As we have seen, especially in Chapters 6 and 7, the earth has provided anything but a constant environment for its inhabitants. Over the last 500 million years, continents and oceans have changed position, continental seas and glaciers have covered extensive areas of land, mountains have risen and rivers have changed course, and climatic regimes have changed drastically. Over just the last 2 million years the earth has experienced repeated cycles of glaciation at high latitudes and elevations. We also know from the fossil record that there have been several episodes of widespread, catastrophic extinctions (see Chapter 8), which may have been caused by some of these changes, but may also have been caused by a more sudden, cataclysmic event, such as an asteroid striking the earth. It is certainly reasonable to expect that many extant patterns of diversity might be attributable to such historical perturbations, especially since the most recent of them occurred only 10,000 years ago, at the end of the Pleistocene.

There can be no doubt that these events in earth history changed the geographic distribution of diversity by causing the extinction or range contraction of many species while enabling others to invade and speciate in newly available regions. The clearest evidence that the legacies of these perturbations affect extant diversity comes from insular habitats that were created or isolated since the end of the Pleistocene (see Chapter 13). It is tempting to attribute latitudinal diversity gradients to similar causes. There may have been insufficient time for species to colonize and adapt to the extensive high-latitude environments that became available with the retreat of the vast continental glaciers and the associated climatic changes. This is even more likely if there were few refuges where arctic and temperate species could have survived glacial periods. Thus the lower diversity of temperate forest trees in Europe than in North America has been attributed to the fact that in Europe the major mountain ranges, such as the Pyrenees and the Alps, run east to west, whereas in North America the major ranges, the Appalachian and Rocky Mountains and the Sierra Nevada, run north to south. In Europe, the trees, and perhaps other kinds of organisms, may have been "trapped" against the mountains and become extinct, whereas in North America they may have been able to retreat southward in front of advancing glaciers and cooling climates, and thus survived.

There are several problems, however, with invoking Pleistocene changes as a general explanation for latitudinal and other patterns of species diversity. First, although it was once generally believed that Pleistocene climatic changes were confined to high latitudes, there is now excellent evidence that major shifts in climate and habitats occurred worldwide (see Chapter 7). For example, it appears that drier climatic regimes during glacial periods caused the contraction of rain forest and the expansion of savanna in the Amazon basin and in other tropical areas around the world. If tropical habitats have maintained their high diversity despite such perturbations, this weakens the argument that comparable changes can account in large part for the low diversity of species in the temperate zones.

Second, marine species richness shows the same qualitative trend from the tropics to the poles as terrestrial species richness, yet marine organisms have almost certainly been less affected by Pleistocene changes. Water temperatures changed less on average than terrestrial climates (see Chapter 7; CLIMAP 1976), and the northern oceans apparently were not covered with much more ice than at present, so marine habitats were not greatly restricted or fragmented. In any event, high-latitude forms could have remained in relatively constant environmental conditions simply by shifting their ranges toward the equator during colder periods.

Third, low diversity is not confined to the cooler regions at high latitudes, but is also characteristic of cooler high-elevation environments at all latitudes. Although high mountains in the tropics have been subjected to climatic changes and even glaciation during the Pleistocene (Simpson 1974, 1975), the resulting shift of climatic and vegetation zones to lower levels would have increased the size and decreased the isolation of these areas, having the opposite effect of the continental glaciers and associated climatic changes at high latitudes. Taken together, these patterns suggest that some common influence of low temperatures per se may be the simplest explanation for the low species diversity of cold regions.

Other historical perturbations.   Other historical explanations for diversity patterns push the time of important events back beyond the Pleistocene, invoking one or more of a number of geological and climatic changes, including shifting poles, long-term climatic shifts, the formation of mountains and other geological features, and so on. Although the postulated events may indeed have occurred, there are two major problems with using them to account for contemporary patterns of species diversity. First, the best estimates of the total number of species inhabiting the earth at various times in the past (Sepkoski et al. 1981; see also Raup 1972, 1976; Boucot 1975a,b; Sepkoski 1976; Flessa and Sepkoski 1978) suggest that there has been only a modest increase in total species richness over the last 500 million years (Figure 15.20). Even the

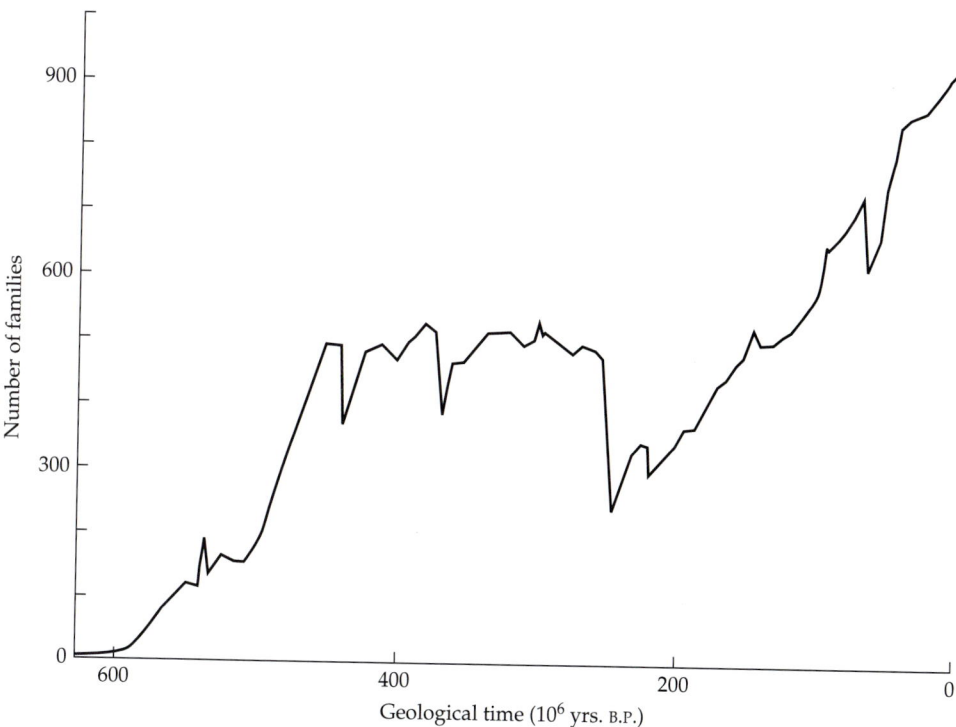

Figure 15.20   Diversity of families of benthic marine organisms over the Phanerozoic. These data, like those in Figures 8.18 and 8.19, suggest that diversity has increased over evolutionary time. Note, however, that the rate of change has been far from constant; most notably, diversity was drastically reduced several times by mass extinction events and did not recover for several million years. (After Sepkoski 1984.)

catastrophic extinctions at the end of the Permian and Cretaceous, the former of which was estimated to have exterminated over 88% of all marine species (see Chapter 8; Raup 1979; Raup and Sepkoski 1982), did not prevent the rapid reattainment of high diversity within a few million years. The history of many taxonomic groups is characterized by rapid early radiation and then a long period during which diversity changes relatively slowly. In Chapter 8 we show examples of such patterns for mammals (Figure 8.18; Lillegraven 1972) and land plants (Figure 8.19; Niklas et al. 1980; Knoll 1986).

Second, as evidenced by the numerous species that have originated during or since the Pleistocene (see the fish examples in Chapter 8 and the Hawaiian examples in Chapter 12), 1 or 2 million years is sufficient evolutionary time for adaptation and speciation. If some groups have been able to colonize and adapt to cold environments at high latitudes and elevations, why have not others been able to do the same? Why have the cold-adapted forms not speciated and radiated to fill those environments with more species? In the case of some Pleistocene perturbations, it may be true that there has not been sufficient time, but even in these cases, the historical explanation should be advanced with caution. Dov Sax (personal communication) has shown that several kinds of exotic organisms, including European plants and North American freshwater fishes, exhibit both a latitudinal diversity gradient and Rapoport's rule in their new environments. Since these species have been introduced by modern humans from other landmasses, one cannot invoke any influence of Pleistocene climatic changes or other historical perturbations for their patterns of diversity on the newly colonized continents. One cannot discount, however, the possibility that some lingering legacy of historical perturbations that occurred on their continents of origin was "carried over" to affect their diversity on the continents to which they were introduced (see Figure 15.21; Strong 1974; Strong et al. 1977).

To make clear the distinction between historical perturbation hypotheses and the equilibrial hypotheses that will be described below, one final point warrants discussion. Many investigators have attempted to explain diversity

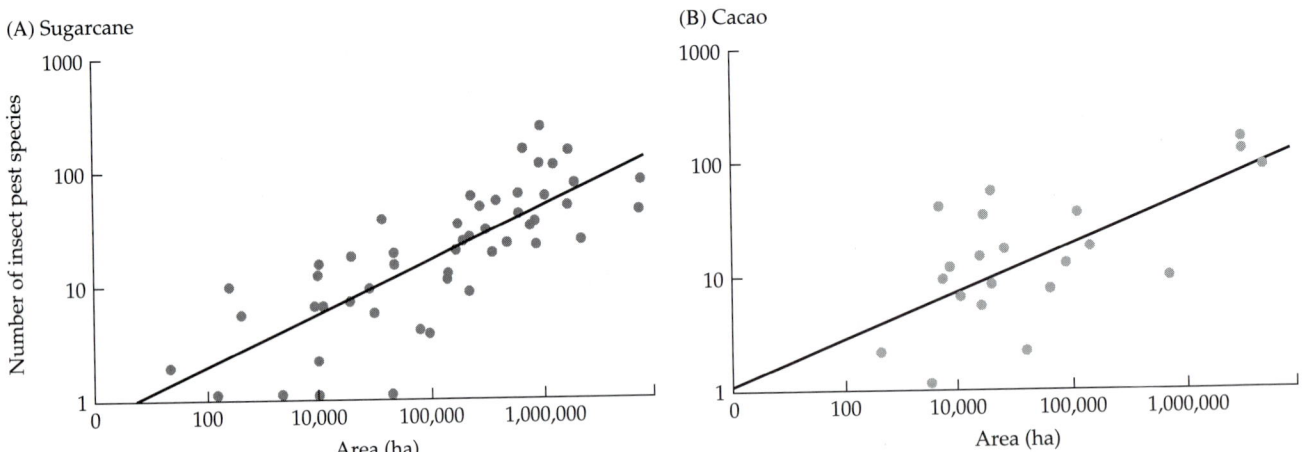

Figure 15.21 Number of species of insect pests recorded on two crop plants, (A) sugarcane and (B) cacao, as a function of the area of the crop under cultivation in various geographic regions. The rapid increase in herbivore diversity as these crops have been planted around the world suggests that even potentially specialized insect species can rapidly make the behavioral and physiological changes necessary to exploit an abundant new host plant. (After Strong et al. 1977.)

gradients in terms of differences in speciation and extinction rates, suggesting that speciation rates must be higher or extinction rates lower in regions of high diversity, such as the tropics. The rates of these processes are important, but the crucial question is how they change with the number of species already present (e.g., Rosenzweig 1975, 1995). Stanley (1979) perceptively points out that speciation and extinction rates are somewhat analogous to birth and death rates, so that an increase in species richness is analogous to population growth (see also Brown 1995). Because speciation is a branching or multiplicative process, the number of species has the potential to increase exponentially, just as the number of individuals in a population does (see Chapter 3). For growing populations, however, some ecological resource eventually becomes scarce, and growth levels off. Just as populations are limited ultimately by the availability of resources and other environmental conditions, the number of species that the earth, or any part of it, can support is also ultimately set by environmental limiting factors.

The crucial question for evaluating the relative importance of historical and equilibrial processes concerns the frequency and magnitude of historical perturbations relative to the proliferation of species. If environmental changes are sufficiently frequent and severe that most of the time species numbers are increasing (or decreasing) at nearly exponential rates, then clearly historical perturbations must be assigned a primary role in the determination of diversity patterns (Figure 15.22A). If, on the other hand, environmental changes are sufficiently small and infrequent, then most of the time species richness will be limited primarily by resource availability and other characteristics of contemporary environments (Figure 15.22B). During the nearly exponential increase phase, the rate of species increase will be a positive function of the number of species already present, and as the equilibrium point is approached, it will be a decreasing function.

The actual temporal pattern of species diversity probably includes both exponential and environmentally limited phases (Figure 15.22C). Life on earth has been decimated several times by catastrophic extinctions, but the overall patterns of diversity through time (see Figure 15.20; see also Figures 8.18 and 8.19) suggest that only for a brief period immediately following these catastrophes has there been a nearly exponential increase in species richness. It is unlikely, however, that species richness has ever attained equilibrial levels, for two reasons: first, the environment is always changing, and second, organisms are continually evolving, improving their capacity to use the resources, tolerate the biotic interactions, and exploit the abiotic environments present on earth. This continual evolution is suggested by the continual increase in diversity over time (see Figures 15.20–15.22). It can probably be attributed to both individual and species selection operating on the phyletic lineages that have persisted since the origin of life to increase the number of ways in which species can make a living and pack into ecological communities.

Equilibrial Mechanisms

Productivity.   It is probably not a coincidence that most of the regions and habitats that contain the greatest numbers of species are also highly productive. Productivity and biomass are much greater in rain forests and coral reefs in the tropics than on tundra and continental shelves at high latitudes. Although not all productive environments are rich in species, the general correspondence between geographic variation in productivity and the major latitudinal, elevational, aridity, and depth gradients of diversity is striking.

A causal relationship between productivity and diversity has been suggested by Hutchinson (1959), Connell and Orias (1964), MacArthur (1965,

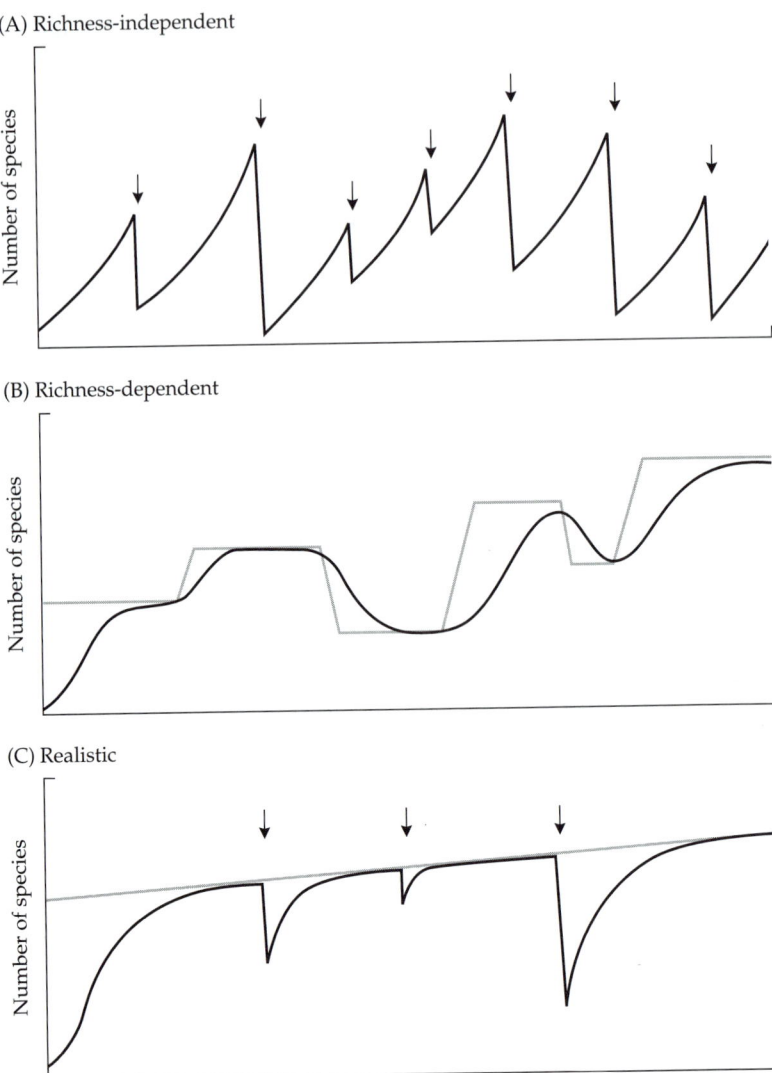

Figure 15.22 Three graphical representations of possible scenarios for the pattern of biological diversity over time. (A) A richness-independent model in which the number of species tends to increase exponentially, but is periodically reduced by extinction events of varying magnitudes (represented by arrows). (B) A richness-dependent model in which diversity tends to approach some equilibrial value (a "carrying capacity for species"; gray line), but the equilibrium shifts with changing environmental conditions. (C) A more realistic representation in which the number of species shows a gradually increasing trend (gray line), which is interrupted by episodic extinction events.

1972), Brown (1981, 1988), Wright (1983), Currie (1991), and others. MacArthur (1972) presented a graphical model of resource use that shows the processes likely to be involved in the way in which species pack together to form communities (Figure 15.23). Basically, this model suggests that more productive environments support more species because, on average, species can be more specialized and still maintain sufficiently large populations to avoid extinction. It assumes implicitly that there is a trade-off between being specialized and being generalized, so that specialists use their narrow range of resources more efficiently than generalists do, and thus are superior competitors in the overlapping portions of their niches. However, specialists are constrained to have smaller populations, and hence are more susceptible to extinction than generalists. Because, other things being equal, the areas under the resource utilization curves in Figure 15.23 are proportional to population size, at an equilibrium between rates of origination and extinction, a more productive environment would contain more specialist species. These species would each have about the same average population size as each of the fewer species of generalists living an environment where resources are scarce.

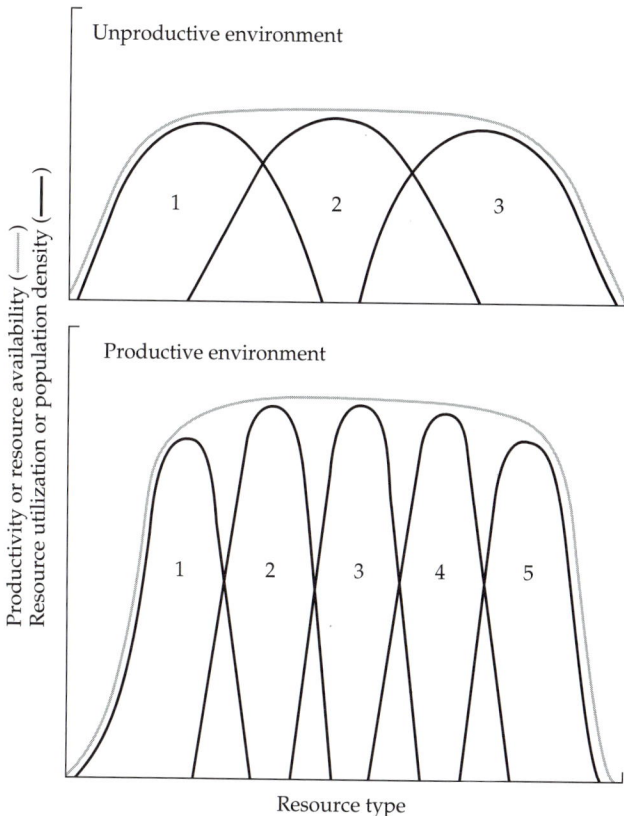

Productivity or resource availability (———)
Resource utilization or population density (——)

Unproductive environment

1 2 3

Productive environment

1 2 3 4 5

Resource type

Figure 15.23   A simple graphical model showing how a more productive environment can support a greater number of more specialized species. The gray lines depict the production of energy or some other essential resource, and the black lines show how those resources are utilized by coexisting species (numbered). If the population size of any species is a function of the area under its resource utilization curve, the more productive environment can accommodate a larger number of more specialized species, each of which can be sufficiently abundant to have a low probability of extinction.

Our considerations of limiting factors and community energetics in Chapters 4 and 5 bolster the reasoning developed above (see Hutchinson 1959; Van Valen 1973b; Brown 1981, 1988). Usable energy is the primary limiting resource required to sustain life. It follows logically that the more energy available, the more species can subdivide the supply, with each still acquiring a sufficient quantity to maintain a viable population size. Thus one might expect close correlations between species diversity and rates of energy supply: rates of solar energy input for plants, rates of primary productivity for heterotrophic organisms, and rates of food supply for particular animal groups. However, other limiting factors influence the extent to which different kinds of organisms can use various energy sources. For example, much of the solar energy in desert ecosystems cannot be used by plants because of limited water supplies. Shortages of soil nutrients limit primary production in some other habitats. Similarly, nitrogen and other nutritional factors besides energy content limit the capacities of many heterotrophs to use apparently suitable food resources.

Although the logic supporting a positive correlation between productivity and diversity seems irrefutable, the empirical relationships and causal mechanisms involved remain poorly understood. Increasingly accurate measures of both productivity and diversity in various plant and animal groups show a very close correlation on a continental scale (Figure 15.24; Currie and Paquin 1987; Currie 1991). The productivity of surface waters and its influence on resource availability seems to be the factor most likely to explain the latitudinal gradient of diversity in the abyssal depths of the oceans (see Figure 15.6; Stuart and Rex 1994). At smaller scales, however, there are many obvious exceptions. Small, shallow eutrophic lakes have fewer species of algae, zooplankton, and probably of other aquatic organisms than large, deep oligo-

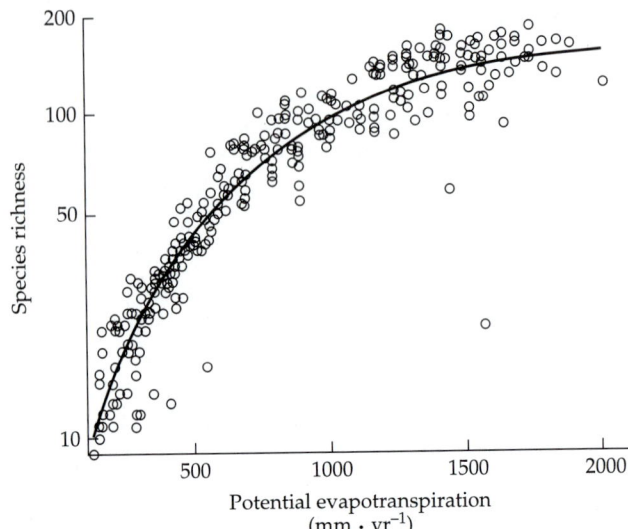

Figure 15.24   Species richness of nonvolant terrestrial vertebrates (amphibians, reptiles, and mammals) in North America as a function of potential evapotranspiration, an estimate of productivity. In these taxa, diversity increases continuously with increasing productivity. (After Currie 1991.)

trophic lakes. Salt marshes and hot springs are among the most productive ecosystems, but they typically contain many individuals and a high biomass of a very few species. In studies in which there has been an effort keep some of the potentially confounding factors more constant, the relationship between diversity and productivity often appears to be hump-shaped, with diversity peaking at intermediate productivity levels (Figure 15.25; Tilman 1988; Rosenzweig 1995). Clearly, productivity alone is an inadequate explanation for geographic variation in species richness.

Harshness and abiotic stress.   Many of the habitats that support a small number of species are ones that we would intuitively think of as harsh or stressful—that is, having conditions inimical to life in general. A list of such environments would include not only unproductive habitats such as deserts, arctic tundra, caves, and abyssal depths, but also highly productive ones such as salt marshes, estuaries, hot springs, eutrophic lakes, and temporary ponds. Is it possible to make the concept of **harshness** sufficiently precise that it becomes a useful tool for explaining patterns of species richness?

There are actually at least two concepts of harshness. The first suggests that certain conditions and environments are harsh or stressful in some absolute sense. Life is constrained by its most basic physical and chemical features. Low temperatures, for example, constitute a major environmental stress for all living organisms. Because the rates of chemical reactions are temperature-dependent, the biochemical and physiological performance of all organisms tends to increase with increasing temperature up to a certain point—typically about 37° C, the body temperature of endothermic birds and mammals. At higher temperatures, proteins tend to denature, imposing another stress. Because biochemical reactions occur in an aqueous medium, desiccation and freezing are also major stresses. Other environmental conditions, such as low oxygen levels and extremes of pH and salinity, can be shown to have negative effects on biochemical and physiological performance.

Presumably these abiotic stresses affect diversity because progressively fewer species can produce and maintain the adaptations required to persist in increasingly stressful environments. At a proximate level, avoiding or counteracting stress usually requires the expenditure of energy that otherwise could be allocated to other activities, such as reproduction and biotic interactions. Freezing, for example, creates a major problem for plants. Nearly all tropical plants are highly intolerant of frost. Many of the temperate plants that live in macroenvironments where frosts occur avoid freezing temperatures altogether by having life history stages, such as seeds, bulbs, and tubers, that survive in protected subterranean microenvironments. Those plants that do experience freezing temperatures, such as trees and shrubs, have special adaptations that protect their tissues from the damage that would otherwise be caused by ice crystals. Similarly, plants in arid environments have adaptations for going dormant to avoid drought stress, for acquiring water during rainy periods and storing it for use during droughts, and for minimizing water loss from their tissues. The energy costs of these adaptations are evidenced by the fact that plants from cold and arid environments typically have lower growth rates than tropical plants when grown under warm, moist conditions (see Chapter 4).

The second concept of harshness is a relative one. Terborgh (1973b; see also Brown 1981, 1988) has suggested that harshness, and its opposite, **favorableness**, be used to reflect the relative rates of species origination and extinction in different habitats or geographic regions. By this definition, harsh environments would be those in which extinction rates are high or colonization and speciation rates are low, or both. Several characteristics of environments determine these rates, and hence their relative harshness. Geographic isolation and physical conditions that are very different from those of the surrounding areas reduce the rate of successful colonization. Small and ephemeral habitats have high extinction rates and concomitantly low speciation rates.

To better understand this concept of harshness, consider the example of hypersaline lakes. A few landlocked bodies of water in desert regions, such as the Great Salt Lake and the Dead Sea, have salt concentrations several times that of seawater. As mentioned in Chapter 10, only a few kinds of animals and plants occur in these unusual habitats, and these organisms, such as the brine shrimp *Artemia* and the alga *Dunaliella*, tend to have wide distributions, occurring in many, if not all, of the widely scattered hypersaline lakes around the world. It is not difficult to imagine why this low diversity persists. The lakes are difficult to colonize, not only because they are small and isolated, but also because most potential colonists cannot tolerate the high salt concentrations. Extinction rates must be high in the lakes, not only because they are small, but also because they are ephemeral. Great Salt Lake, for example, was a giant freshwater lake during the pluvial periods of the Pleistocene, most recently only 10,000 years ago. Most of the inhabitants of hypersaline lakes, including *Artemia* and *Dunaliella*, are relatives of marine taxa that have specialized to occupy temporarily landlocked hypersaline pools and embayments along seacoasts. The fact that some species can inhabit these environments suggests that their harshness cannot be attributed to high salinity per se. If all the oceans of the world were as saline as Great Salt Lake, and their salt concentrations had increased gradually over hundreds of millions of years, we would expect organisms to have adapted and diversified in those seas pretty much as they have in our present oceans.

Used in this way, the concept of relative harshness, together with productivity, promises to provide a potentially powerful equilibrial explanation for many contemporary patterns of diversity (Brown 1981). Most habitats with high species richness are both productive and favorable, whereas habitats with low species richness are either unproductive (e.g., deserts, tundra, abyssal depths), harsh (e.g., salt marshes, hot springs, and temporary ponds), or both (e.g., caves and alpine lakes). The productivity and harshness mechanisms exert their effects on diversity proximately by influencing population

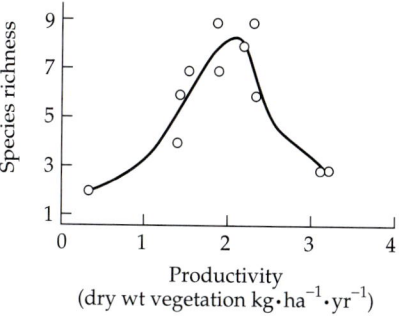

Figure 15.25 The relationship between species richness and productivity for rodents in the Gobi Desert region of central Asia. Some researchers have suggested that such hump-shaped diversity-productivity relationships are the general rule. (After Rosenzweig 1995.)

sizes and ultimately by affecting the balance between origination and extinction rates. The interaction of these mechanisms may even help to explain some of the more perplexing diversity patterns, such as the high diversity of plankton in the central North Pacific and of benthic marine invertebrates on the continental slopes and perhaps on the abyssal plain. These environments are not highly productive, but they cover extensive areas. Therefore, although the local densities of most species populations are low, the species have wide geographic ranges and thus large total population sizes. Similarly, the inverse relationship between the diversity of zooplankton and productivity in temperate lakes is resolved if one supposes that the critical factor is the total population size of each species, and that this is determined by the total productivity of the lake, rather than by the productivity per unit of surface area. If total productivity is calculated by multiplying measured productivity per unit of surface area by the surface area of the lake, the relationship between diversity and productivity becomes positive rather than hump-shaped (see Figure 15.11).

Climatic stability.   Some investigators have suggested that the short-term, especially seasonal, climatic fluctuations at high latitudes are a primary cause of latitudinal diversity gradients. We include such fluctuations under equilibrial processes because they do not limit diversity by causing wholesale extinctions. They occur with sufficient regularity that the organisms that live in these fluctuating environments are presumably able to tolerate them.

There are several possible mechanisms by which climatic fluctuations can affect diversity. They may be thought of as one kind of absolute harshness or abiotic stress. Special adaptations may be required to deal with them, and these may involve the expenditure of energy that could otherwise be used for reproduction, foraging, defense, and other fitness-enhancing activities. Seasonal and other short-term fluctuations may also select for generalist phenotypes, thereby preventing the development of assemblages containing many highly specialized species.

There are two problems with climatic stability as a general explanation for global patterns of species diversity. First, there are numerous exceptions. Many regions with low species richness, such as tropical mountaintops and the abyssal depths, do not experience highly seasonal physical environments. On the other hand, many habitats with high species richness do experience seasonal climates. For example, as pointed out in Chapters 3 and 5, most tropical regions have highly seasonal rainfall regimes. In some tropical areas the dry season is so severe that deciduous forest or savanna vegetation takes the place of evergreen tropical forest. Although these habitats are typically less productive and less rich in species than rain forests, they are more productive and diverse than most temperate and arctic regions. Second, there is the same argument that can be made against historical perturbation hypotheses: if some species can adapt to these habitats by becoming dormant during unfavorable seasons, or by other means, why cannot other species invade, and why cannot the present inhabitants speciate to generate greater diversity?

Nevertheless, as pointed out above, the seasonality of solar radiation input is one of the few features of the earth that is directly related to latitude. And because of the special adaptations required to cope with any kind of wide environmental variation, seasonality can be thought of as a kind of abiotic stress. Furthermore, seasonality, whether of temperature regime at high latitudes or of precipitation regime in the tropics, affects total annual productivity. Not only does productivity vary on an annual cycle, but total annual productivity is almost always lower than for otherwise comparable, but less seasonal, environments. Thus, because it is inextricably linked with abiotic stress, productivity, or both, seasonality is almost certainly an important factor influencing global patterns of diversity, especially along gradients of latitude and aridity.

Habitat heterogeneity.   In a pioneering study that had great influence on community ecology, MacArthur and MacArthur (1961) showed that the number of bird species in different habitats in the eastern United States was highly correlated with a measure of the number of vertical layers of vegetation present, which they called **foliage height diversity** (Figure 15.26). MacArthur and his associates have shown that the general relationship between bird species diversity and foliage height diversity holds for a variety of mesic habitats in North America, from Panama to the northeastern United States (MacArthur et al. 1966). Many other close correlations between species richness and the complexity of vegetation structure have been documented, not only for birds in a wide range of habitats (e.g., MacArthur et al. 1966; Recher 1969; Cody 1968, 1974, 1975; Abbott 1978), but also for a variety of other organisms in different habitats, such as desert rodents (Rosenzweig and Winakur 1969) and desert lizards (Pianka 1967). Clearly, coral reefs, probably the most diverse marine habitats, also exhibit a great deal of structural complexity.

It is important to ask whether these patterns reflect a direct causal role for structural habitat heterogeneity in promoting species richness, or whether they reflect correlations between habitat structure and some third factor, such as historical perturbation, productivity, or harshness. For many taxa, it is clear that the relationship holds, and that it reflects the tendency of coexisting species to apportion essential resources, such as food, along niche dimensions of habitat structure, such as foliage layers in birds (e.g., MacArthur 1958; Cody 1974) and distance from shrub cover in desert rodents (e.g., Rosenzweig and Winakur 1969; Brown 1975). This is not good evidence, however, that physical habitat heterogeneity by itself promotes species richness. Productivity also varies in concert with vegetation heterogeneity: tall, multilayered tropical rain forests are more productive than temperate forests, which in turn are more productive than temperate old fields and shrublands. In relatively arid regions of the western United States, Mexico, and Central America, there are pine forests with tall but widely spaced trees and sparse understory vegetation. These pine forests have a foliage height diversity comparable to that of eastern deciduous forests, but substantially lower bird species diversity, and probably lower productivity as well (E. W. Stiles and G. H. Orians, personal communication).

Particularly intriguing are the patterns found among some aquatic organisms, such as the benthic marine invertebrates of the continental slope and abyssal plain and the freshwater and marine plankton. These organisms attain extremely high diversity in habitats that seemingly lack much in the way of structural heterogeneity. Hutchinson (1961) was so impressed by the numbers of plankton species coexisting in seemingly uniform habitats that he termed this "the paradox of the plankton." Recent oceanographic and limnological studies indicate that in fact, oceans, and even lakes, are by no means uniform water masses (e.g., Wiebe 1976, 1982), but it is not yet clear to what extent the spatial heterogeneity that is being revealed on different scales can resolve Hutchinson's paradox.

Another problem is that structural habitat heterogeneity is often a product of the organisms themselves. Two obvious examples are tropical rain forests and coral reefs. Another possible example, mentioned above, is the suggestion that the abyssal plain may actually be extremely heterogeneous owing to biogenic processes, such as disturbance and clumping of food materials. So is

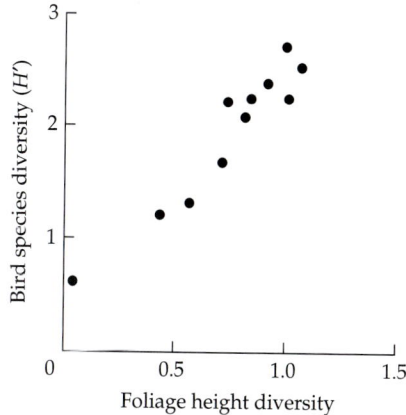

Figure 15.26   The relationship between bird species diversity, measured using H' (one of the widely used diversity indexes), and foliage height diversity, measured as the number of equally dense layers of foliage at different heights above ground, for deciduous forest habitats in eastern North America. This positive relationship, which holds well within forested habitats, is an example of how species diversity often varies with the structural heterogeneity of the habitat. (After MacArthur and MacArthur 1961.)

structural complexity a cause or a consequence of diversity? Almost certainly the answer is both. Much of biotic structural heterogeneity is indeed a consequence of diverse, productive ecosystems. But structural complexity facilitates the existence and coexistence of many species with special requirements and adaptations for finding food, avoiding enemies, and coping with abiotic stress. In addition, environmental heterogeneity at large spatial scales certainly promotes global diversity by facilitating the evolution and persistence of different endemic taxa in isolated areas with similar environmental conditions (see Chapter 10). The much higher diversity of freshwater than marine fish species provides an excellent example. Although only about 1% of the earth's unfrozen water surface is fresh water, more than half of all fish species occur there. While other factors also play some role in this diversity, the most important is simply that freshwater habitats are highly fragmented and isolated by biogeographic barriers. As a consequence, many lake and river systems have developed distinct fish faunas characterized by high degrees of endemism.

Area.   The earth is, in a sense, "fatter" at the equator, because the lines of longitude converge toward the poles. Of course, the earth is a sphere, so it is only because latitude is defined in terms of the earth's relation to the sun that the earth has more surface area at low latitudes. Nevertheless, the tropics are spatially extensive, whereas the temperate and polar zones have more restricted areas and are divided between the Northern and Southern Hemispheres. Rosenzweig (1992, 1995; see also Terborgh 1973; Colwell and Hurtt 1994) has suggested that this feature of geography has importantly influenced global patterns of species diversity (see Figure 6.24).

As we have seen from the description of insular species-area relationships in Chapter 13, large areas do indeed support more species than smaller ones. Similar patterns hold at larger scales. Flessa (1975; see also Brown 1986) found contemporary mammalian diversity to be closely correlated with the areas of continents. Raup (1972, 1976) noted that the diversity of benthic marine organisms tended to be high during periods in earth history when there were extensive shallow seas on the continental plates (see Figure 6.22). He suggested that changes in habitat due to shifting sea levels and tectonic events might explain, at least in part, the fluctuations in the diversity of marine invertebrates over the Phanerozoic. The extensive areas of certain tropical lands and oceans, such as the tropical forests of the Amazon and Zaire basins and the warm, shallow seas of the Indo-Pacific region, undoubtedly contribute importantly to their high diversity. On the other hand, area cannot be more than one part of the explanation for the latitudinal diversity gradient. The fact that tropical islands support many more species than temperate or polar islands of the same size shows that area alone cannot account for the pattern. Other factors, such as evolutionary history, productivity, and harshness, must also play important roles.

Recently, Rosenzweig (1995) put together a large body of data and theory to clarify the relationship between area and species diversity. He proposed a general theory of species-area relationships for both islands and continents. He showed that larger and nonisolated regions have not only more species in total (higher gamma diversity), but also more species per habitat (higher alpha diversity) than smaller and insular areas (Figure 15.27). Rosenzweig's theory explains how speciation and ecological specialization contribute to the buildup of biotas that at all scales are more diverse on larger land masses. We will return to these ideas at the end of this chapter.

Biotic interactions.   It is often suggested that the high species diversity of the tropics can be attributed to the fact that one or more of the classes of inter-

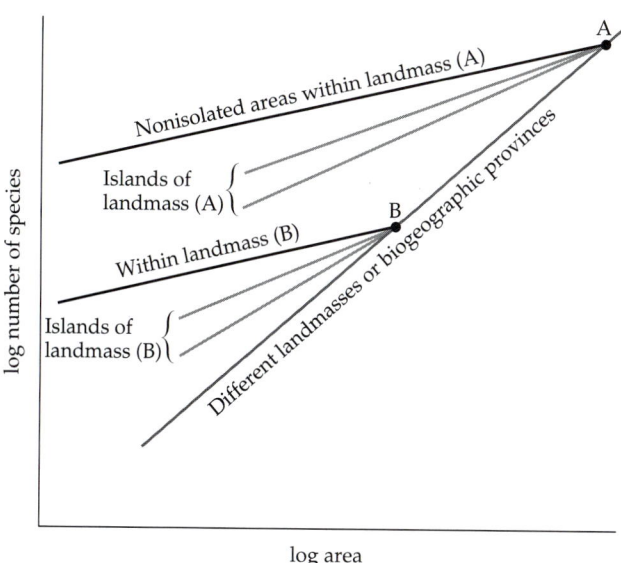

Figure 15.27 A schematic representation of how species-area relationships vary among large landmasses or geographic provinces, nonisolated sample areas within landmasses, and isolated islands or habitat patches whose biotas are derived from those landmasses. When data are plotted on logarithmic axes, typical z-values or slopes of the lines are: different landmasses, 0.5–1.0; nonisolated areas within a landmass, 0.1–0.2; islands or isolated habitat patches, 0.2–0.5. (After Rosenzweig 1995.)

specific interactions is more important at low than at high latitudes. One often hears that the intensity of interspecific competition, the relative effect of interspecific as compared with intraspecific competition, the intensity of predation, the complexity of food webs, or the development of mutualistic relationships is greater in the tropics and perhaps also in other species-rich regions (e.g., Dobzhansky 1950; Paine 1966; MacArthur 1972; Janzen 1970; Connell 1975; Menge and Sutherland 1976). Some or all of these assertions may be true. As the number of species increases, one would expect, from purely statistical considerations, more frequent interspecific interactions, and more opportunities to coevolve special relationships with other species. As in the case of habitat heterogeneity, the real challenge is to distinguish the correlates and consequences of diversity from its primary causes.

In one sense this is not difficult, because all general patterns of diversity must ultimately be explained in terms of physical causes: some combination of historical perturbations and contemporary variation in the abiotic environment. This does not mean, however, that interspecific interactions play no role in producing the geographic patterns of diversity. The abiotic environment can influence the rules and the outcomes of biotic interactions. Competition, predation, and mutualism determine in large part how the physical resources of a region are allocated among species to produce the observed species diversity and community structure (MacArthur 1958, 1972; Brooks and Dodson 1965; Paine 1966; Harper 1969; Cody 1974, 1975; Brown 1975, 1981; Connell 1975, 1978; Menge and Sutherland 1976; Lubchenco 1978, 1980; Lubchenco and Menge 1978; Vermeij 1978). For reasons mentioned above and in Chapter 4, the most important of these resources is solar energy, either in its original form or in its converted form as the organic molecules that constitute the bodies of living and dead organisms.

Speciation and extinction rates. Like biotic interactions, speciation and extinction processes affect the way in which the abiotic template of the earth has been divided up among species. It has been suggested that in the tropics, extinction rates are lower, speciation rates are higher, or both, than at high latitudes. Of course, everything else being equal, we would expect the absolute rate of both extinction and speciation to be higher in the tropics, because there are more species there. The critical question is whether the per species rate of either origination or extinction shows a clear latitudinal trend. Some authors have suggested that the more constant (less seasonal) environment of the tropics results in lower extinction rates, facilitating the persistence of rare species with small geographic ranges. This hypothesis is consistent with the trends exemplified in Rapoport's rule and the dominance-diversity relationship mentioned above.

These patterns might alternatively be explained, at least in part, by higher speciation rates in the tropics. Even if extinction rates were comparable, the tropics would tend to have higher diversity if rare, geographically restricted species were formed at a higher rate. As mentioned in Chapter 9, Janzen (1967) has suggested one reason why speciation rates per species might be higher in the tropics: because of the reduced seasonality, barriers to dispersal are more severe. As Janzen points out, "mountain passes are (effectively) higher in the tropics." In the temperate zones, a lowland species could cross a mountain pass at, say, 2000 m elevation in summer and encounter no more stressful abiotic conditions than it would experience in the lowlands in the winter. This is not true in the tropics. No matter what time of year it crossed, a tropical species would encounter lower temperatures at 2000 m than it had ever experienced in the lowlands (see Figure 9.11). The effectiveness of barriers formed by mountain ranges, and perhaps by other geological and climatic conditions, may promote the formation and persistence of species endemic to small areas. Thus, Janzen's hypothesis suggests how abiotic stress and speciation rates might vary with latitude in a way that would generate the pattern of smaller geographic ranges in the tropics described by Rapoport's rule.

Toward a Synthetic Explanation?

So how do we account for geographic patterns of species diversity? And, in particular, why are there so many species in the tropics? Do biogeographers and community ecologists have anything more to say about one of the most pervasive features of life on earth? Can the explanation(s) for the latitudinal diversity gradient be generalized to explain the elevational, aridity, depth, and other gradients?

We believe that the answers to the above questions are "yes." We can definitely say that all the factors discussed above affect diversity. They obviously operate on different temporal and spatial scales, and some are more direct or proximate than others. We believe, however, that we are approaching the point at which it will be possible to compile the information needed to arrive at a synthetic understanding of the variables and processes that determine the main patterns of diversity. We suspect that this synthesis will have two main components.

Time and space.   The first component is the influence of time and space on the historical buildup of diversity. While the effects of these two factors are not identical, there are many similarities between them. Longer periods of time increase the opportunities for both speciation and colonization. They also allow for evolution and coevolution, so that the accumulating species can adapt both to the abiotic environment and to one another. Larger spaces, similarly, provide larger targets for colonization, and increase opportunities for speciation because they are more likely to contain barriers to dispersal and gene flow, isolated habitat patches, and steep environmental gradients. Larger areas support larger populations, at least of some generalized species, and their greater range of environmental conditions is more likely to include the particular kinds of habitats and combinations of niche variables required by specialized species. So the overall effect of both longer time periods and larger areas is to increase colonization and speciation and to decrease extinction, with the resulting accumulation of a larger number of more differentiated and often more specialized species.

Furthermore, the effects of colonization, speciation, adaptation, coevolution, and specialization tend to feed back so as to facilitate one another. The consequence is that with increasing time and area, species diversity is higher on all scales. Not only do more species occur in larger areas (increasing gamma diversity), as expected from the species-area relationship, but increased specialization also leads to both higher diversity in local areas (increasing alpha diversity) and greater turnover of species between local areas (increasing beta diversity: see Rosenzweig 1995).

Time and space tend to differ in the way in which they mediate the influence of temporal environmental variations, especially large perturbations with the potential to cause extinctions. Long time periods increase the probability that such perturbations will occur. Episodic extinctions set back the accumulation of diversity (see Chapter 8). On the other hand, large spaces increase the probability that there will be local areas that are not affected by even the most severe perturbations, or at least that there will be local combinations of environmental conditions that will permit a greater number of species to survive.

Productivity, abiotic stress, and biotic interactions. The second component of a synthetic explanation for geographic patterns of diversity is a deeper understanding of the relationships among productivity, abiotic stress, and biotic interactions. These three factors are conceptually distinct, but functionally interrelated. As mentioned above, productivity is the rate at which photosynthesis in plants converts the energy of sunlight into the organic molecular forms of energy that power all biological processes. It is not surprising, therefore, that productivity varies inversely with latitude, reflecting the supply of solar energy to the earth's surface. It is also not surprising that species diversity generally increases with increasing productivity. The more solar energy that is made available for biological processes, the more different kinds of organisms are likely to be able to obtain a sufficient share of energy to become established and persist.

Because productivity is a biological process, however, it is affected by abiotic stress. In particular, productivity is limited by abiotic conditions that restrict the number, diversity, and performance of plants. Such stressors include aridity, low temperatures, high salinity, high concentrations of heavy metals or other toxicants, and many others. These stressful conditions usually result in low plant productivity, which leads straightforwardly to low species diversity in other organisms as well, because they are limited both by the harsh physical conditions and by low energy availability. Examples of stressful, low-productivity, low-diversity environments include deserts, mountaintops, areas of serpentine and saline soils, and hypersaline lakes. Some physically stressful but nutrient-rich environments, such as salt marshes and shallow lakes, are exceptions to this general pattern. A few kinds of plants have been able to adapt to the harsh physical conditions in these environments and attain high abundances and high rates of photosynthesis. Even though productivity is high, however, only a few kinds of other organisms can tolerate the stressful conditions, so species diversity remains low. The fact that these physically harsh, highly productive environments tend also to be small, isolated, and ephemeral (see above) contributes to their low diversity by limiting colonization, adaptation, and the buildup of a stress-tolerant biota.

Now, returning to the more usual situation, we can bring in biotic interactions. As productivity increases and abiotic stress decreases, interactions with increasing numbers of competitors, predators, parasites, and pathogens play an increasing role in limiting the abundance and distribution of species (Dobzhansky 1950; MacArthur 1972; Brown 1995; Kaufman 1995). Biotic interactions have several interrelated effects on diversity. They tend to limit geographic ranges at the low-stress, high-productivity ends of ecological gradients, such as closer to the equator or at lower elevations. Competitors and predators are also able to exclude species from otherwise habitable areas within their geographic ranges. Because the effects of interspecific interactions on populations are density-dependent, they also tend to prevent the numerical dominance of a few species, allowing the large energy supply to be divided among a greater number of rarer species.

Therefore, the interacting influences of productivity, abiotic stress, and biotic interactions appear to offer a synthetic explanation for the syndrome of phenomena associated with the gradient in diversity from the poles to the equator. Toward the poles, abiotic stress plays a much larger role than biotic interactions in limiting abundances and distributions, and a few stress-adapted, wide-ranging species divide up the little energy available. In the tropics, by contrast, abiotic conditions are much less harsh, but the large number of predators and competitors limits the abundance and distribution of each species, thereby allowing the large supply of energy to be divided among many relatively rare and geographically restricted species. MacArthur (1972) used the example of botanical gardens to illustrate the trade-off between abiotic stresses and biotic interactions in regulating diversity. He pointed out that the vast majority of plant species from throughout the world can be grown outdoors in the wet lowland tropics, provided that competitors are weeded out, herbivores are excluded, and bacterial and fungal pathogens are controlled. On the other hand, in order to grow these same plant species in the temperate zone, the primary requirement is a greenhouse to protect them from frost.

Coda.   We are still some way from having a synthetic explanation for geographic patterns of biological diversity that is both logically sound and consistent with the empirical evidence. For one thing, we need to incorporate into a single, unified framework the influences of a host of factors that vary across spatial and temporal dimensions—factors that have direct effects that may be straightforward and consistent, but which may also interact in complex ways to influence species diversity. Rosenzweig (1995) has made an important start by clarifying the interrelationships among area, isolation, and productivity.

It still remains, however, to incorporate the influences of other important factors, especially evolutionary history, abiotic stress, and biotic interactions. We are optimistic that this can be done, and we look forward to a day when biogeographers will have a new answer to the question, "Why are there so many species in the tropics?" Instead of saying, "Well, we have about 25 different competing, overlapping, and still untested hypotheses," they will be able to say, "We have a general, synthetic, well-supported theory to explain tropical species diversity."

CHAPTER *16*

Continental Patterns and Processes

In Chapters 13 and 15 we examined patterns in the diversity of species on a variety of scales, from small, isolated patches of habitat to vast oceans and continents. In Chapter 14 we considered other characteristics of insular species and biotas that reflect the influence of evolutionary and ecological processes operating in isolated habitats with low overall biodiversity. Now we will consider how these kinds of evolutionary and ecological processes have shaped the characteristics of species and assemblages occurring in much less isolated habitats on continents and in oceans. As noted in the previous chapter, the correlates and causes of biogeographic patterns on continents are much more complicated than those on islands. In part for this reason, from the time of Darwin and Wallace, islands have been used as empirical models of evolutionary and ecological phenomena. It is important to ask, however, to what extent the patterns and processes known to occur on islands also hold in geographically less isolated and biologically more diverse settings.

As we shall see, the answer to this question is mixed. Often the same kinds of processes are operating. Abiotic and biotic factors are limiting the abundances and distributions of species, selecting for characteristics of organisms that enhance their abilities to survive and reproduce, and affecting the composition of local ecological communities, regional assemblages, and continental biotas. The processes that determine the dynamics of populations and lineages are also operating: demography, dispersal, speciation, and extinction. But the outcome is often distinctly different on continents than on islands, largely because these processes are operating in a dramatically different environmental setting. An ancient and complicated legacy of earth history, a greater diversity of species and lineages, and less isolated but more extensive habitats all combine to produce a rich variety of biogeographic patterns. As in the case of the major geographic gradients of species diversity, while the species-level patterns are often clear and well documented, their causes are usually much less well understood.

Single-Species Patterns

Ecogeographic Rules

Bergmann's rule.   Some of the oldest and best-known biogeographic patterns are regular geographic clines of variation within species (see also Chapter 4). Among the first such patterns to be described were those in the morphology of terrestrial animals, especially vertebrates. In 1847 Carl Bergmann noted that among closely related kinds of mammals and birds, the largest forms often occurred at higher latitudes. Bergmann realized that larger animals have lower surface-to-volume ratios than their smaller relatives, which he suggested must be advantageous to endotherms in conserving body heat in colder climates.

The history of the study of latitudinal size variation is typical of many questions in continental biogeography. After Bergmann, many studies documented a positive correlation between body size and latitude among populations of a single species or among closely related species. The pattern proved to be sufficiently general that it came to be called **Bergmann's rule** (e.g., Allee et al. 1949; Rensch 1960; Mayr 1963). The link between climate and body size was strengthened by studies showing that patterns of geographic size variation were often closely correlated with environmental temperature (Brown and Lee 1969; James 1970; Smith et al. 1995; Hadley 1997). But as more studies were reported, the picture became less clear. Enough exceptions to the rule were noted that some authors began to question its empirical generality and its presumed adaptive basis in temperature regulation (e.g., Scholander 1955; McNab 1971). McNab made the point that organisms do not live on a per gram (or per volume) basis. Larger organisms have less surface area per mass or volume, but their *total* surface area, and their food requirements, are higher than those of smaller organisms. This might suggest that high-latitude environments, where temperatures and productivity are chronically low, would favor traits, such as small body size, that would minimize individual energy requirements.

Thus, surface-to-volume ratio may be an unlikely causal mechanism for Bergmann's rule. This, of course, does not preclude the operation of other physiological or ecological mechanisms that may account for the general increase in body size as we move poleward. For example, lower critical temperatures (the temperatures below which endotherms must increase metabolic energy expenditures to maintain a constant body temperature) decrease with increasing body size. In addition, because energy and water stores increase more rapidly with body size than do energy requirements, larger animals can endure food and water shortages for longer periods than can smaller individuals.

Nevertheless, thermoregulatory mechanisms may still be involved. Recent studies of intraspecific geographic size variation in the bushy-tailed woodrat (*Neotoma cinerea*) and the pocket gopher (*Thomomys talpoides*) support both the empirical existence of Bergmann's rule and its thermoregulatory basis—at least for these particular species of small mammals. The study of woodrats made three critical observations:

1. The capacity of individuals to tolerate stressfully high environmental temperatures, as indicated by upper lethal ambient temperature, decreases with increasing body size.
2. Geographic variation among contemporary populations shows a very strong negative correlation between body size and maximum July temperature in the habitat.

3. Historical changes in body size (determined from the sizes of fecal pellets preserved in middens: see Chapter 7) over the last 20,000 years show a similarly close negative relationship with environmental temperature (Smith et al. 1995).

As the climate changed—most spectacularly, as it warmed at the end of the Pleistocene—body size evolution in woodrats closely tracked the changes. As predicted by Bergmann's rule, each time the climate exhibited a warming trend, woodrats throughout the southwestern United States became smaller (Figure 16.1). Similar patterns of geographic and temporal variation in body size were documented in pocket gophers (Hadley 1997).

Thus, a thermoregulatory basis for the pattern, albeit different from the one first proposed by Bergmann, appears to apply to woodrats, pocket gophers, and probably some other endothermic vertebrates. The inverse correlation between body size and environmental temperature is not always so easily explained, however. One issue still to be resolved is how much of the observed geographic or temporal variation is genetic. James (1970) performed an important experiment on red-winged blackbirds (*Agelaius phoeniceus*), which show a pattern of increasing size in cooler, more northerly regions. When she transferred eggs between nests in different areas, she found that the offspring grew to a size characteristic of their rearing environment, not their ancestral population. While this does not disprove the adaptive significance of Bergmann's rule for blackbirds and other endotherms, it suggests that phenotypic plasticity, rather than genetic differences, may account for much of the observed variation.

The thermoregulatory consequences of being large is not the only hypothesis to account for increasing size in colder climates. Bergmann's rule may in fact be causally linked to the latitudinal gradient of decreasing species diversity from the equator to the poles. Low alpha diversity means reduced interspecific interactions, including competition. In the absence of their larger competitors, smaller members of the same guild often exhibit ecological release, increasing in size in species-poor environments—such as those at high latitudes (McNab 1971; see also Dayan 1990; Dayan et al. 1989; Iriarte et al. 1990).

The thermoregulatory basis of Bergamnn's rule is also challenged by repeated documentation that in many groups of ectothermic organisms (including nematodes, aquatic invertebrates, bees, ants, and reptiles), intraspecific variation in body size is also positively correlated with latitude and inversely correlated with environmental temperature (e.g., Ray 1960; Cushman et al. 1993; Hawkins 1995; Kaspari and Vargo 1995; Van Voorhies 1996). Van Voorhies (1996) showed that nematodes grown at colder temperatures exhibited extended growth and delayed maturation, ultimately attaining

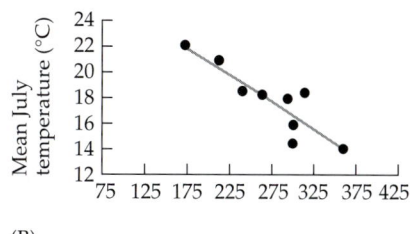

(B)

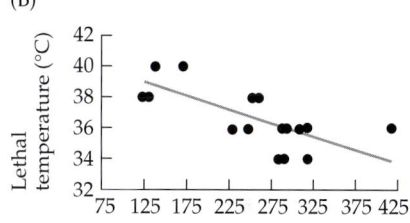

(C)

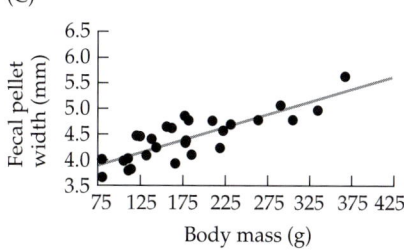

Body mass (g)

(D)

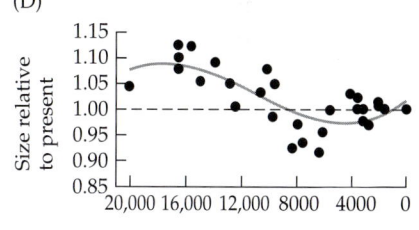

(E)

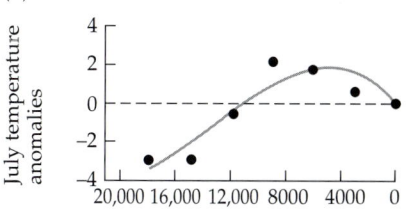

(F)

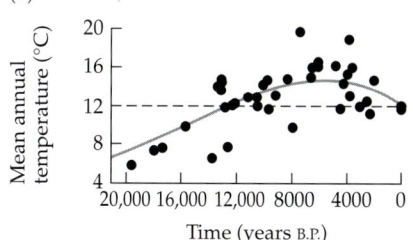

Time (years B.P.)

Figure 16.1 Bergmann's rule and adaptation to temperature in the bushy-tailed woodrat, *Neotoma cinerea*, over both geographic space and evolutionary time. (A) Geographic variation in body size among contemporary populations as a function of July temperature, showing that larger individuals occur in cooler climates. (B) Upper lethal temperature as a function of body size, showing that larger individuals are more susceptible to heat stress. (C) Fecal pellet size as a function of body size, showing that larger animals produce larger feces and allowing body sizes of past populations to be estimated from fossilized fecal pellets. (D) Relative sizes of past populations, estimated from the sizes of radiocarbon-dated fecal pellets, over the last 20,000 years. (E, F) Two measures of environmental temperature over the last 20,000 years, showing a pattern that almost exactly mirrors the changes in body size. So, over both space and time, there are adaptive shifts toward smaller sizes in warmer environments. (From Smith et al. 1995.)

larger adult body sizes. The generality of this developmental phenomenon remains to be explored (see Lonsdale and Levington 1985; Sand et al. 1995), but it might explain observations of Bergmann's rule not only in other ectotherms, but also in blackbirds and some other endotherms as well.

In summary, Bergmann's rule may represent still another case in which a general pattern in nature results not from a single mechanism, but from a combination of convergent forces, all of which vary in a predictable manner across the geographic template. Whatever unifying explanation emerges, it must account for the many exceptions to and variations on the rule: endotherms and ectotherms that fail to exhibit the pattern (mammals: Fuentes and Jaksic 1979; Erlinge 1987; Alacantra 1991; Thurber and Peterson 1991; butterflies: Hawkins and Lawton 1995; Barlow 1994) and related but conflicting size trends along elevational gradients (e.g., mammals: Zimmerman 1950; Taylor et al. 1985; butterflies: Roff 1990).

Allen's and Gloger's rules.   Other ecogeographic rules have been coined to describe other widespread patterns of morphological variation. Allen (1877) noted that among closely related endothermic vertebrates, those forms living in hotter environments often tend to have longer appendages. Again, the hypothesized mechanism invoked thermoregulatory advantages: reducing the surface area of appendages promotes heat conservation in cold environments, whereas increasing the surface area facilitates heat dissipation when ambient temperatures reach stressfully high levels. Some of the longest appendages occur in organisms living in hot deserts, and physiological research has shown that some of these, such as the ears of rabbits and elephants and the legs of camels, do play important roles in heat dissipation (Schmidt-Nielsen 1963, 1964). In North America, hares of the genus *Lepus* show a dramatic reduction in the length of their ears with decreasing temperature, from the black-tailed and white-sided jackrabbits of the hot, arid Southwest to the varying and arctic hares of the perpetually cold north (Figure 16.2).

At present, however, the empirical evidence supporting Allen's rule still consists of a limited number of isolated cases (Hesse et al. 1951; Mayr 1963; Johnston and Selander 1971). The number of exceptions has led some authors

Figure 16.2 Two classic examples of Allen's rule: variation in the relative sizes of the ears of hares (A, from left: *Lepus alleni, L. californicus, L. americanus,* and *L. arcticus*) and foxes (B, from left: *Fennecus zerda, Vulpes vulpes,* and *Alopex lagopus*). In each case, the species on the left occurs in a hot desert habitat, the one on the right in a cold tundra habitat, and the one or two in the middle in intermediate, temperate environments.

to question whether there is really any general rule. For example, it is questionable whether North American rabbits of the genus *Sylvilagus* exhibit the expected pattern (Stevenson 1986). The presumed mechanistic basis, heat dissipation through enlarged appendages, can also be questioned. While long ears can be excellent heat dissipators, the short, well-insulated tails of rabbits and hares can play little role in dissipating heat. In fact, heat exchange can be readily regulated by changing insulation (adding to or fluffing up fur or feathers) and/or by altering blood flow to the extremities. Some relatively large, well-insulated endotherms use their long appendages for heat dissipation in cool environments (e.g., the legs of caribou, the tails of beavers, the flippers of seals). Aside from thermoregulatory considerations, long ears may be advantageous in warm environments because they aid in sound detection (sound attenuates more rapidly at higher temperatures: Erulkar 1972; Harris 1996). Thus, while Allen's rule has received less thorough study than Bergmann's, the empirical and mechanistic status of the two rules is quite similar.

Still another ecogeographic rule is attributed to Gloger (1883), who noted that the coloration of related forms was often correlated with the humidity of their environments: darker colors occurred in more humid environments. In this case, while some physiological mechanisms have been proposed, they have received little support. Dark colors might be thought to facilitate basking to absorb solar radiation in cold environments, and light colors to reflect solar radiation and prevent overheating in hot environments. But much radiant heat exchange occurs in the nonvisible wavelengths, and the thermally effective absorbance and reflectance of fur, feathers, scales, and other body surfaces is often not well correlated with their color. In fact, many animals tend to be white in cooler environments (e.g., polar bears, arctic foxes, arctic hares, and ptarmigans), just the opposite of what we would expect based on thermoregulatory arguments.

The most likely explanation for Gloger's rule invokes selection for crypsis to avoid visual detection by predators or prey. Humid climates are usually associated with dark soils and dense vegetation that casts deep shadows, whereas arid environments (hot or cold) often have light-colored soils and relatively sparse vegetation. Animals that match the color of their environment should have an advantage in avoiding visually hunting predators, such as insectivorous birds and lizards, hawks, owls, and mammalian carnivores. Kettlewell's (1961) famous study of industrial melanism showed that moths matching the soot-darkened tree trunks and other surfaces in industrial areas were less likely to be discovered and eaten by birds. Less well publicized, but perhaps even more thorough and convincing, are the studies of Sumner (1932), Blair (1943), and Dice (1947) on the adaptive basis of geographic variation in coat color in deer mice (*Peromyscus maniculatus*). They found three key pieces of evidence:

1. The pattern of geographic variation is such that dorsal pelage of local and regional populations tends to closely match background coloration in the environment.
2. Most of this variation is heritable, controlled by both simple Mendelian genes and more complex polygenic systems.
3. Individuals that do not match their backgrounds are selectively captured by predators (owls) in controlled experiments.

Crypsis can also help predators avoid detection by their prey, potentially explaining why polar bears are white and lions are the color of dry grass. The extensive variation in pelage color of North American black bears also appears to be cryptic, with a higher frequency of light-colored forms in relatively dry habitats (Rounds 1987).

This large body of integrated evidence is reminiscent of the support for a thermoregulatory advantage of Bergmann's rule in bushy-tailed woodrats. Here, too, however, there remain questions about the extent to which both the empirical pattern and the proposed mechanism can be generalized to other kinds of organisms. Many species, even ones that are taken by visually hunting predators—such as many songbirds—are obviously not colored to match their backgrounds. The usual explanation is that other selective pressures, such as the need for bright, contrasting colors to attract mates or to convey other signals to conspecific or heterospecific individuals, outweigh the advantages of cryptic coloration. This case reminds us once again that while it is possible to document the adaptive basis of specific cases of geographic variation in morphology, it is hazardous to extrapolate from such studies to formulate general explanations for ecogeographic rules.

Geographic Variation in Life History and Population-Level Characteristics

Interesting patterns of geographic variation are not confined to the morphological traits of individuals. There is an extensive literature on geographic variation in clutch size in birds and other egg-laying animals (e.g., Moreau 1944; Lack 1947; Skutch 1949; Cody 1966; Moll and Legler 1971; MacArthur 1972) and litter size in mammals (Lord 1960; Glazier 1985; Cockburn et al. 1983; Millar 1989; Cockburn 1991). Often the observed patterns are related to latitude, with larger clutches or litters at higher latitudes being the general rule (e.g., Figures 8.6 and 16.3). Again, however, there is more consensus about the pattern than about its mechanism(s).

Several of the studies listed above suggest that clutch size is influenced by the combined effects of climate, food availability, and interspecific interactions. Cody (1966) explained geographic variation in clutch size within the frame-

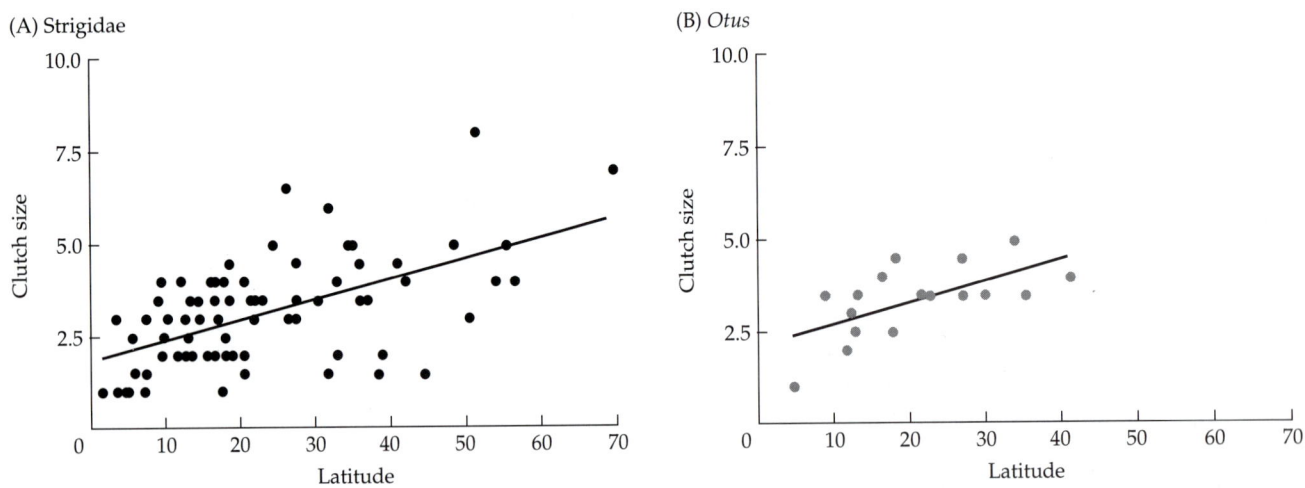

Figure 16.3 Clutch size as a function of latitude in owls. Such patterns have been found at three levels of relationship: (A) among the entire owl family, Strigidae, and (B) among the genus *Otus*, which contains the screech owls and their relatives; similar, though perhaps not statistically significant, trends have been found among populations of several widely distributed owl species. The strong positive relationships in these graphs are obvious, although the mechanisms underlying these typical patterns are much debated. (Courtesy of J.-L. Cartron and J. Kelly.)

work of *r*- and *K*-selection (essentially, selection favoring higher population growth rates and productivity versus selection favoring more efficient utilization of resources, respectively; see also MacArthur and Wilson 1967). Birds inhabiting seasonally variable mainland environments with short growing seasons should be subject to *r*-selection, producing relatively few, but large, clutches. Conversely, on temperate islands, where climates tend to be milder and less variable, selective pressures may shift toward more efficient use of available resources and clutches of fewer, but higher-quality, offspring (i.e., *K*-selection; see Cody 1966; Isenmann 1982). Other studies have reported the same trend for insular salamanders (Anderson 1960) and for microtine rodents (Tamarin 1977; Gliwicz 1980; Ebenhard 1988). Similarly, in the mainland tropics, where the even more stable climates and the diversity of competitors and predators put a premium on efficient resource use, specialization, and predator avoidance, clutch size should again be relatively low.

While the *r*/*K* paradigm has been considered too simplistic and has fallen into disfavor in recent decades, Cody (1966) offered some interesting data in support of his hypothesis (Table 16.1). Other patterns of geographic variation in clutch and litter size may require somewhat different explanations. Why, for example, do clutch and litter sizes sometimes decrease from the tropical lowlands to the cool highlands of equatorial mountains (see Lack and Moreau 1965; Flux 1969)? Others have offered alternatives to Cody's explanation for geographic variation in clutch and litter size (e.g., see Lack 1947; Skutch 1949, 1967; Ashmole 1963; Charnov and Krebs 1974). For the most part, these hypotheses are based on trade-offs associated with climatic fluctuations, the risks of migration, seasonal pulses in productivity, risk of predation, and other factors. As is often the case, these hypotheses are not mutually exclusive, and some combination of them may be necessary to explain the observed geographic clines in clutch and litter size within any single species as well as the pattern for terrestrial vertebrates in general.

Table 16.1

Differences between clutch sizes (numbers of eggs per nest) of the same bird species on islands and nearby mainland localities of the same latitude

For temperate islands off the coast of New Zealand [a]			**For tropical islands in the Caribbean** [b]		
Species or genus	Average mainland clutch	Average island clutch	Species or genus	Average mainland clutch	Average island clutch
Anas spp.	8 (3 sp.)	3.5 (1 sp.)	*Saltator albicollis*	2.0	2.5
Bowdleria punctata	3.1	2.5	*Tangara gyrola*	2.0	2.0
Gerygone spp.	4.5 (1 sp.)	4.0 (1 sp.)	*Habia rubica*	2.3	2.0
Petroica macrocephala	3.5	3.0	*Cacicus cela*	2.0	2.0
Miro australis	2.6	2.5	*Coereba flaveola*	2.5	2.5
Anthornis melanura	3.5	3.0	*Empidonax euleri*	2.0	3.0
Cyanoramphus novaezelandeae	6.5	4.0	*Elaenia flavogaster*	2.3	2.5

Source: MacArthur and Wilson 1967; Cody 1966.

Note: Clutch size averages smaller on the islands, as opposed to the mainland, in the temperate region but not in the tropical region.

[a]Species considered were either indigenous to the offshore islands and had a mainland relative in the same genus, or were subspecifically distinct on the island. The difference between the mainland and island means is 90% significant, by *t*-test.

[b]Species on isolated West Indian islands are compared to the mainland, and also the extreme southerly Lesser Antillean islands (isolated) are compared to Trinidad, which is very close to the mainland. The means are not significantly different.

Multispecies Assemblages

Areography: Sizes, Shapes, and Overlaps of Ranges

Sizes of ranges.   There is enormous variation in the sizes of geographic ranges of individual species. Among the smallest ranges are the natural distributions of the Soccoro isopod (*Thermosphaeroma thermophilum*) and the Devil's Hole pupfish (*Cyprinodon diabolis*), each of which occurs in a single freshwater spring with a surface area of less than 100 m^2. Among the largest ranges are those of several marine organisms, such as the blue whale (*Balaenoptera musculus*), which include most of the world's unfrozen oceans, areas on the order of 300,000,000 km^2. Among terrestrial organisms, species with very large native ranges include the peregrine falcon, barn owl, and osprey, which are widely distributed over all of the continents except Antarctica. Of course, modern *Homo sapiens* is now one of the most widely distributed species, and humans have carried several species of symbionts and exotics with them as they have spread over the entire earth.

1982 saw the publication of the English language edition of a fascinating little book entitled *Areography* by the Argentine ecologist and biogeographer Eduardo Rapoport. Working largely in isolation in Latin America, Rapoport showed that simple quantitative analyses of spatial distributions of organisms—such as the maps of geographic ranges in Hall's (1981) *Mammals of North America* or Critchfield and Little's (1966) *Geographic Distribution of Pines of the World*—could reveal fascinating patterns and suggest hypotheses about the mechanisms that limit distributions and influence species diversity. One such pattern is the inverse relationship between range size and latitude that is called Rapoport's rule (see Figures 15.15 and 15.16). Rapoport's book inspired a number of ecological biogeographers to undertake conceptually and methodologically similar studies. The result was a renewed emphasis on comparative continental biogeography that was similar in some ways to the emphasis on insular biogeography stimulated by MacArthur and Wilson's (1967) book about 15 years earlier.

Another interesting areographic pattern is the distinctively shaped frequency distribution of the sizes (areas) of geographic ranges, first documented by Willis (1922) and subsequently confirmed by Rapoport (1982) and others (e.g., Gaston 1991, 1996; Pagel et al. 1991; Brown 1995; Brown et al. 1996). Two features of this frequency distribution are of particular interest. First, within most large taxonomic groups, the majority of species have restricted ranges, and only a few species occur widely over most of a continent. Thus, for North American land birds and terrestrial mammals, the majority of species are confined to a range that encompasses less than one-fourth and one-fifth of the land area of the continent, respectively. Second, when the data are plotted on either arithmetic or logarithmic axes (Figure 16.4), distinctively shaped frequency distributions are observed in several different taxa (Brown et al. 1996; Gaston 1996). Taken together, these two features suggest that the boundaries of ranges are set by similar kinds of environmental limiting factors, and that the dynamics of colonization, speciation, and extinction processes have similar influences on the relative sizes of ranges, in different taxa. The fact that the patterns are so similar across so many different kinds of organisms, from butterflies and conifers to birds and mammals, encourages the search for unifying principles of biogeography.

Shapes of ranges.   There are also interesting areographic patterns in the shapes of ranges. Rapoport (1982) noted that despite the orders-of-magnitude

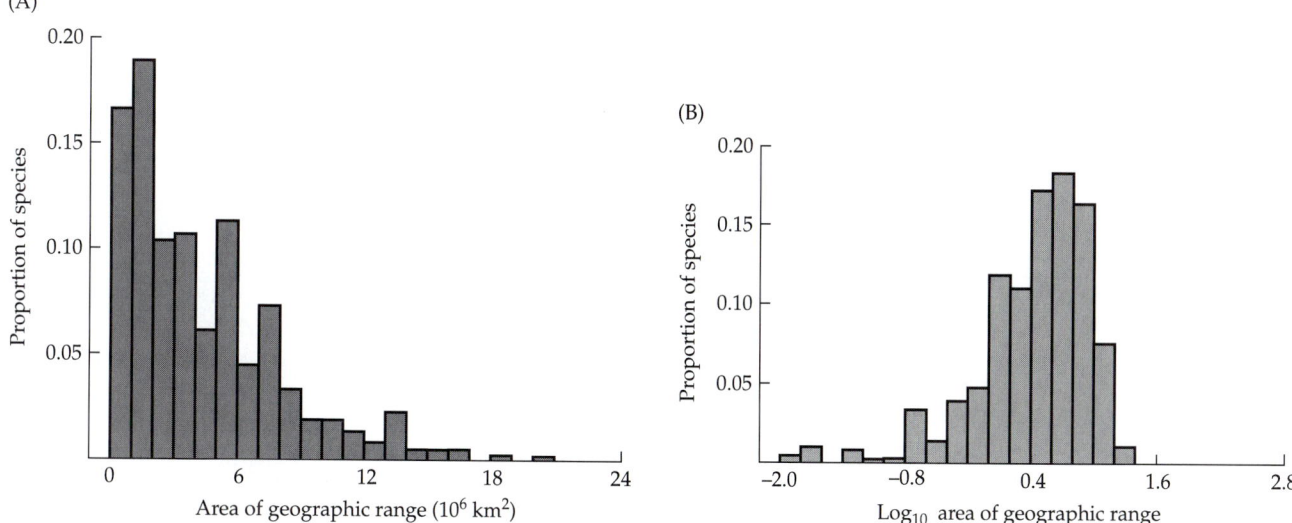

(A)

(B)

Figure 16.4 Frequency distribution of the sizes of geographic ranges of North American land mammals, plotted on (A) arithmetic and (B) logarithmic scales. The message of this pattern, which holds for other kinds of organisms and on other continents, is that the majority of species have relatively small ranges. (After Brown 1995.)

variation in the areas of the ranges of North American mammals, their perimeter-to-area ratio remained relatively constant. That is, when he measured the length of a range boundary and the area encompassed within that boundary, he found that the ratio of the two variables was a nearly constant 10. This is surprising, because for similarly shaped ranges, perimeter-to-area ratios should decrease with range size.

This apparent invariance of perimeter-to-area ratios suggests that, when considered at the same spatial scale, large ranges have more convoluted boundaries than small ranges do. While this pattern might suggest something interesting about colonization-extinction dynamics, ecological limiting factors, or some combination of these, it is first important to evaluate an alternative hypothesis: that it simply reflects an unintentional bias of mapmakers. We have an inherent tendency to draw maps with a fractal structure, including more detail about boundaries (and other features) as the spatial scale decreases. In many cases, including Hall's (1981) treatise on North American mammals, small ranges are mapped at greater magnifications than large ones. Thus, the apparent constancy of perimeter-to-area ratios may be an artifact of the cartographer's tendency to include detail at different spatial scales.

On the other hand, if the fractal inclinations of mapmakers could be taken into account, the periphery-to-area ratio could serve as a simple measure of range shape that warrants further study. Maurer (1994) has considered the fractal nature of distributional patterns and has offered some techniques for quantifying the shapes of range boundaries. Brown and Maurer (1989) used a simple technique to quantify both the shapes and the orientations of ranges. They plotted maximum north-south distance across a range as a function of maximum east-west distance, thus referencing variation in shape to geography. The result is the kind of graph shown in Figure 16.5, in which the line of equality has been plotted for reference. In such a graph, ranges with equal dimensions, such as circles or squares, would fall along the diagonal line, with small ranges to the lower left and large ones to the upper right. Ranges longer

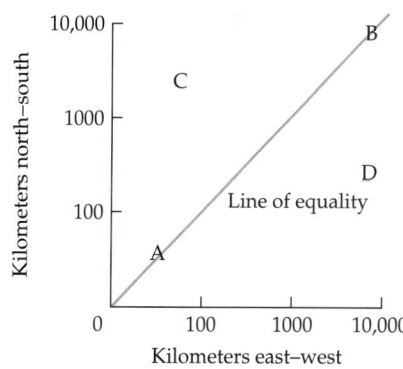

Figure 16.5 A simple way of showing both the size and shape of a geographic range on the same graph is to plot its greatest north-south and east-west dimensions. If the range has equal dimensions, as it would if it were circular, it will plot on the diagonal "line of equality": near the origin for a small range (point A) or far from it for a large one (point B). A range that is elongated in a north-south dimension will plot above the equality line (point C), whereas one that is elongated east-west will plot below the line (point D).

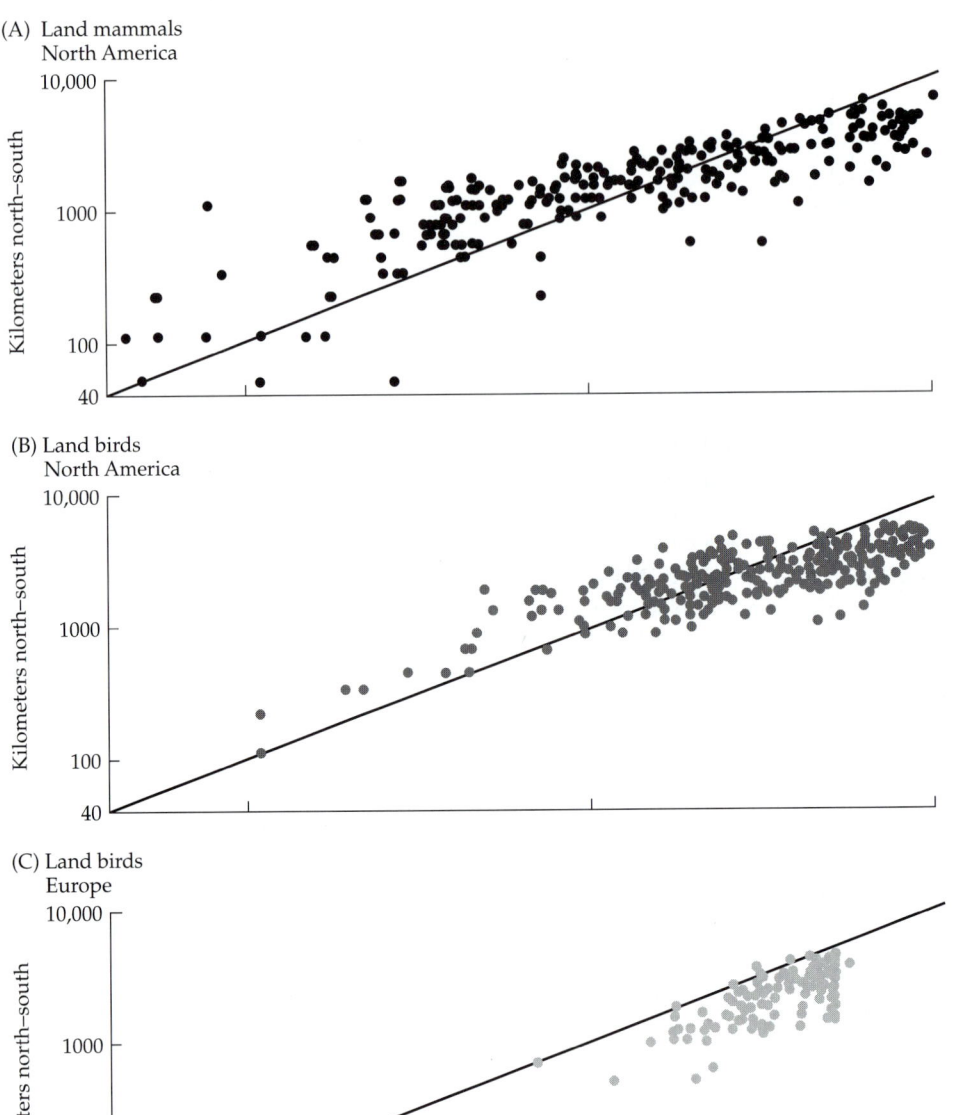

Figure 16.6   Greatest north-south and east-west dimensions of geographic ranges, plotted as in Figure 16.5, for three groups of organisms: (A) North American terrestrial mammals; (B) North American land birds; and (C) European land birds. Note that in North America the small ranges tend to be oriented north-south (above the equality line), whereas in Europe they are oriented east-west (below the line). Presumably this reflects the orientation of geographic features on the two continents. The larger ranges tend to be oriented east-west on both continents, presumably due to the influence of broad climatic zones. (From Brown and Maurer 1989.)

in a north-south direction would fall above the line of equality, whereas those longer in an east-west direction would fall below it.

The few such graphs that have been compiled show some interesting patterns (Figure 16.6; Brown and Maurer 1989; Brown 1995). Although there is considerable scatter, North American mammals, birds, and reptiles all show a consistent trend: small ranges tend to be oriented north-south, whereas large ones tend to be oriented east-west. European mammals and birds present an interesting contrast, with both small and large ranges oriented east-west. Brown and Maurer suggested that these patterns of range orientation reflect the physical geography of the continents. The east-west orientation of large ranges on both continents reflects the influence of the major east-west-oriented belts of climate and vegetation on species with large ranges. Brown and Maurer hypothesized that the difference between the continents in the orientation of small ranges reflects the influence of geographic features, such as mountain ranges, river valleys, and coastlines, in determining the boundaries of small ranges. For example, in North America, the Appalachians, Rocky Mountains, Sierra Nevada, Pacific Coast ranges, Mississippi River, and Central Valley of California all run north-south, whereas in Europe the Alps, Pyrenees, and Mediterranean Sea are all oriented east-west. These ideas could be pursued further with analyses of range shapes in relation to abiotic and biotic environmental variables in other kinds organisms and in other geographic and ecological settings.

Overlap of ranges. Knowledge of the extent to which the ranges of species overlap is central to understanding geographic patterns of species richness. It can also provide at least indirect evidence about the environmental factors that limit distributions and set range boundaries. In Chapter 4 we presented examples of pairs of closely related species whose ranges are adjacent but almost completely nonoverlapping, suggesting that at least one of the species is limited by interspecific competition. Such competitive exclusion has been clearly demonstrated by experimental studies on organisms as different as barnacles (Connell 1961) and chipmunks (Brown 1971). In some other organisms, such as desert rodents, these patterns of nonoverlapping distributions are so striking that they scarcely need statistical analyses to show that they are not due to chance (see Figure 4.25 and below). Such isolated examples, however, are hardly sufficient to show a causal relationship between competition and distributions. Indeed, we know of many examples of species that compete intensely, yet overlap in their geographic ranges and coexist in the same local habitats. And of course the majority of species with nonoverlapping geographic ranges are not closely related, do not use similar resources, and do not compete to any significant extent.

So, can we find any general patterns in the extent to which the geographic ranges of different species overlap? Specifically, do the ranges of close relatives, such as congeners, overlap either more or less than one might expect on the basis of chance? There are good a priori reasons for predicting either outcome. On the one hand, closely related forms are likely to occur in the same general region because they share a fairly recent common ancestor, possess similar dispersal capabilities, and have similar requirements for environmental conditions and resources. On the other hand, if species have originated by allopatric speciation, we might expect the vicariant pattern to be preserved, and if they are too similar in their requirements, we might expect competition to inhibit or preclude their coexistence.

Is it possible to determine which of these patterns predominates by quantitative analyses of the geographic ranges of well-studied taxa? It is simplest to consider the overlap of pairs of species. Pielou (1978) analyzed the distributions of algae along the coasts of North and South America. Within each of three groups, Rhodophyta (red algae), Phaeophyta (brown algae), and Chlorophyta (green algae), on both Atlantic and Pacific coasts, she considered all pairs of species. Each pair was scored as to whether it was congeneric or more distantly related, and whether its distribution was overlapping or disjunct. Results for one group, the Rhodophyta on the Pacific coast, are shown in Table 16.2.

Statistical analyses of these data provide no grounds for rejecting a null hypothesis of independent (i.e., random) distributions of these species. The probability that this pattern could be due to chance is high ($P = 0.2$). There is, however, a qualitative tendency for congeneric pairs to overlap more than distantly related ones. This same trend is present in each of the six data sets

Table 16.2
Patterns of overlap in the geographic ranges of red algae (Rhodophyta) on the Pacific coasts of North and South America

Species pairs	Number of species pairs studied	Percentage with overlapping ranges
Congeneric	37	57%
More distantly related	2664	44%

Source: Pielou 1978.

Pielou examined. Because this in itself is statistically unlikely ($P < 0.02$), Pielou concluded that closely related species tend to overlap in their ranges, and that interspecific competition does not usually result in mutually exclusive geographic ranges.

Pielou's results were similar to those of Simberloff (1970), who used somewhat different methods to analyze the coexistence of congeneric pairs of bird and plant species on islands. The fact that these two studies, performed independently using different techniques, gave similar results for widely divergent groups of organisms is significant. It suggests that closely related species often have similar dispersal capabilities and ecological requirements, and that these tend to result in overlapping distributions. No influence of either allopatric speciation or competitive exclusion is apparent at this level of analysis.

It would be premature, however, to conclude that the effects of vicariance and interspecific competition cannot be detected in the distributions of closely related species. In fact, recent statistical studies of North American seed-eating desert rodents suggest that interspecific competition does indeed limit the overlap of both local and geographic ranges, as might be expected from our discussion in Chapter 4. Kelt and Brown (in press) have reviewed the many studies and reanalyzed the data on geographic overlap in these species. They found that congeneric and morphologically similar pairs of species overlapped in their geographic ranges less than expected by chance (Table 16.3). While quantitative analyses remain to be done, other kinds of organisms appear to exhibit similar patterns. In North America, these include chipmunks (*Eutamias*; see Figure 4.26), ground squirrels (*Spermophilus, Amospermophilus*), marmots (*Marmota*), prairie dogs (*Cynomys*), and pocket gophers (*Thomomys*,

Table 16.3
Comparison of observed and expected frequencies of overlapping geographic ranges for pairs of desert rodent species in the same or different genus or of similar or different body size

	Observed	Expected
Same genus	111	146.3
Different genus	785	749.7
Similar size	217	242.2
Different size	697	653.8

Source: After Kelt and Brown, in press.

Note: Closely related and morphologically similar species overlap less frequently than expected by chance. In both cases the differences are statistically significant ($p < 0.05$).

Geomys) among mammals, and jays (*Aphelocoma, Cyanocitta*) and chickadees and titmice (*Parus*) among birds. The last genus is especially interesting, because in Eurasia the exact opposite pattern obtains: several species of *Parus* have broadly overlapping ranges and forage together in mixed-species flocks.

At least with respect to overlap of congeners, then, desert rodents exemplify patterns that are just the opposite from those observed in Pielou's and Simberloff's studies. How do we account for this discrepancy? We can suggest several hypotheses. First, compared with some other kinds of organisms, competition among desert rodents may be more frequent and more intense. In experimental studies, several species of desert rodents doubled or tripled their local populations when competitors were removed (e.g., Brown and Munger 1985; Heske et al. 1997; Valone and Brown 1995). The issue of scale may also be important. Species may compete strongly and exclude each other at small spatial scales and still have broadly overlapping distributions at geographic scales. Many intertidal algae and land plant species show clearly segregated patterns of local elevational zonation (e.g., Pielou 1978; Yeaton 1981). As Connell (1961; see Figure 4.11) demonstrated for barnacles, such zonation may reflect local competitive exclusion among species with extensively overlapping geographic ranges.

A second and related hypothesis is that intensity of competition is more closely correlated with taxonomic affinity in desert rodents than in some other organisms. Desert rodent species in the same genus tend to be similar in many characteristics that probably influence competitive interactions, such as diet, body size, and morphological specialization for locomotion, foraging, and predator avoidance (Kelt and Brown, in press). The term **guild** (see Root 1967) has been coined to describe sets of species that use similar resources in similar ways. By this criterion, most congeneric desert rodents are clearly members of the same guild.

A third hypothesis is that the nonoverlapping ranges of congeneric desert rodents reflect in part the enduring influence of allopatric speciation. Such a historical legacy would be difficult to disentangle from the influence of contemporary ecological interactions. Indeed, if speciation has occurred as a result of geographic isolation, we would expect the most closely related species to be most similar, to compete most intensely, and as a result, to maintain largely allopatric ranges. There is abundant evidence to support such a scenario in desert rodents. There are many species that are morphologically and ecologically nearly indistinguishable, but recognizable as distinct species by clear differences in their chromosomal or molecular genetic characteristics (e.g., Patton et al. 1984; Sullivan et al. 1986; Lee et al. 1996). Such species are often called **sibling species** because of their close phylogenetic relationship, or **cryptic species** because of their great similarity. They often appear to have formed fairly recently, during the Pleistocene or late Tertiary, as a result of geographic isolation of populations due to geological and climatic changes. The predominant pattern is for pairs of such species to exhibit abutting, but largely nonoverlapping, distributions, even at very small scales (Figure 16.7). Thus, what was once thought to be a single species with a relatively continuous geographic range is often shown to consist of two or more very closely related species that come into close proximity, but rarely co-occur in the same habitat. Of course, competitive exclusion could play a major role by preventing overlap in geographic ranges and thus maintaining the biogeographic legacy of allopatric speciation events. In some other organisms, if speciation has occurred by sympatric mechanisms, or occurred at a much earlier time in the geological record (as is likely in the algae, for example), then the relationships between speciation, competition, and geographic isolation may not be so clear.

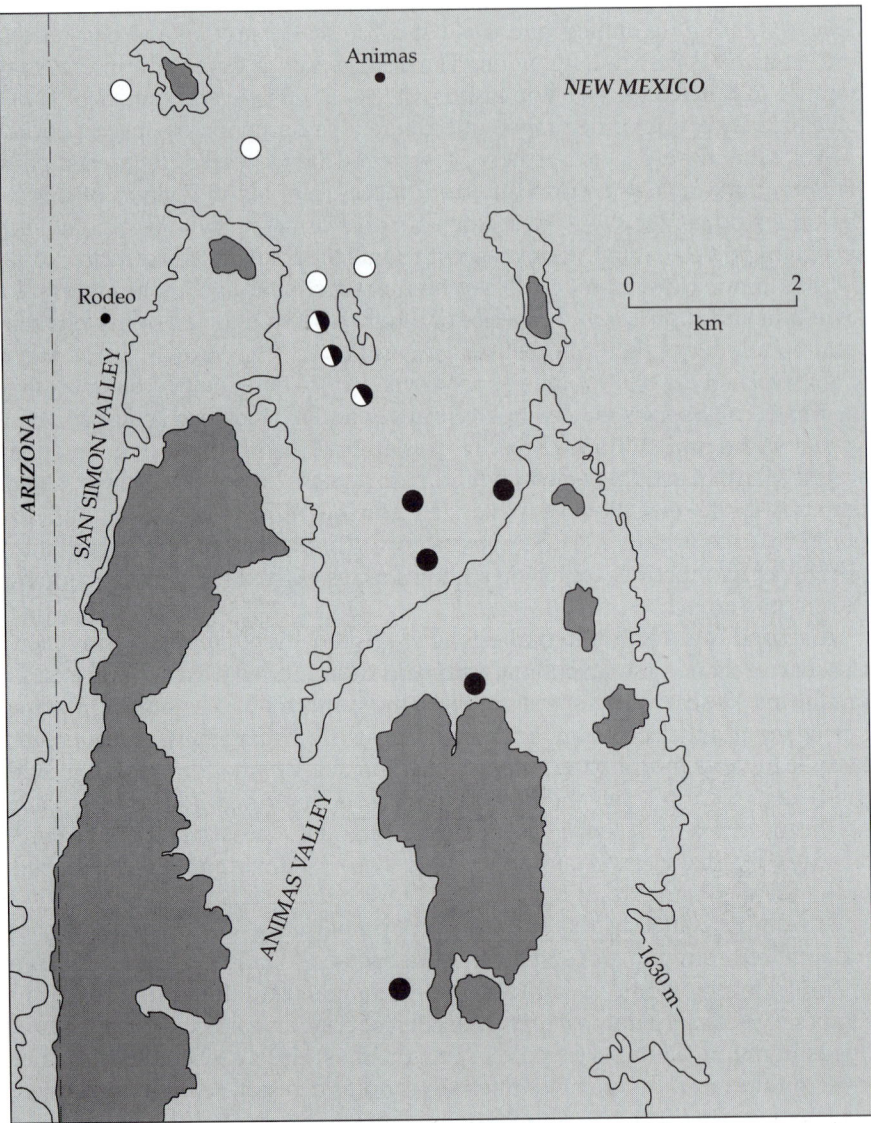

Figure 16.7 Map of a small region of southwestern New Mexico, showing the adjoining but minimally overlapping geographic ranges of two very closely related and ecologically similar species of North American grasshopper mice, *Onychomys arenicola* (open circles) and *O. torridus* (solid circles). These species are so similar that they were identified only recently by differences in karyotype (number and kinds of chromosomes). Individuals of both species have been found together at only three nearby localities. Presumably, competitive exclusion prevents local coexistence of these forms. (From Sullivan et al. 1986.)

Macroecology: Assembly of Continental Biotas

In this chapter and in Chapters 4, 5, 7, 10, and 12, we have looked at the patterns of range limitation exhibited by individual species or small sets of closely related or ecologically similar species in response to both contemporary ecological factors and historical events. In Chapter 12 we considered the role of ecological and historical processes in the generation and maintenance of biogeographic provincialism: the tendency of continents, oceans, and large regions within them to contain biotas of distinctive composition. Between these two levels lie some of the most interesting but difficult questions in biogeography. What are the patterns and processes of assembly of continental and regional biotas? What historical and ecological factors limit the composition of biotas to a restricted number of species and phylogenetic lineages? Given the species and higher taxonomic groups that make up the overall species pool, how are these organisms distributed within a continent or smaller region? What are the patterns of co-occurrence and diversity at different spatial scales, and what are the roles of historical events and ecological conditions in generating and maintaining these patterns?

The statistical macroecological approach.   One approach to investigating the assembly and structure of biotas is **macroecology** (Brown and Maurer 1987, 1989; Brown 1995). Macroecology is a quantitative and statistical approach that overlaps and shares many questions and methods with areography. The major difference is that macroecology does not focus so exclusively on geographic ranges and related spatial patterns and processes. Because the macroecological approach has been advanced even more recently than the areographic one, it is hard to evaluate its potential applications and long-term utility.

The essence of macroecology is that it tries to identify general ecogeographic patterns and to understand the underlying mechanisms by focusing on the statistical distributions of variables among large numbers of equivalent (but not identical) ecological particles. These particles can be almost anything: individual organisms within a local population or entire species, species within local communities or larger biotas, or even replicated sample areas or patches of habitat (for examples, see Brown 1995). The premise of macroecology is that the discovery of repeated patterns in the statistical distributions of ecologically relevant variables can lead to the development and testing of hypotheses about the general mechanistic processes involved. Let's consider an example.

The distribution of range sizes.   In our discussion of areography above, we noted that the sizes of the geographic ranges of species within a continent exhibit a distinctively shaped frequency distribution (see Figure 16.4). What are the mechanisms that account for the preponderance of relatively small ranges and the relatively few very large ones? Additional insights come from plotting the sizes of ranges as a function of body size (Figure 16.8). While such a plot, on logarithmic axes, does not suggest a nice linear or curvilinear relationship that could be well fit with a regression equation, neither is it a random shotgun blast of data points. Conspicuous in the plots for both North American birds and mammals in Figure 16.8 is the dearth of large species with small geographic ranges. The points tend to fall within a roughly triangular space, with small to medium-sized species exhibiting a wide range of range sizes, but large species restricted to relatively large ranges.

These distinctively shaped distributions can be viewed as reflecting the constraints of continental area on the maximum size of ranges and of extinction on the minimum size of ranges (Brown 1981; Brown and Maurer 1987). While the former hypothesis is straightforward, the latter is of considerable interest. Because population density varies inversely with body size (Brown and Maurer 1987; Damuth 1991; Silva and Downing 1995), and probability of extinction varies inversely with absolute population size (e.g., MacArthur and Wilson 1967; Leigh 1975; Gilpin and Hanski 1991), species with large body sizes tend to be rare and extinction-prone. Only if they have large geographic ranges are they likely to have sufficient numbers of individuals to persist. This hypothesis is supported by the observation that there are few large mammals with small ranges, and also by the differential extinction of Pleistocene megafauna in North America, Australia, Madagascar, and New Zealand (see Chapter 7). It follows from this reasoning that any species with a small range relative to its body size might be extinction-prone. Many such species, such as the Morro Bay kangaroo rat, black-footed ferret, lesser prairie chicken, and Kirtland's and golden-cheeked warblers, are considered threatened or endangered.

Still further insights into the factors influencing the sizes of geographic ranges can be obtained by plotting average local abundance against geographic range size (Figure 16.9). Again, the distribution of data points has a triangular shape. For example, there are very few species of North American birds with

(A) Land mammals, North America

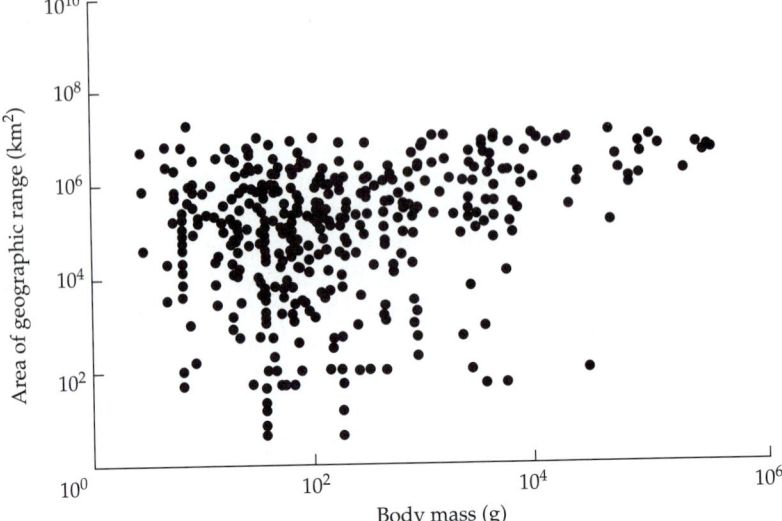

(B) Land birds, North America

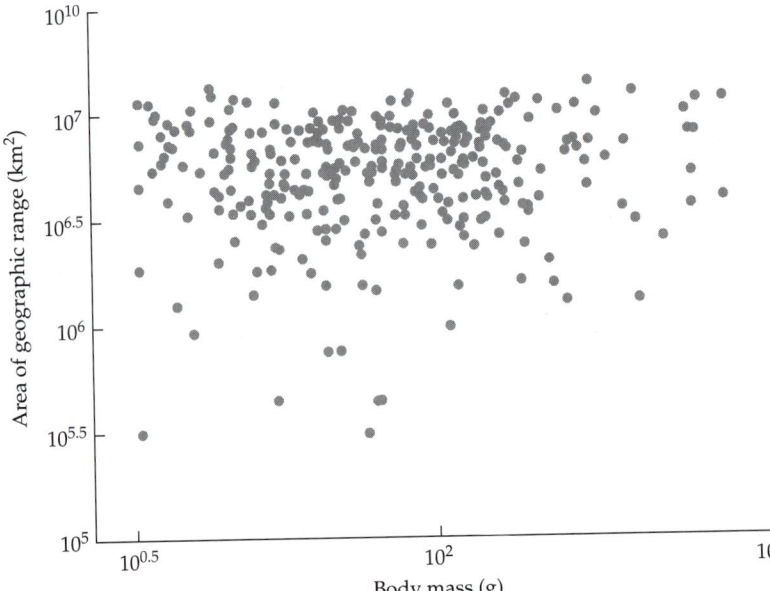

Figure 16.8   Relationship between area of geographic range and body size on logarithmic axes for North American terrestrial mammals (A) and birds (B). Note that while there is much variation, it is far from random. In particular, there are few large species with small geographic ranges. (After Brown 1995.)

small geographic ranges that also have high local population densities, but many species with large geographic ranges that have both high and low average local abundances. This pattern supports the relationship between niche breadth, abundance, and distribution developed in Chapter 4. Species with less restrictive resource requirements and broad tolerances for abiotic and biotic limiting factors are able to be both locally abundant and widely distributed. Furthermore, in part for these reasons, these species are also likely to have a low probability of extinction, even in the face of major environmental change.

In the above examples, the particles of analysis are individual species within a continental biota, and the variables of interest are area of geographic range, average abundance, and body size. The limits to the variation in the bivariate plots suggest constraints on range size due to continental area, body

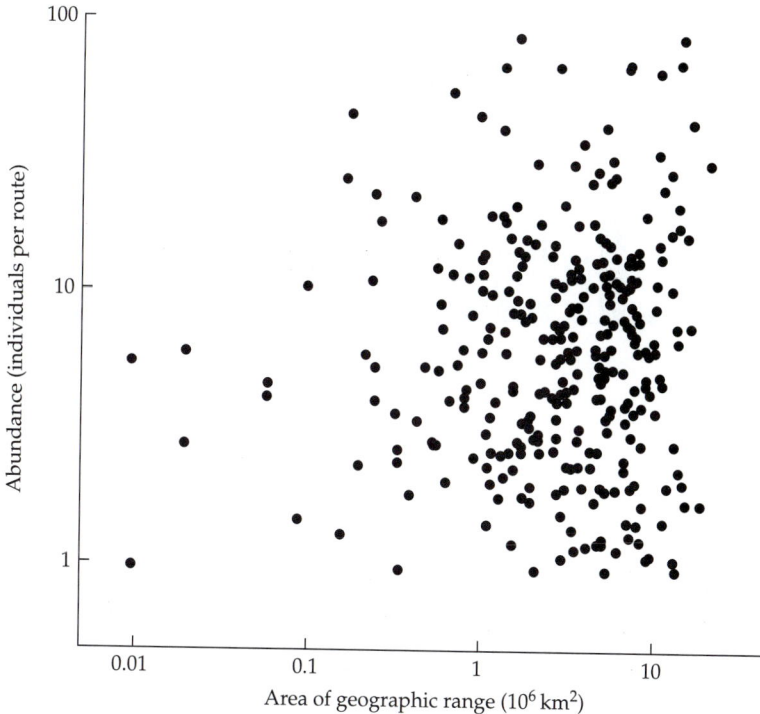

Figure 16.9 Relationship between area of geographic range and local population density for North American land birds. Note that while there is extensive variation, it is far from random. In particular, there are few species with high abundances and small geographic ranges. (After Brown 1995.)

size, and abundance. These insights into the correlates and causes of variation in range size obviously have important implications for conservation biology (see Chapters 17 and 18).

Body size and species diversity.   Additional insights into the processes that govern the assembly of continental biotas come from considering macroecological patterns on different spatial scales. We will focus on the body sizes of North American mammals, which have been studied by Brown and Nicoletto (1991) and Holling (1992). If we plot the frequency distribution of the body sizes of all species of North American terrestrial mammals on a logarithmic scale, we obtain the pattern shown in Figure 16.10A. The distinctive shape of this distribution appears to be very general, and we shall examine its occurrence in other groups in more detail below.

For the moment, however, let's focus on how the shape of this distribution varies with spatial scale. Brown and Nicoletto (1991) assembled complete species lists of the mammals from localities across the North American continent and at two smaller scales: (1) within biomes, large areas of relatively homogeneous climate and vegetation (see Chapter 5); and (2) within small patches of uniform habitat. Comparison of these frequency distributions (Figure 16.10) reveals a distinctive phenomenon: they become progressively flatter from continental to biome to local scales. Thus, while the continental fauna contains predominantly small to medium-sized mammals, local habitats contain an approximately equal representation of species in each logarithmic size category. A similar pattern has been documented for Australian mammals (Blackburn and Gaston 1994).

How do we explain this pattern? In order for the relative number of small to medium-sized species to decrease with decreasing spatial scale, it is necessary for them to turn over more rapidly across the continental landscape. That is, species with extreme body sizes (largest or smallest) tend not only to have

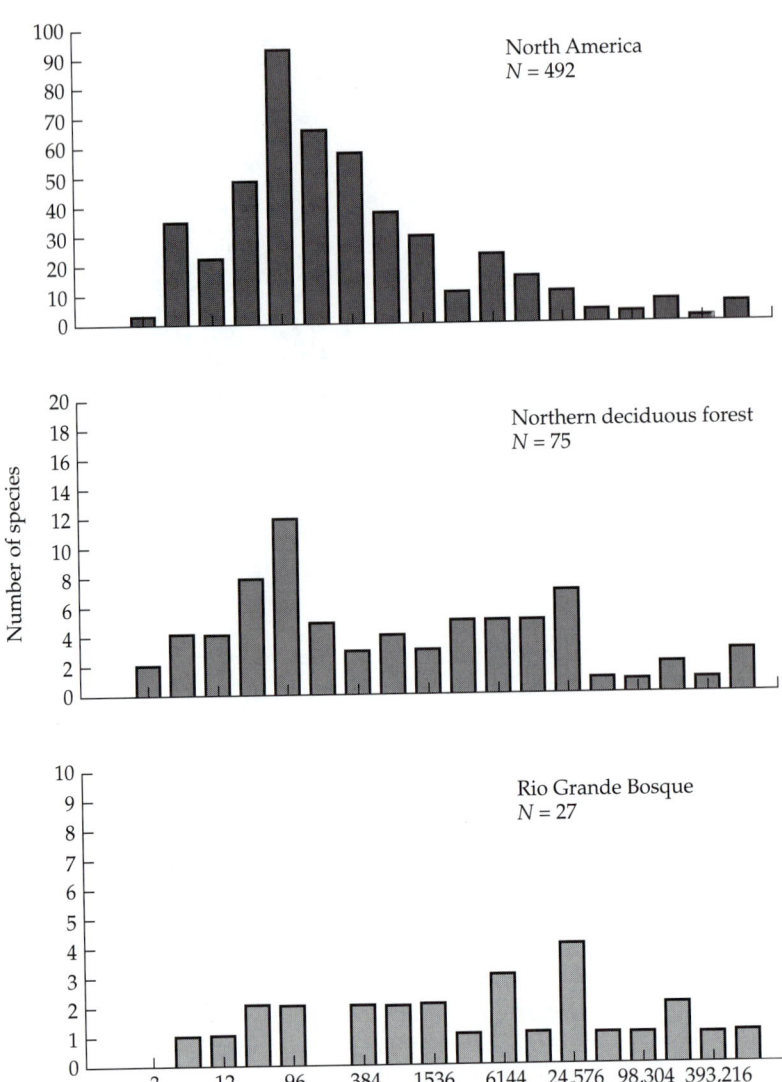

Figure 16.10   Frequency distributions of body sizes on a logarithmic axis for North American terrestrial mammals at three spatial scales: (A) the entire continent (492 species); (B) the deciduous forest biome (75 species); and (C) a small patch of deciduous forest habitat along the Rio Grande in New Mexico (27 species). Note that as the spatial scale and number of species decrease from continent to local habitat patch, the range of body sizes remains nearly identical, but the distribution becomes much more uniform. (From Brown and Nicoletto 1991.)

large geographic ranges (see above and Figure 16.8), but also to occur in a broad variety of habitats and local communities within their ranges. In contrast, small to medium-sized species have smaller geographic ranges and occur in a smaller proportion of the local communities within their ranges. This obviously reflects variation in degree of ecological specialization with body size: the small to medium-sized species that replace each other from one habitat or biome to the next are more specialized in their ecological requirements. Degree of specialization is hard to measure, but this spatial pattern indicates that it plays a role in the assembly of continental biotas.

How general is this phenomenon? If ecological specialization in any group of organisms were greatest at the modal body size for the group, we would expect flattening of body size frequency distributions with decreasing spatial scale, as we have observed in mammals. However, we would not necessarily expect to always observe completely flat distributions at the smallest spatial scales. Indeed, data on insects support these predictions: the size distributions do flatten in samples from decreasing areas, but there is still a distinct mode of small to medium-sized species even within small patches of local habitat (R. Harris, pers. comm.; Siemann et al. 1996).

Holling (1992) has called attention to another feature of the distributions of body sizes. He has analyzed the distributions of sizes of North American terrestrial mammals and land birds within distinct biome or habitat types, and has found a distinctive "lumpy/gappy" distribution. In other words, the species tend to be aggregated in certain size categories—"lumps"—separated by "gaps"—size categories containing few, if any, species (Figure 16.11). Holling suggests that the lumps reflect a match between the ecological characteristics of species of particular sizes and spatial or temporal characteristics of the habitat that favor species with these traits.

Both the flattening of body size frequency distributions and the lumpy/gappy distributions described by Holling imply that local ecological communities are highly nonrandom samples of regional or continental species pools. And indeed, this interpretation is confirmed by statistical analyses. Interspecific competition is an obvious process that could generate, or at least contribute to, both phenomena, preventing the local coexistence of many species in the small to medium body size categories and causing those species that do coexist to be separated by gaps in body size. Holling's (pers. comm.) analysis supports this interpretation: the "lumps" contain fewer species in the same guild than would be expected by chance.

But competition by itself seems an incomplete explanation. Why should there be more competition and geographic turnover among species in some body size categories? Holling (1992) suggests that species of different sizes interact with the environment at different temporal and spatial scales. This idea is reminiscent of Hutchinson and MacArthur's (1959; see also Hutchinson 1959) conjecture to explain the high frequency of small to medium-sized species in the faunas of large continental regions. They noted that the environment appears to be inherently more heterogeneous, or "grainy," at smaller scales, and hypothesized that the higher diversity of small organisms reflects their specialization to exploit this heterogeneity. Similar hypotheses, couched in modern concepts of fractal geometry, have been advanced by May (1978, 1986), Morse et al. (1988), and Lawton (1990). While this idea has intuitive appeal, it is presently difficult to formulate as a hypothesis and test rigorously. In particular, it is difficult to define "specialization" and "environmental het-

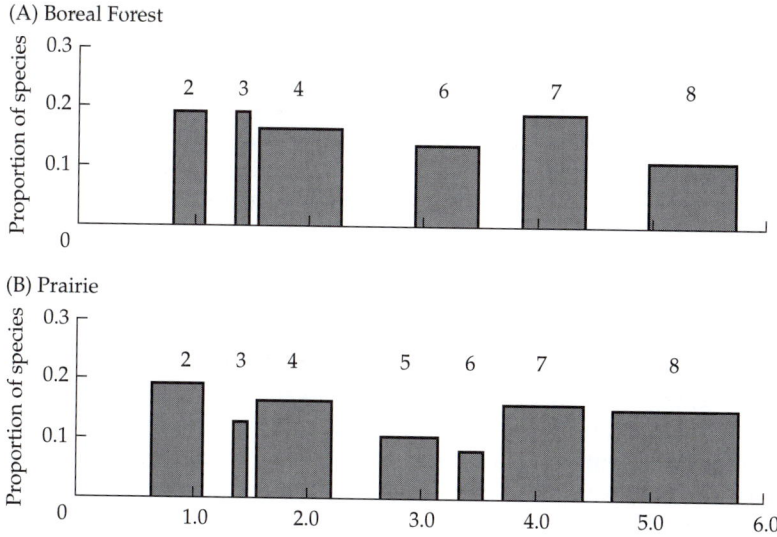

(A) Boreal Forest

(B) Prairie

Figure 16.11 Frequency distributions of mammal species in different size classes in two biomes of North America: boreal forest (A) and prairie (B). The data have been plotted to call attention to the seemingly lumpy/gappy distribution of sizes, which may reflect the responses of organisms of different sizes to distinctive spatial or temporal features of their environment. (From Holling 1992.)

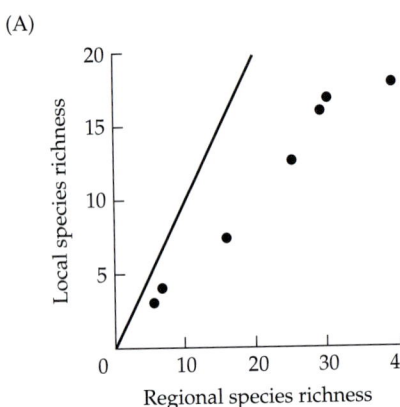

(A)

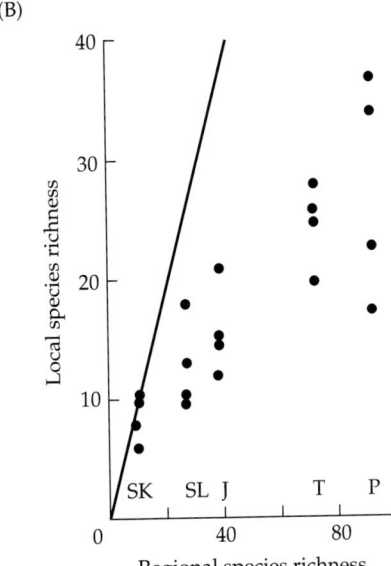

(B)

Figure 16.12 A plot of local versus regional species richness for (A) cynipine gall wasps on oaks in California and (B) Caribbean island land birds. The straight line, drawn for reference, indicates values for which local diversity would equal regional diversity. Note that nearly all of the local communities contain fewer species than the regional species pool, and that the proportion of species that coexist in local habitats increases, but with a decreasing slope, with regional diversity. While this pattern presumably represents the outcome of some combination of small-scale ecological interactions and larger-scale regional and historical processes, the exact nature of these processes is far from clear. (From Ricklefs 1987.)

erogeneity" and thus to assess how they vary with body size and affect geographic distributions.

If these difficulties can be overcome, the ecological and biogeographic correlates and consequences of body size should be a fruitful area of investigation. Body size is one of the most important axes of biological diversity. The sizes of currently living organisms span more than 21 orders of magnitude, from 10^{-13} g *Mycoplasma* and other tiny microorganisms to 10^8 g whales! Size influences the scaling of all characteristics of organisms, from the movements of individuals to the geographic ranges of species, from the life spans of individuals to the durations of species in the fossil record (McMahon and Bonner 1983; Peters 1983; Calder 1984; Schmidt-Nielsen 1984; May 1986; Brown 1995).

Relationships between Local and Regional Diversity

Some investigators have taken a more explicitly ecological approach to questions about the assembly of continental biotas by investigating the relationship between local and regional diversity (e.g., Ricklefs 1987; Schluter and Ricklefs 1993; Cornell 1993; Caley and Schluter 1997). While this approach has much in common with both areography and macroecology, it differs in several important respects. It focuses on how local biotas are assembled from the regional species pool. It assumes that the composition of local communities is determined primarily by ecological interactions that usually occur on small spatial and temporal scales. These local interactions determine which of the species potentially able to colonize a habitat are actually able to tolerate the abiotic environmental conditions and coexist with the other species that occur there. The composition of the pool of potential colonists, on the other hand, is assumed to reflect regional and historical processes such as barriers to dispersal and speciation and extinction events.

One line of investigation is to ask whether, and in what respects, local assemblages differ from "null" communities constructed by drawing species at random from the regional species pool. Statistical demonstration of such nonrandom assembly is usually taken as evidence that a community is strongly influenced by local ecological interactions. Much of the emphasis in this line of research has been on the role of interspecific competition, which led to an often acrimonious debate in the 1980s over its importance (e.g., Strong et al. 1984; Diamond and Case 1986)—a debate that has continued to some extent right up until the present (e.g., Ricklefs and Schluter 1993; Stone et al. 1996; Weiher and Keddy, in press). With some trepidation, we summarize this huge and still controversial body of work by saying that competition is one of the processes that influences the assembly of local communities—but it is not the only such process. Indeed, it should be apparent from the well-documented examples in Chapter 4 that any limiting niche variable—including not only competition but also other biotic interactions such as predation, parasitism, disease, and mutualism, as well as abiotic conditions—can prevent a species from occurring in a local area that it otherwise would be able to colonize.

A second approach to investigating the relationship between local and regional processes is to plot local diversity as a function of regional diversity. In Figure 15.18 we show the results of a somewhat similar analysis in which we plot the fraction of the species in the regional pool that coexist in local habitats. In that case, relative local diversity decreases as species diversity increases from high latitudes toward the equator. When the data are plotted as in Figure 16.12, somewhat different inferences are often drawn. For example, a leveling off in the slope of the relationship between local and regional diversity may be claimed to imply "saturation" of communities and limitation of community membership by local interactions, whereas a continuously increasing relationship may suggest that local communities are "unsaturated" and are limited primarily by regional and historical processes (Ricklefs 1987; Schluter and Ricklefs 1993; Cornell 1993; but see Caley and Schluter 1997).

We applaud these efforts to characterize community patterns quantitatively, either by comparing real and null communities or by analyzing plots of local versus regional diversity. We are concerned, however, that interpreting the results in terms of a simplistic dichotomy between local or regional, ecological or historical phenomena obscures more complex relationships between patterns and processes. Several authors have assumed that "local" ecological processes, such as interspecific competition, do not influence the composition of the regional species pool. This is clearly incorrect. As pointed out in Chapter 4, the edges of nearly all geographic ranges are limited by local ecological niche variables that prevent populations of species from expanding into adjacent areas. If the species pool, by definition, consists of those species whose geographic ranges overlap some local site, and if local interactions can limit geographic ranges, then local interactions influence the composition of the pool. Put another way, no species would occur in the regional pool unless its local ecological relationships enabled it to occur within some local assemblages of the region. So we expect the composition of the regional species pool to be strongly influenced, not just by regional and historical processes, but by local ecological interactions, including competition (see Colwell and Winkler 1984).

We also see problems in deciding when or if a local community is "saturated." A decreasing slope in the plot of local versus regional diversity is not sufficient to show saturation. Indeed, so long as the relationship is increasing at all, species obviously are being added to local communities, so they cannot be considered saturated. We question whether communities are ever truly saturated. This would imply that they could resist invasion of all exotics. At best, this is a virtually impossible proposition to test empirically. Furthermore, it represents a very rigid equilibrial view of ecology and biogeography. As pointed out in Chapters 4, 13, and 19, this view is increasingly being challenged by nonequilibrial concepts, models, and theories that reflect the reality of a constantly changing world.

Biotic Interchange

The fate of artificially introduced species provides some limited insight into the kinds of interactions that can occur when representatives of one formerly isolated biota come into contact with those of another. This has happened naturally in the past, not only when individual species have managed to disperse across barriers, but also when the barriers themselves have been reduced or abolished, bringing two distinct, previously isolated biotas into direct contact. Such contacts have occurred many times as continents have drifted over the earth and new land and water connections have been formed (Vermeij 1978). Unfortunately, the record of their results is often poor, because it must be pieced together largely from limited fossil evidence. For example, when the Indian plate collided with southern Asia, it presumably brought a distinct Gondwanan biota into contact with a large Eurasian biota. You may recall from our discussion of taxa that were originally distributed on Gondwanaland that little mention was made of relictual Gondwanan populations in India. Was the Indian biota derived from Gondwanaland eliminated by interactions with the more diverse biota of the larger Eurasian landmass? This is a plausible explanation, but unfortunately, the fossil inhabitants of the Indian subcontinent are so poorly known that there is little evidence to support or refute this hypothesis. The interchange that took place when North and South America joined is much better documented, and is the focus of the following discussion.

The Great American Interchange

> During the age of mammals, the Cenozoic in geological terms, South America was an island continent. Its land mammals then evolved in almost complete isolation, as in an experiment with a closed population. To make the experiment even more instructive, the isolation was not quite complete, and while it continued two alien groups of mammals were nevertheless introduced, as might be done to study perturbation in a laboratory experiment. Finally, to top the experiment off the isolation was ended, and there was extensive mixture and interaction between what had previously been quite different populations, each with its own ecological variety and balance. (G. G. Simpson, 1980)

The ecological experiment that Simpson was referring to was the exchange of mammals between North and South America following the formation of the Central American landbridge about 3.5 million years ago. This event, known as the **Great American Interchange**, provides a fascinating case study of the combined effects of dispersal, interspecific interaction, extinction, and evolution on biological diversity. Most of the available information concerns mammals, in part because they left a rich fossil record. Interestingly, it seems that the effects of the interchange were quite different in other terrestrial organisms, such as amphibians, reptiles, birds, and plants.

Historical background.   Mammals evolved from reptilian ancestors during the Mesozoic, beginning about 220 million years ago, when the continents were united in the single great landmass of Pangaea. By the time Pangaea began to break up, mammals had spread over the continents and diversified somewhat. Compared with the ruling reptiles, however, they remained a small component of the biota until the mass extinctions at the end of the Cretaceous, about 65 million years B.P. This biotic upheaval, which saw the extinction of many terrestrial and marine lineages, including the dinosaurs (see Chapter 8), was followed by rapid diversification of birds, mammals, and some other surviving groups.

The most important geological, climatic, and biotic events associated with the history of South American mammals are summarized in Table 16.4 (see also Stehli and Webb 1985). South America was a part of Gondwanaland until about 160 million years B.P., but then drifted apart and remained a giant island continent until about 3.5 million years B.P. During this period of what Simpson called "splendid isolation," a distinct endemic land mammal fauna evolved. At least one monotreme (a platypus), several groups of marsupials, and at least one lineage of eutherian mammals speciated and differentiated to produce a morphologically and ecologically diverse fauna that included large carnivores (even a marsupial saber-toothed "tiger") and giant herbivores. This radiation was in some ways comparable to, but even more diverse than, that which occurred on the other island continents formed by the breakup of Gondwanaland—namely, Australia and Madagascar.

The biogeographic isolation of South America was not complete, however. As Simpson noted in the above quote, it was interrupted by one or two transient periods of limited exchange, presumably across the Proto-Antillean (140–120 million years B.P.) and Central American (10–5 million years B.P.) archipelagoes. During these periods the ancestors of the present South American primates, edentates (armadillos, sloths, and anteaters), and caviomorph rodents (porcupines, capybaras, pacas, agoutis, guinea pigs, chinchillas, and

Table 16.4
Summary of tectonic, climatic, and biotic events associated with the Great American Interchange

220–160 million years B.P.	South America remains connected the rest of Gondwanaland, and to North America. The origin, spread and diversification of mammals and birds across the landmasses.
140–75 million years B.P.	South America becomes isolated. It's biota evolves in isolation.
140–120 million years B.P.	Proto-Antilles serves as a transient stepping stone, sweepstakes route. Limited biotic exchange among Neartic and Neotropical regions.
10–5 million years B.P.	Central American Archipelago serves as a transient stepping stone, sweepstakes route. Limited biotic exchange among Neartic and Neotropical regions.
~3.5 million years B.P.	Emergence of the Central American Landbridge (closure of the Bolivar Trench). Provides a filter dispersal route for terrestrial forms, but a barrier for marine organisms.
2–0 million years B.P.	Lowering of sea-level and extension of savanna and other open-habitat biomes during glacial maxima opens a corridor or filter route for biotic exchange. Subsequent invasions and diversification of invaders, extinctions of invaders and natives.

other forms) colonized the continent. In each of these groups, speciation and adaptive radiation subsequently produced multiple families and genera from a few founding lineages.

The isolation of South America ended dramatically about 3.5 million years B.P. when the Central American uplift formed the land connection to North America that still exists today. Since its formation, however, the Central American landbridge has served more as a filter than as a highway (Figure 16.13). David Webb (1991) has suggested that dispersal across the landbridge was strongly influenced by the climatic cycles of the Pleistocene (see Chapter 7). Interchange was greater during glacial periods, when savanna habitats expanded to cover much of Central and northern South America, than during interglacial periods, such as the present, when most of the landbridge was covered with tropical forest.

Patterns and consequences of mammalian faunal exchange.   The fossil records of North and South America over the last 10 million years reveal the magnitude of the Great American Interchange. The diversity of the North American fauna, at the generic level, remained virtually unchanged despite the invasion of some South American forms such as the extant porcupines, opossums, and armadillos and the now extinct hippo-sized glyptodonts and bear-sized giant ground sloths. The overall diversity of the South American fauna increased due to the invasion and establishment of lineages from North America. The many North American groups that crossed the Isthmus of Panama and became established in South America included not only the extant shrews, rabbits, squirrels, dogs, bears, raccoons, weasels, cats, deer, peccaries, tapirs, and camels, but also the now extinct mastodons and horses. The magnitude of change was even greater than the final figures for generic or familial diversity suggest, however, because several endemic South American forms went extinct. These included several kinds of large marsupial carnivores and even larger eutherian herbivores.

The imbalance of the interchange can be seen in the fact that about half of contemporary South American species are derived from North American ancestors, whereas only about 10 percent of North American species are of

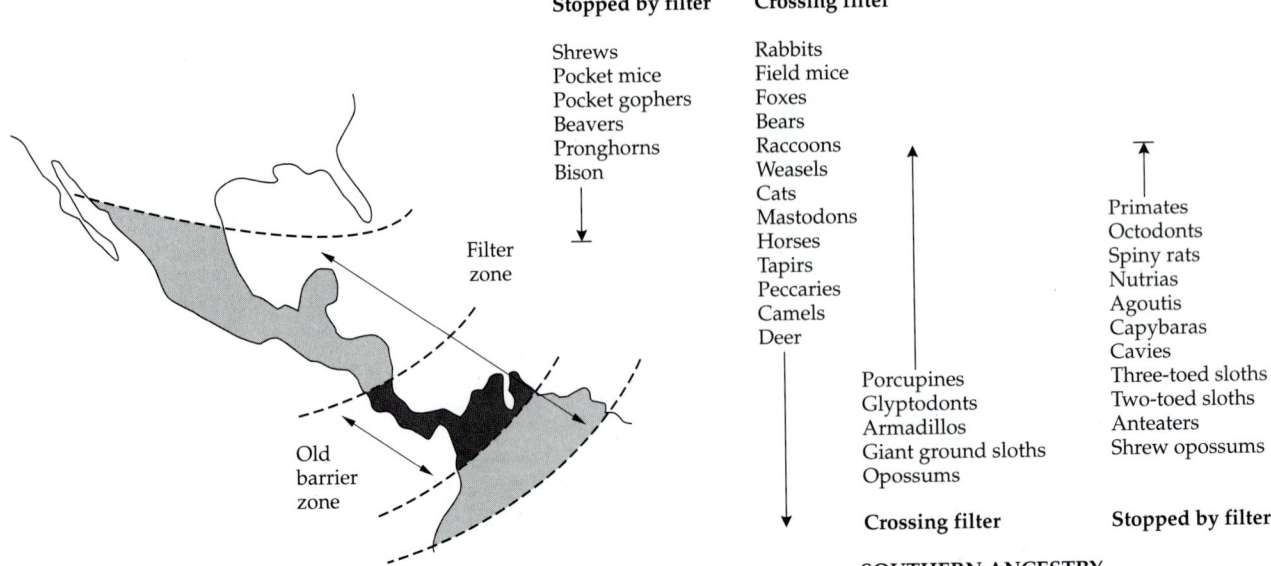

Figure 16.13 Map showing the location of the Central American landbridge, with lists of the mammalian families of both North and South American origin that either crossed through the filter of tropical lowland habitats in Central America during the Great American Interchange to colonize temperate regions of the other continent, or were stopped in or near the filter. Note the asymmetry, with more groups of North American origin passing through the filter and more families of South American origin stopped by the filter.

South American origin (Figure 16.14). The northern forms appear to have had three advantages that contributed to this imbalance:

1. *They were better migrators.* While many northern forms (listed above and in Figure 16.13) crossed the landbridge and invaded deep into South America, only three species (porcupine, opossum, and armadillo) of ancient South American ancestry managed to colonize and persist to the present in temperate North America. Interestingly, however, several other groups of South American ancestry (primates, sloths, anteaters, and other kinds of marsupials and caviomorph rodents) have colonized northward across the Isthmus of Panama, but extend no farther than the tropical forests of southern Mexico.

2. *They were better survivors and speciators.* Most of the northern forms that invaded South America not only survived (mastodons and horses were exceptions), but also speciated and diversified there, giving rise to such distinctive forms as kinkajous, coatimundis, giant otters, swamp deer, guanacos, and vicuñas. The mouselike sigmodontine rodents, which probably colonized South America by island hopping before the completion of the landbridge, are now the most diverse group of mammals on the continent, comprising more than 50 genera and 250 species (Reig 1989; Marshall 1979; Webb and Marshall 1982). Furthermore, over the last 3.5 million years, while lineages of North American ancestry have been diversifying, the ancient South American lineages have been dwindling away because speciation has not kept pace with extinction.

3. *They were better competitors.* In the face of the differential colonization, survival, and speciation of North American forms, it is hard to avoid the conclusion that they have been, on average, superior competitors. Clearly the North American forms not only have increased in generic and species diversity at the expense of their more ancient South American counterparts, but also have radiated to usurp their ecological roles. For example, all of the large carnivores and herbivores and the vast majority of the mouselike rodents in South America today are descendants of northern invaders.

Some authors have questioned whether such differential ecological and evolutionary success can be attributed to "competition." Indeed, not all of the interactions are strictly competitive. Predation, parasitism, and disease probably play a significant role. Furthermore, much of the competition that does occur may be diffuse and involve many species rather than simple pairwise interactions. It is also clear that not all mammals of South American ancestry are competitively inferior; witness the success of porcupines, opossums, and armadillos in invading North America. The latter two species have actually been expanding their ranges northward in recent decades (see Figure 9.3D).

Returning to Simpson's analogy, the Great American Interchange constituted a vast natural experiment. By the movement of continents, the mammalian fauna of South America were first allowed to evolve in "splendid isolation," and then brought it into contact with a more diverse fauna—one that had evolved on the larger North American landmass and had had frequent interchanges with the fauna of the Old World. The outcome of the experiment is clear. Taken as a whole, the North American mammalian fauna proved superior, and differentially "replaced" much, though by no means all, of the original South American fauna.

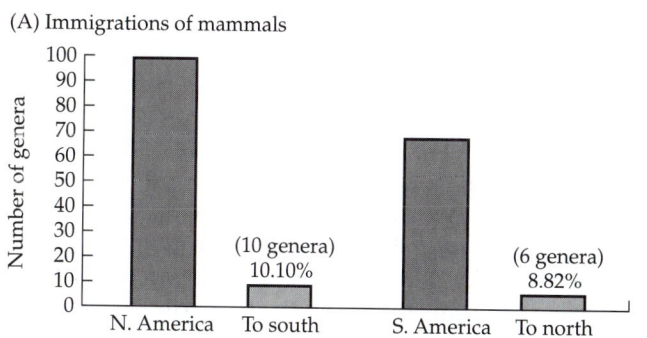

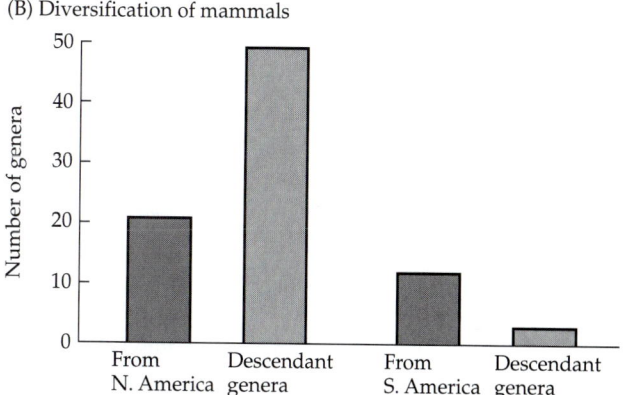

Figure 16.14 The imbalance of the Great American Interchange is indicated by the numbers of genera that invaded each continent and that are derived from these invaders. (A) Numbers of genera in the fauna of each continent that did not cross the Isthmus of Panama during the Pliocene, between 2 and 5 million years B.P. (left bars) and that did invade the other continent (right bars). Note that movement across the isthmus was approximately balanced, with about 10% of the fauna crossing in each direction. (B) Post-Pliocene diversification of mammalian genera in North and South America with respect to their continent of origin. The left bar gives the number of genera derived by direct immigration, and the right bar, the number that differentiated in situ from immigrant ancestors. Note that many of the genera now endemic to South America were derived from North American immigrants. (After Webb and Marshall 1982.)

Inter-American exchange in other vertebrates.   Information on historical exchanges between North and South America in vertebrate groups other than nonvolant mammals is scanty, because the fossil record of these groups is relatively poor. It is also more difficult to interpret, because some of these groups, such as reptiles and especially birds and bats, are better over-water colonists than nonvolant mammals. This means that some level of faunal interchange between the two continents probably occurred continuously throughout the Cenozoic, rather than being concentrated in the last 3.5 million years following the completion of the Central American landbridge (Vuilleumier 1984, 1985; Vanzolini and Heyer 1985; Estes and Baez 1985).

Two patterns are noteworthy. First, there is not a clear dichotomy between an ancient South American fauna, which dates back to the isolation of Gondwanaland, and relatively recent invaders. Instead, the South American herpetofauna and avifauna appear to have been assembled more gradually. They are made up of a mixture of ancient Gondwanan forms, lineages that colonized across water during the early and mid-Cenozoic, and forms that crossed the Central American landbridge (or the island chain that preceded it) during the last several million years. Thus, at least for terrestrial reptiles, amphibians, and birds, it is difficult to distinguish easily between South American "natives" and North American "invaders."

Second, the late Tertiary invasion of South America by North American forms, so clearly seen in mammals, simply did not occur in many other groups. The exchange of birds appears to have been much more balanced than that of mammals. While some groups, such as pigeons, owls, woodpeckers, and jays, colonized South America from the north, others, such as hummingbirds, tyrant flycatchers, vireos, wood warblers, blackbirds, orioles, tanagers, and emberizine buntings (grosbeaks and sparrows), moved in the opposite direction. An interesting feature of the North American avifauna is that the Neotropical migrants, which make up the majority of breeding passerines in most temperate habitats, are virtually all of South American ancestry.

The dominance of the North American fauna by lineages of South American origin is even more pronounced in reptiles and amphibians. Thus South America has no salamanders and perhaps only one species of frog that colonized from North America. In contrast, of the North American frog fauna, only three families (Ascaphidae, Pelobatidae, and Ranidae) are of northern origin, whereas four (Bufonidae, Hylidae, Leptodactylidae, and Microhylidae) are of South American origin. Similarly, in reptiles, the predominant movement has been from South America to North America, although this pattern has sometimes been complicated by secondary centers of speciation and diversification in Central America and Mexico. Thus, for example, the two large families of New World lizards, Iguanidae and Teidae, and most of the snakes have dispersed northward from South America to North America.

The completion of the Central American landbridge resulted in some limited interchange of freshwater fishes (Miller 1966; Rosen 1975; Bussing 1985). Again, the predominant direction of dispersal was from south to north. Because of their requirement for freshwater connections in order to disperse, however, the invasion of primary freshwater forms has been limited. Several South American lineages have reached as far north as central Mexico (Miller 1966), whereas no North American group has made it farther south than Costa Rica (Bussing 1985).

To summarize, the Great American Interchange, that wonderful natural experiment so thoroughly documented by Simpson and later researchers, shows a clear pattern: terrestrial mammals of North American ancestry were better dispersers, survivors, speciators, and competitors. Over the last 3.5 million years, since the completion of the Central American landbridge, they have invaded South America and largely supplanted the ancient South American groups. Completely different patterns, but usually also with unbalanced exchange, occurred in the other vertebrates.

Lessepsian Exchange: The Suez Canal

Continuing with Simpson's analogy of biotic interchange as a natural experiment, humans have unintentionally performed one such manipulation by bringing into contact two previously long-isolated marine biotas. This experiment began in 1869 with the completion of the Suez Canal between the Mediterranean Sea and the Red Sea. The resulting biotic exchange has been called the **Lessepsian migration** or **exchange** (Por 1971, 1977). Although the hypersaline lakes through which the canal passes constitute an impassable barrier for many marine forms, an increasing number of taxa have been able to disperse between the two seas. Again, the biotic exchange has been very unbalanced. More than 50 species of fish, 20 species of decapod crustaceans, and 40 species of mollusks have colonized the eastern Mediterranean from the Red Sea, but it is hard to document cases of migration in the opposite direction.

Three explanations for this unidirectional dispersal have been proposed, and all three factors probably contribute to successful colonization by Red Sea forms. First, the Gulf of Suez, at the southern end of the canal, is itself more saline than ocean water, so the species that occur there may have been preadapted to cross the barriers formed by the hypersaline lakes. Second, most of the successful migrants inhabit shallow sandy or muddy bottom habitats in the Indian Ocean. Such habitat affinities also appear to be preadaptations, because they facilitate movement through similar habitats in the Suez Canal. Finally, species of the Indo-Pacific biota may be more resistant to predation than their Mediterranean counterparts, competitively superior to them, or both. The Red Sea is an arm of the Indian Ocean, which contains a far more diverse biota than the Mediterranean. Biotic interactions are implicated by observations of declining populations of some endemic Mediterranean species in the eastern part of the sea (e.g., along the coast of Israel) where Red Sea colonists have become well established. (For a much more complete discussion of the history and results of the Lessepsian exchange, see Por 1971, 1975, 1978, 1977; Golani and Ben-Tuvia 1990; Golani 1993; and Vermeij 1978, 1991b.)

One final point is warranted. The differential extinctions of South American mammals in the face of North American invaders and the unbalanced exchange of marine forms through the Suez Canal are both consistent with a general pattern noted by many biogeographers. As early as 1915, W. D. Matthew (see also Willis 1922) noted that organisms from diverse biotas on large landmasses are best able to successfully invade smaller areas and replace the native organisms. Darlington (1957, 1959) reemphasized this point, although he argued that the successful forms usually originated in tropical regions, whereas Matthew had thought they came from temperate climates. Because the climates of regions that are now temperate and tropical have changed greatly over their geological history (see Chapters 5 and 14), this point may be difficult to resolve. Nevertheless, the success of organisms from large, diverse biotas in colonizing small, isolated regions containing fewer native species seems to be a consistent phenomenon in biogeography. The interactions that have been seen on a continental scale fit the pattern of invasions of the island continent of Australia from larger continents, as well as the natural and human-assisted colonization of many islands by continental forms (see Chapters 15, 17, and 18).

Maintenance of Distinct Biotas

The opposite of biotic interchange is biotic segregation or biogeographic provincialism. In Chapter 2 we noted that Buffon, Sclater, Wallace, and other early biogeographers called attention to the distinct assemblages of plants and animals that occurred on the different landmasses, even in regions with similar climates, soils, and other abiotic environmental conditions. In Chapters 8, 10, and 12 we discussed the historical processes responsible for the differentiation of endemic biotas. Here we consider briefly how this distinctness is maintained. Given the present landbridges connecting Africa and Eurasia as well as North and South America, and the frequent Pleistocene connections between North America and Eurasia, why hasn't biotic interchange been more complete? What processes are responsible for the preservation of biogeographic provincialism, especially in organisms that are good dispersers?

At one level, the answer is fairly straightforward. Since biotic interchange is due to dispersal and subsequent ecological success, the maintenance of provincialism must be due largely to some combination of continued isolation and resistance to invasion. Both factors are important.

Barriers between Biogeographic Regions

Most biogeographic regions are isolated in the sense that they are separated by ecological barriers to dispersal. For example, Figure 2.5 immediately reveals that even where the terrestrial zoogeographic realms are connected by landbridges, these are either narrow isthmuses, harsh deserts, or high mountains. Furthermore, the geographic ranges of the majority of species within a region do not extend to the boundary and, therefore, fall short of the landbridge. In order to move between regions, the majority of species, including distinctive endemic forms, would have to disperse long distances through unfavorable habitats. Consequently, the opportunity for interchange between regions is limited to a small fraction of the biota. For example, even though the Bering landbridge was exposed by the lowered sea levels that prevailed during most of the last 2 million years, it allowed only limited exchange between North America and Eurasia. It did provide a dispersal corridor, but only for those organisms whose ranges extended into steppe, coniferous forest, and tundra habitats at high latitudes. Some species, such as bears, wolves, ermines, caribou, moose, and lemmings among the mammals and ravens, hawks, owls, ptarmigan, siskins, and crossbills among the birds, dispersed freely, and now occur on both continents. In contrast, many groups of North American and Eurasian amphibians, reptiles, birds, and mammals that did not inhabit these high-latitude environments did not disperse across the Bering landbridge.

Resistance to Invasion

Although it is tricky to document, it also appears that the biotas of large landmasses are relatively resistant to invasion. When faced with a diverse native biota that has evolved to tolerate the abiotic environment and coevolved to withstand the existing biotic interactions, it is difficult for invading species to become established. Evidence in support of this hypothesis comes from the fate of exotic species (for individual case histories, see Chapter 18 and Elton 1958; Udvardy 1969; Hengeveld 1989: Drake et al. 1989). Numerous Old World plant and animal species have become locally abundant and geographically widespread in North America within the last four centuries. Although the success of these invaders is impressive, they represent only a small fraction of the species that have been intentionally or accidentally introduced. The vast majority of introduced populations have gone extinct, and many others have not spread far beyond the site of their introduction. An example is the European tree sparrow (*Passer montanus*): more than a century after its establishment, it is still confined to a small area near St. Louis, Missouri.

Furthermore, of the hundreds of Eurasian plants established in North America, most can be classified as weeds, species that occur in successional habitats created primarily by human disturbance. Of the many introduced insects, the majority of successful species are crop pests, associated with introduced plant species, or confined largely to disturbed habitats. This pattern holds even for vertebrates. Many Eurasian birds have been introduced into North America, but only a few have become established. The two amazingly successful introduced bird species that have spread to cover most of the North American continent, the house sparrow (*Passer domesticus*) and the starling (*Sturnus vulgaris*; see Figure 9.3A,B), are largely commensals with humans, using artificial structures for nesting sites and urban and agricultural habitats for food resources. The success of these Eurasian exotics in North America and the much lower success of New World species in the Old World suggest that most of the Eurasian forms are adapted to occupy niches that are dependent on human activity, and exploit similar niches in North America. As evidence of this we note not only that the exotics have been generally unsuccessful at invading undisturbed native habitats, but also that it is difficult to point with confidence to the extinction of a native continental species owing to replacement in its niche by an introduced competitor.

A quantitative macroecological analysis of resistance to invasion was performed for North American freshwater fishes by Gido and Brown (see Brown 1995). They plotted the number of colonizing exotic species as a function of the number of native species for 135 watersheds in temperate North America. The result is a triangular-shaped distribution (Figure 16.15). Watersheds with few native species show wide variation in the number of exotics present. Some of them have been colonized by few invaders, perhaps because few species have been introduced or because stressful abiotic conditions have prevented their establishment, but others have been invaded by scores of exotic species. In contrast, watersheds with large numbers of native species have uniformly low numbers of exotics. This pattern is consistent with Elton's (1958) suggestion that diverse biotas that have radiated to fill many niches may indeed be relatively resistant to invasion.

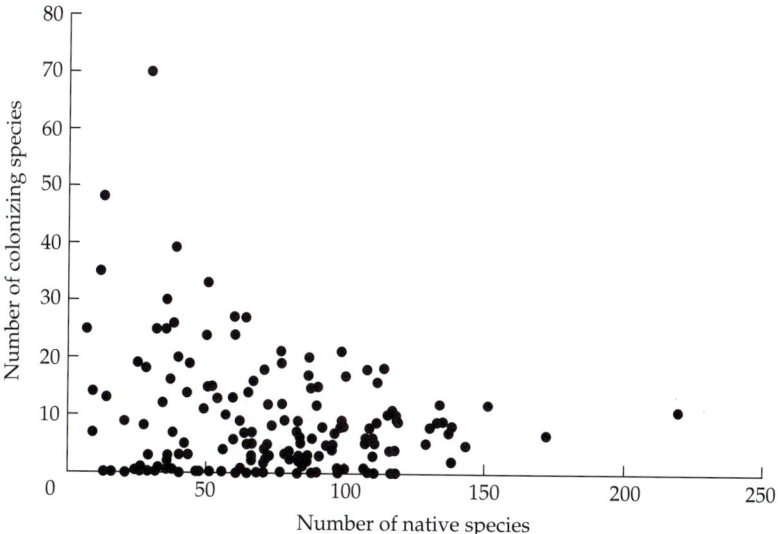

Figure 16.15 Number of exotic fish species that have become established as a function of the number of native species for 135 large watersheds in temperate North America. Note that all of the points fall within a roughly triangular space, showing that watersheds with diverse native fish faunas have been relatively resistant to invasion, whereas some of the watersheds with few native species have been colonized by many invading species. (Courtesy of K. Gido.)

Avian Migration and Provincialism

Birds, with their capacity to disperse by flight over great distances and formidable geographic barriers, would seem to be one of the groups of organisms least likely to exhibit biogeographic provincialism. Indeed, a few bird species, such as the peregrine falcon, osprey, and barn owl, are virtually cosmopolitan, and birds have managed to colonize even the most remote of oceanic islands (see Chapter 14). Remember, however, that Sclater (1858) originally divided the world into biogeographic provinces based on the distributions of birds. This seems counterintuitive.

An examination of avian distributions on a global scale reveals that the observed provincialism is due largely to two very different "functional groups" of birds. On the one hand, birds with very limited powers of dispersal and specialized adaptations for particular environments often exhibit high degrees of endemism. Such birds include the flightless ratites (ostriches, emus, cassowaries, and rheas) of the southern continents and the weakly flying tinamous, hoatzins, and curassows of tropical America, secretary birds and turacos of Africa, megapodes, lyrebirds, bowerbirds, and birds of paradise of Australia, and turkeys of North America. They also include highly specialized birds such as the nectar-feeding hummingbirds of the New World, sunbirds of Africa and southern Asia, and honeyeaters of Australasia, as well as the wood creepers, ovenbirds, antbirds, and manakins of South America, the honeyguides, wood hoopoes, and weaver finches of Africa, and the lorikeets, cockatoos, wood swallows, whistlers, and logrunners of Australia. Two additional comments are worth making. First, many of these groups are endemic to one of the southern continents, presumably reflecting their long history of geographic isolation. Second, other groups of birds that might seem to be equally poor fliers or equally specialized are surprisingly widespread. Examples include the large, rarely flying bustards (Africa, Europe, Asia, and Australia) and the marsh-living, rarely flying rails (cosmopolitan).

The other "functional group" that contributes importantly to biogeographic provincialism consists of small land birds that are long-distance migrants (Bohning-Gaese et al., in press). These include the hummingbirds, tyrant flycatchers, vireos, wood warblers, tanagers, orioles, blackbirds, and emberizine buntings and sparrows, all families or subfamilies endemic to the New World, and also the true flycatchers and warblers, groups restricted to the Old World. These subfamilies and families of birds include many species that migrate twice each year between widely separated breeding grounds in the subarctic and temperate regions of either North America or Eurasia and wintering grounds in tropical areas of South and Central America or Africa, southern Asia, and Australia (Figure 16.16). With very few exceptions, representatives of these groups do not occur in both the New and Old Worlds. The exceptions prove the rule: a few species of Old World warblers breed in the Aleutian Islands and even on the Alaskan mainland, but migrate to winter in tropical Asia or Australasia (Figure 16.17).

Also surprising is the fact that similarly small land birds that are either nonmigratory or only short-distance migrants show exactly the opposite pattern: many families and subfamilies contain genera that occur in both North America and Eurasia. These groups include woodpeckers, nuthatches, creepers, wrens, crows, tits, finches, thrushes, and waxwings (Figure 16.18). These relatively sedentary avian taxa have distributions very similar to those of land mammals, insects, and both coniferous and angiosperm plants: closely related species occur across Arctic, subarctic, and cool temperate regions of both Eurasia and North America. Apparently such Holarctic distributions reflect a long

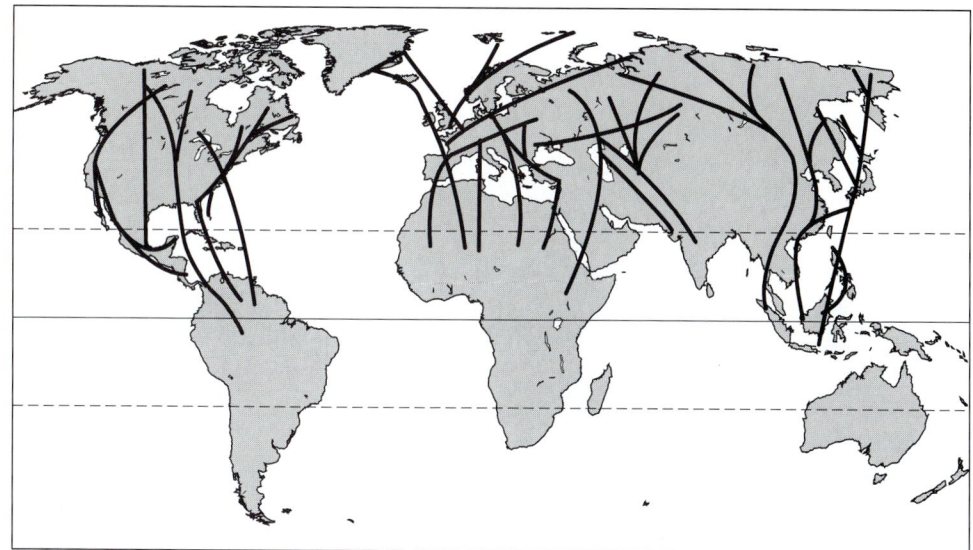

Figure 16.16 Major avian migratory flyways. As the plotted routes suggest, in both the New and Old Worlds, many bird species travel hundreds and even thousands of kilometers twice each year, commuting between breeding grounds in Arctic and temperate regions to wintering grounds that are located mostly in the tropics. Very few of these birds, however, cross between the New and Old Worlds. (After Mc-Clure 1974 and Baker 1978.)

history of close proximity, including repeated land connections and resulting biotic exchanges, most recently across the Bering landbridge during the Pleistocene (see above and Chapters 7 and 9).

Why are the long-distance migrants so different? What has prevented interchange between the New and Old Worlds? At least two interrelated factors seem to be involved. First, traits associated with long-distance migration may actually make it difficult for migrants to colonize a new continent. An initially small founding population not only must become established in a suitable breeding area, it must also find a new, distant wintering ground and a route there and back. The difficulty of developing new migratory routes appears to be so severe as to make exchange between New and Old Worlds highly improbable. Support for this hypothesis comes from the observation that the land bird faunas of isolated islands are constituted primarily of species derived from nonmigrants and short-distance migrants, rather than long-distance migratory ancestors (L. Gonzalez-Guzman, pers. comm.). Long-distance migrants can and do reach distant places, but they are unlikely to stay there or find their way back to breed.

Second, this pattern seems to have a historical component. The lineages that contain most of the long-distance migrants appear to be of tropical origin. Migration seems to be an adaptation that allowed highly mobile birds to rear their young using the seasonal pulse of productivity that is available for only a few months at higher latitudes. Thus, most of the long-distance migrants feed their nestlings on insects that are active, abundant, and accessible only during the warm months. This is especially true in the New World, where, as noted above, the families and subfamilies containing most of the long-distance migrants are of South American origin.

The special case of birds illustrates how evolutionary constraints and ecological factors can interact to maintain the historical legacy of biogeographic provincialism. Even though the continents may be connected by landbridges that would seem to provide dispersal routes, and even though organisms may possess traits that would seem to permit long-distance dispersal, the actual interchange and mixing of long-isolated biotas has been relatively limited. The result is that the influence of ancient earth history, especially of tectonic events, is preserved in the distributions of contemporary forms, even such vagile ones as migratory birds.

(A) Arctic Warbler (*Phylloscopus borealis*)

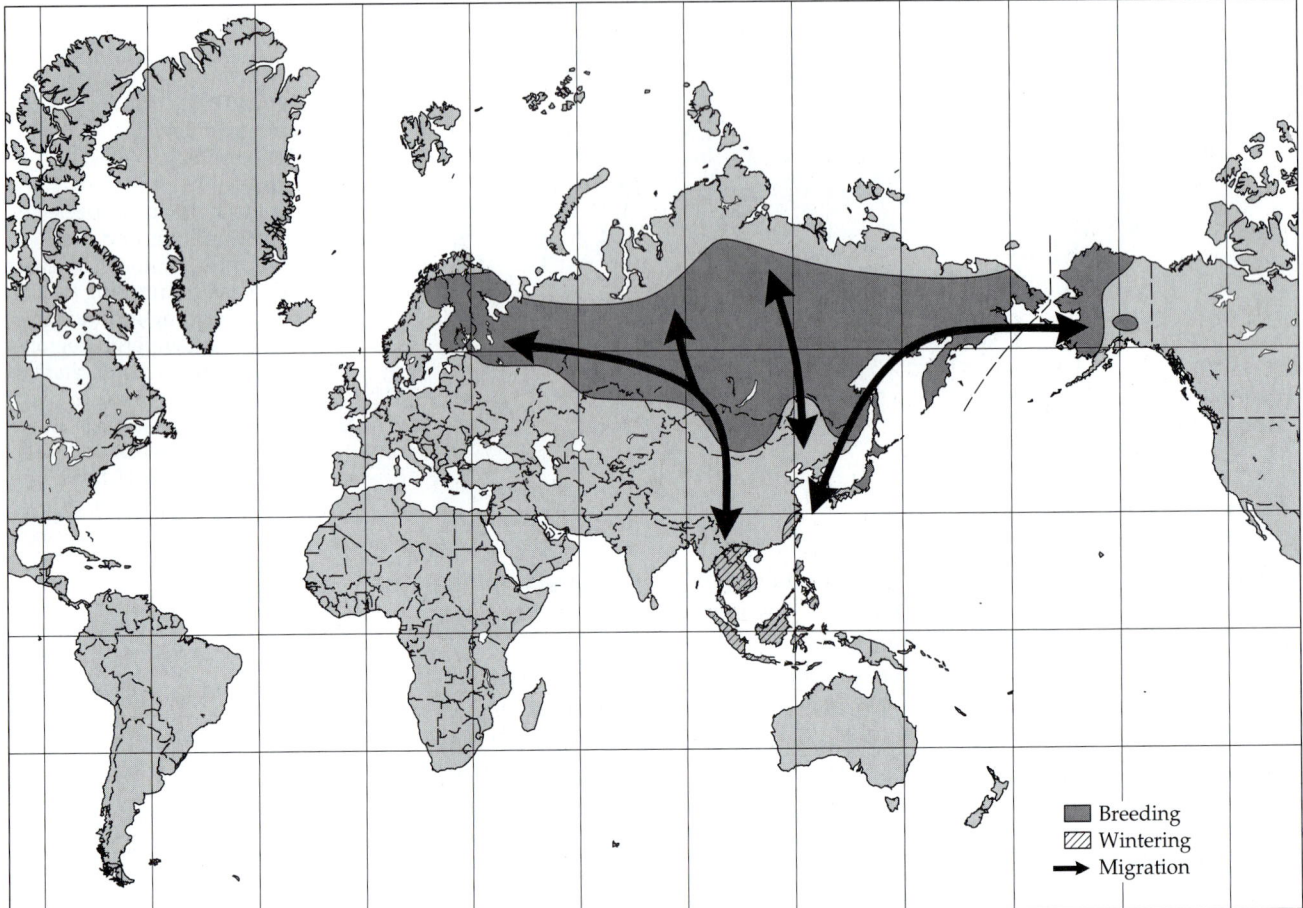

Figure 16.17   Map showing the breeding ranges (dark shaded areas), winter ranges (hatched areas), and migratory routes (arrows) of two passerine bird species: (A) the arctic warbler, *Phylloscopus borealis*, which breeds in western Alaska, and (B) the northern wheatear, *Oenanthe oenanthe*, which breeds in Alaska, northern Canada, and Greenland. Both of these species have colonized the New World, but show their Eurasian origins by migrating to winter in the Old World tropics.

Divergence and Convergence of Isolated Biotas

We have considered what happens when long-isolated biotas are brought into contact. Now let's examine the opposite situation. What happens when a previously continuous region with an undifferentiated biota is fragmented by the creation of a new biogeographic barrier?

Divergence

When the Central American landbridge was formed, it not only permitted the interchange of formerly isolated terrestrial organisms between North and South America, but also created a barrier that completely isolated the formerly continuous tropical Atlantic and Pacific Oceans. This event was as important for marine organisms as it was for the terrestrial forms we discussed above. It is especially important in view of the fragile nature of the isthmus as a barrier.

This fragility became particularly apparent in the 1960s, when the possibility of constructing a sea-level canal across the Isthmus of Panama was being

(B) Northern Wheatear (*Oenanthe oenanthe*)

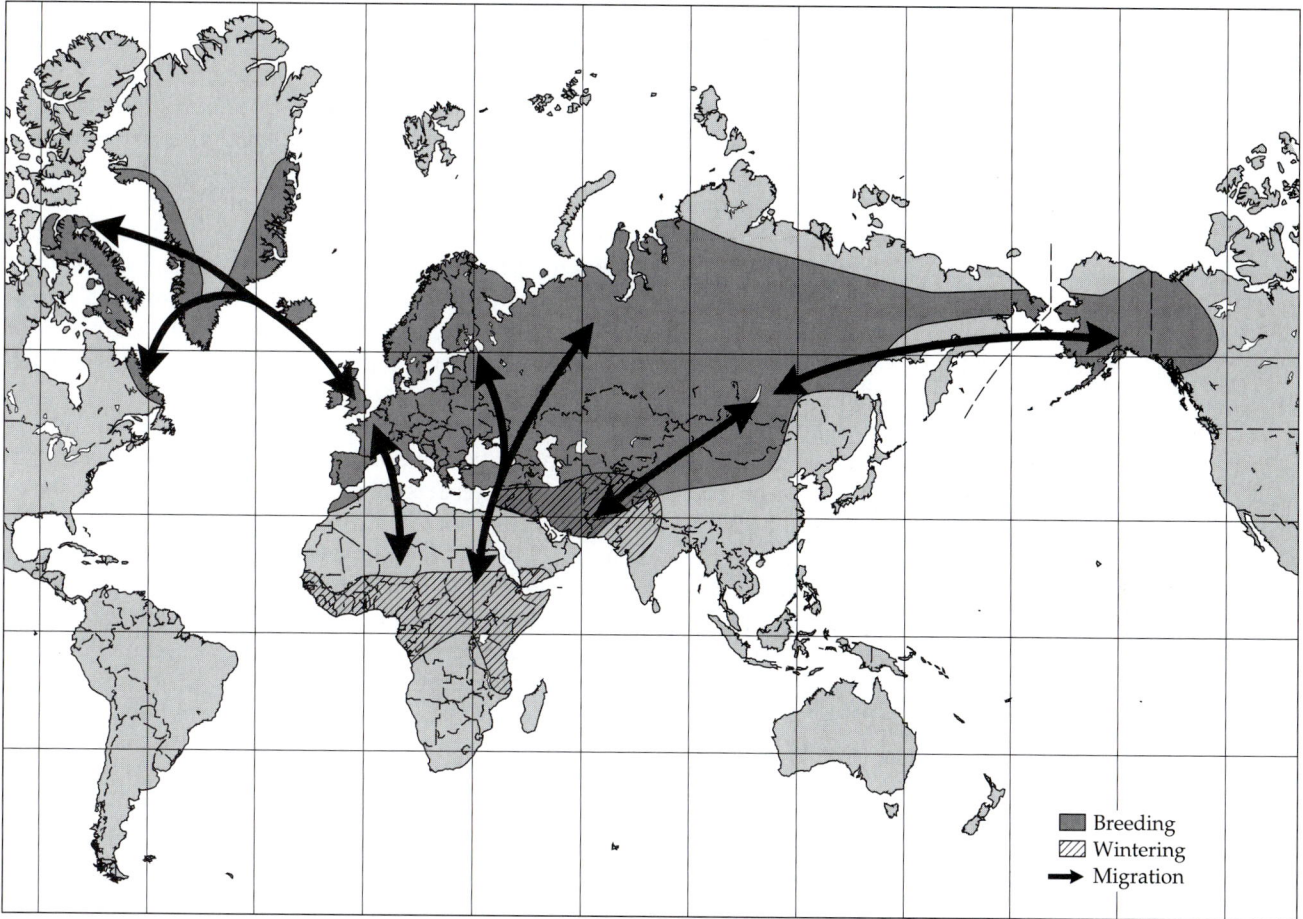

seriously considered. The present Panama Canal, constructed in the early part of the twentieth century, incorporates a large body of fresh water, Gatun Lake, and uses a series of locks to raise and lower ships as they traverse the isthmus. The fresh water effectively prevents interchange between most elements of the Pacific and Caribbean tropical marine biotas. The construction of a sea-level canal would constitute a biogeographic experiment of gigantic proportions. Its effect on marine biotas would be analogous to the influence of the original establishment of the isthmus on terrestrial forms, with the exception that the Caribbean and Pacific biotas have been isolated for only about 3.5 million years, whereas North and South America had been separated for at least 135 million years.

Controversy surrounding the possible ecological effects of an interchange of species as a result of a sea-level canal stimulated much research on the similarities and differences between the Caribbean and eastern tropical Pacific biotas. These studies revealed major differences, especially in species richness, between the marine faunas of the two regions (Briggs 1968, 1974; Rubinoff 1968; Porter 1972, 1974; Vermeij 1978). Most groups are more diverse in the Pacific; examples include most major taxa of mollusks, crabs, and echinoderms. There are exceptions, however. The Caribbean has about 900 species of shallow-water and coral reef fish, compared with only about 650 species in the eastern Pacific. Sea grasses and their specialized animal fauna are abundant, widespread, and diverse in the Caribbean, but virtually absent from the east-

(A)

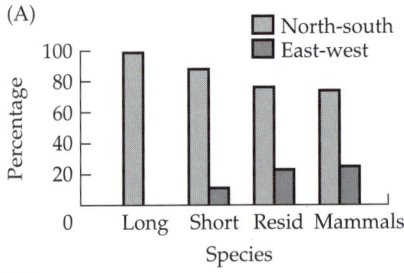

(B)

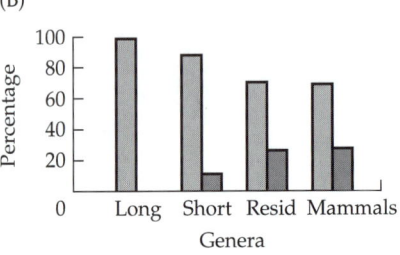

(C)

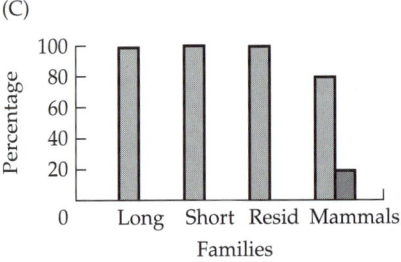

Figure 16.18 Comparison of the distributions between hemispheres of long-distance migrant (long), short-distance migrant (short) and nonmigratory (resid) birds and terrestrial nonvolant mammals at three levels of taxonomic classification: (A) species, (B) genera, and (C) families. North-south (light gray bars) indicates that the geographic range of the taxon includes either both North America and South America in the New World or both Europe and Africa in the Old World; east-west (dark gray bars) indicates that the range of the taxon includes both North America and Eurasia in the Northern Hemisphere. Note that the long-distance migrant birds have exclusively north-south ranges, indicating that they have not colonized between the Old World and the New World; in contrast, resident birds resemble nonvolant mammals in having a substantial number of taxa with east-west distributions, indicating colonization between North America and Eurasia. (After Bohning-Gaese et al., in press.)

ern Pacific, where suitable, highly productive shallow-water habitats are not extensive.

In many groups there are closely related sister species on either side of the isthmus (Jordan 1908; Vermeij 1978). The rates of divergence of some of these forms have been of considerable interest to systematists and evolutionists (e.g., Rubinoff and Rubinoff 1971), because the time of isolation is known quite accurately and is presumably the same for all groups. Although there has been some differentiation, most of these species pairs remain similar in morphology and presumably in their ecological niches. This suggests that competition between such species pairs might prevent much interchange across a sea-level canal. On the other hand, if some forms were superior competitors and able to invade the other ocean, competitive exclusion might result in the extinction of some sister species without greatly affecting the diversity of the biotas on each side of the isthmus.

The absence of complementary species in a few exceptional groups has caused more concern. Two Pacific taxa in particular that do not have close relatives or obvious ecological counterparts in the Caribbean are the sea snake *Pelamis*, and the crown of thorns starfish, *Acanthaster planci*. The former, of course, is highly venomous, whereas the latter feeds voraciously on certain corals and occasionally devastates reefs in its native region (Chester 1969). It is possible that *Acanthaster* might wreak even greater havoc if it were able to colonize the rich Caribbean reefs, where the coral species have had no opportunity to evolve resistance to it (Porter 1972; Vermeij 1978).

Convergence

Most of what we have said thus far would cause one to expect that geographically isolated organisms should diverge. Much of phylogenetic systematics and vicariance biogeography is based on the assumption that genetically and geographically isolated lineages tend to become more different as they evolve independently of each other. This is certainly true of many attributes of organisms, including most of those that are used in classification. It need not be true of ecological characteristics, however. If groups are isolated in regions of different area, geology, and climate, these differences in the physical environment will tend to promote ecological divergence. If the physical environments are similar, however, distantly related organisms may independently evolve similar adaptations. We call this phenomenon **convergent evolution,** and recognize that it can occur on many levels. Within species, it may be restricted to a few traits, or it can involve essentially the entire organism, resulting in convergence in morphology, physiology, and behavior as unrelated forms specialize for similar niches. Convergence can also occur at the level of entire biotas, resulting in geographically isolated ecological communities with similar structures and functions.

Convergence at the species level.   It is possible to find examples of geographically isolated, distantly related pairs of species that are spectacularly similar. Some of the best examples occur in plants, especially those living in regions where similarly stressful abiotic environments have selected for similar form and function. In Chapters 3 and 4 we discussed the Mediterranean climates that occur at about 30° latitude where cold ocean currents flow down the west coasts of continents, in the Mediterranean region, southwestern Africa, western Australia, Chile, and California. The latter two regions were the sites of extensive comparative ecological studies by the International Biological Program (IBP) in the early 1970s (Mooney 1977). Analyses of the anatomy and physiology of the woody plants revealed many similarities. The dominant plants in both matorral habitat in Chile and chaparral in California are shrubs with small to medium-sized, thick (sclerophylous), evergreen leaves. It is possible to identify many pairs of species that are extremely similar in growth form and leaf morphology (Figure 16.19). In addition, these species usually have similar physiological and life history adaptations to photosynthesize and grow during the cool winter rainy season, to minimize water loss and survive through the hot summer dry season, and to regenerate rapidly—usually from vegetative sprouts from the stumps, but sometimes by seed—after the frequent wildfires (Mooney et al. 1977).

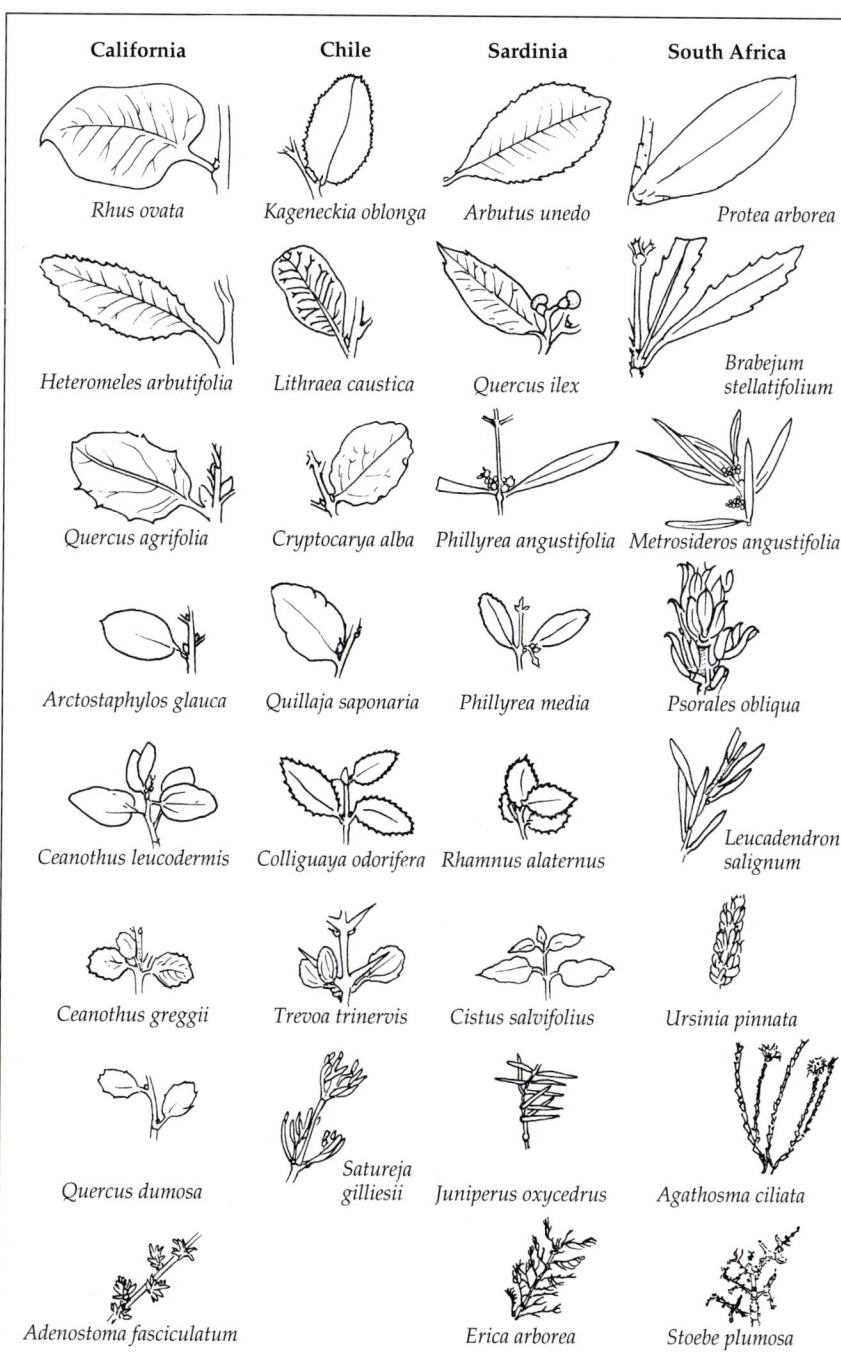

California | **Chile** | **Sardinia** | **South Africa**

Rhus ovata — *Kageneckia oblonga* — *Arbutus unedo* — *Protea arborea*

Heteromeles arbutifolia — *Lithraea caustica* — *Quercus ilex* — *Brabejum stellatifolium*

Quercus agrifolia — *Cryptocarya alba* — *Phillyrea angustifolia* — *Metrosideros angustifolia*

Arctostaphylos glauca — *Quillaja saponaria* — *Phillyrea media* — *Psorales obliqua*

Ceanothus leucodermis — *Colliguaya odorifera* — *Rhamnus alaternus* — *Leucadendron salignum*

Ceanothus greggii — *Trevoa trinervis* — *Cistus salvifolius* — *Ursinia pinnata*

Quercus dumosa — *Satureja gilliesii* — *Juniperus oxycedrus* — *Agathosma ciliata*

Adenostoma fasciculatum — *Erica arborea* — *Stoebe plumosa*

Figure 16.19 Leaf morphology of distantly related plant species from evergreen shrub habitats in Mediterranean climates in four widely separated regions: California, Chile, Sardinia (Mediterranean), and South Africa. Presumably the similarities, not only in the sizes and shapes of the leaves shown here, but also in their physiological characteristics, reflect convergence: the independent evolution of similar traits in response to natural selection for similar adaptations to similar environments. (After Cody and Mooney 1978.)

Plants from desert regions throughout the world provide equally spectacular examples of convergent form and function. Examples are exhibited in several botanical gardens, including the Arizona Sonora Desert Museum near Tucson, Arizona, and Kew Gardens outside London, England. Different genera of succulent thorny cacti (family Cactaceae) from North and South America are remarkably similar not only to each other, but also to the much more distantly related euphorbias (family Ephorbiaceae) of Africa. Another convergent theme is succulents with whorls of tough, pointed leaves: agaves in North America, terrestrial bromeliads in South America, and aloes in Africa (each representing a different family). And the similarity does not stop at superficial resemblances among life forms, but extends to details of cellular anatomy, physiology, and biochemistry. For example, these desert succulents all share a special form of photosynthesis, called crassulacean acid metabolism (CAM). This adaptation enables them to conserve water by opening their stomates at night when temperatures and rates of evaporative water loss are low, taking up CO^2 and storing carbon in the form of organic acids, and then completing photosynthesis with their stomates closed during the day when the sun shines.

These examples of convergence are convincing. Some geographically isolated plant species in different families that live in similar environments are much more similar to one another than to more closely related species that occur on the same continents, but in different environments. Furthermore, their similarities clearly represent evolutionary adaptations to deal in similar ways with similar kinds of abiotic environmental conditions. It should be emphasized, however, that these convergent plants are not strikingly similar in all of their characteristics. Naturally, they each retain distinctive traits indicating their divergent ancestry. In addition, they often differ considerably in reproductive biology, exhibiting divergent forms and functions of flowers, fruits, and seeds that reflect adaptations for different agents of pollination and seed dispersal.

Examples of convergence among geographically isolated animal species are equally spectacular. As in plants, many come from desert regions. The mammalian order Rodentia has a virtually cosmopolitan distribution, but many rodent families have much more restricted distributions. Each of the biogeographic realms has large areas of desert and semiarid habitat, and each of these has a distinct, highly specialized desert rodent fauna made up of different families that have independently evolved to fill a variety of niches. The most striking case of convergence is perhaps the independent evolution of forms with elongated hind legs, long, tufted tails, and bipedal hopping (saltatorial or ricochetal) locomotion in different families on several continents (Figure 16.20; Mares 1976, 1993a). Some of these rodents also share other adaptations, including light-colored pelages to match their backgrounds, enlarged ear cavities (auditory bullae) to detect predators, short, long-clawed front legs for digging burrows and collecting food, and urine-concentrating kidneys for maintaining water balance on a dry diet. On the other hand, as in the plants, their convergence does not extend to all traits. Thus, the North American kangaroo rats and kangaroo mice are highly specialized seed eaters, whereas the similar-looking hopping mice in Australia and jerboas in North Africa, the Middle East, and central Asia eat some seeds, but also include substantial amounts of other items, such as insects, tubers, and leaves, in their diets. Furthermore, some of the rodent species in these different deserts that are not very similar in external morphology are much more similar in diet and, as a consequence, in digestive and excretory physiology (Mares 1976; Mares and Lacher 1987; Mares 1993a,b; Kelt et al., in press).

Similar generalizations could be drawn for other taxa, such as the superficially similar snakes and lizards from arid regions of North America, Africa, Asia, and Australia. The pronghorn or "antelope" of the plains of North America is the sole representative of an endemic family (Antilocapridae) that is convergent in morphology and behavior with the true antelopes (family Bovidae) of the grasslands of Africa. Several species of toucans (family Rhamphastidae) of the New World tropics are superficially similar to some species of hornbills (family Bucerotidae) of the Old World. Representatives of both families have striking black and white plumage and very large, light, keel-shaped bills adapted for feeding on fruit. Toucans and hornbills differ conspicuously, however, in their reproductive biology, with only hornbills showing the distinctive behavior of males building mud walls to imprison their mates in a nesting cavity. The European fire salamander (*Salamandra salamandra*) and North American tiger salamander (*Ambystoma tigrinum*) are similar in their large size, robust body shape, poison glands in the skin, and striking yel-

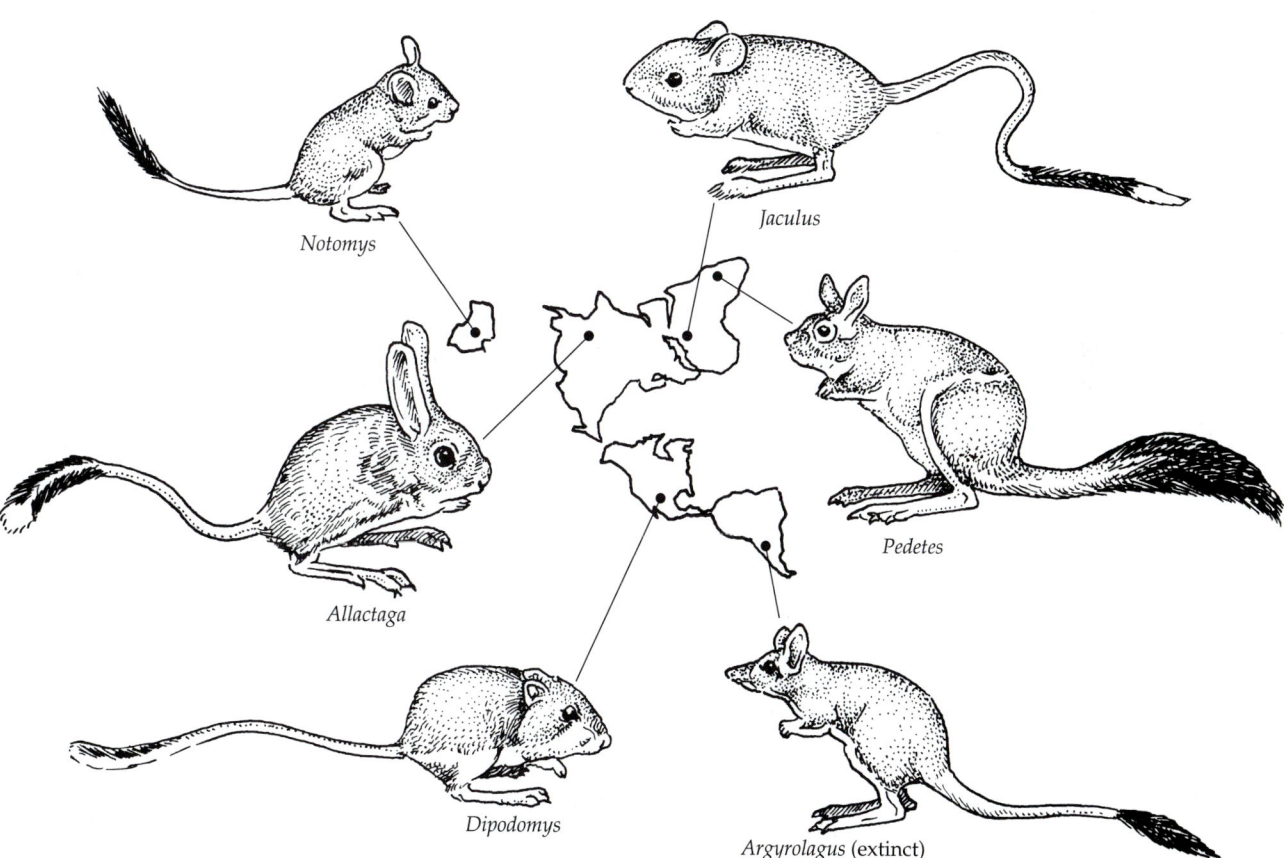

Notomys

Jaculus

Allactaga

Pedetes

Dipodomys

Argyrolagus (extinct)

Figure 16.20 Apparent convergence, at least in morphology, of five genera of rodents and one extinct marsupial from deserts throughout the world. All of the rodents are derived from unspecialized mouselike ancestors, and have independently evolved long hindlimbs, short forelimbs, long tufted tails, light brown dorsal and white ventral pelage, and bipedal hopping locomotion. Some of them share other morphological, physiological, and behavioral characteristics, but some of them also differ conspicuously in ear length, diet, and other characteristics. (After Mares 1993a.)

low on black warning coloration, but they differ in other aspects of their ecology and behavior.

Convergence of entire assemblages?   While such convergence between isolated species and groups may be extremely precise, some investigators have suggested that there can be at least as much convergence at the level of entire biotas. In the literature there are many illustrations purporting to show pairs of similar species that make up the biotas of isolated regions with similar environments. The mammals of Australia and North America, shown in Figure 16.21, are perhaps the most frequently cited example, but others are the mammals inhabiting tropical forests of Africa (or Asia) and South America (e.g., Bourliere 1973; Eisenberg 1981) and the plants and birds living in semiarid (including Mediterranean) or desert climates on different continents (e.g., Cody 1968, 1973; Mooney 1977; Solbrig 1972).

Such illustrations are not intended to be misleading. Nevertheless, they often exaggerate the degree of overall resemblance between the biotas. For one thing, the pairs of species that are depicted in the drawings may not be as similar in body size, physiology, behavior, and ecology as in general body form. Thus, for example, in Figure 16.21, the marsupial "cat" is smaller, more insectivorous, and less arboreal than the ocelot; the flying phalanger is much larger than and differs in diet from the flying squirrel; the wombat is much larger than the groundhog; the numbat or marsupial anteater is much smaller than the giant anteater; and the marsupial "mole" inhabits sandy deserts, unlike any true mole. In addition, such figures rarely show the majority of the species in the two biotas that are not similar. Australia, for example, has no close ecological equivalents of many North American mammals, including mountain lions, bison, weasels, skunks, prairie dogs, and beavers; similarly, North America has no species that are really similar to duck-billed platypuses, spiny anteaters, bandicoots, koalas, and the numerous small and medium-sized wallabies. And finally, other elements of the North American and Australian biotas exhibit even less convergence than the mammals. North America has no real equivalents of Australia's parrots and emus among the birds or the eucalypts and acacias among the plants, and its lizard fauna is much less diverse (Pianka 1986). Australia has no arid-zone plants that are similar to the cacti and agaves of the North American deserts. In fact, after visiting both North America and Australia, many naturalists begin to question the dogma of convergence.

It is not really surprising that the differences in these two biotas far outweigh their similarities. Comparison of the geography, ecology, and geology of the two continents reveals that their environments are in fact very different in several respects. First, as illustrated in Figure 16.22, Australia is much more tropical than North America. The Tropic of Capricorn passes through the very center of Australia just north of Alice Springs, whereas the Tropic of Cancer passes just north of Mazatlan, Mexico. Compared with North America, Australia has extensive older geological formations (Paleozoic and Mesozoic rather than Cenozoic), little recent tectonic activity, more eroded landscapes, and hence much less elevational relief (the highest mountain is only 2229 m compared with 6194 m in North America). As a result of their long exposure to erosion and leaching, Australian soils are extremely low in nutrients. The low availability of essential resources has strongly influenced many aspects of the ecology and evolution of Australian plants, and the plants in turn have affected the animals. Given these major differences in their geological and geographic histories and current environments, it seems naive of biogeographers, evolutionary biologists, and ecologists to expect to find any substantial degree

Placentals

Marsupials

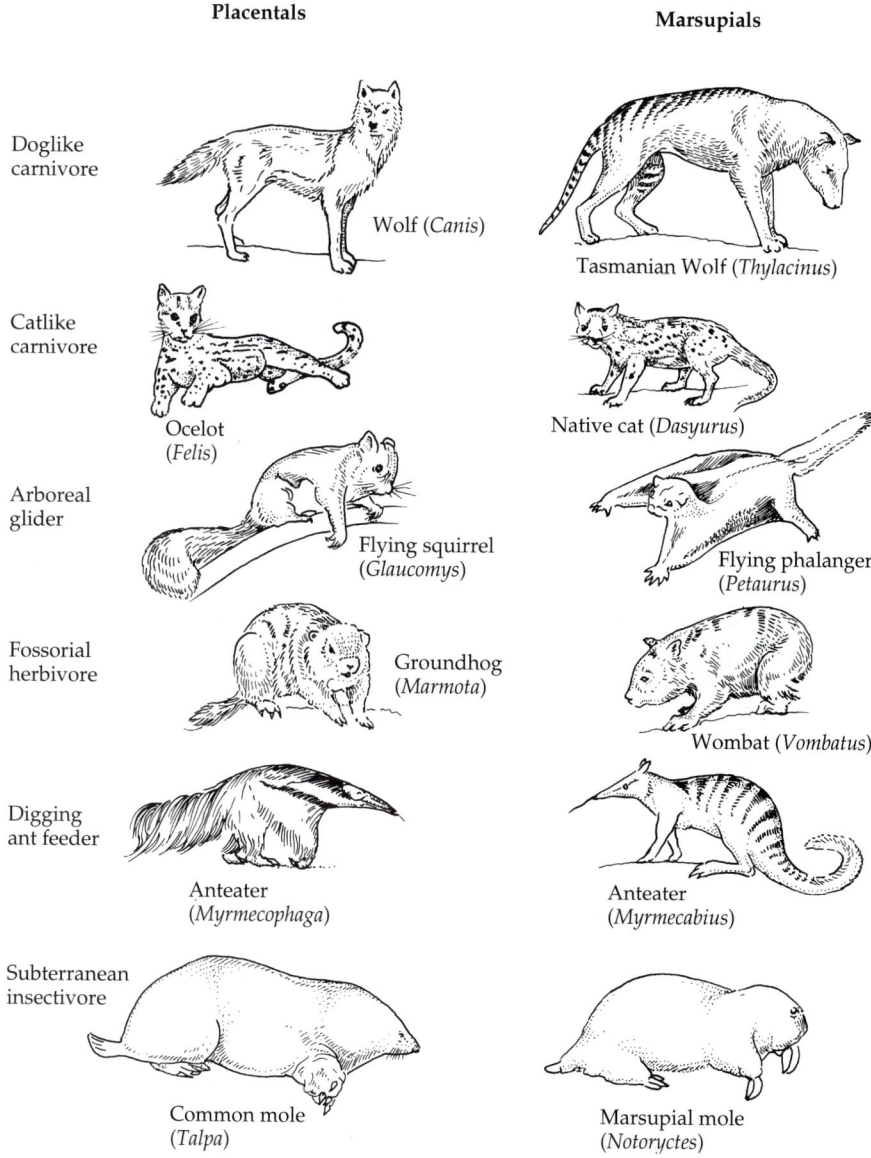

Doglike carnivore

Wolf (*Canis*)

Tasmanian Wolf (*Thylacinus*)

Catlike carnivore

Ocelot (*Felis*)

Native cat (*Dasyurus*)

Arboreal glider

Flying squirrel (*Glaucomys*)

Flying phalanger (*Petaurus*)

Fossorial herbivore

Groundhog (*Marmota*)

Wombat (*Vombatus*)

Digging ant feeder

Anteater (*Myrmecophaga*)

Anteater (*Myrmecabius*)

Subterranean insectivore

Common mole (*Talpa*)

Marsupial mole (*Notoryctes*)

Figure 16.21 Drawings of pairs of species of North American and Australian mammals purporting to show convergence. Figures such as this one are somewhat misleading. As described in more detail in the text, the species paired here are often not drawn to the same scale, and some of those that look alike do not have similar ecological niches. (After Begon et al. 1986.)

of convergence between the biotas of these two continents. Similar caution should be exercised before uncritically accepting other examples of expected or claimed convergence at the level of whole floras or faunas (e.g., Cody and Mooney 1978; Lomolino 1993a; Kelt et al., in press).

A somewhat different approach to convergence has been taken by some ecologists (e.g., Cody 1973, 1974, 1980, 1985; Fuentes 1976; Schluter 1986). Although it may be difficult to identify pairs of species that are close ecological equivalents, some faunas or floras show evidence of having diversified so that similar resources are used in similar ways by a similar number of taxonomically unrelated species (Figure 16.23). In one example of this approach,

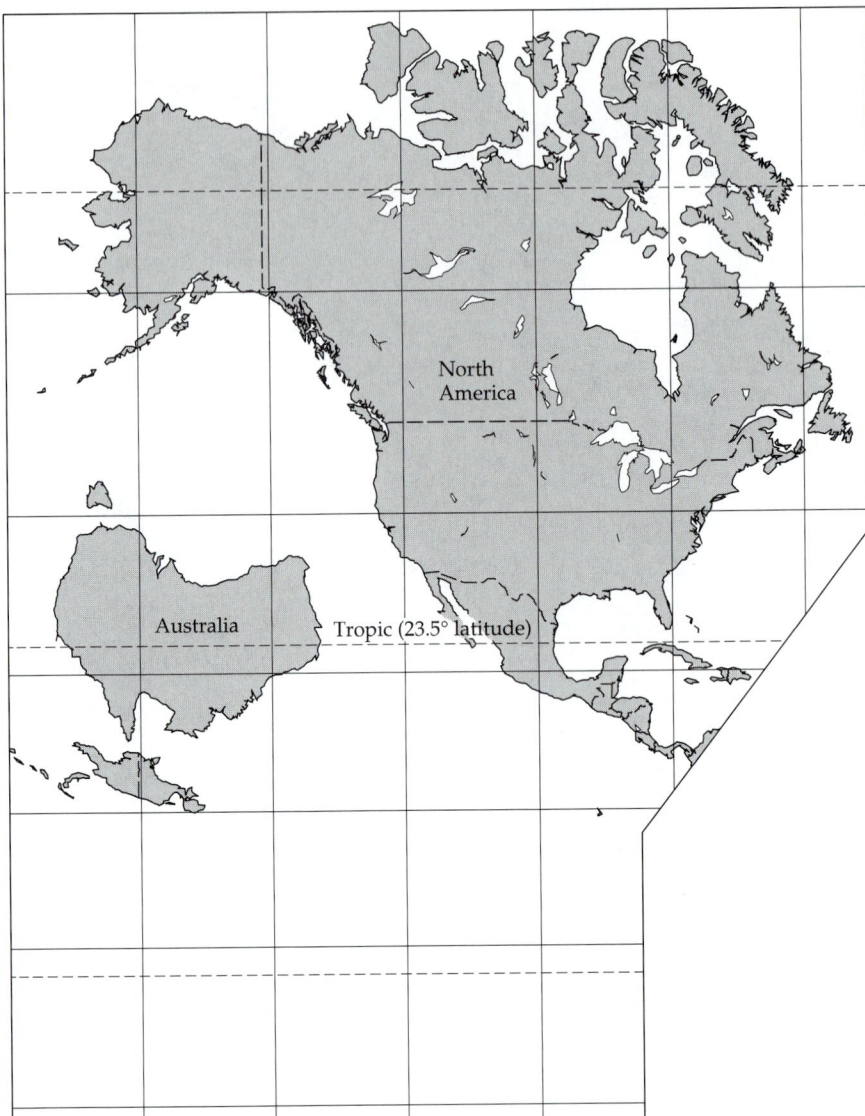

Figure 16.22   A map in which the locations of Australia and North America have been juxtaposed while maintaining their latitudinal positions. Note that while Australia is about the same size as the United States, its latitudinal position overlaps more with that of Mexico. The locations of the arid habitats that are often suggested to contain convergent species or ecological communities are, on average, much more tropical in Australia than in North America.

Cody divided birds into several major feeding niches or foraging guilds. When he compared four different shrub communities in Mediterranean climates, he found that their avifaunas were much more similar to one another in their organization than they were to those of two "control" communities from other kinds of nearby habitats (Table 16.5). Nevertheless, even this overall convergence was far from perfect, so that in certain regions some resources were not used, at least by birds (e.g., there were no nectarivores in Sardinian macchia). Lizards have been studied in California, Chile, and Sardinia, where there are four, five, and three species, respectively (Cody and Mooney 1978). Careful comparisons of two of these three communities indicate that it is difficult to document obvious convergence in the characteristics of particular pairs of species, but the communities are organized so that the major substrates and food items available to lizards are used quite similarly by the different combinations of species on each continent (Figure 16.24; Fuentes 1976; but see Schluter 1988).

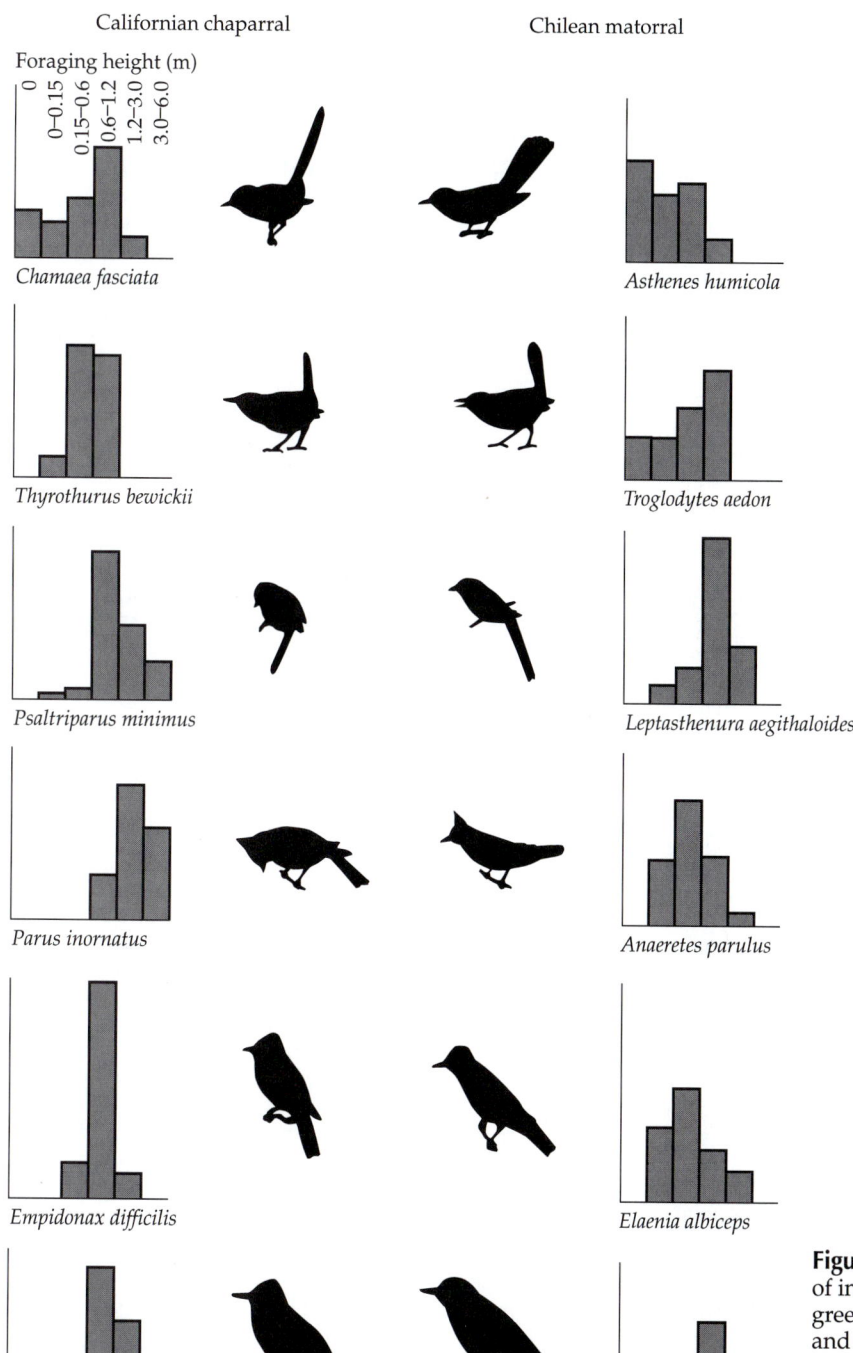

Californian chaparral   Chilean matorral

Foraging height (m)

Chamaea fasciata   *Asthenes humicola*

Thyrothurus bewickii   *Troglodytes aedon*

Psaltriparus minimus   *Leptasthenura aegithaloides*

Parus inornatus   *Anaeretes parulus*

Empidonax difficilis   *Elaenia albiceps*

Myiarchus cinerascens   *Pyrope pyrope*

Figure 16.23 Ecological convergence of pairs of insectivorous birds species inhabiting evergreen shrub habitats in regions of California and Chile with similar Mediterranean climates. The silhouettes show that the paired species are of similar size and shape, and the histograms show that they forage at similar heights in the vegetation. (After Cody and Mooney 1978.)

Overview

So we end by emphasizing the complexity evident in the assembly of continental biotas. On the one hand, as noted by such early biogeographers as Buffon, de Candolle, Sclater, and Wallace and confirmed by many more recent investigators, the biotas of the major landmasses are each made up of distinct kinds of plants and animals. To some extent these differences reflect geological

Table 16.5

Bird community organization for four sites with Mediterranean shrub habitats

Foraging guild	California chaparral		Chilean matorral		Sardinian macchia	
	Number of species by family	Total density pairs/hectare	Number of species by family	Total density pairs/hectare	Number of species by family	Total density pairs/hectare
Foliage insectivores	Chamaeidae (1) Troglodytidae (1) Paridae (2) Vireonidae (1) Sylviidae (1) Parulidae (1)	6.61	Tyrannidae (1) Troglodytidae (1) Furnariidea (2)	5.75	Sylviidae (5) Paridae (3) Troglodytidae (1)	7.74
Sallying flycatchers	Tyrannidae (2)	0.39	Tyrannidae (2)	1.51	Musicapidae (1)	0.30
Nectarivores	Trochilidae (2)	1.18	Trochilidae (2)	1.11		0.00
Ground foragers	Emberizidae (2) Corvidae (1) Phasianidae (1) Columbidae (1) Cuculidae (1) Mimidae (1)	5.31	Emberizidae (2) Mimidae (1) Rhinocryptidae (3) Icteridae (1) Phasianidae (1) Tinamidae (1) Columbidae 91) Tryannidae (1)	7.06	Turdidae (2) Fringillidae (1) Corvidae (1) Laniidae (1) Phasianidae (1)	4.36
Seed- and fruit-eaters	Fringillidae (2) Emberizidae (1)	0.60	Phytotomidae (1) Fringillidae (1)	1.05	Fringillidae (3)	0.94
Trunk and bark foragers	Picidae (2)	0.35	Picidae (2)	0.08		0.00
Aerial insectivores	Apodidae(1)	common	Hirundinidae (1)	common	Apodidae(1)	rare
TOTAL	17 families 23 species	14.4	15 families 24 species	16.56	10 families 20 species	13.34

Source: Cody and Mooney 1978.

Note: Data are for sites on different continents and include two control sites in different habitat types. Note that although representatives of different families fill the niches on the different continents, the number of species and densities are relatively similar among the Mediterranean shrub habitats.

histories of isolation and the presence of different lineages, each with their unique constraints and potentials to diversify and adapt to the abiotic and biotic environment. The pattern and process of diversification has been additionally influenced by differences, both small and large, in the environmental settings, which have favored the differential diversification of some lineages and the evolution of certain traits.

On the other hand, the biotas of the major continents are also similar in some respects. They share some taxonomic groups—and even some species are cosmopolitan, or nearly so. Some of these shared lineages have persisted from the time before the continents were isolated. Others have spread more recently, either colonizing across long-standing barriers or dispersing at times

Table 16.5 *(continued)*
Bird community organization for four sites with Mediterranean shrub habitats

Foraging guild	South African fynbos		Californian oak woodland		British successional scrub	
	Number of species by family	Total density pairs/hectare	Number of species by family	Total density pairs/hectare	Number of species by family	Total density pairs/hectare
Foliage insectivores	Zosteropidae (1) Sylviidae (4)	2.18	Paridae (2) Parulidae (2) Emberizidae (1) Troglodytidae (3) Vireonidae (2) Chamaeidae (1) Sylviidae (1) Iceteridae (1)	9.43	Sylviidae (5) Paridae (4) Troglodytidae (1)	6.42
Sallying flycatchers	Musicapidae (2)	0.79	Tyrannidae (3)	2.00	Musicapidae (1)	0.10
Nectarivores	Nectariniidae (2) Promeropidae (1)	2.66	Trochilidae (2)	1.25		0.00
Ground foragers	Turdidae (2) Laniidae (1) Phasianidae (1) Ploceidae (1) Columbidae (2) Sturnidae (1)	2.56	Emberizidae (2) Carvidae (1) Columbidae (1) Mimidae (1) Phasianidae (1) Turdidae (1)	4.60	Prunellidae (1) Emberizidae (1) Turdidae (4) Corvidae (2) Phasianidae (1) Motacillidae (1) Columbidae (1) Fringillidae (1)	7.90
Seed- and fruit-eaters	Fringillidae (3) Pycnonotidae (1)	0.93	Emberizidae (1) Fringillidae (4) Ptilogonatidae (1) Columbidae (1)	2.53	Fringillidae (4) Columbidae (1)	4.08
Trunk and bark foragers	Picidae (1)	0.20	Picidae (3)	0.93	Picidae (1)	0.15
Aerial insectivores	Hirundinidae (1)	common	Apodidae (1)	rare	Hirundinidae (1) Apodidae (1)	common
TOTAL	15 families 24 species	9.32	19 families 36 species	20.74	15 families 31 species	18.65

in the past when bridges of habitat permitted biotic interchange. Similar environments on different continents have facilitated the colonization and persistence of closely related organisms with similar requirements, as well as the convergent evolution of distantly related forms to use similar environments in similar ways.

The diversity of life on earth reflects the outcome of opposing forces promoting both similarities and differences. Evolutionary conservatism, phylogenetic constraints, gene flow, and similar environments limit the rates and directions of diversification, and thus tend to maintain similarities among biotas. Different environmental conditions, geographic isolation, evolutionary innovations, adaptation, and speciation enhance the rates and directions of divergence and diversification, and thus tend to promote differences among biotas. Nowhere is the complex interplay of these opposing forces more evident than in the diversity of the plants and animals inhabiting different continents.

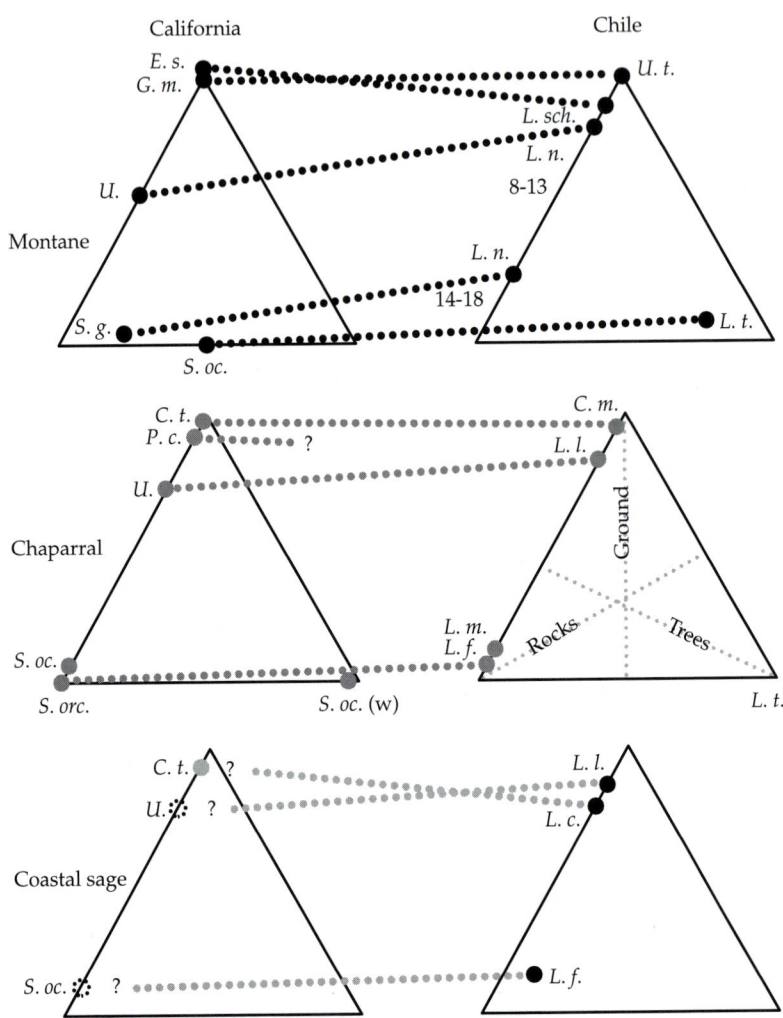

Figure 16.24 Ecological convergence in the use of foraging substrates by communities of lizards in three habitats (montane, chaparral, and coastal sage shrub) with similar climate and vegetation in California and Chile. The triangles indicate proportional use of three substrates (ground, rocks, and trees) by the lizard species (circles labeled with letters). Pairs of "ecologically equivalent" species are connected by dotted lines. Note the tendency for a similar number of species to use substrates in similar ways in each habitat. (After Fuentes 1976.)

UNIT 5

Biogeography and Conservation

CHAPTER 17

The Status of Biodiversity

Biodiversity is, in the simplest terms, the variety of life. It encompasses the variation among species or other biological elements, including alleles and gene complexes, populations, guilds, communities, ecosystems, landscapes, and biogeographic regions. Biodiversity can be expressed as the variation within a given location, or the variation among elements across geographic and temporal scales. This variation can include the number of different types of species or elements, their relative frequencies, the degree of variation among these elements, or variation in key processes such as gene flow, interspecific interactions, or ecological succession.

Given such a broad and inclusive concept, our hope of conserving biodiversity may at first seem like an impossible dream. Yet, in more pragmatic terms, conserving biodiversity distills down to one simple, yet still challenging, goal: maintaining species diversity. As Aldo Leopold put it, "the first requisite of intelligent tinkering is to save all the pieces." By "pieces," of course, he meant native species. With the loss of a species, biodiversity is diminished at all levels, from genetic and local scales to regional and global ones. Charles Elton's (1958) views were similar to Leopold's, but with a more explicit emphasis on the geographic dimensions of what we now term biodiversity. In his discussion of a "wilderness in retreat," Elton stated that "conservation should mean the keeping or putting in the landscape of the greatest possible ecological variety—in the world, every continent or island, and so far as practicable in every district."

Our goals in this chapter are fourfold. First, we will summarize the current status of biodiversity. Second, we will review some of the geographic dimensions of biodiversity—patterns we discussed in earlier chapters, especially those in species diversity and endemicity across geographic gradients. Third, we will explore the geography of extinction, or the tendency for extinctions to be nonrandomly distributed across the globe. In the final section we will discuss potential applications of biogeography for designing nature reserves and for predicting the effects of anthropogenic changes in regional landscapes and global climates.

The Biodiversity Crisis and the Linnaean Shortfall

Few fully informed and objective students of biological diversity question that the current loss of biodiversity has reached crisis proportions. Elton, Leopold, and other ecologists of this and the previous century repeatedly warned of the ongoing losses of native species and landscapes. The existence of a biodiversity crisis is documented in the record of historical extinctions and the vulnerable and imperiled status of many extant species (Figure 17.1A,B; World Conservation Monitoring Centre 1992). During the past four centuries, biologists have tallied some 115 extinctions of birds, including such wondrous forms as the great auk (*Alca impennis*), dodo (*Raphus cucullatus*), and passenger pigeon (*Ectopistes migratorius*). Over the same time period, 58 species of mammals, 100 reptiles, and 64 amphibians have vanished, including one amphibian family and 38 genera (most amphibian extinctions have occurred during the last two decades; see Duellman and Trueb 1986).

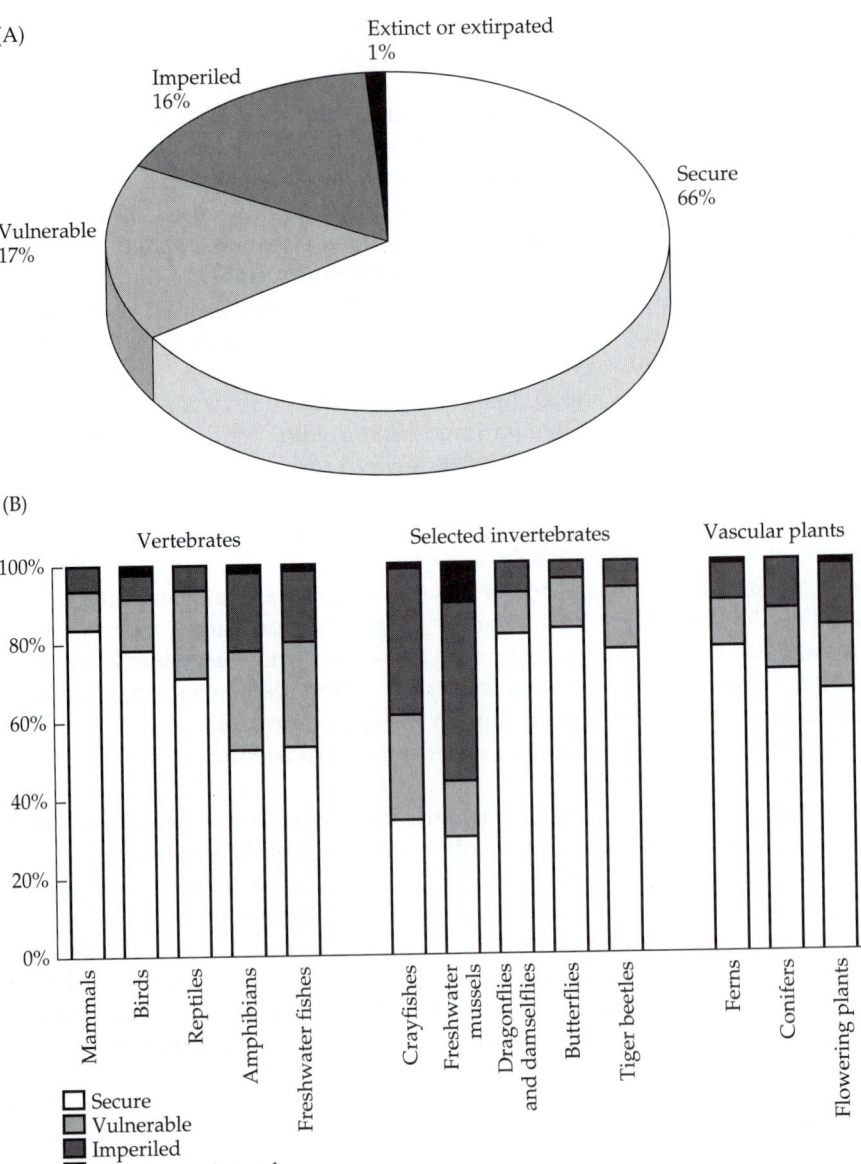

Figure 17.1   (A) Status of plant and animal species in the United States. (B) The same information, broken down by taxon. (Source: Natural Heritage Databases maintained by The Nature Conservancy, 1996.)

Aquatic species, including freshwater fishes, crayfishes, and mussels, have also been affected by this wave of extinctions. The rate of extinctions of freshwater fishes in North America has steadily increased over the past 100 years, rising to some 40 species or subspecies lost during the past decade (Williams and Miller 1990). In the Great Lakes of East Africa, the diversity of native fishes has plummeted just in the past few decades. In Lake Victoria alone, cichlid diversity has declined by as much as 200 species. Worldwide, as much as 20% of all freshwater fish species (about 1800 species) are now extinct or in serious jeopardy.

Loss in diversity of mollusks has been almost equally severe. Since 1600, invertebrate biologists have recorded 191 extinctions of mollusks—including freshwater, marine, and terrestrial forms. The prospects for the near future are just as grave. In the freshwater systems of the United States, 7% of the 297 recognized species of mussels are presumed extinct, while another 65% are either endangered, threatened, or candidates for federal protection (Williams and Neves 1995).

This wave of extinctions has by no means been restricted to the animal kingdom. Worldwide, plant diversity has declined by over 600 species during the past 400 years (Smith et al. 1993). In the United States alone, 176 recognized species and subspecies are now extinct, while another 2465 species are officially listed as imperiled (Natural Heritage Databases 1995). The status of insular plants is especially grave. On the Hawaiian Islands, some 108 endemic plant taxa are now extinct, and another 175 are listed as endangered or vulnerable. On Saint Helena Island in the Atlantic Ocean, 7 endemic plant species are extinct, and all of the remaining 46 endemic species of plants are endangered or threatened.

Even the fungi, a sometimes neglected yet diverse and fascinating kingdom, have experienced a startling pulse of extinctions during recent decades. For example, between 1930 and 1990, the species diversity of European fungi dropped by 40 to 50% (Jaenike 1991).

As alarming as these statistics may seem, it is likely that they underestimate the actual decline in biodiversity. While scientists have now described over 1.4 million species, this represents only a fraction of the total species thought to occur on earth. Estimates of total diversity vary, but typically range between 5 and 50 million species. By almost all estimates, most of what is out there is unknown to us. This deficiency, sometimes called the **Linnaean shortfall**, represents a glorious opportunity for field biologists. There are many species waiting to be discovered, some of them so distinct that they become the sole known representatives of new families, orders, or even phyla (Table 17.1; see also Finlay et al.'s review on the global diversity of microbes).

Given current trends in the specialization and training of taxonomists, however, the Linnaean shortfall is likely to remain with us for some time. While a great majority of the undiscovered animals are thought to be insects, spiders, and other invertebrates, only 30% of today's taxonomists specialize in these groups (Figure 17.2). The geographic distribution of taxonomists—or, at least, their home institutions—is no more reassuring. Over three-fourths of today's taxonomists are trained in temperate areas of the Nearctic and Palearctic regions—which, as we have seen, are not the world's most species-rich areas. Similarly, a quick review of Table 17.1 reveals that the marine realm is still a great biological frontier. Less than 10% of the world's oceans have been adequately sampled for biological diversity, and even moderately rare species are easily missed (Culotta 1994).

One of the most serious downsides of the Linnaean shortfall is the likelihood that many of the undiscovered species will go extinct before they are known to

Table 17.1
The Linnaean shortfall

1. Eleven of the 80 extant cetaceans have been discovered in this century, one in 1991.

2. One of the largest shark species, "megamouth," was discovered in 1976.

3. Three new families of flowering plants were discovered in Mexico during the past decade.

4. Two new *phyla* were discovered during the 1990s, one (Loricefera) in the marine benthos and the other (Cyliophora) clinging to the mouthparts of lobsters.

5. Perhaps 90% of tropical forest insects remain unknown.

6. Almost 1.5 million fungi remain to be discovered.

7. About 4000 bacterial species have been described; a gram of soil may contain 4000 to 5000 species.

(Modified from Raven and Wilson 1992.)

science. E. O. Wilson (1992) calls these **Centilenan extinctions**, and they no doubt represent a significant component of the actual loss in biodiversity.

Whatever the extent of this unknown component, it is certain that we are witnessing a major surge in extinctions, one perhaps beginning to approach some of the mass extinctions of the fossil record. The avian extinction rate during the early Quaternary period, for example, is estimated to have been one species every 83 years. In recent years, this rate may have risen to one bird species every 4 years, and may well reach one species every 6 months by the year 2000 (Temple 1986). Overall, the global extinction rate of plants and animals combined may now be 100 species per year.

One species—*Homo sapiens*–now dominates the earth's life support systems. We utilize some 20 to 40% of the total primary production of terrestrial ecosystems, and exploit or overexploit approximately 80% of marine fish production (Figure 17.3). Our activities account for over half of terrestrial nitrogen fixation, and we utilize over half of all accessible fresh surface water. There can

Figure 17.2   A comparison between the relative number of species worldwide in each of the principal taxonomic groups and the number of taxonomists who specialize in them. (From Barrowclough 1992.)

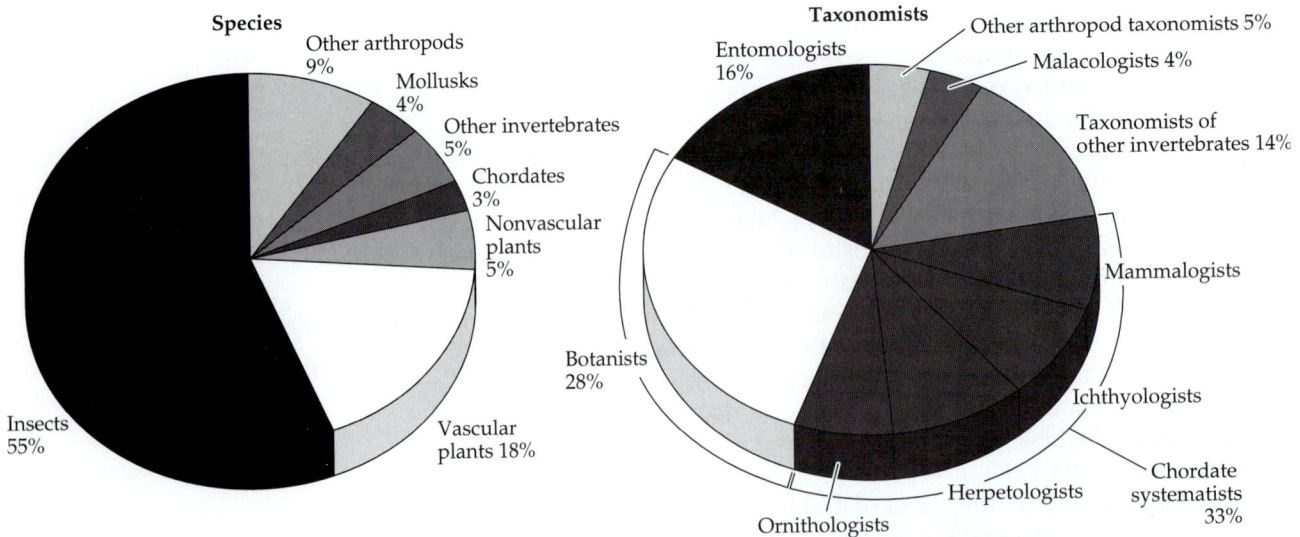

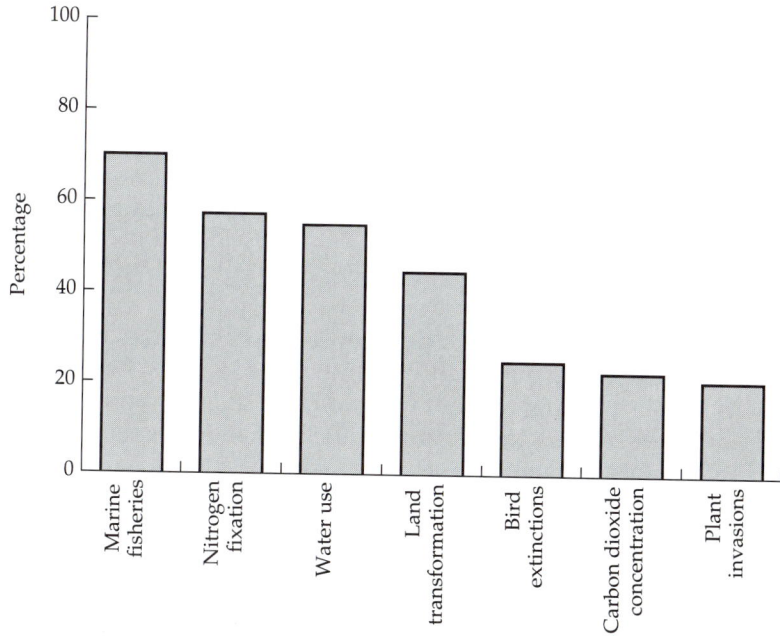

Figure 17.3 Measures of human dominance and exploitation of several major components of the earth's life support systems. Marine fisheries = percentage of major marine fisheries that are fully exploited, overexploited, or depleted by humans; nitrogen fixation = percentage of terrestrial nitrogen fixation that is human-caused; water use = percentage of accessible surface fresh water used by humans; land transformation = percentage of the land surface transformed by humans; bird extinctions = percentage of bird species on earth that have become extinct in the past two millennia, almost all as a consequence of human activities; carbon dioxide = percentage of the current atmospheric carbon dioxide concentration that results from human activities; plant invasions = percentage of plant species in Canada that humanity has introduced from elsewhere. (After Vitousek et al. 1997.)

be little doubt that this human domination of the earth's life support systems has caused the current biodiversity crisis.

As scientists, we are called upon to identify the specific causes of the biodiversity crisis and objectively apply our knowledge to develop effective strategies to minimize the losses. The task is great, and it requires a pluralistic approach, one involving many disciplines and methods. Yet biogeography has much to offer. More than any other scientific discipline, ours focuses on biological variation at local to global scales. In addition to studying patterns in the sizes, locations, and dynamics of geographic ranges, we study patterns in species diversity and endemicity across time and space. As we shall see below, the geographic signature of historical extinctions has allowed us to identify some of their principal causes and to develop some alternative strategies for abating the ongoing wave of extinctions.

Geographic Variations in Biodiversity

Geographic patterns of biodiversity are central to both biogeography and conservation biology. We have discussed many of these patterns in earlier chapters of this book (Chapters 13 through 16). Three of the most relevant are

1. Species diversity tends to increase as we move toward the equator. Systems at lower latitudes tend to have both higher local, or alpha, diversity and higher between-system, or beta, diversity.
2. On islands, on mountaintops, and in other isolated systems, species diversity tends to increase with area and decrease with isolation.
3. **Endemicity**, or the relative number of unique species, tends to be higher for larger and more isolated regions.

Each of the above patterns derives from more fundamental patterns of variation in the characteristics of geographic ranges. It is clear from areographic studies that geographic ranges not only vary markedly among species, but vary in a systematic and predictable manner along geographic gradients. For example, as we move toward the equator, geographic ranges tend to decrease in size, allowing tighter packing of species and higher species richness. In addition, geographic ranges are not randomly distributed across the globe, but tend to concentrate and overlap in particular regions, called **hot spots**. The term "hot spot" can refer either to the simple geographic co-occurrence of many species or, more specifically, to a site or region with an unusually high number of local endemics, also termed **restricted-area species**. It would seem that the latter—hot spots of high endemicity—would be most relevant to conserving biodiversity.

Terrestrial Hot Spots

While patterns of diversity and endemicity provide important clues for locating and ultimately protecting many rare and endangered species, two important questions remain: (1) How well can we predict the intensity and location of hot spots for a particular taxonomic group? (2) To what degree do different taxon-specific hot spots overlap?

Our task in conserving biodiversity would, of course, be easier if hot spots were few, intense, and restricted to the same locations for most taxa. Again, most of the earth's diversity has yet to be catalogued, so we do not know to what extent this is the case. We have, however, developed reasonably reliable estimates of diversity for some groups, especially the more easily observable taxa such as birds.

Under the assumption that extinctions are more likely for species with smaller ranges, conservation of restricted-area species should be of paramount importance. Recent analyses conducted by the International Council for Bird Preservation (Bibby et al. 1992) indicated that 2609 land bird species, or 27% of all birds, have breeding ranges restricted to less than 50,000 km². Restricted ranges tend to be relatively common among some avian families, such as kingfishers, white-eyes, barn and grass owls, and megapodes, but rare in other

---

Table 17.2

The relative number of restricted-range species of birds varies dramatically among families

Family[a]		Restricted-range species	Percent of family
Alcedinidae	Alcedinid kingfishers	20	84%
Zosteropidae	White-eyes	75	78%
Tytonidae	Barn and grass owls	10	59%
Megapodidae	Megapodes	11	58%
Rhinocryptidae	Tapaculos	16	57%
Formicariidae	Ground antbirds	30	56%
Scolopacidae	Sandpipers, curlews, etc.	7	8%
Paridae	Titmice, chickadees, etc.	5	8%
Anatidae	Ducks, swans and geese	11	7%
Falconidae	Falcons and caracaras	6	6%
Threskiornithidae	Ibises and spoonbills	2	6%
Ardcidae	Herons, bitterns and egrets	3	5%
Meropidae	Bee-eaters	1	4%

Source: Bibby et al. 1992.

Note: The families listed above have significantly more (above the dashed line) or fewer (below the line) restricted range species than would be expected by chance. Restricted range species include those with breeding ranges less than 50,000 km².

[a]Following taxonomy in Sibley and Monroe (1990).

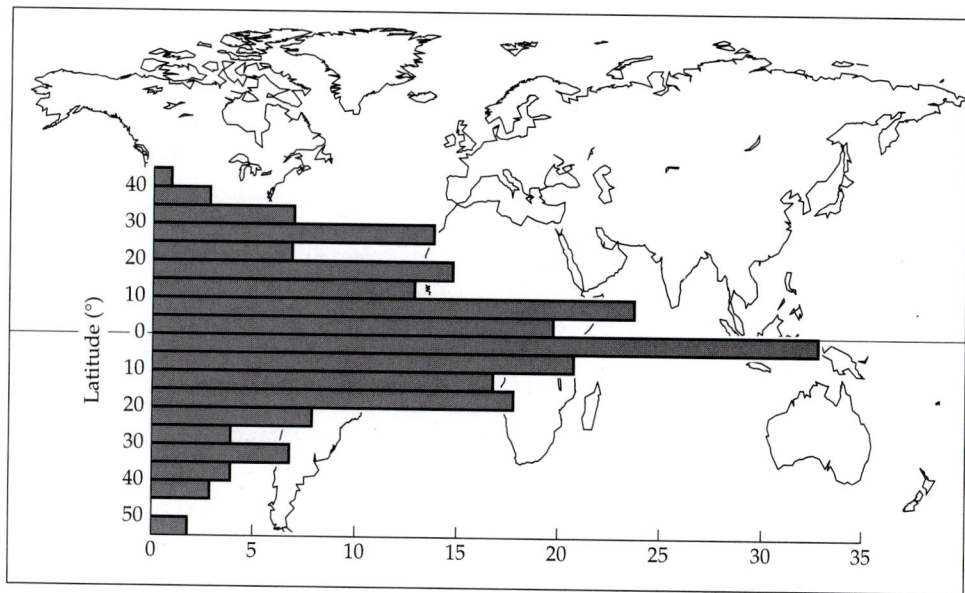

Figure 17.4 The latitudinal trend in the distribution of Endemic Bird Areas (i.e., areas containing the breeding ranges of at least two restricted-range species) follows that of species richness in general. (From Bibby et al. 1992.)

families composed of larger, migratory or nomadic species, such as ducks, falcons, herons, and egrets (Table 17.2).

The International Council for Bird Preservation refers to areas containing the breeding ranges of at least two restricted-range species as **endemic bird areas** (EBAs). As we might expect, EBAs are not randomly distributed, but tend to be concentrated in tropical regions (Figure 17.4). In addition, a highly disproportionate number of EBAs occur on islands. Although islands cover less than 10% of the earth's land area, nearly half of all EBAs are insular (Figure 17.5). Furthermore, whether insular or continental, the richness of EBAs increases with their area (Figure 17.6). Thus we have answered our first question, at least for this one well-studied group of terrestrial vertebrates: the intensity of avian hot spots is quite high. The total area occupied by all EBAs combined is 6.5 million km². That means that just under 5% of the world's total land area provides breeding habitat for a majority of its most threatened bird species.

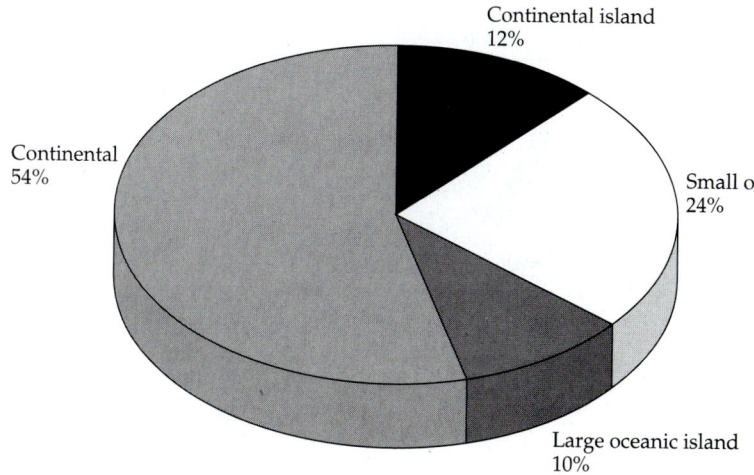

Figure 17.5 Endemic Bird Areas are not randomly distributed across the globe, but tend to be disproportionately common on islands. While islands cover less than 10% of the total land surface, nearly half of the world's EBAs are insular. (From Bibby et al. 1992.)

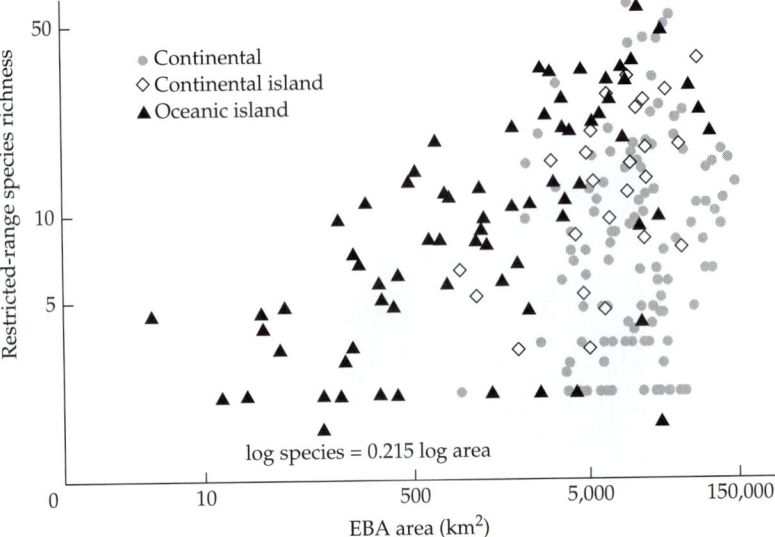

Figure 17.6 Consistent with the species-area relationship for most taxa, the number of restricted-range species increases with the area of endemic bird areas. The slope for the log-log relationship (i.e., the z value for the power model) is 0.215, which falls within the range observed for most insular systems (see Chapter 13). (From Bibby et al. 1992.)

Now to our second question: To what degree do avian hot spots overlap with those of other animals and plants? The preliminary findings are equivocal, and depend on both the particular taxa and regions in question. Areas of endemism in Central America, for example, correspond very closely for birds, reptiles and amphibians, but much less so when butterflies are included (Fig-

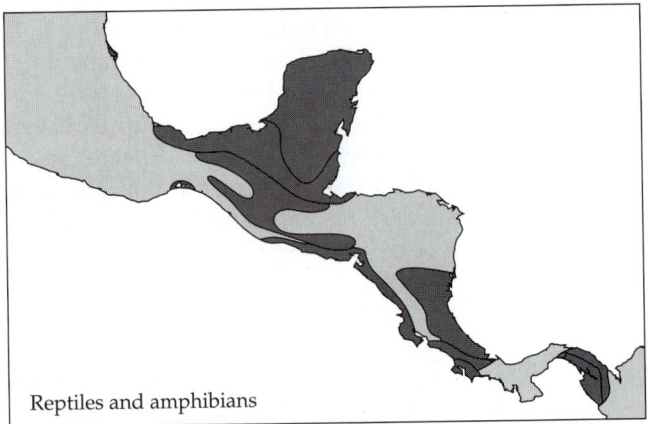

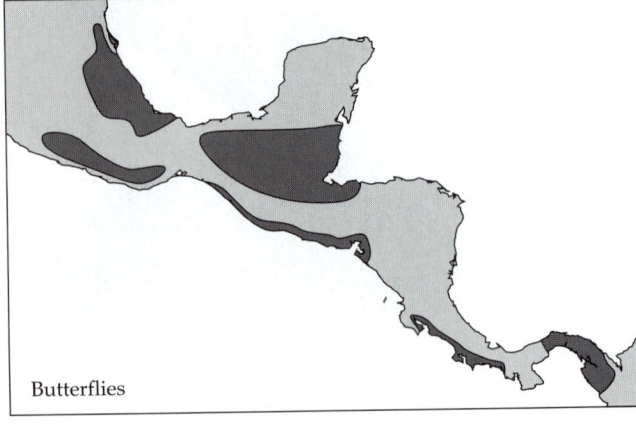

Figure 17.7 A critical question in conservation biology is to what degree hot spots or areas of endemism for one taxon overlap with those for other taxa. In Central America, areas of endemism for birds show relatively high overlap with those for reptiles and amphibians, but considerably less overlap with areas of endemism of butterflies. (From Bibby et al. 1992.)

ure 17.7). On the African continent, areas of endemism are strikingly similar for amphibians and mammals, but birds and plants have additional hot spots exclusive of these groups (Figure 17.8). Similar results have been obtained for other regions and on a global scale for most major groups (Figure 17.9; see Vaisanen and Heliovaara 1994; Dobson et al. 1997). Species within each of the

Figure 17.8 Just as we have seen for Central American faunas (Figure 17.7), overlap of areas of endemism in Africa are quite high for some groups (e.g., amphibians and mammals), but relatively low for others (e.g., plants and mammals). (A–C from Bibby et al. 1992; D from World Conservation Monitoring Centre 1992.)

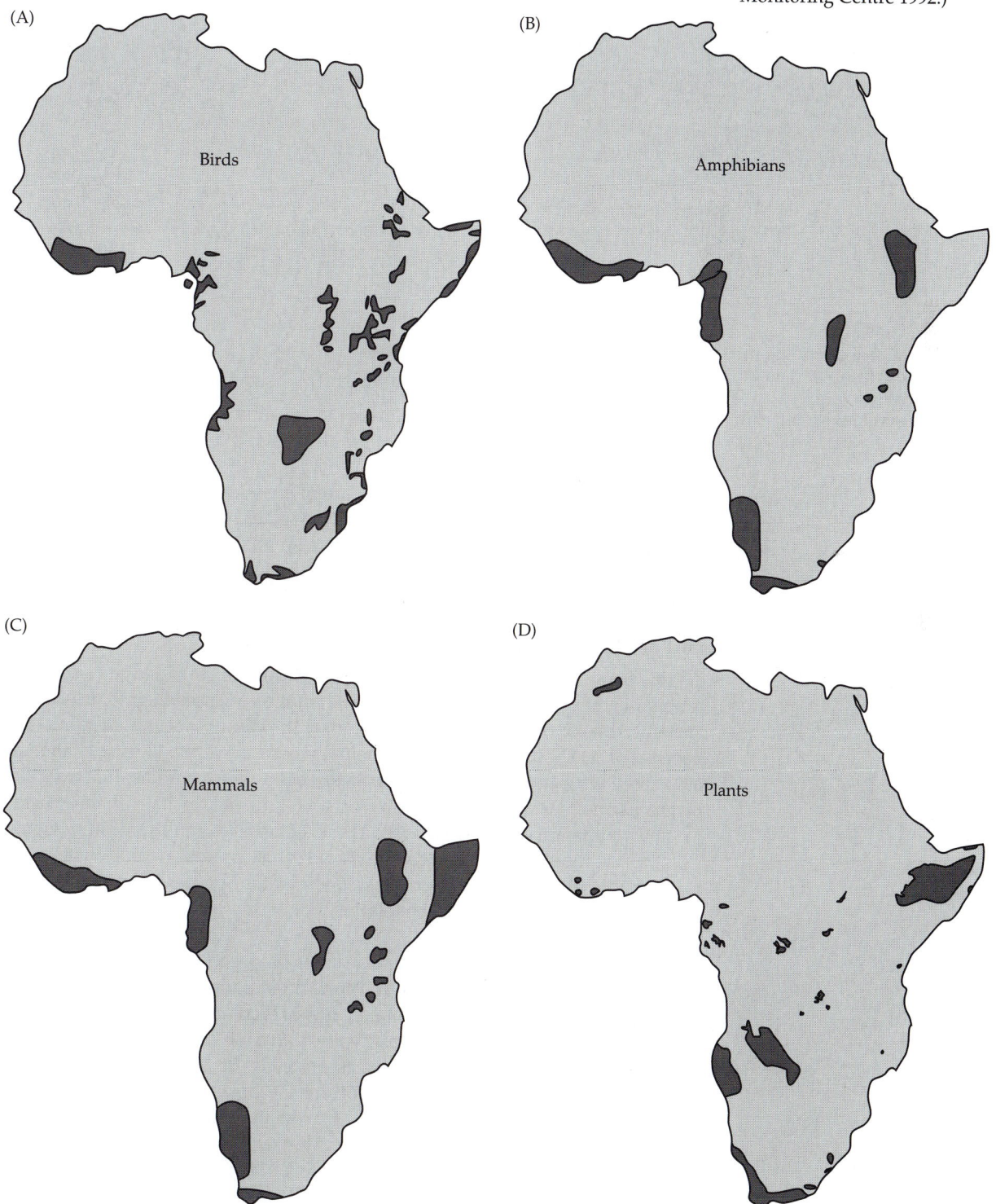

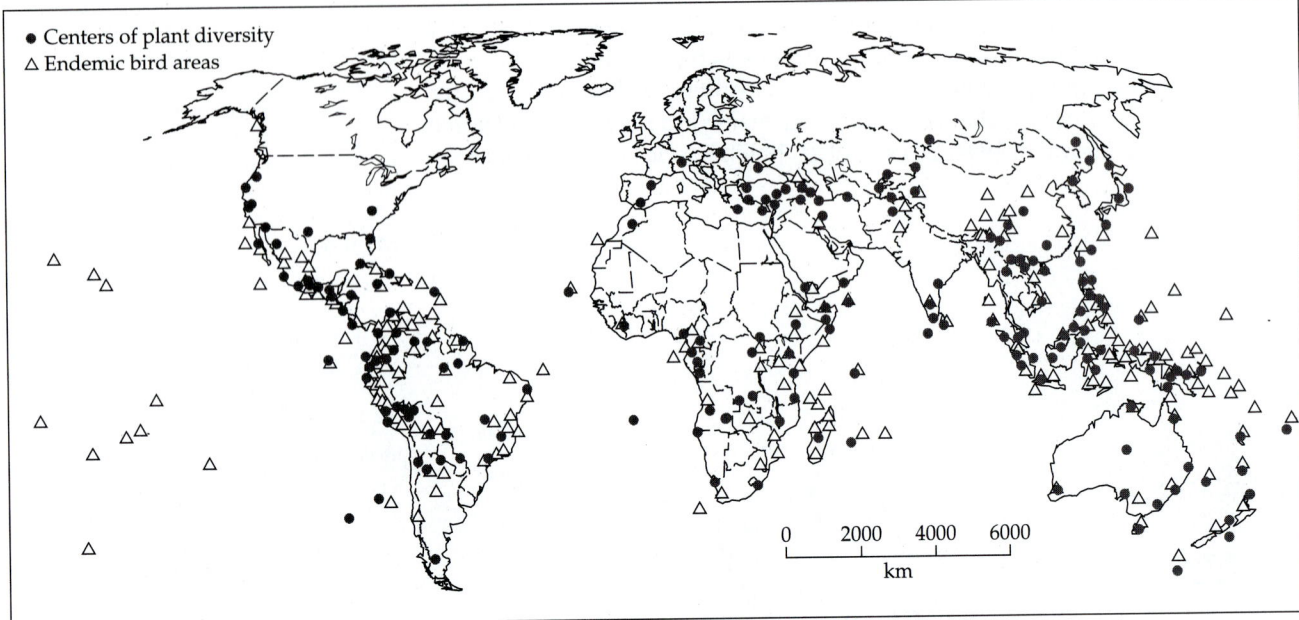

Figure 17.9 The global distribution of endemic bird areas (triangles) and centers of plant diversity (circles). (After World Conservation Monitoring Center 1992.)

well-studied taxa tend to be nonrandomly distributed, with definite and intense hot spots of diversity and endemicity. Although hot spots for different taxa do overlap much more than would be expected by chance, the overlap is not perfect. In addition, hot spots of diversity may show little overlap with hot spots of endemicity. Still, hot spots of endemicity—those of paramount importance for conserving rare species—are fairly well known for some groups, and even if all were to be made reserves, only a small fraction of the earth's land surface would need to be protected to maintain a great majority of these species.

Hot Spots in the Marine Realm

As this book has pointed out repeatedly, most of our knowledge as biogeographers comes from studies of terrestrial and, to a lesser degree, freshwater systems. In contrast, we know precious little about marine systems—not because of a lack of interest, but principally because of the logistic challenges associated with underwater and especially deep-water studies. Just over 70% of the earth's surface is covered by oceans, and nearly half of the world's marine waters are over 3000 m deep (World Conservation Monitoring Centre 1992). Coral reefs, mostly found in shallow coastal waters, are famous for their productivity and diversity. In contrast, deep-water benthic communities were until very recently thought to be relatively depauperate and inhabited largely by cosmopolitan species.

Technological advances in recent decades have provided some fascinating new insights into the marine realm. While most patterns in marine biodiversity remain tentative due to the infancy of research on deep-sea communities, marine faunas often exhibit biogeographic patterns similar to their mainland counterparts. For example, the species diversity of the few marine groups studied tends to increase as we move from the poles to the equator (Figures 15.5 and 15.6; Buzas and Culver 1991). Marine benthic diversity also tends to be strongly correlated with depth, increasing to reach a maximum at depths between 2000 and 3000 m (see Rex 1983). Below these depths, diversity tends to decrease, but endemicity may increase.

One of the key reasons for the high endemicity of deep-water communities derives from the isolated nature of the deepest areas. Indeed, trench, seep, and hydrothermal vent communities are marine analogues of isolated oceanic islands. These **hadal** communities, defined as those below 6000 m, account for just 1% of the ocean's benthic area and are distributed among highly disjunct sites, including those of oceanic trenches (Figure 17.10A). Trenches harbor highly distinct communities with endemicities ranging between 50% and 90%.

In a similar fashion, hydrothermal vents are tectonically derived, island-like systems (Figure 17.10B). In comparison to other deep-sea communities, the

Figure 17.10 Global distributions of (A) the principal oceanic trenches and (B) hydrothermal vents and cold seep communities. (From World Conservation Monitoring Center 1992.)

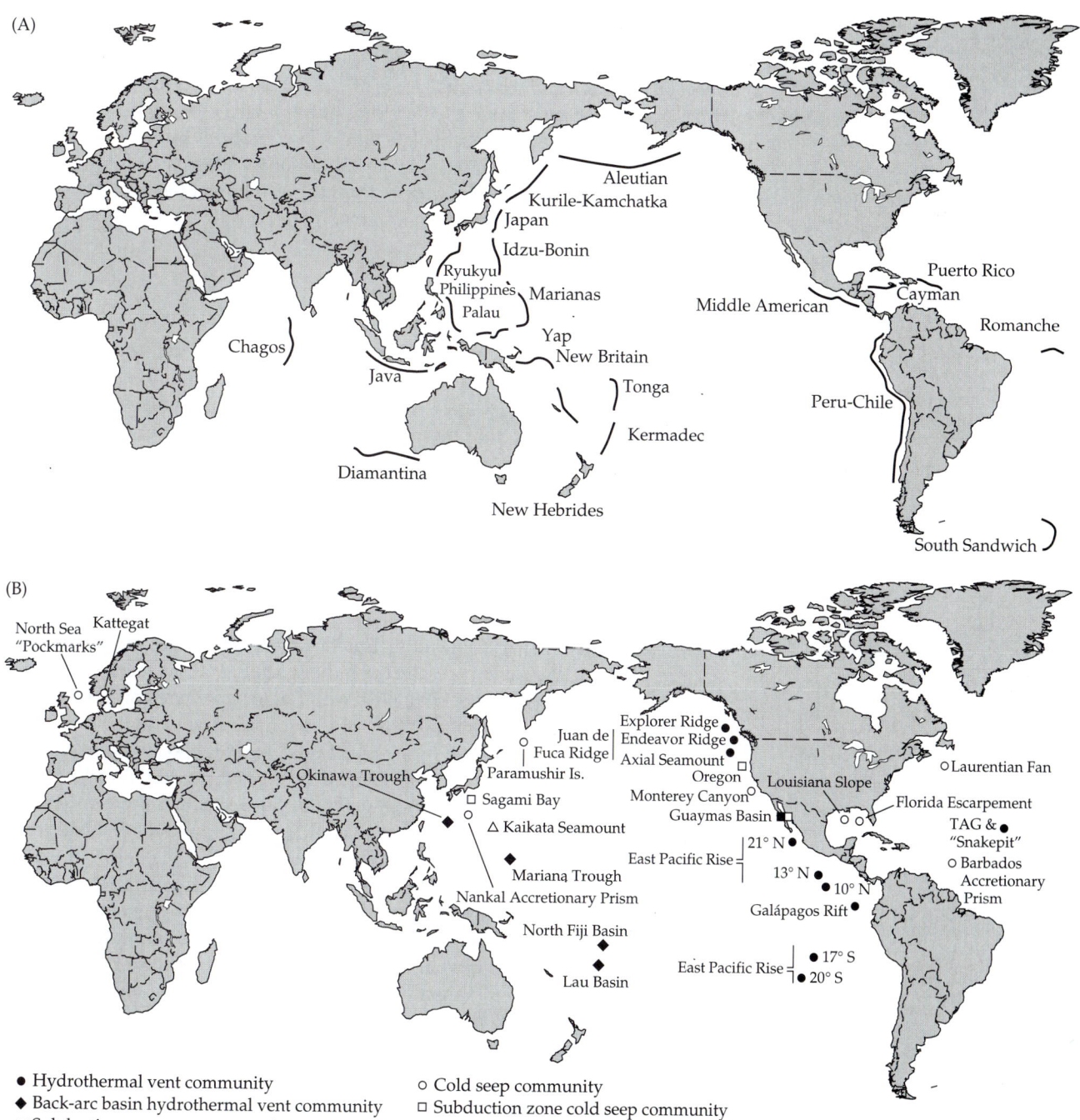

- ● Hydrothermal vent community
- ◆ Back-arc basin hydrothermal vent community
- ■ Subduction zone vent community
- ○ Cold seep community
- □ Subduction zone cold seep community
- △ Volcanic front cold seep community

diversity of hydrothermal vent communities appears to be relatively low, but their endemicity is extremely high. Although discovered only in 1977, surveys of vent communities have yielded one new phylum, over 14 new families, and 50 new genera—most of which are monospecific (Grassle 1989; Gage and Tyler 1991). Mollusks, polychaetes, and arthropods account for approximately 93% of the 375 species described from all vents worldwide; 90% are restricted to vent habitats. The high distinctness and endemicity of vent communities derives largely from three characteristics of hydrothermal vents: their isolation, their antiquity (vent habitats have existed throughout the Phanerozoic, providing ample time for evolutionary divergence), and the special adaptations required to live in a high-pressure, chemically reducing and metal-rich environment where the food chain is based on chemosynthetic rather than photosynthetic producers (see Tunnicliffe and Fowler 1996).

We have only begun to scratch the surface here, but it is obvious that these deep-water communities are hot spots of biological diversity. In contrast to terrestrial and freshwater species, the biotas of deep-water hot spots may be less in jeopardy because they are isolated from hot spots of human activity. Yet even these, perhaps the most remote biotas on earth, may be affected by anthropogenic *global* change, which may trigger the next big wave of extinctions in marine and terrestrial ecosystems.

The Geography of Extinctions

Given that diversity varies in a nonrandom manner across geographic gradients, we might expect that extinctions will also exhibit a definite geographic signature. Do extinctions vary with latitude, elevation, depth, area, or isolation? Do different taxa exhibit the same geographic trends in susceptibility to anthropogenic extinctions? These are no longer just academic questions. Given the limited time, energy, and financial support available for preserving biodiversity and preventing extinctions, it is imperative that we direct our efforts where they will do the most good.

The Prehistoric Record of Extinctions

Studies of the fossil record provide some useful clues. The mass extinctions that occurred at the end of the Cretaceous period are marked by a 60 to 80% drop in species diversity at a global scale. David Raup and David Jablonski have searched for hot spots of local extinction among the rich fossil record of marine bivalves (Raup and Jablonski 1993). When all 340 genera were considered, they found that both the absolute number of end-Cretaceous extinctions and the proportion of local genera suffering extinctions were highest for the equatorial regions. The region with the highest extinction rate was the New World tropics, which includes the site of the Chicxulub crater where the infamous asteroid is thought to have impacted (see Sharpton et al. 1992). Raup and Jablonski warn, however, that their results were strongly influenced by the rudist bivalves: large, attached or recumbent filter feeders that were primarily restricted to tropical waters. When rudists were excluded, the loss of bivalve genera was uniform across the globe—that is, regardless of latitude, the proportion of local assemblages suffering extinctions was approximately constant (about 52%). Still, the total number of extinctions, an index perhaps more relevant to conserving biodiversity, was highest in the richer tropical waters.

Other paleontological evidence sheds additional light on the geographic correlates of natural extinctions. Plate tectonics and continental drift may have played an important role in many of the extinction events recorded in the fossil record. The Great American Interchange is an important case in point (see

Chapter 16). With the formation of the Central American landbridge approximately 3.5 million years B.P., the Nearctic and Neotropical regions were connected after a long period of isolation. The ensuing migrations and radiations of immigrants caused a wave of extinctions among native species, especially in South America. As we shall see, this lesson in paleobiogeography is especially relevant to the biodiversity crisis. Throughout the fossil record, whether due to the formation or breakup of Pangaea, the Great American Interchange, or climate-driven range shifts associated with glacial cycles of the Pleistocene, exchange among biotas that evolved in isolation from one another has typically resulted in waves of extinctions among the native assemblages.

It is only in the most recent slice of this great record, the last 2 million years, that our own species has become a significant component of environmental change, biotic exchange, resource depletion, and resultant extinctions. Throughout this period, prehistoric human societies colonized almost all corners of the globe, preying on native species, introducing exotics, and transforming landscapes. Throughout the world, as human societies colonized and developed advanced hunting and agricultural practices in new regions, native species suffered. Technological advances also spread among human populations, often with devastating effects on native biotas. The extinction of the Pleistocene megafauna is an important case in point. The mass extinction of mammoths, ground sloths, giant raptors such as teratorns, and other large mammals and birds was not synchronous across the globe, but almost invariably followed on the heels of human colonization, or more precisely, the spread of advanced hunting techniques and intensive agricultural practices, including fire (see Chapter 18). Humans colonized Australia over 35,000 years ago, but it wasn't until the aborigines used fire on a grand scale that they caused massive extinctions of native wildlife. Hunting and gathering societies migrated across Beringia to colonize North America between 15,000 and 30,000 years ago. Again, it wasn't until they began more intensive use of fire and developed advanced hunting skills that they had a significant effect on the diversity of the native biota. As we remarked in Chapter 7, Africa is the exception that proves the rule. Unlike other biotas that evolved in isolation from humans, Africa's rich wildlife had the opportunity to gradually adapt to the activities of aboriginal societies. It is only within the twentieth century that human population densities and their transformation of African landscapes have become so intense that they now threaten its megafauna as well as many other native animals and plants (see the discussion on the biogeography of humanity in Chapter 18).

The Historical Record of Extinctions

Reviews of the historical record have provided us with important clues for understanding the ongoing biodiversity crisis. Like the prehistoric extinctions of the Pleistocene megafauna, many historical extinctions occurred when waves of human colonization and development advanced over previously pristine hot spots of endemicity. The thousands of recorded extinctions have a definite geographic signature—one bearing witness to the high endemicity and fragility of insular communities. As Figure 17.11 reveals, a highly disproportionate number of the animal extinctions recorded since 1600 have occurred on islands. This insular bias applies to plants as well. Although just one in six species of plants are insular, one in three threatened species are endemic to islands (Schemske et al. 1994).

These insular extinctions have resulted from a combination of factors, most of which are related to the isolation and ecological naïveté of insular biotas. Many insular plants and animals, especially those endemic to more isolated

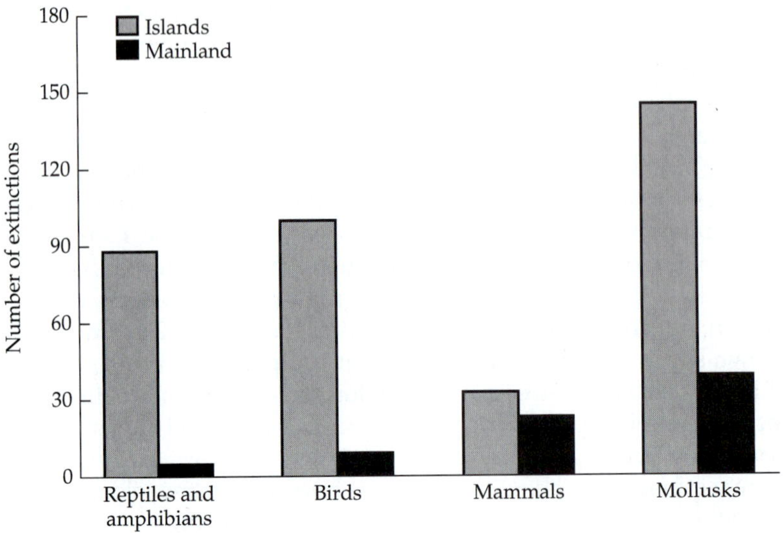

Figure 17.11   Patterns in recorded extinctions (1600–1990) of terrestrial vertebrates and mollusks. Extinctions have been far more common among animals inhabiting islands than among their mainland counterparts. (After World Conservation Monitoring Centre 1992.)

and relatively depauperate islands, have been subjected to the effects of scores of exotic species introduced after humans colonized their islands. On Chiloe Island, Chile, Darwin found the native foxes—later named Darwin's fox (*Pseudalopex fulvipes*)—so naive that he was able to collect the type specimen by hitting it over the head with his geological hammer. On visiting the Galápagos, he was amazed that the avifauna was so tame that he could prod a hawk with the butt of his rifle.

> A gun here would be superfluous. What havoc the introduction of any beast of prey must cause in a country before the inhabitants have become adapted to the stranger's craft of power. (Darwin 1860)

Of course, this was one of the many cases in which Darwin's words proved prophetic. Two-thirds of the historical extinctions of insular birds have been caused either directly or indirectly by introduced mammals.

Throughout their history, most insular biotas evolved in the absence of large ground-dwelling homeotherms. Many species of insular plants, for example, were able to flourish without evolving defenses against mammalian herbivores. In a very real sense, the biological novelty and fragility of insular biotas derive from their isolation. With the introduction of exotic species, whether planned or accidental, this "splendid isolation" was lost. Just as the formation of the Central American landbridge triggered the great faunal interchange and the subsequent extinctions of many species endemic to South America, anthropogenic introductions of exotic species onto islands effectively connected them to the mainland and triggered a wave of extinctions.

Actually, there have been multiple waves of insular extinctions. Europeans were not the only civilization to colonize islands and subsequently devastate the native biota. Subfossil evidence now indicates that many insular extinctions were caused by "native" islanders centuries, and sometimes millennia, before Europeans ever set foot on the islands. These groups include the aboriginals of Malaysia, Indonesia, Australia, and Tasmania, the Micronesians, Melanesians, and Polynesians of the Pacific Islands, and the Amerindians of the West Indies (Morgan and Woods 1986). For example, Olson and James (1982a) estimate that the Polynesians may have caused the extinctions of nearly half the native avifauna of Hawaii. In New Zealand, the moas, which included some 11 species of immense flightless birds, were extirpated by the Maori people and their introduced dogs and rats before European colonization (Halliday 1978). The much smaller kiwis (*Apteryx* sp.) are New Zealand's only surviving ratites.

Because introduced species have played such an important role in historical extinctions, let us review the ecology and geography of species introductions.

Species Introductions: The Ecology and Geography of Invasions

Magnitude of the problem.   The frequency and diversity of anthropogenic introductions of exotic species are staggering. More than any other events, including those associated with plate tectonics or glacial cycles, they have had a homogenizing effect on the world's biota. Species introductions, whether intentional or accidental, have touched and transformed all types of ecosystems, from oceanic islands and isolated mountaintops to tropical rain forests and the far reaches of the Antarctic (see Table 17.3 and Figure 17.12).

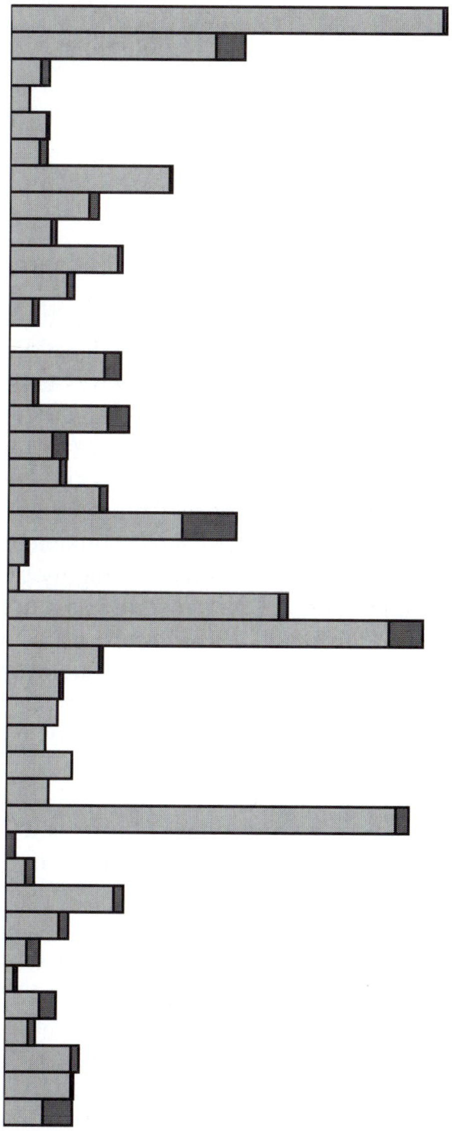

Figure 17.12   The number of introduced, or non-native, species of plants varies substantially among geographic regions and tends to be disproportionately high for islands such as Bermuda, Hawaii, the Cook Islands, and New Zealand, where exotics may comprise roughly half of the extant flora. (From Vitousek et al. 1996.)

Region	Number of native species	Number of non-native species	Percentage of non-native species
Mediterranean	23,000	250	1.1
Europe	11,000	1,568	12.5
Germany	1,718	429	20.0
Finland	1,006	221	18. 0
Poland	1,958	293	13.0
Britain (+Ireland)	1,623	442	21.4
Turkey	8,575	79	0.9
France	4,200	438	9.4
Czech Republic	2,288	194	7.8
Italy	5,599	294	5.0
Switzerland	3,030	280	8.5
Denmark	1,250	239	16.1
USA	--	2,000	--
California	4,844	1,025	17.5
Alaska	1,229	144	10.5
Florida	4,994	1,210	19.5
New England	1,995	877	30.5
Great Plains	2,495	394	13.6
Texas	4,498	492	9.9
Canada	9,028	2,840	23.9
Newfoundland	906	292	24.4
Greenland	427	86	16.8
Mexico	14,140	639	4.3
Australia	15–20,000	2,000	10.0
Uganda	4,848	152	3.0
Egypt	2,185	86	3.0
Rwanda	2,500	93	3.6
Tanzania	1,940	19	1.0
Namibia	3,159	60	1.9
Senegal	1,980	120	5.7
South Africa	20,263	824	3.9
Bermuda	165	303	64.7
Bahamas	1,104	246	18.2
Cuba	5,790	376	6.1
Puerto Rico	2,741	356	11.5
Hawaii	956	861	47.4
Cook Islands	284	273	49.0
Fiji	1,628	1,000	38.1
French Polynesia	959	560	36.9
New Caledonia	3,250	500	13.3
Solomon Islands	3,172	200	5.9
New Zealand	1,790	1,570	46.7

Table 17.3
The number of introduced, or non-native, species of birds and freshwater fishes in selected regions

Region	Breeding birds		Freshwater fishes	
	Native	Non-native	Native	Non-native
Europe	514	27	—	74
California			76	42
Alaska			55	1
Canada			177	9
Mexico			275	26
Australia		32	145	22
South Africa	900	14	107	20
Peru			—	12
Brazil	1635	2	517	76
Bermuda	—	6		
Bahamas	288	4		
Cuba	—	3	—	10
Puerto Rico	105	31	3	32
Hawaii	57	38	6	19
New Zealand	155	36	27	30
Japan	248	4	—	13

Source: Vitousek et al. 1996.

Note: The relative number of exotic species is especially high on islands such as Puerto Rico, Hawaii and New Zealand, where the number of non-native fishes may rival or exceed that of native species.

Although it may be impossible to estimate the total number and diversity of species introductions that have occurred even within historic times, we know that many were conducted as part of "naturalization" programs—deliberate attempts by Europeans to surround themselves with familiar species in exotic lands. While no doubt devastating to many native species, such deliberate programs do at least provide us with useful information on the magnitude and diversity of introductions. In 1988 Torbjorn Ebenhard reviewed all available records of historical introductions of birds and mammals. Again, this has to be viewed as an incomplete record: an inestimable number of introductions were unreported. Yet Ebenhard's survey is highly instructive. He found published reports documenting 788 mammalian introductions, including 118 species, 30 families, and 8 orders. The most frequently introduced mammals included common rabbits (*Oryctolagus cuniculus*), domestic cats (*Felis catus*), rats (including *Rattus norvegicus, R. rattus*, and *R. exulans*), house mice (*Mus musculus*), and domestic pigs, cattle, goats, and dogs. Not surprisingly, the pool of introduced species was not a random sample of the mainland faunas. Carnivores and even-toed ungulates constituted 19% and 31% of the introductions, respectively, whereas they represent less than 7% of the mainland species pool. Ebenhard also tallied 771 introductions of birds, including 212 species from 46 families and 16 orders. Again, species introductions were not random: waterfowl, gallinaceous birds, pigeons, and parrots were overrepresented in the sample of introduced species (46% of the introduced species versus just 12% of all avian species belong to these orders).

Species introductions of mammals and birds also had a strong geographic bias (Figure 17.13). On the continents, the number of species introductions appears to be roughly a function of the area of the region, with more introductions in the Palearctic region than in any other continental region. Australia,

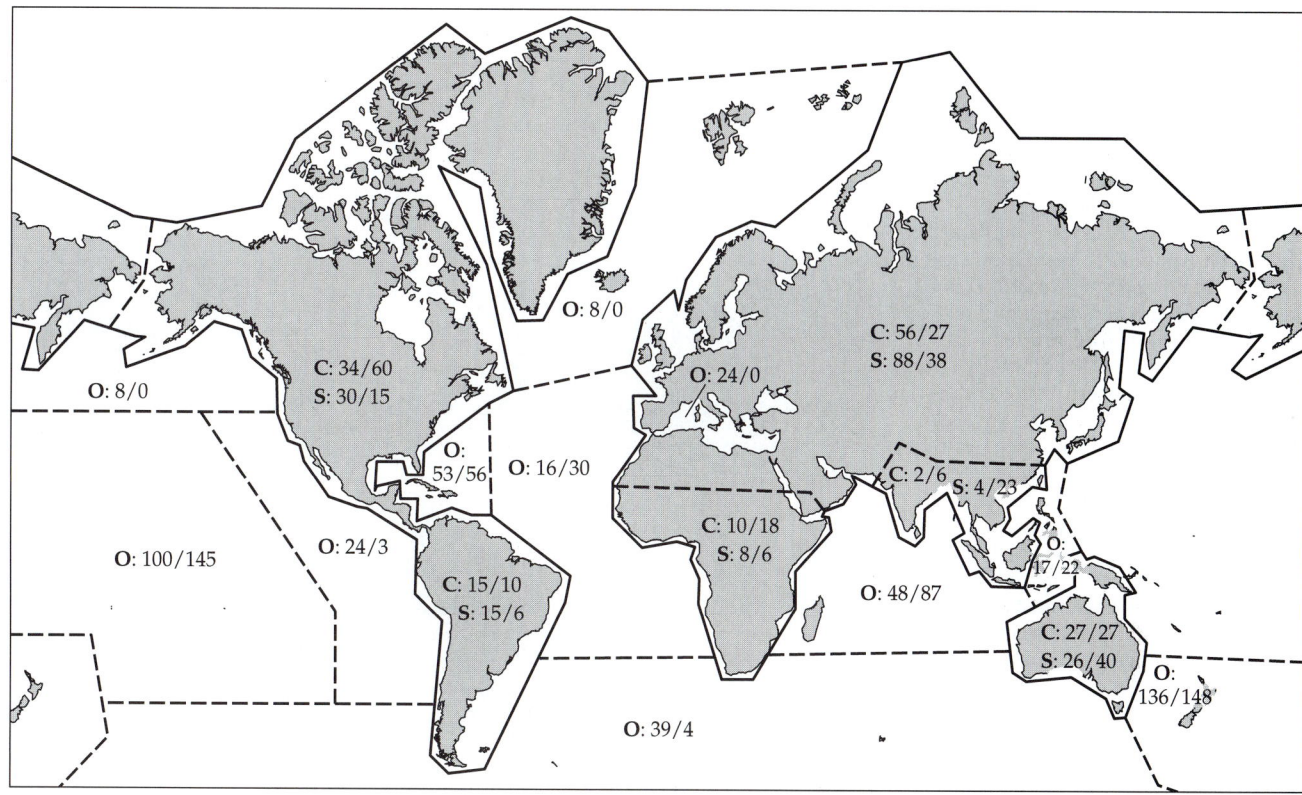

Figure 17.13 The geographic pattern of recorded introductions of mammals and birds. C = continental; S = continental shelf islands; O = oceanic islands. The numbers before and after the slash represent numbers of mammal and bird species respectively. (From Ebenhard 1988b.)

however, appears to be an important exception to this species-area relationship, receiving many more introductions per area than any other continental landmass. The most striking feature of the geographic distribution of introductions is that, despite their isolation and relatively small area, oceanic islands have received the bulk of species introductions. Sixty percent of all mammal and bird introductions have occurred on oceanic islands. In another important monograph on species introductions in birds, Long (1981) catalogued 162 species introductions to the Hawaiian Islands, 133 to New Zealand, and 56 to Tahiti, and well as over 500 to some 90 other islands or archipelagoes across the globe. In addition to receiving a greater number of introductions, islands have also received a greater diversity of invasive species. More species of mammals and many more species of birds have been introduced onto oceanic islands than to the continents or islands of the continental shelves.

Introductions of exotic plants, although of greater magnitude, were similar in their geographic bias. Again, Australia and oceanic islands received a highly disproportionate number of introductions (Drake et al. 1989). It is estimated that there are now between 1500 and 2000 exotic plants in Australia (see also MacDonald et al. 1986). Nearly half the plants of New Zealand are invaders (1790 natives versus 1570 exotics; Heywood 1989; Atkinson and Cameron 1993). Before human colonization, over 90% of the flowering plants of the Hawaiian Islands were endemic. Now nearly half of Hawaii's flowering plants are invaders (over 850 out of about 2000 species).

Aquatic systems have also been significantly affected by introductions of exotic species (see Box 17.1). In the state of California alone, 48 out of 137 species of fish are exotics. Forty-six of these are either known or suspected to have negatively affected native species. Of the 95 species that have bred in Arizona, 67 (71%) are not native to the state. Many exotic species established in freshwater

Box 17.1
Spread and impact of zebra mussels in North American freshwaters
by D. M. Lodge

In about 1985, the zebra mussel (*Dreissena polymorpha*), a filter-feeding Eurasian bivalve, became established in the North American Great Lakes system, probably having been introduced in ballast water. Since then, zebra mussels have continued to spread into all five American Great Lakes, and many rivers and lakes of the eastern United States. In the nineteenth century, zebra mussels spread with human commerce from the Caspian and Black Seas region into much of Europe, but the ecological impacts were not well documented. In North America, the spread of zebra mussels has been far more rapid and its impact far greater than most exotics.

The characteristics that distinguish a successful invader are often difficult to recognize. Zebra mussels, however, are doubly unique among freshwater bivalves in North America. First, larvae are free-swimming, and second, adults (which grow to 4–5 cm) attach themselves to firm substrates, e.g. boat bottoms, with byssal threads. These attributes allow larval and adult zebra mussels to disperse easily, and allow adults to occupy space like no other freshwater species in North America.

In lakes St. Clair and Erie, where zebra mussels occur at densities of

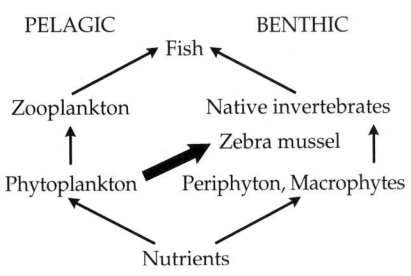

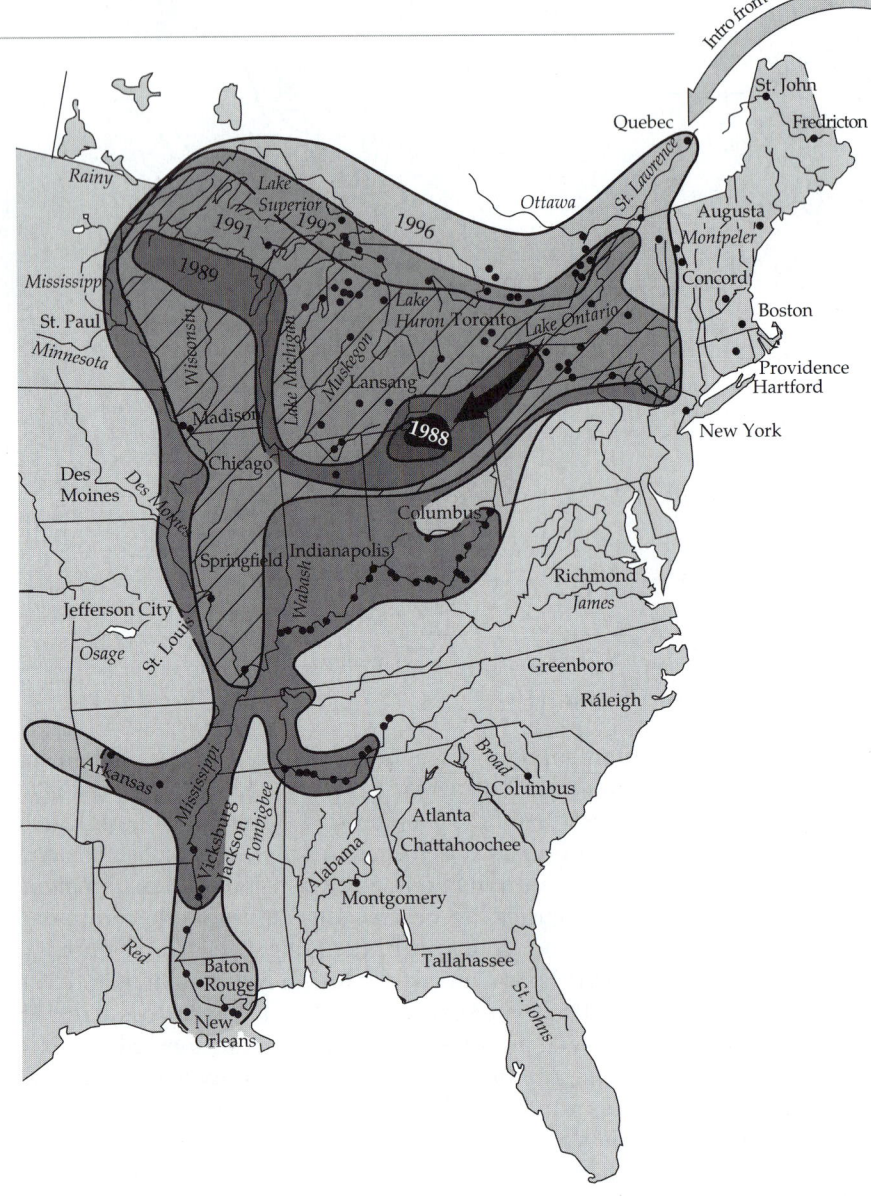

systems of the United States subsequently invaded Canadian waters (Crossman 1991). In Africa, introduction of the Nile perch (*Lates niloticus*) into Lake Victoria appears to have been one of the primary causes of the extinction of as many as 200 endemic cichlids (see Goldschmidt et al. 1993; Baskin 1992).

Perhaps the most troubling feature of the history of species introductions is that most have occurred in those systems with the greatest endemicity and vulnerability to invading species: oceanic islands. The prospect for the future is no less troubling. The proportion of threatened terrestrial vertebrates that are affected by introduced species is much higher on oceanic islands (31%) and on the "island continent" of Australia (51%) than in other continental areas (0 to 10%; Table 17.4). Biological invasions also threaten the diversity of

Box 17.1 *(continued)*

100,000-400,000 m^{-2}, their filtration of phytoplankton has greatly increased water transparency (see simplified lake food web). Now light reaches more of the bottom, allowing colonization by rooted vascular plants. Native unionid clams, many species of which are endangered, become covered by zebra mussels, and populations have declined. In contrast, total macroinvertebrate abundance is ten times greater in the presence of zebra mussels. Negative impacts on fish populations, while predicted, have not been documented. Ducks have reduced some local European populations of zebra mussels. However, adult mussels appear to have few predators, and the extent to which zebra mussels are a trophic dead end is an important question.

Overall, current evidence suggests that zebra mussels profoundly change pelagic and benthic communities, and shunt carbon and nutrients from the pelagic to the benthic zone. Clearly, zebra mussels will eventually inhabit many, if not most, of the lakes and streams of temperate North America. The magnitude and long-term impacts on community structure and ecosystem function are important topics of research.

Source: Lodge, D. M. 1993. Biological invasions: Lessons for ecology. *TREE* 8:133-137. Reprinted with permission from Elsevier Science.

nature reserves. While we are wise to invest much time and energy in managing and protecting nature reserves, most of them have already been significantly impacted by exotic species (MacDonald et al. 1989). On the continents, introduced species represent from 3% to 28% of all vascular plant species found in nature reserves. Again, the situation is much worse on islands, where

Table 17.4
Percentage of threatened terrestrial vertebrate species that are affected by introduced species

Taxanomic group	Continental areas within biogeographic realms								All mainland areas	All insular areas
	Eurasia	N. America	Africa	Indo-Malaya	Oceania	Antarctica	S. America	Australia		
Mammals	16.7 (42)	3.4 (29)	8.0 (100)	12.7 (55)	0.0 (8)	0.0 (0)	10.0 (60)	64.4 (45)	19.4 (283)	11.5 (61)
Birds	4.2 (24)	13.3 (15)	2.5 (118)	0.0 (30)	0.0 (1)	0.0 (0)	4.2 (71)	27.3 (11)	5.2 (250)	38.2 (144)
Reptiles	5.9 (17)	16.7 (24)	25.0 (16)	4.3 (23)	14.3 (7)	0.0 (0)	14.3 (28)	22.2 (9)	15.5 (84)	32.9 (76)
Amphibians	0.0 (8)	6.3 (16)	0.0 (3)	0.0 (0)	0.0 (0)	0.0 (0)	0.0 (1)	0.0 (2)	3.3 (30)	30.8 (13)
Total for all groups considered	9.9 (91)	9.5 (84)	6.3 (237)	7.4 (108)	6.3 (16)	0.0 (0)	8.1 (160)	50.7 (67)	12.7 (647)	31.0 (294)

Source: MacDonald et al. 1989.

Note: The total number of threatened species in each realm is given in parenthesis.

between 31% and 64% of all vascular plants species in nature reserves are exotics. Similarly, while introduced birds and mammals tend to be minor components of mainland reserves (averaging 1 to 5% for most reserves), they represent significant, if not the dominant, elements of the biota in reserves on oceanic islands (20% and 81% of the total species for birds and mammals, respectively). In many reserves that include lakes and streams, well over half the species of freshwater fish are exotic.

Susceptibility of insular biotas to invasion.   As we have seen, isolated oceanic islands tend to be inhabited by relatively depauperate communities that are often composed of novel, yet ecologically naive species. Wilson's taxon cycle theory (see Chapter 14) may be especially relevant to understanding the susceptibility of insular biotas to the effects of introduced species. According to this theory, one of the most striking patterns in insular biogeography is that insular populations may be evolutionary dead ends, doomed to eventual extinction. This results in the taxon cycle, the net movement of species from continents to islands over evolutionary time. Systematic studies show that insular biotas are derived from taxa that originated on larger landmasses. Sometimes there is a stepping-stone effect, so that small or distant islands are colonized by species that have come immediately from other islands that are usually larger or nearer to continents (see Chapter 13). Some island species are formed by endemic speciation within an island or an archipelago, as in the Hawaiian *Drosophila*, but continents remain the ultimate source of insular forms. This appears to be a special case of the general tendency of organisms originating in large areas with diverse biotas to invade small areas with lower organic diversity whenever the opportunity arises, and ultimately to replace the native species (see Chapters 9 and 15).

Even the exceptions to this pattern support the general rule. The region around Miami in southern Florida has been colonized successfully by several species from the Caribbean islands, including about six *Anolis* species that apparently were imported either intentionally or accidentally by humans during the twentieth century. However, southernmost Florida is essentially an island, an isolated region of tropical habitat attached to temperate North

Table 17.5
Numbers of native, endemic, and introduced species of various taxonomic groups in the Hawaiian Islands

Taxonomic Group	No. of native species	No. (%) of endemic species	No. (%) of introduced species
Flowering plants	970	883 (91%)	800 (45%)
Ferns and allies	143	105 (73%)	21 (13%)
Hepaticae (liverworts)	168	ca. 112 (67%)	2 (0.01%)
Musci (mosses)	233	112 (48%)	3 (0.01%)
Lichens	678	268 (38%)	0 (0%)
Resident birds	57	44 (77%)	38 (42%)
Mammals (on land)	1	0 (0%)	18 (94%)
Reptiles (on land)	0	0	13 (100%)
Amphibians	0	0	4 (100%)
Freshwater fishes	6	6 (100%)	19 (76%)
Arthropods	6000–10,000	(98%)	ca. 2000 (20–30%)
Mollusks	ca. 1060	(99%)	9 (<1%)

Source: Loope and Mueller-Dombois, 1989.

America by a low-lying peninsula that was frequently inundated by seawater during the Pleistocene. Although the Caribbean species seem to thrive in the lush tropical gardens in the Miami region, they have not extended their ranges northward to invade more temperate habitats.

The extreme susceptibility of islands to invasion is dramatically illustrated by two case studies. The avifauna of the Hawaiian archipelago is currently composed of 95 species of birds. Of these, 57 are native, while 38 (40%) are exotics that have been introduced since 1800. During the same period 14 native species have become extinct, and several others are rare and endangered. Of 94 species known to have been introduced prior to 1940, 53 became established, at least locally and temporarily, and only 41 failed completely. Most of the introduced species were deliberately released to augment the local avifauna with game birds and species of beautiful plumage and song. A strange mixture of species derived from different taxonomic groups and biogeographic provinces now coexists on the archipelago, including the mockingbird (*Mimus polyglottos*), cardinal (*Cardinalis cardinalis*), and California quail (*Callipepla californica*) from North America; the Indian mynah (*Acridotheres tristis*), lace-necked dove (*Streptopelia chinensis*), and Pekin robin (*Leiothrix lutea*—a babbler, not a thrush) from Asia; the house sparrow (*Passer domesticus*), skylark (*Alauda arvensis*), and ring-necked pheasant (*Phasianus colchicus*) from Europe; and the Brazilian cardinal (*Paroaria cristata*) from South America (Elton 1958). A quick review of Table 17.5 reveals that assemblages of the less vagile vertebrates, including nonvolant mammals, amphibians, reptiles, and freshwater fishes, are dominated by exotic species.

The fate of the Hawaiian archipelago is by no means unique. Consider the birds and mammals of New Zealand (Table 17.6). The potential effect of these exotic species on the insular biota can be better appreciated when one realizes that the only native terrestrial mammals of New Zealand are two species of bats: all of the nonvolant mammals and over half of the bird species have been introduced (Wodzicki 1950; Elton 1958; Falla et al. 1966; Atkinson and Cameron 1993). It is little wonder that these exotics have had a tremendous impact on the native species of lizards and ground-nesting birds, which evolved in the complete absence of mammalian predators.

Given the magnitude and diversity of introductions, it should not be surprising that their effects on native species, while often negative, have involved many different mechanisms. Introduced species can directly compete with or prey on native species. Rats and mongooses are perhaps the most notorious of the predators introduced to oceanic islands. By preying on ground-nesting birds and reptiles, they have decimated many endemic species (Figures 17.14

Table 17.6

Native and introduced land birds and mammals of New Zealand

	Birds	Mammals	Total
Native			
Families	19	2	21
Species	39	2	41
Introduced			
Families	13	11	24
Species	26	26	52
Source of introduced species			
Europe	16	15	31
Australia	6	2	8
Asia	3	4	7
North America	1	3	4
Polynesia	0	2	2

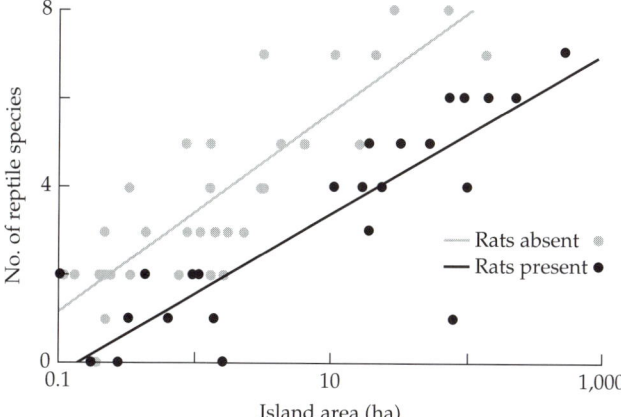

Figure 17.14 The effect of introduced rats on the species diversity of native insular reptiles is revealed by consistent differences in species-area relationships for reptiles on islands off New Zealand with and without rats. (From Whitaker 1978.)

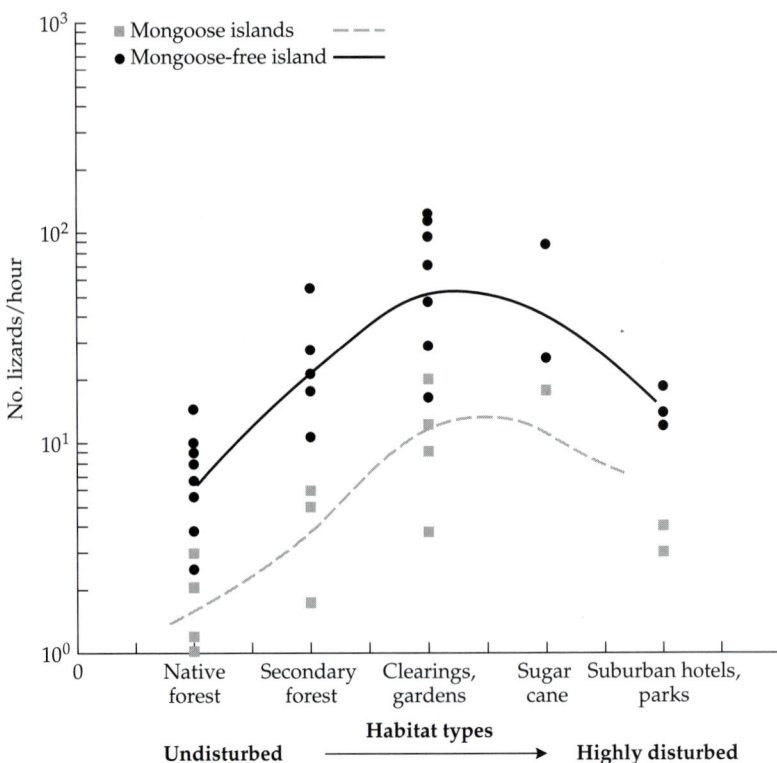

Figure 17.15 The effects of introduced mongooses on density and activity (as measured by number of lizards observed per hour) of native lizards inhabiting tropical islands of the Pacific Ocean. (From Case and Bolger 1991.)

and 17.15). Overgrazing by pigs, goats, sheep, and other mammalian herbivores has affected many insular plants and has indirectly contributed to the demise of insular vertebrates by destroying their food supply or habitat. Other introduced species have served as vectors for pathogens, such as avian malaria and bird pox virus, Dutch elm disease (*Ophiostoma ulmi*), and chestnut blight (*Castanea dentata*; see Jackson 1978; Berger 1981; von Broembsen 1989), which have caused range contractions and extinctions in many animals and plants. Hybridization of exotics with native species is another factor threatening biological diversity, especially in plants and fishes.

In most cases, however, native species have suffered from a combination of extinction forces. For example, four insular subspecies of the Galápagos tortoise were extirpated on islands that were infested with a combination of introduced mammals. Introduced rats preyed on their eggs, cats and dogs preyed on the young tortoises, and goats competed with the adults for vegetation. In still other cases, the effects of introduced species may be even more complex, involving indirect effects that cascaded throughout the native community. Mongooses, which were introduced onto the Virgin Islands to control introduced rats, turned out to prey most heavily on ground-nesting birds and ground lizards (e.g., *Ameiva exsul*). The extirpation of these insectivores on some small islands was followed by outbreaks of insects, which in turn caused widespread damage to native plants.

Regardless of the specific causes of historical extinctions, they all speak to the susceptibility of insular biotas to invasion. These phenomena are consistent with theories of both general ecology and island biogeography. As Elton (1958) suggested, invasibility tends to be higher for disturbed and species-poor communities (see also Hobbs and Huenneke 1992). For example, the proportion of exotic birds and mammals tends to decrease with increasing richness of native species (Figure 17.16). Greenway's (1967) studies of extinct and

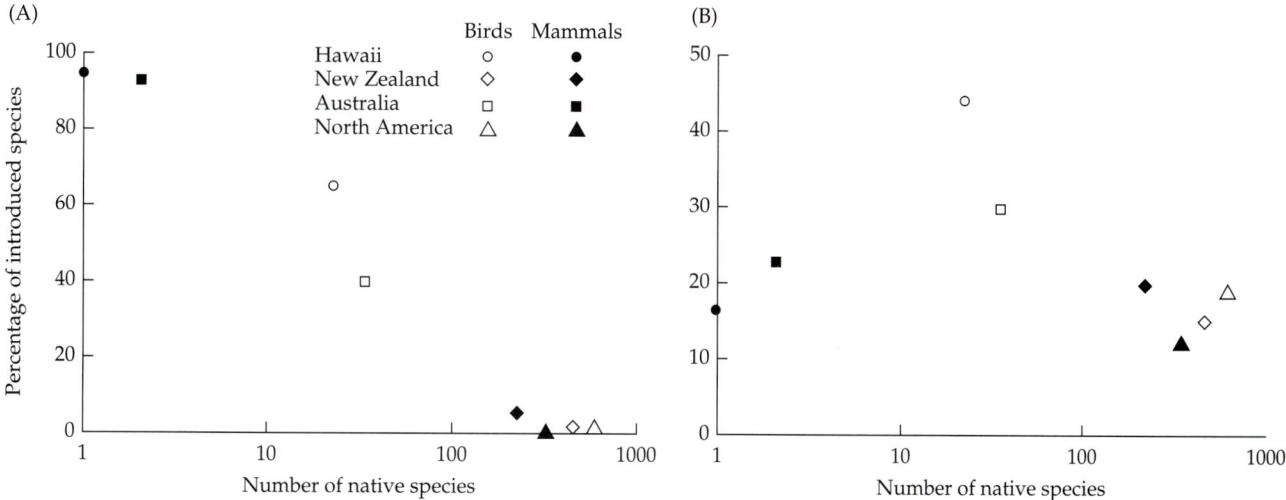

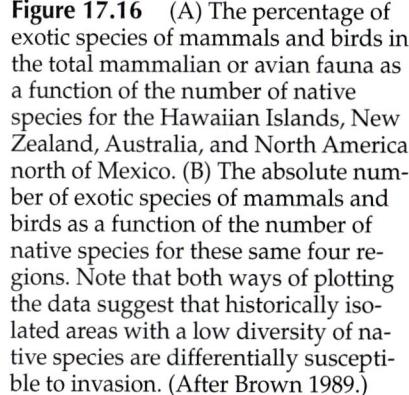

Figure 17.16 (A) The percentage of exotic species of mammals and birds in the total mammalian or avian fauna as a function of the number of native species for the Hawaiian Islands, New Zealand, Australia, and North America north of Mexico. (B) The absolute number of exotic species of mammals and birds as a function of the number of native species for these same four regions. Note that both ways of plotting the data suggest that historically isolated areas with a low diversity of native species are differentially susceptible to invasion. (After Brown 1989.)

endangered birds of the Pacific islands revealed that the avifauna of the more isolated archipelagoes were more prone to extinctions and harbored a higher proportion of threatened species (Figure 17.17). Within archipelagoes, avian extinction rates decreased with increasing area (Figure 17.18). This pattern appears to be a general phenomenon and one consistent with the equilibrium theory of island biogeography: extinction rates are higher on smaller and more isolated islands (see also Case and Bolger 1991).

Moreover, habitat alteration for agriculture and related anthropogenic disturbances tend to reduce the carrying capacity of islands for their native species while increasing the likelihood that introduced species can gain a foothold. This hypothesis is supported by Case's (1996) recent review of global patterns in the invasion success of exotic birds. Across a wide variety of archipelagoes and continental regions, invasion success increases with the number of native avian species already extirpated (Figure 17.19). In fact, Case found that the number of exotic species of birds established on an island was close to the number of species lost to extinctions.

While these results seem to support Elton's earlier contention that species-poor systems are more easily invaded, Case suggested that the invasion success of introduced birds was more directly linked to habitat disturbance. On islands, exotic species are largely restricted to disturbed sites, while native birds inhabit

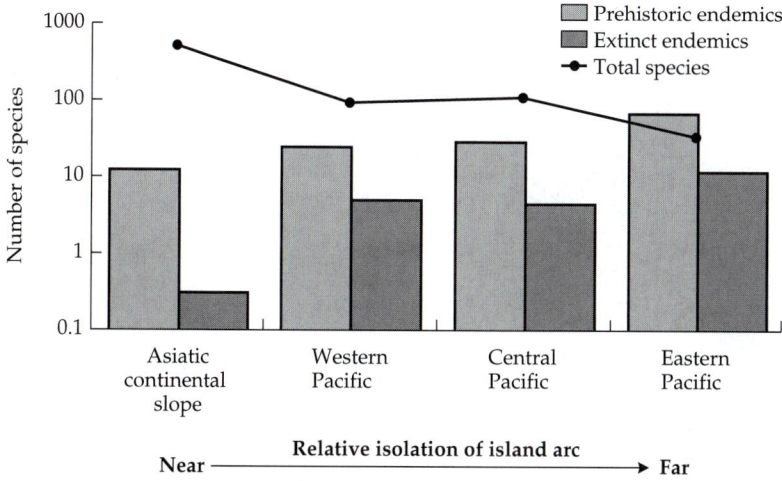

Figure 17.17 While the species richness of land birds decreases with the isolation of island arcs in the Pacific Ocean, endemicity and susceptibility to extinction increase. (These figures do not include prehistoric extinctions, which in some cases included as much as half of the native diversity of these islands prior to European colonization.) (After Greenway 1967.)

Figure 17.18 Extinction rates among land birds of the Pacific islands tend to decrease with island area. (After Greenway 1967.)

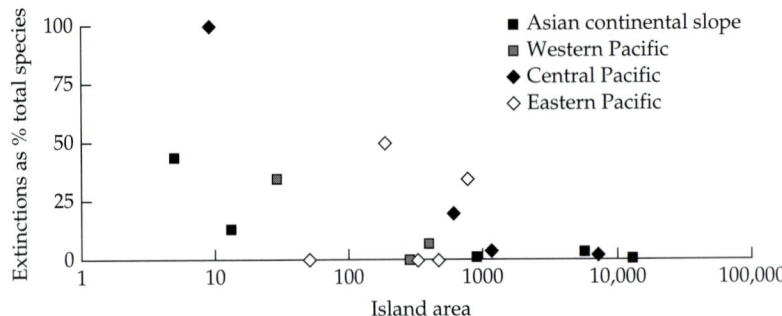

undisturbed refugia away from human activities. Thus, as humans and their commensals have converted native habitats into urban, agricultural, or otherwise disturbed sites, native birds have lost habitat and suffered extirpations, while exotic species have been provided with the habitat they require to establish and maintain populations. Of course, Case's and Elton's hypotheses are not mutually exclusive, and it is possible, if not likely, that habitat disturbance and reduced competition from native species combine to facilitate invasions by exotics. Although this debate will likely remain unresolved for some time, clearly once they have established populations on islands, exotic species have often devastated natives in both historic and recent times.

Current Patterns of Endangerment

The historical record shows clearly that exotic species and habitat alteration have been important factors in causing extinctions, particularly among island endemics. Hunting, collection, and other forms of overexploitation have also taken a heavy toll, accounting for nearly a third of the avian extinctions and half the mammalian extinctions worldwide since 1600 (Figure 17.20). Extinction patterns in mollusks, perhaps the hardest-hit group of animals, present a somewhat different picture. Most of their historical extinctions were associated with habitat destruction and pollution (60%), while introduced species and collecting accounted for a smaller, yet still significant, portion (24% and 14%, respectively).

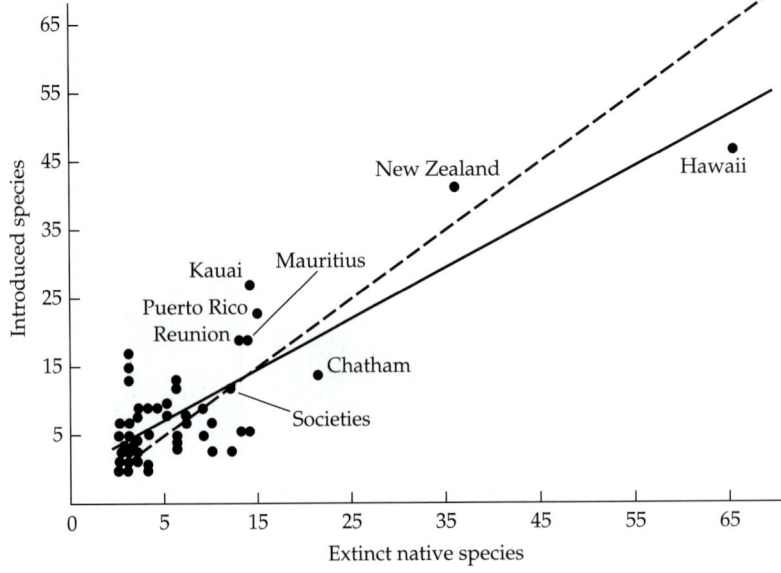

Figure 17.19 The relationship between the number of introduced bird species and the number of extinct native bird species for several insular and continental regions. Invasion success appears to increase with the number of native species already extinct. Dashed line is line of equality; solid line is regression line. (From Case 1996.)

(A) Causes of historical extinctions in animals

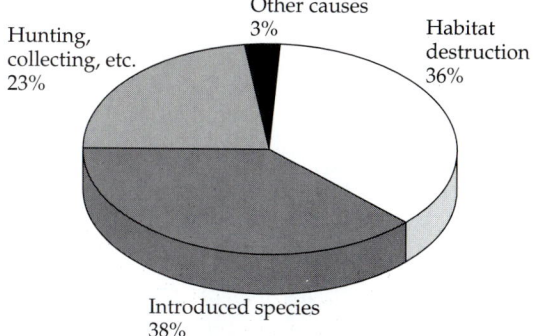

Other causes 3%
Habitat destruction 36%
Hunting, collecting, etc. 23%
Introduced species 38%

(B) Causes of endangerment in animals

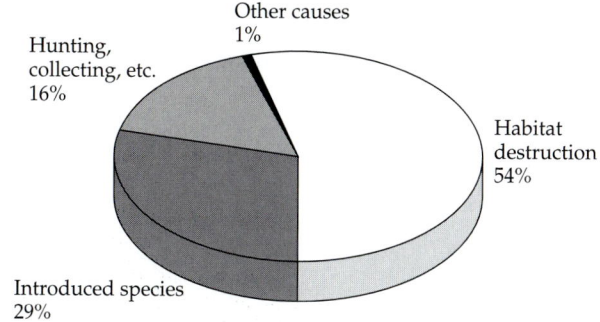

Other causes 1%
Hunting, collecting, etc. 16%
Habitat destruction 54%
Introduced species 29%

(C) Causes of historical extinctions in birds

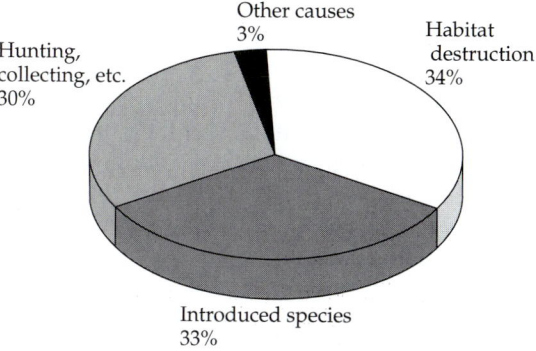

Other causes 3%
Habitat destruction 34%
Hunting, collecting, etc. 30%
Introduced species 33%

(D) Causes of endangerment in birds

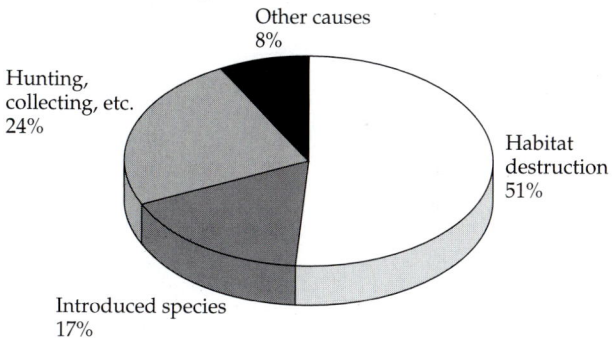

Other causes 8%
Hunting, collecting, etc. 24%
Habitat destruction 51%
Introduced species 17%

(E) Causes of historical extinctions in mammals

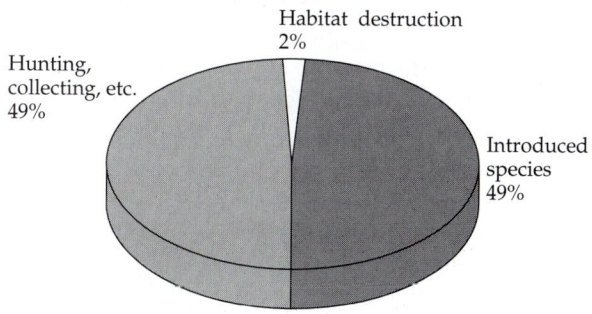

Habitat destruction 2%
Hunting, collecting, etc. 49%
Introduced species 49%

(F) Causes of endangerment in mammals

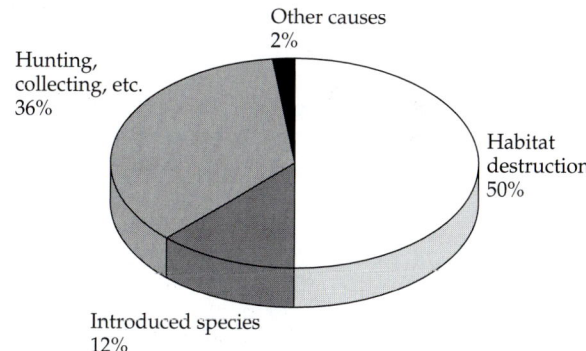

Other causes 2%
Hunting, collecting, etc. 36%
Habitat destruction 50%
Introduced species 12%

(G) Causes of historical extinctions in mollusks

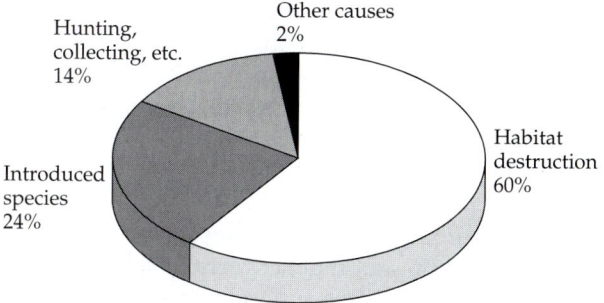

Other causes 2%
Hunting, collecting, etc. 14%
Introduced species 24%
Habitat destruction 60%

(H) Causes of endangerment in mollusks

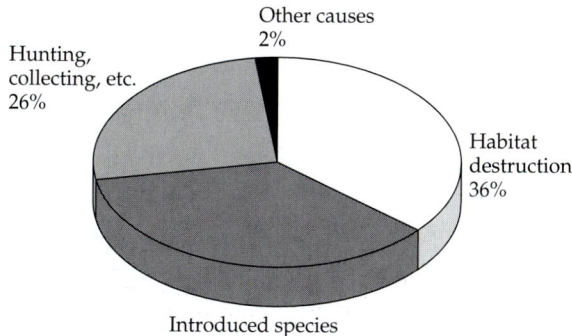

Other causes 2%
Hunting, collecting, etc. 26%
Habitat destruction 36%
Introduced species 36%

Figure 17.20 Causes of historical extinctions (1600–1980) and of current endangerment in selected animal taxa. (After Flather et al. 1994.)

Figure 17.21 Historical trends in the relative numbers of insular and mainland extinctions of animals worldwide. (After World Conservation Monitoring Centre 1992.)

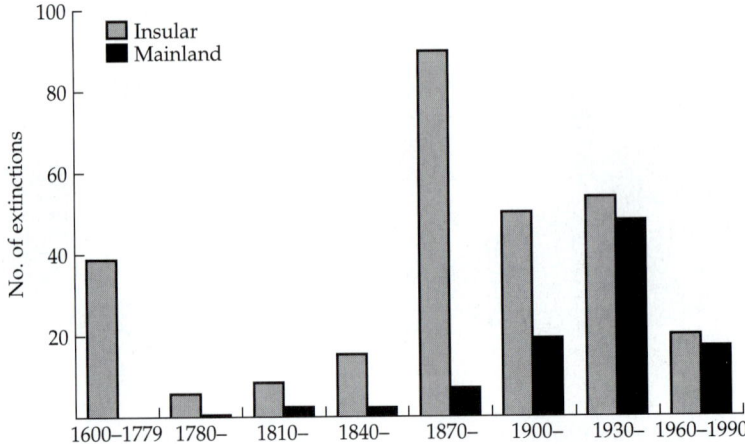

While we have learned much by examining the patterns of past extinctions, it appears that the geographic signature of extinctions and the predominant causal forces are changing. Figure 17.21 reveals an important temporal-spatial trend in animal extinctions: the extinction front has expanded to include the mainland. Although nearly all animal extinctions recorded between 1600 and 1980 were insular, the number of extinctions of mainland species now rivals that of insular forms.

Ironically, the ongoing wave of extinctions may still be an insular phenomenon. Habitat destruction and fragmentation on the continents have reduced once expansive stands of native habitats to ever-shrinking archipelagoes of habitat islands (Figure 17.22). Terrestrial biotas stranded in these remnants of native habitats may undergo a process of ecological relaxation similar to that described in Chapter 13. In contrast to the climatic and tectonic changes that have isolated habitats in the past, these anthropogenic habitat changes are occurring much more rapidly and may, therefore, take a heavier toll on the stranded species.

Whereas anthropogenic introductions and hunting were the primary causes of historical extinctions, habitat loss and fragmentation have now become the primary causes of endangerment in terrestrial animals (see Figure 17.20). In mollusks, species introductions and habitat alteration contribute equally to species endangerment. Patterns of endangerment in plants are similar to those reported for terrestrial animals (Figure 17.23). Habitat alteration, including development for housing, water control, mining, oil and gas, logging, off-road vehicles, and agriculture, represents the primary cause of endangerment for plant species listed by the U.S. Fish and Wildlife Service under the Endangered Species Act. The effects of introduced species and the effects of grazing and trampling by domestic livestock account for another 25% of species endangerment. Perhaps the most sobering finding is that natural causes account for just 1% of the endangerment of plants.

Habitat Loss and Fragmentation

Magnitude of the problem. According to many conservation biologists, habitat loss and fragmentation now represent the most serious threat to biological diversity (Wilcox and Murphy 1985). This is evident from the patterns of endangerment described above. Yet we need to avoid overstating and simplifying what really is a very complex problem. The relative importance of different extinction forces varies markedly among taxa (see Figures 17.20 and

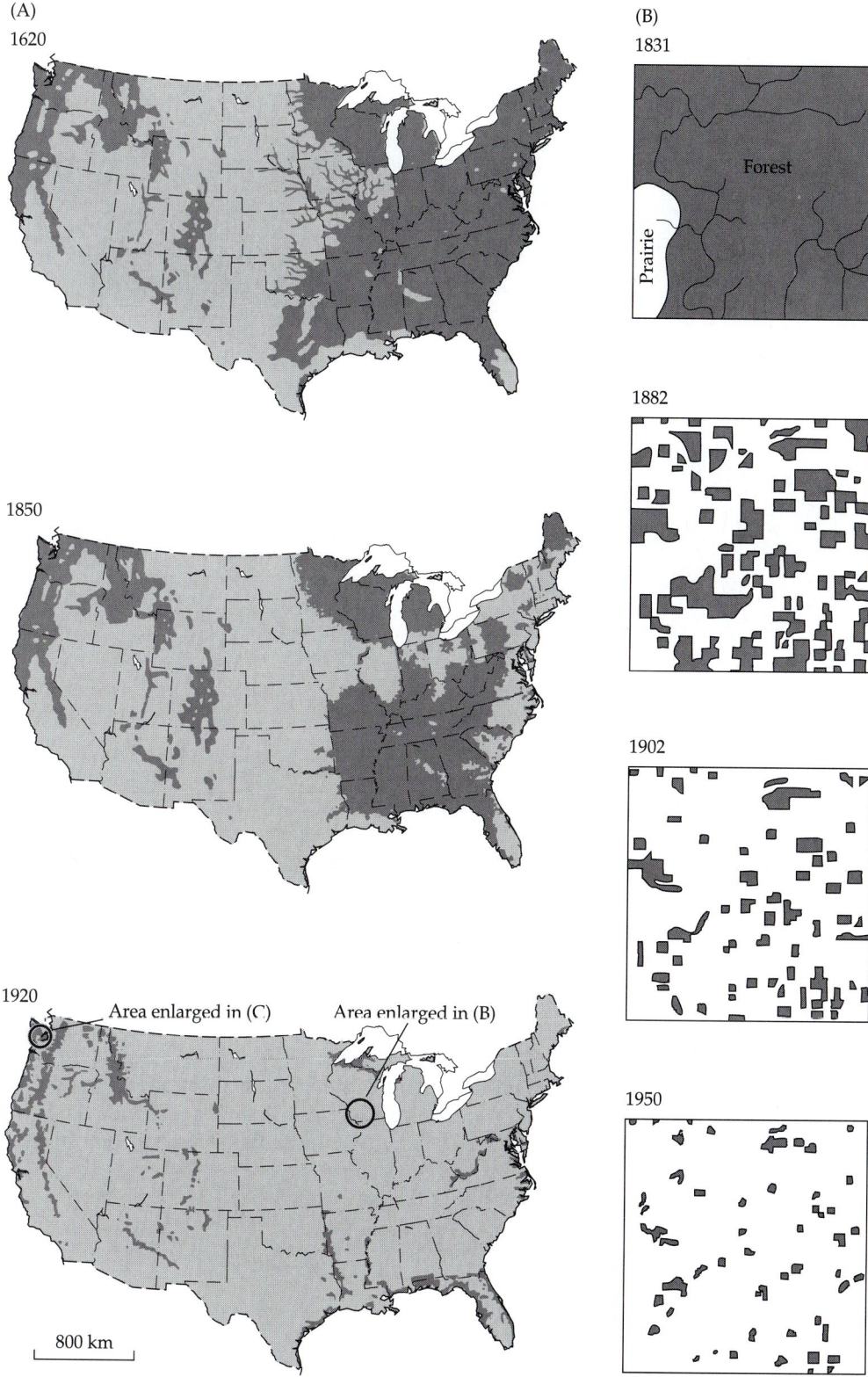

Figure 17.22 The clearing of native North American forests (darker shaded areas) by European settlers has rapidly reduced once expansive stands of native forest habitat to archipelagoes of habitat islands. (A) United States. (B) Cadiz Township, Wisconsin. (C, next page) Olympic National Forest, Washington. (A from Greely 1925, Williams 1990; B from Thomas 1956; C from D. R. Perault, unpublished report.)

Figure 17.22
continued

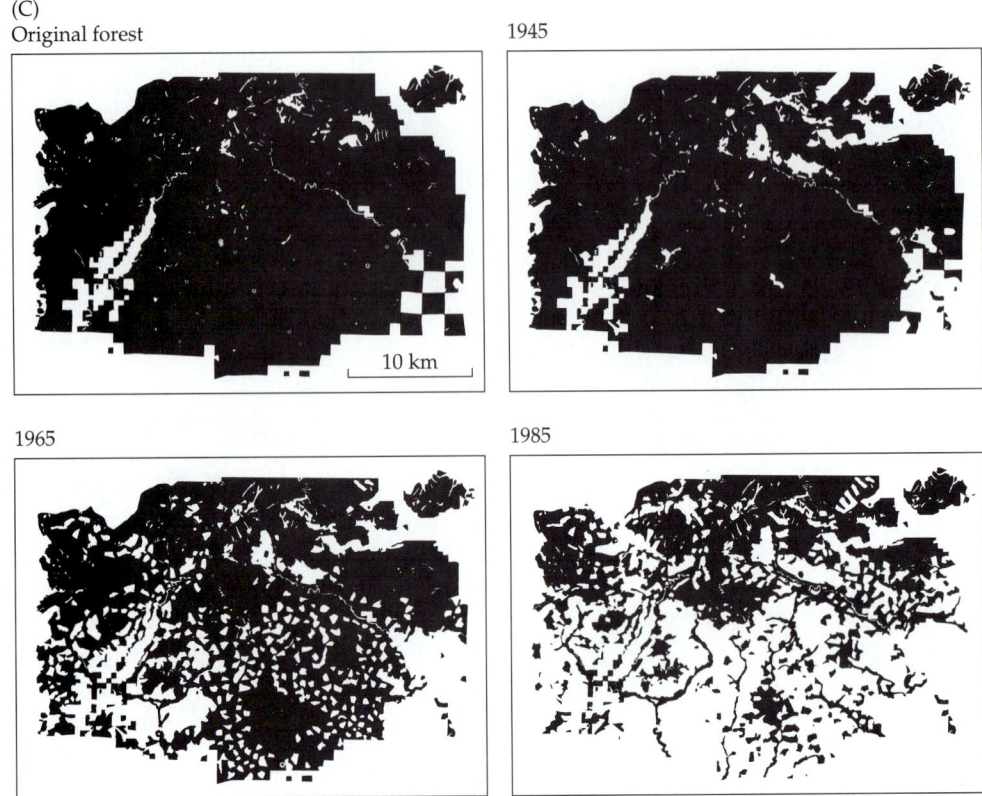

(C)
Original forest   1945

10 km

1965   1985

17.23). Species introductions, hunting, pollution, climatic change, and other effects of human activities continue to pose serious threats to most endangered species. Moreover, almost every endangered species is threatened by a combination of these factors, many of them interacting with the effects of habitat loss and fragmentation. Even on oceanic islands such as Hawaii, however, where half the plants are already exotics, habitat destruction by humans and their domestic and feral livestock has become the primary threat to ferns and other endangered plants (Wagner 1995). Here, we shall first discuss the intensity and magnitude of habitat loss and fragmentation before examining the specific effects of these landscape dynamics on biological diversity.

The earth's natural ecosystems are being transformed at an unprecedented rate. These include some of the world's most diverse systems, including the tropical rain forest biome, in which just 7% of the earth's surface is home to over

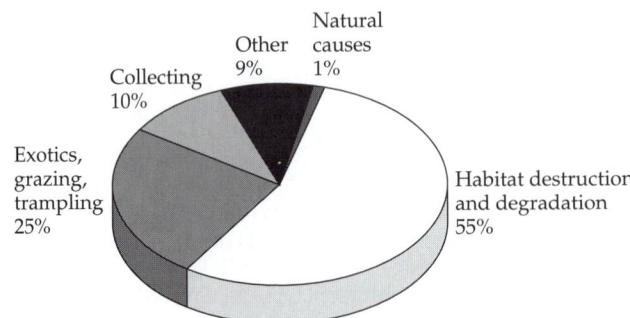

Other 9%

Natural causes 1%

Collecting 10%

Exotics, grazing, trampling 25%

Habitat destruction and degradation 55%

Figure 17.23   Causes of endangerment for plants of the United States. (After Schemske et al. 1994.)

50% of its species (Wilson 1988). A single rain forest tree can contain over 40 species of ants from 26 genera, approximately equal to the entire ant fauna of the British Isles. Yet tropical rain forests are being destroyed at a rate of approximately 76,000 km² per year. Madagascar retains only 7% of its tropical forests, and only 1% of Brazil's Atlantic Coastal forest remains uncut. Since 1819, the primary forest of Singapore has been all but completely removed, with less than 1% remaining (Corlett 1992). Similar patterns of deforestation have severely impacted tropical forest communities across the globe (Figure 17.24).

Loss of natural ecosystems has by no means been restricted to tropical rain forests, however (Figure 17.25). Prior to the impact of humans, grasslands covered approximately 40% of the earth's surface (Clements and Shelford 1939). Now, just half of this amount remains (see World Conservation Monitoring Centre 1992). Pristine prairies once covered approximately 1 million km² of North America; now only 10% of that remains. Less than 1% of tallgrass prairie remains, nearly all of it on shallow, rocky soils or in tiny isolated reserves. The loss of wetlands has been no less severe. Estimates are that as much as 50% of the world's wetlands have been lost, most of this within the twentieth century, and that 65% of biologically significant tropical wetlands

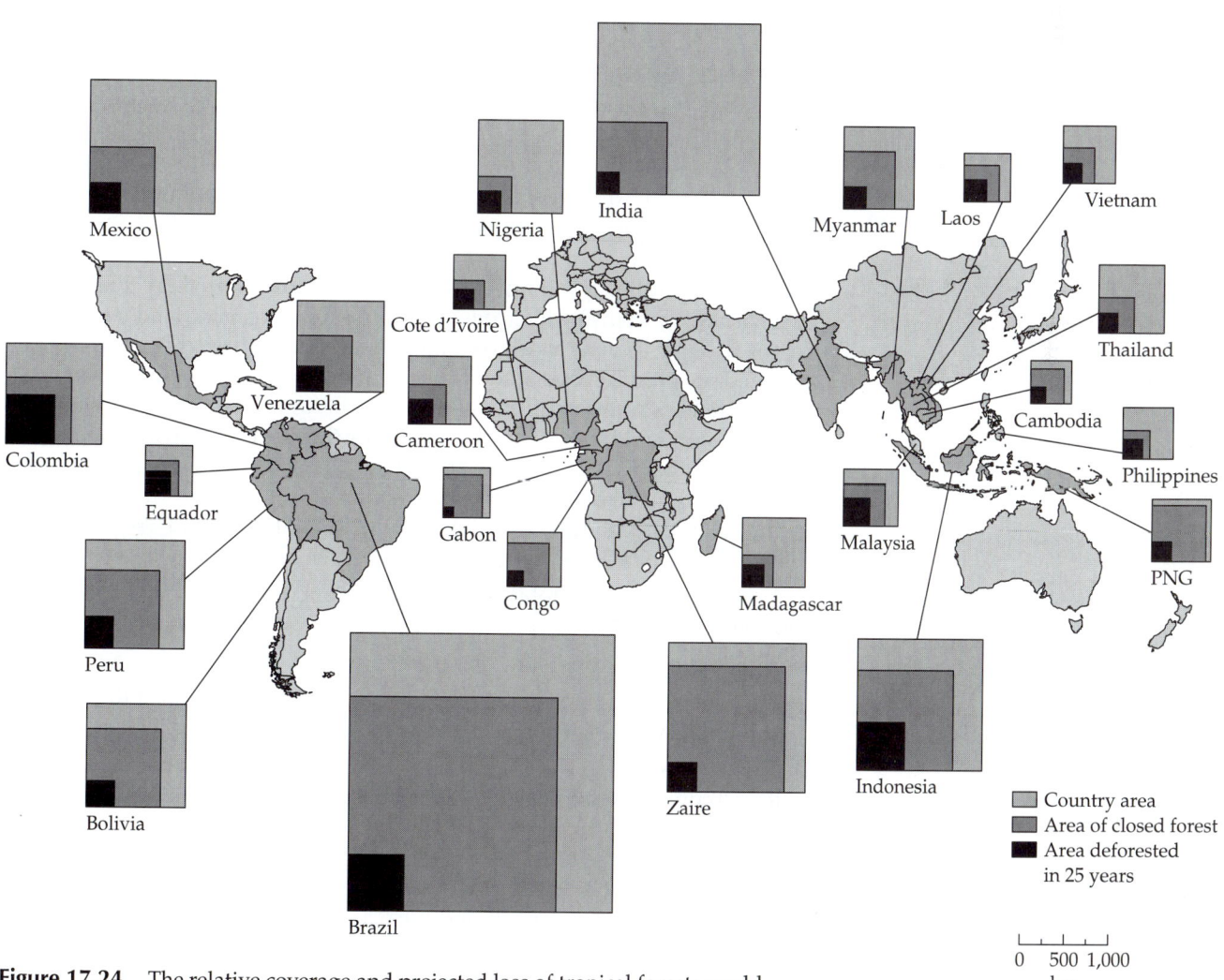

Figure 17.24 The relative coverage and projected loss of tropical forests worldwide. (From World Conservation Monitoring Centre 1992.)

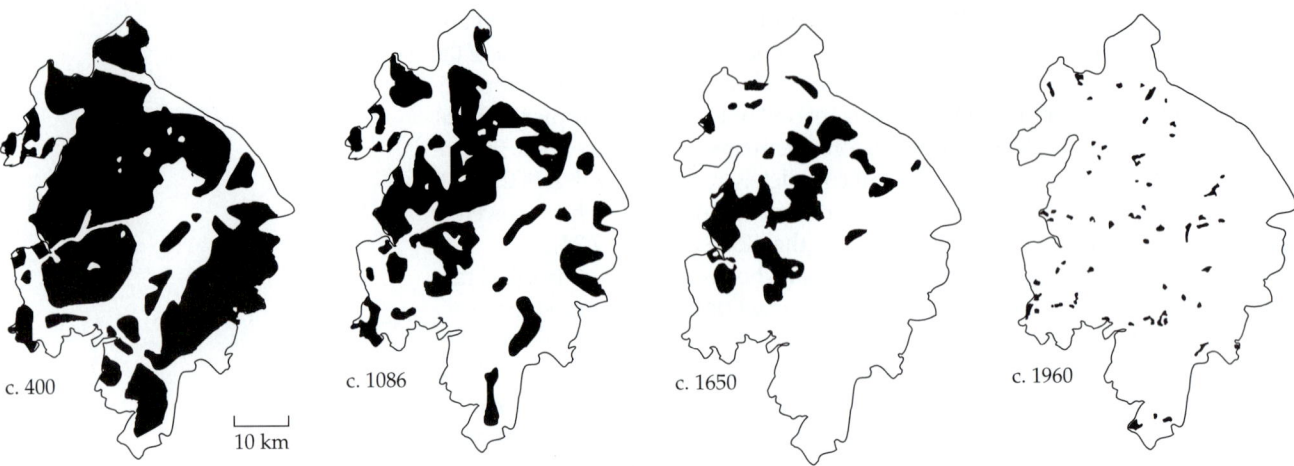

Figure 17.25 Tropical rain forests are not the only habitats that have suffered extensive losses. This map shows the deforestation and fragmentation of temperate native forests in Warwickshire, England, between 400 and 1960 A.D. (From Wilcove et al. 1986.)

are currently threatened (Moyle and Leidy 1992). Freshwater systems are heavily affected by dams and other water control projects, which effectively fragment streams and rivers in most developed regions. In the marine realm, coral reefs are the aquatic equivalent of tropical rain forests in both diversity and productivity. Yet many are threatened by a variety of human activities, including pollution, sedimentation, and overexploitation.

Effects of habitat loss and fragmentation.   Few of us can deny that modern civilizations are transforming the very nature of the earth's land surface to a degree rivaling, and perhaps exceeding, the climate-driven changes that characterized the glacial cycles of the Pleistocene. There are, however, a number of key differences. Anthropogenic transformations are occurring at a much more rapid rate, too rapid for many species to track their optimal environments. Just as important, natural landscapes are not just being lost, but those that remain are being fragmented and degraded in quality.

The physical changes associated with habitat loss and fragmentation include the following (see review by Saunders et al. 1991):

1. A reduction in the total area, resources, and productivity of native habitats.
2. Increased isolation of remnant fragments and their local populations.
3. Significant changes in the environmental characteristics of the fragments, including changes in solar radiation, wind, and water flux.

Especially worrisome is the possibility that such chronic changes in local climate will trigger feedback effects, which in turn can accelerate the rate of habitat loss or prevent the regeneration of native ecosystems. Deforestation, for example, often leads to drier conditions in remnant patches of forest, which makes them susceptible to fires of unnaturally high intensity and frequency. In Borneo, deforestation, the slash-and-burn agriculture that replaced the forests, and seasonal droughts associated with El Niño events combined to trigger massive forest fires, which destroyed a large portion of the island's remaining forests. Although fire is a natural and essential component of many ecosystems, it is now becoming a much more prevalent and serious threat to conserving the diversity of rain forests and other ecosystems not normally subjected to its effects.

As a result of these anthropogenic changes, natural systems may undergo a type of ecological relaxation similar to that described by Brown (1971) and Diamond (1972) for landbridge islands (see Chapter 13). In most cases, relaxation or biotic collapse after habitat fragmentation is expected to proceed in a sequence of stages (after Wilcove 1987):

Stage 1: Initial exclusion. Some species will be lost from the landscape simply because their original ranges did not include any of the remnant patches.

Stage 2: Extirpation due to lack of essential resources. Species vary tremendously in their resource requirements; many require very large areas and/or very rare resources. The likelihood that all of a species' resource requirements can be met decreases as the remaining area decreases.

Stage 3: Perils associated with small populations. Small populations are much more susceptible to a host of genetic, demographic, and stochastic problems (see Soulé 1987). As the total area of the remnant patches decreases, these problems become increasingly severe.

Stage 4: Deleterious effects of isolation. As we discussed in Chapter 13, some populations may be rescued from extinction by migration and recruitment of individuals from other populations. The likelihood of such rescue effects decreases as isolation increases.

Stage 5: Ecological imbalances. While communities may not respond as superorganisms or units in the Clementsian sense (see Chapter 5), most species are strongly influenced by interactions with other species. Loss of one species during any of the above stages of relaxation may result in the subsequent loss of its predators, parasites, mutualists, or commensals. In addition, habitat disturbance and reductions in community diversity during the earlier stages of relaxation may facilitate the establishment of introduced species, triggering a cascade of subsequent extirpations.

Ecological imbalances and the cascading effects of initial losses in biological diversity are fairly well documented by the historical record. For example, as a result of the fragmentation of native forests in North America, populations of brown-headed cowbirds (*Molothrus ater*) underwent dramatic increases in density and expanded their ranges into newly cleared areas and edges of remnant forests. As a result, nest parasitism by cowbirds now takes a heavy toll on many Neotropical songbirds, especially those breeding within 500 m of the forest edge (see Brittingham and Temple 1983; O'Conner and Faaborg 1992). These songbirds suffer from indirect as well as direct effects of deforestation in both their breeding and wintering ranges. In the United States, deforestation has caused significant range contractions among large mammalian predators, including the cougar (*Puma concolor*), gray wolf (*Canis lupus*), and red wolf (*Canis rufus*). The loss of these top carnivores has resulted in compensatory increases in the densities of smaller carnivores (especially coyotes) and omnivores (raccoons, opossums, and skunks), which prey heavily on smaller animals—namely, songbirds and their eggs or nestlings. Fragmentation may also influence community structure by limiting the densities of native decomposers. In the fragmented tropical rain forests north of Manaus, Brazil, the diversity and density of carrion- and dung-feeding beetles was found to be markedly lower in clear-cuts and fragments than in adjacent stands of contiguous forest (Klein 1989). As a result, decomposition and recycling of dead organic material were greatly reduced.

Although ecological imbalances and secondary extinctions represent the final stage of relaxation, this stage may be of long duration. Modeling studies by David Tilman and his colleagues (Tilman et al. 1994) suggested that even moderate intensities of habitat destruction may well cause time-delayed, secondary extinctions even among currently dominant species. The cascading effects we have just discussed often take generations to move from one trophic level to another. Because any realistic scenario may involve many steps, today's habitat losses are likely to result in a substantial "extinction debt," a future ecological cost that may take many generations, possibly centuries, before it is fully realized.

Applications of Biogeographic Theory

Our purpose in this chapter is not to present a gloom and doom scenario on the status of biodiversity. Yes, the challenge of conserving biodiversity in the face of expanding human populations is a great one, but we are guardedly optimistic that conservation biologists can make some real progress toward understanding the problems and reducing the losses. As we stated earlier, effective strategies for conserving biodiversity require insights from many disciplines of science. Yet, of all the many fields that must be called upon to help develop conservation strategies, biogeography is especially relevant. Biological diversity, and the many human activities that threaten it, have definite geographic signatures. They exhibit clear patterns across geographic gradients as well as hot spots—sites of exceptionally high diversity and endemicity of native species or sites of intense human activity. Hot spots of anthropogenic disturbance are not stable, but exhibit a strong tendency to intensify and spread from one region to another.

Designing Nature Reserves

Waves of human colonization and advances in technology have overwhelmed hot spots of diversity and generated surges in extinction rates, both locally and globally. Thus, anthropogenic extinctions are case studies in dynamic biogeography. As we have seen, historical extinctions have been especially severe on oceanic islands, and the ongoing wave of extinctions appears to be concentrated on archipelagoes of fragmented remnant habitats. It is only logical, therefore, to believe that island biogeography may offer some important lessons for conserving biodiversity.

Not surprisingly, E. O. Wilson, the author of the taxon cycle theory and co-author of the equilibrium theory of island biogeography, was one of the first persons to make this argument. In a symposium volume dedicated to Robert MacArthur, Wilson and his colleague Ed Willis (1975) presented a set of recommendations and geometric "rules" for designing nature reserves and networks of reserves. These rules were based largely on the principles and patterns of island biogeography: extinction rates decrease with increasing area; immigration rates (and rescue effects) decrease with increasing isolation. Wilson and Willis urged planners and managers of nature reserves to follow four basic recommendations:

1. Individual reserves should be made as large as possible. Extinction rates are lower on larger reserves.

2. Unique habitats and biotas should be contained in multiple reserves, and these should be located as close to one another as possible. While

each isolated population has a real risk of extinction, it is much less likely that extinctions will occur simultaneously among all the isolated populations. Given that the isolation is not too great, reserves can be recolonized by populations that persist on other reserves.

3. Reserves of fixed area should be as round in shape and as continuous as possible. This recommendation was based on the peninsula effect (the tendency for peninsulas to have relatively low diversities), but it also follows from edge effects (the potential negative effects of exotic species and disturbances that act along the edges of habitat fragments). The relative size of edges is minimized in circular reserves.

4. Managers and conservation biologists should give highest priority to those biotas with the highest degrees of endemicity and vulnerability.

Jared Diamond (1975) and John Terborgh (1974) presented nearly identical sets of rules for designing nature reserves (Figure 17.26).

Each of these recommendations seems perfectly sound, based on a wealth of empirical studies and biogeographic theory. Yet these papers—or more specifically, their geometric rules for designing reserves—set off one of the most contentious debates ever to visit the fields of biogeography and conservation biology. Unfortunately, they were taken too literally and often applied out of context. The most infamous case in point is what conservation biologists call the **SLOSS debate**. Basically, this debate focused on which was best, a *Single Large* reserve *Or Several Small* ones (i.e., the two alternatives in Figure 17.26A). Best for what—conserving a maximum variety of species or preserving populations of a particular endangered species for as long as possible? Conservation biologists and biogeographers alike locked horns in the SLOSS debate for over a decade (Abele and Connor 1979; Gilpin and Diamond 1980; Simberloff and Abele 1982; Soulé and Simberloff 1986). The outcome was a stalemate at best. Wilson and Willis's recommendations were sound, but neither they nor the geometric "rules" were meant to be generic prescriptions for all environmental problems.

More to the point, the optimal strategy for designing reserves and networks of reserves depends on the scale of the problem and the particulars of the species, communities, and landscapes in question. Where most species are ubiquitous, or where communities are strongly nested (i.e., when species-rich communities include all the species found in species-poor communities), one large reserve should indeed capture more species then several small reserves of the same total area (Patterson 1987). Yet the SLOSS debate considered just two of a great many possible combinations of reserves (see Lomolino 1994a). It turns out that for a variety of taxa and ecological systems, neither of the two SLOSS strategies is optimal (Figure 17.27). To include all focal species, either of the two SLOSS strategies would require much more area than the optimal combination of reserves (based on computer simulations of all possible combinations of reserves) and, in fact, tend to require more area than just a random combination of reserves. Yes, bigger is better, and many reserves are better than one or just a few, but on the question of whether one big reserve is better than many small ones of the same total size, biogeographic theory cannot provide a definitive answer.

Putting the SLOSS debate aside, biogeographers have developed a very useful set of tools for designing networks of nature reserves. Computer simulations are now readily available to reserve planners faced with the challenge of selecting reserves or prioritizing their efforts among many potential reserves. Given specific conservation objectives, these tools can be used to select

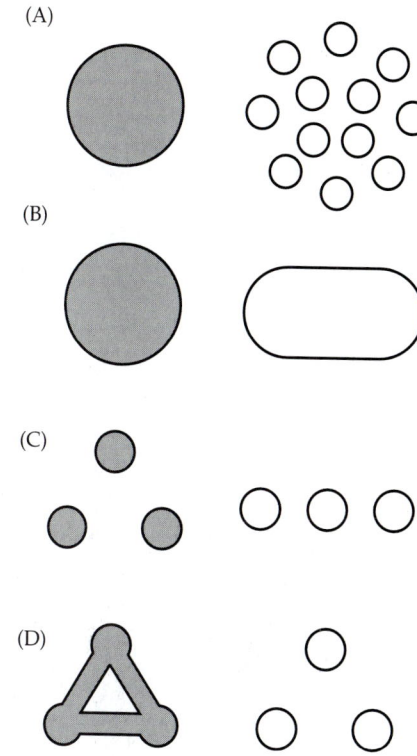

Figure 17.26 The application of biogeographic principles to the optimal design of nature reserves results in several recommendations. In each pair of figures, the configuration on the left is to be preferred over that on the right, even though both incorporate the same total area. (A) A continuous reserve is preferable to a fragmented one. (B) The ratio of area to edge should be maximized. (C) Distance between reserves should be minimized. (D) Dispersal corridors should be provided between isolated fragments. (From Wilson and Willis 1975.)

(A) Large mammals of Western U.S. Parks (S = 15)

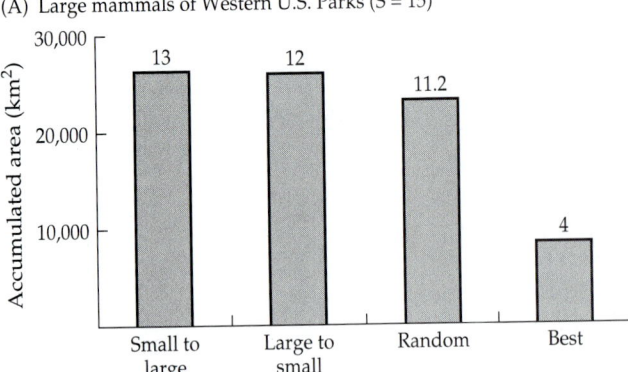

(B) Amphibians of Lake Michigan Islands (S = 10)

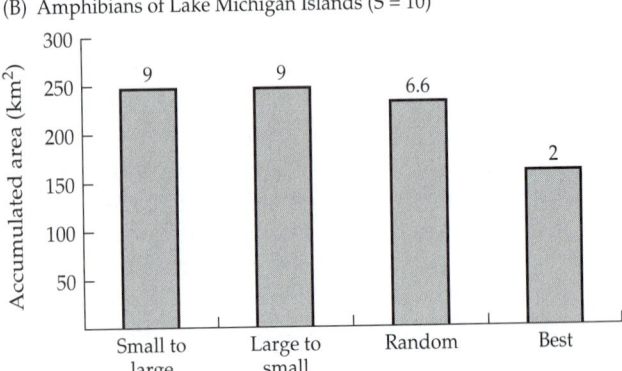

(C) Ants of Frisian Islands (S = 25)

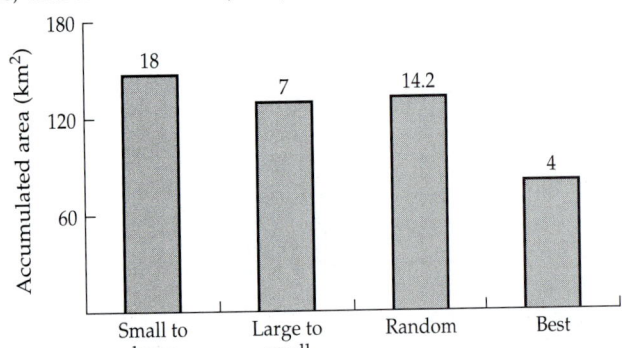

(D) Arctic and alpine plants of Adirondack Mountains (S = 18)

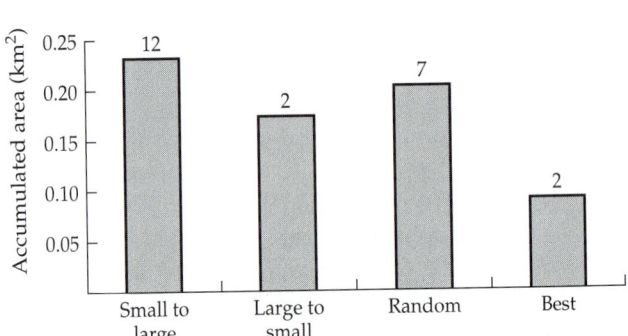

Figure 17.27 Given a large set of natural areas to be included in developing a reserve network, which ones should be selected? If the goal of reserve design is to include a maximal number of target species in a limited total area of protected lands, then the two accumulation strategies considered in the SLOSS debate (accumulating natural areas from the smallest to the largest, or largest to the smallest) perform poorly, requiring a much greater area and number of reserves (numbers above bars) to conserve the same total number of species (S, number of target species) than the optimal collection of reserves (BEST, determined by computer simulations). In fact, the SLOSS strategies perform no better than if reserves were selected at random. (From Lomolino 1994a.)

reserves most likely to achieve particular goals. The critical limiting factor here is information on the distributions of the focal species among prospective reserves. Given these data, reserve planners can select sets of reserves that maximize the number of species included in a limited number or total area of reserves.

Yet we know that the inclusion of a species in a reserve will not guarantee its preservation over a significant time period. Biotas are not static, but are strongly influenced by recurrent extinctions and immigrations. Much of the current emphasis on establishing dispersal corridors stems from this observation. Indeed, Wilson and Willis included the role of immigration and corridors in their design principles. In order to maintain diversity over time, landscapes should be designed to facilitate immigration among reserves. Isolation among reserves should be decreased (see Figure 17.26C), dispersal corridors should be established between reserves (see Figure 17.26D), and the intervening matrix of habitats should be managed to facilitate dispersal of focal species.

From these suggestions, a rich literature on landscape design and the potential importance of corridors has developed (e.g., see Soulé and Gilpin 1991; Nicholls and Margules 1991; Newmark 1993). It is not our purpose here to review all of this important information, but we do caution that along with their obvious potential benefits, corridors also may have some negative effects (see Simberloff et al. 1992; Hobbs 1992). For example, corridors may facilitate the spread of exotic species and disease, or, by increasing dispersal and gene flow among populations, they may promote swamping of local adaptation. In a very real sense, corridors may violate Wilson and Willis's second recommendation by putting all our eggs in one basket. Fortunately, conservation biologists are well aware of these potential problems and are actively evaluating their relevance. Just as important, a variety of studies are assessing the extent to which species utilize corridors and the various geographic and ecological factors influencing dispersal along corridors (e.g., corridor width, length, and quality and characteristics of the surrounding matrix).

Moreover, a number of conservation biologists have made the sobering observation that if we had just reserves and corridors, we would lose the majority of native species. Most individuals of a great majority of species live outside reserves, in the intervening matrix of habitats. If the habitat matrix is degraded or destroyed, all of the above consequences of habitat loss and fragmentation will ensue.

Predicting the Effects of Global Climatic Change

Humanity's influence on the biosphere goes even beyond our ability to transform landscapes. It is now clear that we can cause climatic changes at magnitudes rivaling those associated with the glacial cycles of the Pleistocene. If current trends continue, we will be heating up an interglacial period by perhaps as much as 3° C over the next century. Most ecologists agree that such a change, if it does occur, will dramatically influence the distributions and diversity of life. Some communities, including desert mammals of the American Southwest and the tropical rain forests of Amazonia, may already be signaling the early effects of global warming. But surely we can be more specific. Which species and which ecosystems will be most affected? How much, and in what ways, will the geographic ranges of species expand or contract, and which species will go extinct?

Biogeographers, ecophysiologists, landscape ecologists, and many other scientists are presently tackling this problem (e.g., see Gates 1993; Abrahamson 1989; Karieva et al. 1993; Peters and Lovejoy 1992). Here we illustrate just one of many approaches to predicting the ecological effects of global climate change. This particular approach is based on the patterns of biotic relaxation described in Chapter 13. Using the fairly extensive data on the distributions of small boreal mammals inhabiting montane forests of the Great Basin (see Figure 13.18), McDonald and Brown (1992) were able to predict how many and which species should be lost from these forests if global climates warm by 3° C. Such a change would cause an upward shift in the lower elevational limits of woodlands and forests by approximately 500 m. Because there is less area at the higher elevations of a mountain, such elevational shifts would reduce the habitat area available to boreal mammals (Figure 17.28). McDonald and Brown used topographic maps and planimetry to predict the future area of woodlands and forests in each mountain range. Then, using the species-area relationships of extant communities, they were able to calculate how many species should be lost from each mountain range (Figure 17.29).

Their results are sobering. They predict a total of 45 extirpations across the nineteen mountain ranges, with four ranges losing at least half their species.

(A)

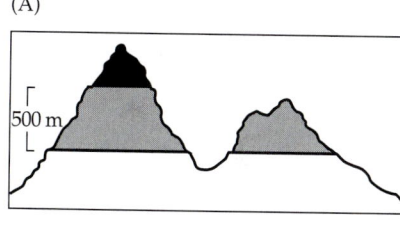

(B)

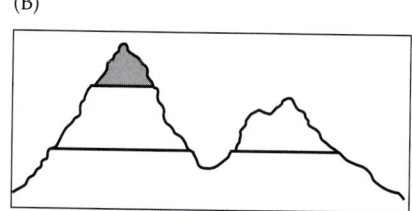

Figure 17.28 The approximate elevational boundaries of the vegetation types on the isolated mountain ranges of the Great Basin (A) today and (B) in the future after postulated climatic warming of approximately 3° C. Unshaded = desert shrub; lightly shaded = piñon-juniper woodland; darkly shaded = mixed coniferous forest. An elevational shift of 500 m would decrease the area of woodland on all mountain ranges in the region and eliminate coniferous forest from some of them. (From McDonald and Brown 1992.)

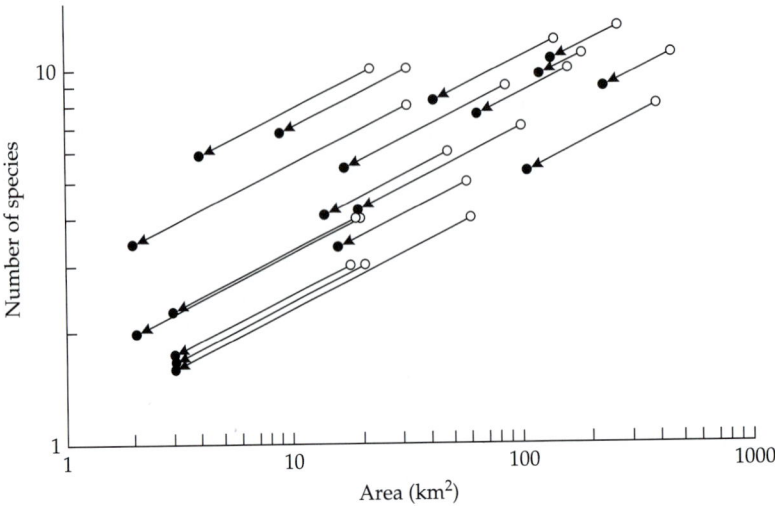

Figure 17.29 The species-area relationship can be used to predict changes in boreal mammal species richness among the isolated mountain ranges of the Great Basin as a result of climatic warming. Numbers of species were plotted as a function of the area above 2280 m elevation. The arrows show the changes in area and numbers of species predicted to be caused by climatic warming: the open circle at the base of each arrow indicates the present number of species in each mountain range, and the solid circle at the point of the arrow indicates the number predicted to remain after a 3° C increase in average temperature. (From McDonald and Brown 1992.)

Assuming that the extinctions would be selective, McDonald and Brown were able to go one step further and predict not just how many, but which species would be lost from each mountain range. Selective extinction in these communities is thought to have produced the nested structure illustrated in Table 17.7. Given this structure, extirpations should be most likely for the populations along the diagonal running from the lower left to the upper right of the table—that is, those marked with an E. Of the fourteen species considered in this study, only two were predicted to survive on all mountain ranges, whereas three were predicted to be lost from the entire system.

As McDonald and Brown (see Brown 1995) acknowledged, their predictions should be viewed with caution. In order to make them, they had to make several assumptions about the magnitude of regional temperature change, elevation shifts in boreal habitats, and the responses of small mammals to these changes. They also assumed that the effects of increased fragmentation and isolation on these habitats and interactions among species were unimportant. Thus, while they provided a sobering thought experiment, McDonald and Brown's approach was probably too simplistic to reliably predict the biotic effects of global warming.

Biodiversity Surveys

Although we have presented just two examples of applied biogeography, it should be clear that biogeographic theory can play an important role in conserving biological diversity. It can help us identify the key factors and processes to study, which data we need, and how to use it. Yet, if we have learned anything from the above case studies, it is that theory alone won't solve any of these problems. Without fundamental information on the ecology, systematics, and biogeography of species, even the most sophisticated theory

Table 17.7

Distribution of fourteen small boreal mammal species among nineteen isolated mountain ranges in the Great Basin at present and after predicted extinctions due to the effects of global warming

Species	Mountain Ranges																			Number of ranges occupied	
	4	8	5	1	15	19	16	13	3	18	12	9	11	7	6	14	10	2	17	Present	Predicted
Eutamias umbrinus	X	X	X	X	X	X	X	X	X	X	X	X	X	X	X	X	X			17	17
Neotoma cinerea	X	X	X	X	X	X	X	X	X	X	X	X		X		X	X	X	X	17	17
Eutamias dorsalis	X		X	X	X	X	X	X	X	X	X	X	E	E		E	E	X	X	17	14
Spermophilus lateralis	X	X	X	X	X		X	X	E		X	X	X	E		E			E	14	10
Microtus longicaudus	X	X	X	X	X	X	X	X	E	X	E		E		X					13	10
Sylivilagus nuttalii	X	X	X	X	X		E	E	E		E	E	E					E		12	5
Marmota flaviventris	X	X	X	X	X	X	E	E	E	X		E								11	7
Sorex vagrans	X	X	X	X	X	X	E	E		X	E									10	7
Sorex palustris	X	X	X	X	E	X				E						E				8	5
Mustella erminea	X		E	E	E	E				E										6	1
Ochotona princeps	X	E	E	E					E											5	1
Zapus princeps	E	E	E			E										E				5	1
Spermophilus beldingi	E	E																		2	0
Leups townsendii		E				E														2	0
Present number of species	13	12	11	11	10	10	9	8	8	8	7	6	5	4	4	4	3	3	3		
Predicted number of species	11	8	9	10	8	7	5	5	3	6	4	4	3	2	2	2	2	2	2		

Source: From McDonald and Brown 1992.

Note: X = species now present and expected to persist; E = species now present and predicted to go extinct.

will be rendered useless. Optimal strategies for selecting nature reserves can be used only when we know which species occur among the prospective reserves. Similarly, McDonald and Brown's application of biogeographic theory would have been impossible without a good record of species distributions on the mountain ranges of the Great Basin. In short, while we should continue to look to theoretical advances in all relevant disciplines, a critical factor for conserving biological diversity is biogeographic data—the distributions of the focal species. Reflect on the Linnaean shortfall. We have described just a small fraction of what is out there, and field biologists continue to discover new species, or rediscover those long thought to be extinct (see Patterson 1994; Sunquist et al. 1994; Ferrari and Queiroz 1994; Pine 1994; Morell 1996; Finlay et al. 1996). But we need more than accurate lists of which species occur *somewhere* on earth. We need global biodiversity surveys to describe species distributions and identify hot spots of diversity.

Some systematists and conservation biologists suggest that a global biodiversity survey could be completed over the course of 50 years (Raven and Wilson 1992). While this may be a bit optimistic, efforts toward this goal would certainly go a long way toward reducing the great gaps in our knowledge. We are encouraged by the increasing number of biological surveys that are being undertaken in the meantime. Many of these, such as Conservation International's Rapid Assessment Program (RAP), have targeted their limited resources at the sites in greatest need of surveys and conservation: the tropical hot spots of diversity. RAP is essentially a SWAT team of some of the world's best field biologists, who can, within the span of weeks, conduct first-cut surveys to document some of the biological diversity of sites facing imme-

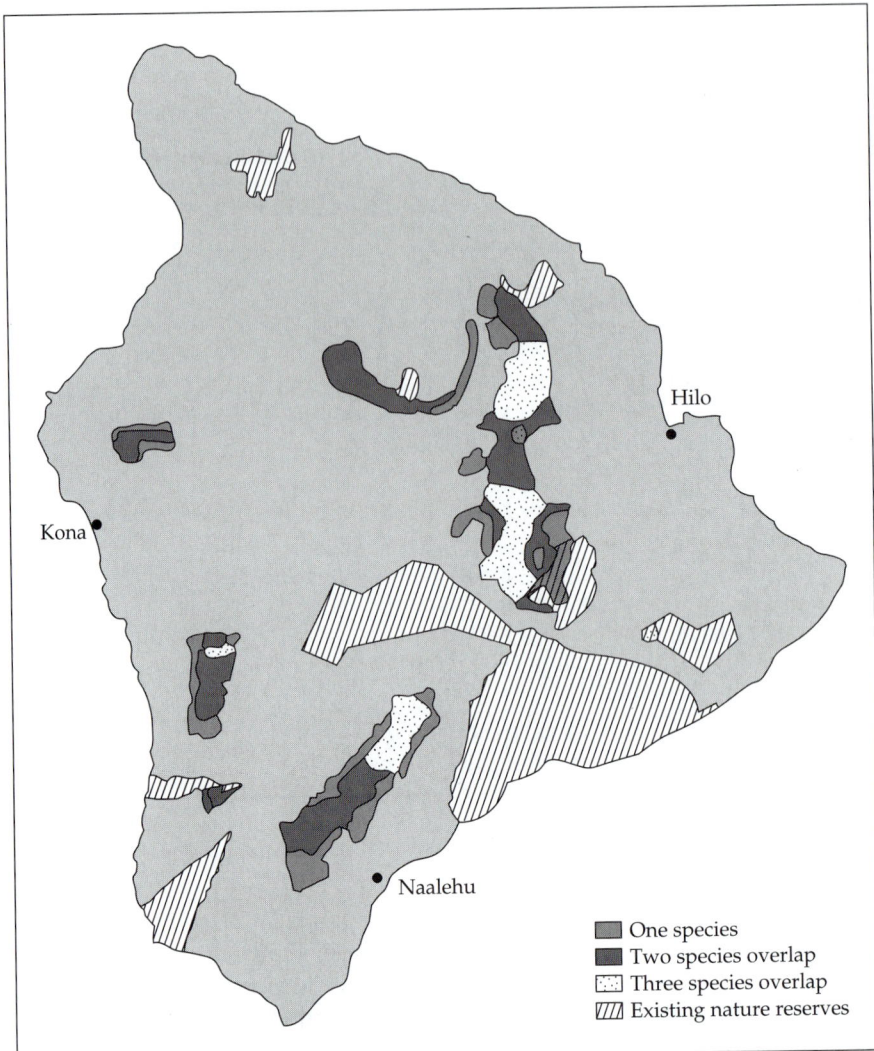

One species

Two species overlap

Three species overlap

Existing nature reserves

Figure 17.30   The U.S. Gap Analysis Program's approach for identifying sites in need of protection is illustrated here in its first application: developing conservation strategies for the avifauna of the island of Hawaii. Maps of species distributions are overlaid on one another to identify hot spots (i.e., sites with a significant number of priority species). Hot spots that are not located within existing nature reserves (hatched areas) are identified as "gaps" in the protection of biodiversity, and are thus targeted for future nature reserves. (From Scott et al. 1993.)

diate danger. RAP selects sites based on four criteria: high total diversity, high endemicity, uniqueness of the ecosystem, and high risk of extinctions. Other efforts, such as the U.S. Gap Analysis Program, utilize remote sensing and GIS (Geographic Information Systems) technology along with extensive existing information on species distributions at regional scales to map hot spots of diversity and locations inhabited by rare and endangered species (Scott et al. 1993). Such maps or GIS images can then be compared with maps depicting the locations of national parks and other protected lands to identify gaps in the protection of biodiversity (i.e., hot spots not falling into a protected area; Figure 17.30). These gaps can then be given top priority by environmental planners and conservation biologists.

Many countries have instituted biodiversity surveys and programs similar to those described above. As Daniel Simberloff (1988) once urged, it is now up to us to employ our waiting army of "unemployed and underemployed" ecologists to go out and collect the data. Perhaps the great wealth of biogeographic patterns we have discussed in this book, many of them documented for just a handful of terrestrial taxa, will prove to be equally applicable to the poorly studied groups—the undiscovered majority. We strongly doubt it: biological diversity cannot be so simple. Indeed, we hope that there will be many exceptions, each one telling us a bit more about the forces structuring the natural world and the great diversity of organisms that inhabit it.

CHAPTER *18*

Applied Biogeography: Single Species

The Biogeography of Humanity

The biogeography of humanity is a subject that must intrigue nearly all of us, scientist and layperson alike. Where did our species originate, and how did we come to occupy and dominate nearly all points of land? These questions are as challenging as they are captivating in that they require a synthesis of information from many disciplines, ranging from genetics and evolutionary biology to comparative and functional anatomy, systematics, physical and cultural anthropology, archaeology, paleoecology, and paleoclimatology.

The geographic history of humanity provides biogeographers with an exceptional data base with which to test the generality of many biogeographic principles. Has our own species exhibited patterns of dispersal and modification similar to those of most other animals? Were the diversification and range expansion of our human ancestors strongly influenced by tectonic and climatic events? Were our colonizations of remote continental regions and islands the result of unique immigration events involving a limited number of founders, or did they depend on recurrent waves of immigration consistent with MacArthur and Wilson's equilibrium theory? Once they colonized remote oceanic islands, were human populations then locked into a taxon cycle, destined to undergo changes (ecological and otherwise) that would eventually lead to their extinction, perhaps at the hands of a new wave of human colonists?

While the biogeographic dynamics of our own species is a compelling subject in its own right, it is also one of profound relevance to the prehistoric and historical waves of global extinctions. We may never be able to tease apart all the potential causal factors, but we are certain that the great majority of these extinctions have been anthropogenic, or at least hastened by the activities of anatomically modern man, *Homo sapiens sapiens*. Here we present an overview of the origin and early expansion of our species (see the more extensive discussions in Howells 1973; Terrell 1986; Fagan 1990; Irwin 1992; Bell and Walker 1992; Clark 1992; Gamble 1994). We acknowledge that our account is far from complete, but we have tried to summarize the principal features and events of the geographic history of humanity, especially those most relevant to the late Holocene surge in extinction rates.

Human Origins and Colonization of the Old World

The origin of Primates, the order to which we and all other hominids belong, dates at least as far back as the early Tertiary, some 65 million years B.P. Over the next 20 or 30 million years, the primates diversified and expanded their ranges to include much of both the New and Old Worlds. These early prosimian primates comprised a collection of relatively primitive forms related to extant species, including galagos of tropical Africa, lemurs of Madagascar, lorises of Southeast Asia, and tarsiers of the Philippines and Indonesia. By 35 million years B.P. the prosimians gave rise to a second, more derived group of primates, the **anthropoids**, which includes the platyrrhines (New World monkeys) and the catarrhines (Old World monkeys and apes). Humans are descended from a subgroup of the catarrhines that is often referred to as the great apes, or **hominoids**, and includes gorillas, orangutans, chimpanzees, and **hominids**—australopithecines and members of the genus *Homo* (including *H. habilis*, *H. ergaster*, *H. erectus*, and *H. sapiens*).

It appears that the hominid lineage diverged from gorillas and chimpanzees between 5 and 7 million years B.P. in Africa. While these dates are, and will continue to be, the subject of much speculation and refinement, it does seem certain that Africa was the site of human origins (Guilaine 1991). Fossils of the oldest known hominid, *Australopithecus afarensis*, found at sites in Ethiopia and Tanzania, date back to about 3.7 million years B.P. Over the next 1 to 2 million years, it appears that the evolutionary and cultural development of our ancestors was confined to savannas of eastern and southern Africa. The first representative of the genus *Homo* was *H. habilis*, which means "handy person." This was apparently the first hominid to make its own tools—relatively simple flaked stones, which it used to prepare animal carcasses and cut vegetation. The tools were rudimentary and unspecialized, but they mark a fundamental advance in our ability to exploit and modify our environment.

By approximately 1.4 million years B.P., a new and by most standards more "advanced" hominid, *Homo erectus*, had evolved. In comparison to *H. habilis*, *H. erectus* was much larger (5 to 6 ft), larger-brained, more dexterous, and more advanced in its ability to manufacture tools and presumably to communicate, at least in a rudimentary fashion, with conspecifics. *H. erectus* was able to develop effective strategies for hunting big game, and may have been able to capture and use fire. This latter advance cannot be overemphasized. With fire, *H. erectus* could greatly expand its niche and modify its environment to the extent that it could influence the geographic ranges of other species (Fagan 1990). Fire could be used for protection against predators, for hunting game, and for preparing smaller prey such as rodents and insects. Fire could also be used to render toxins, common to many vegetable foods, harmless. At first, *H. erectus* just captured natural fires, but eventually developed the ability to create as well as preserve it. With this important tool, in combination with those fashioned from rock, bone, and wood, *H. erectus* could expand its range out of the savannas and woodlands and into a variety of other habitats. In fact, because of their relatively large body size and more developed social systems, populations of *H. erectus* may have needed to expand their home ranges and colonize new habitats to meet their heightened energy demands.

By 1.5 million years B.P., and perhaps as early as 2 million years B.P., *H. erectus*, the apparent predecessor of *H. sapiens*, may have been capable of expanding its geographic range out of Africa and into Asia and Europe, but these migrations did not occur for many more hundreds of thousands of years. Other catarrhine primate groups (including the ancestors of the Asian great apes) had migrated out of Africa when the African and Arabian plates collided some 17 million years ago. The convergence of these plates closed the Tethys

Sea and provided a landbridge to Eurasia across the Arabian Peninsula. The range expansions of hominids, however, did not immediately follow the formation of a terrestrial dispersal route out of Africa. Instead, these emigrations seem to have been strongly influenced by shifts in climate and habitat associated with the glacial cycles of the Pleistocene. With each cycle, the tropical savanna habitats of hominids, and many other mammals for that matter, split, shifted, and rejoined, setting the stage for the vicariant processes that often drive evolutionary and biogeographic change. Each wave of hominid expansion from Africa, from *H. habilis* to *H. ergaster*, *H. erectus*, and finally *H. sapiens*, occurred in at least two stages and probably required repeated migrations of many populations.

First, hominid populations long adapted to mesic savannas and woodlands had to penetrate and eventually cross the great xeric barrier formed by the Sahara Desert. These expansions into the Sahara region most likely occurred during the relatively brief, but wet, interglacial periods. The second stage of hominid migration out of Africa was probably triggered by the return of glacial conditions that again brought a period of great aridity. Populations of *H. erectus* migrated out of the Sahara in all directions, retreating to the savannas and grassland to the south, but also emigrating northward and then eastward across the Sinai and into western Asia (Figure 18.1). Just as it had served as a landbridge earlier for the catarrhine primates, the Arabian Peninsula now served as a migration route for hominids into Asia and Europe. Once each wave of hominids passed beyond this barrier, they rapidly expanded their geographic range to include most areas of Eurasia below 45° N latitude.

H. erectus first colonized the Arabian Peninsula around 800,000 B.P., then spread to establish populations from the eastern edge of Asia to western Europe by 500,000, and perhaps as early as 600,000 B.P. (Figure 18.2; see Larick and Ciochon 1996). Yet with all its skills and advances, and despite its relatively extensive geographic range, *H. erectus* was soon to be overshadowed and replaced by another wave of hominid migrations, again originating in Africa. A few anthropologists have hypothesized that *H. sapiens* resulted from the independent, parallel evolution of isolated subspecies of *H. erectus* in Europe, Africa, and Asia. The general consensus, however, holds that *H. sapiens* evolved in the savannas of Africa prior to 300,000 B.P. as a descendant of the African form of *H. erectus*. Then, following in the footsteps of its primate

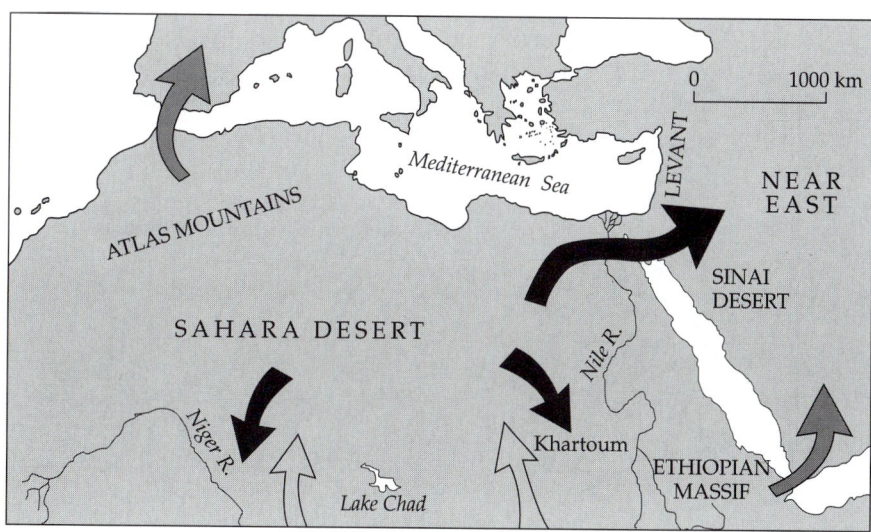

Figure 18.1 Stages in the colonization of the Old World by hominids. This two-stage migration was strongly associated with glacial cycles and was probably repeated during recurrent waves of hominid emigrations out of Africa during the Pleistocene. Open arrows = penetration of the Sahara Desert from the savanna "homeland" during interglacial periods; solid arrows = emigration from the Sahara during relatively arid glacial maxima; shaded arrows = possible migration routes during periods of low sea level during glacial maxima. (After Fagan 1990.)

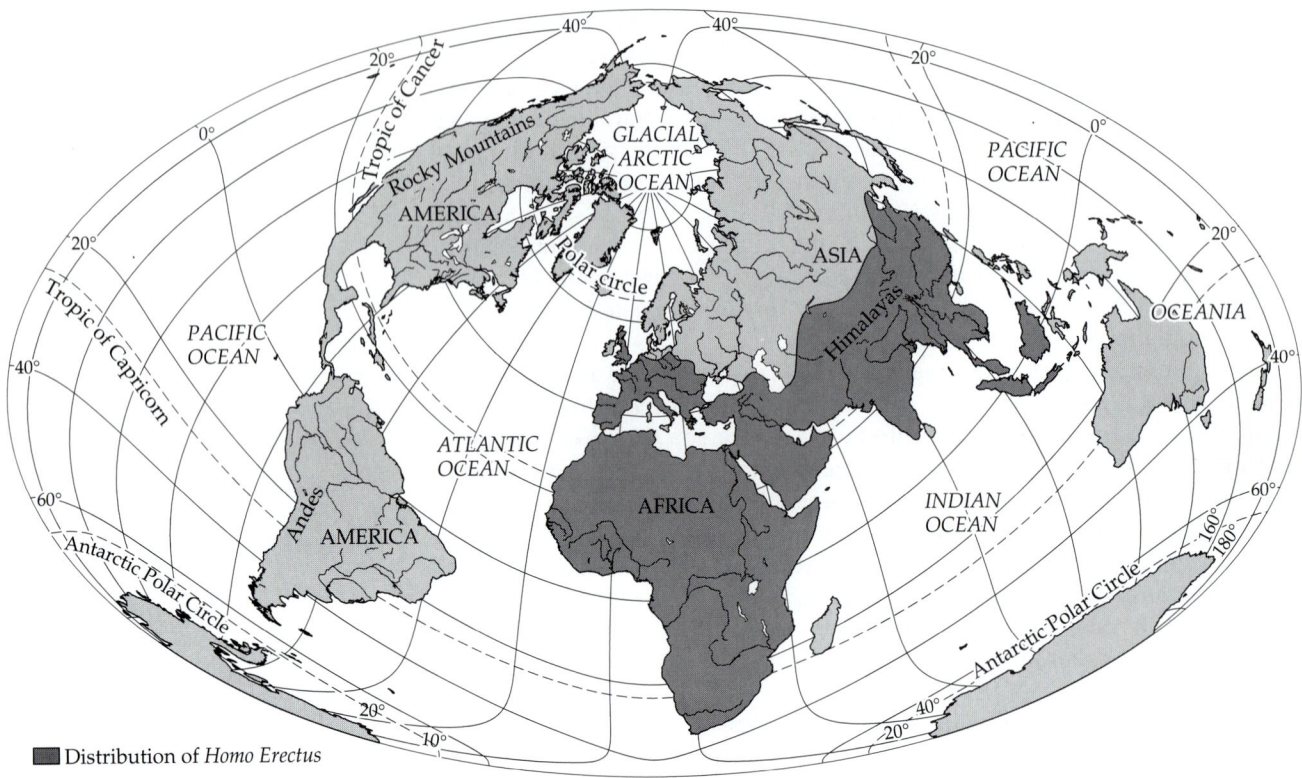

Distribution of *Homo Erectus*

Figure 18.2 Estimated geographic range of *Homo erectus* between 500,000 and 600,000 B.P. (After Guilaine 1991.)

and hominid ancestors, *H. sapiens* repeated the two-stage journey into the Sahara and then across the Sinai to colonize Eurasia. By 100,000 B.P., *H. sapiens* had expanded its range out of Africa and then, either by assimilation or by direct interaction, replaced *H. erectus* (see Gibbons 1996).

In comparison to *H. erectus*, *H. sapiens* was empowered by a larger brain, more sophisticated language, and superior abilities to plan, organize, and coordinate group activities (Figure 18.3). It manufactured diverse tool kits, each with a variety of implements specialized for different local environments, from forests and open savannas to coastal habitats. It continued to develop the

Figure 18.3 Relative brain size (as measured by encephalization quotient, EQ) of hominoids has steadily increased over the past 3.5 million years. (From Gamble 1994.)

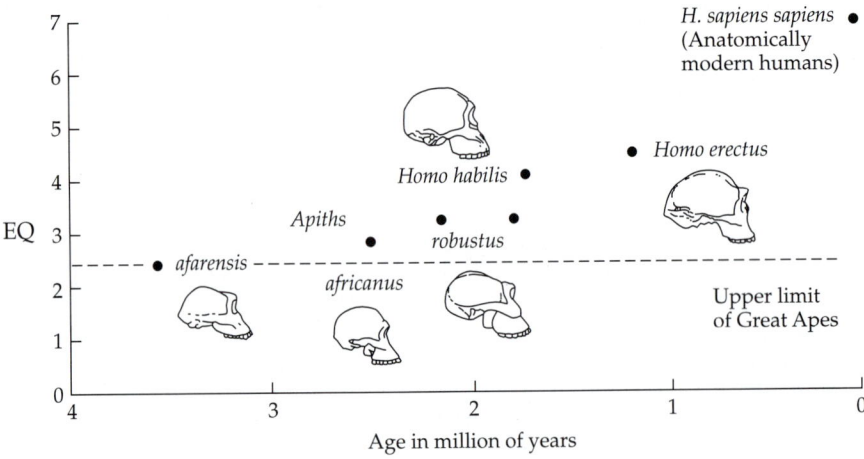

ability to manufacture and use fire for protection, warmth, hunting, and preparing foods. It also built shelters and used caves for protection against the elements. In short, *H. sapiens* possessed an unrivaled ability to adapt to, modify, and eventually dominate a variety of environments. Thus, once *H. sapiens* migrated out of Africa, its range expansion was extremely rapid, overtaking and eventually extending far beyond the ultimate range of *H. erectus*. The probable pathways and timing of the aboriginal migrations of *H. sapiens* across the globe are summarized in Figure 18.4.

Conquering the Cold: Expansion to the New World

The global expansion of *H. sapiens* beyond the range of its hominid predecessor posed two principal challenges: conquering the cold and navigating the oceans. In many ways, *H. sapiens* was preadapted for meeting these challenges, or at least for rapidly developing the ability to master them. With its ability to create fire and to fashion tools for new purposes, along with its relatively advanced intelligence and ability to communicate, *H. sapiens* achieved an unrivaled ability to adapt to new environmental challenges. Yet its expansion into Europe and the cooler climates of the higher latitudes was a relatively slow process, perhaps occurring over some 40,000 to 50,000 years after its first appearance in the Near East.

Again, glacial cycles seem to have had a strong influence on this phase of range expansion by *H. sapiens*. Full glacial conditions persisted throughout Europe for most of the period between 75,000 and 40,000 B.P., except for a relatively brief period of warming around 50,000 B.P. As Fagan (1990) noted, it

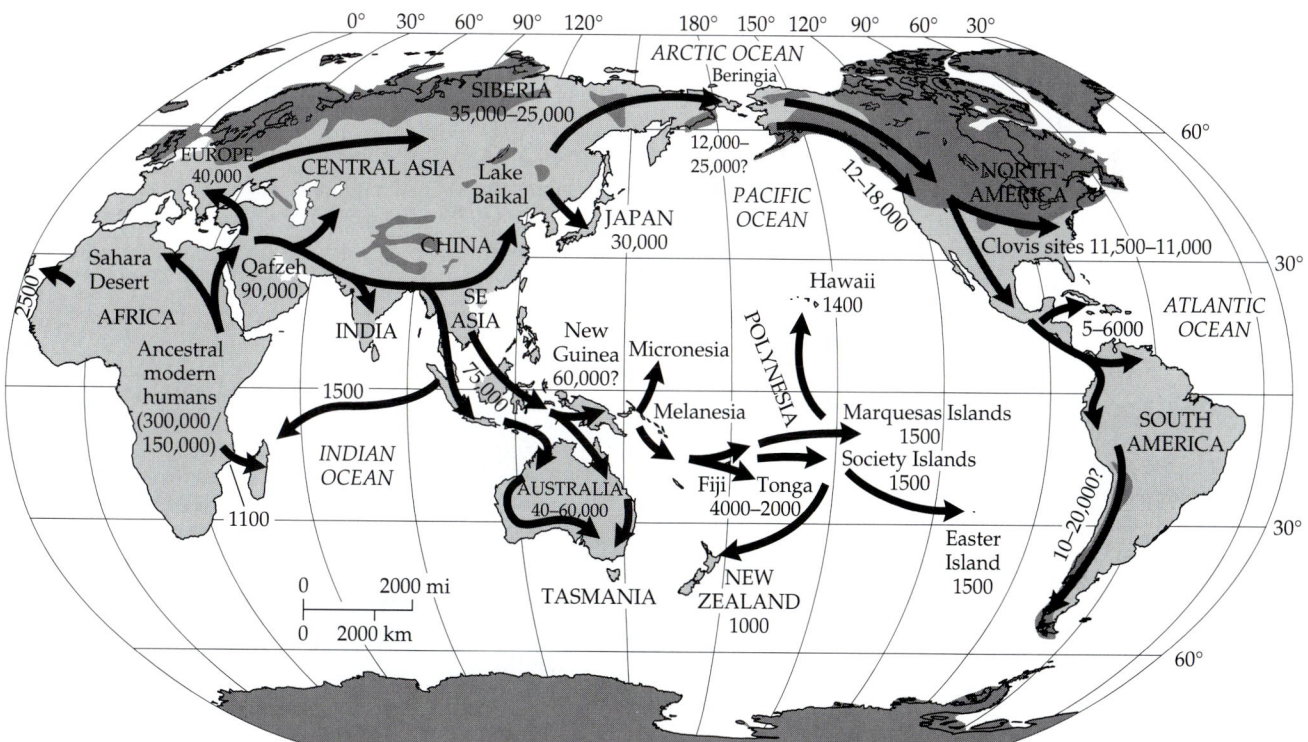

Figure 18.4 Colonization of the globe by anatomically modern humans (*Homo sapiens sapiens*). The exact dates of these migrations, given here in years before present, are the subject of continuing debate and frequent revision. Shading indicates extent of glaciers during last glacial maximum (~18,000 years B.P.). (After Gamble 1994.)

may not be just a coincidence that modern humans first appeared in Europe during this brief interglacial. Tools, shelters, and social organization associated with cold adaptation were developed rapidly after 40,000 B.P., allowing populations of *H. sapiens* to extend their ranges far into the higher latitudes of Eurasia and then into North America. During their migrations, they developed the projectile weapons and group strategies necessary for taking large game. They were then able to use the bones of mammoths and other large mammals to build shelters and fashion tools. Their cultural and ecological adjustments to the cold may have been augmented by morphological and physiological adjustments. Now populations of *H. sapiens* could migrate along with the game and other species that tracked the retreat of the glaciers and shrub-steppe communities with each glacial cycle.

By 40,000 to 50,000 B.P., populations of *H. sapiens* had expanded northward to the edge of the glaciers in Europe and Siberia. Their colonization of northeastern Siberia between 30,000 and 40,000 B.P. set the stage for the great leap into the Western Hemisphere. Again, this major event in human colonization was strongly influenced by climatic cycles and may have required two fundamental steps: colonization of Alaska, and southward dispersal through the remainder of North America and into South America (Figure 18.5; see Roosevelt et al. 1996; Batt and Pollard 1996; Gibbons 1996). Between 25,000 and 18,000 B.P. glacial conditions prevailed, and with the lowering of sea levels by roughly 100 m, Beringia formed a 1500 km wide landbridge between the Palearctic and Nearctic regions. During this period, Beringian environments probably varied substantially, ranging from a relatively productive cold steppe that supported a diversity of large mammals to a polar desert (see Chapter 7 and Hoffecker et al. 1993). Recent evidence suggests that *H. sapiens* may have migrated across Beringia along with their game species to colonize Alaska before 25,000 B.P. Following this first stage, subsequent expansion through the remainder of the Americas was probably blocked by the glaciers that persisted until about 16,000 B.P. It is possible that human populations migrated southward along the ice-free coasts or along the chain of nunataks between the Laurentian and Cordilleran glaciers prior to 16,000 B.P.. Colonization of the New World prior to 15,000 B.P. is still controversial, however, and a number of paleoecologists maintain that colonization occurred only after the last glaciation (see Batt and Pollard 1996).

Figure 18.5 The colonization of North America by anatomically modern humans may have occurred in two stages and was closely associated with glacial cycles. During glacial maxima, lowered sea levels exposed Beringia, which served as a corridor for the dispersal of many terrestrial animals, including humans. Southward migration from Alaska, however, may have been blocked by continuous ice sheets that formed an east-west barrier across northern North America until glacial recession (after 16,000 B.P.). Recent claims of earlier colonization of North America by humans (before 25,000 B.P.) imply that intermittent corridors may have existed between the Cordilleran and Laurentian glaciers, or along a relatively thin strip of unglaciated land along the Pacific coast. (After Guilaine 1991.)

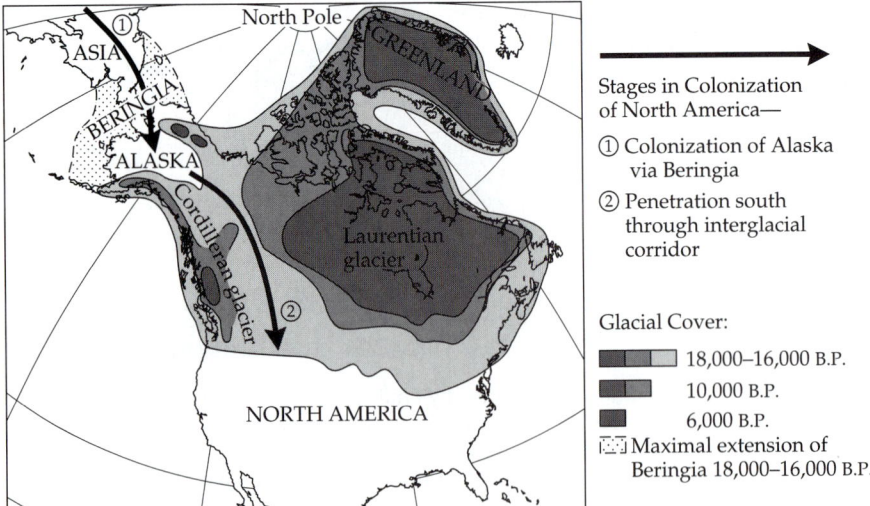

Either way, following their penetration south of the ice sheets, human populations expanded southward through North America and then South America with perhaps unprecedented rapidity (Mosiman and Martin 1975). Barring some taphonomic artifact, it appears that the dates of the earliest known sites of ecologically significant human occupation across the extent of the Americas (from North America to Tierra del Fuego in South America) span just a few thousand years. In contrast to the continents of the Eastern Hemisphere, where most barriers (e.g., the Sahara, the Mediterranean, the Alps) run east to west, these features tend to run north to south in the Western Hemisphere, and thus may have actually facilitated human migrations southward from Alaska to the remainder of the Nearctic and Neotropical regions. Certainly, future paleontological studies will revise many of these dates, but the peopling of the Americas may continue to represent one of the most rapid and dramatic episodes of range expansion evidenced for any animal taxon. This rapid spread of ecologically significant hunting societies between 12,000 and 10,000 B.P. seems to be one of the prime causes of coincident megafaunal extinctions in North America (see Figure 7.30).

Conquering the Oceans: The Island Biogeography of Humanity

The last stage of human colonization represents one of the most important with respect to losses in biodiversity. Again, patterns in the colonization of isolated archipelagoes by *H. sapiens* provide an excellent opportunity to test the generality of many principles and hypotheses of island biogeographic theory. Moreover, because the prehistoric and historical waves of extinctions occurred most strongly on islands, this stage of human colonization is likely to shed some light on coincident losses in biodiversity.

Our key focus here is on the colonization of Indonesia, Australia, and the islands of the Pacific, all of which lie between 40° N and S latitudes. Yet even in these now tropical and subtropical regions, human colonization was strongly influenced by the glacial cycles of the Pleistocene. Populations of anatomically modern humans were established along the coasts of East Asia by about 80,000 to 100,000 B.P. Their migrations across much of what is now the islands of Indonesia paradoxically required only modest seafaring abilities. In fact, most of this area may have been colonized across ancient landbridges. During glacial maxima of the Pleistocene, this area comprised four main regions: Sunda, Wallacea, Sahul, and Oceania (Figure 18.6). As a result of the lowering of sea levels, the Asian mainland extended as a continuous landmass from the Malay Peninsula northeast to the Philippines and eastward to Bali and Borneo—that is, to the western border of Wallacea. Thus the colonization of most of what is present-day Indonesia probably took place by walking, and it probably occurred soon after *H. sapiens* first colonized Southeast Asia, perhaps by 75,000 B.P. (Fagan 1990). The jump from Sunda to Sahul (the united landmasses of New Guinea, Australia, and Tasmania) required crossing the Wallacean archipelago, with a maximum interisland distance of perhaps just 60 km. The crossing to New Guinea may have occurred as early as 70,000 B.P., and from there humans spread rapidly to occupy most of Australia and Tasmania by 40,000 to 60,000 B.P.

The final step in the peopling of the Pacific islands was the most demanding because, even during periods of lowered sea levels, it required true navigation and seafaring skills. The archipelagoes closest to Sahul were colonized first, including the Bismarck Archipelago (colonized around 32,000 B.P.) and the Solomon Islands (20,000 to 28,000 B.P.; see Fagan 1990). Colonization of the more isolated archipelagoes, however, took thousands of years longer and required the discovery of new islands hundreds to thousands of kilometers

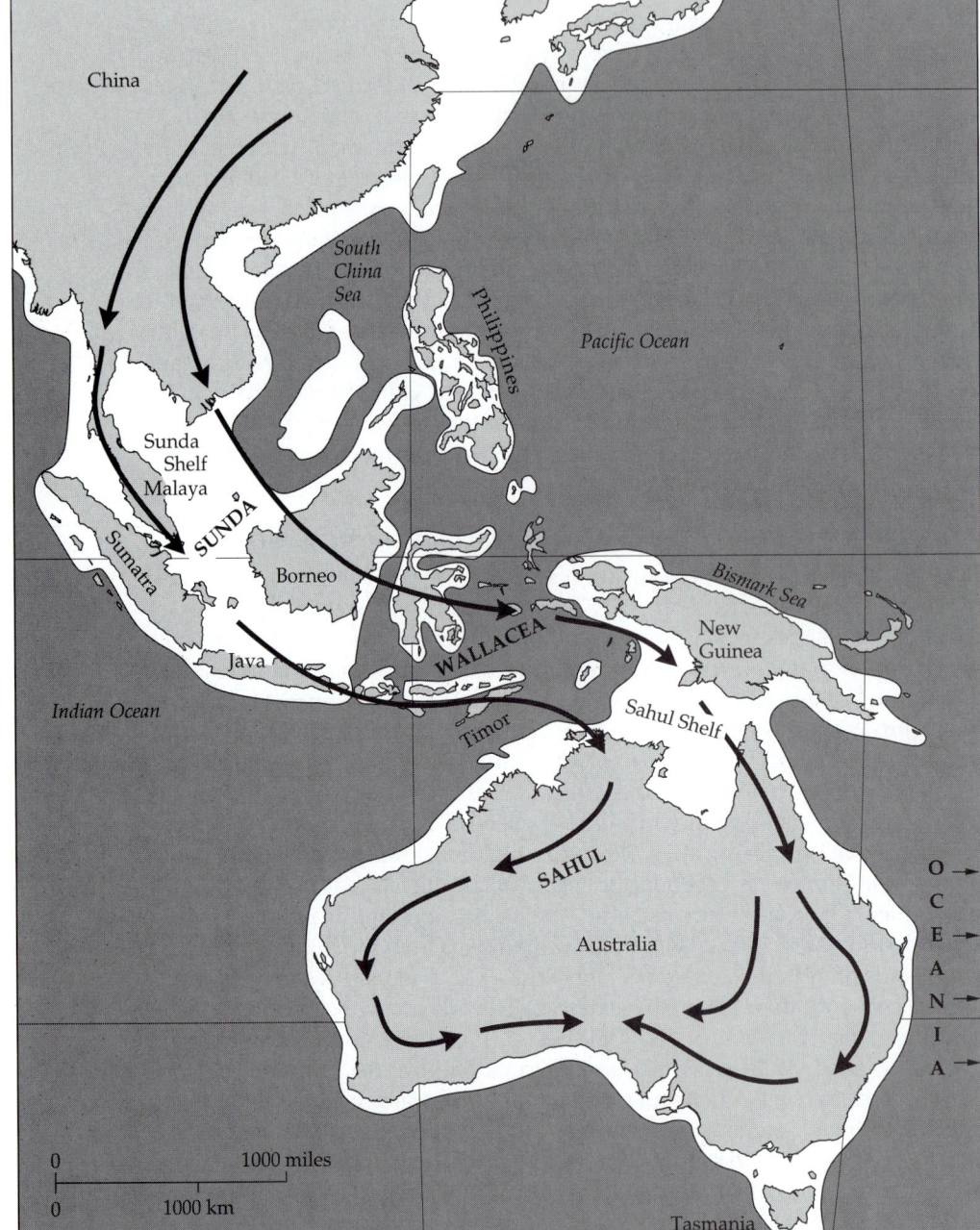

Figure 18.6   The lowering of sea levels during glacial maxima of the Pleistocene caused the exposure of continental shelves and the formation of dispersal routes across four regions of the eastern Pacific: Sunda, Wallacea, Sahul, and Oceania. (White areas = land exposed during glacial maxima; dark shading = deep water (> 200 m); possible dispersal routes are indicated by arrows). (After Fagan 1990; Guilaine 1991.)

from colonized sites. (Interestingly, flocks of native colonial seabirds may have provided important clues for locating some of the most remote oceanic islands.) These migrations by early seafaring people were indeed explorations, purposeful movements against the prevailing winds and currents, requiring sophisticated navigational skills and the development and construction of vessels specially designed for long voyages (Figure 18.7). These prehistoric Pacific voyagers carried with them essential resources, including domesticated plants and animals, along with the men and women required to establish viable populations. But the cultural shift from hunter-gatherers to sophisticated seafarers took millennia to evolve. Thus, Fiji wasn't colonized until about 4000 B.P., over

(A)

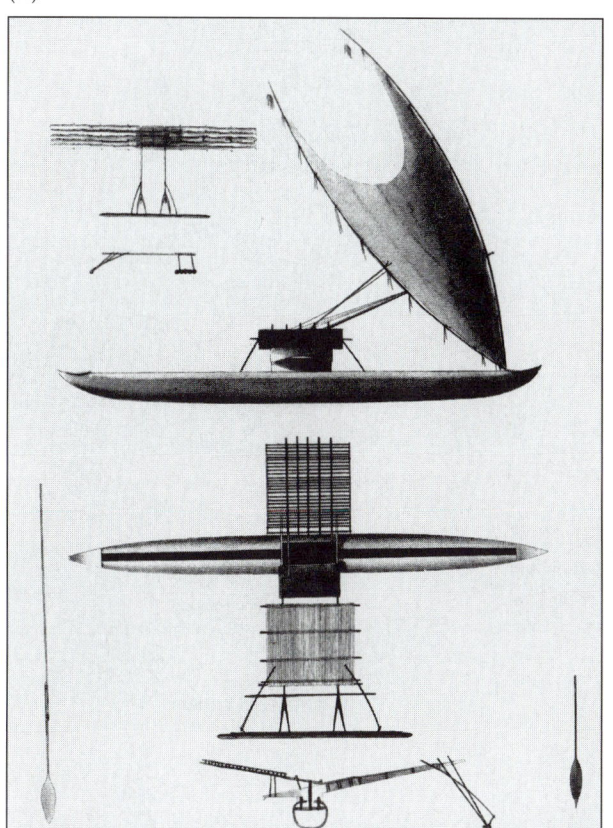

(B)

(C)

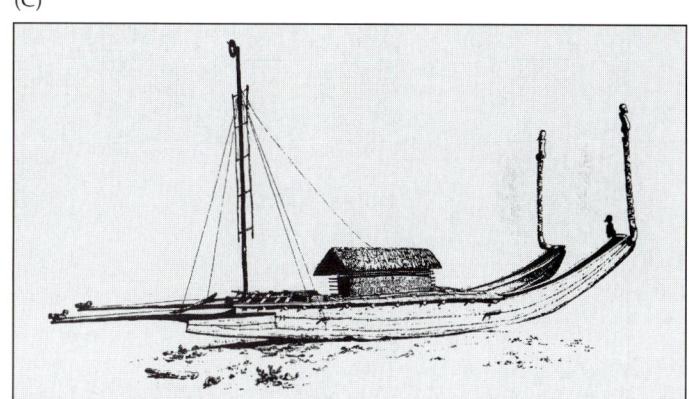

Figure 18.7 Colonization of the Pacific realm by humans required the development of sophisticated, seaworthy sailing vessels many centuries before the Europeans began their "age of exploration." Shown here are (A) detailed plans for a voyaging canoe, (B) an artist's rendering of one of the Fijian vessels spotted by D'Urville's ship on its voyage between 1826 and 1829, and (C) a Tahitian double-canoe sketched by Webber in 1777 during Cook's third voyage. (From Irwin 1992.)

35,000 years after human populations were firmly established across Sahul. Once people had acquired the necessary technology, however, the colonization of what is now Polynesia was extremely rapid. The leap from Fiji to Tonga took approximately 2000 years, and in just another 500 to 600 years Polynesians reached the distant outposts of the Marquesas, Societies, Hawaii, and Easter Island (see Figure 18.4).

Yet despite the Polynesians' mastery of the seas, it appears that some isolated islands were never colonized, perhaps never even visited, before the age of European expansions. These include many isolated islands of central and eastern Polynesia. Many other islands were apparently visited and perhaps briefly colonized by the Polynesians, but they failed to maintain populations for significant periods (Figure 18.8).

The implications of these distribution patterns with respect to the equilibrium theory of island biogeography are intriguing. In an insightful account of the colonization of the Pacific islands, John Terrell (1986) analyzed patterns in the distribution and diversity of humans using many of the principles and models of island biogeographic theory. He found, for example, that just as the theory predicts, humans first colonized the less isolated islands, and were

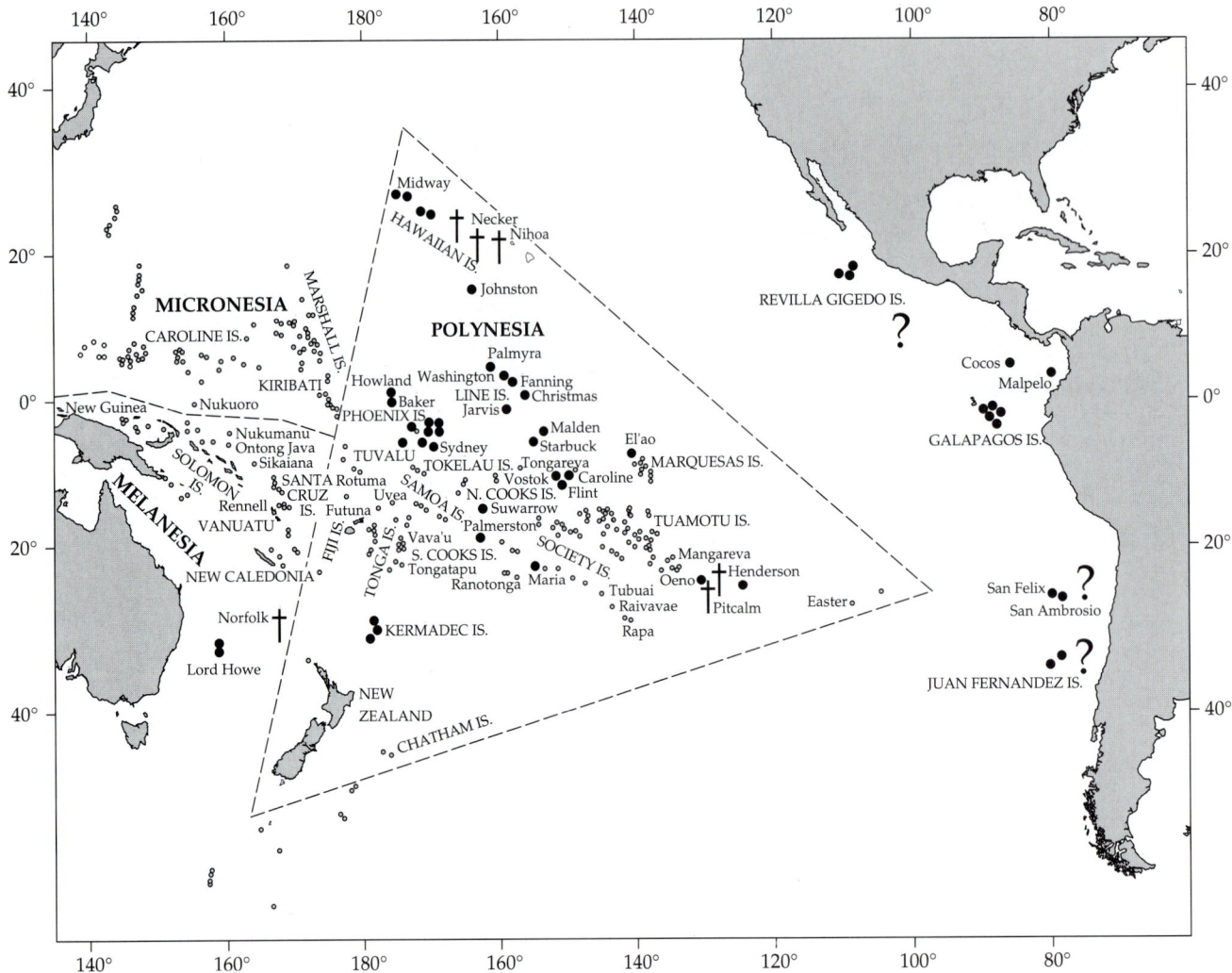

Figure 18.8 At the time of European discovery, many islands in the central and eastern Pacific were uninhabited (large dots). Crosses mark the islands that were colonized by Pacific Islanders prior to 1500 A.D., but were then abandoned in historic times, usually after extinctions of many native species. Question marks indicate those islands where pre-European colonization is suspected, but not clearly documented. (After Terrell 1986.)

likely to maintain populations on the more isolated islands only if they also were relatively large. That is, it appears that the minimum island area required to maintain populations of *Homo sapiens* increased with isolation (see the discussion of "rescue effects" in Chapter 13). In the Atlantic, prehistoric civilizations occupied many small islands relatively close to continents (e.g., the Scilly Isles, Sottish, and Channel Islands) by, if not many millennia before, the Bronze Age (5500 to 3000 B.P.). In contrast, the very isolated islands of the South Atlantic (e.g., Ascension, Saint Helena, Tristan da Cunha, and South Georgia) were uninhabited when Europeans first reached them. Other isolated islands of Oceania and elsewhere were colonized by repeated waves of colonists that either supplemented existing insular populations or replaced those that went extinct.

As observed for other insular faunas, persistence of human populations on islands appears to have been positively correlated with island area. For example, during glacial recession and the associated rise in sea levels during the early Holocene, the landbridge connecting Tasmania to the Australian mainland was fragmented. As a result, insular populations of humans were extirpated on all but the largest islands of Bass Strait. That is, just as Brown (1971) and Diamond (1972) reported for the relaxation of other insular biotas, human populations were subject to nonrandom extinctions following fragmentation of once expansive landmasses. All of these observations are of course consistent with the fundamental tenets of the equilibrium theory of island biogeography: immigrations and extinctions tend to be recurrent processes and are strongly influenced by island characteristics (i.e., isolation and area).

Terrell and his colleagues took this analogy one step further (see Terrell 1986). Not only does the similarity of languages and other cultural and morphological characteristics among insular populations decrease with isolation, but linguistic diversity (i.e., number of different cognate words) increases with island area and decreases with island isolation (Figure 18.9). Interestingly, and consistent with patterns of species diversity, linguistic diversity on the conti-

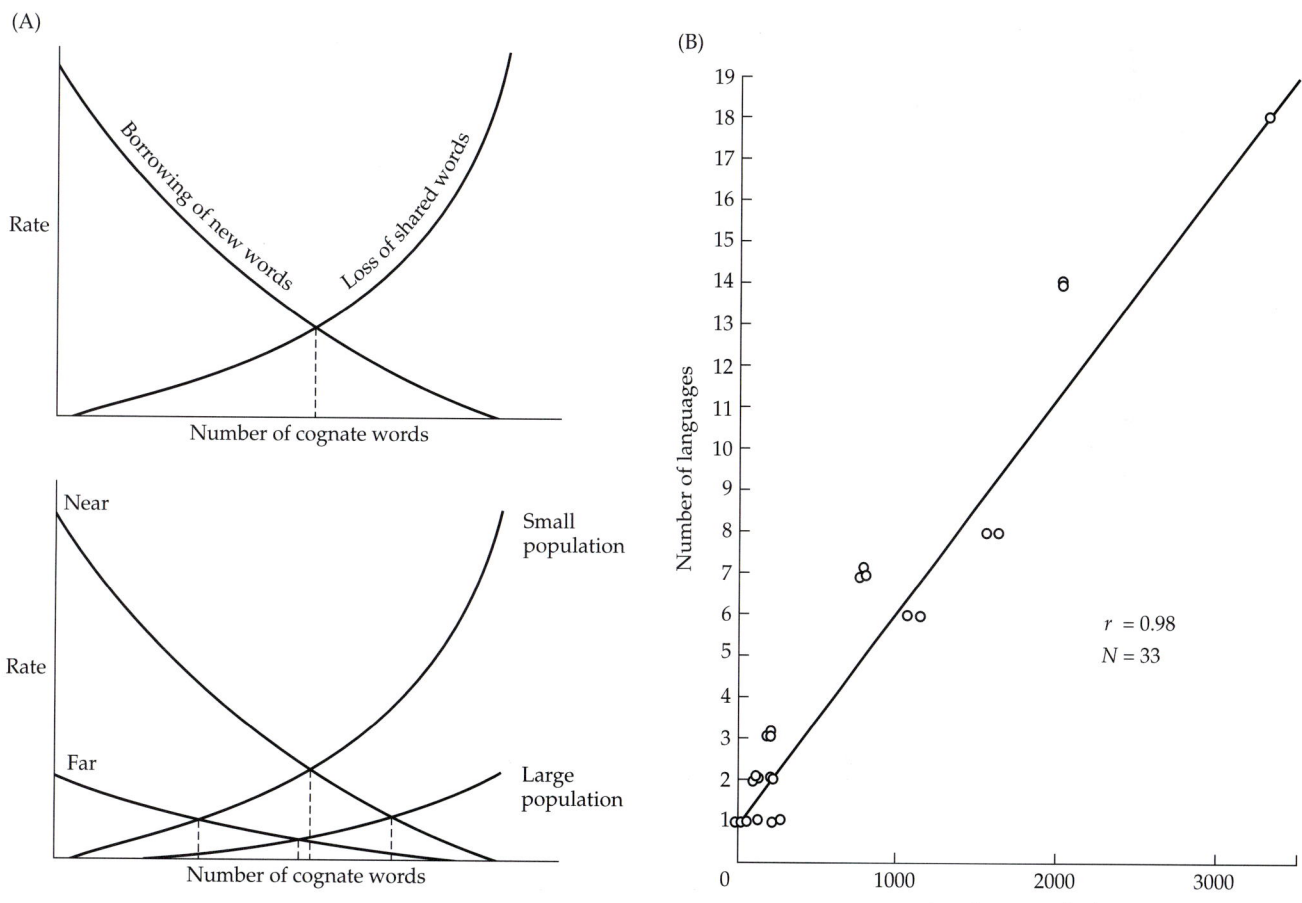

Figure 18.9   (A) Borrowing from island biogeographic theory, Terrell hypothesized that linguistic diversity (number of cognate words) results from a balance between the rate of borrowing of new words between insular societies (which should decrease with increasing distance between islands) and the rate of loss of shared words (which should be higher for smaller islands with smaller populations). Accordingly, linguistic diversity (number of cognate words and number of languages) on islands decreases with isolation and increases with area. (B) The relationship between island area and the number of languages spoken on a given island for the Solomon Islnds supports this hypothesis. (From Terrell 1986.)

nents also exhibits a marked latitudinal gradient, increasing from the poles toward the equator (Mace and Pagel 1995). According to Terrell, the diversity of languages on islands can be explained by a linguistic version of MacArthur and Wilson's equilibrium theory. He hypothesized that linguistic diversity results from a balance between rates of borrowing ("immigration") of new words from other insular populations and the loss ("extinction") of shared, extant words. These rates are functions of both the existing number of cognate words on an island and its physical characteristics (isolation and area; see Figure 18.9). Similarly, the development and distribution of native linguistic groups in North America appears to have been strongly influenced by the distributions of physical barriers and ecogeographic zones of the late Pleistocene and early Holocene (Rogers et al. 1990).

Finally, Jared Diamond (1977) suggested that insular human populations may undergo subsequent transformations much like those associated with Wilson's (1961) taxon cycles (see Chapter 13; see also Diamond 1997; Flannery 1994). Like taxon cycles, the colonization cycles of humans tend to be unidirectional, with most movements from larger landmasses to smaller ones, but not the reverse. After the initial colonists arrive, human populations often expand, not just geographically but ecologically as well, exploiting a broader range of resources and habitats. As population densities increase and resources become saturated, the initial founding population may enter a stage of local adaptation in which selective pressures switch from those favoring colonizing abilities (high dispersal ability, high reproductive rate, and generalized niches) to those favoring efficiency of resource use and competition with conspecifics. Increased competition from newly arriving human populations may then trigger a "retreating stage" in which the niche and distributional range of the initial colonists contracts. Finally, just as envisioned by Wilson in his taxon cycle hypothesis, Diamond's colonization cycle of man ends in extinction—that is, with the more ancient and specialized civilizations being replaced by waves of invading populations of generalists.

The biotas of the systems that were the last to be colonized by humans—oceanic islands—were the most severely affected by human activities (Figure 18.10). Recent studies by David Steadman and his col-

(A)

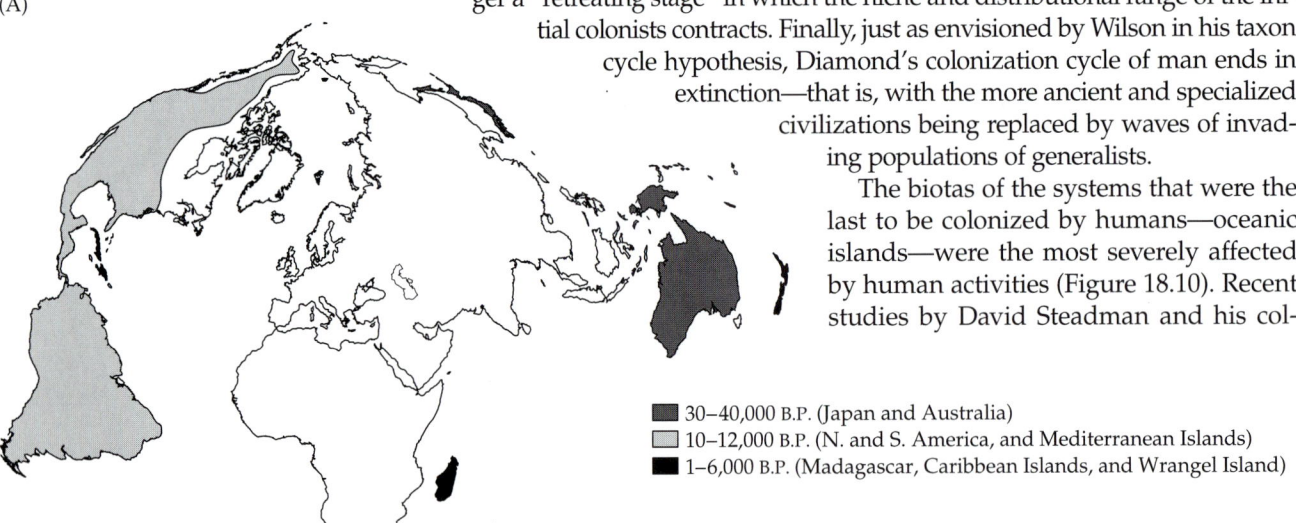

- ■ 30–40,000 B.P. (Japan and Australia)
- ☐ 10–12,000 B.P. (N. and S. America, and Mediterranean Islands)
- ■ 1–6,000 B.P. (Madagascar, Caribbean Islands, and Wrangel Island)

(B)

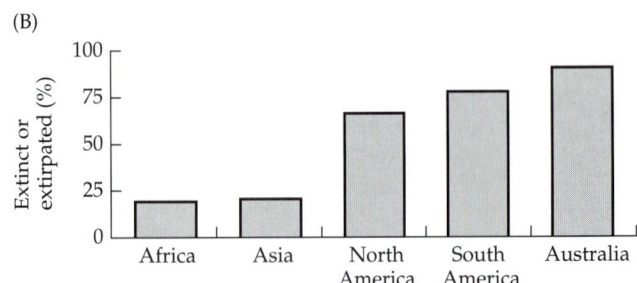

Figure 18.10 (A) Waves of extinctions of large mammals and birds during the Pleistocene seem to be coincident with waves of colonization of environmentally significant humans (see Figures 18.4 and 7.30). (B) The relative number of late Pleistocene/early Holocene (40,000 to 10,000 B.P.) extinctions or extirpations of the mammalian megafauna (species heavier than 44 kg). Note that the continents long occupied by humans (Africa and Asia) were relatively immune to this wave of extinctions. (A after Martin 1984; B after Martin 1990.)

leagues reveal that the extinctions of as many as two-thirds of the native birds and bats of Oceania's islands were coincident with colonization by humans (Steadman 1995; Koopman and Steadman 1995). These extinctions included an estimated average loss of 10 species or populations per island over 800 major islands, yielding 8000 extinctions or extirpations. Steadman (1993) suggests that human colonization of oceanic islands influenced insular avifauna more strongly than any tectonic, climatic, or biological event of the past 10,000 years (Figure 18.11).

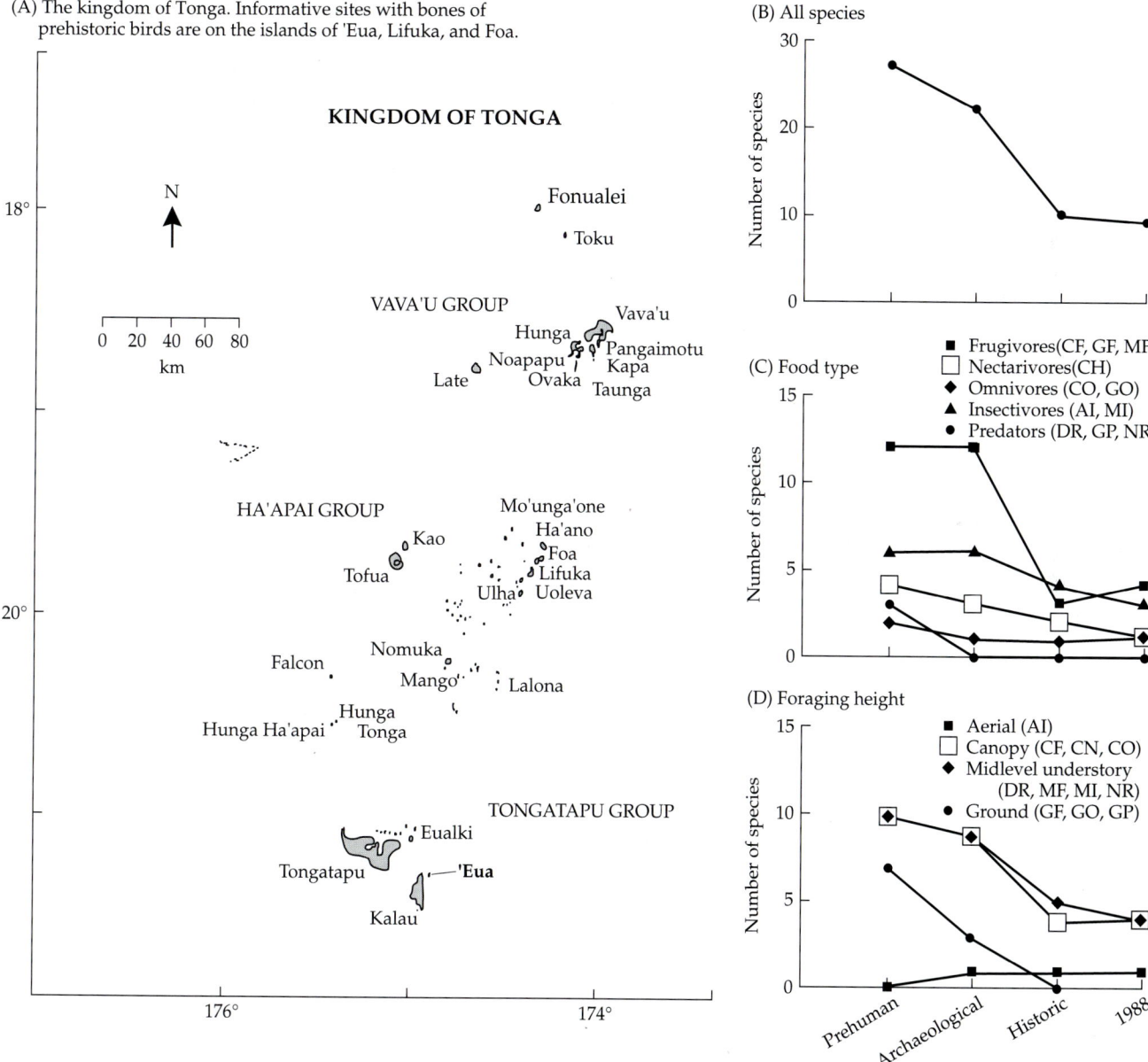

Figure 18.11   The diversity of native forest birds on 'Eua Island in the Kingdom of Tonga (A) has decreased dramatically since the island was colonized by Polynesians (B, C, D). Hardest hit were the frugivorous and ground-dwelling species, which were especially susceptible to predation by humans and their ground-dwelling commensals. (After Steadman 1995.)

Lessons from the Biogeography of Humanity

Before turning our attention to applied biogeography, it is worth summarizing some of the general lessons of the foregoing discussion:

1. On a global scale, human expansions were not gradual and continual, but episodic. Long periods of geographic stasis were followed by rapid expansions once landscapes changed or hominids developed technologies necessary to cross persistent barriers.
2. Early expansions of hominids were strongly influenced by the same types of geographic barriers, filters, and corridors that influenced our mammalian relatives. Thus, the geographic expansions of our species were strongly influenced by tectonic events and especially by the climatic cycles that were prevalent during the Pleistocene.
3. Our range expansions on continents and across archipelagoes were the result of recurrent bouts of colonization and expansion, often followed by collapse and extinction at the hands of other, "more advanced" peoples, technologies, and cultures.
4. Throughout our evolutionary history, the ability of humans to colonize distant sites and to modify and eventually dominate a greater variety of ecosystems has continually increased.

The global range expansion of humans has been followed by exponential increases in human population densities, especially during the last few decades. While some other species may have benefited from these changes, many more have been negatively affected and have suffered reduced populations, range contractions, and extinctions. Predicting which species will be negatively affected and developing effective programs for conserving their populations will require contributions from many scientific disciplines. In the remainder of this chapter we discuss some of the important insights that biogeographers can provide.

Applied Biogeography: Focal Species Patterns and Approaches for Conserving Biodiversity

Given the limited resources allocated for conserving endangered species, it is essential that we focus our efforts where they will do the most good. Toward this end, we can learn a great deal from past extinction events. Reviews of records of prehistoric and historical extinctions reveal several common characteristics of extinction-prone species (see Marshall 1988; Diamond 1984):

1. Specialized habitat requirements or diet
2. Position in the higher trophic levels (e.g., top carnivores)
3. Small and variable populations
4. Large body size
5. Long generation time (age to reproductive maturity) or low reproductive potential
6. Poor dispersal ability or limited opportunities for dispersal
7. Restricted geographic range

While each of these characteristics may represent useful clues for identifying endangered species, the latter two are fundamental to biogeography. That species with poor dispersal abilities and restricted ranges should be more prone to extinctions throughout the fossil record, as well as in recent times, should not be surprising. The previous chapter discussed the relatively high vulnerability of insular biotas, groups of species that by default have limited opportunities for dispersal and restricted geographic ranges. On the other hand, species with greater dispersal abilities and broader geographic ranges are more likely to survive environmental disturbances. As range size increases, there is less chance that disturbances such as floods, fires, or hurricanes will fully encompass all populations. And even in those cases in which environmental disturbances are pervasive and affect all populations, the more vagile species can adapt by shifting their ranges.

The relationship between susceptibility to extinction and geographic range size is well evidenced by the fossil record. During periods of "normal" or background extinction, most of the above listed factors have influenced species' persistence, whereas during mass extinction events, range size has been especially important (Figure 18.12; Jablonski 1986, 1989; Rosenzweig 1995). It appears that the forces of mass extinction operate on such a grand scale that they often overwhelm the influence of traits that are honed by natural selection and operate at much finer scales (e.g., population size, body size, and trophic strategy). As Jablonski (1991) put it, many species suffered extinctions not because they "were poorly adapted by the standards of background extinctions, but because they occurred in lineages lacking the environmental tolerances or geographic distributions necessary for surviving the mass extinction." Jablonski predicts that the ongoing extinction crisis will also "impinge most heavily on rare, geographically restricted species."

These extrapolations from the fossil record to the present are substantiated by studies of more recent extinctions and extirpations. Hanski (1982) reexamined the results of Simberloff and Wilson's (1969 and 1970) classic defaunation experiments on mangrove islands in Florida (see Chapter 13) and found that, as predicted, the more widespread invertebrate species were less likely to be extirpated from the islands (Figure 18.13). Hanski found the same relationships for mollusks inhabiting ponds in England and for leafhoppers from Finnish meadows: species that occurred across more sites were less prone to extirpation from any particular site. Not surprisingly, lists of species currently threatened with extinction are composed of a highly disproportionate number of species with relatively small geographic ranges (Figure 18.14).

The tendency for extinction-prone species to have small geographic ranges is a very general one, but it is far from a rule. It would be dangerous to assume that all species with broad geographic ranges were immune to extinction and, therefore, to ignore these species when setting priorities for conservation biology. Even abundant and widespread species can go extinct. The passenger pigeon (*Ectopistes migratorius*) and the American chestnut tree (*Castanea dentata*) provide two dramatic examples. Both of these species were broadly dis-

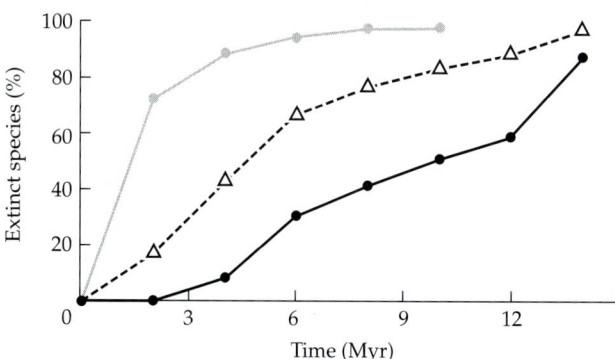

Figure 18.12   Extinction rates of marine bivalves and gastropods living along the Atlantic and Gulf coastal plains of North America during the late Cretaceous were consistently higher for species with relatively small geographic ranges. Species ranges: gray circles = < 1000 km; triangles = 1000–2500 km; black circles = > 2500 km. (After Jablonski 1986.)

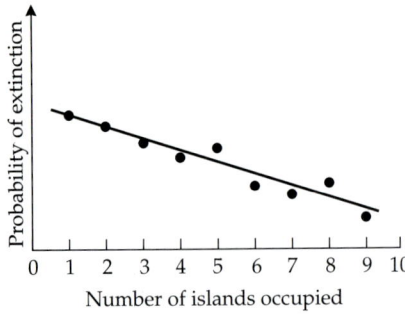

Figure 18.13   The probability of extirpation for insects inhabiting mangrove islands of coastal Florida decreases as the number of islands occupied by the species increases (From Hanski 1982; based on analyses of Simberloff's [1976] data.)

tributed across most of eastern North America, and both were dominant members of their communities, yet both suffered dramatic range collapse and extinction (or virtual extinction in the case of the American chestnut) in less than a century (Rosenzweig 1995). The passenger pigeon was the most common bird in eastern North America during the eighteenth and early nineteenth centuries. Total population estimates vary, but range between 2 and 5 billion birds. If these estimates are accurate, then before its decline and ultimate extinction in 1914, one out of every two or three birds in eastern North America was a passenger pigeon. Similarly, before their decline, chestnut trees may have accounted for one-fourth of all trees in their range. Yet deforestation and overhunting wiped out the passenger pigeon, and the introduced chestnut blight (*Endothia parasitica*) destroyed chestnut trees across all of their historical range (Figure 18.15; see Roane et al. 1986).

Unfortunately, these are not the only exceptions to the rule. Some of the most broadly distributed species of North America, including bison (*Bison bison*), trumpeter swans (*Olor buccinator*), whooping cranes (*Grus americana*), and sandhill cranes (*Grus canadensis*), have suffered substantial range reductions and near extinction. Species with broad ranges may be better off on average, but the power of anthropogenic disturbance is such that it can overwhelm even the biogeographic factors that were correlated with survival during earlier waves of extinctions. Species with historically restricted geographic ranges still constitute a highly disproportionate number of all endangered species, but their ranks are now joined by many species that, until recent decades, enjoyed the benefits of very large ranges.

Patterns of Range Collapse

When a species' range collapses, does the process exhibit a consistent or predictable pattern? For example, do populations in particular portions of the range tend to fare better than others? Classic biogeographic theory, along with more recent patterns emerging from areography and macroecology, suggests some answers. Before discussing these, however, it is important to point out again that this is not just an interesting academic question, but one with important implications for guiding conservation strategies. Efforts to reintroduce or translocate populations of endangered species, or to search for undiscovered remnants of a once widespread species, would be greatly facilitated if we could develop a spatial search image for sites most likely to yield favorable results.

In fact, such a search image has been developed. Biogeographic theory suggests that reintroductions, translocations, and searches for remnant populations should be conducted near the core of a species' historical geographic

Figure 18.14   The relationship between geographic range size and endangerment in waterfowl (Anseriformes). "Endangered" refers to species listed as vulnerable or endangered based on three criteria: population size, rate of population decline, and estimated impact and frequency of disturbances. (After Mace 1993.)

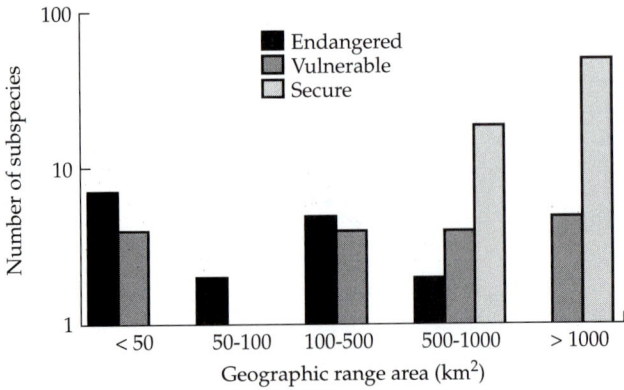

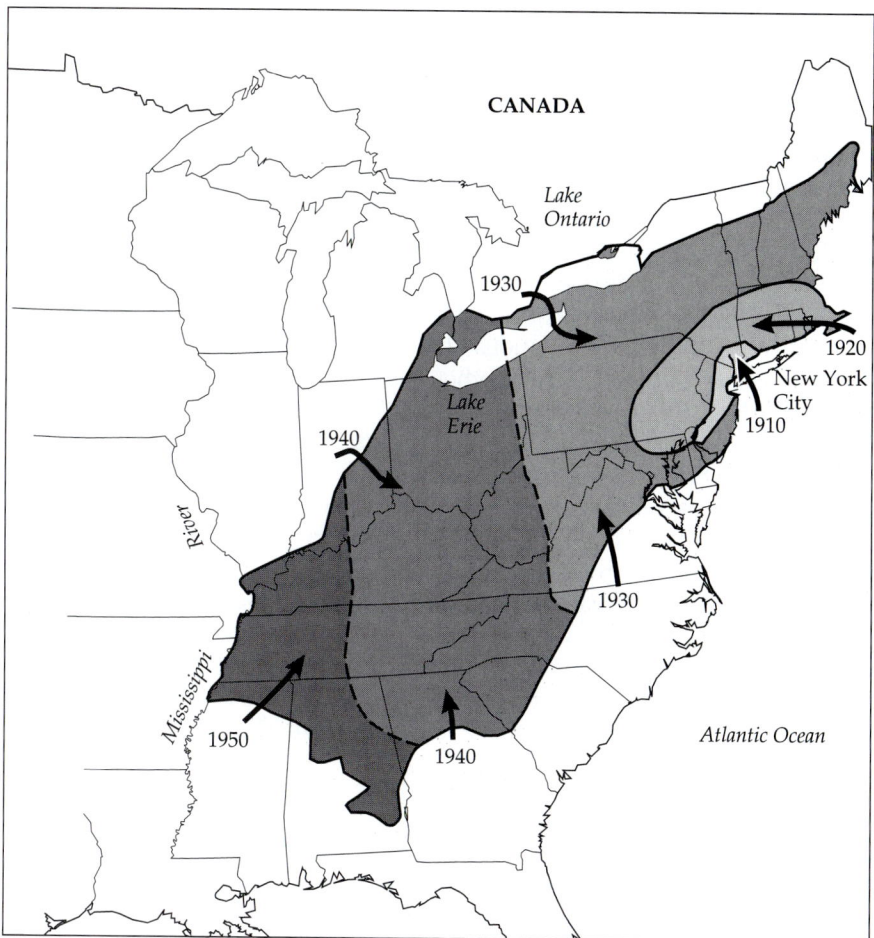

Figure 18.15 The American chestnut (*Castanea dentata*) was once one of the most common trees in the eastern deciduous forests of the United States. The chestnut blight, first introduced near New York City around 1910, spread rapidly to cause the decline and range collapse of the chestnut. The solid outline indicates the natural range of the American chestnut prior to 1910; dashed lines and shaded regions indicate the sequential collapse of chestnut populations between 1910 and 1950. (After Anderson 1974; Bell and Walker 1992.)

range. After all, the range boundaries are just that, sites where environmental conditions are so poor that the species cannot maintain its populations without recruitment from other populations. Areas along the periphery of a species' range are often thought to represent sink habitats (sensu Pulliam 1988; see Chapter 4) for individuals emigrating out from more central source populations. Biogeographers have long predicted that when a species' range collapses, it should implode, collapsing first along the periphery and then toward the center of the historical range.

> On a continent, the process of extinction will generally take effect on the circumference of the area of distribution, because it is there that the species comes into contact with such adverse conditions or competing forms as prevent it from advancing further. (A. R. Wallace, 1876)

> The actual limits of range will then be determined not by the limits of favorable ground but by a constantly fluctuating equilibrium between tendency to spread at the center of the range and tendency to recede at the margins. (P. J. Darlington, Jr., 1957)

Today's conservation biologists have followed the lead of Wallace and Darlington, often viewing the range periphery as the domain of the "living dead," sites with little value for conserving endangered species. Emerging patterns in macroecology and areography seem to substantiate these claims. As we saw in Chapter 16, surface maps of population densities for a great variety of species follow the predicted patterns very closely. As we move from the periphery to the more central areas of a species' geographic range, densities tend to increase, while variability in densities tends to decrease (Figure 18.16). As a result, the more central populations should be less prone to extinction, and sites closer to the center of a species' range should be optimal for locating or reestablishing breeding populations of an endangered species.

Unfortunately, this may be another case in which generic prescriptions, even those based on very sound theory and well-documented empirical patterns, can prove misleading. Recent analyses of range collapse in rare and endangered species of vertebrates, invertebrates, and plants indicate that, contrary to the above predictions, relictual populations of these now geographically restricted species occur along the periphery, not near the core of the species' historical geographic ranges (Lomolino and Channell 1995, 1998; Channell 1998). Two hundred-forty (98%) of the 245 species studied maintained at least one population along the periphery of their historical range, and 167 species (68%) maintained more populations along the range periphery than expected by chance. In fact, populations of 91 of these species were exclusively limited to the range periphery while only 5 species maintained popula-

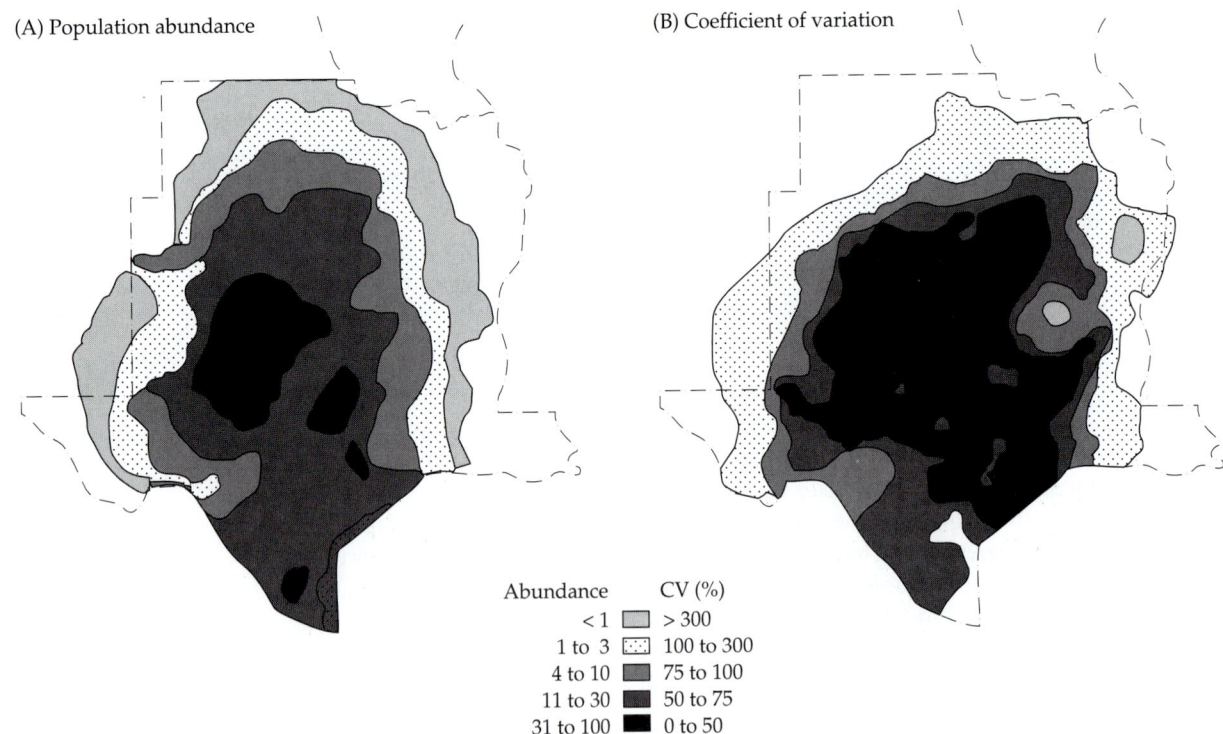

(A) Population abundance

(B) Coefficient of variation

Abundance		CV (%)	
< 1		> 300	
1 to 3		100 to 300	
4 to 10		75 to 100	
11 to 30		50 to 75	
31 to 100		0 to 50	

Figure 18.16 Abundance and population variability for the scissor-tailed flycatcher (*Muscivora forficata*) in Texas and Oklahoma. This species illustrates the general tendency for populations of many terrestrial animals to be highest and least variable near the centers of their geographic ranges. Abundance = the average count in Breeding Bird Surveys; CV = coefficient of variation of abundance over a 25-year period. (From D. Certain, unpublished report.)

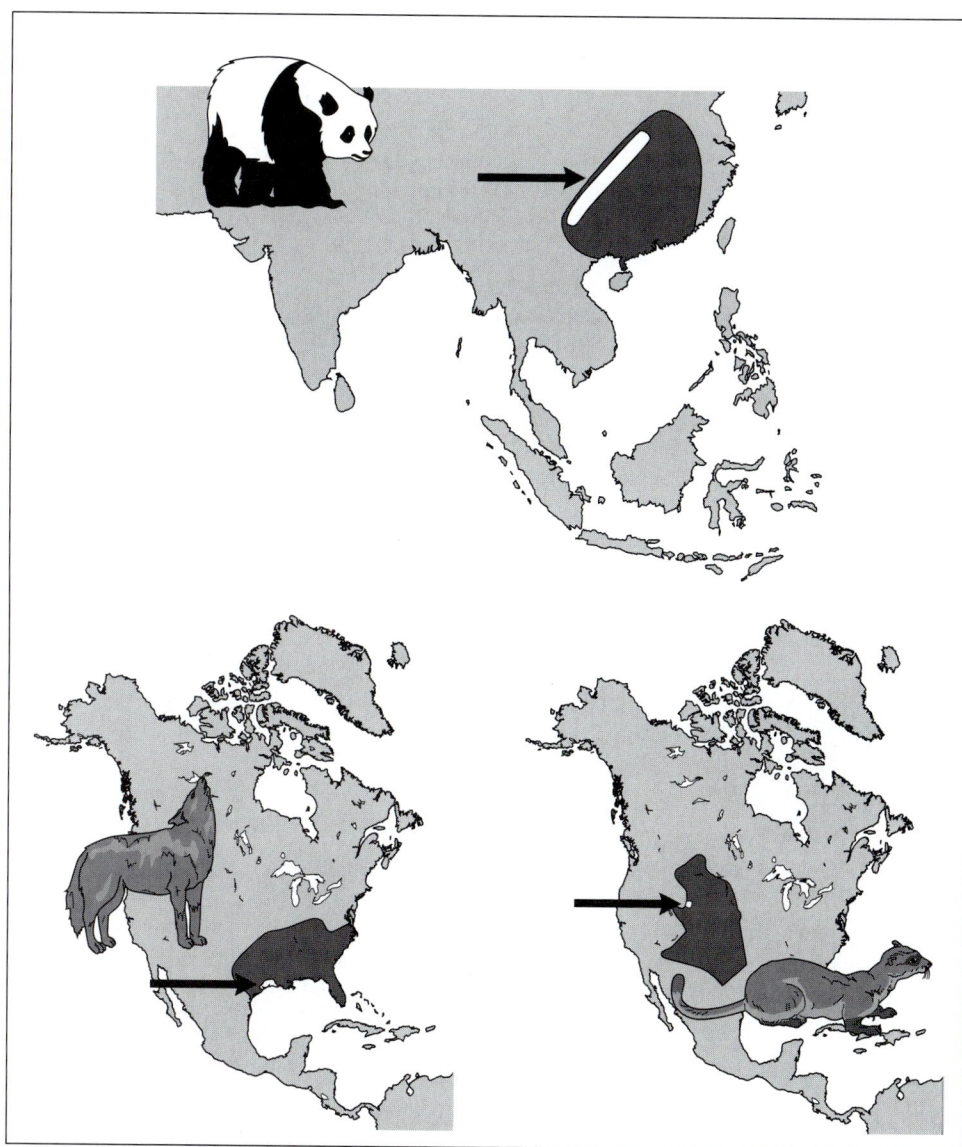

Figure 18.17 Patterns of range collapse in three species of terrestrial mammals: the giant panda (*Ailuropoda melanoleuca*), red wolf (*Canis rufus*), and black-footed ferret *(Mustela nigripes).* The locations of remnant populations of these endangered species are indicated in white; the dark gray polygons represent the historical ranges of the species. (After Lomolino and Channell 1995.)

tions exclusively within the range center. Among the species persisting along the range periphery are at least five "poster species" for the conservation movement: giant panda (*Ailuropoda melanoleuca*), red wolf (*Canis rufus*), black-footed ferret (*Mustela nigripes*), California condor (*Gymnogyps californianus*), and whooping crane (*Grus americana*), (see Figures 18.17 and 18.18). Analyses of range collapse in other taxa are concordant with the results (Figures 18.19 and 18.20).

Why do these patterns of range collapse appear so anomalous? The answer may well lie in the nature of the extinction forces to which these species have been subjected. Even though the macroecological patterns of population density and variability appear to be very general ones, they may be overwhelmed by anthropogenic disturbance, which tends to spread across the landscape like a contagion. Anthropogenic activities, including agriculture, fragmentation, introduced species, diseases, and advanced hunting technologies, tend to move across the landscape like fire across a burning leaf: the last piece to burn will be along an edge. In these circumstances, isolation may be the key to per-

(A)

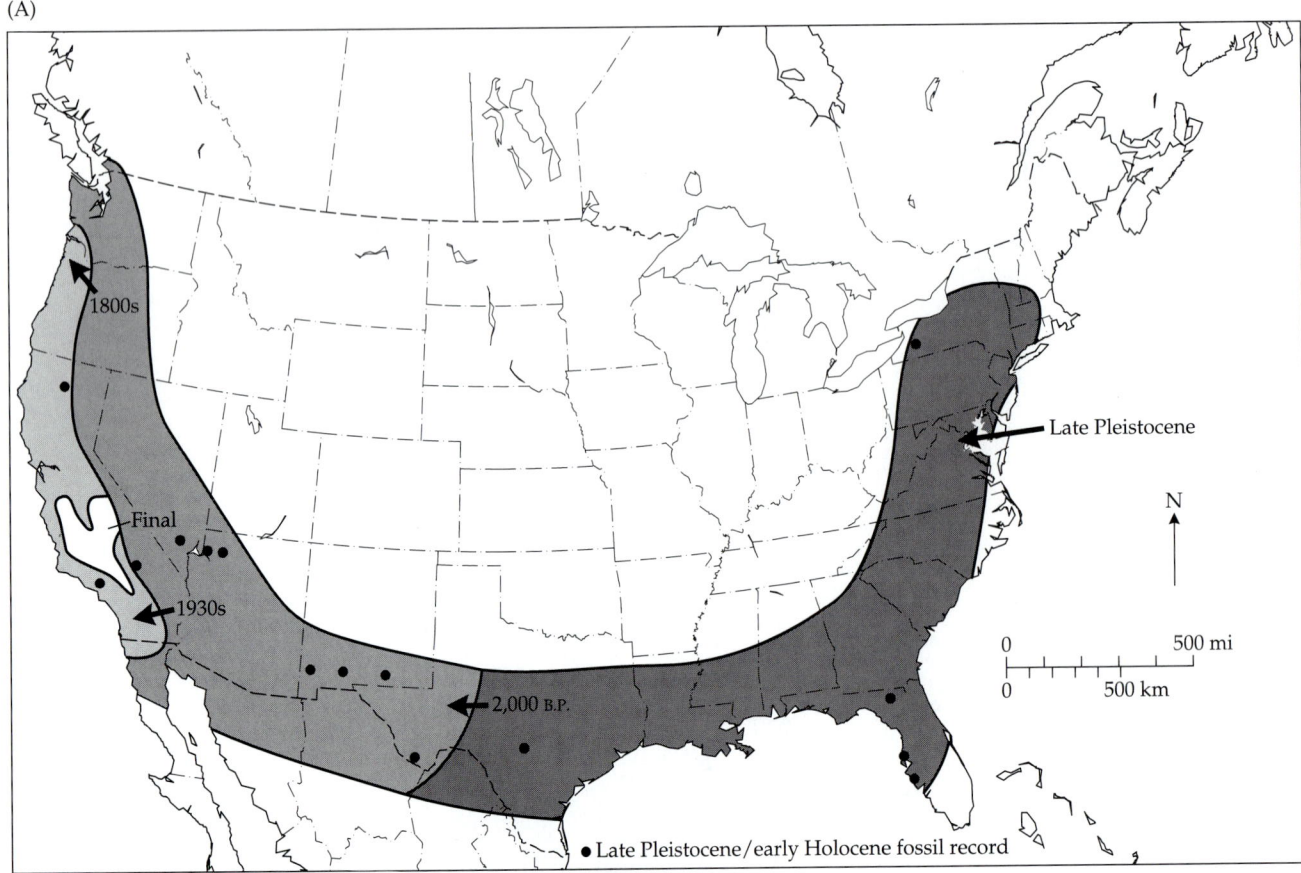

Figure 18.18   Patterns of range collapse in four endangered or extinct North American bird species: (A) California condor (*Gymnogyps californianus*), and (B) the ivory-billed woodpecker (*Campephilus principalis*), Kirtland's warbler (*Dendroica kirtlandii*), and whooping crane (*Grus americana*). The locations of the remnant or last known populations of these species are indicated in white; the dark gray polygons represent the historical ranges of the species. (After R. Channell, pers. comm.)

sistence. Thus, remnant populations should occur in isolated regions, either on an island, an isolated fragment of the historical range, or along the periphery of the range.

One last result of Lomolino and Channell's (1995) initial study of range collapse in 31 species of mammals seems especially relevant to this hypothesis. Of the 18 species of mammals whose ranges tended to collapse toward just one of the cardinal directions, a disproportionate number (11) collapsed toward the west. We know of no such western bias in the density profiles of mammals, or of any other animal for that matter. On the other hand, human populations and their activities tend to spread westward, especially in North America, Eurasia, and Australia. The geographic ranges of endangered and extinct species may have receded with the westward-advancing front of human disturbances.

It is important to note that this study represents one of the first articulations of empirical patterns in range collapse, and we have no doubt that many surprises await those pursuing similar questions in dynamic biography. Because taxonomic groups vary in their vagilities and their susceptibilities to anthropogenic disturbance, we expect that patterns of collapse will vary

(B)

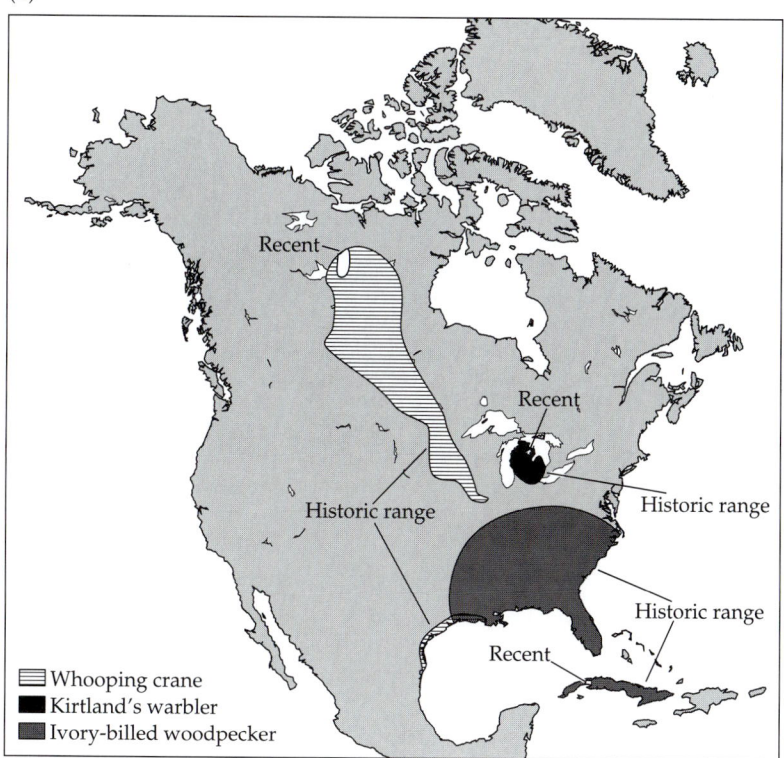

among taxa. Even within the same taxonomic group, we expect to find different geographic trends in range collapse among regions. Because humans coevolved with the native plants and animals of Africa, for example, range collapse on this continent may fail to exhibit any noticeable directionality. In fact, Channell (1998) found that while over 75% of the species of the Australian, North American and European/Asian regions collapsed to the range periphery, only 58% of the African species exhibited peripheral collapse.

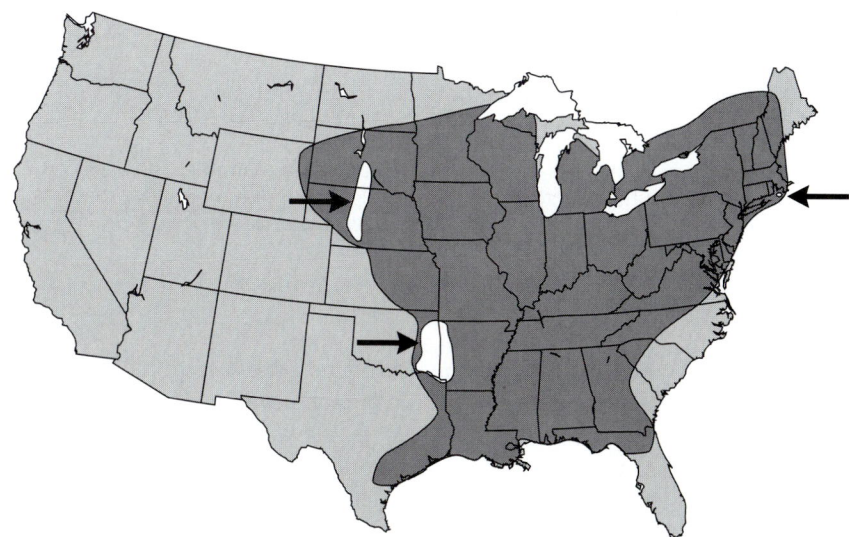

Figure 18.19 Range collapse of the American burying beetle (*Nicrophorus americanus*). The historical range of this species (ca. 1900–1940) is indicated by shading, while extant range is indicated by white polygons. (After Lomolino et al. 1995, and references therein.)

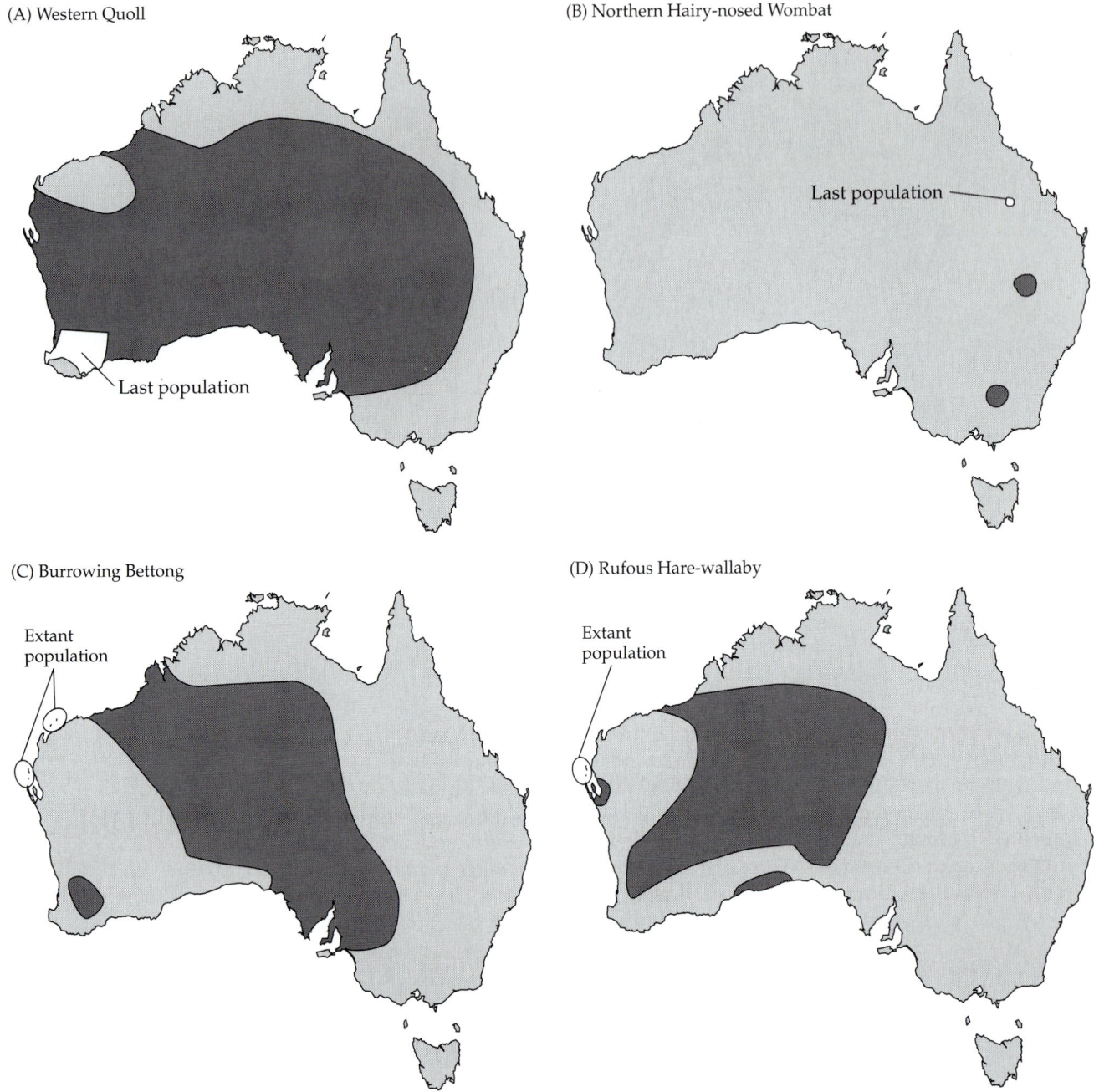

(A) Western Quoll

Last population

(B) Northern Hairy-nosed Wombat

Last population

(C) Burrowing Bettong

Extant population

(D) Rufous Hare-wallaby

Extant population

Figure 18.20 Range collapse in four species of Australian mammals (dark gray denotes historical range, white, current or final range). Note that remnant populations are typically found along the periphery of the species' historical ranges, especially on islands. (From Lomolino and Channell 1995.)

Human colonization of oceanic islands exhibited little directional bias, but spread from the coastal lowlands toward the higher elevations. In Hawaii, the ranges of native birds collapsed away from lower elevations and other areas of anthropogenic disturbance (M. Scott, pers. comm.). In another insular case study, ranges of native birds on the island of Guam collapsed not along an elevational gradient, but with the well-documented expansion of the range of the

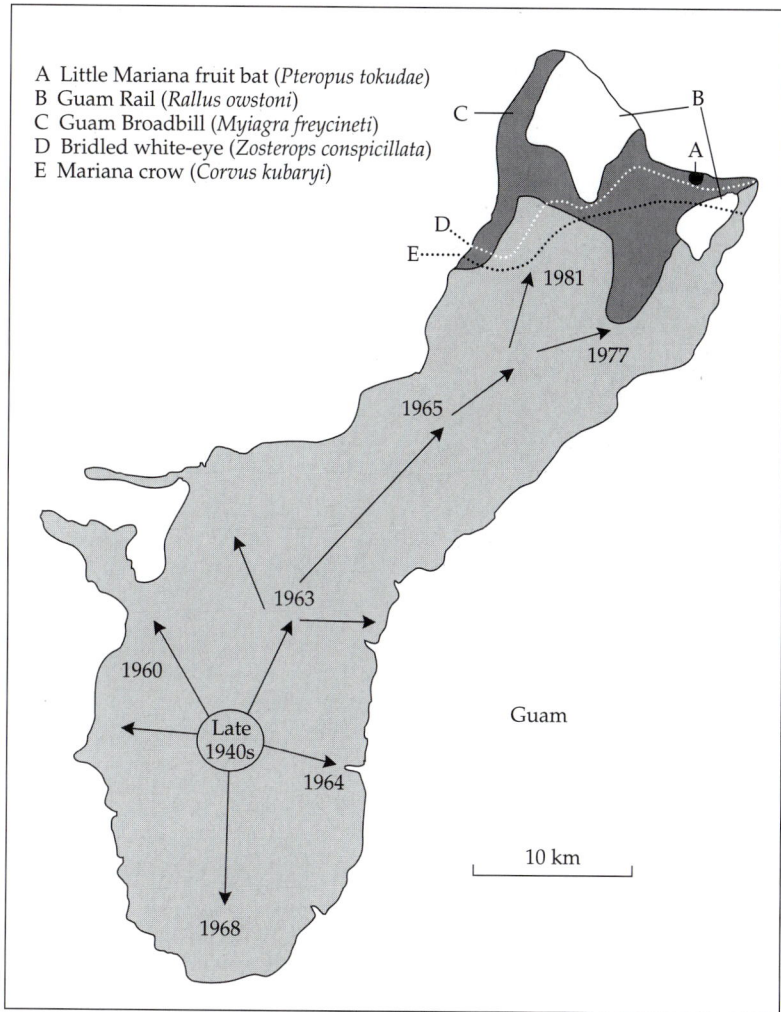

A Little Mariana fruit bat (*Pteropus tokudae*)
B Guam Rail (*Rallus owstoni*)
C Guam Broadbill (*Myiagra freycineti*)
D Bridled white-eye (*Zosterops conspicillata*)
E Mariana crow (*Corvus kubaryi*)

Figure 18.21   Following its introduction onto Guam in the late 1940s, the spread of the brown tree snake (*Boiga irregularis*) caused range contractions of numerous species of native birds and bats whose remnant populations retreated to the northern portion of the island, i.e., the last area to be colonized by *Boiga*. (After Savidge 1987; Channell 1998.)

introduced brown tree snake (*Boiga irregularis*), which preyed heavily on native birds (Figure 18.21). In all these cases, whether populations persist along the edges of once expansive ranges, on offshore islands, or at high altitudes, the common factor is that these sites are isolated and are therefore the last to be affected by the spread of anthropogenic disturbances.

Finally, the putative cause of range collapse in these endangered species—human civilization—may itself serve as a subject for parallel studies. As the ranges of aboriginal civilizations collapsed, did they exhibit spatial trends similar to those of other mammals? We have looked at only a few such maps, but the results are tantalizing. The range of the Anasazi of North America has collapsed toward the periphery of their historical range (Figure 18.22A). The Incas, whose expansive empire once encompassed much of western South America, present another interesting case of range collapse, one that seems similar to the patterns exhibited by Guamanian and Hawaiian birds. With the invasion of the Spanish Conquistadores, which spread southward from Ecuador, the Inca empire collapsed to its final stronghold of Vilcabamba high in the Andes Mountains (Figure 18.22B).

These studies of range collapse have obvious applications for conserving endangered species. In developing effective strategies, conservation biologists must consider the geographic dynamics of extinction forces as well as the ecological and biogeographic characteristics of the species in question. Just as

(A)

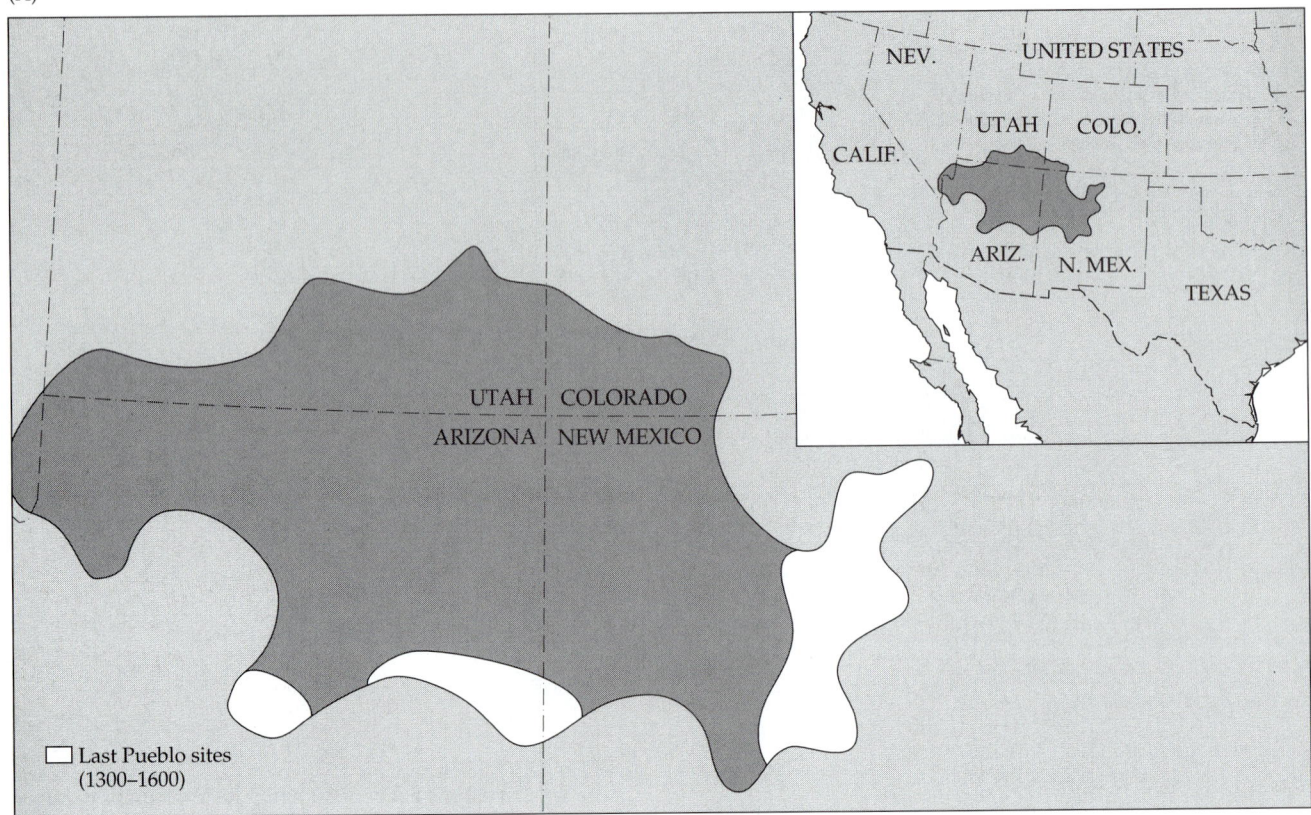

Figure 18.22   (A) Range collapse of the Anasazi of North America. As the shaded regions indicate, the last pueblo sites of these native Americans were situated along the periphery of their historical range. (B) With the invasions of the Spanish conquistadores (arrows), the once vast range of the Incas collapsed to their final stronghold in the high and isolated sites of Machu Picchu and Vilcabamba. (A courtesy of National Geographic Society; B after Davies 1995.)

important, these studies show that sites along the periphery of a species' historic range should no longer be dismissed as sites with little conservation value. Instead, reintroductions, translocations, and surveys for relictual populations should be conducted wherever favorable sites can be located throughout the historical range of an endangered species, and perhaps outside the known range as well. Of course, conservation biologists must also consider the possibility that a species translocated outside its native range would, as many other exotics have, negatively affect species native to the site in question.

In a practical sense, we should realize that geographic range boundaries are subjective summaries, open to interpretation or misinterpretation. For example, the last population of black-footed ferrets was actually found not just along the western edge of the species' historic range, but outside its described range (see Figure 18.17; see also Blackburn and Gaston 1995; Ferrari and Queiroz 1994). Moreover, sites outside a species' historic range may actually prove to be more favorable under the present regime of extinction forces. The plight of the American chestnut, discussed above, is an excellent case in point. The last field population of American chestnut trees is composed of individuals transplanted outside their historic range, and outside the range of the blight that wiped out other populations (Rosenzweig 1995). Conservation biologists of New Zealand, Guam, and Australia are using the same strategy to

(B)

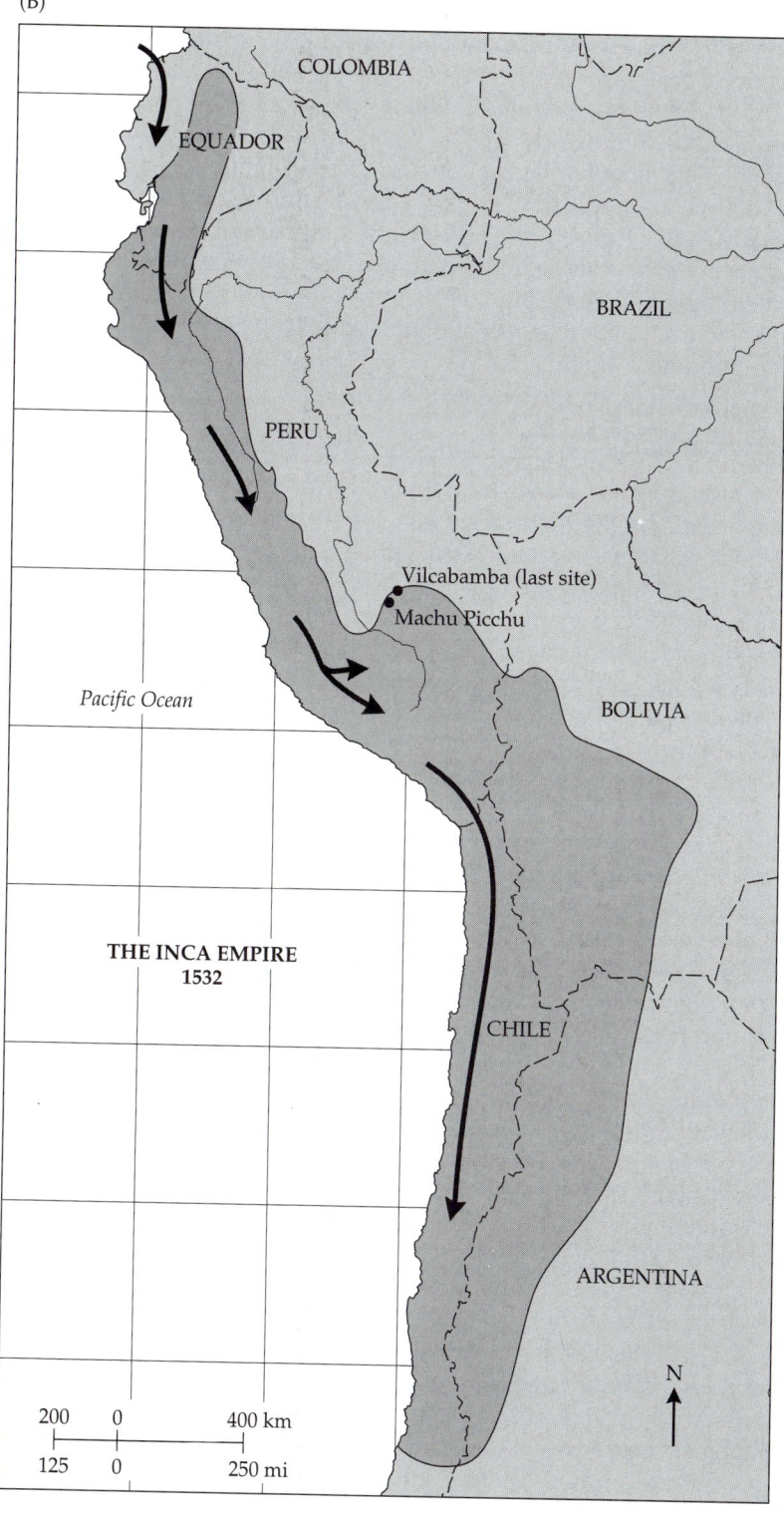

establish breeding populations of endangered birds and reptiles on offshore islands (Towns and Ballantine 1993; Atkinson 1988). The islands may be outside the historic ranges of these species, but they also lack the introduced cats, stoats (weasels), goats, and rats that apparently destroyed their conspecifics in less isolated areas. In an ironic twist of fate, it seems that islands, the site of a highly disproportionate number of prehistoric and historical extinctions, may now represent critical refugia for many endangered species.

Given that the fundamental unit of biogeography is the geographic range, studies on patterns of range dynamics would appear to be of fundamental importance to biogeography. The patterns of dispersal, range expansion, and invasions that we discussed in Chapters 9 and 17, and the patterns of range collapse that we have focused on here, all have obvious applications for conserving endangered species. Yet, it is only very recently that biogeographers have acquired adequate mapping and analytical technologies to investigate these patterns. Beyond the many exciting questions that await the biogeographers who study these patterns, the results of their work will have immediate applications for addressing the biodiversity crisis.

Patterns of Distribution among Insular or Fragmented Habitats

As many conservation biologists have noted, conservation has now become the challenge of preserving populations of endangered species on dwindling fragments of their native habitats. As we saw in the previous chapter, early attempts to apply island biogeographic theory to the development of strategies for designing networks of nature reserves sometimes failed, not because the theory was unsound, but because it was applied with little regard for the particulars of the species and systems in question. Relevant theory can, however, provide some important tools for developing effective strategies for conserving endangered species.

Metapopulation theory and the species-based model of island biogeography, presented in Chapter 14, provide some important insights for predicting the minimal area required to conserve populations of an endangered species. Both of these models are derivatives of MacArthur and Wilson's model, and both predict that, contrary to conventional wisdom, the minimum area or minimum reserve size required to maintain populations of a species increases with isolation (Figure 18.23). Basically, because immigration rates decrease with isolation, populations on more isolated reserves must be larger to prevent extinctions from occurring before immigrants can rescue the population. These models also predict that minimum area requirements will be higher for less vagile species and for populations isolated by less hospitable matrix habitats. Distributions of hazel grouse (*Bonasa bonasia*) in isolated habitat fragments in Sweden (Aberg et al. 1995) and of small mammals on forested islands in the Great Lakes region of North America (Lomolino 1994) are consistent with the latter prediction (Figure 18.24).

Beyond these qualitative predictions, the species-based model summarized in Figure 18.23 can serve as a spatially explicit tool for designing and managing nature reserves or for predicting the effects of future anthropogenic disturbances. Given information on the distribution of an endangered species across a sample of islands, fragments, or prospective reserves, we can calculate the relationship between minimum area requirements and isolation—that is, the biogeographic space where the species should occur. The distribution of red squirrels (*Tamiasciurus hudsonicus*) among isolated montane forests of the American Southwest provides an illustrative case study. As Figure 18.25 reveals, the area requirements of red squirrel populations tend to increase with isolation. Montane site *a* appears to be one with a high potential for natural colonization by red squirrels, while sites *b*, *c*, and *d* may be good candidates for translocation and reestablishment of additional populations. Again, we caution against indiscriminately translocating species outside their native range without evaluating the potential effects of the focal species on the native

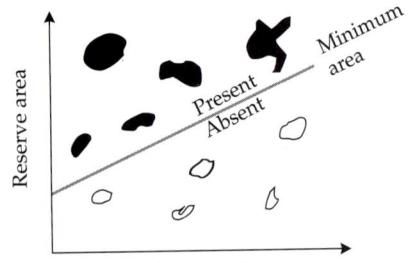

Figure 18.23 This species-based empirical model can serve as a spatially explicit tool for predicting reserve areas required to maintain populations of endangered or other high-priority species in isolated reserves. Solid symbols indicate species presence; open symbols indicate absence.

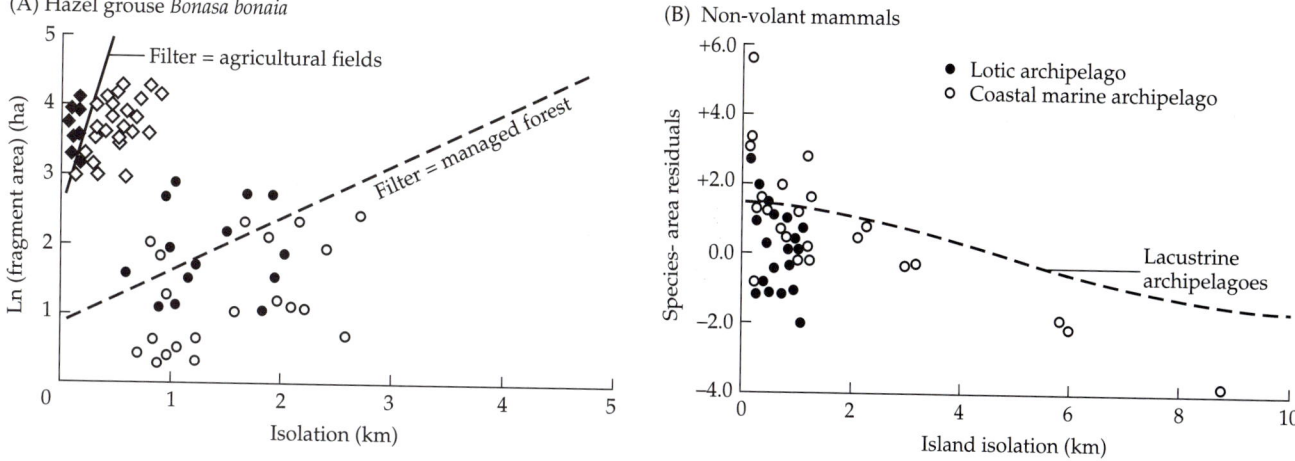

Figure 18.24 Species distribution patterns can reveal the influence of intervening habitats and immigration filters. (A) Distributions of hazel grouse (*Bonasa bonasia*) in isolated forest fragments in Sweden. Although the grouse are widely distributed across forest fragments that are isolated by managed forests, in regions where the habitat matrix is dominated by agricultural fields, populations of grouse are restricted to the largest and least isolated fragments. Black symbols indicate presence, open symbols indicate absence. (B) Distributions of small mammals on forested islands in the Great Lakes region, North America. In comparison to that of lacustrine systems (archipelagoes in Lakes Huron and Michigan), the species richness of nonvolant mammals inhabiting forested islands of lotic and coastal marine systems (the Thousand Island region of the Saint Lawrence River and islands off the coast of Maine), declines much more rapidly with increasing isolation, presumably because immigrations (against currents or across ice in winter) are more difficult in the latter systems. (A after Aberg et al. 1995; B after Lomolino 1994b.)

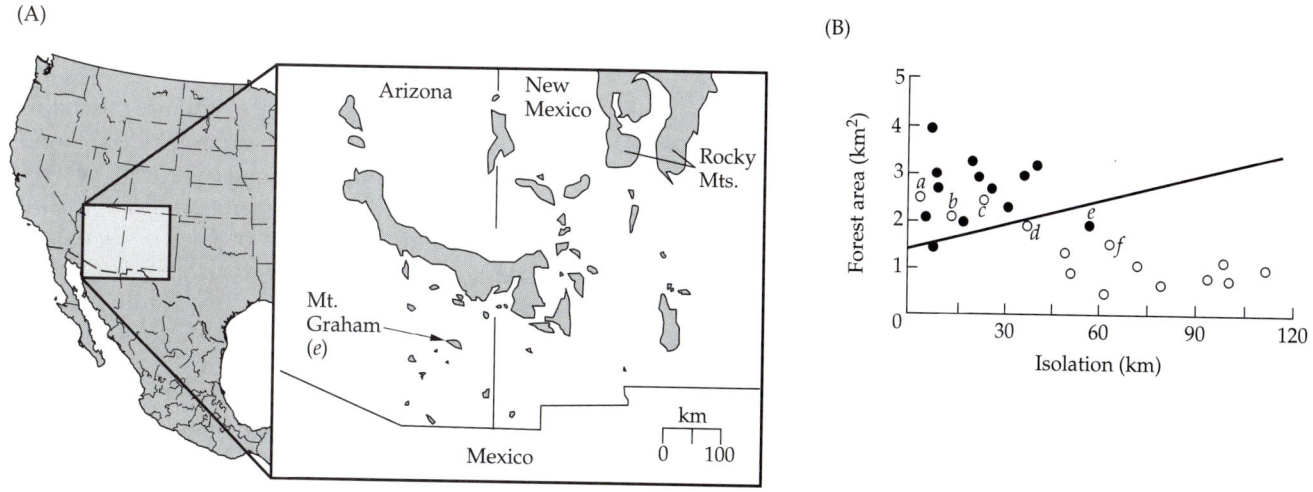

Figure 18.25 (A) Montane forest islands of the American Southwest. (B) Distribution of red squirrels (*Tamiasciurus hudsonicus*) among these forests (solid symbols indicate presence; open symbols indicate absence). Montane forest *a* appears to be a site with a high potential for natural colonization by red squirrels, while sites *b, c, d,* and possibly *f* may be good candidate sites for translocations and reestablishment of additional populations. The graph also indicates that the most tenuous and isolated population of red squirrels in this region inhabits site *e*, in the Pinaleno Mountains, the only forest inhabited by the endangered subspecies *T. h. grahmensis*.

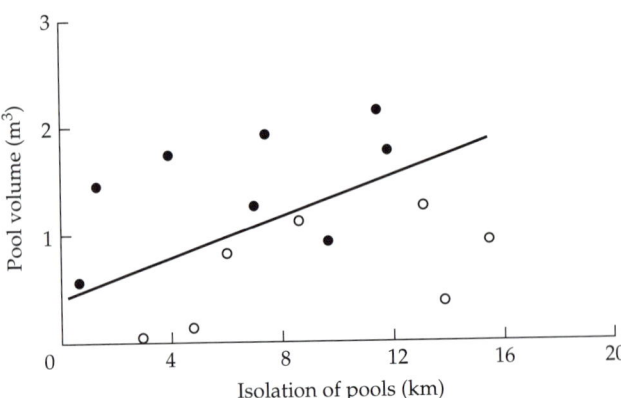

Figure 18.26   Just as for many terrestrial animals, the minimum habitat requirements (here, pool volume) of the endangered Ouachita Mountain shiner (*Lythurus snelsoni*) of eastern Oklahoma increase with the isolation of stream pools (solid symbols indicate presence; open symbols indicate absence). (From Taylor 1997.)

biota. In addition to these concerns, it is important to note that the most tenuous and most isolated population of red squirrels in this region inhabits site *e*. This is Mount Graham, the only montane forest inhabited by the endangered subspecies *T. h. grahamensis*. Given this fact, we should modify our translocation strategy. If we wish to establish additional populations of this genetically distinct subspecies (see Sullivan and Yates 1995; Brett Riddle, pers. comm. 1995), we should translocate a founding population to montane forests that are both large enough to promote population establishment and survival, and isolated enough such that intermountain migration and gene flow are unlikely. Thus, site *d*, and possibly site *f*, may be the optimal sites for establishing additional populations of the Mount Graham red squirrel.

The island model presented here may also have some relevance for conserving biodiversity in aquatic systems. The distribution of the Ouachita Mountain shiner (*Lythurus snelsoni*) among isolated stream pools is qualitatively identical to that of red squirrels in montane forests (Figure 18.26; Taylor 1997). This species is a candidate for listing as endangered by the U.S. Fish and Wildlife Service. As isolation increases, the fish requires larger pools to maintain its populations. We will return to this important case study later in this chapter.

The bog copper butterfly (*Lycaena epixanthe*) provides another illustrative example of the utility of spatially explicit tools in conservation biology. Populations of this species inhabit patchily distributed peatlands of New England. Joanne Michaud (1995) has shown that populations in southern Rhode Island are strongly limited by both the area and isolation of patches of their host plants (cranberries: *Vaccinium* spp.). As Figure 18.27 reveals, populations of this butterfly may persist in some relatively small cranberry patches if they are located near other occupied patches, whereas even some of the largest patches may be unoccupied if they are isolated from other populations.

In a similar manner, the distribution of the endangered bay checkerspot butterfly of California (*Euphydryas editha bayensis*) reveals the importance of biogeographic variables (Harrison et al. 1988). In this case, however, the distribution of the butterflies appears to be most strongly associated with isolation, and populations occur on some of the smallest patches of their optimal habitat (serpentine grasslands) as long as those patches lie within 5 km of source populations (Figure 18.28). Prior to 1986, checkerspots occupied all but four patches of grasslands within this critical isolation distance. In 1986, they colonized two of those patches (labeled *a* and *b* in the figure), including the largest unoccupied patch within 5 km of a source population. Thus, even when populations do not exhibit the predicted relationship between minimum area and

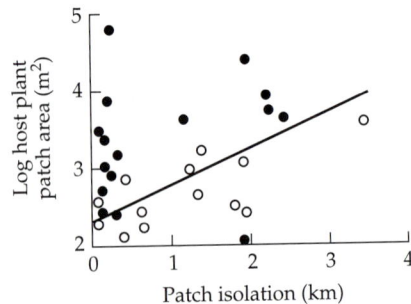

Figure 18.27   The area of cranberry patches required to maintain populations of the bog copper butterfly (*Lycaena epixanthe*) increases with patch isolation. (Solid symbols indicate presence; open symbols indicate absence.) (After Michaud 1995.)

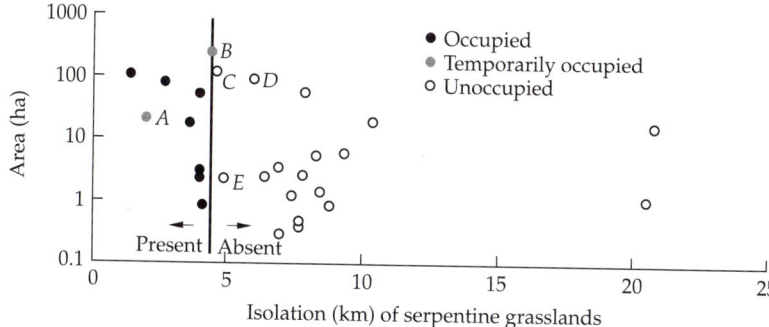

Figure 18.28   The distribution of the endangered bay checkerspot butterfly (*Euphydryas editha bayensis*) is strongly associated with the isolation of serpentine grassland patches, but not with area of those habitats. (Solid symbols indicate presence; open symbols indicate absence.) (After Harrison et al. 1988.)

isolation, this spatially explicit approach still can serve to assist conservation strategies. For example, of the remaining unoccupied sites, those labeled *c, d,* and *e* would seem to be good candidates for establishing viable populations of this endangered butterfly.

Many recent advances in spatial modeling and geostatistics, GIS technology, metapopulation theory, and landscape ecology hold great promise for assessing the effects of anthropogenic disturbances and ameliorating some of those effects. We conclude this chapter with a brief summary of how some researchers have applied these advances to predict the effects of impending changes in global climate and associated disturbances, including habitat loss and fragmentation.

Species Responses to Global Climatic Change

Magnitude of the problem.   The consensus among most meteorologists is that the already high and still increasing levels of carbon dioxide (CO_2), methane (CH_4), chlorofluorocarbons (CFCs), sulfur dioxide (SO_2), nitrous oxide (NO_2), and other anthropogenically derived **greenhouse gases** in the atmosphere have increased average global temperatures by approximately 0.5° C since 1860 (Figure 18.29). The term "greenhouse gas" refers to the fact that these gases allow solar radiation in the form of sunlight to penetrate the atmosphere, but retard the outward movement of infrared radiation from the earth's surface, thereby trapping heat and increasing global surface temperatures. Human activities over the past two centuries have been responsible for an estimated 25% increase in atmospheric levels of CO_2 and a 100% increase in levels of methane (see Karieva et al. 1993; Peters and Lovejoy 1992; Vitousek 1992). Nitrous oxides have also increased dramatically, especially during the past three or four decades. Of all the greenhouse gases, CFCs, which are solely anthropogenic and are the primary agents in the erosion of the protective ozone layer in the upper atmosphere, have increased most rapidly since their first detection during the early 1950s. Although these gases vary in their origin, concentration, and ability to absorb infrared radiation, they all are increasing exponentially, reflecting the exponential increases in human population densities and industrial technologies during the past two centuries.

While predicting future changes in global climate is an exceptionally complex task, projected concentrations of greenhouse gases can serve as a useful yardstick. If the current trends continue over the next century, levels of these gases are likely to increase so as to effectively double their present capacity to trap heat in the lower atmosphere (referred to as **CO_2 doubling**). Schneider (1993) has summarized the climatic changes predicted under a CO_2-doubling scenario, which include changes in a variety of environmental factors in addition to average global temperatures (Table 18.1). For the moment, however,

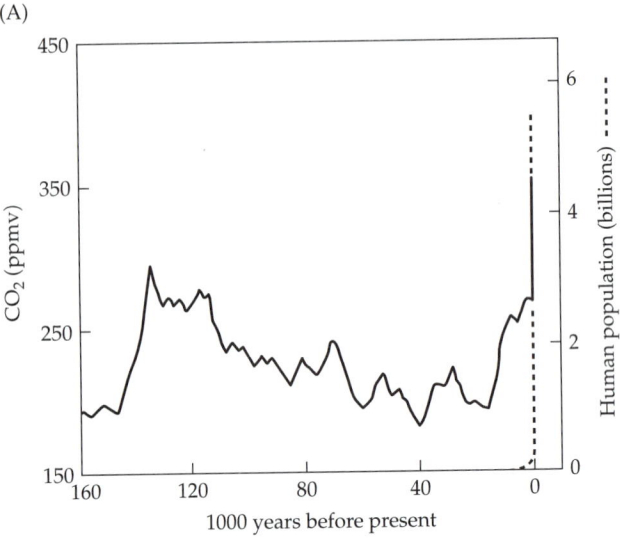

Figure 18.29   (A) Natural variation in atmospheric concentrations of carbon dioxide (solid line) over the past 160,000 years (based on ice core analysis). Recent anthropogenic increases in carbon dioxide concentrations are coincident with exponential growth of the human population (dashed line). (B) Increases in the concentrations of relatively stable greenhouse gases since the onset of the industrial age. (From Vitousek 1992.)

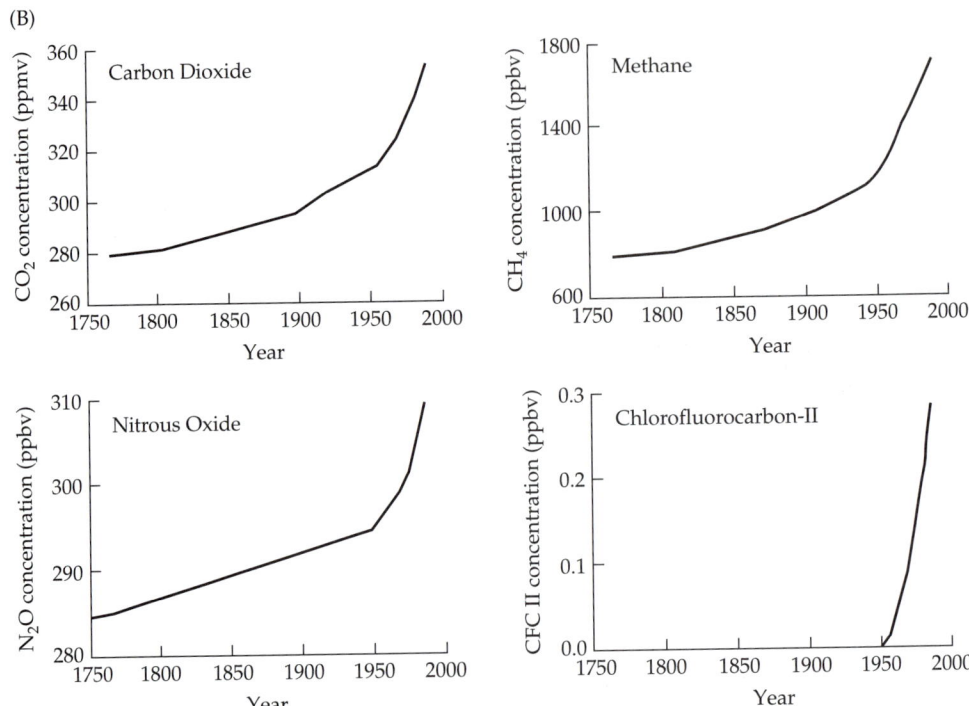

let's focus on temperature change. The projected climatic changes include a 1.5° to 4.5° C increase in average global temperatures over the next century, with a "best guess" of 2.5° C. It is important that we put this increase in context. Since the peak of the last glacial maximum, average global temperatures have increased by approximately 4.5° C, most of this taking place over a period of about 5000 years (see Chapter 7). This change was sufficient to melt the great continental glaciers and cause sea levels to rise by approximately 100 m. Now contrast this rate of warming—less than 1° C per thousand years—with the predicted increase in average global temperatures of 2.5° C in just one century. We know that even the relatively slow, natural warming of global climates had an enormous impact on biogeography and the biodiversity of late

Table 18.1

Predicted changes in climatic conditions as a result of doubling in atmospheric concentrations of CO_2

Degree of confidence[a]	Predicted change
	Temperature
*****	The lower atmosphere and earth's surface warm
*****	The stratosphere cools
***	Near the earth's surface, the global average warming lies between +1.5° C and +4.5° C, with a "best guess" of 2.5° C
***	The surface warming at high latitudes is greater than the global average in winter but smaller than in summer (in time-dependent simulations with a deep ocean, there is little warming over the high-latitude ocean)
***	The surface warming and its seasonal variation are least in the tropics
	Precipitation
****	The global average increases (as does that of evaporation); the larger the warming, the larger the increase
***	Increases at high latitudes throughout the year
***	Increases globally by 3–15% (as does evaporation)
**	Increases at mid-latitudes in winter
**	The zonal mean value increases in the tropics, although there are areas of decrease. Shifts in the main tropical rain bands differ from model to model, so there is little consistency between models in simulated regional changes.
**	Changes little in subtropical arid areas
	Soil Moisture
***	Increases in high latitudes in winter
**	Decreases over northern mid-latitude continents in summer
	Snow and Sea Ice
****	The area of sea ice and seasonal snow cover diminish

Source: Schneider 1993.

[a] The number of asterisks indicates the degree of confidence determined subjectively from the amount of agreement between models, and confidence in the representation of the relevant process in the model. Five stars indicate virtual certainties: one star indicates low confidence in the prediction.

Quaternary biotas. If the predicted surge in global temperatures actually occurs, its effects may be even more dramatic.

In the previous chapter we discussed some community-level approaches for predicting the effects of global climatic change. Reviews of biogeographic dynamics during the last glacial recession, however, reveal that species often shift their ranges individualistically, independent of other species (see Chapter 7). Even within the same taxonomic group, species differ markedly in their ability to adapt to new environmental conditions or to disperse to other sites when local conditions deteriorate. Thus, during the last glacial recession, species ranges shifted in different directions as well as at different rates, with some species migrating so slowly that their range expansion continues today, some 5000 to 10,000 years following the last major climatic shifts (see Figure 7.19).

Effects on species distributions.   If the projected changes in global climates do occur, they will no doubt alter the distributions of most terrestrial plants and animals. A starting point for predicting climate-driven change in the geographic range of a species is to define the species' **climate space**—the climatic dimensions of its niche, or the set of climatic conditions required by a particular species (Figure 18.30). Once we have defined the limits of a species' cli-

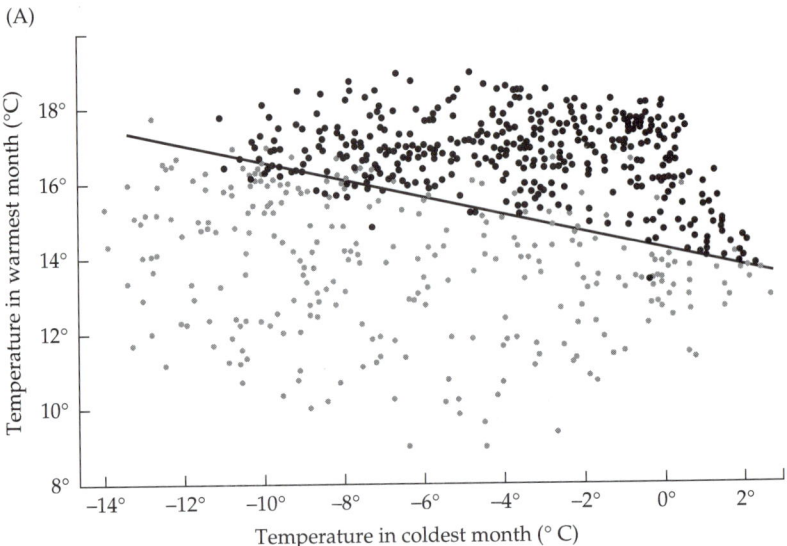

(A)

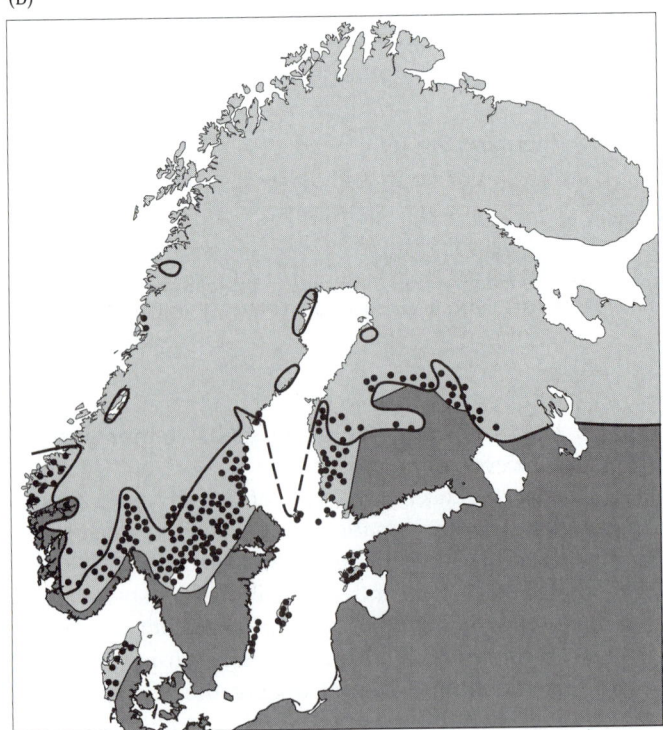

(B)

Figure 18.30   (A) A depiction of the climate space of the basswood tree, *Tilia cordata*, based on average temperatures during the warmest and coldest months. (Solid dots indicate locations within the range of this species, gray dots depict those outside the range.) (B) The actual distribution (solid dots). The range margin (solid line), as translated from the straight line in A, defines the marginal climate space for this species. (From Hengeveld 1992.)

mate space, we can apply them to maps of projected climates to predict changes in the species' geographic range. But this, of course, is too simplistic. We know that species do not occupy all regions within their climate space. Much like the ecological concepts of realized and fundamental niches, the geographic range realized by a species will always be less than that defined by its climate space.

We can think of at least two reasons for this to be so. First, realized geographic ranges are influenced by species' interactions with all features of the environment, including abiotic factors (e.g., soil type and disturbance history) and other species. Reproduction, mortality, immigration, and emigration are influenced by other niche dimensions, both abiotic (requirements for soil, salinity, and nutrient conditions) and biotic (interactions with predators, competitors, parasites, and mutualists). Second, as we noted above, range shifts and expansions may be quite slow, and, at least in some—perhaps many— instances, species may never jump across all barriers to colonize otherwise suitable habitats. Thus, we need at least four types of information to predict the effects of climatic change on geographic ranges of native species.

1. The rate and direction of climatic shifts, and the projected distribution of climatic zones across geographic regions.
2. Estimates of the climate space for focal species, with special emphasis on climatic variables typically included in climatic change models (e.g., average annual precipitation and mean monthly temperatures).
3. Predicted distributions of habitats and of landscape features (e.g., rivers, mountains, inhospitable habitats) that might serve as immigration barriers, filters, or corridors.
4. Estimates of the habitat affinities of the species and their capacities to disperse across future landscapes.

Below we provide several examples of how such information can be used to predict climate-driven changes in the geographic ranges of organisms.

Effects on terrestrial ecosystems. Climate space is a well-established concept among plant ecologists. The environmental limits of this space can be determined by laboratory growth experiments or by noting environmental conditions that coincide with the geographic range limits—both past and present—of the focal species (Figure 18.31). Zabinski and Davis (1989) used three climatic variables (mean January temperature, mean July temperature, and annual precipitation) to predict range changes in four tree species of deciduous forests in the Great Lakes region of North America. The northern distributional limits of all four species coincide strongly with the mean January isotherm of $-15°$ C. On the other hand, the southern limits of yellow birch and hemlock coincide closely with mean July temperatures, and the growth of all four species is strongly limited by annual precipitation. Zabinski and Davis used the output of a global circulation model (GCM) to map the projected steady-state climate under the scenario of CO_2 doubling. They assumed that these tree species could migrate at a rate of 100 km per century. Yet, as Gates (1993) noted, these rates are probably much too high, perhaps by a factor of five or ten. In her earlier work on range shifts of trees during the last glacial recession, Davis (1981) estimated that dispersal rates ranged between 10 and 40 km per century (20 km/century for hemlock, beech, and maple: see Table 7.2). Given the more fragmented nature of current and projected landscapes, future rates of tree dispersal may be much lower than those during the last glacial recession.

Eastern Hemlock (*Tsuga canadensis*)

Beech (*Fagus grandifolia*)

Yellow Birch (*Betulla alleghaniensis*)

Sugar Maple (*Acer saccharum*)

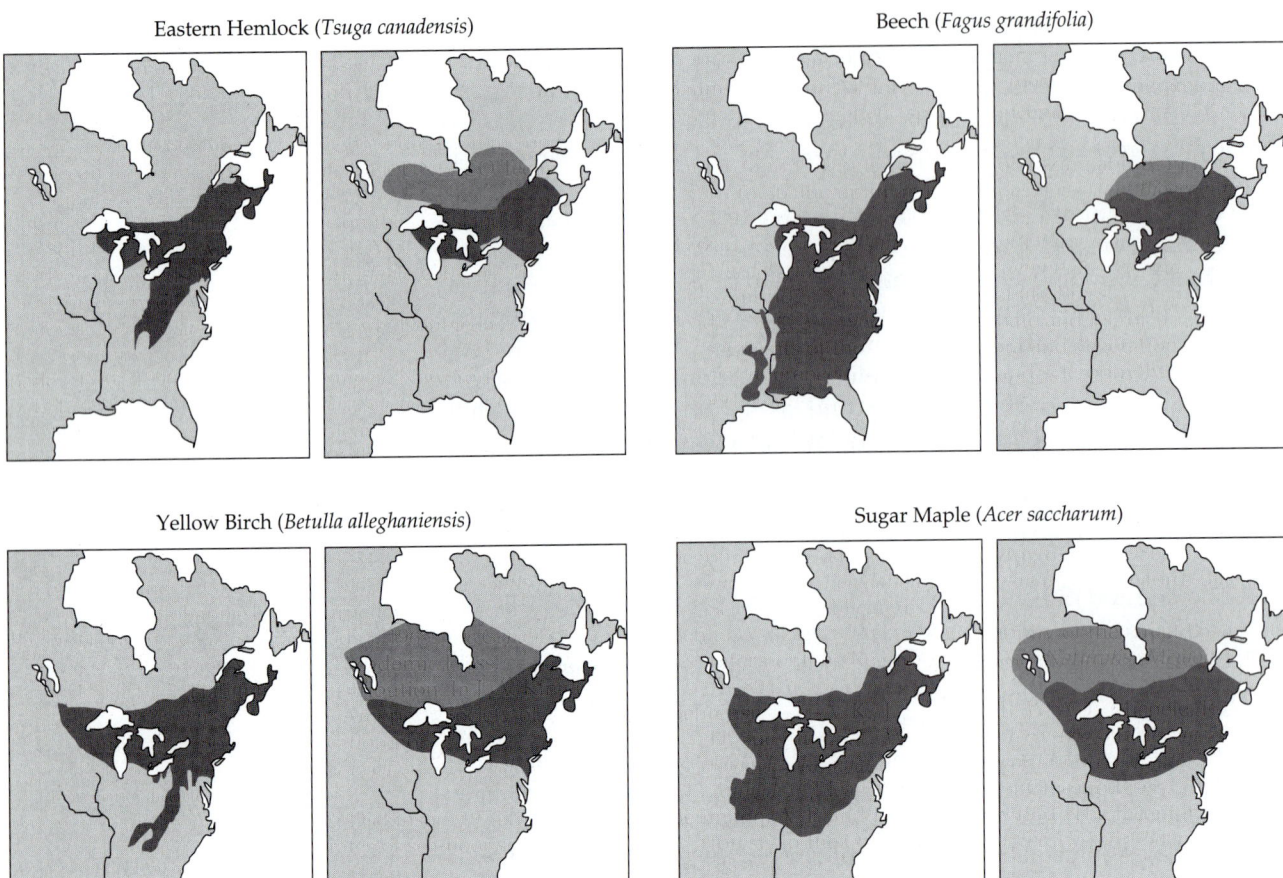

400 km

Figure 18.31 Present (left) and predicted (right) geographic ranges for four species of trees common to eastern regions of North America. The predicted ranges are based on the assumption of a doubling of current concentrations of CO_2; the black area depicts the projected range in 2090 considering estimated species-specific rates of migration, while the gray area depicts the ultimate potential ranges provided enough time for species to migrate to all regions within their climate space. (After Gates 1993.)

Nevertheless, Zabinski and Davis's (1989) results are informative. The differences between the current ranges of the four tree species and their projected ranges are presented in Figure 18.31. Even with the relatively high dispersal rate used in this exercise, the projected ranges after 100 years of migration are much less than the potential ranges based on the species' climate space. In addition, the projected realized ranges for all four species are much less than the extant ranges of these species.

The cascading ecological effects of range shifts by these species, while extremely difficult to predict, may be no less important than the more direct effects of climatic change (Gates 1993). With reductions in the ranges of these now dominant tree species, the geographic ranges and cover of aspen and other early successional species are likely to increase, in turn triggering substantial increases in populations of white-tailed deer. Overgrazing by deer, compounded by a reduction in beech seed production, may contribute to significant declines in birds and small mammals. Gates (1993) also suggests that fire may become a more prevalent component of these ecosystems, and that trees persisting in marginal environments will be stressed and likely to suffer from a higher incidence of insect pests and disease. In fact, even with the high rates of dispersal assumed by Zabinski and Davis, they concluded that many hardwood species of the Great Lakes region would be threatened with extinction under this climatic scenario.

In a second illustrative case study, Terry Root modeled the effects of climate on the winter ranges of songbirds in North America (Root 1988a,b,c, 1993; Root and Schneider 1993, 1995; see also Repasky 1991). Both Hutchinson and MacArthur had hypothesized that, while the southern range limits of terrestrial animals of North America are largely set by interactions with other species, their northern range limits are more strongly influenced by abiotic factors, especially temperature. Following their lead, Root based her projections primarily on the locations of isotherms during the most stressful time of year: winter. She assumed that during winter, these diurnal birds must survive the night by tapping their energy supplies (stored as fat and other metabolizable tissues). The cooler the temperature, the more energy a bird must use for thermoregulation. Using known or estimated metabolic rates for a variety of species, Root was able to estimate the minimum winter temperature at which a bird must exhaust all its energy supplies to survive through the night. This critical winter temperature represents one dimension of the climate space of these birds. For most of the species she studied, it lies between 0 and –5° C. Root then compared the actual winter ranges of songbird species with minimum January isotherms across North America, and found that the northern range limits of 51 species of passerines were strongly associated with minimum winter temperatures (Figure 18.32). She also found that among these 51 species, larger birds had ranges that extended farther north, consistent with bioenergetic considerations (cold tolerance in homeotherms generally increases with body size).

Using maps of projected shifts in winter isotherms under climatic change, Root's model can be applied to predict climate-induced shifts in the winter ranges of birds (see Root and Schneider 1995). But the shifting of winter isotherms is just one of the many factors, abiotic and biotic, that can influence range dynamics. For example, Kirtland's warbler, which is narrowly restricted

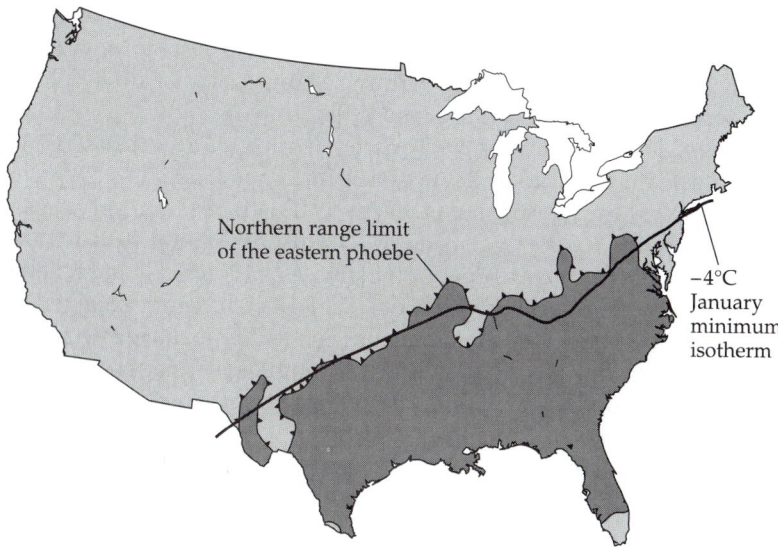

Figure 18.32 The northern limit of the winter range of the eastern phoebe (*Sayornis phoebe*) corresponds very closely with the –4° C minimum January isotherm (heavy solid line). Bioenergetic models also indicate that below this temperature energy stores of these small birds are insufficient to maintain a constant body temperature during their nightly fasts. (From Root 1993.)

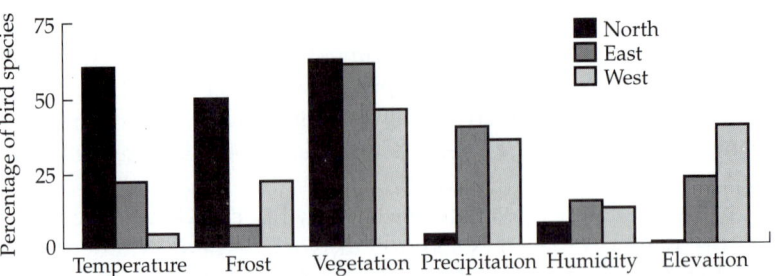

Figure 18.33   The percentage of North American bird species whose northern, eastern, and western range boundaries during winter were associated with various environmental factors. Temperature = average minimum January temperature; frost = mean length of frost-free period; vegetation = potential vegetation; precipitation = average annual precipitation; humidity = average relative humidity; elevation = elevation above sea level. Southern boundaries of winter ranges were not analyzed because they often extended beyond the study area. (From Root 1988c.)

during the breeding season to jack pine habitats of Michigan, is unlikely to survive even a small climatic change (Root 1993; Root and Schneider 1993; see Botkin et al. 1991). Jack pines require relatively well drained, sandy soils for survival, soils that do not occur directly north of the birds' current range. Therefore, even if Kirtland's warblers could potentially track their climate space as it shifts northward, they would probably be stranded because of their other niche requirements.

In an earlier study, Root (1988c) investigated spatial associations between the northern, eastern, and western range boundaries of songbird species and a variety of climatic conditions, as well as potential vegetation (Figure 18.33; southern range boundaries were not analyzed because they often extended beyond the region sampled in Root's study). Overall, range boundaries were most closely associated with potential vegetation, which is not surprising given that it reflects the combined variation in climatic and many other abiotic and biotic variables. The association of range boundaries with the climatic variables varied substantially. For example, northern range boundaries were strongly associated with average minimum January temperatures and mean length of the frost-free period. In contrast, eastern range boundaries were most closely associated with annual precipitation, while western range boundaries were associated with these two variables plus elevation. Elevational limits along western range boundaries are coincident with the Rocky Mountains, which run north to south and mark the range boundaries of many North American species (see also Harte and Shaw 1995).

Again, just as McDonald and Brown's model (Figure 17.29) was based on a number of untested assumptions, those of Root and her colleagues may also be too simplistic. Yet their results again serve to illustrate the multifactorial nature of ecology and biogeography, and to identify information essential to predicting the effects of global climatic change. Our ability to make reliable predictions of the biogeographic and ecological effects of climatic change will require a better understanding of the independent and interactive effects of a variety of environmental factors, including geology, soils, vegetation, and topography as well as other plant and animal species, in addition to anticipated changes in temperature and precipitation.

Effects on freshwater and marine ecosystems.   The geographic ranges of marine and freshwater species will also be affected by global warming if precipitation and evaporation potentials change according to current projections (see Table 18.1). For example, Nash and Gleick (1991) estimate that a 4° C increase in average annual temperature will result in a 9–21% increase in mean annual runoff into streams of the Colorado Basin of western North America. The anticipated anthropogenic changes in climates over the next century may also cause seasonal shifts in periods of peak runoff and streamflow (e.g., see Gleick 1989). The predicted increased ratio of rain to snowfall and the anticipated elevation of the snowline may increase streamflow during fall and winter, while decreasing it during spring and summer. These shifts could cause changes in water quality and in the levels and circulation patterns of lakes and other freshwater systems, which in turn could result in significant and widespread changes in the composition and diversity of aquatic communities (see also Grimm 1993). Because of the limited size of freshwater systems compared with terrestrial ones, it is much more likely that freshwater species will become stranded, unable to persist by dispersing to other streams and lakes or across watersheds.

In contrast to freshwater systems, it may seem logical to assume that, because of their massive volume, marine systems will be buffered against most effects of anthropogenic changes in global and regional climates. Yet marine communities actually may be severely affected by a combination of environmental changes associated with warming climates. It appears that surface waters will respond to global climatic change with increased warming and evapotranspiration. Ocean currents, including major ones such as the Gulf Stream and Humboldt, will then shift substantially. As Norse (1993) observed, the atmosphere and the ocean are two tightly linked parts of the same system, exchanging huge amounts of gases and heat. Oceanographers have identified several changes in marine environments that may directly or indirectly affect native species. The effects of CO_2 doubling on the marine environment are likely to include the following (see Fields et al. 1993; Titus 1990):

1. Latitudinal and vertical shifts in water temperatures will occur. In general, warm waters will shift poleward and to greater depths.
2. Currents and their effects on horizontal and vertical mixing of waters will be altered significantly.
3. Surface water temperatures will increase in many regions, but decrease in others, especially areas along the west coasts of continents, where upwelling may increase in magnitude.
4. Salinity of surface waters near the equator and at higher latitudes will decrease due to increased regional precipitation.
5. pH of surface waters is expected to drop by 0.3, equivalent to a doubling in acidity.
6. Sea levels are expected to rise by 0.5 to 2.0 meters globally due to thermal expansion of seawater and additional ice melt.

While the precise magnitude and geographic profile of these disturbances are impossible to predict with any reasonable degree of certainty, they will dramatically affect species distributions and biodiversity in marine communities. For example, the predicted rise in global sea levels will substantially reduce the size of coastal wetlands and possibly coral reefs as well (Norse 1993). The projected rate of increase in sea levels is likely to overwhelm the ability of wetlands to re-form farther inland and of reef growth to track the rising waters. As we learned in earlier chapters, decreases in habitat area translate into decreases in species diversity. In the United States alone, some 80 species of endangered plants and animals occur only in the narrow 3-meter band above sea level (Reid and Trexler 1991). These and many other rare intertidal species worldwide are likely to perish unless they can migrate apace with rising sea levels. Not surprisingly, insular biotas will be especially threatened by the projected increase in sea levels. For example, the Maldives Archipelago, which includes 1190 small islands rising just 2 meters above the surface of the Indian Ocean, would be submerged, thus destroying natural habitat for five endemic species of plants (Norse 1993).

As we also learned in the foregoing chapter, mixing of biotas often results in surges in the extinction rates of native species. This fact is especially relevant for predicting the effects of climatic change on marine communities. If surface waters warm substantially, the geographic ranges of cold-stenothermal species should contract, while marine species now restricted to tropical and subtropical regions may expand their ranges far into temperate and possibly subarctic regions (see Figure 7.25). If this mixing of long-isolated biotas occurs, it will result in the loss of many species due to competitive exclusion, predation, parasitism, or other interspecific interactions.

Studies of some recent transient shifts in ocean circulation may provide a sample of the ecological and biogeographic effects of more intense and persistent changes in global climates. El Niño phenomena are caused by fluctuations in the strength of trade winds. As a result, warm and nutrient-poor subtropical waters flow farther poleward along the western continental margins. The combined effects of increased water temperatures, reduced nutrient availability, decreased primary productivity, and the influx of subtropical invaders have severely affected marine communities. For example, the 1982–1983 El Niño event increased surface temperatures by as much as 2° C along the western coasts of North America, reduced primary productivity off the coast of California to its lowest recorded level, and triggered northward range expansions of at least 18 species of invertebrates and 38 vertebrates (McGowan 1985; Pearcy and Schoener 1987). Such effects no doubt cascaded through marine communities, affecting trophic levels from zooplankton to seabirds (Fields et al. 1993).

The high frequency and intensity of El Niño events during recent decades may actually be an early signal that global climatic change is already affecting marine communities. Other potential signals are also apparent. In particular, the massive die-offs, or "bleaching," of coral reefs that occurred during the 1980s may be connected with increases in global temperatures. Coral reefs, the most productive and diverse systems in the marine realm, arise from a complex yet fragile mutualistic relationship between dinoflagellate algae and scleractinian corals. This mutualism is obligatory, and is easily disrupted by environmental disturbances, including relatively slight but abrupt changes in water temperature (see Glynn 1991). If current trends in levels of greenhouse gases continue, they may cause massive bleaching of coral reefs, drastically reducing their geographic ranges and triggering cascading effects on the overall diversity of the marine realm.

Changes in connectivity and isolation.   How well can we now predict the effects of climatic change on the distributions of terrestrial and aquatic species? Let us review our list of requisite information for predicting the effects of climatic change on the geographic ranges of species. It should be clear by now that the modeling approaches discussed earlier in this chapter, while offering many potential insights, are based on many untested assumptions. Climatologists, however, continue to make great progress in their ability to model climate change at global to regional levels (item 1 on our list). Climate space (item 2)—the climatic limits of a species' fundamental niche—has been estimated for a variety of species, both plants and animals. Shifts in landscape features and habitats (item 3) must be estimated by integrating information from items 1 and 2: climatic shifts and estimates of climate space for plants, which, in turn, constitute a major habitat component for many other organisms. Recent climatic shifts, such as El Niño events, along with the more prolonged shifts associated with glacial cycles, can serve as valuable yardsticks for evaluating the assumptions used to model projected climates. They can also provide estimates of the abilities of some species to migrate across relatively continuous habitats (item 4).

Most approaches to modeling the biogeographic effects of climatic change, however, share one shortcoming: they do not explicitly address the effects of changes in habitat connectivity and isolation on the abilities of species to shift their ranges or survive within fragments of their current ranges. Real landscapes are heterogeneous mixtures of discontinuous habitats and topographic features that serve as dispersal barriers, filters, or corridors, depending on the focal species. Here again is a case in which biogeographic data may provide some valuable information. As we discussed in the previous chapter, McDonald and Brown (1992) used such data to estimate the effects of climate-driven reductions in forest cover on boreal mammals of the Great Basin. Unfortunately, their approach cannot be used to assess the effects of changes in habitat isolation on species distributions. Such effects, however, can be predicted by using spatially explicit models, such as the mechanistic models developed by Hanski and others working in metapopulation theory (e.g., see Hanski et al. 1995; Wahlberg et al. 1996; Hanski and Thomas 1994). Empirically based, spatially explicit models such as the one illustrated in Figure 18.23 also can be used to estimate the effects of fragmentation on the distributions of endangered species.

Let us illustrate this approach with a hypothetical case study: hypothetical in that we arbitrarily assume values for changes in habitat area and isolation due to climatic change, but real in that the projections are based on the actual, extant distribution of a rare fish species—the Ouachita Mountain shiner (*Lythrurus shelsoni*)—in stream pools in eastern Oklahoma. For simplicity, we have assumed that changes in area and isolation would be constant for all stream pools in this freshwater system. If regional precipitation decreases, then stream pools should become smaller and more isolated. Using arbitrarily chosen values for changes in volume (-0.5 m^3) and isolation ($+2$ km), we can now predict the new distribution of the shiner. Our hypothetical changes in stream habitats would cause five of the eight extant populations of this threatened fish species to become extinct (Figure 18.34). Note also that because population persistence should be influenced by the combined effects of immigration (isolation) and extinction (area), even the population now occupying the largest stream pool (labeled 1 in Figure 18.34B) would be extirpated, while those inhabiting smaller but less isolated pools (2–4) should survive.

Again, we emphasize that this is just a hypothetical example used for illustrative purposes. Without more direct information on the dispersal abilities of focal species, these types of biogeographic models can apply only to steady-state conditions. That is, they cannot be used to estimate *rates* of relaxation, only the final distributions after climates and landscapes (or seascapes) have stabilized. In fact, none of the approaches discussed in this chapter can stand alone. The challenge of predicting the ecological and biogeographic effects of global climatic change may well pose a problem of unrivaled complexity. It calls for temporally and spatially explicit models that span a range of scales from individual to global levels, and include the many potential interactions and feedback loops among model components, including hydrological and

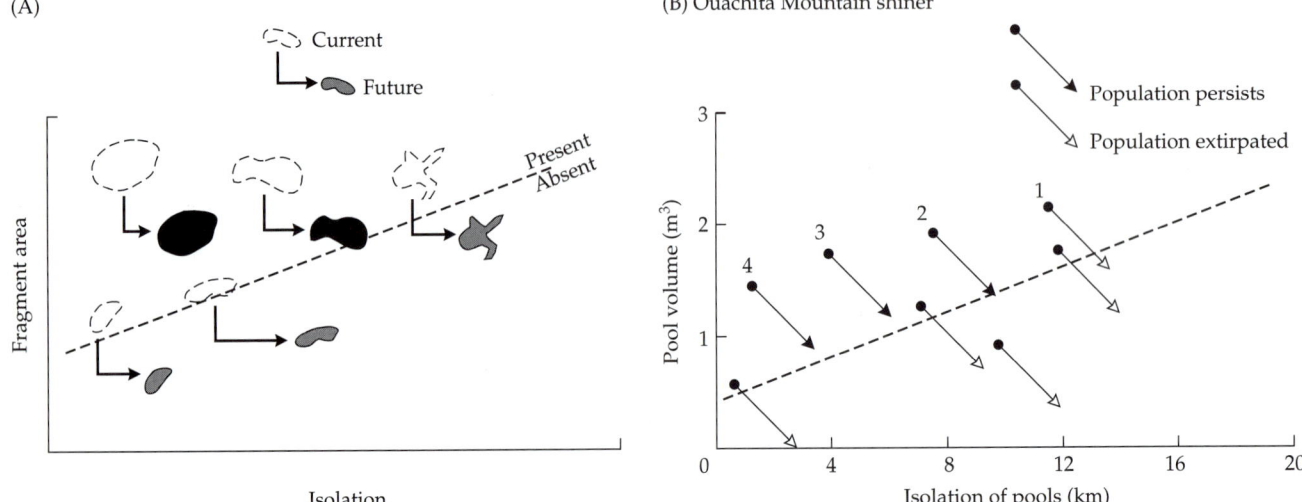

Figure 18.34   Given accurate information on the distribution of a focal species and estimates of changes in the area and isolation of its principal habitat, the species-based model of Figure 18.23 can be used to predict the effects of anthropogenic climatic change. (A) A general model for predicting the effects of changes in the size and isolation of habitat patches for a hypothetical species. (B) An illustration of how the model can be used to predict changes in distribution and site occupancy, based on the actual current distribution of the Ouachita Mountain shiner (*Lythrurus snelsoni*) (see Figure 18.26). This example arbitrarily assumes that a decline in regional precipitation would reduce pool volumes by 0.5 m³ and increase their isolation by 2 km. Solid symbols indicate presence, open symbols indicate absence; arrows indicate changes in patch or pool characteristics. (Data from Taylor 1996.)

nutrient cycles and interspecific interactions. Yet this complexity is not cause for apoplexy. To echo our conclusions in the previous chapter, we now possess an impressive arsenal of technological tools and sophisticated theory. While we need to continue to develop these tools, our greatest handicap lies in our limited knowledge of the diversity and distributions of the earth's species. With each newly described life form and each new map of a geographic range, we not only advance relevant theory, but we also enhance our ability to understand and perhaps mitigate the ongoing crisis in biological diversity.

CHAPTER 19

Biogeography for the Twenty-first Century

What does the future hold for the discipline of biogeography? What will the field be like a decade or a century from now? Obviously, many uncertainties limit any attempt to make such a prediction. A large asteroid colliding with the earth might cause another mass extinction, including the disappearance of *Homo sapiens*. A world war or global depression might curtail the progress of all science, including biogeography. On the other hand, revolutionary techno-logical innovations and the discovery of additional energy sources to replace declining global oil supplies might lead to rates and directions of scientific progress that defy imagination.

Nevertheless, it is fun to speculate about the future of biogeography. Some limited speculation may lead to fairly accurate predictions, at least in the short term. Several trends are apparent in the current direction of the field, and we can expect them to continue, at least for the next several years. We can also see the potential for current advances in other fields to influence the future devel-opment of biogeography. And we can engage in some wishful thinking, because there are areas in which the current progress of the discipline appears to disappointingly slow—either because critical empirical information or new theoretical insights are needed, or because there is dispute and controversy where there should be cooperation and synthesis.

Technological Advances

New Data

The last few decades have seen an explosion of information in every field. As indicated in many places throughout this book, this has had an enormous influence on biogeography. New technologies have given scientists new win-dows through which to view the natural world. In the period between World War II and the 1980s, many of these advances took the form of better "micro-scopes"—better ways of obtaining data about small-scale phenomena. Particle accelerators, electron microscopes, automated amino and nucleic acid sequencers, and other advances in particle physics and cellular and molecular biology come to mind. Since the 1980s, however, technology has also provided powerful new "macroscopes" that have permitted new insights into large-scale patterns and processes. Examples include technology for mapping submarine and subsurface geology and the resulting advances in plate tectonics (although, as indicated in Chapter 6, this capability began to be developed in World War II); satellite imagery and other forms of automated remote and ground-based sensing and the resulting advances in climatology, oceanography, and geography; and computer-related technology, such as digitizing, scanning, image analysis, visualization, communication, and data storage and management, and the resulting compilation and analysis of data in many fields on an unprecedented scale.

The net result of these macroscopic technologies has been a great increase in the data on the earth that inform the physical earth sciences (geology, climatology, oceanography, and physical limnology), the larger-scale biological sciences (systematics, ecology, and biogeography), and the social sciences (geography). We are compiling increasingly detailed and accurate maps of the spatial distributions and dynamic temporal patterns of many important physical, biotic, and human-related variables and processes on scales from local to global. These maps depict such things as shifting ocean currents and climatic patterns; seasonal and interannual variation in vegetation, plankton, and terrestrial and marine productivity; and changes in the human population and its effects, such as habitat conversion, fragmentation, and global climatic change. We can not only see how the earth is changing in response to both natural fluctuations and human influences, we can also readily visualize the consequences of these changes for the distributions of organisms and the conservation of biodiversity.

Analytical Methods

Along with ever-increasing quantities of higher-quality data have come advances in techniques for compiling, storing, managing, distributing, and analyzing those data, and for depicting and interpreting the spatial patterns and dynamic processes they reveal. Most of these methodological advances are a direct result of computer technology. The last two decades have seen a revolution in computing in which large mainframes—which were slow, expensive, and accessible to only a few trained users—have given way to small, inexpensive personal computers, which have greater speed, memory, and computing capacity than the mainframes of the 1970s. Today's computer systems require relatively little specialized expertise in operating systems, languages, and programming, and are available in every scientific laboratory and many households.

Along with these improvements in hardware have come amazing advances in software. The pace of development has been awesome, both for "relatively simple" packages for word processing, graphics, data management, and statistical analysis, and for enormously complex applications such as the Internet and the World Wide Web communication network, geographic information systems (GIS), multipurpose platforms for simulation modeling, and forms of "adaptive computing" such as robotics, genetic algorithms, and artificial intelligence. This may be hard for student readers to visualize, but the first edition of this book, written between 1979 and 1982, was typed on mechanical typewriters, and the original illustrations were all drawn by hand. Only the index, done in 1983, used the then-new technology of *computerized word processing!*

Contributions of Technology

Many of these developing technologies promise to have important effects on the field of biogeography, although it is difficult to predict just what those effects will be. The sheer spatial and temporal scales of biogeographic patterns and processes make it invaluable to obtain, analyze, and visualize previously unavailable kinds and quantities of data. Thus the already great influence of technology on the field can only increase. Satellite imagery and other kinds of remote sensing technology promise to provide a huge amount of information on the geology and topography of the earth's surface and on the dynamic changes in climate, ocean circulation, vegetation, land use, and other human and animal activities that are occurring. Some applications are already in operation: networks of satellite and ground-based sensors to monitor weather, ground-penetrating radar for mapping soils and other subsurface structures, newly declassified high-resolution satellite images useful for mapping individual plants and comparably sized objects, satellite-based global positioning systems (GPS) for precisely locating and mapping places and objects, high-resolution spectral analyses for making fine discriminations among soil and vegetation types, satellite-based biotelemetry for mapping the long-distance movements of ocean fish, migratory birds, and marine and terrestrial mammals, and radar for quantifying numbers, sizes, and directions of nocturnally migrating birds are just a few examples. Other, more powerful data collecting systems will soon be operational. One of the most promising is the new generation of EOS (Earth Observing System) satellites, designed specifically to provide new and better data on the physical, biological, and anthropogenic features of the earth's surface.

Given what is already being done, the amounts and kinds of data that should soon be available are truly awesome. For example, we soon will have the capability to monitor individual trees in temperate and tropical forests using remote sensing technology: to map their locations; measure their heights, canopy areas, and trunk diameters; identify individual trees to species; and record their seasonal patterns of leafing, flowering, and fruiting. It should also be possible to monitor many physical, chemical, and biological characteristics of the microenvironment, not just of the canopy, but also of the different layers of the forest and the forest floor. Similarly, we should be able to follow a pod of whales or a school of tuna as they migrate thousands of kilometers, keeping track of the horizontal and vertical movements of each individual, and for selected individuals with implanted transmitters, using satellite receivers to monitor feeding and mating behaviors, heart rates, energy consumption, and blood chemistry and hormone levels. Similar data could be collected from mixed-species herds of grazing mammals on the Serengeti Plain of central Africa or from groups of kangaroos in the arid interior of Australia.

Ironically, our capacity to obtain data may exceed our ability to manage, analyze, interpret, synthesize, and productively use the information. Remote, automated sensors, most of them controlled by computers, have the potential of supplying more kinds of data more rapidly than they can be assimilated. Here again, however, computer technology will play a major role. Advances in hardware are making it possible to store and transfer enormous quantities of data at amazing speeds. Developments in software are making possible ever more sophisticated analyses, including representations of complex two- and three-dimensional relationships among multiple variables, and animations to visualize dynamic processes that occur on time scales from microseconds to millennia.

The ultimate constraint appears to be human limitations, especially the capacity of our minds to process and use all of this information. The new technologies, and the data that they can supply, will be scientifically useful only to the extent that they can address important basic and applied questions. In order to increase our understanding of the natural world and to apply this knowledge to practical issues of management and conservation, it is still essential to formulate the right questions and to know when they have been answered. In few disciplines are these human limitations so important as in biogeography and related areas of the environmental and earth sciences. Until recent decades, our human senses allowed us to perceive our surroundings on only a very limited spatial and temporal scale: from millimeters to kilometers in space, and from milliseconds to decades in time. Technology has widened our spatial and temporal windows by many orders of magnitude in all dimensions. But it is extremely difficult for the human mind, unaided by computers, to visualize and think about phenomena that occur on scales from nanometers and picoseconds to light-years and billions of years.

Consider just one example: the region containing the greatest concentration of life on the planet. It is hard enough to visualize the more than 7,000,000 km^2 of the Amazon basin, with its over 3000 m of elevational relief, its varying hydrology and geology, and the complex, dynamically changing abundances and distributions of its millions of plant, animal, and microbial species. Add to this the ever-increasing effects of human activities, which include not only a spatial kaleidoscope of fires, forest clearing, agriculture, and settlement, but also local and regional shifts in climatic and other conditions due to global changes. Maps of vegetation types, species distributions, hot spots of endemism and diversity, fires, and clear-cuts can depict the spatial and dynamic patterns of the system in a snapshot of time and help us visualize the kinds and magnitudes of both its natural components and human influences. It is harder, however, to understand the ecological and historical processes that have produced these patterns, and to develop management plans and social policies that will prevent the degradation of the Amazon basin ecosystem and the extinction of a large fraction of its biota. How does one use even a fraction of the available information to do basic or applied science on such complex systems?

Yes, new technologies have produced an explosive increase in information, and more is yet to come. But the great challenge to humankind during the coming years and decades will be to use this information to better understand our natural world so that we can maintain essential ecological processes, preserve biodiversity, and enhance the quality of human life.

Conceptual Advances

One could make the case that progress in theoretical science has not kept pace with the explosion of empirical data. In fact, it is hard to point to many conceptual breakthroughs in the last two decades. Since the mid-1970s, many sciences, including biogeography and related disciplines, appear to have been in a descriptive phase. The technological advances mentioned above have enabled scientists to produce increasingly precise and detailed descriptions of natural phenomena. Thus, for example, we have highly detailed, multicolored maps of vegetation, soils, and geology for much of North America, Europe, Australia, and other regions of the world—maps prepared using satellite imagery, computer-based classification systems, multivariate and geostatistics, and geographic information systems. We also have maps, compiled from museum collections and other records, of the distributions of many plant and animal species. Some of these data have a substantial temporal dimension, supplied by fossils, museum specimens, and other historical records, so that we can document changes in the distributions of vegetation types and some plant and animal species over time periods ranging from a few years to several millennia.

What we lack, however, are new ideas and syntheses to place these empirical findings in a general conceptual framework. For example, we still have no widely accepted theories to explain some of the most pervasive patterns in biogeography, such as the latitudinal and other major gradients of species diversity (see Chapter 15) or the distributions of body sizes and abundances among species within local communities and on larger spatial scales (see Chapter 16). We have little ability to predict whether an immigration event will result in the establishment of an exotic species (see Chapter 18), and we remain puzzled by the apparent paradox of the taxon cycle, in which immigrant species, with no previous experience in an environment, are able to colonize it and eventually replace differentiated endemic forms that we would expect to be better adapted to that environment (see Chapter 14).

New Theory

We would be surprised to see this apparent imbalance between rates of empirical and theoretical advances continue for much longer. As mentioned above, technological innovations have not only made possible the collection of large quantities of increasingly accurate data, but have also facilitated the manipulation, analysis, and interpretation of those data. While some of the details and complexities that have been revealed may seem daunting, in other cases clear patterns of seemingly wide generality have begun to emerge; we have just mentioned several examples. The increasing quantity and quality of data will no doubt assist our search for causal mechanisms, permitting rigorous evaluation of existing hypotheses and stimulating development and testing of new models. Yet it seems highly unlikely that the compilation of data for so many disparate purposes, no matter how massive, will by itself lead to any great scientific revolutions. As Poincaré once said, "Science is built up of facts, as a house is built of stones, but an accumulation of facts is no more a science than a heap of stones is a house." It seems imperative, then, during this phase of exponential accumulation of facts, that biogeographic research continue to be guided by relevant theory and by questions that challenge accepted views and stretch the conceptual envelope of the discipline.

It is difficult, of course, to predict just where and when conceptual breakthroughs will occur. It is a bit easier to say something about what they may be like and how they will differ from current theories. Much of the existing theory of biogeography and related disciplines dates back at least to the 1960s and 1970s. Much of this theory was based on the reductionist idea that the natural world can be understood by taking complex systems apart, identifying their basic components, studying how these function in isolation and interact with one another, and then putting the components back together to build up subsystems of incrementally increasing complexity. This is the approach that underlay the attempt to understand the organization and species diversity of ecological communities by considering first the environmental requirements of single species in isolation, then the basic competitive, predator-prey, and mutualistic interactions between pairs of species, and finally the roles of multiple and indirect interactions (see Chapters 4 and 5). In addition, most current theory is fundamentally equilibrial in that it assumes the preservation of a balance between opposing processes whose rates are set by unchanging, extrinsic environmental conditions. A classic example is MacArthur and Wilson's theory of island biogeography, which assumes a balance between immigration and extinction, which are functions of the constant, extrinsic conditions of island size and isolation, and which counteract each other to maintain a fixed equilibrium number of species (see Chapter 13).

We suspect that whatever the new theoretical advances in biogeography may be, they will be increasingly holistic and nonequilibrial. They will attempt to account for the complexity of systems less by taking them apart and studying their components in isolation than by trying to explain the emergent properties of the entire intact systems. They will not assume that there is a balance of nature, but rather that nature is continually changing in response to both its own intrinsic dynamics and to perturbations in its extrinsic environment.

Consider how such an alternative perspective might change the equilibrium theory of island biogeography. The original theory assumed that different species were nearly identical "particles," similar in this way to the molecules of a gas. MacArthur and Wilson actually realized that both the species already present on an island and those in the pool of potential colonists differed in important ways, and that these differences led to some important features of island biogeography, such as the concave-upward shapes of immigration and extinction rate curves (MacArthur 1972; see Figure 13.8). Their original model, however, did not take explicit account of the variation among species in such characteristics as population size, resource requirements, competitive ability, susceptibility to predation, dispersal ability, or vulnerability to extinction. Yet we know that it is possible to incorporate such complexities into more realistic models of island biogeography (e.g., Gilpin and Diamond 1981; Lomolino 1998). MacArthur and Wilson's equilibrium model also implicitly assumed that the environment of each island remains constant over time, so that the mathematical functions (shapes of the curves) characterizing the rates of colonization and extinction do not vary over time. Thus, the original model cannot account directly for the large changes in species richness due to events such as the eruption of Krakatau (Bush and Whittaker 1991), the within-island and within-archipelago speciation in Hawaii (Wagner and Funk 1995), the post-Pleistocene isolation of Great Basin mountaintops (Brown 1971) or land-bridge islands (Diamond 1972), or the settlement of first Polynesians and then Europeans on Polynesian islands (Olson and James 1982a, 1984; Steadman 1995). These shifts in species distributions and diversity could be much better explained by a nonequilibrial framework that still invokes the opposing forces of origination and extinction, but allows these processes to vary over time with changes in the abiotic environment and biotic composition of the islands.

Synthetic and Interdisciplinary Studies

Throughout this book we have emphasized that biogeography is inherently an interdisciplinary science. Some would divide the field into two subdisciplines: historical biogeography and ecological biogeography. This separation has been counterproductive, because *all* patterns of distribution and diversity have been influenced by both historical and ecological processes. Interactions with the environment have influenced the abundance, range boundaries, long-distance dispersal, adaptation, speciation, and extinction of every species throughout its evolutionary history (see Chapters 3, 4, 5, and 9). In turn, the characteristics of every species have been constrained and molded by its history: both the lineage of ancestry and descent that constitutes its phylogenetic history (see

Chapters 8, 10, 11, and 12) and the changes in the geology, climate, and other features of the earth that constitute its environmental history (see Chapters 6 and 7). Any really complete and satisfying explanation for the distribution of any species, higher taxon, or functional group must include both ecological and evolutionary processes, and both phylogenetic and earth history.

This conclusion is so obvious, and our understanding of ecological and evolutionary processes, phylogenetic relationships, and historical events is advancing so rapidly, that the distinction between ecological and historical biogeography can be expected to diminish. It is too much to hope that it will disappear entirely. Like the organisms that they study, all biogeographers are specialists, at least to some degree. Their backgrounds and interests, tools, and concepts have been shaped and constrained by their own unique environments and histories. They cannot be expected to assimilate and integrate all of the enormous quantities of environmental and historical information that are becoming available. Despite these limitations, we expect biogeography and biogeographers to become increasingly synthetic.

Because of the explosion of data, many empirical patterns of distribution and diversity have been well documented for a variety of different kinds of organisms. What is still lacking is an equally clear understanding of the processes that have produced these patterns, interactions among those processes, and feedback effects between communities and environmental conditions. The latitudinal gradient of species diversity is a case in point. While compilation and analyses of new data have provided increasingly detailed information on geographic variation in species richness among many kinds of organisms, there has not been a corresponding increase in our understanding of the ecological and historical processes that have produced these patterns, of the interactions among geological, climatic, and other physical processes, or of the manner in which species influence local environmental conditions and one another. In fact, the number of hypotheses proposed to explain the latitudinal gradient has increased about fourfold from the 1960s to the 1990s (see Chapter 15). It is hard to imagine that this trend will continue for another three decades! Surely we can hope to make substantial progress in explaining one of the pervasive features of life on our planet. It is equally certain that such advances are most likely to come from a broad, synthetic approach that takes account of the empirical patterns across multiple groups of organisms and invokes some combination of ecological and historical processes.

There seems to be an increasing realization that many of the greatest challenges to modern science lie at the interfaces between the traditional disciplines. We have seen how this is true of the ecological and historical subdisciplines of biogeography. It is equally true of the interfaces between biogeography and other fields, such as systematics, ecology, geology, and climatology. Some of these interdisciplinary syntheses are well under way. Excellent examples are provided in Chapters 6, 7, 11, and 12, in which we discuss how recent advances in earth history and phylogenetic systematics have contributed to our understanding of past and present distributions.

Another example, which is only beginning to be pursued intensively, is the interface between biogeography and climatology. Stimulated in part by practical concerns about anthropogenic climatic changes, there has recently been great progress in understanding how the earth's climate works (see Chapters 2 and 18). Climatologists have produced increasingly refined models that explain how the climate is influenced by characteristics of the atmosphere, oceans, and lands, and how it changes in response to changes in the earth's orbit, fluxes of solar energy, and human-caused changes in landscapes and atmospheric chemistry. There is still much to be learned, however, before we will be able to predict with any certainty how anthropogenic changes in global and regional climates will affect the distributions and diversity of organisms. Useful insights can come from knowledge of the environmental factors and ecological and historical processes that limit the current geographic ranges of species (see Chapter 3), and of the shifts in patterns of distribution and diversity that were caused by major climatic changes in the past, such as at the end of the Pleistocene (see Chapter 7). But the interdisciplinary synthesis goes both ways. The climate of the earth is powerfully influenced by organisms. For example, vegetation cover strongly influences the absorption of solar radiation, and photosynthesis by plants and respiration by plants, animals, and microbes influence the concentration of CO_2 in the atmosphere. Thus, accurate predictions of future climates depend on knowledge of changes in the geographic distributions of organisms, structures and functional types of vegetation, and kinds of human land use.

Biogeography: Past, Present, and Future

It may be instructive to reflect on the persistent themes in biogeography (see Table 2.1) that have formed the conceptual framework of the field since long before Darwin and Wallace began to publish their seminal works. It may be a great lesson in humility to realize that over the past two centuries we have not added to this list, only modified the emphasis that we have placed on different themes at different periods. Theme 1, classifying geographic regions based on their biotas, was once the dominant theme in biogeography, but now receives only passing attention from most of today's scientists. Either we are satisfied with the existing systems of geographic classification, or such systems are viewed to be trivial descriptive exercises. In either case, this seems unfortunate, as there remains much interesting and important work here, especially in classifying oceanic regions and subregions of terrestrial systems.

In contrast, aided by advances in evolutionary biology and computer science, reconstructing the historical development of biotas (Theme 2) may have become the most popular theme of today's biogeography. Phylogenies have been developed for many species groups, which in turn have provided important insights for both historical and ecological biogeography. Yet even here, some of the most interesting theories of the past—perhaps most notably the taxon cycle—seem to have been largely ignored. Few papers deal with this theory, yet it remains one of the most general explanations for a suite of ecological and evolutionary characteristics of insular biotas.

On a similar note, we have made surprisingly little progress in explaining patterns of species richness among geographic areas (Theme 3), although we have made great progress in describing patterns of diversity across a great variety of regions, ecosystems, and geographic dimensions (e.g., isolation, area, elevation, and depth). Granted, theories such as MacArthur and Wilson's have had great heuristic value and have stimulated an impressive body of research on species diversity, but even these theories have not achieved general acceptance. Moreover, the second part of Theme 3, explaining differences in species composition among geographic areas, has received very little attention indeed, yet it is no less fundamental to the field. Community convergence, ecologically equivalent biotas, invasions of exotic species, disharmonic communities, and nested subsets are no less intriguing now then when these patterns were first described, but general explanations for most of them remain largely ad hoc and problematic.

Finally, Theme 4, explaining geographic variation in the characteristics of individuals, seems to have been dormant throughout most of the history of biogeography, with only brief episodes of activity evidenced by the works of distinguished biogeographers such as Bergmann, Allen, and Gloger. However, it now appears that this beast has awakened. With the earlier works of Rapoport (1982) and the more recent developments in the burgeoning field of macroecology, biogeographers and ecologists have now begun to explore the inner structure of geographic ranges. This new generation of biogeographers is marshaling the knowledge garnered from over 200 years of biological surveys, the collective intelligence of some of humanity's greatest minds, and an unparalleled tool kit of technological advances to explore geographic patterns in nature. This is an exciting time to be a biogeographer, and we will no doubt have many fascinating discoveries to add to the next update of this textbook.

Applications

It is obvious that many areas of future biogeographic research will be heavily influenced by the need to understand and manage the increasingly large impact of our own species. Within the last few decades, we have realized that modern humans are causing profound changes on earth, including dramatic changes in local and global climates, cosmopolitization of many human commensal species, and expansion of anthropogenic landscapes with concomitant reduction and fragmentation of native habitats. Many of these changes pose severe threats to the continued survival of our own species as well as to the other organisms that share the planet with us. In order to appreciate and hopefully to avert these threats, there is an urgent need to understand the nature of our own species and to manage its population and its environmental impacts.

Human Ecology

In Chapter 18 we summarized current information on the phylogenetic and geographic history of our own species, which has been studied extensively by anthropologists. Equally important, but largely unstudied, is the ecology of *Homo sapiens*. It is ironic that we probably know more about the ecology of prairie grasses, pitcher plant mosquitoes, Caribbean *Anolis* lizards, and desert rodents than about our own species. To a large extent this appears to be due to the view, widespread in Western culture, that nature is something apart from humanity. Efforts to study natural science have gone to great lengths to exclude human beings and all their influences. This tendency is exemplified by ecologists, who have traditionally chosen to work in reserves and wild areas where the effects of humans have been minimal.

Fortunately, this situation is changing in response to an inescapable reality. Ecologists can no longer pretend that nature is separate from humanity when the influences of humans are apparent in even the most remote and seemingly unspoiled places on earth. Particles of industrial waste are being deposited in the polar icecaps and sinking to the ocean depths. It has become clear not only that humans are the single main cause of the drastic recent decrease in global biodiversity, but also that active human intervention will be required to save endangered species and habitats (see Chapter 17). Respected ecologists, once nearly universally unwilling to study anything "applied," are now doing research aimed at preserving biological diversity, wild habitats, and ecosystem processes. Some are even beginning to study the ecology of such human-dominated systems as agricultural fields, suburban landscapes, and inner cities.

While these changes are praiseworthy, there is still a long way to go. Still largely missing is a perspective that considers *Homo sapiens* and all of its activities as an integral part of the system. Thus, for example, studies in "urban ecology" are still more likely to focus on the animals and plants living in the parks and streams than on the rats in the sewers, the weeds in the sidewalk cracks and vacant lots, and the pets and houseplants in the apartments. While there are increasing numbers of studies on the effects of agricultural, suburban, and urban areas on the surrounding lands and waters, there are few on the ecological resources and interactions that influence the human populations in these densely settled areas (Cohen 1995; Kates 1996).

We foresee a greatly increased emphasis on human ecology, driven by environmental and social necessity. We believe that this will be accomplished in two ways. First, ecologists and other natural scientists will increasingly study humans beings, and the social and ecological systems that contain them, in the same ways, and using similar techniques, as they have done for other species and relatively pristine ecosystems. There is much to be learned from such approaches. Consider, for example, a comparative study of the human food supply of a large city and a small rural settlement. What quantities of which species constitute the diet, where do they come from, and what are the ecological effects of their exploitation and consumption? Similar studies could measure the flows of fossil fuel energy, water, and other resources through human-dominated ecosystems. We know, but usually only in a general, qualitative way, that consumption in the developed countries is depleting these resources, causing environmental degradation, and contributing to the loss of biodiversity in developing countries. If we are serious about reducing and ultimately reversing these impacts, we must study the ecology and biogeography of the most cosmopolitan species, *Homo sapiens*. Recently, Jared Diamond (1997) and Tim Flannery (1994) have produced two exemplary works using just this approach: insightful and provocative case studies that interweave the ecology and biogeography of humans with those of the ecosystems they have colonized.

Second, ecologists and other natural scientists will increasingly collaborate with social scientists to span the conceptual and empirical gulf that has separated the study of things natural and human. While this interdisciplinary effort will be fraught with difficulty and dissention, it is inevitable that it will go forward. The historical prejudice of natural scientists—that studies of our own species inherently lack scientific rigor and objectivity—has caused them to ignore humans. This prejudice is largely responsible for the lack of a scientific approach to humanity—except, of course, in the medical applications of biology. On the other hand, the historical prejudice of social scientists—that the natural laws of physics, chemistry, and biology have little relevance for human behavior and social systems—has caused many of them to ignore the implications of these laws for the present and future welfare of humanity. Until we understand human behaviors, social and political systems, and economies from the perspective of both the natural and social sciences, our ability to understand and manage a world in which *Homo sapiens* is the dominant species will be critically compromised.

Management and Conservation

Finally, as the population and environmental impact of humans continue to increase, it seems inevitable that there will be increasing demand for scientific applications to understand and cope with these changes. It is encouraging that so many of the students entering graduate programs in biogeography, ecology, evolution, and related disciplines are interested in conservation biology and anxious to make a tangible contribution to preserving biological diversity and wild areas.

To date, much of conservation biology has been focused on two themes: preservation of endangered species and protection of unspoiled habitats. These efforts have been effective. Recovery actions have stabilized the populations of many once declining species, and in some cases have led to dramatic recoveries. The Pacific gray whale and American alligator have been removed from the Endangered Species List, and peregrine falcons and bald eagles have made excellent progress toward recovery. Reserves of unique and threatened habitats have been established throughout the world. We are learning how to use fire, animal disturbance, and other ecological processes to manage reserves—from tiny patches of tallgrass prairie in the central United States to immense Kakadu National Park in northern Australia—so as to maintain healthy ecosystems and native species. These traditional conservation efforts call for a great deal of applied ecology, but not so much biogeography.

Increasingly, however, conservationists are realizing that these traditional approaches will not by themselves be sufficient to accomplish their goals of slowing and eventually halting ecological degradation and loss of biodiversity. The problem is simply too large in magnitude and spatial scale to rectify solely by saving species and habitats one at a time. An inestimable number of species, especially plants, insects, and other invertebrates, are going extinct before they can be collected, identified, and described. In many parts of the world there are no more "relatively unspoiled" habitats to preserve, and the effects of encroaching human populations are threatening existing reserves. Problems of such magnitude require big and drastic solutions. The focal species approach of the U.S. Endangered Species Act and similar legislation in other countries should be complemented by legislation that addresses other levels of biological diversity, from local communities to regional and global systems and processes.

Biogeography has much to contribute to this effort. Only a tiny fraction of the earth's surface—less than 2% worldwide—has been set aside in nature reserves. Most global biodiversity is supported by the "semi-natural matrix" of habitats, often somewhat degraded by human use, that occurs between these reserves. When species can live in the intervening matrix, they will tend to thrive in the reserves, but when the matrix becomes uninhabitable, the reserves become isolated, and survival in the reserves becomes tenuous. Increasing awareness of the importance of the matrix on the part of conservationists is leading to new conservation strategies. In particular, because the existing matrix is often used by local people to harvest renewable natural resources, such as water, timber, fish, game, and grass for domestic livestock, it is leading to an emphasis on sustainable use of these lands and their resources. Implicit in such strategies is the knowledge that a consideration of the geographic ranges of species can be critical to their survival. In Chapter 17 we pointed out that contraction and fragmentation of the geographic range is often an integral part of the syndrome of endangerment and extinction. The applied message, then, is that in order to prevent losses of species, it is important to understand both the biogeography of native biotas and the biogeographic dynamics of humans and anthropogenic extinction forces.

As we saw in the early chapters of this book, biogeography has a long and distinguished history. After all, this was the science of the world's most prominent early biologists, from Foster, Bates, Candolle, and Linnaeus to Sclater, Darwin, Hooker, and Wallace. In the twentieth century, the torch was passed to some of our most distinguished early ecologists and evolutionary biologists, including Schimper, Good, Merriam, Simpson, Mayr, Whittaker, MacArthur, and Wilson. Biogeography will be no less relevant for future generations of scientists attempting to explore, understand, and preserve the diversity of life on earth.

Implicit in this goal is the recognition that humans and all their activities are an integral part of the natural world. As much as we might wish that *Homo sapiens* could live on the earth without affecting every area, habitat, and species, with 5.5 billion people, this is obviously no longer possible. Any realistic effort to prevent further ecological degradation and loss of biodiversity will have to take into account this fact of nature. The fate of our species is inextricably linked with the survival of other species, because both are dependent on the preservation of a healthy, productive environment at all scales, from local to global. Any effort to preserve what remains of "nature" must necessarily include managing the population growth and environmental impacts of our own species. Let us hope that we are up to the task.

GLOSSARY

abiotic Pertaining to the nonliving components of an ecosystem, such as water, heat, solar radiation, and minerals.

abscissa On a graph, the horizontal (x) axis or the horizontal coordinate of a point.

abyssal plain The relatively flat floor of a deep ocean, mostly between 4 and 6 km beneath the surface.

abyssal zone In deep bodies of water, the zone between 4 and 6 km through which solar radiation does not penetrate (aphotic zone) and in which temperature remains at or slightly below 4° C year-round.

acid rain Precipitation with an extremely low pH. The acid condition is caused by the combination of water vapor in the atmosphere with chemicals such as hydrogen sulfide vapor released from the burning of fossil fuels, producing sulfuric acid.

active dispersal (vagility) The movement of an organism from one point to another by its own motility, such as by active swimming, walking, or flying, rather than by being carried along by some other force; compare with *passive dispersal*.

actualism The philosophical assumption that the physical processes now operating are timeless, and therefore that the fundamental laws of nature have remained unchanged; also called *uniformitarianism*.

adaptation Any feature of an organism that substantially improves its ability to survive and leave more offspring over that of other ancestral forms or coexisting phenotypes.

adaptive radiation The evolutionary divergence of a monophyletic taxon (from a single ancestral condition) into a number of very different forms and lifestyles (adaptive zones).

adaptive zone A way of life, including such properties as ecological preference and mode of feeding, that has been adopted by a group of organisms.

adiabatic cooling The decrease in air temperature as a result of a decrease in air pressure (not a loss of heat to the outside) as warm air rises and expands. The rate of cooling is about 1° C per 100 m for dry air and 0.6° C per 100 m for moist air.

aerial Occurring in the air.

aestivation A specialized type of animal behavior and physiology in which the organism lives through the summer in a dormant condition.

age and area hypothesis According to Willis, a hypothesis stating that the greater the age of a taxon, the larger its distributional range.

allele One of two or more alternative forms of a gene located at a single point (locus) on a chromosome.

Allen's rule Among homeotherms, the ecogeographic (morphogeographic) trend for limbs and extremities to become shorter and more compact in colder climates than in warmer ones.

allochthonous Having originated outside the area in which it now occurs.

allometry The manner in which the relative size of one part of an organism increases in relation to the size increase of the entire organism; also known as *scaling*.

allopatric Occurring in geographically different places; i.e., ranges that are mutually exclusive.

allopatric speciation The formation of new species that occurs when populations are geographically separated.

allopolyploid A hybrid polyploid formed following the union of two gametes, usually from distantly related species, with nonhomologous chromosomes.

alpha diversity The number of species within a small area of a relatively homogeneous habitat.

amphitropical Occurring in subtropical or temperate areas on opposite sides of the tropics.

anagenesis The process of evolution that produces entirely new levels of structural organization (grades).

ancestor The individual or population that gave rise to some subsequent individuals or populations with different features.

aneuploidy The formation of a new chromosomal arrangement resulting in an increase or decrease of the chromosome number by one pair; often caused by an uneven meiotic division.

anthropochory The intentional transport of disseminules by humans.

aphotic zone The lower zone in a water column, usually below 50 to 100 m, in which the intensity of solar radiation is too low to permit photosynthesis by plants.

apomixis Reproduction without the union of sexual cells (gametes).

apomorphy In cladistics or phylogenetic systematics, a derived character state.

apterous Wingless; often used to contrast these forms with their primitive ancestors that had wings and the ability to fly.

aquatic Living exclusively or for most of the time in water.

arboreal Living predominantly or entirely in the canopies of trees.

arborescent Treelike.

arctic Pertaining to all nonforested areas north of the coniferous forests of the Northern Hemisphere, especially everything north of the Arctic Circle.

area cladogram A line diagram that hypothesizes the historical relationships of the areas concerned based on the hypothesized relationships of the disjunct taxa living in those areas.

arid Exceedingly dry; strictly defined as any region receiving less than 10 cm of annual precipitation.

asthenosphere A fluid, viscous zone of the upper mantle on which the continental and oceanic plates float (ride) and over which they move.

athenosphere See *asthenosphere*.

austral Pertaining to the temperate and subtemperate zones of the Southern Hemisphere.

Australasia The continental fragments of the original Australian Plate, including Australia, New Zealand, New Guinea, Tasmania, Timor, New Caledonia, and several smaller islands.

Australian Region The biogeographic region including Australia, New Zealand, and New Guinea, other nearby islands, and the Indonesian islands lying east of Wallace's line.

autochthonous Having originated in the area in which it presently occurs.

autopolyploid A polyploid possessing more than two sets of homologous chromosomes.

autosome A chromosome that is not a sex chromosome; a somatic chromosome.

autotroph An organism that uses carbon dioxide occurring in the environment as its primary source of cellular carbon.

avifauna All the species of birds inhabiting a specified region.

balanced biota See *harmonic biota*.

barrier Any abiotic or biotic feature that totally or partially restricts the movement (flow) of genes or individuals from one population or locality to another.

basin-and-range topography In geology, a geographic region characterized by wide, parallel valleys periodically interrupted by small mountain ranges.

bathyal zone The deep sea; in particular, that portion within the aphotic zone but less than 4 km deep.

bathymetry The depth and configuration of the bottom of a body of water.

Benioff zones Zones of high earthquake activity located on the back sides of trenches. The earthquakes, which are caused by the subduction of a plate, are shallow near the trench and progressively deeper at greater distances.

benthic Living at, in, or associated with structures on the bottom of a body of water.

Bergmann's rule Among homeotherms, the ecogeographic (morphogeographic) trend for populations from cooler climates to have larger body sizes, and hence smaller surface-to-volume ratios, than related populations living in warmer climates.

Beringia The geographic area of western Alaska, the Aleutians, and eastern Siberia that was connected in the Cenozoic by a landbridge when the Bering Sea and adjacent shallow waters receded.

beta diversity The rate of change (turnover) in species composition with relatively small changes in habitat.

biogeography The study of the geographic distributions of organisms, both past and present. More generally, the study of variation in biological characteristics (e.g., body size or species richness) on a geographical scale.

biological species A group of potentially interbreeding populations that are reproductively isolated from all other populations.

biomass The total body mass of an organism, population, or community.

biome A major type of natural vegetation that occurs wherever a particular set of climatic and soil conditions prevail, but that may contain different taxa in different regions; e.g., temperate grassland.

biosphere Collectively, all the living things of the earth and the areas they may inhabit.

biota All species of plants, animals, and microbes inhabiting a specified region.

biotic Pertaining to the components of an ecosystem that are living or came from a once-living form.

bipedal Using two hindlimbs for locomotion, usually by hopping or jumping, such as a kangaroo or a kangaroo rat.

bipolar Occurring at both poles, in the cold or sub-temperate zones.

bivoltine Breeding twice per year.

boreal Occurring in that portion of the temperate and subtemperate zones of the Northern Hemisphere that characteristically contains coniferous (evergreen) forests and some types of deciduous forests.

bottleneck In evolutionary biology, any stressful situation that greatly reduces the size of a population.

brackish Having a salt concentration greater than fresh water ($> 0.5‰$) and less than seawater ($35‰$)

breeding area In migratory land animals, the area in which populations mate and produce offspring.

browser An animal that feeds on plant materials, especially on woody parts of trees and shrubs.

calcareous In soil biology, pertaining to a soil whose horizons are rich in calcium carbonate and have a basic reaction.

calcification The formation of a soil under continental climatic conditions of relatively low moisture and hot to cool temperatures, resulting in a soil rich in calcium carbonate ($CaCO_3$) because rainfall is not sufficient to leach calcium from the upper soil horizons.

caliche A hard, often rocklike layer of calcium carbonate that forms in soils of arid regions at the level to which the leached calcium salts from the upper soil horizon are precipitated.

canonical distribution A lognormal distribution of the number of individuals or species according to the mathematical formulation of Preston (1962).

carnivore An animal that feeds mostly or entirely on animal prey.

carrying capacity The total number of individuals of a species that the resources of a habitat can support.

catadromous Living in fresh water but breeding in seawater.

catastrophic extinction See *mass extinction*.

chaparral A type of sclerophyllous scrub vegetation occurring in the southwestern region of North America with a Mediterranean climate.

character Any feature or attribute of an organism.

character displacement The divergence of a feature of two similar species where their ranges overlap so that each uses different resources.

character state One of several alternative forms of a character; e.g., the ancestral form or one of several derived forms.

chromosome An organelle consisting of genetic material on long-stranded deoxyribonucleic acid (DNA) wrapped with proteins.

circumboreal Occurring in the temperate or subtemperate zones of the New and Old World portions of the Northern Hemisphere.

clade Any evolutionary branch in a phylogeny, especially one that is based on genealogical relationships.

cladistic biogeography A sub-branch of vicariance biogeography that reconstructs the events that led to observed distributions of biotas, based primarily on phylogenetic methods.

cladistics The method of reconstructing the evolutionary history (phylogeny) of a taxon by identifying the branching sequence of differentiation through analysis of shared derived character states. Also called *phylogenetic systematics*.

cladogenesis The process of evolution that produces a series of branching events.

cladogram A line diagram derived from a cladistic analysis showing the hypothesized branching sequence (genealogy) of a monophyletic taxon and using shared derived character states (synapomorphies) to determine when each branch diverged.

climax Pertaining to a community that perpetuates itself under the prevailing climatic and soil conditions; therefore, the last stage in secondary succession.

cline A change in one or several heritable characteristics of populations along a geographic transect, attributable to changes in the frequencies of certain alleles and often correlated with a gradual change in the environment.

coevolution The simultaneous, interdependent evolution of two unrelated species that have strong ecological interactions, such as a flower and its pollinator or a predator and its prey.

coexistence Living together in the same local community.

colonization The immigration of a species into new habitat followed by the successful establishment of a population.

commensalism An interspecific relationship in which one species draws benefits from the association and the other is unaffected.

community An assemblage of organisms that live in a particular habitat and interact with one another.

community ecology The study of interactions among co-occurring organisms living in a particular habitat.

competition Any interaction that is mutually detrimental to both participants. Interspecific competition occurs among species that share requirements for limited resources.

competitive exclusion The principle that when two species with similar resource requirements co-occur, one eventually outcompetes and causes the extinction of the other.

congeners Species belonging to the same genus.

continental drift A model, first proposed by Alfred Wegener, stating that the continents were once united and have since become independent structures that have been displaced over the surface of the globe.

continental island An island that was formed as part of a continent and that has a nucleus of continental (sialic) rocks.

convergent evolution The development of two or more species with strong superficial resemblances from totally unlike and unrelated ancestors.

Cope's rule A trend in directional evolution (orthogenesis) toward increased body size.

coprolite Fossil excrement.

Coriolis effect A physical consequence of the law of conservation of angular momentum whereby, as a result of the earth's rotation, a moving object appears to veer to the right in the Northern Hemisphere and to the left in the Southern Hemisphere.

corridor A dispersal route that permits the direct spread of many or most taxa from one region to another.

cosmopolitan Occurring essentially worldwide, as on all habitable landmasses or in all major oceanic regions.

craton The stable crustal nucleus of a continent or continental island, which is older than 600 million years; also called *Precambrian shield*.

crust The outermost rock layer of the earth, covering the mantle.

cryptic species Species within a genus that are morphologically so similar that they cannot be visually distinguished by superficial features.

Curie point The temperature at which remnant magnetism develops in cooling minerals; e.g., 680° C for hemitite and 580° C for magnetite.

deciduous In plants, having leaves that are shed for at least one season, usually in response to the onset of cold or drought.

decomposer An organism (usually a bacterium or fungus) capable of metabolically breaking down organic materials into simple organic and inorganic compounds and releasing them into the ecosystem.

deductive reasoning A method of analysis in which one reasons from general constructs to specific cases.

defaunation The elimination of animal life from a particular area.

dehiscence The act of splitting along a natural line to discharge the contents, such as that of an anther to release pollen grains or of a capsule to release seeds.

delta diversity A measure of the dissimilarity of species pools among large areas, each including a variety of habitats and ecosystems. Similar to beta diversity, but applied at much larger scales, such as comparisons in species composition between continents or biogeographic regions.

density compensation   In island biogeography, an increase in the density of a species inhabiting an island habitat when one or more taxonomically similar competitors are absent.

density overcompensation   In island biogeography, a case in which the total densities of a few species inhabiting a small island exceed the combined densities of a much greater number of species of the same taxon occupying similar habitats on a large island or continent.

desert   A general term for an extremely dry habitat, especially one where water is unavailable for plant growth most of the year; in particular, a habitat with long periods of water stress and sparse coverage by plants, often with perennials covering less than 10% of the total area.

deterministic   Determined or controlled by some regulatory force, such as natural selection; compare with *stochastic*.

detritivore   An organism that feeds solely on detritus.

detritus   Freshly dead or partially decomposed organic matter.

diadromous   Referring to an aquatic animal that must migrate between fresh water and seawater to complete its life cycle, such as certain lampreys and eels.

diapause   An arrested state of development in the life cycle, especially of many insects, during which the organism has reduced metabolism and is more resistant to stressful environmental conditions, such as cold, heat, or drought.

diaspore   Any part or stage in the life cycle of an organism that is adapted for dispersal.

diffuse competition   A type of competition in which one species is negatively affected by numerous other species that collectively cause a significant depletion of shared resources.

diffusion   A form of range expansion that is accomplished over generations by individuals spreading out from the margins of the species' range.

dimorphic   Having two distinct forms in a population.

dioecious   Having individuals with only male or only female reproductive systems.

diploid   Having two sets of chromosomes (2*N*).

disharmonic biota   A biota that is biased toward those groups with good dispersal abilities and lacking those with poor dispersal abilities; therefore, a disharmonic biota is not a random subset of the source pool from a biological viewpoint.

disjunct   A taxon whose range is geographically isolated from that of its closest relatives.

disjunction   A discontinuous range of a monophyletic taxon in which at least two closely related populations are separated by a wide geographic distance.

dispersal   The movement of organisms away from their point of origin.

dispersal biogeography   A branch of historical biogeography that attempts to account for present-day distributions based on the assumption that they resulted from differences in the dispersal abilities of individual lineages.

dispersion   The spatial distribution of individual organisms within a local population.

disseminule   Any part of a plant that is used for dispersal; occasionally restricted to include only seeds and seed-bearing structures. See also *diaspore*.

doldrums   A narrow equatorial zone characterized by long periods of calm or light shifting winds, caused by the upward movement of air masses from this region of high atmospheric pressure to higher latitudes with relatively lower pressure.

dominant   A species having great influence on the composition and structure of a community by virtue of its abundance, size, or aggressive behavior.

ecogeographic (morphogeographic) rule   Any generalized statement that accounts for a regular change in the morphology of organisms along a geographic (and climatic) gradient; e.g., Allen's rule.

ecological biogeography   The study of the ecological factors influencing the distributions of organisms.

ecological niche   See *niche*.

ecological time   The period during which a population can interact with its environment and respond to environmental fluctuations without undergoing substantial evolutionary modification; compare *evolutionary time*.

ecology   The study of the abundance and distribution of organisms and of the relationships between organisms and their biotic and abiotic environments.

ecosystem   The set of biotic and abiotic components in a given environment.

ecotone   A zone of transition between two habitats or communities.

ectoparasite   A parasite that lives on the exterior of its host, such as a louse.

ectotherm   An animal whose body temperature is determined largely by the temperature of the environment, not by its own metabolism; compare with *endotherm*.

edaphic   Pertaining to soil.

emigration   Dispersal of organisms away from a region of interest.

endemic   Pertaining to a taxon that is restricted to the geographic area specified, such as a continent, lake, biome, or island.

endotherm   An animal whose body temperature is maintained largely by its own metabolic heat production; compare with *ectotherm*.

entropy   A measure of the unavailable energy in a closed thermodynamic system.

epeiric sea   A large but relatively shallow body of salt water that lies over a part of a continent.

epibiotic   A taxon that was once wide-ranging but is now narrowly restricted following the extinction of most populations.

epicenter   The point on the earth's surface directly above the origin (focus) of an earthquake.

epicontinental sea   See *epeiric sea*.

epifaunal   Living on a substrate, such as a barnacle or coral.

epipelagic   Living in open water, mostly within the upper 100 m.

epiphyllous   Living on a leaf.

epiphyte A plant that usually lives on another plant (is not rooted in soil) and derives its moisture and nutrients from atmospheric precipitation and whatever materials are released by the organisms in the immediate vicinity.

equilibrium A condition of balance between opposing forces, such as birth and death rates or immigration and extinction rates.

equilibrium theory of island biogeography The theory proposed by MacArthur and Wilson stating that the number of species on an island results from a dynamic equilibrium between the opposing rates of immigration and extinction.

equilibrium turnover rate The change in species composition per unit of time when immigration equals extinction.

equinox Either of two times (March 21 and September 22) in a year when the sun passes the equator so that day and night are the same length everywhere on earth.

establishment The successful start or founding of a population.

Ethiopian Region The portion of Africa south of the Sahara Desert plus Madagascar and other nearby islands.

euryhaline Having a tolerance to an extremely wide range of salt concentrations.

eurytopic Having a tolerance to an extremely wide range of habitats and environmental conditions.

eutherian A placental mammal.

eutrophic lake A lake that is rich in dissolved nutrients and highly productive, but that is usually shallow and seasonally deficient in oxygen.

evapotranspiration The sum total of water lost through evaporation from land and transpiration by plants.

evolution In the strictest sense, any irreversible change in the genetic composition of a population.

evolutionary species A discrete cluster of individuals and populations, evolving separately from other clusters, that exhibits a clear pattern of ancestry and descent; compare with *biological species*.

evolutionary systematics A method of reconstructing the evolutionary history (phylogeny) of a taxon by analyzing the evolution of major features along with the distribution of both shared primitive and shared derived characteristics; compare with *cladistics*.

evolutionary time The period during which a population can evolve and become adapted to an environment by means of genetic changes; compare *ecological time*.

excess density compensation See *density overcompensation*.

exoskeleton An external skeleton of an animal, such as that of a clam or insect.

extant Living at the present time.

extinct No longer living.

extratropical Occurring outside the tropics.

facultative relationship An interaction between two organisms that is not essential to the survival of either.

family A taxonomic category above the level of genus and below the level of order.

fault In geology, a weakness in the earth's crust along which there can be crustal motion and displacement.

filter A geographic or ecological barrier blocks the passage of some forms, but not others.

filter feeder An aquatic animal that feeds on plankton or other minute organic particles by using one of a variety of filtering mechanisms.

fitness The ability of a genotype to leave offspring in the next generation or succeeding generations as compared with that of other genotypes.

flyway An established air route used by vast numbers of migratory birds.

forest Any of a variety of vegetation types dominated by trees and usually having a fairly well developed or closed canopy when the trees have leaves.

fossil A remnant, impression, or other trace of a living organism from the past.

fossorial Referring to an animal that lives in and forages on plants from a burrow.

founder effect Genetic drift that occurs when a newly isolated population is founded, such as on an island, by one or a few colonists (founders). The features of the new population may be markedly different from those of the ancestral population because the gene pool of the founders may be a biased and small sample of the source population.

fragment In plate tectonics, a portion of a former landmass.

fresh water In the strictest sense, water that has a salt concentration of less than 0.5‰.

frugivore An animal that feeds mainly on juicy fruits.

fundamental niche The total range of environmental conditions in which a species can survive and reproduce.

fynbos A type of sclerophyllous scrub vegetation occurring in the region of South Africa with a Mediterranean climate.

gamete One of two cells, usually from different parents, that fuse to form a zygote.

gamma diversity The total species richness of a large area comprising a number of communities. Gamma diversity is a function of both alpha (within-habitat) diversity and beta (between-habitat) diversity.

gene The small unit of a DNA molecule that codes for a specific protein to produce one of the chemical, physiological, or structural attributes of an organism.

genealogy The study of the exact sequence of descent from an ancestor to all derived forms.

gene flow The movement of alleles within a population or between populations caused by the dispersal of gametes or offspring.

gene frequency (allelic frequency) The proportions of gene forms (alleles) in a population.

generalized track In panbiogeography, a line drawn on a map representing the coincident distributions of numerous disjunct taxa.

genetic drift Changes in gene frequency within a population caused solely by chance—i.e., which individuals happen to mate and leave offspring in the next generation—without any influence of natural selection.

genome   A full set of chromosomes.

genotype   The total genetic message found in a cell or an individual.

genus   A taxonomic category for classifying species derived from a common ancestor; a level below that of family and tribe.

geographic isolation   Spatial separation of two potentially interbreeding populations; allopatry.

geographic speciation   See *allopatric speciation*.

gleization   The formation of a soil under moist and cool or cold conditions, resulting in an acidic soil with a large amount of organic matter and iron present in a reduced state (FeO).

Gondwanaland   The southern half of the supercontinent Pangaea, consisting of all the southern continental landmasses and India, which were united for at least 1 billion years but broke up during the late Mesozoic.

granivore   An animal that feeds mainly on dry seeds and fruits; a subtype of herbivore.

grassland   Any of a variety of vegetation types composed mostly of grasses and other herbaceous plants (forbs) but few if any trees and shrubs.

greenhouse effect   The retention of heat in the atmosphere when clouds (water vapor) and carbon dioxide absorb the infrared (heat) radiation reradiated from the earth rather than permitting the heat to escape.

guide fossil   A fossil used to date the age of a sedimentary stratum; also called an *index fossil*.

guild   Groups of species that exploit the same class of environmental resources in a similar manner (e.g., desert granivores or foliage-gleaning birds).

guyot   A flat-topped submarine volcano.

habitat island   A geographically isolated patch of habitat, such as a pond, mountaintop, or cave, that can be studied in the same ways as oceanic islands for patterns of colonization and extinction.

habitat selection   The preference of an organism for a particular habitat type.

hadal   Of or pertaining to the deepest zones of the ocean, below 6 km, which have nearly constant environments with year-round temperatures near 4° C and no light penetration; e.g., in the oceanic trenches.

half-life   The amount of time needed for half of the radioactive material in a rock to decay to a stable element.

haploid   Having one set of chromosomes (N).

harmonic biota (balanced biota)   A biota that is similar to a random subset of its source pool.

herbivore   An animal that feeds mostly or entirely on plants.

hermaphroditic   Having both male and female reproductive structures in the same individual.

heterogeneity   The state of being mixed in composition, as in genetic or environmental heterogeneity.

heterotroph   An organism that uses organic carbon (compounds made by living organisms) as its source of cellular carbon.

historical biogeography   The study of the relationship of present and past distributions of organisms to the physical history of the earth.

Holarctic region   The extratropical zone of the Northern Hemisphere, which includes both the Nearctic and Palearctic regions.

homeostasis   The maintenance of a constant internal state despite fluctuations in the external environment.

homeotherm   An animal that maintains a fairly constant body temperature.

homologous chromosomes   Two chromosomes (in a diploid organism) that have essentially the same gene sequence and that are similar enough to pair during meiosis.

horizon   In soils, the major stratifications or zones, each of which has particular structural and chemical characteristics.

horse latitudes   The zones of dry descending air between 30° and 40° N and S latitude, where many deserts of the world are located.

hot spot   In plate tectonics, a stationary weak point in the upper mantle that discharges magma (molten rock) as a plate passes over it, producing a narrow chain of islands. In biodiversity, an area with a relatively high number of species or high number of endemic species.

hybridization   The production of offspring by parents of two different species or dissimilar populations or genotypes.

hyperosmotic   Referring to an environment in which water will diffuse from an organism because the external solution has a salt concentration higher than its internal concentration.

hypersaline   Having a higher salt concentration than normal seawater.

hypothetico-deductive reasoning   A method of analysis in which one starts with a new, tentative hypothesis and then tests the predictions and assumptions following from it one by one in an attempt to falsify it.

ichthyofauna   All the species of fishes inhabiting a specified region.

immigration   The arrival of new individuals to an isolated site.

included niche   A niche of a specialized species characterized by a narrow range of conditions and lying entirely within the larger fundamental niche of a more generalized species.

index fossil   See *guide fossil*.

inductive reasoning   A method of analysis in which one uses specific observations to derive a general principle.

infaunal   Living within a substrate, such as clams that bury themselves in sand or mud and that feed by means of a long siphon.

insectivore   An animal that feeds mainly on insects.

interglacial   During the Quaternary, a phase when glacial ice sheets retreated and the climate became more equable.

intertidal zone   The zone above the low tide mark and below the high tide mark of a body of water; the littoral zone.

inversion   A complete reversal in the orientation of a portion of a chromosome from the ancestral condition; a type of chromosomal mutation.

isolating mechanism Any structural, physiological, ecological, or behavioral mechanism that blocks or strongly interferes with hybridization or gene exchange between two populations.

isostasy In geology, the term that describes how continental blocks float on a viscous layer of the mantle. While doing this the block may rise or fall to achieve an equilibrium.

isotherm A line on a map connecting all locations with the same mean temperature.

*K***-selection** Selection favoring a more efficient utilization of resources, which is more pronounced when the species is at or near carrying capacity (*K*).

karyotype The morphological appearance of a set of chromosomes in the first metaphase of meiosis.

laterite In tropical soils, a hard, rocklike layer composed principally of ferric oxide, produced when this compound accumulates in high concentration in the soil.

laterization The formation of a soil under conditions of abundant moisture, warm temperatures, and high decomposer activity, resulting in a soil from which bases and silica have been removed (leached), leaving behind a clay rich in ferric and aluminum oxides.

Laurasia The northern half of the supercontinent Pangaea, including North America, Europe, and parts of Asia.

leaching In soil science, the removal of soluble substances by water.

lentic Referring to standing freshwater habitats, such as ponds and lakes.

leptokurtic A mathematical distribution characterized by a sharply peaked curve with a long tail.

liana A climbing woody or herbaceous vine that is especially common in wet tropical forests.

Liebig's law of the minimum An early ecological generalization, now discredited or greatly modified, which stated that abundance and distribution are limited by the single factor in the shortest supply. See also *limiting factor.*

life zones The characteristic changes in vegetation composition and form that occur along an elevational or latitudinal gradient.

limiting factor The resource or environmental parameter that most limits the abundance and distribution of a population.

lithosphere The earth's crust, exclusive of water (hydrosphere) and living organisms (biosphere).

littoral zone The marginal zone of the sea; i.e., the intertidal zone. In fresh water, the shallow zone that may contain rooted plants.

log-transformed Referring to the change of values in a data set by taking the logarithm of each.

long-distance (jump) dispersal The movement of an organism across inhospitable environments to colonize a favorable distant habitat.

lotic Referring to running freshwater habitats, such as brooks and rapids.

macchia A type of sclerophyllous scrub vegetation occurring in regions of the Old World with a Mediterranean climate.

macroevolution A general term for evolution above the population level.

magnetic reversals In geology, episodes during which the direction of earth's magnetic field has been reversed, occurring approximately twice every million years.

magnetic stripes In geology, long alternating stripes of normally and reversely magnetized basaltic rock on the ocean floor.

malacofauna All the species of mollusks inhabiting a specified region.

mantle The second and thickest layer of the earth.

maquis A type of sclerophyllous scrub vegetation occurring in portions of the Mediterranean region with its characteristic climate.

marginal population A population that has difficulty in surviving as a result of limiting abiotic or biotic factors.

marine Living in salt water.

maritime climate In general, a coastal or island environment with little or no freezing, much cloud cover and fog, and less variance in temperature, and thus a milder year-round climate, than nearby inland or mainland localities.

mass extinction A major episode of extinction for many taxa, occurring fairly suddenly in the fossil record.

matorral A type of sclerophyllous scrub vegetation occurring in the region of Chile with a Mediterranean climate.

Mediterranean climate A semiarid climate characterized by mild, rainy winters and hot, dry summers.

megafauna A general term for the large terrestrial vertebrates inhabiting a specified region.

mesic Relatively moist and equable.

Messinian crisis The sudden drainage of the Mediterranean Sea in the Cenozoic.

metabolism The sum total of the positive and negative chemical reactions in an organism or a cell that provide the energy and chemical substances necessary for its existence.

metamorphosis A major change in the form of an individual animal during development; e.g., a caterpillar becomes a moth or a butterfly and a tadpole becomes a frog or a toad.

metatherian A marsupial mammal.

microclimate The fine-scale environmental regime.

microcosm A small community that represents in miniature the components and processes of a larger ecosystem.

microhabitat The fine-scale environment that often determines the presence or absence of each kind of organism.

microphyllous Having small, narrow leaves; characteristic of many plants in very dry habitats.

midden A solid mass of collected organic debris left by an animal such as pack rats (Rodentia).

midoceanic ridge In plate tectonics, a submarine mountain chain, within which seafloor spreading of the oceanic plates occurs.

migration   Dispersal from and return to and from an area, typically a breeding site, by an individual or its immediate descendants.

mimicry   The marked resemblance of an organism to another organism or a background (e.g., a leaf, tree bark, or sand) to deceive predators or prey by "disappearing" (crypsis) or by causing the predator or prey to confuse the mimic with something it is not.

mobility   The ability to move or be moved.

monoecious   In botany, having separate male and female flowers on the same plant.

monomorphic   Having only one form in a population.

monophyletic   Having arisen from one ancestral form; in the strictest sense, from one initial population.

monospecific   Having only one species in the genus.

monotypic   Having only one species in the taxon.

morphogeographic rule   See *ecogeographic rule*.

motility   The ability to move under one's own power, as by wings or a flagellum.

mutation   Any change in the genetic information that results in either the alteration in a gene (point mutation) or a major modification in the karyotype (chromosomal mutation), neither of which is usually reversible in the strictest sense.

mutualism   An interspecific relationship in which both species receive positive benefits from their interaction.

mycorrhiza   A symbiotic relationship between a fungus and a plant root that benefits the plant by providing a source of useful nitrogen, which is manufactured by the fungus.

natural selection   The process of eliminating from a population through differential survival and reproduction those individuals with inferior fitness.

Nearctic region   The extratropical region of North America.

nectarivore   An animal that feeds on plant nectar.

nekton   Organisms that are free-swimming in the upper zone of open water and strong enough to move against currents.

neoendemic   An endemic that evolved in fairly recent times.

Neotropical Region   The region from southern Mexico and the West Indies to southern South America.

neritic zone   The shallow water adjoining a seacoast, especially the zone over a continental shelf.

nestedness   The tendency for the species lists of relatively depauperate communities to form regular subsets of those from more species-rich communities.

niche   The total requirements of a population or species for resources and physical conditions.

niche breadth   The range of resources and physical conditions used by a particular population relative to that of other populations.

niche expansion   An increase in the range of habitats or resources used by a population, which may occur when a potential competitor is absent.

nomadic   Having no fixed pattern of migration.

null hypothesis   A statistical hypothesis stating what would be expected by chance alone, which can be tested in order to determine whether an observation could be a result of chance or is instead the result of some directing force.

nunatak   An area in a glaciated region that was not covered by an ice sheet; hence, a refugium.

obligate relationship   An interaction between species in which at least one of the species cannot survive or reproduce without the other; e.g., many host-parasite relationships, in which a single species of host is required.

oceanic   In marine ecology, of or pertaining to open ocean with very deep water.

oceanic island   An island that was formed de novo from the floor of the ocean through volcanic activity and that has never been attached to a continent.

oligotrophic lake   A deep lake with low primary productivity.

omnivore   An animal that feeds on both plants and animals.

order   A taxonomic category above the level of family and below the level of class.

ordinate   On a graph, the vertical (y) axis or the vertical coordinate of a point.

Oriental Region   The tropical zone of Southeast Asia eastward to the margin of the continental shelf (Wallace's line).

orthogenesis   The supposed intrinsic tendency of organisms to evolve steadily in a particular direction, e.g., to become larger or smaller.

osmoconformer   An organism that does not osmoregulate, but instead has internal salt concentrations in osmotic balance with its environment.

osmoregulation   The process of maintaining homeostasis by maintaining a constant internal concentration of body fluids in changing external solutions.

outbreak area   In organisms that irrupt, the area into which populations expand during peak population densities.

outcrossing   Having gametes that are exchanged between different genotypes.

overturn   Vertical mixing of the water column in a lake caused by temperature changes over the seasons.

Pacifica   A hypothesized ancient continent that may have existed somewhere in the Pacific basin.

Palearctic region   The region of extratropical climates in Eurasia and in the coastal area of northernmost Africa.

paleocirculation   The ocean currents of the past.

paleoclimatology   The study of past climates, as elucidated mainly through the analysis of fossils.

paleoecology   The branch of paleontology that attempts to reconstruct the structure of and the processes affecting ancient populations and communities.

paleoendemic   An endemic that evolved in the distant past; compare with *neoendemic*.

paleoflora   All the species of plants inhabiting a specified region in the past.

paleomagnetism The magnetism or magnetically induced orientation of microstructures that has existed in a rock since its origin.

paleontology The field of study devoted to describing, analyzing, and explaining the fossil record.

paleotropical Occurring in the Old World tropics and subtropics; i.e., in Africa, Madagascar, India, and Southeast Asia.

panbiogeography A sub-branch of vicariance biogeography, developed by Croizat, that attempts to reconstruct the events leading to observed distributions of taxa by drawing lines on a map (tracks) connecting known distributions of related taxa. Unlike cladistic vicariance biogeography, panbiogeography does not require phylogenies of the focal taxa.

Pangaea In plate tectonics, the supercontinent of the Permian that was composed of essentially all the present continents and major continental islands.

panmixis The condition whereby interbreeding within a large population is totally random.

Panthalassa The great global ocean that existed during the Permian coincident with the supercontinent Pangaea.

pantropical Occurring in all major tropical areas around the world.

parallel evolution The evolution of species with strong resemblances from fairly closely related ancestors; hence the evolution of two or more taxa in the same direction from related ancestors.

paramo Tropical alpine vegetation, characteristic of high mountains at the equator, that has a low and compact perennial cover as a response to the perpetually wet, cold, and cloudy environment.

parapatric Having contiguous but nonoverlapping distributions.

parapatric speciation A mode of speciation in which differentiation occurs when two populations have contiguous but nonoverlapping ranges, often representing two distinct habitat types.

paraphyletic In cladistics, referring to taxa that are classified chiefly on the basis of shared primitive character states.

parasitism An interspecific relationship in which one species (the parasite) draws nutrition from or is somehow dependent for survival on the other species (the host), which is negatively affected by the interaction.

parsimony The logical principle or rule of thumb that the simplest solution (i.e., involving the fewest logical steps or conditions) should be chosen from two or more conflicting explanations; also call Ockham's razor.

parthenogenesis The development of eggs without fertilization by a male gamete.

passive dispersal (pagility) The movement of an organism from one location to another by means of a stronger force, such as wind or water or via a larger animal.

pattern Nonrandom, repetitive organization.

pedogenic regimes The soil-forming processes; e.g., laterization, podzolization, calcification, and gleization.

pelagic Occurring in open water and away from the bottom.

peninsula effect The hypothesized tendency for species richness to decrease along a gradient from the axis to the most distal point of a peninsula.

peripheral population Any population of a species that occurs along or near the edges of a range, around either the perimeter or the elevational limit. Not synonymous with *marginal population*.

permanence theory The view, widely held before continental drift was accepted, that the distribution pattern of ocean basins and continents remained relatively constant over the history of the earth.

perturbation Any event that greatly upsets the equilibrium of or alters the state or direction of change in a system.

phenetics The study of the overall similarities of organisms; compare with *phylogeny*.

phenotype The expression of the genetic message of an individual in its morphology, physiology, and behavior.

phoresy The transportation of organisms by other organisms.

photic (euphotic) zone The uppermost zone in a water column where solar radiation is adequate to permit photosynthesis by plants.

photoautotroph An organism that uses light as the energy source and carbon dioxide as the carbon source for its basic metabolism; hence, an organism that is photosynthetic.

photosynthesis The chemical process of using pigments to capture sunlight and then using that energy, along with water and carbon dioxide, to make organic compounds (sugars), releasing oxygen in the process.

phyletic gradualism An evolutionary process whereby a species is gradually transformed over time into a different organism; compare with *punctuated equilibrium*.

phylogenetic systematics See *cladistics*; compare with *evolutionary systematics*.

phylogeny The evolutionary relationships between an ancestor and all of its known descendants.

physiognomy The external aspect of a landscape; e.g., the topography and other physical characteristics of a land form and its vegetation.

physiological ecology The study of how the physiological characteristics of organisms relate to the abundances and distributions of those organisms in their natural habitats.

phytogeography The study of the distribution of plants.

phytosociology The quantitative study of the composition of plant communities and how these relate to environmental factors.

placental A mammal that has a placenta connecting the mother to the fetus.

plankton Small organisms (especially tiny plants, small invertebrates, and juvenile stages of larger animals) that inhabit water and are transported mainly by water currents and wave action rather than by individual locomotion.

planktotrophic Referring to aquatic larvae that have no long-term storage of nutrients and so must feed on small organisms in the plankton during their development.

plate In plate tectonics, a portion of the earth's upper surface, about 100 km thick, that moves over the asthenosphere during seafloor spreading.

plate tectonics The study of the origin, movement, and destruction of plates and how these events have been involved in the evolution of the earth's crust.

plesiomorphy In cladistics, a primitive or ancestral character state.

pluvial Referring to periods of high rainfall and water runoff.

podzolization The formation of a soil under conditions of adequate moisture and low decomposer activity, resulting in a soil in which the bases, humic acids, colloids, and ferric and aluminum oxides have been removed (leached) from the upper horizon.

poikilotherm An animal with a relatively variable body temperature, often ectothermic; i.e., relying on the external environment to control its body temperature.

pollination The transfer of pollen grains to a receptive stigma, usually by wind or flower-visiting animals.

polymorphic Having several distinct forms in a population.

polyphagous Feeding on a variety of different kinds of food.

polyploid Any organism or cell that has three or more sets of chromosomes.

Precambrian shield See *craton*.

predation The act of feeding on other organisms; an interspecific interaction that has negative effects on the species that is consumed or used.

predator In a predatory relationship, the species that consumes other organisms.

prey In a predatory relationship, the species that is consumed or used by another organism.

primary division freshwater fish Any fish that is totally intolerant of salt water.

primary production The production of biomass by green plants.

primary succession The gradual transformation of bare rock or another sterile substrate into a soil that supports a living ecological community.

progression rule The idea that a primitive form remains in the center of origin, whereas derived forms are found at greater distances from the center of origin.

propagule Any part of an organism, or group of organisms, or stage in the life cycle that can reproduce the species and thus establish a new population.

provincialism The coincident occurrence of large numbers of well-differentiated endemic forms in an area; regional or provincial distinctiveness.

punctuated equilibrium The hypothesis that evolution occurs during periods of rapid differentiation (often accompanying speciation), which are followed by long periods in which few if any characters evolve.

quadrupedal Having four limbs for locomotion.

r-selection Selection that favors high population growth rate ("r"), which is more prominent when population size is far below carrying capacity.

radiation In evolutionary biology, a general term for the expansion of a group, implying that many new species have been produced.

rafting The transport over water of living organisms on large floating mats of debris.

realized niche The actual environmental conditions in which a species survives and reproduces in nature; a subset of the fundamental niche.

Red Queen hypothesis A hypothesis that states that a species must continually evolve in order to keep pace with an environment that is perpetually changing because all other species are also evolving, altering the availability of resources and the nature of biotic interactions.

refugium An area in which climate and vegetation type have remained relatively unchanged while areas surrounding it have changed markedly, and which has thus served as a refuge for species requiring the specific habitat it contains.

relict A surviving taxon from a group that was once widespread (epibiotic relict) or diverse (taxonomic relict).

reproductive isolation Inability of individuals from different populations to produce viable offspring.

rescue effect On islands near continents, the tendency of a continual influx of individuals belonging to species already present to prevent the disappearance of those species.

resident A species that lives year-round in a particular habitat or location.

reticulate evolution The formation of new species by the hybridization of dissimilar populations; e.g., interspecific hybridization.

saline Having high concentrations of salts and especially ions of chloride and sulfate.

scaling See *allometry*.

scavenger An animal that feeds on carrion.

sclerophyllous Having tough, thick, evergreen leaves.

scrub Any of a wide variety of vegetation types dominated by low shrubs; in exceedingly dry locations, scrub vegetation has few or no trees and widely spaced low shrubs, but in areas of fairly high rainfall, scrub has trees and grades into either woodland or forest.

seafloor spreading In plate tectonics, the process of adding crustal material at a midoceanic ridge and thus displacing older rocks, usually on both sides, away from their point of origin.

seamount A peaked submarine volcano.

secondary division freshwater fish A fish that prefers fresh water but can live for short periods in salt water.

secondary succession A series of changes in the vegetational composition of an environment in response to disturbance, involving the gradual and regular replacement of species and ending, at least hypothetically, with the return to a stable state (climax).

self-compatible Having the ability to produce offspring without requiring gametes from another individual.

self-incompatible Requiring two individuals to exchange gametes in order to produce offspring.

semiaquatic Living partly in or adjacent to water.

semiarid Having a fairly dry climate with low precipitation, usually 25 to 60 cm per year, and a high evapotranspiration rate, so that potential loss of water to the environment exceeds the input.

semidesert A semiarid habitat characterized by low vegetation; e.g., small, widely spaced shrubs.

seral stage One of the stages in the ecological succession of communities.

sere A series of stages in community transformation during ecological succession.

serpentine A rock or soil type rich in magnesium but deficient in calcium.

sessile Remaining fixed in the same spot throughout adulthood; e.g., most plants and certain benthic aquatic invertebrates.

shrubland Any of a wide variety of vegetation types dominated by shrubs, which may form a fairly solid cover; e.g., sclerophyllous scrub (chaparral), or sparse cover; e.g., desert scrub.

sial Rock rich in silica-aluminum; the principal component of continental rocks.

siliceous Containing silica.

sima Rock rich in silica-magnesium; the principal component of basalt and of oceanic plates.

similarity index An estimate of the similarity or relatedness of two communities, biota, or taxa.

sister taxa In cladistics, the two taxa that are most closely (and therefore most recently) related.

solstice Either of two times in a year (June 22, December 22) when the sun reaches its highest latitude (23.5° N and S, respectively).

speciation The process in which two or more contemporaneous species evolve from a single ancestral population.

species The fundamental taxonomic category for organisms; a group of organisms that are morphologically and reproductively more similar to one another than to other populations and that share a singular ancestor-descendent heritage.

species-area relationship A plot of the numbers of species of a particular taxon against the area of islands or other biogeographic regions. The relationship is often linearized by log-transforming one or both variables.

species composition The types of species that constitute a given sample.

species pool All the organisms present in neighboring source areas that are theoretically available to colonize a particular habitat or island.

species richness The number of species in a given sample.

species selection An analogue of natural selection at the species level, in which some species with certain characteristics increase while others decrease or become extinct.

stasipatric speciation See *parapatric speciation*.

stenohaline Having a tolerance for only a narrow range of salt concentrations.

stenothermal Having a tolerance for only a narrow range of temperatures.

stochastic Random, expected (statistically) by chance alone; compare with *deterministic*.

stratigraphy The branch of geology dealing with the sequence of deposition of rocks and fossils as well as their composition, origin, and distribution.

stream capture The interception and incorporation of a stream or part of a stream by another stream or watershed.

subduction In plate tectonics, the movement of one plate beneath another, leading to the heating and subsequent remelting of the lower plate.

subfamily A taxonomic category used for grouping genera within a family.

sublittoral zone The coastal marine zone below the intertidal zone and therefore below the point at which the sea bottom is periodically exposed to the atmosphere.

subspecies A taxonomic category used by some systematists to designate a genetically distinct set of populations with a discrete range.

subterranean Living underground.

supercontinent An ancient landmass formed by the collision and connection of most, if not all, of the present global landmasses (e.g., Pangaea).

superfamily A taxonomic category used for grouping families within a suborder.

superparamo A vegetation type of high elevations in the equatorial Andes mountains.

superspecies A group of closely related species.

supertramp A species that has excellent colonizing abilities but is a poor competitor in diverse communities.

sweepstakes route A severe barrier that results in the partly stochastic dispersal of some elements of a biota, resulting in the establishment of a disharmonic biota.

symbiosis A long-term interspecific relationship in which two unrelated and unlike organisms live together in close association so that each receives some adaptive benefit.

sympatric In the strictest sense, living in the same local community, close enough to interact; in the more general sense, having broadly overlapping geographic ranges.

sympatric speciation The differentiation of two reproductively isolated species from one initial population within the same local area; hence, speciation that occurs under conditions in which much gene flow potentially could or actually does occur. Compare with *allopatric speciation; parapatric speciation*.

symplesiomorphy In cladistics, a primitive character state shared between taxa.

synapomorphy In cladistics, a derived character state shared between taxa.

systematics The study of the evolutionary relationships between organisms.

taxon (pl. **taxa**) A convenient and general term for any taxonomic category; e.g., a species, genus, family, or order.

taxon cycle A proposed series of ecological and evolutionary changes in a newly arrived colonist on an isolated oceanic island, from the state of being indistinguishable from its mainland relatives to that of a highly differentiated endemic, to extinction and replacement by new colonists.

taxonomy In the strictest sense, the study of the names of organisms, but often used for entire process of classification; see *systematics*.

tectonic Referring to any process involved in the production or deformation of the earth's crust.

terrestrial Living on land.

tetrapod Any four-legged vertebrate, including amphibians, reptiles, and mammals.

thermocline In a water column, the subsurface zone in which the temperature drops sharply.

tillites Glacial rock deposits.

timberline The uppermost limit of forest vegetation at high elevations.

track In panbiogeography, a line drawn on a map connecting the geographically isolated ranges of the species in a taxon that are closest relatives (vicariants).

trade winds Winds blowing toward the equator between the horse latitudes and the doldrums in the Northern and Southern Hemispheres.

transform fault A fault in an oceanic plate, perpendicular to the midoceanic ridge, that divides the plate into smaller units.

translocation A kind of chromosomal mutation in which a segment of one chromosome becomes attached to a different chromosome.

transpiration The loss of water vapor from plants through pores called stomates.

trench In plate tectonics, an exceedingly deep cut in the ocean floor where the subduction of an oceanic plate is occurring.

triple junction In plate tectonics, a point at which three oceanic plates meet; the position of the junction shifts because each of the plates drifts at a different rate.

trophic status The position or role of an organism in the nutritional structure of a community; e.g., primary producer, herbivore, or top carnivore.

turnover The rate of replacement of species in a particular area as some taxa become extinct but others immigrate from outside.

undersaturated Having fewer taxa of a particular kind than expected on the basis of an equilibrium between colonization, speciation, and extinction.

uniformitarianism See *actualism*.

upwelling The vertical movement of deep water, containing dissolved nutrients from the ocean bottom, to the surface.

vagility The ability to move actively from one place to another.

vicariance biogeography A branch of historical biogeography that attempts to reconstruct the historical events that led to observed distributional patterns based largely on the assumption that these patterns resulted from the splitting (vicariance) of areas and not long-distance dispersal. Compare *dispersal biogeography*.

vicariants Two disjunct species that are closely related to each other and that are assumed to have been created when the initial range of their ancestor was split by some historical event.

Viking funeral ship A term coined by McKenna for a landmass, such as a fragment, containing fossils that were laid down when the land was in one location, but that were transported to a completely different locality via continental drift.

waif In dispersal biogeography, a diaspore or any type of individual that is carried passively by waves or air currents to a distant place; e.g., most colonizers of oceanic island beaches.

Wallace's line The most famous biogeographic line, running between Borneo and Celebes and between Bali and Lombok, which marks the boundary between the Oriental and Australian Regions.

Wegenerism The general idea of continental drift according to Alfred Wegener.

wintering area In migratory land animals, the area where populations spend the cold season and feed, but do not breed.

woodland Any of a variety of vegetation types consisting of small, widely spaced trees with or without substantial undergrowth.

xerophyte A plant with physiological and structural features adapting it for a habitat that undergoes strong diurnal or seasonal water stress.

zoogeography The study of the distributions of animals.

BIBLIOGRAPHY

Abbott, I. 1978. Factors determining the number of land bird species on islands around southwestern Australia. *Oecologia* 33: 221–233.

Abbott, I. 1983. The meaning of z in species/area regressions and the study of species turnover in island biogeography. *Oikos* 41: 385–390.

Abbott, I., L. K. Abbott and P. R. Grant. 1977. Comparative ecology of Galapagos ground finches (*Geospiza* Gould): Evaluation of the importance of floristic diversity and interspecific competition. *Ecological Monographs* 47: 151–184.

Abele, L. G. and E. F. Connor. 1979. Application of island biogeography theory to refuge design: Making the right decision for the wrong reasons. In R. M. Linn (ed.), *Proceedings of the First Conference on Scientific Research in the National Parks, New Orleans, LA*, 89–94. Washington, DC: Department of the Interior (National Park Service).

Abele, L. G. and W. Kim. 1989. The decapod crustaceans of the Panama canal. *Smithsonian Contributions to Zoology*: 482. Washington, DC: Smithsonian Institution Press.

Aberg, J., G. Jansson, J. E. Swenson and P. Angelstam. 1995. The effect of matrix on the occurrence of hazel grouse (*Bonasa bonasia*) in isolated habitat fragments. *Oecologia* 103: 265–269.

Abrahamson, D. E. 1989. *The Challenge of Global Warming*. Washington, DC: Island Press.

Ackerly, D. D. and M. J. Donoghue. 1995. Phylogeny and ecology reconsidered. *Journal of Ecology* 83: 730–733.

Adams, C. C. 1902. Southeastern United States as a center of geographical distribution of flora and fauna. *Biological Bulletin* 3: 115–131.

Adams, C. C. 1909. *An Ecological Survey of Isle Royale, Lake Superior*. Michigan Biological Survey, Lansing, MI.

Adams, C. G. and D. V. Ager. 1967. *Aspects of Tethyan Biogeography*. The Systematics Association Publications, no. 7, Wetteren, Universa.

Adler, G. H. 1994. Avifaunal diversity and endemism on tropical Indian Ocean islands. *Journal of Biogeography* 21: 85–95.

Alacantra, M. 1991. Geographical variation in body size of the wood mouse *Apodemus sylvaticus* L. *Mammalian Reviews* 21: 143–150.

Alatalo, R. V. 1982. Bird species distributions in the Galapagos and other archipelagoes: Competition or chance? *Ecology* 63: 881–887.

Albrecht, F. O. 1967. *Polymorphisme phasaire et biologie des acridiens migrateurs*. Paris: Masson.

Allard, G. O. and V. J. Hurst. 1969. Brazil-Gabon geologic link supports continental drift. *Science* 163: 528–532.

Allee, W. C., O. Park, A. E. Emerson, T. Park, and K. P. Schmidt. 1949. *Principles of Animal Ecology*. Philadelphia: W. B. Saunders.

Allen, J. A. 1878. The influence of physical conditions in the genesis of species. *Radical Review* 1: 108–140.

Alvarez, L. W., W. Alvarez, F. Asaro and H. V. Michel. 1980. Extraterrestrial cause for the Cretaceous-Tertiary extinction. *Science* 208: 1095–1108.

Alvarez, W., E. G. Kauffman, F. Surlyk, L. W. Alvarez, F. Asaro and H. V. Michel. 1984. Impact theory of mass extinctions and the invertebrate fossil record. *Science* 223: 1135–1141.

Amadon, D. 1950. The Hawaiian honeycreepers (Aves, Drepaniidae). *Bulletin of the American Museum of Natural History* 95: 153–262.

Amiran, D. H. K. and A. W. Wilson (eds.). 1973. *Coastal Deserts: Their Natural and Human Environments*. Tucson: University of Arizona Press.

Anderson, A. B. and W. W. Brown. 1980. On the number of tree species in Amazonian forests. *Biotropica* 12: 235–237.

Anderson, I. W. 1974. The chestnut pollen decline as a time horizon in lake sediments in eastern North America. *Canadian Journal of Earth Sciences* 11: 678–685.

Anderson, P. K. 1960. Ecology and evolution in island populations of salamanders in San Francisco Bay region. *Ecological Monographs* 30: 359–385.

Andrewartha, H. G. and L. C. Birch. 1954. *The Distribution and Abundance of Animals*. Chicago: University of Chicago Press.

Arnold, E. N. 1979. Indian Ocean giant tortoises: Their systematics and island adaptations. *Philosophical Transactions of the Royal Society of London*, Series B 286: 127–145.

Arnold, S. J. 1972. Species densities of predators and their prey. *American Naturalist* 106: 220–236.

Arrhenius, O. 1921. Species and area. *Journal of Ecology* 4: 68–73.

Ashmole, A. P. 1963. The regulation of numbers of tropical ocean birds. *Ibis* 103b: 458–473.

Athias-Binche, F. 1993. Dispersal in varying environments: The case of phoretic ceropodid mites. *Canadian Journal of Zoology-Revue Canadienne de Zoologie* 71: 1793–1798.

Atkinson, I. A. E. 1988. Presidential address: Opportunities for ecological restoration. *New Zealand Journal of Ecology* 11: 1–12.

Atkinson, I. A. E. and E. K. Cameron. 1993. Human influence on terrestrial biota and biotic communities of New Zealand. *Trends in Ecology and Evolution* 8: 447–451.

Atmar, W. and B. D. Patterson. 1993. The measure of order and disorder in the distribution of species in fragmented habitats. *Oecologia* 96: 373–382.

Auffenberg, W. 1971. A new fossil tortoise, with remarks on the origin of South American testudines. *Copeia* (1): 106–117.

Auffenberg, W. 1981. *The Behavioral Ecology of the Komodo Monitor*. Gainesville: University of Florida Presses.

Avise, J. C. 1994. *Molecular Markers, Natural History and Evolution*. New York: Chapman and Hall.

Axelrod, D. I. 1967. Quaternary extinctions of large mammals. *University of California Publications in Geological Science* 74: 1–42.

Azzaroli, A. 1982. Insularity and its effects on terrestrial vertebrates: Evolutionary and biogeographic aspects. In E. M. Gallitelli (ed.), *Paleontology, Essentials of Historical Geology*, 193–213. Mucchi, Modena, Italy: S.T.E.M.

Backus, R. H. 1986. Biogeographic boundaries in the open ocean. In *Pelagic Biogeography*, 9–14. UNESCO Technical Papers in Marine Science, 49.

Bahre, C. J. 1995. Human impacts on the grasslands of Southwestern Arizona. In M. P. McClaran and T. R. Van Devender (eds.), *The Desert Grassland*, 230–264. Tucson: University of Arizona Press.

Bailey, E. 1963. *Charles Lyell*. New York: Doubleday and Company.

Bailey, R. G. 1996. *Ecosystem Geography*. New York: Springer.

Bak, P. 1996. *How Nature Works: The Science of Self-Organized Criticality*. New York: Copernicus.

Baker, A. J., C. H. Daugherty, R. Colbourne and J. L. McLennan. 1995. Flightless brown kiwis of New Zealand possess extremely subdivided population structure and cryptic species like small mammals. *Proceedings of the National Academy of Sciences, USA* 92: 8254.

Baker, H. A. and E. G. H. Oliver. 1967. *Ericas in Southern Africa*. Capetown/Johannesburg: Purnell & Sons.

Baker, H. G. and G. L. Stebbins (eds.). 1965. *The Genetics of Colonizing Species*. New York: Academic Press.

Baker, R. H. 1968. Habitats and distribution. In J. A. King (ed.). *Biology of Peromyscus*, 98–126. American Society of Mammalogists Special Publication no. 2.

Baker, R. R. 1978. *The Evolutionary Ecology of Animal Migration*. London: Hodder & Stoughton.

Baldwin, B., P. J. Coney and W. R. Dickinson. 1974. Dilemma of a Cretaceous time scale and rates of sea-floor spreading. *Geology* 2: 267–270.

Banfield, A. W. F. 1954. The role of ice in the distribution of mammals. *Journal of Mammalogy* 35: 104–107.

Banfield, A. W. F. 1961. A revision of the reindeer and caribou, genus *Rangifer*. *Bulletin of National Museum of Canada* 177: 1–137.

Banfield, A. W. F. 1974. *The Mammals of Canada*. Toronto: University of Toronto Press.

Banks, H. P. 1975. Early vascular land plants: Proof and conjecture. *Bioscience* 25: 730–737.

Barbour, C. D. and J. H. Brown. 1974. Fish species diversity in lakes. *American Naturalist* 108: 473–489.

Barlow, N. D. 1994. Size distributions of butterfly species and the effect of latitude on species sizes. *Oikos* 71: 326–332.

Barnes, R. K. and K. H. Mann (eds.). 1980. *Fundamentals of Aquatic Ecosystems*. Oxford: Blackwell.

Barron, E. J., C. G. A. Harrison, J. L. Sloan and W. W. Hay. 1981. Paleogeography, 180 million years ago to the present. *Ecologae Geologicae Helveticae* 74: 443–470.

Barrowclough, G. F. 1992. Biodiversity and conservation biology. In N. Eldredge (ed.), *Systematics, Ecology and the Biodiversity Crisis*, 121–143. New York: Columbia University Press.

Barton, N. H. and B. Charlesworth. 1984. Genetic revolutions, founder effects, and speciation. *Annual Review of Ecology and Systematics* 15: 133–164.

Baskin, Y. 1992. Africa's troubled waters: Fish introductions and a changing physical profile muddy Lake Victoria's future. *BioScience* 42: 476–481.

Bates, J. M., S. J. Hackett and J. Cracraft. In press. Area-relationships in the Neotropical lowlands: An hypothesis based on raw distributions of passerine birds. *Journal of Biogeography*.

Batt, C. M. and A. M. Pollard. 1996. Radiocarbon calibration and the peopling of North America. *American Chemical Society Symposium Series* 625: 415–433.

Baum, D. A. and M. J. Donoghue. 1995. Choosing among alternative phylogenetic species concepts. *Systematic Botany* 20: 560–573.

Baur, B. and J. Bengtsson. 1987. Colonizing ability in land snails on the Baltic uplift archipelagoes. *Journal of Biogeography* 14: 329–341.

Bazzaz, F. A. 1996. *Plants in Changing Environments: Linking Physiological, Population, and Community Ecology*. Cambridge: Cambridge University Press.

Beadle, N. C. W. 1966. Soil phosphate and its role in molding segments of the Australian flora and vegetation, with special reference to xeromorphy and sclerophylly. *Ecology* 47: 991–1007.

Beadle, N. C. W. 1981. *The Vegetation of Australia*. Stuttgart: Gustav Fischer Verlag.

Beckwith, S. L. 1954. Ecological succession on abandoned farm lands and its relationship to wildlife management. *Ecological Monographs* 24: 349–376.

Begon, M., J. L. Harper and C. R. Townsend. 1986. *Ecology: Individuals, Populations, and Communities*. Oxford: Blackwell Scientific Publications.

Behle, W. H. 1978. Avian biogeography of the Great Basin and Intermontane Region. *Great Basin Naturalist Memoirs* 2: 55–80.

Behrensmeyer, A. K., J. D. Damuth, W. A. DiMichele, R. Potts, H. Sues and S. L. Wing. 1992. *Terrestrial Ecosystems Through Time: Evolutionary Paleoecology of Terrestrial Plants and Animals*. Chicago: University of Chicago Press.

Beilmann, A. P. and L. G. Brenner. 1951. The recent intrusion of forests in the Ozarks. *Annals of the Missouri Botanical Garden* 38: 261–282.

Bell, M. and M. J. C. Walker. 1992. *Late Quaternary Environmental Change: Physical and Human Perspectives*. New York: John Wiley and Sons.

Benioff, H. 1954. Orogenesis and deep crustal structure: Additional evidence from seismology. *Bulletin of the Geological Society of America* 65: 385–400.

Benson, L. and R. S. Thompson. 1987. The physical record of lakes in the Great Basin. In W. F. Ruddiman and H. E. Wright, Jr. (eds.), *Geology of North America Volume K-3: North America and Adjacent Oceans during the Last Deglaciation*, 241–260. Boulder, CO: Geological Society of America.

Berger, A. J. 1981. *Hawaiian Birdlife*. 2nd edition. Honolulu: University of Hawaii Press.

Bergmann, C. 1847. Über die Verhältnisse der Wärmeökonomie der Thiere zu ihren Grösse. *Göttinger Studien* 1: 595–708.

Bernabo, J. C. and T. Webb. 1977. Changing patterns in the Holocene pollen record from northeastern North America: A mapped summary. *Quaternary Research* 8: 64–96.

Berra, T. M. 1981. *An Atlas of Distribution of the Freshwater Fish Families of the World*. Lincoln: University of Nebraska Press.

Berry, R. J. 1964. The evolution of an island population of the house mouse. *Evolution* 18: 468–483.

Berry, W. B. N. 1973. Silurian-Early Devonian graptolites. In A. Hallam (ed.), *Atlas of Palaeobiogeography*, 81–87. Amsterdam: Elsevier Scientific Publishing Co.

Bertness, M. D. 1984. Habitat and community modification by an introduced herbivorous snail. *Ecology* 65: 370–381.

Bertness, M. D. 1985. Fiddler crab regulation of *Spartina alterniflora* production on a New England salt marsh. *Ecology* 66: 1042–1055.

Bertness, M. D. 1989. Competitive and facilitative interactions and acorn barnacle populations in a sheltered habitat. *Ecology* 70: 257–268.

Bertness, M. D. 1991. Interspecific interactions among high marsh perennials in a New England salt marsh. *Ecology* 72: 125–137.

Bertness, M. D. and S. W. Shumway. 1993. Competition and facilitation in marsh plants. *American Naturalist* 142: 718–724.

Betancourt, J. L., T. R. Van Devender and P. S. Martin. 1990. *Packrat Middens: The Last 40,000 Years of Biotic Exchange.* Tucson: University of Arizona Press.

Beven, S., E. F. Conner and K. Beven. 1984. Avian biogeography in the Amazon basin and the biological model of diversification. *Journal of Biogeography* 11: 383–399.

Bibby, C. J., N. J. Collar, M. J. Crosby, M. F. Heath, Ch. Imboden, T. H. Johnson, A. J. Long, A. J. Stattersfield and S. J. Thirgood. 1992. *Putting Biodiversity on the Map: Priority Areas for Global Conservation.* Cambridge: International Council for Bird Preservation.

Bierhorst, D. W. 1971. *Morphology of Vascular Plants.* New York: Macmillan.

Biju-Duval, B. and L. Montadert (eds.). 1977. Structural history of the Mediterranean basins. International Symposium of the 25th Plenary Congress Assembly of the International Commission for the Scientific Exploration of the Mediterranean. Paris: *Société des Éditions Technip.*

Biological Sciences Curriculum Study. 1963. *Biological Science: Molecules to Man.* Boston: Houghton Mifflin.

Birge, E. A. and C. Juday. 1911. The inland lakes of Wisconsin. *Bulletin of the Wisconsin Geological Natural History Survey* 22: 1–259.

Birks, H. J. 1989. Holocene isochrone maps and patterns in tree-spreading in the British Isles. *Journal of Biogeography* 16: 503–540.

Birks, H. J. B. 1987. Recent methodological developments in quantitative descriptive biogeography. *Annales Zoologici Fennici* 24: 165–178.

Bishop, I. R. and M. J. Delaney. 1963. The ecological distribution of small mammals in the Channel Islands. *Mammalia* 27: 99.

Blackburn, T. M. and K. J. Gaston. 1994. Animal body size distributions change as more species are described. *Proceedings of the Royal Society of London,* Series B 257: 293–297.

Blackburn, T. M. and K. J. Gaston. 1995. What determines the probability of discovering a species? A study of South American oscine passerine birds. *Journal of Biogeography* 22: 7–14.

Blair, W. F. 1943. Activities of the Chihuahuan deer-mouse in relation to light intensity. *Journal of Wildlife Management* 7: 92–97.

Bock, C. E. and L. W. Lepthian. 1976. Synchronous eruptions of boreal seed-eating birds. *American Naturalist* 110: 559–571.

Bohning-Gaese, K., L. I. Gonzalez-Guzman and J. H. Brown. 1998. Constraints on dispersal and the evolution of the avifauna of the Northern Hemisphere. *Evolutionary Ecology* 12: 767–783.

Borodin, A. M., A. G. Bannikov and V. E. Sokolov (eds.). 1984. *Red Data Book of the USSR: Rare and Endangered Species of Animals and Plants.* 2nd edition. Moscow: Forest Industry Publishers.

Botkin, D. P., D. A. Woodby and R. A. Nisbet. 1991. Kirtland's warbler habitat: A possible early indicator of climate warming. *Biological Conservation* 56: 63–78.

Boucher, D. H. 1985. *The Biology of Mutualism: Ecology and Evolution.* London: Croom Helm.

Boucher, D. H., S. James and K. Kesler. 1984. The ecology of mutualism. *Annual Review of Ecology and Systematics* 13: 315–347.

Boucot, A. J. and J. G. Johnson. 1973. Silurian brachiopods. In A. Hallam (ed.), *Atlas of Paleobiogeography,* 59–65. Amsterdam, Elsevier Scientific.

Boucot, A. J. 1975a. Standing diversity of fossil groups in successive intervals of geologic time in the light of changing levels of provincialism. *Journal of Paleontology* 49: 1105–1111.

Boucot, A. J. 1975b. *Evolution and Extinction Rate Controls.* New York: Elsevier North-Holland.

Bourliere, F. 1973. The comparative ecology of rain forest mammals in Africa and tropical America: Some introductory remarks. In B. J. Meggers, E. S. Ayensu and W. D. Ducksworth (eds.), *Tropical Forest Ecosystems in Africa and South America: A Comparative Review,* 279–292. Washington, DC: Smithsonian Institution Press.

Bowers, M. 1982. Insular biogeography of mammals in the Great Salt Lake. *Great Basin Naturalist* 42: 589–596.

Bowers, M. A. and J. H. Brown. 1982. Body size and coexistence in desert rodents: Chance or community structure. *Ecology* 63: 391–400.

Bowers, M. A. and C. H. Lowe. 1986. Plant-form gradients on Sonoran Desert bajadas. *Oikos* 46: 284–291.

Boyd, E. M. and S. A. Nunneley. 1964. Banding records substantiating the changed status of 10 species of birds since 1900 in the Connecticut Valley. *Bird-Banding* 35: 1–8.

Braun, E. L. 1950. *Deciduous Forests of Eastern North America.* New York: Hafner Press.

Bremer, K. 1993. Intercontinental relationships of African and South American Asteraceae: A cladistic biogeographic analysis. In P. Goldblatt (ed.), *Biological Relationships Between Africa and South America,* 105–135. New Haven, CT: Yale University Press.

Brenner, W. 1921. Vaxtgeografiska studien: Barosunds skargard. I. Allman del och floran. *Acta Societatis pro Fauna et Flora Fennica* 49: 1–151.

Briggs, D. and S. M. Walters. 1984. *Plant Variation and Evolution.* Cambridge: Cambridge University Press.

Briggs, J. C. 1968. Panama sea-level canal. *Science* 162: 511–513.

Briggs, J. C. 1970. *Tropical Shelf Zoogeography.* Publications of the California Academy of Sciences.

Briggs, J. C. 1974. *Marine Zoogeography.* New York: McGraw-Hill.

Briggs, J. C. 1987. *Biogeography and Plate Tectonics.* Amsterdam: Elsevier.

Briggs, J. C. 1994. The genesis of Central America: Biology versus geophysics. *Global Ecology and Biogeography Letters* 4: 169–172.

Brittingham, M. C. and S. A. Temple. 1983. Have cowbirds caused forest songbirds to decline. *BioScience* 33: 31–35.

Brooks, D. R. 1985. Historical ecology: A new approach to studying the evolution of ecological associations. *Annals of the Missouri Botanical Garden* 72: 660–680.

Brooks, D. R. 1988. Macroevolutionary comparisons of host and parasite phylogenies. *Annual Review of Ecology and Systematics* 19: 235–259.

Brooks, D. R. and S. M. Bandoni. 1988. Coevolution and relicts. *Systematic Zoology* 37: 19–33.

Brooks, D. R. and D. A. McLennan. 1991. *Phylogeny, Ecology, and Behavior.* Chicago: University of Chicago Press.

Brooks, J. L. and S. I. Dodson. 1965. Predation, body size, and composition of plankton. *Science* 150: 28–35.

Brower, L. P. 1977. Monarch migration. *Natural History* 86(6): 40–53.

Brower, L. P. and S. B. Malcolm. 1991. Animal migrations: Endangered phenomena. *American Zoologist* 31: 265–276.

Brown, D. E. and R. Davis. 1995. One hundred years of vicissitude: Terrestrial bird and mammal distribution changes in the American Southwest. In *Biodiversity and Management of the Madrean Archipelago: The sky Islands of Southwestern*

United States and Northwestern Mexico, 231–244. USDA Forest Service, General Technical Report RM-GTR-264.

Brown, J. H. 1968. Adaptation to environmental temperature in two species of woodrats, *Neotoma cinerea* and *N. albingula*. *University of Michigan Museum of Zoology Miscellaneous Publications* 135: 1–48.

Brown, J. H. 1971a. Mechanisms of competitive exclusion between two species of chipmunks (*Eutamias*). *Ecology* 52: 306–311.

Brown, J. H. 1971b. Mammals on mountaintops: Non-equilibrium insular biogeography. *American Naturalist* 105: 467–478.

Brown, J. H. 1971c. The desert pupfish. *Scientific American* 225(5): 104–110.

Brown, J. H. 1975. Geographical ecology of desert rodents. In M. L. Cody and J. M. Diamond (eds.), *Ecology and Evolution of Communities*, 315–341. Cambridge, MA: Belknap Press.

Brown, J. H. 1978. The theory of insular biogeography and the distribution of boreal birds and mammals. *Great Basin Naturalist Memoirs* 2: 209–227.

Brown, J. H. 1981. Two decades of homage to Santa Rosalia: Toward a general theory of diversity. *American Zoologist* 21: 877–888.

Brown, J. H. 1984. On the relationship between abundance and distribution of species. *American Naturalist* 124: 255–279.

Brown, J. H. 1986. Two decades of interaction between the MacArthur-Wilson model and the complexities of mammalian distributions. *Biological Journal of the Linnean Society* 28: 231–251.

Brown, J. H. 1988. Species diversity. In A. Myers and R. S. Giller (eds.), *Analytical Biogeography*, 57–89. London: Chapman and Hall.

Brown, J. H. 1989. Patterns, modes and extents of invasions in vertebrates. In J. A. Drake et al. (eds.), *Biological Invasions: A Global Perspective*, 85–110. New York: John Wiley and Sons.

Brown, J. H. 1995. *Macroecology*. Chicago: University of Chicago Press.

Brown, J. H. and M. A. Bowers. 1984. Patterns and processes in three guilds of terrestrial vertebrates. In D. R. Strong, D. Simberloff, L. G. Abele and A. B. Thistle (eds.), *Ecological Communities: Concepts, Issues and the Evidence*, 282–296. Princeton, NJ: Princeton University Press.

Brown, J. H. and M. A. Bowers. 1985. On the relationship between morphology and ecology: Community organization in hummingbirds. *Auk* 102: 251–269.

Brown, J. H. and C. R. Feldmeth. 1971. Evolution in constant and fluctuating environments: Thermal tolerances of desert pupfish (*Cyprinodon*). *Evolution* 25: 390–398.

Brown, J. H. and A. C. Gibson. 1983. *Biogeography*. St. Louis, MO: Mosby.

Brown, J. H. and A. Kodric-Brown. 1977. Turnover rates in insular biogeography: Effect of immigration on extinction. *Ecology* 58: 445–449.

Brown, J. H. and A. Kodric-Brown. 1993. Highly structured fish communities in Australian desert springs. *Ecology* 74: 1847–1855.

Brown, J. H. and A. K. Lee. 1969. Bergmann's rule and climatic adaptation in woodrats (*Neotoma*). *Evolution* 23: 329–338.

Brown, J. H. and M. V. Lomolino. 1989. On the nature of scientific revolutions: Independent discovery of the equilibrium theory of island biogeography. *Ecology* 70: 1954–1957.

Brown, J. H. and B. A. Maurer. 1986. Body size, ecological dominance, and Cope's rule. *Nature* 324: 248–250.

Brown, J. H. and B. A. Maurer. 1987. Evolution of species assemblages: Effects of energetic constraints and species dynamics on the diversification of the North American avifauna. *American Naturalist* 130: 1–17.

Brown, J. H. and B. A. Maurer. 1989. Macroecology: The division of food and space among species on continents. *Science* 243: 1145–1150.

Brown, J. H. and J. C. Munger. 1985. Experimental manipulation of a desert rodent community: Food addition and species removal. *Ecology* 66: 1545–1563.

Brown, J. H. and P. F. Nicoletto. 1991. Spatial scaling of species assemblages: Body masses of North American land mammals. *American Naturalist* 138: 1478–1512.

Brown, J. H., D. W. Mehlman and G. C. Stevens. 1995. Spatial variation in abundance. *Ecology* 76: 2028–2043.

Brown, J. H., G. C. Stevens and D. M. Kaufman. 1996. The geographic range: Size, shape, boundaries, and internal structure. *Annual Review of Ecology and Systematics* 27: 597–623.

Brown, J. M. and D. S. Wilson. 1992. Local specialization of phoretic mites on sympatric carrion beetle hosts. *Ecology* 73: 463–478.

Brown, J. W. 1987. The peninsular effect in Baja California: An entomological assessment. *Journal of Biogeography* 14: 359–365.

Brown, J. W. and P. A. Opler. 1990. Patterns of butterfly species density in peninsular Florida. *Journal of Biogeography* 17: 615–622.

Brown, K. S. Jr. 1982. Paleoecology and regional patterns of evolution in neotropical forest butterflies. In G. T. Prance (ed.), *Biological Differentiation in the Tropics*, 255–308. New York: Columbia University Press.

Brown, M. and J. J. Dinsmore. 1988. Habitat islands and the equilibrium theory of island biogeography: Testing some predictions. *Oecologia* 75: 426–429.

Brown, W. H., E. D. Merrill and H. S. Yates. 1919. The revegetation of Volcano Island, Luzon, Philippine Islands, since the eruption of Taal Volcano in 1911. *Philippine Journal of Science*, Sect. C, vol. 12.

Brown, W. L. and E. O. Wilson. 1956. Character displacement. *Systematic Zoology* 5: 49–64.

Browne, R. A. 1981. Lakes as islands: Biogeographic distribution, turnover rates, and species composition in the lakes of central New York. *Journal of Biogeography* 8: 75–83.

Browne, R. A. and G. H. MacDonald. 1982. Biogeography of the brine shrimp, *Artemia*: Distribution of parthenogenetic and sexual populations. *Journal of Biogeography* 9: 331–338.

Bruhnes, B. 1906. Recherches sur la direction d'aimentation des roches volcaniques (1). *Journal Physique*, 4e Sér., 5: 705–724.

Brundin, L. 1965. On the real nature of transantarctic relationships. *Evolution* 19: 496–505.

Brundin, L. 1966. Transantarctic relationships and their significance, as evidence by chironomid midges. *Kungliga Svenska Vetenskapsakademiens Handlingar*, 4th ser., II, no. 1: 1–472.

Brundin, L. 1967. Insects and the problem of austral disjunctive distribution. *Annual Review of Entomology* 12: 149–168.

Brundin, L. 1972. Phylogenetics and biogeography. *Systematic Zoology* 21: 69–79.

Brundin, L. 1988. Phylogenetic biogeography. In A. A. Myers and P. S. Gillers (eds.), *Analytic Biogeography: An Integrated Approach to the Study of Animal and Plant Distributions*. London: Chapman and Hall.

Buckley, R. C. and S. B. Knedlhans. 1986. Beachcomber biogeography: Interception of dispersing propagules by islands. *Journal of Biogeography* 13: 69–70.

Buffon, G. L. L., Comte de. 1761. *Histoire Naturelle, Generale et Particuliere*, vol. 9. Paris: Imprimerie Royale.

Bullard, E. C., J. E. Everett and A. G. Smith. 1965. Fit of the continents around the Atlantic. In P. M. S. Blackett, E. C. Bullard and S. K. Runcorn (eds.), *A Symposium on Continental Drift*, 41–75. *Philosophical Transactions of the Royal Society of London*, Series A, 248.

Bunt, J. S. 1975. Primary productivity of marine ecosystems. In H. Lieth and R. H. Whitaker, *Primary Productivity of the Biosphere*, 169–202. Washington, DC: National Academy of Sciences.

Bureau de Recherches Géologiques et Minières. 1980a. Colloque C5. Géologie des chaines alpines issues de la Téthys. *Mémoirs*, no. 115.

Bureau de Recherches Géologiques et Minières. 1980b. Colloque C6. Géologie de l'Europe du Précambrien aux bassins sedimentaires post-hercyniens. *Mémoirs*, no. 108.

Burke, K., J. F. Dewey and W. S. F. Kidd. 1977. World distribution of sutures: The sites of former oceans. *Tectonophysics* 40: 69–99.

Burr, B. M. and R. L. Mayden. 1992. Phylogenetics and North American freshwater fishes. In R. L. Mayden (ed.), *Systematics, Historical Ecology, and North American Freshwater Fishes*. Stanford, CA: Stanford University Press.

Bush, G. L. 1975. Modes of animal speciation. *Annual Review of Ecology and Systematics* 6: 339–364.

Bush, M. B. 1994. Amazonian speciation: A necessarily complex model. *Journal of Biogeography* 21: 5–17.

Bush, M. B. and R. J. Whittaker. 1991. Krakatau: Colonization patterns and hierarchies. *Journal of Biogeography* 18: 341–356.

Bush, M. B. and R. J. Whittaker. 1993. Non-equilibrium in island theory and Krakatau. *Journal of Biogeography* 20: 453–457.

Bussing, W. A. 1985. Patterns of distribution of the Central American Icthyofauna. In G. G. Stehli and S. D. Webb (eds.), *The Great American Interchange*, 453–473. New York: Plenum Press.

Butler, R. 1995. When did India hit Asia? *Nature* 373: 20–21.

Buzas, M. A. and S. J. Culver. 1991. Species diversity and dispersal of benthic foraminifera. *BioScience* 41(7): 483–489.

Cabrera, A. L. and A. Willink. 1973. *Biogeografia de America Latina*. Washington, DC: Programa Regional de Desarrollo Cientifico y Tecnologico, Departamento Asuntos Cientificos, Secretario General de la Organizacion de los Estados Americanos.

Cain, S. A. 1944. *Foundations of Plant Geography*. New York: Harper and Brothers.

Calder, W. A. III. 1984. *Size, Function and Life History*. Cambridge, MA: Harvard University Press.

Caley, M. J. and D. Schluter. 1997. The relationship between local and regional diversity. *Ecology* 78: 70–80.

Calvert, A. J., E. W. Sawyer, W. J. Davis and J. N. Ludden. 1995. Archaean subduction inferred from seismic images of a mantle suture in the Superior Province. *Nature* 375: 670–674.

Candolle, A. P. de. 1820. *Essai Elementaire de Geographie Botanique*. De l'imprimerie de F. G. Levrault.

Candolle, A. P. de. 1855. *Géographie Botanique Raisonnée*. 2 vols. Paris: Masson.

Carey, S. W. 1955. The orocline concept of geotectonics, part I. *Proceedings of the Royal Soicety of Tasmania, Papers* 89: 255–288.

Carey, S. W. 1958. A tectonic approach to continental drift. In *Continental Drift: A Symposium*, 177–355. University of Tasmania, Hobart.

Carlquist, S. 1965. *Island Life*. Garden City, NY: Natural History Press.

Carlquist, S. 1974. *Island Biology*. New York: Columbia University Press.

Carlquist, S. 1981. Chance dispersal. *American Scientist* 69: 509–515.

Carothers, S. W. and R. R. Johnson. 1974. Population structure and social organization of Southwestern riparian birds. *American Zoologist* 14: 97–108.

Carpenter, F. M. 1977. Geological history and evolution of the insects. In D. White (ed.), *Proceedings of the 15th International Congress of Entomology, Washington*, 63–70. College Park, MD: Entomological Society of America.

Carpenter, S. R. 1988. *Complex Interactions in Lake Communities*. New York: Springer-Verlag.

Carpenter, S. R., J. F. Kitchell and J. R. Hodgson. 1985. Cascading tropic interactions and lake ecosystem productivity. *BioScience* 35: 635–639.

Carpenter, S. R., J. F. Kitchell, J. R. Hodgson, P. A. Cochran, J. J. Elser, M. M. Elser, D. M. Lodge, D. Kretchmer, X. He and C. von Ende. 1987. Regulation of lake primary productivity by food web structure. *Ecology* 68: 1867–1876.

Carson, H. L. 1971. Speciation and the founder principle. *University of Missouri, Stadler Symposium* 3: 51–70.

Carson, H. L. 1981. Microevolution in insular ecosystems. In D. Mueller-Dombois, K. W. Bridges and H. L. Carson (eds.), *Island Ecosystems: Biological Organization in Selected Hawaiian Communities*, 471–482. Stroudsburg, PA: Hutchinson Ross.

Carson, H. L. and A. R. Templeton. 1984. Genetic revolutions in relation to speciation phenomena: The founding of new populations. *Annual Review of Ecology and Systematics* 15: 97–131.

Carson, H. L. and K. Y. Kaneshiro. 1976. *Drosophila* of Hawaii: Systematics and ecological genetics. *Annual Review of Ecology and Systematics* 7: 311–346.

Carson, H. L., D. E. Hardy, H. T. Spieth and W. S. Stone. 1970. The evolutionary biology of the Hawaiian Drosophilidae. In M. K. Hecht and W. C. Steere (eds.), *Essays in Evolution and Genetics in Honor of Theodosius Dobzhansky*, 437–543. New York: Appleton-Century-Crofts.

Case, T. J. 1975. Species numbers, density compensation and the colonizing ability of lizards on islands in the Gulf of California. *Ecology* 56: 3–18.

Case, T. J. 1978. A general explanation for insular body size trends in terrestrial vertebrates. *Ecology* 59: 1–18.

Case, T. J. 1987. Testing theories of island biogeography. *American Scientist* 75: 402–411.

Case, T. J. 1996. Global patterns in the establishment and distribution of exotic birds. *Biological Conservation* 76: 69–96.

Case, T. J. and D. T. Bolger. 1991. The role of interspecific competition in the biogeography of island lizards. *Trends in Ecology and Evolution* 6(4): 135–139.

Case, T. J. and T. D. Schwaner. 1993. Island/mainland body size differences in Australian varanid lizards. *Oecologia* 94: 102–109.

Case, T. J. and R. Sidell. 1983. Pattern and chance in the structure of model and natural communities. *Evolution* 37: 832–849.

Case, T. J., J. Faaborg and R. Sidell. 1983. The role of body size in the assembly of West Indian bird communities. *Evolution* 37: 1062–1074.

Case, T. J., D. T. Bolger and K. Petren. 1994. Invasion and competitive displacement among house geckos in the tropical Pacific. *Ecology* 75: 464–477.

Caughley, G. 1987. The distribution of eutherian body weights. *Oecologia* 74: 319–320.

Channell, R. 1998. *A Geography of Extinction: Patterns in the Contraction of Geographic Ranges*. Ph.D. dissertation, University of Oklahoma, Norman, OK.

Chappell, M. A. 1978. Behavioral factors in the altitudinal zonation of chipmunks (*Eutamias*). *Ecology* 59: 565–579.

Charnov, E. L. and J. R. Krebs. 1974. On clutch size and fitness. *Ibis* 116: 217–219.

Cheatham, A. H. and J. E. Hazel. 1969. Binary (presence-absence) similarity coefficients. *Journal of Paleontology* 43: 1130–1136.

Chester, R. H. 1969. Destruction of Pacific corals by the sea star *Acanthaster planci. Science* 165: 280–283.

Clagg, H. B. 1966. Trapping of air-borne insects in the Atlantic-Antarctic area. *Pacific Insects* 8: 455–466.

Clark, G. 1992. *Space, Time and Man: A Prehistorian's View*. New York: Cambridge University Press.

Clarke, A. 1992. Is there a latitudinal cline in the sea? *Trends in Ecology and Evolution* 7: 286–287.

Clausen, J., D. D. Keck and W. M. Hiesey. 1940. *Experimental Studies on the Nature of Species*. I. *Effect of Varied Environments on Western North American Plants*. Carnegie Institute Publication 520. Washington, DC: Carnegie Institute.

Clausen, J., D. D. Keck and W. M. Hiesey. 1947. Heredity of geographically and ecologically isolated races. *American Naturalist* 81: 114–133.

Clausen, J., D. D. Keck and W. M. Hiesey. 1948. *Experimental Studies on the Nature of Species*. III: *Environmental Responses of Climatic Races of* Achillea. Carnegie Institute Publication 581: 1–129. Washington, DC: Carnegie Institute.

Clegg, J. A., M. Almond and P. H. S. Stubbs. 1954. The remnent magnetism of some sedimentary rocks in Britain. *Philosophical Magazine* 45: 583–598.

Clements, F. E. 1916. *Plant Succession: An Analysis of the Development of Vegetation*. Carnegie Institute Publication no. 242. Washington, DC: Carnegie Institute.

Clements, F. E. and V. E. Shelford. 1939. *Bio-Ecology*. New York: John Wiley & Sons.

CLIMAP project members. 1976. The surface of the ice-age earth. *Science* 191: 1131–1137.

Cockburn, A. 1991. *An Introduction to Evolutionary Ecology*. London: Blackwell Scientific Publications.

Cockburn, A., A. K. Lee and R. W. Martin. 1983. Macrogeographic variation in litter size in *Antechinus* spp. (Marsupialia: Dasyuridae). *Evolution* 37: 86–95.

Cody, M. L. 1966. The consistency of intra- and inter-continental grassland bird species counts. *American Naturalist* 100: 371–376.

Cody, M. L. 1968. On the methods of resource division in grassland bird communities. *American Naturalist* 102: 107–137.

Cody, M. L. 1973. Parallel evolution and bird niches. In F. di Castri and H. A. Mooney (eds.), *Mediterranean-Type Ecosystems: Origin and Structure*, 307–338. Ecological Studies 7. New York: Springer-Verlag.

Cody, M. L. 1974. *Competition and the Structure of Bird Communities*. Monographs in Population Biology, no. 7. Princeton, NJ: Princeton University Press.

Cody, M. L. 1975. Towards a theory of continental species diversity. In M. L. Cody and J. M. Diamond (eds.), *Ecology and Evolution of Communities*, 214–257. Cambridge, MA: Belknap Press.

Cody, M. L. 1980. Evolution of habitat use: Geographic perspectives. In R. Nöhring (ed.), *Acta XVII Congressus Internationalis Ornithologici*, vol. 2, 1013–1018. Berlin: Verlag der Deutschen Ornithologen-Gesellschaft.

Cody, M. L. 1985. *Habitat Selection in Birds*. Orlando, FL: Academic Press.

Cody, M. L. and J. M. Diamond (eds.). 1975. *Ecology and Evolution of Communities*. Cambridge, MA: Belknap Press.

Cody, M. L. and H. A. Mooney. 1978. Convergence versus non-convergence in Mediterranean-climate ecosystems. *Annual Review of Ecology and Systematics* 9: 265–321.

Cody, M. L. and J. M. Overton. 1996. Short-term evolution of reduced dispersal ability in island plant populations. *Journal of Ecology* 84: 53–61.

Cogger, H. G. 1975. Sea snakes of Australia and New Guinea. In W. A. Dunson (ed.), *The Biology of Sea Snakes*, 59–139. Baltimore, MD: University Park Press.

Cohen, J. E. 1995. *How Many People Can the Earth Support?* New York: W. W. Norton.

Cole, K. L. 1982. Late Quaternary zonation of vegetation in the eastern Grand Canyon. *Science* 217: 1142–1145.

Coleman, B. D., M. A. Mares, M. R. Willig and Y.-H. Hsieh. 1982. Randomness, area, and species richness. *Ecology* 63: 1121–1133.

Colinvaux, P. A. 1981. Historical ecology in Beringia: The south-land bridge coast at St. Paul Island. *Quaternary Research* 16: 18–36.

Colinvaux, P. A. 1989. Ice-age Amazonia revisited. *Nature* 340: 188–189.

Colinvaux, P. A. 1996. Low-down on a landbridge. *Nature* 382: 21–23.

Colinvaux, P. A., P. E. De Oliveira, J. E. Moreno, M. C. Miller and M. B. Bush. 1996. A long pollen record from lowland Amazonia: Forest and cooling in glacial times. *Science* 274: 86–88.

Collins, N. M. 1989. Termites. In H. Lieth and M. J. A. Werger (eds.), *Ecosystems of the World*, vol. 14B, *Tropical Rain Forest Ecosystems: Biogeographical and Ecological Studies*, 455–471. Amsterdam: Elsevier.

Collins, N. M. and J. A. Thomas. 1991. *The Conservation of Insects and Their Habitats*. London: Academic Press.

Colwell, R. K. 1973. Competition and coexistence in a simple tropical community. *American Naturalist* 107: 737–760.

Colwell, R. K. 1979. The geographical ecology of hummingbird flower mites in relation to their host plants and carriers. In J. S. Rodrigues (ed.), *Recent Advances in Acarology*, vol. 2, 461–468. New York: Academic Press.

Colwell, R. K. and G. C. Hurtt. 1994. Non-biological gradients in species richness and a spurious Rapoport effect. *American Naturalist* 144: 570–595.

Colwell, R. K. and D. W. Winkler. 1984. A null model for null models in biogeography. In D. R. Strong, D. Simberloff, L. G. Abele and A. B. Thistle (eds.), *Ecological Communities: Conceptual Issues and the Evidence*, 344–359. Princeton, NJ: Princeton University Press.

Coney, P. J. 1982. Plate tectonic constraints on biogeographic connections between North and South America. *Annals of the Missouri Botanical Garden* 69: 432–443.

Coney, P. J., D. L. Jones and I. W. H. Monger. 1980. Cordilleran suspect terrains. *Nature* 288: 329–333.

Connell, J. H. 1961. The influence of interspecific competition and other factors on the distribution of the barnacle *Chthamalus stellatus. Ecology* 42: 710–723.

Connell, J. H. 1975. Some mechanisms producing structure in natural communities: A model and evidence from field experiments. In M. L. Cody and J. M. Diamond (eds.), *Ecology and Evolution of Communities*, 460–490. Cambridge, MA: Belknap Press.

Connell, J. H. 1978. Diversity in tropical rain forests and coral reefs. *Science* 199: 1301–1310.

Connell, J. H. and E. Orias. 1964. The ecological regulation of species diversity. *American Naturalist* 98: 399–414.

Connell, T. H. and R. O. Slatyer. 1977. Mechanisms of succession in natural communities and their role in community stability and organization. *American Naturalist* 111: 1119–1144.

Connor, E. F. and E. D. McCoy. 1975. The statistics and biology of the species-area relationship. *American Naturalist* 113: 791–833.

Connor, E. F. and D. Simberloff. 1978. Species number and compositional similarity of the Galápagos flora and avifauna. *Ecological Monographs* 48: 219–248.

Cook, R. E. 1969. Variation in species density of North American birds. *Systematic Zoology* 18: 63–84.

Cook, R. R. 1995. The relationship between nested subsets, habitat subdivision, and species diversity. *Oecologia* 101: 204–210.

Cook, R. R. and J. F. Quinn. 1995. The influence of colonization in nested species subsets. *Oecologia* 102: 413–424.

Cooke, H. B. S. 1972. The fossil mammal fauna of Africa. In A. Keast, F. C. Erk and B. Glass (eds.), *Evolution, Mammals, and Southern Continents*, 89–139. Albany: State University of New York Press.

Cooper, W. S. 1913. The climax forest of Isle Royale, Lake Superior and its development. *Botanical Gazette* 15: 1–44, 115–140, 189–235.

Corlett, R. T. 1992. The ecological transformation of Singapore, 1819–1990. *Journal of Biogeography* 19: 411–420.

Cornelius, J. M. and J. F. Reynolds. 1991. On determining the statistical significance of discontinuities with ordered ecological data. *Ecology* 72: 2057–2070.

Cornell, H. V. 1993. Unsaturated patterns in species assemblages: The role of regional processes in setting local species richness. In R. E. Ricklefs and D. Schluter (eds.), *Species Diversity in Ecological Communities: Historical and Geographical Perspectives*, 243–252. Chicago: University of Chicago Press.

Cotgreave, P. and P. Stockley. 1994. Body size, insectivory and abundance in assemblages of small mammals. *Oikos* 71: 89–96.

Cowles, H. C. 1889. The ecological relations of the vegetation of the sand dunes of Lake Michigan. *Botanical Gazette* 27: 95–117, 167–202, 281–308, 361–391.

Cowles, H. C. 1901. The physiographic ecology of Chicago and vicinity: A study of the origin, development and classification of plant societies. *Botanical Gazette* 31: 73–108, 145–182.

Cox, C. B. 1973a. The distribution of Triassic terrestrial tetrapod families. In D. H. Tarling and S. K. Runcorn (eds.), *Implications of Continental Drift to the Earth Sciences*, vol. 1, 369–371. New York: Academic Press.

Cox, C. B. 1973b. Triassic tetrapods. In A. Hallam (ed.), *Atlas of Palaeobiogeography*, 213–223. Amsterdam: Elsevier Scientific.

Cox, C. B. 1990. New geological theories and old biogeographical problems. *Journal of Biogeography* 17: 117–130.

Cox, C. B. and P. D. Moore. 1985. *Biogeography: An Ecological and Evolutionary Approach*. Boston: Blackwell Scientific Publications.

Cox, G. W., L. C. Contreras and A. V. Milewski. 1994. Role of fossorial animals in community structure and energetics of Mediterranean ecosystems. In M. T. Kilinin de Arroyo, P. H. Zedler and M. D. Fox (eds.), *Ecology of Convergent Ecosystems: Mediterranean-Climate Ecosystems of Chile, California and Australia*, 383–398. New York: Springer-Verlag.

Cracraft, J. 1980. Avian phylogeny and intercontinental biogeographic patterns. In R. Nöhring (ed.), *Acta XVII Congressus Internationalis Ornithologici*, vol. 2, 1302–1308. Berlin: Verlag der Deutschen Ornithologen-Gesellschaft.

Cracraft, J. 1981. Toward a phylogenetic classification of the Recent birds of the world (class Aves). *Auk* 98: 681–714.

Cracraft, J. 1982. Geographic differentiation, cladistics, and vicariance biogeography: Reconstructing the tempo and mode of evolution. *American Zoologist* 22: 411–424.

Cracraft, J. 1988. Deep-history biogeography: Retrieving the historical pattern of evolving continental biotas. *Systematic Zoology* 37: 221–236.

Cracraft, J. 1989. Speciation and its ontology: The empirical consequences of alternative species concepts for understanding patterns and processes of differentiation. In D. Otte and J. A. Endler (eds.), *Speciation and Its Consequences*, 27–59. Sunderland, MA: Sinauer Associates.

Cracraft, J. and R. O. Prum. 1988. Patterns and processes of diversification: Speciation and historical congruence in some Neotropical birds. *Evolution* 42: 603–620.

Crame, J. A. 1993. Latitudinal range fluctuations in the marine realm through geological time. *Trends in Ecology and Evolution* 8: 162–166.

Crane, P. R. and S. Lidgard. 1989. Angiosperm diversification and paleolatitudinal gradients in Cretaceous floristic diversity. *Science* 246: 675–678.

Craw, R. C. 1982. Phylogenetics, areas, geology and the biogeography of Croizat: A radical view. *Systematic Zoology* 31: 304–316.

Craw, R. C. 1988. Panbiogeography: Method and synthesis in biogeography. In A. A. Myers and P. S. Giller (eds.), *Analytical Biogeography: An Integrated Approach to the Study of Animal and Plant Distributions*, 405–435. London: Chapman and Hall.

Craycraft, J. and R. O. Prum. 1988. Patterns and processes of diversification: Speciation and historical congruence in some neotropical birds. *Evolution* 42: 603–620.

Creer, K. M., E. Irving and S. K. Runcorn. 1954. The direction of the geomagnetic field in remote epochs in Great Britain. *Journal of Geomagnetism and Geoelectricity* 6: 163–168.

Creer, K. M., E. Irving and S. K. Runcorn. 1957. Geophysical interpretation of palaeomagnetic directions from Great Britain. *Philosophical Transactions of the Royal Society of London*, Series A 250: 144–156.

Cressie, N. 1991. *Statistics for Spatial Data*. New York: John Wiley and Sons.

Crisci, J. V., M. M. Cigliano, J. J. Morrone and S. Roigjunent. 1991. Historical biogeography of Southwestern South America. *Systematic Zoology* 40: 152–171.

Critchfield, W. B. and E. J. Little. 1966. *Geographic Distributions of Pines of the World*. USDA Forest Service, Miscellaneous Publication 991.

Croizat, L. 1952. *Manual of Phytogeography*. The Hague: Dr. W. Junk.

Croizat, L. 1958. *Panbiogeography*. 2 vol. Caracas: Published by the author.

Croizat, L. 1960. *Principia Botanica*. Caracas: Published by the author.

Croizat, L. 1964. *Space, Time, Form: The Biological Synthesis*. Caracas: Published by the author.

Croizat, L., G. J. Nelson and D. E. Rosen. 1974. Centers of origin and related concepts. *Systematic Zoology* 23: 265–287.

Crosby, G. T. 1972. Spread of the cattle egret in the Western Hemisphere. *Bird-Banding* 43: 205–212.

Crossman, E. J. 1991. Introduced freshwater fishes: A review of the North American perspective with emphasis on Canada. *Canadian Journal of Fisheries and Aquatic Sciences* 48 (suppl. 1): 46–57.

Crowell, K. L. 1962. Reduced interspecific competition among the birds of Bermuda. *Ecology* 43: 75–88.

Crowell, K. L. 1973. Experimental zoogeography: Introductions of mice to small islands. *American Naturalist* 107: 535–558.

Crowell, K. L. 1986. A comparison of relict versus equilibrium models for insular mammals of the Gulf of Maine. *Biological Journal of the Linnean Society* 28: 37–64.

Crowell, K. L. and S. L. Pimm. 1976. Competition and niche shifts of mice introduced onto small islands. *Oikos* 27: 251–258.

Cruden, R. W. 1966. Birds as agents of long-distance dispersal for disjunct plant groups of the temperate Western Hemisphere. *Evolution* 20: 517–532.

Cuellar, O. 1977. Animal parthenogenesis. *Science* 197: 837–843.

Culotta, E. 1994. Is marine diversity at risk? *Science* 263: 918–920.

Culver, D. C. 1970. Analysis of simple cave communities. I. Caves as islands. *Evolution* 29: 463–474.

Cumber, R. A. 1953. Some aspects of the biology and ecology of bumblebees bearing upon the yields of red clover seed in New Zealand. *New Zealand Journal of Science and Technology* 34: 227–240.

Currie, D. J. 1991. Energy and large-scale patterns of animal- and plant-species richness. *American Naturalist* 137: 27–49.

Currie, D. J. and V. Paquin. 1987. Large-scale biogeographical patterns of species richness of trees. *Nature* 329: 326–327.

Curtis, J. T. 1959. *The Vegetation of Wisconsin*. Madison: University of Wisconsin Press.

Cushman, J. H., J. H. Lawton and B. F. J. Manly. 1993. Latitudinal patterns in European ant assemblages: Variation in species richness and body size. *Oecologia* 95: 30–37.

Cutler, A. 1991. Nested faunas and extinction in fragmented habitats. *Conservation Biology* 5: 496–505.

Dalrymple, G. B., E. A. Silver and E. D. Jackson. 1973. Origin of the Hawaiian Islands. *American Scientist* 61: 294–308.

Dammermann, K. W. 1948. The fauna of Krakatau, 1883–1933. *Koninklijke Nederlandsche Akademie Wetenschappen Verhandelingen* 44: 1–594.

Damuth, J. 1991. Of size and abundance. *Nature* 351: 268–269.

Daniels, R. J. R. 1992. Geographic distribution patterns of amphibians in the western Ghats, India. *Journal of Biogeography* 19: 521–529.

Dansereau, P. M. 1957. *Biogeography: An Ecological Perspective*. New York: Ronald Press.

Darlington, P. J. Jr. 1938. The origin of the fauna of the Greater Antilles, with discussion of dispersal of animals over water and through the air. *Quarterly Review of Biology* 13: 274–300.

Darlington, P. J. Jr. 1943. Caribidae of mountains and islands: Data on the evolution of isolated faunas, and on atrophy of wings. *Ecological Monographs* 13: 37–61.

Darlington, P. J. Jr. 1957. *Zoogeography: The Geographical Distribution of Animals*. New York: John Wiley & Sons.

Darlington, P. J. Jr. 1959a. Darwin and zoogeography. *Proceedings of the American Philosophical Society* 103: 307–319.

Darlington, P. J. Jr. 1959b. Area, climate and evolution. *Evolution* 13: 488–510.

Darlington, P. J. Jr. 1965. *Biogeography of the Southern End of the World: Distribution and History of Far Southern Life and Land, with an Assessment of Continental Drift*. Cambridge, MA: Harvard University Press.

Darwin, C. 1839. *Journal of the Researches into the Geology and Natural History of Various Countries Visited by H.M.S. Beagle, under the Command of Captain Fitzroy, R.N. from 1832 to 1836*. London: Henry Colburn.

Darwin, C. 1859. *On the Origin of Species by Means of Natural Selection or the Preservation of Favored Races in the Struggle for Life*. London: John Murray.

Darwin, C. 1860. *The Voyage of the Beagle*. New Jersey: Doubleday.

Daubenmire, R. 1978. *Plant Geography with Special Reference to North America*. New York: Academic Press.

Davies, N. 1995. *The Incas*. Niwat, CO: University Press of Colorado.

Davis, M. B. 1969. Palynology and environmental history during the Quaternary period. *American Scientist* 57: 317–332.

Davis, M. B. 1976. Pleistocene geography of temperate deciduous forests. *Geoscience and Man* 13: 13–26.

Davis, M. B. 1981. Quaternary history and the stability of forest communities. In D. C. West, H. H. Shugart and D. B. Botkin (eds.), *Forest Succession: Concepts and Application*, 132–153. Springer-Verlag, New York.

Davis, M. B. 1986. Climatic stability, time lags and community disequilibrium. In J. M. Diamond and T. Case (eds.), *Community Ecology*, 269–284. New York: Harper and Row.

Davis, M. B. 1994. Ecology and paleoecology begin to merge. *Trends in Ecology and Evolution* 9: 357–358.

Davis, R. 1996. The Pinalenos as an island in a montane archipelago. In C. A. Istock and R. S. Hoffmann (eds.), *Storm over a Mountain Archipelago: Conservation Biology and the Mt. Graham Affair*, 123–134. Tuscon: University of Arizona Press.

Davis, R. and D. E. Brown. 1989. Role of post-Pleistocene dispersal in determining the modern distribution of Albert's squirrel. *Great Basin Naturalist* 49: 425–434.

Davis, R. and J. R. Callahan. 1992. Post-Pleistocene dispersal in the Mexican vole (*Microtus mexicanus*): An example of an apparent trend in the distribution of southwestern mammals. *Great Basin Naturalist* 52: 262–268.

Davis, R. and C. Dunford. 1987. An example of contemporary colonization of montane islands by small, nonflying mammals in the American Southwest. *American Naturalist* 129: 398–406.

Dawson, W. R. and C. Carey. 1976. Seasonal acclimation to temperature in cardueline finches. *Journal of Comparative Physiology* 112: 317–333.

Day, D. S. and W. G. Pearcy. 1968. Species associations of benthic fishes on the continental shelf and slope off Oregon. *Journal of the Fisheries Research Board of Canada* 25: 2665–2675.

Dayan, T. 1990. Feline canines: Community-wise character displacement in the small cats of Israel. *American Naturalist* 136: 39–60.

Dayan, T., E. Tchernov, Y. Yom-Tov and D. Simberloff. 1989. Ecological character displacement in Saharo-Arabian *Vulpes*: Outfoxing Bergmann's rule. *Oikos* 55: 263–272.

Dayton, P. K. 1971. Competition, disturbance and community organization: The provision and subsequent utilization of space in a rocky intertidal community. *Ecological Monographs* 41: 351–389.

DeGraaf, R. M. and J. H. Rappole. 1995. *Neotropical Migratory Birds: Natural History, Distribution, and Population Change*. Ithaca, NY: Comstock Press (Cornell University Press).

Delany, M. J. 1972. The ecology of small rodents in tropical Africa. *Mammalian Reviews* 2: 1–41.

Delcourt, H. R. and P. A. Delcourt. 1994. Postglacial rise and decline of *Ostrya virginiana* (Mill.) K. Koch and *Carpinus caroliniana* Walt. in eastern North America: Predictable responses of forest species to cyclic changes in seasonality of climate. *Journal of Biogeography* 21: 137–150.

Delcourt, P. A. and H. R. Delcourt. 1977. The Tunica Hills, Louisiana-Mississippi: Late glacial locality for spruce and deciduous forest species. *Quaternary Research* 7: 218–237.

Delcourt, P. A. and H. R. Delcourt. 1987. Late-quarternary dynamics of temperate forests: Applications of paleocology to issues of global environmental change. *Quaternary Science Review* 6: 129–146.

Delcourt, P. A. and H. R. Delcourt. 1996. Quaternary paleoecology of the lower Mississippi Valley. *Engineering Geology* 45: 219–242.

Dell, B., G. Muir and M. Palmer. 1980. Lizard assemblages and reserve size and structure in the Western Australian wheatbelt—some implications for conservation. *Biological Conservation* 17: 25–61.

Den Boer, P. J. 1977. *Dispersal Power and Survival: Carabids in a Cultivated Countryside*. Wageningen, the Netherlands: H. Veenman and Zonen B. V.

Desmond, A. 1965. How many people have ever lived on earth? In L. K. Y. Ng and S. Mudd (eds.), *The Population Crisis: Implications and Plans for Action*. Bloomington: Indiana University Press.

de Vries, A. L. 1971. Freezing resistance in fishes. In W. S. Hoar and D. J. Randall (eds.), *Fish Physiology*, vol. 6, 157–190. New York: Academic Press.

de Wet, J. M. J. 1979. Origins of polyploids. In W. H. Lewis (ed.), *Polyploidy: Biological Relevance*, 3–15. New York: Plenum Press.

Dewey, J. F. 1977. Suture zone complexities: A review. *Tectonophysics* 40: 53–67.

Dexter, R. W. 1978. Some historical notes on Louis Agassiz's lectures on zoogeography. *Journal of Biogeography* 5: 207–209.

D'Hondt, S. and M. A. Arthur. 1996. Late Cretaceous oceans and the cool tropic paradox. *Science* 271: 1838–1841.

Diamond, J. M. 1969. Avifaunal equilibria and species turnover rates on the Channel Islands of California. *Proceedings of the National Academy of Sciences, USA* 64: 57–63.

Diamond, J. M. 1970a. Ecological consequences of island colonization by Southwest Pacific birds. I. Types of niche shifts. *Proceedings of the National Academy of Sciences, USA* 67: 529–536.

Diamond, J. M. 1970b. Ecological consequences of island colonization by Southwest Pacific birds. II. The effect of species diversity on total population density. *Proceedings of the National Academy of Sciences, USA* 67: 1715–1721.

Diamond, J. M. 1972. Biogeographic kinetics: Estimation of relaxation times for avifaunas of Southwest Pacific islands. *Proceedings of the National Academy of Sciences, USA* 69: 3199–3203.

Diamond, J. M. 1973. Distributional ecology of New Guinea birds. *Science* 179: 759–769.

Diamond, J. M. 1974. Colonization of exploded volcanic islands by birds: The supertramp strategy. *Science* 184: 803–806.

Diamond, J. M. 1975a. The island dilemma: Lessons of modern biogeographic studies for the design of natural reserves. *Biological Conservation* 7: 129–146.

Diamond, J. M. 1975b. Assembly of species communities. In M. L. Cody and J. M. Diamond (eds.), *Ecology and Evolution of Communities*, 342–444. Cambridge, MA: Belknap Press.

Diamond, J. M. 1977. Colonization cycles in man and beast. *World Archaeology* 8: 249–261.

Diamond, J. M. 1984. "Normal" extinctions of isolated populations. In M. N. Nitecki (ed.), *Extinctions*, 191–246. Chicago: University of Chicago Press.

Diamond, J. M. 1991a. Did Komodo dragons evolve to eat pygmy elephants? *Nature* 326: 832.

Diamond, J. M. 1991b. A new species of rail from the Solomon Islands and convergent evolution of insular flightlessness. *Auk* 108: 461–470.

Diamond, J. M. 1997. *Guns, Germs and Steel: The Fates of Human Societies*. New York: W. W. Norton.

Diamond, J. M. and T. J. Case (eds.). 1986. *Community Ecology*. New York: Harper and Row.

Diamond, J. M. and M. E. Gilpin. 1983. Biogeographic umbilici and the origin of the Philippine avifauna. *Oikos* 41: 307–321.

Diamond, J. M. and R. M. May. 1976. Island biogeography and the design of natural preserves. In R. M. May (ed.), *Theoretical Ecology: Principles and Applications*, 163–186. Philadelphia: W. B. Saunders Co.

Diamond, J. M. and S. L. Pimm. 1993. Survival times of bird populations: A reply. *American Naturalist* 142: 1030–1035.

Dice, L. R. 1947. Effectiveness of selection by owls of deer mice (*Peromyscus maniculatus*) which contrast in color with their background. *Contributions, Laboratory Vertebrate Biology, University of Michigan* 50: 1–15.

Dice, L. R. and P. M. Blossom. 1937. *Studies of Mammalian Ecology in Southwestern North America with Special Attention to the Colors of Desert Mammals*. Carnegie Institute Publication 485: 1–29. Washington, DC: Carnegie Institute.

Dickerson, J. E. Jr. and J. V. Robinson. 1986. The controlled assembly of microcosmic communities: The selective extinction hypothesis. *Oecologia* 71: 12–17.

Dickman, C. R. 1996. The impact of exotic generalist predators on the native fauna of Australia. *Wildlife Biology* 2: 185–195.

Dietz, R. S. 1961. Continental and ocean basin evolution by spreading of the sea floor. *Nature* 190: 854–857.

Dixon, J. R. 1979. Origin and distribution of reptiles of low-land tropical rain forest of South America. In W. E. Duellman (ed.), *The South American Herpetofauna: Its Origin, Evolution, and Dispersal*, 217–240. Monograph no. 7. Lawrence: Museum of Natural History, University of Kansas.

Doak, D. F. and L. S. Mills. 1994. A useful role for theory in conservation. *Ecology* 75: 615–626.

Dobson, A. P. 1996. *Conservation and Biodiversity*. New York: Scientific American Library.

Dobson, A. P., J. P. Rodriguez, W. M. Roberts and D. S. Wilcove. 1997. Geographic distribution of endangered species in the United States. *Science* 275: 550–553.

Dobzhansky, Th. 1950. Evolution in the tropics. *American Scientist* 38: 209–221.

Dobzhansky, Th. 1951. *Genetics and the Origin of Species*. 3rd ed. New York: Columbia University Press.

Docters van Leeuwen, W. M. 1936. Krakatau, 1883–1933. *Ann. Jard. Bot. Buitenzorg* 46–47: 1–506.

Dodd, A. P. 1959. The biological control of prickly pear in Australia. In A. Keast (ed.), *Biogeography and Ecology in Australia*, 565–577. Monographiae Biologicae 8. The Hague: Dr. W. Junk.

Dorst, J. 1962. *The Migrations of Birds*. Boston: Houghton Mifflin.

Doube, B. M., A. MacWueen, T. J. Ridsill-Smith and T. A. Weir. 1991. Native and introduced dung beetles in Australia. In I. Hanski and Y. Camberfort (eds.), *Dung Beetle Ecology*, 255–278. Princeton, NJ: Princeton University Press.

Doyle, J. A. 1978. Origin of angiosperms. *Annual Review of Ecology and Systematics* 9: 365–392.

Drake, J. A., H. A. Mooney, F. di Castri, R. H. Groves, F. J. Kruger, M. Rejmanek and M. Williamson (eds.). 1989. *Biological Invasions: A Global Perspective*. New York: John Wiley & Sons.

Dressler, R. L. 1981. *The Orchids: Natural History and Classification*. Cambridge, MA: Harvard University Press.

Drude, O. 1887. *Atlas der Pflanzenverbreitung*. 5 Abt. Gotha, Berghaus Physikalischer Atlas.

Drury, W. H. and I. C. T. Nisbet. 1973. Succession. *Journal of the Arnold Arboretum* 54: 331–368.

Due, A. D. and G. A. Polis. 1986. Trends in scorpion diversity along the Baja California Peninsula. *American Naturalist* 128: 460–468.

Duellman, W. E. and L. Trueb. 1986. *Biology of Amphibians*. London, New York: McGraw-Hill.

Dunmire, W. W. 1960. An altitudinal survey of reproduction in *Peromyscus marliculatus*. *Ecology* 41: 174–182.

Dunn, C. P. and C. Loehle. 1988. Species-area parameter estimation: Testing the null model of lack of relationships. *Journal of Biogeography* 15: 721–728.

Dunson, W. A. (ed.) 1975. *The Biology of Sea Snakes*. Baltimore, MD: University Park Press.

Dunson, W. A. and F. J. Mazotti. 1989. Salinity as a limiting factor in the distribution of reptiles in Florida Bay: A theory for the estuarine origin of marine snakes and turtles. *Bulletin of Marine Science* 44: 229–244.

Durham, J. W. and F. S. MacNeil. 1967. Cenozoic migrations of marine invertebrates through the Bering Strait region. In D. M. Hopkins (ed.), *The Bering Land Bridge*, 326–349. Stanford, CA: Stanford University Press.

D'Urville, M. J. D. 1833. Voyage de la corvette L'Atrolable execute pendant les annees 1826–1827–1828–1829 sans le commandement de M. Jules Dummont D'Urville, capitaine de vaisseau. *Atlas Historique Paris, J. Tartu.*

du Toit, A. L. 1927. *A Geological Comparison of South America with South Africa.* Publication no. 381, 1–157. Washington, DC: Carnegie Institute.

du Toit, A. L. 1937. *Our Wandering Continents.* Edinburgh: Oliver & Boyd.

Dyer, R. A. 1975. *The Genera of Southern African Flowering Plants,* vol. 1. Pretoria: Department of Agricultural Technical Services, Botanical Research Institute.

Ebenhard, T. 1988a. *Demography and Island Colonization of Experimentally Introduced and Natural Vole Populations.* Ph.D. dissertation, Uppsala University, Sweden.

Ebenhard, T. 1988b. Introduced birds and mammals and their ecological effects. *Swedish Wildlife Research* 13: 1–107.

Ebenhard, T. 1989. Bank vole (*Clethrionomys glareolus*) propagules of different sizes and island colonization. *Journal of Biogeography* 16: 173–180.

Ebenhard, T. 1991. Colonization of metapopulations: A review of theory and observations. *Biological Journal of the Linnean Society* 42: 105–121.

Echelle, A. A. and I. Kornfield (eds.). 1984. *Evolution of Fish Species Flocks.* Orono: University of Maine Press.

Edwards, P. J., R. M. May and N. R. Webb (eds.). 1994. *Large-Scale Ecology and Conservation Biology.* London: Blackwell.

Egyed, L. 1956. The change of earth's dimensions determined from paleographical data. *Geofisica Pura e Applicata* 33: 42–48.

Egyed, L. 1957. A new dynamic conception of the internal constitution of the Earth. *Geologische Rundschau* 46: 101–121.

Ehrlich, P. R. 1961. Intrinsic barriers to dispersal in the checkerspot butterfly. *Science* 134: 108–109.

Ehrlich, P. R. 1965. The population biology of the butterfly *Euphydryas editha.* II. The structure of the Jaspar Ridge colony. *Evolution* 19: 327–336.

Ehrlich, P. R. and P. H. Raven. 1965. Butterflies and plants: A study in coevolution. *Evolution* 18: 586–608.

Ehrlich, P. R. and R. W. Holm. 1963. *The Process of Evolution.* New York: McGraw-Hill, Inc.

Ehrlich, P. R., R. R. White, M. C. Singer, S. W. McKechnie and L. E. Gilbert. 1975. Checkerspot butterflies: A historical perspective. *Science* 188: 221–228.

Eisenberg, J. F. 1981. *The Mammalian Radiations: An Analysis of Trends in Evolution, Adaptation, and Behavior.* Chicago: University of Chicago Press.

Eisenberg, J. F. and E. Gould. 1970. The Tenrecs: A study in mammalian behavior and evolution. *Smithsonian Contributions to Zoology* 27: 1–137.

Ekman, S. 1953. *Zoogeography of the Sea.* London: Sidgwick & Jackson.

Eldredge, N. and I. Cracraft. 1980. *Phylogenetic Patterns and the Evolutionary Process.* New York: Columbia University Press.

Eldredge, N. and S. J. Gould. 1972. Punctuated equilibria: An alternative to phyletic gradualism. In T. J. M. Schopf (ed.), *Models in Paleobiology,* 82–115. San Francisco: Freeman, Cooper & Co.

Elena, S. F., V. S. Cooper and R. E. Lenski. 1996. Punctuated evolution caused by selection of rare beneficial mutations. *Science* 272: 1802–1804.

Elias, S. A. 1997. *The Ice-Age History of Southwestern Parks.* Washington, DC: Smithsonian Institution Press.

Elias, S. A., S. K. Short and C. H. Nelson. 1996. Life and times of the Bering landbridge. *Nature* 382: 60–63.

Elliot, D. H., E. H. Colbert, W. J. Breed, I. A. Jensen and T. S. Powell. 1970. Triassic tetrapods from Antarctica: Evidence for continental drift. *Science* 169: 197–201.

Elliot, G. F. 1951. On the geographical distribution of terebratelloid brachiopods. *Annals and Magazines of Natural History,* ser. 12, 4: 305–334.

Ellson, J. A. 1969. Late Quaternary marine submergence of Quebec. *Review of Geography, Montreal* 23: 247–258.

Elmberg, J., P. Nummi, H. Paysa and K. Sjoberg. 1994. Relationship between species number, lake size and resource diversity in assemblages of breeding waterfowl. *Journal of Biogeography* 21: 75–84.

Elton, C. 1927. *Animal Ecology.* New York: Macmillan.

Elton, C. S. 1958. *The Ecology of Invasions by Animals and Plants.* London: Methuen & Co.

Elton, C. S. 1966. *The Patterns of Animal Communities.* London: Methuen & Co.

Embleton, B. J. J. 1973. The palaeolatitude of Australia through Phanerozoic time. *Journal of the Geological Society of Australia* 19: 475–482.

Emlen, J. T. 1978. Density anomalies and regulation mechanisms in land bird populations on the Florida peninsula. *American Naturalist* 112: 265–286.

Emlen, J. T. 1979. Land bird densities on Baja California islands. *Auk* 96: 152–167.

Enckell, P. H., S. A. Bengtson and B. Wiman. 1987. Serf and waif colonization, distribution, and dispersal of invertebrate species in Faroe Island settlement areas. *Journal of Biogeography* 14: 89–104.

Endler, J. A. 1977. *Geographic Variation, Speciation, and Clines.* Monographs in Population Biology, no. 10. Princeton, NJ: Princeton University Press.

Endler, J. A. 1982. Problems in distinguishing historical from ecological factors in biogeography. *American Zoologist* 22: 441–452.

Enquist, B. J., M. A. Jordan and J. H. Brown. 1995. Connections between ecology, biogeography, and paleobiology: relationship between local abundance and geographic distribution in fossil and recent molluscs. *Evolutionary Ecology* 9: 586–604.

Erickson, R. O. 1945. The *Clematis fremontii* var. *riehlii* population in the Ozarks. *Annals of the Missouri Botanical Garden* 32: 413–460.

Erlinge, S. 1987. Why do European stoats *Mustela erminea* not follow Bergmann's rule? *Holarctic Ecology* 10: 33–39.

Erulkar, S. D. 1972. Comparative aspects of spatial localization of sound. *Physiological Reviews* 52: 237–360.

Erwin, D. H. 1993. *The Great Paleozoic Crisis: Life and Death in the Permian.* New York: Columbia University Press.

Erwin, T. L. 1983. Tropical forest canopies: The last biological frontier. *Bulletin of the Entomological Society of America* 19: 14–19.

Erwin, T. L. 1988. The tropical forest canopy: The heart of biotic diversity. In E. O. Wilson (ed.), *Biodiversity,* 123–129. Washington, DC: National Academy Press.

Estes, J. A. and D. O. Duggins. 1995. Sea otters and kelp forests in Alaska: Generality and variation in a community ecology paradigm. *Ecological Monographs* 65: 75–100.

Estes, R. and A. Baez. 1985. Herptofaunas of North and South America during the Late Cretaceous and Cenozoic: Evidence for interchange? In F. G. Stehli and S. D. Webb (eds.), *The Great American Interchange,* 140–200. New York: Plenum Press.

Estes, R. and O. A. Reig. 1973. The early fossil record of frogs: A review of the evidence. In J. L. Vial (ed.), *Evolutionary Biology of the Anurans,* 11–63. Columbia: University of Missouri Press.

Facelli, J. and S. T. A. Pickett. 1990. Markovian chains and the role of history in succession. *Trends in Ecology and Evolution* 5: 27–30.

Faeth, S. H. 1984. Density compensation in vertebrates and invertebrates: A review and an experiment. In D. R. Strong, D. Simberloff, L. G. Abele and A. B. Thistle (eds.), *Ecological Communities: Conceptual Issues and the Evidence*, 491–509. Princeton, NJ: Princeton University Press.

Fagan, B. M. 1990. *The Journey from Eden: The Peopling of Our World*. London: Thames and Hudson.

Falconer, D. S. 1960. *Introduction to Quantitative Genetics*. Edinburgh: Oliver and Boyd.

Falla, R. A., R. B. Sibson and E. G. Turbott. 1966. *A Field Guide to the Birds of New Zealand and Outlying Islands*. London: William Collins Sons & Co.

Fallaw, W. C. 1979. Trans-North Atlantic similarity among Mesozoic and Cenozoic invertebrates correlated with widening of the ocean basin. *Geology* 7: 398–400.

Farrell, T. M. 1991. Models and mechanisms of succession: An example from a rocky intertidal community. *Ecological Monographs* 61: 95–113.

Feduccia, A. 1980. *The Age of Birds*. Cambridge, MA: Harvard University Press.

Feduccia, A. 1995. Explosive evolution in Tertiary birds and mammals. *Science* 267: 637–638.

Feduccia, A. 1996. *The Origin and Evolution of Birds*. New Haven, CT: Yale University Press.

Felsenstein, J. 1985. Phylogenies and the comparative method. *American Naturalist* 125: 1–15.

Fernandes, G. W. and A. C. F. Lara. 1993. Diversity of Indonesian gall-forming herbivores along altitudinal gradients. *Biodiversity Letters* 1: 186–192.

Ferrari, S. F. and H. L. Queiroz. 1994. Two new Brazilian primates discovered, endangered. *Oryx* 28: 31–36.

Fichman, M. 1977. Wallace: Zoogeography and the problem of landbridges. *Journal of the History of Biology* 10: 45–63.

Fields, P. A., J. B. Graham, R. H. Rosenblatt and G. N. Somero. 1993. Effects of expected global climate change on marine faunas. *Trends in Ecology and Evolution* 8: 361–366.

Findley, J. S. 1969. Biogeography of southwestern boreal and desert mammals. In J. K. Jones Jr. (ed.), *Contributions in Mammalogy*. Lawrence: University of Kansas, Museum of Natural History Publication no. 51.

Findley, J. S. and C. Jones. 1964. Seasonal distribution of the hoary bat. *Journal of Mammalogy* 45: 461–470.

Finerty, J. P. 1980. *The Population Ecology of Cycles in Small Mammals: Mathematical Theory and Biological Fact*. New Haven, CT: Yale University Press.

Finlay, B. J., J. O. Corliss, G. Esteban and T. Fenchel. 1996. Biodiversity at the microbial level: The number of free-living ciliates in the biosphere. *Quarterly Review of Biology* 71: 221–237.

Firth, R. and J. W. Davidson. 1945. *Pacific Islands*, vol. 1. General Survey. Naval Intelligence Division, London, 20.3.5.

Fischer, A. G. 1960. Latitudinal variation in organic diversity. *Evolution* 14: 64–81.

Fischer, A. G. 1984. The two Phanerozoic supercycles. In W. A. Berggren and J. A. Van Couvering (eds.), *Catastrophes and Earth History*, 129–150. Princeton, NJ: Princeton University Press.

Fischer, A. G. and M. A. Arthur. 1977. Secular variations in the pelagic realm. *Society for Econonomic Paleontology and Mineralogy Special Publications* 25: 19–50.

Flannery, T. F. 1994. *The Future Eaters: An Ecological History of the Australasian Lands and People*. Kew, Victoria, N.S.W., Australia: Reed International Books.

Flather, C. H., L. A. Joyce and C. A. Bloomgarden. 1994. *Species Endangerment Patterns in the United States*. USDA Forest Service, General Technical Report RM-241.

Fleming, T. H. 1973. Numbers of mammal species in North and Central American forest communities. *Ecology* 54: 555–563.

Fleming, T. H. and A. Estrada (eds.). 1993. *Frugivory and Seed Dispersal: Ecological and Evolutionary Aspects*. Dordrecht: Kuwer Academic.

Flenley, J. R. 1979a. *The Equatorial Rain Forest: A Geological History*. London: Butterworth.

Flenley, J. R. 1979b. The Late Quaternary vegetational history of the equatorial mountains. *Progress in Physical Geography* 3: 488–509.

Flessa, K. W. 1975. Area, continental drift and mammalian diversity. *Paleobiology* 1: 189–194.

Flessa, K. W. 1980. Biological effects of plate tectonics and continental drift. *Bioscience* 30: 518–523.

Flessa, K. W. 1981. The regulation of mammalian faunal similarity among the continents. *Journal of Biogeography* 8: 427–438.

Flessa, K. W. and J. J. Sepkoski Jr. 1978. On the relationship between Phanerozoic diversity and changes in habitable area. *Paleobiology* 4: 359–366.

Flessa, K. W., S. G. Barnett, D. B. Cornue, M. A. Lomaga, N. Lombardi, J. M. Miyazaki and A. S. Murer. 1979. Geologic implications of the relationship between mammalian faunal similarity and geographic distance. *Geology* 7: 15–18.

Flint, R. F. 1971. *Glacial and Quaternary Geology*. New York: John Wiley & Sons.

Flohn, H. 1969. *Climate and Weather*. New York: World University Library, McGraw-Hill.

Flux, J. E. C. 1969. Current work on the African hare, *Lepus capensis* L. in Kenya. *Journal of Reproduction and Fertilization* (Suppl.) 6: 225–227.

Forbes, E. 1856. Map of the distribution of marine life. In A. K. Johnston (ed.), *The Physical Atlas of Natural Phenomena*. Philadelphia: Lea and Blanchard.

Forbes, E. 1859. *The Natural History of European Seas*. London: John Van Voorst.

Forman, R. T. T. (ed.). 1979. *Pine Barrens: Ecosystem and Landscape*. New York: Academic Press.

Forster, J. R. 1778. *Observations Made during a Voyage Round the World, on Physical Geography, Natural History, and Ethic Philosophy*. London: G. Robinson.

Foster, J. B. 1963. *The Evolution of Native Land Mammals of the Queen Charlotte Islands and the Problem of Insularity*. Ph.D. dissertation, University of British Columbia, Vancouver.

Foster, J. B. 1964. Evolution of mammals on islands. *Nature* 202: 234–235.

France, R. 1992. The North American latitudinal gradients in species richness and geographic range of freshwater crayfish and amphipods. *American Naturalist* 139: 342–354.

Free, J. B. 1970. *Insect Pollination of Crops*. New York: Academic Press.

Friis, E. M., W. G. Chaloner and P. R. Crane (eds.). 1987. *The Origins of Angiosperms and Their Biological Consequences*. Cambridge: Cambridge University Press.

Fritts, H. C. 1976. *Tree Rings and Climate*. New York: Academic Press.

Fryer, G. and T. D. Iles. 1972. *The Cichlid Fishes of the Great Lakes of Africa: Their Biology and Evolution*. Edinburgh, Oliver & Boyd.

Fuentes, E. R. 1976. Ecological convergence of lizard communities in Chile and California. *Ecology* 57: 3–17.

Fuentes, E. R. and F. M. Jaksic. 1979. Latitudinal size variation of Chilean foxes: Tests of alternative hypotheses. *Ecology* 60: 43–47.

Funk, V. A. and W. L. Wagner. 1995. Biogeographic patterns in the Hawaiian islands. In W. L. Wagner and V. A. Funk (eds.), *Hawaiian Biogeography: Evolution on a Hot Spot Archipelago*. Washington, DC: Smithsonian Institution Press.

Futuyma, D. J. 1998. *Evolutionary Biology*. 3rd edition. Sunderland, MA: Sinauer Associates.

Gadow, 1913. *The Wanderings of Animals*. Cambridge: Cambridge University Press.

Gaffney, E. S. 1975. A phylogeny and classification of the higher categories of turtles. *Bulletin of the American Museum of Natural History* 155: 387–436.

Gage, J. G. and P. A. Tyler. 1991. *Deep-Sea Biology: A Natural History of Organisms at the Deep-Sea Floor*. Cambridge: Cambridge University Press.

Gaines, S. D. and M. D. Bertness. 1992. Dispersal of juveniles and variable recruitment in sessile marine species. *Nature* 360: 579–580.

Gamble, C. 1994. *Timewalkers: The Prehistory of Global Colonization*. Cambridge, MA: Harvard University Press.

Gardner, S. L. 1991. Phyletic coevolution between subterranean rodents of the genus *Ctenomys* (Rodentia, Hystricognathi) and nematodes of the genus *Paras*. *Zoological Journal of the Linnean Society* 102: 169–201.

Gaston, K. J. 1991. How large is a species' geographic range? *Oikos* 61: 434–438.

Gaston, K. J. 1994. *Rarity*. London: Chapman & Hall.

Gaston, K. J. 1996. Species-range-size distributions: Patterns, mechanisms and implications. *Trends in Ecology and Evolution* 11: 197–201.

Gaston, K. J. and R. David. 1994. Hot spots across Europe. *Biodiversity Letters* 2: 108–116.

Gaston, K. J. and R. M. May. 1992. Taxonomy of taxonomists. *Nature* 356: 281–282.

Gates, D. M. 1993. *Climate Change and Its Biological Consequences*. Sunderland, MA: Sinauer Associates.

Gause, G. F. 1934. *The Struggle for Existence*. Baltimore, MD: Williams & Wilkins.

Gayet, M., J. C. Rage, T. Sempere and P. Y. Gagner. 1992. Mode of interchanges of continental vertebrates between North and South America during the late Cretaceous and Paleocene. *Bulletin de la Societe Geologique de France* 163: 781–791.

Genoways, H. H. and J. H. Brown (eds.). 1993. *Biology of the Heteromyidae*. American Society of Mammalogists, Special Publication no. 10.

George, T. L. 1987. Greater land bird densities on island vs. mainland: Relation to nest predation level. *Ecology* 68: 1393–1400.

Gibbons, A. 1996a. Did Neandertals lose an evolutionary arms race? *Science* 272: 1586–1587.

Gibbons, A. 1996b. The peopling of the Americas. *Science* 274: 31–33.

Gilbert, F. S. 1980. The equilibrium theory of island biogeography: Fact or fiction? *Journal of Biogeography* 7: 209–235.

Gilbert, L. E. 1984. The biology of butterfly communities. In R. I. Vane-Sright and P. R. Ackery (eds.), *The Biology of Butterflies*, 41–53. London: Academic Press.

Gilpin, M. and I. Hanski. 1991. *Metapopulation Dynamics: Empirical and Theoretical Investigations*. London: Academic Press.

Gilpin, M. E. and J. M. Diamond. 1976. Calculations of immigration and extinction curves from the species-area distance relation. *Proceedings of the National Academy of Sciences, USA* 73: 4130–4134.

Gilpin, M. E. and J. M. Diamond. 1980. Subdivisions of nature reserves and the maintenance of species diversity. *Nature* 285: 567–568.

Gilpin, M. E. and J. M. Diamond. 1981. Immigration and extinction probabilities for individual species: Relation to incidence functions and species colonization curves. *Proceedings of the National Academy of Sciences, USA* 78: 392–396.

Gingerich, P. D. 1976a. Paleontology and phylogeny: Patterns of evolution at the species level in Early Tertiary mammals. *American Journal of Science* 276: 1–28.

Gingerich, P. D. 1976b. Cranial anatomy and evolution of early Tertiary Plesiadapidae. *University of Michigan Papers in Paleontology* no. 15, 1–140.

Gittleman, J. L. 1985. Carnivore body size: Ecological and taxonomic correlates. *Oecologia* 67: 540–554.

Glaessner, M. F. 1969. Decapoda. In R. C. Moore (ed.), *Treatise on Invertebrate Paleontology*, Part R, *Arthropoda*, vol. 2, 400–532. Boulder, CO: Geological Society of America.

Glazier, D. S. 1985. Energetics of litter size in five species of *Peromyscus* with generalizations for other mammals. *Journal of Mammalogy* 66: 629–642.

Gleason, H. A. 1917. The structure and development of the plant association. *Bulletin of the Torrey Botanical Club* 53: 7–26.

Gleason, H. A. 1922. On the relation between species and area. *Ecology* 3: 158–162.

Gleason, H.A. 1926. The individualistic concept of plant associations. *Bulletin of the Torrey Botanical Club* 53: 7–26.

Gleick, P. H. 1989. Climate change, hydrology and water resources. *Reviews of Geophysics* 27(3): 329–344.

Glenn-Lewin, D. C. 1977. Species diversity in North American temperate forests. *Vegetatio* 33: 153–162.

Glick, P. A. 1939. *The Distribution of Insects, Spiders and Mites in the Air*. Washington, DC: U.S. Department of Agriculture.

Gliwicz, J. 1980. Island populations of rodents: Their organization and functioning. *Biological Reviews* 55: 109–138.

Gloger, C. L. 1883. *Das Abandern der Vogel durch Einfluss des Klimas*. Breslau: A. Schulz.

Glynn, P. W. 1991. Coral reef bleaching in the 1980s and possible connections with global warming. *Trends in Ecology and Evolution* 6(6): 175–179.

Godfrey, G. K. and P. Crowcroft. 1960. *The Life of the Mole*. London: Museum Press.

Goin, F. J. and A. A. Carlini. 1995. An early tertiary microbiotherid marsupial from Antarctica. *Journal of Vertebrate Paleontology* 15: 205–207.

Golani, D. 1993. The sandy shore of the Red Sea—launching pad for Lessepsian (Suez Canal) migrant fish colonizers of the eastern Mediterranean. *Journal of Biogeography* 20: 579–585.

Golani, D. and A. Ben-Tuvia. 1990. Two Red Sea flatheads, immigrants in the Mediterranean (Pisces: Platycephalidae). *Cybium* 14: 57–61.

Goldblatt, P. 1978. An analysis of the flora of southern Africa: Its characteristics, relationships, and origins. *Annals of the Missouri Botanical Garden* 65: 369–436.

Goldschmidt, T., F. Witte and J. Wanink. 1993. Cascading effects of the introduced Nile perch on the detritivorous/planktivorous species in the sublittoral areas of Lake Victoria. *Conservation Biology* 7: 686–700.

Good, R. 1974. *The Geography of the Flowering Plants*. 3rd edition. White Plains, NY: Longman. [First published in 1947.]

Goode, G. B. 1896. *Biobliograph of the Published Writings of Philip Lutley Sclater, FRS Secretary of the Zoological Society of London*. Washington, DC: Government Printing Office.

Gordon, K. R. 1986. Insular evolutionary body size trends in *Ursus. Journal of Mammalogy* 67: 395–399.

Gotelli, N. J. 1991. Metapopulation models: The rescue effect, the propagule rain, the core-satellite hypothesis. *American Naturalist* 138: 768–776.

Gotelli, N. J. and G. R. Graves. 1996. *Null Models in Ecology.* Washington, DC: Smithsonian Institution Press.

Gould, S. J. 1965. Is uniformitarianism necessary? *American Journal of Science* 263: 223–228.

Gould, S. J. 1979. An allometric interpretation of species-area curves: The meaning of the coefficients. *American Naturalist* 114: 335–343.

Gould, S. J. and C. B. Calloway. 1980. Clams and brachiopods: Ships that pass in the night. *Paleobiology* 6: 383–396.

Grace, J. B. and R. G. Wetzel. 1981. Habitat partitioning and competitive displacement in cattails (*Typha*): Experimental field studies. *American Naturalist* 118: 463–474.

Graham, R. W. 1986. Response of mammalian communities to environmental changes during the Late Quaternary. In J. M. Diamond and T. J. Case (eds.), *Community Ecology*, 300–313. New York: Harper and Row.

Graham, R. W., E. L. Lundelius Jr., M. A. Graham, E. K. Schroeder, R. S. Toomey III, E. Anderson, A. D. Barnosky, J. A. Burns, C. S. Churcher, C. K. Grayson, R. D. Guthrie, C. R. Harrington, G. T. Jefferson, L. D. Martin, H. G. McDonald, R. E. Morlan, H. A. Semken Jr., S. D. Webb, L. Werdelin and M. C. Wilson. 1996. Spatial response of mammals to late-Quaternary environmental fluctuations. *Science* 272: 1601–1606.

Grant, B. R. and P. R. Grant. 1989. *Evolutionary Dynamics of a Natural Population: The Large Cactus Finch of the Galápagos.* Chicago: University of Chicago Press.

Grant, P. R. 1965. The adaptive significance of some size trends in island birds. *Evolution* 19: 355–367.

Grant, P. R. 1966a. The density of land birds on the Tres Marias Island in Mexico. I. Numbers and biomass. *Canadian Journal of Zoology* 44: 391–400.

Grant, P. R. 1966b. Ecological compatibility of bird species on islands. *American Naturalist* 100: 451–462.

Grant, P. R. 1968. Bill size, body size and the ecological adaptations of bird species to competitive situations on islands. *Systematic Zoology* 17: 319–333.

Grant, P. R. 1971. The habitat preference of *Microtus pennsylvanicus* and its relevance to the distribution of this species on islands. *Journal of Mammalogy* 52: 351–361.

Grant, P. R. 1986. *Ecology and Evolution of Darwin's Finches.* Princeton, NJ: Princeton University Press.

Grant, P. R. and I. Abbott. 1980. Interspecific competition, island biogeography and null hypotheses. *Evolution* 34: 332–341.

Grant, P. R. and B. R. Grant. 1992. Hybridization of bird species. *Science* 256: 93–197.

Grant, V. 1981. *Plant Speciation.* New York: Columbia University Press.

Grassle, J. F. 1986. The ecology of deep-sea hydrothermal vent communities. *Advances in Marine Biology*, vol. 23. New York: Academic Press.

Grassle, J. F. 1989. Species diversity in deep-sea communities. *Trends in Ecology and Evolution* 4(1): 12–15.

Grassle, J. F. 1991. Deep-sea benthic biodiversity. *BioScience* 41: 464–469.

Grassle, J. F. and N. J. Maciolek. 1992. Deep-sea richness: Regional and local diversity estimates from quantitative bottom samples. *American Naturalist* 139: 313–341.

Gray, J. and A. J. Boucot (eds.). 1979. *Historical Biogeography, Plate Tectonics, and the Changing Environment.* Corvallis: Oregon State University Press.

Gray, J., A. J. Boucot and W. B. N. Berry (eds.). 1981. *Communities of the Past.* Stroudsburg, PA: Hutchinson Ross.

Grayson, D. K. 1977. Pleistocene avifaunas and the overkill hypothesis. *Science* 195: 691–692.

Grayson, D. K. 1981. A mid-Holocene record for the heather vole, *Phenacomys* cf. *intermedius*, in the central Great Basin and its biogeographic significance. *Journal of Mammalogy* 62: 115–121.

Grayson, D. K. 1987a. An analysis of the chorology of late Pleistocene extinctions in North America. *Quaternary Research* 28: 281–289.

Grayson, D. K. 1987b. The biogeographic history of small mammals in the Great Basin: Observations on the last 20,000 years. *Journal of Mammalogy* 68: 359–375.

Grayson, D. K. 1993. *The Desert's Past: A Natural Prehistory of the Great Basin.* Washington, DC: Smithsonian Institution Press.

Grayson, D. K. and S. D. Livingston. 1993. Missing mammals on Great Basin mountains: Holocene extinctions and inadequate knowledge. *Conservation Biology* 7: 527–532.

Greely, W. B. 1925. The relation of geography to timber supply. *Economic Geography* 1: 1–11.

Greenway, J. C. Jr. 1967. *Extinct and Vanishing Birds of the World.* New York: Dover Publications.

Greenwood, P. H. 1974. The cichlid fishes of Lake Victoria, East Africa: The biology and evolution of a species flock. *Bulletin of the British Museum of Natural History*, Supplement 6: 1–134.

Greenwood, P. H. 1984. African cichlids and evolutionary theories. In A. A. Echelle and I. Kornfield (eds.), *Evolution of Fish Species Flocks.* Orono: University of Maine Press.

Gressitt, J. L. 1954. *Insects of Micronesia.* Hawaii: Bishop Museum.

Gressitt, J. L. (ed.). 1963. *Pacific Basin Biogeography.* Honolulu: Bishop Museum Press.

Griffiths, M. 1978. *The Biology of the Monotremes.* New York: Academic Press.

Grimm, N. B. 1993. Effects of climate change for stream communities. In P. M. Karieva, J. M. Kingsolver and R. B. Huey. *Biotic Interactions and Global Change*, 293–314. Sunderland, MA: Sinauer Associates.

Grinnell, J. 1917. The niche-relationship of the California thrasher. *Auk* 34: 427–433.

Grinnell, J. 1922. The role of the "accidental." *Auk* 39: 373–380.

Guilaine, J. 1991. *Prehistory: The World of Early Man.* New York: Facts on File Publishers.

Guilday, J. E. 1967. Differential extinction during Late Pleistocene and Recent times. In P. S. Martin and H. E. Wright Jr. (eds.), *Pleistocene Extinctions: The Search for a Cause*, 121–140. New Haven, CT: Yale University Press.

Guthrie, R. D. 1990. *Frozen Fauna of the Mammoth Steppe: The Story of Blue Babe.* Chicago: University of Chicago Press.

Hackett, S. J. 1993. Phylogenetic and biogeographic relationships in the Neotropical genus *Gymnopithys* (Formicariidae). *Wilson Bulletin* 105: 301–315.

Hackett, S. J. and K. V. Rosenberg. 1990. A comparison of phenotypic and genetic differentiation in South American antwrens (Formicariidae). *Auk* 107: 473–489.

Hadley, E. A. 1997. Evolutionary and ecological response of pocket gopher (*Thomomys talpoides*) to late-Holocene climatic change. *Biological Journal of the Linnean Society* 60: 277–296.

Haeckel, E. H. P. A. 1866. *Generelle Morphologie der Organismen.* Berlin: Reimer.

Haffer, J. 1969. Speciation in Amazonian forest birds. *Science* 165: 131–137.

Haffer, J. 1974. *Avian Speciation in Tropical South America, with a Systematic Survey of the Toucans (Ramphastidae) and Jacamars (Galbulidae).* Cambridge, MA: Nuttall Ornithological Club.

Haffer, J. 1978. Distribution of Amazon forest birds. *Bonn. Zool. Beitr.* 29: 38–78.

Haffer, J. 1981. Aspects of Neotropical bird speciation during the Cenozoic. In G. Nelson and D. E. Rosen (eds.), *Vicariance Biogeography: A Critique*, 371–391. New York: Columbia University Press.

Hafner, M. S. and S. A. Nadler. 1991. Cospeciation in host-parasite assemblages: Analysis of rates of evolution and timing of cospeciation events. *Systematic Zoology* 39: 192–204.

Hafner, M. S. and R. D. M. Page. 1995. Molecular phylogenies and host-parasite cospeciation: Gophers and lice as a model system. *Philosophical Transactions of the Royal Society of London*, Series B 349: 77–83.

Hagan, J. M. III and D. W. Johnston, eds. 1992. *Ecology and Conservation of Neotropical Migrant Landbirds*. Washington, DC: Smithsonian Institution Press.

Haila, Y. 1986. On the semiotic dimension of ecological theory: The case of island biogeography. *Biology and Philosophy* 1: 377–387.

Haila, Y. and I. Hanski. 1993. Birds breeding on small British islands and extinction risks. *American Naturalist* 142: 1025–1029.

Haila, Y., I. Hanski, O. Jarvinen and E. Ranta. 1982. Insular biogeography: A Northern European perspective. *Acta Oecologica* 3: 303–318.

Hall, E. R. 1981. *The Mammals of North America*. 2nd edition. 2 vols. New York: John Wiley & Sons.

Hallam, A. 1967. The bearing of certain palaeogeographic data on continental drift. *Palaeogeography, Palaeoclimatology, and Palaeoecology* 3: 201–224.

Hallam, A. 1973a. Provinciality, diversity and extinction of Mesozoic marine invertebrates in relation to plate movements. In D. H. Tarling and S. K. Runcorn (eds.), *Implications of Continental Drift to the Earth Sciences*, vol. 1, 287–294. New York: Academic Press.

Hallam, A. (ed.). 1973b. *Atlas of Palaeobiogeography*. Amsterdam: Elsevier Scientific.

Hallam, A. 1983. Plate tectonics and evolution. In D. S. Bendall (ed.), *Evolution From Molecules to Men*, 367–386. Cambridge: Cambridge University Press.

Hallam, A. 1994. *Outline of Phanerozoic Biogeography*. Oxford Biogeography Series, 10. Oxford: Oxford University Press.

Hallam, A. and B. W. Sellwood. 1976. Middle Mesozoic sedimentation in relation to tectonics in the British area. *Journal of Geology* 84: 301–321.

Halliday, T. 1978. *Vanishing Birds: Their Natural History and Conservation*. New York: Holt, Rinehart and Winston.

Hamilton, T. H. and N. E. Armstrong. 1965. Environmental determination of insular variation in bird species abundance in the Gulf of Guinea. *Nature* 207: 148–151.

Hamilton, T. H., R. H. Barth Jr. and I. Rubinoff. 1964. The environmental control of insular variation in bird species abundance. *Proceedings of the National Academy of Sciences, USA* 52: 132–140.

Hansen, T. A. 1980. Influence of larval dispersal and geographic distribution on species longevity in neogastropods. *Paleobiology* 6: 193–207.

Hanski, I. 1982. Dynamics of regional distribution: The core and satellite species hypothesis. *Oikos* 38: 210–221.

Hanski, I. 1986. Population dynamics of shrews on small islands accord with the equilibrium model. *Biological Journal of the Linnean Society* 28: 23–36.

Hanski, I. 1992. Inferences from ecological incidence functions. *American Naturalist* 139: 657–662.

Hanski, I. and Y. Cambefort. 1991. Spatial processes. In I. Hanski and Y. Camberfort (eds.), *Dung Beetle Ecology*, 283–304. Princeton, NJ: Princeton University Press.

Hanski, I. and A. Peltonen. 1988. Island colonization and peninsulas. *Oikos* 51: 105–106.

Hanski, I. and C. D. Thomas. 1994. Metapopulation dynamics and conservation: A spatially explicit model applied to butterflies. *Biological Conservation* 68: 767–786.

Hanski, I., T. Pakkala, M. Kuussaari and G. Lei. 1995. Metapopulation persistence of an endangered butterfly in a fragmented landscape. *Oikos* 72: 21–28.

Hardin, G. 1960. The competitive exclusion principle. *Science* 131: 1292–1297.

Hardisty, M. V. 1979. *Biology of the Cyclostomes*. London: Chapman & Hall.

Harper, J. L. 1969. The role of predation in vegetational diversity. *Brookhaven Symposia in Biology* 22: 48–62.

Harris, C. M. 1996. Absorption of sound in air versus humidity and temperature. *Journal of Acoustic Society of America* 40: 148–159.

Harris, P., D. Peschken and J. Milroy. 1969. The status of biological control of the weed *Hypencum perforatum* in British Columbia. *Canadian Entomologist* 101: 1–15.

Harris, V. T. 1952. An experimental study of habitat selection by prairie and forest races of the deer mouse, *Peromyscus maniculatus*. Contributions to the Laboratory of Vertebrate Biology, University of Michigan 56: 1–53.

Harrison, S. 1989. Long-distance dispersal and colonization in the Bay checkerspot butterfly, *Euphydryas editha bayensis*. *Ecology* 70: 1236–1243.

Harrison, S., D. D. Murphy and P. R. Ehrlich. 1988. Distribution of the bay checkerspot butterfly, *Euphydryas editha bayensis*: Evidence for a metapopulation model. *American Naturalist* 132: 360–382.

Harte, J. and R. Shaw. 1995. Shifting dominance within a montane vegetation community: Results of a climate-warming experiment. *Science* 267: 876–879.

Hartnett, D. C., K. R. Hickman and L. E. Fischer Walter. 1996. Effects of bison grazing, fire and topography on floristic diversity in tallgrass prairie. *Journal of Range Management* 49: 413–420.

Harvey, P. H. and M. D. Pagel. 1991. The comparative method in evolutionary biology. Oxford: Oxford University Press.

Hastings, J. R. and R. M. Turner. 1965. *The Changing Mile*. Tucson, AZ: University of Arizona Press.

Hastings, J. R., R. M. Turner and D. K. Warren. 1972. *An Atlas of Some Plant Distributions in the Sonoran Desert*. Tucson, AZ: University of Arizona Institute of Atmospheric Physics, Technical Reports of the Meteorology of Arid Lands.

Hatt, R. T. 1928. The relation of the meadow mouse *Microtus pennsylvanicus p.* to the biota of a Lake Champlain island. *Ecology* 9: 88–93.

Hawkins, B. A. 1994. *Parasitoid Community Ecology*. Oxford: Oxford University Press.

Hawkins, B. A. 1995. Latitudinal body-size gradients for the bees of the eastern United States. *Ecological Entomology* 20: 195–198.

Hawkins, B. A. and J. H. Lawton. 1995. Latitudinal gradients in butterfly body sizes: Is there a general pattern? *Oecologia* 102: 31–36.

Hawkins, L. K. and P. F. Nicoletto. 1992. Kangaroo rat burrow structure and the spatial organization of ground-dwelling animals in a semiarid grassland. *Journal of Arid Environments* 23: 199–208.

Hay, O. P. 1927. *The Pleistocene of the Western Region of North America and its Vertebrate Animals*. Washington, DC: Carnegie Institute.

Hayward, T. L. and J. A. McGowan. 1979. Pattern and structure in an oceanic zooplankton community. *American Zoologist* 19: 1045–1055.

Heads, M. 1984. *Principia Botanica*: Croizat's contribution to botany. *Tuatara* 27(1): 8–13

Heaney, L. R. 1978. Island area and body size of insular mammals: Evidence from the tri-colored squirrel (*Calliosciurus prevosti*) of Southwest Africa. *Evolution* 32: 29–44.

Heaney, L. R. 1984. Mammalian species richness on islands of the Sunda Shelf, Southwest Asia. *Oecologia* 61: 11–17.

Heaney, L. R. 1985. Zoogeographic evidence for middle and late Pleistocene land bridges to the Philippine Islands. *Modern Quaternary Research in Southeast Asia* 9: 127–143.

Heaney, L. R. 1991. Distribution and ecology of small mammals along an elevational transect in southeastern Luzon, Philippines. *Journal of Mammalogy* 72: 458–469.

Heaney, L. R. 1991. An analysis of patterns of distribution and species richness among Philippine fruit bats (Pteropodidae). *Bulletin of the American Museum of Natural History* 206: 145–167.

Heaney, L. R. and E. A. Rickart. 1990. Correlations of clades and clines: Geographic, elevational, and phylogenetic distribution patterns among Philippine mammals. In G. Peters and R. Hutterer (eds.), *Vertebrates in the Tropics*, 321–332. Bonn: Museum Alexander Koenig.

Heaney, L. R., E. A. Rickart, R. B. Utzurrum and J. S. F. Klompen. 1989. Elevational zonation of mammals in the central Philippines. *Journal of Tropical Ecology* 5: 259–280.

Heatwole, H. 1987. *Sea Snakes*. Kensington, NSW: New South Wales University Press.

Hedgpeth, J. W. 1957. Classification of marine environments. In J. W. Hedgpeth (ed.), *Treatise on Marine Ecology and Paleoecology*, vol. 1, *Ecology*, 17–18. Memoirs of the Geological Society of America, 67.

Hemmingsen, A. M. 1960. *Energy Metabolism as Related to Body Size and Respiratory Surfaces, and Its Evolution*. Reports of the Steno Memorial Hospital and the Nordisk Insulin Laboratorium, Copenhagen, vol. 9(2).

Hendrickson, J. A. Jr. 1981. Community-wide character displacement reexamined. *Evolution* 35: 794–809.

Hengeveld, H. G. 1990. Global climate change: Implications for air-temperature and water supply in Canada. *Transactions of the American Fisheries Society* 119: 176–182.

Hengeveld, R. 1989. *Dynamics of Biological Invasions*. London: Chapman and Hall.

Hengeveld, R. 1992. *Dynamic Biogeography*. Cambridge: Cambridge University Press.

Hennig, W. 1950. *Grundzüge einer theorie der phylogenetischen Systematik*. Berlin: Deutscher Zentralverlag.

Hennig, W. 1966. *Phylogenetic Systematics*. 3rd edition. Trans. D. D. Davis and R. Zanderl. Urbana: University of Illinois Press.

Herbertson, A. J. 1905. The major natural regions: An essay in systematic geography. *Geography Journal* 25: 300–312.

Heske, E. J., D. L. Rosenblatt and D. W. Sugg. 1997. Population dynamics of small mammals in an oak woodland-grassland-chaparral habitat mosaic. *Southwestern Naturalist* 42: 1–12.

Hespenheide, H. A. 1978. Are there fewer parasitoids in the tropics? *American Naturalist* 112: 766-769.

Hess, H. H. 1962. History of ocean basins. In A. E. J. Engel, H. L. James and B. F. Leonard (eds.), *Petrological Studies: A Volume in Honor of A. F. Buddington*, 599–620. New York: Geological Society of America.

Hesse, R., W. C. Allee and K. P. Schmidt. 1951. *Ecological Animal Geography*. 2nd edition. New York: John Wiley & Sons.

Heyer, W. R. and L. R. Maxson. 1982. Distributions, relationships, and zoogeography of lowland frogs/the *Leptodactylus* complex in South America, with special reference to Amazonia. In G. T. Prance (ed.), *Biological Differentiation in the Tropics*. New York: Columbia University Press.

Heywood, V. H. 1989. Patterns, extents and modes of invasions by terrestrial plants. In J. A. Drake et al. (eds.), *Biological Invasions: A Global Perspective*, 31–60. New York: John Wiley and Sons.

Hildebrand, S. F. 1939. The Panama canal as a passageway for fishes, with lists and remarks of the fishes and invertebrates observed. *Zoologica* 24: 15–46.

Hilgard, E. W. 1860. *Report on the Geology and Agriculture of State of Mississippi*. Mississippi: E. B. Jackson Printers.

Hillis, D. M. and C. Moritz. 1990. *Molecular Systematics*. Sunderland, MA: Sinauer Associates.

Hillis, D. M., C. Moritz and B. K. Mable. 1996. *Molecular Systematics*. 2nd edition. Sunderland, MA: Sinauer Associates.

Hintikka, V. 1963. Ueber das grossklima einiger pflanzenareale, durch zwei limakoordinatensystemen dargestellt. *Annales Botanici Societatis Zoologicae Botanicae Fennicae "Vanamo"* 34: 1–65.

Hnatiuk, S. H. 1979. A survey of germination of seeds of some vascular plants found on Aldabra Atoll. *Journal of Biogeography* 6: 105–114.

Hobbs, R. J. 1992. The role of corridors in conservation: Solution or bandwagon. *Trends in Ecology and Evolution* 7(11): 389–391.

Hobbs, R. J. and L. F. Huenneke. 1992. Disturbances, diversity, and invasion: Implications for conservation. *Conservation Biology* 6: 324–337.

Hocker, H. W. Jr. 1956. Certain aspects of climate as related to the distribution of loblolly pine. *Ecology* 37: 824–834.

Hockin, D. C. 1982. Experimental insular zoogeography: Some tests of the equilibrium theory using meiobenthic harpacticoid copepods. *Journal of Biogeography* 9: 487–497.

Hocutt, C. H. and E. O. Wiley (eds.). 1986. *The Zoogeography of North American Freshwater Fishes*. New York: John Wiley & Sons.

Hocutt, C. H., R. F. Denoncourt and J. R. Stauffer Jr. 1978. Fishes of the Greenbrier River, West Virginia, with drainage history of the Central Appalachians. *Journal of Biogeography* 5: 59–80.

Hoffecker, J. F., W. R. Powers and T. Goebel. 1993. The colonization of Beringia and the peopling of the New World. *Science* 259: 46–52

Hoffmann, R. S. 1971. Relationships of certain Holarctic shrews, genus *Sorex*. *Zeitschrift fur Saugetierkunde* 36: 193–200.

Hoffmann, R. S. 1981. Different voles for different holes: Environmental restrictions on refugial survival of mammals. In G. E. Scudder and J. L. Reveal (eds.), *Evolution Today: Proceedings of Systematic and Evolutionary Biology*, 24–45. Pittsburgh: Hunt Institute for Botanical Documentation, Carnegie-Mellon University.

Hoffmann, R. S., J. W. Koeppl and C. F. Nadler. 1979. The relationships of the amphiberingian marmots (Mammalia: Sciuridae). *Occasional Papers, Museum of Natural History, University of Kansas* 83: 1–56

Holdgate, M. W. 1960. The fauna of the mid-Atlantic islands. *Proceedings of the Royal Society of London*, Series B 152: 550–567.

Holdridge, L. R. 1947. Determination of world plant formations from simple climatic data. *Science* 105: 367–368.

Holland, R. F. and S. K. Jain. 1981. Insular biogeography of vernal pools in the central valley of California. *American Naturalist* 117: 24–37.

Holling, C. S. 1992. Cross-scale morphology, geometry and dynamics of ecosystems. *Ecological Monographs* 62: 447–502.

Holloway, T. D. and N. Jardine. 1968. Two approaches to zoogeography: A study based on the distributions of butterflies, birds and bats in the Indo-Australian area. *Proceedings of the Linnean Society of London* 179: 153–188.

Holt, R. E. 1994. Simple rules for interspecific dominance in systems with exploitative and apparent competition. *American Naturalist* 144: 741–771.

Hooijer, D. A. 1976. Observations on the pygmy mammoths of the Channel Islands, California. Athlon, Festschrift Loris Russel, Royal Ontario Museum, 220–225.

Hooker, J. D. 1853. *The Botany of the Antarctic Voyage of H.M.S. Discovery Ships "Erebus" and "Terror" in the Years 1839–1843.* London: Lovell Reeve.

Hooker, J. D. 1867. *Lecture on Insular Floras.* London. Delivered before the British Association for the Advancement of Science at Nottingham, August 27, 1866.

Hooper, E. T. 1942. An effect on the *Peromyscus maniculatus* Rassenkreis of land utilization in Michigan. *Journal of Mammalogy* 23: 193–196.

Hopkins, D. M. and P. A. Smith. 1981. Dated wood from Alaska and the Yukon: Implications for forest refugia in Beringia. *Quaternary Research* 15: 217–249.

Hopkins, D. M., J. W. Mathews Jr., C. E. Schweger and S. B. Young. 1982. *Paleoecology of Beringia.* New York: Academic Press.

Horn, H. S. 1974. The ecology of secondary succession. *Annual Review of Ecology and Systematics* 5: 25–37.

Horn, H. S. 1975. Markovian processes in forest succession. In M. L. Cody and J. M. Diamond (eds.), *Ecology and Evolution of Communities*, 196–211. Cambridge, MA: Belknap Press.

Horn, H. S. 1981. Succession. In R. M. May (ed.), *Theoretical Ecology*, 253–271. Oxford: Blackwell Scientific Publications.

Horn, M. H. and L. G. Allen. 1978. A distributional analysis of California coastal marine fishes. *Journal of Biogeography* 5: 23–42.

Horner, E. 1954. Arboreal adaptations of *Peromyscus* with special reference to use of the tail. *Contributions, Laboratory of Vertebrate Biology, University of Michigan* 61: 1–84.

Howell, A. B. 1917. *Birds of the Islands off the Coast of Southern California.* Pacific Coast Avifauna, no. 12. Hollywood, CA: Cooper Ornithological Club.

Howells, W. 1973. *The Pacific Islanders.* London: Weidenfeld and Nicolson.

Hsu, K. J. 1972. When the Mediterranean dried up. *Scientific American* 227: 26–36.

Hubbell, S. P. 1979. Tree dispersion, abundance, and diversity in a dry tropical forest. *Science* 203: 1299–1309.

Hubbell, S. P. 1995. Towards a theory of biodiversity and biogeography on continuous landscapes. In G. R. Carmichael, G. E. Folk and J. L. Schnoor (eds.), *Preparing for Global Change: A Midwestern Perspective*, 171–199. Amsterdam: SPB Academic Publishing.

Hubbell, S. P. and R. B. Foster. 1986. Biological chance, history and the structure of tropical rainforest tree communities. In J. M. Diamond and T. J. Case (eds.), *Community Ecology*, 314–329. New York: Harper and Row.

Hubbs, C. L. and R. R. Miller. 1948. The zoological evidence: Correlation between fish distribution and hydrographic history in the desert basins of western United States, with emphasis on glacial and postglacial times. In *The Great Basin*, 17–166. *Bulletin of the University of Utah, Biology*: 107.

Huffaker, C. B. and C. E. Kennett. 1959. A ten-year study of vegetational changes associated with biological control of Klamath weed. *Journal of Range Management* 12: 69–82.

Hugueny, B. 1989. West African rivers as biogeographic islands: Species richness of fish communities. *Oecologia* 79: 236–243.

Hultén, E. 1937. *Outline of the History of Arctic and Boreal Biota during the Quaternary Period.* Stockholm: Bokforlags Aktiebolaget Thule.

Humboldt, A. von. 1805. *Essai sur la geographie des plantes accompagne d'un tableau physique des regions equinoxiales, fonde sur des mesures executees, depuis le dixieme degre de latitude boreale jusqu'au dixieme degre de latitude australe, pendant les annees 1799, 1800, 1801, 1802 et 1803.* Paris: Levrault Schoell.

Humphrey, S. R. 1974. Zoogeography of the nine-banded armadillo (*Dasyopus novemcinctus*) in the United States. *BioScience* 24: 457–462.

Humphries, C. J. 1981. Biogeographical methods and the southern beeches (Fagaceae: *Nothofagus*). In V. A. Funk and D. R. Brooks (eds.), *Advances in Cladistics: Proceedings of the First Meeting of the Willi Hennig Society*, 177–207. New York: New York Botanical Survey.

Humphries, C. J. 1993. Cladistic biogeography. In P. Forey, C. Humphries, I. Kitching, R. Scotland, D. Siebert and D. Williams (eds.), *Cladistics: A Practical Course in Systematics*, 137–159. The Systematics Association, Publication no. 10. Oxford, Clarendon Press.

Humphries, C. J. and L. R. Parenti. 1986. *Cladistic Biogeography.* Oxford: Clarendon Press.

Humphries, C. J., P. Y. Ladiges, M. Roos and M. Zandee. 1988. Cladistic biogeography. In A. A. Myers and P. S. Giller (eds.), *Analytical Biogeography: An Integrated Approach to the Study of Animal and Plant Distributions.* London: Chapman and Hall.

Humphries, J. M. 1984. Genetics of speciation in pupfishes from Laguna Chichancanab, Mexico. In A. A. Echelle and I. Kornfield (eds.), *Evolution of Fish Species Flocks*, 129–140. Orono: University of Maine Press.

Humphries, J. M. and R. R. Miller. 1981. A remarkable species flock of pupfishes, genus *Cyprinodon*, from Yucatan, Mexico. *Copeia* 1981: 53–64.

Huntley, B. and H. J. B. Birks. 1983. *An Atlas of Past and Present Pollen Maps for Europe: 0–12000 Years Ago.* Cambridge: Cambridge University Press.

Hurley, P. M. 1968. The confirmation of continental drift. *Scientific American* 218(4): 52–62.

Hurley, P. M. and J. R. Rand. 1969. Pre-drift continental nuclei. *Science* 164: 1229–1242.

Hutchinson, G. E. 1957. *A Treatise on Limnology*, vol. 1. New York: John Wiley & Sons.

Hutchinson, G. E. 1958. Concluding remarks. *Cold Spring Harbor Symposia on Quantitative Biology* 22: 415–427.

Hutchinson, G. E. 1959. Homage to Santa Rosalia, or why are there so many kinds of animals? *American Naturalist* 93: 145–159.

Hutchinson, G. E. 1961. The paradox of the plankton. *American Naturalist* 95: 137–145.

Hutchinson, G. E. 1967. *A Treatise on Limnology*, vol. 2. New York: John Wiley & Sons.

Hutchinson, G. E. 1975. *A Treatise on Limnology*, vol. 3. New York: John Wiley & Sons.

Hutchinson, G. E. 1978. *An Introduction to Population Ecology.* New Haven, CT: Yale University Press.

Hutchinson, G. E. 1993. *A Treatise on Limnology*, vol. 4. *The Zoobenthos.* New York: John Wiley & Sons.

Hutchinson, G. E. and R. H. MacArthur. 1959. A theoretical ecological model of size distributions among species of animals. *American Naturalist* 93: 117–125.

Hutton, J. 1795. *Theory of the Earth with Proofs and Illustrations.* Edinburgh.

Huxley, J. S. 1932. *Problems of Relative Growth.* New York: Dial

Inger, R. F. 1954. Systematics and zoogeography of Philippine Amphibia. *Fieldiana Zoology* 33: 181–531.

Inouye, R. S., N. J. Huntly, D. Tilman, J. R. Tester, M. A. Stillwell and K. C. Zinnel. 1987. Old field succession on a Minnesota sand plain. *Ecology* 68: 12–26.

Iriarte, J. A., W. L. Franklin, W. E. Johnson and K. H. Redford. 1990. Biogeographic variation of food habits and body size of the American puma. *Oecologia* 85: 185–190.

Irving, E. 1956. Paleomagnetic and paleoclimatological aspects of polar wandering. *Pure and Applied Geophysics* 33: 23–41.

Irving, E. 1959. Paleomagnetic pole positions. *Journal of the Royal Astronomy Society Geophysics* 2: 51–77.

Irwin, G. 1992. *The Prehistoric Exploration and Colonisation of the Pacific*. Cambridge: Cambridge University Press.

Isenmann, P. 1982. The influence of insularity on fecundity in tits (Aves, Paridae) in Corsica. *Acta Oecologia* 3: 295–301.

Ivantsoff, W., P. Unmack, B. Saeed and L. E. L. M. Crowley. 1991. A redfinned blue-eye, a new species and genus of the family Pseudomugilidae from central western Queensland. *Fishes of Sahul* 6: 277–282.

Jablonski, D. 1982. Evolutionary rates and modes in Late Cretaceous gastropods: Role of larval ecology. In B. Mamet and M. J. Copeland (eds.), *Proceedings of the Third North American Paleontological Convention, Toronto*.

Jablonski, D. 1986. Background and mass extinctions: The alternation of macroevolutionary regimes. *Science* 231: 129–133.

Jablonski, D. 1989. The biology of mass extinction: A paleontological view. *Philosophical Transactions of the Royal Society of London*, Series B 325: 357–368.

Jablonski, D. 1991. Extinctions: A paleontological perspective. *Science* 253: 754–757.

Jablonski, D. and D. J. Bottjer. 1990. Onshore-offshore trends in marine invertebrate evolution. In R. M. Ross and W. D. Allmon (eds.), *Causes of Evolution: A Paleontological Perspective*, 21–75. Chicago: University of Chicago Press.

Jablonski, D. and D. J. Bottjer. 1991. Environmental patterns in the origins of higher taxa: The post-Paleozoic fossil record. *Science* 252: 1831–1833.

Jablonski, D. and R. A. Lutz. 1980. Molluscan shell morphology: Ecological and paleontological applications. In D. C. Rhoads and R. A. Lutz (eds.), *Skeletal Growth of Aquatic Organisms*, 323–377. New York: Plenum Press.

Jablonski, D., J. J. Sepkoski Jr., D. J. Bottjer and P. M. Sheehan. 1983. Onshore-offshore patterns in the evolution of Phanerozoic shelf communities. *Science* 222: 1123–1125.

Jaccard, P. 1902. Etude comparative de la distribution florale dans une portion des Alpes et du Jura. *Bulletin de la Societe Vaudoise de la Science Naturelle* 37: 547–579.

Jaccard, P. 1908. Nouvelles recherches sur la distribution florale. *Bulletin de la Societe Vaudoise de la Science Naturelle* 44: 223–276.

Jackson, E. D., E. A. Silver and G. B. Dalrymple. 1972. Hawaiian-Emperor chain and its relation to Cenzoic circumpacific tectonics. *Bulletin of the Geological Society of America* 83: 601–618.

Jackson, H. H. T. 1919. An apparent effect of winter inactivity upon the distribution of mammals. *Journal of Mammalogy* 1: 58–64.

Jackson, J. A. 1978. Alleviating problems of competition, predation, parasitism, and disease in endangered birds: A review. In S. A. Temple (ed.), *Endangered Birds: Management Techniques for Preserving Threatened Species*, 75–112. Madison: University of Wisconsin Press.

Jackson, J. B. C. 1974. Biogeographic consequences of eurytopy and stenotopy among marine bivalves and their evolutionary significance. *American Naturalist* 108: 541–560.

Jackson, S. T. and D. R. Whitehead. 1991. Holocene vegetation patterns in the Adirondack Mountains. *Ecology* 72: 641–653.

Jaenike, J. 1978. Effect of island area on *Drosophila* population densities. *Oecologia* 36: 327–332.

Jaenike, J. 1991. Mass extinction of European fungi. *Trends in Ecology and Evolution* 6(6): 174–175.

James, F. C. 1970. Geographic size variation in birds and its relationship to climate. *Ecology* 51: 365–390.

James, F. C., R. F. Johnston, N. O. Wamer, G. J. Niemi and W. J. Boecklen. 1984. The Grinellian niche of the wood thrush. *American Naturalist* 124: 17–47.

James, H. F. 1995. Prehistoric extinctions and ecological changes on oceanic islands. In P. M. Vitousek, L. L. Loope and H. Andsersen (eds.), *Islands: Biological Diversity and Ecosystem Function*, 87–102. New York: Springer Verlag.

Jannasch, H. W. and C. O. Wirsen. 1980. Chemosynthetic primary production at East Pacific sea floor spreading center. *Bioscience* 29: 592–598.

Janzen, D. H. 1966. Coevolution of mutualism between ants and acacias in Central America. *Evolution* 20: 249–275.

Janzen, D. H. 1967. Why mountain passes are higher in the tropics. *American Naturalist* 101: 233–249.

Janzen, D. H. 1970. Herbivores and the number of tree species in tropical forests. *American Naturalist* 104: 501–528.

Janzen, D. H. 1973. Sweep samples of tropical foliage insects: Effects of seasons, vegetation types, elevation, time of day, and insularity. *Ecology* 54: 687–708.

Janzen, D. H. 1981. The peak in North American ichneumonid species richness lies between 38° and 42° N. *Ecology* 62: 532–557.

Janzen, D. H. 1985. The natural history of mutualisms. In D. H. Boucher (ed.), *The Biology of Mutualisms*, 40–99. London: Croom Helm.

Jarrard, R. D. and D. A. Clague. 1977. Implications of Pacific island and seamount ages for the origin of volcanic chains. *Review of Geophysics and Space Physics* 15: 57–76.

Johansson, M. E. and P. A. Keddy. 1991. Intensity and asymmetry of competition between two plant pairs of different degrees of similarity: An experimental study on two guilds of wetland plants. *Oikos* 60: 27–34.

Johnson, D. L. 1978. The origin of island mammoths and the Quaternary land bridge history of the Northern Channel Islands, California. *Quaternary Research* 10: 204–225.

Johnson, D. L. 1980. Problems in the land vertebrate zoogeography of certain islands and the swimming powers of elephants. *Journal of Biogeography* 7: 383–398.

Johnson, D. L. 1981. More comments on the Northern Channel Island mammoths. *Quaternary Research* 15: 105–106.

Johnson, G. A. L. 1973. Closing of the Carboniferous sea in western Europe. In D. H. Tarling and S. K. Runcorn (eds.), *Implications of Continental Drift to the Earth Sciences*, vol. 2, 843–850. New York: Academic Press.

Johnson, L. A. S. and B. G. Briggs. 1975. On the Proteaceae: The evolution and classification of a southern family. *Botanical Journal of the Linnean Society* 70(2): 83–182.

Johnson, L. E. and J. T. Carlton. 1996. Post-establishment spread in large-scale invasions: Dispersal mechanisms of the zebra mussel *Dreissena polymorpha*. *Ecology* 77: 1688–1690.

Johnson, M. P. and P. H. Raven. 1973. Species number and endemism: The Galapagos Archipelago revisited. *Science* 179: 893–895.

Johnson, M. P., L. G. Mason and P. H. Raven. 1968. Ecological parameters and species diversity. *American Naturalist* 102: 297–306.

Johnson, N. K. 1975. Controls of the number of bird species on montane islands in the Great Basin. *Evolution* 29: 545–567.

Johnson, T. C., C. A. Scholz, M. R. Talbot, K. Kelts, R. D. Ricketts, G. Ngobi, K. Beuning, I. Ssemmanda and J. W. McGill. 1996. Late Pleistocene desiccation of Lake Victoria and rapid evolution of cichlid fishes. *Science* 273: 1091–1093.

Johnston, M. C. 1963. Past and present grasslands of southern Texas and northeastern Mexico. *Ecology* 44: 456–466.

Johnston, R. F. and W. J. Klitz. 1977. Variation and evolution in a granivorous bird: The house sparrow. In J. Pinowski and S. C. Kendeigh (eds.), *Granivorous Birds in Ecosystems*, 15–51. Cambridge: Cambridge University Press.

Johnston, R. F. and R. K. Selander. 1964. House sparrows: Rapid evolution of races in North America. *Science* 144: 548–550.

Johnston, R. F. and R. K. Selander. 1971. Evolution in the house sparrow. II. Adaptive differentiation in North American populations. *Evolution* 25: 1–28.

Jones, C. G. and J. H. Lawton. 1994. *Linking Species and Ecosystems*. London: Chapman & Hall.

Jones, D. L., A. Cox, P. Coney, and M. Beck. 1982. The growth of western North America. *Scientific American* 247(5): 70–84.

Jones, H. L. and J. M. Diamond. 1976. Short-time-base studies of turnover in breeding bird populations on the California Channel Islands. *Condor* 78: 526–549.

Jones, J. K. Jr. and H. H. Genoways. 1970. Chiropteran systematics. In B. H. Slaughter and D. W. Walton (eds.), *About Bats*, 3–21. Dallas: Southern Methodist University Press.

Jones, J. R. E. 1949. A further study of calcareous streams in the "Black Mountain" district of South Wales. *Journal of Animal Ecology* 18: 142–159.

Jordan, D. S. 1891. *Temperature and Vertebrae: A Study in Evolution*. New York: Wilder-Quarter Century Books.

Jordan, D. S. 1908. The law of geminate species. *American Naturalist* 42: 73–80.

Jordan, P. 1971. *The Expanding Earth: Some Consequences of Dirac's Gravitational Hypothesis*. New York: Pergamon Press.

Kadmon, R. 1995. Nested species subsets and geographic isolation: A case study. *Ecology* 76: 458–465.

Karieva, P. M., J. G. Kingsolver and R. B. Huey (eds.). 1993. *Biotic Interactions and Global Change*. Sunderland, MA: Sinauer Associates.

Karl, D. M., C. O. Wirsen and H. W. Jannasch. 1980. Deep-sea primary production at the Galápagos hydrothermal vents. *Science* 207: 1345–1347.

Karr, J. R. 1982 Avian extinction on Barro Colorado Island, Panama: A reassessment. *American Naturalist* 119: 220–239.

Karr, J. R. 1990. Avian survival rates and the extinction process on Barro Colorado Island, Panama. *Conservation Biology* 4: 391–397.

Kaspari, M. and E. L. Vargo. 1995. Colony size as a buffer against seasonality: Bergmann's rule in social insects. *American Naturalist* 145: 610–632.

Kates, R. W. 1996. Population, technology, and the human environment: A thread through time. *Daedalus* 125: 43–71.

Kattan, G. H., H. Alvarez-Lopez and M. Giraldo. 1994. Forest fragmentation and bird extinctions: San Antonio eighty years later. *Conservation Biology* 8: 138–146.

Kaufman, D. M. 1995. Diversity of New World mammals: Universality of the latitudinal gradient of species and bauplans. *Journal of Mammalogy* 76: 322–334.

Kaufman, D. M. and M. R. Willig. In press. Latitudinal patterns of mammalian species richness in the New World: The effects of sampling method and faunal group. *Journal of Biogeography*.

Kaufman, L. and P. Ochumba. 1993. Evolutionary and conservation biology of cichlid fishes as revealed by faunal remnants in northern Lake Victoria. *Conservation Biology* 7: 719–730.

Keast, A. 1971. Adaptive evolution and shifts in niche occupation in island birds. In W. L. Stern (ed.), *Adaptive Aspects of Insular Evolution*, 39–53. Pullman: Washington State University Press.

Keast, A. 1972a. Continental drift and the biota of the mammals on southern continents. In A. Keast, F. C. Erk and B. Glass (eds.), *Evolution, Mammals, and Southern Continents*, 23–87. Albany: State University of New York Press.

Keast, A. 1972b. Australian mammals: Zoogeography and evolution. In A. Keast, F. C. Erk and B. Glass (eds.), *Evolution, Mammals, and Southern Continents*, 195–246. Albany: State University of New York Press.

Keast, A. 1972c. Comparisons of contemporary mammal faunas of southern continents. In A. Keast, F. C. Erk and B. Glass (eds.), *Evolution, Mammals, and Southern Continents*, 433–501. Albany: State University of New York Press.

Keast, A. 1977a. Zoogeography and phylogeny: The theoretical background and methodology to the analysis of mammal

and bird fauna. In H. J. Frith and J. H. Calaby (eds.), *Proceedings of the 16th International Ornithological Congress, Australian Academy of Sciences*, 246–312.

Keddy, P. A. 1982. Population ecology on an environmental gradient: *Cakile edentula* on a sand dune. *Oecologia* 52: 348–355.

Keddy, P. A., and P. MacLelan. 1990. Centrifugal organization in forests. *Oikos* 59: 75–84.

Kelt, D. A. and J. H. Brown. In press. Community structure and assembly rules: Confronting conceptual and statistical issues with data on desert rodents. In E. Weiher and P. A. Keddy (eds.), *The Search for Assembly Rules in Ecological Communities*. Cambridge: Cambridge University Press.

Kelt, D. A., J. H. Brown, K. Rogovin and G. Shenbrot. In press. Patterns in the structure of Asian and North American desert small mammal communities. *Ecological Monographs*.

Kennedy, W. J. 1977. Ammonite evolution. In A. Hallam (ed.), *Patterns of Evolution as Illustrated by the Fossil Record*, 251–304. Amsterdam: Elsevier.

Kerr, R. A. 1995. Earth's surface may move itself. *Science* 269: 1214–1216.

Kettlewell, H. B. D. 1961. The phenomenon of industrial melanisms in Lepidoptera. *Annual Review of Ecology and Systematics* 6: 245–262.

Kiester, A. R. 1971. Species density of North American amphibians and reptiles. *Systematic Zoology* 20: 127–137.

Kikkawa, J. and K. Pearse. 1969. Geographical distribution of land birds in Australia: A numerical analysis. *Australian Journal of Zoology* 17: 821–840.

Kikkawa, J. and E. E. Williams. 1971. Altitudinal distribution of land birds in New Guinea. *Search* 2: 64–69.

King, C. M. and P. J. Moors. 1979. On co-existence, foraging strategy and the biogeography of weasels and stoats (*M. nivalis* and *M. erminea*) in Britain. *Oecologia* 39: 129–150.

Kitchener, D. J., A. Chapman, J. Dell, B. G. Muir and M. Palmer. 1980. Lizard assemblage and reserve size and structure in the Western Australian wheatbelt—some implications for conservation. *Biological Conservation* 17: 25–61.

Klein, B. C. 1989. Effects of forest fragmentation on dung and carrion beetle communities in Central Amazon. *Ecology* 70: 1715–1725.

Knoll, A. H. 1986. Patterns of change in plant communities through geological time. In J. Diamond and T. J. Case (eds.), *Community Ecology*, 126–141. New York: Harper and Row.

Kodric-Brown, A. and J. H. Brown. 1979. Competition between distantly related taxa and the co-evolution of plants and pollinators. *American Zoologist* 19: 1115–1127.

Kohn, A. J. 1978. Ecological shift and release in an isolated population: *Conus miliaris* at Easter Island. *Ecological Monographs* 48: 323–336.

Koopman, K. F. and J. K. Jones. 1970. Classification of bats. In B. H. Slaughter and D. W. Walton (eds.), *About Bats*, 22–28. Dallas: Southern Methodist University Press.

Koopman, K. F. and D. W. Steadman. 1995. Extinction and biogeography of bats on 'Eua, Kingdom of Tonga. *American Museum Novitates* 3125: 1–13.

Korobytsina, K. V., D. F. Nadler, N. N. Vorontsov and R. S. Hoffmann. 1974. Chromosomes of the Siberian snow sheep, *Ovis nivicola*, and implications concerning the origin of amphiberingian wild sheep. *Quaternary Research* 4: 235–245.

Krebs, C. J. 1978. *Ecology: The Experimental Analysis of Distribution and Abundance*. New York: Harper & Row.

Krebs, C. J., B. L. Keller and R. H. Tamarin. 1969. *Microtus* population biology: Demographic changes in fluctuating populations of *M. ochrogaster* and *M. pennsylvanicus* in southern Indiana. *Ecology* 50: 587–607.

Kremp, G. O. W. 1992. Earth expansion theory versus statical assumption. In S. Chatterjee and N. Hotton III (eds.), *New*

Concepts in Global Tectonics, 297–309. Lubbock: Texas Tech University Press.

Krzanowski, A. 1967. The magnitude of islands and the size of bats (Chiroptera). *Acta Zoologica Cracoviensia* 15, XI: 281–348.

Kunin, W. E. and K. J. Gaston. 1997. *The Biology of Rarity: Causes and Consequences of Rare-Common Differences.* New York: Chapman and Hall.

Kurtén, B. and E. Anderson. 1980. *Pleistocene Mammals of North America.* New York: Columbia University Press.

Lack, D. 1947. *Darwin's Finches.* Cambridge: Cambridge University Press.

Lack, D. 1970. Island birds. *Biotropica* 2: 29–31.

Lack, D. 1973. The numbers of species of hummingbirds in the West Indies. *Evolution* 27: 326–337.

Lack, D. 1976. *Island Biology Illustrated by the Land Birds of Jamaica.* Studies in Ecology, vol. 3. Berkeley: University of California Press.

Lack, D. and R. E. Moreau. 1965. Clutch size in tropical passerine birds of forest and savanna. *L'Oiseau* 35: 76–89.

La Greca, M. and C. F. Sacchi. 1957. Problemi del popamento animale nelle piccole isole mediterranee. *Ann. Inst. Mus. Zool. Univ. Napoli* 9: 1–189.

Lai, D. Y. and P. L. Richardson. 1977. Distribution and movement of Gulf Stream rings. *Journal of Physical Oceanography* 7: 670–683.

La Marche, V. C. 1973. Holocene climatic variations inferred from treeline fluctuations in the White Mountains, California. *Quaternary Research* 3: 632–660.

La Marche, V. C. 1978. Tree-ring evidence of past climatic variability. *Nature* 276: 334–338.

Larick, R. and R. Ciochon. 1996. The first Asians: A cave in China yields evidence of the earliest migration out of Africa. *Archaeology* 49: 51–53.

Latham, R. E. and R. E. Ricklefs. 1993. Global patterns of tree species diversity in moist forests: Energy-diversity theory does not account for variation in species richness. *Oikos* 67: 325–333.

Laurance, W. F. 1990. Comparative responses of five arboreal marsupials to tropical forest fragmentation. *Journal of Mammalogy* 71: 641–653.

Lawlor, T. E. 1983. The mammals. In T. J. Case and M. L. Cody (eds.), *Island Biogeography of the Sea of Cortez,* 265–289, 482–500. Berkeley: University of California Press.

Lawlor, T. E. 1986. Comparative biogeography of mammals on islands. *Biological Journal of the Linnean Society* 28: 99–125.

Lawton, J. H. 1990. Species richness and population dynamics of animal assemblages. Patterns in body size: Abundance and space. *Philosophical Transactions of the Royal Society of London,* Series B 330: 283–291.

Lawton, J. H., S. Nee, A. J. Letcher and P. H. Harvey. 1994. Animal distributions: Patterns and processess. In P. J. Edwards, R. M. May and N. R. Webb (eds.), *Large-Scale Ecology and Conservation Biology,* 41–58. London: Blackwell.

Lazell, J. D. Jr. 1983. Biogeography of the herptofauna of the British Virgin Islands, with description of a new anole (Sauria: Iguanidae). In A. G. J. Rhodin and K. Miyata (eds.), *Advances in Herpetology and Evolutionary Biology,* pp. 99–117. Cambridge, MA: Museum of Comparative Zoology.

Lee, T. E. Jr., B. R. Riddle and P. L. Lee. 1996. Speciation in the desert pocket mouse (*Chaetodipus penicillatus* Woodhouse). *Journal of Mammalogy* 77: 58–68.

Leigh, E. 1975. Population fluctuations and community structure. In W. H. Van Dobben and R. H. Lowe-McConnel (eds.), *Unifying Concepts in Ecology,* 67–88. The Hague: Dr. W. Junk.

Leigh, E. G. 1981. The average lifetime of a population in a varying environment. *Journal of Theoretical Biology* 90: 213–239.

Leith, H. 1956. Ein Beitrag zur Frage der korrelation zwischen mittleren klimawerten und vegetationsformationen. *Berichte Deutsche Botanische Gesellschaft* 69: 169–176.

Leston, D. 1957. Spread potential and colonization of the islands. *Systematic Zoology* 6: 41–46.

Levin, D. A. (ed.). 1979. *Hybridization: An Evolutionary Perspective.* Stroudsburg, PA: Dowden, Hutchinson, & Ross.

Lewis, W. H. (ed.). 1979. *Polyploidy: Biological Relevance.* New York: Plenum Press.

Lewontin, R. C. and L. C. Birch. 1966. Hybridization as a source of variation for adaptation to new environments. *Evolution* 20: 315–336.

Lidicker, W. Z. Jr. 1988. Solving the enigma of microtine "cycles." *Journal of Mammalogy* 69: 225–235.

Lidicker, W. Z. Jr. 1992. *Animal Dispersal: Small Mammals as a Model.* New York: Chapman and Hall.

Lieth, H. 1973. Primary production: Terrestrial ecosystems. *Human Ecology* 1: 303–332.

Lillegraven, J. A. 1972. Ordinal and familial diversity in Cenozoic mammals. *Taxon* 21: 261–274.

Lillegraven, J. A., Z. Kielan-Jaworowska and W. A. Clemens. 1979. *Mesozoic Mammals: The First Two-Thirds of Mammalian History.* Berkeley: University of California Press.

Lindeman, R. 1942. The trophic-dynamic aspect of ecology. *Ecology* 23: 399–418

Lister, A. M. 1989. Rapid dwarfing of red deer on Jersey in the last interglacial. *Nature* 342: 539–542.

Lister, B. C. 1976a. The nature of niche expansion in West Indian *Anolis* lizards. I. Ecological consequences of reduced competition. *Evolution* 30: 659–676.

Lister, B. C. 1976b. The nature of niche expansion in West Indian *Anolis* lizards. II. Evolutionary consequences. *Evolution* 30: 677–692.

Lithgow-Bertelloni, C. and M. A. Richards. 1995. Cenozoic plate driving forces. *Geophysical Research Letters* 22: 1317–1320.

Livezey, B. C. 1992. Morphological corollaries and ecological implications of flightlessness in the kakapo (Psittaciformes: *Strigops habroptilus*). *Journal of Morphology* 213: 105–145.

Livezey, B. C. 1993. An ecomorphological review of the dodo (*Raphus cucullatus*) and solitaire (*Pezophaps solitaria*), flightless Columbiformes of the Macarene Islands. *Journal of Zoology* (London) 230: 247.

Lockwood, J. L. and M. P. Moulton. 1994. Ecomorphological pattern in Bermuda birds: The influence of competition and implications for nature preserves. *Evolutionary Ecology* 8: 53–60.

Lodge, D. M. 1993. Biological invasions: Lessons for ecology. *Trends in Ecology and Evolution* 8: 133–137.

Lomolino, M. V. 1982. Species-area and species-distance relationships of terrestrial mammals in the Thousand Island Region. *Oecologia* 54: 72–75.

Lomolino, M. V. 1983. *Island Biogeography, Immigrant Selection, and Mammalian Body Size on Islands.* Ph.D. dissertation, Department of Biology, State University of New York at Binghamton.

Lomolino, M. V. 1984. Immigrant selection, predatory exclusion and the distributions of *Microtus pennsylvanicus* and *Blarina brevicauda* on islands. *American Naturalist* 123: 468–483.

Lomolino, M. V. 1985. Body size of mammals on islands: The island rule re-examined. *American Naturalist* 125: 310–316.

Lomolino, M. V. 1986. Mammalian community structure on islands: Immigration, extinction and interactive effects. *Biological Journal of the Linnean Society.* 28: 1–21.

Lomolino, M. V. 1988. Winter immigration abilities and insular community structure of mammals in temperate archipelagoes. In J. F. Downhower (ed.), *Biogeography of the Island Region of Western Lake Erie,* 185–196. Columbus: Ohio State University Press.

Lomolino, M. V. 1989. Interpretation and comparisons of constants in the species-area relationship: An additional caution. *American Naturalist* 133: 71–75.

Lomolino, M. V. 1989. Bioenergetics of cross-ice movements of *Microtus pennsylvanicus*, *Peromyscus leucopus* and *Blarina brevicauda*. *Holarctic Ecology* 12: 213–218.

Lomolino, M. V. 1990. The target area hypothesis: The influence of island area on immigration rates of non-volant mammals. *Oikos* 57: 297–300.

Lomolino, M. V. 1993a. Matching of granivorous mammals of the Great Basin and Sonoran deserts on a species-for-species basis. *Journal of Mammalogy* 74: 863–867.

Lomolino, M. V. 1993b. Winter filtering, immigrant selection and species composition of insular mammals of Lake Huron. *Ecography* 16: 24–30.

Lomolino, M. V. 1994a. An evaluation of alternative strategies for building networks of nature reserves. *Biological Conservation* 69: 243–249.

Lomolino, M. V. 1994b. Species richness patterns of mammals inhabiting nearshore archipelagoes: Area, isolation, and immigration filters. *Journal of Mammalogy* 75: 39–49.

Lomolino, M. V. 1996. Investigating causality of nestedness of insular communities: Selective immigrations or extinctions? *Journal of Biogeography* 23: 699–703.

Lomolino, M. V.1998. A species-based, hierarchical model of island biogeography. In E. Wiher and P.A. Keddy (eds.), *The Search for Assembly Rules in Ecological Communities*. New York: Cambrige University Press.

Lomolino, M. V. and R. Channell. 1995. Splendid isolation: Patterns of range collapse in endangered mammals. *Journal of Mammalogy* 76: 335–347.

Lomolino, M. V. and R. Channell. 1998. Range collapse, reintroductions, and biogeographic guidelines for conservation. *Conservation Biology* 12: 481–484.

Lomolino, M. V. and R. Davis. 1997. Biogeographic scale and biodiversity of mountain forest mammals of western North America. *Global Ecology and Biogeography Letters* 6: 57–76.

Lomolino, M. V., J. H. Brown and R. Davis. 1989. Island biogeography of montane forest mammals in the American Southwest. *Ecology* 70: 180–194.

Lomolino, M. V., J. C. Creighton, G. D. Schnell and D. L. Certain. 1995. Ecology and conservation of the endangered American burying beetle (*Nicrophorus americanus*). *Conservation Biology* 9: 605–614.

Long, J. 1981. *Introduced Birds of the World*. London: David and Charles.

Lonsdale, D. J. and J. S. Levington. 1985. Latitudinal differentiation in copepod growth: An adaptation to temperature. *Ecology* 66: 1397–1407.

Loope, L. L. and D. Mueller-Dombois. 1989. Characteristics of invaded islands, with special reference to Hawaii. In J. A. Drake et al. (eds.), *Biological Invasions: A Global Perspective*, 257–280. New York: John Wiley and Sons.

Lord, R. D. Jr. 1960. Litter size and latitude in North American mammals. *American Midland Naturalist* 64: 488–499.

Losos, J. B. 1992. A critical comparison of the taxon cycle and character displacement models of size evolution in *Anolis* lizards in the lesser Antilles. *Copeia* 1991: 279–288.

Losos, J. B. and H. W. Greene. 1988. Ecological and evolutionary implications of diet in monitor lizards. *Biological Journal of the Linnean Society* 35: 379–407.

Losos, J. B., J. C. Marks and T. W. Schoener. 1993. Habitat use and ecological interactions of an introduced and a native species of Anolis lizard on Grand Cayman, with a review of the outcomes of anole introductions. *Oecologia* 95: 525–532.

Lovejoy, T. E., R. O. Birregaard, Jr., A. B. Rylands, J. R. Malcolm, C. E. Quintela, L. H. Harper, K. S. Brown, Jr., A. H. Powell, G. V. N. Powell, H. O. R. Schubart and M. B. Hays. 1986. Edge and other effects of isolation on Amazon forest fragments. In M. E. Soulé (ed.), *Conservation Biology: The Science of Scarcity and Diversity*, 257–285. Sunderland, MA: Sinauer Associates.

Lowther, P. E. and C. L. Cink. 1992. *Passer domesticus*: House sparrow. The Birds of North America. *American Ornithologists' Union*, no. 12.

Lubchenco, J. 1978. Plant species diversity in a marine intertidal community: Importance of herbivore food preferences and algal competitive abilities. *American Naturalist* 112: 23–39.

Lubchenco, J. 1980. Algal zonation in the New England rocky intertidal community: An experimental analysis. *Ecology* 61: 333–344.

Lubchenco, J. and B. A. Menge. 1978. Community development and persistence in a low rocky intertidal zone. *Ecological Monographs* 48: 67–94.

Ludwig, J. A., and J. F. Reynolds. 1988. *Statistical Ecology: A Primer on Methods and Computing*. New York: John Wiley & Sons.

Lundberg, J. G. and B. Chernoff. 1992. A Miocene fossil of the Amazonian fish *Arapaima* (Teleostei, Arapaimidae) from the Magdalena river region of Colombia: Biogeographic and evolutionary implications. *Biotropica* 24: 2–14.

Lundberg, J. G., A. Machado-Allison and R. F. Kay. 1986. Miocene characid fishes from Columbia: Evolutionary stasis and extirpation. *Science* 234: 208–209.

Lyell, C. 1969. *Principles of Geology*. New York. Johnson Reprint Company. [First published in 1830.]

Lynch, J. D. 1979. The amphibians of the lowland tropical forests. In W. E. Duellman (ed.), *The South American Herpetofauna: Its Origin, Evolution and Dispersal*, 189–215. Monograph no. 7. Lawrence: Museum of Natural History, University of Kansas.

Lynch, J. D. 1988. Refugia. In A. A. Myers and P. S. Giller (eds.), *Analytical Biogeography: An Integrated Approach to the Study of Animal and Plant Distributions*. London: Chapman and Hall.

Lynch, J. D. and N. V. Johnson. 1974. Turnover and equilibria in insular avifaunas, with special reference to the California Channel Islands. *Condor* 76: 370–384.

MacArthur, R. H. 1958. Population ecology of some warblers of northeastern coniferous forests. *Ecology* 39: 599–619.

MacArthur, R. H. 1965. Patterns of species diversity. *Biological Review* 40: 510–533.

MacArthur, R. H. 1972. *Geographical Ecology: Patterns in the Distributions of Species*. New York: Harper & Row.

MacArthur, R. H. and T. H. Connell. 1966. *The Biology of Populations*. New York: John Wiley & Sons.

MacArthur, R. H. and R. Levins. 1967. The limiting similarity, convergence, and divergence of coexisting species. *American Naturalist* 101: 377–385.

MacArthur, R. H. and J. W. MacArthur. 1961. On bird species diversity. *Ecology* 42: 594–598.

MacArthur, R. H. and E. O. Wilson. 1963. An equilibrium theory of insular zoogeography. *Evolution* 17: 373–387.

MacArthur, R. H. and E. O. Wilson. 1967. *The Theory of Island Biogeography*. Monographs in Population Biology, no. 1. Princeton, NJ: Princeton University Press.

MacArthur, R., H. Recher and M. Cody. 1966. On the relation between habitat selection and species diversity. *American Naturalist* 100: 319–332.

MacArthur, R. H., J. M. Diamond and J. Karr. 1972. Density compensation in island faunas. *Ecology* 53: 330–342.

MacArthur, R. H., J. MacArthur, D. MacArthur and A. MacArthur. 1973. The effect of island area on population densities. *Ecology* 54: 657–658.

MacDonald, I. A. W., F. J. Kruger and A. A. Ferrar. 1986. *The Ecology and Management of Invasions in Southern Africa*. Cape Town: Oxford University Press.

MacDonald, I. A. W., L. L. Loope, M. B. Usher and O. Hamann. 1989. Wildlife conservation and the invasion of nature reserves by introduced species: A global perspective. In J. A. Drake et al. (eds.), *Biological Invasions: A Global Perspective*, 215–256. New York: John Wiley & Sons.

Mace, G. M. 1993. An investigation into methods for categorizing the conservation status of species. In P. J. Edwards, R. M. May and N. R. Webb (eds.), *Large-Scale Ecology and Conservation Biology*, 293–312. London: Blackwell Scientific Publications.

Mace, R. and M. Pagel. 1995. A latitudinal gradient in the density of human languages in North America. *Proceedings of the Royal Society of London*, Series B 261: 117–121.

MacMahon, J. A. 1987. Disturbed lands and ecological theory: An essay about a mutualistic association. In W. R. Jordan, M. E. Gilpin and J. D. Aber, (eds.), *Restoration Ecology: A Synthetic Approach to Ecological Research*. New York: Cambridge University Press.

MacPhee, R. D. E. and P. A. Marx. 1997. The 40,000-year plague: Humans, hyperdisease, and first-contact extinctions. In S. Goodman and B. D. Patterson (eds.), *Human Impact and Natural Change in Madagascar*, 169–217. Washington, DC: Smithsonian Institution Press.

Magurran, A. E. 1988. *Ecological Diversity and Its Measurement*. Princeton, NJ: Princeton University Press.

Maloney, B. K. 1980. Pollen analytical evidence for early forest clearance in North Sumatra. *Nature* 287: 324–326.

Manly, B. F. J. 1991. *Randomization and Monte Carlo Methods in Biology*. London: Chapman and Hall.

Mares, M. A. 1976. Convergent evolution in desert rodents: Multivariate analysis and zoogeographic implications. *Paleobiology* 2: 39–63.

Mares, M. A. 1993a. Desert rodents, seed consumption, and convergence. *BioScience* 43: 373–379.

Mares, M. A. 1993b. Heteromyids and their ecological counterparts: A pandesertiv view of rodent ecology and evolution. In H. H. Genoways and J. H. Brown (eds.), *Biology of the Heteromyidae*, 652–719. American Society of Mammalogists, Special Publication no. 10.

Mares, M. A. and T. E. Lacher Jr. 1987. Ecological, morphological and behavioral convergence in rock-dwelling mammals. In H. H. Genoways (ed.), *Current Mammalogy*, vol. 1, 307–347. New York: Plenum.

Markgraf, V., M. McGlone and G. Hope. 1995. Neogene paleoenvironmental and paleoclimatic change in southern temperate ecosystems: A southern perspective. *Trends in Ecology and Evolution* 10: 143–147.

Marks, G. and W. K. Beatty. 1976. *Epidemics*. New York: Charles Scribner's Sons.

Marshall, L. G. 1979. Evolution of metatherian and eutherian (mammalian) characters: A review based on cladistic methodology. *Zoological Journal of the Linnean Society* 66: 369–410.

Marshall, L. G. 1988. Extinction. In A. A. Meyers and P. S. Giller (eds.), *Analytical Biogeography: An Integrated Approach to the Study of Animal and Plant Distributions*, 219–254. New York: Chapman and Hall.

Marshall, L. G. and R. S. Corruccini. 1978. Variation, evolutionary rates and allometry in dwarfing lineages. *Paleobiology* 4: 101–119.

Marshall, L. G. and J. G. Lundberg. 1996. Miocene deposits in the Amazonian Foreland Basin. *Science* 273: 123–124.

Martin, T. E. 1981. Species-area slopes and coefficients: A caution on their interpretation. *American Naturalist* 118: 823–837.

Martin, A. R. 1990. *The Illustrated Encyclopedia of Whales and Dolphins*. New York: Portland House.

Martin, J. L. 1992. Niche expansion in an insular bird community: An autecological perspective. *Journal of Biogeography* 19: 375–381.

Martin, J. and P. Gurrea. 1990. The peninsular effect in Iberian butterflies (Lepidoptera: *Papilionoidea* and *Hesperioidea*). *Journal of Biogeography* 17: 85–96.

Martin, P. S. 1967. Prehistoric overkill. In P. S. Martin and H. E. Wright Jr. (eds.), *Pleistocene Extinctions: The Search for a Cause*, 75–120. New Haven, CT: Yale University Press.

Martin, P. S. 1973. The discovery of America. *Science* 180: 969–974.

Martin, P. S. 1984. Prehistoric overkill. In P. S. Martin and R. G. Klein (eds.), *Quaternary Extinctions: A Prehistoric Revolution*, 354–403. Tucson: University of Arizona Press.

Martin, P. S. 1990. 40,000 years of extinction on the "planet of doom." *Palaeogeography, Palaeoclimatology, Palaeoecology* 82: 187–201.

Martin, P. S. 1995. Mammoth extinction: Two continents and Wrangel Island. *Radiocarbon* 37: 1–6.

Martin, P. S. and R. G. Klein (eds.). 1984. *Quaternary Exctinctions: A Prehistoric Revolution*. Tucson: University of Arizona Press.

Martin, P. S. and H. E. Wright Jr. (eds.). 1967. *Pleistocene Extinctions: The Search for a Cause*. New Haven, CT: Yale University Press.

Mason, H. L. 1954. Migration and evolution of plants. *Madroño* 12: 161–192.

Matthew, W. D. 1915. Climate and evolution. *Annals of the New York Academy of Sciences* 24: 171–318.

Maurer, B. A. 1994. *Geographic Population Analysis: Tools for Analysis of Biodiversity*. London: Blackwell Scientific Publications.

Maurer, B. A., J. H. Brown and R. D. Rusler. 1992. The micro and macro of body size evolution. *Evolution* 46: 939–953.

May, R. M. 1978. The dynamics and diversity of insect faunas. In L. A. Mound and N. Waloff (eds.), *Diversity of Insect Faunas*. 188–204. New York: Blackwell Scientific Publications.

May, R. M. 1986. The search for patterns in the balance of nature: Advances and retreats. *Ecology* 67: 1115–1126.

May, R. M. 1988. How many species are there on earth? *Science* 241: 1441–1449.

Mayden, R. L. 1987a. Historical ecology and North American highland fishes: A research program in community ecology. In W. J. Matthews and D. C. Heins (eds.), *Community and Evolutionary Ecology of North American Stream Fishes*, 210–222. Norman: University of Oklahoma Press.

Mayden, R. L. 1987b. Pleistocene glaciation and historical biogeography of North American Central Highland fishes. In W. C. Johnson (ed.), *Quaternary Environments of Kansas*, 141–151. Lawrence: Kansas Geological Survey.

Mayden, R. L. 1992a. Explorations into the past and the dawn of systematics and historical ecology. In R. L. Mayden (ed.), *Systematics, Historical Ecology, and North American Freshwater Fishes*. Stanford, CA: Stanford University Press.

Mayden, R. L. (ed.) 1992b. *Systematics, Historical Ecology, and North American Freshwater Fishes*. Stanford, CA: Stanford University Press.

Mayr, E. 1942. *Systematics and the Origin of Species*. New York: Columbia University Press.

Mayr, E. 1944a. Wallace's Line in the light of recent zoogeographic studies. *Quarterly Review of Biology* 19: 1–14.

Mayr, E. 1944b. The birds of Timor and Sunda. *Bulletin of the American Museum of Natural History* 83: 127–194.

Mayr, E. 1963. *Animal Species and Evolution*. Cambridge, MA: Harvard University Press.

Mayr, E. 1965a. The nature of colonization in birds. In H. G. Baker and G. L. Stebbins (eds.), *The Genetics of Colonizing Species*, 3047. New York: Academic Press.

Mayr, E. 1965b. Avifauna: Turnover on islands. *Science* 150: 1587–1588.

Mayr, E. 1969. *Principles of Systematic Zoology*. New York: McGraw-Hill Book Co.

Mayr, E. 1974. Cladistic analysis or cladistic classification? *Z. Zool. Syst. Evolut.-forsch.* 12: 94–128.

McAtee, W. L. 1947. Distribution of seeds by birds. *American Midland Naturalist* 38: 214–223.

McAuliffe, J. R. 1994. Landscape evolution, soil formation, and ecological processes in Sonoran Desert bajadas. *Ecological Monographs* 64: 111–148.

McClure, H. E. 1974. *Migration and Survival of the Birds of Asia*. Bangkok: Applied Scientific Research Corporation of Thailand.

McCook, L. J. 1994. Understanding ecological succession: Causal models and theories, a review. *Vegetatio* 110: 115–147.

McCoy, E. D., S. S. Bell and K. Walters. 1986. Identifying biotic boundaries along environmental gradients. *Ecology* 67: 749–759.

McDonald, K. A. and J. H. Brown. 1992. Using montane mammals to model extinctions due to global change. *Conservation Biology* 6: 409–415.

McDowell, S. B. 1969. Notes on the Australian sea snake *Ephalophis greyi* M. Smith (Serpentes: Elapidae, Hydrophiinae) and the origin and classification of sea snakes. *Zoological Journal of the Linnean Society* 48: 333–349.

McDowell, S. B. 1972. The genera of sea snakes of the *Hydrophis* group (Serpentes: Elapidae). *Transactions of the Zoological Society of London* 32: 189–247.

McDowell, S. B. 1974. Additional notes on the rare and primitive sea snake, *Ephalophis greyi*. *Journal of Herpetology* 8: 123–128.

McElhinny, M. W. 1973a. *Paleomagnetism and Plate Tectonics*. Cambridge: Cambridge University Press.

McElhinny, M. W. 1973b. Paleomagnetism and plate tectonics of eastern Asia. In P. J. Coleman (ed.), *The Western Pacific: Island Arcs, Marginal Seas, Geochemistry*, 407–414. New York: Crane, Russak & Co.

McFarlane, D. A. 1989. Patterns of species co-occurrence in the Antillean bat fauna. *Mammalia* 53: 59–66.

McGowan, J. A. 1985. El Niño effects in the Eastern Subarctic Pacific Ocean. In W. S. Wooster and D. L. Fluharty (eds.), *El Niño North*, 166–184. Seattle: Washington Sea Grant, University of Washington Press.

McGowan, J. A. and P. W. Walker. 1979. Structure in the copepod community of the North Pacific central gyre. *Ecological Monographs* 49: 195–226.

McGuinness, K. A., 1984. Equations and explanations in the study of species-area curves. *Biological Review* 59: 423–440.

McIntosh, R. P. 1967. The continuum concept of vegetation. *Botanical Review* 33: 130–187.

McIntosh, R. P. 1981. Succession in ecological theory. In D. C. West, H. H. Shugart and D. B. Botkin (eds.), *Forest Succession: Concepts and Applications*, 10–23. New York: Springer-Verlag.

McKenna, M. C. 1972a. Eocene final separation of the Eurasian and Greenland-North American landmasses. *24th International Geological Congress, Section 7*: 275–281.

McKenna, M. C. 1972b. Was Europe connected directly to North America prior to the middle Eocene? In Th. Dobzkansky, M. K. Hecht and W. C. Steere (eds.), *Evolutionary Biology 6*, 179–188. New York: Appleton-Century-Crofts.

McKenna, M. C. 1973. Sweepstakes, filters, corridors, Noah's Arks, and beached Viking funeral ships in paleogeography. In D. H. Tarling and S. K. Runcorn (eds.), *Implications of Continental Drift to the Earth Sciences*, vol. 1, 295–308. New York: Academic Press.

McKenzie, D. P. and W. J. Morgan. 1969. The evolution of triple junctions. *Nature* 226: 239–243.

McKinney, H. L. 1972. *Wallace and Natural Selection*. New Haven, CT: Yale University Press.

McLaughlin, J. F. and J. Roughgarden. 1989. Avian predation on *Anolis* lizards in the northeastern Caribbean: An inter-island contrast. *Ecology* 70: 617–628.

McLaughlin, S. P. 1986. Floristic analysis of the southwestern United States. *Great Basin Naturalist* 46: 46–65.

McLaughlin, S. P. 1989. Natural floristic area of the western United States. *Journal of Biogeography* 16: 239–248.

McMahon, T. and J. T. Bonner. 1983. *On Size and Life*. New York: Scientific American Books.

McNab, B. K. 1963. Bioenergetics and the determination of home range size. *American Naturalist* 97: 113–140.

McNab, B. K. 1971. On the ecological significance of Bergmann's rule. *Ecology* 52: 845–854.

McNab, B. K. 1994a. Energy conservation and the evolution of flightlessness in birds. *American Naturalist* 144: 628–642.

McNab, B. K. 1994b. Resource use and the survival of land and freshwater vertebrates on oceanic islands. *American Naturalist* 144: 643–660.

McNeill, W. H. 1976. *Plagues and Peoples*. Garden City, NY: Anchor Press.

McPhail, J. D. 1994. Speciation and the evolution of reproductive isolation in the sticklebacks (*Gasterosteus*) of southwestern British Columbia. In M. A. Bell and S. A. Foster (eds.), *Evolutionary Biology of the Threespine Stickleback*, 399–437. Oxford: Oxford University Press.

McPhail, J. D. and C. C. Lindsey. 1986. Zoogeography of the freshwater fishes of Cascadia (the Columbia rivers north of the Stikine). In C. H. Hocutt and E. O. Wiley (eds.). *Zoogeography of North American Freshwater Fishes*, 615–637.

Means, D. B. and D. Simberloff. 1987. The peninsula effect: Habitat-correlated species decline in Florida's herptofauna. *Journal of Biogeography* 14: 551–568.

Melhop, P. and J. F. Lynch. 1978. Population characteristics of *Peromyscus leucopus* introduced to islands inhabited by *Microtus pennsylvanicus*. *Oikos* 31: 17–26.

Melville, R. 1981. Vicariance plant distributions and paleogeography of the Pacific region. In G. Nelson and D. E. Rosen (eds.), *Vicariance Biogeography: A Critique*, 238–274. New York: Columbia University Press.

Menge, B. A. and J. P. Sutherland. 1976. Species diversity gradients: Synthesis of the roles of predation, competition, and temporal heterogeneity. *American Naturalist* 110: 351–369.

Menge, B., E. L. Berlow, C. Blanchette, A. Carol, S. A. Navarrete and S. B. Yamada. 1994. The keystone species concept: Variation in interaction strength in a rocky intertidal community. *Ecological Monographs* 64: 249–285.

Merriam, C. H. 1890. Results of a biological survey of the San Francisco Mountain region and the desert of the Little Colorado, Arizona. *North American Fauna* 3: 1–136.

Merriam, C. H. 1894. Laws of temperature control of the geographic distribution of terrestrial animals and plants. *National Geographic* 6: 229–238.

Mertens, R. 1934. Die Inseleidenchsen des Golfes von Salerno. *Senckenbergiana Biologica* 42: 31–40.

Metcalf, H. and J. F. Collins. 1911. The control of the chestnut bark disease. *Farmer's Bulletin of the U. S. Deptartment of Agriculture*: 467: 1–24.

Meyer, A. 1993. Phylogenetic relationships and evolutionary processes in East African cichlid fishes. *Trends in Ecology and Evolution* 8: 279–284.

Meyer, A., T. D. Kocher, P. Basasibwaki and A. C. Wilson. 1990. Monophyletic origin of Lake Victoria cichlid fishes suggested by mitochondrial DNA sequences. *Nature* 347: 550–553.

Meyers, G. S. 1953. Ability of amphibians to cross sea barriers, with especial reference to Pacific zoogeography. *Proceedings of the Seventh Pacific Science Congress* 4: 19–17

Meyers, N. 1980. *The Sinking Ark*. Oxford: Pergamon Press.

Michaud, J. M. 1995. Biogeography of the bog copper butterfly (*Lycaena epixanthe*) in southern Rhode Island peatlands: A metapopulation perspective. Master's thesis, University of Rhode Island.

Miles, D. B. and A. E. Dunham. 1996. The paradox of the phylogeny: Character displacement of analyses of body size in island *Anolis*. *Evolution* 50: 594–603.

Miles, J. 1987. Vegetation and succession: Past and present perceptions. In A. J. Gray, M. J. Crawley and P. J. Edwards (eds.), *Colonisation, Succession, and Stability*, 1–30. Oxford: Blackwell Scientific Publications.

Millar, J. S. 1989. Reproduction and development. In G. L. Kirkland Jr. and J. N. Layne (eds.), *Advances in the Study of* Peromyscus *(Rodentia)*, 169–232. Lubbock, TX: Texas Tech University Press.

Miller, A. I. 1989. Spatio-temporal transitions in Paleozoic Bivalvia: An analysis of North America fossil assemblages. *Historical Biology* 1: 251–273.

Miller, R. R. 1948. The cyprinodont fishes of the Death Valley system of eastern California and southwestern Nevada. *University of Michigan Museum of Zoology Miscellaneous Publications* 42: 1–80.

Miller, R. R. 1961a. Man and the changing fish fauna of the American Southwest. *Papers of the Michigan Academy of Science, Arts and Letters* 46: 365–404.

Miller, R. R. 1961b. Speciation rates in some freshwater fishes of western North America. In W. F. Blair (ed.), *Vertebrate Speciation*, 537–560. Austin: University of Texas Press.

Miller, R. R. 1966. Geographical distribution of Central American freshwater fishes. *Copeia* (4): 773–802.

Miller, R. S. 1964. Ecology and distribution of pocket gophers (Geomyidae) in Colorado. *Ecology* 45: 256–272.

Miller, R. S. 1967. Pattern and process in competition. *Advances in Ecological Research* 4: 1–74.

Milne, B. T. and R. T. Forman. 1986. Peninsulas in Maine: Woody plant diversity, distance, and enviromental patterns. *Ecology* 67: 967–974.

Moll, E. O. and J. M. Legler. 1971. The life history of a Neotropical slider turtle, *Pseudomys scripta* (Schoepff), in Panama. *Bulletin of the Los Angeles County Museum of Natural History of Science*, no. 11.

Molles, M. C., Jr. 1978. Fish species diversity on model and natural reef patches: Experimental insular biogeography. *Ecological Monographs* 48: 289–305.

Monk, C. D. 1968. Successional and environmental relationships of the forest vegetation of north central Florida. *American Midland Naturalist* 79: 441–457.

Mooney, H. A. (ed.). 1977. *Convergent Evolution in Chile and California: Mediterranean Climate Ecosystems*. US/IBP Synthesis Series 5. Stroudsburg, PA: Dowden, Hutchinson, & Ross.

Mooney, H. A., J. Kummerow, A. W. Johnson, D. J. Parsons, D. Kelley, A. Hoffman, R. I. Hays, J. Giliberto and C. Chu. 1977. The producers: Their resources and adaptive responses. In H. A. Mooney (ed.), *Convergent Evolution in Chile and California*, 85–143. Stroudsburg, PA: Dowden, Hutchinson, and Ross.

Moore, H. E. Jr. 1973. Palms in the tropical forest ecosystems of Africa and South America. In B. J. Meggers, E. S. Ayensu and W. D. Duckworth (eds.), *Tropical Forest Ecosystems in Africa and South America: A Comparative Review*, 63–88. Washington, DC: Smithsonian Institution Press.

Moreau, R. E. 1944. Clutch size: A comparative study with special reference to African birds. *Ibis* 86: 286–347.

Morell, V. 1996. New mammals discovered by biology's new explorers. *Science* 273: 1491.

Morgan, G. S. and C. A. Woods. 1986. Extinction and zoogeography of West Indian land mammals. *Biological Journal of the Linnean Society* 28: 167–203.

Morgan, W. J. 1972a. Deep mantle convection plumes and plate motion. *Bulletin of the American Association of Petroleum Geologists* 56: 203–213.

Morgan, W. J. 1972b. Plate motions and deep mantle convection. *Memoirs of the Geological Society of America* 132: 7–22.

Morrone, J. J. and J. V. Crisci. 1995. Historical biogeography: Introduction to methods. *Annual Review of Ecology and Systematics* 26: 373–401.

Morse, D. R., N. E. Stock and J. H. Lawton. 1988. Species numbers, species abundance, and body length relationships of arboreal beetles in Bornean lowland rain forest trees. *Ecological Entomology* 13: 25–37.

Morton, E. S. 1978. Avian arboreal folivores: Why not? In G. G. Montgomery (ed.), *The Ecology of Arboreal Folivores*, 123–130. Washington, DC: Smithsonian Institution Press.

Mosiman, J. E. and P. S. Martin. 1975. Simulating overkill by paleoindians. *American Scientist* 63: 304–313.

Moulton, M. P. 1985. Morphological similarity and co-existence of congeners: An experimental test with introduced birds. *Oikos* 44: 301–305.

Moulton, M. P. 1993. The all-or-none pattern in introduced Hawaiian passeriforms: The role of competition sustained. *American Naturalist* 141: 105–119.

Moulton, M. P. and J. L. Lockwood. 1992. Morphological dispersion of introduced Hawaiian finches: Evidence for competition and the Narcissus effect. *Evolutionary Ecology* 6: 45–55.

Moulton, M. P. and S. L. Pimm. 1983. The introduced Hawaiian avifauna: Biogeographical evidence for competition. *American Naturalist* 121: 669–690.

Moulton, M. P. and S. L. Pimm. 1986. The extent of competition in shaping an introduced avifauna. In J. M. Diamond and T. Case (eds.), *Community Ecology*, 80–97. New York: Harper and Row.

Moulton, M. L. and S. L. Pimm. 1987. Morphological assortment and introduced Hawaiian passerines. *Evolutionary Ecology* 1: 113–124.

Moyle, P. B. and R. A. Leidy. 1992. Loss of biodiversity in aquatic ecosystems: Evidence from fish faunas. In P. L. Fiedler and S. K. Jain (eds.), *Conservation Biology: The Theory and Practice of Nature Conservation, Preservation, and Management*, 127–170. New York: Chapman & Hall.

Muller, J. 1970. Palynological evidence on early differentiation of angiosperms. *Biological Reviews of the Cambridge Philosophical Society* 45: 417–450.

Muller, P. 1974. *Aspects of Zoogeography*. The Hague: Dr. W. Junk.

Muller, R. A. and G. J. MacDonald. 1995. Glacial cycles and orbital inclination. *Nature* 377: 107–108.

Munroe, E. G. 1948. *The Geographical Distribution of Butterflies in the West Indies*. Ph.D. dissertation, Cornell University, Ithaca, NY.

Munroe, E. G. 1953. The size of island faunas. In *Proceedings of the Seventh Pacific Science Congress of the Pacific Science Association*, vol. IV, Zoology, 52–53. Auckland, New Zealand: Whitcome and Tombs.

Murray, J. 1895. A summary of scientific results. In *Challenger Report Summary*. 2 volumes. London: Nell and Company.

Myers, A. A. and P. S. Giller (eds.). 1989. *Analytical Biogeography*. London: Chapman and Hall.

Nadler, C. F., N. N. Vorontosov, R. S. Hoffmann, I. I. Formichova and C. F. Nadler Jr. 1973. Zoogeography of Transferins in arctic and long-tailed ground squirrel populations. *Comparative Biochemistry and Physiology* 44B: 33–40.

Nadler, C. F., M. Zhurkevich, R. S. Hoffmann, A. I. Kozlovskii, L. Deutsch and D. F. Nadler Jr. 1978. Biochemical relatedness of some Holarctic voles of the genera *Microtus, Arvicola* and *Clethrionomys* (Rodentia: Arvicolae). *Canadian Journal of Zoology* 56: 1564–1575.

Nalepa, T. F. and D. W. Schlosser. 1993. *Zebra Mussels: Biology, Impacts and Control.* Boca Raton, FL: CRC Press.

Nash, L. L. and P. H. Gleick. 1991. Sensitivity of streamflow in the Colorado Basin to climate changes. *Journal of Hydrology* 125: 221–241.

Neill, W. T. 1958. The occurrence of amphibians and reptiles in saltwater areas, and the bibliography. *Bulletin of Marine Science of the Gulf and Caribbean* 8: 1–97.

Neill, W. T. 1969. *The Geography of Life.* New York: Columbia University Press.

Nelson, G. J. 1969a. The problem of historical biogeography. *Systematic Zoology* 18: 243–246.

Nelson, G. J. 1969b. Gill arches and the phylogeny of fishes, with notes on the classification of vertebrates. *Bulletin of the American Museum of Natural History* 141: 475–552.

Nelson, G. J. 1978. From Candolle to Croizat: Comments on the history of biogeography. *Journal of the History of Biology* 11: 269–305.

Nelson, G. J. and N. Platnick. 1981. *Systematics and Biogeography: Cladistics and Vicariance.* New York: Columbia University Press.

Nelson, G. J. and D. E. Rosen (eds.). 1981. *Vicariance Biogeography: A Critique.* New York: Columbia University Press.

Neumayr, M. 1887. Uber klimatische Zonen wahrend der Jura- und Kreidzeit. *Koniglische Akademie der Wissenschaft Wien Denkschrift* 47: 277–310.

Nevo, E. and H. Bar-El. 1976. Hybridization and speciation in fossorial mole rats. *Evolution* 30: 831–840.

Nevo, E., M. Corti, G. Heth, A. Beiles and S. Simpson. 1988. Chromosomal polymorphisms in subterranean mole rats: Origins and evolutionary significance. *Biological Journal of the Linnean Society* 33: 309–322.

Newmark, W. D. 1993. The role and design of wildlife corridors with examples from Tanzania. *Ambio* 22: 500–504.

Nicholls, A. O. and C. R. Margules. 1991. The design of studies to demonstrate the biological importance of corridors. In D. A. Saunders and R. J. Hobbs (eds.), *The Role of Corridors in Nature Conservation,* 49–61. Chipping Norton, NSW, Australia: Surrey Beatty.

Niering, W. A. 1963. Terrestrial ecology of Kapingamarangi Atoll, Caroline Islands. *Ecological Monographs* 33: 131–160.

Niethammer, G. 1958. Tiergeographie (Bericht uber die Jahren 1950–56). *Fortschr. Zool.* 11: 35–141.

Niklas, K. J., B. H. Tiffney and A. H. Knoll. 1983. Apparent changes in the diversity of fossil plants. In M. C. Hecht, W. C. Steere and B. Wallace (eds.), *Evolutionary Biology* 12, 1–89. New York: Plenum Press.

Nilsson, S. G. and I. N. Nilsson. 1978. Species richness and dispersal of vascular plants to islands in Lake Mockeln, Southern Sweden. *Ecology* 59: 473–480.

Nobel, P. S. 1978. Surface temperarures of cacti: Influences of environmental and morphological factors. *Ecology* 59: 986–996.

Nobel, P. S. 1980a. Morphology, nurse plants, and minimum apical temperatures for young *Carnegiea gigantea. Botanical Gazette* 141: 188–191.

Nobel, P. S. 1980b. Morphology, surface temperatures, and northern limits of columnar cacti in the Sonoran Desert. *Ecology* 61: 1–7.

Nores, M. 1995. Insular biogeography of birds on mountaintops in north Argentina. *Journal of Biogeography* 22: 61–70.

Norse, E. A. 1993. *Global Marine Biological Diversity: A Strategy for Building Conservation into Decision Making.* Washington, DC: Island Press.

Nur, A. and Z. Ben-Avraham. 1977. Lost Pacifica continent. *Nature* 270: 41–43.

Oberdorff, T., J. Guegan and B. Hugeny. 1995. Global scale patterns in fish species diversity in rivers. *Ecography* 18: 345–352.

Ochumba, P. B. O., M. Gophen and L. S. Kaufman. 1993. Changes in oxygen availability in the Kenyan portion of Lake Victoria: Effects on fisheries and biodiversity. Abstract. In *People, Fisheries, Biodiversity, and the Future of Lake Victoria.* Report 93–3. Boston: New England Aquarium.

O'Conner, R. J. and J. Faaborg. 1992. The relative abundance of the brown-headed cowbird (*Molothrus ater*) in relation to exterior and interior edges in forests of Missouri. *Transactions of the Missouri Academy of Science* 26: 1–9.

Odening, W. R., B. R. Strain and W. C. Oechel. 1974. The effects of decreasing water potential on net CO_2 exchange of intact desert shrubs. *Ecology* 55: 1086–1095.

Odum, E. P. 1969. The strategy of ecosystem development. *Science* 164: 262–270.

Odum, E. P. 1971. *Fundamentals of Ecology.* 3rd edition. Philadelphia: W. B. Saunders.

Odum, H. T. 1957. Trophic structure and productivity of Silver Springs, Florida. *Ecological Monographs* 27: 55–112.

Olson, S. L. 1973. Evolution of the rails of the South Atlantic Islands. *Smithsonian Contributions to Zoology* 152: 1–43.

Olson, S. L. 1976. Oligocene fossils bearing on the origins of the Totidae and Momotidae (Aves: Coraciiformes). *Smithsonian Contributions to Paleobiology* 27: 111–119.

Olson, S. L. and H. F. James. 1982a. Fossil birds from the Hawaiian Islands: Evidence for wholesale extinction by man before Western contact. *Science* 217: 633–635.

Olson, S. L. and H. F. James. 1982b. Promodromus of the fossil avifauna of the Hawaiian Islands. *Smithsonian Contributions to Zoology* 365.

Olson, S. L. and H. F. James. 1984. The role of Polynesians in the extinction of the avifauna of the Hawaiian Islands. In P. S. Martin and R. G. Klein (eds.), *Quarternary Extinctions,* 768–780. Tucson: University of Arizona Press.

Omi, P. N., L. C. Wensel and J. L. Murphy. 1979. An application of multivariate statistics to land-use planning: Classifying land units into homogeneous zones. *Forest Science* 25: 399–414.

Ortmann, A. E. 1896. *Grundzuge der marinen tiergeographie.* Jena: G. Fisher.

Ospovat, D. 1977. Lyell's theory of climate. *Journal of the History of Biology* 10: 317–339.

Otte, D. 1976. Species richness patterns of New World desert grasshoppers in relation to plant diversity. *Journal of Biogeography* 3: 197–209.

Otte, D. and J. A. Endler (eds.). 1989. *Speciation and Its Consequences.* Sunderland, MA: Sinauer Associates.

Owen, D. F. and J. Owen. 1974. Species diversity in temperate and tropical Ichneumonidae. *Nature* 249: 583–584.

Owen, H. G. 1992. Has the earth increased in size? In S. Chatterjee and N. Hotton III (eds.), *New Concepts in Global Tectonics,* 289–296. Lubbock: Texas Tech University Press.

Owen-Smith, N. 1987. Pleistocene extinctions: The pivotal role of megaherbivores. *Paleobiology* 13: 351–362.

Owen-Smith, N. 1989. Megafaunal extinctions: The conservation message from 11,000 years B.P. *Conservation Biology* 3: 405–412.

Owen-Smith, R. N. 1988. *Megaherbivores: The Influence of Very Large Body Size on Ecology.* Cambridge: Cambridge University Press.

Pacala, S. and J. Roughgarden. 1985. Population experiments with the *Anolis* lizards of St. Maarten and St. Eustatius. *Ecology* 66: 128–141.

Page, R. D. M. 1990. Tracks and trees in the antipodes: A reply. *Systematic Zoology* 39: 288–299.

Page, R. D. M. 1991. Clocks, clades, and cospeciation: Comparing rates of evolution and timing of cospeciation events in host-parasite assemblages. *Systematic Zoology* 40: 188–198.

Page, R. D. M. 1993a. Genes, organisms, and areas: The problem of multiple lineages. *Systematic Biology* 42: 77–84.

Page, R. D. M. 1993b. Parasites, phylogeny and cospeciation. *International Journal for Parasitology* 23: 499–506.

Page, R. D. M. 1994. Parallel phylogenies: Reconstructing the history of host-parasite assemblages. *Cladistics* 10: 155–173.

Pagel, M. D., R. M. May and A. R. Collie. 1991. Ecological aspects of the geographical distribution and diversity of mammalian species. *American Naturalist* 137: 791–815.

Paine, R. T. 1966. Food web complexity and species diversity. *American Naturalist* 100: 65–76.

Paine, R. T. 1974. Intertidal community structure: Experimental studies on the relationship between a dominant competitor and its principal predator. *Oecologia* 15: 93–120.

Parenti, L. R. 1981. Discussion [of C. Patterson, *Methods of Paleobiogeography*]. In G. Nelson and D. E. Rosen (eds.), *Vicariance Biogeography: A Critique*, 490–497. New York: Columbia University Press.

Parmesan, C. 1996. Climate and species range. *Nature* 382: 765–766.

Pascual, R., M. Archer, E. O. Jaureguizar, J. L. Prado, H. Godthelp and S. J. Hand. 1992. First discovery of monotremes in South America. *Nature* 356: 704–706.

Paterson, H. E. H. 1982. Perspective on speciation by reinforcement. *South African Journal of Science* 78: 53–57.

Patrick, R. 1961. A study of the numbers and kinds of species found in rivers in eastern United States. *Proceedings of the Academy of Natural Sciences of Philadelphia* 13: 215–258.

Patrick, R. 1966. The Catherwood Foundation Peruvian Amazon Expedition: Limnological and systematic studies. *Monographs of the Academy of Natural Sciences of Philadelphia* 14: 1–495.

Patterson, B. D. 1980. Montane mammalian biogeography in New Mexico. *Southwestern Naturalist* 25: 33–40.

Patterson, B. D. 1984. Mammalian extinction and biogeography in the southern Rocky Mountains. In M. H. Nitecki (ed.), *Extinctions*, 247–294. Chicago: University of Chicago Press.

Patterson, B. D. 1987. The principle of nested subsets and its implications for biological conservation. *Conservation Biology* 1: 323–334.

Patterson, B. D. 1990. On the temporal development of nested subset patterns of species composition. *Oikos* 59: 330–342.

Patterson, B. D. 1994. Accumulating knowledge on the dimensions of biodiversity: Systematic perspectives on Neotropical mammals. *Biodiversity Letters* 2: 79–86.

Patterson, B. D. 1995. Local extinctions and the biogeographic dynamics of boreal mammals in the southwest. In C. A. Istock and R. S. Hoffman (eds.), *Storm over a Mountain Island: Conservation Biology and the Mount Graham Affair*. Tucson: University of Arizona Press.

Patterson, B. D. and W. Atmar. 1986. Nested subsets and the structure of insular mammalian faunas and archipelagoes. *Biological Journal of the Linnean Society* 28: 65–82.

Patterson, B. D. and J. H. Brown. 1991. Regionally nested patterns of species composition in granivorous rodent assemblages. *Journal of Biogeography* 18: 395–402.

Patterson, B. D. and R. Pascual. 1972. The fossil mammal fauna of South America. In A. Keast, F. C. Erk and B. Glass (eds.), *Evolution, Mammals, and Southern Continents*, 247–309. Albany: State University of New York Press.

Patterson, B. D., V. Pacheco and S. Solari. 1996. Distribution of bats along an elevational gradient in the Andes of south-east Peru. *Journal of the Zoological Society of London* 240: 637–658.

Patterson, B. D., D. F. Stotz, S. Solari, J. W. Fitzpatrick and V. Pacheco. In press. Contrasting patterns of elevational zonation for birds and mammals in the Andes of southeastern Peru. *Journal of Biogeography*.

Patterson, C. 1981. Methods of paleobiogeography. In G. Nelson and D. E. Rosen (eds.), *Vicariance Biogeography: A Critique*, 446–489. New York: Columbia University Press.

Patton, J. L. 1969. Chromosomal evolution in the pocket mouse, *Perognathus goldmani* Osgood. *Evolution* 23: 645–662.

Patton, J. L. 1972. Patterns of geographic variation in karyotype in the pocket gopher, *Thomomys bottae* (Eydoux and Gervais). *Evolution* 25: 574–586.

Patton, J. L. 1985. Population structure and the genetics of speciation in pocket gophers, genus *Thomomys*. *Acta Zoologica Fennica* 170: 109–114.

Patton, J. L., M. F. Smith, R. D. Price and R. A. Hellenthal. 1984. Genetics of hybridization between the pocket gophers *Thomomys bottae* and *Thomomys townsendii* in northeastern California. *Great Basin Naturalist* 44: 431–440.

Patton, J. L., M. N. F. da Silva and J. R. Malcom. 1994. Gene genealogy and differentiation among arboreal spiny rats (Rodentia: Echymidae) of the Amazon Basin: A test of the riverine barrier hypothesis. *Evolution* 48: 1314–1323.

Pearcy, W. G. and A. Schoener. 1987. Changes in the marine biota coincident with the 1982–1983 El Niño in the northeastern subarctic Pacific Ocean. *Journal of Geophysical Research* 92, C13: 14417–14428.

Pearson, T. G. (ed.). 1936. *Birds of America*. Garden City, NY: Garden City Publishing Co.

Peattie, D. C. 1922. The Atlantic coastal plain element in the flora of the Great Lakes. *Rhodora* 24: 57–70, 80–88.

Peltonen, A. and I. Hanski. 1991. Patterns of island occupancy explained by colonization and extinction rates in shrews. *Ecology* 72: 1698–1708.

Peltonen, A., S. Peltonen, P. Vilpas and A. Beloff. 1989. Distributional ecology of shrews in three archipelagoes in Finland. *Annales Zoologici Fennici* 26: 381–387.

Pendergast, J. R., R. M. Quinn, J. H. Lawton, B. C. Eversham and D. W. Gibbons. 1993. Rare species, the coincidence of diversity hot spots and conservation strategies. *Nature* 365: 335–337.

Peng, C. H., J. Guiot, E. VanCamp and R. Cheddar. 1995. Temporal and spatial variations of terrestrial biomes and carbon storage since 13,000 yr BP in Europe: Reconstruction from pollen data and statistical models. *Water, Air and Soil Pollution* 82: 375–390.

Perault, D. R. 1998. *Landscape Heterogeneity and the Role of Corridors in Determining the Spatial Structure of Insular Mammal Populations*. Ph.D. dissertation, University of Oklahoma, Norman, OK.

Peters, R. H. 1983. *The Ecological Implications of Body Size*. Cambridge: Cambridge University Press.

Peters, R. L. and T. E. Lovejoy. 1992. *Global Warming and Biological Diversity*. New Haven, CT: Yale University Press.

Peterson, C. H. 1991. Intertidal zonation of marine invertebrates in sand and mud. *American Scientist* 79: 236–249.

Peterson, R. L. 1955. *North American Moose*. Toronto: University of Toronto Press.

Petren, K., D. T. Bolger and T. J. Case. 1993. Mechanisms in the competitive success of an invading sexual gecko over an asexual native. *Science* 259: 354–358.

Petuch, E. J. 1995. Molluscan diversity in the late Neogene of Florida: Evidence for a two-staged mass extinction. *Science* 270: 275–277.

Pianka, E. R. 1966. Latitudinal gradients in species diversity: A review of concepts. *American Naturalist* 100: 33–46.

Pianka, E. R. 1967. On lizard species diversity: North American flatland deserts. *Ecology* 48: 331–351.

Pianka, E. R. 1978. *Evolutionary Ecology*. 2nd edition. New York: Harper & Row.

Pianka, E. R. 1986. *Ecology and Natural History of Desert Lizards*. Princeton, NJ: Princeton University Press.

Pianka, E. R. 1988. *Evolutionary Ecology*. New York: Harper and Row.

Pianka, E. R. 1995. Evolution of body size: Varanid lizards as a model system. *American Naturalist* 146: 398–414.

Pickett, S. T. A. and P. S. White. 1985. *The Ecology of Natural Disturbance and Patchiness*. New York: Academic Press, Inc.

Pielou, E. C. 1975. *Ecological Diversity*. New York: John Wiley & Sons.

Pielou, E. C. 1977a. *Mathematical Ecology*. New York: John Wiley & Sons.

Pielou, E. C. 1977b. The latitudinal spans of seaweed species and their patterns of overlap. *Journal of Biogeography* 4: 299–311.

Pielou, E. C. 1978. Latitudinal overlap of seaweed species: Evidence for quasi-sympatric speciation. *Journal of Biogeography* 5: 227–238.

Pielou, E. C. 1979. *Biogeography*. New York: John Wiley & Sons.

Pielou, E. C. 1991. *After the Ice Age*. Chicago: University of Chicago Press.

Pijl, L. van den. 1972. *Principles of Dispersal in Higher Plants*. 2nd edition. New York: Springer-Verlag.

Pimm, S. L. 1991. *The Balance of Nature? Ecological Issues in the Conservation of Species and Communities*. Chicago: University of Chicago Press.

Pimm, S. L., H. L. Jones and J. M. Diamond. 1988. On the risk of extinction. *American Naturalist* 132: 757–785.

Pine, R. H. 1994. New mammals not so seldom. *Nature* 368: 593.

Platnick, N. I. and G. Nelson. 1978. A method of analysis for historical biogeography. *Systematic Zoology* 27: 1–16.

Platt, I. 1969. What we must do. *Science* 166: 1115–1121.

Platt, W. J. 1975. The colonization and formation of equilibrium plant species associations on badger disturbances in a tall-grass prairie. *Ecological Monographs* 45: 285–305.

Poinar, G. O. and D. A. Grimaldi. 1990. Fossil and extant macrochelid mites (Ascari: Macroelidae) phoretic on Drosophilid flies (Diptera: Drosophilidae). *Journal of the New York Entomological Society* 98: 88.

Poinar, G. O., G. M. Thomas and B. Lighthart. 1990. Bioassay to determine the effect of commercial preparations of bacillus-thuringiensis on entomogeneous rhabditoid nematodes. *Agriculture Ecosystems and Environment* 30: 195–202.

Popper, K. R. 1968a. *The Logic of Scientific Discovery*. 2nd edition. New York: Harper & Row.

Popper, K. R. 1968b. *Conjectures and Refutations*. New York: Harper & Row.

Por, F. D. 1971. One hundred years of Suez Canal: A century of Lessepsian migration: Retrospect and viewpoints. *Systematic Zoology* 20: 138–159.

Por, F. D. 1975. Pleistocene pulsation and preadaptation of biotas in Mediterranean seas: Consequences for Lessepsian migration. *Systematic Zoology* 24: 72–78.

Por, F. D. 1978. *Lessepsian Migration: The Influx of Red Sea Biota into the Mediterranean by Way of the Suez Canal*. Ecological Studies 23. Berlin: Springer-Verlag.

Por, F. D. 1990. Lessepsian migration: An appraisal and new data. *Bulletin, Institut Oceanographique (Monaco)* Special 7: 1–10.

Porter, J. W. 1972. Ecology and species diversity of coral reefs on opposite sides of the Isthmus of Panama. *Bulletin of the Biological Society of Washington* 2: 89–116.

Porter, J. W. 1974. Community structure and coral reefs on opposite sides of the Isthmus of Panama. *Science* 186: 543–545.

Poulson, T. L. and W. B. White. 1969. The cave environment. *Science* 165: 971–981.

Power, D. M. 1972. Numbers of bird species on the California Islands. *Evolution* 26: 451–463.

Prance, G. T. 1973. Phytogeographic support for the theory of Pleistocene forest refugia in the Amazon basin, based on evidence from distribution patterns in Caryocaraceae, Chrysobalanaceae, Dichapetalaceae, and Lecythidaceae. *Acta Amazonica* 3(3): 5–28.

Prance, G. T. 1978. The origin and evolution of the Amazon flora. *Interciencia* 3: 207–222.

Prance, G. T. (ed.). 1982. *The Biological Model of Diversification in the Tropics*. New York: Columbia University Press.

Preston. F. W. 1957. Analysis of Maryland statewide bird counts. *Maryland Birdlife [Bulletin of the Maryland Ornithological Society]* 13: 63–65.

Preston, F. W. 1962a. The canonical distribution of commonness and rarity: part I. *Ecology* 43: 185–215.

Preston, F. W. 1962b. The canonical distribution of commonness and rarity: part II. *Ecology* 43: 410–432.

Price, P. W. 1980. *Evolutionary Biology of Parasites*. Monographs in Population Biology, no. 15. Princeton, NJ: Princeton University Press.

Primack, R. 1998. *Essentials of Conservation Biology*. 2nd edition. Sunderland, MA: Sinauer Associates.

Pruvot, G. 1896. *Essai sur les fonds et la faune de la Manche occidentale (cotes de Bretagne) compares a ceux du golfe du Lion*. Paris: Schleicher Frères.

Pulliam, H. R. 1988. Sources, sinks, and population regulation. *American Naturalist* 132: 652–661.

Purves, W. K., G. H. Orians, H. C. Heller and D. Sadava. 1997. *Life: The Science of Biology*, 5th edition. Sunderland, MA: Sinauer Associates, Inc.

Putnam, R. J. 1994. *Community Ecology*. New York: Chapman and Hall.

Rabinowitz, D., S. Cairns and T. Dillon. 1986. Seven forms of rarity and their frequency in the flora of the British Isles. In M. E. Soulé (ed.), *Conservation Biology: The Science of Scarcity and Diversity*, 182–204. Sunderland, MA: Sinauer Associates.

Raffi, S., S. M. Stanley and R. Marasti. 1985. Biogeographic patterns and Plio-Pleistocene extinction of Bivalvia in the Mediterranean and southern North Sea. *Paleobiology* 11: 368–388.

Rahbek, C. 1995. The elevational gradient of species richness: A uniform pattern? *Ecography* 18: 200–205.

Raikow, R. J. 1976. The origin and evolution of the Hawaiian honeycreepers (Drepaniidae). *Living Bird* 15: 95–117.

Rapoport, E. H. 1982. *Areography: Geographical Strategies of Species*. New York: Pergamon Press.

Rapoport, E. H., M. E. Diaz Betancourt and I. R. Lopez Moreno. 1983. *Aspectos de la Ecologia urbana en la Cuidad de Mexico: Flora y de las calles y baldios*. Mexico, D.F.: Editorial Limusa.

Rass, T. S. 1986. Vicariance icthyogeography of the Atlantic ocean pelagial. In *Pelagic Biogeography*, 237–241. UNESCO Technical Papers in Marine Science, 49.

Raunkiaer, C. 1934. *The Life Forms of Plants and Statistical Plant Geography*. Oxford: Clarendon Press.

Raup, D. M. 1972. Taxonomic diversity during the Phanerozoic. *Science* 177: 1065–1071.

Raup, D. M. 1976. Species diversity in the Phanerozoic: An interpretation. *Paleobiology* 2: 289–297.

Raup, D. M. 1979. Size of the Permo-Triassic bottleneck and its evolutionary implications. *Science* 206: 217–218.

Raup, D. M. and D. Jablonski. 1993. Geography of end-Cretaceous marine bivalve extinctions. *Science* 260: 971–973.

Raup, D. M. and J. J. Sepkoski Jr. 1982. Mass extinctions in the marine fossil record. *Science* 215: 1501–1503.

Raup, D. M., S. J. Gould, T. J. M. Schopf and D. Simberloff. 1973. Stochastic models of phylogeny and the evolution of diversity. *Journal of Geology* 81: 525–542.

Raven, P. H. and E. O. Wilson. 1992. A fifty-year plan for biodiversity surveys. *Science* 258: 1099–1100.

Ray, C. 1960. The application of Bergmann's and Allen's rule to the poikilotherms. *Journal of Morphology* 106: 85–109.

Ray, C., M. Gilpin and A. T. Smith. 1991. The effect of conspecific attraction on metapopulation dynamics. *Biological Journal of the Linnean Society* 42: 123–134.

Real, L. A. and J. H. Brown (eds.). 1991. *Foundations of Ecology: Classic Papers with Commentaries*. Chicago: University of Chicago Press.

Recher, H. 1969. Bird species diversity and habitat diversity in Australia and North America. *American Naturalist* 103: 75–80.

Reichman, O. J. 1985. Impact of pocket gopher burrows on overlying vegetation. *Journal of Mammalogy* 66: 720–725.

Reichman, O. J. and S. C. Smith. 1985. Impact of pocket gopher burrows on overlying vegetation. *Journal of Mammalogy* 66: 720–725.

Reichman, O. J., D. T. Wicklow and C. Rebar. 1985. Ecological and mycological characteristics of caches in the mounds of *Dipodomys spectabilis. Journal of Mammalogy* 66: 643–651.

Reid, W. V. and M. C. Trexler. 1991. *Drowning the National Heritage: Climate Change and U.S. Coastal Biodiversity*. Washington, DC: World Resources Institute.

Reig, O. A. 1989. Karyotypic repatterning as one triggering factor in cases of explosive speciation. In A. Fondevila (ed.), *Evolutionary Biology of Transient Unstable Populations*, 246–289. Berlin: Springer-Verlag.

Rensch, B. 1960. *Evolution Above the Species Level*. New York: Columbia University Press.

Repasky, R. R. 1991. Temperature and the northern distribution of wintering birds. *Ecology* 72: 2274–2285.

Rex, M. A. 1981. Community structure in the deep-sea benthos. *Annual Review of Ecology and Systematics* 12: 331–354.

Rex, M. A. 1983. Geographic patterns of species diversity in the deep-sea benthos. In G. T. Rowe (ed.), *Deep Sea Biology*, vol. 8, *The Sea*, 452–472. New York: John Wiley and Sons.

Rex, M. A., C. T. Stuart, R. R. Hessler, J. A. Allen, H. L. Sanders and G. D. F. Wilson. 1993. Global-scale patterns of species diversity in the deep sea benthos. *Nature* 365: 636–639.

Rey, J. R. 1981. Ecological biogeography of arthropods on *Spartina* islands in northwest Florida. *Ecological Monographs* 51: 237–265.

Rice, D. W. and A. A. Wolman. 1971. *The Life History and Ecology of the Gray Whale* (Eschrichtius roboustus). American Society of Mammalogists, Special Publication.

Rich, T. H. 1985. *Megalania prisca* (Owen, 1859), the giant goanna. In P. V. Rich and F. F. van Tets (eds.), *Kadimakara: Extinct Vertebrates of Australia*, 152–155. Lildale, Victoria: Pioneer Design Studio.

Richards, P. W. 1957. *The Tropical Rainforest*. Cambridge: Cambridge University Press.

Richman, A. D., T. J. Case and T. D. Schwaner. 1988. Natural and unnatural extinction rates of reptiles on islands. *American Naturalist* 131: 611–630.

Richter-Dyn, N. and N. S. Goel. 1971. On the extinction of a colonizing species. *Theoretical Population Biology* 3: 406–433.

Ricklefs, R. E. 1987. Community diversity: Relative roles of local and regional processes. *Science* 235: 167–171.

Ricklefs, R. E. and D. Schluter (eds.). 1993. *Species Diversity in Ecological Communities: Historical and Geographical Perspectives*. Chicago: University of Chicago Press.

Ricklefs, R. E. and G. W. Cox. 1972. Taxon cycles of the West Indian avifauna. *American Naturalist* 106: 295–219.

Ricklefs, R. E. and G. W. Cox. 1978. Stage of taxon cycle, habitat distribution, and population density in the avifauna of the West Indies. *American Naturalist* 112: 875–895.

Riddle, B. R. 1995. Molecular biogeography in the pocket mice (*Perognathus* and *Chaetodipus*) and grasshopper mice (*Onychomys*): The late Cenozoic development of a North America aridlands rodent guild. *Journal of Mammalogy* 76: 283–301.

Riddle, B. R. 1996. The molecular phylogenetic bridge between deep and shallow history in continental biotas. *Trends in Ecology and Evolution* 11: 207–211.

Ridley, H. N. 1930. *The Dispersal of Plants throughout the World*. Ashford, England: L. Reeve & Co.

Roane, M. K., G. J. Griffin and J. R. Elkins. 1986. *Chestnut Blight, other* Endothia *Diseases and the Genus* Endothia. St. Paul, MN: American Phytopathological Society.

Robinove, C. J. 1979. *Integrated Terrain Mapping with Digital Landsat Images in Queensland, Australia*. Professional Paper 1102. Washington, DC: U.S. Geological Survey. 39 pp.

Robinson, E. and J. F. Lewis. 1971. Field guide to aspects of the geology of Jamaica. *An International Field Institute Guidebook to the Caribbean Island-Arc System*, 2–29. American Geological Institute (Jamaica section).

Robinson, G. R., R. D. Holt, M. S. Gaines, S. P. Hamburg, M. L. Johnson, H. S. Fitch, E. A. Martinko. 1992. Diverse and contrasting effects of habitat fragmentation. *Science* 257: 524–526.

Roff, D. A. 1986. The evolution of wing dimorphism in insects. *Evolution* 40: 1009–1020.

Roff, D. A. 1990. The evolution of flightlessness in insects. *Ecological Monographs* 60: 389–421.

Rogers, R. A., L. D. Martin and T. D. Nicklas. 1990. Ice-age geography and the distribution of native North American languages. *Journal of Biogeography* 17: 131–143.

Rogers, R. A., L. A. Rogers, R. S. Hoffmann and L. D. Martin. 1991. Native American biological diversity and the biogeographic influence of Ice Age refugia. *Journal of Biogeography* 18: 623–630.

Rohde, K. 1992. Latitudinal gradients in species diversity: The search for the primary cause. *Oikos* 65: 514–527.

Rohde, K. 1996. Rapoport's rule is a local phenomenon and cannot explain latitudinal gradients in species diversity. *Biodiversity Letters* 3: 10–13.

Rohde, K. and M. Heap. 1996. Latitudinal ranges of teleost fish in the Atlantic and Indo-Pacific oceans. *American Naturalist* 147: 659–665.

Rohde, K., M. Heap and D. Heap. 1993. Rapoport's rule does not apply to marine teleosts and cannot explain latitudinal gradients in species richness. *American Naturalist* 142: 1–16.

Romer, A. S. 1966. *Vertebrate Paleontology*. Chicago: Chicago University Press.

Roosevelt, A. C., M. Lima deCosta, C. Lopes Machado, M. Michab, N. Mercier, H. Vallada, J. Feathers, W. Barnett, M. Imazio da Silveira and K. Schick. 1996. Paleoindian cave dwellers in the Amazon: The peopling of the Americas. *Science* 272: 373–384.

Root, R. B. 1967. The niche exploitation pattern of the blue-grey gnat-catcher. *Ecological Monographs* 37: 317–350.

Root, T. 1988a. *Atlas of Wintering North American Birds: An Analysis of Christmas Bird Count Data*. Chicago: University of Chicago Press.

Root, T. 1988b. Energy constraints on avian distributions and abundances. *Ecology* 69: 330–339.

Root, T. 1988c. Environmental factors associated with avian distributional boundaries. *Journal of Biogeography* 15: 489–505.

Root, T. L. 1993. Effects of global climate change on North American bird communities. In P. M. Karieva, J. M. King-solver and R. B. Huey (eds.), *Biotic Interactions and Global Change*, 280–292. Sunderland, MA: Sinauer Associates.

Root, T. L. and S. H. Schneider. 1993. Can large-scale climatic models be linked with multiscale ecological studies. *Conservation Biology* 7: 256–270.

Root, T. L. and S. H. Schneider. 1995. Ecology and climate: Research strategies and implications. *Science* 269: 334–431.

Rosen, B. R. and A. B. Smith. 1988. Tectonics from fossils? Analysis of reef coral and sea urchin distributions from late Cretaceous to Recent, using a new method. In M. G. Audley-Charles and A. Hallam (eds.), *Gondwana and Tethys*, 275–306. Special publication. Oxford: Geological Society of London.

Rosen, D. E. 1975. A vicariance model of Caribbean biogeography. *Systematic Zoology* 24: 431–464.

Rosen, D. E. 1978. Vicariant patterns and historical explanations in biogeography. *Systematic Zoology* 27: 159–188.

Rosen, D. E. 1979. Fishes from the uplands and intermontane basins of Guatemala: Revisionary studies and comparative geography. *Bulletin of the American Museum of Natural History* 162: 267–376.

Rosenzweig, M. L. 1966. Community structure in sympatric Carnivora. *Journal of Mammalogy* 47: 602–612.

Rosenzweig, M. L. 1968. Net primary productivity of terrestrial communities: Predictions from climatological data. *American Naturalist* 102: 67–74.

Rosenzweig, M. L. 1975. On continental steady states of species diversity. In M. L. Cody and J. M. Diamond (eds.), *Ecology and Evolution in Communities*, 121–141. Cambridge, MA: Belknap Press.

Rosenzweig, M. L. 1978. Competitive speciation. *Biological Journal of the Linnaean Society* 10: 275–289.

Rosenzweig, M. L. 1992. Species diversity gradients: We know more and less than we thought. *Journal of Mammalogy* 73: 715–730.

Rosenzweig, M. L. 1995. *Species Diversity in Space and Time*. New York: Cambridge University Press.

Rosenzweig, M. L. and J. Winakur. 1969. Population ecology of desert rodent communities: Habitats and environmental complexity. *Ecology* 50: 558–572.

Roth, V. L. 1990. Insular dwarf elephants: A case study in body mass estimation and ecological inference. In J. Damuth and B. J. MacFadden (eds.), *Body Size in Mammalian Paleobiology: Estimation and Biological Implications*, 151–179. New York: Cambridge University Press.

Roth, V. L. and M. S. Klein. 1986. Maternal effects of body size of large insular *Peromyscus maniculatus*: Evidence from embryo transfer experiments. *Journal of Mammalogy* 67: 37–45.

Roughgarden, J. 1974. Niche width: Biogeographic patterns among *Anolis* lizard populations. *American Naturalist* 108: 429–442.

Roughgarden, J. 1992. Comments on the paper by Losos: Character displacement versus taxon loop. *Copeia* 1992: 288–295.

Roughgarden, J. 1995. Anolis *Lizards of the Caribbean: Ecology, Evolution and Plate Tectonics*. Oxford: Oxford University Press.

Roughgarden, J. and E. R. Fuentes. 1977. The environmental determinants of size in solitary populations of West Indian *Anolis* lizards. *Oikos* 29: 44–51.

Roughgarden, J. and S. Pacala. 1989. Taxon cycle among *Anolis* lizard populations: Review of the evidence. In D. Otte and J. Endler (eds.), *Speciation and Its Consequences*, 403–442. Sunderland, MA: Sinauer Associates.

Roughgarden, J., D. Heckel and E. R. Fuentes. 1983. Coevolutionary theory and the biogeography and community structure of *Anolis*. In R. B. Huey, E. R. Pianka and T. W. Schoener (eds.), *Lizard Ecology: Studies of a Model Organism*, 371–410. Cambridge, MA: Harvard University Press.

Roughgarden, J., S. D. Gaines and S. Pacala. 1987. Supply side ecology: The role of physical transport processes. In P. Giller and J. Gee (eds.), *Organization of Communities: Past and Present*, 491–518. Oxford: Blackwell Scientific Publications.

Roughgarden, J., R. M. May and S. A. Levin. 1989. *Perspectives in Ecological Theory*. Princeton, NJ: Princeton University Press.

Rounds, R. C. 1987. Distribution and analysis of colourmorphs of the black bear (*Ursus americanus*). *Journal of Biogeography* 14: 521–538.

Rowe, J. S. 1980. The common denominator in land classification in Canada: An ecological approach to mapping. *Forestry Chronicle* 56: 19–20.

Roy, K., D. Jablonski and J. W. Valentine. 1994. Eastern Pacific molluscan provinces and latitudinal diversity gradient: No evidence for Rapoport's rule. *Proceedings of the National Academy of Sciences, USA* 91: 8871–8874.

Rubinoff, I. 1968. Central American sea-level canal: Possible biological effects. *Science* 161: 857–861.

Rubinoff, R. W. and I. Rubinoff. 1971. Geographic and reproductive isolation in Atlantic and Pacific populations of Panamanian *Bathygobius*. *Evolution* 25: 88–97.

Runcorn, S. K. 1956. Paleomagnetic comparisons between Europe and North America. *Proceedings of the Geological Association of Canada* 8: 77–85.

Runcorn, S. K. 1962. Paleomagnetic evidence for continental drift and its geophysical cause. In S. K. Runcorn (ed.), *Continental Drift*, 1–40. International Geophysics Series 3. New York: Academic Press.

Sand, H., G. Cederlund and Kjell Danell. 1995. Geographical and latitudinal variation in growth patterns and adult body size of Swedish moose (*Alces alces*). *Oecologia* 102: 433–442.

Sanders, H. L. 1968. Marine benthic diversity: A comparative study. *American Naturalist* 102: 243–282.

Sanders, H. L. 1969. Benthic marine diversity and the stability-time hypothesis. *Brookhaven Symp. Biol.* 22: 71–81.

Sauer, J. 1969. Oceanic islands and biogeographic theory: A review. *Geographical Review* 59: 582–593.

Sauer, J. D. 1988. *Plant Migration: The Dynamics of Geographic Patterning in Seed Plants*. Berkeley: University of California Press.

Saunders, D. A., R. J. Hobbs and C. R. Margules. 1991. Biological consequences of ecosystem fragmentation: A review. *Conservation Biology* 5: 18–32.

Savage, J. M. 1973. The geographic distribution of frogs: Patterns and predictions. In J. L. Vial (ed.), *Evolutionary Biology of the Anurans*, 351–445. Columbia: University of Missouri Press.

Savidge, J. A. 1987. Extinction of an island forest avifauna by an introduced snake. *Ecology* 68: 660–668.

Savilov, A. I. 1961. The distribution of the ecological forms of the by-the-wind sailor, *Velella lata*, Ch. and Eys., and the Portugese man-of-war *Physalia utriculus* (La Martiniere) Esch., in the North Pacific. *Transactions Institute Okenologyi, Akademis Nauk*. SSSR 45: 223–239.

Schall, J. J. and E. R. Pianka. 1978. Geographical trends in numbers of species. *Science* 201: 679–686.

Scheiner, S. M. and J. M. Rey-Benayas. 1994. Global patterns of plant diversity. *Evolutionary Ecology* 8: 331–347.

Schemske, D. W., B. C. Husband, M. H. Ruckelshaus, C. Goodwillie, I. M. Parker and J. G. Bishop. 1994. Evaluating approaches to the conservation of rare and endangered plants. *Ecology* 75: 584–606.

Schimper, A. F. W. 1898. *Pflanzengeographie auf physiologischer Grundlage*. 1st ed. Jena.

Schimper, A. F. W. 1903. *Plant-Geography upon a Physiological Basis*. Trans. W. R. Fisher. Oxford: Clarendon Press.

Schlinger, E. I. 1974. Continental drift, *Nothofagus*, and some ecologically associated insects. *Annual Review of Entomology* 19: 323–343.

Schluter, D. 1986. Tests of similarity and convergence of finch communities. *Ecology* 67: 1073–1086.

Schluter, D. 1988. The evolution of finch communities on islands and continents: Kenya vs. Galápagos. *Ecological Monographs* 58: 229–249.

Schluter, D. 1993. Adaptive radiation in sticklebacks: Size, shape and habitat use efficiency. *Ecology* 74: 699–709.

Schluter, D. 1996. Ecological speciation in postglacial fishes. *Philosophical Transactions of the Royal Society of London* 351: 807–814.

Schluter, D. and P. R. Grant. 1984. The distribution of *Geospiza difficilis* in relation to *G. fuliginosa* in the Galápagos Islands: Tests of three hypotheses. *Evolution* 36: 1213–1226.

Schluter, D., T. D. Price and P. R. Grant. 1985. Ecological character displacement in Darwin's finches. *Science* 227: 1056–1059.

Schmidt-Nielsen, K. 1963. Osmotic regulation in higher vertebrates. *Harvey Lectures*, ser. 58: 53–93.

Schmidt-Nielsen, K. 1964. *Desert Animals: Physiological Problems of Heat and Water*. New York: Oxford University Press.

Schmidt-Nielsen, K. 1984. *Scaling: Why Is Animal Size So Important?* New York: Cambridge University Press.

Schmitt, R. J. 1987. Indirect interactions between prey: Apparent competition, predator aggregation, and habitat segregation. *Ecology* 68: 1887–1897.

Schneider, S. H. 1993. Scenarios of global warming. In P. M. Karieva, J. G. Kingsolver and R. B. Huey (eds.), *Biotic Interactions and Global Change*, 9–23. Sunderland, MA: Sinauer Associates.

Schoener, T. W. 1968a. The *Anolis* lizards of Bimini: Resource partitioning in a complex fauna. *Ecology* 49: 704–726.

Schoener, T. W. 1968b. Sizes of feeding territories among birds. *Ecology* 49: 123–141.

Schoener, T. W. 1970. Size patterns in West Indian *Anolis* lizards. II. Correlations with the sizes of particular sympatric species: Displacement and divergence. *American Naturalist* 104: 155–174.

Schoener, T. W. 1974. The species-area relationship within archipelagos: Models and evidence from island land birds. In H. J. Firth and J. H. Calaby (eds.), *Proceedings of the 16th International Ornithological Congress*, 629–642. Canberra, Australian Academy of Science.

Schoener, T. W. 1975. Presence and absence of habitat shift in some widespread lizard species. *Ecological Monographs* 45: 233–258.

Schoener, T. W. 1976. The species-area relationship within archipelagoes: Models and evidence from island birds. *Proceedings of the XVI International Ornithological Congress* 6: 629–642.

Schoener, T. W. 1983. Rate of species turnover decreases from lower to higher organisms: A review of the data. *Oikos* 41: 372–377.

Schoener, T. W. 1986. Patterns in terrestrial vertebrate vs. arthropod communities: Do systematic differences in regulation exist? In J. Diamond and T. J. Case (eds.), 556–586. *Community Ecology*. New York: Harper and Row.

Schoener, T. W. 1988. The ecological niche. In J. M. Cherrett (ed.), *Ecological Concepts*, 79–113. Oxford: Blackwell.

Schoener, T. W. 1989. Food webs from the small to the large. *Ecology* 70: 1559–1589.

Schoener, T. W. and G. H. Adler. 1991. Greater resolution of distributional complementarities by controlling for habitat affinities: A study of Bahamian lizards. *American Naturalist* 137: 669–692.

Schoener, T. W. and A. Schoener. 1983. Distribution of vertebrates on some very small islands. I. Occurrence sequences of individual species. *Journal of Animal Ecology* 52: 209–235.

Schoener, T. W. and D. A. Spiller. 1987. High population persistence in a system with high turnover. *Nature* 330: 474–477.

Scholander, P. F. 1955. Evolution of climatic adaptation in hometherms. *Evolution* 9: 15–26.

Schopf, J. M. 1970a. Gondwana paleobotany. *Antarctic Journal of the U.S.* 5: 62–66.

Schopf, J. M. 1970b. Relation of floras of the Southern Hemisphere to continental drift. *Taxon* 19: 657–674.

Schopf, J. M. 1974. Permo-Triassic extinctions: Relation to seafloor spreading. *Journal of Geology* 82: 129–144.

Schopf, J. M. 1975. Precambrian paleobiology: Problems and perspectives. *Annual Review of Earth and Planetary Sciences* 3: 213–249.

Schopf, J. M. 1976. Morphologic interpretation of fertile structures in glossopterid gymnosperms. *Review of Palaeobotany and Palynology*. 21: 25–64.

Schouw, F. 1823. *Grunzüge einer algemeinen Pflanzengeographie*. Berlin.

Schreiber, H. 1978. *Dispersal Centres of Sphingidae (Lepidoptera) in the Neotropical Region*. Biogeographica 10. The Hague: Dr. W. Junk.

Schuchert, C. 1932. Gondwana landbridges. *Bulletin of the Geological Society of America* 43: 875–916.

Schuster, M. F. and S. McDaniel. 1973. A vegetative analysis of a Black Belt prairie relict site near Aliceville, Alabama. *Journal of the Mississippi Academy of Science* 19: 153–159.

Schwaner, T. D. and S. D. Sarre. 1990. Body size and sexual dimorphism in mainland and island tiger snakes. *Journal of Herpetology* 24: 320–322.

Schwarzbach, M. 1980. *Alfred Wegener: The Father of the Continental Drift*. Madison, WI: Science Technology.

Sclater, J. G. and C. Tapscott. 1979. The history of the Atlantic. *Scientific American* 240: 156–174.

Sclater, P. L. 1858. On the general geographical distribution of the members of the class Aves. *Journal of the Linnean Society, Zoology* 2: 130–145.

Sclater, P. L. 1897. On the distribution of marine mammals. *Proceedings of the Zoological Society of London* 41: 347–359.

Scotese, C. R., L. M. Gahagan and R. L. Larson. 1988. Plate tectonic reconstructions of the Cretaceous and Cenozoic ocean basins. *Tectonophysics* 155: 27–48.

Scott, J. M., S. Mountainspring, F. L. Ramsey and C. B. Kepler. 1986. Forest bird communities of the Hawaiian Islands: Their dynamics, ecology and conservation. *Studies in Avian Biology* 9: 1–431.

Scott, M. J., F. Davis, B. Csuiti, B. Butterfield, C. Groves, H. Anderson, S. Caicco, F. D'Erchia, T. C. Edwards Jr., J. Ulliman and R. G. Wright. 1993. GAP analysis: A geographical approach to protection of biological diversity. *Wildlife Monographs* 123: 1–41.

Scriber, J. M. 1973. Latitudinal gradients in larval feeding specialization of the world Papilionidae (Lepidoptera). *Psyche* 80: 355–373.

Seastedt, T. R., O. J. Reichman and T. C. Todd. 1986. Microarthropods and nematodes in kangaroo rat burrows. *Southwestern Naturalist* 31: 114–116.

Sepkoski, J. J. Jr. 1976. Species diversity in the Phanerozoic: Species-area effects. *Paleobiology* 2: 298–303.

Sepkoski, J. J. Jr. 1984. A kinetic model of Phanerozoic taxonomic diversity. III. Post-Paleozoic families and mass extinctions. *Paleobiology* 10: 246–267.

Sepkoski, J. J. Jr., R. K. Bambach, D. M. Raup and J. W. Valentine. 1981. Phanerozoic marine diversity and the fossil record. *Nature* 293: 435–437.

Shane, C. K. and E. J. Cushing. 1991. *Quaternary Landscapes.* Minneapolis: University of Minnesota Press.

Sharpton, V. L., G. B. Dalrymple, L. E. Marin, G. Ryder and B. C. Schuraytz. 1992. New links between the Chicxulub impact structure and the Cretaceous/Tertiary boundary. *Nature* 359: 819.

Shipley, B. and P. A. Keddy. 1987. The individualistic and community-unit concepts as falsifiable hypothesis. *Vegetatio* 69: 47–55.

Short, J. and A. Smith. 1994. Mammal decline and recovery in Australia. *Journal of Mammalogy* 75: 288–297.

Shreve, F. 1921. Conditions indirectly affecting vertical distribution on desert mountains. *Ecology* 3: 269–274.

Shrode, J. B. 1975. Developmental temperature tolerance of a Death Valley pupfish (*Cyprirodon nevadensis*). *Physiological Zoology* 48: 378–389.

Shuntov, V. P. 1974. *Sea Birds and the Biological Structure of the Ocean.* Washington, DC: National Information Service, U. S. Department of Commerce.

Sibley, C. G. 1970. A comparative study of the egg-white proteins of passerine birds. *Bulletin of the Peabody Museum of Natural History, Yale University* 32: 1–131.

Sibley, C. G. and J. E. Ahlquist. 1990. *Phylogeny and Classification of Birds.* New Haven, CT: Yale University Press.

Sibley, C. G. and B. L. Monroe. 1990. *Distribution and Taxonomy of Birds of the World.* New Haven, CT: Yale University Press.

Siemann, E., D. Tilman and J. Haarstad. 1996. Insect species diversity, abundance and body size relationships. *Nature* 380: 704–706.

Silva, M. and J. A. Downing. 1995. The allometric scaling of density and body size: A non-linear relationship for terrestrial mammals. *American Naturalist* 141: 704–727.

Silva, M. N. F. and J. L. Patton. 1993. Amazonian phylogeography: mtDNA sequence variation in arboreal echimyid rodents (Caviomorpha). *Molecular Phylogenetics and Evolution* 2: 243–255.

Silvertown, J. 1985. History of a latitudinal diversity gradient: Woody plants in Europe. *Journal of Biogeography* 12: 519–525.

Simberloff, D. S. 1970. Taxonomic diversity of island biotas. *Evolution* 24: 23–47.

Simberloff, D. S. 1974a. Equilibrium theory of island biogeography and ecology. *Annual Review of Ecology and Systematics* 5: 161–182.

Simberloff, D. S. 1974b. Permo-Triassic extinctions: Effects of area on biotic equilibrium. *Journal of Geology* 82: 267–274.

Simberloff, D. S. 1976. Trophic structure determination and equilibrium in an arthropod community. *Ecology* 57: 395–398.

Simberloff, D. S. 1978. Using island biogeographic distributions to determine if colonization is stochastic. *American Naturalist* 112: 713–726.

Simberloff, D. S. 1988. Contribution of population and community biology to conservation science. *Annual Review of Ecology and Systematics* 19: 473–511.

Simberloff, D. S. and L. G. Abele. 1982. Refuge design and island biogeographic theory: Effects of fragmentation. *American Naturalist* 120: 41–50.

Simberloff, D. S. and W. Boecklen. 1991. Patterns of extinction in the introduced Hawaiian avifauna: A re-examination of the role of competition. *American Naturalist* 138: 300–327.

Simberloff, D. S. and J. Martin. 1991. Nestedness of insular avifaunas: Simple summary statistics masking complex species patterns. *Ornis Fennica* 68: 178–192.

Simberloff, D. S. and E. O. Wilson. 1969. Experimental zoogeography of islands: The colonization of empty islands. *Ecology* 50: 278–296.

Simberloff, D. S. and E. O. Wilson. 1970. Experimental zoogeography of islands: A two-year record of colonization. *Ecology* 51: 934–937.

Simberloff, D. S., K. L. Heck, E. D. McCoy and E. F. Conner. 1981a. There have been no statistical tests of cladistic biogeographical hypothesis. In G. Nelson and D. E. Rosen (eds.), *Vicariance Biogeography: A Critique*, 40–64. New York: Columbia University Press.

Simberloff, D. S., K. L. Heck, E. D. McCoy and E. F. Conner. 1981b. Response. In G. Nelson and D. E. Rosen (eds.), *Vicariance Biogeography: A Critique*, 85–93. New York: Columbia University Press.

Simberloff, D. S., J. Cox and D. W. Mehlman. 1992. Movement corridors: Conservation bargains or poor investments? *Conservation Biology* 6: 493–504.

Simpson, B. B. 1974. Glacial migrations of plants: Island biogeographical evidence. *Science* 185: 697–700.

Simpson, B. B. 1975. Pleistocene changes in the flora of the high tropical Andes. *Paleobiology* 1: 273–294.

Simpson, G. G. 1936. Data on the relationships of local and continental mammal faunas. *Journal of Paleontology* 10: 410–414.

Simpson, G. G. 1940. Mammals and land bridges. *Journal of the Washington Academy of Science* 30: 137–163.

Simpson, G. G. 1943a. Mammals and the nature of continents. *American Journal of Science* 241: 1–31.

Simpson, G. G. 1943b. Turtles and the origin of the fauna of Latin America. *American Journal of Science* 241: 413–429.

Simpson, G. G. 1945. The principles of classification and a classification of mammals. *Bulletin of the American Museum of Natural History* 85: i–xvi, 1–350.

Simpson, G. G. 1952a. Periodicity in vertebrate evolution. *Journal of Paleontology* 26: 359–370.

Simpson, G. G. 1952b. Probabilities of dispersal in geologic time. *Bulletin of the American Museum of Natural History* 99: 163–176.

Simpson, G. G. 1953. *The Major Features of Evolution.* New York: Columbia University Press.

Simpson, G. G. 1956. Zoogeography of West Indian land mammals. *American Museum Novitates* no. 1759.

Simpson, G. G. 1961a. *Principles of Animal Taxonomy.* New York: Columbia University Press.

Simpson, G. G. 1961b. Historical zoogeography of Australian mammals. *Evolution* 15: 413–446.

Simpson, G. G. 1964. Species density of North American Recent mammals. *Systematic Zoology* 13: 57–73.

Simpson, G. G. 1965. *The Geography of Evolution.* Philadelphia: Chilton Book Co.

Simpson, G.G. 1969. South American mammals. In E. J. Fittkau, J. Illies, H. Klinge, G. H. Schwabe and H. Sioli (eds.), *Biogeography and Ecology in South America*, vol. 2, Monographiae Biologicae 20: 879–909. The Hague: Dr. W. Junk.

Simpson, G. G. 1970. Uniformitarianism: An inquiry into principle, theory, and method in geohistory and biohistory. In M. K. Hecht and W. C. Steere (eds.), *Essays in Evolution and Genetics in Honor of Theodosius Dobzhansky*, 43–96. New York: Appleton-Century-Crofts.

Simpson, G. G. 1978. Early mammals in South America: Fact, controversy, and mystery. *Proceedings of the American Philosophical Society* 122: 318–328.

Simpson, G. G. 1980a. *Why and How: Some Problems and Methods in Historical Biology.* Oxford: Pergamon Press.

Simpson, G. G. 1980b. *Splendid Isolation: The Curious History of Mammals in South America.* New Haven, CT: Yale University Press.

Sinclair, W. A. 1964. Comparisons of recent declines of white ash, oaks and sugar maple in northeastern woodlands. *Cornell Plantations.* 20: 62–67.

Skutch, A. F. 1949. Do tropical birds rear as many young as they can nourish? *Ibis* 91: 430–455.

Skutch, A. F. 1967. Adaptive limitation of the reproductive rate of birds. *Ibis* 109: 579–599.

Slatkin, M. 1973. Gene flow and selection in a cline. *Genetics* 75: 733–756.

Slaughter, B. H. 1967. Animal ranges as a cue to Late-Pleistocene extinctions. In P. S. Marion and H. E. Wright Jr. (eds.), *Pleistocene Extinctions: The Search for a Cause*, 155–167. New Haven, CT: Yale University Press.

Slobodkin, L. B. and H. L. Sanders. 1969. On the contribution of environmental predictability to species diversity. In G. M. Woodwell and H. H. Smith (eds.), *Diversity and Stability in Ecological Systems*, 82–93. Brookhaven Symp. Biol. 22.

Smith, A. T. 1974. The distribution and dispersal of pikas: Consequences of insular population structure. *Ecology* 55: 1112–1119.

Smith, A. T. 1980. Temporal changes in insular populations of the pika (*Ochotona princeps*). *Ecology* 61: 8–13.

Smith, C. H. 1983. A system of world mammal faunal regions I: Logical and statistical derivation of regions. *Journal of Biogeography* 10: 455–466.

Smith, F. A. 1992. Evolution of body size among woodrats from Baja California, Mexico. *Functional Ecology* 6: 265–273.

Smith, F. A., J. L. Betancourt and J. H. Brown. 1995. Evolution of body size in the woodrat over the past 25,000 years of climate change. *Science* 270: 2012–2014.

Smith, F. D. M., R. M. May, R. Pellew, T. H. Johnson and K. R. Walter. 1993. How much do we know about the current extinction rate? *Trends in Ecology and Evolution* 8(10): 375–377.

Smith, G. B. 1979. Relationships of eastern Gulf of Mexico reef-fish communities to the species equilibrium theory of insular biogeography. *Journal of Biogeography* 6: 49–61.

Smith, G. R. 1978. Biogeography of intermountain fishes. *Great Basin Naturalist Memoirs* 2: 17–42.

Smith, G. R. 1981. Late Cenozoic freshwater fishes of North America. *Annual Review of Ecology and Systematics* 12: 163–193.

Smith, J. M. B. 1994. Patterns of disseminule dispersal by drift in the northwest Coral Sea. *New Zealand Journal of Botany* 30: 57–67.

Smith, J. M. B., H. Heatwole, M. Jones and B. M. Waterhouse. 1990. Drift disseminules on cays of the Swain Reefs, Great Barrier Reef, Australia. *Journal of Biogeography* 17: 5–17.

Smith, M. H. and J. T. McGinnis. 1968. Relationships of latitude, altitude and body size and mean annual production of offspring in *Permoyscus*. *Research on Population Ecology* 10: 115–126.

Snead, R. E. 1980. *World Atlas of Geomorphic Features*. Huntington, NY: Krieger.

Sneath, P. H. A. and R. R. Sokal. 1973. *Numerical Taxonomy*. San Francisco: W. H. Freeman.

Snelgrove, P. V. R., J. F. Grassle and R. F. Petrecca. 1992. The role of food patches in maintaining deep-sea diversity: Field experiments with hydrodynamically unbiased colonization trays. *Limnology and Oceanography* 37: 1543–1550.

Snelgrove, P. V. R., J. F. Grassle and R. F. Petrecca. 1996. Experimental evidence for aging food patches as a factor contributing to deep-sea macrofaunal diversity. *Limnology and Oceanography* 41: 605–614.

Solbrig, O. T. 1972. The floristic disjunctions between the "Monte" in Argentina and the "Sonoran Desert" in Mexico and the United States. *Annals of the Missouri Botanical Garden* 59: 218–223.

Solem, A. 1959. Zoogeography of the land and freshwater mollusca of the New Herbrides. *Fieldiana Zoology* 43: 239–343.

Solem, A. 1981. Land-snail biogeography: A true snail's pace of change. In G. Nelson and D. E. Rosen (eds.), *Vicariance Biogeography: A Critique*, 197–221. New York: Columbia University Press.

Soltz, D. L. and R. J. Naiman. 1978. *The Natural History of Native Fishes in the Death Valley System*. Los Angeles, CA: Natural History Museum of Los Angeles County.

Sondaar, P. Y. 1977. Insularity and its effect on mammal evolution. In M. K. Hecht, P. C. Goody and B. M. Hecht (eds.), *Major Patterns in Vertebrate Evolution*, 671–707. New York: Plenum.

Soulé, M. E. 1966. Trends in insular radiation of a lizard. *American Midland Naturalist* 100: 47–64.

Soulé, M. E. 1987. *Viable Populations for Conservation*. New York: Cambridge University Press.

Soulé, M. E. and M. E. Gilpin. 1991. The theory of wildlife corridor capability. In D. A. Saunders and R. J. Hobbs (eds.), *The Role of Corridors in Nature Conservation*, 3–8. Chipping Norton, NSW, Australia: Surrey Beatty.

Soulé, M. E. and D. S. Simberloff. 1986. What do genetics and ecology tell us about the design of nature reserves? *Biological Conservation* 35: 19–40.

Southern California Coastal Water Research Project. 1973. *The Ecology of the Southern California Bight: Implications for Water Quality Management*. Technical Report 104: 1–531. El Segundo, CA.

Souza, W. P., S. C. Schroeter and S. D. Gaines. 1981. Latitudinal variation in intertidal algal community structure. *Oecologia* 48(3): 297–303.

Spaulding, W. G. and L. J. Graumlich. 1986. The plast pluvial climatic episodes in the deserts of Southwestern North America. *Nature* 320: 441–444.

Spencer, A. W. and H. W. Steinhoff. 1968. An explanation of geographical variation in litter size. *Journal of Mammalogy* 49: 281–286.

Stanley, S. M. 1975. A theory of evolution above the species level. *Proceedings of the National Acadedmy of Sciences, USA* 72: 646–650.

Stanley, S. M. 1979. *Macroevolution: Pattern and Process*. San Francisco: W. H. Freeman.

Stanley, S. M. 1984. Marine mass extinctions: a dominant role for temperature. In R. H. Nitecki, *Extinctions*, 69–117. Chicago, IL: University of Chicago Press.

Stanley, S. M. 1987. *Extinction*. New York: Scientific American Books, Inc.

Steadman, D. W. 1986. Two new species of rails (Aves: Rallidae) from Mangaia, southern Cook Islands. *Pacific Science* 40: 27–43.

Steadman, D. W. 1987. California condor associated with spruce-jack pine woodland in the late-Pleistocene of New York. *Quaternary Research* 28: 415–426.

Steadman, D. W. 1989. Extinction of birds in eastern Polynesia: A review of the record, and comparisons with other Pacific islands groups. *Journal of Archaeological Science* 16: 177–205.

Steadman, D. W. 1993. Biogeography of Tongan birds before and after human contact. *Proceedings of the National Academy of Sciences* 90: 818–822.

Steadman, D. W. 1995. Prehistoric extinctions of Pacific island birds: Biodiversity meets zooarchaeology. *Science* 267: 1123–1131.

Steadman, D. W. and P. S. Martin. 1984. Extinctions of birds in the late Pleistocene of North America. In P. S. Martin and R. J. Klein (eds.), *Quaternary Extinctions*, 466–477. Tucson: University of Arizona Press.

Steadman, D. W. and S. L. Olson. 1985. Bird remains from an archaeological site on Henderson Island, South Pacific. *Acta XX Congresus Internationalis Ornithologici* 2: 361–382.

Stebbins, G. L. 1971a. Adaptive radiation of reproductive characteristics in angiosperms. 2. Seeds and seedlings. *Annual Review of Ecology and Systematics* 2: 237–260.

Stebbins, G. L. 1971b. *Chromosomal Evolution in Higher Plants.* London: Edward Arnold.

Steenbergh, W. F. and C. H. Lowe. 1976. Ecology of the saguaro. I. The role of freezing weather in a warm desert plant population. In *Research in the Parks: National Park Service Symposium,* ser. no. 1: 49–92. Washington, DC: U.S. Government Printing Office.

Steenbergh, W. F. and C. H. Lowe. 1977. Ecology of the saguaro. II. Reproduction, germination, establishment, growth, and survival of the young plant. *National Park Service Scientific Monographs,* ser. no. 8. Washington, DC: U.S. Government Printing Office.

Stehli, F. G. 1968. Taxonomic diversity gradients in pole locations: The recent model. In E. T. Drake (ed.), *Evolution and Environment,* 163–227. Peabody Museum Centennial Symposium. New Haven, CT: Yale Unversity Press.

Stehli, F. G. and S. D. Webb (eds.). 1985. *The Great American Biotic Interchange.* New York: Plenum Press.

Stehli, F. G. and J. W. Wells. 1971. Diversity and age patterns in hermatypic corals. *Systematic Zoology* 20: 115–126.

Stehli, F. G., R. G. Douglas and N. D. Newell. 1969. Generation and maintenance of gradients in taxonomic diversity. *Science* 164: 947–949.

Steinauer, E. M. and S. L. Collins. 1996. Prairie ecology: The tallgrass prairie. In F. B. Samson and F. L. Knopf (eds.), *Prairie Conservation,* 39–52. Washington, DC: Island Press.

Stenseth, N. C. and W. Z. Lidicker, Jr. 1992. *Animal Dispersal: Small Mammals as a Model.* New York: Chapman and Hall.

Stenthouarskis, S. 1992. Altitudinal effect on species richness of *Oniscidea* (Crustacea; Isopoda) on three mountains in Greece. *Global Ecology and Biogeography Letters* 2: 157–164.

Sterba, G. 1966. *Freshwater Fishes of the World.* London: Studio Vista.

Stevens, G. C. 1989. The latitudinal gradients in geographic range: How so many species coexist in the tropics. *American Naturalist* 132: 240–256.

Stevens, G. C. 1992. The elevational gradient in altitudinal range: An extension of Rapoport's latitudinal rule to altitude. *American Naturalist* 140: 893–911.

Stevens, G. C. 1996. Extending Rapoport's rule to Pacific marine fishes. *Journal of Biogeography* 23: 149–154.

Stevens, G. C. and B. J. Enquist. 1998. Macroecological limits to the abundance and distribution of *Pinus.* In D. M. Richardson (ed.), *Ecology and Biogeography of the Genus* Pinus, 183–190. Cambridge: Cambridge University Press.

Stevens, G. C. and J. F. Fox. 1991. The causes of treeline. *Annual Review of Ecology and Systematics* 22: 177–191.

Stevenson, R. D. 1986. Allen's rule in North American rabbits (*Sylvilagus*) and hares (*Lepus*) is an exception, not a rule. *Journal of Mammalogy* 67: 312–316.

Stodart, E. and I. Parer. 1988. Colonization of Australia by the rabbit, *Oryctolagus cunicularis.* Project Report no. 6, *CSIRO Division of Wildlife and Ecology.*

Stone, L., T. Dayan and D. Simberloff. 1996. Community-wide assembly patterns unmasked: The importance of species' differing geographical ranges. *American Naturalist* 148: 997–1015.

Storey, B. C. 1995. The role of mantle plumes in continental breakup: Case histories from Gondwanaland. *Nature* 377: 301–308.

Stouffer, P. C. and R. O. Birregaard, Jr. 1995. Use of Amazonian forest fragments by understory insectivorous birds. *Ecology* 76: 2429–2445.

Stout, J. and J. Vandermeer. 1975. Comparison of species richness for stream-inhabiting insects in tropical and mid-latitude streams. *American Naturalist* 109: 263–280.

Strahler, A. N. 1975. *Physical Geography,* 4th edition. New York: John Wiley & Sons, Inc.

Strauss, S. Y. 1991. Indirect effects in community ecology: Their definition, study and importance. *Trends in Ecology and Evolution* 6: 206–210.

Strayer, D. L. 1991. Projected distribution of zebra mussel, *Dreissena polymorpha,* in North America. *Canadian Journal of Fisheries and Aquatic Sciences* 48: 1389–1395.

Strecker, U., C. G. Meyer, C. Sturmbauer and H. Wilkens. 1996. Genetic divergence and speciation in an extremely young species flock in Mexico formed by the genus *Cyprinodon* (Cyprinodontidae, Teleostei). *Molecular Phylogenetics and Evolution* 6: 143–149.

Strong, D. R. Jr. 1974. Rapid asymptotic species accumulation in phytophagous insect communities: The pests of cacao. *Science* 185: 1064–1066.

Strong, D. R. Jr. and J. R. Rey. 1982. Testing for MacArthur-Wilson equilibrium with the arthropods of the miniature *Spartina* archipelago at Oyster Bay, Florida. *American Zoologist* 22: 355–360.

Strong, D. R., E. D. McCoy and J. R. Rey. 1977. Time and the number of herbivore species: The pests of sugarcane. *Ecology* 58: 167–175.

Strong, D. R., D. Simberloff, L. G. Abele and A. B. Thistel (eds). 1984. *Ecological Communities: Conceptual Issues and the Evidence.* Princeton, NJ: Princeton University Press.

Stuart, C. T. and M. A. Rex. 1994. The relationship between developmental pattern and species diversity in deep-sea prosobranch snails. In C. M. Young and K. J. Eckelbarger (eds.), *Reproduction, Larval Biology, and Recruitment of the Deep-Sea Benthos,* 119–136. New York: Columbia University Press.

Stute, M., M. Forster, H. Frischkorn, A. Serejo, J. F. Clark, P. Schlosser, W. S. Broeker and G. Bonani. 1995. Cooling of tropical Brazil: 5°C during the last glacial maximum. *Science* 269: 379–380.

Stutz, H. C. 1978. Explosive evolution of perennial *Atriplex* in western America. In K. T. Harper and J. L. Reveal (eds.), *Intermountain Biogeography: A Symposium,* 161–168. Great Basin Naturalist Memoirs. Provo, UT: Brigham Young University.

Sugihara, G. 1981. $S = CAz$, $z = 1/4$: A reply to Connor and McCoy. *American Naturalist* 117: 790–793.

Sullivan, R. M. and T. L. Yates. 1995. Population genetics and conservation biology of relict populations of red squirrels. In C. A. Istock and R. S. Hoffman (eds.), *Storm over a Mountain Island: Conservation Biology and the Mt. Graham Affair,* 193–208. Tucson: University of Arizona Press.

Sullivan, R. M., D. J. Hafner and T. L. Yates. 1986. Genetics of a contact zone between two chromosomal forms of the grasshopper mouse (genus *Onychomys*): A reassessment. *Journal of Mammalogy* 67: 640–659.

Sumner, F. B. 1932. Genetic, distributional and evolutionary studies of the subspecies of deer mice (*Permoscus*). *Bibliographia Genetica* 9: 1–106.

Sunquist, M., C. Leh, F. Sunquist, D. M. Mills and R. Rajaratnam. 1994. Rediscovery of the Bornean bay cat. *Oryx* 28: 67–69.

Swisher, C. C. III, J. M. Grajales-Nishimura, A. Montanari, S. V. Margolis, P. Claeys, W. Alvarez, P. Renne, E. Cedillo-Pardo, F. J.-M. R. Maurrasse, G. H. Curtis, J. Smit and M. O. McWilliams. 1992. Coeval 40AR/39AR ages of 65.0 million years ago from Chicxulub crater melt rock and Cretaceous-Tertiary boundary tektites. *Science* 257: 954–958.

Swofford, D. L. 1993. *PAUP: Phylogenetic Analysis Using Parsimony, Version 3.1.1.* Champaign: Illinois Natural History Survey.

Taitt, M. J. and C. J. Krebs. 1985. Population dynamics and cycles. In R. H. Tamarin (ed.), *Biology of New World Microtus,* 567–620. American Society of Mammalogists, Special Publication no. 8.

Tamarin, R. H. 1977. Dispersal in island and mainland voles. *Ecology* 58: 1044–1054.

Taper, M. L. and T. J. Case. 1992. Models of character displacement and the theoretical robustness of taxon cycles. *Evolution* 46: 317–333.

Tarling, D. H. 1962. Tentative correlation of Samoan and Hawaiian Islands using "reversals" of magnetism. *Nature* 196: 882–883.

Tarr, C. L. and R. C. Fleischer. 1995. Evolutionary relationships of the Hawaiian honeycreepers (Aves, Drepanidinae). In W. L. Wagner and V. A. Funk (eds.), *Hawaiian Biogeography: Evolution on a Hot Spot Archipelago.* Washington, DC: Smithsonian Institution Press.

Taulman, J. F. and L. W. Robbins. 1996. Recent range expansion and distributional limits of the nine-banded armadillo (*Dasypus novemcinctus*) in the United States. *Journal of Biogeography* 23: 635–648.

Taylor, B. 1991. Investigating species incidence over habitat fragments of different areas: A look at error estimation. *Biological Journal of the Linnean Society* 42: 477–491.

Taylor, C. M. 1996. Abundance and distribution within a guild of benthic stream fishes: Local processes and regional patterns. *Freshwater Biology* 36: 385–396.

Taylor, C. 1997. Fish species richness and incidence patterns in isolated and connected stream pools: Effects of pool volume and spatial position. *Oecologia* 110: 560–566.

Taylor, C. M. and N. J. Gotelli. 1994. The macroecology of *Cyprinella*: Correlates of phylogeny, body-size, and geographical range. *American Naturalist* 144: 549–569.

Taylor, D. W. 1996. *Flowering Plant Origin, Evolution, and Phylogeny.* New York: Chapman and Hall.

Taylor, F. B. 1910. Bearing of the Tertiary mountain belt on the origin of the earth's plan. *Geological Society of America Bulletin* 21: 179–226.

Taylor, J. D. and C. N. Taylor. 1977. Latitudinal distribution of predatory gastropods on the eastern Atlantic shelf. *Journal of Biogeography* 4: 73–81.

Taylor, J. M., S. C. Smith and J. H. Calaby. 1985. Altitudinal distribution and body size among New Guinean *Rattus* (Rodentia, Muridae). *Journal of Mammalogy* 66: 353–358.

Taylor, R. J. and P. J. Regal. 1978. The peninsular effect on species diversity and the biogeography of Baja California. *American Naturalist* 112: 583–593.

Teal, J. M. 1957. Community metabolism in a temperate cold spring. *Ecological Monographs* 27: 283–302.

Teal, J. M. 1962. Energy flow in the salt marsh ecosystem of Georgia. *Ecology* 43: 614–624.

Tegelstrom, H. and L. Hansson. 1987. Evidence for long distance dispersal in the common shrew (*Sorex araneus*). *Sonderdunk aus Zeitschrift fur Saugetierkunde* 52: 52–54.

Temple, S. 1986. The problem of avian extinctions. *Current Ornithology* 3: 453–485.

Templeton, A. R. 1980a. The theory of speciation via the founder principle. *Genetics* 94: 1011–1038.

Templeton, A. R. 1980b. Modes of speciation and inferences based on genetic distances. *Evolution* 34: 719–729.

Templeton, A. R. 1981. Mechanisms of speciation: A population genetic approach. *Annual Review of Ecology and Systematics* 12: 23–48.

Terborgh, J. 1973a. Chance, habitat, and dispersal in the distribution of birds in the West Indies. *Evolution* 27: 338–349.

Terborgh, J. 1973b. On the notion of favorableness in plant ecology. *American Naturalist* 107: 481–501.

Terborgh, J. 1974. Preservation of natural diversity: The problem of extinction prone species. *Bioscience* 24: 715–722.

Terborgh, J. 1977. Bird species diversity on an Andean elevational gradient. *Ecology* 58: 1007–1019.

Terborgh, J. 1986. Keystone plant resources in tropical forest. In M. E. Soulé (ed.), *Conservation Biology: The Science of Scarcity and Diversity,* 330–344. Sunderland, MA: Sinauer Associates.

Terborgh, J. 1992. *Diversity and the Tropical Rain Forest.* New York: Scientific American Library.

Terborgh, J. and J. Faaborg. 1973. Turnover and ecological release in the avifauna of Mona Island, Puerto Rico. *Auk* 90: 759–779.

Terborgh, J. and B. Winter. 1980. Some causes of extinction. In M. E. Soulé and B. A. Wilcox, (eds.), *Conservation Biology: An Ecological-Evolutionary Perspective,* 119–134. Sunderland, MA: Sinauer Associates.

Terborgh, J., L. Lopez and J. Tellos. 1997. Bird communities in transition: The Lago Guri Islands. *Ecology* 78: 1494–1501.

Terrell, J. 1986. *Prehistory in the Pacific Islands: A Study of Variation in Language, Customs, and Human Biology.* Cambridge: Cambridge University Press.

Thomas, W. L. 1956. *Man's Role in Changing the Face of the Earth.* Chicago: Univeristy of Chicago Press.

Thompson, D. B. 1990. Different spatial scales of adaptation in the climbing behavior of *Permyscus maniculatus*: Geographic variation, natural selection, and gene flow. *Evolution* 44: 952–965.

Thompson, D. Q., R. L. Stuckey and E. B. Thompson. 1987. *Spread, Impact, and Control of Purple Loosestrife* (Lythrum saclicaria) *in North American Wetlands.* Washington, DC: U.S. Fish and Wildlife Service, Research 2.

Thompson, R. S. and J. I. Mead. 1982. Late Quaternary environments and biogeography in the Great Basin. *Quaternary Research* 17: 39–55.

Thornton, I. W. B. 1992. K. W. Dammerman: Forerunner of island theory? *Global Ecology and Biogeography Letters* 2: 145–148.

Thornton, I. 1996. *Krakatau: The Destruction and Reassembly of an Island Ecosystem.* Cambridge, MA: Harvard University Press.

Thornton, I. W. B., T. R. New, R. A. Zann and P. A. Rawlinson. 1990. Colonization of the Krakatau Islands by animals: Perspective from the 1980s. *Philosophical Transactions of the Royal Society of London* Series B 328: 131–165.

Thornton, I. W. B., R. A. Zann and S. van Balen. 1993. Colonization of Rakata (Krakatau Is.) by non-migrant land birds from 1883 to 1992 and implications for the value of island biogeography theory. *Journal of Biogeography* 20: 441–452.

Thrower, N. J. W. and D. E. Bradbury (eds.). 1977. *Chile-California Mediterranean Scrub Atlas: A Comparative Analysis.* Stroudsburg, PA: Dowden, Hutchinson & Ross.

Thurber, J. M. and R. O. Peterson. 1991. Changes in body size associated with range expansion in the coyote (*Canis latrans*). *Journal of Mammalogy* 72: 750–755.

Tilman, D. 1988. *Plant Strategies and the Dynamics and Structure of Plant Communities.* Princeton, NJ: Princeton University Press.

Tilman, D., R. M. May, C. L. Lehman and M. A. Nowak. 1994. Habitat destruction and the extinction debt. *Nature* 371: 65–66.

Titus, J. G. 1990. Greenhouse effects, sea level rise and land use. *Land Use Policy* 7: 138–153.

Tivy, J. 1993. *Biogeography: A Study of Plants in the Ecosphere.* New York: John Wiley and Sons.

Tonn, W. M. and J. J. Magnunson. 1982. Patterns in the species composition and richness of fish assemblages in northern Wisconsin lakes. *Ecology* 63: 1149–1166.

Towns, D. R. and W. J. Balantine. 1993. Conservation and restoration of New Zealand island ecosystems. *Trends in Ecology and Evolution* 8: 452–457.

Tracy, C. R. and T. L. George. 1992. On the determinants of extinction. *American Naturalist* 139: 102–122.

Tracy, C. R. and T. L. George. 1993. Extinction probabilities of British island birds: A reply. *American Naturalist* 142: 1036–1037.

Trewick, S. A. 1996. Morphology and evolution of two takahe: Flightless rails of New Zealand. *Journal of Zoology* (London) 238: 221.

Tull, D. S. and K. Bohning-Gaese. 1993. Patterns of drilling predation on gastropods of the family Turritellidae in the Gulf of California. *Paleobiology* 19: 476–486.

Tunnicliffe, V. and C. M. R. Fowler. 1996. Influence of sea-floor spreading on the global hydrothermal vent fauna. *Nature* 379: 531–533.

Turner, J. R. G. 1981. Adaptation and evolution in *Heliconius:* A defense of neo-Darwinism. *Annual Review of Ecology and Systematics* 12: 99–121.

Turner, R. M., J. E. Bowers and T. L. Burgess. 1995. *Sonoran Desert Plants: An Ecological Atlas.* Tucson: University of Arizona Press.

Turrill, W. B. 1953. *Pioneer Plant Geography: The Phytogeography Researches of Sir Joseph Dalton Hooker.* The Hague: Martinus Nijhoff.

Udvardy, M. D. F. 1969. *Dynamic Zoogeography.* New York: Van Nostrand Reinhold.

Upton, G. and B. Fingleton. 1990. *Spatial Data Analysis by Example.* New York: John Wiley & Sons.

Urquhart, F. A. 1960. *The Monarch Butterfly.* Toronto: University of Toronto Press.

Usher, M. B. 1979. Markovian approaches to ecological succession. *Journal of Animal Ecology* 48: 413–426.

Vagvolygi, J. 1975. Body size, aerial dispersal, and origin of the Pacific land snail fauna. *Systematic Zoology* 24: 465–488.

Vaisanen, R. and K. Heliovaara. 1994. Hot-spots of insect diversity in northern Europe. *Annales Zoologici Fennici.*

Valentine, J. W. 1961. Paleoecologic molluscan geography of the Californian Pleistocene. *University of California Publications in Geological Science* 34: 309–442.

Valentine, J. W. 1966. Numerical analysis of marine molluscan ranges on the extratropical northeastern Pacific shelf. *Limnology and Oceanography* 11: 198–211.

Valentine, J. W. 1989. Phanerozoic marine faunas and the stability of the Earth system. *Global and Planetary Change* 1: 137–155.

Valentine, J. W. and D. Jablonski. 1982. Larval strategies and patterns of brachiopod diversity in space and time. *Geological Society of America Abstracts* 14: 241.

Valentine, J. W. and D. Jablonski. 1993. Fossil communities: Compositional variation at many time scales. In R. E. Ricklefs and D. Schluter (eds.), *Species Diversity in Ecological Communities: Historical and Geographical Perspectives,* 341–349. Chicago: University of Chicago Press.

Valone, T. J. and J. H. Brown. 1995. Effects of competition, colonization, and extinction on rodent species diversity. *Science* 267: 880–883.

Van Den Bosch, F., R. Hengeveld and J. A. J. Metz. 1992. Analyzing the velocity of animal range expansion. *Journal of Biogeography* 19: 135–150.

Van der Hammen, T. 1982. Paleoecology of tropical South America. In G. T. Prance (ed.), *Biological Diversification in the Tropics,* 60–66. New York: Columbia University Press.

Van der Hammen, T., T. A. Wijmstra and W. H. Zagwijn. 1971. The floral record of the late Cenozoic of Europe. In K. K. Turekian (ed.), *The Late Cenozoic Glacial Ages,* 391–424. New Haven, CT: Yale University Press.

Van Devender, T. R. 1977. Holocene woodlands in the southwestern deserts. *Science* 198: 189–192.

Van Devender, T. R. and W. G. Spaulding. 1979. Development of vegetation and climate in the southwestern United States. *Science* 204: 701–710.

Vanni, M. J. and D. L. Findlay. 1990. Trophic cascades and phytoplankton community structure. *Ecology* 71: 921–937.

Van Riper, C., S. G. Van Riper, M. L. Goff and M. Laird. 1986. The epizootiology and ecological significance of malaria in Hawaiian islands. *Ecological Monographs* 56: 327–344.

Van Valen, L. 1970. Late Pleistocene extinctions. In *Proceedings of the North American Pelontologist Conference,* 469–485. Lawrence, KS: Allen Press.

Van Valen, L. 1973a. Body size and the number of plants and animals. *Evolution* 27: 27–35.

Van Valen, L. 1973b. A new evolutionary law. *Evolutionary Theory* 1: 1–33.

Van Voorhies, W. A. 1996. Bergmann size clines: A simple explanation for their occurrence in ectotherms. *Evolution* 50: 1259–1264.

Vanzolini, P. E. and W. R. Heyer. 1985. The American herptofauna and the interchange. In F. G. Stehli and S. D. Webb (eds.), *The Great American Interchange,* 475–483. New York: Plenum Press.

Vanzolini, P. E. and E. E. Williams. 1970. South American anoles: The geographic differentiation and evolution of the *Anolis chtysolepis* species group (Sauria, Iguanidae). *Arq. Zool.* (São Paulo) 19: 1–298.

Varley, G. C. 1970. The concept of energy flow applied to a woodland community. In A. Watson (ed.), *Animal Populations in Relation to Their Food Resources,* 389–405. London: Blackwell Scientific Publications.

Vartanyan, S. L., V. E. Garutt and A. V. Sherr. 1993. Holocene dwarf mammoths from Wrangel Island in the Siberian Arctic. *Nature* 362: 337–340.

Vartanyan, S. L., K. A. Arslanov, T. V. Tertychnaya and S. B. Chernov. 1995. Radiocarbon dating evidence for mammoths on Wrangel Island, Arctic Ocean, until 2000 B.C. *Radiocarbon* 37: 1–6.

Vaughan, T. A. 1967. Two parapatric species of pocket gophers. *Evolution* 21: 148–158.

Vaughan, T. A. and R. M. Hansen. 1964. Experiments on interspecific competition between two species of pocket gophers. *American Midland Naturalist* 72: 444–452.

Vermeij, G. J. 1974. Marine faunal dominance and molluscan shell form. *Evolution* 28: 656–664.

Vermeij, G. J. 1978. *Biogeography and Adaptation: Patterns of Marine Life.* Cambridge, MA: Harvard University Press.

Vermeij, G. J. 1991a. Anatomy of an invasion: The trans-Arctic interchange. *Paleobiology* 17: 281–307.

Vermeij, G. J. 1991b. When biotas meet: Understanding biotic interchange. *Science* 253: 1099–1103.

Vine, F. J. and D. H. Matthews. 1963. Magnetic anomalies over a young oceanic ridge. *Nature* 199: 947–949.

Vinogradova, N. G. 1962. Vertical zonation in the distribution of deep-sea benthic fauna in the ocean. *Deep-Sea Research* 8: 245–250.

Vitousek, P. M. 1992. Global environmental change: An introduction. *Annual Review of Ecology and Systematics* 23: 1–14.

Vitousek, P. M., P. R. Ehrlich, A. H. Ehrlich and P. A. Matson. 1986. Human appropriation of the products of photosynthesis. *BioScience* 36: 368.

Vitousek, P. M., L. L. Loope and H. Andersen. 1995. *Islands: Biological Diversity and Ecosystem Function*. Berlin: Springer Verlag.

Vitousek, P. M., C. M. D'Antonio, L. L. Loope and R. Westbrooks. 1996. Biological invasions as a global environmental challenge. *American Scientist* 84: 468–478.

Vitousek, P. M., H. A. Mooney, J. Lubchenco and J. M. Melillo. 1997. Human domination of earth's ecosystems. *Science* 277: 494–499.

von Broembsen, S. L. 1989. Invasions of natural ecosystems by plant pathogens. In J. A. Drake et al. (eds.), *Biological Invasions: A Global Perspective*, 77–84. New York: John Wiley & Sons.

Voris, H. K. 1977. A phylogeny of the sea snakes (Hydrophiidae). *Fieldiana Zoology* 70(4): 79–166.

Vuilleumier, F. 1970. Insular biogeography in continental species. I. The Northern Andes of South America. *American Naturalist* 104: 373–388.

Vuilleumier, F. 1973. Insular biogeography in continental regions. II. Cave faunas from Tesin, southern Switzerland. *Systematic Zoology* 22: 64–76.

Vuilleumier, F. 1975. Zoogeography. In D. S. Farner and J. R. King (eds.), *Avian Biology*, vol. 5, 421–469. New York: Academic Press.

Vuilleumier, F. 1984. Faunal turnover and development of fossil avifaunas in South America. *Evolution* 38: 1384–1396.

Vuilleumier, F. 1985. Fossil and recent avifaunas and the Interamerican interchange. In F. G. Stehli and S. D. Webb (eds.), *The Great American Biotic Interchange*, 387–424. New York: Plenum Press.

Vuilleumier, F. and D. Simberloff. 1980. Ecology versus history as determinants of patchy and insular distributions in high Andean birds. In M. K. Hecht, W. C. Steere and B. Wallace (eds.), *Evolutionary Biology* 12, 235–379. New York: Plenum Press.

Wagner, W. H. Jr. 1995. Evolution of Hawaiian ferns and fern allies in relation to their conservation status. *Pacific Science* 49: 31–41.

Wagner, W. L. and V. A. Funk (eds.). 1995. *Hawaiian Biogeography: Evolution on a Hot Spot Archipelago*. Washington, DC: Smithsonian Institution Press.

Wahlberg, N., A. Moilanen and I. Hanski. 1996. Predicting the occurrence of endangered species in fragmented landscapes. *Science* 273: 1536–1538.

Wahrman, R., R. Goitein and E. Nevo. 1969. Mole rat *Spalax*: Evolutionary significance of chromosome variation. *Science* 164: 82–84.

Wake, D. B. 1966. Comparative osteology and evolution of the lungless salamanders, family Plethodontidae. *Memoirs of the Southern California Academy of Sciences* 4: 1–111.

Wallace, A. R. 1857. On the natural history of the Aru Islands. *Annals and Magazine of Natural History*, Supplement to Volume 20, December.

Wallace, A. R. 1860. On the zoological geography of the Malay Archipelago. *Journal of the Linnaean Society of London* 4: 172–184.

Wallace, A. R. 1869. *The Malay Archipelago: The Land of the Orangutan, and the Bird of Paradise*. New York: Harper.

Wallace, A. R. 1876. *The Geographical Distribution of Animals*. 2 vols. London: Macmillan.

Wallace, A. R. 1878. *Tropical Nature and Other Essays*. New York: Macmillan.

Wallace, A. R. 1880. *Island Life, or the Phenomena and Causes of Insular Faunas and Floras*. London: Macmillan.

Waloff, Z. 1966. The upsurges and recessions of the desert locust plague: An historical survey. *Anti-Locust Memoirs* 8: 1–111.

Warming, E. 1895. *Plantesamfund*. Copenhagen.

Warming, E. 1905. *Oecology of Plants: An Introduction to the Study of Plant Communities*. Oxford: Oxford University Press.

Warner, R. E. 1969. The role of introduced diseases in the extinction of the endemic Hawaiian avifauna. *Condor* 70: 101–120.

Waser, N. M., L. Chittka, M. V. Price, N. M. Williams and J. Ollerton. 1996. Generalization in pollination systems, and why it matters. *Ecology* 77: 1043–1060.

Watson, H. C. 1859. *Cybele Britannica, or British Plants and their Geographical Relations*. London: Longman and Company.

Watters, G. T. 1992. Unionids, fishes, and the species-area curve. *Journal of Biogeography* 19: 481–490.

Watts, D. 1979. The new biogeography and its niche in physical geography. *Geography* 78: 324–337.

Webb, S. D. 1985. Late Cenozoic mammal dispersals between the Americas. In F. G. Stehli and S. D. Webb (eds.), *The Great American Biotic Interchange*, 357–386. New York: Plenum Press.

Webb, S. D. 1991. Ecogeography and the great American interchange. *Paleobiology* 17: 266–280.

Webb, S. D. and A. D. Barnosky. 1989. Faunal dynamics of Pleistocene mammals. *Annual Reviews of Earth and Planetary Sciences* 17: 413–438.

Webb, S. D. and L. G. Marshall. 1982. Historical biogeography of recent South American land mammals. In M. A. Mares and H. H. Genoways (eds.), *Mammalian Biology in South America*, 39–52. Special Publication Series 6, Pymatuning Laboratory of Ecology, University of Pittsburgh.

Webb, T. III. 1987. The appearance and disappearance of major vegetational assemblages: Long-term vegetational dynamics in eastern North America. *Vegetatio* 69: 177–187.

Weber, F. R., T. D. Hamilton, D. M. Hopkins, C. A. Repenning and H. Haas. 1981. Canyon Creek: A Late Pleistocene vertebrate locality in interior Alaska. *Quaternary Research* 16: 167–180.

Wecker, S. C. 1963. The role of early experience in habitat selection by the prairie deer-mouse, *Peromyscus maniculatus bairdi*. *Ecological Monographs* 33: 307–325.

Wecker, S. C. 1964. Habitat selection. *Scientific American* 211: 109–116.

Wegener, A. 1912a. Die Entstehung der Kontinente. *Petermanns Geogr. Mitt.* 58: 185–195, 253–256, 305–308.

Wegener, A. 1912b. Die Entstehung der Kontinente. *Geol. Rundsch.* 3: 276–292.

Wegener, A. 1915. *Die Entstehung der Kontinente und Ozeane*. Braunschweig: Vieweg. [Other editions 1920, 1922, 1924, 1929, 1936.]

Wegener, A. 1966. *The Origin of Continents and Oceans*. New York: Dover Publications. [Translation of 1929 edition by J. Biram.]

Weigert, R. G., and D. F. Owen. 1971. Trophic structure, available resources and populations densities in terrestrial ecosystems versus aquatic ecosystems. *Journal of Theoretical Biology* 30: 69–81.

Weiher, E. and P. A. Keddy (eds). In press. *The Search for Assembly Rules in Ecological Communities*. New York: Cambridge University Press.

Weller, M. W. 1980. *The Island Waterfowl*. Ames: Iowa State University Press.

Wells, P. V. 1976. Macrofossil analysis of wood rat (*Neotoma*) middens as a key to the Quaternary vegetational history of arid America. *Quaternary Research* 6: 223–248.

Wells, P. V. 1978. Postglacial origin of the present Chihuahuan Desert less than 11,500 years ago. In R. H. Wauer and D. H. Riskind (eds.), *Transactions of the Symposium on the Biological Resources of the Chihuahuan Desert Region, United States and Mexico,* 67–83. U.S. Department of Interior, National Park Service Transactions and Proceedings, ser. no. 3.

Wells, P. V. 1979. An equable glaciopluvial in the West: Pleniglacial evidence of increased precipitation on a gradient from the Great Basin to the Sonoran and Chihuahuan Deserts. *Quaternary Research* 12: 311–325.

Wells, P. V. 1983. Paleobiogeography of montane islands in the Great Basin since the last glaciopluvial. *Ecological Monographs* 53: 341–382.

Wells, P. V. and R. Berger. 1967. Late Pleistocene history of coniferous woodland in the Mohave Desert. *Science* 155: 1640–1647.

Wenner, A. M. and D. L. Johnson. 1980. Land vertebrates on the California Channel Islands: Sweepstakes or bridges? In D. M. Power (ed.), *The California Islands: Proceedings of a Multi-Disciplinary Symposium,* 497–530. Santa Barbara Museum of Natural History.

Werger, M. J. A. and A. C. van Bruggen (eds.). 1978. *Biogeography and Ecology of Southern Africa.* 2 vols. Monographiae Biologicae 31. The Hague: Dr. W. Junk.

Werner, P. A. 1975 A seed trap for determining patterns of seed deposition in terrestrial plants. *Canadian Journal of Botany* 58: 810–813.

West, R. G. 1977. *Pleistocene Geology and Biology.* 2nd edition. London: Longman Group.

Westing, A. H. 1966. Sugar maple decline: An evaluation. *Economic Botany* 20: 196–212.

Wetzel, R. G. 1975. *Limnology.* Philadelphia: W. B. Saunders.

Whitaker, A. H. 1978. The effects of rodents on reptiles and amphibians. In P. R. Dingwall and I. A. Atkinson (eds.), *The Ecology and Control of Rodents in New Zealand's Nature Reserves,* 75–88. Department of Lands and Survey Information Series no. 4.

White, M. J. D. 1973. *Animal Cytology and Evolution.* 3rd edition. Cambridge: Cambridge University Press.

White, M. J. D. 1978. *Modes of Speciation.* San Francisco: W. H. Freeman.

White, T. C. R. 1976. Weather, food, and plagues of locusts. *Oecologia* 22: 119–134.

Whitehead, D. R. and C. E. Jones. 1969. Small islands and the equilibrium theory of insular biogeography. *Evolution* 23: 171–179.

Whiteside, M. C. and R. V. Harmsworth. 1967. Species diversity in chydorid (Cladocera) communities. *Ecology* 48: 664–667.

Whitford, P. C. 1979. An explanation of altitudinal deviations from Bergmann's rule as applied to birds. *The Biologist* 61: 1–10.

Whittaker, R. H. 1956. Vegetation of the Great Smoky Mountains. *Ecological Monographs* 22: 1–44.

Whittaker, R. H. 1960. Vegetation of the Siskiyou Mountains, Oregon and California. *Ecological Monographs* 30: 279–338.

Whittaker, R. H. 1967. Gradient analysis of vegetation. *Biological Review* 42: 207–264.

Whittaker, R. H. 1975. *Communities and Ecosystems.* 2nd edition. New York: Macmillan.

Whittaker, R. H. 1977. Evolution of species diversity in land communities. In M. K. Hecht, W. C. Steere and B. Wallace (eds.), *Evolutionary Biology* 10, 1–67. New York: Plenum Press.

Whittaker, A. H. 1978. The effects of rodents on reptiles and amphibians. In P. R. Dingwall and I. A. Atkinson (eds.), *The Ecology and Control of Rodents in New Zealand's Nature Reserves,* 75–88. Department of Lands and Survey Information Series no. 4.

Whittaker, R. J. and S. H. Jones. 1994. The role of frugivorous bats and birds in the rebuilding of a tropical forest ecosystem, Krakatau, Indonesia. *Journal of Biogeography* 21: 245–258.

Whittaker, R. H. and G. E. Likens. 1973. Carbon in the biota. In G. M. Woodwell and E. V. Pecan (eds.), *Carbon and the Biosphere,* 281–300. Conf. 72501. Springfield, VA: National Technical Information Service.

Whittaker, R. H. and W. A. Niering. 1965. Vegetation of the Santa Catalina Mountains, Arizona: A gradient analysis of the south slope. *Ecology* 46: 429–452.

Whittaker, R. H. and W. A. Niering. 1968. Vegetation of the Santa Catalina Mountains, Arizona. IV. Limestone and acid soils. *Journal of Ecology* 56: 523–544.

Whittaker, R. H. and W. A. Niering. 1975. Vegetation of the Santa Catalina Mountains, Arizona. V. Biomass, production and diversity along the elevation gradient. *Ecology* 56: 771–790.

Wiebe, P. H. 1976. The biology of cold-core rings. *Oceanus* 19: 69–76.

Wiebe, P. H. 1982. Rings of the gulf stream. *Scientific American* 246: 60–70.

Wiens, H. J. 1962. *Atoll Environment and Ecology.* New Haven, CT: Yale University Press.

Wilcove, D. S. 1987. From fragmentation to extinction. *Natural Areas Journal* 7: 23–29.

Wilcove, D. S., C. H. McLellan and A. P. Dobson. 1986. Habitat fragmentation in the temperate zone. In M. E. Soulé (ed.), *Conservation Biology: The Science of Scarcity and Diversity,* 237–256. Sunderland, MA: Sinauer Associates.

Wilcox, B. A. 1978. Supersaturated island faunas: A species-age relationship for lizards on post-Pleistocene land-bridge islands. *Science* 199: 996–998.

Wilcox, B. A. 1980. Insular ecology and conservation. In M. E. Soulé and B. A. Wilcox, (eds.), *Conservation Biology: An Ecological-Evolutionary Perspective,* 95–117. Sunderland, MA: Sinauer Associates.

Wilcox, B. A. and D. D. Murphy. 1985. Conservation strategy: The effects of fragmentation on extinction. *American Naturalist* 125: 879–887.

Wildenow, K. L. 1792. *Nomenclator botanicus sistens plantas omnes in Caroli a Linne Speciebus plantarum.* Halae Magdeb: Typis Io. Christ. Hendelii.

Wiley, E. O. 1981. *Phylogenetics: The Theory and Practice of Phylogenetic Systematics.* New York: Wiley-Interscience.

Wiley, E. O. 1988. Vicariance biogeography. *Annual Review of Ecology and Systematics* 19: 513–542.

Wiley, E. O. and R. L. Mayden. 1985. Species and speciation in phylogenetic systematics, with examples from the North American fish fauna. *Annals of the Missouri Botanical Garden* 72: 596–635.

Wilkinson, D. M. 1993. Equilibrium island biogeography: Its independent invention and the marketing of scientific theories. *Global Ecology and Biogeography Letters* 3: 65–66.

Williams, C. B. 1953. The relative abundance of different species in a wild animal population. *Journal of Animal Ecology* 22: 14–31.

Williams, C. B. 1964. *Patterns in the Balance of Nature and Related Problems in Quantitative Ecology.* New York: Academic Press.

Williams, E. E. 1976. West Indian anoles: A taxonomic and evolutionary summary. I. Introduction and a species list. *Breviora* no. 440.

Williams, E. E. 1983. Ecomorphs, faunas, island size and diverse end points in island radiations of *Anolis.* In R. B. Huey, E. R. Pianka and T. W. Schoener (eds.), *Lizard Ecology: Studies on a Model Organism.* Cambridge, MA: Belknap Press.

Williams, J. D. and R. J. Neves. 1995. Freshwater mussels: A neglected and declining aquatic resource. In E. T. Laroe (ed.), *Our Living Resources: A Report to the Nation on the Distribution, Abundance, and Health of U.S. Plants, Animals and Ecosystems*, 177–179. Washington, DC: U.S. Department of the Interior, National Biological Service.

Williams, J. E. and R. R. Miller. 1990. Conservation status of North American fish fauna in fresh water. *Journal of Fish Biology* 37: 79–85.

Williams, M. 1990. Clearing of the forests. In M. P. Conzen (ed.), *The Making of the American Landscape*, 146–168. Boston: Unwin Hyman.

Williamson, M. 1981. *Island Populations*. Oxford: Oxford University Press.

Williamson, M. 1989. The equilibrium theory today: True but trivial. *Journal of Biogeography* 16: 3–4.

Willig, M. R. and E. R. Sandlin. 1989. Gradients of species density and species turnover in New World bats: A comparison of quadrat and band methologies. In M. A. Mares and D. J. Schmidly (eds.), *Latin American Mammals: Their Conservation, Ecology, and Evolution*, 81–96. Norman, OK: University of Oklahoma Press.

Willig, M. R. and K. W. Selcer. 1989. Bat species density gradient in the New World: A statistical assessment. *Journal of Biogeography* 16: 189–195.

Willis, B. 1932. Isthmian links. *Bulletin of the Geological Society of America* 43: 917–952.

Willis, E. O. 1974. Populations and local extinctions of birds on Barro Colorado Island, Panama. *Ecological Monographs* 44: 153–169.

Willis, J. C. 1922. *Age and Area*. Cambridge: Cambridge University Press.

Wilson, D. S. 1975. The adequacy of body size as a niche difference. *American Naturalist* 109: 769–784.

Wilson, E. O. 1959. Adaptive shift and dispersal in a tropical ant fauna. *Evolution* 13: 122–144.

Wilson, E. O. 1961. The nature of the taxon cycle in the Melanesian ant fauna. *American Naturalist* 95: 169–193.

Wilson, E. O. 1975. *Sociobiology*. Cambridge, MA: Belknap Press.

Wilson, E. O. (ed.) 1988. *Biodiversity*. Washington, DC: National Academy Press.

Wilson, E. O. 1992. *The Diversity of Life*. Cambridge, MA: Belknap Press.

Wilson, E. O. and D. S. Simberloff. 1969. Experimental zoogeography of islands: Defaunation and monitoring techniques. *Ecology* 50: 267–278.

Wilson, E. O. and E. O. Willis. 1975. Applied biogeography. In M. L. Cody and J. M. Diamond (eds.), *Ecology and Evolution of Communities*, 522–534. Cambridge, MA: Belknap Press.

Wilson, J. T. 1963a. A possible origin of the Hawaiian Islands. *Canadian Journal of Physics* 41: 863–870.

Wilson, J. T. 1963b. Evidence from islands on the spreading of the ocean floors. *Nature* 197: 536–538.

Wilson, J. W. III. 1974. Analytical zoogeography of North American mammals. *Evolution* 28: 124–140.

Wilson, M. V. and A. Schmida. 1984. Measuring beta diversity with presence-absence data. *Journal of Ecology* 72: 1055–1064.

Windley, B. F. 1977. *The Evolving Continents*. London: John Wiley & Sons.

Winemiller, K. O. 1990. Spatial and temporal variation in tropical fish trophic networks. *Ecological Monographs* 60: 331–367.

Wisheu, I. C. and P. A. Keddy. 1989. Species richness: Standing crop relationships along 4 lakeshore gradients: constraints on the general model. *Canadian Journal of Botany* 67: 1609–1617.

Wodzicki, K. A. 1950. Introduced mammals of New Zealand: An ecological and economic survey. *Bulletin Department of Scientific and Industrial Research* 98: 1–255.

Wolfe, J. A. 1979. *Temperature Parameters of Humid to Mesic Forests of Eastern Asia and Relation to Forests of Other Regions of the Northern Hemisphere and Australasia*. U.S. Geological Survey Professional Paper no. 1106. Washington, DC: U.S. Government Printing Office.

Wollaston, T. V. 1854. *Insecta Maderensia*. London: John Van Voorst.

Wollaston, T. V. 1877. *Coleoptera Sanctae-Helenae*. London: John van Voorst.

Woodburne, M. O. and J. A. Case. 1996. Dispersal, vicariance, and the late Cretaceous to early Tertiary land mammal biogeography from South America to Australia. *Journal of Mammalian Evolution* 3: 121–161.

Woodburne, M. O. and W. J. Zinmeister. 1982. Fossil land mammals from Antarctica. *Science* 218: 284–286.

Woodring, W. P. 1966. The Panama canal landbridge as a sea barrier. *Proceedings of the American Philosophical Society* 110: 425–433.

Woodroffe, C. D. 1986. Vascular plant species-area relationship on Nui Atoll, Tuvalu, Central Pacific: A reassessment of the small island effect. *Australian Journal of Ecology* 11: 21–31.

Woods, K. D. and M. D. Davis. 1989. Paleoecology of range limits: Beech in the upper peninsula of Michigan. *Ecology* 70: 681–696.

Wootton, J. T. 1992. Indirect effects, prey susceptibility and habitat selection: Impacts of birds on limpets and algae. *Ecology* 73: 981–991.

World Conservation Monitoring Centre. 1992. *Global Biodiversity: Status of the Earth's Living Resources*. New York: Chapman and Hall.

Wright, D. H. 1983. Species-energy theory: An extension of species-area theory. *Oikos* 41: 496–506.

Wright, D. H. 1987. Estimating human effects on global extinction. *International Journal of Biometereology* 31: 293.

Wright, D. H. and J. H. Reeves. 1992. On the meaning and measurement of nestedness of species assemblages. *Oecologia* 92: 416–428.

Wright, D. H., B. D. Patterson, G. M. Mikkelson, A. Cutler and W. Atmar. 1996. A comparative analysis of nested subset patterns of species composition. *Oecologia* 113: 1–20.

Wright, H. E. 1976. Ice retreat and revegetation of the western Great Lakes area. In W. C. Mahaney (ed.), *Quaternary Stratigraphy of North America*, 119–132. Stroudsburg, PA: Dowden, Hutchison and Ross.

Wright, S. 1978. *Variability Within and Among Natural Populations*. Chicago, IL: University of Chicago Press.

Wright, S. J. 1980. Density compensation in island avifaunas. *Oecologia* 45: 385–389.

Wright, S. J. 1981. Inter-archipelago vertebrate distributions: The slope of the species-area relation. *American Naturalist* 118: 726–748.

Wright, S. J. 1985. How isolation affects rates of turnover of species on islands. *Oikos* 44: 331–340.

Wulff, E. V. 1943. *An Introduction to Historical Plant Geography*. Trans. E. Brissenden. Waltham, MA: Chronica Botanica.

Wyatt, R. E. 1992. *Ecology and Evolution of Plant Reproduction*. New York: Chapman and Hall.

Yeaton, R. I. 1974. An ecological analysis of chaparral and pine forest bird communities on Santa Cruz Island and mainland California. *Ecology* 55: 959–973.

Yeaton, R. I. 1981. Seedling morphology and the altitudinal distribution of pines in the Sierra Nevada of central California: A hypothesis. *Madroño* 28: 67–77.

Yeaton, R. I., R. W. Yeaton and J. E. Horenstein. 1981. The altitudinal replacement of digger pine by ponderosa pine on the western slopes of the Sierra Nevada. *Bulletin of the Torrey Botanical Club* 107: 487–495.

Yoda, K. 1967. A preliminary survey of the forest vegetation of eastern Nepal. II. General description, structure and floristic composition of sample plots chosen from different vegetation zones. *Journal of the College of Art and Science. Chi'ba University National Science* 5: 99–140.

Zabinski, C. and M. B. Davis. 1989. Hard times ahead for Great Lakes forests: A climate threshhold model predicts responses to CO_2-induced climate change. In J. B. Smith and D. Tirpak (eds.), *The Potential Effects of Global Climate Change on the United States*, Appendix D. Washington, DC: U.S. Environmental Protection Agency.

Zaret, T. M. 1980. *Predation and Freshwater Communities*. New Haven, CT: Yale University Press.

Zaret, T. M. and R. T. Paine. 1973. Species introduction in a tropical lake. *Science* 182: 449–455.

Zeh, D. W. and J. A. Zeh. 1992. Failed predation or transportation? Causes and consequences of phoretic behavior in the pseudoscorpion *Dinocheirus arizonensis* (Pseudoscorpionida: Chernetidae). *Journal of Insect Behavior* 5: 37–50.

Zeh, D. W., J. A. Zeh and G. Tavakilian. 1992. Sexual selection and sexual dimorphism in the harlequin beetle *Acrocinus longimanus*. *Biotropica* 24: 86–96.

Zimmerman E. C. 1948. *The Insects of Hawaii*, vol. 1. Honolulu: University of Hawaii Press.

Zimmerman, K. 1950. Die randformen der mitteleuropaischen Wuhlmause. In A. Jordans and F. Peus (eds.), *Syllegomena Biologica Festschrift*, 454–471. Leipzig: Kleinschmidt Verlag.

INDEX

Abert's squirrel, 398
Abies, 121
Abiotic factors, 72
Abiotic stress, 91
 diversity and, 478–480,
 485–486
Abundance, distribution and,
 66, 70–72
Abyss, species diversity in,
 467–468, 479–482
Abyssal zone, 126
Acacia spp., 87, 114, 248, 320
Acanthaster planci, 520
Accipiter brachyurus, 421–422
Acer negundo, 276
Acer saccharym, 605–607
Achillea millefolium, 93
Acridotheres tristis, 553
Active dispersal, 273–276
Actualism, 9
Adaptation, 223
 gene flow and, 91–94
Adaptive radiation, 246–249
Adelaide warbler, 446
Adiabatic cooling, 42
Aerial photographs, 64
Aerial plankton, 277
Africa
 convergence in, 524
 endemics, 295–296, 541
 formation of, 163–164
 glaciation in, 179
 human origins and, 574
 human role in extinctions
 in, 216, 545
 precipitation, 44
 range collapse in, 593
 savanna, 115–116
 shift in orientation since
 Triassic, 149
 species introductions, 550
 species-area relationships,
 400
African migratory locust, 71
African Plate, 167–168
African Rift Valley, 241
Afroeurasia, Pleistocene
 extinctions, 218
Agassiz, Louis, 21, 34
Agathis, 120
Agave, 116, 522
Age
 in classifying endemics,
 300–301

of earth, 18, 136
Agelaius phoeniceus, 489
Aggregated distribution pat-
 tern, 70
Agoseris spp., 322
Ailuropoda melanoleuca, 591
Air temperature, mean glob-
 al, 183
Alauda arvensis, 553
Albedo, 181
Alca impennis, 534
Alces alces gigas, 210
Aldabra Atoll, 282
Aleutian Islands, 414
Aleutian Trench, 169
Algae
 distributions, 30–31
 range overlap, 497–498
Alleles, 232
Allelopathy, 83
Allen, J. A., 27
Allen's rule, 28, 490–492
Allochthonous endemic, 300
Allometry, 28
Allopatric speciation, 32–33,
 235–239
 Galápagos Islands,
 237–239
 Hawaiian Islands, 364
 range overlap and, 499
Allopolyploidy, 240
Aloes, 522
Alopex lagopus, 490
Alpha diversity, 450, 469–470,
 482, 485, 489
Alpine tundra, 121–122
Alsinidendron, 362–363
Alvarez, Walter, 254
Amazon basin, mapping, 616
Amazon River, 167, 308,
 457–458
Amazonia, 205–206
Ambrosia ambrosioides, 62
Ambrosia dumosa, 276
Ambystoma tigrinum, 523
Ameghino, C., 32
Ameiva exsul, 554
Ameiva spp., 424
American burying beetle, 593
American chestnut, 195, 252,
 587–588, 589, 596
American elm, 252
American Museum of
 Natural History, 352

American muskrat, 267–269
Americas, Pleistocene extinc-
 tions, 218
Ammonites, 172–173
Ammophila breviligulata, 188
Amospermophilus, 498
Amphibians
 endemism in Africa, 541
 endemism in Central
 America, 540
 long-distance dispersal,
 409
 patterns in recorded
 extinctions of, 546
 species-area relationship,
 374
 status of species in United
 States, 534
Amphiprion spp., 89
Amphitropical distributions,
 154
Anaeretes parulus, 527
Anas spp., 493
Anasazi, range collapse of,
 595–596
Anax, 275
Andes, 172, 191–192, 397,
 399, 464
Aneuploidy, 239
Angiosperms
 endemic to Neotropical
 regions, 296
 oldest known fossil, 338
 radiation in, 257
Anglerfishes, 248
Animals. *See also* Mammals
 causes of historical extinctions, 557
 classification schemes, 303
 convergence in, 522–524
 desert, 116
 disjunctions, 321–322
 dispersal, 409, 432–434
 response to Pleistocene
 climatic cycles, 190
 status of species in United
 States, 534
 terrestrial vertebrates,
 278–279
Annelids, 455
Anolis spp., 85, 424–426, 427,
 447, 552
Anseriformes, 588
Antarctic Circle, 41

Antarctic ocean region, 124
Antarctic
 diversity patterns,
 450–451
 formation of, 164, 170–171
 fossil record, 339–340
 ocean region, 124
Antarctica Plate, 157
Anteater, 323, 524
Antelopes, 523
Antennanus, 248
Anthornis melanura, 493
Anthropogenic islands,
 turnover on, 387–388
Anthropoids, 574
Antillea, Pleistocene extinc-
 tions on, 218
Antilles, history of, 164–166.
 See also Greater Antilles;
 Lesser Antilles
Ants, 87, 267, 375, 378, 411,
 415, 444–445
Anurans, fossil, 338. *See also*
 Amphibians; Frogs
Aphelocoma, 499
Aphotic zone, 125–126
Apis mellifera, 267
Aplonis metallica, 421
Apomictic plant, 292
Apomorphic character states,
 328–329
Apomorphic characteristic,
 225
Appalachian Mountains,
 191–192
Apteryx, 547
Aquatic communities,
 122–123
 biomass, 129
 food web, 101
 freshwater. *See* Freshwater
 communities
 marine. *See* Marine communities
 productivity, 129
Aquatic environments
 cosmopolitanism in, 299
 diversity patterns,
 465–468
 latitudinal gradient and,
 454–458
 microenvironmental vari-
 ation and, 59
 oceanic circulation, 55–57

pressure and salinity, 57
  stratification, 53–55
  tides and intertidal zones, 57–58
Aquatic organisms, habitat heterogeneity and, 481
Aquatic systems, postglacial and pluvial lakes, 200–203
Aquilegia triternata, 88
Arabian Peninsula, human origins and, 575
Arabian Plate, 167–168
Arabian subcontinent, 285–286
Arafura Basin, 285
Arafura Sea, 204
Aral Sea, 127
Arceuthobium americanum, 276
Archaeopteryx, 337–338
Archerfishes, 295
Archilochus colubris, 274
Archilochus spp., 88
Archipelago, 170. *See also* Island biogeography; Island patterns; Islands
Arctic, diversity patterns, 450–451
Arctic Circle, 41
Arctic climatic region, 124
Arctic ocean region, 124
Arctic warbler, 518
Area, diversity and, 482
Area cladogram, 353–354, 356, 357, 359–360
  Hawaiian Islands, 361–364
Area hypothesis, diversity patterns and, 460
Arecacae, 451
Areography, 494
Areography, 494–500
  macroecology and, 501
Argyroxiphium spp., 333
Aridity, diversity patterns and, 464–465. *See also* Desert
Aristotle, 13, 261
Arizona Sonora Desert Museum, 522
Armadillos, 267–269, 509, 510
Arrhenius equation, 373
Artamidae, 306
Artemia salina, 79–80
Artemia spp., 277–278, 479
Arthropods, 544
Artiodactyls, 438
Asexual reproduction, colonization and, 292–293
Asia
  distribution patterns in, 71
  formation of, 163
  landbridges to, 187
Asprotilapia leptura, 241
Assembly rules, 422–423

Assortative equilibrium, 389–390
Asteroid, as cause of extinctions, 253–254
Asthenes humicola, 527
Asthenosphere, 150
Atacama Desert, El Niño and, 46–47
Atmar, Wirt, 416
Atriplex, 301
Aulonocara nyassae, 242
Australia, 32, 303
  adaptive radiation in, 248
  biogeographic provinces, 303–304
  body size and species diversity in, 503
  convergence in, 524–526
  diffusion of species, 268–270
  dwarfism in, 441
  endemics, 295–296, 306
  exotic fishes and birds, 548–549, 555
  extinctions, 216–217, 545
  formation of, 163–164, 170–171
  glaciation in, 179
  human colonization of, 579
  insect distribution, 65, 71
  landbridges to, 187
  marsupial radiation, 256–257
  range collapse in, 593–594
  sea snakes and, 347–348
  shift in orientation since Triassic, 149
  target area effect, 392–393
  temperate rain forest, 119
  vegetation zone shifts during Pleistocene, 193–194
Australian fruit fly, 91
Australian Region, 303
Australopithecus afarensis, 574, 576
Autochthonous endemic, 300
Autocorrelation, 73
Autopolyploidy, 240
Autotrophs, 95
Avian malaria, 554
Azolla, 299
Azonal soils, 51

Bacon, Sir Francis, 138
Badgers, 82
Baetis, 101
Bahamas, 424–425, 548
Bailey's pocket mouse, 307
Baja California, 462
Bajadas, 52
Balaenoptera musculus, 494
Balanced biotas, 408
Balanus balanoides, 68–69
Bald eagle, 386

Bali, 25
Banks, Sir Joseph, 16
Barnacle, niche variables, 68
Barro Colorado Island, 252, 284, 387, 413
Basal meristems, 120
Basal metabolic rate, 98–99
Basin-and-range topography, 203
Basins, 203
Bass, 281
Bass Straits, 204
Bassian Province, 303
Basswood, 604
Bathyal zone, 126
Bathybates ferox, 241
Bathymetry, 126, 313–314
Bats, 274, 299, 319–320, 338, 409, 439, 454
Bay checkerspot butterfly, 600–601
Bay of Bengal, 455
Beech, 109, 195, 605–607. *See also* Southern beeches
Beetles, 99, 146, 356, 431
Bejerinck's law, 262
Bendire's thrasher, 307
Benioff, Hugo, 148
Benioff zones, 148
Benthic organisms, 126
Bergmann, Carl, 27, 488
Bergmann's rule, 27, 488–490
Bering landbridge. *See* Beringia
Bering Strait, 204
Beringia, 6, 163, 187, 204, 208–209, 285, 578
Bermuda, 422, 423, 427–428, 547, 548
Beta diversity, 450, 469–470, 485
Betulla alleghaniensis, 605–607
Bidens pilosa, 276, 433
Bighorn sheep, 213
Binomial nomenclature, 14
Biodiversity, 533
  crisis in, 534–537
  geographic variations, 537–538
  local vs. regional, 506–507
  marine hot spots, 542–544
  surveys, 568–571
  terrestrial hot spots, 538–542
Biogeographic implications of fossil record, 337–343
Biogeographic lines, 308–312
Biogeographic provinces, 303
Biogeographic regions, 24–27
  barriers between, 514
  marine, 31
  oceans, 124–125
Biogeographic relicts, 300
Biogeography. *See also* Historical biogeography

applications, 564–571, 621–624
  basic principles, 7–9
  conceptual advances, 616–621
  contemporary, 9–12
  defined, 3–5
  number of publications in, 35
  relation to other sciences, 5–7, 618–619
  technological advances, 613–616
  themes, 14
Biogeography, history of, 13–14
  Darwin, 19–23
  eighteenth century, 14–17
  Hooker, 19, 23–24
  nineteenth century, 17–31
  Sclater, 19, 24–25, 31
  twentieth century, 32–35
  Wallace, 19, 25–26
Biological species concept, 32, 224–225
Biomass, 101, 129–132
  diversity and, 475–478
Biomass pyramid, 102
Biomes, 96. *See also* Terrestrial biomes
  global comparison, 128–133
Biosis, 11
Biosphere, 97
Biotas
  convergence in, 524–527
  historical development of, 620
  quantifying similarity among, 317–320
Biotic exchange, 285–290
  glacial cycles and, 203–205
Biotic factors, 72
Biotic interactions, 91, 482–483
  diversity and, 485–486
Biotic interchange, 507–508
  Great American Interchange, 508–513
  Lessepsian Exchange, 513
Biotic regions, 16, 17
Biotic segregation
  avian migration and provincialism, 515–518
  barriers between biogeographic regions, 514
  resistance to invasion, 514–515
Bird census, 375
Bird extinctions, 534, 546, 556, 557
  human impact, 537
Bird pox virus, 554
Birds. *See also* Bird extinctions; *individual species names*

biogeographic regions
and, 24–25, 319–320
center of origin and, 349
clutch size and latitude
and, 492–493
colonization and, 411, 413
community organization
in, 528–529
competition in, 422
consensus area clado-
gram, 360
convergence in, 527
density overcompensa-
tion, 427–428
dispersal, 408
distribution, 128
endemism, 306, 540–541
evolution of flightlessness,
430–431
exotics, 515, 548, 549
foliage height diversity
and, 481
geographic ranges, 496
geographic variation in
North America, 453
Great American
Interchange, 512
on Krakatau, 379, 402–403
local vs. regional species
richness, 507
migration and provincial-
ism, 516–519
on montane forest islands,
399
oldest known fossil, 338
Pleistocene relicts,
396–397
relation between body
size and range, 99, 501,
502–503
restricted-range species,
538–540
selective immigration and,
415
species introductions,
548–549, 552–553
species richness, 378
status of species in United
States, 534
in Tonga, 585
trends in insular body
size, 443–444
turnover, 385–386
variation in range with
latitude, 469–471
variation in richness with
elevation, 464
Birth rates, 69
extinction and, 250
Bismarck Archipelago,
419–421
human colonization, 579
Bison, 82, 210–211, 588
Bison bison, 588
Bison latifrons, 210–211
Bivalves, extinction, 544, 587

Black walnut, 83
Blackett, P. M. S., 149
Blackfinch, 446
Blackfly, 101
Black-footed ferret, 591, 596
Blarina brevicauda, 423–424
Blechnum, 120
Blue jay, 63
Blue whale, 494
Body mass, community organization and, 98–100
Body size, 28, 29
evolution of flightlessness
in birds and, 430–431
evolution on islands,
434–444
latitude and, 488–490
population density and,
501
reduced dispersal ability
and, 432
relation to extinction, 412,
413
species diversity and,
503–506
Body weights, extinction
and, 213, 215
Bog copper butterfly, 600
Boiga irregularis, 595
Bombus spp., 87
Bonasa nonasia, 598
Boreal forest, 110, 121
mammal distribution, 505,
568–569
in North America during
Pleistocene, 192
productivity and biomass,
129, 130, 132
Boulengerochromis microlepis,
241
Bowdleria punctata, 493
Bowlesia incana, 322
Brachiopods, 258–259
Brain size, of hominoids, 576
Branta spp., 122
Brazilian cardinal, 553
Bridled white-eye, 595
Briggs, John, 31
Brine fly, 80
Brine shrimp, 79–80, 277–278,
479
Bristlecone pine, 76, 77,
300–301
Bromeliads, 295, 522
Brongniart, Adolphe, 18
Brooks, Daniel, 356
Brown bears, 210
Brown lemmings, 197
Brown tree snake, 595
Brown-headed cowbirds, 563
Bruhnes, Bernard, 149
Bubulcus ibis, 267
de Buffon, Comte, 15–16, 35,
346, 349
vs. Linnaeus, 15
Buffon's law, 16, 17

Bufo marina, 409
Bufo olivarius, 307
Bulinus truncatus, 230
Bullfinch, 446
Bumblebee, 87
Burrowing bittong, 594
Bushy-tailed woodrat, 60,
488, 489
Butterflies. *See also*
Checkerspot butterfly
in Baja California, 462–463
biogeographic regions in
Indo-Australian area,
319–320
endemism in Central
America, 540
on Krakatau, 402

Cacao, 474
Cacicus cela, 493
Cacti, 52, 114, 116, 249, 295
Cactoblastis cactorum, 86
Cactospiza spp., 238
Caddis, 101
Caelatura spp., 230
Cain, Stanley A., 33, 346
Cakile edentula, 69, 188
Calcification, 48, 49–50
Caliche, 50
California condor, 591, 592
California quail, 553
Callipepla californica, 553
Callosciurus prevosti, 439
Calypte spp., 88
CAM. *See* Crassulacean acid
metabolism
Camarhynchus spp., 238
Camels, 273
fossil localities, 339–340
Campephilus principalis, 592,
593
Canary Islands, 409
de Candolle, Augustin, 17,
29, 302
Canis lupus, 103, 563
Canis rufus, 563, 591
Canoes, 581
Carabidae, 146
Carabids, flightless, 432
Carbon dioxide concentra-
tion, human impact on,
537
Carbon dioxide doubling,
601–603, 609
Carboniferous, continental
drift, 158
Cardinalis cardinalis, 74, 553
Carduelis flammea, 72
Carey, S. W., 142
Caribbean islands, bird
clutch size and latitude,
493
Caribbean Plate, 166, 170
Caribou, 122, 213
Carlquist, Sherwin, 33, 291

Carnegiea gigantea, 73–74, 75,
76, 307
Carnivores, 100, 438
Carolina parakeet, 252
Carolina wren, 71
Carrying capacity, 102–103,
250
Cascade Range, 155
Cassowaries, 296, 320–321,
443
Castanea dentata, 195, 252,
554, 587–588, 589
Castelleja spp., 88
Casuarius spp., 443
Catalina ironwood, 298
Catastrophic death assem-
blages, 341
Catfish, 417
Catfish eels, 295
Catharacta spp., 283
Cats, domestic, 548
Cattails, 85
Cattle egret, 267
Causal historical biogeogra-
phy, 23
Caves, 127–128
Cenozoic, 137
Cenozoic radiation, 255–256
Centers of origin, 32, 346–350
Centilenan extinctions, 536
Central America
endemism, 540
Great American
Interchange, 508–513
history of, 164–166
landbridge, 545
Ceratophyllum, 299
Cercidium microphyllum, 116
Cercocarpus traskae, 298
Cerdidium, 321
Certhidea olivacea, 238
Chaetodipus baileyi, 307
Chaetodipus intermedius, 53
Chagras River, 252, 387
Chamaea fasciata, 527
Champlain Sea, 188
Channel Islands, 298, 385
Chaparral, 6, 117–118, 521
birds and, 528
convergence in, 527
lizards and, 530
Character displacement, 227,
246
Character states, 328–329
Checkerboards, 419–423
Checkerspot butterfly, 321
predation and, 85
Chestnut blight, 554, 588
Chewing lice, 356–357
Chickadees, 499
Chihuahuan Desert, 44, 307
Chiloe Island, 546
Chilopsis linearis, 79, 276
Chipmunks, 84, 210, 245, 498
Chironomidae, 101
Chlamydogoblius, 417

Chlorophyta, 497
Chorology, 24
Chromosomal changes, 239–240
Chronospecies, 243
Chrysolina quadrigemina, 86
Chthamalus stellatus, 68–69
Chuckwallas, 53
Cichla ocellaris, 86
Cichlids, 240–242, 246, 400, 457, 535
Cincinnati Zoo, 251
Cirsium neomexicanum, 390
Clade, 61, 328
Cladistic biogeography, 262, 358
Cladogenesis, 223
   example, 330
Cladogram, 328–333
   area, 353–354, 356
   limitations of, 334–335
Clams, 258–259, 308
Clematis fremontii, 66
Clements, F. E., 95
Cleopatra ferruginea, 230
Clethrionomys spp., 97
Climate, 16, 18
   body size and, 488
   continental drift and, 171–176
   diversity patterns and, 472–473
   influence of elevation on, 46
   predicting effects of, 567–568
   role in extinctions, 214–215, 253
   soil types and, 50
   solar energy and temperature regimes, 40–42
   in Wallace's theory, 28
   winds and rainfall, 42–47
Climate change. *See* Global climate change
Climate space, 604–605
Climatic cycles, Pleistocene, 190
Climatic events
   Great American Interchange and, 509
   of Phanerozoic, 174–175
Climatic patterns, 30
Climatic regions, 45
   oceans, 124–125
Climatic stability, 458
   diversity and, 480–481
Climatic zones
   geographic shifts in, 184–186
   relation to soil type and vegetation communities, 113
Climatology, biogeography and, 619–620
Climax vegetation, 48

Cline, 235
Clownfishes, 89
Clumped distribution pattern, 70–71
Clutch size, as function of latitude, 492–493
Coastline
   abundance and, 72, 73
   temperate rain forest, 119
Cobaea scandens, 248
Cobb Seamount, 170
Cocos Plate, 166, 170
Coereba flaveola, 493
Colbert, E. H., 32, 146
Cold seep communities, global distribution, 543
Coleopterans, 432
Collared lemmings, 197
Collared lizards, 53
Collision zones, 151
Colonization, 291–293
   of microenvironments, 59–60
   in vicariance biogeography, 354
Colony, establishing, 291–293
Colorado River, 167
Coloration, humidity and, 491
Colossoma macropomum, 338
Columbia River, 167
Commander Islands, 218
Common redpoll, 72
Communities, 95–98
   defined, 96
   global comparison, 128–133
Community distribution
   spatial patterns, 103–107
   temporal patterns, 107–110
Community function, 96
Community organization, 98–103
Community structure, 96
Competition, 34, 82–85, 106–107
   body size, species diversity, and, 505
   density compensation and niche shifts on islands, 423–427
   density overcompensation, 427–429
   distribution, checkerboards, and incidence functions, 419–423
   diversity and, 483, 506–507
   diversity patterns and, 458, 460
   experiments, 84–85
   Great American Interchange and, 511
   range overlap and, 499
   in Wallace's theory, 28

Competitive exclusion, 106, 245
Congo River, 308
Conifers, 117, 121, 451–452
Connectivity, global climate changes and, 610–612
Connell, Joseph, 68
Consensus area cladogram, 359–360
Conservation
   range collapse and, 595–596
   of species, 533
Conservation biology, 622–624
Conservation International, 569
Contact, 237
Continental biotas, 500–506
Continental drift, 7, 8, 18, 33, 137–139. *See also* Plate tectonics
   early opposition to, 141–142
   evidence for, 142–157
   role in extinctions, 253
   role in prehistoric extinctions, 544–545
   vicariance and, 261–262
   Wegener's theory, 137–141
Continental extinction, vs. insular extinction, 558, 560
Continental islands, 312–314, 398–399
Continental patterns, 487, 527–530. *See also* Biotic interchange; Isolated biotas; Multispecies assemblages; Single-species patterns
Continents, formation of. *See also individual continents*
   breakup of Gondwanaland, 163–164
   breakup of Laurasia, 163
   breakup of Pangaea, 162–163
   diversity and, 477
   Gondwanaland, Laurasia and Pangaea, 157–162
   latitudinal gradient and, 451–454
   in Wallace's theory, 28
Contour maps, 63–64, 73
Convergence
   of assemblages, 524–527
   at the species level, 520–524
Convergent evolution, 520
Cook, Captain James, 16
Cook Islands, exotics, 547
Cope's rule, 29, 441
Copernicia alba, 64–65
Coprolites, 219
Coral reef

diversity and, 316, 481–482, 542
   global climate changes and, 610
   latitudinal variation in, 457
   loss of, 562
   productivity and biomass, 102, 129, 132
Corbicula consobrina, 230
Cordillera, 170
Cordilleran ice sheet, 207–208, 578
Corematodus shiranus, 242
Coriolis effect, 42–43, 55
Corridors, 285
Corrus kubaryi, 595
Cosmopolitan species, 16, 39
Cosmopolitanism, 299–300, 315
Costa Rica
   tropical deciduous forest, 114
   tropical rain forest, 112
Cougar, 563
Crabs, oldest known fossil, 338
Cranberries, 600
Crassulacean acid metabolism, 522
Craterocephalus, 417
Cratons, 143
Creosote bush, 79, 321
Cretaceous, 137
   continental drift, 160
   extinctions, 253–254, 544
Crickets, 364
Crocodilians, 326
Croizat, Leon, 34, 350
Crossbills, 71–72
Crotophytus collaris, 53
Crown of thorns starfish, 520
Cryogenic lakes, 200–202
Crypsis, 491–492
Cryptic species, 245, 499
Ctenotus spp., 248
Cuba, exotic fishes and birds, 548
Cuckoo doves, 420–421
Cunningtonia longiventralis, 241
Curie point, 148
Currents, ocean, 43
Curtis, J. L., 96
Cutthroat trout, 203
Cyanocitta, 499
Cyanocitta cristata, 63
Cyanoramphus novaezelandeae, 493
Cyathochromis obliquidens, 242
Cyclical vicariance, 205
Cynomys, 498
Cynotilapia afra, 242
Cyprinodon diabolis, 298, 494
Cyprinodon spp., 77, 78, 203, 242–243

Cyprinodon variegatus, 281
Dacus tryoni, 91
Dall sheep, 122, 210, 213
Danaus plexipus, 275
Dandelion, 292
Daphnia, 299
Darlington, Phillip J., 33, 349
Darters, 248, 308, 359
Darwin, Charles, 8, 10, 19, 20–23, 29, 142, 236, 261, 370, 429, 430
Darwin's finches, 226–227, 237–239, 370, 422
Darwin's fox, 546
Dasypus novemcinctus, 267–269
Data, in biogeography, 10
Davis, Margaret, 195
Davis's line, 287
Day length, seasonal variation in, 41
Dead Sea, 127, 278, 479
Death rate, 69
  extinction and, 250
Death Valley, 6, 45, 77, 203, 399
Deciduous forest, mammal distribution, 504
Deductive reasoning, 8
Deer, 107, 213
Deer mice, 92, 210, 233–235, 291–292, 423, 491
Defaunation, experimental, 388–390
Deforestation, 559–562
Deglaciation, 181
Deinacrida rugosa, 431
Deinogalerix koenigswaldi, 435
Demideserts, 116
Dendroica adelaidae, 446
Dendroica kirtlandii, 592, 593
Dendroica petechia, 89–90
Density compensation, 423–427
Density enhancement, 428–429
Density overcompensation, 427–429
Density stasis, 428–429
Derived characteristic, 225
Desert, 110, 112, 116–117
  convergence in, 522
  as dispersal barrier, 281
  as isolating mechanisms, 397
  pluvial lakes, 202–203
  precipitation, 44
  productivity and biomass, 129, 130, 132
  provinces, 307
Desert hopping mice, 257
Desert mountains, diversity patterns, 463–465
Desert pupfishes, 77, 78
Desert scrub, 116
Desert willow, 79

Desertification, 81
Deterministic mechanisms, 34
Detritivores, 100
Detritus, 102
  in temperate rain forest, 120
  in tropical rain forest, 113–114
Devil's Hole pupfish, 298, 494
Devil's Lake, 202
Devonian, 148
  continental drift, 158, 162
Dew point, 43–44
Diacamma rugosum, 415
Diamesine, 356
Diamond, Jared, 215, 385, 391, 565, 584, 622
Diaspores, 276–277
Dice, Lee R., 32, 92
Dice similarity index, 318
Dicksonia, 120
Dicrostonyx, 197
Didelphis virginiana, 267
Dietz, R. S., 150
Diffuse competition, 106
Diffusion, 266–273
Digger pine, 106
Dinocras, 101
Dinosaurs, 253
Diornis giganteus, 435
Dipodomys deserti, 53
Dipodomys spp., 83, 257, 297–298
Discontinuous variety, 226
Discrete communities, 104–105
Disharmonic biotas, 408
Disjunction, 34, 296
  patterns, 320–322
  processes, 322–323
Dispersal, 21, 23, 261–265, 299–300, 408
  diffusion, 266–273
  of endemics, 301–302
  evolution of reduced ability on islands, 429–434
  glacial cycles and, 203–205
  jump dispersal, 265–266
  in microenvironments, 59–60
  secular migration, 273
  vs. vicariance, 262–263
  in Wallace's theory, 28
Dispersal barriers, 279–281
  ecological and psychological, 283–284
  physiological, 281–283
Dispersal biogeography, 353–354
Dispersal curves, 289–290
Dispersal events, 236–237
Dispersal mechanisms
  active dispersal, 273–276
  passive dispersal, 276–279

Dispersal routes, 289
  corridors, 285, 565, 567
  dispersal curves, 289
  filters, 285–287
  glacial events and, 189
  sweepstakes routes, 287–289
Dispersion, 263
Disruptive selection, 239
Distance, in Wallace's theory, 28
Distribution, 419–423
  abundance and, 70–72
  of individuals, 64–65
  past, 339–340
  of populations, 65–72
  of single species, 61–64, 418–419
Districts, 302
Disturbance, 81–82
Divergence, 518–520
Diversification
  adaptive radiation, 246–249
  ecological differentiation, 244–246
Diversity. *See* Biodiversity; Diversity patterns; Species diversity
Diversity indexes, 449–450
Diversity patterns, 471–472
  aquatic environments, 465–468
  aridity, 464–465
  dominance-diversity relationships, 470–471, 484
  elevation, 463–464
  equilibrial mechanisms, 475–484
  latitudinal gradient, 450–461
  nonequilibrial mechanisms, 472–475
  peninsulas, 461–463
  Rapoports' rule, 468–470
  synthetic explanation, 484–486
Diving petrels, 283
Docimodus johnstoni, 242
Dodo, 430, 534
Dodonaea viscosa, 276
Dokuchaev, V. V., 30
Dominance-diversity relationship, 470–471, 484
Donoghue, Michael, 229
Dot maps, 61–63
Double-canoe, 581
Doyle, Sir Arthur Conan, 215
Dragonflies, 275, 402
Dreissena polymorpha, 271, 550–551
Driftless areas, 207
Dromaeius novaehollandiae, 443
Dromiciops gliroides, 300
Drosophila, 32, 346, 362–363

Drude, O., 29
Duckweeds, 299
Ducula, 293
Dunaliella, 479
Dung beetles, 411
Dutch elm disease, 554
Dwarfism, 434–435
Dynamic equilibrium, 371

Earth
  age of, 18, 136
    geological time scale, 135–137
    history of, 135
    solar radiation and orbit of, 180–181
Earth Observing System, 615
Earthquakes, 151
East African Rift Valley, 152
Easter Island, 170, 581
Easterly trade winds, 55
Eastern chipmunks, 197
Eastern hemlock, 605–607
Eastern phoebe, 607
EBAs. *See* Endemic Bird Areas
Ebenhard, Torbjorn, 548
Eccentricity, of earth's orbit, 180–181
Ecdyonurus, 101
Echinocereus triglochidiatus, 88
Ecogeographic rules
  Allen's rule, 490–492
  Bergmann's rule, 488–490
  Gloger's rule, 490–492
Ecological barriers, to dispersal, 283–284
Ecological biogeography, 4, 33, 34, 618–619
Ecological differentiation, 244–246
Ecological niche, 67
  geographic range and, 68–70
Ecological pyramids, 102
Ecological release, 419
Ecology, 5, 10, 13, 24
Ecoregions, 96
Ecosystems, 96–98
Ecotone, 104
Ecotypes, 227–228
Ectopistes migratorius, 195, 251, 534, 587–588
Edentata, 454
Edge effects, 565
Egretta intermedia, 421–422
Eichhornia crassipes, 271
Ekman, Sven, 33
El Niño, 43, 46–47
  global climate changes and, 610, 611
El Niño Southern Oscillation, 46
Elaenia albieceps, 527
Elaenia flavogaster, 493
Elaphus, 275–276, 435

Eldredge, Niles, 229
Elephant, 275–276, 435
Elephant birds, 313, 320–321, 430
Elevation
 cooling effect of, 41–42
 as dispersal barrier, 282
 diversity patterns and, 463–464, 473
 effect on bird ranges, 608
 influence on climate, 46
Elliot, D. H., 146
Elm, 271–272
Elton, Charles, 533, 534
Emigration, 69
Emperor seamount chain, 169
Empidonax difficilis, 527
Empidonax euleri, 493
Empirical observation, 8
Emus, 296, 320–321, 443
Endangered species, management of, 623. *See also* Focal species patterns
Endangerment, current patterns, 556–558
Endeavor, 16
Endemic Bird Areas, 539–540
 global distribution, 542
Endemicity, 537
Endemism, 295–299. *See also* Provincialism
 classifying endemics, 300–302
 cosmopolitanism, 299–300
Endothia parasitica, 252, 588
Energy, 40–41
Energy flow pyramid, 102
Energy supply, species diversity and, 477
Enhydra lutris, 252
ENSO. *See* El Niño Southern Oscillation
Entosphenus, 86
Environmental gradients, distribution and, 104–105
Environmental variation, small-scale, 59
Eocene, continental drift, 160
Eons, 137
EOS, 615
Epeiric seas, 166–167
Ephalophis, 348
Epibiotics, 300
Epicontinental seas, 166–167
Epiphylls, 112
Epiphytes, 112, 119
Epochs, 137
Equator
 extinction in, 544
 geographic ranges and, 537–538
Equatorial countercurrent, 46
Equatorial region, 124
Equilibria, series, 390

Equilibrium theory of island biogeography, 34, 379–382, 618
 additional patterns of insular species richness, 390–394
 history of, 370–373
 human colonization and, 581–582
 nonequilibrium biotas, 394–401
 strengths and weaknesses, 382–384
 tests of model, 384–390
Equinoxes, 41
Eras, 137
Erica, 346
Eriocaulon aquaticum, 263
Ermine, 210
Eschrichtius robustus, 252
Essay on the Principle of Population, 67
Establishment
 vs. immigration, 403–404
 of insular populations, 410–411
Estuaries, 128
 productivity and biomass, 129, 130, 132
Etheostoma, 248, 308
Etheostoma variatum spp., 359
Ethiopian, 303
Eubalaena glacialis, 317
Eucalyptus, 52, 120, 248, 303
Eupera ferruginea, 230
Euphorbia polygonifolia, 188
Euphorbias, 114, 116, 522
Euphydryas editha, 321
 predation and, 85
Euphydryas editha bayensis, 600–601
Eurasia, 303
 formation of, 163
 origin of life and, 16
Eurasian Plate, 152
Europe
 distribution patterns in, 71
 formation of, 163
 geographic ranges of birds, 496
 glacial-interglacial cycles, 184
 range collapse in, 593
European fire salamander, 523
Euryhaline organisms, 57
Eustatic changes, 186–188
Eutamias spp., 84, 498, 569
Eutrophic lakes, 127
Evapotranspiration, 478
Evolution, 21. *See also* Macroevolution; Microevolution
 of body size on islands, 434–444

branching vs. reticulate, 223–224
 effects on species richness, 405
 of flightlessness and reduced dispersal ability, 429–434
 in fossil record, 229–231
Evolutionary biology, 5, 10
Evolutionary classifications, 326–327
Evolutionary equilibrium, 390
Evolutionary species concept, 225
Excess density compensation, 427–429
Exotics. *See* Invasion; Species introductions
Expanding earth hypothesis, 154–155
Exploitative interactions, 82–83
Exponential growth, 67
Extensionists, 21–23
Extinction, 18, 34, 223, 534–537. *See also* Equilibrium theory of island biogeography; Focal species patterns; Mass extinctions
 current pattern of endangerment, 556–558
 differential immigration and, 414–419
 diversity patterns and, 460–461, 473–475
 ecological processes, 249–251
 extinction-prone species, 586–587
 in fossil record, 253–255
 glacial cycles and, 210–219
 habitat loss and fragmentation, 562–564
 historical record, 545–547
 human colonization and, 584
 in insular communities, 411–414, 585
 on Krakatau, 401, 402
 overkill hypothesis and, 212–214
 prehistoric, 413, 544–545
 range size and, 501
 recent, 251–253
 species introductions and, 547–556
 in vicariance biogeography, 354
 in Wallace's theory, 28
Extinction events, in marine realm, 172
Extinction rates, diversity and, 483–484
Eyrean Subregion, 303

Fagus grandifolia, 605–607
Fagus spp., 109, 356
 consensus area cladogram, 360
Falco peregrinus, 299, 386
Family, 228
Farallon Plate, 170
Farnes Islands, 386
Favorableness, 479
Felis catus, 548
Fence effect, 428
Fennecus zerda, 490
Ferns, 120, 299
Fieldwork, 10
Fiji, human colonization, 580–581
Filters, 285–287
 oceanic, 414–415
Finger Lakes, 202, 399, 466
Fir, 121
Fire
 in chapparal, 118
 habitat loss and, 562
 human origins and, 574
 role in distribution, 81
 in temperate rain forest, 120
 in tropical savanna, 115
Fire ants, 267
First Kulczynski similarity index, 318
First law of thermodynamics, 101
Fischer, Alfred, 451
Fish, 28, 146. *See also individual species names*
 biogeographic provinces, 309
 consensus area cladogram, 360
 disjunctions, 323
 dispersal, 410
 dispersal barriers, 281
 distributional lines, 311
 endemism, 308
 exotics, 548
 in Great American Interchange, 512
 latitudinal variation, 456–457
 nestedness in, 417
 post-Pleistocene dynamics, 399–401
 resistance to invasion, 515
 species introductions, 549–550
 species richness as function of depth, 466
 species-area relationship, 400
 status of species in United States, 534–535
Fitchia spp., 433
Fitzroy, Robert, 20
Flannery, Tim, 441, 622

Flightlessness, evolution of on islands, 429–434
Florida Keys, 389
Floristic belts, 17
Flycatcher, 233, 419, 421, 451, 454
Flying phalanger, 524
Flying squirrel, 524
Focal species patterns, 586–588
  insular and fragmented habitat patterns, 598–601
  patterns of range collapse, 588–598
  species responses to global climate change, 601–612
Foliage height diversity, 481
Food chains, 100
Food supply, range boundaries and, 74–75
Food webs, 90, 100–101
Forbes, Edward, 21, 31, 261
Forest, 112. *See also* Temperate deciduous forest; Temperate forest; Temperate rain forest; Tropical; Tropical deciduous forest; Tropical rain forest
  deciduous forest, 504
  deforestation, 559–562
  subtropical evergreen forest, 110, 118
  subtropical forest, 193
  summer-green deciduous forest, 119
  swamp forest, 121
Forster, Johann Reinhold, 16
Fossil pollen, 108–109
Fossil record, 11, 18, 135–137, 335–336
  biogeographic implications of, 337–343
  cladistic reconstruction and, 334–335
  evolution in, 229–231
  extinctions in, 253–255
  Great American Interchange and, 509
  latitudinal gradient and, 458
  limitations of, 336–337
  lineages and, 337–339
  overkill hypothesis and, 213
  phyletic speciation and, 243–244
  susceptibility to extinction and geographic range in, 587
  in Wallace's theory, 28
Fossils
  defined, 336
  living, 300
Fossorial species, 121
Foster, J. Bristol, 437

Founder effect, 233
Founder events, 236–237
Fouquieria splendens, 76
Fragmentation, habitat loss and, 562–564
Fragmented habitats, patterns of distribution, 598–601
Freshwater communities, 126–128
  effect of global climate changes, 608–610
  exotics, 548
  latitudinal gradient and, 454–458
  productivity and biomass, 129, 130, 132
Freshwater fauna, post-Pleistocene dynamics, 399–401
Freshwater organisms, endemism, 308
Frogfishes, 248
Frogs, 146, 512. *See also* Amphibians
Fruit bats, 451
Fruit pigeons, 293
Fungi, extinctions, 535
Furnariidae, 295
Fynbos, 6, 118
  birds and, 529

Gabbiella senaariensis, 230
Galápagos finches, 246
Galápagos Islands
  allopatric speciation in, 21, 239, 442
  El Niño and, 47
  extinctions, 216–217
  Pleistocene extinctions, 218
  quantifying similarities, 318–320
Galápagos tortoise, extinctions of, 554
Galaxioidea, 146
Gall wasps, 506
Gamma diversity, 450, 482, 485
Gastropods, 587
Gatun Lake, 86, 252, 284, 286, 288, 387, 519
Gaussian function, 289–290
GCM. *See* Global circulation model
Geckos, 293
Geese, 122
Gene flow, 233
  adaptation and, 91–94
Genera, 228
Genetic differentiation, mechanisms, 232–235
Genetic drift, 232, 236–237
Genus, 224
Genyochromis mento, 242
Geoclimatic cycles, 138

Geographic Distribution of Pines of the World, 494
Geographic information systems, 11, 35, 64, 570, 614
Geographic races, 227
Geographic range, 39
  endangerment and, 587–589
  of single species, 61–65
Geographic speciation. *See* Allopatric speciation
Geographic variation, 233
The Geographical Distribution of Animals, 25
Geography, relation to biogeography, 5, 10
Geological periods, fossil record and, 337
Geological time scale, 135–137
Geology, 10
  marine, 143–144, 148
Geomydoceus spp., 357
Geomys spp., 245, 237–239, 357, 498–499
Gerygone spp., 493
Giant bison, 210–211
Giant clams, 451
Giant deer, 435
Giant panda, 591
Giant shrew, 435
Gigantism, 434–437, 442
Gilia spp., 248
Gilpin, Michael, 391
Ginkgo biloba, 300
Ginkgoads, oldest known fossil, 338
GIS. *See* Geographic information systems
Glacial cycles
  biotic exchange and, 203–205
  extinctions and, 210–219
  Homo sapiens range expansion and, 577–578
Glacial Lake Agassiz, 200
Glacial lakes, 200–202
Glacial refugia, 207–210
Glacial-interglacial cycles, 178–181
  diversity patterns and, 459
Glaciation
  aquatic systems and, 200–203
  biotic exchange and glacial cycles, 203–205
  diversity patterns and, 472–473
  evolutionary response to Pleistocene refugia, 205–210
  extent and causes of, 177–181
  nonglaciated areas, 181–189

plant communities in southwestern United States and, 198–200
  terrestrial biotas and, 189–198
Glaciers, 108
  Cordilleran, 207–208, 578
  Laurentide, 185, 207–208, 578
  in North America, 6
Glacio-pluvial periods, 178
Gleason, H. L., 95, 107–108
Gleization, 48, 50
Global circulation model, 605
Global climate change, 567–568
  changes in connectivity and isolation, 610–612
  effect on distributions, 604–605
  effect on freshwater and marine ecosystems, 609–610
  effect on terrestrial ecosystems, 605–608
  species responses, 601–604
Global Ecology and Biogeography Letters, 11
Global positioning systems, 615
Global temperatures, 601–603
Global warming, 181
Globicephala melas, 317
Gloger, C. L., 27
Gloger's rule, 27, 490–492
Glossopteris, 145, 339
Gobi Desert, species diversity, 479
Goby, 417
God, diversity of life and role of, 14
Golden plover, 274
Gondwanaland, 6, 23, 172
  breakup of, 145–146, 148, 163–164
  formation of, 157–162
  Great American Interchange and, 508
  history of, 345
  vicariance biogeography and, 355–356
Gondwanaland relicts, 297, 507
Good, R., 33
Gorda Plate, 151, 155, 170
Gould, Stephen J., 229
GPS. *See* Global positioning systems
Grasses, in tropical savanna, 115
Grasshopper, 244
Grasshopper mice, 500
Grassland, 112
  loss of, 561
  response to climatic cycles in Pleistocene, 190

Gray, Asa, 21, 23, 34, 261
Gray kingbird, 446
Gray whale, 252
Gray wolf, 563
Grazing, tropical savanna
    and, 115
Great American Interchange,
    165–166, 205, 273,
    286–287, 508–513
    extinction and, 544–545
Great auk, 252, 534
Great Basin, 167, 307,
    395–396
    boreal mammals in,
        568–569
Great Lakes, 5–6, 188,
    399–400, 459, 466
    zebra mussel invasion,
        550–551
Great Lakes region, mam-
    malian distribution, 599
Great Plains, soil map, 49
Great Rift Valley, 167–168
Great Salt Lake, 79–80, 127,
    203, 278, 301, 479
    fossil record, 336
Great Smoky Mountains, 46
Greater Antilles, 422, 425
Greenhouse effect, 42
Greenhouse gases, 181,
    601–602
Grinnell, Joseph, 32, 274
Ground lizards, 554
Ground sloths, 210–211, 509
Ground squirrels, 498
Groundhog, 524
Grus americana, 588, 592, 593
Grus canadensis, 588
Guam, range collapse,
    595–596
Guam broadbill, 595
Guam rail, 595
Guanaco, 273
Gudgeon, 417
Guide fossils, 135
Guild, 499
Gulf of Mexico, soil along
    coast, 53
Gulf Stream rings, 56–57
Guyot, 143–144, 152–153, 155
Gymnogyps californicus, 591,
    592
Gyres, 56

Habia rubica, 493
Habitat destruction, as cause
    of extinction, 556, 558
Habitat heterogeneity, 458
    diversity and, 481–482
Habitat loss, fragmentation
    and, 558–562
Habitat selection, 283–284,
    291
Habitations, 17
Hadal communities, 543
Hadley cells, 42

Haeckel, Ernst, 24
Half-life, 136
Haliaeetus leucocephalus, 386
Halomorphic soil, 51–52
Halophytic plants, 52
Haplochromis cyaneus, 242
Haplochromis spp., 242
Hardyhead, 417
Harmonic biotas, 408
Harshness, 458, 460
    diversity and, 478–480
Hawaii
    hot spots, 169, 570
    range collapse in, 594
Hawaiian honeycreepers,
    246, 247, 362, 364
Hawaiian Islands
    adaptive radiation on, 247
    biogeographic history of,
        360–365
    colonization of, 32, 409,
        413
    competition among birds,
        422
    evolution of flightless
        birds, 430
    exotics, 547, 548, 549
    extinctions, 413–414, 535,
        546, 555, 556
    human colonization, 581
    species introductions of
        birds, 552–553
Hawaiian silverswords, 333
Hawaiian birds, avian malar-
    ia, 87
Hawks, 421–422
Hazel grouse, 598
Heat energy, 40
Hedgpeth, J. W., 33
Hemitilapia oxyrhynchus, 242
Hemlock, 109, 605–607
Hennig, Willi, 34, 328, 352
Hennig's paradigm, 328–333,
    360
Herbivores, 100
Herons, 421–422
Herrings, 240
Hess, Herman H., 143, 149,
    150
Heteractis malu, 89
Heterotheca latifolia, 290
Heterotrophs, 95
Hibiscadelphus, 362–363
Hierarchical classification
    system, 228–229
Hilgard, E. W., 30
Himalayas, 152–152, 172, 463
Hispaniola, 425–426
Historical biogeography, 4,
    345, 618–619
    centers of origin, 346–350
    Croizat's panbiogeogra-
        phy, 350–352
    Hennig's progression rule,
        352
    modern, 357–365

    vicariance biogeography,
        352–357
Historical perturbation, 458
    diversity patterns and,
        473–475
History of lineage, 325–327,
    343, 358–359
History of place, 325,
    358–359
Histosols, 48
Histrio histrio, 248
HMS *Beagle*, 20–21
Hoatzin, 295
Holarctic, 163, 303, 346
Holmes, Sherlock, 215
Holocene
    Australian extinctions, 216
    geographic range shifts of
        rodents during, 197
    pluvial lake disappear-
        ance, 203
    range extension of trees,
        196
    temperature, 182–184
Hominids, 574
Hominoids, 574
Homo erectus, 574–576
    geographic range, 576
Homo ergaster, 574, 575, 576
Homo habilis, 574, 575, 576
Homo sapiens
    biodiversity crisis and,
        536–537
    classification of, 228
    as cosmopolitan species,
        299
    ecology of, 621–622
    effect on natural world,
        624
    estimated population
        growth of, 67
    origin, 574–577
    range, 39, 494
    role in extinctions, 190,
        212–214, 218, 253, 441
Homo sapiens sapiens, 573, 576
    colonization of globe, 577
Homoizoic belts, 31
Honeyeaters, 306, 421
Hooker, Joseph Dalton, 8, 10,
    19, 21, 23, 140, 261, 370
Hopping mice, 522
Hornbills, 523
Hornwort, 299
Horse latitudes, 42
    precipitation in, 44
Horses, 339–340
Hot spots, 169–170
    marine, 542–544
    terrestrial, 538–542
Hot springs, 128
House mice, 548
House sparrows, 232,
    267–269, 386, 515, 553
Hultén, E., 208–209
Human ecology, 621–622

Humanity, biogeography of,
    573, 586
    expansion to New World,
        577–579
    human origins and colo-
        nization of Old World,
        574–577
    island biogeography,
        579–585
Humans. *See also Homo sapi-
    ens; Homo sapiens sapiens*
    estimated population
        growth, 67
    habitat destruction by,
        559–560
    role in extinctions,
        212–214, 218, 251, 253,
        545
Humboldt Current, 46
Humidity, coloration and,
    491
Hummingbirds, 422
    North American distribu-
        tion, 88
Humus, 48–49
Hunting, 212–214, 556
Hurricanes, 47, 404
    role in distribution, 81
Hutchinson, Evelyn, 67
Hutchinson, G. E., 34, 370
Huxley's line, 310
Hybridization, 225
    cladistic reconstruction
        and, 334
    of exotics, 554
Hydrilla verticilata, 271
Hydromantes, 321
Hydromys, 257
Hydrophis, 348
Hydropsyche, 101
Hydrothermal vents, 543–544
Hypericum canadense, 263
Hypericum perforatum, 86
Hypersaline lakes, 127, 485
    harshness and, 479
Hypothetico-deductive rea-
    soning, 8
Hypsithermal, 183

Ibis, 24
IBP. *See* International
    Biological Program
Ice plants, 52
Ichneumonid wasps, 451–452
Immigrant pattern, 416
Immigration, 34, 69. *See also*
    Equilibrium theory of
        island biogeography
    differential, 414–419
    equilibrium theory and,
        372, 375, 379
    in insular communities,
        409–410
    on Krakatau, 401, 402–403
    nature reserve design and,
        566

Incas, range collapse of, 595–597
Incidence function, 419–423
Incubator bird, 421
Index fossils, 135
Indian mynah, 553
Indian Ocean, 167
Indian Plate, 152
Individual selection, 255
Individuals
  distribution of, 64–65
  geographic variation in, 621
  vs. species, 223
Indonesia
  human colonization, 579
  selective immigration in, 415
Inductive reasoning, 8
Industrial melanism, 231, 232, 491
Insects
  distribution in Australia, 65, 71
  evolution of flightlessness in, 431–432
  long-distance dispersal, 409
  oldest known fossil, 338
  probability of extirpation, 587–588
  sympatric speciation in, 239
Insular communities, 407
  differential migration and extinction, 414–419
  establishing populations, 410–411
  extinction in, 411–414, 556, 558
  immigration in, 408–410
  interspecific interaction, 419–429
  nestedness of, 414–418
Insular distribution, 418–419
Insular habitats, patterns of distribution, 598–601
Insular mammals, 436–440
Insular species richness, 390–394
Interactive equilibrium, 389–390
Interdisciplinary studies, biogeography and, 5–7, 618–619
Interference, 83
Interglacial periods, 178
International Biological Program, 520
International Council for Bird Preservation, 538–539
Interspecific interactions, 458, 460
Intertidal zone, 57–58, 104, 126

Intertropical convergence zone, 115
Invasion, 547–548
  resistance to, 514–515
Inversion, 239
Invertebrates
  evolution of flightlessness in, 431–432
  propagules, 277–278
Ipomopsis aggregata, 88, 248
Ipomospsis longiflora, 248
Iridium, 254
Irvingtonian extinction, 215
Island arc, 170, 312
Island biogeography, 16, 23–24, 33. *See also* Equilibrium theory of island biogeography
  conservation and, 598–600
  history of, 369–372
  humans and, 579–585
Island chains, tectonic development of, 166–171
Island Life, 23, 25
Island patterns
  species turnover, 378–379
  species-area relationship, 372–376
  species-isolation relationship, 376–378
Island rule, 437–438
Island size, equilibrium theory and, 381, 384
Islands. *See also* Insular communities; *individual island names*
  adaptive radiation on, 246–247
  classifying, 312–314
  as endangered species refugia, 598
  Endemic Bird Areas on, 539
  evolution of body size on, 434–444
  evolution of flightlessness and reduced dispersal ability on, 429–434
  extinctions, 545–547
  founder events and, 236–237
  influence on biogeography, 9
  ridges and, 153
  species introductions, 549–558
  sweepstakes routes and, 287–289
  in Wallace's theory, 28
Isle of Cumbrae, 68–69
Isolated biotas
  convergence, 520–527
  divergence, 518–520
Isolating mechanisms, 237
Isolation, 21, 23–24, 236–237
  biodiversity and, 537–538

equilibrium theory and, 381, 388
  global climate changes and, 610–612
  habitat loss and, 563
Isostatic changes, 186–188
Isthmus of Panama, 6, 310, 509, 518–519
Isthmus of Tehuantepec, 310
Ivory-billed woodpecker, 592, 593

Jablonski, David, 544
Jaccard similarity index, 318
Jack pines, 81
Jaegers, 283
Jaguar, 103
Jamaica, 387
Janzen, Daniel, 281
Java, 265
Jays, 499
Jellyfish, 278
Jerboas, 522
Jet stream, 42, 185
  passive dispersal and, 277
Jordan, D. S., 28
Jordan, Pascual, 142
Jordan's law of vertebrae, 28
Journal of Biogeography, 11
Juan de Fuca Plate, 170
Juglans nigra, 83
Julidochromis transcriptus, 241
Jump dispersal, 265–266
Juncus spp., 263
Juniper tree, 64
Juniperous osteosperma, 64
Juniperus, 117
Juniperus californica, 276
Jurassic, 137, 143
  continental drift, 159

Kangaroo mice, 53, 297
  convergence, 522
Kangaroo rat, 53, 83, 245, 257, 297–298
  in Baja California, 462
  convergence, 522
Kangaroos, 441
Kapingamarangi Atoll, 393–394
Kettle lakes, 201–202
Kew Gardens, 522
Keystone species, 90, 106, 109
Killer bees, 267
Kinglets, 96, 97
Kirtland's warbler, 592, 593, 607–608
Kiwis, 297, 313, 320–321, 547
Klamath weed, 86
Krakatau, 377–379
  jump dispersal and, 265–266
  recolonization, 401–405
  soil formation, 48
Krameria grayi, 276
Kriging, 64

K-selection, 493
K-strategists, 410
Kula Plate, 170
Kurtén, B., 32

Labeotropheus fuelleborni, 242
Labidochromis vellicans, 242
Lace-necked dove, 553
Lack, David, 370
Lago Chicancanab, 242–243
Lago Guri, 388
Lagomorphs, 438
Lagopus, 122
Lake Baikal, 308, 400, 466
Lake Bonneville, 203
Lake Chad, 203
Lake Champlain, 188
Lake Chichancanab, 459
Lake Erie, antecedent, 6
Lake Eyre Basin, 303
Lake Gatun, 86, 256, 284, 286, 288, 387, 519
Lake Malawi, 241–242, 457, 466
Lake Mendota, 202
  temperature profile, 64
Lake Michigan, antecedent, 6
Lake Missoula, 202
Lake Tanganyika, 457, 466
Lake trout, 86
Lake Turkana Basin, 230
Lake Victoria, 86, 241, 400, 457, 535, 550
Lakes, sympatric speciation in, 240–243
Lama guanicoe, 273
Lampreys, 86
  fossil, 338
Land snails, 409, 410–411, 432
Land tortoises, 442
Land transformation, human impact, 537
Landbridge islands, 398–399
  estimates of turnover on, 385–387
Landbridges, 22–23, 187, 189, 261–262, 285–286. *See also* Beringia
Landmasses, productivity and, 131
Lapland rosebay, 207
Larch, 121, 209
Largemouth black bass, 86
Larix spp., 121, 209
Larrea, 321
Larrea tridentata, 79
Lasiodiamesa, 332
Lasiorhinus spp., 441
Lasiurus, 274
Laterization, 48, 49, 50
Lates niloticus, 86, 550
Laticauda laticauda, 348
Latitude. *See also* Latitudinal gradient
  body size and, 488–490
  clutch size and, 492–493

effects of biotic and abiotic factors, 91

geographic ranges and, 537–538

Rapoport's rule and, 468–470

solar radiation and, 40–41

variation in diversity and, 315–316

Latitudinal diversity, abyssal plain and, 468

Latitudinal gradient, 450–451

patterns, 451–458

processes, 458–461

Laurasia, 306

breakup of, 163

formation of, 157–162

Laurels, 118

Laurentide ice sheet, 185, 207–208, 578

Leaching, 49

Leiopotherapon unicolor, 417

Leiothrix lutea, 553

Lemmings, 71, 122

Lemmus, 197

Lemurs, 246

Lentic habitats, 126

Leopold, Aldo, 533, 534

Lepomis, 281

Leporillus, 257

Leptasthenura aegithaloides, 527

Leptodactylon californicum, 248

Lepus spp., 490

Lessepsian migration, 513

Lesser Antilles, 422, 425, 426, 447

Lesser Sunda Islands, 286

Lethrinops brevis, 242

Leups townsendii, 569

Lianas, 112

Lichens, 122

Liebig's law of the minimum, 74

Life history characteristics, geographic variation in, 492–493

Life zones, 29, 30, 96

Limestone, 336, 341

global distribution, 342

Limnetic zone, 127

Limnochromis leptosoma, 241

Limnologists, 123

Linanthus dianthiflorus, 248

Line Islands, 170

Line of equality, 495

Lineages. *See also* Fossil record; Systematics; Vicariance biogeography

fossil histories of, 337–339

history of, 325–327, 343

Linguistic diversity, island area and, 583–584

Linnaean classification system, 228

Linnaean shortfall, 534–537

Linnaean Society of London, 24

Linnaeus, Carl, 14–15, 228, 326

Liquidambar, 321

Liriodendron tulipfera, 81

Lithosphere, 150

Little Ice Ages, 183–184

Little Mariana fruit bat, 595

Littoral zone, 126

Living fossils, 300

Lizards, 53, 248, 442, 530

competition and, 424–426, 427

convergence in, 523, 526

disjunction in, 321, 323

species richness on land-bridge islands, 378

trends in insular body size, 443–444

Lobelia cardinalis, 88

Loblolly pine, 75–76

Lobochilotes labiatus, 241

Local diversity, relation to regional diversity, 506–507

Locusta migratoria, 71

Locusts, distribution pattern, 71

Lodgepole pine, 81, 195–196

Long-finned pilot whale, 317

Lonicera arizonica, 88

Loriidae, 306

Lorikeets, 306

Lotic habitats, 126

Lowe's line, 287

Loxia leucoptera, 71–72

Loxigilla noctis, 446

Lungfishes, 340

Lungless salamanders, 321

Lycaena epixanthe, 600

Lyddekker's line, 308–310

Lyell, Charles, 9, 18, 21–22, 30–31, 138, 261

Lystrosaurus, 146, 339

Lythrum salicaria, 271–272

Lythrurus snelsoni, 600, 611–612

MacArthur, Robert, 8, 11, 34, 370, 373, 379–397, 407

Macchia, 6, 118

birds and, 528

Macrodactyla doreensis, 89

Macroecology, 29, 500–506

Macroevolution, 229–232

Macropus spp., 441

Macropygia, 420

Madagascar, 303, 312, 442

adaptive radiation on, 246–248

endemics and, 295, 297, 298

evolution of flightless birds, 430

extinctions, 216, 218, 413–414

formation of, 163–164

Madeira Islands, 431–432

Magma, 150

Magnetic reversals, 149

Magnetic stripes, 149

Magnolia, 118

Mahogany, 298

The Malay Archipelago, 25

Maldives Archipelago, 610

Malthus, Thomas, 67

Mammals

body size and range, 99, 100, 501, 502

body size and species diversity, 503–505

Cenozoic radiation of, 256–257

convergence in, 524–526

dispersal, 408–409

endemism in Africa, 541

extinctions, 214–215, 217, 219, 412, 546, 557

geographic ranges in North America, 452, 495, 496

Great American Interchange and, 509–511

insular, 436–440

matrix of similarity coefficients, 319

nonvolant, 266

oldest known fossil, 338

placental, 256–257

Pleistocene relicts, 396

range limits of Neotropical, 311

response to Pleistocene climatic cycles, 190

species introductions, 548–549

species-area relationship on Krakatau, 403

status of species in United States, 534

target area effect and, 392–393

Mammals of North America, 494

Mammoths, 210–211

Mammuthus primigenius, 436–437

Mammuthus spp., 210–211

Management, conservation, 622–624

Mangrove, 52, 263, 389

selective immigration and, 415–416

Mangrove swamps, 128

Mantle drag, 150–151

Maple, 605–607

Maps, 614

range shapes and, 495–496

range of species, 61–64

Maquis, 6, 118

Mariana crow, 595

Marine basins, tectonic development of, 166–171

Marine biogeography, 30–31

Marine communities, 123–126

effect of global climate changes, 608–610

productivity and biomass, 129, 130, 132

Marine fisheries, human impact, 537

Marine geology, continental drift theory and, 143–144, 148

Marine hot spots, 542–544

Marine invertebrates, 210

Marine organisms, glacial cycles and, 203–204

Marine realms and provinces, 314–317

Marine zoogeography, 33

Marls, 127

Marmota flaviventris, 45, 70

Marmots, 210, 498

Marquesas, human colonization, 581

Marshes, 127

Marsilea, 299

Marsupials, 303, 339

body size, 441, 438

disjunctions, 321

phylogenetic hypothesis of relationships among, 335

radiation, 256–257

Martin, Paul, 253

Mascarenes, Pleistocene extinctions, 218

Mass extinctions, 108–109, 175–176, 253. *See also* Extinction

Mathematical models, equilibrium theory, 372–373

Mathematical theory, 11

Matorral, 6, 118, 521, 527, 528

Matrix, 623

Matthew, W. D., 32

Matthews, Drummond, 149

Maurer, Brian, 438

Mayfly, 101

Mayr, Ernst, 11, 32, 229, 235, 240, 370

Meadow voles, 423–242

Mealybugs, 278

Mediterranean, Pleistocene extinctions, 218

Mediterranean climates, 44, 118

convergence in, 520–521, 526–527

distribution, 6–7

Mediterranean Sea, 285

formation of, 167

Lessepsian migration, 513

Mediterranean shrub habitats, bird community organization, 528–529
Megaceros, 435
Megalonyx spp., 210–211
Meganesia, Pleistocene extinctions, 218
Megapodius freycinet, 421–422
Melaleuca, 248
Melanospiza richardsoni, 446
Meliphagidae, 306
Merriam, C. Hart, 29, 30
Mesophytes, 79
Mesozoic, 137, 140
Mesozoic radiation, 255–256
Mesquite, 321
Metallicine, 356
Metapopulation, 70
Metapopulation theory, 418
 species conservation and, 598
Meteorologica, 13
Mice, 423, 435, 462, 500, 522. *See also* Deer mice; Kangaroo mice
Michaud, Joanne, 600
Micoceanic island chains, 312
Microbial biogeography, 4
Microclimates, 59
Microcosms, 97
Microdipodops, 297
Microdipodops pallidus, 53
Microenvironments, 59–60
Microevolution, 229, 231–232
Micropterus salmoides, 86
Micropterus spp., 281
Microtus agrestis, 440
Microtus longicaudus, 569
Microtus pennsylvanicus, 423–424
Middens, 198–199
Midges, 101, 355–356
 cladogram, 332, 361
 consensus area cladogram, 360
Midway Island, 169
Migration
 avian, 516–519
 Pleistocene, 195
Milankovitch cycles, 180–181
Mimus polyglottos, 553
Miocene, continental drift, 160
Miro australis, 493
Mirouga angustirostris, 252
Mississippi River, 167, 308
Moas, 313, 320–321, 430, 435
Mockingbird, 553
 range boundaries, 74
Mogumda mogumda, 417
Moisture, range boundaries and, 78
Moisture gradients, 105
Mojave Desert, 44, 298, 307
Mole, 524

Molecular systematics, 333–334
Mollusks, 316
 endemism, 308
 evolution of, 230–231, 257–258
 extinctions, 535, 546, 556, 557, 558
 in hydrothermal vents, 544
 latitudinal gradient and, 455, 457
 Lessepsian migration, 513
 long-distance dispersal, 409
 Rapoport's rule and, 469
 species richness, 466–467
Molothrus ater, 563
Mona, 387
Monarch butterflies, 275
Monarcha casteneoventris, 233
Monarcha cinerascens, 421
Mongooses, 553, 554
Monito, 300
Monitor lizard, 443
Monophyletic, 328
Monotremes, 256–257
 fossil, 338
Monsoon forest, 114
Monsoons, 185–186
Montane forest, 118, 530, 567
Monte Desert, 45
Moon, effect on tides, 57–58
Moors, 127
Moose, 210, 213
Morphological species concept, 224
Moss, 122
Moth, 86, 375
Mount Ararat, 15
Mount Kilimanjaro, 41
Mount Lemmon, 45–46
Mount Vesuvius, 341
Mountain lion, 103
Mountain ranges, orientation, 497
Mountains
 as barriers, 484
 diversity patterns, 463–464
Multiple interactions, 89–91
Multispecies assemblages
 areography, 494–500
 local and regional diversity, 506–507
 macroecology, 500–506
Munroe, Eugene Gordon, 371, 372
Munzothamnus sp., 298
Murray, John, 31
Mus musculus, 548
Muscivora forficata, 73, 590
Musk oxen, 122
Mussel, 81
Mustela nigripes, 591
Mustella erminea, 569

Mutela nilotica, 230
Mutualism, 34, 87–89, 109
 diversity and, 483
Myiagra freycineti, 595
Myiarchus cierascens, 527
Myopopone castanea, 415
Myriophyllum, 299
Mytilus californianus, 81
Myzomela pammelaena, 421

Nama dichotomum, 322
Nannodectes spp., 244
Natural selection, 21, 232–233
Nature reserve design, 564–567
Nazca Plate, 166, 170
Neap tides, 58
Nearctic, 165, 303
Nekton, 126
Nelson, G., 352, 353
Neoceratodus fosteri, 340
Neoendemic, 300
Neofiber, 257
Neogene, 137. *See also* Holocene; Pleistocene
Neosilurus, 417
Neotoma cinerea, 60, 488, 489, 569
Neotoma spp., 198, 257
Neotropical mammals, range limits, 311
Neotropical Pleistocene refugia, 205–207
Neritic zone, 126
Nestedness, 414–418
Net primary productivity, 128–130
New Caledonia, 313, 409
 endemics and, 295, 296
New Guinea, 375, 378
 landbridge islands, 398–399
 sea snakes and, 347–349
 selective immigration in, 415
 variation in bird species richness, 464
 vegetation zone shifts during Pleistocene, 193
New synthesis, 229
New World, colonization of, 577–579
New Zealand, 312, 313, 409, 410
 bird clutch size and latitude, 493
 endemics, 295–297, 298–299
 evolution of flightless birds, 430
 exotics, 547, 548, 549, 553, 555, 556
 extinctions, 218, 546–547
 formation of, 163–164
 vegetation zone shifts during Pleistocene, 194

Newton, Isaac, 14, 57
Niche, 67–70
Niche shifts, on islands, 423–427
Nicrophorus americanus, 593
Nile perch, 86, 550
Niobrara River, 341
Nitrogen fixation, human impact on, 537
Node, 330
Nomadacris septemfasciata, 71
Nonequilibrium biotas
 landbridge islands, 398–399
 Pleistocene refugia, 394–398
 post-Pleistocene dynamics of freshwater faunas, 399–401
Nonglaciated areas, 181–189
Non-interactive equilibrium, 389–390
Nonvolant mammals, 266
North America, 303
 biogeographic provinces for freshwater fish, 309
 bird range boundaries, 74
 body size and range, 100, 501, 502–503
 body size and species diversity, 503–505
 convergence in, 524–526
 current and Holocene range shifts, 197, 199
 density compensation, 423
 deserts, 53, 307, 318
 diffusion of species, 267–272
 distribution patterns in, 71, 311, 323, 399–400
 diversity patterns, 451–456
 extinctions, 210–214, 217–219, 251–253, 412, 535, 545
 forest clearing, 559
 formation of, 163
 fossil localities of horse and camels, 339–340
 geographic ranges of land mammals, 495, 496
 glacial recession in, 178–179
 glacial-interglacial cycles, 184
 global climate effects, 605–608
 Great American Interchange, 508–513
 history of, 5–6
 hummingbird distribution, 88
 latitudinal boundaries of marine provinces, 316
 life zones, 30

lungless salamander distribution, 321
Monarch butterflies, 275
neoendemics, 301
patterns of range collapse, 592, 593
percentage of exotics of mammals and birds, 555
Pleistocene refugia, 395
pluvial lakes, 202
recolonization after Pleistocene glaciation, 108–109
role of fire in species distribution, 81–82
secular migration, 273
songbird distribution pattern, 71
species richness of non-volant vertebrates, 478
spruce-fir forests, 96, 97
stages in human colonization, 578
temperature cycles in Pleistocene, 184
timberline, 77
variation in bird ranges with latitude, 469
variation in vegetative communities, 195
vegetation zone shifts during Pleistocene, 192
North American Breeding Bird Survey, 73
North American muskrats, 257
North American plants, diaspores, 276
North American Plate, 170
Northern boreal region, 124
Northern elephant seal, 252
Northern hairy-nosed wombat, 594
Northern Hemisphere
glacial-interglacial cycles in, 184
glaciation of, 178
landmasses, 303
Northern notal region, 124
Northern pocket gophers, 197
Northern subtropical ocean region, 124
Northern temperate ocean region, 124
Northern wheatear, 519
Nothofagus spp., 118, 120, 296–297, 303, 306, 313, 339, 356
consensus area cladogram, 360
Notomys, 257
Notropis, 308
NPP. *See* Net primary productivity

Numerical taxonomy, 327
Nunataks, 207–210
Nyctea scandiaca, 71–72

Oak, 117, 118, 271
Oak-laurel forest, 118
Obliquity, of earth's orbit, 180–181
Ocean temperature
mean global, 183
in Pleistocene, 185
Oceania, 579, 580
Oceanic circulation, 55–57
Oceanic islands, 312–314
species introductions, 549–558
Oceanographers, 123
Oceans, 123. *See also* Aquatic communities; Aquatic environments; Marine communities; Panthalassa
currents, 43
diversity and, 477–478
latitudinal gradient and, 454–455
productivity, 131–132
productivity and biomass, 129, 130, 132
properties of ocean floor, 149
Ocelot, 524
Ochotona princeps, 250–251, 391, 569
Ocotillo, 76
Odocoileus spp., 107
Oenanthe oenanthe, 519
Old Sandstone Continent, 162, 172
Old World, colonization of, 574–577
Oligotrophic lakes, 127
Olor buccinator, 588
Olson, E. C., 32
Olympic Mountains, 172
Ondatra zibethica, 267–269
Onychomys spp., 500
Oparanthus spp., 433
Ophiostoma ulmi, 554
Opisthocomus hoazin, 295
Opossums, 267, 509, 510
Opththalmochromis nasutus, 241
Opuntia stricta, 86
Orchids, 463
Ordovician, 148
Origin, place of in classifying endemics, 300
The Origin of Continents and Oceans, 139
The Origin of Species, 21, 23
Orthogenesis, 29
Orthogeomys spp., 357
Ortmann, Arnold, 31

Oryctalagus cuniculus, 268–270, 548
Osmorhiza spp., 322
Osprey, 386
Ostrich, 296, 320–321
Ottawa River, 188
Otus, 492
Ouachiata Mountain shiner, 600, 611–612
Outbreak areas, 71
Outgroup, 329
Outline maps, 61–63
Ovenbirds, 295
Overdispensed species, 104
Overkill hypothesis, 212–214, 218, 253, 441
objections to, 216
Overturn, 65
Ovibos moshatus, 122
Ovis dallii, 122
Owls, 492

Pachycepahlinae, 306
Pachycephala spp., 419–420
Pachyornis spp., 430
Pacific basin, inferred history, 170
Pacific Ocean, dynamics of, 167, 169
Pacific Plate, 151, 170
Pacifica, 154
Pack rats, 198, 257
Pagility, 273
Palearctic, 303
Paleobotanists, 108
Paleocirculations, 170–171
Paleoclimates, 170–171
Paleoecologists, 108
Paleoecology, 4, 340–343
Paleoendemic, 300
Paleogene, 137
Paleomagnetism, 148–150
Paleontology, 5, 10, 32
Paleothermometer, 182
Paleozoic, 148
Palms
distribution of, 64–65
pantropical distribution, 451
Palo verde, 116, 321
Pampas, 120
Panama Canal, 288
Panbiogeography, 350–352
Pandion haliaetus, 386
Pangaea, 140, 142, 145–146, 157, 171, 508
breakup of, 162–163
formation of, 157–162
Pangolins, 323
Panthalassa, 162, 166, 170–171
Panthera onca, 103
Paradox of the plankton, 481
Paramecium, 245
Paramo, 122

Parapatric speciation, 239. *See also* Sympatric speciation
Parasites, 87, 356
dispersal, 279
energy flow, 103
sympatric speciation in, 239
Paroaria cristata, 553
Parus inornatus, 527
Passenger pigeon, 195, 251, 534, 587–588
Passer domesticus, 267–269, 386, 515, 553
Passer montanus, 515
Passive dispersal, 276–279
Pastoralists, 115–116
Pattern, defined, 7
Patterson, Bruce, 416
Pedogenic regimes, 48
Pekin robin, 553
Pelagic organisms, 126
Pelamis, 520
Pelamis platurus, 347–348
Penguins, 283, 451
Peninsula effect, 461–462
nature reserve design and, 565
Peninsulas, diversity patterns, 461–463
Penstemon barbatus, 88
Peppered moths, 231, 232, 491
Perch, 417
Peregrine falcon, 299, 386
Periods, 137
Perla, 101
Permafrost, 121
Permian, 137, 145–146, 155
continental drift, 159
extinctions, 253
Peromyscus maniculatus, 92, 233–235, 491
Peromyscus spp., 291–292, 423, 440
Petrochromis polyodon, 241
Petroica macrocephala, 493
Petromyzon, 86
Petrotilapia tridentiger, 242
Pezophaps solitariaon, 430
Phacelia spp., 322
Phaeophyta, 497
Phalacrocorax harrisi, 430
Phanerozoic, 137, 157
taxonomic diversity, 473
tectonic, climatic, and biotic events, 174–175
Phasianus colchicus, 553
Philippines, 314, 315, 347
Philopotamus, 101
Philosophy of science, 7–8
Phlox, 248
Phoca sibirica, 308
Phoenix Plate, 170
Phoresy, 279
Photic zone, 53, 125

Photosynthesis, 96
  in marine communities, 125–126
  in water, 53
Phrynosoma solare, 307
Phyletic speciation, 243–244
Phylloscopus borealis, 518
Phylogenetic cladogram, 359
Phylogenetic classification, 33–34. *See also* Systematics
  limitations of, 334–335
Phylogenetic species concept, 225–227
Phylogenetic systematics, 228, 328–333
Phylogeography, 358
Physiological barriers, to dispersal, 281–283
Phytogeographers, 4
Phytogeography, 16–17, 24, 29
Phytoplankton, 53–55, 102, 126
  species richness, 466
Phytosociology, 29
Picea glauca, 195–196
Picea spp., 96, 97, 121
Pickleweeds, 52
Pikas, 250–251, 391
Pinaceae, 451–452
Pine barrens, 52
Pines, 106. *See also individual species names*
  latitudinal variation in, 456
Piñon pine, 199
Pinus banksiana, 81, 195–196
Pinus contorta, 81
Pinus contorta var. *latifolia*, 195–196
Pinus longaeva, 76, 300–301
Pinus ponderosa, 106
Pinus sabiniana, 106
Pinus spp., 106, 117
Pinus taeda, 75–76
Pipidae, 146
Pisaster ocraceous, 81
Pitcher plants, 52
Placental mammals, radiation of, 256–257
Plankton, 126, 481
Plant biogeography, 16–17
Plant communities, in southwestern United States, 198–200
Plant diversity centers, 542
Plant extinctions, 210, 534–535
Plant physiological ecology, 29
Plantago hookeriana, 85
Plants, 33
  adaptations to unusual soil types, 51–53

classification schemes, 303–305
colonization and, 411
community and, 108
convergence, 522, 526
desert, 116
as disjuncts, 321–322
effect of physical factors on range, 78–79
endemism, 306, 541
exotics, 537, 547
hybridization and, 225
oldest known fossils, 338
patterns of endangerment, 558, 560
polyploidy in, 240
productivity, abiotic stress, and, 485
reduced dispersal ability, 432–434
response to Pleistocene climatic cycles, 190
species introductions, 549
status of species in United States, 534
variation in species richness with elevation, 464–465
Plasmodium, 87
Plastids, 334
Plate tectonics, 11, 18, 33, 137, 150–153, 155–157. *See also* Continental drift
  climatic and biogeographic consequences, 171–176
  role in prehistoric extinctions, 544–545
Platnick, N., 352, 353
Platypus, fossil, 340
Platyspiza crassirostris, 238
Pleidon, 230
Pleistocene. *See also* Glaciation
  biogeographic responses to climatic cycles of, 190
  effect on human colonization, 579–580
  extinctions, 210–219, 253, 584
  glaciation, 178–181
  island formation, 312, 314, 315
  lake formation, 200–203
  nonglaciated areas, 181–189
  sea level changes in, 186–189
Pleistocene assemblages, 341
Pleistocene changes, diversity patterns and, 472–473
Pleistocene refugia, 205–210, 394–398
Pleistocene refugium hypothesis, 206, 358
Plesiadapis spp., 244

Plesiomorphic character states, 328
Pliocene, Great American Interchange, 511
Pliocene Blancan extinction, 214–215
Plotosidae, 295
Plumbago scandens, 276
Plunge pools, 201–202
Pluvial lakes, 202–203
Pluvialis dominica, 274
Pocket gophers, 82, 245, 356–357, 488, 498
Pocket mice, in Baja California, 462
Podocarpus, 120
Podochlus, 361
Podonomopsis, 361
Podonomus, 361
Podzolization, 48–49, 50
Polemonium spp., 248
Polychaete worms
  in hydrothermal vents, 544
  species richness, 466–467
Polynesia, human colonization, 581–583
Polyploidy, 93, 239–240
Polytypic species, 226
Pomerine ants, 375, 378
Ponderosa pine, 106, 199
Pools, 126
Population biology, 5
Population density, body size and, 501
Population distribution
  distribution and abundance, 70–72
  geographic range as reflection of the niche, 68–70
  Hutchinson's multidimensional niche concept, 67–68
  population growth and demography, 65–67
Population growth
  demography and, 65–67
  species richness and, 475
Population-level characteristics, geographic variation in, 492–493
Porcupines, 323, 509, 510
Postglacial lakes, 200–203
Potassium-argon method, 136–137
Power model, 371, 373–374, 394
Prairie, 120
  mammal distribution, 505
  response to climatic cycles in Pleistocene, 190
Prairie dogs, 498
Precambrian shields, 143, 145
Precession, 180–181
Precinctiveness, 431

Precipitation
  biome classification and, 111
  in desert, 116
  effect on bird ranges, 608
  effect on diversity, 463–464
  formation of soil types and, 50
  net primary productivity and, 130
  patterns, 43–45
  in sclerophyllous woodland, 118
  in subtropical evergreen forest, 118
  in temperate rain forest, 119, 120
  in tropical rain forest, 112
  in tropical savanna, 115
  in tundra, 121
Predation, 34, 85–87, 106–107
  crypsis and, 491–492
  diversity and, 483
  diversity patterns and, 458, 460
  in Wallace's theory, 28
Prehistoric extinction, 413, 544–545
Pre-Pangaean supercontinent, 154
Pressure, salinity and, 57
Preston, Frank, 374–376
Prickly pear cactus, 86
Primary consumers, 100
Primary cordillera, 77
Primary producers, 100
Primary productivity, 96
Primary succession, 47–48, 107
Primates, 454
Principles of Geology, 18, 19, 20, 30, 138
Process, 7
Procolophodon, 339
Productivity, 128–132, 458, 460
  in communities, 98–103
  diversity and, 475–478, 485–486
Profundal zone, 127
Prognathogryllus, 364
Progression rule, 346, 352
  Hawaiian Islands and, 361–362
Pronghorn, 523
Propagule, 291–293
Prosopis, 321
Proteads, 303
Protonemura, 101
Provinces, 302
Provincialism, 295–296, 514–518
  avian, 516–519
  biogeographic lines, 308–312

classifying islands, 312–314

marine regions and provinces, 314–317

quantifying similarity among biotas, 317–320

terrestrial regions and provinces, 302–308

Pruvot, G., 31

Psaltriparus minimus, 527

Pseudalopex fulvipes, 546

Pseudobovaria, 230

Pseudomyrmex, 87

Pseudotropheus spp., 242

Psychological barriers, to dispersal, 283–284

Ptarmigan, 122

Pteropus tokudae, 595

Puerto Rico, 46, 548, 556

Puma concolor, 103, 563

Punctuated equilibrium, 229–231

Pupfishes, 203, 240, 242–243, 281, 459

Purple loosestrife, 271–272

Puszta, 120

Pyrope pyrope, 527

Quaternary, 137

Queen Charlotte Archipelago, 210

Quercus spp., 117, 118, 271

Rabbits, 268–270, 548

Radiant energy, 40–41

Radiation, 223–224

Radiocarbon dating, 137

Rafinesquin neomexicana, 276

Raillus owstoni, 595

Rain forest. *See* Temperate rain forest; Tropical rain forest

Rain shadow deserts, 44–45

Rainfall. *See* Precipitation

Rain-green forest, 114

Rana cancrivora, 409

Range boundaries competition and, 82–85

disturbance and, 81–82

multiple interactions and, 89–91

mutualism and, 87–89

physical limiting factors, 72–80

predation and, 85–87

Range collapse, patterns of, 588–598

Range expansion diffusion, 266–273

jump dispersal, 265–266

secular migration, 273

Range measurement and mapping, 61–64

Range sizes, distribution, 501–503

Ranges

overlap, 497–500

shapes of, 494–497

sizes of, 494

Rangifer tarandus, 122

RAP. *See* Rapid Assessment Program

Raphus cucullatus, 430, 534

Rapid Assessment Program, 569

Rapids, 126

Rapoport, Eduardo, 494

Rapoport's rule, 468–470, 484, 494

Ratite birds, 296–297, 306, 313, 320–321, 516

consensus area cladogram, 360

Rats, 548. *See also* Kangaroo rat

as invaders, 553

Rattus spp., 548, 553

Raup, David, 544

Realistic diversity model, 476

Realms, 302

Red algae, range overlap, 497–498

Red clover, 87

Red fox, 268–270

Red locust, 71

Red Queen hypothesis, 249

Red Sea formation of, 167

Lessepsian migration, 513

Red squirrels, 598–600

Red wolf, 563, 591

Red-backed voles, 97

Red-eyed vireo, 70–71

Redfinned blue-eye, 295

Red-winged blackbirds, 489

Refugia, 205–210, 394–398

Regal horned lizard, 307

Regional biotas, 17

Regional diversity, relation to local diversity, 506–507

Regions, 302

Regulus spp., 96, 97

Reinforcement, 237

Reinwardtoena, 420

Relaxation, 387–388, 565–566 island area and, 397

Relaxation model, 416

Relict pattern, 416

Relicts, 300

Remnant magnetism, 148

Remote sensing, 615

Rensch, B., 32

Reproductive isolation, 225

Reptiles, 374. *See also* Lizards endemism in Central America, 540

on Krakatau, 402

long-distance dispersal, 409

patterns in recorded extinctions, 546

species-area relationship, 374

status of species in United States, 534

trends in insular body size, 443–444

Rescue effect, 390–391, 392

Respiration, 96

Restricted-area species, 538–540

Reticulate cladogenesis, 223–224

Reticulate evolution, 223–224

Rhamphochromis macrophthalmus, 242

Rheas, 296, 320–321

Rheochlus, 361

Rhinoplocephalus, 348

Rhizophora mangle, 389

Rhododendron lapponicum, 207

Rhodophyta, 497–498

Rhyacophia, 101

Richness-dependent diversity model, 476

Richness-independent diversity model, 476

Ridge push, 150–151

Riffles, 126

Rift Valley, 308, 400

Rift zones, 152

Rifting, 167

Right whale, 317

Ring of fire, 148

Ring-necked pheasant, 553

Riparian deciduous woodland, 119

Rivera Plate, 170

Rock pocket mice, 53

Rocky Mountains, 49, 192, 287, 395

Rodents. *See also* Deer mice; Kangaroo mice; Kangaroo rat; Mice convergence, 522–523

geographic range shifts during Holocene, 197

gigantism in, 437–439

range overlap, 498–499

Romer, A. S., 32, 327

Root, Terry, 607

Rosen, D., 352

Ross, Sir James Clark, 23

Rotifers, 299

Royal Geographic Society of London, 24

r-selection, 493

r-strategists, 410–411

Rubus, 276

Ruby-throated hummingbird, 274

Rufous hare-wallaby, 594

Sabertooth cats, 210–211

Saguaro cactus, 73–74, 75, 76, 307

Saguaro National Monument, 75

Sahara Desert, human origins and, 575

Sahul, 579, 580

Sailing vessels, 581

Saint Helena Island, 414 extinctions, 535

Saint Lawrence River, 167, 189, 284, 378 target area effect and, 392–393

Saint Lawrence Valley, 188

Salamanders, 321, 493, 523

Salamandra salamandra, 523

Salinity, 123

as dispersal barrer, 282

ecological niche and, 68

effect on range, 78–80

oceanic distribution and, 409

pressure and, 57

productivity and, 485

water density and, 55

Salmo clarki, 203

Salsola iberica, 292

Salt marshes, 128, 485

Saltator albicollis, 493

Saltbush, 301

Salvelinus namaycush, 86

Salvinia, 299

Salvinia molesta, 271

San Andreas fault, 157

San Clemente, 298

Sandhill cranes, 588

Sandpipers, 451 geographic variation in North America, 455

Sanicula crassicaulis, 322

Santa Catalina Island, 298

Santa Catalina Mountains, 105, 465

Santa Cruz Island, 298

Santa Rosa Island, 298

Sargassum fish, 248

Sauromalus obesus, 53

Savanna, 110, 115–116 productivity and biomass, 129, 130, 132

response to climatic cycles in Pleistocene, 190

response to glaciation, 189

wildfire effects, 80

Sax, Dov, 474

Sayornis phoebe, 607

Scaptomyza frustulifera, 431

Scaturiginichthys, 295

Schiedea, 362–363

Schimper, A. F. W., 29

Schoener, Thomas, 388

Science, philosophy of, 7–8

Scientific names, 224

Scissor-tailed flycatcher, 73, 590

Sciurus aberti, 398

Sclater, Philip Lutley, 19, 24–25, 31, 302, 303
Sclater-Wallace classification scheme, 302–308
Sclerophyllous leaves, 117
Sclerophyllous woodland, 110, 117–118
Scolopacidae, 455
Scorpions, 462
Scotia Plate, 157
Scrub, 112
Sculpins, 240
Sea. *See* Marine communities; Oceans
Sea anemones, 89
Sea level, changes in Pleistocene, 186–189
Sea of Cortez, tide calendar, 58
Sea otters, 107, 252
Sea snakes, 520
    center of origin, 346–349
Sea surface temperatures during Pleistocene, 191
Sea urchins, 107
Seafloor spreading, 143–144, 150, 289
    North Atlantic, 173
Seals, 308, 451
Seamounts, 143–144, 152–153, 155
Seasonality, 480–481
Seasons, 41
Second law of thermodynamics, 101
Secondary consumers, 100
Secondary succession, 48, 81, 107
Secular migration, 273
Sedge, 115, 122
Seismic refraction techniques, 143
Selasphorus spp., 88
Selection. *See* Species selection
Selenidera spp., 206
Semi-logarithmic model, 374, 376, 394
Senecio, 299
Seres, 107
Sexual reproduction, colonization and, 291–293
Shiners, 308
Short-tailed shrews, 423–424
Shrews, 209–210, 423, 435
Shrubland, 112
Shyok Ice Lake, 201
Sial, 140
Sibling species, 245, 499
Sierra Madre, 287
Sierra Nevada, 45, 93, 106, 395
Silene laciniata, 88
Silphid, 356
Silurian
    continental drift, 158

cosmopolitanism in, 315
Sima, 140
Simberloff, Daniel, 388, 571
Similarity indexes, 318
Simpson, George Gaylord, 9, 11, 32, 33, 141, 262–263, 287, 451, 508
Simpson similarity index, 318–319
Simulium, 101
Single-species patterns
    ecogeographic rules, 488–492
    life history and population-level characteristic variation, 492–493
Sink habitats, 69, 589
Siskiyou Mountains, 105, 464
Skuas, 283
Skylark, 553
Slab pull, 150–151
SLOSS debate, 565–566
Small island effect, 393–394
Smilodon spp., 210–211
Smithsonian Institution, 252
Snakes. *See also* Sea snakes
    convergence in, 523
    trends in insular body size, 443
Snider-Pelligrini, Antonio, 33, 139, 142
Snowy owls, 71–72
Soccoro isopod, 494
Societies, human colonization and, 581
Soil chemistry, range boundaries and, 75–76, 78
Soils
    formation of, 48
    primary succession, 47–48
    relation to climatic zone and vegetation communities, 113
    requiring special plant adaptations, 51–53
    types, 48–51
    world distribution, 51
Solanum douglasii, 276
Solar radiation
    diversity and, 459–460, 480–481
    glaciation and, 179–180
    latitude and, 40–41
    in Pleistocene, 186
    water and, 53
Solitaires, 430
Solomon Islands, 233, 347
    human colonization, 579
    island area and linguistic diversity, 583
Solstice, 41
Somali Plate, 168
Songbirds, distribution pattern, 71
Sonoran Desert, 44, 307

range boundaries, 74
range of *Ambrosia ambrosioides*, 62
Sonoran Desert toad, 307
Sorex obscurus, 440
Sorex spp., 569
Soulé, Michael, 442
Source habitats, 69
South America, 303
    biogeographic provinces, 304–305
    convergence in, 524
    diffusion of species, 267
    distributional lines for freshwater fish, 311
    endemics, 296, 306
    extinctions, 218, 253, 545
    formation of, 163–164
    Great American Interchange, 508–513
    marsupials, 256–257, 335
    overkill hypothesis, 213
    secular migration, 273
    temperature cycles in Pleistocene, 184
    vegetation zone shifts during Pleistocene, 193–194
Southern beeches, 118, 296–297, 306, 313, 356
    consensus area cladogram, 360
Southern boreal region, 124
Southern Hemisphere
    disjunctions, 320–321, 323
    distributions, 355
    glaciation in, 178–179
Southern notal region, 124
Southern subtropical ocean region, 124
Southern temperate ocean region, 124
Space and time, diversity and, 484–485
Spalax ehrenbergi, 93
Spathodus marlieri, 241
Spatial patterns
    abundance and, 72
    of community, 103–107
Spatial scale, effect on perception of species distribution, 65
Spatial variation, small-scale, 45–47
Speciation, 32, 223, 229–230
    allopatric speciation, 235–239
    diversity and, 483–484
    diversity patterns and, 460–461
    equilibrium theory and, 384, 404
    genetic differentiation mechanisms, 232–235
    on Hawaiian Islands, 364

phyletic speciation, 243–244
    sympatric speciation, 239–243
    in Wallace's theory, 28
Speciation pump model, 205
Species
    conservation of, 533
    convergence, 520–524
    cosmopolitan, 16, 39
    geographic range, 61–65
    geographic shifts during Pleistocene, 194–198
    interactions among, 106
Species concepts
    biological species concept, 224–225
    evolutionary species concept, 225
    morphological species concept, 224
    phylogenetic species concept, 225–227
    subspecies and ecotypes, 227–228
Species distribution, 39
    effect of global climatic change, 604–605
Species diversity. *See also* Diversity patterns; Focal species patterns; Single-species patterns; Species richness
    body size and, 503–506
    in tropical rain forest, 112
Species diversity theory, 16
Species introductions, 547–548
    as cause of extinctions, 556
Species Plantarum, 16
Species richness, 17–18, 449–450. *See also* Island patterns; Latitudinal gradient; Species diversity
    effects of evolution, 405
    equilibrium theory and, 388
    insular, 390–394
    on Krakatau, 401–402
Species selection, 229–231
    examples, 255–259
    processes, 255
Species turnover, 377–379
Species-area relationship, 372–376, 397, 482–483
    boreal mammals, 567
    constants, 376
    on Krakatau, 403
Species-isolation relationship, 376–378, 393, 396–397
Spermophilus spp., 498, 569
Sphenodon punctatus, 298, 340
Spiders, 278, 424–425

long-distance dispersal, 409
Spreading zones, 151–152
Spring tides, 57–58
Spruces, 96, 97, 121
Stanley, Steven M., 229
Starfish, 81
Starlings, 267–269, 386, 421, 515
Stasipatric speciation, 239. *See also* Sympatric speciation
Stasis, 229–230
Static theory, 371
Stations, 17
Statistical macroecology, 501
Steadman, David, 584–585
Steller's sea cow, 252
Stellula calliope, 88
Stephen Island wren, 298–299
Steppe, 120, 189
Stercorarius spp., 283
Sternus vulgaris, 515
Stick nest builders, 257
Sticklebacks, 240, 242, 459
Stochastic mechanisms, 34
Stochasticity, 266
Stonefly, 101
Straits of Gibraltar, 167
Stratification, in aquatic environments, 53–55
Stratigraphic evidence, for continental drift, 145
Streptopelia chinensis, 553
Sturnus vulgaris, 267–269, 386
Subantarctic ocean region, 124
Subarctic ocean region, 124
Subduction zones, 150, 155
Sublittoral zone, 126
Subpopulation, 70
Subregions, 302
Subspecies, 227–228
Subtropical evergreen forest, 110, 118
Subtropical forests, during Pleistocene, 193
Succession, 107, 404. *See also* Primary succession; Secondary succession
Successional scrub, birds and, 529
Suez Canal, 513
Sugar maple, 605–607
Sugarcane, 474
Sumatra, 265
Summer-green deciduous forest, 119
Sun, effect on tides, 57–58
Sunda, 579, 580
Sunda Shelf, 187, 204, 285, 314, 315
Sundews, 52
Sunfishes, 281
Sunflower, fossil, 338

Supertramps, 420–421
Survival, in new habitat, 293
Suture zones, 157
Swamp forest, 121
Swamps, 127
Sweepstakes routes, 287–289
Sweet gum tree, 321
Sylivilagus nuttalii, 569
Sylvilagus, 491
Sympatric speciation, 239–243
Symplesiomorphies, 329
Synapomorphies, 329
Systematics, 5, 10
    cladogram construction, 331
    evolutionary classification, 326–327
    limitations of phylogenetic classifications, 334–335
    molecular systematics, 333–334
    phylogenetic systematics, 328–333

Tahiti, 422, 549
Taiga, 121
Tallgrass prairie, 120, 128
Tamias striatus, 197
Tamiasciurus hudsonicus, 598–599
Tamiasciurus hudsonicus grahamensis, 600
Tangara gyrola, 493
Taphonomy, 336
Taraxacum, 292
Tardigrades, 299
Target area effect, 391–393
Tasman Basin, 285
Taxa, centers of origin and, 346–350
Taxon cycle, 444–447
    insular human populations and, 584
Taxon cycle theory, invasions and, 552–553
Taxonomic categories, endemics and, 297
Taxonomic relicts, 300
Taxonomists, species specialization and, 536, 537
Taxonomy
    classifying endemics, 300
    defined, 224
    higher classifications, 228
    species concepts, 224–228
Taylor, F. B., 33, 139
Tectonic events, Great American Interchange, 509
Tectonic history
    continents, 157–166
    marine basins and island chains, 166–171
Tectonic plates, 152
Telmatochromis spp., 241

Temperate deciduous forest, 110, 118–119
    productivity and biomass, 129, 130, 132
Temperate forest
    deforestation, 562, 563
    pyramid of numbers, 102
Temperate grassland, 110, 120
    productivity and biomass, 129, 130, 132
Temperate lakes, overturn in, 65
Temperate rain forest, 110, 113, 119–120
    productivity and biomass, 129, 130, 132
Temperature
    aquatic communities and, 123
    biome classification and, 111
    as dispersal barrier, 282–283
    ecological niche and, 68
    effect on bird ranges, 608
    effect on diversity, 463–464
    formation of soil types and, 50
    glacial-interglacial cycles and, 182–184
    net primary productivity and, 130
    range boundaries and, 73–78
    of sea surface during Pleistocene, 191
    of subtropical evergreen forest, 118
    of temperate rain forest, 119
    of tropical rain forest, 112
    of tundra, 121
    water density and, 64
Temporal patterns, of community, 107–110
Temporal variation, small-scale, 45–47
Tenrecs, 246–247
Teratornis spp., 210–211
Teratorns, 210–211
Terborgh, John, 388, 565
Terella model, expanding earth hypothesis, 154–155
Terranes, 155–156
Terrell, John, 581, 583
Terrestrial biomes, 110–112
    boreal forest, 110, 121
    desert, 110, 116–117
    sclerophyllous woodland, 110, 117–118
    subtropical evergreen forest, 110, 118

temperate deciduous forest, 110, 118–119
temperate grassland, 110, 120–121
temperate rain forest, 110, 119–120
thorn woodland, 110, 114–115
tropical deciduous forest, 110, 114
tropical rain forest, 110, 112–114
tropical savanna, 110, 115–116
tundra, 110, 121–122
Terrestrial biotas, biogeographic responses to glaciation, 189–198
Terrestrial communities, productivity and biomass, 129, 130, 132
Terrestrial ecosystems, effects of global climatic change, 605–609
Terrestrial environments, microenvironmental variation and, 59
Terrestrial hot spots, 538–542
Terrestrial regions and provinces, 302–308
Terrestrial vertebrates, 278–279
Tertiary, 137
Tethyan Seaway, 170–171, 285
Tethys Sea, 163, 167, 170, 286, 574–575
Tetramolopium, 362–363
Thais lapillus, 68–69
Thermal stratification, 53–55
Thermocline, 64
Thermodynamics, 40
Thermosphaeroma thermophilum, 494
Thistle, 390
Thomomydoecus spp., 357
Thomomys spp., 245, 357, 498
Thomomys talpoides, 197, 488
Thorn scrub, 115
Thorn woodland, 110, 114–115
Thousand Islands, 599
Thrinaxodon, 339
Thryothurus ludovicianus, 71
Thyrosthurus bewickii, 527
Tidal cycles, 123
Tides, 57–58
Tiger salamander, 523
Tilia cordata, 604
Tillites, 140
Tilman, David, 564
Timberline, 76–77
Time and space, diversity and, 484–485
Time dwarfing, 441
Titmice, 499

Todidae, 340
Todies, 298
du Toit, Alexander, 141
Tonga, 585
Torres Strait, 303
Torresian Province, 303
Tortoises, 237–239, 443
Toucans, 523
Toxostoma bendirei, 307
Toxotidae, 295
Trace fossils, 336
Trachymesopus spp., 415
Trade winds, 43
Transform zones, 151, 155, 157
Transition zones, 414–415
Translocation, 239
Tree sparrow, 515
Trees
    growth rings, 137
    latitudinal trends in species richness, 452–453
    rates of migration, 195–196
    species diversity, 459
    variation in range with latitude, 470–471
Trenches, 143–144, 312
    global distribution, 543
Triassic, 137, 149
    extinctions, 253
Trichotanypus, 332
Tri-colored squirrel, 439
Trifolium pratense, 87
Triple junctions, 169–170
Troglodytes aedon, 527
Trophic levels, 100
Trophic status, community organization and, 98, 100
Tropic of Cancer, 41, 44
Tropic of Capricorn, 41
Tropical alpine scrubland, 122
Tropical deciduous forest, 110, 114
    productivity and biomass, 129, 130, 132
Tropical forest
    biomass pyramid, 102
    during Pleistocene, 193
Tropical ocean region, 124
Tropical rain forest, 110, 112–114
    diversity and, 481–482
    productivity, 128–129
    productivity and biomass, 129, 130, 132
    projected losses, 561
    response to glaciation, 189
Tropical savanna. *See* Savanna
Tropical waters, 124
Trumpeter swans, 588
Tsuga canadensis, 605–607
Tsuga, 109

Tuamoto Archipelago, 170
Tuatara, 298, 300, 340
Tufted titmouse, 74
Tulip tree, 81
Tumbleweed, 292
Tundra, 110, 121–122
    in North America during Pleistocene, 192
    productivity and biomass, 129, 130, 132
    response to climatic cycles in Pleistocene, 190
Turgai Sea, 167
Turnover, 371, 377–379
    on anthropogenic islands, 387–388
    of arthropod species on thistle, 390–391
    on landbridge islands, 385–387
Turtles, 338
Typha spp., 85
Tyrannidae, 454
Tyrannus dominicensis, 446

U.S. Endangered Species Act, 558, 623
U.S. Endangered Species List, 252
U.S. Fish and Wildlife Service, 558
U.S. Gap Analysis Program, 570
Ulmus spp., 271–272
Uma spp., 53, 321, 323
Ungulates, 209, 435
Uniformitarianism, 9, 18–19
United States. *See also* North America
    causes of endangerment for plants, 560
    plant communities in southwestern, 198–200
Upwelling, 55–56
Urals, 172
Uta stansburiana, 442
Utricularia purpurea, 189

Vaccinium spp., 600
Vagility, 273
Valvata, 230
Vandiermenella, 244
Vanga shrikes, 247
Varanus, 248
Varieties, 227
Vascular plants, 338
Vegetation
    biome classification and, 111
    during Pleistocene, 190–198
    relation to soil type and climatic zone, 113
Veil line, 375
Veldt, 120
Vellela, 278

Venus's flytraps, 52
Vertebrates, 25, 32, 33
    cladogram, 329
    evolutionary classification, 327
    extinctions, 210
    insular, 436–444
    terrestrial, 278–279
Vicariance, vs. dispersal, 262–263
Vicariance biogeography, 23–24, 262, 352–357
Vicariant events, 236–237
Vicariants, 34
Vicugna vicugna, 273
Vicuña, 273
Viking funeral ships, 289
Vine, Frederick, 149
Vireo olivaceus, 70–71
Viruses, 334
Volant species, 275
Volcanism, 151–152, 404
Voles, 71, 122, 210
Volga River, 308
von Humboldt, Alexander, 17
von Liebig, Justus, 74
Vulpes vulpes, 268–270, 490

Wallace, Alfred Russel, 8, 10, 19, 21, 24, 25–26, 135–136, 261, 302
    biogeographic principles, 28
Wallace's line, 25, 187, 308, 310
Wallacea, 286, 579, 580
Warming, E., 29
Wasps, 451–452
Water availability, range boundaries and, 75–76
Water density
    salinity and, 55
    temperature and, 64
Water depth
    marine communities and, 125
    variation in species richness, 466–467
Water dispersal, 278–279
Water fern, 271
Water flea, 299
Water hyacinths, 271
Water milfoil, 299
Water rats, 257
Water use, human impact on, 537
Waterfowl, 588
Webb, David, 509
Wecker, Stanley, 291
Wegener, Alfred, 7, 8, 33, 139
West Indies, 374
    taxon cycle in birds, 445–447
Westerlies, 43, 185
Western quoll, 594

Westralia, 303
Wetas, flightless, 431
Wetlands, loss of, 561–562
Whales, 317
White spruce, 195–196
Whitefishes, 240, 459
White-winged crossbill, 71–72
Whittaker, R. H., 96
Whittaker, Robert, 104–105
Whooping crane, 588, 592, 593
Wilkesia spp., 333
Willdenow, Karl, 16
Williams, E. E., 424
Williamson, P. G., 230
Willis, Ed., 564
Wilson, Edward O., 11, 34, 370, 373, 379–397, 407, 444, 536, 564
Wilson, J. Tuzo, 169
Wind, passive dispersal and, 277
Wind patterns, 42–43
Wolf, 103, 107
Wombat, 441, 524, 594
Wood swallows, 306
Woodland, 112
    birds and, 529
    productivity and biomass, 129, 130, 132
Woodrats, 60
Woods Hole Oceanographic Institution, 56
Wooly mammoth, 436–437, 442
Wrangel Island, 218, 436–437, 442
Würm glacial maximum, 191

Xenicus lyalli, 298–299
Xenotilapia spp., 241
Xerophytes, 78–79
Xyris caroliniana, 188

Yellow birch, 605–607
Yellow warbler, 89–90
Yosemite National Park, 106
Yucca, 116

Zagros Mountains, 167
Zapus princeps, 569
Zebra mussels, 271, 550–551
Zelandochlus, 361
Zonal soils, 51
Zoogeographers, 4
Zoogeography, 24–25
Zoological Society of London, 24
Zooplankton, 53, 126, 299, 465
Zosterops conspicillata, 595

ABOUT THE AUTHORS

James H. Brown is a Regents' Professor of Biology at the University of New Mexico. He received his B.A. from Cornell University, and his Ph.D. from the University of Michigan, where he worked with Emmet T. Hooper and William R. Dawson. Dr. Brown is the author or coauthor of four books and numerous scientific articles, and a past President of the American Society of Mammalogists, the American Society of Naturalists, and the Ecological Society of America. His diverse research has been united by long-standing interests in natural history and biodiversity. Current research projects include: long-term experimental studies of both rodents and plants in the Chihuahuan Desert; "macroecological" explorations of the interface between ecology and biogeography; theoretical investigations of the influence of body size and biological scaling on biodiversity; and applied efforts to preserve ranching livelihoods and biodiversity along the border between the United States and Mexico.

Mark V. Lomolino is an Associate Professor in the Department of Zoology and the Oklahoma Natural Heritage Inventory of the Oklahoma Biological Survey at the University of Oklahoma. He received his B.S. from the State University of New York at Cortland, an M.S. degree from the Department of Wildlife Biology at the University of Florida, with Katherine C. Ewel, and his Ph.D. from Binghamton University, with John Jaenike. Dr. Lomolino's research combines empirical and theoretical approaches to explore patterns and processes in nature, advance biogeography and community ecology theory, and develop effective strategies for conserving biological diversity.

ABOUT THE BOOK

Editor: Andrew D. Sinauer
Project Editor: Nan Sinauer
Production Manager: Christopher Small
Electronic Book Production: Wendy Beck, Jefferson Johnson, and Janice Holabird
Illustration Program: Precision Graphics, Inc., and Nancy J. Haver
Copy Editor: Norma S. Roche
Book Design: Janice Holabird
Cover Design: Jefferson Johnson
Book Manufacturer: Courier Companies, Inc.